Vassil N. Alexandrov
Geert Dick van Albada Peter M.A. Sloot
Jack Dongarra (Eds.)

Computational Science – ICCS 2006

6th International Conference
Reading, UK, May 28-31, 2006
Proceedings, Part IV

AF678522

UTD LIBRARY
2601 N. FLOYD RD.
RICHARDSON TX 75083

Springer

UT-DALLAS LIBRARIES
3 1863 007 376 423

Volume Editors

Vassil N. Alexandrov
University of Reading
Centre for Advanced Computing and Emerging Technologies
Reading RG6 6AY, UK
E-mail: v.n.alexandrov@rdg.ac.uk

Geert Dick van Albada
Peter M.A. Sloot
University of Amsterdam
Department of Mathematics and Computer Science
Kruislaan 403, 1098 SJ Amsterdam, The Netherlands
E-mail: {dick,sloot}@science.uva.nl

Jack Dongarra
University of Tennessee
Computer Science Department
1122 Volunteer Blvd., Knoxville, TN 37996-3450, USA
E-mail: dongarra@cs.utk.edu

Library of Congress Control Number: 2006926429

CR Subject Classification (1998): F, D, G, H, I, J, C.2-3

LNCS Sublibrary: SL 1 – Theoretical Computer Science and General Issues

ISSN 0302-9743
ISBN-10 3-540-34385-7 Springer Berlin Heidelberg New York
ISBN-13 978-3-540-34385-1 Springer Berlin Heidelberg New York

This work is subject to copyright. All rights are reserved, whether the whole or part of the material is concerned, specifically the rights of translation, reprinting, re-use of illustrations, recitation, broadcasting, reproduction on microfilms or in any other way, and storage in data banks. Duplication of this publication or parts thereof is permitted only under the provisions of the German Copyright Law of September 9, 1965, in its current version, and permission for use must always be obtained from Springer. Violations are liable to prosecution under the German Copyright Law.

Springer is a part of Springer Science+Business Media

springer.com

© Springer-Verlag Berlin Heidelberg 2006
Printed in Germany

Typesetting: Camera-ready by author, data conversion by Scientific Publishing Services, Chennai, India
Printed on acid-free paper SPIN: 11758549 06/3142 5 4 3 2 1 0

Lecture Notes in Computer Science 3994

Commenced Publication in 1973
Founding and Former Series Editors:
Gerhard Goos, Juris Hartmanis, and Jan van Leeuwen

Editorial Board

David Hutchison
Lancaster University, UK

Takeo Kanade
Carnegie Mellon University, Pittsburgh, PA, USA

Josef Kittler
University of Surrey, Guildford, UK

Jon M. Kleinberg
Cornell University, Ithaca, NY, USA

Friedemann Mattern
ETH Zurich, Switzerland

John C. Mitchell
Stanford University, CA, USA

Moni Naor
Weizmann Institute of Science, Rehovot, Israel

Oscar Nierstrasz
University of Bern, Switzerland

C. Pandu Rangan
Indian Institute of Technology, Madras, India

Bernhard Steffen
University of Dortmund, Germany

Madhu Sudan
Massachusetts Institute of Technology, MA, USA

Demetri Terzopoulos
University of California, Los Angeles, CA, USA

Doug Tygar
University of California, Berkeley, CA, USA

Moshe Y. Vardi
Rice University, Houston, TX, USA

Gerhard Weikum
Max-Planck Institute of Computer Science, Saarbruecken, Germany

RECEIVED

OCT 1 2 2006

UTD LIBRARY
ACQUISITION

Preface

The Sixth International Conference on Computational Science (ICCS 2006) was held in Reading, United Kingdom, May 28-31 and continued the traditions of previous conferences in the series: ICCS 2005 in Atlanta, Georgia, USA; ICCS 2004 in Krakow, Poland; ICCS 2003 held simultaneously at two locations in, Melbourne, Australia and St. Petersburg, Russia; ICCS 2002 in Amsterdam, The Netherlands; and ICCS 2001 in San Francisco, California, USA.

Since the first conference in San Francisco, rapid developments in Computational Science as a mainstream area facilitating multi-disciplinary research essential for the advancement of science have been observed. The theme of ICCS 2006 was "Advancing Science through Computation", marking several decades of progress in Computational Science theory and practice, leading to greatly improved applications science. The conference focused on the following major themes: tackling Grand Challenges Problems; modelling and simulations of complex systems; scalable algorithms and tools and environments for Computational Science. Of particular interest were the following major recent developments in novel methods and modelling of complex systems for diverse areas of science, scalable scientific algorithms, advanced software tools, computational grids, advanced numerical methods, and novel application areas where the above novel models, algorithms and tools can be efficiently applied such as physical systems, computational and systems biology, environmental systems, finance, and others.

Keynote lectures were delivered by Mateo Valero (Director, Barcelona Supercomputing Centre) - "Tackling Grand Challenges Problems"; Chris Johnson (Distinguished Professor, University of Utah) - "Visualizing the Future"; José Moreira (IBM, Chief Architect, Commercial Scale Out) - "Achieving Breakthrough Science with the Blue Gene/L Supercomputer"; Martin Curley (INTEL, Global Director of Innovation and IT Research) - "IT Innovation: A New Era"; Vaidy Sunderam (Samuel Candler Dobbs Professor of Computer Science, Emory University, USA) - "Metacomputing Revisited: Alternative Paradigms for Distributed Resource Sharing"; and Ron Bell (AWE plc.) - "The AWE HPC Benchmark".

In addition, two special sessions were held - one by industry and one by the funding bodies. Three tutorials preceded the main technical program of the conference: "Tools for Program Analysis in Computational Science" by Dieter Kranzlmüller; "P-GRADE Portal" by P. Kascuk, T. Kiss and G. Sipos; and "Scientific Computing on Graphics Hardware" by Dominik Göddeke. We would like to thank all the keynote, the invited, and the tutorial speakers for their inspiring talks.

Apart from the plenary sessions and tutorials the conference included twelve parallel oral sessions and two poster sessions. Since the first ICCS in San

Francisco the conference has grown steadily attracting increasing numbers of researchers in the field of Computational Science. For ICCS 2006 we received over 1,400 submissions, around 300 for the main track and over 1,100 for the originally proposed workshops. Of these submissions, 98 were accepted as a full papers and 29 as posters for the main track; and 500 were accepted as full papers, short papers or posters for the 32 workshops. This selection was possible due to the tremendous work done by the Program Committee and the 720 reviewers. The author index contains over 1,000 names and over 600 participants from all the major continents. The papers cover a wide variety of topics in Computational Science, ranging from Grand Challenges problems and modelling of complex systems in various areas to advanced numerical algorithms and new scalable algorithms in diverse application areas and software environments for Computational Science. The ICCS 2006 Proceedings consist of four volumes, 3991 to 3994, where the first volume contains the papers from the main track and all the posters; the remaining three volumes contain the papers from the workshops. ICCS this year is primary published on a CD and we would like to thank Springer for their cooperation and partnership. We hope that the ICCS 2006 Proceedings will be a major intellectual resource for many computational scientists and researchers for years ahead. During the conference the best papers from the main track and workshops as well as the best posters were nominated and commended on ICCS 2006 website. A number of selected papers will also be published in special issues of relevant mainstream journals.

We would like to thank all workshop organisers and the program committee for the excellent work, which further enhanced the conference's standing and led to very high quality event with excellent papers. We would like to express our gratitude to Advanced Computing and Emerging Technologies Centre staff, postgraduates and students for their wholehearted support of ICCS 2006. We would like to thank the School of Systems Engineering, Conference Office, Finance Department and various units at the University of Reading for different aspects of the organization and for their constant support in making ICCS 2006 a success. We would like to thank the Local Organizing Committee for their persistent and enthusiastic work towards the success of ICCS 2006. We owe special thanks to our sponsors: Intel, IBM, SGI, Microsoft Research, EPSRC and Springer; and to ACET Centre and the University of Reading for their generous support. We would like to thank SIAM, IMACS, and UK e-Science programme for endorsing ICCS 2006.

ICCS 2006 was organized by the Advanced Computing and Emerging Technologies Centre, University of Reading, with support from the Section Computational Science at the Universiteit van Amsterdam and Innovative Computing Laboratory at the University of Tennessee, in cooperation with the Society for Industrial and Applied Mathematics (SIAM), the International Association for Mathematics and Computers in Simulation (IMACS), and the UK Engineering and Physical Sciences Research Council (EPSRC). We invite you to visit the ICCS 2006 website (http://www.iccs-meeting.org/iccs2006/) and ACET Centre website (http://www.acet.reading.ac.uk/) to recount the events leading up to the conference, to view the technical programme, and to recall memories of three and a half days of engagement in the interest of fostering and advancing Computational Science.

June 2006

Vassil N. Alexandrov
G. Dick van Albada
Peter M.A. Sloot
Jack J. Dongarra

Organisation

ICCS 2006 was organised by the Centre for Advanced Computing and Emerging Technologies (ACET), University of Reading, UK, in cooperation with the University of Reading (UK), the Universiteit van Amsterdam (The Netherlands), the University of Tennessee (USA), Society for Industrial and Applied Mathematics (SIAM), International Association for Mathematics and Computers in Simulation (IMACS) and Engineering and Physical Sciences Research Council (EPSRC). The conference took place on the Whiteknights Campus of the University of Reading.

Conference Chairs

Scientific Chair - Vassil N. Alexandrov (ACET, University of Reading, UK)
Workshops Chair - G. Dick van Albada (Universiteit van Amsterdam, The Netherlands)
ICCS Series Overall Chair - Peter M.A. Sloot (Universiteit van Amsterdam, The Netherlands)
ICCS Series Overall Co-Chair - Jack J. Dongarra (University of Tennessee, USA)

Local Organising Committee

Vassil N. Alexandrov
Linda Mogort-Valls
Nia Alexandrov
Ashish Thandavan
Christian Weihrauch
Simon Branford
Adrian Haffegee
David Monk
Janki Dodiya
Priscilla Ramsamy
Ronan Jamieson
Ali Al-Khalifah
David Johnson
Eve-Marie Larsen
Gareth Lewis
Ismail Bhana
S. Mehmood Hasan
Sokratis Antoniou

Sponsoring Institutions

Intel Corporation
IBM
SGI
Microsoft Research
EPSRC
Springer
ACET Centre
University of Reading

Endorsed by

SIAM
IMACS
UK e-Science Programme

Program Committee

D. Abramson - Monash University, Australia
V. Alexandrov - University of Reading, UK
D.A. Bader - Georgia Tech, USA
M. Baker - University of Portsmouth, UK
S. Belkasim - Georgia State University, USA
A. Benoit - Ecole Normale Superieure de Lyon, France
I. Bhana - University of Reading, UK
R. Blais - University of Calgary, Canada
A. Bogdanov - Institute for High Performance Computing and Information Systems, Russia
G. Bosilca - University of Tennessee, USA
S. Branford - University of Reading, UK
M. Bubak - Institute of Computer Science and ACC Cyfronet - AGH, Poland
R. Buyya - University of Melbourne, Australia
F. Cappello - Laboratoire de Recherche en Informatique, Paris Sud, France
T. Cortes - Universitat Politecnica de Catalunya, Spain
J.C. Cunha - New University of Lisbon, Portugal
F. Desprez - INRIA, France
T. Dhaene - University of Antwerp, Belgium
I.T. Dimov - University of Reading, UK
J. Dongarra - University of Tennessee, USA
C. Douglas - University of Kentucky, USA
G.E. Fagg, University of Tennessee, USA
M. Gerndt - Technical University of Munich, Germany

Y. Gorbachev - Institute for High Performance Computing and Information Systems, Russia
A. Goscinski - Deakin University, Australia
A. Haffegee - University of Reading, UK
L. Hluchy - Slovak Academy of Science, Slovakia
A. Hoekstra - Universiteit van Amsterdam, The Netherlands
A. Iglesias - University of Cantabria, Spain
R. Jamieson - University of Reading, UK
D. Johnson - University of Reading, UK
J. Kitowski - AGH University of Science and Technology, Poland
D. Kranzlmüller - Johannes Kepler University Linz, Austria
A. Lagana - Universita di Perugia, Italy
G. Lewis - University of Reading, UK
E. Luque - University Autonoma of Barcelona, Spain
M. Malawski - Institute of Computer Science AGH, Poland
M. Mascagni - Florida State University, USA
E. Moreno - Euripides Foundation of Marilia, Brazil
J. Ni The - University of Iowa, Iowa City, IA, USA
G. Norman - Russian Academy of Sciences, Russia
S. Orlando - University of Venice, Italy
B. Ó Nulláin - UUniversiteit van Amsterdam, The Netherlands
M. Paprzycki - Computer Science Institute, SWSP, Warsaw, Poland
R. Perrott - Queen's University of Belfast, UK
R. Renaut - Arizona State University, USA
A. Rendell - Australian National University, Australia
D. Rodriguez-García - University of Reading, UK
P. Roe Queensland - University of Technology, Australia
S.L. Scott - Oak Ridge National Laboratory, USA
D. Shires - U.S. Army Research Laboratory, USA
P.M.A. Sloot - Universiteit van Amsterdam, The Netherlands
G. Stuer - University of Antwerp, Belgium
R. Tadeusiewicz - AGH University of Science and Technology, Poland
A. Thandavan - University of Reading, UK
P. Tvrdik - Czech Technical University, Czech Republic
P. Uthayopas - Kasetsart University, Thailand
G.D. van Albada - Universiteit van Amsterdam, The Netherlands
J. Vigo-Aguiar - University of Salamanca, Spain
J.A. Vrugt - Los Alamos National Laboratory, USA
J. Wasniewski - Technical University of Denmark, Denmark
G. Watson - Los Alamos National Laboratory, USA
C. Weihrauch - University of Reading, UK
Y. Xue - Chinese Academy of Sciences, China
E. Zudilova-Seinstra - Universiteit van Amsterdam, The Netherlands

Reviewers

A. Adamatzky
A. Arenas
A. Belloum
A. Benoit
A. Bielecki
A. Bode
A. Cepulkauskas
A. Chkrebtii
A. Drummond
A. Erzan
A. Fedaravicius
A. Galvez
A. Gerbessiotis
A. Goscinski
A. Griewank
A. Grösslinger
A. Grzech
A. Haffegee
A. Hoekstra
A. Iglesias
A. Jakulin
A. Janicki
A. Javor
A. Karpfen
A. Kertész
A. Knuepfer
A. Koukam
A. Lagana
A. Lawniczak
A. Lewis
A. Li
A. Ligeza
A. Mamat
A. Martin del Rey
A. McGough
A. Menezes
A. Motter
A. Nasri
A. Neumann
A. Noel
A. Obuchowicz
A. Papini
A. Paventhan
A. Pieczynska
A. Rackauskas
A. Rendell
A. Sánchez
A. Sánchez-Campos
A. Sayyed-Ahmad
A. Shafarenko
A. Skowron
A. Sosnov
A. Sourin
A. Stuempel
A. Thandavan
A. Tiskin
A. Turan
A. Walther
A. Wei
A. Wibisono
A. Wong
A. Yacizi
A. Zelikovsky
A. Zhmakin
A. Zhou
A.N. Karaivanova
A.S. Rodinov
A.S. Tosun
A.V. Bogdanov
B. Ó Nualláin
B. Autin
B. Balis
B. Boghosian
B. Chopard
B. Christianson
B. Cogan
B. Dasgupta
B. Di Martino
B. Gabrys
B. Javadi
B. Kahng
B. Kovalerchuk
B. Lesyng
B. Paternoster
B. Payne
B. Saunders
B. Shan
B. Sniezynski
B. Song
B. Strug
B. Tadic
B. Xiao
B.M. Rode
B.S. Shin
C. Anthes
C. Bannert
C. Biely
C. Bischof
C. Cotta
C. Douglas
C. Faure
C. Glasner
C. Grelck
C. Herrmann
C. Imielinska
C. Lursinsap
C. Mastroianni
C. Miyaji
C. Nelson
C. Otero
C. Rodriguez Leon
C. Schaubschläger
C. Wang
C. Weihrauch
C. Woolley
C. Wu
C. Xu
C. Yang
C.-H. Huang
C.-S. Jeong
C.G.H. Diks
C.H. Goya
C.H. Kim
C.H. Wu
C.K. Chen
C.N. Lee
C.R. Kleijn
C.S. Hong
D. Abramson

D. Brinza
D. Brown
D. Che
D. Déry
D. Donnelly
D. Evers
D. Göddeke
D. Johnson
D. Kim
D. Kranzlmüller
D. Laforenza
D. Li
D. Luebke
D. Maringer
D. Pfahl
D. Plemenos
D. Rodriguez-García
D. Shires
D. Stoffer
D. Stokic
D. Szczerba
D. Taniar
D. Thalmann
D. Vasuinin
D. Wang
D. Xu
D.A. Bader
D.B. Davies
D.B.D. Birkbeck
D.C. Ghosh
D.C. Lee
D.J. Roberts
D.M. Chiu
D.M. Tartakovsky
D.R. Green
D.S. Kim
D.S. Perry
E. Atanasov
E. Grabska
E. Huedo Cuesta
E. Jaeger-Frank
E. Lee
E. Luque
E. Macias
E. Moreno
E. Nawarecki
E. Puppo
E. Roanes-Lozano
E. Valakevicius
E. Zeng
E. Zotenko
E. Zudilova-Seinstra
E.A. Castro
E.N. Huh
E.S. Quintana-Orti
F. Capkovic
F. Cappello
F. Desprez
F. Gava
F. Hirata
F. Iavernaro
F. Kiss
F. Lamantia
F. Lee
F. Loulergue
F. Markowetz
F. Melendez
F. Perales
F. Rogier
F. Terpstra
F. Zuccarello
F.-X. Roux
F.J. Keil
G. Alexe
G. Allen
G. Bosilca
G. Chen
G. Cheng
G. Dobrowolski
G. Dong
G. Erlebacher
G. Farin
G. Felici
G. Frenking
G. Gheri
G. Jeon
G. Kolaczek
G. Kou
G. Lewis
G. Lin
G. Mauri
G. Messina
G. Mounié
G. Narasimhan
G. Norman
G. Pavesi
G. Rojek
G. Slusarczyk
G. Stuer
G. Szabó
G. Tempesti
G. Volkert
G. Watson
G. Zheng
G.-L. Park
G.D. van Albada
G.D. Vedova
G.E. Fagg
G.J. Rodgers
H. Bungartz
H. Choo
H. Diab
H. Fangohr
H. Jin
H. Kaltenbach
H. Kosina
H. Labiod
H. Lee
H. Moradkhani
H. Müller
H. Munakata
H. Oh
H. Sarafian
H. Stockinger
H. Suzuki
H. Umeo
H. Wang
H. Yanami
H.-K. Choi
H.-K. Lee
H.C. Chojnacki
H.F. Schaefer III
H.K. Kim
H.P. Luehi
H.S. Nguyen

H.Y. Lee
I. Bhana
I. Boada
I. Kolingerova
I. Lee
I. Mandoiu
I. Moret
I. Navas-Delgado
I. Podolak
I. Schagaev
I. Suehiro
I. Tabakow
I. Taylor
I.T. Dimov
J. Abawajy
J. Aroba
J. Blower
J. Cabero
J. Cai
J. Cao
J. Chen
J. Cho
J. Choi
J. Davila
J. Dolado
J. Dongarra
J. Guo
J. Gutierrez
J. Han
J. He
J. Heo
J. Hong
J. Humble
J. Hwang
J. Jeong
J. Jurek
J. Kalcher
J. Kang
J. Kim
J. King
J. Kitowski
J. Koller
J. Kommineni
J. Koo
J. Kozlak
J. Kroc
J. Krueger
J. Laws
J. Lee
J. Li
J. Liu
J. Michopoulos
J. Nabrzyski
J. Nenortaite
J. Ni
J. Owen
J. Owens
J. Pang
J. Pjesivac-Grbovic
J. Quinqueton
J. Sanchez-Reyes
J. Shin
J. Stefanowski
J. Stoye
J. Tao
J. Utke
J. Vigo-Aguiar
J. Volkert
J. Wang
J. Wasniewski
J. Weidendorfer
J. Wu
J. Yu
J. Zara
J. Zhang
J. Zhao
J. Zivkovic
J.-H. Nam
J.-L. Koning
J.-W. Lee
J.A. Vrugt
J.C. Cunha
J.C. Liu
J.C. Teixeira
J.C.S. Lui
J.F. San Juan
J.H. Hrusak
J.H. Lee
J.J. Alvarez
J.J. Cuadrado
J.J. Korczak
J.J. Zhang
J.K. Choi
J.L. Leszczynski
J.M. Bradshaw
J.M. Gilp
J.P. Crutchfield
J.P. Suarez Rivero
J.V. Alvarez
J.Y. Chen
K. Akkaya
K. Anjyo
K. Banas
K. Bolton
K. Boryczko
K. Chae
K. Ebihara
K. Ellrott
K. Fisher
K. Fuerlinger
K. Gaaloul
K. Han
K. Hsu
K. Jinsuk
K. Juszczyszyn
K. Kubota
K. Li
K. Meridg
K. Najarian
K. Ouazzane
K. Sarac
K. Sycara
K. Tai-hoon Kim
K. Trojahner
K. Tuncay
K. Westbrooks
K. Xu
K. Yang
K. Zhang
K.-J. Jeong
K.B. Lipkowitz
K.D. Nguyen
K.V. Mikkelsen
K.X.S. Souza
K.Y. Huang

L. Borzemski
L. Brugnano
L. Cai
L. Czekierda
L. Fernandez
L. Gao
L. Gonzalez-Vega
L. Hascoet
L. Hluchy
L. Jia
L. Kotulski
L. Liu
L. Lopez
L. Marchal
L. Neumann
L. Parida
L. Taher
L. Xiao
L. Xin
L. Yang
L. Yu
L. Zheng
L. Zhigilei
L.H. Figueiredo
L.J. Song
L.T. Yang
M. Aldinucci
M. Baker
M. Bamha
M. Baumgartner
M. Bhuruth
M. Borodovsky
M. Bubak
M. Caliari
M. Chover
M. Classen
M. Comin
M. Deris
M. Drew
M. Fagan
M. Fras
M. Fujimoto
M. Gerndt
M. Guo
M. Hardman
M. Hobbs
M. Houston
M. Iwami
M. Jankowski
M. Khater
M. Kim
M. Kirby
M. Kisiel-Dorochinicki
M. Li
M. Malawski
M. Mascagni
M. Morshed
M. Mou
M. Omar
M. Pérez-Hernández
M. Palakal
M. Paprzycki
M. Paszynski
M. Polak
M. Rajkovic
M. Ronsse
M. Rosvall
M. Ruiz
M. Sarfraz
M. Sbert
M. Smolka
M. Suvakov
M. Tomassini
M. Verleysen
M. Vianello
M. Zhang
M.A. Sicilia
M.H. Zhu
M.J. Brunger
M.J. Harris
M.Y. Chung
N. Bauernfeind
N. Hu
N. Ishizawa
N. Jayaram
N. Masayuki
N. Murray
N. Navarro
N. Navet
N. Sastry
N. Sundaraganesan
N.T. Nguyen
O. Beckmann
O. Belmonte
O. Habala
O. Maruyama
O. Otto
O. Yasar
P. Alper
P. Amodio
P. Balbuena
P. Bekaert
P. Berman
P. Blowers
P. Bonizzoni
P. Buendia
P. Czarnul
P. Damaschke
P. Diaz Gutierrez
P. Dyshlovenko
P. Geerlings
P. Gruer
P. Heimbach
P. Heinzlreiter
P. Herrero
P. Hovland
P. Kacsuk
P. Li
P. Lingras
P. Martineau
P. Pan
P. Praxmarer
P. Rice
P. Roe
P. Sloot
P. Tvrdik
P. Uthayopas
P. van Hooft
P. Venuvanalingam
P. Whitlock
P. Wolschann
P.H. Lin
P.K. Chattaraj
P.R. Ramasami
Q. Deng

R. Aspin
R. Blais
R. Buyya
R. Dondi
R. Drezewski
R. Duran Diaz
R. Jamieson
R. Jothi
R. Kakkar
R. Katarzyniak
R. Kobler
R. Lambiotte
R. Liu
R. Marcjan
R. Mikusauskas
R. Nock
R. Perrott
R. Ramaroson
R. Rejas
R. Renaut
R. Rizzi
R. Ruiz
R. Sander
R. Schaefer
R. Simutis
R. Strzodka
R. Tadeusiewicz
R. Walentynski
R. Westermann
R. Wismüller
R. Wolff
R.G. Giering
R.Q. Wu
S. Abe
S. Aluru
S. Ambroszkiewicz
S. Balla
S. Bandini
S. Belkasim
S. Bhowmick
S. Böcker
S. Branford
S. Chen
S. Chiu
S. Cho
S. Dong
S. El Yacoubi
S. Forth
S. Gilmore
S. Gimelshein
S. Gorlatch
S. Green
S. Gremalschi
S. Han
S. Jhang
S. Kawano
S. Kim
S. Lee
S. Lightstone
S. Maniccam
S. Olariu
S. Orlando
S. Pal
S. Rahmann
S. Rajasekaran
S. Sanchez
S. Thurner
S. Tsunekawa
S. Turek
S. Valverde
S. Yi
S. Yoon
S.-B. Scholz
S.-R. Kim
S.-Y. Han
S.C. Lo
S.H. Cho
S.J. Han
S.K. Ghosh
S.L. Gargh
S.L. Scott
S.S. Manna
T. Angskun
T. Atoguchi
T. Cortes
T. Dhaene
T. Dokken
T. Ezaki
T. Fahringer
T. Hu
T. Ida
T. Korkmaz
T. McKenzie
T. Milledge
T. Politi
T. Przytycka
T. Recio
T. Strothotte
T. Suzudo
T. Takahashi
T. Tsuji
T. Wang
T. Ward
T. Worsch
T.-J. Lee
T.B. Ho
T.C. Lu
T.L. Zhang
T.N. Troung
T.V. Gurov
T.W. Kim
U. Ruede
U. Ufuktepe
U. Vaccaro
U.N. Naumann
V. Alexandrov
V. Aquilanti
V. Debelov
V. Hargy
V. Korkhov
V. Parasuk
V. Rafe
V. Robles
V. Srovnal
V. Weispfenning
V.A. Emanuele II
V.C. Chinh
V.V. Krzhizhanovskaya
V.V. Shakhov
W. Alda
W. Bronsvoort
W. Choi
W. Dou
W. Funika
W. Lee

W. Miller
W. Rachowicz
W. Yan
W. Yin
W. Zhang
W. Zheng
W.K. Tai
X. Huang
X. Liao
X. Wan
X. Wang
X. Zhang
X.J. Chen
X.Z. Cheng
Y. Aumann
Y. Byun
Y. Cai
Y. Cotronis
Y. Cui
Y. Dai
Y. Li
Y. Liu
Y. Mun
Y. Pan
Y. Peng
Y. Shi
Y. Song
Y. Xia
Y. Xue
Y. Young Jin
Y.-C. Bang
Y.-C. Shim
Y.B. Kim
Y.E. Gorbachev
Y.J. Ye
Y.Q. Xiong
Y.S. Choi
Y.Y. Cho
Y.Z. Cho
Z. Cai
Z. Hu
Z. Huang
Z. Liu
Z. Pan
Z. Toroczkai
Z. Wu
Z. Xin
Z. Zhao
Z. Zlatev
Z.G. Sun
Z.M. Zhou

Workshop Organisers

Third International Workshop on Simulation of Multiphysics Multiscale Systems

V.V. Krzhizhanovskaya - Universiteit van Amsterdam, The Netherlands and St. Petersburg State Polytechnical University, Russia
Y.E. Gorbachev - St. Petersburg State Polytechnic University, Russia
B. Chopard - University of Geneva, Switzerland

Innovations in Computational Science Education

D. Donnelly - Department of Physics, Siena College, USA

Fifth International Workshop on Computer Graphics and Geometric Modeling (CGGM 2006)

A. Iglesias - University of Cantabria, Spain

Fourth International Workshop on Computer Algebra Systems and Applications (CASA 2006)

A. Iglesias - University of Cantabria, Spain
A. Galvez - University of Cantabria, Spain

Tools for Program Development and Analysis in Computational Science

D. Kranzlmüller - GUP, Joh. Kepler University, Linz, Austria
R. Wismüller - University of Siegen, Germany
A. Bode - Technische Universität München, Germany
J. Volkert - GUP, Joh. Kepler University, Linz, Austria

Collaborative and Cooperative Environments

C. Anthes - GUP, Joh. Kepler University, Linz, Austria
V.N. Alexandrov - ACET, University of Reading, UK
D.J. Roberts - NICVE, University of Salford, UK
J. Volkert - GUP, Joh. Kepler University, Linz, Austria
D. Kranzlmüller - GUP, Joh. Kepler University, Linz, Austria

Second International Workshop on Bioinformatics Research and Applications (IWBRA'06)

A. Zelikovsky - Georgia State University, USA
Y. Pan - Georgia State University, USA
I.I. Mandoiu - University of Connecticut, USA

Third International Workshop on Practical Aspects of High-Level Parallel Programming (PAPP 2006)

A. Benoît - Laboratoire d'Informatique du Parallélisme, Ecole Normale Supérieure de Lyon, France
F. Loulergue - LIFO, Université d'Orléans, France

Wireless and Mobile Systems

H. Choo - Networking Laboratory, Sungkyunkwan University, Suwon, KOREA

GeoComputation

Y. Xue - Department of Computing, Communications Technology and Mathematics, London Metropolitan University, UK

Computational Chemistry and Its Applications

P. Ramasami - Department of Chemistry, University of Mauritius

Knowledge and Information Management in Computer Communication Systems (KIMCCS 2006)

N.T. Nguyen - Institute of Control and Systems Engineering, Wroclaw University of Technology, Poland

A. Grzech - Institute of Information Science and Engineering, Wroclaw University of Technology, Poland
R. Katarzyniak - Institute of Information Science and Engineering, Wroclaw University of Technology, Poland

Modelling of Complex Systems by Cellular Automata (MCSCA 2006)

J. Kroc - University of West Bohemia, Czech Republic
T. Suzudo - Japan Atomic Energy Agency, Japan
S. Bandini - University of Milano - Bicocca, Italy

Dynamic Data Driven Application Systems (DDDAS 2006)

F. Darema - National Science Foundation, USA

Parallel Monte Carlo Algorithms for Diverse Applications in a Distributed Setting

I.T. Dimov - ACET, University of Reading, UK
V.N. Alexandrov - ACET, University of Reading, UK

International Workshop on Intelligent Storage Technology (IST06)

J. Shu - Department of Computer Science and Technology, Tsinghua University, Beijing, P.R. China

Intelligent Agents in Computing Systems

R. Schaefer - Department of Computer Science, Stanislaw Staszic University of Science and Technology in Kraków
K. Cetnarowicz - Department of Computer Science, Stanislaw Staszic University of Science and Technology in Kraków

First International Workshop on Workflow Systems in e-Science (WSES06)

Z. Zhao - Informatics Institute, University of Amsterdam, The Netherlands
A. Belloum - University of Amsterdam, The Netherlands

Networks: Structure and Dynamics

B. Tadic - Theoretical Physics Department, J. Stefan Institute, Ljubljana, Slovenia
S. Thurner - Complex Systems Research Group, Medical University Vienna, Austria

Evolution Toward Next Generation Internet (ENGI)

Y. Cui - Tsinghua University, P.R. China
T. Korkmaz - University of Texas at San Antonio, USA

General Purpose Computation on Graphics Hardware (GPGPU): Methods, Algorithms and Applications

D. Göddeke - Universität Dortmund, Institut für Angewandte Mathematik und Numerik, Germany
S. Turek - Universität Dortmund, Institut für Angewandte Mathematik und Numerik, Germany

Intelligent and Collaborative System Integration Technology (ICSIT)

J.-W. Lee - Center for Advanced e-System Integration Technology, Konkuk University, Seoul, Korea

Computational Methods for Financial Markets

R. Simutis - Department of Informatics, Kaunas Faculty, Vilnius University, Lithuania
V. Sakalauskas - Department of Informatics, Kaunas Faculty, Vilnius University, Lithuania
D. Kriksciuniene - Department of Informatics, Kaunas Faculty, Vilnius University, Lithuania

2006 International Workshop on P2P for High Performance Computational Sciences (P2P-HPCS06)

H. Jin - School of Computer Science and Technology, Huazhong University of Science and Technology, Wuhan, China
X. Liao - Huazhong University of Science and Technology, Wuhan, China

Computational Finance and Business Intelligence

Y. Shi - Graduate School of the Chinese Academy of Sciences, Beijing, China

Third International Workshop on Automatic Differentiation Tools and Applications

C. Bischof - Inst. for Scientific Computing, RWTH Aachen University, Germany
S.A. Forth - Engineering Systems Department, Cranfield University, RMCS Shrivenham, UK
U. Naumann - Software and Tools for Computational Engineering, RWTH Aachen University, Germany
J. Utke - Mathematics and Computer Science Division, Argonne National Laboratory, IL, USA

2006 Workshop on Scientific Computing in Electronics Engineering

Y. Li - National Chiao Tung Univeristy, Hsinchu City, Taiwan

New Trends in the Numerical Solution of Structured Systems with Applications

T. Politi - Dipartimento di Matematica, Politecnico di Bari, Itali
L. Lopez - Dipartimento di Matematica, Universita' di Bari, Itali

Workshop on Computational Science in Software Engineering (CSSE'06)

D. Rodríguez García - University of Reading, UK
J.J. Cuadrado - University of Alcalá, Spain
M.A. Sicilia - University of Alcalá, Spain
M. Ruiz - University of Cádiz, Spain

Digital Human Modeling (DHM-06)

Y. Cai - Carnegie Mellon University, USA
C. Imielinska - Columbia University

Real Time Systems and Adaptive Applications (RTSAA 06)

T. Kuo - National Taiwan University, Taiwan
J. Hong - School of Computer Science and Engineering, Kwangwoon University, Seoul, Korea
G. Jeon - Korea Polytechnic University, Korea

International Workshop on Grid Computing Security and Resource Management (GSRM'06)

J.H. Abawajy - School of Information Technology, Deakin University, Geelong, Australia

Fourth International Workshop on Autonomic Distributed Data and Storage Systems Management Workshop (ADSM 2006)

J.H. Abawajy - School of Information Technology, Deakin University, Geelong, Australia

[illegible] - University Carlos III of Madrid, Spain

Table of Contents – Part IV

Evolution Toward Next Generation Internet (ENGI)

A New Energy Efficient Target Detection Scheme for Pervasive Computing
Thanh Hai Trinh, Hee Yong Youn 1

A Load Balance Based On-Demand Routing Protocol for Mobile Ad-Hoc Networks
Liqiang Zhao, Xin Wang, Azman Osman Lim, Xiangyang Xue 9

Handover Control Function Based Handover for Mobile IPv6
Guozhi Wei, Anne Wei, Ke Xu, Hui Deng 17

Unified Error Control Framework with Cross-Layer Interactions for Efficient H.264 Video Transmission over IEEE 802.11e Wireless LAN
Jeong-Yong Choi, Jitae Shin 25

A Novel Control Plane Model of Extensible Routers
Kun Wu, Jianping Wu, Ke Xu 33

AM-Trie: A High-Speed Parallel Packet Classification Algorithm for Network Processor
Bo Zheng, Chuang Lin 41

Speedup Requirements for Output Queuing Emulation with a Sliding-Window Parallel Packet Switch
Chia-Lung Liu, Woei Lin, Chin-Chi Wu 49

Combining Cross-Correlation and Fuzzy Classification to Detect Distributed Denial-of-Service Attacks
Wei Wei, Yabo Dong, Dongming Lu, Guang Jin 57

Convergence of the Fixed Point Algorithm of Analytical Models of Reliable Internet Protocols (TCP)
Debessay Fesehaye Kassa, Sabine Wittevrongel 65

A Peer-to-Peer Approach to Semantic Web Services Discovery
Yong Li, Sen Su, Fangchun Yang 73

Multicast Routing Protocol with Heterogeneous and Dynamic Receivers
Huimei Lu, Hongyu Hu, Quanshuang Xiang, Yuanda Cao 81

Using Case-Based Reasoning to Support Web Service Composition
Ruixing Cheng, Sen Su, Fangchun Yang, Yong Li 87

Secure OWL Query
Baowen Xu, Yanhui Li, Jianjiang Lu, Dazhou Kang 95

Efficient Population Diversity Handling Genetic Algorithm for QoS-Aware Web Services Selection
Chengwen Zhang, Sen Su, Junliang Chen 104

A New Algorithm for Long Flows Statistics—MGCBF
Mingzhong Zhou, Jian Gong, Wei Ding 112

Estimating Original Flow Length from Sampled Flow Statistics
Weijiang Liu, Jian Gong, Wei Ding, Yanbing Peng 120

Easily-Implemented Adaptive Packet Sampling for High Speed Networks Flow Measurement
Hongbo Wang, Yu Lin, Yuehui Jin, Shiduan Cheng 128

Multi-layer Network Recovery: Avoiding Traffic Disruptions Against Fiber Failures
Anna Urra, Eusebi Calle, Jose L. Marzo 136

An Algorithm for Estimation of Flow Length Distributions Using Heavy-Tailed Feature
Weijiang Liu, Jian Gong, Wei Ding, Guang Cheng 144

Performance Evaluation of Novel MAC Protocol for WDM/Ethernet-PON
Bokrae Jung, Hyunho Yun, Jaegwan Kim, Mingon Kim, Minho Kang 152

An Efficient Mobility Management Scheme for Two-Level HMIPv6 Networks
Xuezeng Pan, Zheng Wan, Lingdi Ping, Fanjun Su 156

Analysis of Packet Transmission Delay Under the Proportional Fair Scheduling Policy
Jin-Hee Choi, Jin-Ghoo Choi, Chuck Yoo 160

Precise Matching of Semantic Web Services
Yonglei Yao, Sen Su, Fangchun Yang 164

Evolving Toward Next Generation Wireless Broadband Internet
Seung-Que Lee, Namhun Park, Choongho Cho, Hyongwoo Lee, Seungwan Ryu 168

A Decision Maker for Transport Protocol Configuration
Jae-Hyun Hwang, Jin-Hee Choi, Chuck Yoo 172

On the Generation of Fast Verifiable IPv6 Addresses
Qianli Zhang, Xing Li 176

A MAC Protocol to Reduce Sleep Latency and Collisions in Wireless Sensor Network
Jinsuk Pak, Jeongho Son, Kijun Han 180

IC Design of IPv6 Routing Lookup for High Speed Networks
Yuan-Sun Chu, Hui-Kai Su, Po-Feng Lin, Ming-Jen Chen 184

General Purpose Computation on Graphics Hardware (GPGPU): Methods, Algorithms and Applications

GPU Accelerated Smith-Waterman
Yang Liu, Wayne Huang, John Johnson, Sheila Vaidya 188

A Graphics Hardware Accelerated Algorithm for Nearest Neighbor Search
Benjamin Bustos, Oliver Deussen, Stefan Hiller, Daniel Keim 196

The Development of the Data-Parallel GPU Programming Language CGiS
Philipp Lucas, Nicolas Fritz, Reinhard Wilhelm 200

Spline Surface Intersections Optimized for GPUs
Sverre Briseid, Tor Dokken, Trond Runar Hagen, Jens Olav Nygaard 204

A GPU Implementation of Level Set Multiview Stereo
Patrick Labatut, Renaud Keriven, Jean-Philippe Pons 212

Solving the Euler Equations on Graphics Processing Units
Trond Runar Hagen, Knut-Andreas Lie, Jostein R. Natvig 220

Particle-Based Fluid Simulation on the GPU
Kyle Hegeman, Nathan A. Carr, Gavin S.P. Miller 228

Spiking Neurons on GPUs
Fabrice Bernhard, Renaud Keriven ... 236

Intelligent and Collaborative System Integration Technology (ICSIT)

SONA: An On-Chip Network for Scalable Interconnection of AMBA-Based IPs
Eui Bong Jung, Han Wook Cho, Neungsoo Park, Yong Ho Song ... 244

Semi-automatic Creation of Adapters for Legacy Application Migration to Integration Platform Using Knowledge
Jan Pieczykolan, Bartosz Kryza, Jacek Kitowski ... 252

A Self-configuration Mechanism for High-Availability Clusters
Hocheol Sung, Sunyoung Han, Bok-Gyu Joo, Chee-Wei Ang, Wang-Cho Cheng, Kim-Sing Wong ... 260

Development of Integrated Framework for the High Temperature Furnace Design
Yu Xuan Jin, Jae-Woo Lee, Karp Joo Jeong, Jong Hwa Kim, Ho-Yon Hwang ... 264

A Distributed Real-Time Tele-operation System Based on the TMO Modeling
Hanku Lee, Segil Jeon ... 272

A Sharing and Delivery Scheme for Monitoring TMO-Based Real-Time Systems
Yoon-Seok Jeong, Tae-Wan Kim, Chun-Hyon Chang ... 280

An Algorithm for the Generalized k-Keyword Proximity Problem and Finding Longest Repetitive Substring in a Set of Strings
Inbok Lee, Sung-Ryul Kim ... 289

A Grid-Based Flavonoid Informatics Portal
HaiGuo Xu, Karpjoo Jeong, Seunho Jung, Hanku Lee, Segil Jeon, KumWon Cho, Hyunmyung Kim ... 293

Computational Methods for Financial Markets

Computer Construction of Quasi Optimal Portfolio for Stochastic Models with Jumps of Financial Markets
Aleksander Janicki ... 301

A New Computational Method of Input Selection for Stock Market Forecasting with Neural Networks
Wei Huang, Shouyang Wang, Lean Yu, Yukun Bao, Lin Wang 308

Short-Term Investment Risk Measurement Using VaR and CVaR
Virgilijus Sakalauskas, Dalia Kriksciuniene 316

Computational Asset Allocation Using One-Sided and Two-Sided Variability Measures
Simone Farinelli, Damiano Rossello, Luisa Tibiletti 324

Stock Trading System Based on Formalized Technical Analysis and Ranking Technique
Saulius Masteika, Rimvydas Simutis 332

Deriving the Dependence Structure of Portfolio Credit Derivatives Using Evolutionary Algorithms
Svenja Hager, Rainer Schöbel 340

Stochastic Volatility Models and Option Prices
Akvilina Valaitytė, Eimutis Valakevičius 348

Extraction of Interesting Financial Information from Heterogeneous XML-Based Data
Juryon Paik, Young Ik Eom, Ung Mo Kim 356

A Hybrid SOM-Altman Model for Bankruptcy Prediction
Egidijus Merkevicius, Gintautas Garšva, Stasys Girdzijauskas 364

Learning and Inference in Mixed-State Conditionally Heteroskedastic Factor Models Using Viterbi Approximation
Mohamed Saidane, Christian Lavergne 372

2006 International Workshop on P2P for High Performance Computational Sciences (P2P-HPCS06)

Constructing a P2P-Based High Performance Computing Platform
Hai Jin, Fei Luo, Xiaofei Liao, Qin Zhang, Hao Zhang 380

LDMA: Load Balancing Using Decentralized Decision Making Mobile Agents
M. Aramudhan, V. Rhymend Uthariaraj 388

A Hybrid Scheme for Object Allocation in a Distributed Object-Storage System
Fang Wang, Shunda Zhang, Dan Feng, Hong Jiang, Lingfang Zeng, Song Lv 396

Survive Under High Churn in Structured P2P Systems: Evaluation and Strategy
Zhiyu Liu, Ruifeng Yuan, Zhenhua Li, Hongxing Li, Guihai Chen 404

Analyzing Peer-to-Peer Traffic's Impact on Large Scale Networks
Mao Yang, Yafei Dai, Jing Tian 412

Analyzing the Dynamics and Resource Usage of P2P File Sharing by a Spatio-temporal Model
Riikka Susitaival, Samuli Aalto, Jorma Virtamo 420

Understanding the Session Durability in Peer-to-Peer Storage System
Jing Tian, Yafei Dai, Hao Wang, Mao Yang 428

Popularity-Based Content Replication in Peer-to-Peer Networks
Yohei Kawasaki, Noriko Matsumoto, Norihiko Yoshida 436

Computational Finance and Business Intelligence

A New Method for Crude Oil Price Forecasting Based on Support Vector Machines
Wen Xie, Lean Yu, Shanying Xu, Shouyang Wang 444

Credit Risk Evaluation Based on LINMAP
Tai-yong Mou, Zong-fang Zhou, Yong Shi 452

Logic Mining for Financial Data
G. Felici, M.A. Galante, L. Torosantucci 460

Mining Both Associated and Correlated Patterns
Zhongmei Zhou, Zhauhui Wu, Chunshan Wang, Yi Feng 468

A New Multi-criteria Convex Quadratic Programming Model for Credit Analysis
Gang Kou, Yi Peng, Yong Shi, Zhengxin Chen 476

Multiclass Credit Cardholders' Behaviors Classification Methods
Gang Kou, Yi Peng, Yong Shi, Zhengxin Chen 485

Hybridizing Exponential Smoothing and Neural Network for Financial Time Series Predication
Kin Keung Lai, Lean Yu, Shouyang Wang, Wei Huang 493

Assessment the Operational Risk for Chinese Commercial Banks
Lijun Gao, Jianping Li, Jianming Chen, Weixuan Xu 501

Pattern Recognition for MCNs Using Fuzzy Linear Programming
Jing He, Wuyi Yue, Yong Shi 509

Comparisons of the Different Frequencies of Input Data for Neural Networks in Foreign Exchange Rates Forecasting
Wei Huang, Lean Yu, Shouyang Wang, Yukun Bao, Lin Wang 517

Third International Workshop on Automatic Differentiation Tools and Applications

Automatic Differentiation of C++ Codes for Large-Scale Scientific Computing
Roscoe A. Bartlett, David M. Gay, Eric T. Phipps 525

A Sensitivity-Enhanced Simulation Approach for Community Climate System Model
Jong G. Kim, Elizabeth C. Hunke, William H. Lipscomb 533

Optimal Checkpointing for Time-Stepping Procedures in ADOL-C
Andreas Kowarz, Andrea Walther 541

On the Properties of Runge-Kutta Discrete Adjoints
Adrian Sandu 550

Source Transformation for MATLAB Automatic Differentiation
Rahul V. Kharche, Shaun A. Forth 558

The Data-Flow Equations of Checkpointing in Reverse Automatic Differentiation
Benjamin Dauvergne, Laurent Hascoët 566

Linearity Analysis for Automatic Differentiation
Michelle Mills Strout, Paul Hovland 574

Hybrid Static/Dynamic Activity Analysis
Barbara Kreaseck, Luis Ramos, Scott Easterday, Michelle Strout, Paul Hovland 582

Automatic Sparsity Detection Implemented as a Source-to-Source Transformation
Ralf Giering, Thomas Kaminski 591

2006 Workshop on Scientific Computing in Electronics Engineering

Lattice Properties of Two-Dimensional Charge-Stabilized Colloidal Crystals
Pavel Dyshlovenko, Yiming Li 599

Self-consistent 2D Compact Model for Nanoscale Double Gate MOSFETs
S. Kolberg, T.A. Fjeldly, B. Iñiguez 607

Neural Network Based MOS Transistor Geometry Decision for TSMC 0.18μ Process Technology
Mutlu Avci, Tulay Yildirim 615

Vlasov-Maxwell Simulations in Singular Geometries
Franck Assous, Patrick Ciarlet Jr. 623

Fast Rigorous Analysis of Rectangular Waveguides by Optimized 2D-TLM
Ayhan Akbal, Hasan H. Balik 631

A New Approach to Spectral Domain Method: Functional Programming
Hasan H. Balik, Bahadir Sevinc, Ayhan Akbal 638

Optimized Design of Interconnected Bus on Chip for Low Power
Donghai Li, Guangsheng Ma, Gang Feng 645

A Conservative Approach to SystemC Parallelization
B. Chopard, P. Combes, J. Zory 653

Modular Divider for Elliptic Curve Cryptographic Hardware Based on Programmable CA
Jun-Cheol Jeon, Kee-Won Kim, Jai-Boo Oh, Kee-Young Yoo 661

New Trends in the Numerical Solution of Structured Systems with Applications

A General Data Grid: Framework and Implementation
Wu Zhang, Jian Mei, Jiang Xie 669

Path Following by SVD
Luca Dieci, Maria Grazia Gasparo, Alessandra Papini 677

Comparing Leja and Krylov Approximations of Large Scale Matrix Exponentials
L. Bergamaschi, M. Caliari, A. Martínez, M. Vianello 685

Combined Method for Nonlinear Systems of Equations
Peng Jiang, Geng Yang, Chunming Rong 693

A General Family of Two Step Runge-Kutta-Nyström Methods for $y'' = f(x, y)$ Based on Algebraic Polynomials
Beatrice Paternoster 700

Schur Decomposition Methods for the Computation of Rational Matrix Functions
T. Politi, M. Popolizio 708

Piecewise Constant Perturbation Methods for the Multichannel Schrödinger Equation
Veerle Ledoux, Marnix Van Daele, Guido Vanden Berghe 716

State Dependent Symplecticity of Symmetric Methods
Felice Iavernaro, Brigida Pace 724

On the Solution of Skew-Symmetric Shifted Linear Systems
T. Politi, A. Pugliese 732

Workshop on Computational Science in Software Engineering (CSSE'06)

Search Based Software Engineering
Mark Harman 740

Modular Monadic Slicing in the Presence of Pointers
Zhongqiang Wu, Yingzhou Zhang, Baowen Xu 748

Modified Adaptive Resonance Theory Network for Mixed Data Based on Distance Hierarchy
Chung-Chian Hsu, Yan-Ping Huang, Chieh-Ming Hsiao 757

Checking for Deadlock, Double-Free and Other Abuses in the Linux Kernel Source Code
Peter T. Breuer, Simon Pickin 765

Generating Test Data for Specification-Based Tests Via Quasirandom Sequences
Hongmei Chi, Edward L. Jones, Deidre W. Evans, Martin Brown 773

Support Vector Machines for Regression and Applications to Software Quality Prediction
Xin Jin, Zhaodong Liu, Rongfang Bie, Guoxing Zhao, Jixin Ma 781

Segmentation of Software Engineering Datasets Using the M5 Algorithm
D. Rodríguez, J.J. Cuadrado, M.A. Sicilia, R. Ruiz 789

A Web User Interface of the Security Requirement Management Database Based on ISO/IEC 15408
Daisuke Horie, Shoichi Morimoto, Jingde Cheng 797

Domain Requirements Elicitation and Analysis - An Ontology-Based Approach
Yuqin Lee, Wenyun Zhao 805

Digital Human Modeling (DHM-06)

Integrative Computational Frameworks for Multiscale Digital Human Modeling and Simulation
Richard C. Ward, Line C. Pouchard, James J. Nutaro 814

Multi-scale Modeling of Trauma Injury
Celina Imielinska, Andrzej Przekwas, X.G. Tan 822

Investigation of the Biomechanic Function of Cruciate Ligaments Using Kinematics and Geometries from a Living Subject During Step Up/Down Motor Task
Luigi Bertozzi, Rita Stagni, Silvia Fantozzi, Angelo Cappello 831

Optimization Technique and FE Simulation for Lag Screw Placement in Anterior Column of the Acetabulum
Ruo-feng Tong, Sheng-hui Liao, Jin-xiang Dong 839

Model of Mechanical Interaction of Mesenchyme and Epithelium in Living Tissues
Jiří Kroc 847

Three-Dimensional Virtual Anatomic Fit Study for an Implantable Pediatric Ventricular Assist Device
Arielle Drummond, Timothy Bachman, James Antaki 855

Soft Computing Based Range Facial Recognition Using Eigenface
Yeung-Hak Lee, Chang-Wook Han, Tae-Sun Kim 862

A Privacy Algorithm for 3D Human Body Scans
Joseph Laws, Yang Cai 870

The Study of the Detection and Tracking of Moving Pedestrian Using Monocular-Vision
Hao-li Chang, Zhong-ke Shi, Qing-hua Fu 878

An Implementation of Real Time-Sentential KSSL Recognition System Based on the Post Wearable PC
Jung-Hyun Kim, Yong-Wan Roh, Kwang-Seok Hong 886

Patient Modeling Using Mind Mapping Representation as a Part of Nursing Care Plan
Hye-Young Ahn, Eunja Yeon, Eunmi Ham, Woojin Paik 894

Real Time Systems and Adaptive Applications (RTSAA 06)

A Technique for Code Generation of USN Applications Based on Nano-Qplus
Kwangyong Lee, Woojin Lee, Juil Kim, Kiwon Chong 902

A Study on the Indoor Real-Time Tracking System to Reduce the Interference Problem
Hyung Su Lee, Byunghun Song, Hee Yong Youn 910

A Task Generation Method for the Development of Embedded Software
Zhigang Gao, Zhaohui Wu, Hong Li 918

Active Shape Model-Based Object Tracking in Panoramic Video
Daehee Kim, Vivek Maik, Dongeun Lee, Jeongho Shin, Joonki Paik 922

Interworking of Self-organizing Hierarchical Ad Hoc Networks and the Internet
Hyukjoon Lee, Seung Hyong Rhee, Dipankar Raychaudhuri, Wade Trappe 930

A Dependable Communication Network for e-Textiles
Nenggan Zheng, Zhaohui Wu, Lei Chen, Yanmiao Zhou, Qijia Wang 938

EAR-RT: Energy Aware Routing with Real-Time Guarantee for Wireless Sensor Networks
Junyoung Heo, Sangho Yi, Geunyoung Park, Yookun Cho, Jiman Hong 946

A Design of Energy-Efficient Receivers for Cluster-Head Nodes in Wireless Sensor Networks
Hyungkeun Lee, Hwa-sung Kim 954

An Error Control Scheme for Multicast Video Streaming on the Last Hop Wireless LANs
Junghoon Lee, Mikyung Kang, Gyungleen Park, Hanil Kim, Choelmin Kim, Seongbaeg Kim 962

Design of a Fast Handoff Scheme for Real-Time Media Application on the IEEE 802.11 Wireless LAN
Mikyung Kang, Junghoon Lee, Jiman Hong, Jinhwan Kim 970

Accuracy Enhancement by Selective Use of Branch History in Embedded Processor
Jong Wook Kwak, Seong Tae Jhang, Chu Shik Jhon 979

A Novel Method of Adaptive Repetitive Control for Optical Disk Drivers
Kyungbae Chang, Gwitae Park 987

A Real Time Radio Link Monitoring Using CSI
Hyukjun Oh, Jiman Hong 991

Adaptive Encoding of Multimedia Streams on MPSoC
Julien Bernard, Jean-Louis Roch, Serge De Paoli, Miguel Santana 999

International Workshop on Grid Computing Security and Resource Management (GSRM'06)

A Mechanism to Make Authorization Decisions in Open Distributed Environments Without Complete Policy Information
Chiu-Man Yu, Kam-Wing Ng 1007

A Reputation-Based Grid Information Service
J.H. Abawajy, A.M. Goscinski 1015

Transparent Resource Management with Java RM API
Arkadiusz Janik, Krzysztof Zieliński 1023

Resource Discovery in Ad-Hoc Grids
Rafael Moreno-Vozmediano 1031

JIMS Extensions for Resource Monitoring and Management of Solaris 10
Krzysztof Zieliński, Marcin Jarząb, Damian Wieczorek, Kazimierz Balos 1039

An Agent Based Semi-informed Protocol for Resource Discovery in Grids
Agostino Forestiero, Carlo Mastroianni, Giandomenico Spezzano 1047

Fourth International Workshop on Autonomic Distributed Data and Storage Systems Management Workshop (ADSM 2006)

Replica Based Distributed Metadata Management in Grid Environment
Hai Jin, Muzhou Xiong, Song Wu, Deqing Zou 1055

Data Replication Techniques for Data-Intensive Applications
Jaechun No, Chang Won Park, Sung Soon Park 1063

Managing Data Using Neighbor Replication on Triangular-Grid Structure
Ali Mamat, M. Mat Deris, J.H. Abawajy, Suhaila Ismail 1071

Author Index 1079

A New Energy Efficient Target Detection Scheme for Pervasive Computing*

Thanh Hai Trinh and Hee Yong Youn**

School of Information and Communication Engineering,
Sungkyunkwan University, 440-746, Suwon, Korea
hai9381@skku.edu, youn@ece.skku.ac.kr

Abstract. Energy is one of the critical constraints for the applications of sensor network. In the earlier target detection and tracking algorithm power saving is achieved by letting most of the non-border sensor nodes in the cluster stay in hibernation state. However, the border sensor nodes consume a significant amount of energy since they are supposed to be on all the time for target detection. In this paper we propose a new target detection scheme which lets the border sensor nodes be on shortly one after another in a circular fashion to minimize the energy consumption. Computer simulation shows that the proposed scheme can significantly reduce the energy consumption in target detection and tracking compared to the earlier scheme.

Keywords: Cluster, energy saving, target detection, pervasive computing, wireless sensor network.

1 Introduction

With advances in computation, communication, and sensing capabilities, large scale sensor-based distributed environments are emerging as a predominant pervasive computing infrastructure. One of the most important areas where the advantages of sensor networks can be exploited is tracking mobile targets. The scenarios where the network may be deployed can be both military (tracking enemy vehicles and detecting illegal border crossings) and civilian [1]. In developing the sensor networks for target tracking, battery power conservation is a critical issue.

In the sensor network a large number of sensor nodes carry out a given task. In [2], a distributed, randomized clustering algorithm was given to organize the sensors in the wireless sensor network into a hierarchy of clusters with an object of minimizing the energy spent for the communication with the information center. The problem of tracking the targets using the sensor network has received attention from various angles. In [5], the authors consider the case where a set of k targets need to be tracked with three sensors per target from the viewpoint of resource requirement. They identified that the probability that all targets can be assigned three unique sensors shows

* This research was supported in part by the Ubiquitous Autonomic Computing and Network Project, 21st Century Frontier R&D Program in Korea and the Brain Korea 21 Project in 2006.

** Corresponding author.

V.N. Alexandrov et al. (Eds.): ICCS 2006, Part IV, LNCS 3994, pp. 1 – 8, 2006.
© Springer-Verlag Berlin Heidelberg 2006

phase transition properties as the level of communication between the sensors increases. In [7], the information driven sensor collaboration mechanism is proposed. In this mechanism the measures of information utility are used to decide future sensing actions. Collaborative signal processing aspects for target classification in sensor networks is addressed in [8]. The techniques for locating targets using a variety of mechanisms have been proposed in [9-11]. In [3], the authors present a prediction based approach, called localized predictive, for power efficient target tracking. In [4], tracking based on a distributed and scalable predictive algorithm called the distributed predictive tracking algorithm (DPT) is proposed. The DPT algorithm divides the sensor nodes in a cluster into border sensor nodes and non-border sensor nodes, and achieves power saving by letting most of the non-border sensor nodes stay in hibernation state. The border sensor nodes are awake all the time to detect the targets.

In this paper we propose a scheme detecting the targets in the sensor network, which can save the energy of border sensor nodes as well. One after another in the circular fashion, the border sensor nodes are activated to detect the target moving into the sensing area and then returns to hibernation mode. Consequently, the energy spent for detecting the target with the proposed scheme is much smaller than that of the DPT algorithm. If there exist several targets simultaneously moving into the sensing area, several messages will be sent to the cluster head. Therefore, we consider the relation between the processing time in the cluster head and velocity of the target movement to model the target detectability and tracking capability of a sensor network. Computer simulation shows that the proposed scheme can significantly reduce the energy consumption compared to the DPT algorithm for the targets of typical speeds.

The rest of the paper is organized as follows. In the following section we present a review of the DPT algorithm to identify the power problem. Section 3 presents the proposed detection scheme, and Section 4 discusses the simulation results. Finally, Section 5 presents the concluding remarks and future work.

2 The Related Work

The distributed predictive tracking (DPT) [4] algorithm employs the distributed and scalable prediction based approach to accurately track mobile targets using a sensor network. The fundamental guideline followed throughout the design of the DPT algorithm was to keep it as simple as possible. The DPT algorithm uses a cluster-based architecture for scalability and robustness. Given a target to track, the protocol provides a distributed mechanism for locally determining an optimal set of sensors for tracking. Only the nodes are then activated to minimize the energy consumption. Most of the sensors stay in hibernation mode until they receive an activation message from their cluster head. This is made possible by predicting the target's next location.

The DPT algorithm distinguishes the border sensor nodes from the non-border sensor nodes in terms of their operation. While the border sensors are required to be awake all the time in order to detect any target entering the sensing region, the non-border sensors hibernate unless they are asked to wake up by their cluster head. Hence, the energy of border sensor nodes will decrease quickly, while the task of border sensor nodes is critical for target detection. Therefore, an approach saving the energy of border sensor nodes as much as possible without reducing the target detectability is needed.

3 The Proposed Scheme

In this section we present the proposed scheme allowing energy efficient target detection and tracking with a sensor network. The scheme allows the border sensor nodes to minimize the energy consumption on target detection.

3.1 Assumptions

We first discuss the assumptions made by the proposed scheme. It assumes a cluster-based architecture for the sensor network [2,6]. Initially, all the border sensor nodes have information on their neighboring sensor nodes for their identity, location, and energy level. Each cluster head also has such information on the sensor nodes belonging to its cluster. When tracking a moving target, it decides which border sensor nodes detect the presence of a target. The assumptions on the sensor nodes are given below.

- Each border sensor is activated for ΔT time, and then returns to hibernation mode.
- The border sensor nodes are assumed to be uniformly distributed over the sensing region and the number of sensor nodcs is large.
- Let d be the distance between two neighboring border sensors. d is smaller or equal to the radius of the sensible area of a sensor, r. If d is larger than r, the intersecting area between two neighboring sensors will become small and the detection probability decreases accordingly. (Refer to Figure 1.)
- The targets originate outside the sensing area and then move into the area with constant velocity.

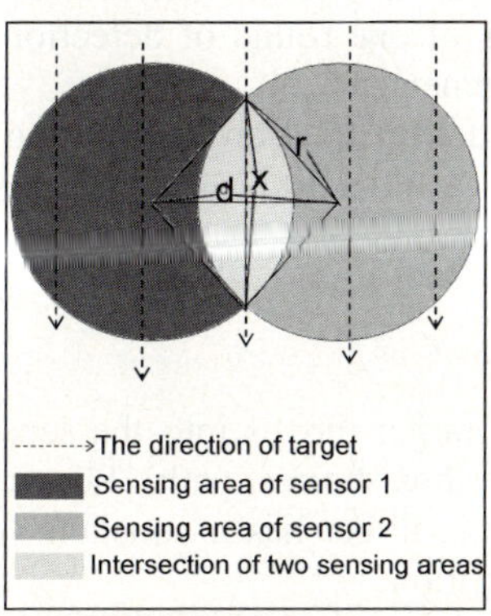

Fig. 1. Intersection of two sensing areas

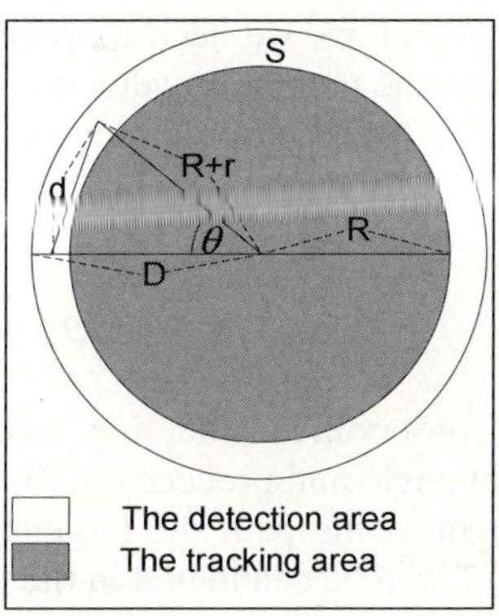

Fig. 2. The structure of a cluster

3.2 The Detection Algorithm

When targets move into the sensing area, the border sensors detect the presence of them and then inform the cluster head on it. The proposed mechanism is explained using the parameters listed in Table 1.

Table 1. The parameters used in the proposed mechanism (Refer to Figure 2)

Parameter	Description
V	The velocity of target
T	The period a border sensor is turned on and off
N	The number of border nodes
r	The radius of the sensing area of a sensor node
d	The distance between two neighboring border sensor nodes
R	The distance from the cluster head to the edge of detection area
D	The radius of the cluster, which is $R+2r$
θ	The angle at the cluster head made by two neighboring border sensors
S	The detection area covered by the border nodes
λ	The sensor density

The proposed scheme divides the detection period, T, into equal interval of ΔT given by: $\Delta T = T/N$. In other words, every border node turns on for ΔT time in every T time unit. In the DPT algorithm, all the border sensor nodes are activated to detect the targets. If there is only one target moving into the sensing area, a significant amount of energy will thus be wasted. Therefore, we propose to let only one border sensor is activated during ΔT and others are put in hibernation mode. Once a border sensor wakes up during ΔT, it then stays in hibernation mode during $(N-1)\Delta T$. This process repeats continuously with a period of T. By letting the border sensor nodes on and off fast enough, any target penetrating the border can be detected regardless of the entering direction. However, if a target moves faster than a certain threshold, it cannot be detected since the border sensor node in charge of it might still be in hibernation mode. Therefore, we need to decide the time T of one round of detection period according to the maximum velocity of target movement, V_{max}.

Assume that the border sensors are deployed with Poisson distribution with a parameter λS. In Figure 2, since the number of border sensors is very large, θ is very small and

$$\sin\theta \approx \theta = \frac{2\pi}{N} = \frac{d}{r+R} \tag{1}$$

For conservative modeling, assume that the target moves into the sensing area in the right angle and proceeds in the straight line (thus shortest path). Let us denote x_{min} the length of the path the target takes. Since the target has a constant velocity, the period, T, can be calculated in the following equation.

$$T = \frac{x_{\min}}{V_{\max}} \tag{2}$$

As shown in Figure 1, the minimum length x_{min} results in when the target moves through the intersection area of two neighboring border sensors. If the intersection area is small, the target detection probability will be low accordingly. To guarantee a reasonably high detection probability, we assume that the length of intersection, x_{min}, is equal to the radius of the sensing area of a sensor. Thus, $d \leq r\sqrt{3}$. According to Equation (1),

$$d = \frac{2\pi(r+R)}{N} \tag{3}$$

$$N = \lambda S = \lambda\pi\left(D^2 - R^2\right) = \lambda\pi\left((R+2r)^2 - R^2\right) = \lambda\pi\left(4Rr + 4r^2\right) = \lambda\pi 4r(R+r) \tag{4}$$

Putting Equation (4) into Equation (3),

$$\lambda \geq \frac{1}{2r^2\sqrt{3}} \quad \text{and} \quad \lambda_{\min} = \frac{1}{2r^2\sqrt{3}} \tag{5}$$

If all the border sensor nodes are activated during T, the consumed energy, E, is given by

$$E = N \times T \times E_{active} \tag{6}$$

Here E_{active} is the energy consumed by an active sensor and given by *727.5 mW* [12]. In the proposed scheme, the energy consumed, E_p, is given by

$$E_p = N \times \lfloor \Delta T \times E_{active} + (N-1) \times \Delta T \times E_{sleep} \rfloor \tag{7}$$

E_{sleep} is the energy consumed by a sensor in sleep mode and given by *416.3mW* [12]. The energy saved by the proposed scheme compared to the DPT algorithm becomes

$$\Delta E = E - E_p = (N-1) \times T \times \left(E_{active} - E_{sleep}\right) \tag{8}$$

Here N is given by Equation (4) and T is given by Equation (2) with $x_{min} = r$, and thus

$$\Delta E = [\lambda \times \pi \times 4 \times r \times (R+r) - 1] \times \frac{r}{V_{\max}} \times \left(E_{active} - E_{sleep}\right) \tag{9}$$

From Equations (5)

$$\Delta E_{\min} = \left[\frac{2 \times \pi \times (R+r) - r \times \sqrt{3}}{V_{\max} \times \sqrt{3}}\right] \times \left(E_{active} - E_{sleep}\right) \tag{10}$$

In Equation (10), the difference between E and E_p is the function of target velocity, V_{max}, radius of the sensing area of a sensor, r, and radius of tracking area, R. In Section 4, the relation between V_{max} and the detection probability, p, will be identified.

We consider only one target above. There might be several targets, however, moving into the sensing area with different velocities and directions. We assume that distribution of the number of targets is Poisson with the rate of η. The rate the targets are detected in time T is given by

$$\mu = p \times \eta \tag{11}$$

If several border sensors detect the presence of a target, several messages are sent to the cluster head simultaneously. The cluster head then has to put the messages into its queue for processing them sequentially. Table 2 summarizes the parameters used in the model for the process of multiple targets.

Table 2. The parameters used in the process of multiple targets

Parameter	Description
η	The rate of targets moving into the sensing area
μ	The rate of targets detected during T
γ	The target processing rate
M	The average number of messages in the system
t	The average target processing time
t_q	The waiting time in the queue for each message

Assume that arrival rate of the messages in the queue is constant and it is equal to the target detection rate, μ. Since only one cluster head processes the target information, this model represents a single server. Using a birth-death system of M/M/1 queue, the birth rate is μ and the death rate is the processing rate γ. We also assume that the death rate is constant and target population is infinite for the arrival process. To calculate the probability that the queue is in state-k (i.e., has k messages waiting including the one in service), we use the general birth-death formulas.

$$p_k = \left(1 - \frac{\mu}{\gamma}\right) \times \left(\frac{\mu}{\gamma}\right)^k \qquad (\mu < \gamma) \qquad (12)$$

The average number of messages in the system, M, is

$$M = \frac{\delta}{1-\delta}, \ (\delta = \frac{\mu}{\gamma}), \ \ M = \mu \times t, \ t = \frac{\frac{1}{\gamma}}{1-\delta}, \ t_q = \frac{\frac{\mu}{\gamma^2}}{1-\delta} \qquad (13)$$

4 Performance Evaluation

In this section we present the simulation results evaluating the performance of the proposed scheme. The simulation study mainly focuses on the energy saved by the proposed scheme compared to the DPT algorithm, the detection probability, and the message processing time (delay time) as the number of targets varies.

We simulated a scenario where a target moves with random velocity. The simulator was developed to detect the target, considering the detection probability and energy consumed. The simulation program is a discrete-event simulator developed using C language. In the simulation we distribute the sensors uniformly over a cluster with D = 120*m* and r = 10*m*. The velocity of target movement varies randomly from 0*m/s* to 60*m/s* in order to evaluate its impact on the performance of the proposed scheme.

Figure 3 shows the energy saved by the proposed scheme compared to the DPT scheme obtained using Equation (10). The amount of energy saved decreases as the velocity of target increases. This is because the border sensor nodes need to fastly on and off if the target moves fastly, and thus energy consumption increases compared to the case of slow target. Note, however, that the velocity here is up to 100 meter per second. For typical targets of below the 20 or 30*m/s* speed, the energy saving is very significant.

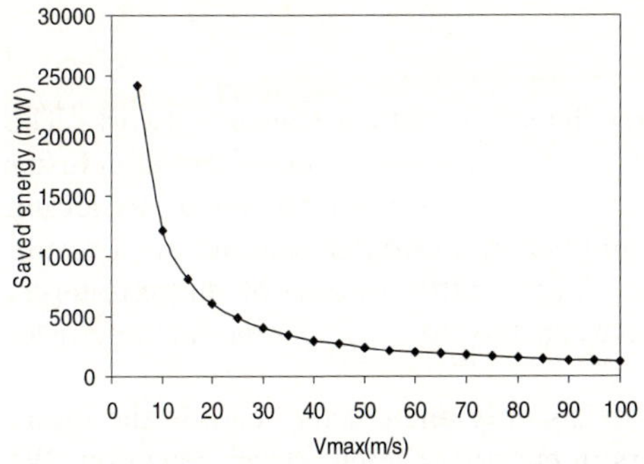

Fig. 3. The amount of energy saved by the proposed scheme as the target speed varies

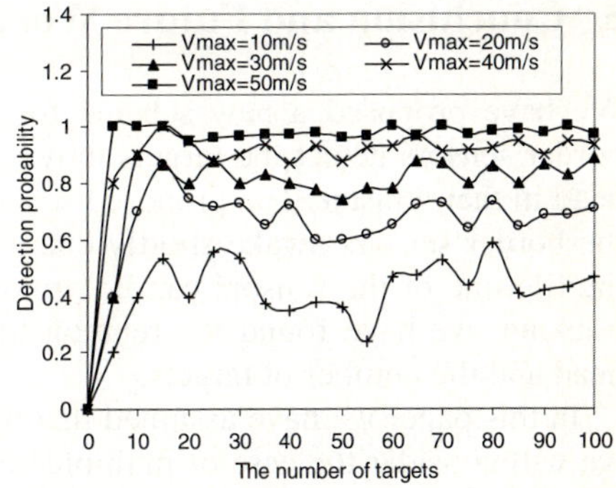

Fig. 4. Detection probability vs. number of targets

Figure 4 presents the relationship between the detection probability and the number of targets. As the threshold velocity of target increases from 10*m/s* to 50*m/s*, the detection probability increases. When the velocity of the target is smaller than the threshold velocity, the number of missed targets decreases since the rotation time *T* is small and the period that the border sensors are activated is short. However, the amount of energy saved becomes small for high threshold velocity. Hence, we need to choose a proper period so that the detection probability is high while energy saving at the border sensor nodes is substantial.

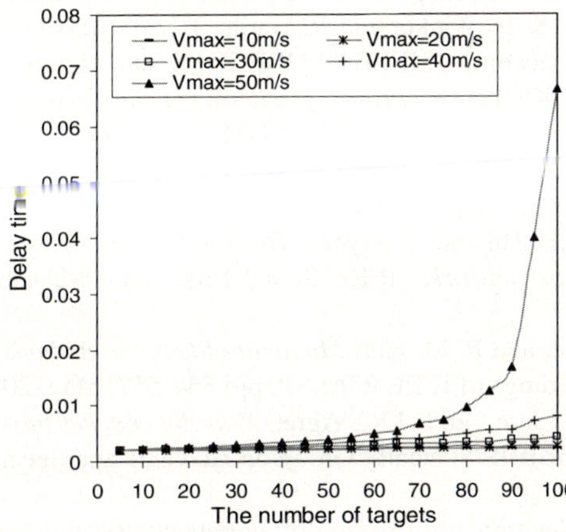

Fig. 5. The relationship between the delay time and the number of targets

In Figure 5, the relationship between the delay (processing time) and velocity of the target is shown. As the velocity changes from 10*m/s* to 40*m/s*, the delay increases slowly. For *V*=50*m/s*, the delay increases rapidly when the number of targets becomes larger than 90. The reason for this is the ratio between the death rate and birth rate, $\delta=\mu/\gamma$, becomes close to 1.

5 Conclusion and Future Work

We have proposed a new scheme for target detection with a sensor network. The border sensors detect the targets moving into the sensing area and inform the cluster head in their cluster. The proposed scheme allows significant energy saving by letting the border sensors awake shortly one after another in a circular fashion. As a result, the lifetime of the sensors can be prolonged. Additionally, in case of multiple-target tracking, we have found the relationship between the processing time at the cluster head and the number of targets.

In this paper we have assumed that there exists only one cluster head. In the future we will consider the case of multiple clusters in energy efficient target detection. We will also investigate the impact of other parameters on the target detectability.

References

[1] J. Nemeroff, L. Garcia, D. Hampel, and S. DiPierro. *Application of sensor network communications.* In military Communications Conference, 2001.

[2] S. Bandyopadhyay and E.J. Coyle. *An Energy Efficient Hierarchical Clustering Algorithm for Wireless Sensor Networks.* INFOCOM 2003. Twenty-Second Annual Joint Conference of the IEEE Computer and Communications Societies. IEEE.

[3] Y. Xu and W. C. Lee. *On localized prediction for power efficient object tracking in sensor networks.* In Proceeding 1. st. International Workshop on. Mobile Distributed Computing (MDC), Providence, Rhode Island, May 2003.

[4] H. Yang and B. Sikdar. *A protocol for tracking mobile targets using sensor networks.* Proceedings of IEEE Workshop on Sensor Network Protocols and Applications, 2003.

[5] B. Krishnamachari, S. B. Wicker and R. Berjar. *Phase transition phenomena in wireless ad-hoc network*, proceeding of IEEE GLOBECOM, San Atonio, TX, November 2001.

[6] S. Banerjee and S. Khuller. *A clustering scheme for hierarchy control in multi-hop wireless networks.* Proceedings of IEEE INFOCOM, April 2001.

[7] F. Zhao, J. Shin and J. Reich. *Information-driven dynamic sensor coloration for target tracking*, IEEE signal processing maganize, vol. 19, no. 2, pp 61-77. March 2002.

[8] D. Li, K. Wong, Y.H. Hu and A. Sayeed. *Detection, classification and tracking of targets in distributed sensor networks.* IEEE Signal Processing Maganize, vol 9, no.2, March 2002.

[9] L.M. Kaplan, Q. Le, and P. Molnar. *Maximum likehood methods for bearing-only target localization.* Proceedings of IEEE ICASSP, pp. 554-557, May, 2001.

[10] J.S. Scholl, L.P. Clare, and J.R. Agre. *Wavelet packet-based target classification schemes.* Meeting of IRIS Specialty Group on Acoustic and Seismic Sensing, laurel, MD, September, 1998.

[11] K. Yao, et. al., *Estimation and tracking of an acoustic/seismic source using a beamforming array based on residual minimizing methods.* Proceedings of IRIA-IRIS, pp. 153-163, January 2001.

[12] V. Raghnathan, et. al., *Energy aware wireless microsensor networks.* IEEE Signal Processing Magazine, March 2002.

A Load Balance Based On-Demand Routing Protocol for Mobile Ad-Hoc Networks

Liqiang Zhao[1], Xin Wang[2], Azman Osman Lim[3], and Xiangyang Xue[2]

[1] School of Software, Fudan University, Shanghai, China
[2] Dept. of Computer Science and Engineering, Fudan University, Shanghai, China
{zhaolq, xinw, xyxue}@fudan.edu.cn
[3] Dept. of Computer Science, Kyoto University, Kyoto, Japan
alvin1973@hotmail.co.jp

Abstract. Since the existing ad hoc routing protocols lack of load balance capabilities, they often fail to provide satisfactory performance in the presence of a large volume of traffic. In this paper, we propose a new load balance mechanism and a novel bandwidth estimation method for ad hoc on-demand routing protocols. Destination chooses the optimal path via the route information carried by RREQ, whilst congested intermediate nodes dropped RREQs to avoid RREQ storm. Simulation shows that our new scheme improves packet delivery ratio, reduces end-to-end latency and decreases routing overhead.

1 Introduction

A Mobile Ad hoc Network (MANET) is a self-configuring network of mobile hosts connected by wireless links. Due to its unique character of self-organization, quick deployment and infrastructure-free, MANET has a wide range of applications which include battlefield command and control, emergency disaster relief, mine site operations.

Currently, on-demand routing protocol is the dominant routing protocol in MANET. As the typical on-demand routing protocols, AODV [7] and DSR [2] select the shortest routes as the optimum routes. However, due to the special characters of ad hoc network, many researches realize that the shortest route may not be the best criteria of route selection [9,14].

Meanwhile, researchers have been focusing on network load balance. Due to the constrained ad hoc network bandwidth resource, routing protocols are required to properly distribute the traffic flow over the network. Otherwise, the overloaded nodes may cause network congestion and long delay. In considering the significance of network traffic balance, we propose a load balance mechanism, which can virtually be applied to any on-demand routing protocol. In this paper, we combine this mechanism with AODV, forming a new routing protocol called load balance based AODV (LBB-AODV).

The rest of this paper is organized as follows: Section 2 introduces some related work about load balance routing in ad hoc network. In section 3, we propose a novel

V.N. Alexandrov et al. (Eds.): ICCS 2006, Part IV, LNCS 3994, pp. 9 – 16, 2006.
© Springer-Verlag Berlin Heidelberg 2006

residual bandwidth estimation model and a new load balance mechanism, both of which are incorporated into the routing protocol LBB-AODV. Section 4 is the simulation and analysis. Finally, section 5 concludes this paper.

2 Related Work

Multiple paths schemes, such as [10] and [13], are fully exploited to provide better load-balancing capabilities. The underlying idea of the scheme is to distribute the traffic among multiple paths, which are maintained at nodes and used for routing. However, maintaining alternative paths requires more routing table space and computational overhead while selecting the best route among several discovered paths [11]. Moreover, multiple routes schemes are effective only if the alternate multi-path are disjoint, which is difficult to achieve in MANET [6]. Ganjali et al. [8] further demonstrates that the load distribution of multi-path routing is almost the same as that of single path routing.

DLAR [3], a single-path load balance mechanism, is proposed. The RREQ records queue occupancy information of each node it traverses. The destination chooses the optimal path based on the load information kept in RREQs. However, to utilize the most up-to-date load information, DLAR prohibits intermediate nodes from replying to RREQ, which may result in RREQ storm when there are a large number of nodes in the mobile ad hoc network.

In [4], a new scheme is presented that each node forward RREQs selectively according to the load status of the node, preventing the new routes set up through overloaded nodes. This may alleviate the possibility of RREQ storm, but since it works in a fully distributed manner, the route selected may not be the best one due to the lack of comparison of candidate routes.

3 Load Balance Based-AODV

LBB-AODV is an enhanced version of AODV based on load balance, which enable nodes to forward RREQ selectively during route discovery stage. When a node receives route request, it first checks its available bandwidth. The node forwards the RREQ only if it is not CONGESTED. After the destination receives all the route information via RREQ, it picks the best route. The detail description of bandwidth estimation and LBB-AODV are presented as follows.

3.1 Bandwidth Estimation

To guarantee the network load balance, it is necessary to know each node's available bandwidth. The end-to-end throughput is a concave parameter [5], which is determined by the bottleneck bandwidth of the intermediate nodes along the route. Therefore, estimating available bandwidth of a specific route can be simplified into finding the minimal residual bandwidth available among the nodes in the route.

We propose a simple and effective available bandwidth estimation model in ad hoc network. In the new scheme, each the node is set to promiscuous mode, so it can perceive any frame sent by its neighboring nodes. Every node accumulates the size of the perceived frames within a predefined period of time, and adds it to the size of frames the node sends itself during the time, the throughput of the small area centering the node within the time window is gained. By examining the activities of both the node itself and its surrounding neighbors, we are able to obtain a good approximation of the bandwidth usage. The available bandwidth of a node is:

$$B_{avail} = B_{raw} - B_{tran} - B_{recv} \tag{1}$$

B_{avail} is the available bandwidth of the node, while B_{raw} is ad hoc network pure channel bandwidth. B_{tran} is the consumed bandwidth of the node by sending frames, including data and routing control frames and 802.11MAC layer RTS/CTS/ACK control frames. B_{recv} is the bandwidth took up by neighboring nodes.

B_{tran} and B_{recv} can be represented as follows:

$$B_{tran} = TH_{tran} / t \tag{2}$$

$$B_{recv} = TH_{recv} / t \tag{3}$$

Here, t is the predefined period of time , when TH_{tran} and TH_{recv} are the size of frames transmitted and received, respectively, within t by the node.

Taking into consideration the channel fading, physical error, the frame collision , as well as the burst nature of control packets overhead, factor$\theta \in (0,1)$ is used to adjust $B_{avail.}$ Combine function (1), (2) and (3), we get:

$$B_{avail} = \left(B_{raw} - (TH_{tran} + TH_{recv}) / t\right) \cdot \theta \tag{4}$$

At the beginning of the simulation, TH_{tran} and TH_{recv} of every node are set to 0 while the timeout is set to t. When it timeouts, the available bandwidth can be calculated by (4). Store it in the cache. Then reset TH_{tran} and TH_{recv} to 0, calculate the bandwidth of next period.

The calculation of the period time t is critical; it should not be too big or too small. If it is set too small, there might be no node to send packet during that short period. In this case, the function only reflects transit bandwidth utilization, and it might not be the stable bandwidth consumption condition. Moreover, short period means frequent calculation, which will consume a batch of CPU time and battery power. If t is set too big, because of the movement of the node and change of link state, the function might not be able to reflect the real bandwidth usage. Moreover, t should be proportional to density of the whole network, since the higher the node's density is, the higher chance of packet collision, and the larger t should be [12].

To smooth bandwidth estimation, we define a smoothing constant $\beta \in (0,1)$. Suppose the last bandwidth is $B_{avail(n-1)}$ and the bandwidth measured in the current sampling time window is B_{avail}. Then, the current bandwidth $B_{avail(n)}$ is given as:

$$B_{avail(n)} = \beta \cdot B_{avail(n-1)} + (1-\beta) \cdot B_{avail} \tag{5}$$

3.2 LBB-AODV

For our purpose, a new field B_{min} is added in RREQ format to record the minimum bandwidth along the path it traverses. When a source node decides to launch a RREQ, it first estimates its bandwidth using mechanism described above, and records it in B_{min}.

When an intermediate node first receives a RREQ, it will collect its load information in order to define present network condition. According to the network condition, routing layer determines whether to forward or just drop the RREQ. The network has two conditions: CONGESTION and NORMAL, which is defined by ratio of the current available bandwidth (B_{avail}) and the total channel bandwidth (B_{raw}) withφ being the congestion factor:

$$\text{CONGESTION: } B_{avail} / B_{raw} \leq \phi \tag{6}$$

$$\text{NORMAL: } B_{avail} / B_{raw} > \phi \tag{7}$$

When an intermediate node is in the CONGESTION mode, the node will no longer process any route request; it simply discards RREQ, making itself impossible to be an intermediate node for other traffic. Each node begins to allow additional traffic flows whenever its congested status is dissolved.

A new field B_t is added to the route table of each node to record the so-far-discovered-biggest route bandwidth from the source to the node. When the node is in NORMAL condition, if it is the first time receiving RREQ from the source node with the broadcast id, B_t will be set as B_{min} for the corresponding entry in the route table. Then compare B_{min} and the node's available bandwidth, keep the smaller one in the B_{min} and forward RREQ. If the node has already received RREQ from the same source node with the same broadcast id, it compares B_{min} with the corresponding B_t. If the former is not bigger, simply discard the RREQ. Otherwise, update the route table entry to replace B_t with B_{min} and redirect the previous node index to the node where the RREQ is from. Finally, discard the RREQ rather than forwarding it to avoid RREQ storm.

After receiving the first RREQ, the destination waits for an appropriate amount of time to learn all possible routes. After receiving duplicate RREQ from different previous nodes, the destination chooses the route with the largest B_{min} and sends RREP.

By rejecting route request at the individual congested nodes, the network is able to reduce the large volume of routing requests and alleviate the burden brought by the broadcasting. On the other hand, choosing the best route at the destination node provide the best route along light-loaded nodes.

When the local congestion occurs, since the congested nodes refuse to forward routing request, it is possible to fail the route discovery. To cope with this scenario, after the first round route discovery failure, the original AODV will be effective to find the route from the second round.

4 Numerical Simulation

In this section, we evaluate the performance of LBB-AODV compared to the conventional AODV. The simulation environment, metrics and results are presented.

4.1 Simulation Environment

We evaluate LBB-AODV and AODV by using the network simulator ns-2. In ns-2 simulator, we use the default value of AODV to demonstrate a fair comparison of LBB-AODV. The distribution coordination function (DCF) of IEEE 802.11 standard is used as the MAC protocol with a data rate of 2Mbps.

A network consists of 100 nodes with 250 m transmission range spread randomly over an area of 2400 m × 800 m. The nodes uniformly choose velocity from 0 m/s to 5 m/s following the random-waypoint model. Each simulation runs for 200 s with pause time set to 50 s.

Source and destination node pairs are randomly selected and each source generates CBR packets with packet size of 512 bytes. We use 40 traffic flows and gradually increase the data rate of each flow from 3 to 9 packets/second (or 480 to 1440 Kb/s). The traffic load represents the combined throughput of the data sources. By adjusting the frequency of sources sending packets, we eventually control the traffic load of the network.

After several simulation comparisons, we choose the following parameter values used for LBB-AODV, which are specified in Table 1.

Table 1. Parameters Used for LBB-AODV

Parameter	Value
t	1s
θ	0.9
β	0.2
φ	0.1

4.2 Simulation Metrics

The following three metrics are considered to evaluate the performance of the protocols:

1. Packet delivery ratio – the ratio of the total number of data packets received by destinations over the total number of data packets transmitted by sources. It captures the network throughput.
2. End-to-end delay –the average of delays for all received data packet from the sources to destinations. It includes all possible delays caused by buffering during route discovery, queuing at the interface queue, retransmission delays at the medium access control layer, and propagation and transfer time.
3. Normalized routing overhead – the total number of routing control packets (RREQ, RREP and RERR) normalized by the total number of received data packets. It evaluates the efficiency of the routing protocol.

4.3 Simulation Results

Fig.1 plots the packet delivery ratio versus traffic load for both AODV and LBB-AODV protocols. Numerical results reveal that LBB-AODV achieves a larger packet delivery ratio compared to AODV, especially when the traffic load increases. This is because LBB-AODV chooses the least congested route for data transmission, which indirectly increases the overall network throughput. On the other hand, AODV without load-balancing mechanism suffers the high probability of the route failure and the data retransmission as the network traffic load becomes larger.

As can be seen in Fig.2, LBB-ADOV outperforms AODV in terms of average end-to-end delay. The largest fall of average end-to-end delay is up to 17.3%. Delay is mainly due to queuing in heavily loaded buffers. Even though the route path may be longer than that of AODV, LBB-ADOV tries to route packets along a less congested path to avoid overloading nodes, thereby minimizing path congestion and in turn reducing the overall queuing delay.

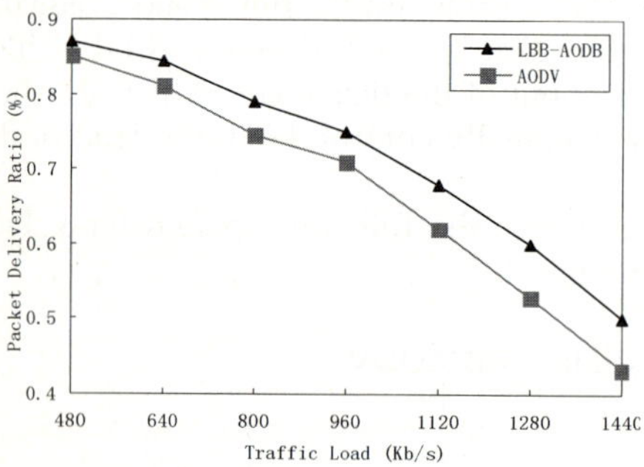

Fig. 1. Packet delivery ratio

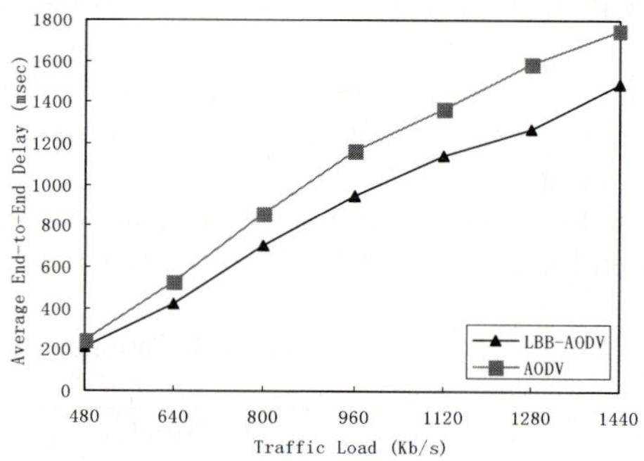

Fig. 2. Average end-to-end delay

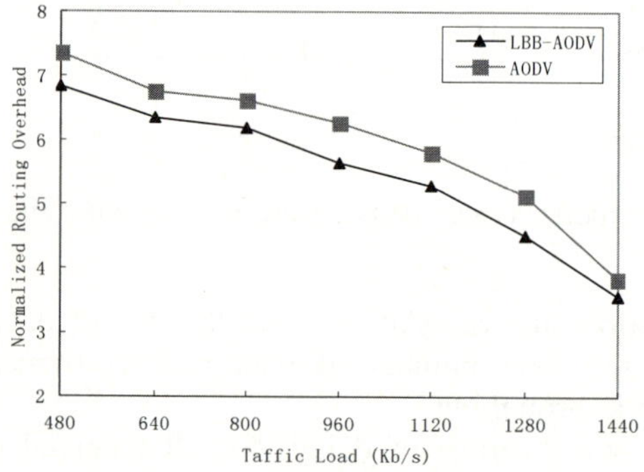

Fig. 3. Normalized routing overhead

In Fig.3, LBB-AODV demonstrates up to 16.2% reduction of the routing overhead against AODV. This is mainly due to the suppression of forwarding RREQs at the intermediate nodes, as the forwarding RREQs make up about 90% of the routing packets in AODV [8]. Because LBB-AODV prevents those over-loaded nodes from forwarding RREQs, the overall routing overhead is dramatically decreased. Moreover, LBB-AODV can establish routes more stable than AODV. As a result, LBB-AODV reduces link breakage and routing overhead, which is mainly caused by the route reconstruction. When the traffic load increases, the overall normalized routing overhead of both protocols decreases because of the source can send more data packets to its corresponding destination by using the discovered route.

5 Conclusions

In this paper, we proposed the load balance mechanism for the conventional AODV by considering the available bandwidth at each node. According to the local bandwidth status, each node has the right to re-broadcast the RREQ message in order to avoid network congestion and RREQ broadcast storm. Beside that, we also proposed that the RREQ message should contain an additional field to allow nodes to exchange the route bandwidth information among themselves. Moreover, we proposed a simple and effective bandwidth estimation approach based on the perceived bandwidth consumption of a node and its neighboring nodes.

Numerical simulations reveal that LBB-AODV significantly improves the packet delivery ratio and reduces the average end-to-end latency as well as the routing overhead in the presence of large traffic volume. The new scheme successfully balances the network load among nodes, and it can easily be incorporated in the existing on-demand routing protocol.

To facilitate practical implementation of our proposal, further research is required to investigate the optimal parameter under different network settings for LBB-AODV. Furthermore, we will look into how the proposed load balance mechanism can be incorporated in other on-demand routing protocols and make a fair comparison with other load balance schemes, like [3] and [4], in order to further improve the performance of LBB-AODV.

Acknowledgement

This work was supported in part by Shanghai Municipal R&D Foundation under contracts 035107008, 04DZ15021-5A, 055115009, Shanghai Key Laboratory of Intelligent Information Processing (IIPL) and LG Electronics.

References

1. Y. Ganjali, A. Keshavarzian. Load balancing in ad hoc networks: single-path routing vs. multi-path routing. INFOCOM 2004. Twenty-third Annual Joint Conference of the IEEE Computer and Communications Societies Volume 2, Page(s):1120 - 1125 vol.2., March 2004.
2. D. Johnson and D. Maltz. Dynamic source routing in ad hoc wireless networks. Mobile Comput., pp. 153–181, 1996.
3. S. J. Lee, M. Gerla. Dynamic load-aware routing in ad hoc networks. Communications, 2001. ICC 2001. IEEE International Conference on Volume 10, Page(s):3206 - 3210 vol.10., June 2001.

4. Y. J. Lee, G. F. Riley. A workload-based adaptive load-balancing technique for mobile ad hoc networks. Wireless Communications and Networking Conference, 2005 IEEE Volume 4, Page(s):2002 - 2007 Vol. 4., March 2005.
5. P. Mohapatra, J. Li, and C. Gui. QoS in mobile ad hoc networks, IEEE Wireless Commun. Mag. (Special Issue on QoS in Next-Generation Wireless Multimedia Communications Systems), pp. 44–52, 2003.
6. A. Nasipuri, R. Castaneda, and S. Das. Performance of multipath routing for on-demand protocols in ad hoc networks. ACM/Kluwer Mobile Networks and Applications (MONET) Journal, Vol. 6, No. 4, pp. 339-349, August 2001.
7. C. Perkins and E. Royer. Ad-Hoc On-Demand Distance Vector Routing (AODV). July 2003. RFC 3561.
8. C. Perkins, E. Royer, S. Das, M. Marina. Performance comparison of two on-demand routing protocols for ad hoc networks. Personal Communications, IEEE. Volume 8, Issue 1, Page(s):16 – 28., Feb. 2001.
9. D. D. Perkins, H. D. Hughes, C. B. Owen. Factors affecting the performance of ad hoc networks. Communications, 2002. ICC 2002. IEEE International Conference on Volume 4, Page(s):2048 - 2052 vol.4, May 2002.
10. P. Pham and S. Perreau, Multi-path routing protocol with load balancing policy in mobile ad hoc network. IFIP Int'l Conference on Mobile and Wireless Communications Networks (MWCN 2002), September 2002.
11. P. Pham and S. Perreau. Performance analysis of reactive shortest path and multi-path routing mechanism with load balance. IEEE Conference on Computer Communications (INFOCOM 2003), March 2003.
12. Liang Zhang, Yantai Shu, Yan Liuand Guanghong Wang, Adaptive Tuning of Distributed Coordination Function (DCF) in the IEEE 802.11 to Achieve Efficient Channel utilization. Future Telecommunication Conference 2003. Beijing, china, Dec. 2003.
13. Lingfang Zhang, Zenghua Zhao, Yantai Shu, Lei Wang, Yang, O.W.W. Load balancing of multipath source routing in ad hoc networks. Communications, 2002. ICC 2002. IEEE International Conference on Volume 5, Page(s):3197 - 3201 vol.5., May 2002.
14. Xiaofeng Zhong, Youzheng Wang, Shunliang Mi, Jing Wang. An Experimental Performance Study of Wireless ad hoc System Utilizing 802.11a Standard Base on different Routing Protocols. Asia Pacific Optical and Wireless Communications, October 2002.

Handover Control Function Based Handover for Mobile IPv6

Guozhi Wei[1], Anne Wei[1], Ke Xu[2], and Hui Deng[3]

[1] Université de Paris XII 122 Rue Paul,
Armangot 94400 Vitry-sur-Seine, France
{Guozhi.Wei, Wei}@univ-paris12.fr
[2] Department of Computer Science and Technology,
Tsinghua University, China
Xuke@csnet1.cs.tsinghua.edu.cn
[3] Hitachi (China) Investment, Ltd, Beijing, China
Hdeng@hitachi.cn

Abstract. IEEE 802.11 Wireless LAN (WLAN) has been enthusiastically adopted in business offices, homes, and other spaces, for both public and private wireless local network connection. The users would like to deploy Voice IP (VoIP) and Video Phone based on Mobile IPv6 Protocol over Wireless LAN network. However, the frequent change in the mobile node's location causes evident signaling overhead, handover latency and packet loss, which in turn leads to the service degradation of real time traffic in Mobile IPv6. In this paper, we propose a scheme based on Wireless LAN by adding a new component called Handover Control Function (HCF) in MIPv6, which records all APs's MAC address, backend ARs' address and network prefix of those AP's. By the means that all Mobile Nodes (MNs) report periodically all AP's MAC address and signal strength information to HCF which MN can probe, HCF decides whether or which AP MN shall associate with and notifies MN about the new AP/AR's information, meanwhile, a bi-cast mechanism shall be applied to further improve handover performance by reducing the packet loss during handover.

1 Introduction

Wireless LAN (WLAN) technologies, especially the IEEE 802.11 standards [1], have got great attention in recent years. A growing number of WLANs have been set up in public buildings or corporate environments as access networks to the Internet. These WLANs are connected to the Internet through layer 2 Access Points (APs) and layer 3 Access Routers (ARs). In WLAN, users could freely change their places when they are communicating with other users. However the real-time applications (such as VoIP and Video Phone) couldn't be supported due to long handover delays and high packet losses brought by the handover process. The handover process occurs when MNs moves from one AP/AR to another.

In order to support the mobility of MNs, Mobile IPv6 (MIPv6) [2] is proposed by the Internet Engineering Task Force (IETF), which describes the protocol operations

V.N. Alexandrov et al. (Eds.): ICCS 2006, Part IV, LNCS 3994, pp. 17 – 24, 2006.
© Springer-Verlag Berlin Heidelberg 2006

for MN to maintain connectivity with the Internet as it moves between subnets. These operations involve movement detection, IP address configuration, and location update. Many approaches are submitted to improve the performance of handover and to meet the requirements of real-time applications.

In Consideration of Wireless LAN's special feature, we propound a scheme to achieve MIPv6 fast handover in WLAN by introducing a new component called Handover Control Function (HCF) in Hierarchical Mobile IPv6 (HMIPv6) [3].

The remainder of the paper is organized as follows. Section 2 gives a brief resume to previous works related to handover management in Mobile IPv6. Section 3 presents our HCF Based Handover for MIPv6 scheme and the detailed protocol operation. Analysis and comparison are shown in section 4. Finally, conclusion and future works are mentioned in section 5.

2 Background and Related Works

Recently, several extensions to MIPv6 have been proposed aimed to reduce the handover latency and packet loss.

Actually, the main proposals accepted by IETF are Hierarchical Mobile IPv6 (HMIPv6) and Fast Handover for MIPv6 (FHMIPv6). HMIPv6 [3, 4] introduces Mobility Anchor Point (MAP) (a special node located in the network visited by MN) who acts somewhat like a local Home Agent (HA) for the visiting MN. Moreover, HMIPv6 separates MN mobility into micro-mobility (within one domain or within the same MAP) and macro-mobility (between domains or between MAPs). With this hierarchical network structure, MAP can limit the amount of signaling required outside the MAP's domain. Therefore, the amount and latency of signaling between a MN, its HA and one or more Correspondence Nodes (CNs) decrease. Consequently, the layer 3 handover delays are reduced.

FHMIPv6 [5] reduces packets loss by providing fast IP connectivity as soon as a new link is established. The network uses layer 2 triggers to launch either Pre-Registration or Post-Registration handover scheme [6]. In Pre-Registration scheme, the network provides support for preconfiguration of link information (such as the subnet prefix) in the new subnet while MN is still attached to the old subnet. By reducing the preconfiguration time on the new subnet, it enables IP connectivity to be restored at the new point of attachment sooner than would otherwise be possible. In Post-Registration scheme, by tunneling data between the previous AP/AR and new AP/AR, the packets delivered to the old Care of Address (CoA) are forwarded to the new CoA during link configuration and binding update. So it is possible to provide IP connectivity in advance of actual Mobile IP registration with the HA or CN.

Besides the main proposals, there have been numerous approaches for providing lossless handover and minimizing the handover delay. H.Chaouchi [7] propounded a Pre-Handover Signaling (PHS) protocol to support the triggering of a predictive handover and to allow the network to achieve accurate handover decisions considering different constraints such as QoS, the user profile and the mobile node service requirements. Y.Bi [8] submitted a Hierarchical Network-layer Mobility Management (HNMM) framework in which an integrated IP-layer handover solution is proposed to provide optimized network connectivity. The solution includes Enhanced IP-layer Handover mechanism (EIHO) with combined policies and Adaptive IP-layer Handover Control scheme (AIHC). Based on the IPv6 Soft Handover mechanism proposed by E.Perkins [9], the Competition based Soft Handover Management (CSHM) protocol of J. Kristiansson [10], the Multi-path Transmission Algorithm of S. Kashihara [11] are also proposed to decrease packet loss during handover.

3 Handover Control Function Based Handover for Mobile IPv6

In our paper, we focus on optimizing handover performance in MIPv6 over WLAN. When being used as link layer, WLAN doesn't permit users to employ FHMIPv6. Indeed, in WLAN, when MN moves towards a new AP, it must disconnect from its previous AP and scan for others APs in range. Based on the result of these scans, it will choose its new AP. Hence, there is no possibility for MN to use neither Pre-registration nor Post-registration mechanism as MN doesn't know in advance where to go. Furthermore, once a new AP is selected, MN is no longer connected with its previous AP.

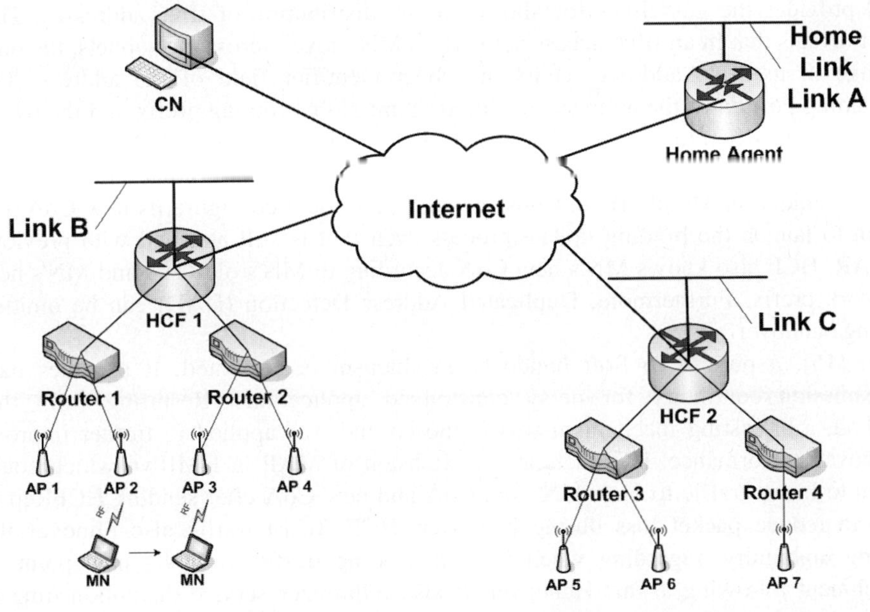

Fig. 1. HCF Based Handover for Mobile IPv6

To resolve this problem, IEEE 802.11f standard, known as Inter-Access Point Protocol (IAPP) [12] [13] has been proposed. IAPP enables the WLAN's APs to communicate with each other. By using IAPP, FHIMPv6 could be applied in WLAN. However, IAPP can neither reduce greatly the layer 2 delay nor avoid the packet losses. Furthermore, it couldn't provide the global management, such as dynamic load balancing on APs and ARs.

Therefore, we introduce a new component Handover Control Function (HCF) and add two new messages --- HCFReq and HCFRep messages in Mobile IPv6 in order to resolve issues mentioned above. The architecture of the HCF Based Handover for Mobile IPv6 is shown in figure 1.

In this network scenario, one AR might connect to multiple APs. Those APs might have same network prefix or different network prefix. While MN moves from AP2 to AP3, the attached Access Router also changes from AR1 to AR2 as well.

MN reports periodically to HCF all APs' MAC addresses and signal strengths that MN can probe. Based upon those reported information such as AP's loading and MN's movement, etc, by using a predefined algorithm, HCF decides whether or which AP MN shall associate with and notifies MN about the new AP/AR's information, such as AP's MAC address, AR interface address, and network prefix. HCF decides which AR's interface MN should move to as well. Consequently, the new network prefix of MN will be notified by HCF through HCFRep message accordingly.

The "IPv6 address allocation and assignment policy" draft issued by RIPE NCC [14] provides the guidelines for allocation and distribution of IPv6 addresses. This draft reveals that in an IPv6 access network as MN moves across the subnets, the only change in its IPv6 address occurs in subnet identifier field of the address. The remaining portion of the address, including 48 bit global routing prefix and the 64 bit interface identifier remains unchanged. Moreover, in our proposal, MN's interface identifier is allocated according to the norm of EUI-64. It ensures that the MN's new CoA is unique in Mobile IPv6. Consequently, MN could configure its new CoA and begin to launch the binding update process even if it is still attached with previous AP/AR. HCF also knows MN's new CoA according to MN's old CoA and MN's new network prefix. Furthermore, Duplicated Address Detection (DAD) can be omitted during handover.

In [15], a pure IPv6 Soft handover mechanism is presented. It provides data transmission continuity for delay constrained applications. Respiring from this method, a bi-casting mechanism was proposed and was applied to further improve handover performance. HCF acts as an extension of MAP in HMIPv6 which could begin to bicast traffic to both MN's old CoA and new CoA after sending HCFRep to MN to reduce packet loss during handover. HCF Bicast traffic also removes the timing ambiguity, regarding when to start sending traffic to MN's new point of attachment following a Fast Handover. It also minimizes service disruption time in the case of ping-pong movement.

Figure 2 shows messages exchange in the handover procedure.

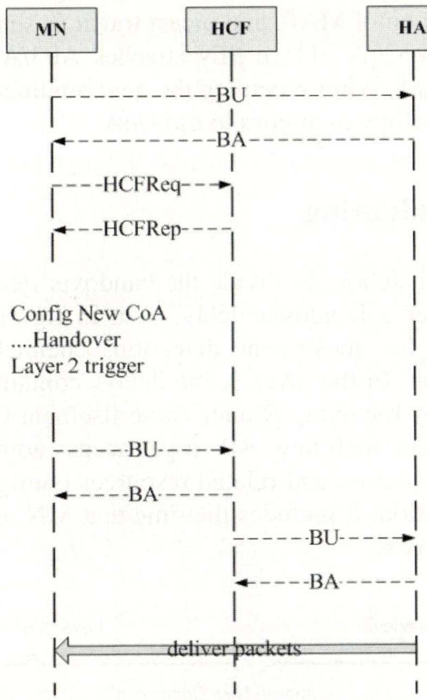

Fig. 2. Protocol for HCF based handover

1. When MN registers to HA at the first time, it will send Binding Update message to HA, and HA will response with Binding Acknowledgement.
2. MN probes all neighbor AP's information, including signal strength. Once the signal strength threshold is gotten over, MN shall send HCFReq message directly to HCF to report the information of its neighbor AP.
3. After HCF receives the HCFReq message, it will decide whether or which AR/AP MN shall associate with.
4. Detail algorithm for HCF judgment of MN handover mostly is based mostly on the signal strength that MN received from neighbor APs and HCF's network administrating policy.
5. HCF judge MN in some sense that can also help loading balance among different ARs and APs, if the number of registered MNs in one AR or AP have reached a limit, HCF will not approve MN to move to that network.
6. After HCF makes decision that which AR/AP MN shall move to, HCF will notify MN about new AR/AP's information, such as link prefix and AR's address. The information will help MN make a new CoA before it handover. This address also has been known by HCF since new CoA is made based on EUI-64.
7. After MN receive the HCFRep message, it knows that which AR/AP it will associate with and will configure its new CoA based on HCFRep message about new AP/AR.
8. When MN moves from AP2 to AP3, the attached Access Router also changes from AR1 to AR2 accordingly. MN will intensively disassociate with AP2 and associate with AP3.

9. HCF works as an extension of MAP, and bicast traffic is sent out from HCF to both MN's old CoA and new CoA. Once MN attaches AP3/AR2, that traffic will go directly to MN's new CoA. After receiving the new binding update from new CoA, HCF will remove the traffic which goes to old CoA.

4 Analyzing and Evaluating

To better analyze handover delay, we divide the handover delay into two parts: layer 2 handover delay and layer 3 handover delay, seen in figure 3. In the layer 2, the delays include the time for movement detection, channel scan, authentication, detection and re-association. In the layer 3, the delays contain: (1) the time that MN detects the AR changing by receiving Router Advertisement (RA) from new AR. (2) The time that MN negotiates with new AR to get access admission. It includes MN authentication, CoA configuration and related resources configurations in AR. (3) the time for Mobile IP registration. It includes the time that MN sends Binding Update to CN, HA and receives BU ACK.

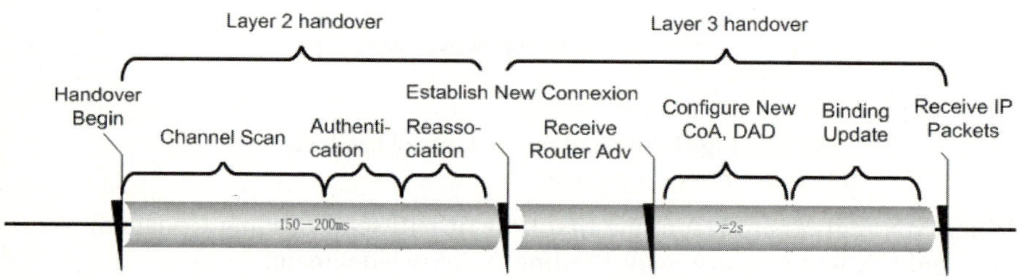

Fig. 3. Layer 2 and Layer 3 Handover Process

As shown in figure 3, the handover delay is too high to meet a real-time application need. We try to reduce the processing time or to omit unnecessary process in our approach. Compared with other proposals, HCF decides whether or which AP MN shall attach to before MN disconnects with its previous AP. By this means, MN reduces the time of channel scanning and switching phase running - authentication and re-association requests - with new AP. Being notified of the new network prefix by HCF, MN is informed of the new AR/AP's information, and gets its new CoA. So that it doesn't need either to wait for Router Advertisement message or to execute the new CoA configuration process. Besides, since unique IP address is no longer necessary to be verified again, DAD process execution can be omitted accordingly. Moreover, Pre-registration mechanism is permitted to be used in our proposal, hence, MN could carrying on binding update with its HA and CNs before leaving its previous AP. In conclusion, handover delay could be greatly minimized, and our proposal will be proved by the future simulation.

Packet Loss Rate (PLR) is another element which influences the handover performance. Packet Loss Rate is influenced not only by L3 handover delay, but also by L2 handover delay. The low packet loss rate can be realized by minimizing the handover delays. As we mentioned that our scheme could cut down either the layer 2 or layer 3 handover delay. In addition to decrease the handover delays, there are some other means to reduce the packet loss. Some proposals want to reduce the packet loss by buffering packets in old AP/AR when L2 handover is detected and forwarding packets to new AP/AR as soon as L2 handover ends. While this increases the charge of AP/AR and the signaling exchange in the system. Some proposals reduce the packet loss by using the IPv6 Soft handover mechanism, which sends the traffic to both MN's old CoA and new CoA while it needs the precise time control. In our scheme, HCF decides which AR/AP MN shall attach to, and HCF decides exactly how and when to send and end bicast traffic to which AR/AP.

5 Conclusion and Prospects

In this paper, we propose a simple but effective scheme, which could reduce the handover latency and packet losses without modifying AP/AR of WLAN. Our proposal permits MN to get the new CoA and to lance binding update process before moving to the new AR/AP. Moreover, the omission of DAD process optimizes greatly handover performance. Furthermore, by the means of sending bicast traffic from HCF to both old and new CoA, the packet loss could be minimized.

This paper expounds our primary concept. In the next step, our proposal will be simulated and evaluated by using OPNET (a discrete network simulation tool) [16]. In our future study, we are about to deal with the other MIPv6 issues, such as AAA, QoS and to enable HCF to better manage the mobility and the network resource in the WLAN.

References

1. "IEEE 802.11b: Wireless LAN Medium Access Control (MAC) and Physical Layer (PHY) Specifications," IEEE Standard, 1999.
2. D.Johnson, C.Perkins, and J.Arkko, "Mobility Support in IPv6", RFC 3775, June 2004.
3. H.Soliman, C.Castelluccia, K.Malki, and L.Bellier, "Hierarchical Mobile IPv6 mobility management (HMIPv6)", RFC 4140, August 2005.
4. Wei Kuang Lai, Jung Chia Chiu, "Improving Handoff Performance in Wireless Overlay Networks by Switching Between Two-Layer IPv6 and One-Layer IPv6 Addressing," IEEE JOURNAL ON SELECTED AREAS IN COMMUNICATIONS, VOL. 23, NO. 11, NOVEMBER 2005.
5. R. Koodli, Ed."Fast Handovers for Mobile IPv6", RFC 4068, July 2005.
6. C.Blondia, O.Casals, et al. "Performance Evaluation of Layer 3 Low Latency Handoff Mechanisms," Mobile Network and Application 9, pp.633-645, 2004
7. H.Chaouchi, P.Antunes, "Pre-handover signaling for QoS aware mobility management," INTERNATIONAL JOURNAL OF NETWORK MANAGEMENT 14, pp.367–374, 2004;
8. Y.Bi, P.Iyer et al. "An Integrated IP-layer Handover Solution for Next Generation IP-based Wireless Network", Vehicular Technology Conference, 2004 Vol. 6 pp.3950 - 3954
9. E.Perkins, "IP mobility support for IPv4", RFC 3220, January 2002.
10. J. Kristiansson and P. Parnes, "Application-layer Mobility support for Streaming Real-time Media," Wireless Communications and Networking Conference, Vol.1 pp.268-273, 2004.

11. S. Kashihara, K. Iida, H. Koga, Y. Kadobayashi, and S. Yamaguchi, "End-to-End Seamless Handover using Multi-path Transmission Algorithm," In Internet Conference 2002
12. “IEEE 802.11f: Recommended Practice for Multi-Vender Access Point Interoperability via an Inter-Access Point Protocol Access Distribution Systems Supporting IEEE 802.11 Operation,” IEEE Standard 802.11f/D1, Jan. 2002 (draft).
13. Chun-Ting Chou and Kang G. Shin, "An Enhanced Inter-Access Point Protocol for Uniform Intra and Intersubnet Handoffs," IEEE TRANSACTIONS ON MOBILE COMPUTING, VOL. 4, NO. 4, JULY/AUGUST 2005
14. IPv6 Address Allocation and Assignment Policy, ripe-267, January 2003. http://www.ripe.net/ripe/docs/ipv6policy.html
15. F.Belghoul, Y.Moret, C.Bonnet, “IP-Based Handover Management over Heterogeneous Wireless Networks”, Local Computer Networks, 2003. LCN '03. Proceedings. 28th Annual IEEE International Conference on 20-24 Oct. 2003 pp.772 - 773
16. www.opnet.com

Unified Error Control Framework with Cross-Layer Interactions for Efficient H.264 Video Transmission over IEEE 802.11e Wireless LAN

Jeong-Yong Choi and Jitae Shin

School of Information and Communication Engineering,
Sungkyunkwan University, Suwon, 440-746, Korea
{eldragon, jtshin}@ece.skku.ac.kr

Abstract. Efficient H.264 video transmission, as a dominant video coding standard used in Digital Multimedia Broadcasting (DMB) and other advanced video conferencing, over wireless/mobile networks becomes dominant. However, wireless video transmission suffers from deficient wireless channel conditions such as high bit error rate, error bursts due to channel fading and bandwidth limitations. In this paper, a novel design of unified error-control with cross-layer interaction over IEEE 802.11e Wireless LAN, in order to improve error protection performance, is proposed. This framework combines cross-layer error protection techniques, i.e., error correction code in the link layer, erasure code in the application layer and automatic repeat retransmission across the link layer and the application layer.

1 Introduction

In Korea, Satellite Digital Multimedia Broadcasting (S-DMB) and Terrestrial Digital Multimedia Broadcasting (T-DMB) services have been widely standardized on a national basis. S-DMB was launched as a test-service in January and as a main-service in May 2005, respectively, and T-DMB commenced service in December 2005. The Korean DMB system [1] adopts the European Digital Audio Broadcasting (DAB) system known as Eureka-147 [2] as its base, and adds various coding, networking and error correcting tools to process multimedia content. While S-DMB and T-DMB services have a different system, the associated video coding techniques are based on the H.264/Advanced Video Coding (AVC) baseline profile [3]. The H.264/AVC video coding standard provides for coding efficiency and network friendliness, splitting the Video Coding Layer (VCL) and Network Adaptation Layer (NAL).

In the trend of broadcasting communication convergence, multimedia content is supplied via various communication channels. IEEE 802.11-based Wireless LANs (WLANs) can be one a great communication channel, supplying users with portability. IEEE 802.11e [4] has been developed and standardized to support QoS of Audio/Visual (AV) data. In contrast to wired networks, since wireless

V.N. Alexandrov et al. (Eds.): ICCS 2006, Part IV, LNCS 3994, pp. 25–32, 2006.
© Springer-Verlag Berlin Heidelberg 2006

channels suffer from restrictions such as bit-error and bandwidth, video transmission over wireless channels is a challenging task.

There has been considerable research invested, to overcome such inferior channel conditions. Many hybrid ARQ (HARQ) techniques combine various error-control schemes [5][6]. They include error detection code, Forward Error Correction (FEC) and Automatic Repeat reQuest (ARQ). Although aforementioned error-control-related research proposes efficient error protection schemes, there have been few concrete cross-layer frameworks in WLANs.

In this paper, a novel cross-layer error-control framework in IEEE 802.11e wireless LAN is proposed to improve error protection performance. This framework combines cross-layer error protection schemes, i.e., error correction code (Reed-Solomon code) in the link layer, erasure code in the application layer and ARQ across both the link layer and the application layer.

The remainder of this paper is organized as follows. A brief overview of H.264 video transmission in wireless channels and a review of error-control techniques are provided in Section 2. Section 3 describes the proposed unified cross-layer error-control framework. In Section 4, the performance of proposed framework is evaluated. Lastly, conclusions and further works are presented in Section 5.

2 Related Works on Key Error-Control Components

2.1 Error-Controls for Video Transmission

FEC and delay-constrained ARQ are the main error-control techniques in video transmission. FECs, which are based both on the error correction code and on the erasure code, realize a proactive error-control mechanism. The sender generates redundant data and transmits these data with the original data. Then, the receiver recovers the erroneous bits or packets with redundant data. In order to prevent confusion, FECs are classified into two categories, i.e., error correction code-based FEC (namely Symbol-level FEC, S-FEC) and erasure code-based FEC (namely Packet-level FEC, P-FEC). While both S-FEC and P-FEC provide proactive solutions, ARQ is driven by reactive feedback information. ARQ spends less overhead in comparison with constantly rate-wasteful FECs, because ARQ is driven only when the received packet is erroneous, ARQ tends to result in greater delay than FECs.

In order to alleviate error resilience, hybrid techniques were attempted, and type-I [7] and type-II HARQ [8] were introduced. H. Liu et al. proposed a HARQ technique that combines the benefit of type-I and type-II techniques, and showed its efficacy for video transmission over wireless networks [5]. Y. Shan et al. presented a priority-based error protection scheme in the sense of the cross-layer approach [6]. Lastly, P. Ferré et al. attempted to modify the IEEE 802.11 legacy MAC to improve the throughput efficiency [9]. The article includes composition of an MAC frame and modification of MAC-level retransmission.

This paper is dominantly motivated by [9], to move the wireless channel to IEEE 802.11e, and makes use of the FEC option and BlockAck mechanism of IEEE 802.11e MAC.

2.2 IEEE 802.11e MAC-Level Error Control Techniques

Symbol-Level Forward Error Correction (S-FEC) [10]. Fig. 1 shows the MAC Protocol Data Unit (MPDU) format defined in the draft specification of IEEE 802.11e. Basically, a (224, 208) shortened Reed Solomon (RS) code, defined in GF(256), is used. Since an MAC Service Data Unit (MSDU), from the higher layer, can be much larger than 208 octets, an MSDU may be split into (up to 12) multiple blocks, and then each block is separately encoded by the RS encoder. The final RS block in the frame body can be shorter than 224 octets, using a shortened code. A (48, 32) RS code, which is also a shortened RS code, is used for the MAC header, and CRC-32 is used for the Frame Check Sequence (FCS). It is important to note that any RS block can correct up to 8 byte errors. The outer FCS allows the receiver to skip the RS decoding process if the FCS is correct. The inner FCS (or FEC FCS) allows the receiver to identify a false decoding by the RS decoder.

MAC Header		Frame Body (N Blocks)							FCS
Header	Header FEC	$MSDU_1$	FEC	$MSDU_2$	FEC	....	$MSDU_N$ + FEC FCS	FEC	FCS
32	16	208	16	208	16	....	208	16	4

Fig. 1. IEEE 802.11e MPDU format with the optional FEC

ARQ [11]. The legacy MAC of IEEE 802.11 is based on the simple Stop-and-Wait ARQ (SW-ARQ) scheme. This involves a lot of overheads due to the immediate transmissions of acknowledgements (ACKs). In 802.11e, a new Selective-Repeat ARQ (SR-ARQ) mechanism named block acknowledgement (BlockAck) is introduced. In this mechanism, a group of data frames can be transmitted one by one with SIFS interval between them. Then, a single BlockAck frame is transmitted back to the sender to inform ACKs how many packets have been received correctly. Obviously, this scheme can improve channel efficiency. There are two kinds of BlockAck mechanisms used in 802.11e: *immediate* and *delayed*. Details of each modes are presented in [11]. In this paper, the delayed Block-Ack mechanism will be adopted between two mechanisms, so as to satisfy the requirements of delayed operations across layers.

3 Unified Cross-Layer Cooperating Error-Control Framework

In this section, a unified error-control framework with cross-layer interactions is proposed, and the role of each layer and cross-layer cooperation are described. The overall structure of the proposed framework is depicted in Fig. 2.[1] The

[1] Since, in this paper, the range of the implementation is restricted, the implemented components are marked as shadow in Fig. 3.

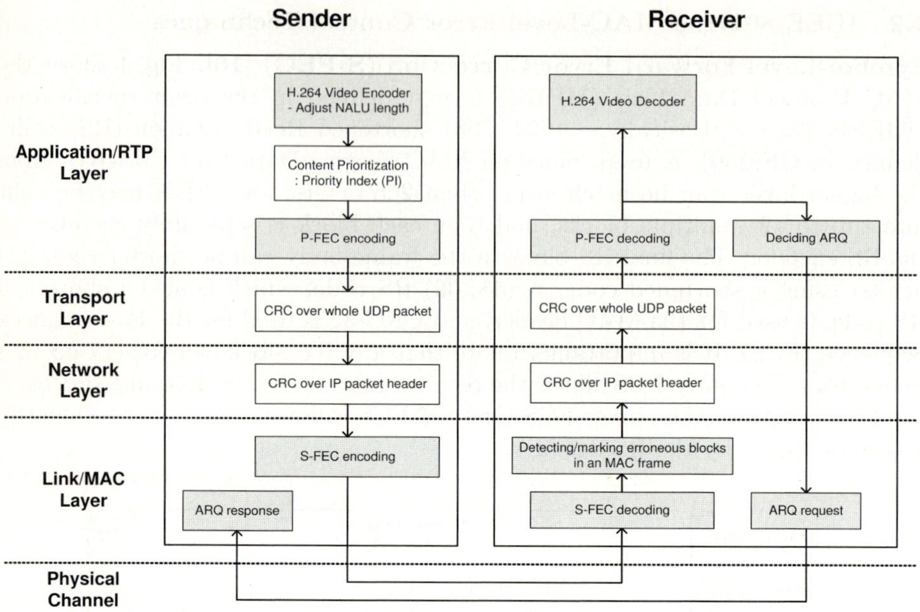

Fig. 2. Overall structure of the unified error-control framework with cross-layer interactions

techniques and schemes previously explained in Section 2 are combined, in order to achieve an efficient framework.

3.1 Link/MAC Layer

As explained in 2.2, the link/MAC layer performs S-FEC. On the sender side, the MAC header and all MSDUs are encoded using (48, 32) and (224, 208) RS codes, respectively, and then composed into an MAC frame. After S-FEC encoding, the whole MAC frame is calculated and the CRC-32 and FCS field are generated.

On the receiver side, prior to S-FEC decoding, the FCS check is performed. If the FCS is correct, the S-FEC decoding procedure is unnecessary, and is therefore not performed. However, if the FCS is incorrect, all S-FEC blocks are decoded and erroneous blocks are corrected. If MAC header is erroneous and uncorrectable, all of the MSDUs belonging to the MAC header will be dropped. If an FEC block of MSDUs is uncorrectable, the block is marked as erroneous and then passed to a higher layer. The uncorrectable S-FEC block is determined whether to be retransmitted in the application layer.

In addition to S-FEC, the link/MAC layer provides ARQ scheme for the erroneous MSDUs, as explained in 2.2. Since the uncorrectable S-FEC block is passed to higher layers, not immediately requested for retransmission, the delayed BlockAck scheme is used, rather than the immediate BlockAck scheme.

The sequence numbers of the uncorrectable S-FEC blocks determined to be retransmitted are bitmapped in the BlockAck frame.

3.2 Application Layer

In the proposed cross-layer framework, the application layer is responsible for *packet-level adaptation* and *cross-layer interaction*. Packet-level adaptation includes H.264 encoding, priority indexing (PI), RTP packetization and P-FEC. In this procedure, the H.264 encoder should consider the maximum length of an MSDU at the link/MAC layer. This implies that if one RTP packet is not fragmented into greater than one MSDU at the link/MAC layer, so as not to deteriorate error resilience, the maximum length of an RTP packet, including 2-byte P-FEC header, should not exceed 180 bytes, as presented in Fig. 3. Size-adjusted H.264 NAL Units (NALUs) are P-FEC-encoded and packetized into an RTP packet. Priorities of each NALUs are estimated and indexed, while NALUs are simultaneously encoded. The PI method which is used in [12] may be extended for H.264.[2] The PI-values can be criteria for UEP strategy.

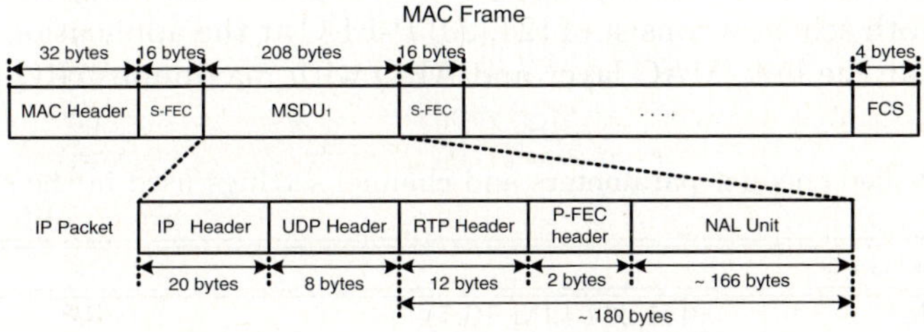

Fig. 3. Structure of an MAC frame

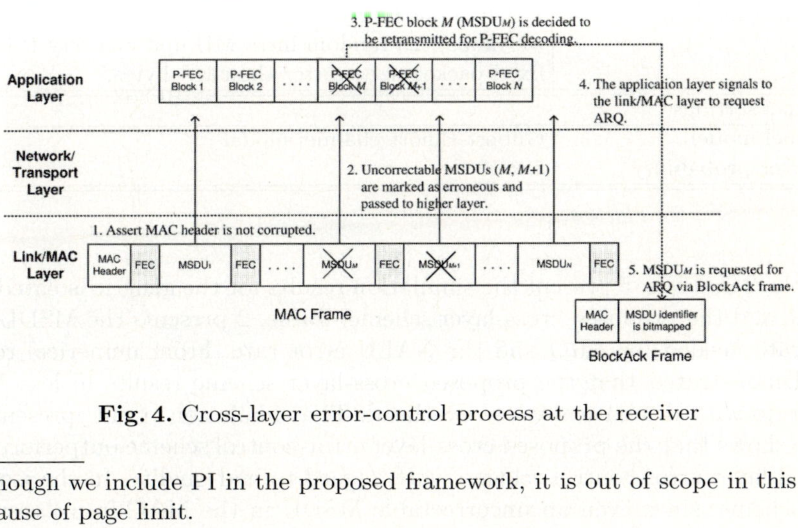

Fig. 4. Cross-layer error-control process at the receiver

[2] Although we include PI in the proposed framework, it is out of scope in this paper because of page limit.

Next, consider the cross-layer interaction. The proposed framework complements S-FEC with P-FEC and ARQ. First, S-FEC protects all symbols in a packet. Next, if S-FEC proves to be uncorrectable, P-FEC recovers the erroneous data blocks. Lastly, if P-FEC fails, the application layer signals to the link layer to request for retransmission of the necessary data blocks, to ensure that cross-layer cooperation is completed. The detailed error-control process is depicted and explained in Fig. 4. To conclude, the application layer adapts the video content to the underlying structure of the link/MAC layer, and decides the error-control strategy, whereas the link/MAC layer provides the application layer with the transport information of the packets and a means for retransmission.

4 Simulation

In this section, the performance of the proposed cross-layer error-control framework is demonstrated over IEEE 802.11e WLAN, by performing simulations under several BER conditions. The "Foreman" sequence (CIF, 352x288) was encoded using H.264 with encoding parameters, as presented in Table 1, and then transmitted through the Gilbert-Elliott channel model. The performance of the generic error-control scheme and the proposed error-control scheme are compared. Both schemes consist of (24, 20) P-FEC at the application layer, (224, 208) S-FEC at the link/MAC layer and ARQ with maximum ARQ count of 2.

Table 1. Video encoder parameters and channel settings used in the simulation

Parameter settings	
Video encoder	H.264 (JM 10.1)
Sequence name	Foreman
Image format	CIF (352×288)
Number of encoded frames	1200
Encoding method	1 IDR-frame followed by 1199 P-frames with CBR (384kbps), 20 random intra MB update every P-frame, fixed packet size shorter than 180 bytes.
Channel settings	
Channel model	Gilbert-Elliott channel model
Bit error probability	5×10^{-3}

Table. 2 and Fig. 5 present the simulation results for the generic isolated-layer scheme and the proposed cross-layer scheme. Table. 2 presents the MSDU overhead rate needed for ARQ and the NALU error rate. From numerical results, it is demonstrated that the proposed cross-layer scheme results in less NALU error rate with dramatically less overhead. The simulation result, presented in Fig. 5, shows that the proposed cross-layer error-control scheme outperforms the isolated-layer error-control scheme, with regard to end-quality. In the isolated-layer scheme, since even an uncorrectable MSDU in the MAC frame results in

Table 2. MSDU overhead rate for ARQ and NALU error rate

	Isolated-layer scheme	Proposed cross-layer scheme
MSDU overhead rate for ARQ (%)	32.84	0.29
NALU error rate (%)	0.40	0.33

retransmission of the whole MAC frame, the isolated-layer scheme experiences higher bit error probability than the cross-layer scheme. Thus, greater retransmission overhead is required, and high MAC frame error probability deteriorates throughput, eventually, producing degradation of objective video quality.

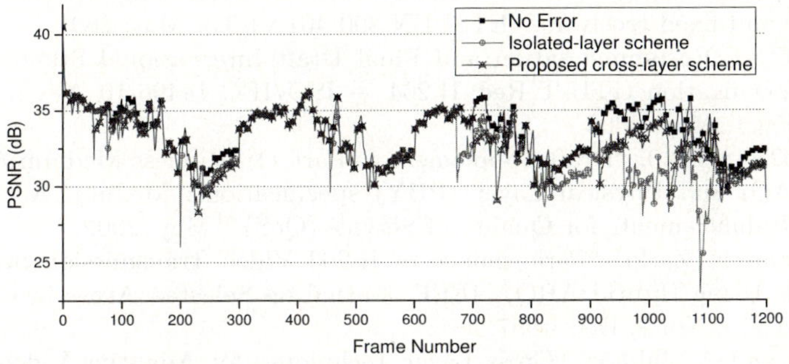

Fig. 5. PSNR distribution of Y-component of transmitted sequences: (a) No error, (b) Isolated-layer error-control scheme, and (c) Proposed cross-layer error-control scheme

From simulation results, it can be inferred that the proposed cross-layer error-control scheme presents excellent performances from the viewpoint of both end-quality and transmission efficiency.

5 Conclusions and Future Works

In this paper, a unified cross-layer error-control framework was proposed, and the performance of the framework demonstrated. In the proposed error-control framework, each layer does not perform its dedicated role but also considers overall layers, thus unnecessary operations can be omitted. From simulation results, it can be inferred that the proposed cross-layer error-control scheme demonstrates excellent performance from the viewpoint of both end-quality and transmission efficiency.

FEC and ARQ have been the most powerful error-control tools in data communication. In order to maximize the performance of FEC and ARQ, combined with packet-level interleaving and symbol-level interleaving can be considered. In current work, we cannot present the mathematical analysis of the proposed scheme, because of page limit. Mathematical analysis under certain parameters of Gilbert-Elliott channel helps to estimate the simulation result.

Acknowledgements

This research was supported by the Ministry of Information and Communication (MIC), Korea, under the Information Technology Research Center (ITRC) support program supervised by the Institute of Information Technology Assessment (IITA) (IITA-2005-(C1090-0502-0027)).

References

1. "Digital Multimedia Broadcasting," Telecommunications Technology Association, 2003SG05.02-046, 2003.
2. "Radio Broadcasting System: Digital Audio Broadcasting (DAB) to mobile, portable and fixed receivers," ETSI EN 300 401 v1.3.3, May, 2001.
3. Draft ITU-T Recommendation and Final Draft International Standard of Joint Video Specification (ITU-T Rec. H.264 — ISO/IEC 14496-10 AVC), Document: JVT-G050r1, May. 2003.
4. IEEE 802.11e/D3.0, "Draft Supplement to Part 11: Wireless Medium Access Control (MAC) and Physical Layer (PHY) specifications: Medium Access Control (MAC) Enhancements for Quality of Service (QoS)," May, 2002.
5. H. Liu and E. Zarki, "Performance of H.263 Video Transmission over Wireless Channels Using Hybrid ARQ," IEEE Journal on Selected Areas in Communications, Vol. 15, No. 9, Dec. 1997.
6. Y. Shan and A. Zakhor, "Cross Layer Techniques for Adaptive Video Streaming over Wireless Networks," International Conference on Multimedia and Expo, Lausanne, Switzerland, Aug. 2002, pp. 277-280.
7. H. Deng and M. Lin, "A Type I Hybrid ARQ System with Adaptive Code Rates," IEEE Trans. Communications, vol. 46, pp. 733-737, Feb. 1995.
8. S. Lin and D. Costello, "Error Control Coding: Fundamentals and Applications," Englewood Cliffs, NJ: Prentice-Hall, 1983.
9. P. Ferré, A. Doufexi, A. Nix, D. Bull, and J. Chung-How, "Packetisation Strategies for Enhanced Video Transmission over Wireless LANs," Packet Video Workshop (PV2004), Irvine, CA, USA, 13 14 Dec. 2004.
10. S. Choi, "IEEE 802.11e MAC-Level FEC Performance Evaluation and Enhancement," IEEE Global Telecommunications Conference, 2002.
11. Q. Ni, L. Romdhani and T. Turletti, "A Survey of QoS Enhancements for IEEE 802.11 Wireless LAN," RFC 3550, IETF, Jul. 2003.
12. J. Shin, J. G. Kim, J. W. Kim and C.-C.J. Kuo, "Dynamic QoS mapping control for streaming video in relative service differentiation networks," European Transactions on Telecommunications, Vol. 12, No. 3, May-June 2001, pp. 217-230.

A Novel Control Plane Model of Extensible Routers*

Kun Wu, Jianping Wu, and Ke Xu

Department of Computer Science and Technology,
Tsinghua University, Beijing 100084, P.R. China
{kunwu, xuke}@csnet1.cs.tsinghua.edu.cn,
jianping@cernet.edu.cn

Abstract. The extensible router is the next generation router architecture. This paper studies the extensibility of the router control plane, which contains complicated modules such as routing protocol processing, router information management, etc. It is difficult to decide when to extending the control plane, and how many nodes are needed. Moreover, the module dispatching configurations is another issues. This paper describes the extensible architecture of the router system and the control plane. It defines three variables to evaluate the extensibility. And a two-step extensible model is provided to achieve the maximum extensibility under these evaluations. Finally, the experiments show that the model is correct and efficient.

1 Introduction

1.1 Motivation

The next generation router will concentrate in the extensible architecture[1][2][3]. The single chassis router, which has been widely used in the conventional router system, cannot catch up with the Internet explosion [4]. A fully extensible structure is borrowed from computer clusters[5], which can help in meeting the expanding requirements of the future Internet.

In the recent years, research on the scalability in the data plane has grown enormously. However, very few solutions can be found by far on how to make a control plane extensible enough, which is as critical as that of the the data plane. While the system is becoming larger and larger by clustering more small nodes, the routing protocol processing and other service routines can cause bottlenecks in both of the processing power and the memory storage.

1.2 Previous Work

The router extensibility is an interesting topic. Some demos or products under the extensible architectures have been released. The systems can be extended

* This work was supported by the National Key Fundamental Research Plan (973) of China (No 2003CB314801).

V.N. Alexandrov et al. (Eds.): ICCS 2006, Part IV, LNCS 3994, pp. 33–40, 2006.
© Springer-Verlag Berlin Heidelberg 2006

to multiple chassis with specific inter-connecting networks. Many manufactures announced their systems, such as the CRS from CISCO, T640 from Juniper and Avici's TSR system. The interconnecting styles embedded are very different. CRS uses the multi-stage inter-connecting network. And, TSR employs a 3D-torus architecture. However, the majority research and improvement are located in the extensibility on the data plane. Very few evolutions can be found on the control plane, which can not keep up with the extension on the date plane.

Nevertheless, some research groups are active on the extensible control plane, especially to provide an extensible router software platform. The Scout[1] is an configurable operating system, which can provide the functional extensibility. Click[6] is a module-oriented router model, which is composed by simple elements. Router Plugins[7] is also a way to extend the router functions by import new plugins onto the processing path. However, all of these models are mainly focusing on the functional extensibility, other than the structure extending. Zebra is an excellent module-based routing software platform. It is well designed and implemented under a structural manner. It can contribute much on the extensible router system although it is not extensible itself. The XORP[8] is an active project on providing a flexible and extensible software router platform for network reach. It provides a flexible well-defined structure. However it is not focus on the performance-driven extensibility.

In this paper, we analyze the evolution of the modern router architecture, and provide a feasible model for extensible router systems. Section2 give a overview of the architecture of router and the control plane. Section3 describes the extensibility requirements and analyze the key issues. Section4 presents the extensible model in details. Section5 evaluates the performance of the model. Section6 gives a conclusion of the paper.

2 Architecture Overview

In this section, we will describe why an extensible model is necessary, and what the structure it looks like. Fig1 depicts the conceptual extensible router architecture. The forward entities are the components located in the data plane. They are connected by the interconnecting network. This is beyond this paper's discussion, which focus on the control plane.

As shown in Fig. 1, the control entities are the control part of each node. The system running control and routing processing are located in them, which constitute the control plane. The control entities are connected together by a communication network. This communication network can be implemented in different ways, such as a switch, or even a virtual network provided by the data plane. The main problem now is how to distribute the processing modules to different control entities in the control plane.

The control plane architecture is shown in Fig. 2. The concrete processes on each control entity are represented as *modules*, such as OSPF processing. The lines in Fig. 2 represent the relationships between each pair of nodes.

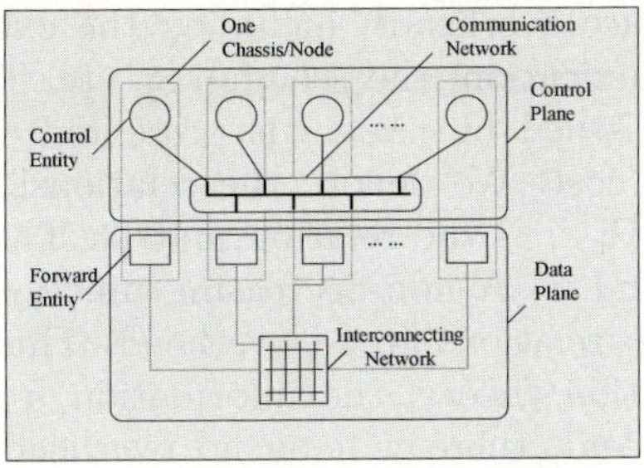

Fig. 1. Extensible Router Architecture

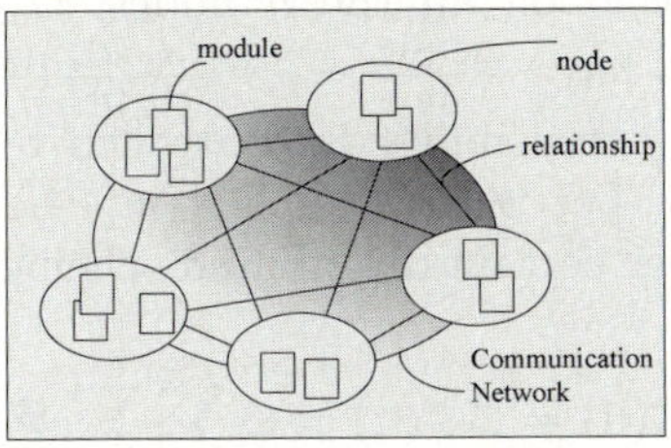

Fig. 2. Control Plane Architecture

3 Extensibility Requirements

Benefited from the optical technology improvements[9][10], the interface speed and density reach to an extreme high level. The core router will provide carrier class high availability to achieve non-stop forwarding[11]. Moore's Law[12] shows that the computer performance would double every 18 months. But the data traffic in Internet has grown with a tremendous speed[13][14], which is approximately doubling each year. It is not possible to support the traffic and processing in one box.

One of the key issues for extensibility is to limit the cost increasing during the extending. The overall cost is denoted as $C(n)$ for a n-nodes router. The effect function is $E(n)$, which is a integrated measurement for that the extension can benefit. We define the derivative $\frac{\Delta C(n)}{\Delta n}$ as the **Extending Cost Rate (ECR)**. The $\frac{\Delta E(n)}{\Delta n}$ shows the efficiency, which is defined as the **Extending Efficiency Rate (EER)**. The requirements for the extensibility is to minimize the ECR, and maximize the EER respectively. Moreover, we define **Extending Balance Level (EBL)** to reflect the balancing result, which can be measured by the geometric mean for the utilities on each node.

Under the current philosophy of parallel computing, it is impossible to reduce ECR to a level lower than zero. We define the inherent cost for one problem, i.e. routing processing in router, as $\hat{C}$. In the extended system with n nodes, $\hat{C}$ can be distributed as $\hat{c}_i$, where $i = 0, 1, \cdots, n-1$. The communications among these nodes can introduce overheads, including the extra temporary memory, the inter-process communication, the scheduling, etc. The overhead on node j caused by information issued by node i is presented as o_{ij}. These overheads will add to the overall cost. We denote the relationship between node i and node j as r_{ij}, and the $\mathbf{R}_{i\times j}$ is the relation matrix. The sets of overheads and relation values are $\mathbb{O}$ and $\mathbb{R}$. A non-decreasing function $\mathscr{F}(r_{ij}) : \mathbb{R} \mapsto \mathbb{O}$ can be given to convert the relationship to overhead. That is $o_{ij} = \mathscr{F}(r_{ij})$. According to the information theory, the information from node i to j cannot be negative, $r_{ij} \geq 0$. And, there will be no overhead on node j caused directly by node i, if node j is not depended directly on node i. Then we have $\mathscr{F}(0) = 0$. So for $\forall i, j = 1, 2, \cdots, n$, we have $o_{ij} = \mathscr{F}(r_{ij}) \geq \mathscr{F}(0) = 0$. That is $o_{ij} \geq 0$. While being extended to n nodes, the total cost is the accumulate cost and overhead on all the n nodes. $C_{all}(n) = \Sigma c_i + \Sigma\Sigma o_{ij} = \hat{C} + \Sigma\Sigma\mathscr{F}(r_{ij})$. Then we have $\frac{\Delta C(n)}{\Delta n} = \frac{C_{all}(n) - C_{all}(n-1)}{n-(n-1)} = \Delta\Sigma\Sigma\mathscr{F}(r_(ij))$. It shows that the potential extensibility is controlled by the relationship matrix and the convert function. Let $\mathscr{G}(n) = \Sigma\Sigma\mathscr{F}(r_{ij})$ be the integrated overhead for a n-nodes system under certain distributed scheme. Hence, the ECR can be written as

$$ECR(n) = \Delta\mathscr{G}(n) = \Delta\sum_{i=1}^{n}\sum_{j=1}^{n}\mathscr{F}(r_{ij}) \tag{1}$$

For node i, we write the resources it can provided as a_i. The total resources in the extensible system will be $A(n) = \sum_{i=1}^{n} a(i)$. After cutting off the overhead, we get the available resources $A(n) - \mathscr{G}(n)$. Therefore, the EER can be written as

$$EER(n) = a_n - \Delta\mathscr{G}(n) = a_n - ECR(n) \tag{2}$$

And, by definition, we can write the extending balance level as

$$EBL(n) = \left(\prod_{i=1}^{n}\frac{c_i}{a_i}\right)^{\frac{1}{n}} \tag{3}$$

From these equations, we can point out that the extending result is dominated by the correlation between each pair of the nodes, the contribution of the relation to the cost, the capability of the new participant node, and the system-wise balance level.

4 Extensibility Model

Consider the situation that the information dependence of the tasks, which are separated onto the new coming node against those on the rest nodes, increase to a large value, so that the overhead will overwhelm the capacity. In the router systems, many high cost tasks are also too tightly coupled to be separated. At some point when the the overhead for dispatching a new module is greater than the incremental capacity, the extending upper-bound is reached. By the definitions, we can calculate the extension limitation for a system by the ECR and EER, so that the system can still be extended to more nodes. But it is not enough. Because it has no senses to assign a trivial module to a new node, although it will contribute to the ECR and EER. The balance level (EBL) decides which and how many modules should be dispatched. So, the EBL is used to evaluate if an extension scheme is good or bad while the node number is given.

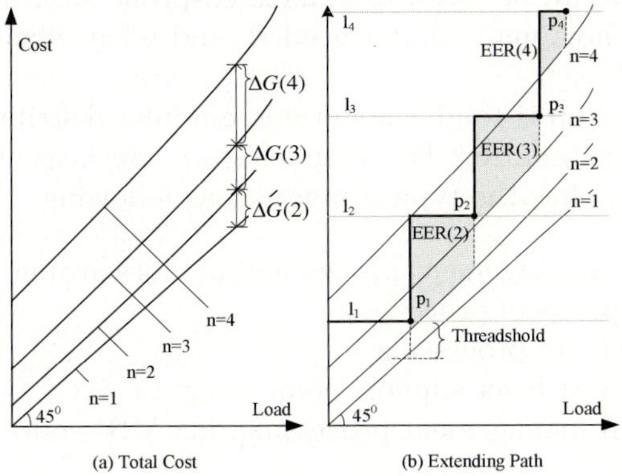

Fig. 3. Extending Path

The extensibility model contains two parts. One is to decide how many nodes are needed for the given system load. The other is to assign the modules to these separated nodes.

The first step is depicted in Fig. 3. The left fig shows the cost increasing with the load under different amount of nodes. The number of nodes is denoted as n. $n = 1$ means the single chassis system.

The function in Equation1 is a convertor from information to overhead. $\mathscr{F}(r_{ij})$ is the overhead on node j caused by the information issued by node i. While considering a real router system employ a multi-tasks operating system (OS), the job scheduling, bus arbitration, and buffer management can introduce the side effect of OS overhead. Long term measurement can show us that the processing overhead at node j , with the attached management overhead is counted in, can increase faster than the increment of information from i to j. Hence, the conversion function $\mathscr{F}(r_{ij})$ is roughly a convex function. This is why the curve with a larger node number is above that with a smaller one. And the space between every two adjacent curves has the same trend.

While the total node capacity is considered, we will have the relationship between system resource and overall cost along with variety of the load. This is illustrated in right part of Fig. 3. The horizontal lines from l_1 to l_4 is the capabilities with node number $n = 1, 2, 3, 4$. Each horizontal line is a resource level the system can provide. The space from the line to the corresponding cost curve describes the resource available. This space reduces with the load increasing. A threshold is introduced here to preserve some resources for the current configuration. While the threshold is reached, a decision must make if it will transfer to higher resource level. The decision points are shown as p_1, p_2, p_3, p_4. And, for each value of n, we will have the EER as the shadow areas.

If we connect each decision point and the next horizontal lines, which is depicted as the bold line in the right figure, a decision border is given. We name it as the *Extending Path*. The area above the path is all the permitted conditions. And the area below can be resource exhausted-prone situations. From this figure, we can decide how many nodes needed, and what efficiency the extension can achieve.

Now, the second step is to give a suitable modules distributing configuration so that the EBL is maximized. For simplicity, we will inspect the situation in a simple system, which has the typical modules as following.

- $RTMGT$: route management processing, i.e. forwarding table maintaining.
- $OSPF$: OSPF protocol processing
- BGP: BGP protocol processing
- $TRANS$: transport layer support computing, i.e. TCP/UDP/IP.
- $MGMT$: system management processing, i.e. MIB collecting.

And the information dependencies among them are listed below:

- MIB: Management Information Base.
- RT_{BGP}: BGP routing table changes.
- RT_{OSPF}: OSPF routing table changes.
- UPD_{BGP}: BGP update information.
- UPD_{OSPF}: OSPF update information.

We can sort the information dependencies in a more general sense.
$MIB \gg \Delta RT_{BGP} > \Delta RT_{OSPF} \gg \Delta UPD_{BGP} > \Delta UPD_{OSPF}$
So, we can arrange the dependencies into the following table.

Table 1. Information Dependencies

Source	$RTMGT$	$OSPF$	BGP	$TRANS$	$MGMT$
$RTMGT$	0	ΔRT_{BGP}	ΔRT_{OSPF}	ε	MIB
$OSPF$	ΔRT_{OSPF}	0	0	ΔUPD_{OSPF}	MIB
BGP	ΔRT_{BGP}	0	0	ΔUPD_{BGP}	MIB
$TRANS$	ε	ΔUPD_{OSPF}	ΔUPD_{BGP}	0	MIB
$MGMT$	ε	ε	ε	ε	0

The source of the information flow is list in the first column. Each one of the rest columns is the information dependencies to the first item in the column. A *small* value ε is introduced for the dependencies that can be ignored. From the table we can classify the modules into three levels. The *Level I* is the closely associated module, i.e. the $MGMT$, is tightly coupled with every other modules.

The *Level II* is the partly coupled module, such as *OSPF* or *BGP*, which is much depend on some of the other partly coupled modules. The *Level III* is the weakly coupled service module, i.e. *TRANS*, which depends loosely on other modules.

To balance the load on each node, we have the following guidelines for modules dispatching schemes.

1. Distribute Level I modules evenly to each node.
2. Distribute Level II modules in the descent order.
3. Distribute Level III modules with the corresponding modules in Level II.

For a router system, the task granularity cannot small enough to achieve a fully balanced configuration, because many resource-consuming tasks are composed by modules tightly-coupled. However, the guidelines can achieve the maximum EBL under this condition.

5 Model Evaluations

We simulate the 2-step extensible model under one PC. Each router node is implemented as a process, and the modules are programmed as resources-consuming threads. The router system is abstracted as a typical router system containing the modules BGP, OSPF, route management, Transport layer and management routines. The information dependencies are assigned as the dynamic traffics among the modules under the same manner as Table 1. The results are shown in Fig. 4.

The experiment for extending decision points shows that the simulating results are almost the same as the decision path in Fig. 3, except that the simulating decision points come some earlier before the calculating. That's an error introduced by the dynamic randomness of the modules.

The experiment for the module dispatching configuration shows that the system load is balanced to some extend. The balance level is not far beyond the theoretical lower bound.

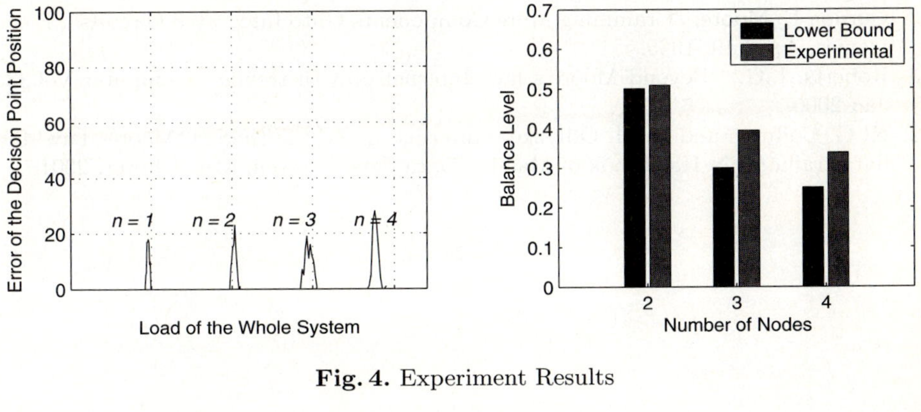

Fig. 4. Experiment Results

6 Conclusion

The extensible router is the next generation core router architecture. This paper focuses on the extensibility of the control plane. The control plane structure is deeply inspected. EER, EFR, and EBL are given to evaluate the extension schemes. We present a two-steps extensible model in this paper. The first step is to decide how many nodes are needed for the extension and when to extend. The next step contains a list of rules to find out a modules configuration scheme so that the extension is balanced.

References

1. S.Karlin, L.Peterson. "VERA: An extensible router architecture", Computer Networks, Vol. 38, Issue 3 (February 2002).
2. Lukas Ruf, Ralph Keller, Bernhard Plattner, "A Scalable High-performance Router Platform Supporting Dynamic Service Extensibility On Network and Host Processors", ICPS'04, 2004.
3. H.J.Chao, "Next Generation Routers", Proceedings of the IEEE, vol.90, no.9, Sep. 2002, pp.1518-1558.
4. H.Jonathan Chao, Kung-li Deng, Zhigang Jing, "A Petabit Photonic Packet Switch (P^3S)", IEEE INFOCOMM'03, San Francisco, April 1-3, 2003.
5. Buyya, R. (ed.), "High Performance Cluster Computing: Architectures and Systems", Volume 1 and 2, Prentice Hall PTR, NJ, USA, 1999.
6. E. Kohler, R. Morris, B. Chen, J. Jonnotti, M. F. Kaashoek, "The Click Modular Router", ACM Transactions on Computer Systems 18(3), Aug. 2000.
7. D Decasper, Z Dittia, G Parulkar, B Plattner, "Router Plugins: A Software Architecture for Next-Generation Routers" ACM Transactions on Networking, 2000.
8. M Handley, O Hodson, E Kohler, "XORP: An open platform for network research", Proceedings of HotNets-I Workshop, October, 2002.
9. F.Masetti, D.Zriny, etc, "Design and Implementation of a Multi-Terabit Optical Burst/Packet Router Prototype", OFC'2002, 2002.
10. Yang Chen, Chunming Qiao, Xiang Yu, "Optical Burst Switching(OBS): A New Area in Optical Networking Research", IEEE Network, 2004.
11. Avici Inc. "Non-Stop Routing Technology", white paper, 2002.
12. Gordon E. Moore, "Cramming More Components Onto Integrated Circuits", Electronics, April 19, 1965.
13. Roberts, L.G., "Beyond Moore's law: Internet growth trends", Computer Vol.33, Jan 2000.
14. K. G. Coffman and A. M. Odlyzko, "Internet growth: Is there a 'Moores Law' for data traffic?", in Handbook of Massive Data Sets. Norwell, MA: Kluwer, 2001

AM-Trie: A High-Speed Parallel Packet Classification Algorithm for Network Processor

Bo Zheng and Chuang Lin

Dept. of Computer Science and Technology, Tsinghua University,
Beijing 100084, P.R. China
{bzheng, clin}@csnet1.cs.tsinghua.edu.cn

Abstract. Nowadays, many high-speed Internet services and applications require high-speed multidimensional packet classification, but current high-speed classification often use expensive and power-slurping hardware (such as TCAM and FPGA). In this paper, we present a novel algorithm, called AM-Trie (Asymmetrical Multi-bit Trie). Our algorithm creatively use redundant expression to shorten the height of Trie; use compression to reduce the storage cost and eliminate the trace back to enhance the search speed further. Moreover, AM-Trie is a parallel algorithm and very fit for the "multi-thread and multi-core" features of Network Processor; it has good scalability, the increase of policy number influences little to its performance. Finally, a prototype is implemented based on Intel IXP2400 Network Processor. The performance testing result proves that AM-Trie is high-speed and scalable, the throughput of the whole system achieves 2.5 Gbps wire-speed in all situations.

1 Introduction

In recent years, the Internet traffic increases dramatically and the users of Internet increase exponentially. At the same time, the services of Internet turn from traditional best-effort service to quality of service (QoS). Hence, network devices should evolve from the traditional packets forwarding to content awareness, and packet classification is one of the most important basic functions. Many network technology such as VPN, NAT, Firewall, IDS, QoS, and MPLS require packet classification. So the speed of packet classification algorithm will affect the performance of these network devices directly, and it also takes great effect on the next generation Internet. The high-speed packet classification is a hot topic in today's network research, there are many papers about classification [1, 2, 3, 4] published in the top conference, such as Sigcomm and Infocom.

By now, hardware (such as TCAM, FPGA, etc.) is the most common way to achieve high-speed classification, but the hardware is expensive and power-slurping, so the industry hopes to find some high-speed software methods. Network Processor Unit (NPU) is a kind of programable chip which is optimized for network applications processing. Most NPUs are "Multi Thread and Multi-core" architecture [5], they have multiple hardware threads in one microengine and multiple microengines in one NPU chip; NPUs usually have optimized memory

V.N. Alexandrov et al. (Eds.): ICCS 2006, Part IV, LNCS 3994, pp. 41–48, 2006.
© Springer-Verlag Berlin Heidelberg 2006

management unit for packet forwarding; and some NPUs provide special-purpose coprocessors for some specific network operations. With the well designed architecture, NPU combines the flexibility of software and the high-speed of hardware together, it can also greatly shorten the product development cycle. So it is considered as the core technology which impels the development of next generation Internet. Network Processor has a good prospect and becomes a new direction and hot topic [6, 7, 8] in network research after ASIC.

But the traditional software classification algorithms are not fit for NPU, they usually based on the data structure of binary tree or hash table. The complexity of these algorithms is $O(dW)$, because they are not parallel algorithm, where d is the classification dimension (field number of the policy table) and W is the bits of the field. They cannot satisfy current high-speed network, not to mention the high-speed IPv6 applications in the near future. Hence, researches on high-speed packet classification algorithm based on NPU platform are very important.

In this paper, we present a parallel high-speed multidimensional classification algorithm, called AM-Trie (Asymmetrical Multi-bit Trie), which has the following features:

- Use redundant expression to express prefix with arbitrary length and build AM-Trie, which is much lower than normal Trie.
- Compress AM-Trie to eliminate trace back. This increases the lookup speed and reduces the storage cost. The time complexity of AM-Trie is $O(d+h)$ and the space complexity is $O(N^2)$, where d is the dimension of the classification, h is the height of AM-Trie and N is the number of classification policy.
- AM-Trie is a parallel algorithm and very fit for the "multi-thread and multi-core" feature of Network Processor.
- Our algorithm has good scalability, there is almost no influence to the performance when the number of policy increase much.

The remainder of this paper is structured as follows. We describe the detail of AM-Trie algorithm and analyze the algorithm complexity in Section 2. Performance evaluation can be found in Section 3. Finally, Section 4 gives a further discussion and summarizes the paper.

2 AM-Trie Classification Algorithm

The performance metrics of packet classification include time complexity, space complexity, scalability, flexibility, worst case performance, update complexity, preprocessing complexity, etc., and for different applications the importance priority of these metrics are different. At present, the speed of Internet increases dramatically, many new kinds of services appear. Hence, in order to adapt packet classification algorithm to the evolution of next generation Internet, we should consider time complexity, scalability and flexibility first.

However, the the search complexity of traditional tree/hash based classification algorithm is direct proportion of the classified field length. For example, to classify a IPv4 address we have to build a 32-level Binary Trie or establish

32 Hash tables; and 128-level Binary Trie needed for IPv6 address, the search efficiency is too low. Therefore, we hope to construct $m$bits-Trie to reduce the height of Trie, and speed up the search. But because of the widely used Classless Inter-Domain Routing (CIDR) technology, the prefix length may be any length between 0 to field width. Such policy table cannot transform to m-bits-Trie conveniently, because the prefix length may not right suit with the multi-bit Trie (for example, 2bits-Trie, which takes 2-bit as one unit, can not express prefix length for odd number situation).

The first idea of AM-Trie is: express arbitrary length of prefix using the redundancy. Considering a k-bits field, all possible prefixes are $*, 0*, 1*, 00*, 01*, \ldots,$ $\overbrace{11\ldots1}^{k-1}*, \overbrace{00\ldots0}^{k}, \ldots, \overbrace{11\ldots1}^{k}$, totally $2^{k+1}-1$ prefixes. We can number them as $1, 2, \ldots, 2^{k+1}-1$. They may be expressed using one $(k+1)$bits number which we call *Serial Number*. If the prefix is $P*$, and P is a p-bits number, then

$$SerialNumber(P*) = 2^p + P \tag{1}$$

For example, * is the 1st prefix and 1111* is the 31st prefix. Using this redundant expression, we can use one $(k+1)$-bit number to express all k-bit prefix. Then we can use such expression to build appropriate multi-bit Trie.

2.1 AM-Trie Creation

The redundant expression can express arbitrary length of prefix. Therefore, we can cut a classification field in to several variable length sections, and then construct multi-bit Trie – AM-Trie. Without loss of generality, assume a W-bit-wide field is divided into l parts (and constructed as a l-level AM-Trie), the width of each part is $h_1, h_2, \ldots, h_l$ bits, and $\sum_{i=1}^{l} h_i = W$, i.e. the ith level of constructed AM-Trie has 2^{h_i} nodes (How to optimize the l and $h_i, i = 1 \ldots l$ will be mentioned in section 2.3).

Each node of AM-Trie have two pointer, *pNext* points to its children level and *pRule* points to the related policy rule. When creating the AM-Trie of a policy table, we add the policy one by one. Firstly, divide the policy into predefined parts $h_1, h_2, \ldots, h_l$, and calculate the serial number of each section use the redundant expression. Then calculate the offset value of each level according to the serial number. If the related leaf node does not exist, add the policy rule node to the right place, or compare the existing node and the new one and add the higher priority rule. After that, add the next policy. The pseudocode for add a node into AM-Trie is:

```
int AddNode(struct AMTrieNode *root, Rule newrule)
    current=root;
    divide newrule into predefined parts;
    offset=Serial Number(first part);
    while(not the last part)
        current<--current[offset];
        if(current->pNext == NULL) create children under current node;
        offset=Serial Number(next part);
```

```
current<--current[offset];      //handle the last part
if(current->pRule==NULL) current->pRule=this rule; return 1;
else current->pRule=the higher priority rule; return 0;
```

We use an example to give the direct-viewing of AM-Trie creation. Figure 1 shows the AM-Trie derived from Table 1, here we chose l=2, $h_1 = h_2 =$ 2bits. Firstly, we add the the first policy R0 which has only one part (*), it's serial number is 1. So R0 is add to the first node under root. Similarly, R1, R2 is the 2nd and the 5th child of root. As for R3, its first part (01) has serial number 5, under the 5th child node of root we continually process its second part (1*), so R3 is joined to the 3rd child node in the second level. The situation of R4 is similar to R3. If there is a rule R5 (011*), when it joins to the AM-Trie, its position is already occupied by R3, then we needed to compare the priority of R3 and the R5 to judges whom takes this position.

From the generated AM-Trie, we can find that only the nodes in the right part of the Trie may have children nodes, because the first $2^{k_i} - 1$ nodes map to the prefix with a wildcard '*', which is the end of a prefix, so they cannot have child. Therefore this Trie looks asymmetrical, we name this Trie Asymmetrical Multi-bit Trie.

Table 1. A simple policy table

Rule	Field
R0(Default)	*
R1	0*
R2	01*
R3	011*
R4	1111

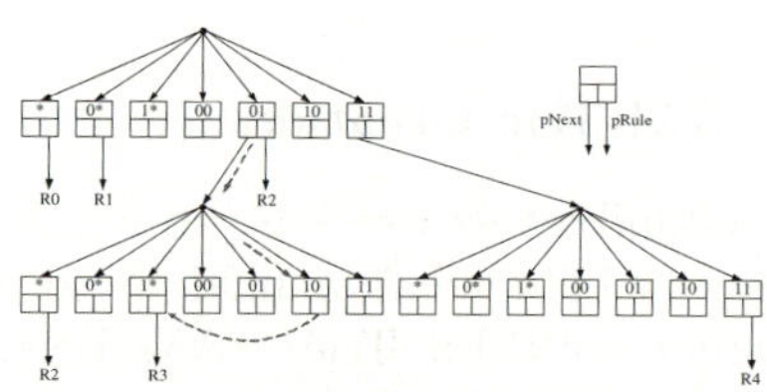

Fig. 1. AM-Trie Creation and Search

2.2 AM-Trie Search, Optimization and Update

Finding a policy in the AM-Trie is similar to search a common Trie except that if a leaf node doesn't contain a policy we should go on searching its "prefix node"[1]. For instance, the red broken line in Figure 1 shows the search progress in the same AM-Trie. If the classification field of the incoming packet is 0110, firstly we calculate the serial number of its first part '01' and go to the 5th child of the root, and then calculate the serial number of the second part '10' and go to the 6th node in the second level. This leaf node has no policy, so we lookup the prefix node of '10' (i.e. '1*'), and go to the 3rd node in this level and find the result is R3.

This example shows that when a leaf node has no rule, we have to trace back to its prefix nodes. We can use a compression algorithm to avoid the trace back and optimize the AM-Trie. We find that when the policy table determined, the

[1] We call a prefix with wildcard '*' *covers* all the prefixes match it, for example, in a 3bits field, 0* covers 00*, 01*, 000, 001, 010 and 011. If A covers B, A is the *prefix node* of B.

longest rule-matched prefix of any prefix (with wildcard or not) determined too. Hence, we can "push" the policy in a prefix to the nodes covered by it and without policy in the period of preprocessing. Then all the leaf nodes contain a policy and no trace back in searching a AM-Trie now. At the same time, the prefix with wildcard '*' in the left part of AM-Trie will never be accessed because their information has been pushed to the nodes they covers in the right part. So, we can delete these node to compress the storage of AM-Trie. Figure 2 illustrates the "push" procedure and the result of compression. The broken line with arrow indicate the push operation, and the red nodes can be delete to compress the AM-Trie. We can save about one half storage after compressing. And the search complexity of compressed AM-Trie is $O(h)$ now.

AM-Trie data structure supports incremental update. The Add operation is just the same as AddNode in AM-Trie creation. The Delete operation is also very simple. Call the search function to locate the policy we want to delete, and then release the storage; if this level contains no policy after deletion, release the whole level. Both Add and Delete operation has the complexity of $O(h)$.

2.3 Multidimensional Packet Classification Using AM-Trie

In order to perform multidimensional classification using AM-Trie, we profit from the idea of Aggregated Bit Vector (ABV) [1]. For a d-dimension classification problem we build d AM-Tries, each AM-Trie maps to a dimension and its leaf nodes stores the ABV. Because the search operations in different dimension are independent, they can be parallel execution. AM-Trie algorithm is very fit for the architecture of Network Processor, it can run in multiple microengines or hardware threads concurrently and increase the performance enormously. When we get the ABVs of all the dimensions, we AND them together and come out the final classification result.

For a d-dimensional classification policy table which has N rules, if the height of generated AM-Trie is h (usually $h \ll W$, where W is the bits of the field), the complexity of AM-Trie algorithm is:

1. The initialization complexity of AM-Trie algorithm is $O(N \log_2 N)$.
   The creation time complexity of AM-Trie is $O(Nh)$, but the initialization of ABV need to sort the policy table. Hence, the initialization complexity $O(N \log_2 N)$.

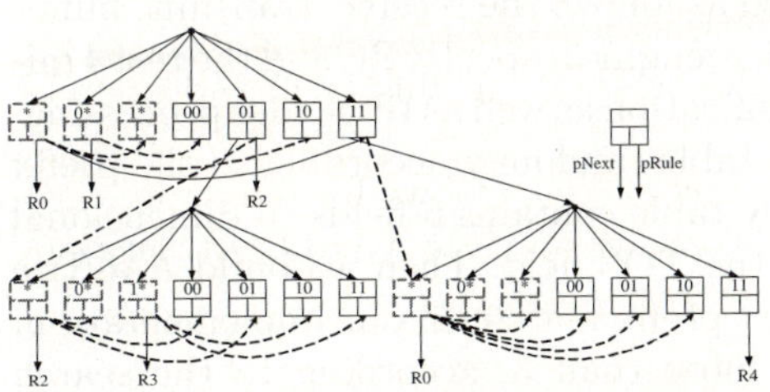

Fig. 2. AM-Trie "push" and compression

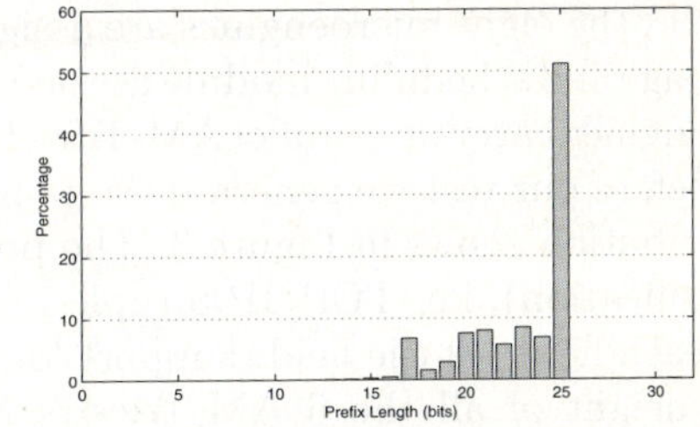

Fig. 3. Prefix distribution in a BGP table

2. The search complexity of AM-Trie algorithm is $O(h + d)$.
In worst case, each AM-Trie needs h comparisons to get the ABV address, different AM-Trie can search concurrently; after that still need to read the d ABVs and AND them to get the final result. So the search complexity of AM-Trie algorithm is $O(h + d)$.
3. The space complexity of AM-Trie algorithm is $O(N^2)$.
Because there are total N policies, the number of prefix in any field $\leq N$. If a level is m-bits, the nodes which have children cost $2^m * sizeof(Node)$ in the child level. In the worst case (all the nodes in the $h-1$ level have children and all the children do not duplicate), the required storage is less than

$$[(h-1) * N + 1] * 2^m * sizeof(Node) + N * sizeof(ABV). \quad (2)$$

Where $sizeof(Node)$ is the size of AM-Trie Node structure (8 bytes), and $sizeof(ABV)$ is the size of aggregated bit vector (N-bits). So the space complexity of AM-Trie algorithm is $O(N^2)$ (Assume m is small enough that 2^m not a big number compare to N).

But how to choose the the number of level and the stride of each level to gain a better performance? The lower the AM-Trie the faster the algorithm, but the more storage it will cost. If we build a AM-Trie with only one level, the 2^m in Equation (2) will be 2^{32}, we can not afford such a huge memory cost.

In order to optimize the space complexity we can take the distribution of prefix length into account. In the practical policy table, the prefix length distribution is very non-uniformity, Figure 3 is the prefix length distribution of a BGP table which we randomly select from [9]. This table contains 59351 policies, most of them belong to a few kind of prefix length, the 24-bit policies even more than half. A straightforward idea is if we choose the prefix length with high probability as the dividing point, the generated AM-Trie will cost less storage. There are some related work on this topic, we can use a similar way as Srinivasan proposed in [10] to optimize the AM-Trie space complexity.

3 Performance Evaluation

We have implemented the AM-Trie Algorithm in single Intel IXP2400 network processor, the eight microengines are assigned as follow: the receive, transmit, buffer manage and scheduling module use one microengine respectively, and the rest 4 microengines carry on parallel AM-Trie classification as well as IP header processing.

Before our test we generate the policy table randomly according to the prefix distribution shows in Figure 3. The policy table contains 6 fields (6-dimensional classification), i.e. TCP/IP 5 tuple plus the TOS field. Then we build AM-Trie for each field, all the fields support longest prefix match. In our implementation, the height of all the 6 AM-Tries is no more than 3, according to the search complexity analysis in section 2.3, only 15 memory access needed to classify one packet. In other words, even if implemented with low speed DDR SDRAM

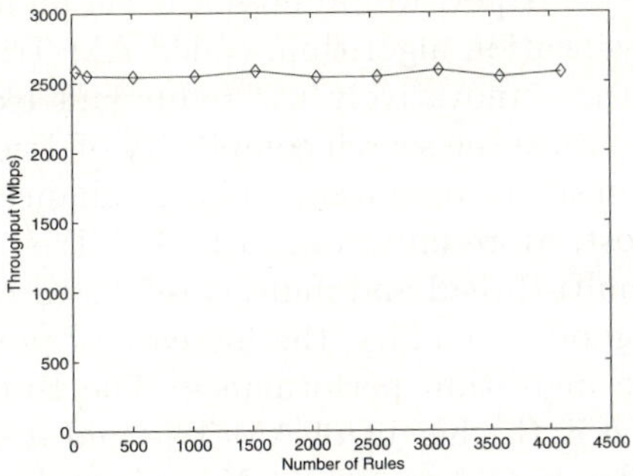

Fig. 4. The throughput of AM-Trie on IXP2400, the system achieves 2.5Gbps wire-speed in all condition (number of policies increases from 32 to 4096)

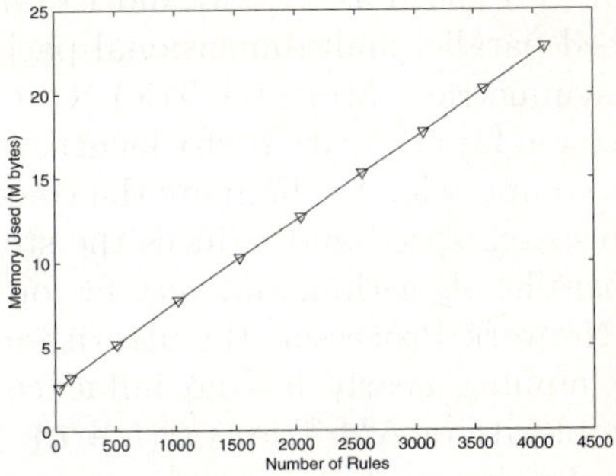

Fig. 5. The cost of memory increase linearly as a function of policy number. AM-Trie only used 23M RAM when the size of policy table is 4096.

(150MHz), AM-TRIE may also achieve fast wire-speed (10Mpps). Finally we begin our test using IXIA 1600 traffic generator to generate traffic to IXP2400 (all the packets are 64 bytes).

Figure 4 shows the throughput of the whole system. The system throughput does not decrease when the number of policies increases from 32 to 4096, the throughput maintain above 2.5Gbps (5Mpps). The reasons are: 1) AM-Trie algorithm has good scalability, it can easily be extend to support 256K policies using hierarchical structure if necessary, and still maintain such high speed; 2) IXP2400 is oriented to OC48 (2.5Gbps) network application, the AM-Trie algorithm already makes the most of IXP2400 capability, by the restriction of other system modules (such as receive, transmit, scheduling, etc.), the system performance is unable to enhance more.

Figure 5 demonstrated the memory cost of AM-Trie algorithm as a function of the number of policy (in order to simplify the implementation, we fix the length of ABV to 4096 bits, therefore the memory cost appear linear growth, but not the theoretically space complexity $O(N^2)$). AM-Trie belongs to algorithm of "trade space for time", it use a lower speed but large capacity and cheap RAM to achieve comparative performance as the expensive and power-slurping hardware (such as FPGA, TCAM, etc.). IXP2400 support 2x64M SRAM and 2G DDR SDRAM, it's very ample for AM-Trie algorithm.

We are implementing this algorithm on Intel high end network processor IXP2800, which designed for OC192 network applications. The preliminary test result is closed to 10Gbps, anticipated throughput will achieve 10Gbps (20M pps) after code optimization.

4 Conclusion and Further Discussion

The rapid growth of Internet needs high-speed packet classification algorithm, but current high-speed classification often use expensive and power-slurping hardware (such as TCAM and FPGA). In this paper, we propose a novel high-speed parallel multidimensional packet classification algorithm, called AM-Trie (Asymmetrical Multi-bit Trie). Our algorithm innovatively use redundant expression for arbitrary prefix length, greatly reduce the search complexity of Trie; use compression to eliminate the trace back in search operation, further enhances the search speed and reduces the storage cost. More important, the AM-Trie is a parallel algorithm, and very fit for the "multi-thread and multi-core" features of Network Processor; the algorithm have good scalability, the increase of policy number nearly has no influence to the algorithm performance. The time complexity of AM-Trie is $O(h + d)$, where h is the height of AM-Trie and d is the dimension of classification; space complex is $O(N^2)$, where N is the policy number.

We also implemented AM-Trie algorithm based on Intel IXP2400. The performance analysis and the test result show that the AM-Trie algorithm can achieves 2.5Gbps wire-speed (all the packets size are 64 bytes, i.e. 5Mpps) in the TCP/IP 6 tuple classification, and it has big performance promotion space in more powerful Network Processor(such as IXP2800 or above). At the same time, our experimental result also indicate that "trade space for time" algorithms on Network Processor using lower speed but cheap and large capacity RAM to obtain high-speed classification is a feasible way.

References

1. F. Baboescu and G. Varghese, Scalable packet classification, in SIGCOMM, 2001, pp. 199–210.
2. G. V. Sumeet Singh, Florin Baboescu and J. Wang, Packet classification using multidimensional cutting, in SIGCOMM, 2003, pp. 213–224.
3. Florin Baboescu and G. Varghese, Packet classification for core routers: Is there an alternative to cams? in INFOCOM, 2003, pp. 53–63.
4. A. R. Karthik Lakshminarayanan and S. Venkatachary, Algorithms for advanced packet classification with ternary cams, in SIGCOMM, 2005.
5. N. Shah, Understanding network processors, Tech. Rep., 2001.
6. Network processing forum (npf). http://www.npforum.org/
7. Network systems design conference. http://www.networkprocessors.com/
8. T. Wolf and M. A. Franklin, Design tradeoff for embedded network processors, in ARCS, 2002, pp. 149–164.
9. BGP routing table analysis reports. http://bgp.potaroo.net/
10. V. Srinivasan and G. Varghese, Faster IP lookups using controlled prefix expansion, in Measurement and Modeling of Computer Systems, 1998, pp. 1–10.

Speedup Requirements for Output Queuing Emulation with a Sliding-Window Parallel Packet Switch

Chia-Lung Liu[1], Woei Lin[1], and Chin-Chi Wu[2]

[1] Department of Computer Science, National Chung-Hsing University, 250, Kuo Kuang Road, Taichung, Taiwan
{s9056005, wlin}@cs.nchu.edu.tw
[2] Nan Kai Institute of Technology
wcc007@nkc.edu.tw

Abstract. This investigation uses an approximate Markov chain to determine whether a sliding window (SW) parallel packet switch (PPS), only operating more slowly than the external line speed, can emulate a first-come first-served (FCFS) output-queued (OQ) packet switch. A new SW packet switching scheme for PPS, which is called SW-PPS, was presented in the authors' earlier study [1]. The PPS class is characterized by deployment of distributed center-stage switch planes with memory buffers that run slower than the external line speed. Given identical Bernoulli and Bursty data traffic, the proposed SW-PPS provided substantially outperformed typical PPS, in which the dispatch algorithm applies a round-robin method (RR) [1]. This study develops a presented Markov chain model that successfully exhibits throughput, cell delay and cell drop rate. A simulation reveals that the chains are accurate for reasonable network loads. Major findings concerning the presented model are that: (1) the throughput and cell drop rates of a SW-PPS can theoretically emulate those of a FCFS-OQ packet switch when each slower packet switch operates at a rate of around R/K (Eq. 19); and, (2) this investigation also proves the theoretical possibility that the cell delay of a SW-PPS can emulate that of an FCFS-OQ packet switch, when each slower packet switch operates at a rate of about (R/cell delay of FCFS-OQ switch) (Eq. 20).

1 Introduction

For efficient data deliver in high-capacity switching systems, the PPS is a good choice in a high-speed communication [1]–[3]. However, traditional PPS which uses a round-robin method cannot effectively use the memory space of center-stage switches, and so requires much memory [3]. Our previous study describes a novel SW packet switching method for PPS, called SW-PPS [1]. This novel switching scheme overcomes the shortcomings of traditional PPS, and uses memory space more effectively than traditional PPS. Accordingly, previous experimental results demonstrate that SW-PPS outperformed traditional PPS [1].

Work conserving [2] refers to a situation in which a switch's output port is working whenever there is a cell in the switch for it. FCFS-OQ switches are work conserving. However, FCFS-OQ switches require buffer memories running at N times

V.N. Alexandrov et al. (Eds.): ICCS 2006, Part IV, LNCS 3994, pp. 49 – 56, 2006.
© Springer-Verlag Berlin Heidelberg 2006

external line rates. A necessary condition for a switch to emulate the performance of FCFS-OQ switch is that it be work conserving. Under identical inputs, if switch A emulates switch B, the performance of switch A equals or transcends that of the switch B. Therefore, this study reveals that the throughput and drop rate of the SW-PPS can emulate those of a FCFS-OQ switch with $S \geq 1$ (Eq. 19), and that cell delay of the SW-PPS can emulate that of a FCFS-OQ switch with $S \geq$ K/delay of FCFS-OQ switch (Eq. 20), where S is the speedup of the internal line speed (R/K).

2 SW-PPS Architecture

The SW-PPS is divided into the following independent stages: 1) the self-routing parameter assignment circuit; 2) the demultiplexers; 3) the slower speed center-stage packet switches; and, 4) the multiplexers (shown in figure 1). The destined output port of the incoming cell, extracted by applying header processing circuits, is indicated by *d*. The incoming cell's destination address *d* is delivered to a self-routing parameter assignment circuit. In processing incoming cells, the self-routing parameter assignment circuit employs the output port *d* and a parameter assignment algorithm to create an additional group of self-routing parameters (*i, j, and d*). These self-routing parameters (*i, j, and d*) are attached to incoming cells as a self-routing tags. Incoming cells then use the attached tags to navigate through the demultiplexers and center-stage switches. Parameters (*i, j, and d*) are defined as follows: the variable *i* in parameters informs the center-stage switch where the cell will be stored; variable *d* indicates which memory module in the *i*th center-stage switch the cell will be stored in; and, variable *j* designates the memory location in the *d*th memory module where the cell will be stored. During the cell WRITE cycle for an incoming cell, the cell is written to *j*th memory location in a given memory module *d* and a given center-stage switch *i*. During the cell READ cycle, cells are sent to multiplexers according to the location of the SW. The number of center-stage switches refers to the reference [2].

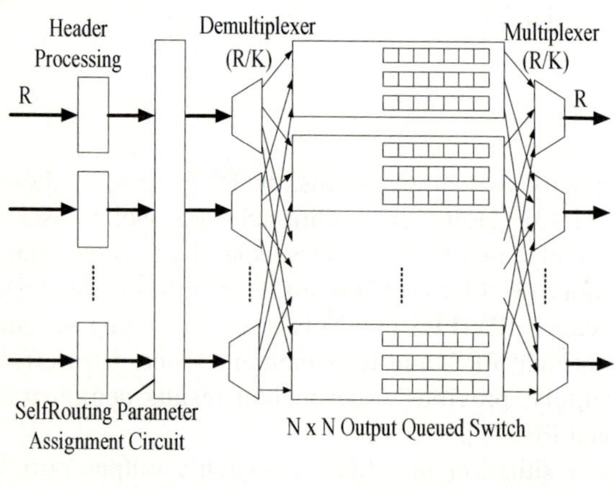

Fig. 1. Architecture of the SW-PPS

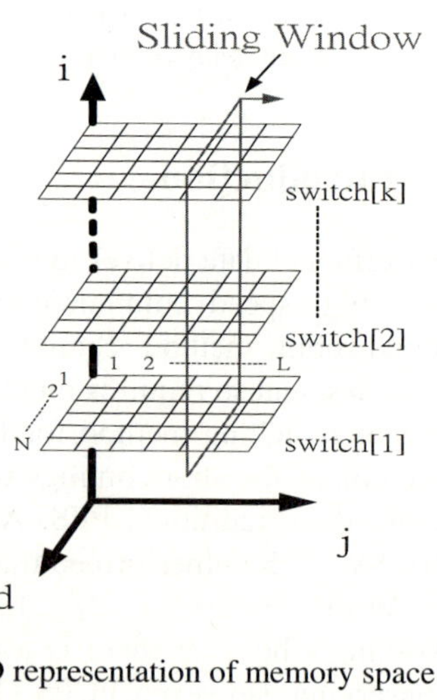

Fig. 2. A 3-D representation of memory space

3 SW Switching Scheme

According to the SW-PPS switching schedule, the overall memory space, including all cell memory locations in all of the center-stage switches, is represented as a three-dimensional (3-D) memory space (*i, j, and d*) (shown in figure 2). The memory locations in the global memory space are represented by a 3-D coordinate system (*i, j, and d*), where the *i*th coordinate represents the center-stage switch; *i*= [1…K], where K is the number of center-stage switches. The *d*th coordinate indicates the memory module; *d*= [1…N], where N is the size of PPS (or the size of the center-stage switch); *j*th coordinate designates the memory location in the memory module; *j*= [1…L], where L is queue size. In other words, L represents the number of memory locations in each memory module. The SW is regarded as a pointer to a group of cells in the 3-D memory space (shown in figure 2). The SW advances one step during each switch cycle. The location of the SW in the global memory space is recorded by one variable: SW.*j*.

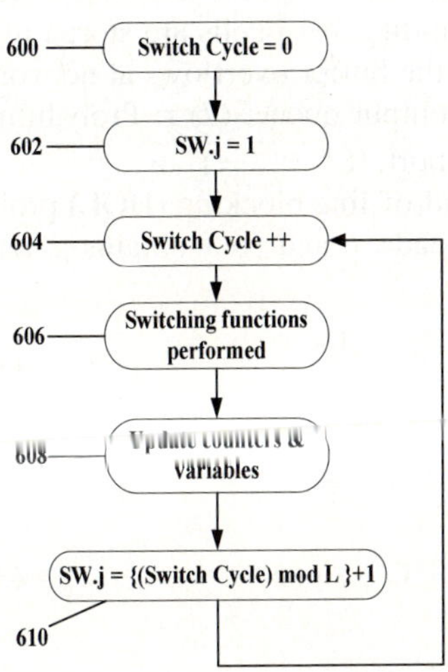

Fig. 3. Flowchart depicting traversal of SW

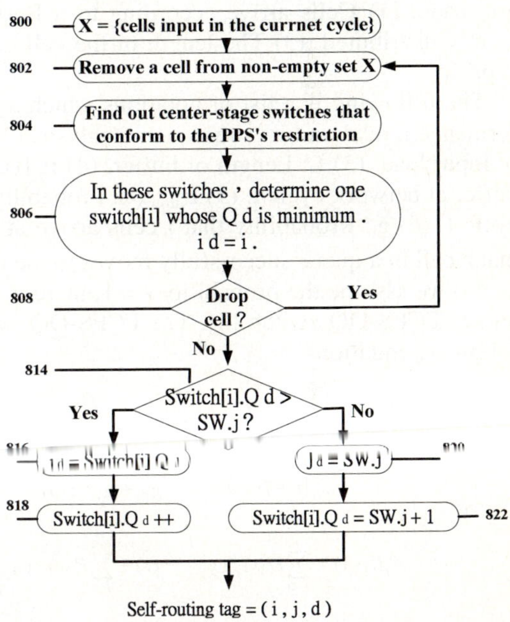

Fig. 4. Assignment process for self-routing parameters (i, j, and d)

The flowchart in figure 3 shows SW traversal across the entire 3-D memory space and its connection to the switch cycle and switching operation. After switching functions in step 606, counters and variables are updated in 608 to account for changes. The SW is then advanced to next location in step 610.

Assigning self-routing parameters (*i, j, and d*) to incoming cells is computed by the parameter assignment circuit. An additional routing tag carrying the self-routing parameters (*i, j, and d*) is attached to incoming cells. Before exploring the process the self-routing parameters (*i, j, and d*), the PPS's restrictions first be understood [1][2]. Determination of self-routing parameters (*i, j, and d*) by the assignment circuit to an incoming cell is shown by the flowchart in figure 4. The symbols used therein are described as follows, (1) d is the cell's output-port destination (2) (i_d, *and* j_d) are the parameters (*i, and j*) of the incoming cell sent to output port d (3) m is the number of the fanout, when it is in the multicast traffic (4) *Switch[i].*Q_d is the memory location that is written next time inside the dth memory module and ith center-stage switch for the cells destined to output port d. (5) X is a set of cells input during a given external cycle, $0 \le |X| \le N$, where N is the number of input ports. According to step 808 (shown in Fig. 4), if the number of cells in the dth queue of the *switch[i]* is greater than L, the cell destined to d is dropped.

4 Analysis for FCFS-OQ Switch

The performance of a FCFS-OQ switch is first analyzed. The following assumptions are made: (1) Cells arrive according to a Bernoulli process. (2)Traffic load is uniformly distributed. (3) The length of the cell is constant. (4) Each queue has a finite capacity.

The following is a list of notations which are used in the development of the performance model and its subsequent analysis. (1) N: Size of the FCFS-OQ switch. (2) ρ: Input load. (3) L: Length of buffer. (4) P_j (t): Probability that j cells are stored in a buffer at network cycle t. (5) P_{drop} (t): Probability that the buffer overflows at network cycle t. (6) g_i: Probability that i cells arrive at same output queue. (7) r: Probability that a cell in a queue successfully moves to the output port. (8) $\bar{r}$: $\bar{r} = 1 - r$.

So we assume the probability r = 1, there is no head-of-line blocking (HOL) problem in FCFS-OQ switch [5]. The FCFS-OQ switch load= ρ and r= 1 obtains g_i and following equations.

$$g_i = C_i^N * \left(\frac{\rho}{N}\right)^i * \left(1 - \frac{\rho}{N}\right)^{N-i}, \qquad 0 \le i \le N \cdot \tag{1}$$

$$P_0(t+1) = P_0(t) * (g_0 + g_1 * (r)) + P_1(t) * (g_0) * (r) \cdot \tag{2}$$

$$P_j(t+1) = \sum_{n=0}^{j} P_n(t) * (g_{j-n}) * (\bar{r}) + \sum_{n=0}^{j+1} P_n(t) * (g_{j+1-n}) * (r), \; 1 \le j \le L-1 \cdot \tag{3}$$

$$P_L(t+1) = \sum_{n=0}^{L} P_n(t) * (g_{L-n}) * (\bar{r}) + \sum_{n=0}^{L} P_n(t) * (g_{L+1-n}) * (r) + P_{drop}(t) \cdot \tag{4}$$

$$P_{drop}(t+1) = \sum_{n=0}^{L} \sum_{i=L+1-n}^{N} P_n(t) * (g_i) * (\bar{r}) + \sum_{n=0}^{L} \sum_{i=L+2-n}^{N} P_n(t) * (g_i) * (r) \cdot \tag{5}$$

The three primary performance measures are given as follows.

$$FCFS\_OQ\_Switch\_Drop \;\; Rate(\rho, S, t, N, K, L) = P_{drop}(t) \cdot \tag{6}$$

$$FCFS\_OQ\_Switch\_Throughput \;\; (\rho, S, t, N, K, L) = (\rho - P_{drop}(t)) \cdot \tag{7}$$

$$FCFS\_OQ\_Switch\_Delay(\rho,S,t,N,K,L)=\frac{[\sum_{i=1}^{L}(i)*(P_i(t))]+L*(P_{drop}(t))}{1-P_0(t)}. \tag{8}$$

5 SW-PPS Analysis

The SW-PPS was further simplified to an output buffer that is described by a Markov chain model. Finally, four equations of performance evaluations are derived. The assumptions made herein are the same as those in section 4 with one additional assumption: Low speed center-stage switches are OQ switches.

For clarity, we list notations that are employed in the development of the performance model and analysis. (1) N: Size of the SW-PPS. (2) K: Number of center-stage switches. (3) ρ: Input load. (4) L_{sw}: Size of buffer. (5) P_{sw_j} (t): Probability that j cells are stored in a buffer at network cycle t. (6) P_{sw_drop} (t): Probability that buffer overflows at network cycle t. (7) g_{sw_i}: Probability that i cells arrive at the same output buffer. (8) r_{sw}: Probability that a cell in a buffer successfully moves to the multiplexer. (9) $\bar{r}_{sw}$: $\overline{r_{sw}}=1-r_{sw}$. (10) S: The speedup of the internal link.

Because (1) External load= ρ, (2) number of center-stage switches= K, (3) SW scheme will choose one of the switches that Q_d is minimum (shown in steps 804 and 806 in the figure 4), we obtained internal load of the SW-PPS (ρ'), g_{sw_i} and r_{sw}.

$$\rho'=\frac{\rho}{C_1^{K(1-\rho)+1}}. \tag{9}$$

$$g_{sw\_i}=C_i^N*(\frac{\rho'}{N}*\frac{1}{S})^i*(1-\frac{\rho'}{N}*\frac{1}{S})^{N-i},\qquad 0\le i\le N. \tag{10}$$

We assume the probability r_{sw} = 1, because there is no head-of-line blocking (HOL) problem in center-stage switches that are OQ switches.

The resulting equations are similar to those for a Markov chain of the FCFS-OQ switch (section 4), because the center stage switches use FCFS-OQ switches. We have following equations.

$$P_{sw\_0}(t+1)=P_{sw\_0}(t)*(g_{sw\_0}+g_{sw\_1}*r_{sw})+P_{sw\_1}(t)*(g_{sw\_0})*(r_{sw}). \tag{11}$$

$$P_{sw\_j}(t+1)=\sum_{n=0}^{j}P_{sw\_n}(t)*(g_{sw\_j-n})*(\bar{r}_{sw})+\sum_{n=0}^{j+1}P_{sw\_n}(t)*(g_{sw\_j+1-n})*(r_{sw}),\ 1\le j\le L_{sw}-1. \tag{12}$$

$$P_{sw\_L_{sw}}(t+1)=\sum_{n=0}^{L_{sw}}P_{sw\_n}(t)*(g_{sw\_L_{sw}-n})*(\overline{r_{sw}})+\sum_{n=0}^{L_{sw}}P_{sw\_n}(t)*(g_{sw\_L_{sw}+1-n})*(r_{sw})+P_{sw\_drop}(t) \tag{13}$$

$$P_{sw\_drop}(t+1)=\sum_{n=0}^{L_{sw}}\sum_{i=L_{sw}+1-n}^{N}P_{sw\_n}(t)*(g_{sw\_i})*(\overline{r_{sw}})+\sum_{n=0}^{L_{sw}}\sum_{i=L_{sw}+2-n}^{N}P_{sw\_n}(t)*(g_{sw\_i})*(r_{sw}). \tag{14}$$

The four performance measures are as follows.

$$SW\_PPS\_Drop\ \ Rate(\rho,S,t,N,K,L_{sw})=P_{sw\_drop}(t). \tag{15}$$

$$SW\_PPS\_Throughput(\rho,S,t,N,K,L_{sw}) = \rho - P_{sw\_drop}(t) \cdot \quad (16)$$

$$SW\_PPS\_Internal\ \ Delay(\rho,S,t,N,K,L_{sw}) = \frac{[\sum_{i=1}^{L_{sw}}(i)*(P_{sw\_i}(t))] + L_{sw} * P_{sw\_drop}(t)}{1 - P_{sw\_0}(t)} \cdot \quad (17)$$

$$SW\_PPS\_External\ \ Delay(\rho,S,t,N,K,L_{sw}) = \frac{[\sum_{i=1}^{L_{sw}}(i)*(P_{sw\_i}(t))]*K + L_{sw} * P_{sw\_drop}(t)*K}{S - P_{sw\_0}(t)*S} \quad (18)$$

6 Speedup Requirements for FCFS-OQ Switch Emulation

If the FCFS-OQ Switch Drop Rate (Eq. 6), divided by the SW-PPS Drop Rate (Eq. 15) exceeds one, the drop rate of SW-PPS emulates that of FCFS-OQ switch. Based on the assumption that length of a FCFS-OQ switch queue equals length of center-stage switch queue, $L = L_{sw}$. *FCFS_OQ_Switch_Drop_Rate/SW_PPS_Drop_Rate≥1*. This means that $P_{drop}(t)/P_{sw\_drop}(t) \geq 1$, so that

$$S \geq 1 \cdot \quad (19)$$

Applying $S = 1$ and $L = L_{sw}$ yields $g_i \geq g_{sw\_i}$ and, $P_j \geq P_{sw\_j}$, where $0 \leq i \leq N$ and $0 \leq j \leq L$. Since $S \geq 1$, $P_{drop} \geq P_{sw\_drop}$. Throughput is determined by subtracting drop rate from arrival rate. With a speedup of $S \geq 1$ and internal link rate $\geq R/K$, the SW-PPS can emulate FCFS-OQ switch throughput and drop rate.

A same procedure demonstrates that SW-PPS can match the cell delay of FCFS-OQ switch, if the following inequality is satisfied. If *FCFS_OQ_Switch_Delay /SW_PPS_External_Delay≥1*, we have *FCFS_OQ_Switch_Delay/SW_PPS_Internal_ Delay*(K/S)≥1*. By increasing speedup S, the SW-PPS Internal Delay (Eq. 17) can be effectively reduced. We also obtain

$$S \geq \frac{SW\_PPS\_Internal\_Delay * K}{FCFS\_OQ\_Switch\_Delay} \cong \frac{K*(1-P_0(t))}{[\sum_{i=1}^{L}(i)*(P_i(t))] + L*(P_{drop}(t))} \cdot \quad (20)$$

7 Comparison and Simulation

The evaluations used to measure performance are mean throughput, cell loss ratio, and mean cell delay. A 32*32 FCFS-OQ switch and SW-PPS are established; the number of center-stage FCFS-OQ switches is K= 64; the queue size (L) is 16 and 4. Figures 5 to 10 show a mathematical analysis and simulation results for the OQ (L, x) and SW-PPS (L, x, speedup). If x= S represents the simulation result, then x= M yields the results of the Markov chains analysis (whose results are plotted as hollow points in figures 5 to 10). When the speedup satisfies Eq. 19, figures 5 to 8 compare the throughput and drop rate for OQ (L, x) and SW-PPS (L, x, speedup= Eq. 19). The SW-PPS accurately emulates the FCFS-OQ switch. The cell delay of SW-PPS (L, x,

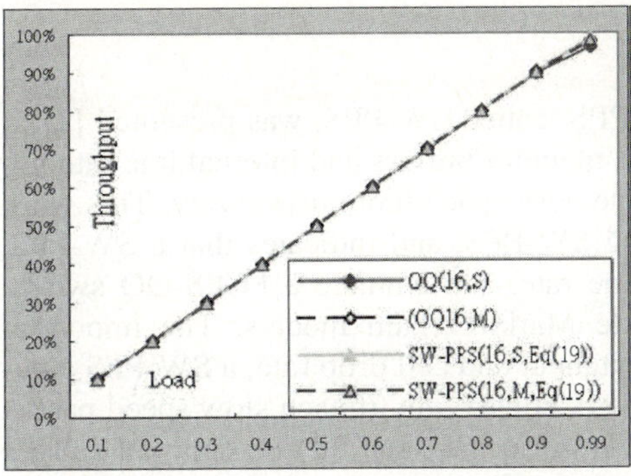

Fig. 5. Emulation of throughput (L=16)

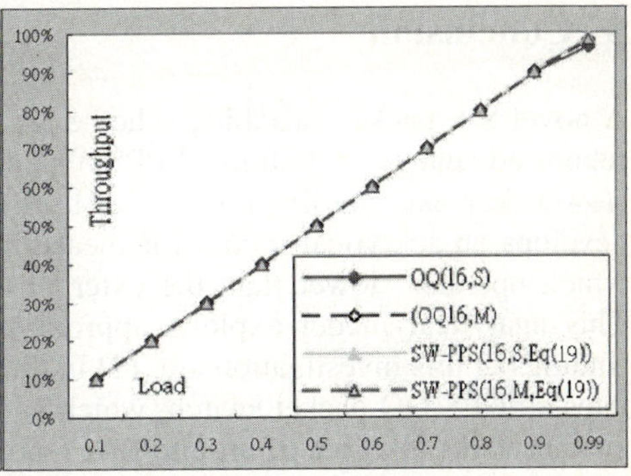

Fig. 6. Emulation of throughput (L=4)

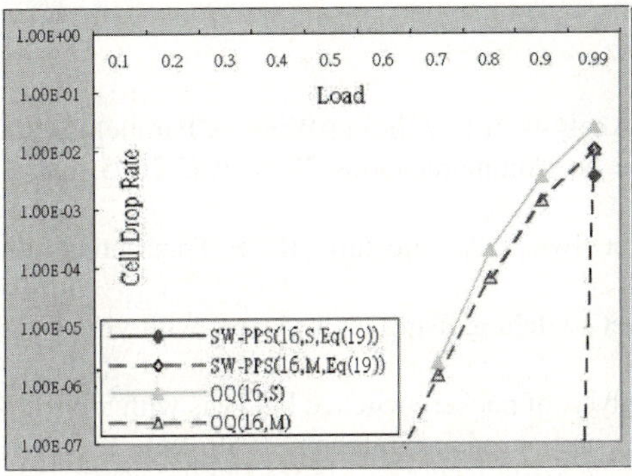

Fig. 7. Emulation of drop rate (L=16)

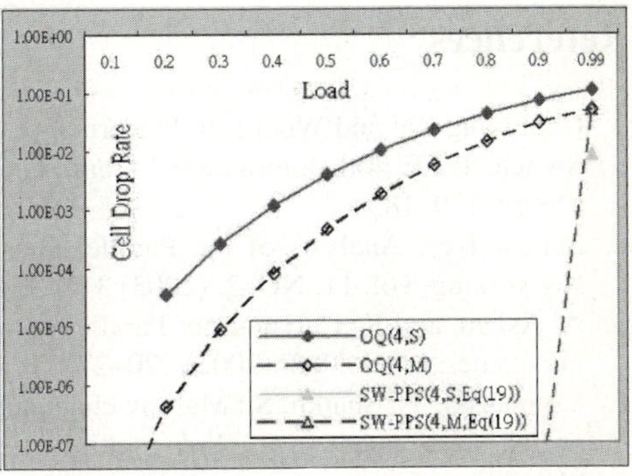

Fig. 8. Emulation of drop rate (L=4)

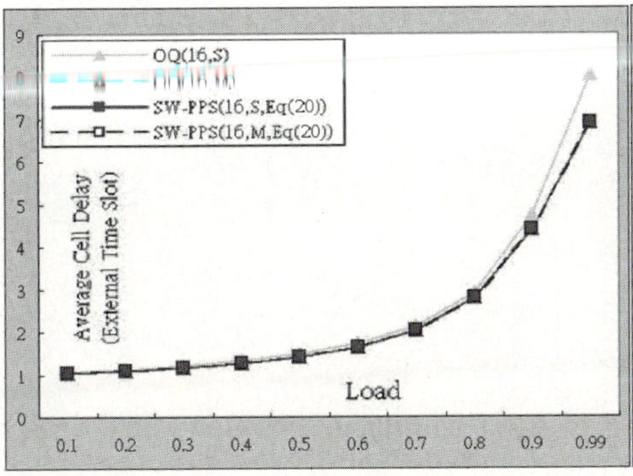

Fig. 9. Emulation of delay (L=16)

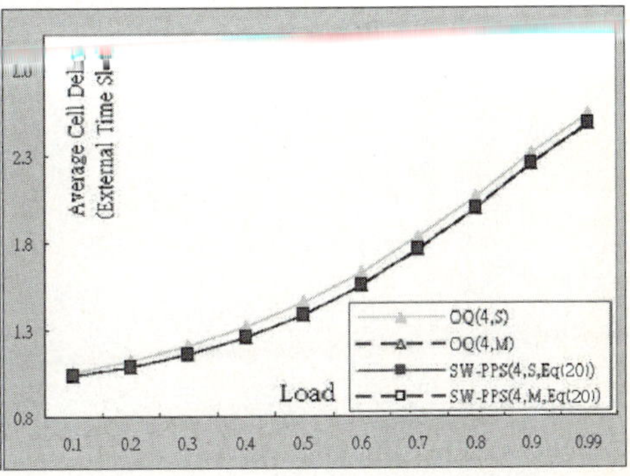

Fig. 10. Emulation of delay (L=4)

speedup= Eq. 20) is very close to that of OQ (L, x) (figures 9 and 10), when the speedup of internal link rate is as given by Eq. 20. Figures 5 to 10 reveal that the FCFS-OQ switch performance is emulated remarkably well closely. The analytical results agree closely with the simulation results.

8 Conclusion

A novel SW packet switching scheme for PPS, called SW-PPS, was presented [1]. It retains advantage of traditional PPS, that all memory buffers and internal line rate run slower than external line rate, and it uses memory space more effectively. This work develops an analytical model for measuring SW-PPS, and indicates that a SW-PPS, which operates slower than the external line rate, can emulate a FCFS-OQ switch. This analytical model exploits approximate Markov chain models. The important findings of this investigation are: (1) in throughput and cell drop rate, a SW-PPS emulates a FCFS-OQ packet switch, which is work conserving, if each slow speed packet switch works at a rate of around R/K (shown in Eq. 19); and, (2) in delay of cells, a SW-PPS can emulate a FCFS-OQ packet switch when each slow speed packet switch works at a rate of about (R/cell delay of FCFS-OQ switch) (shown in Eq. 20).

References

1. Chia Lung Liu and Woei Lin: Performance Analysis of the Sliding-Window Parallel Packet Switch. IEEE 40th International Conference on Communications 2005, ICC 2005. vol. 1 (2005). 179–183
2. Sundar Iyer: Analysis of the Parallel Packet Switch Architecture. IEEE Transactions on Networking, vol. 11, NO. 2. (2003) 314–324
3. A. Aslam, and K. Christensen: Parallel packet switching using multiplexers with virtual input queues. LCN 2002. (2002) 270–277
4. Szymanski, T., Shaikh, S.: Markov chain analysis of packet-switched banyans with arbitrary switch sizes, queue sizes, link multiplicities and speedups. Infocom 1989. vol. 3. (1989) 960–971.
5. Karol, M., Hluchyj, M., Morgan, S.: Input Versus Output Queueing on a Space-Division Packet Switch. Communications, IEEE Transactions on vol. 35. Issue 12. (1987) 1347–1356

Combining Cross-Correlation and Fuzzy Classification to Detect Distributed Denial-of-Service Attacks*

Wei Wei[1], Yabo Dong[1], Dongming Lu[1], and Guang Jin[2]

[1] College of Compute Science and Technology,
Zhejiang University, Hangzhou 310027, P.R. China
[2] College of Information Science and Engineering,
Ningbo University, Ningbo 315211, P.R. China
{weiwei_tc, dongyb, ldm, d05jinguang}@zju.edu.cn

Abstract. In legitimate traffic the correlation exists between the outgoing traffic and incoming traffic of a server network because of the request-reply actions in most protocols. When DDoS attacks occur, the attackers send packets with faked source addresses. As a result, the outgoing traffic to the faked addresses does not induce any related incoming traffic. Our main idea is to find changes in the correlation caused by DDoS. We sample network traffics using Extended First Connection Density (EFCD), and express correlation by cross-correlation function. Because network traffic in DDoS-initiating stage is much similar to legitimate traffic, we use fuzzy classification in order to guarantee the accuracy. Experiments show that DDoS traffic can be identified accurately by our algorithm.

1 Introduction

With the development of the Internet, Distributed Denial of Service (DDoS) becomes a major security threat. In several cases, over ten thousands of attack agents cooperate to send attack packets to flood targets. Attack packets aggregate to block target network or directly bring target host down.

Various methods have been proposed for DDoS detection. J. Mirkovic et al present a detailed conclusion and classification [1] and Q. Li et al give out an analysis on statistical filtering [2]. For statistical detection of DDoS, spectral analysis is adopted. A. Hussain et al use power spectrum to extract fingerprint features from attack stream and DDoS attack scenarios are identified when repeated [3]. Meanwhile, Entropy is used as a summarization tool. A. Lakhina [4] et al get entropy of several packet features and use multi-way subspace method to combine multiple features, then use clustering approach to detect. Besides, L. Feinstein et al give algorithms based on entropy and chi-square [5]. Moreover, many traffic correlation based methods have appeared. S. Jin et al get covariance matrix from several traffic sources and uses the matrix for detection [6]. Wavelet detection methods are also presented. L. Li et al

* Funded by National Natural Science Foundation of China (60503061); Zhejiang Provincial Natural Science Foundation (Y104437); Zhejiang Provincial Science and Technology Program (2005C33034); the Program for New Century Excellent Talents in University (No. NCET-04-0535).

V.N. Alexandrov et al. (Eds.): ICCS 2006, Part IV, LNCS 3994, pp. 57 – 64, 2006.
© Springer-Verlag Berlin Heidelberg 2006

compute the difference of wavelet energy function vector between neighborhood traffic windows [7]. Furthermore, Network Traffic's fractal property is used. M. Li uses Long Range Dependent (LRD) traffic pattern recognition for DDoS detection [8]. Y. Xiang et al identify DDoS through detecting self-similarity changes in traffics [9]. In addition, change point methods are widely used. One is R. Blazek et al's sequential change-point and batch sequential detection [10]. And some pattern classification methods are used, such as support vector machine detection in [11].

As to above methods, the main problems are accuracy and performance. Here we propose a DDoS detection method named Cross-Correlation based Fuzzy Classification (CCFC). The method uses Extended First Connection Density (EFCD) to sample outgoing and incoming traffic, and computes the cross-correlation vector sequence. Then fuzzy classification is applied to the sequence for DDoS detection. Experiments show that CCFC is accurate. The rest of this paper is organized as follows: in section 2 we present the principle of CCFC. In section 3 we focus on the implementation of CCFC while some definitions are provided. In section 4, experiment results are shown and explained. Then conclusion is given in section 5.

2 The Principle of CCFC

2.1 The Definition of EFCD

EFCD is the count of first connections in a time unit. Here connection is defined with 5 fields in IP header, i.e. source address, source port, destination address, destination port, TCP flags (if have). The first 4 fields are usually used for TCP connection identification. In DDoS attacks, these fields may vary a lot and EFCD value changes abnormally. TCP flags field is used to strengthen the cross-correlation of sampled outgoing and incoming EFCD sequences of legitimate TCP traffic.

Table 1. The EFCD value for ten outgoing TCP packets

Packets	Source IP	Source port	Dest IP	Dest port	TCP flag	Conn order
TCP 1	A	80	B	42957	0x26	1
TCP 2	C	23	D	34425	0x13	2
TCP 3	A	80	B	42957	0x26	1
TCP 4	E	443	F	4256	0x13	3
TCP 5	G	42364	H	80	---	4
TCP 6	A	80	B	42957	0x26	1
TCP 7	I	23467	J	23	0x36	5
TCP 8	B	---	A	---	---	6
TCP 9	B	---	A	---	---	6
TCP 10	I	23467	J	23	0x36	5

Table 1 shows 10 TCP packets in 0.1 seconds. As we can see, they belong to 6 different first connections. So with a time unit of 0.1 seconds, the EFCD is 6.

EFCD, as a kind of flow-level sample, is more sensitive than pack-level sample in connection based attack. For EFCD, the time unit should be chosen carefully. Bigger time unit will shorten the computation time while decrease the detection accuracy. Because RTTs (Round Trip Time) of legitimate network connections are mostly less than 1 second, we choose 0.1 seconds as a time unit to balance the accuracy and performance. Fig. 1 is EFCD with the time unit of 0.1 seconds from MIT Lincoln Laboratory outside data [12].

2.2 Principle

The correlation between the traffics of two directions exists because of the request-reply actions in many protocols, such as TCP and many application-level protocols. For example, when a server receives a TCP SYN request, it sends out a TCP ACK packets. And then legitimate client receives the reply and responses data packets with TCP ACK flag set. In above scenario, one EFCD of outgoing traffic contributes to one EFCD of incoming traffic, therefore, there is a positive correlation between values of EFCD of outgoing and incoming traffic and that is shown in Fig. 2.

When DDoS traffic flushes into the server network, most part of EFCD value of incoming attack flows is not caused by EFCD of any earlier outgoing flows. As a result, the correlation between EFCD of the outgoing and incoming traffic is deviated, which can be used for detection. In next section, we will explain our algorithm in details.

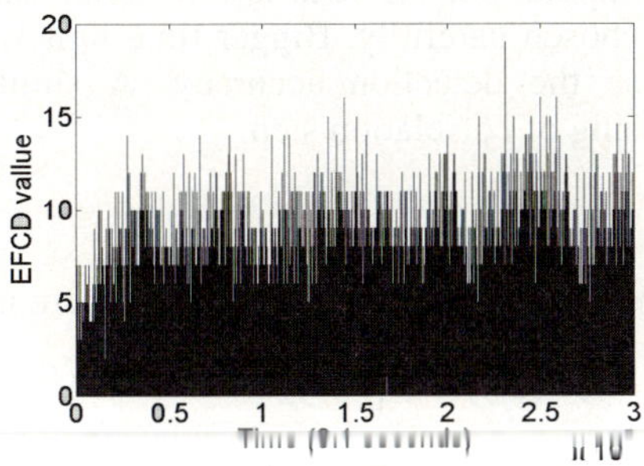

Fig. 1. EFCD sampled values

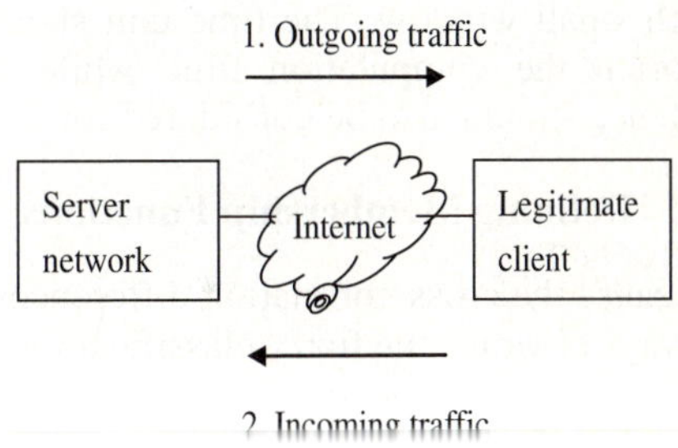

Fig. 2. The request-reply action in normal traffic

3 The Algorithm of CCFC

3.1 Computing Cross-Correlation Vector Sequence

Correlation has been used in network anomaly detection, such as in [6]. It can be well described by cross-correlation function, which is definded as below [13]:

$$R_{xy}(t,t+\tau) = \frac{\operatorname{cov}(X(t),Y(t+\tau))}{\sigma_{x(t)}\sigma_{y(t+\tau)}}$$
$$= \frac{E((X(t)-\mu(X(t)))(Y(t+\tau)-\mu(Y(t+\tau))))}{\sqrt{(E(X^2(t))-(E(X(t)))^2)}\sqrt{(E(Y^2(t+\tau))-(E(Y(t+\tau)))^2)}} . \quad (1)$$

Here X and Y are two random processes and $R_{xy}(t,t+\tau) \in [-1, 1]$ is their cross-correlation function. τ is the correlation step (usually referred to as 'lags') which is an integer. $X(t)$ represents the random variable of X at time t while $Y(t+\tau)$ represents the random variable of Y at time $t+\tau$, $t \in [0,+\infty)$. $\mu(X(t))$ and $\mu(Y(t+\tau))$ are mathematical expectations of $X(t)$ and $Y(t+\tau)$ respectively. $\sigma_{X(t)}$ and $\sigma_{Y(t)}$ are standard deviations of $X(t)$ and $Y(t+\tau)$. For given t, $R_{xy}(t,t+\tau)$ can be written as $R_{xy}(\tau)$. Note that $R_{xy}(\tau)= R_{yx}(-\tau)$.

Because in legitimate traffic the RTT is mostly less than 0.5 seconds, we sample traffic using time unit of 0.1 second. As a result, we compute 1 to 5 step cross-correlation values which are sufficient to represent the cross-correlation of legitimate traffic. For two sequences, their cross-correlation values with different τ compose cross-correlation vector. Here we set $\tau \in \{1,2,3,4,5\}$, the cross-correlation vector of X and Y is $\{R_{xy}(1), R_{xy}(2), R_{xy}(3), R_{xy}(4), R_{xy}(5)\}$.

In this step we will divide EFCD sequence into several subsequences according to a configured time window and compute the cross-correlation between the outgoing traffic and incoming traffic in a same time window. Here a time window of 3 seconds is suitable for accurate computation and detection, i.e. there are 30 EFCD values in a subsequence.

The 1 to 10 step cross-correlation vector of legitimate and DDoS traffic are shown in Fig. 3. The correlation step and time window's width should be chosen carefully. The sampling with big window needs less computation, but it is less accurate than with small window. The time unit should be chosen carefully. Bigger time unit will shorten the computation time while decrease the detection accuracy. A similar balance should also be considered for the choosing of correlation step.

3.2 Defining Membership Functions

Because the cross-correlation differences between legitimate and DDoS traffic are not always obvious, the fuzzy classification can be a good way to differentiate them.

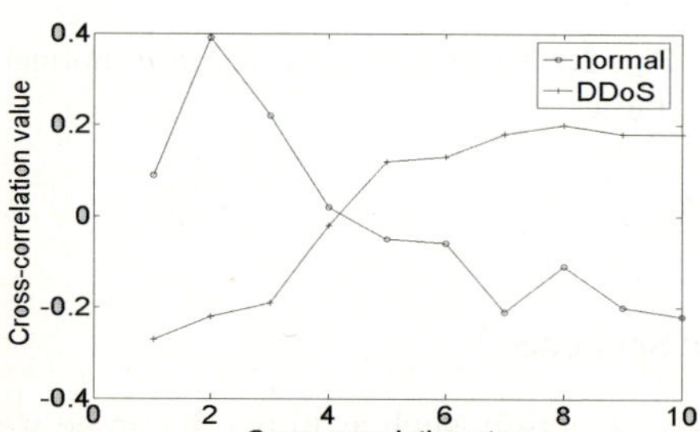

Fig. 3. Typical legitimate and ddos cross-correlation sequence (step 1 to 10)

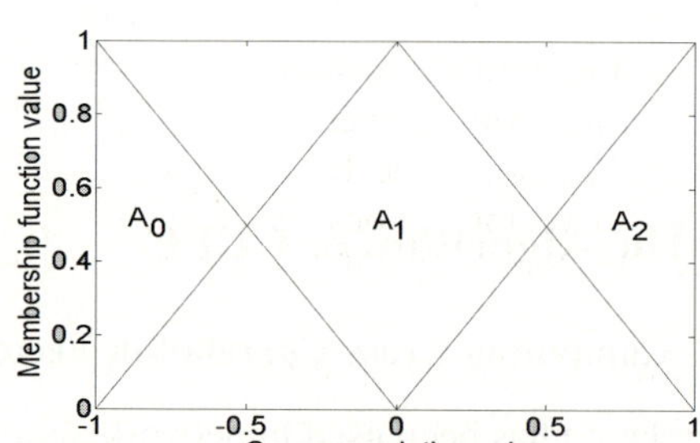

Fig. 4. Fuzzy sets for every axis of pattern space

The sequence of cross-correlation vector is used as the input for fuzzy classification module. Cross-correlation vector is used as the pattern and sequence of cross-correlation vector as the pattern subspace. Dimension number is the length of cross-correlation vector. K fuzzy sets exist in each dimension of the pattern space and here we choose K=3. That is enough for our fuzzy classification. We partition every axis of the pattern space into 3 fuzzy sets A_1, A_2, A_3.

Like usual operation, symmetric triangular membership functions are used and the corresponding result is shown in Fig. 4.

3.3 Training Fuzzy Classifier Using History Data

The input pattern is used as input variables and the output is mapped to result classes. Here the main problem is how to generate fuzzy 'If-Then' rules from history data. In this paper, we use an improved classification method in [14]. Following the definition in (1), the training pattern space is 5-dimension and can be expressed as:

$$[-1,1]\times[-1,1]\times[-1,1]\times[-1,1]\times[-1,1] \quad (2)$$

History data are N patterns, denoted as $V_m=\{(v_{m,1}, v_{m,2}, v_{m,3}, v_{m,4}, v_{m,5})|m=1,2,3\ldots,N\}$, which belong to M output classes, here M=2, i.e. N patterns belong to either legitimate or DDoS class.

The label of rules is shown in Fig 5. 'A_x' represents fuzzy set for one dimension, for each dimension there are three fuzzy sets A_0, A_1, A_2. We just draw the first 2 dimensions here. '$R_{ij\ldots}$' represents fuzzy rule, and its subscript is a 5-character string written as an array S, where $S(\cdot)$ can be '0', '1' and '2'. For one pattern V_m, a rule R_S can be described as bellows:

If $v_{m,1}$ is $A_{S(1)}$, $v_{m,2}$ is $A_{S(2)}$, $v_{m,3}$ is $A_{S(3)}$, $v_{m,4}$ is $A_{S(4)}$ and $v_{m,5}$ is $A_{S(5)}$, then V_m belongs to output class n_S ($n_S \in \{1, 2\}$) with Certainty Factor CF_S, which is the possibility of this rule belonging to the output class.

For each rule, its output class n_S and CF_S are decided in training step. The generations of 'If-Then' rules are given below. Firstly, for each rule R_S, the probability for decision $P_{n,S}$ is computed as follows:

$$P_{n,S} = \sum_{V_m \in \text{class } n} \prod_{t=1}^{L} \rho_{S(t)}(v_{m,t}) \ . \quad (3)$$

In (3), class n is one of the classes the patterns finally belong to. Here $n \in \{1,2\}$. L is the dimension number of input patterns and L=5. $v_{m,t}$ is the value in dimension t of pattern V_m. $\rho_{S(t)}$ is the fuzzy set $A_{S(t)}$'s membership function and $S(t)$ is the value in index t of vector S. For rule R_S, we can get $P_{1,S}$ and $P_{2,S}$. If $P_{1,S} > P_{2,S}$, then the output class n is class 1, else if $P_{1,S} < P_{2,S}$, then the output class n is class 2. If $P_{1,S} = P_{2,S}$, the output class is undecided, so when implementation, this rule is classified into one class with CF_S=0. At last the CF_S of this rule is computed as follows, where the rule belongs to class t'.

$$CF_S = (P_{t',S} - (\sum_{t=1,t\neq t'}^{M} P_{t,S})/(M-1))/\sum_{t=1}^{M} P_{t,S} \ . \quad (4)$$

3.4 Using Fuzzy Classifier to Detect

In this step we use the fuzzy classifier trained in 3.3 to classify unknown patterns. For one input pattern V_m, we compute q_{max} :

$$q_{max} = \max\{(\prod_{t=1}^{L} \rho_{S(t)}(v_{m,t}))\bullet CF_S \mid S \in set\ of\ all\ rules\}\ . \tag{5}$$

With the q_{max} and the corresponding rule R_S, the pattern V_m is assigned to the class n_S that the rule R_S belongs to. If q_{max} is can not be decided and the corresponding rules don't belong to a same output class, the classifier doesn't give a result. And if the result is DDoS class then an alarm is generated.

4 Experiment Results

We used MIT Lincoln Laboratory's data in [12] to train and evaluate our algorithm. The attack events are shown in Fig 6, where x-axis is the time windows sequence number, and in y-axis 0 stands for legitimate while 1 stands for DDoS attack. The DDoS attacks occur at time window 5 to 7 and 19 to 21.

The detection results are shown in Fig. 7. But for the result, there are false positive alarms in time window 11 and 15. Through observing original sampling data, we found at time window 11 and 15 the percentage of TCP traffic is unusually low and UDP traffic has a burst. And in the situation most of outgoing packets are UDP packets and legitimate clients do not reply to what they received, as a result, the correlation is weak.

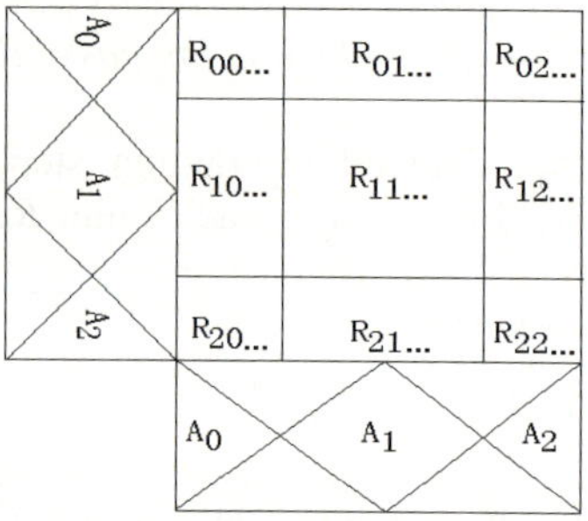

Fig. 5. The label of rules in the 1st and 2nd dimension of 5 dimensions

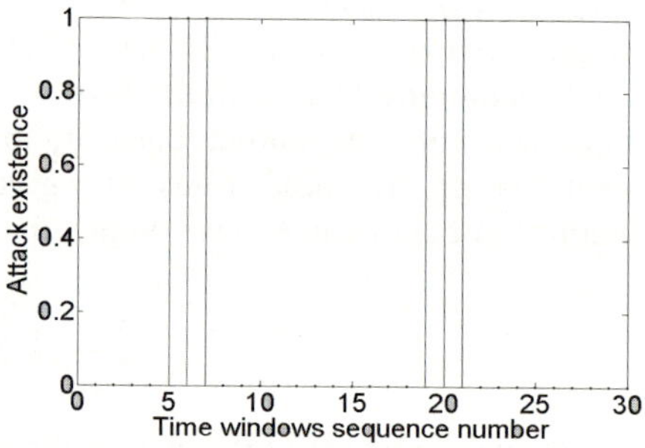

Fig. 6. Attack existence

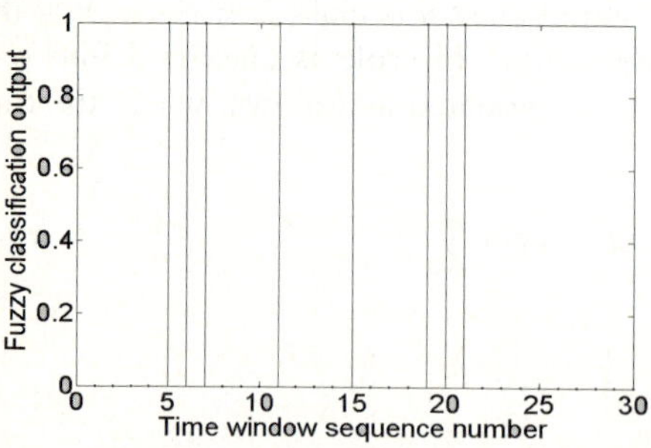

Fig. 7. Detection result

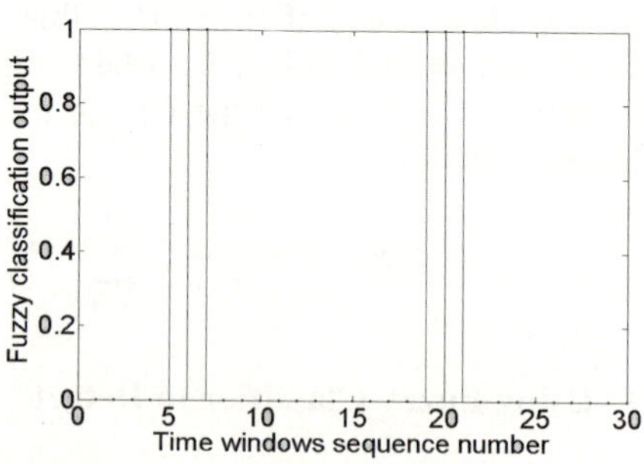

Fig. 8. Detection result after filtering

This type of false positive alarms can be erased through filtering. The main principle is that neighborhood alarms strengthen each other, i.e. alarms are clustered and given a trust value. If the value is greater than a given value, *Tr*, alarms in this cluster are reported. When an alarm rises, if there exists a cluster, the trust value is updated; otherwise, a new cluster is created with an initial trust value. When alarm stops for several time windows, *Tc*, the existing cluster is removed.

The detection result after filtering is accurate as shown in Fig. 8. We evaluate the algorithm with several other data sets [12] and get similar results.

5 Conclusion

In this paper we proposed a new fuzzy classification algorithm for DDoS detection. The algorithm is based on the fact that DDoS attacks can deviate the cross-correlation of EFCD of network traffic.

In this algorithm, we get EFCD sequence of outgoing and incoming traffic of server network and then transform it into 5-step cross-correlation sequences. Then we use legitimate and DDoS data to train fuzzy classification and used it to detect DDoS attacks. The experiment results show that CCFC is accurate enough.

Higher dimension in fuzzy classification could increase the accuracy of detection, while increase the computation consumption. We plan to improve this algorithm with higher dimension pattern space for application in real time.

References

1. Mirkovic, J., Reiher, P.: A taxonomy of DDoS attack and DDoS defense mechanisms. Computer Communication Review, Vol. 34(2), 39-53, 2004.
2. Li, Q., Chang, E. C., Chan, M. C.: On the effectiveness of DDoS attacks on statistical filtering. In Proceedings of IEEE INFOCOM 2005, March 2005.
3. Hussain, A., Heidemann, J., Papadopoulos, C.: Identification of Repeated Denial of Service Attacks. In Proceedings of IEEE INFOCOM 2006, April 2006.
4. Lakhina, A., Crovella, M., Diot, C.: Mining Anomalies Using Traffic Feature Distributions. In Proceedings of ACM SIGCOMM 2005, August 2005.
5. Laura, F. Dan, S. Statistical Approaches to DDoS Attack Detection and Response. In Proceedings of DARPA Information Survivability Conference and Exposition, Vol. 1, 303-314, 2003.
6. Jin, S., Yeung, Y. D.: A covariance analysis model for DDoS attack detection. In Proceedings of IEEE International Conference on Communications, Vol. 4, 1882-1886, 2004.
7. Li, L., Lee, G.: DDoS attack detection and wavelets. Computer Computer Communications and Networks, 421-427, 2003.
8. Li, M.: An approach to reliably identifying signs of DDoS flood attacks based on LRD traffic pattern recognition. Computers and Security, Vol. 23(7), 549-558, 2004.
9. Xiang, Y., Lin, Y., Lei, W. L., Huang, S. J.: Detecting DDoS attack based on network self-similarity. IEEE Proceedings Communications, Vol. 151(3), 292-295, 2004.
10. Blazek, R., Kim, H., Rozovskii, B., Alexander, T.: A novel approach to detection of "denial–of–service" attacks via adaptive sequential and batch–sequential change–point detection methods. In Proceedings of the 2001 IEEE Workshop on Information Assurance and Security United States Military Academy, 2001.

11. Shon, T., Kim, Y., Lee, C.: Jongsub Moon: A machine learning framework for network anomaly detection using SVM and GA. In Proceedings of Systems, Man and Cybernetics (SMC) Information Assurance Workshop, 176-183, 2005
12. http://www.ll.mit.edu/IST/ideval/data/2000/LLS_DDOS_1.0.html
13. Box, G.E.P., Jenkins, G.M., Reinsel, G.C.: Time Series Analysis: Forecasting and Control, Third edition, Prentice Hall, 1994.
14. Ravi, V., Zimmermann, H. J.: Fuzzy rule based classification with FeatureSelector and modified threshold accepting. European Journal of Operational Research, Vol. 123(1), 16-28, 2000.

Convergence of the Fixed Point Algorithm of Analytical Models of Reliable Internet Protocols (TCP)

Debessay Fesehaye Kassa and Sabine Wittevrongel

Department of Telecommunications and Information Processing,
Ghent University, Sint-Pietersnieuwstraat 41
B-9000 Gent, Belgium
debessay@telin.ugent.be

Abstract. Analytical models are important tools for the performance investigation of the Internet. The literature shows that the fixed point algorithm (FPA) is one of the most useful ways of solving analytical models of Internet performance.

Apart from what is observed in experimental literature, no comprehensive proof of the convergence and uniqueness of the FPA is given. In this paper we show how analytical models of reliable Internet protocols (TCP) converge to a unique fixed point. Unlike previous work in the literature the basic principles of our proof apply to both single and multiple bottleneck networks, to short and long-lived TCP connections and to both Drop Tail and Active Queue Management (AQM) routers. Our proof of convergence is based on a well known fixed point theorem and our uniqueness proof exploits the feedback nature of the reliable protocol.

The paper specifies conditions under which the FPA of analytical models of TCP converges to a unique point. The concepts used in the proof can also be extended to analyze the equilibrium, stability and global uniqueness issues of TCP, other reliable protocols and the Internet as a whole.

Keywords: Analytical models, Internet, TCP, fixed point, convergence, equilibrium, stability.

1 Introduction

An equation $\mathbf{f(x)} = \mathbf{x}$ which maps a point $\mathbf{x}$ to itself is called a *fixed point equation*.

Analytical models are important tools for investigating, designing, dimensioning and planning IP (Internet Protocol) networks. The literature (for example see Olsén *et al.* [10]) shows that the fixed point algorithm (FPA) is one of the most useful ways of solving analytical models of Internet performance. The fixed point methods combine the detailed models describing the behavior of the sources with network models resulting in a compound model. Many analytical models such as [3, 4, 5, 9] use the fixed point method.

Each homogeneous group of (TCP) connections (flows with the same path, the same TCP version, the same packet size ...) is represented by a *TCP sub-model*. Each bottleneck link and the associated router traversed by the groups of TCP connections is represented by a *network sub-model*.

V.N. Alexandrov et al. (Eds.): ICCS 2006, Part IV, LNCS 3994, pp. 65–72, 2006.
© Springer-Verlag Berlin Heidelberg 2006

The TCP sub-models calculate the load offered to the respective network sub-models. The network sub-models in turn calculate the loss probability and queueing delay for the corresponding TCP sub-models in an iterative fixed point procedure (FPA). This is analogous to the fact that the packet loss probability and packet delay in the network depend on the sending rate of the sources, and the flow-controlled (TCP) sources adjust their sending rates in response to observed packet loss and packet delay. Hence the usual fixed point elements (entries) used in solving analytical models of TCP performance are the packet loss probability P_L, the average queue length E_N from which the queueing delay and the *RTT* (round trip time – the time from the start of packet transmission until the reception of its acknowledgement) are calculated and the load offered by the TCP connections Λ.

In light of the above discussion, the main contributions of this paper are the following.

- **Convergence and Uniqueness Proof:** Proof of the convergence of the FPA based on a well established theorem (used in Nash Equilibrium theory) and proof of the uniqueness of the fixed points based on the feedback nature of the reliable protocol are given. Unlike some previous works ([9, 3]) our elegant proofs apply to both single and multi-bottleneck networks, to short and long-lived connections and to AQM and Drop Tail routers at different network scenarios.
- **Specification of the Conditions for Convergence and Uniqueness:** While proving the convergence and uniqueness we also specify the conditions under which the FPA of analytical models of reliable Internet protocols like TCP converges to a unique point. None of the proofs in the literature ([9, 3]) has explicitly specified these conditions.
- **Simplicity:** Unlike the few previous works, this paper presents simple and accurate proofs with minimal assumptions.
- **Extendibility:** Unlike the previous works, an extension of our robust proofs which are based on well known concepts allows us to easily analyze the equilibrium, stability and global uniqueness issues of TCP, other reliable Internet protocols and the Internet as a whole.

The rest of the paper is organized as follows. In sections 2 and 3 we show how the FPA converges to a unique fixed point. We give the summary and work in progress in section 4.

2 Existence of the Fixed Point

There are several fixed point theorems based on continuous functions [2] and discontinuous functions [6]. We use the *Brouwer's fixed point theorem* (see [2]) which is based on continuity of the fixed point function to show the existence of the fixed points.

To prove the convergence of the two–dimensional FPA of analytical models of TCP performance for a single bottleneck link using the Brouwer's fixed point theorem it suffices to show that a *continuous* function **f** exists in a *non empty compact convex set* or in a closed n-ball or equivalent such that

$$\mathbf{f}(P_L, E_N) = (P_L, E_N). \tag{1}$$

2.1 The Compact and Convex Set (Region) Where the FPA Is Carried Out

The packet loss probability value P_L is between 0 and 1 and the mean buffer occupancy value E_N is between 0 and K where $K - 1$ is the router buffer capacity. These two intervals form a non empty compact convex set of points (plane) in which the fixed point procedure is carried out.

We next derive the function **f** and show how and why it becomes continuous.

2.2 The Continuous Fixed Point Function

To derive the fixed point Equation 1 and show that it is continuous, we will first use well known formulas for the derivation of the load that the TCP sub-models offer to the network sub-models. We then prove that these formulas are continuous and conclude that the formula for the throughput (load offered by a TCP connection) is continuous. We will also use the packet loss probability and queuing delay as given by the $M/M/1/K$ queuing system for the network sub-model and explain that these functions are continuous. Finally we will construct the vector valued fixed point function given in Equation 1 and show that it is continuous.

The formulas for the TCP sub-models and their continuity. TCP connections adjust the load they offer to the network as a function of the packet loss probability and queueing delay at the bottleneck router. This relationship is given by the well known *PFTK* throughput formula ([11]) for TCP Reno.

Let P_L denote the packet loss probability at a bottleneck router. Let E_N denote the queue length at the bottleneck router. Let Λ denote the load offered by the TCP sub–model(s). Let T_0 denote the TCP initial timeout value in seconds. Let W_m denote the maximum TCP window size expressed in packets. Let $W(P_L)$ denote the expected TCP window size as a function of packet loss probability, P_L. Then [11]

$$W(P_L) = \begin{cases} \frac{2}{3} + \sqrt{\frac{4(1-P_L)}{3P_L} + \frac{4}{9}} & W(P_L) < W_m \\ W_m & W(P_L) \geq W_m . \end{cases} \quad (2)$$

Let $Q(P_L, w)$ denote the probability that a loss in a window of size w is a timeout (TO). Then [11]

$$Q(P_L, w) = \begin{cases} 1 & w \leq 3 \\ \frac{(1-(1-P_L)^3)(1+(1-P_L)^3(1-(1-P_L)^{w-3}))}{1-(1-P_L)^w} & w > 3,\ P_L \neq 0 \\ \frac{3}{w} & w > 3,\ P_L = 0. \end{cases} \quad (3)$$

Let $G(P_L)$ denote a polynomial term used in the *PFTK* formula by

$$G(P_L) = 1 + P_L + 2P_L^2 + 3P_L^3 + 4P_L^4 + 5P_L^5 + 6P_L^6. \quad (4)$$

Let $E[X]$ denote the expected round number when the first packet loss occurs. A round begins when a packet is sent and ends when its ACK arrives. As shown in [11],

$$E[X] = \begin{cases} W(P_L) & W(P_L) < W_m \\ \frac{W_m}{4} + \frac{1-P_L}{P_L W_m} + 1 & W(P_L) \geq W_m . \end{cases} \quad (5)$$

The throughput of a TCP connection is given by the *PFTK* formula

$$\lambda = t(P_L, RTT) = \frac{\frac{1-P_L}{P_L} + \frac{W(P_L)}{2} + Q(P_L, W(P_L))}{RTT(E[X]+1) + \frac{Q(P_L,W(P_L))G(P_L)T_0}{1-P_L}}. \tag{6}$$

Substituting Equation 5 into Equation 6 and multiplying the first part of the resulting equation by $(1-P_L)/(1-P_L)$ and the second part by P_L/P_L yields

$$\lambda = t(P_L, RTT) = \begin{cases} \frac{\frac{(1-P_L)^2}{P_L} + \frac{W(P_L)}{2}(1-P_L) + Q(P_L,W(P_L))(1-P_L)}{RTT(W(P_L)+1)(1-P_L) + Q(P_L,W(P_L))G(P_L)T_0}, \\ W(P_L) < W_m \\ \frac{1-P_L + \frac{W(P_L)}{2}(P_L) + Q(P_L,W(P_L))P_L}{P_L RTT(\frac{W_m}{4} + \frac{1-P_L}{P_L W_m} + 2) + P_L \frac{Q(P_L,W(P_L))G(P_L)T_0}{1-P_L}}, \\ W(P_L) \geq W_m. \end{cases} \tag{7}$$

$W(P_L) < W_m$ implies that

$$\frac{2}{3} + \sqrt{\frac{4(1-P_L)}{3P_L} + \frac{4}{9}} < W_m \tag{8}$$

so that

$$P_L > \frac{1}{\frac{3}{4}W_m^2 - W_m + 1}. \tag{9}$$

The function $W(P_L)$ given in Equation 2 is continuous as

$$\lim_{P_L \to \frac{1}{\frac{3}{4}W_m^2 - W_m + 1}} W(P_L) = W_m = W\left(\frac{1}{\frac{3}{4}W_m^2 - W_m + 1}\right). \tag{10}$$

Similarly it can be shown that $E[X]$ is a continuous function of P_L.

It can be shown using L'Hopital's rule that the function $Q(P_L, W_m)$ described in Equation 3 is a continuous function of P_L. The function $Q(P_L, W(P_L))$ is also continuous as $W(P_L)$ is continuous. The polynomial function $G(P_L)$ given by Equation 4 is also continuous.

Besides the continuity of $\lambda = t(P_L, RTT)$ is not affected by the value of *RTT* which is greater than 0.

Therefore the function $\lambda = t(P_L, RTT)$ of persistent TCP Reno connections given by Equation 7 is continuous since a combination and composition of continuous functions is continuous at all appropriate points. For other TCP implementations with Drop Tail and RED the throughput formulas given in [7] can be used and the continuity can be similarly shown.

Using similar arguments of the above formulas of persistent TCP connections, the continuity of the square root formula for the rate of non–persistent TCP flows given in [1] can be shown. Any point of discontinuity can be removed by re–defining the function.

We next discuss the continuity of the network sub-model formulas.

The formulas for the network sub-models and their continuity. The network sub-model which focuses on the IP network receives the average traffic load Λ packets/sec collectively offered by the TCP sub–model(s). The network sub-model $(M/M/1/K)$ with a router buffer capacity of $K-1$ packets and a link capacity of C packets/sec (the load $\rho = \Lambda/C$) is used to compute the loss probability P_L and the expected number E_N of customers in the queueing system. The queueing delay part of the *RTT* is calculated from E_N for the TCP sub-models.

The $M/M/1/K$ queueing system yields a closed form formula for the packet loss probability and queue length. A simple way of accounting for the burstiness of TCP traffic using the $M/M/1/K$ is shown in [9] so that the closed form expressions of P_L and E_N still hold.

Using L'Hopital's rule the quantities P_L and E_N can be shown to be continuous functions $h_1(\Lambda)$ and $m(\Lambda)$. This implies that *RTT* which is a continuous function $u(E_N)$ is also a continuous function $h_2(\Lambda)$. If there are N TCP connections, the total load offered $\Lambda = N\lambda = Nt(P_L, RTT) = g(P_L, RTT)$ for some continuous function g.

The fixed point formula and its continuity. From the above arguments the fixed point equation used in modeling TCP is given by

$$\begin{aligned}(P_L, E_N) &= (h_1(\Lambda), m(\Lambda)) = (h_1(g(P_L, RTT), m(g(P_L, RTT))) \\ &= (h_1(g(P_L, u(E_N)), m(g(P_L, u(E_N))))) = (f_1(P_L, E_N), f_2(P_L, E_N)) \\ &= \mathbf{f}(P_L, E_N). \qquad (11)\end{aligned}$$

Theorem 2.1 *The function* **f** *given in Equation 11 above is continuous.*

Proof. The functions, h_1, m, u, and g are all shown to be continuous in the preceding sections. Hence the functions f_1 and f_2 which are compositions of continuous functions are also continuous. This implies that the vector valued fixed point function **f** given by Equation 11 is a continuous function.

Now by the Brouwer's fixed point theorem the function **f** given by Equation 11 has a fixed point in the non empty compact convex set explained in section 2.1. We next show that this fixed point is unique.

3 Uniqueness of the Fixed Point of TCP Models

To prove the uniqueness of the fixed point of analytical models of TCP, we first construct a fixed point function of the TCP throughput and show that it is continuous and decreasing. We then state two theorems and prove them. We use these theorems to complete the proof of the uniqueness of the fixed point of the analytical models of TCP.

As shown in [11] and explained in [8] the throughput function given by Equation 7 can be expressed as

$$\lambda = t(P_L, RTT) = \frac{1}{RTT\sqrt{\frac{2P_L}{3}} + 3T_0\sqrt{\frac{3P_L}{8}}P_L(1+32P_L^2)}.$$

This implies that for a single TCP–network sub–model pair with N active TCP connections

$$\Lambda = N\lambda = \frac{N}{h_2(\Lambda)\sqrt{\frac{2h_1(\Lambda)}{3}} + 3T_0\sqrt{\frac{3h_1(\Lambda)}{8}}h_1(\Lambda)(1+32(h_1(\Lambda))^2)} = F(\Lambda) \quad (12)$$

where *RTT* $= h_2(\Lambda)$ and $P_L = h_1(\Lambda)$ as shown in the previous sections.

If there are k TCP sub–models each of which offers λ_i to the same bottleneck link, let N_i denote the number of active TCP connections in TCP sub–model i. Let D denote the queueing delay and c_i refer to other components of *RTT* like the propagation delay which are constant for each TCP sub–model. Since D is a continuous function of E_N which in turn is a continuous function of Λ, D is a continuous function $h_3(\Lambda)$. Now we have

$$\Lambda = \sum_{i=1}^{k} \lambda_i = \sum_{i=1}^{k} N_i\, t(P_L, RTT_i) = \sum_{i=1}^{k} N_i\, t(P_L, D + c_i)$$
$$= \sum_{i=1}^{k} N_i\, t(h_1(\Lambda), h_3(\Lambda) + c_i) = H(\Lambda). \quad (13)$$

The first derivative $F'(\Lambda)$ is

$$F'(\Lambda) = -\left(h_2(\Lambda)\sqrt{\frac{2h_1(\Lambda)}{3}} + 3T_0\sqrt{\frac{3h_1(\Lambda)}{8}}h_1(\Lambda)(1+32(h_1(\Lambda))^2)\right)^{-2} \times D_\Lambda\left(h_2(\Lambda)\sqrt{\frac{2h_1(\Lambda)}{3}} + 3T_0\sqrt{\frac{3h_1(\Lambda)}{8}}h_1(\Lambda)(1+32(h_1(\Lambda))^2)\right).$$

The first derivatives of $h_1(\Lambda) = P_L$ and $h_2(\Lambda) = RTT = u(E_N)$ are positive implying that the functions h_1 and h_2 are increasing. This can be verified by the fact that when the traffic load increases the loss probability P_L and the queuing delay E_N both increase. Hence $F'(\Lambda) < 0$ for all possible values of Λ. This implies that the function $F(\Lambda)$ is continuous and decreasing function for $\Lambda > 0$ ($P_L > 0$). This can also be verified by the fact that when the loss probability and queuing delays increase the TCP throughput decreases.

Similarly it can be shown that the fixed point function H used for the many TCP sub–models case is also a continuous and decreasing function of Λ.

The following statement which is based on continuous and decreasing functions may be a well known fact. However we put it as a theorem in order to easily reference it from the succeeding parts of the paper.

Theorem 3.1. *A continuous decreasing function p of one variable has at most one fixed point.*

Proof. The function $q(x) = x$ is an increasing function. Therefore this function and the decreasing function $p(x)$ intersect at at most one point. This in turn implies that the fixed point function $p(x) = x$ has at most one fixed point.

Hence each of the functions F and H given by Equations 12 and 13 has a unique fixed point as it is also shown from the above statements that a fixed point exists (Brouwer's fixed point theorem).

Theorem 3.2. *The vector valued function of two variables,* **f** *given by Equation 11 has a unique fixed point.*

Proof. Suppose there are two fixed points (P_{L_1}, E_{N_1}) and (P_{L_2}, E_{N_2}). This implies that there are two fixed points $\Lambda = F(\Lambda)$ and $\Lambda' = F(\Lambda')$ where F is defined in Equation 12. But this is a contradiction as the function F has a unique fixed point as shown above. Hence $(P_{L_1}, E_{N_1}) = (P_{L_2}, E_{N_2})$ and the function **f** has a unique fixed point.

4 Summary and Work in Progress

In this paper we have shown how the FPA converges to a unique fixed point. The proof of convergence is based on a well known fixed point theorem and the uniqueness proof exploits the feedback and reliable nature of the protocol (TCP). Unlike the previous works in the literature ([9, 3]), our proof is simple and elegant and its basic principles are applicable to models of both short and long-lived TCP connections for single and multi-bottleneck links with AQM and Drop Tail routers.

We have specified (using different theorems) the conditions under which the FPA of analytical models of reliable Internet protocols like TCP and the performance of the reliable Internet protocol (TCP) itself converge to a unique fixed point.

We are extending the techniques used in this paper to prove the convergence and uniqueness of analytical models of TCP for multi-bottleneck networks with homegenous and heterogeneous (connections with different paths) TCP connections. We will also use these techniques along with some studies in the literature ([12, 13]) to further analyze the equilibrium, stability and global uniqueness issues of TCP, other reliable protocols and the Internet as a whole.

Acknowledgements

This work is supported by grant numbers 2054027 and 2677 from the South African National Research Foundation, Siemens Telecommunications and Telkom SA Limited.

References

1. F. Baccelli and D. McDonald. A square root formula for the rate of non-persistent TCP flows. In *First Conference on Next Generation Internet Networks (NGI 2005)*, pages 171–176, Rome, Italy, April 2005.
2. Kim C. Border. *Fixed Point Theorems with Applications to Economics and Game Theory*. Press Syndicate of the University of Cambridge, The Pitt Building, Trumpington Street, Cambridge, United Kingdom, 1985.
3. T. Bu and D. Towsley. Fixed point approximations for TCP behavior in an AQM network. In *Proceedings of ACM SIGMETRICS*, Cambridge, Massachusetts, USA, June 2001.

4. M. Garetto, R. Lo Cigno, M. Meo, and M. A. Marsan. Closed queueing network models of interacting long-lived TCP flows. *IEEE/ACM Transactions on Networking*, 12(2):300–311, April 2004.
5. R. Gibbens, S. Sargood, C. Van Eijl, F. Kelly, H. Azmoodeh, R. Macfadyen, and N. Macfadyen. Fixed-point models for the end-to-end performance analysis of IP networks. In *Proceedings of 13th ITC Special Seminar:* IP *Traffic Management, Modeling and Management*, Monterey, California, September 2000.
6. J. J. Herings, G. van der Laan, D. Talman, and Z. Yang. A fixed point theorem for discontinuous functions. In *Tinbergen Institute Discussion Papers 05-004/1, Department of Econometrics and Tinbergen Institute*, Vrije Universiteit, De Boelelaan 1105, 1081 HV Amsterdam, The Netherlands, December 2004.
7. I. Kaj and J. Olsén. Stochastic equilibrium modeling of the TCP dynamics in various AQM environments. In *International Symposium on Performance Evaluation of Computer and Telecommunication Systems (SPECTS'02)*, San Diego, USA, July 2002.
8. C. T. Kelly. *Engineering flow controls for the Internet, A dissertation submitted to the Cambridge University for the degree of Doctor of Philosophy*. PhD thesis, LABORATORY FOR COMMUNICATION ENGINEERING, Department of Engineering, Cambridge University, 2004.
9. M. Meo, M.Garetto, M. A. Marsan, and R. Lo Cigno. On the use of fixed point approximations to study reliable protocols over congested links. In *Globecom*, San Francisco, USA, December 2003.
10. Jorgen Olsén. *Stochastic Modeling and Simulation of the* TCP *Protocol*. PhD thesis, Department of Mathematics, Uppsala University, Box 480, SE-751 06 Uppsala, Sweden, 2003.
11. J. Padhye, V. Firoiu, D. Towsley, and J. Kurose. Modeling TCP Reno performance: a simple model and its empirical validation. *IEEE/ACM Transactions on Networking*, 8(2):133–145, April 2000.
12. A. Tang, J. Wang, S. H. Low, and M. Chiang. Network equilibrium of heterogeneous congestion control protocols. In *IEEE INFOCOM*, Miami, FL USA, March 2005.
13. J. Wang, L. Li, S. H. Low, and J. C. Doyle. Cross-layer optimization in TCP/IP networks. *IEEE/ACM Transactions on Networking*, 13(3):568–582, June 2005.

A Peer-to-Peer Approach to Semantic Web Services Discovery*

Yong Li, Sen Su, and Fangchun Yang

State Key Lab. of Networking and Switching,
Beijing University of Posts and Telecommunications
liyong.bupt@gmail.com, {susen, fcyang}@bupt.edu.cn

Abstract. Web Services with distributed and dynamic characteristics need efficient and decentralized discovery infrastructure supporting the semantic descriptions and discovery. In this paper, a novel peer-to-peer indexing system with related P2P registries is proposed to support the completely decentralized discovery capabilities. In the presented system, with the ontology encoding scheme, the semantic service description is distributed into a distributed trie index on the structured P2P network to allow requesters to lookup services with semantic requirements. Finally, experimental result shows that the presented system is efficient and scalable.

1 Introduction

Web Services are emerging the most popular paradigm for distributed computing. Using Web Services technologies the enterprises can interact with each other dynamically. A typical Web Service architecture consists of three entities: service providers that publish Web Services, service brokers that maintain support their discovery, and service requesters that invocate Web Services.

The growing number of Web Services demands for a scalable, flexible and reliable solution to discovery the most appropriate services for the requesters. The mechanisms of service discovery include centralized registry and decentralized approach. Much of work on Web Services discovery is based on the centralized registries, like UDDI [1] or DAML-S matchmaker [2]. However as the number of Web Services grows and become more dynamic, the centralized registries, which lead to a single point failure and performance bottleneck, quickly become impractical.

In order to avoid the disadvantages of the centralized systems, a number of decentralized solutions based on P2P technologies have been proposed. Some systems build on P2P network use ontologies to publish and discover the web services descriptions. The systems depend on classification or metadata routing [4, 5] can offer rich mechanism of query services. However, the unstructured P2P network limits the

* This work is supported by the National Basic Research and Development Program (973 program) of China under Grant No. 2003CB314806; the National Natural Science Foundation project of China under Grant No.90204007; the program for Changjiang Scholars and Innovative Research Team in University (PCSIRT);. National Natural Science Funds for Distinguished Young Scholar(No.60125101).

V.N. Alexandrov et al. (Eds.): ICCS 2006, Part IV, LNCS 3994, pp. 73 – 80, 2006.
© Springer-Verlag Berlin Heidelberg 2006

scalability of these approach. The structured P2P networks based on Distributed Hash Table (DHT) [3] are extremely scalable and lookups can be resolved in logn overlay routing hops for a network of size n nodes. However, DHT overlays support only "exact match" lookups and encounter difficulties in complex queries. Although some DHT-based P2P systems [6, 7] are proposed to enhance capabilities of query, there still lack of open and efficiency solutions for the web services with semantic advertisement and discovery.

In this paper, we propose a scalable system that support semantic service discovery by publishing advertisement of semantic Web services on a structured overlay network. The Web services can be described in semantic method, according to existing standards (e.g. OWL-S [14]), and can be characterized by a set of keywords. We use these keywords to index the Web services descriptions, and store the index at peers in the P2P systems using a DHT approach. In order to support semantic queries, we use multiple ordered keywords sets taken from domain ontologies as index terms for semantic web service descriptions.

The presented system uses the semantic overlay on top of DHT to manage service advertisements. We deploy a distributed trie index on the semantic overlay to support semantic services matchmaking containing subsumes and overlaps. The approach makes it possible to quickly identify the peers containing most likely matched services according to user requests by the combination of the ontology numerical naming scheme.

The rest of this paper is structured as follows. Section 2 compares the presented system to related work. Section 3 introduces our model of Web Services discovery. Section 4 describes the architecture and operation of the presented distributed indexing system. Section 5 shows an experimental evaluation of the system. Last section concludes the paper and presents future work.

2 Related Work

Currents research to Web service discovery can be broadly classified into two categories: centralized and decentralized approaches.. The centralized approaches include UDDI [1], where central registries are used to store Web Service descriptions. Most of the decentralized approaches using p2p structure and combine with ontology to discover the web services. [9, 10] are similar to our method as they are built on top of structured P2P systems. [9] Describe Web Services by a set of keywords and then map the multi-dimensional index to a DHT via linearization. Some service discovery systems adopt ontology-based approach to improve efficiency. [8, 11] make use of ontology to organize web service discovery registries and addresses scalability of the discovery process.

3 System Architecture

Fig 1(a) shows the architecture of our system. It consists of three layers:

The DHT layer is designed as a common communication layer, on which higher level service discovery mechanism can be built. The DHT layer ensures a scalable management and routing for overlay network. The semantic overlay layer is in turn built on top of a scalable distributed hash table, and formed by the semantic description of service.

Distributed trie built on top of the semantic overlay can perform two operations, register a service description or issue a query to search for service. The main task of the distributed trie layer is to determine where to register a service description and where to send a service query. Queries are directly sent to the trie root for resolution according to the keys in the queries, and searching the whole network is not needed.

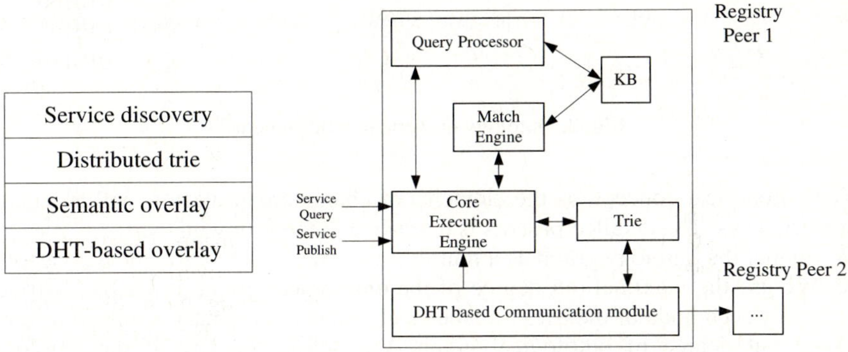

(a) (b)

Fig. 1. Architecture of the system. (a) Layered model of the discovery system and (b) structure of registry node.

Fig .1(b) shows the internal structure of registry node. The core execution engine module connects the other internal components and interacts with users, i.e., publish services, submit queries and receive results. The DHT-based communication module takes charge of underlying P2P network and deals with the connection to other registry peers. The query processor transforms service semantic description into numeric keys that our system can be handled. The numeric key set of service is then sent to the root of distributed trie, and the trie module perform the service publish and query on the distributed trie. After the distributed trie return the service advertisements to the node issuing service query, the match engine with knowledge base will complete ontology-based semantic matching between service query request and service advertisements.

4 Service Description, Registration and Discovery

4.1 Mapping the Semantic Service Description to Keys

In our system, a semantic service description, i.e., a service advertisement, a service request, will be associated with the *characteristic vector*, a set of numeric keys of ontological concepts for the description of the service. Using characteristic vector, service advertisements are assigned to registry peers. Similarly, requesters can discover registries through the vector of the service requests.

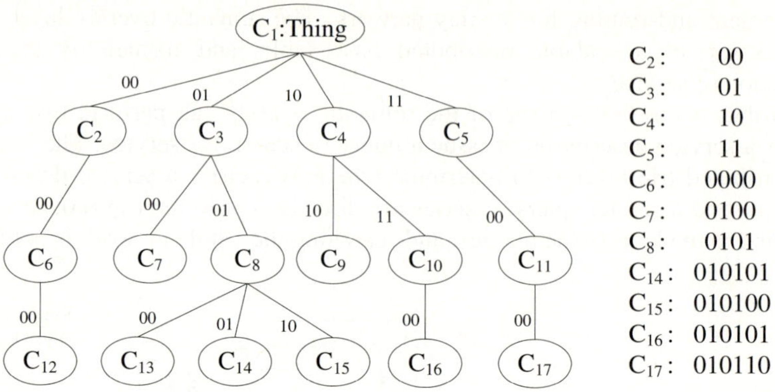

Fig. 2. Ontology numeric naming scheme

The ontological concept, as the elements of characteristic vector, will be encoded to numeric keys. The detailed process is described in the following part.

We model the ontology graph as a multibit-trie where each ontological concept is a node. We use the maximal out degree of the ontological graph to the fixed-stride of the trie. For each node of the trie, we label each branch in a binary string from zero to maximal out degree of ontological graph. We define the key of the ontological concept as a string composed by the branch labels from root to the node representing the concept. This encoding scheme make the code of concepts have the prefix property, which represent the hierarchical relationship among ontological concepts.

The elements of characteristic vector sort in descending order of concept C_i. The descending order of C_is is defined like breadth-first search of tree-like trie structure. A concept C_i have higher order than another concept C_j, if (1) the level of C_i is higher than the level of C_j, (2) both C_i and C_j have the same level and C_i is in the left of C_j in the ontology graph.

An example of ontology numeric encoding scheme is illustrated in Fig. 2, where C_i is an ontological concept. In this graph, the maximal out-degree of nodes is 4, so, from left to right, the branches of C_8 are represented by 00, 01, and 10, and the branches of C_1 are represented by 00, 01, 10 and 11. k_8, the encoding of C_8, is "0101" and k_{15} is "010110". Apparently, according to the encoding of the concepts, we can notice that C_{15} is the child concepts of C_8.

4.2 Publishing Service Advertisements to Peer Registries

Using the numeric encoding of concept as the key of the DHT function, each concept in ontology graph is published on a peer of the object overlay. In this method, the nodes with semantically related concepts form a semantic overlay

We deploy a distributed trie index on the semantic overlay. A distributed trie index is a tree-like index that supports complexity queries on key strings. A trie can be viewed as an m-ary tree where the keys are arranged on extra leaf nodes. Keys with the same k prefix share the same path of k indexing nodes from the root to leaves [13]. Searching a key on a trie index starts from the root and follows the node that meet the query along a trie path until arriving a leaf node.

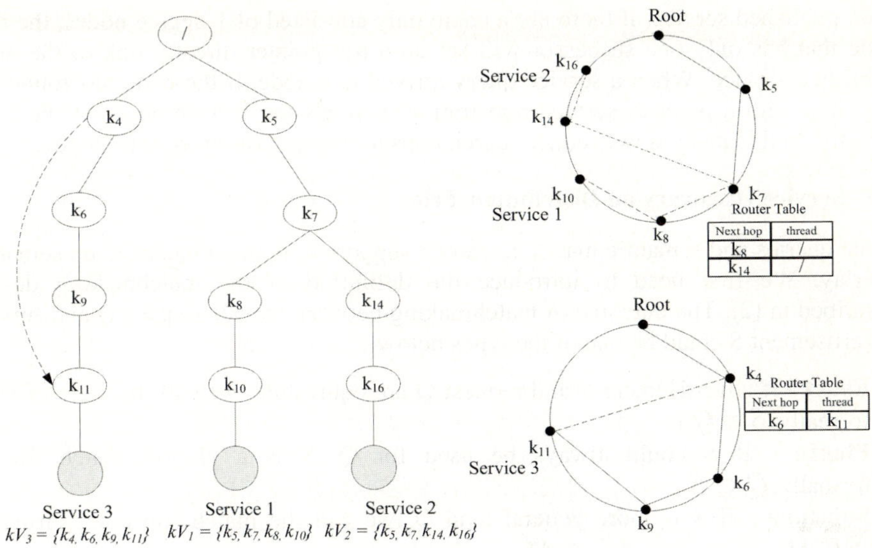

Fig. 3. Service publishing and discovery on distributed trie

We use the characteristic vector of service as the trie path. Publishing a web services is a searching and publishing process of the characteristic vector. As Fig. 3 shows, the method to build a distributed trie and publish the service to overlay works is:

1. For a characteristic vector $CV= \{k_1, k_2... k_n\}$ arriving at root node, if the first key k_1 was not in the router table, insert k_1 , forward service description to the node k_1.
2. For node k_i in CV, when service description arrived, the router table of k_i will be checked. If there is not exists the next node k_{i+1} in router table, insert it; and service description will be forwarded to k_{i+1}.
3. According to 2, a trie path will be formed by the characteristic vector $\{k_1, k_2... k_n\}$, if there exists a partly trie path tp_1 $(k_1\ k_2\ ...k_m)$, build the rest part tp_2 $\{k_{m+1}\ k_2\ ...k_n\}$ of the characteristic vector after the tp_1.
4. Finally, service description arrives at the last node k_m and is stored in this node.

Using distributed trie index, the problem of searching for the registries most likely to store a matched service becomes the problem of finding peers along with trie path. This method of service registration will help us quickly look up the "right" registry peer with a service request in the following discovery process.

The efficiency of a trie index, mainly search hops, is proportional to the average depth of the trie and length of keys. A full trie index has much longer search paths and many hops. Besides the full trie, there are some types of trie to reduce search hops and increase efficiency: pruned trie [13], Patricia trie [13] and compressed trie [12]. However, these trie indexes need to move and adjust existing services from nodes to the others when publishing new services; so, they are not suitable for the distributed environment. We devise a threaded trie index to reduce average search hops and avoid moving existing service advertisement. As the Fig.3 shows, from the node published service, if there are a chain only consisted of 1-degree nodes, the first node that has only one successor will set up a trie pointer directly link to the node published service. When a service query arrived at a node, if the node has found the next hop with a point to service registration, it sends the service query to the node directly. In this method, we reduce search hops to increase discovery efficiency.

4.3 Service Discovery on Distributed Trie

Basic queries and semantic matching can be supported by distributed trie on semantic overlay. We first need to introduce the definition of the matchmaking degree described in [2]. The outcome of matchmaking between service request Q and service advertisement S could be one of the types below:

- **Exact** - If advertisement S and request Q are equivalent, we call the match Exact, formally, $S \equiv Q$ ψ
- **PlugIn** - if S could always be used for Q. S is a plug-in match for Q, formally, $Q \subseteq S$.
- **Subsume** - if S is more general than Q, we call the match subsume. formally, $S \subseteq Q$

In order to query a service in the distributed discovery system, a requester needs to send query to a participating peer, called *introducer*. The introducer first checks local service registrations. If presented, the discovery process terminates. Otherwise, query request will be forwarded to the root of distributed trie and begin a distributed discovery process. To search a service description, we convert the service description to its characteristic vector as mention in 4.1, and then look for it on the trie starting at the root. When the service query arrive at node registering concept k_i, if the next node k_{i+1} is in the router table, the service query can be forwarded to the k_{i+1} for "*Exact*" query. Meanwhile, the service query will be forwarded to the nodes are found to satisfy "*Subsume*" or "*PlugIn*" with k_{i+1} by browsing the router table. Whenever the service requirement reaches the node of registered service, the node will return the service description to service introducer.

Regarding Fig. 4, suppose that we have two services advertisements published in distributed trie: S_1 described with C_5, C_7, C_8, and C_{10} and S_2 described with C_5, C_7, C_{14}, and C_{16}. The characteristic vector CV_1 of S_1 will be $\{k_5, k_7, k_8, k_{10}\}$ and CV_2 will be $\{k_5, k_7, k_{14}, k_{16}\}$. According to our way of distributed service registration, to publish S_1, a *trie path* $<k_5, k_7, k_8, k_{10}>$ *is generated and* S_1 *is registered to the last node* k_{10}, *likewise,* S_2 is attached in k_{16}. In case the characteristic vector of service query is $\{k_5, k_7, k_{14}, k_{16}\}$ and the match requirement is "*Subsume*", after the service query arrive at node k_7, according to matching requirement, the node will choose the next node by browsing the router table and send the service query to the selected nodes: k_8 and k_{14}. The nodes k_8 and k_{14} will return the service description of S_1 and S_2 to the requester.

5 Experimental Evaluation

Accessing a concept encoding requires $O(log_2N)$ hops through the DHT function of Chord overlay where N is the number of physical nodes. Locating a characteristic vector using distributed trie takes $O(W*log_2N)$ hops on Chord, where W is the length of a characteristic vector.

Avg. msg. overhead= total number of *Join-request* msg. sent / total number of join members
Success ratio = number of new receivers accepted / total number of join members
Bandwidth used per member = total bandwidth used by successful receivers / number of new successful receivers.

In our experiment, the bandwidth of each group is uniformly distributed from 2 to 6 Mbps. Video data is encoded into three layers: a base layer, a most significant layer, and an enhancement layer with the bit rates being 30%, 50%, and 20% of the total video respectively. 500 groups are launched gradually in this experiment. The source and the members of each group are selected randomly, however only the simulation results of the middle join rate (10 nodes among 36 nodes join to each group as the member) is shown for the length limitation of this paper. The receiving capability of each receiver is randomly selected among 1, 2, 3 layers.

Fig. 3 gives the simulation results of the three performance metrics. MRDH computes a feasible graft path locally and then sends a *join-request* message along the path to set up the branch. QMRP-LD unicasts a *join-request* message from the DR towards the core. If a router in the unicast path does not satisfy the QoS requirement, the *join-request* message backtracks to the former router and then is flooded to all neighbor routers. As shown in Fig. 3(a), the message overhead of MRDH is much lower than that of QDMR-LD, and almost as low as that of QMRP-LD when the group number is small; but becomes much lower than that of QMRP-LD when group number increases. In Fig. 3(b), with the increase of group number, the percentage of receivers accepted decreases more significantly under QMRP-LD, while decreases gently under MRDH, QDMR-LD and Maxenchuk's algorithm. QMRP-LD relies on unicast routing to find a feasible path without considering the network bandwidth already reserved for a multicast session. Therefore, the success ratio under QMRP-LD is lower than that of MRDH. In Fig. 3(c), the multicast tree constructed under MRDH is more efficient in resource usage compared to that under QDMR-LD and QMRP-LD, and close to that under Maxenchuk's algorithm for one of the most important aims of Maxenchuk's algorithm is to minimize the total bandwidth resources used.

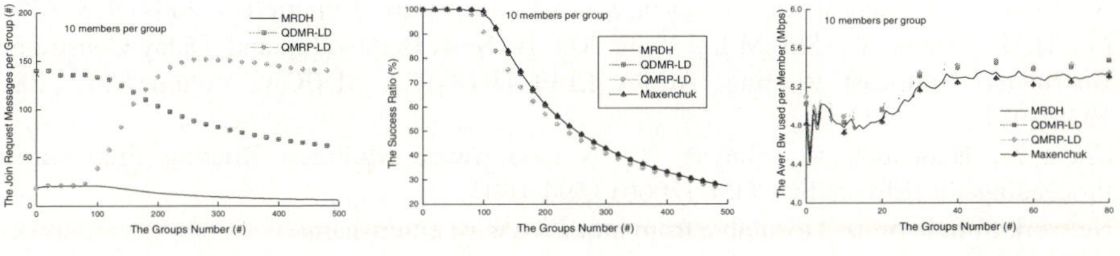

(a) (b) (c)

Fig. 3. Performance comparison of (a) the message overhead; (b) the success routing ratio; (3) the average bandwidth used per member under different algorithms

In conclusion, MRDH has the following prominent advantages: (1) MRDH can calculate in a centralized way by storing two kinds of group-specific information. Feasible paths are calculated locally by DRs, the message overhead for searching feasible paths tends to be very low. (2) If the network state information is up-to-date, the success ratio tends to be high. Unlike others routing, MRDH does not rely on any unicast routing. Hence, every router knows the bandwidth reserved for existing multicast group, which improves the probability of finding an existing feasible path. (3) The network bandwidth usage is optimized. The graft path is the shortest path among all the feasible paths. It leads to efficient use of network resources.

Certainly, MRDH has to maintain a global state, which limits its usage in large networks. But the above pros confirm that MRDH is an effective multicast routing in small range of networks.

5 Conclusions

In this paper, we study the problem of constructing a multicast tree for heterogeneous and dynamic receivers. We demonstrate that by maintaining the distribution of multicast tree and the highest receivable layer of on-tree nodes, feasible paths can be calculated locally. We then describe our protocol in a detailed way. Simulation results shows that our routing tends to have low message overhead, high routing success ratio and optimization usage of network bandwidth.

References

1. Li, B., Liu, J.C.: Multirate Video Multicast over the Internet: An Overview. IEEE Network. Volume 17.(2003)2-6
2. Lui, K.-S., Wang, J., Xiao, L., Nahrstedt K.: QoS Multicast Routing with Heterogeneous Receivers. In: Proceedings of IEEE GLOBECOM. (2003) 3597-3601
3. Maxemchuk, N. F.: Video Distribution on Multicast Networks. IEEE Journal on Selected Areas in Communications. Volume 15. (1997) 357-372
4. Wang, B., Hou, J. C.: QoS-Based Multicast Routing for Distributing Layered Video to Heterogeneous Receivers in Rate-based Networks. In: Proceedings of IEEE INFOCOM. (2000) 480-489
5. Lu, H.M., Xiang Y., Shi M.L., Yang M.: A QoS-Based Dynamic Multicast Routing Algorithm for Streaming Layered Data. Journal of Software. Volume 15. (2004) 928~939
6. Lu, H.M., Xiang Y., Shi M.L., Yang M.: A New Bandwidth and Delay-Constrained Distributed Multicast Routing. ACTA ELECTRONICA SINICA. Volume 20. (2002) 1978-1981
7. Chen, S., Nahrstedt, K., Shavitt, Y.: A QoS-Aware Multicast Routing Protocol. In: Proceedings of IEEE INFOCOM. (2000) 1594-1603
8. Network Simulator ns-2.available from http://www.isi.edu/nsnam/
9. CERNET. Available from http://www.edu.cn/HomePage/cernet_fu_wu/about_cernet/

Using Case-Based Reasoning to Support Web Service Composition*

Ruixing Cheng, Sen Su, Fangchun Yang, and Yong Li

State Key Lab. of Networking and Switching,
Beijing University of Posts and Telecommunications
{Chengruixing, liyong.bupt}@gmail.com,
{susen, fcyang}@bupt.edu.cn

Abstract. With the growing number of Web service, it is necessary to implement web service composition automatically. This paper presents an approach to support large-granularity web service composition accurately and fast according to users' requests. With this approach, it can reduce the cost of web service composition, and improve scalability, reusability and efficiency. Using a proactive well defined service base, web service composition execution engine can gain the logic with Case-Based Reasoning technology. Comparing to other method and qualitative analysis, the approach proposed by this paper can solve the problem of web service composition under the condition of insufficient and ill-defined knowledge, and can reduce the difficulty and cost of web service composition.

1 Introduction

With the emergence of numerous Web services, we need to, according to users' need, compose the current Web services from different environments, platforms and enterprises into a larger-granularity value-added service to satisfy uses' need. It remains an unsolved problem how to efficiently and precisely implement composition of services on a higher level.

This paper proposes a service composition method based on Case-Based Reasoning. First of all, this method builds the service base with structure of two layers, the basic service layer and application service layer, which is constructed according to the different service granularity. And by integrating and sealing cases in basic service layer into cases in application service layer with lager granularity. Then by comparing the similarity of different case, searching and reusing suitable case in the application service layer to obtain the description of composition logics, and realize larger- granularity service composition transparently in a simply, easy-to-use way. Furthermore, by reconfiguring users' individualized restrictive condition, this method can revise the case in the service base so that it can better satisfy users' need.

* This work is supported by the National Basic Research and Development Program (973 program) of China under Grant No. 2003CB314806; the National Natural Science Foundation project of China under Grant No.90204007; the National Natural Science Funds for Distinguished Young Scholar Program of China under Grant No.60125101; the program for Changjiang Scholars and Innovative Research Team in University (PCSIRT).

V.N. Alexandrov et al. (Eds.): ICCS 2006, Part IV, LNCS 3994, pp. 87 – 94, 2006.
© Springer-Verlag Berlin Heidelberg 2006

The second section of this paper introduces the related work. In the third section, the basic idea of using Case-Based Reasoning to support service composition is put forward. The fourth section introduces Case-Based Reasoning and the service-composition framework using Case-Based Reasoning. The last section gives a qualitative analysis of this method and illustrates the future work.

2 Related Work

The as-exist Web service composition methods can be classified into three kinds regarding whether manual tasks are involved in the composition process. They are manual composition method, half-automatic method and automatic method (see [1] for detail).

Manual method provides users interactive interface via graph or text editor. Users are involved in the produce of a script language in the service composition process. The language is then executed by service composition (see [2], [3] for detail).

Based on manual method, half-automatic method introduces the concept of semantic to reduce manual involvement, thus increasing the automatic level of service composition. Sirin and other scholars proposed an interactive service composition framework [4], which is based on semantic analysis. This framework can use semantic to perform filter and selection of service at different stages, produce and execute service composition work flow.

Automatic service composition does not need any manual interference in the whole service composition process. It normally adopts Artificial Intelligent method to realize the automation of the whole process. Among the related research results [5], there is a service composition framework based on Agent, using semantic and universal program to perform service composition. But to some of the service compositions, complicated negotiation and design are needed. The current Artificial Intelligent and work flow methods can hardly satisfy such need [6].

The current research work bases on the as-exist service composition template to realize dynamic service composition. But this method has a flaw in that the creation of service template needs manual interference. For instance, users have to design time constraints and non-function constraints. It also lacks the mechanism of learning from successful experience, making it highly dependent on professionals and experts. Therefore, this method cannot perform well when some users have special demand, or knowledge in certain area is unavailable, or certain domain has mal-definition.

3 Basic Ideas

This paper proposes a highly automated service composition method with large-granularity. The basic idea of this method is to apply Case-Based Reasoning to the stage of service composition discovery, while moving the logic design of the existing service composition engine to Case-Based Reasoning system. This can be done through searching and modifying case. With the well designed definition predetermined by area experts and the reuse of users' past successful experience, this method can establish all sorts of service composition logic set for users before they request. In the service composition process, the method can, according to users requests, search, revise cases, procure satisfactory service composition logic sets and give them to the service execution engine to process.

The structure of service base can be divided into two layers to reduce the complexity of the service composition and to improve the reusability and maintainability. The first layer is basic service layer, which is related with specific domain. The domain experts design case in basic service layer with the interior logic between the services in the domain, and establish service specification and basic infrastructure to map specific physical service to specific domain case. It will not be discussed how to realize the virtualization from physical service to specific domain case in this paper. The second layer is application service layer. By setting some individualized constraints, users can use the services provided by the first layer directly to design application without worrying about how the services in a domain are combined. In this way, users are freed from the complicated service composition process. What they need to do is simply the process of case matching and configuration. The case in the application layer is preconfigured by experts. It can also be a composition of case from a variety of domains based on past successful service composition experience allowing the dynamic configuration of some individualized constraints. An application layer after dynamic configuration can describe a complete application. When a user's request arrives to the application layer of the service base, the suitable case in the application service layer can be found by the process of searching and matching. If no suitable case is found, the pre-constraints will be partly modified and an application description that can satisfy user's need will be formed. The user will evaluate the result from the execution engine. If it is satisfactory, the service base will save the modified case.

When the number application layer case grows into a large one, to increase the efficiency of searching, we first cluster cases in the application service layer. Through clustering, the application service layer cases are divided into different categories. Then we compare user's request with cluster centre, locate the certain category it falls into according to its similarity to cluster centre case, and perform search and match in the very category.

4 Service Composition with Case-Based Reasoning

4.1 Case-Based Reasoning (CBR) and the Design of Web Service Composition

CBR originated from human being's cognition activity. It is a kind of analogism [7]. CBR's basic idea is to form new solutions to new problems and new cases through certain analogy and through appropriate adjustment according to the difference between old problems and new ones. The analogy is made based on the experience and knowledge people have acquired from the past activities dealing with similar problems. To construct service cases base system, sufficient experience of service composition is a must. Long accumulated experience from domain experts and knowledge engineers is needed. On the other hand, direct acquisition of large amount of useful data from past successful experience is also necessary. SB stores substantive service composition cases. To better maintenance and reduce the complexity of service composition, according to the service case granularity, the case base can be divided into two layers: basic service layer and application service layer.

Cases in basic service layer are certain standards summarized based on the analysis of domain experts according to the requirements in the domain, the specific background and the past successful experience. According to these standards and optimized rules, Cases in basic service layer abstract and encapsulate the service logic and composition logic in the domain, and map many atomy service to a Case in basic service layer by service virtualization mechanism.

An Application service layer case consists of one or many basic service layer cases, and can describe composition logics according to successful experience. It can express a larger granularity service.

4.2 Similarity Comparison of Cases

To estimate the similarity between the current case and the new problem, according to the comparison of the cases' similarity, is the key of the case based reasoning and also the essence of showing the correctness of reasoning and the intelligent of performance. By comparing the interface information of cases, the case with the biggest similarity is retrieved. Then the case will be modified according to user's constraints. In the process of case retrieval, to improve the efficiency, we first perform clustering to the case in the case base. Then we decide which kind it belongs to by making similarity comparisons with the case from cluster centre. The algorithms to perform clustering will not be discussed in this paper. The case similarity calculation form adopts attributes-value mode to calculate the similarity in the prototype. The selection of feature weight is significant to CBR system. A weight value of an attribute reflex not only its importance compared to other attributes, but also the level of contribution it will make to problem solving. In prototype, the interface information in the case is for case retrieval. The name of the base case in the interface information in the application layer is essential for problem solving. Therefore the name of a base layer case will be rendered higher weight. The constraint reflexes users' individualized requirement, which often change with the command of different users. Therefore it is rendered lower weight. But when users are very sensitive to a certain constraint, it can also be added higher weight. For example, if a user has a strict requirement to price, it can be given higher weight. CBR system judges to which attributes users are sensitive according to both the past successful experience and the constraints users have input, and renders them corresponding weight. Besides, different attributes have different contribution in solving problems, thus should be rendered different weight value. The selection of algorithms dealing with weight is really important to the optimization of CBR system.

This paper uses a method that performs case retrieval by comparing the parameters in interface information in a case. These parameters can be divided into two kinds - text parameters and data ones. To text parameters, comparisons will be made through the algorithm related with the similarity between concepts. To data parameter, the scope of threshold will be made. The similarity will be zero if the threshold is not in the scope. Otherwise, it will be between 0 and 1.

The algorithm is described below taking service case base in the application layer as an example.

Suppose that the service case base in the application layer has N cases:

$$CBA=\{AC_1,AC_2,\ldots,AC_n\}$$

There are m parameters in the interface information in each application layer case (AC). There parameters' weight vector are $W(W_1,W_2,W_3,\ldots,W_m)$,并且 $\sum_{i=1}^{m} w_i = 1$

The similarity of Case CA_x and Ca_y is: $S(x,y)=1-\sqrt{\sum_{i=1}^{m} w_i \times d(CA_{x_i},CA_{y_i})^2}$.

$\sqrt{\sum_{i=1}^{m} w_i \times d(CA_{x_i},CA_{y_i})^2}$ is the Euclidean distance between CA_x and CA_y, $d(CA_{x_i},CA_{y_i})$ is attribute normal distance.

1) When the type of the parameters is text:
If the text keywords are the same, $d(CA_{x_i},CA_{y_i})=0$. Otherwise $d(CA_{x_i},CA_{y_i})=1$.

2) When the type of the parameters is data: $d(CA_{x_i},CA_{y_i}) = \frac{|CA_{x_i}+CA_{y_i}|}{|T_{max}-T_{min}|}$

$|T_{max} - T_{min}|$ Represents the scope threshold

4.3 The Framework of Service Composition with CBR

The framework of service composition with CBR is shown in figure 1. It can be divided into two parts. The first one realizes this module's ulterior management interface, controlled by domain expert. By managing the view, the experts can manage and build the corresponding service base according to different domain. In addition, they can also test the correctness of the service cases by formal validation tools. Each CBR processor will store local service base and the service base address list scattered in other nodes. The second part deals with users' command. The whole process, from receiving a user's request to producing the result, can be divided into five steps.

1) The user put a request to the user request processor;
2) The processor first divides the request into request parameters and constraints, then passes the result to CBR processor;
3) The CBR processor classifies the user's request according to the request parameters and constrains. To improve the preciseness of the classification, we use a method that supports vector machine. The related work has been stated in Section three. According to the result we search the related service case base to obtain the most similar source case. Then, we can get appropriate service composition description by making comparisons with the source case in similarity. A detailed explanation has been made in the former part of this section.
4) CBR processor translates the description into control logic relations according to the service composition cases found and the user's individualized constraints;
5) Service composition processor takes actions according to the control logic relations;
6) If the user is satisfied with this service, the logic relations will be saved as application case in the service case base. The domain experts will, according to users' different response, fill in corresponding service records, modify and make complements to the service case base.

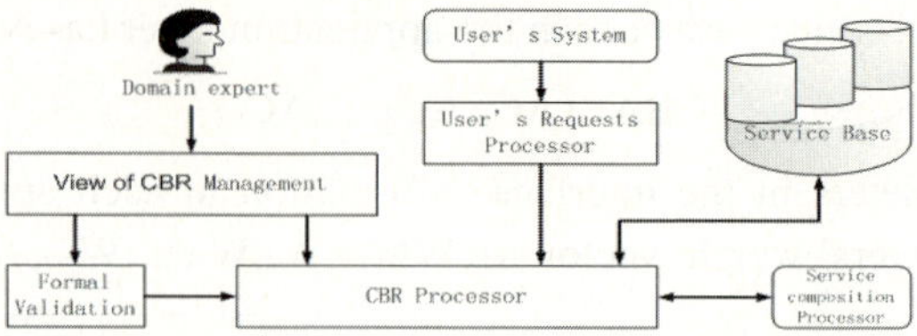

Fig. 1. The Framework of service composition with CBR

5 Experiment and Analysis

5.1 Simulation Experiment

In order to stand the content out, the experiments are based on the hypothesis as follows:

1) All requests can match with cases in the application service layer
2) The time of executing logics will not be taken into account.
3) The time of transmission in the net work will not be considered.

Exp 1. Test of service composition efficiency.

First, 100 cases from the application service layer are selected at random. Then different experiments are made according to the number of sub-services contained by case in application service layer. Comparisons will be made between CBR and the method based on keyword (KC) or semantic (SC).

Fig.2 shows the mean time of service composition of three different method of web service composition. It is shown that the relationship between the number of service and the time spent by the processing of service composition. In addition, it can be shown that the use of the service composition method with Case-Based Reasoning (CBR) can make the efficiency of web service composition higher than that of method based on key word (KC) or semantic (SC).

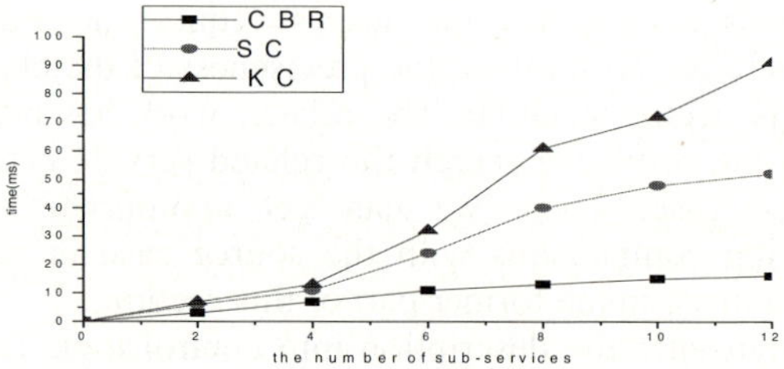

Fig. 2. Experimental results about service composition efficiency, varying the value of three different methods

Exp 2. Test of successful service composition percentage.

First, 100 requests whose similarity is 70%,75%,80%,85%,90%,95%,100% are created respectively. That is, totally 700 request are established. Then comparison are

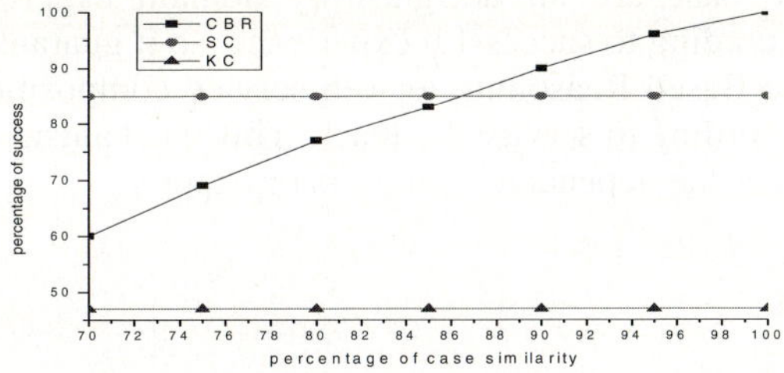

Fig. 3. Experimental results about successful service composition percentage, varying the value of three different methods

made between the Case-Based Reasoning (CBR) method and the method based on key word (KC) or semantic(SC).

Fig.3 shows the mean successful percentage of three different method of web service composition. It can be shown that when the similarity of a case increases, the percentage of successful composition also increases. Under the condition that sample number is 100 and case similarity is greater than 85%, using the service composition method with Case-Based Reasoning (CBR) can make the percentage of success greater than that of the method based on keyword (KC) or semantic(SC).

5.2 Qualitative Analysis and Conclusion

We present a method of using Case-Based Reasoning (CBR) to support web service composition. CBR is applied in the phase of pre-composition to build service based of different service domain. When user's request is received, the process of search and matching cases of application layer in service base is called. And the list of sub-services, parameters and condition of restriction are gained according to the source case. Then they are converted into control logics of web service composition by CBR processor and finished by web service composition processor.

Compare to other methods of web service composition, this method have many advantages shown as follow:

1) Case-Based Reasoning is applied in the process of building service base, before service composition engine deals with user's requests. By searching in the service base, the logic of web service composition is discovered. The result can avoid starting from scratch in designing web service composition logics. It can also simplify the process of reasoning the logic, and improve the efficiency of service composition engine, as well as reduce the cost of service composition.
2) Compare to the method of service composition based on template, ours doesn't need interference of users or experts during the process of composition. It can reuse the successful experience of composition. In addition, it can overcome the difficulties of designing suitable logic of composition under the condition of insufficient knowledge or mal-defined knowledge. So it improves the reliability of applications.
3) With the times of successful service composition increasing, the cases suitable for user's requests in service base also increase. Furthermore, because of applying clustering method in the process of matching and search suitable case in service base, the efficiency of discovering source cases is greatly improved.

4) Cases in service base are all designed by domain experts and validated ,or preconfigured according to successful experience, so it guarantee the correctness.
5) By applying Case-Based Reasoning to web service composition, the service base can be revised according to service feedback. This mechanism improves the ability to learn and reduces the dependence on users or experts.

6 Future Work

There are many related work to be perfected at present. In the future, we intend to take steps to (1) integrate induce algorithm into our method of web service composition with Case-Based Reasoning to improve the ability to learn, (2) realize the virtualization mechanism between physical layer and basic service layer, (3) improve the clustering and similarity algorithm, and (4) add the mechanism of QoS evaluation to our method and use the algorithm based on index of QoS or other recommendations in the process of searching cases in the application service layer.

References

[1] Schahram Dustdar, Wolfgang Schreiner. A survey on web service composition[J]. Web and Grid Service,Vol.1 No.1,2005

[2] Mandell D. J . , McIlrait h S. A. . A bottom up approach to automating Web service discovery , customization , and semantic translation. In : Proceedings of t he 12th International WWW Conference Workshop on E-Services and the Semantic Web ,Budapest , 2003 ,89~96

[3] Taylor I. , Shields M. , Wang I. , Philp R. . Distributed P2P computing within triana : A galaxy visualization test case. In :Proceedings of t he IPDPS 2003 Conference , Nice , France ,2003 , 16~27

[4] E Sirin, B Parsia, J Hendler .Filtering and selecting semantic Web services with interactive composition techniques - IEEE Intelligent Systems, 2004 - ieeexplore.ieee.org

[5] McIlraith S. , Son T. C. . Adapting golog for composition of semantic Web services. In : Proceedings of the 8th International Conference on Knowledge Representation and Reasoning , Toulouse , 2002 , 482~493

[6] J. Koehler and B. Srivastava .Web Service Composition: Current Solutions and Open Problems. ICAPS(The International Conference on Automated Planning & Scheduling) 2003 Workshop on Planning for Web Services

[7] Mario Lenz. Case-based Reasoning: From Foundations to Applications [M]. Berlin: Springer, 1998

Secure OWL Query*

Baowen Xu[1,2], Yanhui Li[1,2], Jianjiang Lu[1,2,3], and Dazhou Kang[1,2]

[1] Department of Computer Science and Engineering,
Southeast University, Nanjing 210096, P.R. China
[2] Jiangsu Institute of Software Quality, Nanjing 210096, P.R. China
[3] Institute of Command Automation,
PLA University of Science and Technology, Nanjing 210007, P.R. China
bwxu@seu.edu.cn

Abstract. With the development of the Semantic Web, the issue on Semantic Web security has received a considerable attention. OWL security is a burgeoning and challengeable sub-area of Semantic Web security. We propose a novel approach about OWL security, especially about inference control in OWL retrieval. We define OWL inference control rules (ICRs) to present the aim of inference control. To achieve ICRs, our approach introduces a novel concept "semantic related view" (SRView) w.r.t an OWL knowledge base, which describes semantic relations in it. We present a construction process of SRViews from OWL knowledge bases and define pruning operations to rewrite them. Based on pruned SRViews, a query framework to achieve inference control is proposed, followed by analysis of correctness and complexity.

1 Introduction

The Semantic Web is an extension of the current Web, which supports machine-understandable semantics to enable intelligent information management. With the increasing efficiency of utilizing data in the Semantic Web, there is a considerable attention for security issues. Thuraisingham pointed out security standards for the Semantic Web, in which security of the whole Semantic Web was divided into security of its components: XML, RDF and OWL security etc. [1].

Various research efforts have been done on XML security, especially about access control in the context of XML. Fan et al. introduced a novel approach that the security restrictions were annotated on the schema structure (DTDs) and queries were corresponding rewritten and optimized [2]. Compared with XML security, RDF security just begins. Reddivari et al. proposed a simple RDF access policy framework (RAP), which contained a policy based access control model to achieve control over various actions possible on an RDF store [3]. Largely dissimilar to XML and RDF cases, OWL security contains a great part of inference control that prevents users from infer-

* This work was supported in part by the NSFC (60373066, 60425206 and 90412003), National Grand Fundamental Research 973 Program of China (2002CB312000), National Research Foundation for the Doctoral Program of Higher Education of China (20020286004), Excellent Ph.D. Thesis Fund of Southeast University, Advanced Armament Research Project (51406020105JB8103) and Advanced Research Fund of Southeast University (XJ0609233).

V.N. Alexandrov et al. (Eds.): ICCS 2006, Part IV, LNCS 3994, pp. 95 – 103, 2006.
© Springer-Verlag Berlin Heidelberg 2006

ring blocked or unallowable information. How to present inference control restrictions and achieve them is still an open problem, and to the best of our knowledge, no published paper is known for it.

In this paper, we will give a novel approach about OWL inference control. Firstly we give a brief introduction to OWL knowledge base (KB) in section 2. Then we define OWL inference control rules (ICRs) to formally present OWL inference control restrictions in section 3. In section 4, a new term "semantic related view" (SRView) is introduced to describe whole semantic relation in an OWL KB. We also give a construction process and pruning operations of SRViews. With pruned SRViews, a query framework to achieve inference control is proposed in section 5, followed by conclusion and future work in section 6.

2 A Brief Introduction to OWL Knowledge Base

The proposed OWL recommendation consists of three languages with increasing expressive power: OWL Lite, OWL DL and OWL Full. For inference in OWL Full is undecidable [4], we focus on inference control in OWL DL. In the following, we use "OWL" instead of "OWL DL". We adopt description logic form to compactly express OWL KB, for OWL DL is description logic SHOIN(D) with RDF syntax.

Definition 1. Let R_A, R_D, C_A, C_D, I_A and I_D be pairwise disjoint sets of atomic abstract roles, atomic concrete roles, atomic concepts, data types, individuals and data values. Let $R \in R_A$, a SHOIN(D) abstract role is either an atomic abstract role R or its inverse role R^-. The set of SHOIN(D) role is defined as $\{R^- \mid R \in R_A\} \cup R_A \cup R_D$. And SHOIN(D) concept constructors are given in Table 1.

Table 1. Syntax and semantics of concept constructors in SHOIN(D)

Syntax	Semantics	Syntax	Semantics
$A \in C_A$	$A^I \subseteq \Delta^I$	$C_1 \sqcap C_2$	$(C_1 \sqcap C_2)^I = C_1^I \cap C_2^I$
$D \in C_D$	$D^I = D^D \subseteq \Delta^D$	$C_1 \sqcup C_2$	$(C_1 \sqcup C_2)^I = C_1^I \cup C_2^I$
$R \in R_A$	$R^I \subseteq \Delta^I \times \Delta^I$	$\neg C$	$(\neg C)^I = \Delta^I \setminus C^I$
$U \in R_D$	$U^I \subseteq \Delta^I \times \Delta^D$	$\{w_1, \cdots, w_n\}$	$\{w_1, \cdots, w_n\}^I = \{w_1^I, \cdots, w_n^I\}$
$i \in I_A$	$i^I \in \Delta^I$	$\forall P.E$	$(\forall P.E)^I = \{d \mid \forall d', (d, d') \in P^I \rightarrow y \in E^I\}$
$v \in I_D$	$v^I = v^D \in \Delta^D$	$\exists P.E$	$(\exists P.E)^I = \{d \mid \exists d', (d, d') \in P^I \wedge y \in E^I\}$
R^-	$(R^-)^I = \{(d, d') \mid (d', d) \in R^I\}$	$\geq n\ P$	$(\geq n P)^I = \{d \mid \#(\{d' \mid (d, d') \in P^I\}) \geq n\}$
$\top$	$\top^I = \Delta^I$	$\leq n\ P$	$(\leq n P)^I = \{d \mid \#(\{d' \mid (d, d') \in P^I\}) \leq n\}$
$\bot$	$\bot^I = \varnothing$		

In table 1, Δ^I and Δ^D are two nonempty sets standing for abstract and concrete domains. $\top$ and $\bot$ are considered as the top and bottom concepts. $\{w_1, \cdots, w_n\}$ is a sub set of I_A or I_D; C_1 and C_2 are abstract concepts; P is a SHOIN(D) role; the expressions $\forall P.E$ and $\exists P.E$ satisfy that if $P \in R_D$, $E \in C_D$; if $P \in \{R^- \mid R \in R_A\} \cup R_A$, E is an abstract concept.

Definition 2. A SHOIN(D) KB Σ (T_Σ, R_Σ, A_Σ) consists of three finite axiom boxes: TBox T_Σ, RBox R_Σ and ABox A_Σ, whose axioms' syntax and semantics are given in table 2.

Table 2. Axioms in a SHOIN(D) KB

	Syntax	Semantics
TBox axiom	$C_1 \sqsubseteq C_2$	$C_1^I \subseteq C_2^I$
RBox axiom	$R_1 \sqsubseteq R_2$	$R_1^I \subseteq R_2^I$
	Trans(R_1)	$R_1^I = (R_1^I)^+$
	$U_1 \sqsubseteq U_2$	$U_1^I \subseteq U_2^I$
ABox axiom	$i : C$	$i^I \in C^I$
	$i_1 = i_2$	$i_1^I = i_2^I$
	$i_1 \neq i_2$	$i_1^I \neq i_2^I$

An interpretation $I(\Delta^I, \cdot^I)$ satisfies an axiom if it satisfies corresponding semantics restriction given in table 2. *I* satisfies a TBox (RBox and ABox respectively), if *I* satisfies any axiom in it. *I* satisfies a SHOIN(D) KB Σ, if *I* satisfies its TBox, ABox and RBox. Such interpretation *I* is called a model of KB Σ.

3 OWL Inference Control Rule

Before discussing OWL ICR, we will give a short introduction of OWL query mechanism. Answers of OWL queries are a collection of logical entailed axioms of the OWL KB. Using semantics discussed in section 2, we define "logical entail": for any axiom α, a KB Σ logical entails α, denoted as $\Sigma \models \alpha$, if for any model $I(\Delta^I, \cdot^I)$ of Σ, *I* satisfies α. With "rolling-up" technique [5], OWL query problems can be simplified to be instance retrievals *Retrieval*( ) of singleton concepts *C*:

$$Retrieval(C) = \{ i \mid \Sigma \models i : C \}$$

Now we give the definition of OWL ICR.

Definition 3. An OWL ICR is a triple (subject, axioms set, sign), where subject is an identification of user, axioms set is a given set of logical entailed ABox axioms of the OWL KB Σ, and sign$\in$ {+(positive), -(negative), ?(unknown)}.

For a positive OWL ICR (*s*, { $\alpha_1,...,\alpha_n$ }, +), it means when the query presenter *s* give a query: *Retrieval*(*C*), if $\alpha_i = i : C$ and *i*$\in$ *Retrieval*(*C*), *i* must be returned to *s*. Negative or unknown ICRs have similar meanings with replacing "must" with "must not" or "maybe". For the set of logical entailed axioms of Σ may be infinite, it is impossible to sign all axioms with +, - or ?. Therefore we set ? as the default value.

Definition 4. For a given OWL ICR set S_R, let $S(s,+)=\bigcup Sa$ for any $(s, Sa, +) \in S_R$ and $S(s,-)=\bigcup Sa$ for any $(s, Sa, -) \in S_R$. S_R is consistent w.r.t an OWL KB Σ, if for any subject *s*, there is a OWL KB Σ^*, where $\Sigma \models \Sigma^*$ and $\Sigma^* \models Sa(s,+)$ and for any axioms α in $Sa(s,-)$, $\Sigma^* \not\models \alpha$. We call such KB Σ^* as a suitable sub-KB of Σ.

For any consistent OWL ICR set S_R w.r.t Σ, perfect inference control aims to find the most general suitable sub-KB Σ^* of Σ for a given subject *s*, where no other suitable sub-KB Σ' satisfying $\Sigma' \models \Sigma^*$. In OWL retrieval, Σ^* will replace Σ as a special view for the subject *s*. However, this problem is as hard as finding most general consistent sub-ontology in an inconsistent big ontology [6], and there is no efficient method to deal with such problem. In this paper, we give a primitive approach to find a biggish suitable sub-KB. The creation of the biggish suitable sub-KB is considered as pruning KB Σ. We use SRView as an abstraction of Σ and define pruning operations for SRView. A pruned SRView will play a role as a suitable sub-KB in OWL query.

4 Semantic Related View of OWL Knowledge Base

Before turning our attention towards SRView, we introduce some common notions. We define the set *sub*(*C*) of sub-concepts of a SHOIN(D) concept *C*:

$$sub(C) = \{C\} \cup \begin{cases} \varnothing & \text{if } C = A \mid \geq n\, P \mid \leq n\, P \mid \{w_1, \cdots, w_n\} \\ sub(C_1) & \text{if } C = \neg C_1 \\ sub(C_1) \cup sub(C_2) & \text{if } C = C_1 \sqcap C_2 \mid C_1 \sqcup C_2 \\ sub(E) & \text{if } C = \exists P.E \mid \forall P.E \end{cases}$$

Let S_Σ be the set of concepts appearing in an OWL KB Σ, and $sub(S_\Sigma) = \bigcup sub(C)$ for any C in S_Σ. After talking about these notions, we present a formal structure of a SRView.

Definition 5. A SRView is a graph *G*(*V*, *E*), where *V* is a set of labels and *E* is a set of edges with connectives. Every edge can be denoted as a triple (l, con_i, l_1), where $l, l_1 \in V$, $con_i \in \{\exists P, \forall P, \neg, \sqcap, \sqcup, \geq n\, P, \leq n\, P, \sqsubseteq\}$. For any label l in V, let $E_{out}(l) = \{(l, con_i, l_1) \mid (l, con_i, l_1) \in E\}$, $E_{in}(l) = \{(l_1, con_i, l) \mid (l_1, con_i, l) \in E\}$.

Additionally, we define a label function L: $sub(S_\Sigma) \rightarrow V$ and it satisfies that for any two concepts (or data types) C_1 and C_2 in $sub(S_\Sigma)$, $L(C_1) \neq L(C_2)$.

Now we will give the detailed creation of a SRView from an OWL KB Σ. At the beginning, we initialize $V = \{L(\top), L(\bot)\}$, $E = \varnothing$, and following creation consists of three ordinal steps:

Developing step: for any concept C in S_Σ, we add a concept tree *Tree*(*C*) in *G*. The addition process *AddTree*(*C*) is described in Figure 1. Obviously, *Tree*(*C*) is a syntax tree to denote the structure of concept *C*.

Connection step: for any axiom $C_1 \sqsubseteq C_2$ in the TBox of Σ, we connect $Tree(C_1)$ and $Tree(C_2)$ by adding $(L(C_1), \sqsubseteq, L(C_2))$ in *E*.

Merging step: if two trees have the same structure, we will merge the two trees into one. For any label l in V, let $C_{out}(l) = \{(con_i, l_1) \mid (l, con_i, l_1) \in E_{out}(l)$ and $con_i \neq \sqsubseteq\}$ An merging rule is presented as follows: for any two labels l and l^*, if $C_{out}(l) = C_{out}(l^*)$, then merge two labels into one label l and add a set of new edges

$\{(l, con_i, l_1)|(l^*, con_i, l_1)\in E_{out}(l^*)\} \cup \{(l_1, con_i, l)|\ (l_1\ con_i, l^*)\in E_{in}(l^*)\}$ in E. When no merging rule can be applied to G, we call G is a SRView of KB Σ and denote it as $G=SRV(\Sigma)$.

Obviously, a SRView $G(V, E)$ expresses syntax structures of concepts by concept tree and "subsumption" relationships by "$\sqsubseteq$" edge. For any edge in E can be seen as a production rule, a SRView explicitly creates a production system to explain complex concepts by containing all restrictions on these concepts.

Procedure *AddTree*(*C*) **begin**
If $L(C)$ is not defined
{add a new label l in V and let $L(C)=l$;
Switch (C) {
 Case $\neg C_1$: *AddTree* (C_1); add a new edge ($L(C), \neg, L(C_1)$) in E;
 Case $C_1 \sqcap C_2$: *AddTree*(C_1); *AddTree*(C_2); add ($L(C), \sqcap, L(C_1)$) and ($L(C), \sqcap, L(C_2)$) in E;
 Case $C_1 \sqcup C_2$: *AddTree*(C_1); *AddTree*(C_2); add ($L(C), \sqcup, L(C_1)$) and ($L(C), \sqcup, L(C_2)$) in E;
 Case $\exists P.E$: *AddTree*(E); add ($L(C), \exists P, L(E)$) in E;
 Case $\forall P.E$: *AddTree*(E); add ($L(C), \forall P, L(E)$) in E;
 Case $\geq n\ P$: add ($L(C), \geq n\ P, L(\top)$) in E;
 Case $\leq n\ P$: add ($L(C), \leq n\ P, L(\top)$) in E; }
}
end

Fig. 1. *AddTree*( ) procedure

After talking about creating SRView, we will go into pruning it. The eight pruning operations are defined as triples (operation, old edge, substituted edge) in table 3.

Table 3. Pruning operations

Operation	Old Edge	Substituted Edge
Del	(l, con_i, l_1)	
$\neg$	$(l, \neg, l_1)$	$(L(\bot), \sqcap, l), (L(\bot), \sqcap, l_1)$
$\sqcap$	$(l, \sqcap, l_1),(l, \sqcap, l_2)$	$(l, \sqsubseteq, l_1), (l, \sqsubseteq, l_2)$
$\sqcup$	$(l, \sqcup, l_1),(l, \sqcup, l_2)$	$(l_1, \sqsubseteq, l),(l_2, \sqsubseteq, l)$
$\exists P$	$(l, \exists P, l_1)$	$(l^*, \exists P^*, l_1), (l^*, \sqsubseteq, l)$, where $P^* \sqsubseteq P$
$\forall P$	$(l, \forall P, l_1)$	$(l^*, \forall P^*, l_1), (l^*, \sqsubseteq, l)$, where $P \sqsubseteq P^*$
$\geq n\ P$	$(l, \geq n\ P, l_1)$	$(l^*, \geq m\ P, l_1), (l^*, \sqsubseteq, l)$, where $m>n$
$\leq n\ P$	$(l, \leq n\ P, l_1)$	$(l^*, \leq m\ P, l_1),(l^*, \sqsubseteq, l)$, where $n>m$

In table 1, l^* is a new label added by pruning operations and "$P^* \sqsubseteq P$" means P^* is a sub-role of P. A pruning operation will replace old edges with substituted ones in E. Pruning operations rewrite SRView to release some restrictions and such rewriting will affect reasoning so that some blocked axioms will not be inferred by the pruned SRView.

Definition 6. An SRView $G^*(V^*, E^*)$ is a pruned view of $G(V, E)$ w.r.t a operation set $S_O=\{O_1, \ldots, O_n\}$, if $G^*(V^*, E^*)$ is the new graph after applying all operations to G.

5 Inference Control with Pruned SRView

Now we give an overview of our query framework (figure 2), which describes data exchange between subject and knowledge base system. The knowledge base system consists of four processors (rectangles) and two stores (rounded rectangles). We will detailedly discuss how knowledge base system deals with a query.

1) When the knowledge base system receives a query $Q=Retrieval(C)$ from subject s, Parser parses C into its concept tree $Tree(C)$ and labels C as $L(C)$.

2) Connector connect $Tree(C)=G_C(V_C, E_C)$ and $SRV(\Sigma)=G_{KB}(V_{KB}, E_{KB})$ into a mixed SRView G, where $V=V_C \cup V_{KB}$ and $E=E_C \cup E_{KB}$. After exhaustively applying merging rules to G, Connector sends G as the final result.

3) Pruner gets G from Connector, and a pruning operation set $So(s)$ from Pruner store. Pruner applies $So(s)$ to prune G and return a pruned SRView G^* w.r.t $So(s)$.

4) Reasoner provides three sub-processors: SRView interpreter, ABox rewriter and RBox preprocessor (Figure 3). SRView interpreter sends an direct production $DP(C_{out}(l))$ and sup-label set $Lsup(l)$ of any label l to Tableau Reasoner.

$$DP(C_{out}(l))=\begin{cases} \neg l^* & \text{if } \#C_{out}(l)=1 \text{ and } (\neg, l^*)\in C_{out}(l) \\ con_i.l^* & \text{if } \#C_{out}(l)=1, (con_i, l^*)\in C_{out}(l) \text{ and } con_i \neq \neg \\ l_1 \sqcup l_2 & \text{if } \#C_{out}(l)=2 \text{ and } (\sqcup, l_1), (\sqcup, l_2)\in C_{out}(l) \\ l_1 \sqcap l_2 & \text{if } \#C_{out}(l)=2 \text{ and } (\sqcap, l_1), (\sqcap, l_2)\in C_{out}(l) \end{cases}$$

$$Lsup(l)= \{l^* | (l, \sqsubseteq, l^*)\in E\}$$

ABox rewriter replaces any concept C with its label $L(C)$ in ABox A_Σ and new ABox is denoted as $L(A_\Sigma)$. For a RBox R_Σ, we introduce $\sqsubseteq_R$ as the transitive-reflexive closure of $\sqsubseteq$ on $R_\Sigma \cup \{S^- \sqsubseteq R^- \mid S \sqsubseteq R \in R_\Sigma\}$. RBox preprocessor computes "$\sqsubseteq_R$" relations that are repeatedly used in reasoning process. Finally Tableau Reasoner will return $Retrieval(L(C))$ as an answer set of the query to subject s.

Now we will prove that any answer in $Retrieval(L(C))$ is a validate answer to $Q=Retrieval(C)$, that means $Retrieval(L(C)) \subseteq Retrieval(C)$.

Definition 7. For a SRView $G=(V, E)$ and any l in V, let $TBox(l)=\{DP(C_{out}(l)) \sqsubseteq l, l \sqsubseteq DP(C_{out}(l))\} \cup \{l \sqsubseteq l^* | l^* \in Lsup(l)\}$, $TBox(G)=\bigcup_{l\in V}(TBox(l))$.

Lemma 1. For a query $Retrieval(C)$, a KB $\Sigma(T, R, A)$ and its SRView G_{KB}, let $G=Merge(G_{KB}, Tree(C))$, we can get that $\Sigma(T, R, A) \models i: C \Leftrightarrow \Sigma^*(TBox(G), R, L(A)) \models i: L(C)$.

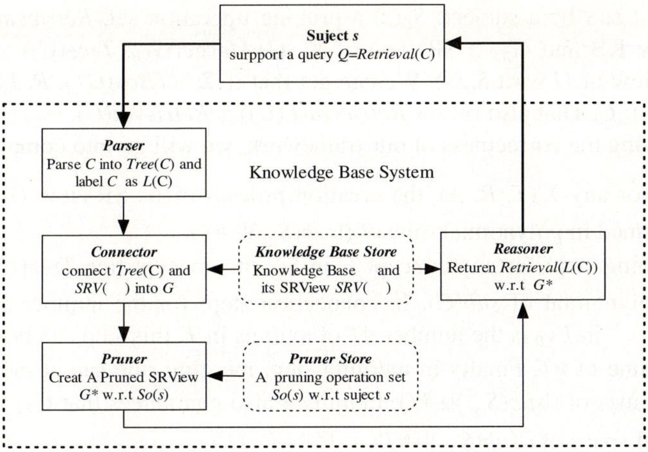

Fig. 2. A query framework with pruned SRViews

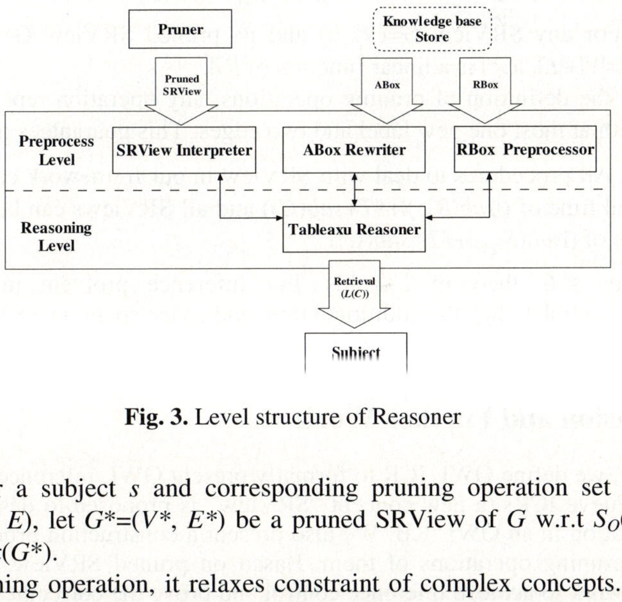

Fig. 3. Level structure of Reasoner

Lemma 2. For a subject s and corresponding pruning operation set $S_O(s)$ and a SRView $G=(V, E)$, let $G^*=(V^*, E^*)$ be a pruned SRView of G w.r.t $S_O(s)$, we have $TBox(G) \models TBox(G^*)$.

For any pruning operation, it relaxes constraint of complex concepts. The pruned SRView can be inferred by original SRView. For example, $\geq n\ P$ rule adds a new label l^* by replacing n with a larger number m and states l^* is subsumed by l, directly $l = \geq n\ P.\top$ $l^* = \geq m\ P.\top$ can infer that $l^* \sqsubseteq l$. Therefore, lemma 2 holds.

Lemma 3. For a query *Retrieval*(C), a KB $\Sigma(T, R, A)$, a pruned SRView $G^*=(V^*, E^*)$ and its label function $L()$, *Retrieval*($L(C)$) w.r.t G^* is the set $\{i \mid \Sigma^*(TBox(G^*), R, L(A)) \models i: L(C)\}$.

This Lemma obviously holds, for Tableau Reasoner considers G^*, R and $L(A)$ as TBox, RBox and ABox of new KB respectively. From above three lemmas, we will have the following theorem.

Theorem 1. Let *s* be a subject, $S_O(s)$ a pruning operation set, *Retrieval*(*C*) a query, $\Sigma(T, R, A)$ a KB and G_{KB} a SRView of Σ. G=*Merge*(G_{KB}, *Tree*(*C*)) and G^* is the pruned SRView of G w.r.t $S_O(s)$. We can get that if $\Sigma^*(TBox(G^*), R, L(A)) \models i: L(C)$, $\Sigma(T, R, A) \models i: C$. That also means *Retrieval*(*L*(*C*)) $\subseteq$ *Retrieval*(*C*).

After proving the correctness of our framework, we will go into complexity issue.

Lemma 4. For any $\Sigma(T, R, A)$, the creation process of its SRView $G_{KB}=(V_{KB}, E_{KB})$ can be performed in polynomial time of $(|sub(S_\Sigma)|+\#T)$.

In developing step: for any concept C in S_Σ, its concept tree *Tree*(*C*) can be performed in polynomial of *sub*(*C*). In connection step: for the number of edges with connective "$\sqsubseteq$" in E_{KB} is the number $\#T$ of axioms in T, this step can be performed in polynomial time of $\#T$. Finally in merging step, merging rule can at most be applied polynomial times of $(|sub(S_\Sigma)|+\#T)$. Lemma 4 also guarantees that G_{KB} can be stored in polynomial space of $(|sub(S_\Sigma)|+\#T)$.

Lemma 5. For a SRView G_{KB} and *Tree*(*C*), G=*Merge*(G_{KB}, *Tree*(*C*)), the creation of G can performed in polynomial time of $(|sub(S_\Sigma)|+\#T+sub(C))$.

Lemma 6. For any SRView $G=(V, E)$ and its pruned SRView $G^*=(V^*, E^*)$ w.r.t $So(s)$, let $|G|=|V|+|E|$, $|G^*|$ is a linear function of $|G|$.

Based on the definition of pruning operations, any operation replaces one or two old edges with at most one new label and two edges. This guarantees that $|G^*| \le 3|G|$.

Theorem 2. All procedures to deal with SRView in our framework can be performed in polynomial time of $(|sub(S_\Sigma)|+\#T+sub(C))$ and all SRViews can be stored in polynomial space of $(|sub(S_\Sigma)|+\#T+sub(C))$.

From Lemma 4-6, theorem 2 holds. For inference problem in SHOIN(D) is NEXPTIME-complete [4], the additional time and space spent on SRView will not be a bottleneck of a knowledge base system.

6 Conclusion and Further Work

In this paper, we define OWL ICR to formally present OWL inference control restrictions. To achieve ICRs, a new concept "SRView" is proposed to describe the whole semantic relation in an OWL KB. We also present a construction process of SRView and define pruning operations of them. Based on pruned SRView, we construct a query framework to achieve inference control and prove the correctness and complexity of it. Creating pruning operation sets w.r.t OWL ICRs is still a manual work in our framework. Future work includes designing some assistant method to semiautomatic or automatic create them, which will make our framework less manually interfered.

References

[1] Thuraisingham, B.: Security standards for the Semantic Web. Computer Standards & Interfaces, vol. 27 (2005) 257-268

[2] Fan, W., Chan, C., Garofalakis, M.: Secure XML Querying with Security Views. In: Proceedings of SIGMOD 2004, (2004) 587-598

[3] Reddivari, P., Finin, T., Joshi, A.: Policy based Access Control for a RDF Store. In: Proceedings of Policy Management for the Web workshop (position paper), 14th Int. World Wide Web Conf, Chiba, (2005) 78-81
[4] Horrocks, I., Patel-Schneider, P.F.: Reducing OWL entailment to description logic satisfiability. In: Proceedings of the 2003 Description Logic Workshop, (2003) 1-8
[5] Horrocks, I., Tessaris, S.: Querying the Semantic Web: a formal approach. In: Proceedings of 2002 Semantic Web Conf. (ISWC 2002), (2002) 177-191
[6] Haase, P., van Harmelen F., Huang Z., Stuckenschmidt, H., Sure Y.: A Framework for Handling Inconsistency in Changing Ontologies. In: Proceedings of the Fourth International Semantic Web Conference (ISWC2005), (2005) 353-367

Efficient Population Diversity Handling Genetic Algorithm for QoS-Aware Web Services Selection*

Chengwen Zhang, Sen Su, and Junliang Chen

State Key Lab of Networking and Switching Technology,
Beijing University of Posts & Telecommunications(BUPT)
187# 10 Xi Tu Cheng Rd., Beijing 100876, China
zwjcbj2007@gmail.com, {susen, chjl}@bupt.edu.cn

Abstract. To maximize user satisfaction during composition of web services, a genetic algorithm with population diversity handling is presented for Quality of Service(QoS)-aware web services selection. In this algorithm, the fitness function, the selection mechanism of the population as well as the competition mechanism of the population are represented. The population diversity and population fitness are used as the primary criteria of the population evolution. By competing between the current population and the historical optimal population, the whole population evolution can be done on the basis of the whole population evolution principle of the biologic genetic theory. Prematurity is overcome effectively. Experiments on QoS-aware web services selection show that the genetic algorithm with population diversity handling can get more excellent composite service plan than the standard genetic algorithm.

1 Introduction

Web service is a software application identified by an URI [1]. How to create robust service compositions becomes the next step work in web services [2] and has attracted a lot of researches [3], [4], [16], [17]. Since the web services with the same functions and different QoS are increasing, and the web services requesters always express their functional requirements as well as their global QoS constraints, a determination needs to be made to select which services are used in a given composite service on the basis of multiple QoS attributes in order to maximize user satisfaction. Hence, QoS-aware web services selection plays an important role in web services composition [5], [6]. In the past years, the researches about web services selection have gained considerable momentums [9], [10], [11], [15], [16], [17], [18], [19].

To figure out QoS-aware web services selection, some approaches are made with the help of semantic web [9], [10], [11], and the others are based on the QoS attributes computation [15], [16], [17], [18], [19]. But it is obvious that the latter approaches are

* The work presented in this paper was supported by the National Basic Research and Development Program (973 program) of China under Grant No. 2003CB314806; the National Natural Science Foundation project of China under Grant No. 90204007; the National Natural Science Funds for Distinguished Young Scholar of China under Grant No. 60125101; the program for Changjiang Scholars and Innovative Research Team in University (PCSIRT); BUPT Education Foundation.

V.N. Alexandrov et al. (Eds.): ICCS 2006, Part IV, LNCS 3994, pp. 104 – 111, 2006.
© Springer-Verlag Berlin Heidelberg 2006

more suitable solutions to satisfy the global QoS requirement. It is a combinatorial optimization issue that the best combination of web services is selected in order to accord with the global constraints.

Genetic Algorithm is a powerful tool to solve combinatorial optimizing problems [12]. The design of a genetic algorithm has the greatest influence on its behavior and performance [20], especially the fitness function and the selection mechanism.

Since keeping the individual diversity is a perfect means to overcome the premature phenomenon, it may be a good way to add the diversity handling into the genetic algorithm. Some diversity studies were in [13], [14], [22], [23], which paid less attention to the population evolution making use of the population diversity.

Following the above analyses, we present a genetic algorithm with population diversity handling, which enables the population to evolve on the basis of the whole population evolution principle of the biologic genetic theory.

The remainder of this paper is organized as follows. After a review of the literature of the QoS-aware web services selection using QoS computation in section 2 and the diversity of genetic algorithm in section 3, Section 4 presents the genetic algorithm with population diversity handling we propose in detail. Section 5 describes the simulations of the proposed algorithm and discusses results aiming to support the work. Finally, section 6 is our conclusions.

2 Quality Computation-Based Selection of Web Services

According to Std. ISO 8402 [7] and ITU E.800 [8], QoS may include a number of nonfunctional properties such as price, response time, availability, reputation. Thus, the QoS value of a composition service can be achieved by the fair computation of QoS of every component web services. There are some techniques in the literatures, including some traditional optimization techniques [15], [16], [17] and strategies based on Genetic Algorithm (GA) [18], [19].

The QoS computation based on the QoS matrix is a representative solution. To normalize the QoS matrix to rank the web services was proposed in [15], however, it was only a local optimization algorithm but not a global one for services selection. Other works in the area of QoS computation include [16], [17], which proposed local optimization and global planning. But, both had limitation to some extent.

The genetic algorithm is well suitable for the QoS-aware services selection belonging to the class of NP-hard [19]. In [18] and [19], the authors proposed only a coding manner of chromosome or the fitness function for the service selection with little further information about the rest parts of the genetic algorithm, such as the selection mechanism.

3 Diversity of Genetic Algorithm

This section describes some diversity control means presented in [13], [14], [22], [23]. In [13], the affinity was defined by means of the information entropy of every gene of chromosome. Finally, the individual with higher fitness and lower affinity had higher priority to evolve. Also, the low diversity of the population was raised by a procedure aiming at maintaining high diversity of the population in [14]. In [22], an individual diversity controlling approach was used to attain the global optimum without getting stuck at a local optimum. And, a fuzzy controller changed the crossover rate and the mutation rate in [23], in order to maintain the proper population diversity during the GA's operation.

According to the biology genetic theory, the evolution of a population is a process during which all of individuals have higher and higher fitness to the environments. From the point of the view of the population, the population diversity evolves from high diversity at initial stage to low one at terminal stage during the population evolution. Hence, comparing population diversity between two populations produced at different generations is a feasible way to evaluate whether the current population has higher evolution extent than the former one, and the population evolution can be handled by the population diversity. Thus, we propose a genetic algorithm with population diversity handling to overcome the local optimal solution.

4 Genetic Algorithm with Population Diversity Handling

In this section, we present a genetic algorithm with diversity handling in order to resolve quality-driven selection. The main difference between our GA and the standard GA is that our GA includes the competition mechanism of the population that does not be expressed by the standard GA.

4.1 Fitness Function

Some QoS models and QoS computation formulas for composite service were available in [17], [19]. But, via comparison in experiments, the optimal solution based on the QoS computation formula in [19] is better than the one based on [17]. Consequently, the QoS computation formula in [19] is adopted, and the objective function is defined in (1):

$$f(g) = \frac{\sum_j (Q_j \times w_j)}{\sum_k (Q_k \times w_k)}. \tag{1}$$

Where $w_j, w_k \in [0,1]$, and w_j, w_k are real positive weight factors, represent the weight of criterion j and k for which end users show their favoritism concerning QoS by providing values respectively. The sum of all of them is 1. Q_j and Q_k denote the sum of QoS values of the individual j and k respectively.

End users usually assert their function requirements as well as the global constraints, e.g. Cost<60, Time<150. The individual whose QoS violates the constraints will be rejected in [19]. However, this approach belongs to the absolute rejection. It influences the diversity of the population seriously and always results in a local optimal solution. The individual disobeying the constraints should be selected proportionally into the next generation population on a basis of a certain technique. The most common method is penalty technique for constrained optimization problems. Hereby, the fitness function with penalty character is defined in (2):

$$Fit(g) = f(g) - \lambda \sum_{j=1}^{n} \left(\frac{\Delta P_j}{R_{jMax} - R_{jMin}}\right)^2 . \tag{2}$$

Where R_{jMax}, R_{jMin} are the maximum value and minimal value of the No.j quality constraint respectively, n is the number of quality constraints, λ is a parameter used to adjust the scale of penalty value. P_j represents the calculation value of a Q_i or some Q_is and these values are limited by a quality constraint. The following is the definition of □P_j:

$$\Delta P_j = \begin{cases} P_j - R_{jMax}; & P_j > R_{jMax} \\ 0 ; & R_{jMin} \le P_j \le R_{jMax} \\ R_{jMin} - P_j; & P_j < R_{jMin} \end{cases}$$

4.2 Design of Diversity

According to the definition of information entropy [14], $H_j(N)$ is the information entropy of the jth locus where N is the number of chromosome and H(N) is the information entropy of all of N chromosomes. Clearly, the smaller the information entropy is, the worse the diversity is.

Firstly, the definition for individuals is (3):

$$d_i = \frac{1}{N} \sum_{j=1}^{N} H(2). \tag{3}$$

Where d_i is used for the diversity of the ith individual, N is the number of chromosomes, H(2) is the information entropy of the ith individual between itself and all of the rest of the population.

Secondly, the population diversity can be acquired from H(N).

4.3 Selection Mechanism of Individuals

The individual with high diversity and with high fitness will survive at higher rate. However, some sets of experiments show that introducing the information entropy to the diversity of the individuals always increases the running time greatly, and can not get obviously more optimal solution than without the diversity of the individuals. Thus, the individuals are selected only according to the fitness value.

4.4 Selection Mechanism of the Population

According to the evolution theory based on natural selection of Darwin, the species evolve in the form of the whole group but not individuals. During the evolution process of the population, in addition to the individuals evolution should be controlled, the whole population evolution should also be handled by means of some policies that prevent the population from degeneration and help to find the optimal solution. In this paper, a selection mechanism of the population is taken into consideration. The historical optimal population is always kept and participates in the competition process between itself and the current population. The population with lower expectation value is known as degeneration and is replaced by the one with higher expectation value.

The expectation value e of the population evolution is calculated in (4):

$$e = \frac{\overline{Fit(g)}}{1 + H(N)}. \tag{4}$$

From (4), we can see that the population with high average fitness and with low diversity will survive at higher rate. So, (4) represents the control mechanism of the diversity because the population with low diversity is promoted and the population with high diversity is suppressed.

Both the maximum of the population fitness and the result of subtracting the minimal fitness of the individuals from the maximal fitness of the individuals are not adopted at the e calculation formula, because the average fitness can cover all of individuals and the others can not reflect the real status of the population, especially in the worst concentricity.

4.5 Structure of Genetic Algorithm with Population Diversity Handling

Hitherto, some important elements in the GA for QoS-driven web services selection have been proposed. Here, the structure of the GA is available in sequence as figure 1.

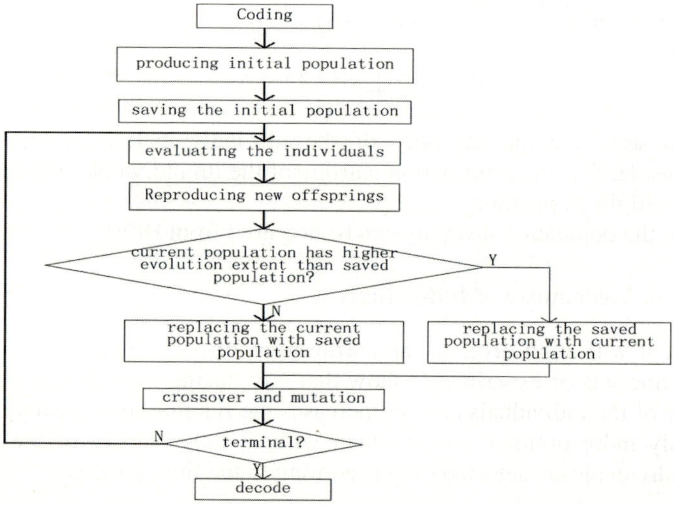

Fig. 1. Structure of the genetic algorithm with population diversity handling

The outstanding point of this paper includes the holding mechanism and competition mechanism of the historical optimal population that ensure the whole population evolution.

5 Experiments

To verify the excellence of the GA of the population diversity handling we have proposed, numerous simulation comparisons between itself and the standard GA had been performed on QoS-aware web services selection. All the experiments were taken on the same software and hardware, which were Pentium 1.6GHz processor, 512MB of RAM, Windows XP Pro, development language JAVA, IDE Eclipse 3.1. The same data were adopted, including workflows of different sizes, 15-50 candidate web services for each task and 5 QoS data for each web service. A simplified representation of web service was used, including an ID number, some QoS data that were retrieved according to some principles in the range of defined values.

The two GAs were set up with the same population size, crossover operation and probability, mutation operation and probability, the fitness function, the selection algorithm of the individuals and the selection mechanism of the individuals. QoS model in [17] was used for both of them. The population diversity handling is the only difference between the two GAs.

The data of the experiments were collected with two methods: statistic data and process data. Figure 2 plotted the fitness function evolution across GA generation. Table 1 presented statistic experiment results of the average fitness and time for the maximal fitness.

Figure 2 was the population size 200, crossover probability 0.7, mutation probability 0.1. The upper curve stood for the GA of population diversity handling and the lower one stood for the standard GA in figure 2. This means the GA of population diversity handling could have larger fitness value than the standard GA.

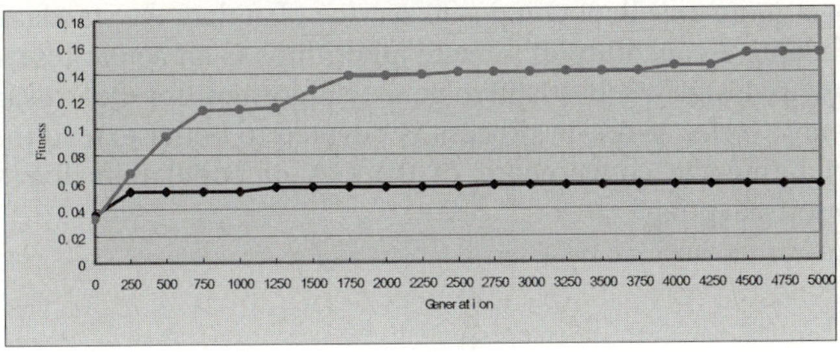

Fig. 2. Fitness comparision when tasks number 30

As shown in table 1, the statistic data were collected after the two GAs running for 50 times with the population size 200, iterations 500, crossover probability 0.7 and mutation probability 0.1. In table 1, the N(ew) represented the GA of population diversity handling and the O(ld) for the standard GA, the time comparison of N and O was the time when the maximal fitness value appears with unit as ms.

Table 1. Statistic Data

Tasks Num	Average fitness (N:O)	Time (N:O)
20	0.125:0.076	7882:2569
30	0.093:0.053	11307:2976

As described above, it is the introduction of the holding mechanism and competition mechanism of the historical optimal population on the basis of the population diversity that ensures that the GA of population diversity handling can get more optimal solution than the standard GA, and to some extent overcome the premature phenomenon of the standard GA. The service composed by the GA of population diversity handling is better than the one composed by the standard GA.

However, it can be seen that the execution time of the GA of population diversity handling is longer than the standard GA at the cost of the holding mechanism and competition mechanism of the historical optimal population. The general indication given by the simulations is that the weaknesses are including the long running time and slow convergence. These would require refining the convergence and shrinking the running time by means of the original population policy and mutation policy, etc.

6 Conclusions

The QoS-aware web services selection is an active research area. In this paper, we present a structure of genetic algorithm characterized by the population competition mechanism. Prematurity is overcome effectively through the conservation of the historical optimal population and the competition between the historical optimal population and the current population.

We also verify the formulas that we use in the genetic algorithm through experiments and the results show that the genetic algorithm with population diversity handling can get more excellent composite service plan than the standard GA.

Providing adaptive capability of genetic algorithms is an active research area [21]. Therefore, how to design a self-adaptive genetic algorithm for QoS-aware selection is one of our future works. Work-in-process is devoted to better extend the approach as follows: accelerating the convergence of the GA of population diversity handling, supplying the self-adaption.

References

1. W3C.Web Services Architecture.http://www.w3.org/TR/2004/NOTE-ws-arch-20040211/ (2004)
2. F. Curbera, R. Khalaf, N. Mukhi, etc.: The Next Step in Web Services. Commnincation of the ACM, 46(10) (2003) 29-34
3. Nikola Milanovic, Miroslaw Malek: Current Solutions for Web Service Composition. IEEE Internet Computing (2004) 51-59
4. B. Orriens, J. Yang, M P Papazoglou: Model Driven Service Composition. In the First International Conference on Service Oriented Computing (ICSOC'03) (2003)
5. D. A. Menascé: QoS Issues in Web Services. IEEE Internet Computing, 6(6) (2002) 72–75
6. D. A. Menascé: Composing Web Services:A QoS View. IEEE Internet Computing (2004) 88-90
7. ISO 8402, Quality Vocabulary
8. ITU-T Recommendation E.800 (1994), Terms and Definitions Related to Quality of Service and Network Performance Including Dependability
9. M. Tian, A. Gramm, H. Ritter, J. Schiller: Efficient Selection and Monitoring of QoS-Aware Web Services with the WS-QoS Framework. IEEE/WIC/ACM International Conference on Web Intelligence (WI'04) (2004)

10. A. Soydan Bilgin, Munindar P. Singh: A DAML-Based Repository for QoS-Aware Semantic Web Service Selection. Proceedings of the IEEE International Conference on Web Services (ICWS'04) (2004)
11. Chen Zhou, Liang-Tien Chia, Bu-Sung Lee: DAML-QoS Ontology for Web Services. IEEE International Conference on Web Services (ICWS'04) (2004)
12. M. Srinivas, L. M. Patnaik: Genetic Algorithm:a Survey. IEEE (1994) 17-26
13. Chun J S, Kim M K, Jung H K. Shape Optimization of Electromagnetic Devices Using Immune Algorithm. IEEE Trans on Magnetics, 33(2) (1997) 1876-1879
14. Yasuhiro TSUJIMURA, Mitsuo GEN: Entropy-Based Genetic Algorithm for Solving TSP.The Second International Conference on Knowledge-Based Intelligent Electronic Systems (1998) 285-290
15. Y. Liu, A. H. Ngu, L. Zeng: QoS Computation and Policing in Dynamic Web Service Selection. In Proceedings of the 13th International Conference on World Wide Web (WWW), ACM Press (2004) 66-73
16. L. Zeng, B. Benatallah, M. Dumas, J. Kalagnanam, Q. Z. Sheng: Quality Driven Web Services Composition. Proc. 12th Int'l Conf. World Wide Web (WWW) (2003)
17. Liangzhao Zeng, Boualem Benatallah, Anne H. H. Ngu,etc.: QoS-Aware Middleware for Web Services Composition. IEEE Transactions on Software Engineering, 30(5) (2004) 311-327
18. LiangJie Zhang, Bing Li, Tian Chao,etc.: On Demand Web Services-Based Business Process Composition.IEEE (2003) 4057-4064
19. G. Canfora, M. Di Penta, R. Esposito, M. L. Villani: A Lightweight Approach for QoS–Aware Service Composition. ICSOC (2004)
20. R. Ignacio, G. Jesús, P. Héctor,etc.: Statistical Analysis of the Main Parameters Involved in the Design of a Genetic Algorithm. IEEE Transactions on Systems, Man, and Cybernetics—Part C: Applications and Reviews, 32(1) (2002) 31-37
21. R. Hinterding, Z. Michalewicz, A. E. Eiben: Adaptation in Evolutionary Computation: a Survey. IEEE EC (1997) 65-69
22. Hisashi Shimodaira: DCGA: A Diversity Control Oriented Genetic Algorithm. The Ninth IEEE International Conference on Tools with Artificial Intelligence (1997) 367-374
23. Kejun Wang:A New Fuzzy Genetic Algorithm Based on Population Diversity.IEEE International Symposium on Computational Intelligence in Robotics and Automation (2001) 108-112

A New Algorithm for Long Flows Statistics—MGCBF

Zhou Mingzhong, Gong Jian, and Ding Wei

Department of Computer Science and Engineering,
Southeast Univercity, Nanjing, Jiangsu, China 210096
{mzzhou, jgong, wding}@njnet.edu.cn

Abstract. Long flows identification and characteristics analysis play more and more important role in modern traffic analysis because long flows take main traffic payload of network. Based on the flows length distribution and long flows characteristics of the Internet, this paper presents a novel long flows' counting and information maintenance algorithm called Multi-granularity Counting Bloom Filter (MGCBF). Using a little fix memory, the MGCBF maintains the counters for all incoming flows with small error probability, and keeps information of long flows whose length are bigger than an optional threshold set by users. This paper builds up an architecture for long flows' information statistics based on this algorithm. And the space used, calculation complexity and error probability of this architecture are also discussed at following. The experiment applied this architecture on the CERNET TRACEs, which indicates that the MGCBF algorithm can reduce the resource usage in counting flows and flows information maintenance dramatically with losing little measurement's accuracy.

1 Introduction

Flow-based measurement is wildly used in network usages just like accounting, bandwidth measurement and network security, etc. A flow is defined as a stream of packets subject to flow specification and timeout. The flow definition can be changed according to its usage, but recent studies show that a very small percentage of flows carry the majority of the packets and bytes [1][2][3][8] regardless of their definition. And so it is very important to improve the network performance by finding out these heavy-hitter flows (called long flows).

In the following of this section, we will introduce the recent related works on long flows identification and statistics, and indicate the main contributions of the Multi-granularity Counting Bloom Filter (MGCBF) algorithm introduced by this paper in flow identification briefly. In the next section, the algorithm is presented in detail, and its performance and error probability are anatomized. And then the optimizations are also expressed. It is proved that the error probability can be controlled through parameters adjusting. In Section 3, some experiments based on the TRACEs from CERNET are employed to illustrate and evaluate the performance and error probability of MGCBF comparing with traditional hash method and CBF method. At last, we discuss the usages of MGCBF in other domains and present the future works.

V.N. Alexandrov et al. (Eds.): ICCS 2006, Part IV, LNCS 3994, pp. 112 – 119, 2006.
© Springer-Verlag Berlin Heidelberg 2006

1.1 Related Works

As the increasing needs of flow-based traffic measurement, Long flows identification and counting are widely studied as one of its main branch [2][3][4][7].

As most widely used way of flow identification, the method presented by IETF RTFM group can gather total or parts of flows information that transmitting by the router. But sampling is widely used in these supports because of the resource restriction of routers in flow identification and exporting according to the RTFM criterions. And sampling satisfies the needs of performance by losing the precision. A. Shaikh, J. Rexford and K. G. Shin[3] applied a method to realize the load balance by keeping the information of every flow, and judged the flow belonging to long or not by the packet number it arrived in a fixed time unit. This is a direct but not efficient way. Smitha I. Kim, and A. L. N.Reddy.[2] provided an algorithm called Least Recent Used (LRU). This algorithm can be used to identify long existing and heavy-hitter flows for load balance, but it can only maintain flows information in a short time and must refresh frequently, and so long flows with large duration will not be kept. C. Estan and G. Varghese [4] used two methods to find out long flows: *sample and hold* and *multistage filters*. And both of them resolve the problem of storing flows information by packet sampling efficiently, but they are only used to find out and keep the information of those long flows, which take a very large ratio of total traffic (i.e. the flow volume is larger than 0.1% of total traffic). A. Kumar, J. Xu, et al.[7] provided an algorithm called SCBF for flow counting, which used limited resource to store all flows' length information with a little errors by sampling. This algorithm applied maximum likelihood estimation (MLE) and mean value estimation (MVE) to estimate the length of every flow. C. Estan, G. Varghese and M. Fisk [10] described a bitmap algorithm to take count of the flows with different length with little storage resource. But all those algorithms and methods do not maintain the detail information for flows, which can satisfy the needs of special applications (i.e. load balance, traffic accounting).

1.2 Main Contribution

This paper proposes a long flows counting and identification algorithm called Multi-granularity Counting Bloom Filter (MGCBF) based on standard bloom filter according to the study of flow distribution and characteristics in detail. This algorithm has the features of adaptability and expansibility because a threshold value can be set according to different usages. And the MGCBF takes advantage of the heavy-tailed distribution of flow length in Internet backbone, and uses a new structure to store flows information, which can save the resource dramatically. This algorithm does not maintain the flow information whose length less than the threshold, and its resource usage is less than other prevalent algorithms [1][11] that kept the flow information.

2 Flow Identification and Statistics Based on MGCBF

This Section provides the prototype of MGCBF based on introducing the standard bloom filter. And the improving model of MGCBF introduced following can decrease its complexity and increase its precision. In the end of this section we will analyze the performance and error rate of this algorithm.

2.1 MGCBF Prototype

For large datasets querying, the time and space complexity are main problems that need to be solved [4][8][9]. MGCBF provided by this paper operates iceberg query to the datasets which items frequencies follow a heavy-tailed distribution. It employs a serial of CBFs (MGCBF={cbf_0,cbf_1, …, cbf_{h-1}}) which use different count spaces (C={1,c_1,c_2, …, c_{h-1}}) to count the frequency of different items in the set.The prototype of this algorithm is introduced as following:

1) When an item *x* wanted to add into MGCBF, the counters at positions $h_1^0(x)$, $h_2^0(x)$, …, $h_{k0}^0(x)$ in vector V_0 increase 1. (V_0 is the vector of MGCBF's first CBF, cbf_0. h_1, h_2, …, h_{k0} are the hash functions in cbf_0. Without the loss of generality, we suppose $h_1^0(x) \le h_2^0(x) \le \ldots \le h_{k0}^0(x)$).
2) Then check the value $h_1^0(x)$. If $h_1^0(x)=c_1$, the counters at positions $h_1^0(x)$, $h_2^0(x)$, …, $h_{k0}^0(x)$ decrease c_1, then values in the counters change to 0, $h_2^0(x)-c_1$, …, $h_{k0}^0(x)-c_1$; the counters at positions $h_1^1(x)$,$h_2^1(x)$, …, $h_{k1}^1(x)$ in V_1 which is the vector of cbf_1, that means $h_1^1(x)+1$, $h_2^1(x)+1$, …, $h_{k1}^1(x)+1$.
3) And then check the value $h_1^1(x)$. If $h_1^1(x)=c_2$, we operate the same action as 2) in cbf_1 and cbf_2. And the check does not stop until there is no carrying or the cbf_h are checked. If we suppose the set of counters' minimum value which is set by an item *x* is M(x)={$min_0(x)$,$min_1(x)$, …, $min_{h-1}(x)$}, then the frequency of *x* in S is:

$$Counter(x) = min_0(x)+min_1(x)*c_1+ \ldots + min_{h-1}(x)*\prod_{i=1}^{h-1} c_i \quad (1)$$

The flow length follows a heavy-tailed distribution in Internet [2][3][4], which means very few long flows take most network traffic. And so it is more efficiency that we use the MGCBF algorithm to replace the traditional counting methods.

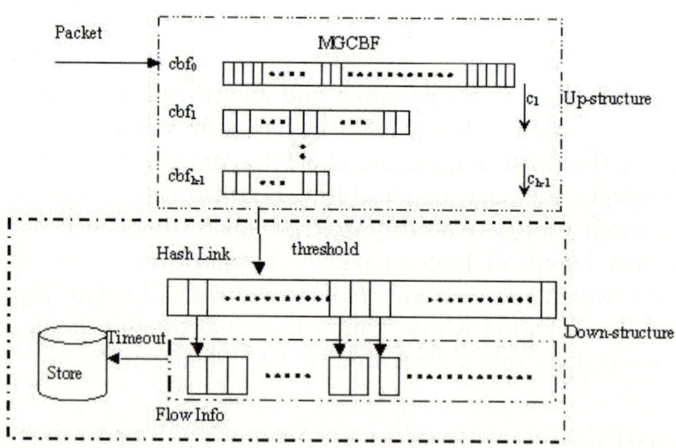

Fig. 1. Long flow information statistics model based on MGCBF

Fig. 1 illustrates the long flow information statistics model based on MGCBF. The data structure of MGCBF in the upside of this figure is used to maintain the packet number of every active flow, and the downside of this figure describes the data struc-

ther more, this model also has excellent expansibility to maintain all flow information whose length is longer than 2 packets. And this model can also be widely used in other related domains if the frequency of items in the datasets follows heavy-tailed distribution.

Acknowledgement

This research is partially support by the National Basic Research Program (called 973 Program), No. 2003CB314803; Jiangsu Province Key Laboratory of Network and Information Security BM2003201 and the Key Project of Chinese Ministry of Education under Grant No.105084.

References

[1] N. Brownlee, C. Mills and G. Ruth. Traffic Flow Measurement: Architecture. RFC 2722. October, 1999.

[2] Smitha, I. Kim, and A. L. N. Reddy. Identifying Long Term High Rate Flows At A Router. In *Proceedings of High Performance Computing*. December, 2001.

[3] A. Shaikh, J. Rexford and K. G. Shin. Load-Sensitive Routing of Long-Lived IP Flows. In *Proceedings of the ACM SIGCOMM 1999*.Cambridge, M.A., USA. August, 1999.

[4] C. Estan and G. Varghese, New Directions in Traffic Measurement and Accounting, In *Proceedings of the ACM SIGCOMM 2002*. Pittsburgh, P.A., USA. August, 2002.

[5] B. Bloom, Space/Time trade-offs in hash coding with allowable errors. In *Commun.ACM*. Vol. 13, no.7, pp. 422-426, July 1970.

[6] S. Cohen, Y. Matias. Spectral Bloom Filters. In *Proceedings of the ACM SIGMOD 2003*. San Diego, C.A., USA. June, 2003.

[7] A. Kumar, J. Xu, et al. Space-Code Bloom Filter for Efficient Per-Flow Traffic Measurement. In *IEEE Infocom 2004*, Hongkong, China. March, 2004.

[8] R. M. Karp, S. Shenker, and C. H. Papadimitriou, A Simple Algorithm For Finding Frequent Elements In Streams And Bags, In *ACM Transactions on Database Systems (TODS)*. Vol. 28, pp. 51-55, 2003.

[9] M. Charikar, K. Chen, and Farach-Colton, Finding Frequent Items In Data Streams, In *ICALP. Lecture Notes in Computer Science, Springer-Verlag, Heidelberg, Germany*. 2002.

[10] C. Estan, G. Varghese and M. Fisk. Bitmap Algorithms For Counting Active Flows On High Speed Links. In *Proceedings of the ACM IMW*, 2003.

[11] L. Fan, P. Cao, J. Almeida, and A. Z. Broder. Summary Cache: A Scalable Wide-Area Web Cache Sharing Protocol. *IEEE/ACM Transactions on Networking*, 8(3):281-293, 2000.

Estimating Original Flow Length from Sampled Flow Statistics

Weijiang Liu, Jian Gong, Wei Ding, and Yanbing Peng

Department of Computer Science and Engineering,
Southeast University, 210096 Nanjing, Jiangsu, China
{wjliu, jgong, wding, ybpeng}@njnet.edu.cn

Abstract. Packet sampling has become an attractive and scalable means to measure flow data on high-speed links. Passive traffic measurement increasingly employs sampling at the packet level and makes inferences from sampled network traffic. This paper proposes a maximum probability method that estimates the length of the corresponding original flow from the length of a sampled flow. We construct the probability models of the original flow length distributions of a sampled flow under the assumptions of various flow length distributions, respectively. Through recovery analyzing with different parameters, we obtain a consistent linear expression that reflects the relationship between the length of sampled flow and that of the corresponding original flow. Furthermore, after using publicly available traces and traces collected from CERNET to do recovery experiments and comparing the experiment outcomes and theoretic values calculated with Pareto distributions, we may conclude that the maximum probability method calculated by using the Pareto distribution with 1.0 can be used to estimate original flow length in the concerned network.

1 Introduction

With the rapid increase of network service types and user number, measuring the volume of network traffic and network performance is becoming more difficult. Packet sampling has become an attractive and scalable means to measure flow data. Sampling entails an inherent loss of information. There are some studies in [1-4] that use statistic inference to recover information as much as possible. However, to our knowledge the above studies are not available to estimate length of the original flow by the given sampled flow length. The original flow length is very important for many applications, e.g., Resources Required for Collecting Flow Statistics, Characterizing Source Traffic and Determining thresholds for setting up connections in flow-switched networks. This paper proposes a Maximum Probability (MP) method that estimates the length of the corresponding original flow from the length of a sampled flow. This method is simple to calculate and easy to implement. The main contributions of this paper include:

1) Maximum Probability (MP) method that estimates the length of the corresponding original flow from the length of a sampled flow is proposed.

V.N. Alexandrov et al. (Eds.): ICCS 2006, Part IV, LNCS 3994, pp. 120–127, 2006.
© Springer-Verlag Berlin Heidelberg 2006

2) A linear expression that reflects the relationship between the length of sampled flow and that of the corresponding original flow is obtained. Although they are different for different Pareto parameters, the differences are very small.
3) We conclude that the value calculated by using the Pareto distribution with 1.0 can be used to estimate original flow length in the concerned network.

The rest of this paper is organized as follows. In the next section, we review some elementary concepts on flow and sampling. In Section 3 we construct the probability models of the original flow length distributions of a sampled flow under the assumptions of various flow length distributions, respectively. In Section 4, we propose the Maximum Probability (MP) method and compare the estimating results of uniform distribution, Pareto distributions and empirical distributions. We conclude with some a summary our contributions in Section 5.

2 Some Elementary Concepts

There are a number of different ways to implement this, e.g., independent sampling of packets with probability $p = 1/N$, and periodic selection of every Nth packet from the full packet stream. In both cases we will call N the sampling period, i.e., the reciprocal of the average sampling rate. Although the length distributions by random and periodic sampling can be distinguished, the differences are, in fact, sufficiently small [2].

Definition 1. A flow is defined as a stream of packets subject to flow specification and timeout.

When a packet arrives, the specific rules of flow specification determine which active flow this packet belongs to, or if no active flow is found that matches the description of this packet, a new flow is created. In this paper, the flow interpacket timeout is 64 seconds. A general flow is a stream of packets subject to timeout and having the same source and destination IP address, same source and destination port numbers (not considering protocol). In this paper, we will use the term original flow to describe the above flow. A flow length is the number of packets in the flow.

Definition 2. A sampled flow is defined as a stream of packets that are sampled at probability $p = 1/N$ from an original flow.

3 Probability Distribution of Original Flow Length

In this paper, sampling probability is $p = 1/N$. For a specific original flow F, let X_F denote the number of packets in F, Y_F denote the number of packets in the sampled flow from F. The conditional distribution of Y_F, given that $X_F = l$, follows a binomial distribution

$$Pr[Y_F = k | X_F = l] = B_p(l, k) = \binom{l}{k} p^k (1-p)^{l-k}.$$

For an original flow F, let $Pr[Y_F = y, X_F = x]$ denote the probability of $X_F = x$ and $Y_F = y$, by the conditional probability formula,

$$Pr[X_F = x|Y_F = y] = \frac{Pr[Y_F = y, X_F = x]}{Pr[Y_F = y]} = \frac{Pr[Y_F = y|X_F = x]Pr[X_F = x]}{Pr[Y_F = y]} \tag{1}$$

and by the complete probability formula, we obtain:

$$Pr[Y_F = y] = \sum_{i=y}^{\infty} Pr[Y_F = y|X_F = i]Pr[X_F = i] = \sum_{i=y}^{\infty} B_p(i,y)Pr[X_F = i] \tag{2}$$

3.1 Uniform Distribution

Suppose original flows lengths satisfy uniform distribution, that is, $Pr[X_F = k] = Pr[X_F = k+1]$, for all $k = 1, 2, \cdots$. Moreover,

$$\sum_{i=k}^{\infty} B_p(l,k) = \sum_{i=k}^{\infty} \binom{l}{k} p^k(1-p)^{l-k} = 1/p = N$$

hence $Pr[Y_F = y] = \sum_{i=y}^{\infty} B_p(i,y)Pr[X_F = i] = Pr[X_F = y] \sum_{i=y}^{\infty} B_p(i,y) = Pr[X_F = y]/p$, so $Pr[Y_F = y|X_F = x] = pB_p(x,y)$ and we obtain the following results:

Lemma 1. *The probability that a sampled flow of length k is sampled from an original flow of length l is $Pr[X_F = l|Y_F = k] = \binom{l}{k} p^{k+1}(1-p)^{l-k}, l = k, k+1, \cdots$.*

Let $a_1 = \frac{B_p(l,k)}{B_p(l-1,k)} = 1 + \frac{kN+1-l}{(l-k-1)N}$. For $l \leq kN$, since $a_1 > 1$, hence $B_p(l,k)$ is increasing as l increases. For $l > kN+1$, since $a_1 < 1$, hence $B_p(l,k)$ is decreasing as l increases. At $l = kN+1$, $a_1 = 1$ means that $B_p(l,k)$ is maximized at $l = kN$ and $l = kN+1$. We have

Lemma 2. *The probability $Pr[X_F = l|Y_F = k]$ is maximized at $l = kN, kN+1$. It is increasing as l increases for $l < kN+1$ and decreasing as l increases for $l > kN+1$.*

3.2 Pareto Distribution

Assume original flow lengths satisfy Pareto distribution. Its probability mass function is

$$Pr[X_F = x] = \beta\alpha^\beta / x^{\beta+1}, \ \ \alpha, \beta > 0, \ \ x \geq \alpha \tag{3}$$

where β is called Pareto parameter. Hence Equation (2) can be written as:

$$Pr[Y_F = y] = \sum_{i=y}^{\infty} B_p(i,y)\beta\alpha^\beta / i^{\beta+1}, \ \ y \geq \alpha$$

Lemma 3. *Under the assumption that original flow lengths satisfy Pareto distribution, the probability that a sampled flow of length $y(\geq \alpha)$ is sampled from an original flow of length x is*

$$Pr[X_F = x|Y_F = y] = \frac{B_p(x,y)/x^{\beta+1}}{\sum\limits_{i=y}^{\infty} B_p(i,y)\beta\alpha^{\beta}/i^{\beta+1}}.$$

4 Maximum Probability Method (MP)

The purpose of the Maximum Probability method (MP) is to estimate the length of the corresponding original flow from the length of a sampled flow. The point we are trying to make is that MP estimates the length of the original flow according to maximum probability. MP contains three steps:

i) Computing probability. Given a sampled flow with fixed length k, compute $Pr[X_F = l|Y_F = k]$ for $l = k, k+1, \cdots$.

ii) Finding maximum probability. Define $mp = \max\limits_{l \geq k}\{Pr[X_F = l|Y_F = k]\}$.

iii) Estimating length. Let $\bar{l} = \min\limits_{l}\{mp = Pr[X_F = l|Y_F = k]\}$, we write our estimate of the length of the original flow as $\bar{l}$.

Below we apply the MP method to different flow length distributions.

4.1 Uniform Distribution

Let the length of original flows be uniform distribution. By Lemma 2, we can use MP to obtain a linear expression as

$$\overline{X}_F = \frac{Y_F}{p} \tag{4}$$

where $\overline{X}_F$ is the estimate of the length of the original flow, Y_F is the length of the sampled flow, p is sampling probability. The estimates of the original flow lengths for $p = 0.1, 0.05, 0.01$ are shown in Figure 1(a). From this figure we can observe the linear relationship between the length of original flow and that of sampled flow under uniform distribution.

4.2 Pareto Distribution

Let the length of original flows be Pareto distribution. By Lemma 3, we can compute the results with $\beta = 0.5, 0.75, 1.0, 1.5$, respectively, and obtain an expression that reflect the relationship between the estimated length of the original flow and the sampled flow length as

$$\overline{X}_F = \frac{Y_F}{p} - n(p,\beta)\ for\ Y_F \geq 1/p \tag{5}$$

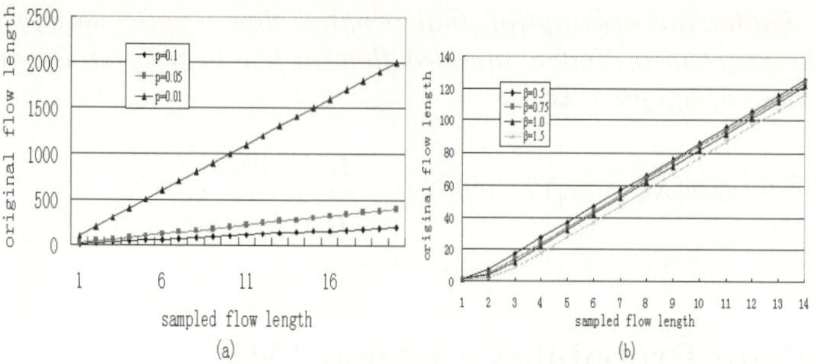

Fig. 1. (a) MP estimate of original flow length under uniform distribution. (b) MP estimate of original flow length under Pareto distribution.

where $\overline{X}_F$ is the estimate of the length of the original flow, Y_F is the length of the sampled flow, p is sampling probability, $n(p, \beta)$ is a positive integer involving p and β as defined in following subsection. In Figure 1(b), with sampling probability $p = 0.1$, and Pareto parameter $\beta = 0.5, 0.75, 1.0, 1.5$, respectively, we can observe that, for a sampled flow with fixed length Y_F, $\overline{X}_F$ is increasing as β decreases. It is minimized at $\beta = 1.5$, maximized at $\beta = 0.5$, medial at $\beta = 1.0$. Though there are differences for different parameter, the estimates show very similar tendencies for all parameters. Therefore, we may conclude that $\overline{X}_F$ calculated by using the parameter $\beta = 1.0$ can be used as approximations under unknown parameter values in the concerned network.

4.3 Finite Pareto Distribution

In the concerned network, due to constraint of measurement time, the lengths and numbers of flows all are finite. Let M denote the maximum original flow length in an original flow distribution, the probability density function in Equation (3) is

$$Pr[X_F = x] = \gamma / x^{\beta+1}, \quad x = \alpha, \cdots, M. \tag{6}$$

where $\gamma = \frac{\beta\alpha^\beta}{\sum_{x=\alpha}^{M} \beta\alpha^\beta / x^{\beta+1}}$, $0 < \gamma < 1$. Equation (6) can be extended so that β can be extended to real field, that is, β can be zero and negative. We call Equation (6) extended Pareto distribution. For $\beta = 0$, Equation (6) is

$$Pr[X_F = x] = \gamma / x, \quad x = \alpha, \cdots, M. \tag{7}$$

For $\beta = -1$, Equation (6) is

$$Pr[X_F = x] = \gamma, \quad x = \alpha, \cdots, M. \tag{8}$$

Equation (8) is uniform distribution, therefore we call uniform distribution as degenerate Pareto distribution.

We now consider how the finite constraint impacts the conditional probability. Due to the constraint of finite number of flows, the conditional probability in

Lemma 3 is written by

$$Pr_M[X_F = x|Y_F = y] = \frac{B_p(x,y)/x^\beta}{\sum_{i=y}^{M} B_p(i,y)/i^{\beta+1}} \tag{9}$$

Obviously $Pr_M[X_F = x|Y_F = y] = \rho(y)Pr[X_F = x|Y_F = y]$, where $\rho(y) = \frac{\sum_{i=y}^{M} B_p(i,y)/i^{\beta+1}}{\sum_{i=y}^{\infty} B_p(i,y)/i^{\beta+1}} < 1$ is a function with variable y. Hence, for a fixed y, the above two probabilities are maximized at the same x. Therefore Equations (4) and (5) are still valid, we rewrite them as consistent form

$$\overline{X}_F = \frac{Y_F}{p} - n(p,\beta) \ for \ Y_F \geq 1/p \tag{10}$$

where $\overline{X}_F$ is the estimate of the length of the original flow, Y_F is the length of the sampled flow, p is sampling probability, $n(p,\beta)$ is a binary function with variables p and β whose value domain is integer set. Here $n(p,\beta)$ is a binary function with variables p and β whose value domain is integer set. Function $n(p,\beta)$ has the following properties:

1) It is a monotone decreasing function on variable p, that is , for fixed β, is decreasing as p increases.

2) It is a monotone increasing function on variable β, that is , for fixed p, is increasing as β increases.

3)$n(p,-1) = 0$, for any $p(0 < p < 1)$

For example, $n(0.1,-1) = 5$, $n(0.1,0) = 10$, $n(0.1,0.5) = 14$, $n(0.1,1.0) = 18$, $n(0.1,1.25) = 21, n(0.1,1.5) = 23, n(0.1,2.0) = 27, n(0.1,3.0) = 36$.

4.4 Comparing for Different Distributions

We use six traces to verify the MP method. The first three traces [5], all containing packets during a 10 minute period, were collected with a Dag3.2E 10/100 MBit/sec Ethernet card at the outside of the firewall servicing researchers at Bell Labs via a 9 MBits/sec link to the Internet in May 2002. The other three Traces, either of which contains packets during a 10-minute period too, were collected at Jiangsu provincial network border of China Education and Research Network (CERNET) in disjoint time interval on April 17, 2004. The backbone capacity is 1000Mbps; mean traffic per day is 587 Mbps. For each trace, we sample at $p = 0.1, 0.05, 0.01$ respectively. Then we use MP to estimate the length of original flow. We find the estimated lengths are very close at same sampling probability for the six traces. For clear display, we only show the estimates in three experiments at sampling probability $p = 0.1$ in Figure 2(a). As shown in Figure 2(a), the estimates are very close.

Figure 2(b) illustrates the estimates for uniform distribution, Pareto distribution ($\beta = 1.0$) and Experiment 1(For first trace, sampling at $p = 0.1$, then we use

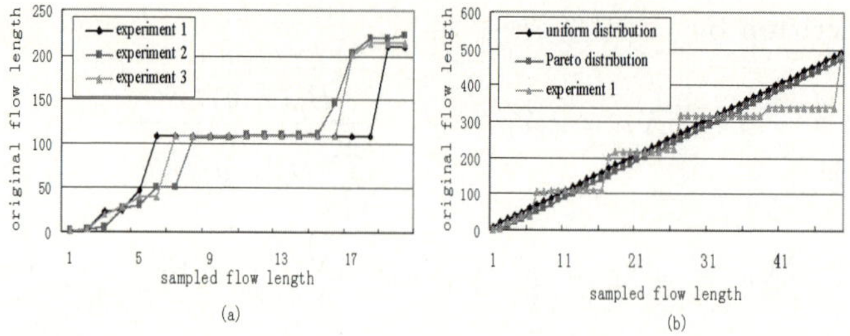

Fig. 2. (a) MP estimate of original flow length under empirical distribution. (b) Comparing MP estimates of original flow length under different distributions.

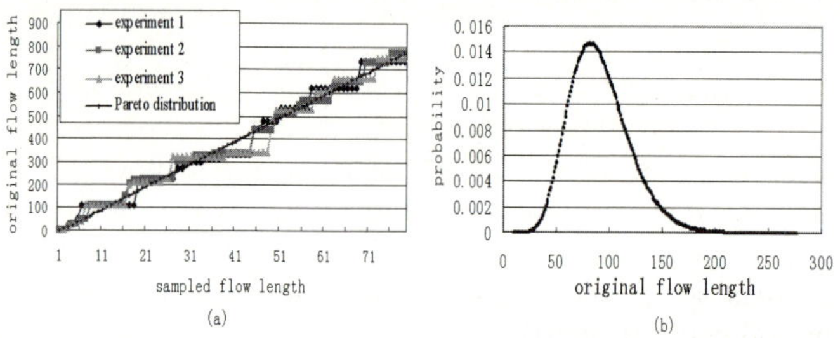

Fig. 3. (a): Comparing MP estimates of original flow length under empirical distributions and Pareto distribution. (b) Probability distribution of original flow length, given length 10 of sampled flow at sampling probability p=0.1.

MP to estimate the length of original flow. Similarly Experiment 2 for the third trace, Experiment 3 for the fifth trace). We can observe that the estimates for uniform are slightly big but the estimates for Pareto distribution with $\beta = 1.0$ fit those for Experiment 1 very well. From Figure 3(a), we can see that the estimate results of several experiments all move up and down at the point of estimate for Pareto distribution $\beta = 1.0$. Therefore we can use the Pareto distribution with $\beta = 1.0$ as theoretical distribution of the three empirical distributions.

4.5 Difficulties and Applications

When we use MP to estimate the original flow length of a sampled flow, we must find the one that makes the probability maximized. However, the maximized probability sometimes is still very small. For example, consider a sampled flow with length 10 at sampling probability 0.1. We calculate the probability by using Pareto distribution with 1.0 to estimate its original flow length. Figure 3(b) displays the probability distribution of the original flow length of the sampled flow. Within it we can observe that the probability at length 82 is maximized with value 0.0147. Although 0.0147 is maximum value, it is too small. It reflects the difficulty and uncertainty associated with an estimate. To improve certainty, we can estimate the confidence interval for original flow length subject to interval width (as small as possible). Suppose that the estimated length may fall into an interval with width 70, we can sum for some values and obtain a maximum probability $Pr[51 \leq x \leq 120|y = 10] = 0.800017$. Therefore, we can use MP to estimate the boundary of length for a specific flow.

5 Conclusions

This paper proposes a naive method (MP) to estimates original flow length from the sampled flow. For different Pareto parameters, we obtain a consistent linear expression that reflects the relationship between the length of sampled flow and that of the corresponding original flow. In the concerned network, the length distributions of flows collected in any time interval do not satisfy Pareto distributions with fixed parameter strictly, but they can follow a Pareto distribution with parameter in interval [0.5, 1.5] approximately. The value 1.0 is the middle value of interval [0.5,1.5] exactly. Theory analysis and experiment results show that it is a reasonable choice using parameter 1.0 to calculate at the condition of unknown parameter value.

Acknowledgement

This work is supported in part by the National Grand Fundamental Research 973 Program of China under Grant No.2003CB314804; the National High Technology Research and Development Program of China (2005AA103001); Jiangsu Planned Projects for Postdoctoral Research Funds.

References

1. Duffield, N.G., Lund, C. , Thorup, M.: Properties and Prediction of Flow Statistics from Sampled Packet Streams. ACM SIGCOMM Internet Measurement Workshop 2002,159-171, November 2002.
2. Duffield, N.G., Lund, C. , Thorup, M.: Estimating Flow Distributions from Sampled Flow Statistics. IEEE/ACM Transation on Networking, **13**(2005) 325-336.
3. Tatsuya Mori, Masato Uchida, Ryoichi Kawahara: Identifying Elephant Flows Through Periodically Sampled Packets. ACM SIGCOMM Internet Measurement Conference 2004,115-120.
4. Noriaki Kamiyama: Identifying High-Rate Flows With Less Memory. IEEE Infocom 2005,2781-2785 ,March 2005.
5. NLANR: Abilene-I data set,http://pma.nlanr.net/Traces/long /bell1.html.

Easily-Implemented Adaptive Packet Sampling for High Speed Networks Flow Measurement*

Hongbo Wang, Yu Lin, Yuehui Jin, and Shiduan Cheng

State Key Laboratory of Networking and Switching Technology,
Beijing University of Posts and Telecommunications, 100876, Beijing, China
{hbwang, linyu, yhjin, chsd}@bupt.edu.cn

Abstract. An Easily-implemented Adaptive Packet Sampling (EAPS) is presented in this paper, which overcomes the shortcoming of NetFlow and Adaptive Netflow. EAPS is easy to be hardware-implemented and scalable for high-speed networks. Additionally, EAPS is automatically adaptive to traffic rate under resource constrains, thus it is convenient to be used by network operators. Experiments are conducted with the real network traces. Results show that EAPS is more accurate than ANF over links with light load or traffic fluctuations.

1 Introduction

Traffic measurement is the basis of IP network monitoring, management and controlling tasks. Particularly, flow-level measurements are widely used for various applications such as traffic profiling, dominant applications or users tracking, and traffic engineering. With the ever increasing speeds of transmission links and volume of network traffic, flow-level measurements face the formidable challenges of the scalability issues. On today's high-speed links, monitoring every packet and recording statistics of every flow consume too much resource at the routers or other network elements. Packet sampling has been suggested[1-5] to address this problem, which is a scalable alternative for network measurement.

Cisco's NetFlow[6] which adopts static packet sampling method are widely deployed by most major ISPs and becomes the most popular flow measurement solution. Though the wide deployment of NetFlow is a proof of its ability to satisfy the important needs of network operators, it has several shortcomings[5]: during flooding attacks, resource consumed by flow records may increase beyond what is available; selecting the right static sampling rate is difficult. Estan et al. proposed Adaptive NetFlow(ANF)[5] to address the problems of NetFlow. When the traffic volume increases, ANF can dynamically decrease the sampling rate until it is low enough for the flow records to fit into flow cache memory. However, the renormalization algorithm adopted by ANF is complex and heuristic, which hinder it from the implementation by hardware. Therefore, ANF itself may become the processing bottleneck when the link speed exceeds OC-48(2.5Gpbs). Furthermore, the maximal sampling rate of ANF

* This work was supported by the NSFC (No. 90204003 and 60502037), CNGI (No. CNGI-04-8-1D) and the 973 project of china (No. 2003CB314806).

V.N. Alexandrov et al. (Eds.): ICCS 2006, Part IV, LNCS 3994, pp. 128 – 135, 2006.
© Springer-Verlag Berlin Heidelberg 2006

(about 1/35) is computed under worst case conditions, but it is suboptimal and less accurate in the light loaded conditions. Additionally, once the sampling rate used by ANF is reduced, it can not increase again in one measurement bin. Consider the cases that the traffic rate fluctuates in the measurement bin: the traffic load is high in the fore-half bin, so ANF adopts small sampling rate; but as the traffic load decrease, ANF can not increase its sampling rate again. Thus ANF becomes less accurate when the traffic load decreases during the measurement bins.

In order to overcome aforementioned shortcomings of ANF, an Easily-implemented Adaptive Packet Sampling EAPS is proposed in this paper for flow measurement. With measurement buffer, EAPS samples a fixed number of packets in every measurement interval by adopting a reservoir sampling method. Since the number of sampled packets is constant in every measurement interval, the sampling rate of EAPS automatically decreases with the increasing of traffic rate. On the contrary, the sampling rate will automatically increase when the traffic rate decreases. Furthermore, the flow cache memory and bandwidth consumption is also constrained by the size of the measurement buffer [1]. Therefore, EAPS is robust during flooding attacks. We also show the upper-limit of the relative standard deviation of EAPS estimation through theoretical analyses. With the experiments of real network traces, the result demonstrates that EAPS is more accurate than ANF over links with light load or traffic fluctuations.

The rest of this paper is organized as follows. Section 2 describes the methodology of EAPS. Section 3 proposes the estimation error. In Section 4, EAPS is compared to ANF using real flow traces. Finally, the paper is concluded in Section 5.

2 Methodology of EAPS

In this section, we will discuss the sampling and estimation methodology of EAPS. NetFlow uses four rules to decide when a flow has ended which then allows the corresponding record to be exported: 1) when indicated by TCP flags (FIN or RST), 2) 15 seconds(configurable) after seeing the last packet with a matching flow ID, 3) 30 minutes (configurable) after the record was created (to avoid staleness) and 4) when the flow cache is full. As shown in [5], most traffic analysis tools divide the traffic stream into fixed analysis bins. Unfortunately, NetFlow records can span bins, causing unnecessary complexity and inaccuracy for traffic analysis. Just as ANF, EAPS divide the NetFlow operation into short bins so that the bins used by traffic analysis are exact multiples of the measurement bins. Differing from ANF, EAPS retains Netflow's four rules during the measurement bins, but terminates all active flow records at the end of each measurement bin. In our experiments we used the more challenging one minute size for the measurement bin.

2.1 Random Sampling Algorithm with a Reservoir

Since the size of measurement buffer n is limited, EAPS can not record all the packets arrived within one measurement bin and then do sampling process. It must do sample at the same time one packet arrives. However, the challenging problem is how to select

[1] Our resource consumption controlling method is like to that of [7]. But our sampling is packet sampling at measurement point in stead of flow sampling at the collector of flow records[7].

a random sample of n packets from N successively arriving packets, where the value of N is unknown beforehand. This can be solved by reservoir random sampling algorithm in literature [8]. The naive algorithm work as follows: the first n packets arrived are stored in the reservoir and became the sample candidates; when the t^{st} ($t>n$) packet is arriving, it has a n/t probability of being a sample candidate, if it became a candidate, it will randomly replace one candidate in the reservoir. It is easy to see that the resulting set of n candidates in the reservoir forms a random sample of the first t packets. The computing complexity of this algorithm is $O(N)$, where N is the total number of packets arrived in one bin. In [8], Vitter proposed algorithm Z with the much less complexity $O(n(1+\log(N/n)))$. EAPS will adopt algorithm Z which is easy to be implemented by hardware with higher efficiency.

2.2 Design of Measurement Buffer

In order to further reduce the size of measurement buffer, we divide the measurement bin into m fixed intervals, and do reservoir sampling in each interval. In our scheme, the measurement buffer is divided into two reservoirs: reservoir A and B. As shown in figure 1, in the ith interval, when flow measurement software reads sampled packets from reservoir A to update flow entries, the packet arriving from network is buffered in another reservoir B. And in the $(i+1)$th interval, the function of reservoir A and B will be swapped ,and so on. The sampling will be done by the hardware implemented Algorithm Z.

In fact, only parts of packet fields (about 21 bytes) are needed to store in the measurement buffer for flow measurement, including TOS, packet length, protocol, source IP address, destination IP, source port, destination port, TCP flags, timestamp, and etc. For a reservoir containing 12000 packets information, a 3.85Mbit SRAM is needed, which is easily implemented by today's semiconductor technology.

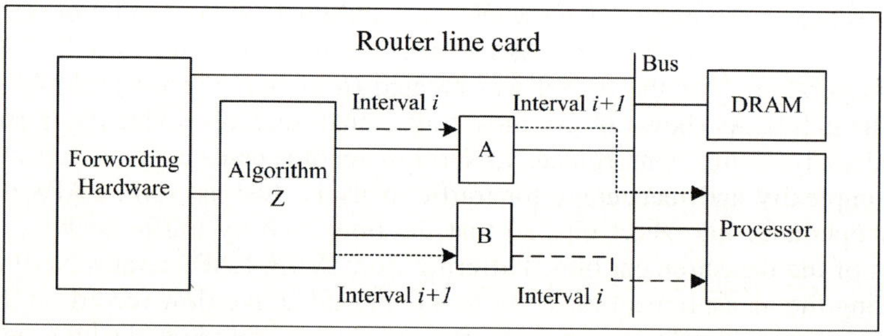

Fig. 1. Design of measurement buffer in router line card

2.3 Estimation Methodology

Denote m intervals in one measurement bin by $t_1, t_2, ..., t_m$. In the interval $t_i (i=1,2,...,m)$, denote a flow appeared by $f_k (k=1,2,...)$ and the total number of packets arrived of all flows by N_i. Since the number of sampled packets in the reservoir is constant n, the sampling rate in this interval is n/N_i. For a flow f_k in interval t_i, let N_i^k be the number of packets belonging to it; and denote the *j*th packet of the flow by $p_{ij}^k (j=1,2,...,N_i^k)$ and its packet size by x_{ij}^k. Similarly, denote the number of sampled packets belonging to flow f_k by n_i^k, and the *j*th packet in the sampled packets by $q_{ij}^k (j=1,2,...,n_i^k)$, its packet size by X_{ij}^k.

The total bytes of flow f_k in the interval t_i, i.e., $x_i^k = \sum_{j=1}^{N_i^k} x_{ij}^k$, is estimated as

$$\hat{x}_i^k = \frac{N_i}{n} \sum_{j=1}^{n_i^k} X_{ij}^k \quad (1)$$

The total packet number of flow f_k in the interval t_i, i.e. N_i^k is estimated as

$$\hat{N}_i^k = \frac{N_i}{n} n_i^k \quad (2)$$

The total bytes of flow f_k in the whole measurement bin, i.e., $x^k = \sum_{i=1}^{m} x_i^k$, is estimated as

$$\hat{x}^k = \sum_{i=1}^{m} \hat{x}_i^k \quad (3)$$

The total packet number of flow f_k in the whole measurement bin, i.e., $N^k = \sum_{i=1}^{m} N_i^k$, is estimated as

$$\hat{N}^k = \sum_{i=1}^{m} \hat{N}_i^k \quad (4)$$

3 Estimation Error

In this section, we firstly prove that $\hat{x}^k$ and $\hat{N}^k$ is the unbiased estimation of x^k and N^k, then give upper-limit of their relative errors. Consider a flow f_k during the measurement interval t_i. Let $\mu_i = \frac{1}{N_i^k} \sum_{j=1}^{N_i^k} x_{ij}^k$, $\sigma_i^2 = \frac{1}{N_i^k} \sum_{j=1}^{N_i^k} (x_{ij}^k - \mu_i)^2$. The average size of packets during the measurement bin is denoted as $\mu^k = \frac{x^k}{N^k}$.

Lemma 1. Let n_i^k be the number of sampled packets, belonging to flow f_k which have N_i^k packets during interval t_i. Then n_i^k is a binomial random variable, and its mean and variance are $E(n_i^k) = N_i^k \frac{n}{N_i}$, $Var(n_i^k) = \frac{n}{N_i}(1 - \frac{n}{N_i}) N_i^k$ respectively.

Proof. Since the buffer size is *n* and the total packet number in interval t_i is N_i, the probability that any packet is sampled is n / N_i. Thus, n_i^k is a binomial random variable with parameters N_i^k and $\frac{n}{N_i}$, so $E(n_i^k) = N_i^k \frac{n}{N_i}$, $Var(n_i^k) = \frac{n}{N_i}(1 - \frac{n}{N_i}) N_i^k$.

Theorem 1. Consider a flow f_k, $\hat{N}^k$ is an unbiased estimation of N^k.

Proof. From equation (2) and Lemma 1, $E(\hat{N}_i^k) = E(\frac{N_i}{n} n_i^k) = \frac{N_i}{n} E(n_i^k) = N_i^k$. Also from equation (4), we have $E(\hat{N}^k) = E(\sum_{i=1}^{m} \hat{N}_i^k) = \sum_{i=1}^{m} E(\hat{N}_i^k) = \sum_{i=1}^{m} N_i^k = N^k$ which proves the theorem.

Theorem 2. Consider a flow f_k, $\hat{x}^k$ is an unbiased estimation of x^k.

Proof. Suppose n_i^k packets belonging a flow f_k are sampled during interval t_i. From the attributes of simple random sampling[9], the sizes of each sampled packets, i.e. $X_{ij}^k (j = 1,2,...,n_i^k)$, is random variable, and $E(X_{ij}^k) = \mu_i$. From equation (1),

$$E(\hat{x}_i^k \mid n_i^k) = E(\frac{N_i}{n} \sum_{j=1}^{n_i^k} X_{ij}^k) = \frac{N_i}{n} n_i^k \mu_i \tag{5}$$

Then $E(\hat{x}_i^k) = E(E(\hat{x}_i^k \mid n_i^k)) = E(\frac{N_i}{n} n_i^k \mu_i) = \frac{N_i}{n} E(n_i^k)\mu_i = N_i^k \mu_i = x_i^k$. From equation (3), we have $E(\hat{x}^k) = E(\sum_{i=1}^{m} \hat{x}_i^k) = \sum_{i=1}^{m} E(\hat{x}_i^k) = \sum_{i=1}^{m} x_i^k = x^k$ which proves the theorem.

Theorem 3. For each flow f_k, the standard deviation of $\hat{N}^k$ is up-bounded by $1 \Big/ \sqrt{n \frac{N^k}{N}}$, where N is the total packet number in one measurement bin.

Proof. From equation (2) and Lemma 1, the variance of $\hat{N}^k$ is $Var(\hat{N}_i^k) = Var(\frac{N_i}{n} n_i^k) = \frac{N_i^2}{n^2} Var(n_i^k) = \frac{N_i^2}{n^2} (\frac{n}{N_i} (1 - \frac{n}{N_i}) N_i^k) = \frac{N_i}{n} (1 - \frac{n}{N_i}) N_i^k < \frac{N_i}{n} N_i^k$

Since the sampling processes of different measurement intervals are independent with each other, $\hat{N}_i^k (i = 1,2...,m)$ are independent random variances. From equation (4), $Var(\hat{N}^k) = Var(\sum_{i=1}^{m} \hat{N}_i^k) = \sum_{i=1}^{m} Var(\hat{N}_i^k) < \sum_{i=1}^{m} \frac{N_i}{n} N_i^k = \frac{1}{n} \sum_{i=1}^{m} N_i N_i^k$. Note that $N_i > 0$ and $N_i^k \geq 0$, then $\sum_{i=1}^{m} N_i N_i^k < \sum_{i=1}^{m} N_i \sum_{i=1}^{m} N_i^k = N N^k$, we thus have $\frac{\sqrt{Var(\hat{N}^k)}}{N^k} < 1 \Big/ \sqrt{n \frac{N^k}{N}}$ which proves the theorem.

Theorem 4. For each flow f_k, the standard deviation of $\hat{x}^k$ is up-bounded by $1 \Big/ \sqrt{n \frac{N^k}{N} \frac{\mu^k}{b_{\max}}}$, where $b_{\max}$ is the maximal size of IP packet, N is the total packet number in an measurement bin.

Proof. From formula of variance of the sample mean in simple random sampling[9],

$$Var(\hat{x}_i^k \mid n_i^k) = Var(\frac{N_i}{n}\sum_{j=1}^{n_i^k} X_{ij}^k) = \left(\frac{N_i}{n}n_i^k\right)^2 Var(\frac{1}{n_i^k}\sum_{j=1}^{n_i^k} X_{ij}^k) = \left(\frac{N_i}{n}n_i^k\right)^2 \frac{\sigma_i^2}{n_i^k}\left(1-\frac{n_i^k-1}{N_i^k-1}\right) < \frac{N_i^2}{n^2}n_i^k\sigma_i^2$$

From equation (5) and Lemma 1, we have

$$\begin{aligned} Var(\hat{x}_i^k) &= E(Var(\hat{x}_i^k \mid n_i^k)) + Var(E(\hat{x}_i^k \mid n_i^k)) < E(\frac{N_i^2}{n^2}n_i^k\sigma_i^2) + Var(\frac{N_i}{n}n_i^k\mu_i) \\ &= \frac{N_i^2}{n^2}E(n_i^k)\sigma_i^2 + \frac{N_i^2}{n^2}Var(n_i^k)\mu_i^2 = \frac{N_i}{n}N_i^k\sigma_i^2 + \frac{N_i}{n}N_i^k(1-\frac{n}{N_i})\mu_i^2 \\ &< \frac{N_i}{n}N_i^k(\sigma_i^2+\mu_i^2) \end{aligned}$$

Since the sampling processes of different measurement intervals are independent with each other, $\hat{x}_i^k (i=1,2...,m)$ are independent random variances,

$$Var(\hat{x}^k) = Var(\sum_{i=1}^{m}\hat{x}_i^k) = \sum_{i=1}^{m}Var(\hat{x}_i^k) < \frac{b_{\max}}{n}\sum_{i=1}^{m}(N_i\sum_{j=1}^{N_i^k}x_{ij}^k) < (\sum_{i=1}^{m}N_i)(\sum_{i=1}^{m}\sum_{j=1}^{N_i^k}x_{ij}^k) = Nx^k$$

. Therefore

$$\frac{\sqrt{Var(\hat{x}^k)}}{x^k} < \sqrt{\frac{b_{\max}N}{nx^k}} = \sqrt{\frac{b_{\max}N}{n\mu^k N^k}} = 1\bigg/\sqrt{n\frac{N^k}{N}\frac{\mu^k}{b_{\max}}}$$

which proves the theorem.

4 Experiment Results and Analysis

In our experiments we use real network traces from OC-48 links collected by CAIDA[10] project[2], and ten one-minute datasets were adopted. We run EAPS over these datasets using different parameters settings, including buffer size of 6k, 12k, 18k and 24k, and measurement interval of 6, 12 and 20 seconds. For same parameters settings, we run EAPS for 27 times.

4.1 Experiment Results of EAPS

Firstly, the relative estimated error of packet number is computed and shown in figure 2.a. We can see that the points are symmetrically distributed around coordinate y. Thus, the estimation is unbiased in accordance with theorem 1.

In practice, it is convenient for network analysis to aggregate individual flows into aggregate flow. In this paper, we define individual flows with same port number as an aggregate flow. As shown in figure 2.b, the upper line is the upper-limit of relative error given by theorem 3, and the spanned points below the line are the results of different experiments. This conforms to theorem 3.

For space limit, we omitted the corresponding traffic bytes results which are similar to figure 2.a. or figure 2.b.

---

[2] We thank CAIDA for providing us the real network traces data.

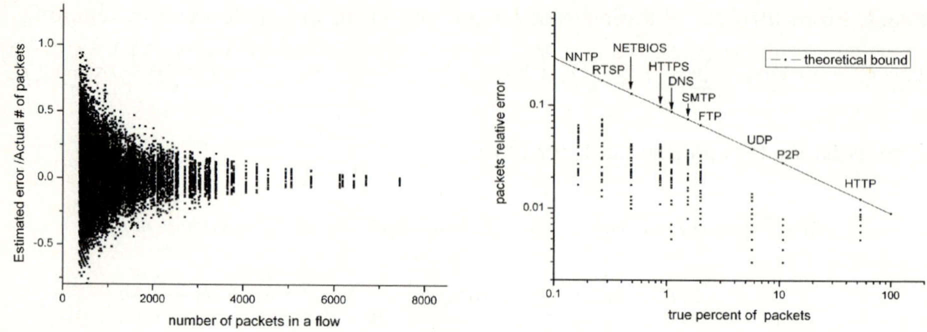

(a) (b)

Fig. 2. Experiment Results of EAPS

4.2 Comparison with ANF

We then compare EAPS with ANF proposed in [5]. For same parameter setting, we run EAPS and ANF respectively.

For space limit, we only show the bytes comparison results of EAPS and ANF based on two typical Datasets. In Dataset1, the traffic load is relatively light. In DataSet2, the traffic load is relatively high, and the traffic rate decreases (fluctuates) during the measurement bin. From table 2, we can see that, for both DataSet1 and DataSet2, EAPS is more accurate than ANF. It can be explained as follows: in the case of DataSet1(on 6s measurement interval), the traffic load is relatively light; and the automatically-adaptive sampling rate of EAPS is about 1/22, while the maximal sampling rate of ANF is 1/35; As to DataSet2(on 12s measurement interval), ANF choose a relatively small sampling rate (about 1/200) when the traffic load is high in the fore-half bin, and can not increase its sampling rate even when the traffic load decreases in the post-half bin; EAPS is automatically adaptive to the changes of traffic rate, it takes a sampling rate about 1/55. Thus EAPS is more accurate than ANF over links with light load or traffic fluctuations.

Table 1. Relative Error Comparison of EAPS with ANF

Aggregate flows		HTTP	P2P	FTP	SMTP	RTSP	HTTPS	DNS
Percent(%)		54.61	12.83	1.74	0.72	0.41	0.33	0.19
Dataset1 (6s)	EAPS	0.003	0.009	0.020	0.036	0.054	0.043	0.026
	ANF	0.007	0.016	0.041	0.070	0.068	0.074	0.048
Dataset2 (12s)	EAPS	0.005	0.014	0.031	0.042	0.073	0.059	0.046
	ANF	0.009	0.022	0.039	0.081	0.077	0.081	0.050

5 Conclusion

Adaptive NetFlow(ANF) has been proposed to overcome the shortcoming of Cisco's NetFlow by dynamically adjusting the sampling rate. However, the renormalization algorithm adopted by ANF is difficult to be implemented by hardware, thus it is not scalable for higher speed links. In this paper, an Easily-implemented Adaptive Packet Sampling (EAPS) is presented. Compared with ANF, EAPS is easier to be hardware-implemented and used, as well as automatically adaptive to traffic rate with certain resource consumption. With the real network traces, the experiments demonstrate that EAPS is more accurate than ANF over links with light load or traffic fluctuations.

References

1. Claffy, K.C., Polyzos, G.C., and Braun, H.-W.: Application of sampling methodologies to network traffic characterization. In Proc. ACM SIGCOMM, (1993) 13-17
2. Hernandez, E.A., Chidester, M.C., George A.D.: Adaptive Sampling for Network Management. Technical Report, University of Florida, (2000)
3. Cheng Guang, Gong Jian, Ding Wei: Network traffic sampling measurement model on packet identification. Acta Electronica Sinica, December, (2002) 1986-1990
4. Choi, B.-Y., Park, J., and Zhang, Z.-L.: Adaptive Random Sampling for Traffic Load Measurement. IEEE International Conference on Communications (ICC '03), May 2003.
5. Estan, C., Keys, K., Moore, D., and Varghese, G.: Building a better netflow. In Proceedings of the ACM SIGCOMM (2004)
6. Cisco netflow. http://www.cisco.com/warp/public/732/Tech/netflow.
7. Duffield, N.G., Lund, C. and Thorup, M.: Flow sampling under hard resource constraints. In Proc. ACM SIGMETRICS 2004 85–96. ACM Press, New York(2004).
8. Vitter, J.S.: Random sampling with a reservoir. ACM Trans. Math. Software. (1985)1137–57.
9. Rice, J.A.: Mathematical Statistics and Data Analysis. Second Edition. Duxbury Press. USA(1995)
10. Cooperative association for internet data analysis.http://www.caida.org

Multi-layer Network Recovery: Avoiding Traffic Disruptions Against Fiber Failures*

Anna Urra, Eusebi Calle, and Jose L. Marzo

Institute of Informatics and Applications (IIiA),
University of Girona, Girona 17071, Spain

Abstract. The next generation backbone networks, optical IP/MPLS networks, enable increasingly higher volumes of information to be transported. In this network architecture, a fiber failure can result in a loss of several terabits of data per second and leads to multiple failures in the upper network layer. Thus, the ability of the network to maintain an acceptable level of reliability has become crucial. In this paper, a dynamic cooperation between packet and wavelength switching domain is considered in order to provide protected paths cost effectively. A new multi-layer routing scheme that incorporates recovery mechanisms in order to guarantee connectivity against any single fiber failure is presented.

1 Introduction

The use of Wavelength Division Multiplexing (WDM) optical network technology in core network combined with IP/Multi-Protocol Label Switching (MPLS) for offering traffic-engineering capabilities has been selected as a suitable choice by many Internet Service Providers (ISPs) [1]. In particular GMPLS offers the instruments for traffic engineering, constraint-based routing and many other services required by future Internet applications in this network architecture [2]. Many of these applications, like e-business critical transactions, require high reliability and QoS guarantees from the network. However, single fiber failures occur frequently causing disruptions in the service of affected applications [3]. Moreover, a fiber failure can result in a loss of several terabits of data per second and leads to multiple failures in the upper network layer at the same time. Recovery techniques are defined in order to avoid these disruptions and reduce the failure impact. Thus, the traffic affected by the failure is switched over from the working path to an alternative/backup path. The selection of both working and backup paths depends on the skill of the routing algorithm applied and the current network state.

In this paper, we propose and analyze a multi-layer routing scheme that incorporates recovery mechanisms against single fiber failures. A dynamic cooperation between packet and wavelength switching domain is taken into account in order to provide protected paths cost effectively. New metrics, such as the equipment cost, switching granularity and resource consumption are also considered.

* This work was supported by the COST293, the Spanish Research Council (TIC2003-05567) and the Ministry Universities, Research and Information Society (DURSI).

V.N. Alexandrov et al. (Eds.): ICCS 2006, Part IV, LNCS 3994, pp. 136–143, 2006.
© Springer-Verlag Berlin Heidelberg 2006

2 Recovery Considerations in Multi-layer Networks

2.1 Photonic MPLS Router: Packet Switching Capabilities

Diverse switching granularity levels exist into the optical IP/MPLS network scenario. From coarser to finer there is fiber, wavelength and packet switching. The new photonic MPLS routers offer packet and wavelength switching [4]. Thus, packet Label Switch Paths (p-LSPs) are routed in the optical network through wavelength paths, called lambda LSPs (λ-LSPs).

For a better utilization of the network resources, p-LSPs should be efficiently multiplexed into λ-LSPs and then, these (λ-LSPs) should be demultiplexed into p-LSPs at some router. This procedure of multiplexing/demultiplexing and switching p-LSPs onto/from λ-LSPs is called traffic grooming [5]. The photonic MPLS routers have the technology to implement traffic grooming. It consists of a p number of Packet-Switching Capable (PSC) ports and w number of wavelengths [6]. The number of PSC indicates how many lambda LSPs can be demultiplexed into this router, whereas the number of wavelengths corresponds to the number of wavelengths connected to the same adjacent router. Based on these parameters, a new resource constraint is added to the network. Three scenarios exists according to p: a) $p = 0$; b) $0 < p < w$ and c) $p = w$. Lets suppose a network scenario where w is equal to 2. If the value of p is equal to 0, then the network does not offer packet switching capability at intermediate nodes. Thus, the protection should be performed either at the optical domain and λ-LSP oriented or at the IP/MPLS domain and p-LSP using only path protection (global). On the other hand, if $0 < p < w$, then not all the wavelengths may be demultiplexed at the intermediate nodes. In this case only one wavelength may be demultiplexed at intermediate nodes. Therefore, not all the p-LSP will be able to perform segment/local protection. Finally, when p is equal to w all the protection strategies, i.e. global, segment and local, are suitable.

2.2 Routing Algorithms in the Multi-layer Architecture

Routing algorithms can be categorized in static or dynamic depending on the type of routing information used for computing LSPs. Static algorithms use network information that does not change with time, meanwhile dynamic algorithms use the current state of the network. In the multi-layer dynamic case, the λ-LSPs are set up, if necessary, whenever a new p-LSP is requested.

A first framework for dynamic multi-layer routing was proposed by Oki [6]. Oki proposed different policies to allocate the packet LSPs to an existing lambda LSP. If the lambda LSP is not available then either 1) a sequence of existing lambda LSPs with two or more hops that connects the source and destination nodes are selected or 2) a new one-hop lambda LSP is established and selected as the new packet LSP. The main drawback of these policies is that the network connectivity is not guaranteed. An example is shown in Fig. 1. Lets suppose that a new packet LSP between the nodes (1,3) is requested and a new lambda LSP (1,2,3), i.e. the λ-LSP$_1$, is set up according to the routing policies presented

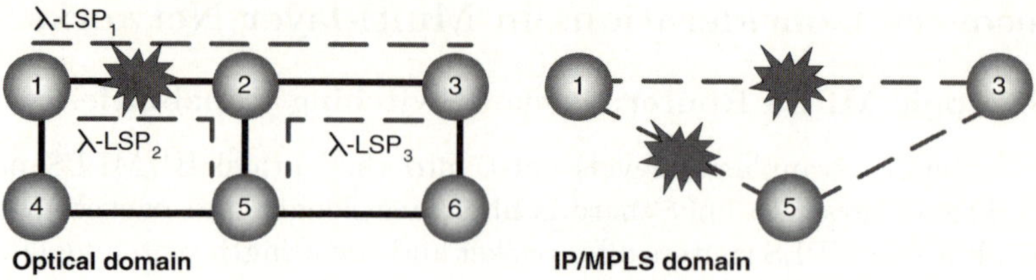

Fig. 1. Loss of connectivity at IP/MPLS domain due to a single link failure

by Oki [6]. In this example, both policies presented by Oki give the same result. The same procedure is applied to set-up two new LSPs between the nodes (1,5) and (3,5) obtaining the λ-LSP$_2$ and λ-LSP$_3$ respectively. Lets consider that the optical fiber (1,2) fails. Automatically the λ-LSPs λ-LSP$_1$ and λ-LSP$_2$ also fail. Considering only the IP/MPLS layer, node 1 is isolated, and the connectivity is lost, whilst the network has still enough resources to recover the failure. For instance, instead of selecting the optical fibers 1-2-5 and 5-2-3 for setting up the λ-LSP$_2$ and λ-LSP$_3$ respectively, the optical fibers 1-4-5 and 5-6-2 should be selected. Thus, the connectivity will remain against any single fiber failure.

In this paper, an on-line dynamic multi-layer routing scheme is proposed. This scheme establishes the λ-LSPs and p-LSPs whenever a new path is requested. Protection resources are reserved at either IP/MPLS or optical layer according to the current network resources, resulting in efficient resource consumption.

3 Reliable and Dynamic Multi-layer Routing

3.1 Network Definition

Let $G_P = (V, E_P)$ and $G_L = (V, E_L)$ represent the physical topology and the logical topology respectively, where V is the set of photonic MPLS routers; E_P and E_L are the set of network physical links and λ-LSPs respectively. Each router has p input and output Packet Switching Capable (PSC) ports, where $PSCi(u)$ input ports and $PCSo(u)$ output ports of node u are already not assigned to any λ-LSP. Each physical link has w wavelengths. When a p-LSP is requested, the proposed routing scheme considers both physical links and λ-LSPs, i.e. $E_P \cup E_L$. In order to univocally identify the physical link and the existing λ-LSPs that connect node pair (i, j) the 3-tuple (i, j, k) is used. Thus, the link (i, j, k), is a physical link if $k = 0$, otherwise $(k > 0)$ it is a λ-LSP.

Each (i, j, k) λ-LSP has an associated B_{ijk} residual bandwidth; total bandwidth reserved to protect physical link $(u, v, 0)$; and T_{ijk} the total shared bandwidth allocated in link (i, j, k). Note that $T_{ijk} = \max_{(u,v,0)\in E_P} S_{ijk}^{uv}$. Each (i, j, k) λ-LSP is a sequence of physical links denoted as a set P_{ijk} and a sequence of wavelengths assigned at each physical link denoted as W_{ijk}.

The p-LSP request is defined by (s, d, r) where (s, d) is the source and destination node pair; and r, specifies the amount of bandwidth required for this

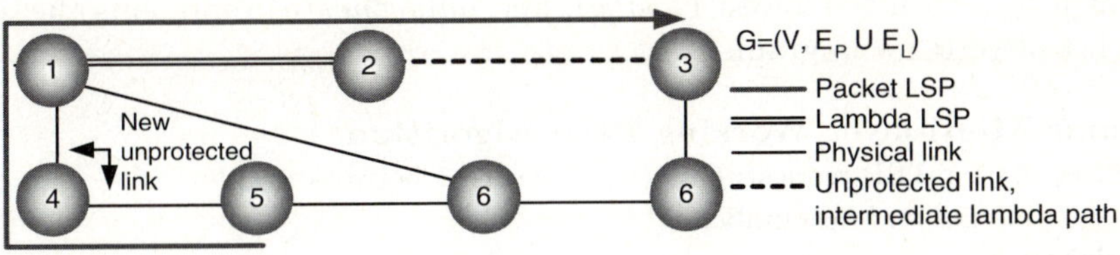

Fig. 2. Working p-LSP computation. Creation of a new λ-LSP using the physical links (5,4) and (4,1).

request. For each setup request, a working p-LSP (WP) has to be set-up and a backup p-LSP (BP) must be also setup, whenever the WP has, at least, one unprotected λ-LSP. If there are not enough resources in the network, for either the WP or the BP for the current request, the request is rejected.

3.2 Working Lambda and Packet LSP Computation

In our proposal, a new procedure to compute the working p-LSP (WP) is presented. In this procedure the following cost parameters are taken into account: 1) the residual bandwidth of the link candidates, B_{ijk}; 2) the maximum number of hops H; and 3) the free packet switching ports of each router, $PCSi$ and $PSCo$. Note that the residual bandwidth of the physical links with free wavelengths is the capacity of the wavelength. The proposed procedure called Dynamic Multi-Layer Working Path (DMWP) algorithm computes the min-hop working path based on a variation of the Dijkstra algorithm. In this case, the number of hops coincides with the number of λ-LSPs. Thus, the consecutive sequence of physical links, that constitutes a λ-LSP, are only considered as one hop. The DMWP procedure uses the network graph composed by the λ-LSPs and physical links, i.e., $G = (V, E_P \cup E_L)$. This procedure ends when it reaches the destination node or there is not any feasible path between source and destination nodes. If a feasible path exists then the procedure may return:

1. A sequence of existing λ-LSPs.
2. A sequence of physical links. In this case, a new λ-LSP is set up between source and destination node.
3. A sequence of lambda LSPs, physical links and intermediate lambda paths (unprotected λ-LSPs). In this case, new intermediate lambda paths are setup for each consecutive sequence of physical links as shown in Fig. 2. In the example, a new intermediate lambda path is set up with the physical links (5,4) and (4,1).

In the Dynamic Multi-Layer Working Path algorithm, $Cost(v)$ is a vector containing the path cost from s to v; $Pred(v)$ contains the v's predecessor node; and $WPlast(v)$ contains the identifier k of link (u, v). Q represents the list of adjacent vertices which are not visited yet. Function $min\_cost(Q)$ returns the element $u \in Q$ with the lowest $Cost(u)$; and $adjacency(u)$ represents the adjacency list of vertex u in graph G.

Dynamic Multi-layer Working Path Algorithm
Input: (s, d, r): p-LSP request; $G = (V, E)$: current network graph;
H: maximum hop number.
Algorithm
For all $(v \in V)$ **do**
  $Cost(v) = \infty$
  $Pred(v) = null$
  $WPlast(v) = 0$
$Cost(s) = 0$
$Q \leftarrow s$
while $(d \notin Q$ **and** $Q \neq \oslash)$ **do**
  $u \leftarrow min\_cost(Q)$
  for all $v \in adjacency(u, G)$ **do**
    for all $(u, v, k) \in E$ **do**
    if $(B_{ijk} \geq b)$ **and** $((k = WPlast(u) = 0)$ or $(Cost(u) + 1 < Cost(v) < H))$ **then**
      if $(PSCi(v) > 0$ **and** $k = 0$ **and** $WPLast(u) > 0)$ **or**
        $(PSCo(v) > 0$ **and** $k > 0$ **and** $WPlast(u) = 0)$ **or**
       $(k = WPlast(u) = 0)$ **or** $(k > 0$ **and** $WPlast(u) > 0)$ **then**
        $Pred(v) = u$
        $WPlast(v) = k$
        $Q \leftarrow v$
        if not $(k = WPlast(u) = 0)$ **then**
          $Cost(v) = Cost(u) + 1$

3.3 Backup Lambda and Packet LSP Computation

Once the WP is known, the backup p-LSP (BP) is computed. Three different procedures could be applied depending on the WP characteristics:

1. If the WP is a sequence of existing λ-LSPs, then each λ-LSP is already protected. In this case, the computation of the BP is not required.
2. If the WP is a new λ-LSP, and exists an available and shareable backup λ-LSP this is used to protect the WP. Otherwise, a new backup λ-LSP is set-up applying DMWP algorithm with $G = (V, E_P)$. A backup λ-LSP is shareable if the new λ-LSP does not belong to the same Shared Rink Link Group (SRLG) [7] of the both backup λ-LSP and the λ-LSPs protected by this backup λ-LSP.
3. If the WP is a combination of λ-LSPs and intermediate lambda paths, then a variation of the Partial Disjoint Path (PDP) presented in [8] is used to compute the BP. The variations are the ones included to the Dijkstra algorithm in order to consider the packet switching ports in the DMWP algorithm. The PDP may overlap with λ-LSPs of the WP, since they are already protected, and the nodes of the WP. Therefore, no extra resource is necessary in the

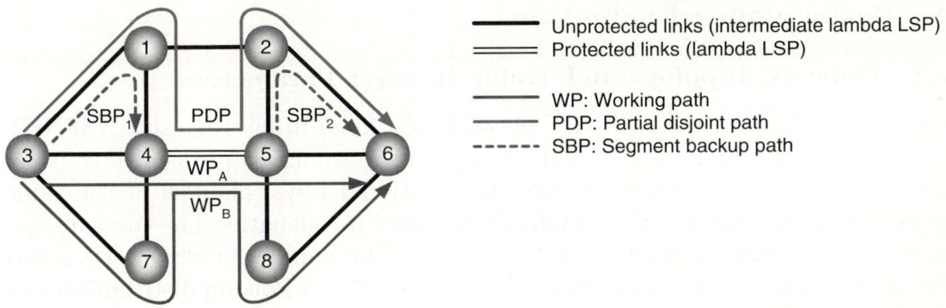

Fig. 3. Partial disjoint path computation and segment backup paths identification

IP/MPLS domain against failure of protected λ-LSPs in the optical layer. When the PDP overlaps the WP, more than one Segment Backup Paths (SBP) are established. An example is shown in Fig. 3 where two WPs are sharing the protected λ-LSP 4-5. In this example, the same PDP is used to protect both the WPs. Two segment backup paths (SBP_1 and SBP_2) are established between the protected segment paths 3-4 and 5-6. Moreover, the SBP bandwidth is shared since the SBP defined at the IP/MPLS layer is not activated against the failure of the λ-LSP 4-5. For more details refer to [8]. The result of the PDP algorithm is a sequence of λ-LSPs, intermediate lambda paths and physical links. With the set of consecutive physical links new intermediate lambda paths are created. Note that in the logical topology λ-LSPs are protected at optical domain and intermediate lambda paths are protected at IP/MPLS domain.

3.4 Multi-layer Routing with Protection Against Single Fiber Failures

We propose the multi-layer routing scheme with protection against single fiber failures (PASFF). PASFF computes the WP using the DMWP algorithm and the BP according to the criteria described in Section 3.3.

In order to compare our proposal, the next two algorithms based on Oki policies [6] are implemented:

- Policy 1 with protection (P1P). The routing policy 1 first tries to allocate the p-LSPs to an existing λ-LSP. If the λ-LSP is not available then a sequence of existing λ-LSPs with two or more hops that connects the source and destination nodes are selected. In order to protect the λ-LSPs, backup λ-LSPs are set up to protect the new λ-LSPs.
- Policy 2 with protection (P2P). The routing policy 2 first tries to allocate the p-LSPs to an existing λ-LSP. If the λ-LSP is not available then a new one-hop λ-LSP is established and selected as the new p-LSP. The same procedure presented in P1P is used to compute the backup λ-LSPs. Note that protection is only applied at optical domain in both P1P and P2P.

4 Performance Evaluation

4.1 Network Topology and Traffic Request Parameters

For the simulations, the NSFNET network described in [6] was used. NSFNET network consists of 14 nodes and 21 physical links. Each physical link is bidirectional i.e., they acted like two unidirectional physical links of the same number of wavelengths. Each physical links has 8 wavelengths. The transmission speed of each wavelength was set to 10 Gbps. The number of PSC ports p was the same in each node. Requests arrived according to a Poisson distribution and exponentially distributed holding times. The required p-LSP bandwidth was set to 500Mbps. When an existing λ-LSP, intermediate lambda path or backup λ-LSP didn't accommodate any p-LSP, then it was disconnected. Ten independent trials were performed over a window of 10.000 requests. The maximum hop number H was set to 2.

4.2 Simulation Results

Figure 4a shows the performance of the proposed algorithm PASFF compared to P1P and P2P in terms of request rejection ratio. All the analyzed algorithms present a sharply decrease of the request rejection ratio as the p factor increases. However, PASFF shows around a 4% of rejected requests. PASFF performs 3 times better than P1P and 4 times than the P2P. This is because, as expected, PASFF is able to find a feasible working and backup p-LSP for most of the p-LSP requests, due to the application of protection at IP/MPLS. So, PASFF provides a better filling of the capacity. PASFF protects the intermediate lambda paths at IP/MPLS layer, whilst, the λ-LSPs are optically protected.

Next two simulated results show the percentage of the network protected at optical and IP/MPLS domain. This is evaluated using two parameters: 1) the rate of backup λ-LSPs respect to the number of logical links (lambda LSPs and lambda paths) shown in Fig. 4b and 2) the rate of spare capacity, i.e. the percentage of bandwidth used as a BP with respect to the bandwidth used as a WP shown at MPLSin Fig. 4c.

In Fig. 4b, P1P and P2P present similar behavior throughout the experiment. Note that protection is applied at optical domain for P1P and P2P algorithm. This means that mostly each λ-LSP has its own backup λ-LSP if not shareable. Therefore, the rate is close to 100% for these algorithms. On the other hand, for our proposed algorithm PASFF the number of backup λ-LSP is much more smaller as shown in Fig. 4b since some logical links are protected at IP/MPLS domain.

Finally Fig. 4c shows the percentage of BP bandwidth used at IP/MPLS domain respect to the bandwidth used as a WP. As the number of PSC, p, increases the amount of bandwidth used to recovery the traffic at IP/MPLS layer increases for PASFF. Moreover, this rate slightly decreases as well as the request rejection ratio, since more resources can be shared. On the other hand, this value is 0 for P1P and P2P since these algorithms only use protection at optical domain.

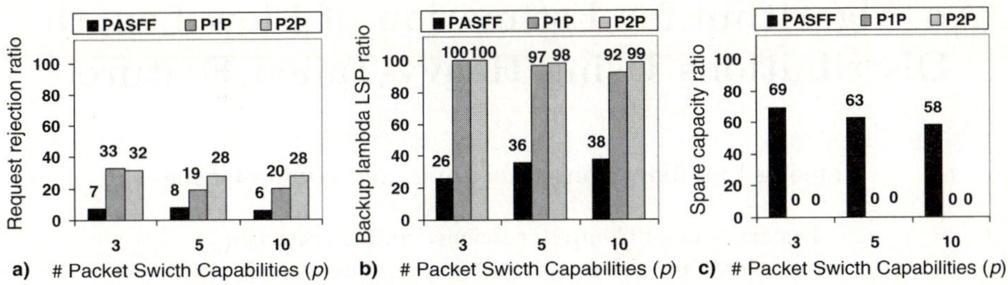

Fig. 4. a) Request rejection b) Backup lambda LSP and c) Spare capacity ratio

5 Conclusions

In this paper a novel dynamic multi-layer routing scheme was introduced for optical IP/MPLS networks. The proposed scheme incorporated protection mechanisms in order to guarantee connectivity against any single fiber failure. As a novelty this scheme takes into account both wavelength and packet switching capabilities in order to provide protected packet LSPs cost effectively. Thus, optical protection and IP/MPLS protection mechanisms are combined. Two kinds of lambda paths were defined: the lambda LSPs that are protected at optical domain and the intermediate lambda paths that are protected at IP/MPLS domain. Shared resources and shared risk link group were also considered in the proposed scheme. Results showed the efficiency of the proposed scheme in terms of resources used to protect the network and the request rejection ratio in different multi-layer network scenarios.

References

1. J. Y. Wei: Advances in the management and control of optical Internet. Selected Areas in Communications, IEEE Journal, vol. 20, no. 4, pp. 768-785, May 2002.
2. E. Mannie: Generalized Multi-Protocol Label Switching (GMPLS) Architecture. RFC 3945, Oct. 2004.
3. W. D. Grover:Mesh-based survivable networks: Options and strategies for optical, MPLS, SONET, and ATM networking. Prentice Hall PTR, 2004.
4. K. Sato et al.: GMPLS-based photonic multi-layer router (Hichari router) architecture: An overview of traffic engineering and signling technology: IEEE Commun. Mag., vol. 40, no. 3, pp. 96-101, Mar. 2002.
5. K. Zhu, H. Zang and B. Mukherjee: A comprehensive study on next-generation optical grooming switches. Selected Areas in Communications, IEEE Journal, vol. 21, no. 7, pp. 1173-1186, Sept. 2003.
6. E. Oki et al.: Dynamic multilayer routing schemes in GMPLS-based IP+optical networks. IEEE Comm. Magazine, vol. 43, pp. 108-114, Jan. 2005.
7. P. Sebos, J. Yates, G. Hjalmtysson, A. Greenberg: Auto-discovery of Shared Risk Link Groups. In Proc. of the Optical Fiber Communication Conference and Exhibit (OFC), pp. WWD3-1-WWD3-3, March 2001.
8. A. Urra, E. Calle, J.L. Marzo: Enhanced multi-layer protection in multi-service GMPLS networks. In Proc. of IEEE Globecom, Dec. 2005.

An Algorithm for Estimation of Flow Length Distributions Using Heavy-Tailed Feature

Weijiang Liu, Jian Gong, Wei Ding, and Guang Cheng

Department of Computer Science and Engineering,
Southeast University, 210096 Nanjing, Jiangsu, China
{wjliu, jgong, wding, gcheng}@njnet.edu.cn

Abstract. Routers have the ability to output statistics about packets and flows of packets that traverse them. Since however the generation of detailed traffic statistics does not scale well with link speed, increasingly passive traffic measurement employs sampling at the packet level. Packet sampling has become an attractive and scalable means to measure flow data on high-speed links. However, knowing the number and length of the original flows is necessary for some applications. This paper provides an algorithm that uses flow statistics formed from sampled packet stream to infer the absolute frequencies of lengths of flows in the unsampled stream. We achieve this through statistical inference and by exploiting heavy-tailed feather. We also investigate the impact on our results of different packet sampling rate. The experiment results show the inferred distributions are accurate in most cases.

1 Introduction

With the rapid increase of network link speed, packet sampling has become an attractive and scalable means to measure flow data. However, knowing the number and lengths of the unsampled flows remains useful for characterizing traffic and the resources required to accommodate its demands. Here are some applications: Resources Required for Collecting Flow Statistics: flow cache utilization and the bandwidth for processing and transmitting flow statistics are sensitive to the sampling rate, the number of flows, and flow lengths and duration; see [1,2]. Characterizing Source Traffic: the measured numbers of flows and the distribution of their lengths have been used to evaluate gains in deployment of web proxies [3], and to determine thresholds for setting up connections in flow-switched networks [4]. Sampling entails an inherent loss of information. We expect use statistic inference to recover information as much as possible. However, more detailed characteristics of the original traffic are not so easily estimated. Quantities of interest include the number of packets in the flow–we shall refer to this as the flow length–and the number of flows with fixed length.

1.1 Related Work

Kumar et al proposed a novel SCBF that performs per-flow counting without maintaining per-flow state in [5] and an algorithm for estimation of flow size

V.N. Alexandrov et al. (Eds.): ICCS 2006, Part IV, LNCS 3994, pp. 144–151, 2006.
© Springer-Verlag Berlin Heidelberg 2006

distribution in [6]. Its disadvantage is that all packet must be processed due to not using sampling. Hohn and Veitch in [7] discussed the inaccuracy of estimating flow distribution from sampled traffic, when the sampling is performed at the packet level.

Although sampled traffic statistics are increasingly being used for network measurements, to our knowledge few studies have addressed the problem of estimating flow size distribution from the sampled packet stream. In [2], the authors studied the statistical properties of packet-level sampling using real-world Internet traffic traces. This is followed by [8] in which the flow distribution is inferred from the sampled statistics. After showing that the naive scaling of the flow distribution estimated from the sampled traffic is in general not accurate, the authors propose an EM algorithm to iteratively compute a more accurate estimation. Scaling method is simple, but it exploits the sampling properties of SYN flows to estimate TCP flow frequencies; EM algorithm does not rely on the properties of SYN flows and hence is not restricted to TCP traffic, but its versatility comes at the cost of computational complexity.

1.2 Some Elementary Concepts

This paper considers sampling some target proportion $p = 1/N$ of the packet stream. There are a number of different ways to implement this. Implementations include independent sampling of packets with probability $p = 1/N$, and periodic selection of every N^{th} packet from the full packet stream. In both cases we will call N the sampling period, i.e., the reciprocal of the average sampling rate. Although the length distributions by random and periodic sampling can be distinguished, the differences are, in fact, sufficiently small [8]. A flow is defined as a stream of packets subject to flow specification and timeout. When a packet arrives, the specific rules of flow specification determine which active flow this packet belongs to, or if no active flow is found that matches the description of this packet, a new flow is created. A TCP flow is a stream of TCP packets subject to timeout and having the same source and destination IP addresses, same source and destination port numbers. Similarly, a UDP flow is a stream of UDP packets associated with above specification. A general flow is a stream of packets subject to timeout and having the same source and destination IP addresses, same source and destination port numbers(not considering protocol). In this paper, we will use the term original flow to describe the above flow. A sampled flow is defined as a stream of packets that are sampled at probability $p = 1/N$ from an original flow.

1.3 Contribution and Outline

This paper presents a novel algorithm for estimation of flow size distributions from sampled flow statistics. Our method is available not only to TCP flows but also to general flows. We complete this work using four approaches. The first formalizes the probability distribution of original flow length of a sampled flow length j. The second classifies two types of flows based on their probability that no packet is sampled. A flow is labeled as small (S) when it's probability that no packet is sampled is more than ε and as large (L) when it's probability that no packet is sampled is less than or equal to ε. The third gives a simple estimation method for large flows. The fourth uses maximum likelihood estimation and EM algorithm to estimate the full distribution of small flows.

The rest of this paper is organized as follows. In Section 2 we analyze the probability models of the original flow length distributions of a sampled flow under the assumptions of Pareto distributions. In Section 3, we classify two types of flows: small flow and large flow. Then we present different estimation methods for small flows and large flows, respectively. In Section 4 we discuss the computational complexity of our method. Furthermore, we compare our method with EM algorithm in estimation accuracy and computational complexity. We conclude in Section 5.

2 Probability Distribution of Original Flow Length

For a specific original flow F, let X_F denote the number of packets in F, Y_F denote the number of packets in the sampled flow from F. The conditional distribution of Y_F, given that $X_F = l$, follows a binomial distribution $Pr[Y_F = k|X_F = l] = B_p(l,k) = \binom{l}{k} p^k(1-p)^{l-k}$. By the conditional probability formula,

$$Pr[X_F = x|Y_F = y] = \frac{Pr[Y_F = y|X_F = x]Pr[X_F = x]}{Pr[Y_F = y]} \tag{1}$$

and by the complete probability formula,

$$Pr[Y_F = y] = \sum_{i=y}^{\infty} B_p(i,y)Pr[X_F = i] \tag{2}$$

We know that flow length distributions have the property of being heavy-tailed. Pareto distribution is the simplest heavy-tailed distribution; its probability mass function is

$$Pr[X_F = x] = \beta\alpha^\beta/x^{\beta+1}, \ \ \alpha, \beta > 0, \ \ x \geq \alpha \tag{3}$$

where β is called Pareto parameter. Hence Equation (2) can be written as:

$$Pr[Y_F = y] = \sum_{i=y}^{\infty} B_p(i,y)\beta\alpha^\beta/i^{\beta+1}, \ \ y \geq \alpha$$

Lemma 1. *Under the assumption that original flow lengths satisfy Pareto distribution, the probability that a sampled flow of length $y(\geq \alpha)$ is sampled from an original flow of length x is*

$$Pr[X_F = x|Y_F = y] = \frac{B_p(x,y)/x^{\beta+1}}{\sum_{i=y}^{\infty} B_p(i,y)\beta\alpha^\beta/i^{\beta+1}}.$$

To describe the properties of the above probability, apply different values of p and β to calculate the probability of Lemma 1. And for fixed p and β, for each $y(\geq \alpha)$, we find x such that the above probability $Pr[X_F = x|Y_F = y]$ is maximized. We have

Lemma 2. *Under the assumption of Lemma 1, for fixed* $p = 1/N, \beta$ *and* $y(\geq \alpha)$, *the probability* $Pr[X_F = x|Y_F = y]$ *is maximized at* $x = Ny - n(p,\beta)$. *It is increasing as* x *increases for* $x < Ny - n(p,\beta)$ *and decreasing as* x *increases for* $x > Ny - n(p,\beta)$.

Here $n(p,\beta)$ is a binary function with variables p and β whose value domain is integer set. Function $n(p,\beta)$ has the following properties:

1) It is a monotone decreasing function on variable p, that is , for fixed β, is decreasing as p increases.

2) It is a monotone increasing function on variable β, that is , for fixed p, is increasing as β increases.

For example, $n(0.1, 0.5) = 14$, $n(0.1, 1.0) = 18, n(0.1, 1.5) = 23$. In the concerned network, the length distributions of flows collected in any time interval do not satisfy Pareto distributions with fixed parameter strictly, but they can follow a Pareto distribution with parameter in interval [0.5,1.5] approximately. The value 1.0 is the middle value of interval [0.5,1.5] exactly. Therefore, to compute the conditional probability we assume that original flow length has a Pareto distribution with parameter 1.0 *a priori* distribution.

3 Estimation Method of Flow Length Distributions

3.1 Flow Classification: Large Flow and Small Flow

Let $y = \{y_j : j = 1, 2, \cdots, n\}$, where y_j is sampled flow frequencies of length j, be a set of sampled flow length frequencies, $f = \{f_i : i = 1, 2, \cdots, n, \cdots\}$ a set of estimated original flow length frequencies. Consider sampling the packets of an original flow of length Nj independently with probability $1/N$, the probability that no packet is sampled is $(1 - 1/N)^{Nj} = ((1 - 1/N)^N)^j$. $\{(1 - 1/N)^N\}$ is increasing in N and $\lim_{N\to\infty} (1 - 1/N)^N = 1/e < 0.37$. Thus for a given error ε,we require $(1-1/N)^{Nj} < (1/e)^j < \varepsilon$ and choose $j_{bord} \geq max(j(\varepsilon) = \lceil log(1/\varepsilon)\rceil, \alpha)$. For example, $j(0.01) = 5, j(0.001) = 7$. We classify two types of flows based on their probability that no packet is sampled. A flow is labeled as small (S) when it's probability that no packet is sampled is more than ε and as large (L) when it's probability that no packet is sampled is less than or equal to ε.

3.2 Estimation for Large Flow

For a sampled flow of length $j > j_{bord}$, by Lemma 2, the original flow length values of the $2N$ relatively large probabilities are $N(j-1)-n(p,\beta)+1, \cdots, N(j+1) - n(p,\beta)$ where $\beta = 1.0$. We estimate the sampled flow is sampled from one of the $2N$ original flows. Then there are $\frac{g_j}{2N}$ sampled flows that are sampled from one of original flows of the above lengths in $g_j(j > j_{bord})$ sampled flows. Therefore, for all large flows of length $i > Nj_{bord}$, we have

$$f_i = \frac{1}{2N}(g_j + g_{j+1}), where\, j = \lfloor (i + n(p, \beta) - 1)/N \rfloor . \tag{4}$$

3.3 Likelihood Function of Small Flows

For all small flows of length $i \leq Nj_{bord}$, we estimate as follows:

$$g_j = \sum_{i=j}^{m} B_p(i, j) f_i \tag{5}$$

where $m = \max\{i : f_i \neq 0\}$. For $i > Nj_{bord}$, substituting (4) into Equation (5):

$$\overline{g}_j = g_j - \sum_{i=Nj_{bord}+1}^{m} B_p(i, j) f_i = \sum_{i=j}^{Nj_{bord}} B_p(i, j) f_i, j = l, \cdots, Nj_{bord}. \tag{6}$$

Because some solutions of Equations (6) may be negative, we don't solve the equations directly. We construct MLE and employ EM algorithm to compute the solutions of Equations (6). For the above some $\overline{g}_j \leq 0$, we replace it with $\delta \overline{g}_{i-1}, 0 < \delta < 1$. Below we only consider all small flows of length $1, \cdots, Nj_{bord}$. Let $\gamma = \sum_{i=1}^{i=Nj_{bord}} \overline{g}_i$, and let ϕ_i denote the frequencies of original flows of length i conditional on at least one of its packets being selected. Our aim is to estimate $\phi = \{\phi_i\}, i = 1, \cdots, Nj_{bord}$ and $\sum_i \phi_i = 1$, from the frequencies $\{\overline{g}_i\}$. We now derive an expression for log-likelihood $L(\phi)$to obtain $\overline{g}_i$ given ϕ. Here, $c_{ij} = B_p(i, j)/(1 - B_p(i, 0))$ is the probability that packets are sampled from a flow of length i,conditional on $j \geq 1$, i.e., that the flow is sampled. For any j , the function is $(\sum_{i=j} \phi_i c_{ij})^{\overline{g}_j}$. Hence we obtain the likelihood function $\prod_{j=1}^{Nj_{bord}} (\sum_{i \geq j} \phi_i c_{ij})^{\overline{g}_j}$. Therefore the logarithm of likelihood function is

$$L(\phi) = \prod_{j=1}^{Nj_{bord}} \overline{g}_j log \sum_{i \geq j} \phi_i c_{ij} \tag{7}$$

where $c_{ij} = B_p(i, j)/(1 - B_p(i, 0))$. We wish to maximize $L(\phi)$ subject to the constraints $\phi \in \Delta = \{\phi : \phi_i \geq 0, \sum_i \phi_i = 1\}$.

3.4 EM Algorithm of Small Flows

Now we adopt a standard iterative approach: the Expectation Maximization (EM) algorithm [9], the standard form is as follows.

Starting with an initial value $\phi^{(0)}$, for example, $\phi^{(0)} = \{\frac{\overline{g}_i}{\gamma}\}$, the algorithm finds $sup\{L(\phi) : \phi \in \Delta\}$, by iterating between the following two steps ($k = 0, 1, \cdots$):

E step. Let f_{ij} denote the frequencies of original flows of length i from which j packets are sampled. Thus $\overline{g}_j = \sum_i f_{ij}$, while $\overline{f}_i = \sum_j f_{ij}$ is the frequency of original flows of length i at least one of whose packets is sampled. Form the complete data likelihood function assuming known f_{ij}

$$L_c(\phi) = \sum_{i \geq j \geq 1}^{Nj_{bord}} f_{ij} \log \phi_i c_{ij} \tag{8}$$

Form the expectation $Q(\phi, \phi^{(k)})$ of $L_c(\phi)$ conditional on the known frequencies $\overline{g}_j$, according to a distribution $\phi^{(k)}$:

$$Q(\phi, \phi^{(k)}) = \sum_{i \geq j \geq 1}^{Nj_{bord}} E_{\phi^{(k)}}[f_{ij}|\overline{g}] \log \phi_i c_{ij} \tag{9}$$

M step. Define $\phi^{(k+1)} = argmax_{\phi \in \Delta} Q(\phi, \phi^{(k)})$. From the Legendre equations in the maximization of $Q(\phi, \phi^{(k)})$ we have: $\phi_i^{(k+1)} = \frac{E_{\phi^{(k)}}[f_{ij}|\overline{g}]}{\gamma}$. Through direct computation of the above conditional expectation we obtain:

$$\phi_i^{(k+1)} = \frac{1}{\gamma} \sum_{i \geq j \geq 1} \frac{\phi_i^{(k)} c_{ij} \overline{g}_j}{\sum_{l \geq j} \phi_l^{(k)} c_{lj}} \tag{10}$$

Iterate steps E and M until some termination criterion is satisfied. Let $\overline{\phi}$ denote the termination point. We write our estimation of original small flows as $f_i = \overline{\phi}_i \gamma / (1 - B_p(i, 0))$, $i = 1, \cdots, Nj_{bord}$.

4 Evaluations and Comparison

Computational complexity. Let j_{max} denote the maximum sampled flow length. The computation for binomial coefficients of Equations (6) is $O(NNj_{bord}j_{max})$. Tabulation of the binomial coefficients for the iteration is $O((Nj_{bord})^2)$. Then for a fixed ϕ_i, each EM iteration is $O((Nj_{bord})^2)$. For all ϕ_i, completing an EM iteration is $O((Nj_{bord})^3)$. We compare the computational complexity of our method against the best known EM algorithm in [8] for estimating flow distribution from sampled traffic. In [8] for all ϕ_i completing an EM iteration is $O(i_{max}^2 j_{size})$. We collect all IP packet heads during a period of 300 minutes at Jiangsu provincial network border of China Education and Research Network (CERNET) (1Gbps) to do offline experiment. For IP header data during a period of 1 minute, sampling packets with $p = 1/10$, in our method let $\varepsilon = 0.01$, then $j_{bord} = 5$, thus $(Nj_{bord})^3 = 50^3$. However, $i_{max} = 2000$, $j_{size} = 200$ in EM algorithm of [8], $i_{max}^2 j_{size} = 6400 * 50^3$.

Estimation accuracy: We adopt Weighted Mean Relative Difference (WMRD) as our evaluation metric. Suppose the number of original flows of length i is n_i and our estimation of this number is $\hat{n}_i$. The value of WMRD is given by: WMRD$= \frac{\sum_i |n_i - \hat{n}_i|}{\sum_i (\frac{n_i + \hat{n}_i}{2})}$.

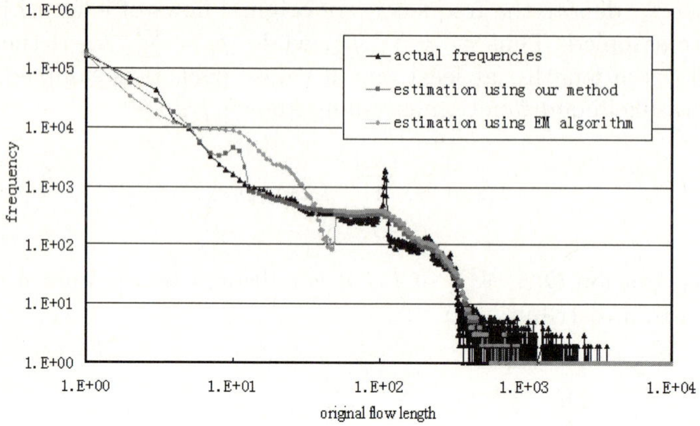

Fig. 1. Comparison of our method and EM algorithm at sampling period $N = 10$ for Jiangsu trace

We use three traces in our comparison experiments. The first trace is the first publicly available 10 Gigabit Internet backbone packet header trace from NLANR: Abilence III data set [10]. In our experiments , we used a minute of traffic from the trace. The second trace, which contains packets during a 5-minute period , was collected at Jiangsu provincial network border of China Education and Research Network (CERNET) on April 17, 2004. The backbone capacity is 1000Mbps; mean traffic per day is 587 Mbps. We call this trace as Jiangsu trace. The third trace, which contains packets during a 10 minute period, was obtained from NLANR: Abilence I [11]. Figure 1 compares the two estimators of Jiangsu trace derived by our method and EM algorithm of [8] at sampling period $N = 10$. Observe that they are so close. Table 1 shows the estimation accuracy of our algorithm is close enough to that of EM algorithm. In most cases, our algorithm is much more accurate.

Table 1. WMRD of our method and EM algorithm

trace	Sampling period	WMRD of our method	WMRD of EM algorithm
Abilence III	10	17%	18%
	30	23%	24%
	100	34%	37%
Jiangsu	10	20%	28%
	30	15%	29%
	100	30%	39%
Abilence I	10	15%	14%
	30	21%	23%
	100	31%	35%

5 Conclusions

Estimating the distribution of flow length is important in a number of network applications. In this paper we present a novel method for estimation of flow length distributions from sampled flow statistics. The main advantage is that it could significantly reduce the computational complexity. The theoretical analysis shows that the computational complexity of our method is well under control. The experimental results demonstrate that our method achieves an accurate estimation for flow distribution.

Acknowledgement

This work is supported in part by the National Grand Fundamental Research 973 Program of China under Grant No.2003CB314804; the National High Technology Research and Development Program of China (2005AA103001); the Key Project of Chinese Ministry of Education under Grant No.105084; the Jiangsu Provincial Key Laboratory of Computer Network Technology No. BM2003201; Jiangsu Planned Projects for Postdoctoral Research Funds.

References

1. Duffield, N.G., Lund, C. , Thorup, M.: Charging from sampled network usage. ACM SIGCOMM Internet Measurement Workshop 2001, 245-256, November 2001.
2. Duffield, N.G., Lund, C. , Thorup, M.: Properties and Prediction of Flow Statistics from Sampled Packet Streams. ACM SIGCOMM Internet Measurement Workshop 2002,159-171, November 2002.
3. Feldmann, A. , Caceres, R. , Douglis, F. , Glass, G., Rabinovich, M.: Performance of Web Proxy Caching in Heterogeneous Bandwidth Environments.IEEE INFOCOM [illegible]
4. Feldmann, A., Rexford, J., and Caceres, R.: Efficient Policies for Carrying Web Traffic over Flow-Switched Networks. IEEE/ACM Transactions on Networking, **6**(1998)673-685.
5. Abhishek Kumar, Jun Xu, Li Li, and Jia Wang: Space Code Bloom Filter for Efficient Traffic Flow Measurement.IEEE INFOCOM 2004 ,1762-1773.
6. Abhishek Kumar, Minho Sung, Jun (Jim) Xu and Jia Wang: Data streaming algorithms for efficient and accurate estimation of flow size distribution, ACM SIGMETRICS 2004,177-188.
7. Nicolas Hohn, Darryl Veitch: Inverting Sampled Traffic. Internet Measurement Conference 2003. October 27-29, Miami Beach ,Florida, USA. 222-233.
8. Duffield, N.G., Lund, C. , Thorup, M.: Estimating Flow Distributions from Sampled Flow Statistics. IEEE/ACM Transation on Networking, **13**(2005) 933-945.
9. Mao shisong, Wang jinglong, and Pu xiaolong: Advanced Mathematical Statistics. China Higher Education Press, Beijing,1998.
10. NLANR: Abilene-III data set, hppt://pma.nlanr.net/Special/ipls3.html.
11. NLANR:Abilene-I data set, http://pma.nlanr.net/Traces/long/bell1.html.

Performance Evaluation of Novel MAC Protocol for WDM/Ethernet-PON

Bokrae Jung[1], Hyunho Yun[2], Jaegwan Kim[1], Mingon Kim[1], and Minho Kang[1]

[1] Information and Communications University (ICU),
119 Munjiro, Yuseong-Gu, Daejeon 305-732, Republic of Korea
{gaole3, tecmania, kmg0803, mhkang}@icu.ac.kr
[2] Electronics and Telecommunications Research Institute (ETRI),
161 Gajeong-Dong, Yuseong-Gu, Daejeon 308-350, Republic of Korea
yhh63129@etri.re.kr

Abstract. This paper proposes a novel MAC protocol adopting time division duplex (TDD) technology for extending total subscribers at the expense of downstream bandwidth in WDM/Ethernet passive optical network (WE-PON). To compensate sacrificed downstream bandwidth, we employ an efficient threshold decision and dynamic bandwidth (DBA) algorithm. Simulation results show that the proposed MAC protocol improves queueing delay, link utilization under asynchronous traffic conditions.

1 Introduction

WDM-PON provides unlimited capacity with the each user enough to accommodate broadband multimedia services. Although WDM-PON is high-performance, it is difficult to be fully accepted today due to its luxurious cost. Therefore, WDM/TDM hybrid PON is a compromising solution guaranteeing cost-effective and flexible upgrade from TDM to WDM [1]. This paper introduces WE-PON architecture as a kind of WDM/TDM hybrid PON. In the WE-PON, single light source is reused at optical network termination (ONT) to modulate upstream signals called loop-back method for the efficient bidirectional transmission. Once adopting such a technology, splitting ratio is considerably limited due to amplitude squeezing effect (ASE) at ONT [2]. To address this problem, we propose an improved MAC protocol applying time division TDD technology [3]. Consequently, maximum 16 subscribers per wavelength can be serviced in the optical layer. But downstream bandwidth is inevitably taken away for upstream traffic in the network layer. To compensate downstream bandwidth used for up-traffic, threshold decision and dynamic bandwidth allocation (DBA) algorithm efficiently response to user request in order to keep tack with almost close performance of ASE-based WE-PON that does not use a proposed method.

2 WE-PON Architecture

Fig. 1 shows the overall architecture of WE-PON network. In the OLT, downstream signals from OLT multiplex 32 wavelengths into single mode fiber (SMF) through

V.N. Alexandrov et al. (Eds.): ICCS 2006, Part IV, LNCS 3994, pp. 152 – 155, 2006.
© Springer-Verlag Berlin Heidelberg 2006

arrayed-waveguide gratings (AWGs). In the RN, downstream signals leaved from 10 km feeder link pass out 1 and port 2 of circulator for the right direction and then demultiplex 32 wavelengths for distributing signal power to 4 ONTs via optical power splitter (OPS). In the ONT, 25 percents of optical power splitted by OPS are fed into receiver. Remaining 75 percents are reused to modulate upstream signals using ASE in RSOA. By using ASE mechanism to regenerate continuous wave (CW) to be modulated for upstream traffic in reflective semiconductor optical amplifiers (RSOA), strictly bounded power range of input signals should be allowed to enter the RSOA. It limits splitting ratio to only four. To solve this problem, we adopt TDD technology to the existing MAC protocol. As this scheme applies to existing frame structure, no more reuse of modulated downstream signals using ASE is need as well as maximum 16 subscribers per wavelength can be achieved.

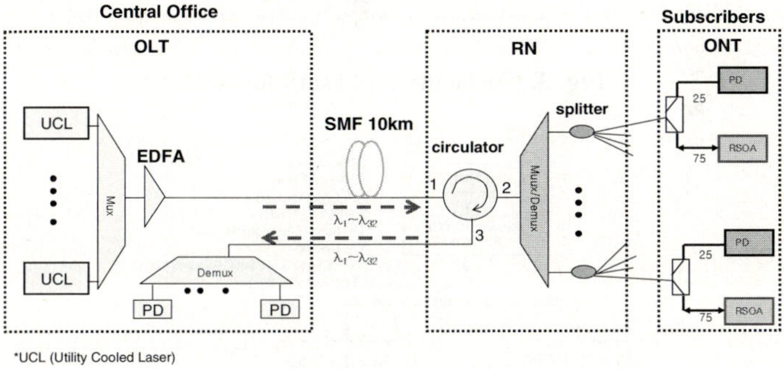

Fig. 1. WE-PON architecture

3 Proposed MAC Protocol

The basic downstream frame structure adopting TDD technology is shown in Fig. 2. In this scheme, a part of the downstream area is shared with upstream bandwidth granted for ONTs. Here the threshold is regarded as the amount of upstream bandwidth. In this case, the length of upstream and downstream is evenly assigned, which call fixed threshold adjustment (FTA). In general, we design all the control message formats and collision avoidance mechanism among ONTs observe 802.3ah EFM. Within the threshold, a GATE message corresponding to the ONT's request, vacant time window for the CW amount to the grant size of the ONT and guard time are consecutively flow as a group to the ONT request. Contrast to FTA, the range of threshold can be adaptively varied from 0 to 2ms according to upstream request for ONTs, called adaptive threshold adjustment (ATA) shown in fig. 3.

Once adopting given MAC protocol in the scheduler, we can not avoiding degrading performance. But well designed threshold decision and DBA algorithm can improve system efficiency and capacity. Scheduling algorithm for determining the threshold and a mount of grant under the DBA policy is shown in fig. 4.

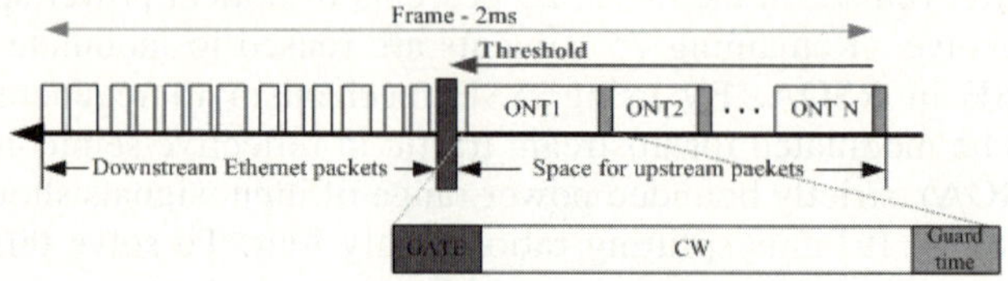

Fig. 2. Fundamental downstream frame structure

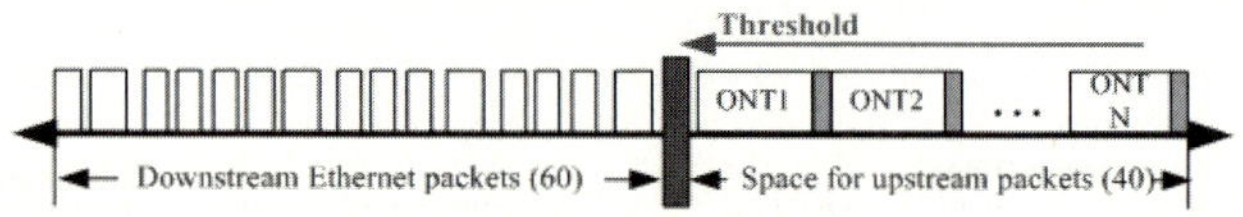

Fig. 3. Elastic threshold shift for ATA

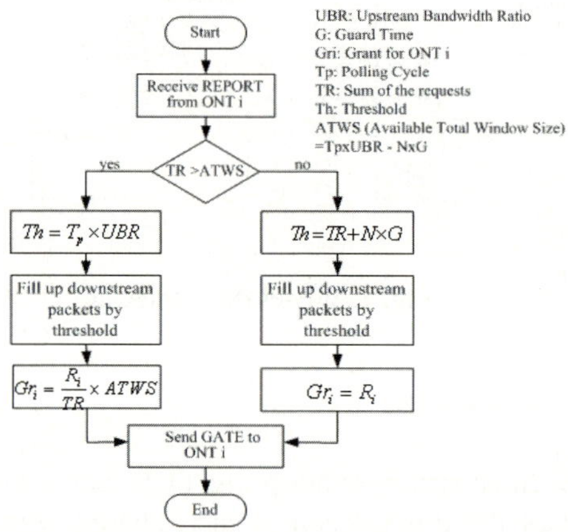

Fig. 4. DBA Algorithm for ATA

4 Simulation Results

We simulate the performance of proposed MAC protocol using OPNET. We assume distance from OLT to ONTs is identically 10km. Link speed for downstream and upstream transmission is set to 1Gbps. Guard time between adjacent frames is let to 5us. Every ONT has a finite queue size, 20 Mbytes. Incoming Ethernet frame uniformly distributed from 64 to 1518 byte. Fig.5 shows average queueing delay, up-link utilization, and traffic scenario-based bandwidth efficiency for proposed MAC protocol.

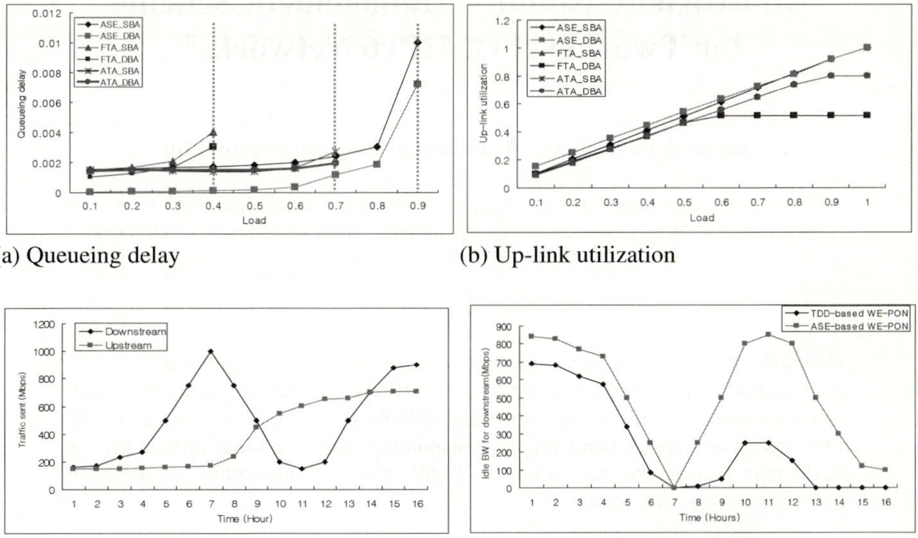

(a) Queueing delay (b) Up-link utilization

(a) Traffic Scenario (b) Unused downstream bandwidth

Fig. 5. Performance evaluation for proposed MAC protocol

5 Conclusion

This paper proposed new MAC protocol adopting TDD technology to improve system efficiency of WE-PON. From the simulation results, the proposed MAC protocol achieved substantial improvement in queueing delay and link utilization under asynchronous traffic conditions by applying threshold decision and DBA algorithm for ATA.

Acknowledgement

This paper was supported in part by KOSEF-OIRC project and ETRI.

References

1. Fu-Tai An and Kyeong Soo Kim, "SUCCESS: a next-generation hybrid WDM/TDM optical access network architecture", J.Lightwave Technol., vol 22, no. 11, pp 2557 – 2569, Nov. 2004.
2. Wooram Lee and Mahn Yong Park, "Bidirectional WDM-PON based on gain-saturated reflective semiconductor optical amplifiers" IEEE Photon. Technol. Lett., vol. 11, no. 11, pp 2460 – 2462, Nov. 2005.
3. Huang, V. and Weihua Zhuang, "Optimal resource management in packet-switching TDD CDMA systems", IEEE Personal Commun., vol. 7, no. 6, pp 26 – 31, Dec. 2000.

An Efficient Mobility Management Scheme for Two-Level HMIPv6 Networks*

Xuezeng Pan, Zheng Wan, Lingdi Ping, and Fanjun Su

College of Computer Science, Zhejiang University, Hangzhou, P.R. China
zhengwan66@yahoo.com.cn, {xzpan, ldping}@zju.edu.cn,
sufanjun@163.com

Abstract. In this paper, we propose an efficient mobility management scheme in a two-level Hierarchical Mobile IPv6 (HMIPv6) architecture. A mobile node chooses a suitable Mobility Anchor Point (MAP) according to its average dwell time and its number of connecting correspondent nodes. Furthermore, a higher MAP may adjust the threshold of MAP selection algorithm periodically according to its traffic load.

1 Introduction

Hierarchical Mobile IPv6 (HMIPv6) [1] was proposed to reduce the amount of registration between the mobile node (MN), its home agent and correspondent nodes (CNs). To improve the performance of HMIPv6 further, many schemes [2-5] proposed that the MN choose a suitable MAP according to its mobility characteristics. For example, Kawano et al. [4-5] introduced a multilevel HMIPv6 architecture and a speed based MAP selection algorithm. However, these algorithms did not consider the influence of traffic parameter of an MN, i.e. the number of connecting CNs. Thus reduction of registration cost that achieved by mobility based algorithm is limited.

In this paper, we propose a mobility and traffic based MAP selection algorithm for two-level HMIPv6 architecture. In addition, a dynamic adjustment mechanism is introduced to reduce the registration cost further, in which a higher MAP may adjust the threshold of MAP selection periodically according to its traffic load.

2 Proposed Scheme

2.1 MAP Selection

We combine the number of CNs (N_c) and dwell time (T_f) as the parameter for MAP selection (called "*combined measure*", referred to as "C"). The following equation is used to compute C.

$$C = (1 + N_c) / T_f \tag{1}$$

* This work was funded in part by Huawei Funds for Science and Technology (YJCB2004025SP) and Science and Technology Plan of Zhejiang Province (2005C21002).

V.N. Alexandrov et al. (Eds.): ICCS 2006, Part IV, LNCS 3994, pp. 156 – 159, 2006.
© Springer-Verlag Berlin Heidelberg 2006

The MAP selection strategy is defined as:

$$MA = \begin{cases} MAP_h & C > T_c \\ MAP_l & C \leq T_c \end{cases} \quad (2)$$

where MA indicates the chosen MAP and T_c denotes the threshold of combined measure. MAP_h and MAP_l denote higher MAP and lower MAP, respectively.

2.2 Dynamic Adjustment

If N_c of each MN often changes, a fixed threshold can not minimize the inter-domain registration cost. Based on this observation we propose the dynamic adjustment mechanism. Each higher MAP periodically checks the number of its serving MNs (N_m) and compares N_m with its capacity (C_{map}). Assume the checking interval is I. We use "*Sat*" to denote the saturation degree of a higher MAP, which is equal to N_m/C_{map}. The higher MAP adjusts T_c according to its current *Sat* as follows:

$$T_{c\_n} = \begin{cases} T_{c\_o} \times (1+\beta) & Sat > Sat_h \\ T_{c\_o} & Sat_h \geq Sat \geq Sat_l \\ T_{c\_o} \times (1-\beta) & Sat < Sat_l \end{cases} \quad (3)$$

where T_{c_n} and T_{c_o} are the new threshold and the old threshold. Sat_h and Sat_l denote two thresholds for adjustment decision. β indicates the degree of adjustment and is a positive constant less than 1.

To deploy proposed scheme, two fields are included into the MAP option:

- *L*: a bit lies after "R" bit, indicating the level of an MAP. "1" refers to a higher MAP and "0" refers to a lower MAP.
- *Threshold*: updated T_c for MAP selection. It is valid when "L" bit is set. It is 64 bits long and lies before "lifetime" field.

3 Performance Evaluation

3.1 Simulation Model

There are 16 higher MAPs distributed in the simulated grid network. Each higher MAP is connected by 16 lower MAPs and each lower MAP covers 4 subnets. There are "high speed MNs" and "low speed MNs", with dwell time T_{f1} and T_{f2} respectively, which follow the uniform distribution with parameters [5s,10s] and [10s,15s]. We set the ratio of high speed MNs to low speed MNs to 1:1. Mobile nodes are also divided into two classes according to their traffic parameters. N_c for low traffic MNs may be 0 and 1, and N_c for high traffic MNs may be 5 and 6. The ratio of these two traffic classes is defined as a simulation parameter (R_T).

We consider three schemes. The first is the traditional mobility based MAP selection scheme, in which high/low speed MNs choose higher/lower MAPs. The second is the combined measure based MAP selection algorithm. And the last scheme deploys dynamic adjustment mechanism on the basis of the second scheme. The three schemes are referred to as *Speed* scheme, *Comb* scheme and *Comb_D* scheme. The simulation time is 5,000 seconds. Each experiment is repeated ten times and the average result is presented. Table 1 summarizes the major parameters.

Table 1. Simulation parameters

Parameter(s)	Description	Value
N_M	The number of simulated MNs	200 to 1000
R_T	Ratio for low to high traffic classes	5:1 to 1:5
β	Adjustment degree for dynamic mechanism	0.2
I	Interval for a higher MAP to check N_m	10
T_c	Initial threshold for MAP selection	0.4
Sat_h	Up threshold for dynamic mechanism	1.0
Sat_l	Down threshold for dynamic mechanism	0.8
C_{map}	Capacity of higher MAP	N_M/16

3.2 Simulation Results

From Fig. 1 we find that the number of binding updates (referred to as N_b) of the three solutions all increases with increasing N_M. The percentage reductions of registration cost obtained by Comb and Comb_D schemes over Speed scheme are approximately 25% and 45% respectively.

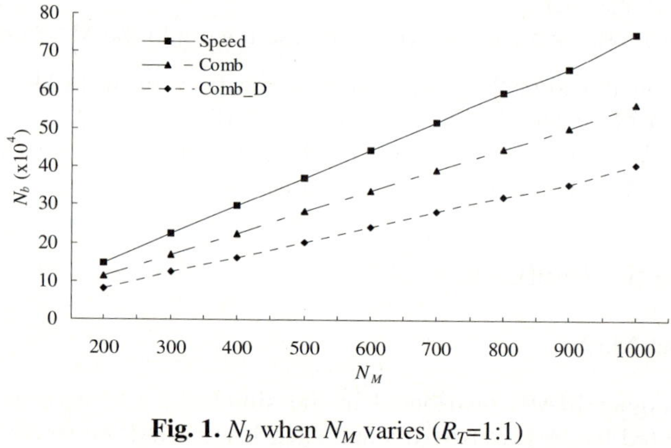

Fig. 1. N_b when N_M varies (R_T=1:1)

We modify R_T to simulate the scenario in which N_c of the MN changes. Served by the same MAP, an MN with more CNs issues more binding updates. Thus more high traffic MNs indicates larger registration cost in Speed scheme, as shown in Fig. 2. On the contrary, the Comb scheme shows its adaptability to various ratios of traffic classes because high traffic MNs try to choose higher MAPs according to our novel MAP selection algorithm. As for Comb_D scheme, when the number of high traffic MNs is low, the higher MAP will reduce its threshold to allow more MNs to be served. As the number of high traffic MNs increases, more MNs are originally served by higher MAPs. Thus performance gain by deploying dynamic adjustment decreases.

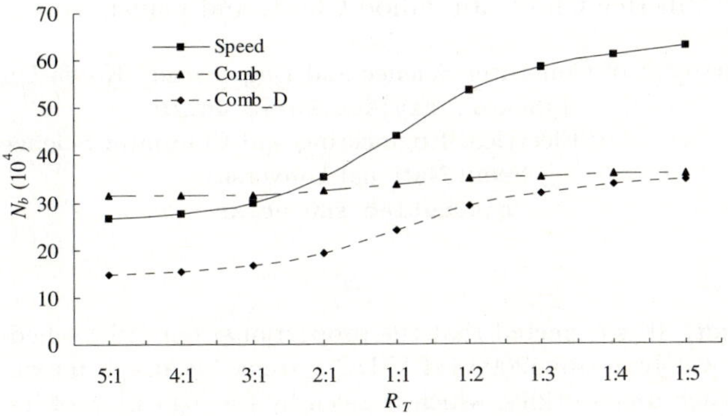

Fig. 2. N_b when R_T varies (N_M=600)

4 Conclusion

In this paper, we propose a mobility and traffic based MAP selection algorithm and a dynamic adjustment mechanism to reduce the registration cost outside MAP domains in a two-level HMIPv6 architecture. Taking traffic parameter into account, the novel MAP selection algorithm achieves better performance in reducing overall registration cost than mobility based MAP selection algorithm. In addition, dynamic adjustment prevents saturation degree of the higher MAP from being a small value so that it reduces the registration cost further.

References

1. H. Soliman, C. Castelluccia, K. El-Malki, and L. Bellier, "Hierarchical Mobile IPv6 mobility management," IETF RFC 4140, Aug. 2005.
2. S.H. Hwang, B.K. Lee, Y.H. Han, and C.S. Hwang, "An adaptive hierarchical mobile IPv6 with route optimization," IEEE Vehicular Technology Conference (VTC), Apr. 2003.
3. T. Kumagai, T. Asaka, and T. Takahashi, "Location management using mobile history for hierarchical mobile IPv6 networks," IEEE GLOBECOM, pp.1585-1589, Nov. 2004.
4. K. Kawano, K. Kinoshita, and K. Murakami, "A multilevel hierarchical distributed IP mobility management scheme for wide area networks," International Conference on Computer Communications and Networks (ICCCN), pp.480-484, Oct. 2002.
5. K. Kawano, K. Kinoshita, and K. Murakami, "A mobility-based terminal management in IPv6 networks," IEICE Transactions on Communications, vol. E85-B, no.10, pp.2090–2099, Oct. 2002.

Analysis of Packet Transmission Delay Under the Proportional Fair Scheduling Policy*

Jin-Hee Choi[1], Jin-Ghoo Choi[2], and Chuck Yoo[1]

[1] Department of Computer Science and Engineering, Korea University
{jhchoi, hxy}@os.korea.ac.kr
[2] School of Electrical Engineering and Computer Science,
Seoul National University
cjk@netlab.snu.ac.kr

Abstract. It is expected that the proportional fair (PF) scheduler will be used widely in cdma2000 1xEV-DO systems because it maximizes the sum of each user's utility, which is given by the logarithm of its average throughput. In this paper, we address an influence of the PF scheduler on the packet transmission delay in base station (BS) and propose an analytic model.

1 Introduction

Recent advances in communication technology make appearance of the packet-based cellular systems such as cdma2000 1xEV-DO [1] and UMTS-HSDPA [2]. Being mainly targeted on high-speed data applications that are tolerant of some packet delay, it is reasonable that their schedulers focus on maximizing the sum of each user's utility. A good way of achieving it is to serve the users with good channel condition in order to utilize the time-varying feature of wireless channels. This approach increases the system throughput significantly. But, some users can be sacrificed since, in wireless environment, users have very different channel condition according to their location.

The proportional fair scheduler [3] is one of the most promising opportunistic schemes that balance system throughput and user fairness. It is very simple to implement, and also it is optimal in the sense of maximizing the sum of each user's utility that is given by the logarithm of average throughput for elastic traffic. However, owing to its reflection on channel state, the scheduler induces some variation on scheduling delay, and the variation may lead to unstable packet transmission delay. Since generally the delay variation makes negative influence on the performance of transport layer protocol and application, it is very important to have an accurate delay model that describes the delay variation. From that reason, we propose a packet transmission delay model in BS with PF scheduler in this paper. Also, we show the comparison of the analytic model and the simulation result using NS-2 [4].

* This work was supported by grant No.R01-2004-000-10588-0 from the Basic Research Program of the Korea Science & Engineering Foundation.

V.N. Alexandrov et al. (Eds.): ICCS 2006, Part IV, LNCS 3994, pp. 160–163, 2006.
© Springer-Verlag Berlin Heidelberg 2006

2 Scheduling Delay Analysis

The wireless networks sometimes have a quite long delay because a base station may have many tasks to reduce the impact of the errors such as Forward Error Correction (FEC), interleaving, retransmission, and so on. In this section, we show an analytic model for BS delay, which is simplified with only the scheduling delay and the retransmission delay.

2.1 1 User Case

First, we build the model with 1 user. The user has the packet size of T bytes and is able to transmit X bytes whenever the scheduling slot is allocated. For example, if T is 1500 and X is 100 in constant, the time for servicing the packet is 15 slots. But when X changes depending on the channel state, the analysis comes to be more difficult.

For convenience of the analysis, we assume that X has an exponential distribution with average m (actually this assumption is exactly correct when the transmission rate linearly increases in proportion as SNR (Signal-to-Noise Ratio) on the Rayleigh channel). When we denote the data size that is successfully transmitted in flow i as X_i, the number of slots that is required to service the packet is $N(T)$. $N(T)$ is minimum N that satisfies $\sum_{i=1}^{N} X_i \geq T$. Analyzing this problem as the Poisson counting process, we can see that $N(T)-1$ has a Poisson distribution with both average and variance $\frac{T}{m}$.

We obtain the required number of slots to service a packet as above. However, owing to wireless channel error, the transmission does not always make a success even if BS successfully transmits the packet. In this model, we denote the error rate of each flow as p and assume that the error rate is independent of the transmission rate. At this time, to transmit the packet successfully in flow i, actually Y_i slots are taken. Because Y_i follows the discrete probability distribution with $Pr(Y_i = n) = p^{n-1}(1-p)$, we get $E(Y_i) = \frac{1}{1-p}$ and $Var(Y_i) = \frac{p}{(1-p)^2}$.

Actual number of slots to transmit a packet is given by $S = \sum_{i=1}^{N(T)} Y_i$, and we can obtain its average and variance as follows.

$$E(S) = E\{N(T)\}E(Y_i) = (\frac{T}{m} + 1)(1-p)^{-1}, \tag{1}$$

$$\begin{aligned} Var(S) &= E\{N(T)\}Var(Y_i) + E^2(Y_i)Var\{N(T)\} \\ &= (\tfrac{T}{m} + 1)p(1-p)^{-2} + (1-p)^{-2}(\tfrac{T}{m} + 1) \\ &= (\tfrac{T}{m} + 1)(1+p)(1-p)^{-2} \end{aligned} \tag{2}$$

2.2 *K* Users Case

Let's consider the case of K users. We assume that each user has a packet to transmit, and the packet size, T, and channel state are same in every user. Also assuming that the scheduler chooses a user and, only after transmitting the user's one packet, selects another user, we analyze the packet transmission time of the

Table 1. Configuration variables for simulation

Configuration Variables	Value
Schedule Interval (SCHED_INTERVAL_)	0.001667 (sec)
Transmission Power (Pt)	10 (Watt)
Bandwidth (BANDWIDTH)	1.25 (MHz)
Distance (DISTANCE)	100 (m)
Noise Density (ND)	2×10^{-14}

last-selected user. When the transmission time of $k-th$ selected user is denoted as D_k, our finding time is $D = D_1 + D_2 + ... + D_k = \sum_{k=1}^{K} D_k$. By applying Central limit theorem [5], we approximate D to a Gaussian distribution with the average $K \cdot E(D_k)$ and the variance $K \cdot Var(D_k)$.

$$\begin{aligned} E(D_k) &= E(S) = (\tfrac{T}{m} + 1)(1-p)^{-1}, \\ Var(D_k) &= Var(S) = (\tfrac{T}{m} + 1)(1+p)(1-p)^{-2}, \end{aligned} \tag{3}$$

Finally D follows the Gaussian distribution with the average $K(\frac{T}{m}+1)(1-p)^{-1}$ and the variance $K(\frac{T}{m}+1)(1+p)(1-p)^{-2}$.

For example, when we consider the case of T=1500, m=100, K=50, and p=0.1, the packet transmission time of the last selected user is as follows[1]. It is necessary to keep in mind that the inter-packet interval of a user comes from the scheduling delay.

- Constant rate with no channel error: 750 slots.
- Variable rate with no channel error: 800 slots with 50%.
- Variable rate with channel error, p: 889 slots with 50%.

In this model, the scheduler services a user's packet sequentially but real PF scheduler services several users' packet little by little, depending on the channel state. Thus, every user finishes the packet transmission at similar time due to their mixed service time while the average rate of the allocated slots is so high as to have a better possibility that reduces the entire transmission time. Consequently every user has a similar finish time with the "last" user.

3 Simulation Result

To validate the model, we perform simulation studies using NS-2. Some key configuration variables are summarized in Table 1.

The configuration values make average m=100, which is used in examples for our analysis. To remove the impact of transport layer protocol, we used UDP and CBR application. Simulation run time is 200 sec, and about 8000 packets are gathered. The following Fig.1 shows the distributions of PF scheduling delay. Respectively, Fig.1(a) is a histogram of "variable rate with no channel error," and Fig.1(b) is a histogram of "variable rate with 10% channel errors". Also, cumulative lines are drawn in both cases.

[1] Note that one slot takes 1.667 ms.

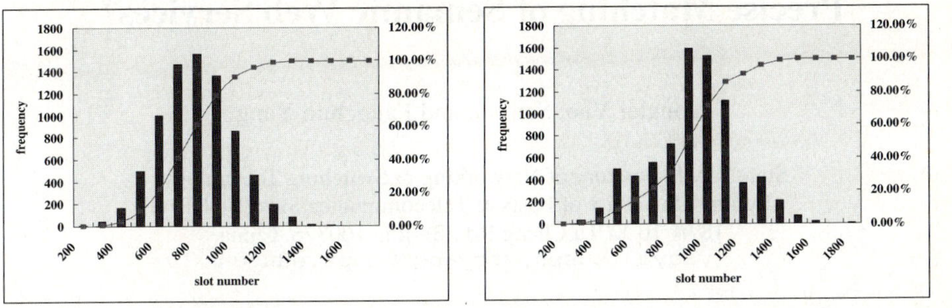

(a) Variable rate with no channel error (b) Variable rate with 10% channel errors

Fig. 1. Distribution of packet transmission delay

In "no channel error" case, we get 775 slots with average, which are taken for about 1.29 sec. And, there are above 800 slots for 48.86% of packets. In "10% channel errors" case, average 880 slots are observed for about 1.46 sec. Also, there are above 889 slots for 48.47% of packets.

4 Conclusion

In this paper, we proposed an analytic delay model for PF scheduler. Although the model includes only the scheduling delay and the retransmission delay, the simplification does not undermine the inter-packet interval of a user. In addition, NS-2 simulation result shows that the analytic model approximates to the simulation model.

References

1. Q. Bi and S. Vitebsky: *Performance Analysis of 3G-1X EVDO High Data Rate System*, Proceedings of WCNC, 2002.
2. R. Love, A. Ghosh, X. Weimin, and R. Ratasuk: *Performance of 3GPP high speed downlink packet access (HSDPA)*, Proceedings of VTC-Fall, 2004.
3. F. Kelly: *Charging and Rate Control for Elastic Traffic*, European Transactions on Telecommunications, volume 8 (1997) pages 33-37.
4. *NS-2 Network Simulator version 2.26*, http://www.isi.edu/nsnam/ns, 2003.
5. K.S. Trivedi: *Probability and Statistics with Reliability, Queuing and Computer Science Applications 2nd edition*, Wiley-Interscience 2002, page 241.

Precise Matching of Semantic Web Services*

Yonglei Yao, Sen Su, and Fangchun Yang

State Key Laboratory of Networking & Switching Technology,
Beijing University of Posts & Telecommunications (BUPT)
187# 10 Xi Tu Cheng Rd., Beijing 100876, China
{yaoyl, susen, fcyang}@bupt.edu.cn

Abstract. Matchmaking is an important aspect of the Web Services interactions, which enables a service requester to locate the most suitable counterpart. However, current service matching algorithms operate on service advertisements, which don't contain enough information to select a service instance that the consumer can immediately interact with. In this paper, we propose a precise matching algorithm, based on WS-Agreement and OWL-S, to deal with this challenge. We show how to combine WS-Agreement with OWL-S to obtain an agreement-style service description, and how the degree of match between two descriptions of this style is calculated[1].

1 Introduction

In recent years, more and more Web Services are becoming available on the Internet, enabling customer to interact with a great number of potential counterparts. On the other hand, this also makes a matchmaking process necessary to locate the most suitable one.

In current matchmaking approaches, the matching algorithms operate on service advertisements and queries [3, 4, 5]. Unfortunately, quality of service and other guarantees cannot simply be advertised. As a result, the matchmaking process can only find a set of services which may possibly meet one's requirements, leaving the task of selecting the most suitable counterpart to the customer.

Against this background, this paper develops a matching algorithm that operates on agreement-style service descriptions, to help a customer to select a specific service that can indeed meets all its requirements. This work advances the state of the art in the following ways. Firstly, it's a novel way to calculate the degree of match between services based on detailed service descriptions which combine WS-Agreement [1] and OWL-S [2]. Secondly, this work develops a matching algorithm which can provide a precise and quantified degree of match between a service requested and a service offered, which in turn can serve as a sound criteria for service selection.

---

* The work presented in this paper was Supported by 973 program of China(2003CB314806), the National Natural Science Foundation project of China(90204007), the program for Changjiang Scholars and Innovative Research Team in University (PCSIRT), and National Natural Science Funds for Distinguished Young Scholar(60125101).

V.N. Alexandrov et al. (Eds.): ICCS 2006, Part IV, LNCS 3994, pp. 164 – 167, 2006.
© Springer-Verlag Berlin Heidelberg 2006

2 Matching Algorithm

In order to select a specific service instance, a service consumer will continuously exchange proposals described in WS-Agreement, in which an OWL-S profile is embedded as service description terms to describe the functionalities provided or required, with potential service providers. Our matching algorithm operates on such agreement-style service descriptions to facilitate the service selection process. Our work presents several matching degree assessment methods as follows:

2.1 Concept Matching Degree

Firstly, we show how to calculate the degree of match between two concepts defined in ontologies. We differentiate between four degrees of concept match in terms of the subsumption relationships, and quantify each degree, as shown following:

```
ConceptMatchDegree(C1, C2){
  if C1≡C2 then return 1;
  else if C2 subsumes C1 then return p;
  else if C1 subsumes C2 then return q;
  else return 0;}
```

Where C_1 and C_2 denote two concepts defined in some ontologies, and p, q $\in(0,1)$ and $p>q$.

2.2 Input Matching Degree

For input matching, we want to show how well the inputs and preconditions required by service providers are satisfied by service consumers.

```
InputMatchDegree(inputsProvider, inputsRequester){
  inputMatchDegree=1;
  for each inputP in inputsProvider do{
  find inputR in inputsRequester that
    m=max(ConceptMatchDegree(inputR, inputP));
  if m>0 then inputMatchDegree=inputMatchDegree*m;
  else return 0;}
  return inputMatchDegree;}
```

2.3 Output Matching Degree

For output matching, we decide how well the outputs and effects required by service consumers are satisfied by those of service providers, as shown following:

```
OutputMatchDegree(outputsProvider, outputsRequester){
  outputMatchDegree=1;
  for each outputR in outputsRequester do{
  find outputP in outputsProvider that
```

```
  m=max(ConceptMatchDegree(outputP, outputR));
 if m>0 then outputMatchDegree=outputMatchDegree*m;
 else return 0;}
 return outputMatchDegree;}
```

2.4 Guarantee Term Matching Degree

Besides functionalities, a service consumer also demands an assurance with respect to the non-functional attributes from the service provider, such as price, delivery time, quality of service, etc. The assurance is described by the guarantee terms of the WS-Agreement document.

Let $t \in \{t_1, t_2, \ldots\ldots, t_n\}$ represent the guarantee terms demanded by the service consumer, and $D_i = [min_i, max_i]$ denotes the intervals of values for quantitative term t_i acceptable for the service consumer. Values for qualitative issues, on the other hand, are defined over a fully ordered domain, i.e., $D_i = \langle q_1, q_2, \ldots\ldots, q_m \rangle$.

For each term t_i, the service consumer has a scoring function V_i: $D_i \rightarrow [0, 1]$ that gives a score to a value of term t_i in the range of its acceptable values. Based on the above model, we can calculate the degree of match between guarantee terms.

```
GuranteeTermMatchDegree(termsRequester, termsProvider){
 guranteeTermMatchDegree=0;
 for each term t_i in termsRequester do{
 if there is a corresponding term in termsProvider with
   value j then GuranteeTermMatchDegree+=W_iV_i(j);}
 return GuranteeTermMatchDegree;}
```

Where W_i denotes the relative importance that the service consumer assigns to term V_i, The weights of all terms are normalized, i.e., $\sum_{1 \le i \le n} W_i = 1$.

2.5 Global Matching Degree

Based on the matching degrees between input parameters, output parameters and guarantee terms, we can determine the degree of match between two agreement-style service descriptions as:

$$\begin{aligned} MatchDegree = {} & W_i \times inputMatchDegree + W_o \times outputMatchDegree \\ & + W_g \times GuaranteeTermMatchDegree \end{aligned} \quad (1)$$

Where W_i, W_o and W_g represent the weights that the service consumer assigns to input match, output match and guarantee term match respectively.

3 Conclusion and Future Work

In this paper, we propose an algorithm for calculating the degree of match between two services based on the agreement-style service descriptions. Given the algorithm we develop, a service consumer can select a specific service which indeed has the ability to meet her requirements.

Future work includes implementing the matching algorithm, applying this technique to various applications, and as the last step, investigating the effectiveness of the algorithm with real world use cases.

References

[1] A. Andrieux, K. Czajkowski, A. Dan, K. Keahey, H. Ludwig, J. Pruyne, J. Rofrano, S. Tuecke, and M. Xu, Web Services Agreement Specification (WS-Agreement): Global Grid Forum, May 2004.

[2] D. Martin, M. Burstein, J. Hobbs, O. Lassila, D. McDermott, S. Mcllraith, S. Narayanan, M. Paolucci, B. Parsia, T. Payne, E. Sirin, N. Srinivasan, and K. Sycara. Owl-s: Semantic markup for web services, http://www.daml.org/services/owl-s/1.1/overview/, 2004.

[3] J. Hau, W. Lee, J. Darlington, A Semantic Similarity Measure for Semantic Web Services, In Proceedings of the fourteenth International World Wide Web Conference (WWW 2005), Chiba, Japan, May 10–14, 2005

[4] L. Li and I. Horrocks, "A software framework for matchmaking based on semantic web technology", In Proceedings of the Twelfth International World Wide Web Conference (WWW 2003), pages 331-339, ACM, 2003.

[5] M. Paolucci, T. Kawamura, T.R. Payne, and K. Sycara., "Semantic matching of web services capabilities", In Proceedings of the 1st International Semantic Web Conference (ISWC), pages 333-348, Sardinia, Italia, June 2002.

Evolving Toward Next Generation Wireless Broadband Internet

Seung-Que Lee[1], Namhun Park[1], Choongho Cho[2], Hyongwoo Lee[3], and Seungwan Ryu[3,⋆]

[1] Mobile Telecommunications Research Division, ETRI, Daejeon, Korea
[2] Department of Computer and Information Science, Korea University, Korea
Department of Electronics and Information Engineering, Korea University, Korea
[3] Department of Information Systems, Chung-Ang University, Korea
{sqlee, nhpark}@etri.re.kr, {chcho, hwlee}@korea.ac.kr, rush2384@cau.ac.kr

Abstract. Recently, much attention has been paid to portable Internet as a solution not only to surpass limits of high-speed Internet, wireless LAN and cellular communications but also to accommodate increasing demands for wireless Internet services. In this paper, we introduce the Wireless Broadband (WiBro) system which has been developed in Korea as a portable Internet system to foster a new home grown wireless system and services.

1 Introduction

Recently, demand for portable Internet service via various mobile terminals such as notebook PC, PDA, and cellular phone is increasing rapidly. As a result, much attention has been paid to portable Internet as a solution not only to surpass limits of high-speed Internet, wireless LAN and cellular communications but also to accommodate increasing demands for Internet services in anytime, anywhere by users in the stationary or mobile environments with a low access cost and high data rate. Portable Internet is aiming to provide wireless Internet services at any time and anywhere via portable wireless devices with high data rate and medium or low mobility.

In this paper, we introduce the wireless Broadband Internet (WiBro) considered as a first system that bridges the wired and wireless realm by giving high speed portable Internet access anywhere at anytime while on moving with a low access cost and high data rate. An overview of the WiBro system architecture and its services are given followed by design concepts and structure of the WiBro Access Terminal which is plays an important role in delivering user friendly wireless Internet access.

⋆ Corresponding author.

V.N. Alexandrov et al. (Eds.): ICCS 2006, Part IV, LNCS 3994, pp. 168–171, 2006.
© Springer-Verlag Berlin Heidelberg 2006

2 The Wireless Broadband Internet (WiBro) System

In Korea, the *Wireless Broadband (WiBro) system* has been developed in 2005 as a portable Internet system based on IEEE802.16 standard [1, 2, 3, 4] to foster a new home grown wireless system and services. The design objectives of the WiBro system is to support seamless multimedia internet services with wider cell coverage than that of the wireless LAN under medium or low mobility. The *WiBro* system deploys OFDMA/TDD to take asymmetric traffic pattern into account for. Commercial WiBro service is scheduled to begin in June, 2006 in Korea.

Three types of WiBro services are *information provisioning type services* such as Internet access, e-mailing, and data searching, *entertainment type services* such as picture transmission, VoD, and gaming, and *business type services* such as remote approval, telemetering, and e-commerce. In order to provide such WiBro services, many requirements are needed. First, the system should be able to support data rate of 2Mbps per user in average, which is almost the same as data rate of the ADSL[1]. It is also required to support up to 50 Mbps data rate under various mobility conditions under indoor or outdoor environments. In addition, it should be able to support various data rate according to radio channel condition with low latency and high reliability.

2.1 WiBro System Architecture

The WiBro system consists of three different components; WiBro-AT (Access Terminal), WiBro-AP (Access-point), and PAR (Packet access router). Many WiBro-AP are connected to a PAR, and PAR is connected to the IP networks which contains many types of servers such as Authorization, Authentication and Accounting (AAA) server, Home Agent (HA) server, and network management and operations (NMO) servers.

The AT is an end point of the radio channel and communicates with AP via OFDMA wireless access. Main functions performed at an AT are radio channel transmission/reception, MAC processing, handover, user authentication and encryption, radio link control. The AP is responsible to deliver information between ATs and a PAR through performing mapping process between radio and wired channels. The AP performs packet retransmission, packet scheduling, bandwidth allocation, ranging, handover, and so on. The PAR manages many APs connected to it, and performs handover control to guarantee high mobility within its control area. In order to perform such functions, a PAR is constructed based on Gigabit Ethernet switch, and connected to APs based on IP protocol.

3 WiBro Access Terminal Subsystem (ATS)

Since the user equipment, i.e., WiBro *Access Terminal (AT)* provides interface with users, it is the most important factor for successful launching and fast

[1] ADSL: Asymmetric Digital Subscriber Line.

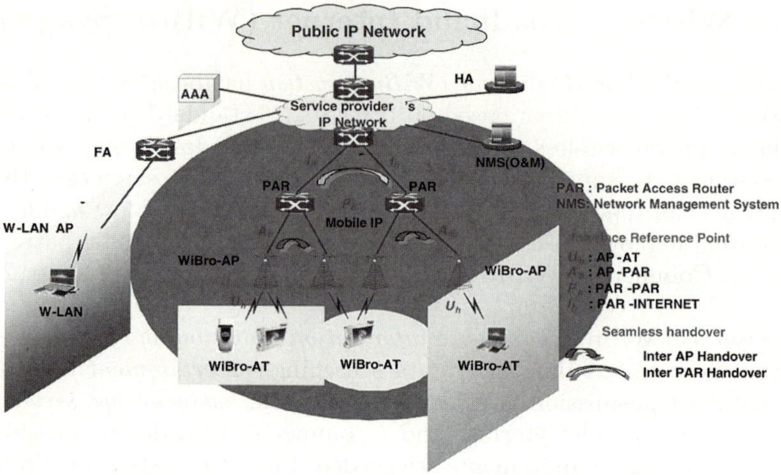

Fig. 1. The WiBro system architecture

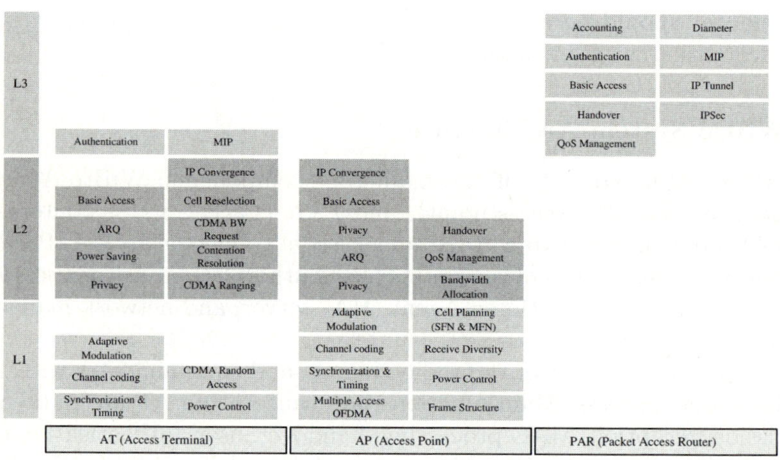

Fig. 2. Functions of each layers of AT, AP and PAR subsystem

expansion of commercial WiBro service scheduled to begin in June 2006 [5]. In this section, we briefly introduce design concepts.

Logical structure of the WiBro ATS shown in figure 3 consists of the high level protocol, the higher MAC, the lower MAC and RF and modem part. The high level protocol is responsible for functions of upper part of MAC protocol such as Mobile IP (MIP), authentication, installation of IP configuration, radio link condition report, and card interface. The higher MAC executes MAC convergence sublayer (MAC-CS) functions and MAC common part sublayer (MAC-CPS) functions. MAC-CS functions include MAC convergence, MAC service flow control, packet classification and header suppression. MAC-CPS functions include

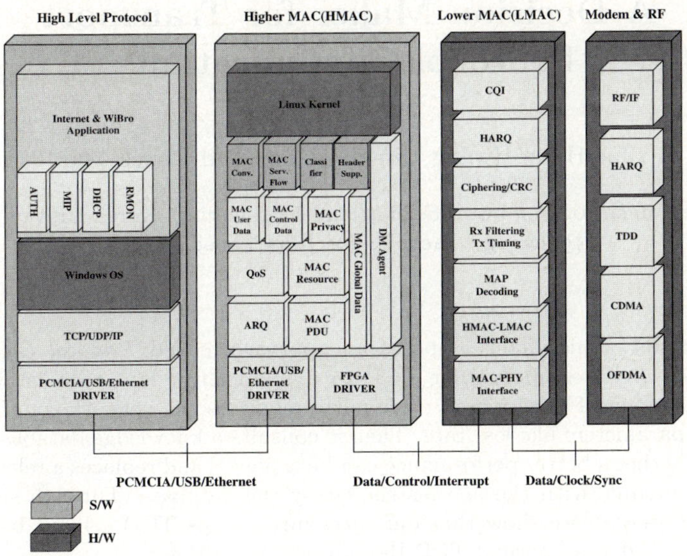

Fig. 3. Logical structure of WiBro access terminal subsystem (ATS)

MAC user and control data management, MAC resource management, MAC privacy, QoS management, MAC PDU, and error control (ARQ).

The lower MAC handles functions having strict time constraint such as channel quality information (CQI) report, HARQ, ciphering, CRC generation/check, Rx filtering, Tx timing, MAP decoding. Modem and RF part is responsible for modulation/demodulation for radio transmission and RF related functions including modulation/demodulation for OFDMA, bit spreading for CDMA-based random access, TDD for data transmission and reception, HARQ for fast error correction, and other functions for radio transmission.

References

1. IEEE Standard, "IEEE Standard for Local and Metropolitan Area Networks; Part 16: Air Interface for Fixed Broadband Wireless Access Systems," May, 2004.
2. IEEE Standard, "IEEE Standard for Local and Metropolitan Area Networks; Part 16: Air Interface for Fixed Broadband Wireless Access Systems, Amendment for Physical and MAC Layers," February, 2005.
3. Telecommunications Technologies Association (TTA), "2.3GHz Portable Internet Standard; Medium Access Control Sublayer," June, 2004.
4. Telecommunications Technologies Association (TTA), "2.3GHz Portable Internet Standard; Physical Layer," June, 2004.
5. S. Lee and N. Park, "An Overview of WiBro Access Terminal," *KICS Journal (in Korean)*, 22(9), pp. 128–137, September, 2005.

A Decision Maker for Transport Protocol Configuration*

Jae-Hyun Hwang, Jin-Hee Choi, and Chuck Yoo

Department of Computer Science and Engineering, Korea University
{jhhwang, jhchoi, hxy}@os.korea.ac.kr

Abstract. This paper proposes an approach, called *Protocol Configuration Decision Maker*, for TCP to dynamically adapt to a network environment. The proposed mechanism monitors the network condition with parameters like loss rate. Then it consults a knowledge database to see whether a better performance can be achieved and replaces a relevant TCP module with the one instructed by the database. Through simulation studies, we show that our mechanism helps TCP achieve better throughput than normal TCP Reno, up to 80∼194%.

1 Introduction

Flexibility and extensibility of network protocol software are getting more important as new networking technologies keep coming up. The configurable and extensible network systems have been proposed in TIP, ADAPTIVE, F-CSS projects[1]. Their goals include: 1) make the extension of new service and protocol easy, and 2) configure the protocol stack based on the application requirements. That is, when an application requests some QoS(Quality of Service)-attributes like performance, reliability, timeliness, security and so forth, the system configures the protocol stack that consists of fine-grained configurable modules in order to satisfy such requirements. However, if the system cannot adapt to dynamically changing environment, the performance degradation would come to be inevitable even though the application requirements are taken into account.

The problem is that the design of existing network systems has never considered network environments as protocol configuration criteria, up to our knowledge. To address the configuration criteria for network environments, this paper proposes a mechanism called Protocol Configuration Decision Maker that chooses the adequate functional module based on network environments. The proposed mechanism classifies the scope of configuration into Slow Start, Congestion Avoidance, Error Recovery like TCP and decides the dominant protocol module affecting the improvement of protocol performance. We evaluate our mechanism through simulation studies, and show that our mechanism helps TCP achieve better throughput than normal TCP Reno significantly.

* This work was supported by grant No.R01-2004-000-10588-0 from the Basic Research Program of the Korea Science & Engineering Foundation and a Korea University grant.

V.N. Alexandrov et al. (Eds.): ICCS 2006, Part IV, LNCS 3994, pp. 172–175, 2006.
© Springer-Verlag Berlin Heidelberg 2006

2 Decision Making for Protocol Configuration

The goal of Protocol Configuration Decision Maker is to dynamically adapt TCP to time-varying networks when the information about network environments is inferred. For the purpose of inferring the network conditions, we use four network parameters: round-trip delay, asymmetric ratio[1], loss rate and loss pattern. Because most of the estimation schemes already have been thoroughly evaluated in the literature, we do not compare or argue the accuracy of them but assume that the network parameters can be estimated precisely.

Now, we explain the process of making the protocol knowledge database for each protocol module. First of all, the throughput of all protocols the system has should be estimated for various network environments. This could be measured by simulations or with experiments in real Internet. Then, we represent the estimated throughput to Expected Throughput as follows.

$$Expected\ Throughput_{(SS\_i,CA\_j,ER\_k)}(RD, AR, LR, LP),$$

where SS_i stands for i-th Slow Start instance, CA_j for j-th Congestion Avoidance instance and ER_k for k-th Error Recovery instance, and RD, AR, LR, LP stand for Round-trip Delay, Asymmetric Ratio, Loss Rate, and Loss Pattern respectively.

Second, we calculate the impact degree of each protocol module on the expected throughput using ANOVA[5]. Let us assume that a system has two different versions of each protocol module. Then we can define three variables x_{SS}, x_{CA} and x_{ER} as follows.

$$x_{SS} = \begin{cases} -1 & for\ SS\_1 \\ 1 & for\ SS\_2 \end{cases},\quad x_{CA} = \begin{cases} -1 & for\ CA\_1 \\ 1 & for\ CA\_2 \end{cases},\quad x_{ER} = \begin{cases} -1 & for\ ER\_1 \\ 1 & for\ ER\_2 \end{cases}$$

y, expected throughput, is regressed to x_{SS}, x_{CA} and x_{ER}. q_0 is the mean performance of y, and q_{SS} is the effect of x_{SS}.

$$y = q_0 + q_{SS}x_{SS} + q_{CA}x_{CA} + q_{ER}x_{ER}$$

We obtain the following four combinations of equation using the above model.

$$\begin{aligned} thru\_1 &= q_0 - q_{SS} - q_{CA} - q_{ER} \\ thru\_2 &= q_0 + q_{SS} - q_{CA} - q_{ER} \\ thru\_3 &= q_0 - q_{SS} + q_{CA} - q_{ER} \\ thru\_4 &= q_0 - q_{SS} - q_{CA} + q_{ER} \end{aligned}$$

By solving the equations, we can get q_0, q_{SS}, q_{CA} and q_{ER}. Finally, the impact rate of each protocol module can be calculated as follows.

[1] The asymmetric ratio is defined as a ratio between forward delay and reverse delay in this paper.

$$x_{SS}'s\ impact = \frac{q_{SS}^2}{q_{SS}^2 + q_{CA}^2 + q_{ER}^2}$$

$$x_{CA}'s\ impact = \frac{q_{CA}^2}{q_{SS}^2 + q_{CA}^2 + q_{ER}^2}$$

$$x_{ER}'s\ impact = \frac{q_{ER}^2}{q_{SS}^2 + q_{CA}^2 + q_{ER}^2}$$

If the system uses [SS_1, CA_1, ER_1] and each protocol module impacts on the throughput improvement by 80%, 15% and 5% respectively, it could get better throughput up to 80% with reconfiguring SS_1 to SS_2.

3 Simulation Result

In this section, we validate our decision maker using NS-2 simulator version 2.26[6]. Assuming that the network system supports partial reconfiguration of

Table 1. TCP variants used in simulation

TCP Variant	Slow Start	Congestion	Error Recovery
Reno	Heuristic	AIMD	Duplicate ACK
Westwood[2]	BW Estimate	AIMD	Duplicate ACK
BI[3]	Heuristic	Binary Search	Duplicate ACK
SACK[4]	Heuristic	AIMD	Selective ACK

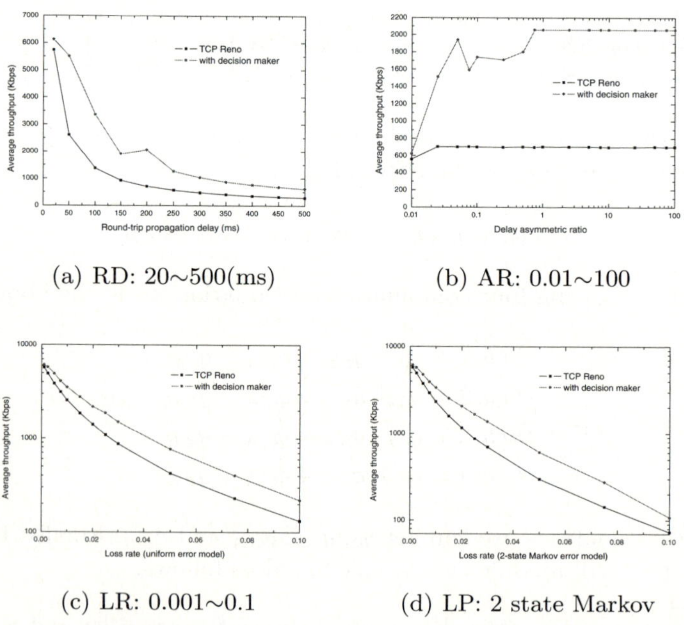

(a) RD: 20∼500(ms) (b) AR: 0.01∼100

(c) LR: 0.001∼0.1 (d) LP: 2 state Markov

Fig. 1. Average throughput graphs: TCP Reno vs TCP with our decision maker the transport protocols, we performed the experiment using TCP variants simply to get same results instead of implementing such system. Table 1 shows TCP variants used in our simulation. According to Table 1, we could emulate the partial reconfiguration of protocol modules in the network system that contains two different instances for each protocol module. Each simulation time is 100 seconds and the throughput of each TCP is averaged by results of 100 iterations for the same environment. The default bandwidth, round-propagation delay, asymmetric ratio and loss rate are 10(Mbps), 20(ms), 1.0 and 0.001 respectively.

Through simulation studies, we estimated the average throughput of all TCP variants and calculated the impact rate of each protocol module changing RD, AR, LR and LP. With the impact rate, we could find out the dominant module for throughput improvement. Then, we compared the throughput of 1) the case only using Reno and 2) the case with applied reconfiguration of the protocol modules properly using our decision maker. The average throughput graphs of both cases with various network parameters are shown in Figure 1. As we know from the results, there are always protocol modules that can take the place of Reno's for an improvement of the throughput.

4 Conclusion

In this paper, we proposed Protocol Configuration Decision Maker for TCP to dynamically adapt to time-varying network environments. While the existing flexible and extensible network systems configure the protocol stack based on application requirements, the proposed mechanism is based on network parameters like loss rate. Through simulation studies, we could show that our mechanism helps TCP achieve better throughput than normal TCP Reno, up to 80∼194%.

References

1. Stenfan Böcking, "Object-Oriented Network Protocols," *In Proceedings of IEEE INFOCOM*, 1997.
2. M. Gerla, M. Y. Sanadidi, R. Wang, A. Zanella, C. Casetti, S. Mascolo, "TCP Westwood: Congestion Window Control Using Bandwidth Estimation," *In Proceedings of IEEE GLOBECOM*, Nov. 2001.
3. L. Xu, K. Harfoush and I. Rhee, "Binary increase congestion control for fast long-distance networks," *In Proceedings of IEEE INFOCOM*, 2004.
4. S. Floyd, M. Mahdavi, M. Mathis and J. Widmer, "An Extension to the Selective Acknowledgement(SACK) option for TCP," *IETF*, RFC 2883, 2000.
5. George W. Cobb, "Introduction to Design and Analysis of Experiments," *Springer*, Mar. 1998.
6. ns2 Network Simulator version 2.26, `http://www.isi.edu/nsnam/ns`, 2003.

On the Generation of Fast Verifiable IPv6 Addresses

Qianli Zhang and Xing Li

Tsinghua University, Beijing 100084, China
zhangql02@mails.tsinghua.edu.cn

Abstract. Many network attacks forge the source address in their IP packets to block traceback. This situation does not change much in IPv6 network since IPSEC is not enabled generally and most IP address spoof attacks have taken effect before packets reached destination. Although ingress filtering can be used to validate source addresses, it could only ensure that the network portion of an address is not spoofed. Since subnets are much larger in IPv6, even with RFC 2827-like filtering an adversary can spoof an enormous range of addresses. In this paper, we propose an IPv6 address assignment scheme to generate verifiable IPv6 addresses in one network. With this scheme, router could validate the IPv6 addresses quickly, thus allow all outgoing packets with improper source addresses and all incoming packets with improper destination addresses to be immediately identified. Apart from the obvious merit to counter denial of service attacks, this scheme also make network audit and pricing easier.

1 Introduction

Attackers commonly forge source addresses to hinder tracing of their malicious packets. Examples include DDoS attacks [1], smurf attacks[2], and TCP SYN flooding attacks[3]. Reliably detecting the attacker is hard because standard routers cannot verify that a packet is indeed sent by the node specified in its source address. Ingress filtering[4] is widely used to validate source addresses. RFC 2827 specifies methods to implement ingress filtering to prevent spoofed traffic at its origin. Unfortunately such filtering lacks of the initiative for the origin network to implement. Also RFC 2827 ensures that only the network portion of an address is not spoofed, not the host portion. For example, for 24-bit subnet 192.0.2.0/24, RFC 2827 filtering ensures that traffic originating from 192.0.3.0 is dropped but does not stop an adversary from spoofing all the hosts within the 192.0.2.0/24. Since subnets are much larger in IPv6, even with RFC 2827-like filtering an adversary can spoof an enormous range of addresses. Currently no techniques are available to mitigate the spoofing of the 64 bits of host address space available in IPv6.

Another approach to the problem of IP spoofing is tracing[5]. Since source addresses are unreliable, tracing requires expensive and complicated techniques to observe traffic as they pass through routers and reconstruct a packets travel

V.N. Alexandrov et al. (Eds.): ICCS 2006, Part IV, LNCS 3994, pp. 176–179, 2006.
© Springer-Verlag Berlin Heidelberg 2006

path at the end. Tracing also becomes ineffective when the volume of attack traffic is small or the attack is distributed. Moreover, tracing is typically performed after an attack is detected, and perhaps the victim has already been damaged.

In this paper, we propose a scheme to assign verifiable IPv6 addresses in a network. With this scheme, router could validate not only the subnet part but also the interface part of the IPv6 addresses quickly. Apart from the obvious value in ingress filtering, this scheme can also ensure that incoming packets with improper destination addresses to be immediately identified and dropped. Thus it provide some initiative for its deployment. With identifier contained in the addresses, it could also be used to identify the possible sources of an attack. Intrusion detection and network problem diagnosis can also be simplified.

This paper is structured as follows: Section 2 presents some background notations and information. Details are provided in section 3. The paper concludes in section 4.

2 Background Notations

An IPv6 address is 128 bits long. It is divided into two parts. The leftmost 64 bits, the subnet prefix, is used for routing IP packets across the Internet to the destination network. The rightmost 64 bits, the interface identifier, identifies an individual node within a local network. The interface identifiers may be chosen in an arbitrary way, e.g. randomly, as long as no two nodes on the same network share the same value.

Two bits of the interface identifier have a special semantics. The 7th bit from the left is the Universal/Local bit or "u" bit. It is usually set to 1 to mean that the interface identifier is configured from an EUI-64 identifier from the interface hardware and, thus, is globally unique. The 8th bit from the left is the Individual/Group or "g" bit, which is set to 1 for multicast addresses.

To better present our scheme, the following notations are used throughout the paper.

- hash: Cryptographic hash function, SHA-1[6] for example.
- hashT: Cryptographic hash function whose output is truncated by taking the T leftmost bits of the output.
- cipher64: 64 bits block cipher, IDEA[7] for example.

3 Verifiable IPv6 Address Generation

The verifiable address is generated by a local authority, DHCP server for example. In figure 1, ID is the identifier generated by authority for tracking, R is not used now and will be set to zero. P is prefix required and used to generate destination specific addresses. Given the destination address D, prefix requirement P_r, subnet prefix N and the correspondent key K_N, the generation procedure is as follows.

1. set $Padding$ and R to 0, P to P_r, ID to the identifier.
2. $SIG = HASH32((D >> (128 - (P_r >> 3)))|N|R|P_r|ID)$, where $|$ is concatenation.

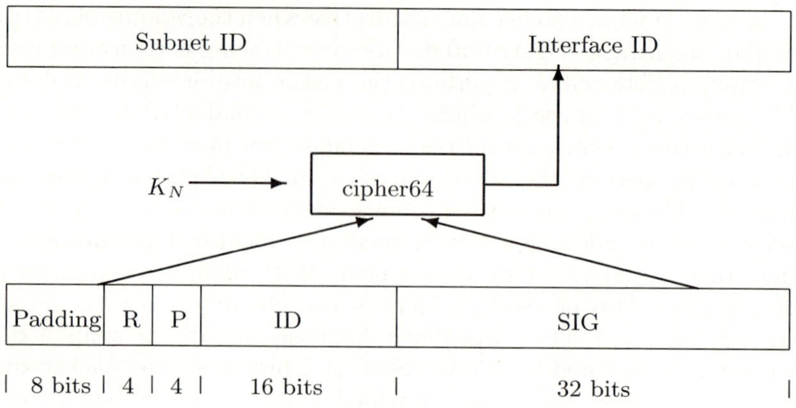

Fig. 1. Verifiable IPv6 address

3. encrypt the 64 bits by *cipher*64 with key K_N. Encryption guaranteed the generated addresses can not be discriminated from randomly generated addresses easily. Encryption also provide a method to keep the ID information confidential.
4. test whether "u" bit is 0 and "g" bit is zero. If not, increase padding by 1 and repeat the last step. Since a total of 256 IPv6 interface ID could be generated, the probability of this 256 IPv6 addresses all have the "u" bit to 1 or "g" bit to 1 is $(\frac{3}{4})^{256}$, which is about 10^{-32} and negligible.

Required prefix mandates when the calculation of SIG also includes the leftmost P_r bytes of destination IPv6 address. When P_r is not zero, the address generated is destination specific and could not communicate to hosts out of the range. If P_r is zero, the generated address is a static IPv6 addresses and can communicate to all IPv6 addresses. Destination specific addresses make IP address spoof even harder since even if an attacker knows a valid IPv6 address, he could not decide whether this address is a static one. The limitation of destination specific addresses is, however, that it requires to extend DHCPv6 protocol. Also, the value of destination specific addresses is limited since servers have to have static addresses.

The verification procedure is similar.

1. decrypt the 64 bits interface ID by cipher64 with key K_N. if R is not zero, discard the packet.
2. set *Padding* to 0, $SIG_v = HASH32((D >> (128 - (P_r >> 3)))|N|R|P_r|ID)$, where $|$ is concatenation, if SIG_v does not equal SIG, discard the packet.

For a large domain, it may be of interest to generate K_N with a master key K_m. For example, $K_N = hash(K_m|N)$. However, this scheme does not mandate the specific method to generate K_N.

In this scheme, only symmetric cryptography is used, which make it scalable for the high speed filtering. Asymmetric cryptographic primitives, such as RSA signatures[8], are computationally expensive: RSA signature verification is about three orders of magnitude slower than one symmetric operation (block cipher or hash function operation), and signature generation is about four orders of magnitude slower. When implemented in hardware, the speed difference is even larger. Thus make this algorithm feasible for high-speed implementation.

4 Conclusion

In this paper, a new scheme to generate verifiable IPv6 addresses is introduced. This scheme make the IPv6 host portion ingress filter feasible. Also since only symmetric cryptography is used, this scheme could be implemented in routers and provide better protection for network bandwidth DOS.

More research is required to resolve the following problems. First of all, for large organizations, it is often desirable to have a key management protocol to deal with the key generation and distribution. Secondly, the process of destination specific addresses generation is worth further research.

References

1. Computer Emergency Response Team. CERT Advisory CA-2000-01 Denial-of-Service Developments, http://www.cert.org/advisories/CA-2000-01.html, January 2000.
2. Computer Emergency Response Team. CERT Advisory CA-1998-01 Smurf IP Denial-of-Service Attacks, http://www.cert.org/advisories/CA-1998-01.html, January 2000.
3. C. L. Schuba, I. V. Krsul, M. G. Kuhn, E. H. Spafford, A. Sundaram, and D. Zamboni. Analysis of a denial of service attack on TCP, Proceedings of IEEE Symposium on Security and Privacy, 1997.
4. P. Ferguson and D. Senie. Network Ingress Filtering: Defeating Denial of Service Attacks Which Employ IP Source Address Spoofing, RFC 2827, May 2000.
5. Stefan Savage, David Wetherall, Anna Karlin, and Tom Anderson, Network Support for IP Traceback, IEEE/ACM TRANSACTIONS ON NETWORKING, VOL. 9, NO. 3, JUNE 2001
6. C. Madson and R. Glenn, The Use of HMAC-SHA-1-96 within ESP and AH, RFC 2404, November 1998.
7. A. J. Menezes, P.C. v. Oorschot, S.A. Vanstone, Handbook of Applied Cryptography, CRC Press New York, 1997, p. 265.
8. R. L. Rivest, A. Shamir, and L. M. Adleman. A Method for Obtaining Digital Signatures and Public-Key Cryptosystems. Communications of the ACM, 21(2):120-126, 1978.

A MAC Protocol to Reduce Sleep Latency and Collisions in Wireless Sensor Network

Jinsuk Pak, Jeongho Son, and Kijun Han*

Department of Computer Engineering, Kyungpook National University,
1370, Sankyuk-dong, Puk-gu, Daegu, 702-701, Korea
{jspak, jhson}@netopia.knu.ac.kr, kjhan@knu.ac.kr

Abstract. This paper presents a MAC protocol which uses separate wakeup slots for each sensor node in sensor networks. Most MAC protocols proposed for sensor network are inefficient under heavy traffic loads, in particular in high density network topology because of frequent collisions and long sleep latency. In this paper, we suggest a MAC protocol in which each node has a different wakeup schedule in the same duty cycle, and it joins the competition only for its own short wakeup slot when the receiver is ready to receive its data. Simulation results indicate that our scheme can reduce energy consumption and minimize idle listening which increases the power efficiency.

1 Introduction

In Wireless Sensor Networks (WSN), energy efficiency is one the most critical issues in order to prolong network lifetime since it is often not feasible to replace or recharge batteries for sensor nodes. Media Access Control (MAC) protocols must minimize the radio energy costs in sensor nodes. Latency and throughput are also important design features for MAC protocols in WSN [3].

The SMAC proposed in [2], which is a modified version of the IEEE 802.11 Distributed Coordinator Function (DCF), provides a tunable periodic active/sleep cycle for sensor nodes. It puts nodes to sleep at certain times to conserve energy [4]. However, periodic sleep may result in a long sleep latency since the sending node has to wait until the receiving node wakes up in its listen period. Timeout MAC (TMAC) alleviated the problems of SMAC by using an adaptive duty cycle. In TMAC, if a node does not observe any activity in the neighborhood for some time, it goes to sleep early. TMAC saves more energy under variable traffic loads, but it still has problems of long sleep latency and low throughput.

We propose a new MAC protocol to solve the sleep delay and collision problems by allocating different listen period to each node in the same duty cycle with a legacy sensor MAC protocol. In our MAC protocol, each sensor node joins the competition only for its own short wakeup slot when the receiver is ready to receive its data. Our MAC protocol can reduce the possibility of collision and decrease sleep delay due to contention. Also, it has a shorter listen period than SMAC, which contributes to reducing energy waste and thus to improving the power efficiency.

* Corresponding author.

V.N. Alexandrov et al. (Eds.): ICCS 2006, Part IV, LNCS 3994, pp. 180 – 183, 2006.
© Springer-Verlag Berlin Heidelberg 2006

2 Our MAC Protocol

In this paper, we propose a MAC protocol to reduce collisions and decrease latency caused by periodic sleeping. As shown in Fig. 1, in our MAC protocol, a superframe is comprised of two parts: a SYNC period for synchronization signal, and a listen/sleep period. The listen/sleep period is again divided into multiple sub-slots. We call these wakeup slots. Each node is assigned a wakeup slot for data transmitting and receiving, and it can wake up only during its own wakeup slot. After its own wakeup slot, it goes to sleep until another wakeup slot is reached in the next superframe. The location of wakeup slot is assigned to each node depending on its ID (SID). For example, the location of wakeup slot can be determined by simply using the residual value of SID divided by the number of wakeup slots (N_{ws}). The number of wakeup slots per listen/sleep period depends on applications used or network deployment. In high dense network, it may need a lot of wakeup slots. At this time, several nodes may share a single wakeup slot.

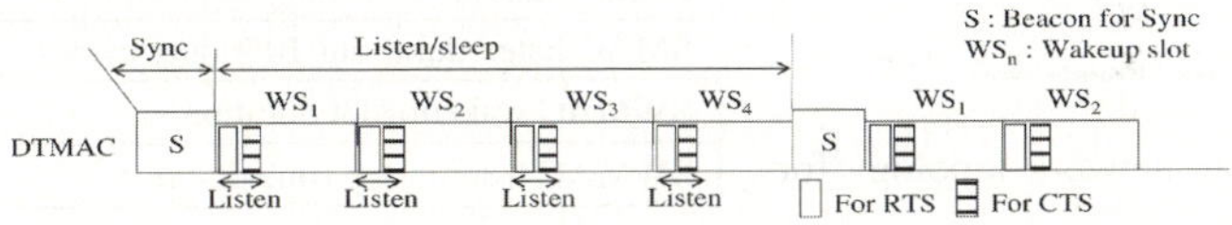

Fig. 1. Frame structure of our MAC protocol

All sensor nodes awake up during SYNC period and join in the contention to broadcast SYNC packet as done similarly in SMAC. The winner sends a SYNC packet to let its neighbor nodes know wakeup schedule information as well as to deal with clock drift [2]. On receiving a SYNC packet, each node knows the location of its own wakeup slots. Each sensor maintains a time synchronization and wakeup schedule information table for its neighboring nodes.

When a node has data to send, it looks up the wakeup schedule information table to find out the location of the wakeup slot assigned to the receiver, and waits until the wakeup slot comes. Upon seeing the receiver's wakeup slot, it starts RTS and CTS handshaking for transmission competition as performed similarly in SMAC. If it fails in the competition of RTS/CTS handshaking, it must wait for the next superframe.

As previously explained, in SMAC, every node joins the competition to transmit its data packet for every listen period. If a node fails in the competition, it must wait for the next listen time. Thus, it becomes the main cause of latency time problems. On the other hand, our MAC protocol can reduce energy consumption and minimize idle listening since it joins the competition only for its own short wakeup slot when the receiver is ready to receive its data.

3 Simulation

We evaluated the performance of our MAC protocol mechanisms through a computer simulation. The simulation parameters are listed in Table 1. To simplify the simulation, we assumed that the radio link propagation delay was zero without transmission error. Energy consumption model is based on real nodes: 0.016mW while sleeping, 12.36mW while idle listening, 12.50mW while receiving, and 14.88mW while transmitting a data packet [1].

The simulation was conducted in a static network with 9 sensor nodes as shown in Fig. 2. Each sending node, modeled as Constant Bit Rate (CBR) traffic source, had 20 packets. The number of wakeup slots was 4. We assumed that the traffic flows to only one way from send nodes to the destination nodes through a unicast path. The simulation ran until every node sent all of its packets.

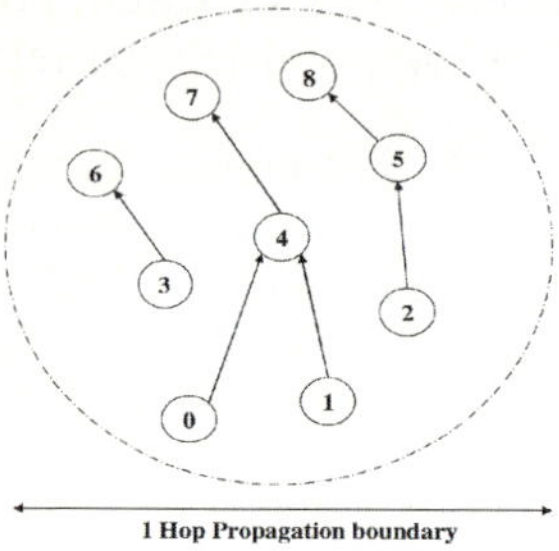

Fig. 2. Two-hop network topology for simulation

Table 1. Simulation parameters

Parameter	Value
Radio bandwidth	20 kbps
Contention window	32 slot
Data packet size	150Bytes
RTS, CTS, ACK size	20 Bytes
Duration of beacon	25ms
Frame interval	625ms
SMAC listen duration(10% duty cycle)	62.5ms
SWMAC wakeup slot duration	150ms
SWMAC listen/sleep time duration	25ms

Fig. 3 shows the average number of collisions until each node sends all packets it has with different traffic load. For comparison, we implement a SMAC with adaptive listening, but we do not consider its synchronization and message passing scheme. In the SMAC with periodic sleep, each node is configured to operate on a 10% duty cycle. Also we implement a Carrier Sense Multiple Access/Collision Avoidance (CSMA/CA) MAC without periodical sleep schedule.

Our MAC protocol causes less collision than SMAC since each sensor node in our MAC protocol has a separate receiving time and it tries to send its packet only when the receiver is ready. In SMAC, on the other hand, if each node has the same duty cycle, they then join in contention of transmission at the same time. Thus, they can choose the same back-off time under heavy traffic load, which causes frequent collisions.

We compare the average packet queuing delay under various traffic loads for three MAC protocols, as illustrated in Fig. 4. In general, the queuing delay depends on the traffic load. In a heavy traffic case, queuing delay becomes a dominant factor in the latency of MAC protocol. In light traffic, there is no queuing delay since few packets are moving through the network. In the MAC protocol without sleeping, it immediately starts carrier sensing and tries to forward packets to the next hop. However, the MAC protocol with periodic sleeping, has an extra delay (called a 'sleep delay'), since when a sender gets a packet to transmit, it must wait until the receiver wakes up. Further, if the sender is defeated in a transmit competition it then must sleep until the next wakeup schedule time. This increases the queuing delay. However, our MAC protocol offers a lower queuing delay even under heavy traffic loads since it distributes competitions over the superframe.

Fig. 5 shows the amount of energy consumed by all nodes in the network until the end of the simulation runs. We compared the total energy consumption of different MAC protocols under different traffic loads [2]. This figure shows that SMAC consumes more energy than our MAC protocol. This is because SMAC produces more retransmissions than our MAC protocol. However, both compared with the CSMA/CA can reduce total energy consumption using periodic listen/sleep schedule.

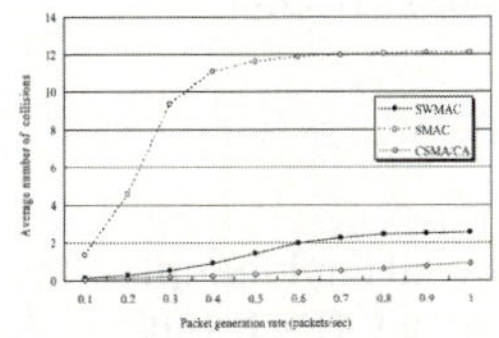

Fig. 3. Average number of collisions

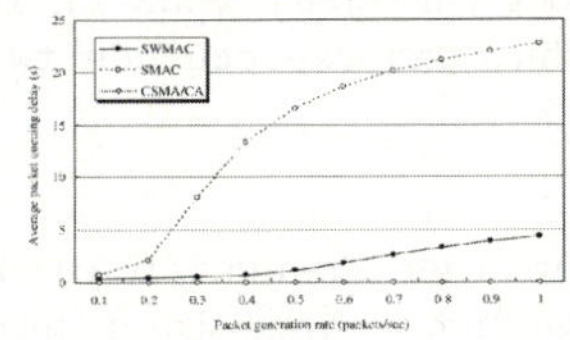

Fig. 4. Average packet queuing delay

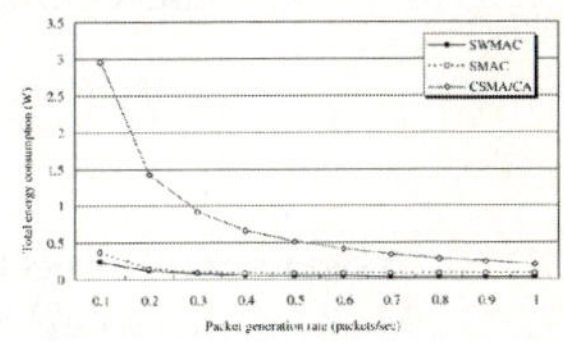

Fig. 5. Total energy consumption

4 Conclusions

We proposed a new MAC protocol, an energy efficient, low collision, and low latency MAC protocol using separate wakeup slots in the same duty cycle for wireless sensor networks. In our MAC protocol, each node joins the competition only for its own short wakeup slot when the receiver is ready to receive its data. Simulation results show that our MAC protocol can reduce probability of collisions and decrease sleep delay, which contributes to enhancing throughput and improving power efficiency.

Acknowledgements. This research is supported by Program for the Training of Graduate Students for Regional Innovation.

References

1. Curt Schurgers, Vlasios Tsiatsis, Saurabh Ganeriwal, Mani Srivastava : Optimizing Sensor Networks in the Energy-Latency-Density Design Space, IEEE Transactions on mobile computing, Vol. 1, No. 1, pp. 70-80, (2002)
2. W. Ye, J. Heidemann, and D. Estrin,: Medium Access Control with Coordinated, Adaptive Sleeping for ireless Sensor Networks, IEEE/ACM Transaction on Networking, Vol. 12, No.3, pp.493-506, (2004)
3. Gang Lu, Bhaskar Krishnamachari, Cauligi S. Raghavendra.: An Adaptive Energy-Efficient and Low-Latency MAC for Data Gathering in Wireless Sensor Networks, WMAN'04, Vol. 13, No. 13, pp. 224a, (2004)
4. Ramakrishnan, S. Huang, H. Balakrishnan, M. Mullen, J.: Impact of sleep in a wireless sensor MAC protocol, VTC2004-Fall, Vol. 7, pp. 4621-4624, (2004)

IC Design of IPv6 Routing Lookup for High Speed Networks

Yuan-Sun Chu, Hui-Kai Su, Po-Feng Lin, and Ming-Jen Chen

Department of Electrical Engineering,
National Chung-Cheng University, Chia-Yi, Taiwan 621, R.O.C
{chu, pat}@ee.ccu.edu.tw

Abstract. In recent years, there are many researches for routing lookup. Most of them can achieve high average search throughput for IPv4, but they are slow in the updating speed and cannot suit to 128 bits IPv6 address even in hardware architecture. This paper proposed a routing lookup system which contains an ASIC of routing lookup table and off-chip memory sets. In the performance analysis, 91.89 % routing entries of the routing table can be searched in one memory access, and the worst case about 10 % needs two memory accesses. The routing lookup system approaches 213.4 Mlps (109.26 Gb/s). It is enough to satisfy the high speed link OC-768 (40 Gb/s) with 150000 routing entries.

1 Introduction

In recent years, there are many researches for routing lookup. [1, 2] create routing lookup table with trie. [3] proposed hierarchical hardware architecture for IPv4 routing lookup. They can achieve high average search throughput for IPv4, but they cannot suit to 128 bits IPv6 address. [4, 5, 6] proposed routing lookup with CAM, and all match action only needs one clock cycle. But it needs special mechanisms to solve the sorting problem and expensive, especially TCAM.

This paper proposes a routing lookup system which contains an ASIC and off-chip RAM for IPv6. The routing scheme is based on the prefix length distribution of 6Net routing tables. In the proposed system, the lookup speed with 281.69 Mlps can satisfy the requirement of OC-768, and only needs 20.04 KB TCAM, 10.24 KB BCAM, and 29.29 MB RAM for 150000 routing entries.

2 System Architecture

Figure 1 shows the system architecture which composed of an ASIC and a memory set stores routing table. The off-chip RAM is with two hierarchical levels. First level is composed of 3 hash tables, and second level is a pure SRAM table. The hash tables store routing entries with prefix equal to 32, 48 and 64 respectively.

The ASIC has the complete functions, including inserting, searching, updating and deleting. Moreover, it focus on 1-64 bits of IPv6 address (network ID).

V.N. Alexandrov et al. (Eds.): ICCS 2006, Part IV, LNCS 3994, pp. 184–187, 2006.
© Springer-Verlag Berlin Heidelberg 2006

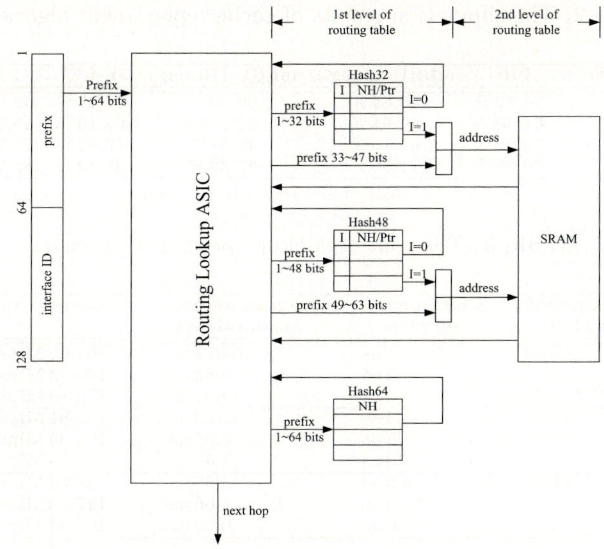

Fig. 1. The routing lookup system

Table 1. The entropy of various hash functions

Hash index	Hash function			
	Bit extraction	Fletch checksum	XOR folding	CRC
12	11.9694	11.9714	11.9753	11.977

80 % hit ratio is guaranteed on the chip. The CAM is used as an on-chip memory for the fast search in the ASIC, and cache replacement algorithm and FIFO are used.

3 Hash Scheme

Table 1 shows the entropy of various hash functions by different schemes. The simulation trace is based on TANET[7] network. It consists of 7.66157 million entries in a period of one hour, and there are 43867 distinct destination addresses. The hash functions are used to simulate Bit Extraction, Fletch Checksum, XOR Folding, and CRC.

Table 1 shows that CRC is the best scheme for the hash function, but it requires complex computation. The XOR folding is also an excellent hash function and simple to be implemented in hardware, so it is used in our scheme.

4 Cache Replacement Algorithm

When a cache misses, the new referenced data needs to be inserted into the cache table. The simulation analyzed five cache replacement algorithms. These

Table 2. The simulation results of cache replacement algorithms

No. of cache entries	LRU	mLRU (4 seg.)	mLRU (16 seg.)	SF-LRU	LFU	FIFO
64	54.72 %	45.84 %	36.20 %	57.15 %	48.65 %	49.21 %
128	61.50 %	49.91 %	42.21 %	64.10 %	55.71 %	56.07 %
512	73.52 %	64.10 %	49.91 %	79.66 %	71.97 %	72.06 %
1024	79.78 %	71.50 %	57.33 %	87.32 %	85.76 %	81.19 %

Table 3. The routing lookup speed in ideal case

Cache hit ratio	Clock period on chip	Clock period off chip	Average routing lookup time	Lookup speed	Provide line rate (64 B)
60 %	3.3 ns	3.3 ns	6.072 ns	164.69 Mlps	84.32 Gb/s
		5 ns	6.82 ns	146.63 Mlps	75.07 Gb/s
		10 ns	9.02 ns	110.86 Mlps	56.76 Gb/s
70 %	3.3 ns	3.3 ns	5.379 ns	185.91 Mlps	95.19 Gb/s
		5 ns	5.94 ns	168.35 Mlps	86.2 Gb/s
		10 ns	7.59 ns	131.75 Mlps	67.46 Gb/s
80 %	3.3 ns	3.3 ns	4.686 ns	213.4 Mlps	109.26 Gb/s
		5 ns	5.06 ns	197.63 Mlps	101.19 Gb/s
		10 ns	6.16 ns	162.34 Mlps	83.12 Gb/s

five algorithms are FIFO [8], LRU [8], mLRU[9], SF-LRU[10], and LFU [11]. The simulation results are shown in Table 2.

In the simulation, SF-LRU and LFU have good performance. But SF-LRU and LFU need counters to record the last reference time and sorting action is too complex in hardware. FIFO has good performance and 81.19 % hit ratio with 1024 entries. It is enough in network traffic. It is simple for hardware design and only needs one register to record the next CAM address.

5 Efficiency Analysis

CCUEE SOC Lab proposes a PF-CDPD CAM [12]. Its clock period can approach 3.3 ns when CAM size is $1024 \times (64 + 8)$. We assume the ideal off-chip clock period can approach the ASIC's speed, i.e., 3.3 ns. We refer to the actual SRAM clock period (5 ns) and CAM clock period (10 ns). The analysis results of lookup speed in ideal case is shown Table 3. The lookup speeds can satisfy the OC-768 requirement. The best ideal lookup speed is 213.4 Mlps with 80 % hit ratio.

Table 4. The comparison with the related works

	Our Scheme	BDD [1]	IPv4/IPv6 dual [5]	Fast TCAM [4]
Implement method	hardware	software	hardware (TCAM)	hardware (TCAM)
Worst search latency	2 memory access	depend on trie depth	7-stage pipeline	3 clock cycle latency, 1 clock is 5 ns
Lookup speed	213.4 Mlps (150000 prefixes)	168.6 Mlps for 29487 prefixes	100 Mlps	200 Mlps
Memory size	TCAM: 20.04 KB, CAM: 10.24 KB, RAM: 29.29 MB	non-available	non-available	21 MB capacity, 21632 entries

6 Conclusion

The IPv6 routing lookup system is proposed with a routing lookup ASIC and a memory set. The scheme is based on the prefix length distribution of 6Net routing tables. The first level in the proposed routing table can cover about 91.89 % routing entries. The ASIC has the complete routing lookup functions: insert, search, update, and delete. A FIFO is used as cache replacement algorithm in the proposed architecture by using a CAM with 1024 entries. It can guarantee 80 % hit ratio, so the speed can approach 213.4 Mlps and satisfy the requirement of OC-768. The system only needs 20.04 KB TCAM, 10.24 KB BCAM, and 29.29 MB RAM for 150000 routing entries. Table 4 shows the comparison with the related works.

References

1. Sangireddy, R., Somani, A.: High-speed IP routing with binary decision diagrams based hardware address lookup engine. IEEE Journal on Selected Areas in Communications **21** (2003) 513 – 521
2. Lampson, B., Srinivasan, V., Varghese, G.: IP lookups using multiway and multicolumn search. IEEE/ACM Transactions on Networking **7** (1999) 324 – 334
3. Huang, N.F., Zhao, S.M.: A novel IP-routing lookup scheme and hardware architecture for multigigabit switching routers. IEEE Journal on Selected Areas in Communications **17** (1999) 1093 – 1104
4. Gamache, B., Pfeffer, Z., Khatri, S.: A fast ternary CAM design for IP networking applications. In: ICCCN 2003. (2003)
5. Wang, Z.X., Wang, H.M., Sun, Y.M.: High-performance IPv4/IPv6 dual-stack routing lookup. In: 18th International Conference on Advanced Information Networking and Applications. Volume 1. (2004)
6. Hayashi, T., Miyazaki, T.: High-speed table lookup engine for IPv6 longest prefix match. In: IEEE Global Telecommunications Conference (GLOBECOM 1999). Volume 2. (1999)
7. The Computer Center of the Ministry of Education in Taiwan: Taiwan academic network (TANET). http://www.edu.tw/EDU_WEB/EDU_MGT/MOECC/EDU0688001/tanet/1.htm (in Chinese) (2005)
8. Tanenbaum, A.S., Woodhull, A.S.: Operating Systems: Design And Implementation, Second Edition. Prentice Hall (1996)
9. Liu, H.: Reducing cache miss ratio for routing prefix cache. In: IEEE Global Telecommunications Conference (GLOBECOM 2002). Volume 3. (2002)
10. Alghazo, J., Akaaboune, A., Botros, N.: SF-LRU cache replacement algorithm. In: Records of the International Workshop on Memory Technology, Design and Testing (MTDT'04). (2004)
11. Shyu, W.L., Wu, C.S., Hou, T.C.: Efficiency analyses on routing cache replacement algorithms. In: IEEE International Conference on Communications (ICC 2002). Volume 4. (2002)
12. Wang, J.S., Li, H.Y., Chen, C.C., Yeh, C.: An AND-type match-line scheme for energy-efficient content addressable memories. In: IEEE International Solid-State Circuits Conference (ISSCC). (2005)

GPU Accelerated Smith-Waterman

Yang Liu[1], Wayne Huang[1,2], John Johnson[1], and Sheila Vaidya[1]

[1] Lawrence Livermore National Laboratory
[2] DOE Joint Genome Institute, UCRL-CONF-218814
{liu24, whuang, jjohnson, vaidya1}@llnl.gov

Abstract. We present a novel hardware implementation of the double affine Smith-Waterman (DASW) algorithm, which uses dynamic programming to compare and align genomic sequences such as DNA and proteins. We implement DASW on a commodity graphics card, taking advantage of the general purpose programmability of the graphics processing unit to leverage its cheap parallel processing power. The results demonstrate that our system's performance is competitive with current optimized software packages.

1 Introduction

Sequence comparison [1] is a fundamental tool for genome scientists to infer biological relationships from large databases of related DNA and proteins sequences. This task cannot be adequately solved by traditional string matching methods because genomic sequences that share the same biological purpose mutate over time when exposed to evolutionary events, and may no longer match identically. Genomic sequence comparison tools such as BLAST and HMMer are based upon approximate string matching principles that measure the overall similarity between strings and are thus more tolerant of mismatches.

This fuzzy string matching problem can be formulated in two different ways. The similarity between two strings can be assessed explicitly, by minimizing an ad hoc cost function (e.g. edit distance) over all possible alignments between the strings. Alternatively, a similarity score can also be computed stochastically, by finding the maximum likelihood path through a hidden Markov model (HMM) trained from an input string. Both of these approaches are optimization problems that require dynamic programming (DP) to solve. The DP step is often very computationally expensive, especially when comparing large strings. Fortunately, the data dependencies in the recurrence relations allow some degree of parallelism and the computation for some cell entries of the DP table can be distributed across a set of processors. This paper describes a parallel hardware implementation of the double affine Smith-Waterman (DASW) alignment algorithm [2][3]. DASW uses DP to find the best local alignment between two genomic sequences by optimizing a scoring function across all possible alignment arrangements, taking into consideration mismatches and gaps in either sequence to maximize the total amount of base pairings.

V.N. Alexandrov et al. (Eds.): ICCS 2006, Part IV, LNCS 3994, pp. 188–195, 2006.
© Springer-Verlag Berlin Heidelberg 2006

We chose to implement DASW on a graphics processing unit (GPU) because GPUs are cheaper and better suited for SIMD computation than conventional CPUs and more commercially available than many other special-purpose processors (i.e. ClearSpeed [4]). In fact, graphics cards can already be commonly found in many desktop and laptop computers. Moreover, we can also leverage existing visualization clusters to accelerate genomic sequence comparison between large databases, which often require days to compute on a single processor.

The NVIDIA GeForce 7800 GTX card in our system contains 24 processors with an aggregated peak compute performance of 313 GFLOPS. Unfortunately, the GPU's internal memory bandwidth of 38.4 GB/s limits the performance of most memory-bound applications (such as DASW) to around 70 GFLOPS. However, this is still very impressive, especially given that the estimated retail cost of the graphics card is only about $500. A dual-core Intel Pentium D 840 processor running at 3.2 GHz achieves roughly 25.6 GFLOPS using the SSE3 extensions and costs around $600. In comparison to CPUs, GPUs have much better cost-performance. However, GPUs are also much more difficult to program since they are designed to accelerate computer games and graphics applications. In practice it is difficult for general computational (i.e. non-graphics) tasks such as DASW to efficiently exploit data parallelism on the GPU architecture due to its strict resource constraints, limited data formats, communication overheads, and restrictive programming models. Furthermore, the GPU cannot efficiently share data with the CPU since memory transactions across the PCI-Express bus are very slow (4 GB/s) relative to the processing power and internal bandwidth of the graphics card. Intermediate data must reside completely in texture memory and be processed entirely on the GPU to avoid the bandwidth bottleneck. However, a typical commodity graphics card has only 256 MB of texture memory, which may be insufficient for applications on large data sets. Nevertheless, despite these practical challenges, many computational problems have been already implemented on the GPU, achieving on average, several times speedup over respective optimized software implementations. These performance gains are encouraging for cheap high performance computing and show the potential of GPUs as versatile math co-processors.

2 Related Work

Since genomic sequence comparisons are very expensive to compute in software, several hardware systems have been proposed to accelerate this task. ClawHMMer [5] is a streaming implementation of hmmsearch on a GPU, reporting performance at least twice as fast as the best optimized software packages. ClawHMMer implements the Viterbi algorithm, which uses DP to find the most likely path through a trained HMM network. In order to maximize throughput, ClawHMMer processes several sequence comparisons in parallel. Our system, in contrast, achieves parallelism but processes sequence comparisons one at a time to also minimize the latency of individual comparisons. TimeLogic's DeCypher card uses field programmable gate array (FPGA) chips to implement the logic for DP used by the DASW and HMMer algorithms. FPGAs are favored for their reconfigurability and fast integer arithmetic, but are also nontrivial to program. Rognes, et al [6] reported a six-fold speedup in their Smith-Waterman implementation using Intels MMX and SSE3 extensions by hand-tuning inline assembly instructions. Unfortunately, a notable drawback is that their implementation store alignment scores in 8-bit, and cannot perform long sequence comparisons where the scores are expected to exceed 255.

3 Smith-Waterman Algorithm

The Smith-Waterman algorithm computes the optimal local alignment for a pair of sequences according to a scoring system defined by a substitution matrix and gap penalty function. The substitution matrix is a symmetric matrix that assigns the cost of pairing bases together. The costs are derived from the observed substitution frequencies in alignments of related sequences. Each potential base pair is given a score representing the observed frequencies of such an occurrence in alignments of evolutionarily related sequences. This score also reflects the sample frequency of each base since some bases occur more frequently in nature than others. Identities are usually assigned the highest positive scores, frequently observed substitutions also receive positive scores, but matches that are observed to be highly unlikely are penalized by negative scores. The two most popular sets of substitution matrices for comparing long sequence are the BLOSUM and PAM matrices [7][8]. Smith-Waterman also supports gaps in the sequences at a penalty to maximize the substitution score. The parameters of the gap penalty function influence the length and frequency of gaps allowed in the alignment. There are generally three types of gap penalty functions:

$$\begin{aligned} \textbf{constant} : g(n) &= bn \\ \textbf{single affine} : g(n) &= a + bn \\ \textbf{double affine} : g(n) &= a + min(n,k)b_0 \\ &\quad +max(0, n-k)b_1 \end{aligned} \tag{1}$$

The constant gap function assigns a fixed cost to each gap space, regardless of its placement in the alignment. The single affine gap function penalizes gap creation to encourage the placement of new gap spaces to extend existing gaps rather than opening new ones. This is a more plausible model for gaps in genomic sequences since a gap of more than one space can be accounted for by a single evolutionary event. The double affine gap function extends this idea by assessing a separate penalty for each gap space that extends a gap beyond the threshold of spaces; is usually set smaller than to encourage longer gaps. We implement this function in our system since it generalizes both the constant and single affine gap functions. The optimal sequence alignment according to this scoring system is found by evaluating a set of recurrence relations over each cell of the DP table. For example, the following relations compute the optimal alignment using the single affine gap penalty function:

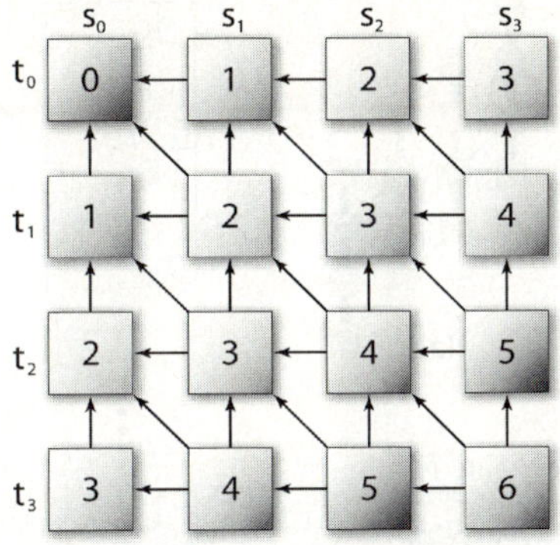

Fig. 1. Data dependency for DP computation. Cells from each diagonal are mutually independent, and depend only on cells from the previous two diagonals.

$$\begin{aligned} E_{0,j} &= E_{i,0} = D_{0,j} = D_{i,0} = I_{0,j} = I_{i,0} = 0 \\ E_{i,j} &= max\{0, E_{i-1,j-1} + match(s_i, t_j), D_{i,j}, I_{i,j}\} \\ D_{i,j} &= max\{E_{i-1,j} + a, D_{i-1,j} + b\} \\ I_{i,j} &= max\{E_{i,j-1} + a, I_{i,j-1} + b\} \\ M_{i,j} &= max\{E_{i,j}, E_{i-1,j}, E_{i,j-1}\} \end{aligned} \tag{2}$$

$\mathbf{s} = s_0 s_1 \ldots s_{m-1}$ and $\mathbf{t} = t_0 t_1 \ldots t_{n-1}$ are the two input sequences, *match* is the substitution cost matrix, and a, b are the single affine penalties for respectively opening and extending a gap. Implementing the double affine gap penalty function requires the intermediate gap lengths to be associated with $D_{i,j}$ and $I_{i,j}$ to select the appropriate penalty for b. The purpose of $M_{i,j}$ is to track the maximum alignment score for all cells (intermediate alignments) in the DP table – the final maximum alignment score will be propagated to the last cell $M_{m-1,n-1}$. Pointers for each $M_{i,j}$ are also maintained to track the cell location with the maximum alignment score and simplify the alignment trace back step.

Figure 1 illustrates the data dependencies involved in computing the recurrence relations over the DP table. In a single processor system, DP cells are processed sequentially, but a multiprocessor system can efficiently exploit the data dependencies by processing independent cells from each DP table diagonal (up to $min(m, n)$ cells) in parallel. Furthermore, the data dependencies also allow opportunities for cache optimization; only two diagonals are accessed during a computation pass so a sufficiently large LRU cache can maximize its cache coherency. With $p = min(m, n)$ processors, the DP table can be computed in $(m + n - 1)$ passes by sequentially processing each diagonal. Unfortunately, there is some efficiency loss since a few processors must stall when processing non-major diagonal. The total number of stalls in the DP computation is $p(p-1)$. However, the query sequence is commonly matched against a database of many target sequences, and the computation of the DP tables can be interleaved together to amortize the processor stalls.

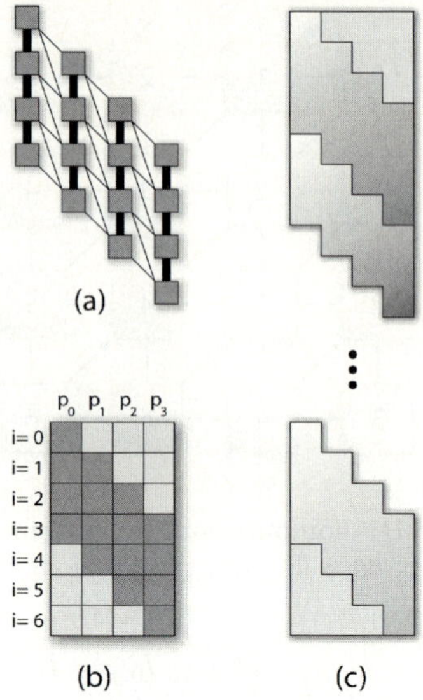

Fig. 2. The data dependency shown in (a) allows cell computations to be assigned to a set of processors as shown in (b). The wasted space can be amortized over several query sequence comparisons by connecting their DP tables together as shown in (c).

4 GPU Implementation

We implement DASW on the NVIDIA GeForce 7800 GTX graphics card using the OpenGL API, and the GL shading language (GLSL). Our implementation only involves two stages from the OpenGL rendering pipeline: geometry transformation and fragment rasterization. The geometry serves as the proxy that initializes the pipeline for the DP computation and defines the area of computation. After copying the query and target sequence data to texture memory, for each diagonal, the geometry transformation stage is passed (the vertices of) a quadrilateral that can compactly contain the DP cells of the diagonal. The dimensions of this quadrilateral must be carefully chosen to minimize wasted cells and to take advantage of any tiling optimizations on the GPU. The geometry transformation stage assigns to each constituent fragment from the quadrilateral a unique texture coordinate address and then engages the fragment rasterization stage, where the DP recurrence relations are evaluated over the fragments by a set of processors. The resulting pixel values are stored into an image buffer in texture memory, to be reused in subsequent passes. This computation loop proceeds until all diagonals have been processed. The last cell in the DP table contains the optimal alignment score $M_{m-1,n-1}$ and is retrieved from texture memory. If alignment generation is desired, then up to $(m + n - 1)$ more pixels are copied out of texture memory, comparing the intermediate values of $E_{i,j}$ with $D_{i,j}$ to build the alignment.

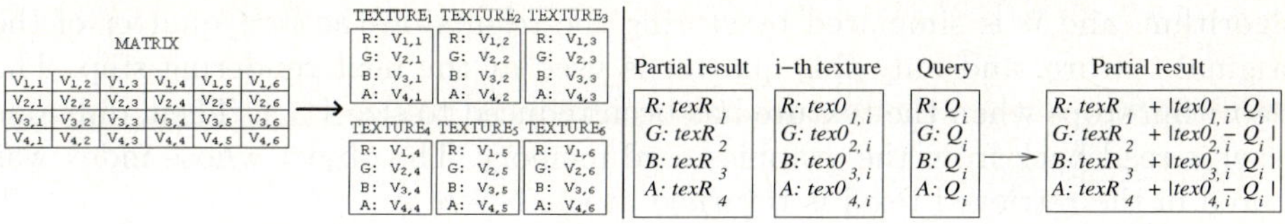

Fig. 1. Data organization in textures (left), and how does FP1 work (right)

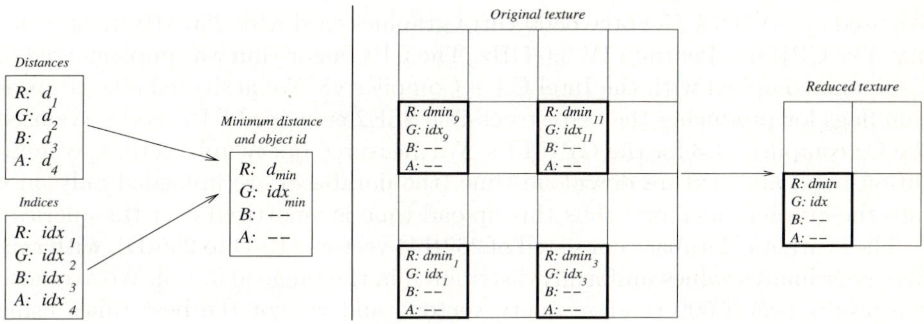

Fig. 2. Texel processing by FP2 (left) and texture reduction by FP3 (right)

distance to q is returned as the NN. We depict now how to efficiently implement this algorithm with the aid of a GPU. The first step of the algorithm is to load the vectors into the graphics card texture memory. For this purpose, we create d 2-D textures. Each of them stores one coordinate value of all vectors. As we use the four color channels to store the data, each texel from the ith texture contains the ith coordinate value of four different vectors (see Figure 1 (left)).

We use three fragment programs (FPs) to implement the linear scan. The first FP computes the distance between each object $u \in \mathbb{D}$ and the query q. As distance δ, we use the *Manhattan distance* (other metrics are possible). To fully exploit the potential of the GPU, we compute the difference between coordinates for several dimensions simultaneously. At each pass, the algorithm processes t textures (dimensions) in parallel, for a total of d/t passes. The results from previous iterations are aggregated with the results of the current pass in an additional texture tex_R (which initially contains zeros). Figure 1 (right) illustrates.

The next rendering pass determines the NN to q. The second FP computes the minimum distance value within the color+alpha channels (d_{min}) and associates this distance value to the index of its correspondent object (idx_{min}) (see Figure 2 (left)). The FP compares the four values stored in each texel, keeping the minimum value in the red channel and storing its associated index (an integer value between 1 and n) in the green channel. The third FP searches the minimum distance between four appropriately selected texels, and iteratively reduces the texture size by a factor of 4 at each pass. The minimum distance and its associated index are stored in the red and green channel, respectively (see Figure 2 (right)). This iterative reduction is the tricky part in the search algorithm, and it is simulated by storing the results on the first quarter of the original texture, and only this quarter is used at the next rendering step. The algorithm stops when the texture has been reduced to size 1×1. Then, only one texel is read back from the graphics card memory. The object whose index was stored in the retrieved texel is returned as the NN of q.

3 Experimental Results

We used an NVIDIA GeForce 6800 Ultra graphics card with 256 MByte of memory. The CPU is a Pentium IV 3.0 GHz. The CPU algorithm was implemented in C++ and compiled with the Intel C++ Compiler v8. We activated all optimization flags for producing the fastest possible SSE2 enhanced CPU code. We used the Cg compiler v1.3 for the GPU FPs. We measured query upload time, computation time, and texture download time (the databases are uploaded only once into the graphics memory, thus this upload time is amortized over the queries).

The synthetic database consisted of 262,144 vectors (16-D to 256-D), with random coordinates values uniformly distributed in the range $[0.0, 1.0]$. We averaged the results over 1,000 random query vectors, and we got the best times using $t = 8$. Figure 3 (left) shows the obtained results: The GPU algorithm achieved an average speed-up factor of 6.4x. Our algorithm also scaled well when using different database sizes. If the data did not fit into one texture, we partitioned the data in blocks of about 1 million objects and run the algorithm on each block iteratively. Figure 3 (right) shows the results for 1 to 7.5 million vectors.

We also tested our GPU algorithm using real-world databases. The first one is the *Forest CoverType* (*UCI-KDD-A*) which contains data about different forest cover types obtained by the U.S. Forest Service (54-D, 250,000 observations). The second one is the *Corel image features* (*UCI-KDD-B*), which contains features of images extracted from a Corel image collection (32-D, 65,000 images). Both sets are available at the *UCI KDD Archive* [6]. The third one is a 3D CAD database (512-D, 16,000 models). For each data set, we selected 1,000 random objects as queries for the NN search. Figure 4 (left) shows the results. We observed speed-up factors of, respectively, 6.4x, 4.5x, and 4.2x over the CPU algorithm.

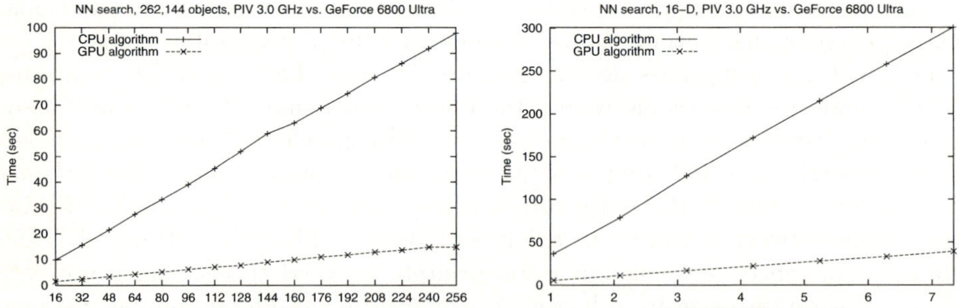

Fig. 3. Results varying dimensionality (left) and database size (right)

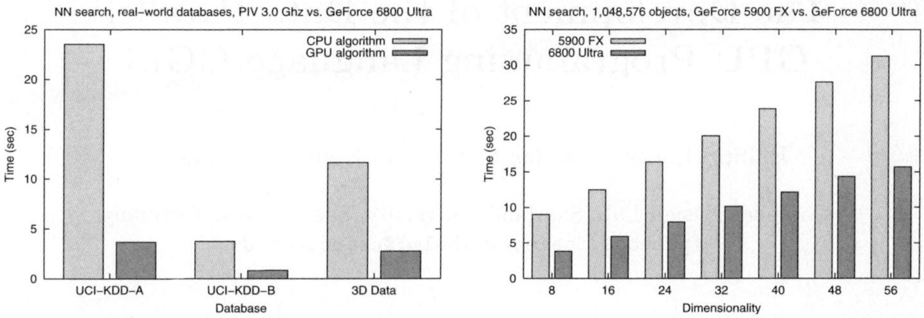

Fig. 4. Results using real data sets (left) and comparison between two GPUs (right)

Finally, we compared the GeForce 6800 Ultra card with one card from the previous generation, namely the GeForce 5900 FX, to estimate what kind of improvements one could expect for the future. With regard to hardware (pixel shaders and memory bandwidth), the GeForce 6800 Ultra should be at least twice as fast as the GeForce FX 5900. Figure 4 (right) shows the obtained results. For the next generation of GPUs, we expect a similar speed-up factor.

4 Conclusions

We presented a GPU accelerated algorithm for NN search in high-dimensional vector spaces. We described how to map vectors into texture data, without restrictions on the dimensionality of the data, and we showed that relatively simple FPs (including a texture reduction process) can efficiently return the NN object. Experimental results using synthetic and real-world data sets showed that our GPU algorithm provide a significant speed improvement over the CPU algorithm, with linear scalability in dimensionality and database size.

References

1. Sun, C., Agrawal, D., Abbadi, A.: Hardware acceleration for spatial selections and joins. In: Proc. ACM Intl. Conf. on Managment of Data. (2003) 455–466
2. Bandi, N., Sun, C., Abbadi, A., Agrawal, D.: Hardware acceleration in commercial databases: A case study of spatial operations. In: Proc. Intl. Conf. on Very Large Databases, Morgan Kaufmann (2004) 1021–1032
3. Govindaraju, N., Lloyd, B., Wang, W., Lin, M., Manocha, D.: Fast computation of database operations using graphics processors. In: Proc. ACM Intl. Conf. on Managment of Data. (2004) 215–226
4. Böhm, C., Berchtold, S., Keim, D.: Searching in high-dimensional spaces: Index structures for improving the performance of multimedia databases. ACM Computing Surveys **33**(3) (2001) 322–373
5. Beyer, K., Goldstein, J., Ramakrishnan, R., Shaft, U.: When is "nearest neighbor" meaningful? In: Proc. 7th Intl. Conf. on Database Theory. (1999) 217–235
6. Hettich, S., Bay, S.: The UCI KDD archive [http://kdd.ics.uci.edu] (1999)

The Development of the Data-Parallel GPU Programming Language CGiS

Philipp Lucas*, Nicolas Fritz*, and Reinhard Wilhelm

Compiler Design Lab, Saarland University, Saarbrücken, Germany
{phlucas, cage, wilhelm}@cs.uni-sb.de

Abstract. In this paper, we present the recent developments on the design and implementation of the data-parallel programming language CGiS. CGiS is devised to facilitate use of the data-parallel resources of current *graphics processing units (GPUs)* for scientific programming.

1 Introduction

The last few years have seen a rapid development in programmable graphics hardware. To exploit the vast computational power of these highly parallel architectures, scientists successfully ported algorithms to GPUs [5]. This is commonly known as *General Purpose Programming on GPUs (GPGPU)*.

While highly tailored solutions could be implemented by specialists, the common programmer without knowledge of the details of GPU programming was left out. For wider access, higher-level languages have emerged, such as Brook for GPUs [1], Sh [4], CGiS [2, 3] and the recent Accelerator [6].

CGiS is designed to improve GPU accessibility by further raising the abstraction level. Scientific programmers not accustomed to programming graphics hardware can transparently use performance enhancing features of the target. Generating efficient GPU code for a general purpose language program is a demanding task. The goal of the CGiS project is to explore the possibilities of high-level data-parallel programming.

This paper presents recent developments of CGiS, its compiler and its applicability with respect to [2]. By example, we also show that, even with the higher level of abstraction, parallel algorithms can be implemented efficiently on GPUs with CGiS.

The remainder of this paper is organised as follows. Section 2 describes CGiS' decisive features, and Section 3 shows a more detailed example. Section 4 concludes the paper with an outlook on future development and work in progress.

2 CGiS

In contrast to other GPU languages, CGiS offers the possibility to write a *whole algorithm* in one language. To achieve portability and performance optimisation possibilities, the compiler has full control over code and data distribution.

* Supported by DFG grant WI 576/10-3.

V.N. Alexandrov et al. (Eds.): ICCS 2006, Part IV, LNCS 3994, pp. 200–203, 2006.
© Springer-Verlag Berlin Heidelberg 2006

Because the programmer never sees the actual GPU kernels, the compiler may reorder functions into kernels for improved code distribution. The compiler can thus use upcoming hardware generations, so that CGIS programs do not have to be rewritten.

2.1 CGIS Compiler Architecture

Figure 1 shows the usage of CGIS. The CGIS compiler takes a CGIS source file as an input and generates GPU and C++ code. Together with the user-provided code which consists of calls to data passing and initialisation routines, and with the runtime library, an executable can be linked. It uses the GPU transparently. The GPU is accessed through OpenGL, and the GPU code is written in standard assembly language. Thus, the compiler supports by design several back-ends. Back-ends for multi-media CPU instruction sets are also planned.

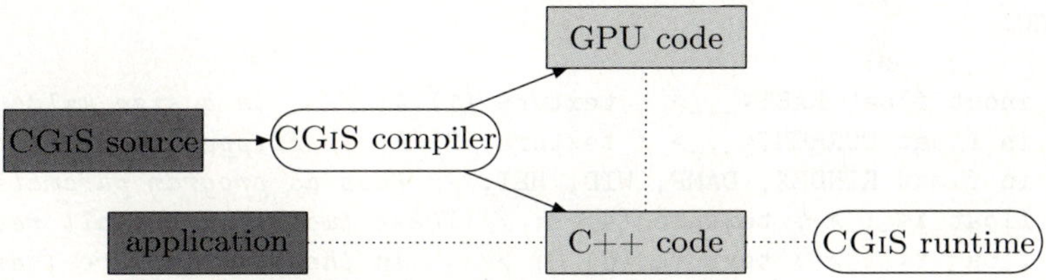

Fig. 1. Basic usage pattern of CGIS. Dotted lines denote linkage, solid arrows denote in- or output. Darker, green rectangles denote user-provided sources, the other rectangles are the output of the CGIS compiler. The ellipses stand for the CGIS base system components. There is no direct connection between the application and the GPU.

2.2 Features

CGIS is a data-parallel programming language, focused on GPUs. It allows parallel and independent computations on streams, as well as reductions, e. g. summing up the elements of a stream into a scalar. For further details, see [2].

The CGIS compiler takes care of optimisations to accommodate inexperienced GPU programmers, but advanced programmers are allowed to specify guidings and hints; for example, whether a certain conditional is to be translated with if-conversion or with native conditional instructions on architectures supporting such. The generated program drives the computation and all necessary rearrangements of data and the data interface to the main application. The GPU stays invisible to the programmer. Textures may be exposed to the outside for visualisation, or shown directly. Like other languages, CGIS excludes recursive functions and pointers because of the hardware's memory constraints.

Tests have shown that for naturally parallel algorithms, CGIS programs can offer performance benefits with respect to CPU code [3], although the quality is less than that of hand-written GPU code.

3 Example: Refraction

In this section we show how CGiS programs differ from programs in other GPU languages. This is exemplified by a program computing wave propagation and refraction on watery surfaces.

CGiS programs consist of three sections: An `INTERFACE` section declares scalar and stream variables, a `CODE` section defines functions working on single elements of streams and a `CONTROL` section defines how these functions work together. Because of the aforementioned hardware constraints and a desired closeness to C, the `CODE` section is similar to other languages. In the following, we will concentrate on the other two sections.

In the `INTERFACE` section, the programmer may provide ids to the compiler for a detailed specification of packing streams into textures. If not, the compiler

```
PROGRAM viswave;

INTERFACE

extern inout float LAST<_,_> : texture (1) A; // _ is a size wildcard.
extern in float CURRENT<_,_> : texture (2) A; // Flipped on each step.
extern in float RINDEX, DAMP, WID, HEI; // Pass as program parameters.
intern float X<_,_> : texture (4) R; // These two streams shall reside
intern float Y<_,_> : texture (4) G; //    in the same texture (id=4).
extern in float3 TEXTURE<_,_>: texture (3) RGB; // Use RGB components
extern out float3 IMAGE<_,_> : texture (5) RGB; //  for visualisation.

CODE
... // Declare kernels called from this section and from CONTROL.

CONTROL

// Single step wave propagation:
forall (float last in LAST; float current in CURRENT){
  propagate (last, current, indexX(last), indexY(last), DAMP, WID, HEI);
}
// Compute refractions in X- and Y-dimension:
forall (float x in X; float y in Y; float height in LAST){
  refractionX (RINDEX, x, height, indexX(height), WID);
  refractionY (RINDEX, y, height, indexY(height), HEI);
}
// Compute refracted image:
forall (float3 pixel in IMAGE; float height in LAST;
        float x in X; float y in Y){
  render (TEXTURE, pixel, height, x, y);
}
// Display image on screen:
show(IMAGE);
```

Fig. 2. Part of a CGiS program for calculating refractions will try to minimise the texture accesses. Another use for the ids is data passing between separately compiled programs, which is implemented by shared textures. For display on the screen, the image should reside in specific colour components, which also is specified. Scalar values are always passed as fragment program parameters.

The `INTERFACE` section gives rise to C++ functions, which the application invokes to pass and retrieve the data. All data transfer is handled by the generated code, and the GPU is invisible.

The `CONTROL` section specifies the calls of kernels which operate on streams, passing index values, stream elements or whole arrays of data. The compiler generates code to upload the kernels, hook necessary textures, run the kernels and copy the data back to textures. Again, the GPU remains invisible. This example also features a `show` statement for interactively displaying the computation results.

4 Conclusion and Future Work

We have described the CGIS language and shown that it is feasible for GPU applications. In special domains, GPU implementations of CGIS programs offer good performance, even at the current development stage of the compiler.

With the design of CGIS finished, we focus our future efforts on the compiler framework. What remains to be done is to implement more program analyses and more optimisations to the code generator, thus removing some of the overhead for more complex programs. The compiler also has to be retargeted to the most recent generation of GPUs and to SIMD-CPUs. When the compiler framework is ready, it will be released as Open Source Software under the BSD license.

References

1. I. Buck, T. Foley, D. Horn, J. Sugerman, K. Fatahalian, M. Houston, and P. Hanrahan. Brook for GPUs: Stream computing on graphics hardware. In *SIGGRAPH*, 2004.
2. N. Fritz, P. Lucas, and P. Slusallek. CGiS, a new language for data-parallel GPU programming. In *"Vision, Modeling, and Visualization" Workshop*, 2004.
3. P. Lucas, N. Fritz, and R. Wilhelm. The CGiS compiler—a tool demonstration. In A. Mycroft and A. Zeller, editors, *Proceedings of the 15th International Conference on Compiler Construction*, LNCS. Springer-Verlag, 2006.
4. M. D. McCool, Z. Qin, and T. S. Popu. Shader metaprogramming. In *Eurographics Workshop on Graphics Hardware*, pages 57–68, 2002. (Revised).
5. J. D. Owens, D. Luebke, N. Govindaraju, M. Harris, J. Krüger, A. E. Lefohn, and T. J. Purcell. A survey of general-purpose computation on graphics hardware. In *Eurographics 2005*, pages 21–51, 2005.
6. D. Tarditi, S. Puri, and J. Oglesby. Accelerator: Simplified programming of graphics processing units for general-purpose uses via data-parallelism. Technical Report MSR-TR-2005-184, Microsoft Research, December 2005.

Spline Surface Intersections Optimized for GPUs

Sverre Briseid[1], Tor Dokken[1,2], Trond Runar Hagen[1,2], and Jens Olav Nygaard[1]

[1] SINTEF, Dept. of Applied Math., P.O. Box 124 Blindern, N-0314 Oslo, Norway
[2] Centre of Mathematics for Applications (CMA), University of Oslo, Norway
{sbr, tdo, trr, jnygaard}@sintef.no
http://www.sintef.no/gpgpu

Abstract. A commodity-type graphics card with its graphics processing unit (GPU) is used to detect, compute and visualize the intersection of two spline surfaces, or the self-intersection of a single spline surface. The parallelism of the GPU facilitates fast and efficient subdivision and bounding box testing of smaller spline patches and their corresponding normal subpatches. This subdivision and testing is iterated until a prescribed level of accuracy is reached, after which results are returned to the main computer. We observe speedups up to 17 times relative to a contemporary 64 bit CPU.

1 Introduction

We can divide most surface intersections into three categories: 1) self-intersections for which the normal degenerates to zero length, 2) self-intersections for which this does not happen, and 3) intersections between two surfaces, resulting in an intersection curve, see Figure 1. There are a number of other kinds of intersections, more or less degenerate, like intersections in points, intersections with coincident parts of the surface(s), etc. See [5] for more information about intersections.

Detecting and finding intersections is particularly important in all computer aided design (CAD) systems, as well as in numerous other applications. In CAD systems, geometries are traditionally represented with the help of spline surfaces and spline curves. For an introduction to splines, see *e.g.*, [2]. With the advent of fast and programmable GPUs offering 32 bit floating point precision, a natural course of action is to use the GPU for such geometric computations.

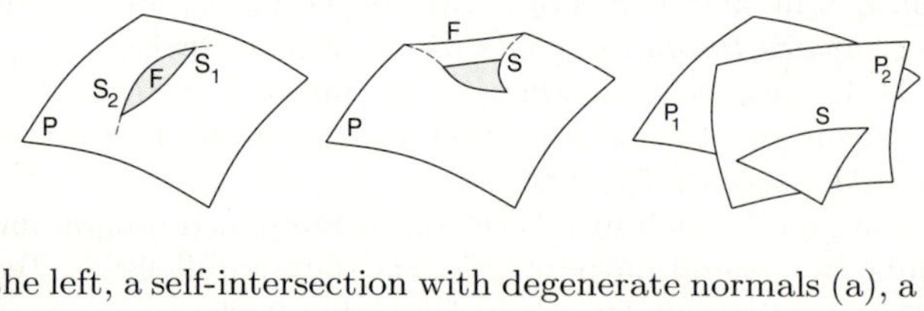

Fig. 1. From the left, a self-intersection with degenerate normals (a), a self-intersection with no degenerate normals (b) and an intersection of two surfaces resulting in a well defined intersection curve (c)

V.N. Alexandrov et al. (Eds.): ICCS 2006, Part IV, LNCS 3994, pp. 204–211, 2006.
© Springer-Verlag Berlin Heidelberg 2006

The main limitations of the GPU are the single precision arithmetics, the limited inter-process communication, and the bottleneck in the passing of results back to the main computer from the graphics card. By adapting our algorithms to the GPU, we have maneuvered around these obstacles. The processing of geometry on the GPU is not farfetched, since the GPU is designed not only to *render* geometry, but also to some degree process it, even though this is focused around the processing of triangles.

2 Background

Current CAD-technology is built on the STEP-standard (ISO 10303) from the early 1990s, and is consequently based on the computer performance at that time. Volume objects in CAD are described by the outer and possibly inner shells limiting the volume. A shell is described by a surface patchwork. Degree 1 and 2 algebraic surfaces, *i.e.*, planes, spheres, cones and cylinders are central in CAD-systems. More complex sculptured shapes are represented by piecewise rational parametric surfaces using non-uniform rational B-spline surfaces (NURBS).

Closed forms can be found for the intersection of surfaces of algebraic degrees 1 and 2. However, for intersections of surfaces of higher algebraic degree, numerical methods have to be used. CAD surface intersection until recently only worked well for transversal intersections, where the normals of the surfaces are non-parallel along the intersection curve [9]. If the normal fields of two surfaces do not intersect then Sinha's theorem [7] states that the surfaces do not have a closed intersection loop. Thus recursive subdivision can be used to create sub-surfaces in regions with potential closed intersection loops and detect these (*loop destruction*) [8]. For singular intersections where the surface normals are parallel along the intersection curve the theorem does not apply. For near singular intersections where the surface normals are near parallel along the intersection curve, the theorem often does not help much, as normal fields are in general approximated before overlap is tested [5, 6].

To better solve the singular and near singular surface intersection problems, as well as attempt to solve the surface self-intersection challenge, the idea of approximate implicitization was introduced [1]. In the EU IST FET-Open GAIA projects (2000-2005) surface intersection and self-intersection algorithms were developed following these ideas. However, performance of these new combined recursive and approximate implicit intersection algorithms is not as good as required for industrial use. Thus the idea was born to use the GPU for "naive" extensive subdivision to establish guaranteed intersection conjectures that can guide the recursion strategy used in the CPU-based algorithms.

3 Numerical Methods and Implementation on the GPU

Given a spline surface, it can contain any number and combination of the intersections in Figure 1. If we find an intersection in a subpart of the surface, this does not exclude other intersections in the same subpart. We will now discuss

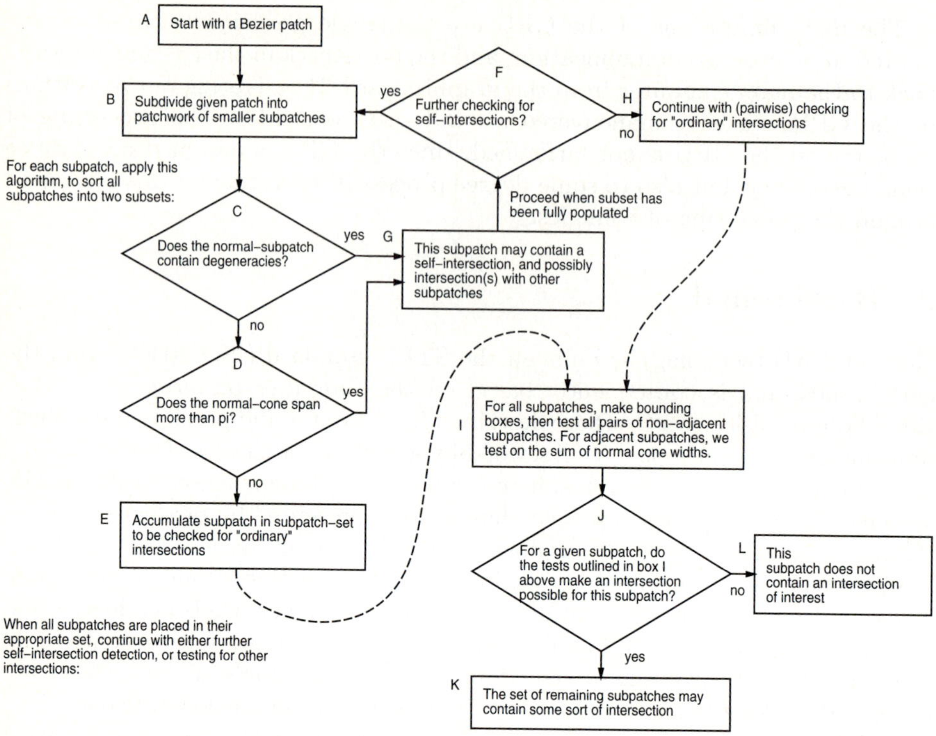

Fig. 2. The flowchart describing the processing of a Bezier-patch. Note that in stages C and D, subpatches can be processed in parallel, and that subpatches are collected in sets in E, G and H. In J, pairs of subpatches are processed in parallel, while sets of subpatches again are accumulated in K and L. Finally, any resulting subpatch in these sets may be further subdivided in A, and the procedure repeated.

how to find the various kinds of intersections, with reference to the schematic flowchart in Figure 2.

3.1 Splines and Spline Surfaces

One special case of a spline surface is the Bezier patch. A Bezier patch of degree d, or order $k=d+1$, is a linear combination of k^2 basis functions, and can therefore be represented by a $k \times k$ matrix of coefficients. A larger spline surface can be subdivided into a set of Bezier patches, each of which can be further subdivided. The coefficient matrix for a patchwork of Bezier patches is a matrix of independent $k \times k$ sub-matrices of coefficients, which makes Bezier patches convenient building blocks in our spline-based GPU-algorithms.

Since we are interested in 3D surfaces, our spline surfaces will have 3D coefficients, or if we use rational spline surfaces (NURBS) 4D coefficients. In this paper, we focus on non-rational splines, but we note that in both cases the coefficients fit nicely into the RGBA-quadruples of GPU fragments.

3.2 Knot Insertion and Subdivision

We have a spline surface $S(u,v) = \sum_{i,j=1}^{p} c_{i,j} N_i(u) N_j(v) = (\boldsymbol{N}^T C \boldsymbol{N})(u,v)$, of degree d, where $\boldsymbol{N}^T = (N_1(\cdot), \ldots, N_p(\cdot))$ consists of p B-spline functions and C is a $p \times p$ matrix of 3D coefficients. By giving all knots multiplicity k by *knot insertion*, we get a set of Bezier subpatches describing the same surface. For simplicity, and without loss of generalization, we use the same knot vectors for the two directions, resulting in the new representation

$$S(u,v) = (\boldsymbol{N}^T C \boldsymbol{N})(u,v) = (\bar{\boldsymbol{N}}^T D \bar{\boldsymbol{N}})(u,v), \tag{1}$$

where the new coefficients D are given by $D = ACA^T$, and we have the relation $\boldsymbol{N} = A\bar{\boldsymbol{N}}$ between the new and old basis functions $\bar{\boldsymbol{N}}$ and $\boldsymbol{N}$. Here, A is the *knot insertion matrix* corresponding to the insertion of the new knots. For more on splines and knot insertion, see again [2].

For S itself a Bezier patch, we would typically subdivide it into 2^{2n} new Bezier subpatches, by splitting the original knot vector interval into 2^n subintervals. We compute the accompanying knot insertion matrix A on the CPU and pass it on as a texture together with C to the GPU. The computation of the new coefficient matrix D can then be done efficiently on the GPU using two passes for the two matrix multiplications in $D = (AC)A^T$. Notice that only the second of these multiplications, AC multiplied with A^T, really does gain from the efficiency of the GPU, since the first multiplication only results in a matrix with order $\mathcal{O}(2^n)$ coefficients, while the latter produces $\mathcal{O}(2^{2n})$ coefficients. These shaders are quite simple, and have good "arithmetic to texture fetch" ratios for our cubic 3D Bezier patches.

We then test each of the subpatches for self-intersections and/or pairs of them for intersections and discard subpatches without any possible (self-)intersections.

This whole procedure can then be iterated. The result is a set of small patches which may contain intersections. These can be used as a starting point for more traditional intersection algorithms which are better suited for the CPU, or used directly for visualization. A subset of the three kinds of intersections may be computed by breaking off proper parts of the algorithm in Figure 2.

3.3 Bounding Box Tests

We look for intersections (not self-intersections) by subdividing patches and discarding subpatches that do not have rectangular bounding boxes that overlap any other than those of their neighbours. Note that the *convex hull property for splines* implies that the subpatches are contained in these boxes. This does not guarantee an intersection in the remaining subpatches, but we will never discard subpatches wrongly.

To avoid making an assumption about adjacent subpatches (in the parameter domain) not intersecting each other even though their bounding boxes always will, we do not use the bounding box test on such pairs, but instead compare the sum of their *normal cone* widths to π. The normal cone is a bounding box

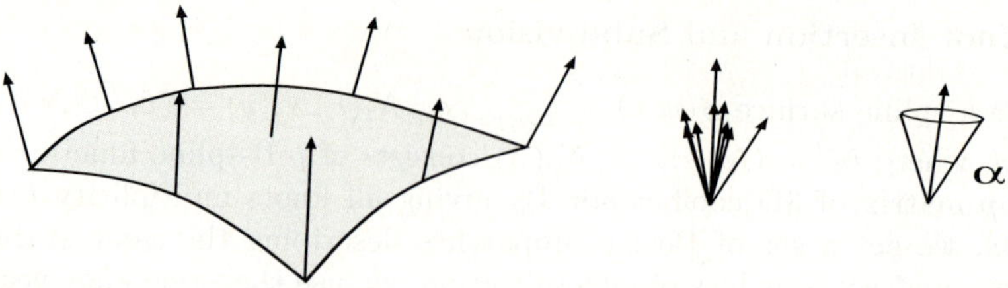

Fig. 3. From the left, a) a Bezier patch with normals indicated, b) the normals collected, and c) the normal cone with the angle/width α indicated

for the normals of a subpatch, specified as a direction and an angle, as shown in Figure 3. If the sum is smaller than π, the adjacent subpatches cannot intersect. (See Figure 2, boxes I and J.)

If we subdivide a Bezier patch into $2^n \times 2^n$ subpatches, we get 2^{4n} ordered pairs of subpatches, including both subpatches paired to themselves and pairs of neighbouring subpatches. Given the subpatches organized in a $2^n k \times 2^n k$ coefficient matrix, as produced by the knot insertion of Section 3.2, we can easily form two matrices of size $2^n \times 2^n$ containing two opposite corners of the bounding boxes of the subpatches, on the GPU.

To compute the bounding boxes, we use the depth buffer to implement a fast maximum and minimum operator to be applied to all coefficients of all the subpatches. The bounding boxes are represented as two $2^n \times 2^n$ textures, that are afterward treated as 1D textures of length 2^{2n}. These are then used to form the tensor product of all pairs, and a new texture of size $2^{2n} \times 2^{2n}$ is used to store the boolean results of all the bounding box tests. Since this matrix is symmetric, we can eliminate half of the tests by rendering a triangle rather than a quad when running the corresponding shader on the GPU.

3.4 Self-intersection Tests

The subpatches making out the full Bezier patch can be independently checked for self-intersections. We do this by checking for degenerate, or "close to degenerate" normals, and checking the width of the normal cone.

Degenerate Normals. If we take a smooth surface without any intersections, and pull a part far away from any edges out and fold it back into itself, we get a self-intersecting surface as shown in Figure 1a. Here, a fold F with the darkened underside of patch P is shown. In the two intersection points S_1 and S_2 we have degenerate normals. Note that on the sides of the protruding fold F we also have intersections like those in Figure 1b and 1c, in which the normals are not degenerate.

To test for degenerate normals, we form the *normal surface* of the patch, and subdivide this like the patch itself. The only difference is that if the surface patch has degree d, the normal surface will have degree $2d-1$. From each subpatch of the normal surface, we find a bounding box for the coefficients, thus getting a bounding box for the normals. If this bounding box does not contain the origin (given some tolerance) the corresponding surface subpatch cannot contain any degenerate normal, and we can dismiss the subpatch as not having this particular kind of self-intersection. This corresponds to the test C in Figure 2.

We compute the normal surface of the d-degree Bezier patch on the CPU, and subdivide it on the GPU as described above in Section 3.2. Since the degree is $2d-1$ we cannot use exactly the same shaders as for the Bezier patch itself, as we do not want the degree to be a variable. Rather, we use specialized shaders for each degree, but the algorithm is the same.

The test of degeneracy is then applied to this hull of all the subpatches, *i.e.*, all the coefficients of the normal subpatches are compared to a specified small tolerance. This will result in a large matrix of boolean values, which can then be treated according to how we use it for specific applications. If further subdivison is to be performed, we can remove the irrelevant subpatches, and start the iteration again on the remaining subpatches. Or these can be returned to the CPU. A third use is to simply detect whether any of the normals at all are degenerate, this can be done very quickly with occlusion culling techniques.

Normal Cone Tests. Note that if the subpatch is to contain any self-intersections, we must have a normal cone with width greater than π. This test corresponds to the one depicted as test D in the algorithm in Figure 2.

For these tests we make approximate normal cones that are not as "tight" as they could be, but faster to construct on the GPU. We average the coefficients of the normal subpatches, *i.e.*, the subdivided normal patches, for the direction of the cones. The width of each cone is computed as the maximal deviation from this average to all normals on the subpatch. This is accomplished by iterating a very simple shader, in effect averaging, or blending, triples (the directions) and storing the largest angle.

3.5 Tying It Together

In Sections 3.1 to 3.4, we have seen how we take a spline surface as input, subdivide it into a larger set of Bezier subpatches, and perform tests on subpatches and pairs of such to determine whether certain intersections can be present. Depending on what tests we want to perform, we may adapt the algorithm in Figure 2 to our purpose. If we are interested in, *e.g.*, only self-intersections of the kind producing degenerate normals, we iterate the sub-loop $ABCGF$. If we are interested in not only detecting the presence of self-intersections for which the normals do not disappear, but also finding them, we must follow a path in the flowchart ending in either E or H, followed by the subpatch-pair testing J, resulting in a set of patches in K that may contain intersections. Further processing of this set can be done either on the CPU, or the subpatches can be subdivided by iterating the algorithm from the top again. In the latter case, the currently slow returning of larger amounts of data from the GPU to the CPU can be avoided, the cost being that one must instead remove the uninteresting subpatches from the pool of such subpatches on the GPU. The authors are currently working on such an addition, following the ideas in [4].

4 Results and Conclusions

We have made use of the GPU to do computations that it was not primarily designed for. This means that the algorithms chosen are somewhat different from what have been used so far on more general computers, for the same purposes.

To fit the GPU architecture, the algorithm has been made less adaptive and more "brute force" than would be natural for a pure-CPU intersection algorithm. For comparison, we have made C-code that mimics the GPU-algorithm. This may seem to favor the GPU over the CPU, but the effect is lessened by the fact that our GPU-algorithm will be even more efficient when combined with proper controlling CPU-based code. Such code will add more adaptivity to the GPU-algorithm. Work is in progress to implement a CPU+GPU-based module that

Table 1. Timings in seconds for a selection of sizes n. For small n, computational overhead is significant, but we see convergence in the speedup factor quickly after. This also testifies to the scalability of the algorithm and code.

n	GPU	CPU	Speedup factor
4	7.456e-03	6.831e-03	0.9
5	1.138e-02	7.330e-02	6.4
6	7.271e-02	1.043e00	14.3
7	9.573e-01	1.607e01	16.8
8	1.515e01	2.555e02	16.9

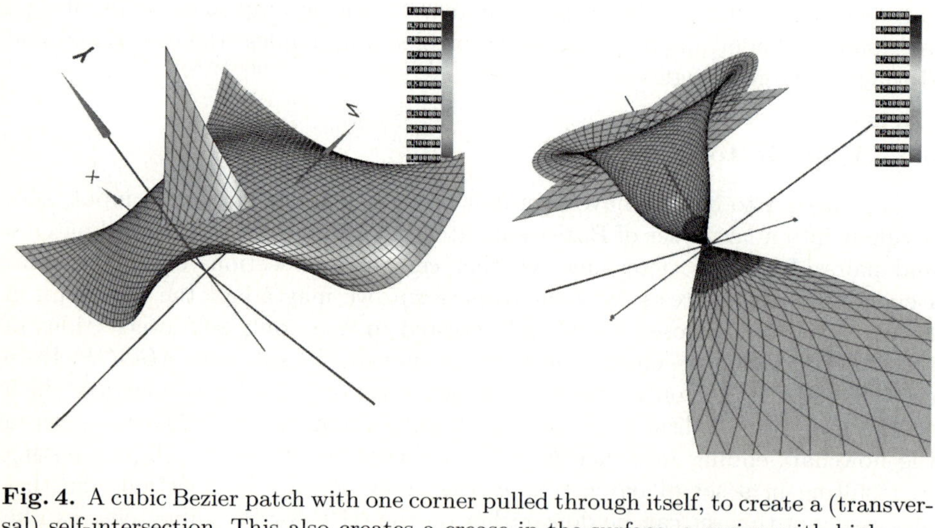

Fig. 4. A cubic Bezier patch with one corner pulled through itself, to create a (transversal) self-intersection. This also creates a crease in the surface, a region with high curvature. If the corner was pulled even more, this fold would produce a self-intersection with degenerate normals. As it is, the normals are only "near degenerate", and the region (darkened) has been detected by our GPU code, given a sufficiently high tolerance for "degeneracy". To the left, the Bezier patch, to the right, the corresponding normal surface, where the near degenerate normals appear as a region around the origin.

can be substituted for a corresponding purely CPU-based module in a widely used CAD system, giving us an even better test bed.

On an NVIDIA GeForce 7800GT graphics card, we have tested the subdivision of cubic Bezier patches into $2^n \times 2^n$ subpatches, followed by tests for degenerate normals, subdivision of the quintic normal patch into the same number of normal subpatches, computation of the approximate normal cones, the bounding boxes, and finally the bounding box pair intersections. The CPU-version (C-code) was run on an AMD X2 4400+ without threading, and is compiled with the GNU gcc compiler with optimization `-O2`. Tests with `-O3`, `-march=k8`, `-mfpmath=sse,387` and `-ffast-math` did not produce significantly different timings. Explicit vector instructions are not used. The corresponding times are listed in Table 1. An illustration of the detection of near degenerate normals in a near-self-intersection situation is shown in Figure 4.

References

1. Dokken T., Aspect of Intersection algorithms and Approximation, Thesis for the doctor philosophias degree, University of Oslo, Norway, 1997, 52-105.
2. Farin, G.: Curves and surfaces for CAGD: a practical guide. Morgan Kaufmann Publishers Inc. San Francisco, CA, USA (2002)
3. Hohmeyer, M. E., Robust and Efficient Surface Intersection for Solid Modelling, Report No. UCB/CSD 92/681, Computer Science Division, University of California, (1992).
4. Horn, D.: Stream Reduction Operations for GPGPU Applications, in *GPUGems 2 : Programming Techniques for High-Performance Graphics and General-Purpose Computation*, Addison-Wesley, (2005) 573–587
5. Patrikalakis, N.M.: Shape Interrogation for Computer Aided Design and Manufacturing. Springer-Verlag New York, Inc. Secaucus, NJ, USA (2002)
6. Sederberg, T.W. and A.K. Zundel, Pyramides that bound surface patches. CVGIP: Graphics Models and Image Processing, (1996), 75-81
7. Sinha, P., E. Klassen and K.K. Wang, Exploiting topological and geometric properties for selective subdivision. In *ACM Symposium on Computational Geometry*, ACM Press, (1985), 39-45.
8. Skytt, V., A recursive approach to surface-surface intersection, in *Mathematical Methods for Curves and Surfaces: Tromsø 2004* , M. Dæhlen, K. Mørken, and L. L. Schumaker (eds.), Nashboro Press, Brentwood, (2005), 327–338.
9. Skytt, V., Challenges in surface-surface intersections, in *Computational Methods for Algebraic Spline Surfaces (COMPASS)*, T. Dokken and B. Jüttler (eds), Springer, (2004), 11-26.

A GPU Implementation of Level Set Multiview Stereo

Patrick Labatut[1], Renaud Keriven[2], and Jean-Philippe Pons[2]

[1] Département d'Informatique, École normale supérieure,
F-75230 Paris Cedex 05, France
patrick.labatut@ens.fr

[2] CERTIS, École Nationale des Ponts et Chaussées,
F-77455 Marne-la-Vallée Cedex 2, France
keriven@certis.enpc.fr, pons@certis.enpc.fr

Abstract. Variational methods that evolve surfaces according to PDEs have been quite successful for solving the multiview stereo shape reconstruction problem since [1]. However just like every other algorithm that tackles this problem, their running time is quite high (from dozens of minutes to several hours). Fortunately graphics hardware has shown a great potential for speeding up many low-level computer vision tasks. In this paper, we present the analysis of the different bottlenecks of the original implementation of [2] and show how to efficently port it to GPUs using well-known GPGPU techniques. We finally present some results and discuss the improvements.

1 Introduction

Three-dimensional shape reconstruction from a set of pictures is one of the oldest problems in computer vision and find its roots back in robotics. Unfortunately the current state-of-the-art algorithms for reconstruction from multiple views are typically very slow and forbid a more widespread use of this technique.

A quite recent idea to improve the running time of many computer vision algorithms consists in using commodity graphics cards not for rendering fancy graphics but for general-purpose computations. We show how this approach was successful for us, allowing quality shape reconstruction within minutes.

The first section of this paper discusses previous work in the field of stereo reconstruction and general-purpose computation on GPUs (GPGPU), the next section describes the shape reconstruction algorithm we used, the following section details our implementation and the final section presents some results.

2 Related Work

2.1 Multiview Stereo Algorithms

Given n (≥ 2) images of the same scene (along with the calibration parameters of the cameras), the goal is to build a 3D model of the scene as close as possible

V.N. Alexandrov et al. (Eds.): ICCS 2006, Part IV, LNCS 3994, pp. 212–219, 2006.
© Springer-Verlag Berlin Heidelberg 2006

to the original. This goal is difficult to reach because occluded parts and lighting can substantially change the appearance of a scene from different viewpoints.

Currently multiview stereo algorithms can be very roughly divided into two classes: on one hand, discrete methods à la *space carving* derivated from [3] which work on an initialy whole discrete volume and incrementally remove chunks of voxels that do not satisfy a photo-consistency condition; on the other hand, variational methods based on the deformation of a surface under a PDE [1, 4]. The algorithm considered here [2] belongs to this latter class.

2.2 General-Purpose Computation on GPUs

In just a few years graphics cards have become heavily parallel processing machines with increased programming capabilities making their use possible for other purposes than standard real-time rendering [5]. A simplistic way to understand what a GPU actually does is to consider it as a stream processor [6] which executes a computational kernel over all the elements of an input stream (possibly accessing other streams) and puts the corresponding results into an output stream.

2.3 Stereo Vision Using Graphics Hardware

Computer vision algorithms are nowadays often GPU-accelerated, as they work on the same kind of data as rendering. Recovering the disparity map of two images has already been thoroughly studied: from simple block matching strategy with a multiscale approach [7], to mixed CPU/GPU approach initializing a graph-cut optimization with crude depth maps computed on GPU [8], and parallel dynamic programming on GPU [9]. Numerical schemes for 2D level sets have also been implemented by brute force [10] and more recently, 3D level sets for segmentation [11] introduced a GPU to CPU message passing system. However, to our knowledge, no multiview stereo algorithm for full shape reconstruction has been adapted to run on GPUs.

3 Shape Reconstruction Method

The variational method of [2] borrows from [1] but formulates the evolution of the surface as an image registration problem. It is thus simpler, more robust than most other methods and also more suitable for a GPU implementation.

3.1 Notations

A surface $S \subset \mathbb{R}^3$ models the shape of the scene being reconstructed. We note $P_i : \mathbb{P}^3 \to \mathbb{P}^2$ the projection matrices and $I_i : \Omega_i \subset \mathbb{R}^2 \to \mathbb{R}^c$ the corresponding images. S_i is the part of the surface S visible in the image I_i. $P_{i,S}^{-1} : P_i(S) \to S_i$ reprojects from the camera P_i to the surface S. Finally $I_j \circ P_j \circ P_{i,S}^{-1} : P_i(S_j) \to \mathbb{R}^c$ is the reprojection of the image I_j in the camera P_i via the surface S.

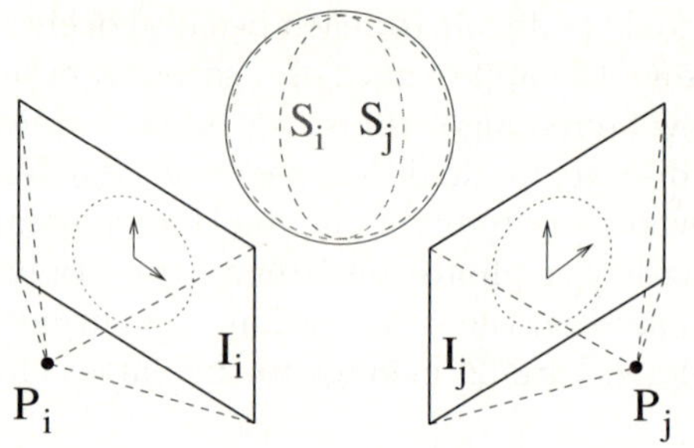

Fig. 1. Cameras setup and notations

3.2 Energy to Minimize

We wish to minimize a sum of dissimilarity terms between pairs of images: each pair is composed of one of the input images and some predicted image obtained by reprojecting another input image into the camera corresponding to the former image. This leads to the following energy (M is the dissimilarity measure between two areas of image):

$$\mathcal{M}(S) = \sum_i \sum_{j \neq i} \mathcal{M}(I_i, I_j)(S) = \sum_i \sum_{j \neq i} M|_{\Omega_i \cap P_i(S_j)} (I_i, I_j \circ P_j \circ P_{i,S}^{-1}) \quad (1)$$

The minimization of this energy results in the evolution of the surface S along its outward normal $\mathbf{N}$, driven by the equation:

$$\frac{\partial S}{\partial t} = \left[-\lambda H + \sum_i \sum_{j \neq i} \delta_{S_i \cap S_j}\, \partial_2 M\, DI_j\, DP_j\, \frac{\mathbf{d}_i}{z_i^3} \right] \mathbf{N} \quad (2)$$

where H is the mean curvature of S (which corresponds to a smoothing term added to the energy), $D\cdot$ is the Jacobian matrix of a function, $\delta_\cdot$ is the Kronecker symbol, $\mathbf{d}_i$ the vector from the camera P_i to the considered point, z_i its depth and λ a smoothing coefficient ($\lambda > 0$).

3.3 Similarity Measure

The described method allows the use of whatever similarity measure we want: cross-correlation, correlation ratio, mutual information, etc... [12]. We limited ourselves to the *local normalized cross-correlation* $\text{cc}(I_i, I_j)$ (which can accomodate even non-lambertian surfaces provided the window size is small enough):

$$\begin{aligned} \mu(I_i) &= \tfrac{G_\sigma * I_i}{\omega} & \text{v}(I_i) &= \tfrac{G_\sigma * I_i^2}{\omega} - \mu^2(I_i) + \tau^2 \\ \text{v}(I_i, I_j) &= \tfrac{G_\sigma * I_i I_j}{\omega} - \mu(I_i)\,\mu(I_j) & \text{cc}(I_i, I_j) &= \tfrac{\text{v}(I_i, I_j)}{\sqrt{\text{v}(I_i)\,\text{v}(I_j)}} \end{aligned} \quad (3)$$

where $\omega(\mathbf{x_0}) = \int_\Omega G_\sigma(\mathbf{x_0} - \mathbf{x})\, d\mathbf{x}$ is the spatial normalization to account for the shape of the correlation window, and G_σ is a gaussian kernel.

The dissimilarity measure $M^{cc}(I_i, I_j)$ between images I_i and I_j is simply the sum of the normalized cross-correlation over the whole domain: $-\int_\Omega \mathrm{cc}(I_i, I_j)(\mathbf{x})\, d\mathbf{x}$ and its partial derivative needed for the minimization is:

$$\partial_2 M^{cc}(I_i, I_j) = \alpha(I_i, I_j)\, I_i + \beta(I_i, I_j)\, I_j + \gamma(I_i, I_j) \tag{4}$$

where:

$$\begin{aligned}
\alpha(I_i, I_j) &= G_\sigma * \frac{-1}{\omega\sqrt{\mathrm{v}(I_i)\,\mathrm{v}(I_j)}} \\
\beta(I_i, I_j) &= G_\sigma * \frac{\mathrm{cc}(I_i, I_j)}{\omega\,\mathrm{v}(I_j)} \\
\gamma(I_i, I_j) &= G_\sigma * \left(\frac{\mu(I_i)}{\omega\sqrt{\mathrm{v}(I_i)\,\mathrm{v}(I_j)}} - \frac{\mu(I_j)\,\mathrm{cc}(I_i, I_j)}{\omega\,\mathrm{v}(I_j)} \right)
\end{aligned} \tag{5}$$

3.4 Energy Minimization

The minimization of the energy by gradient descent is implemented within the *level set* framework [13] and can implicitly cope with surface topology changes. However this comes at a cost and to reduce the computational burden, the *narrow band* algorithm [14] is used to evolve the level sets. As the energy is optimized through a simple steepest gradient descent, it can easily get stuck in a local minimum. The algorithm therefore adopts a multi-scale approach by using the result of the optimization at a coarser scale to initialize the optimization at a finer level.

4 Graphics Hardware Implementation

Whereas other variational methods for multiview stereo are CPU-only, [2] was designed with classical GPU acceleration in mind. We take this one step further by using GPGPU techniques.

4.1 Original Implementation Analysis

The main loop driving the evolution of the surface and executed at each time step can be decomposed as shown in Tab. 1. As all the surface points visible in image I_i should be points from the narrow band, the M^{cc} derivative is computed over the common domain of image I_i and image I_j reprojection, allowing for stream computation. Items 6 and 7 actually spend most of the time doing bilinear interpolations. The depth computations and reprojections were already running on GPU. Finally the level sets computations cover only a fraction of the running time. We thus chose to concentrate our efforts on items 5.2, 6 and 7.

4.2 Reprojection and Visibility Masks Computation

The depth is computed via conventional rendering of the surface and update of the *depth buffer*. The visibility masks $(\Omega_i \cap P_i(S_j))$ and the image reprojections are computed with the *shadow mapping* technique which consists in using the

Table 1. Main loop with typical running time distribution

Time	Step
0 %	1 mesh update
10 %	2 distance function update
5 %	3 mesh download to the GPU
5 %	4 depth computation in every camera
	5 similarity measure derivative update for each camera couple (i, j)
0 %	5.1 reprojection of the image I_j in the camera P_i
20 %	5.2 computation of the similarity measure derivative
20 %	6 computation of band points attributes for each band point / for each camera position / visibility / intensity computation
40 %	7 normal speed computation for each band point / for each camera pair if the point is visible in the two cameras 7.1 corresponding normal speed computation
0 %	8 CFL condition
0 %	9 level sets update

contents of the *depth buffer* we got from the P_j camera as a texture and rendering the surface in the camera P_i. Accessing texels in this special texture triggers a comparison between the current depth and the depth stored in the texture and returns a boolean value. The surface points are used as texture coordinates and the texture matrix (which is applied to the texture coordinates before accessing *texels*) is replaced by the P_j camera matrix. We can thus generate a depth mask using the P_j camera *depth buffer* as a texture. Then the I_j image reprojection is obtained by applying this image as a texture (see Fig. 2(a)).

4.3 Computation of the Similarity Measure Derivative

The convolutions were originally implemented with a recursive filter [15]. IIR filters do not fit very well in the GPU computational model constraints so it was replaced by a simple separable convolution. In order to mask away some of the input records in the stream, we take advantage of the efficient *Z-Culling* technique: it consists in loading a mask in the *depth buffer* that allows masked records to completely skip the execution of the computational kernel. Using this masking technique, the computation of α, β and γ from (5) easily maps to successive kernels: first compute 1, I_i, I_j, $I_i\, I_i$, $I_i\, I_j$, $I_j\, I_j$, then convolve the previous pass result with G_σ, then compute $\omega, \mu(I_i)$, $\mu(I_i)$, $\mathrm{v}(I_i)$, $\mathrm{v}(I_j)$, $\mathrm{v}(I_i, I_j), \mathrm{cc}(I_i, I_j)$ and finally convolve with G_σ to get α, β and γ.

4.4 Computation of the Points Position, Visibility, Intensity and Normal Speed

At the finest scale, the band typically contains many dozens of thousand of points. An input stream containing the coordinates of the band points is first created (as shown in Fig. 2(b)). Computational kernels iterate over the cameras,

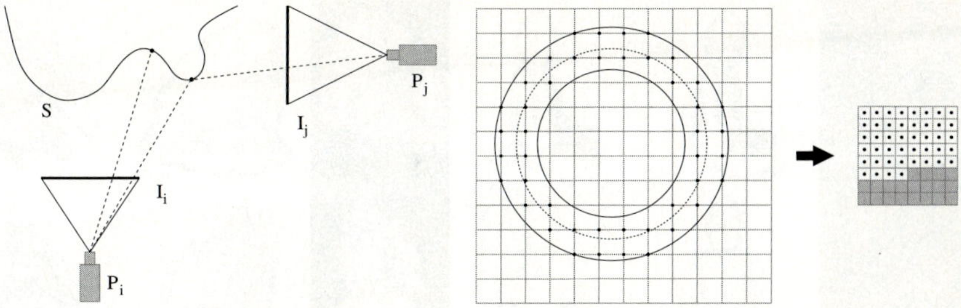

(a) shadow mapping (b) mapping the band to a 2D stream

Fig. 2. Some of the techniques used

Table 2. Input data, parameters and running times

	"buddha"[1]	"dino"[2]
#Images	25	16
Resolution	256 × 256 × 3	256 × 256 × 3
#Pairs	50	32
Level set volume	128 × 128 × 128	192 × 192 × 192
Running time (CPU/GPU)	∼ 780 s	∼ 860 s
Running time (GPU)	∼ 210 s	∼ 240 s

compute the position, visibility and intensity attributes from this input stream, and output corresponding attributes streams. *Z-Culling* is once again used to mask away some parts of the input stream where no computation needs to be done.

For the normal speed we combine visibility streams and the stream mask to generate a mask for *Z-Culling*. The camera pairs are then iterated over while accumulating the normal speeds computed for each points in the narrow band. The level sets are finally updated using this output stream.

5 Results

The graphics hardware was programmed using the OpenGL API and its extensions mechanism. The Cg programming language was also used for prototyping. All the presented results (Tab. 2 and Fig. 3) were obtained on a PC with an Intel Xeon 2.8 GHz CPU and 1 GB of system RAM equipped with an NVIDIA GeForce 7800 GTX graphics card with 256 MB of video RAM.

[1] Intel OpenLF Mapping project: `http://www.intel.com/research/mrl/research/lfm/`

[2] Multiview Stereo Evaluation project: `http://vision.middlebury.edu/mview/`

Fig. 3. "buddha" data set: first column: some input images, second column: ground truth, third column: result; "dino" data set: first column: some input images, second column: evolution (first time step of each scale), third column: result

The overall speed factor is almost about 4 when compared to the already GPU accelerated method from [2]. However the original sections of the algorithms that were using lots of bilinear interpolations observe an elevenfold improvement in general, and the computation of the measure derivative gets a ninefold decrease of its running time.

6 Conclusion

The decrease of the total running time is significative allowing shape reconstruction within minutes and it is even more impressive if we consider the CPU-only method from [1]: the hypothetical overall performance gain would be about 200.

References

[1] Faugeras, O., Keriven, R.: Complete dense stereovision using level set methods. In: European Conference on Computer Vision. Volume 1406. (1998) 379–393

[2] Pons, J.P., Keriven, R., Faugeras, O.: Modelling Dynamic Scenes by Registering Multi-View Image Sequences. International Conference on Computer Vision and Pattern Recognition **2** (2005) 822–827

[3] Kutulakos, K., Seitz, S.: A Theory of Shape by Space Carving. International Journal of Computer Vision **38**(3) (2000) 199–218

[4] Duan, Y., Yang, L., Qin, H., Samaras, D.: Shape Reconstruction from 3D and 2D Data Using PDE-based Deformable Surfaces. In: European Conference on Computer Vision. Volume 3. (2004) 238–251

[5] Owens, J.D., Luebke, D., Govindaraju, N., Harris, M., Krüger, J., Lefohn, A.E., Purcell, T.J.: A Survey of General-Purpose Computation on Graphics Hardware. In: Eurographics 2005, State of the Art Reports. (2005) 21–51

[6] Buck, I., Foley, T., Horn, D., Sugerman, J., Fatahalian, K., Houston, M., Hanrahan, P.: Brook for GPUs: Stream Computing on Graphics Hardware. ACM Transactions on Graphics **23**(3) (2004) 777–786

[7] Yang, R., Pollefeys, M.: Multi-Resolution Real-Time Stereo on Commodity Graphics Hardware. In: IEEE Conference on Computer Vision and Pattern Recognition. Volume 1. (2003) 211–218

[8] Geys, I., Koninckx, T.P., Gool, L.J.V.: Fast Interpolated Cameras by Combining a GPU based Plane Sweep with a Max-Flow Regularisation Algorithm. In: International Symposium on 3D Data Processing, Visualization and Transmission. (2004) 534–541

[9] Gong, M., Yang, Y.H.: Near Real-Time Reliable Stereo Matching Using Programmable Graphics Hardware. In: IEEE International Conference on Computer Vision and Pattern Recognition. (2005)

[10] Rumpf, M., Strzodka, R.: Level Set Segmentation in Graphics Hardware. In: IEEE International Conference on Image Processing. Volume 3. (2001) 1103–1106

[11] Lefohn, A.E., Kniss, J.M., Hansen, C.D., Whitaker, R.T.: A Streaming Narrow-Band Algorithm: Interactive Deformation and Visualization of Level Sets. IEEE Transactions on Visualization and Computer Graphics **10**(40) (2004) 422–433

[12] Hermosillo, G., Chefd'hotel, C., Faugeras, O.: Variational Methods for Multimodal Image Matching. International Journal of Computer Vision **50**(3) (2002) 329–343

[13] Osher, S., Sethian, J.A.: Fronts Propagating with Curvature-Dependent Speed: Algorithms Based on Hamilton-Jacobi Formulations. Journal of Computational Physics **79**(1) (1988) 12–49

[14] Adalsteinsson, D., Sethian, J.A.: A Fast Level Set Method for Propagating Interfaces. Journal of Computational Physics **118**(2) (1995) 269–277

[15] Deriche, R.: Fast Algorithms for Low-Level Vision. IEEE Transactions on Pattern Analysis and Machine Intelligence **1**(12) (1990) 78–88

Solving the Euler Equations on Graphics Processing Units

Trond Runar Hagen[1,2], Knut-Andreas Lie[1,2], and Jostein R. Natvig[1,2]

[1] SINTEF, Dept. Applied Math., P.O. Box 124 Blindern, N-0314 Oslo, Norway
[2] Centre of Mathematics for Applications (CMA),University of Oslo, Norway
{trr, knl, jrn}@sintef.no
http://www.sintef.no/gpgpu

Abstract. The paper describes how one can use commodity graphics cards (GPUs) as a high-performance parallel computer to simulate the dynamics of ideal gases in two and three spatial dimensions. The dynamics is described by the Euler equations, and numerical approximations are computed using state-of-the-art high-resolution finite-volume schemes. These schemes are based upon an explicit time discretisation and are therefore ideal candidates for parallel implementation.

1 Introduction

Conservation of physical quantities is a fundamental physical principle that is often used to derive models in the natural sciences. In this paper we will study one such model, the Euler equations describing the dynamics of an ideal gas based on conservation laws for mass, momentum, and energy. In three spatial dimensions the Euler equations read

$$\begin{bmatrix} \rho \\ \rho u \\ \rho v \\ \rho w \\ E \end{bmatrix}_t + \begin{bmatrix} \rho u \\ \rho u^2 + p \\ \rho uv \\ \rho uw \\ u(E+p) \end{bmatrix}_x + \begin{bmatrix} \rho v \\ \rho uv \\ \rho v^2 + p \\ \rho vw \\ v(E+p) \end{bmatrix}_y + \begin{bmatrix} \rho w \\ \rho uw \\ \rho vw \\ \rho w^2 + p \\ w(E+p) \end{bmatrix}_z = \begin{bmatrix} 0 \\ 0 \\ 0 \\ g\rho \\ g\rho w \end{bmatrix}. \tag{1}$$

Here ρ denotes the density, (u, v, w) the velocity vector, p the pressure, g the acceleration of gravity, and E the total energy (kinetic plus internal energy) given by $E = \rho(u^2 + v^2 + w^2)/2 + p/(\gamma - 1)$. In all computations we use $\gamma = 1.4$. The Euler equations are one particular example of a large class of equations called *hyperbolic systems of conservation laws*, which can be written on the form

$$Q_t + F(Q)_x + G(Q)_y + H(Q)_z = S(Q). \tag{2}$$

This class of PDEs exhibits very singular behaviour and admits various kinds of discontinuous and nonlinear waves, such as shocks, rarefactions, phase boundaries, fluid and material interfaces, etc. Resolving propagating discontinuities accurately is a difficult task, to which a lot of research has been devoted in the

V.N. Alexandrov et al. (Eds.): ICCS 2006, Part IV, LNCS 3994, pp. 220–227, 2006.
© Springer-Verlag Berlin Heidelberg 2006

last 2–3 decades. Today, a successful numerical method will typically be of the high-resolution type (see e.g., [3]) and be able to accurately capture discontinuous waves and at the same time offer high-order resolution of smooth parts of the solution.

Modern high-resolution methods for nonstationary problems are typically based upon explicit temporal discretisation. In explicit methods there is *no* coupling between unknowns in different grid cells, and one therefore avoids the use of linear system solvers, which is a typical bottleneck in many fluid dynamics algorithms. High-resolution methods are therefore relatively easy to parallelise, using e.g., domain decomposition. In this paper we will discuss parallel implementation of gas-dynamics simulations on commodity graphics cards (GPUs) residing in recent desktop computers and workstations. The idea of using GPUs for numerical simulation is far from new—cf. e.g., `http://www.gpgpu.org`—but except for our previous work [1] on shallow water waves, this is the first paper to consider a GPU implementation of high-resolution schemes for models on the form (2).

From a computational point-of-view, a modern GPU can be considered as a single-instruction, multiple-data processor capable of parallel processing of floating-point numbers. Whereas an Intel Pentium 4 CPU has a theoretical performance of at most 15 Gflops, performance numbers as high as 165 Gflops have been reported for the NVIDIA GeForce 7800 cards. The key to this unrivalled processing power is the fact that current GPUs contain up to 24 parallel pipelines that each are capable of processing vectors of length four simultaneously. By exploiting this amazing computational power for 2D and 3D gas-dynamics simulations, we observe speedup factors of order 10—20 on a single workstation.

2 Numerical Methods

To solve the Euler equations in two and three dimensions, we will use a family of semi-discrete finite-volume schemes on a regular Cartesian grid and seek approximations to (2) in terms of the cell-averages $Q_{ijk} = \frac{1}{|\Omega_{ijk}|}\int_{\Omega_{ijk}} Q\,dV$. Integrating (2) over Ω_{ijk}, we obtain an evolution equation for the cell-averages

$$\begin{aligned}\frac{dQ_{ijk}}{dt} &= -\big(F_{i+1/2,jk} - F_{i-1/2,jk}\big) - \big(G_{i,j+1/2,k} - G_{i,j-1/2,k}\big) \\ &\quad - \big(H_{ij,k+1/2} - H_{ij,k-1/2}\big) + S_{ijk},\end{aligned} \tag{3}$$

where $F_{i\pm1/2,jk}$ denote the flux over the surfaces with normal along the x-axis, etc. The fluxes are approximated using a standard Gaussian quadrature (fourth order, tensor product rule):

$$\begin{aligned}F_{i+1/2,jk}(t) &= \frac{1}{|\Omega_{ijk}|}\int_{y_{j-1/2}}^{y_{j+1/2}}\int_{z_{k-1/2}}^{z_{k+1/2}} F\big(Q(x_{i+1/2},y,z,t)\big)\,dydz \\ &\approx \frac{1}{4\Delta x}\sum_{n,m=\{-1,1\}} F\Big(Q\Big(x_{i+1/2},\, y_j + n\frac{\Delta y}{2\sqrt{3}},\, z_k + m\frac{\Delta z}{2\sqrt{3}},\, t\Big)\Big).\end{aligned} \tag{4}$$

To evaluate the integrand, we need to *reconstruct* a continuously defined function from the cell-averages. To this end, we will use a function that is piecewise continuous inside each grid cell. In a first-order method, one would use a piecewise constant function. To obtain second-order (on smooth solutions), we use a piecewise linear reconstruction for each component in Q

$$\begin{aligned}\hat{Q}_{ijk}(x,y,z) = Q_{ij} + L\big(D_x^+ Q_{ijk}, D_x^- Q_{ijk}\big)\frac{x-x_i}{\Delta x} \\ + L\big(D_y^+ Q_{ijk}, D_y^- Q_{ijk}\big)\frac{y-y_j}{\Delta y} + L\big(D_z^+ Q_{ijk}, D_z^- Q_{ijk}\big)\frac{z-z_k}{\Delta z},\end{aligned} \tag{5}$$

where $D_x^{\pm} = \pm(Q_{i\pm 1,jk} - Q_{ijk})$, etc. The so-called limiter L is a nonlinear function of the forward and backward differences, whose purpose is to prevent the creation of overshoots at local extrema. Here we use the family of generalised minmod limiters

$$L(a,b) = \mathrm{MM}(\theta a, \tfrac{1}{2}(a+b), \theta b), \quad \mathrm{MM}(z_1,\ldots,z_n) = \begin{cases}\max_i z_i, & z_i < 0\ \forall i,\\ \min_i z_i, & z_i > 0\ \forall i,\\ 0, & \text{otherwise.}\end{cases} \tag{6}$$

The reconstruction $\hat{Q}(x,y,z)$ is discontinuous across all cell-interfaces, and thus gives a left-sided and right-sided point value, Q_L and Q_R, at each integration point in (4). To evaluate the flux across the interface at each integration point, we use the central-upwind flux [2]

$$\begin{aligned}\mathcal{F}(Q^L,Q^R) &= \frac{a^+F(Q^L) - a^-F(Q^R)}{a^+ - a^-} + \frac{a^+a^-}{a^+ - a^-}(Q^R - Q^L),\\ a^+ &= \max\big(0, \lambda^+(Q^L), \lambda^+(Q^R)\big), \quad a^- = \min\big(0, \lambda^-(Q^L), \lambda^-(Q^R)\big),\end{aligned} \tag{7}$$

where $\lambda^{\pm}(Q)$ are the slow and fast eigenvalues of dF/dQ, given analytically as $u \pm \sqrt{\gamma p/\rho}$.

Finally, we need to specify how to integrate the ODEs (3) for the cell-averages. To this end, we use a second-order TVD Runge–Kutta method [5]

$$\begin{aligned}Q_{ij}^{(1)} &= Q_{ij}^n + \Delta t R_{ij}(Q^n),\\ Q_{ij}^{n+1} &= \frac{1}{2}Q_{ij}^n + \frac{1}{2}\big[Q_{ij}^{(1)} + \Delta t R_{ij}(Q^{(1)})\big],\end{aligned} \tag{8}$$

where R_{ij} denotes the right-hand side of (3). The time step is restricted by a CFL-condition, which states that disturbances can travel at most one half grid cell each time step, i.e., $\max(a^+, -a^-)\Delta t \le \Delta x/2$, and similarly in y and z.

3 GPU Implementation

The data-driven programming model of GPUs is quite different from the instruction-driven programming model most people are used to on a CPU. On a CPU,

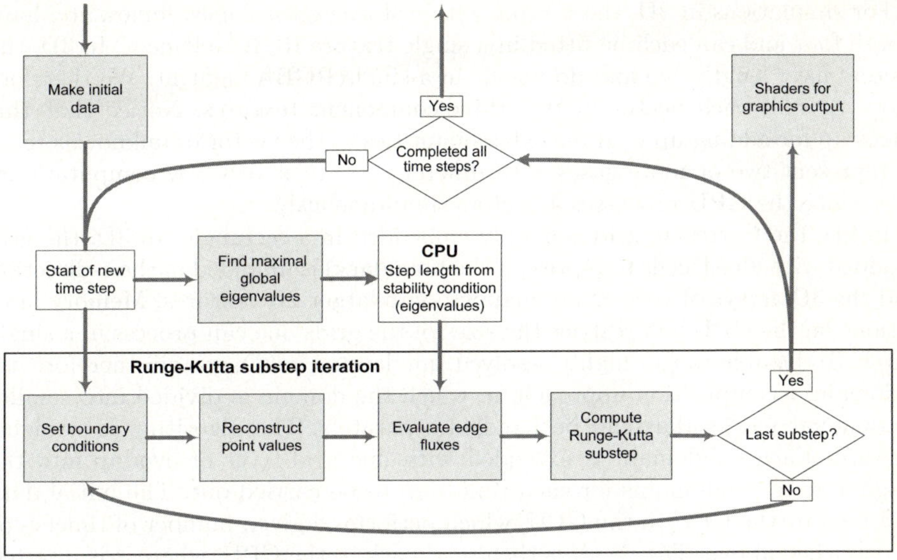

Fig. 1. Flow chart for the GPU implementation of the semi-discrete finite-volume scheme. Gray boxes are executed on the GPU and white boxes on the CPU.

a computer program for the algorithm in Section 2 would consist of a set of arrays and the processing is performed as series of loops that march through all cells to compute reconstructions, integrate fluxes, compute flux differences and evolve the ODEs, etc. On the GPU, each grid cell is associated with a *pixel* (*fragment*) in an off-screen frame buffer. The data stream (cell-averages, fluxes, etc.) is given as *textures* and is invoked by rendering a geometry to a frame buffer. The data stream is processed by a series of kernels (fragment shaders, in graphics terminology) using the fragment-processing capabilities in the rendering pipeline. Writing each computational kernel using Cg (or GLSL) is straightforward for any computational scientist capable of writing C/C++. However, setting up the graphics pipeline requires some familiarity with computer graphics (in our case, OpenGL).

The flow chart for the simulation algorithm is given in Figure 1. Worth noting is the computation of the maximum eigenvalues to determine the time step. Finding the maximum is implemented using an 'all-reduce' operation utilising the depth buffer combined with a read-back to the CPU; see e.g., [1]. In each of the two Runge–Kutta steps, four basic operations are performed. First we set the boundary data. Then we compute the reconstruction using (5) and (6). The most computationally intensive step is the evaluation of edge fluxes and computation of the source term. Before this calculation can start, the time step Δt must be passed to the shader by the CPU. Finally, the step is completed by adding fluxes and the source term to the cell averages. To complete a full time step, the sequence of operations in the Runge–Kutta box are performed twice.

For simulations in 2D, the vectors with cell-averages, slopes, fluxes, etc. have length four and can each be fitted in a single texture RGBA-element. In 3D, the vectors have length five and do not fit in a single RGBA-element. We therefore chose to split each vector in two three-component textures. Notice that this opens up for adding up to three extra quantities in the vector of unknowns (e.g., to represent two or more gases with different γ's) at a very low computational cost, since the GPU processes 4-vectors simultaneously.

In 2D, the Cartesian grid is simply embedded in a rectangle. In 3D, the grid (padded with ghost-cells to represent the boundary) is unfolded in the z-direction and the 3D arrays of vectors are mapped onto larger 2D textures. Memory limitations on the GPU will restrict the sizes of the grids one can process in a single batch. To be able to run highly resolved simulations in 3D, we will therefore use a domain decomposition approach, in which the domain is divided into smaller rectangular blocks that can be handled separately. The algorithm is straightforward: Each subdomain is extended with one grid-layer of overlap into the neighbouring subdomains for each time-step to be carried out. The initial data is passed by the CPU to the GPU, which performs a given number of time-steps as described above. The result is then read back to the CPU, where it is inserted into the corresponding subdomain in the global solution on the next (global) time level. Since passing of initial data and read-back of computational results can be performed asynchronously between the CPU and the GPU, the performance reduction due to stalls on the GPU will be insignificant. Moreover, this algorithm easily extends to multiple CPU-GPU configurations by implementing some kind of message passing and control on the CPU-side.

4 Numerical Examples

In the following we present a few numerical examples to assess the computational efficiency of our GPU implementation. To this end, we compare runtimes on two NVIDIA GeForce graphics cards (6800 Ultra and 7800 GTX) with runtimes on two different CPUs (a 2.8 GHz Intel Xeon CPU and an AMD Athlon X2 4400+, respectively). The timings are averaged over all timesteps and do not include any preprocessing. The CPU reference codes are implemented in C, using a design that has evolved during 6–7 years research on high-resolution schemes. High computational efficiency has been ensured by carefully minimising the number of arithmetic operations, optimal ordering of loops, use of temporary storage, replacing divisions by multiplications whenever possible, etc. Apart from that, our CPU codes contain no hardware-specific hand-optimisation, but rather rely on general compiler optimisation; `icc -O3 -ipo -xP` (version 8.1) for the Intel CPU and `gcc -O3` for the AMD. To ensure a fair comparison, we have used the same design choices for the GPU implementations, trying to retain a one-to-one correspondence of statements in the CPU and GPU computational kernels.

Another important question is accuracy. The numerical methods considered in the paper are stable and the accuracy will therefore not deteriorate significantly due to rounding errors. In fact, all our tests indicate that the difference between

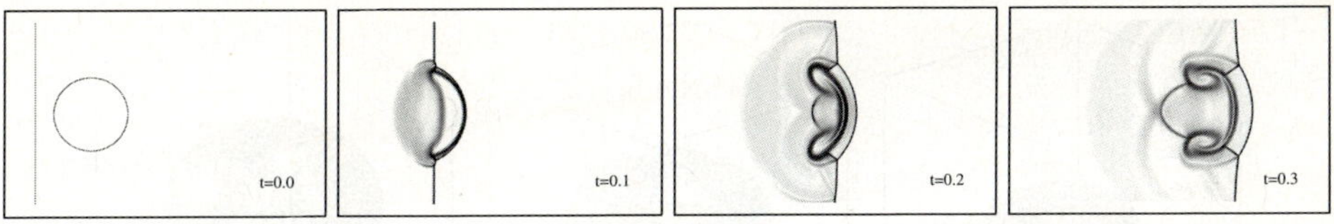

Fig. 2. Emulated Schlieren images of a shock-bubble interaction

Table 1. Runtime per time step in seconds and speedup factor for two CPUs versus two GPUs for the shock-bubble problem run on a grid with $N \times N$ cells for bilinear (upper part) and CWENO reconstruction (lower part)

N	Intel	6800	speedup	AMD	7800	speedup
128	4.37e-2	3.70e-3	11.8	1.88e-2	1.38e-3	13.6
256	1.74e-1	8.69e-3	20.0	1.08e-1	4.37e-3	24.7
512	6.90e-1	3.32e-2	20.8	2.95e-1	1.72e-2	17.1
1024	2.95e-0	1.48e-1	19.9	1.26e-0	7.62e-2	16.5
128	1.05e-1	1.22e-2	8.6	7.90e-2	4.60e-3	17.2
256	4.20e-1	4.99e-2	8.4	3.45e-1	1.74e-2	19.8
512	1.67e-0	1.78e-1	9.4	1.03e-0	6.86e-2	15.0
1024	6.67e-0	7.14e-1	9.3	4.32e-0	2.99e-1	14.4

single precision (GPU) and double precision (CPU) results are of the order ϵ_s, where $\epsilon_s = 1.192 \cdot 10^{-7}$ is the smallest number such that $1 + \epsilon_s - 1 > 0$ in single precision. In other words, for the applications considered herein, the *discretisation* errors dominate errors due to lack of precision.

Example 1 (2D Shock-Bubble Interaction). In this example we consider the interaction of a planar 2.95 Mach shock in air with a circular region of low density. The gas is initially at rest and has unit density and pressure. Inside a circle of radius 0.2 centred at (0.4, 0.5) the density is 0.1. The incoming shock-wave starts at $x = 0$ and has a post-shock pressure $p = 10.0$. Figure 2 shows the evolution of the bubble in terms of emulated Schlieren images (density gradients depicted using a nonlinear graymap) as described by the 2D Euler equations, i.e., (1) with $g = w \equiv 0$.

Table 1 reports a comparison of average runtime per time step for GPU versus CPU implementations of the high-resolution scheme using either the bilinear reconstruction in (5) and (6), or the third-order CWENO reconstruction [4]. The corresponding schemes thus have second-order accuracy in time and second and third order accuracy in space, respectively. For the bilinear reconstruction, the resulting speedup factors of order 20 and 15 for the GeForce 6800 and 7800, respectively, are quite amazing since we did not try to optimise the GPU implementation apart from the obvious use of vector operations whenever appropriate.

The CWENO reconstruction is quite complicated, and we have observed on various Intel CPUs that `icc` gives significantly faster code than `gcc`. We therefore

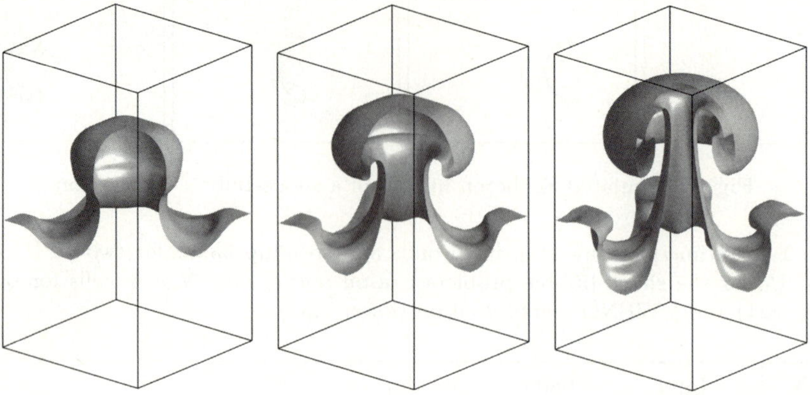

Fig. 3. The Rayleigh–Taylor instability at times $t = 0.5$, 0.6, and 0.7

Table 2. Runtime per time step in seconds and speedup factor for CPU versus GPU for the 3D Rayleigh–Taylor instability run on a grid with $N \times N \times N$ cells

N	AMD	7800	speedup
49	5.23e-1	4.16e-2	12.6
64	1.14e-0	8.20e-2	13.9
81	1.98e-0	1.72e-1	11.5

expect the CWENO code to be suboptimal on the AMD CPU. Moreover, due to the large number of temporary registers required in the CWENO reconstruction, we had to split the computation of edge fluxes in two passes on the GPU: one for the F-fluxes and one for the G-fluxes. This introduces extra computations compared with the CPU code and also extra render target switches and texture fetches. This reduces the theoretical speedup by factor between 25 and 50%, as can be seen in the lower half of Table 1, comparing the 6800 card and the Intel CPU with the `icc` compiler.

Example 2 (3D Rayleigh–Taylor Instability). In the next example we simulate a Rayleigh–Taylor instability, which arises when a layer of heavier fluid is placed on top of a lighter fluid and the heavier fluid is accelerated downwards by gravity. Similar phenomena occur more generally when a light fluid is accelerated towards a heavy fluid. In the simulation, we consider the domain $[-1/6, 1/6]^2 \times [0.2, 0.8]$ with gravitational acceleration $g = 0.1$ in the z-direction. The lower fluid has unit density and the upper fluid density $\rho = 2.0$. Initially the two fluids are at rest, in hydrostatic balance, and separated by an interface located at $z = 1/2 + 0.01\cos(6\pi \min(\sqrt{x^2 + y^2}, 1/6))$. Reflective boundary conditions are assumed on all exterior boundaries. Figure 3 shows the evolution of the instability.

Table 2 reports a comparison of average runtime per time step for a GPU versus a CPU implementation for the high-resolution scheme with the trilinear reconstruction in (5) and (6). Compared with the 2D simulation, the speedup is reduced. There are two factors contributing the the reduced speedup: (i) cache misses on the GPU due to lookup in the unfolded 3D texture, and (ii) use of only three out of four vector components in all basic arithmetic operations. For the 2D solver, all texture fetches are to neighbouring texel locations, whereas the access in the z-direction introduces non-local texture fetches. Similarly, the 2D solver uses a single four-component texture to represent the conserved quantities, whereas the 3D solver needs to use two three-component textures: $(\rho, \rho u, \rho v, \rho w, E)$, plus a passive tracer to distinguish the gases.

5 Concluding Remarks

In this paper we have demonstrated the application of GPUs as high-performance computational engines for compressible gas dynamics simulations in 2D and 3D. Unlike many other fluid dynamics algorithms, the current high-resolution schemes do not involve any linear system solvers, which may be a performance bottleneck in many GPU/parallel implementations. Instead, the schemes are based upon explicit temporal discretisation, for which each cell can be updated independent of the others. This makes the schemes perfect candidates for parallel implementation. Moreover, a high number of arithmetic operations per memory fetch makes these algorithms ideal for high performance data-stream based computer architectures and results in fairly amazing speedup numbers.

The possibility of using a simple domain-decomposition algorithm to handle large simulation models makes it attractive to explore future use of clusters of CPU–GPU nodes for this type of simulations. The communication need between different nodes is low compared with the computations performed on each GPU, and communication bandwidth is therefore not expected to be a major issue.

Acknowledgement

The research is funded by the Research Council of Norway under grants number 158911/I30 (Hagen and Lie) and 139144/431 (Natvig).

References

1. Hagen, T.R., Hjelmervik, J.M., Lie, K.-A., Natvig, J.R., Henriksen, M.O.: Visual simulation of shallow-water waves. Simul. Model. Pract. Theory, **13** (2005) 716–726.
2. Kurganov, A, Noelle, S., Petrova, G.: Semidiscrete central-upwind schemes for hyperbolic conservation laws and Hamilton–Jacobi equations. SIAM J. Sci. Comput. **23 (3)** (2001) 707–740.
3. LeVeque, R: Finite volume methods for hyperbolic problems, Cambridge Texts in Applied Mathematics, Cambridge University Press, Cambridge, 2002.
4. Levy, D., Puppo, G., Russo. G.:Compact central WENO schemes for multidimensional conservation laws. SIAM J. Sci. Comput. **22(2)** (2000) 656–672.
5. Shu, C.-W.: Total-variation-diminishing time discretisations. SIAM J. Sci. Stat. Comput. **9** (1988) 1073–1084.

Particle-Based Fluid Simulation on the GPU

Kyle Hegeman[1], Nathan A. Carr[2], and Gavin S.P. Miller[2]

[1] Stony Brook University
kyhegem@cs.sunysb.edu
[2] Adobe Systems Inc.
{ncarr, gmiller}@adobe.com

Abstract. Large scale particle-based fluid simulation is important to both the scientific and computer graphics communities. In this paper, we explore the effectiveness of implementing smoothed particle hydrodynamics on the streaming architecture of a GPU. A dynamic quadtree structure is proposed to accelerate the computation of inter-particle forces. Our method readily extends to higher dimensions without undue increase in memory or computation costs. We show that a GPU implementation runs nearly an order of magnitude faster than our CPU version for large problem sizes.

1 Introduction

In computer graphics, particles are a popular primitive for the simulation and rendering of numerous effects including cloth, water, steam, smoke, and fire [1, 2, 3]. Producing high quality liquid animations can require hundreds of thousands of particles and take several minutes to compute a single frame [4]. This requirement to scale to large problem sizes has led us to investigate the use of graphics hardware to accelerate the computations. A challenging aspect of this problem is determining all the neighbors within a radius of a given particle, a common operation during the evaluation of inter-particle forces. For CPU based implementations, a simple and effective approach is to place particles in buckets on a uniform grid. Each cell in the grid is the size of the neighborhood radius. To find neighbors for a given particle, only particles within its bucket and buckets of neighboring grid cells need to be considered. Unfortunately, on current GPU architectures it is difficult to maintain a dynamic list of particles. In this paper, we propose an alternative method that uses a hierarchical tree structure to accelerate these queries. We describe algorithms for constructing, maintaining, and evaluating a quadtree data structure in graphics hardware. These quadtree data structures are efficient to traverse and recompute as the particle configuration evolves over time.

2 Previous Work

Recent advances in technology have enabled a host of new applications to profit from the floating point power and bandwidth available in graphics hardware [5]. Several papers [6, 7] have demonstrated that particle systems can be effectively mapped to GPUs. Chiara et al. [8] implement a flocking behavior model with much of the computation being performed on the GPU, the CPU is used to compute neighbor lists for particles. Rather than finding neighbors explicitly, Kolb and Cuntz [9] implement a particle based

V.N. Alexandrov et al. (Eds.): ICCS 2006, Part IV, LNCS 3994, pp. 228–235, 2006.
© Springer-Verlag Berlin Heidelberg 2006

fluid model model by splatting kernels in image space. This method is less accurate than our proposed approach since it involves discretizing kernels onto a grid.

The use of tree data structures such as kd-trees, octrees, and hierarchical bounding volume (HBV) trees in graphics hardware is still relatively new and has been employed for GPU ray-tracing [10, 11], as well as 3D paint [12, 13]. Tree based schemes in graphics hardware are severely limited due to the lack of recursion and register indexing within fragment programs. For this reason image based algorithms have been used on the GPU to accelerate collision detection [14, 15, 16]. Tree based algorithms are a popular means to perform collision detection on the CPU [17, 18]. Existing work with HBV trees on GPUs has focused on efficient tree evaluation, but has relegated tree construction and maintenance to the CPU. We have chosen a fixed topology HBV tree for our implementation. This structure has the nice property that it can be traversed and maintained on the graphics card. We can extend this structure to handle particle-surface interaction by adapting the CPU geometry image collision detection approach of [19] to run entirely in the graphics hardware. Our work is most closely aligned to that of Simonsen et al. [11] who noted that HBV trees map very well onto graphics hardware.

3 Dynamic Quadtrees on the GPU

In this section we describe our algorithms for building and traversing an HBV quadtree on the GPU. A key contribution of this paper is the method for constructing and maintaining the tree in a dynamic simulation. Previous work on GPU acceleration structures rely on the CPU to perform this construction; only the traversal stage is implemented on the GPU.

3.1 Quadtree Storage

Particles are arranged in a 2D array with a resolution of $N \times N$. These particles form the leaves of a quadtree. One level higher in the tree there are $\frac{N}{2} \times \frac{N}{2}$ nodes. Each node has 4 children. The next level is $\frac{N}{4} \times \frac{N}{4}$, etc. until the root of the tree is reached. These levels of the tree form a pyramid. A key observation is that each level of the tree can be computed as a simple reduction operation using data from a lower level. This operation finds the smallest sphere that contains the 4 child spheres. Computation of such reduction pyramids is a common step in graphics algorithms, for example in texture mipmap construction.

Given a tree with fixed depth we can precompute the order of traversal. At each node, we store two links. One link descends deeper into the tree if there is a collision, this is called a "hit" link. The other link is followed if there is no collision, a "miss link". The logical link structure is shown in Figure 1. In memory, the links are stored in 2D arrays that mirror the structure of the tree. There are special links to denote leaf nodes and the end of traversal.

3.2 Traversal

The precomputed traversal links simplify the algorithm for traversing the tree. Traversal of the tree is performed in a top down manner. Starting from the root of the tree, an

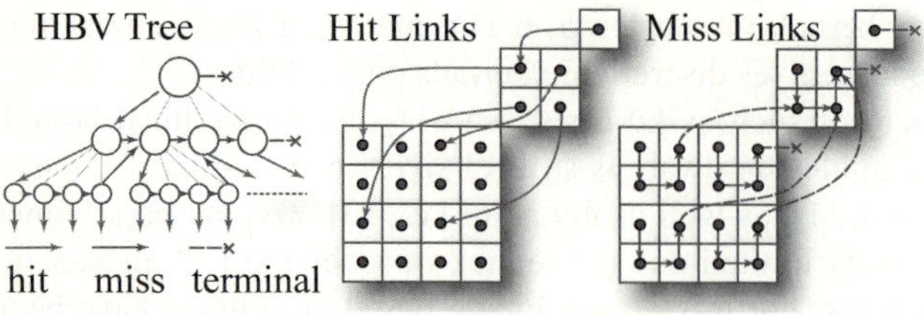


```
while(link != 0){
  s' = getSphere(link)
  if(spheres_overlap(s,s'))
    link = getHitLink(link)
  else
    link = getMissLink(link)
  if(isLeaf(link))
    doAppDependentWork(s,s')
    link = getMissLink(link)
}
```

Fig. 1. Quadtree sphere bounding volume hierarchy with link traversal(*left*), hit links stored on the GPU (*middle*), and miss links stored on the GPU(*right*). The algorithm for GPU link hierarchy traversal is provided(*far right*).

overlap test is performed between the bounding volume being tested and the bounding volume of the node. Based on the result of the overlap test, we follow one of two pre-computed links. If there is overlap the hit link is followed, otherwise the miss link. A link location can be used to determine whether the traversal has hit a leaf node. When this occurs, an application dependent processing step is performed. Traversal is performed until a null *terminal* link is reached, denoting that all overlapping leaf nodes have been visited. Pseudo code for the traversal loop is given in Figure 1(far right).

3.3 Quadtree Optimization

The HBV quadtree is rebuilt each time particle positions are updated in our algorithm. This rebuilding process does not change the tree's topology, however, to construct an efficient tree we require that the particles are stored in a spatially coherent manner. If this is not the case, the reduction algorithm computes large bounding volumes resulting in inefficient traversal. Thus, tree quality only affects performance and not the correctness of the computation. It is easy to initialize the system such that particles are spatially coherent, however, particles are constantly moving in a dynamic system. As shown in Section 5, the degradation in spatial coherency for a particle based liquid simulation is slow. We take advantage of this fact by lazily reoptimizing every α seconds, where α is a user tunable parameter. We can do this optimization asynchronously. For example, particle locations can be read back to the CPU at time t, meanwhile, the GPU continues the simulation process. Once the CPU has completed the tree optimization at some time $t+\Delta t$, the new updated information can be transmitted to the card. In this manner the tree optimized by the CPU is slightly behind the simulation. Nonetheless, this approach results in a tight bounding volume tree when Δt is small.

A top-down recursive bisection algorithm is used to compute an optimized particle configuration. We start by choosing a splitting axis along which to partition our particles into two equal sized sets. A partitioning plane orthogonal to this axis is found such that half the particles fall on either side of the plane. This partitioning process is applied to each of these two sets, resulting in four equal sized sets that form the second level of our quad-tree. By iteratively applying this bisection process, all levels of the tree are built. The choice of splitting axis plays an important role in the bisection process. Our CPU implementation uses principal component analysis (PCA) to determine an axis of greatest variation. Our GPU implementation uses a lower quality approach of choosing

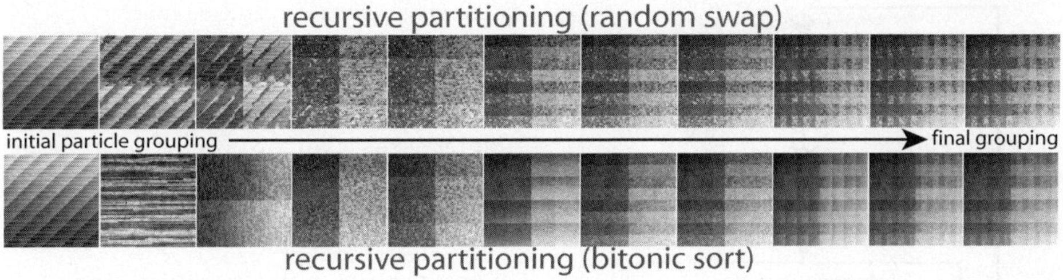

Fig. 2. GPU algorithm for spatial re-ordering of particles to improve tightness of hierarchical bounding tree. Image containing unsorted particles (left), is recursively partitioned and either sorting (bottom) or random swapping(top) is used to place particles into tighter spatial clusters.

axis aligned splitting planes. We alternate between the principal coordinate axes $x, y\,[,z]$. Efficiently re-ordering n particles across a chosen splitting plane requires computation of the ith order statistic which can be done in $O(n)$ time. A less effective choice is to use sorting. To accomplish this task on the GPU, however, we must resort to parallel algorithms such as bitonic sort which has been shown to run in $O(\lg^2 n)$ [20] on graphics hardware. In practice, however, we can use an approximate method and achieve similar results. For this we propose a random swapping algorithm.

Our random swapping algorithm also utilizes recursive bisection to form a complete tree. We start by partitioning our particle image vertically into two rectangular regions. These two regions are partitioned horizontally, forming four square subregions. The process of partitioning subregions vertically and then horizontally is recursively applied until every subregion covers a single particle. Each time we partition a sub-region in half we select a coordinate value (e.g. $x, y[,z]$) to compare against. A random offset is generated and sent to a fragment shader. This shader uses the offset and modula arithmetic to swap particles between partitions based on the magnitude of the selected coordinate value. For example, during the first partitioning we divide the particle image vertically. Particles in the lower partition are compared along the random offset to their corresponding particle in the upper partition. Particles with greater x values are swapped to the top partition(Figure 2). This swapping process between partitions can be iterated a number of times to increase the likelihood that particles are sorted into the correct partition. In practice we have found that applying $\lg(w \times h)$ passes of swapping, where w and h are the width and height of the subregion, produces satisfactory results. Figure 2 shows a comparison of random swapping versus recursively performing GPU bitonic sorting. Note that the random swap algorithm produces nearly the same final particle clustering. By increasing the number of random swaps performed at each level we approach the quality of the tree provided by the more expensive recursive bitonic sorting algorithm. In this way, we can trade off the cost of particle reordering against the quality of the resulting bounding volume tree.

4 Smoothed Particle Hydrodynamics

Smoothed Particle Hydrodynamics (SPH), was developed for use in astrophysics [21, 22]. Müeller et al. [2] use SPH to simulate liquids in interactive computer graphics

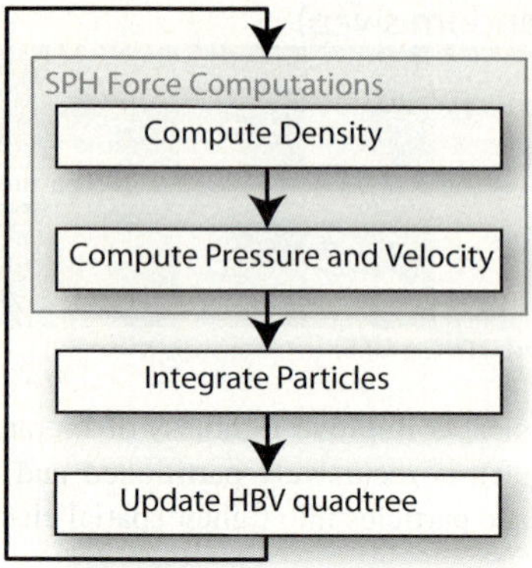


```
// compute density for particle_i
// with neighbors particle_j
density_i = 0;
while(link != 0){
  s' = getSphere(link)
  if(spheres_overlap(s,s'))
    link = getHitLink(link)
  else
    link = getMissLink(link)
  if(isLeaf(link))
    particle_j = getParticle(link)
    dist=|particle_i-particle_j|
    density_i += Wpoly6(dist,h)
    link = getMissLink(link)
}
density_i *= ParticleMass
```

Fig. 3. High level overview of the SPH computation(*left*) and pseudo code for computing density utilizing the quadtree(*right*)

applications. In this section, we describe how we use our quadtree implementation to accelerate the SPH model of Müeller et al. [2], see Figure 3(left) for an overview of the process. Scalar quantities and inter-particle forces in this model are computed using a spatial kernel surrounding each particle with radius h. For example, the equation for density of a particle at position r_i is:

$$\rho(r_i) = \sum_j m_j W(r_i - r_j, h), \quad W_{poly6}(r,h) = \frac{315}{64\pi h^9}\begin{cases}(h^2-r^2)^3 & \text{if } 0 \le r \le h,\\ 0 & \text{otherwise}\end{cases}$$

Where m is mass, h is the smoothing radius and W is the symmetric kernel W_{poly6}. A naive implementation of this model has complexity $O(N^2)$, however the kernels used in the model have finite support over radius h. This means that only particles within a local neighborhood need to be considered. Figure 3(right) contains pseudo code for an algorithm that uses the quadtree to limit the computation to this region. For each particle, the entire tree is traversed using a sphere that represents the smoothing kernel W, e.g. the radius of the sphere is equal to the support of the kernel. This traversal finds all the neighboring particles and evaluates the kernel for each. Pressure and viscosity are computed in a similar fashion using the equations given in [2].

In addition to accelerating force computations, we also use the quadtree for particle to environment collisions. To handle dynamic surfaces using our method, we must store the mesh such that a quadtree can be easily updated. Geometry images [23] encode mesh data in an array where each 2×2 block represents two triangles. This layout has the spatial coherence property required for our reduction method to compute an efficient quadtree. The construction routine is modified slightly such that the first level is built by constructing a bounding volume for the two triangles. After this step, the construction continues in the manner described in Section 3.

To intersect a moving particle with a mesh, we construct a ray from the previous particle position to the new one. We use the quadtree to accelerate the intersection test by traversing using a sphere that encapsulates the ray. If the sphere overlaps a leaf, we perform a ray-triangle intersection with the two triangles bound by the leaf.

The output of the traversal is the nearest point of intersection, if one occurred, and the corresponding surface normal at this point. This information is used to update the position and velocity of particles that collide with the mesh.

5 Results

To evaluate our implementation, we have set up several experiments using the breaking dam configuration as shown in Figure 4(top)[1]. Our test machine is a 3.19 GHz Xeon with 3GB of RAM running a 256MB GeForce FX 7800 GPU. The graph in Figure 4 (bottom right) compares the performance differences as the problem size increases.

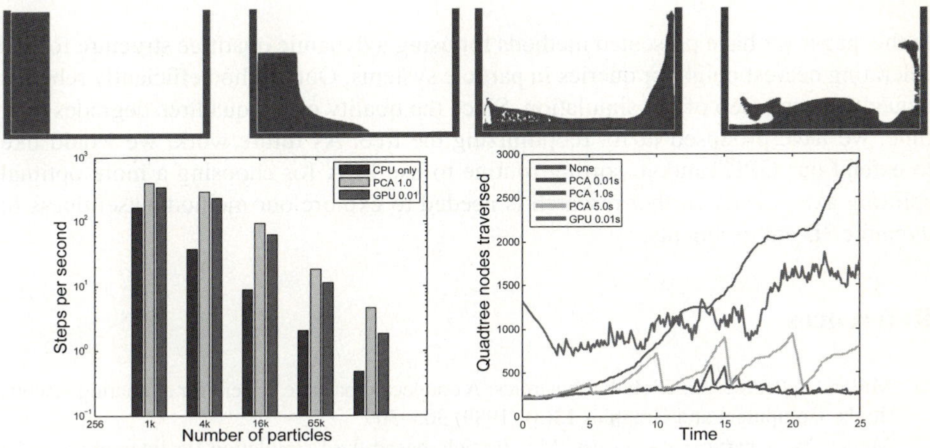

Fig. 4. Images of the breaking dam simulation(*top*). CPU and GPU solvers using both optimization methods for different problem sizes(*bottom left*). Maximum node traversals on the GPU required to resolve particle interactions(*bottom right*).

We compare three solvers: a GPU solver using CPU based PCA tree optimization, a GPU solver using the GPU random sorting method, and a CPU only solver that uses the grid based acceleration strategy. Our CPU implementation is optimized using efficient data structures. Further performance may be achievable by taking advantage of native CPU SIMD instruction sets such as SSE along with cache friendly memory layout [2]. The performance numbers are given as averages over the total simulation run visualization disabled. As the problem size increases, the GPU solvers outperform the purely CPU implementation in some cases by a factor of ten.

As mentioned in Section 3.3, the quality of the quadtree becomes less efficient as the particles move. Since performance is limited by the worst case, a simple measure of this quality is the number of nodes traversed to find the neighboring particles. A second experiment was run with 65k particles using each of the different optimization strategies described in the paper. For each time step, we record the maximum number

[1] A movie is available at `http://www.cs.sunysb.edu/~kyhegem/gpgpu06/sph.mov`

of nodes traversed for a particle(Figure 4 bottom left). As can be seen, the quality of the tree quickly degrades if no optimization is performed, thus the performance of our simulation steadily decreases over time. To achieve the best performance, we have experimented with different optimization frequencies. The quality of the tree produced by the GPU optimization is not as good as the CPU, but is significantly cheaper to run because there is no transfer of data. Because of the cost associated with transferring particle positions to the CPU, it is beneficial to overall performance to reduce the frequency. For this example, optimizing the tree on the CPU once a second produces the best performance.

6 Conclusion and Future Work

In this paper we have presented methods for using a dynamic quadtree structure for accelerating nearest neighbor queries in particle systems. Our method efficiently rebuilds a quadtree each step of the simulation. Since the quality of the quadtree degrades over time, we have proposed lazily reoptimizing the tree. As future work, we would like to extend our GPU random sorting routine to use PCA for choosing a more optimal splitting axis. Lastly further research is needed to explore our method's usefulness in dynamic 3D environments.

References

1. Miller, G., Pearce, A.: Globular dynamics: A connected particle system for animating viscous fluids. Computers and Graphics **13**(3) (1989) 305–309
2. Müller, M., Charypar, D., Gross, M.: Particle-based fluid simulation for interactive applications. In: Proc. of the 2003 ACM SIGGRAPH/Eurographics symposium on Computer animation. (2003) 154–159
3. Baraff, D., Witkin, A.: Large steps in cloth simulation. In: Proc. of SIGGRAPH '98. (1998) 43–54
4. Premože, S., Tasdizen, T., Bigler, J., Lefohn, A., Whitaker, R.T.: Particle-based simulation of fluids. In: Proc. of Eurographics 2003. Volume 22. (2003)
5. Owens, J.D., Luebke, D., Govindaraju, N., Harris, M., Krüger, J., Lefohn, A.E., Purcell, T.J.: A survey of general-purpose computation on graphics hardware. In: Eurographics 2005, State of the Art Reports. (2005) 21–51
6. Kolb, A., Latta, L., Rezk-Salama, C.: Hardware-based simulation and collision detection for large particle systems. In: Proc. of the ACM SIGGRAPH/EUROGRAPHICS conference on Graphics hardware. (2004) 123–131
7. Krüger, J., Kipfer, P., Kondratieva, P., Westermann, R.: A particle system for interactive visualization of 3D flows. IEEE Trans. on Visualization and Computer Graphics **11**(6) (2005)
8. Chiara, R.D., Erra, U., Scarano, V., Tatafiore, M.: Massive simulation using GPU of a distributed behavioral model of a flock with obstacle avoidance. In: Proc. of the Vision, Modeling, and Visualization Conference 2004. (2004) 233–240
9. Kolb, A., Cuntz, N.: Dynamic particle coupling for GPU-based fluid simulation. In: Proc. of 18th Symposium on Simulation Technique. (2005) 722–727
10. Foley, T., Sugerman, J.: KD-tree acceleration structures for a GPU raytracer. In: Proc. of the ACM SIGGRAPH/EUROGRAPHICS conference on Graphics hardware. (2005) 15–22

11. Simonsen, L.O., Thrane, N., Ørbæk, P.: A comparison of acceleration structures for GPU assisted ray-tracing. Masters Thesis (2005)
12. Lefohn, A., Kniss, J.M., Strzodka, R., Sengupta, S., Owens, J.D.: Glift: Generic, efficient, random-access GPU data structures. ACM Transactions on Graphics **25**(1) (2006)
13. Lefebvre, S., Hornus, S., Neyret, F.: Octree textures on the GPU. In Pharr, M., ed.: GPU Gems 2. Addison Wesley (2005) 595–613
14. Govindaraju, N.K., Redon, S., Lin, M.C., Manocha, D.: CULLIDE: Interactive collision detection between complex models in large environments using graphics hardware. In: Proc. of the ACM SIGGRAPH/EUROGRAPHICS conference on Graphics hardware. (2003) 25–32
15. Govindaraju, N.K., Lin, M.C., Manocha, D.: Fast and reliable collision culling using graphics hardware. In: Proc. of the ACM symposium on Virtual reality software and technology. (2004) 2–9
16. Govindaraju, N.K., Knott, D., Jain, N., Kabul, I., Tamstorf, R., Gayle, R., Lin, M.C., Manocha, D.: Interactive collision detection between deformable models using chromatic decomposition. ACM Transactions on Graphics **24**(3) (2005) 991–999
17. Gottschalk, S., Lin, M.C., Manocha, D.: OBBTree: a hierarchical structure for rapid interference detection. In: Proc. of SIGGRAPH '96. (1996) 171–180
18. Klosowski, J.T., Held, M., Mitchell, J.S.B., Sowizral, H., Zikan, K.: Efficient collision detection using bounding volume hierarchies of *k*-dops. IEEE Trans. on Visualization and Computer Graphics **4**(1) (1998) 21–36
19. Beneš;, B., Villanueva, N.G.: GI-COLLIDE: collision detection with geometry images. In: Proc. of the 21st spring conference on Computer graphics. (2005) 95–102
20. Purcell, T.J., Donner, C., Cammarano, M., Jensen, H.W., Hanrahan, P.: Photon mapping on programmable graphics hardware. In: Proc. of the ACM SIGGRAPH/EUROGRAPHICS conference on Graphics hardware. (2003) 41–50
21. Lucy, L.B.: A Numerical Approach to Testing the Fission Hypothesis. The Astronomical Journal **82**(12) (1977) 1013–1924
22. Gingold, R., Monaghan, J.: Smoothed particle hydrodynamics: Theory and application to non-spherical stars. Monthly Notices of the Royal Astronomical Society **181** (1977) 375–389
23. Gu, X., Gortler, S.J., Hoppe, H.: Geometry images. In: Proc. of SIGGRAPH '02. (2002) 355–361

Spiking Neurons on GPUs

Fabrice Bernhard and Renaud Keriven

Projet Odyssée - INRIA/ENS/ENPC
45 rue d'Ulm, 75005 Paris, France
Fabrice.Bernhard@polytechnique.org

Abstract. Simulating large networks of spiking neurons is a very common task in the areas of Neuroinformatics and Computational Neurosciences. These simulations are time-consuming but also often intrinsically parallel. The recent advent of powerful and programmable graphic cards seems to be a pertinent solution to the problem: they offer a cheap but efficient possibility to serve as very fast co-processors for the parallel computing that spiking neural networks need. We describe our implementation of three different problems on such a card: two image-segmentation algorithms using spiking neural networks and one multi-purpose spiking neural-network simulator. Using these examples we show the benefits, the challenges and the limits of such an implementation.

1 Introduction

1.1 Motivations

Considering the power of recent graphic cards in parallel computation, and the possibility to program them, it is no wonder that so many people have tried to divert them from their initial purpose. Simulations of spiking neural networks, very common in Neuroinformatics, are highly parallel and are therefore one of the many tasks that could benefit from such cards.

We found few references to artificial neural networks on graphic cards [1]. However perceptrons are very different from spiking neural networks, and do not serve the same purpose at all. This work is therefore far from what interested us, which we now describe here.

We will briefly introduce spiking neural networks and how they differ from artificial neural networks, and then give details on the implementation of three different problems using such networks: two image-segmentation algorithms and a general-purpose spiking neural network simulator. We will try to show the benefits, the challenges and the limits of such a task and give our conclusion on its pertinence.

1.2 Spiking Neural Networks

Frank Rosenblatt's perceptron [2] in the late fifties was the first famous attempt to simulate the brain on a computer. Although able to learn and useful in some

V.N. Alexandrov et al. (Eds.): ICCS 2006, Part IV, LNCS 3994, pp. 236–243, 2006.
© Springer-Verlag Berlin Heidelberg 2006

simple classification tasks, it is limited, as Minsky and Papert's work showed [3]. More generally, artificial neural networks are too far from a biologically plausible model to interest scientists in Neuroinformatics and computational Neurosciences.

Research in those domains is oriented towards spiking neural networks, for which models also take into account the internal dynamic of each neuron: the membrane potential, the firing rate, the amount of ions released at the synapses, etc. Not only do these models reproduce more closely the dynamics observed in real neural networks, but they also enable to take into account phenomena forgotten in the simple rate-based model, as for example spiking synchronisation. A very simple but common and already powerful spiking model is for example Lapique's integrate-and-fire neuron [4] which describes the neuron's membrane as a leaky condensator. This is the one we are going to use in the three following implementations.

Spiking neural networks' dynamics are however harder to describe mathematically than artificial neural networks'. Simulation is therefore an essential part of their study, but these are also slower: the pertinent time-scale would be roughly for example the average duration between spikes, which can be as long as a hundred cycles of a timestep simulation, a lot more than for an artificial neural network, whose weights are updated at every cycle. Fortunately, since these networks try to mimic the brain's behaviour, where neurons all work independently from one another, they are intrinsically parallel.

1.3 Graphics Processing Unit

The primary goal of 3D graphic cards is to replace the CPU in all the calculations needed to show 3-dimensional objects to the screen: projection of the objects, texturing, lighting, etc.

One of the tasks that graphic cards accelerate very significantly is the calculations of the screen's pixels' colour. After having projected the different objects on the screen according to the position of the viewer, a program (the fragment shader) is run in parallel on all pixels to decide their colour, taking into account objects' textures, lighting, fog, etc. This is accelerated by up to 24 pipelines and is the step that we will divert to serve as a powerful co-processor for our spiking neural networks' simulations.

Basically, we will map a neuron to a pixel and replace the fragment program by one that will calculate the updated state of our neurons' variables. By repeating this multiple times, we will have our timestep simulations. A very good guide on how to efficiently do this can be found on Dominik Göddeke's page [5].

Our technical choice was to use OpenGl to interface with the graphic card, the FrameBufferObject class [6] to render to textures and NVidia's Cg to write the fragment programs. It seemed to us to be the most flexible and efficient solution at the time of the experiments. The GPU card used was mostly a Nvidia 6800 Ultra, and at the end the 7800. We worked on a Bi-Xeon 2.8GHz, 1GB RAM, with Windows. To compare the results, we used all the possible compiling optimisations for the CPU version.

2 Implementation of Three Different Algorithms Using Spiking Neural Networks

2.1 Synchrony and Desynchrony in Integrate-and-Fire Oscillators, S.R. Campbell, D.L. Wang, and C. Jayaprakash

Quick description of the algorithm. This first algorithm [7] is a segmentation of a picture based on a locally-excitatory globally-inhibitory spiking neural network. For each pixel of the picture to segment, there is a spiking neuron associated to it, verifying a Lapicque-like differential equation:

$$\frac{dV}{dt} = -V + I \tag{1}$$

I is defined at the beginning so that neurons in homogeneous regions spike spontaneously, i.e. $I > \theta$ for them, where θ is the threshold of the neuron. Every neuron is connected in an excitatory manner to all its local neighbours (up, down, left and right) if they have a similar colour. If a neuron emits a spike, the potential of its similar neighbours is increased. Every neuron is also globally connected to all the others in an inhibitory manner. The result is that neurons coding a segment of homogeneous colour will excite themselves locally and after a short time will synchronize their spikes. Neurons of different segments will inhibit themselves through the global inhibition and stay desynchronized. Segments of homogeneous colour will therefore appear after a short time by just looking at the spiking times of every neurons: all neurons of a same segment will spike together while different segments will spike at different times.

Implementation on the GPU. All arrays like V and I are put into textures. The evolution of the potential can be implemented as a Euler-timestep, but here in that simple case, we can also observe that:

$$V(t) = I(1 - e^{-t}) \Rightarrow V(t + dt) = I + e^{-dt}\left(V(t) - I\right) \tag{2}$$

The first challenge appears with the spiking. In a graphic card, we have a CREW (concurrent reading, exclusive writing) mechanism: we can read in parallel anywhere in the textures (a process also called "gather" [9]), but we can write only in one place: the destination pixel ("scatter" is not possible). Therefore we cannot, when a neuron fires, update the connected neuron's potential, as we would simply do in sequential mode. To solve this problem we convert the scatter problem to a gather problem by introducing a new array, E, coding the excitatory state of a neuron. At every step, every neuron reads the excitatory state of its connected neurons. If it is equal to 1, we add the appropriate excitation. This method is fast enough for a little number of connections, since reading in a texture is very fast on graphic cards.

For the global inhibition however, this method would mean that every neuron must read at each pass from every other neuron. This would become very

Fig. 1. The original picture followed by the different steps of its segmentation by the algorithm of Campbell, Wang and Jayaprakash

time-consuming. Since the global inhibition here just depends on the number of neurons having fired at the precedent pass we just need to determine that number. The first idea we had, was to implement it through a reduction: For a $n \times n$ square, we calculate

$$E(x,y) = E(x,y) + E(x+2^i,y) + E(x,y+2^i) + E(x+2^i,y+2^i), i \in [|0, log_2(n)|]. \quad (3)$$

We therefore have after $log_2(n)$ cycles $E(0,0) = \sum E(x,y)$. The second idea was to use the hardware acceleration of mipmapping to perform that same reduction. For a texture of size n, the mipmap is a pyramid of $log_2(n)$ textures of sides $\frac{n}{2^i}, i \in [|1, log_2(n)|]$, the pixels of which are the average of the source texture. Therefore the last texture is one pixel large and contains the average of all points of the source texture, in our case: $\frac{1}{n^2}\sum E(x,y)$. However this is not very precise since it is done on 8 bits and is not very flexible. The solution found was to use a trick specific to graphic cards: there is a counter that is able to count the number of pixels written in a pass: the Occlusion Query. We therefore just need to run a pass on the graphic card which writes the pixel (x,y) only if $E(x,y) = 1$, else it exits the fragment program with the CG command `discard()` and then count the number of pixels written with the occlusion query.

Results. Thanks to the occlusion query trick we were able to put the whole algorithm on the GPU and gain a significant increase in speed: it is about 7 times faster on the 6800 Ultra and 10 times faster on the 7800 than on the CPU. The results are identical. The visualisation is also made easier. In figure 1 we represent in red the membrane potential, in green the excitement state. Blue is just there for visualisation reasons: it is set to 1 when the neuron spikes and then decreases exponentially. Neurons of the same segment, which spiked at the same time, have therefore the same blue colour.

Finally it is interesting to see that we are able to determine connexity, since connex neurons are synchronized, a problem that the simple perceptron cannot solve [7]. More generally the use of spiking-synchrony is only possible in spiking neural networks and shows their potential.

2.2 Image Segmentation by Networks of Spiking Neurons, J.M. Buhmann, T. Lange, and U. Ramacher

Quick description of the algorithm. This second algorithm [8] is a segmentation algorithm based on a histogram clustering method: we divide the picture in $N \times N$ little rectangles for which we calculate the luminosity histogram, for M grey levels. The segmentation then tries to assign to a same segment blocs of similar luminosity histogram. For that we use a spiking neural network designed to roughly minimize the objective function

$$H = \sum_{1}^{N \times N} D^{KL}(h_i, \bar{h}_{s(i)}) \tag{4}$$

where D^{KL} is a distance, h_i the histogram of block i and $\bar{h}_{s(i)}$ the average histogram of the segment to which block i belongs. We are therefore looking for segments where the member blocs are the most homogeneous possible.

The spiking neural network is composed of $K \times N \times N$ neurons, where K is the a priori known number of segments: for each bloc (i, j), we will have K neurons that will compete in a winner-take-all mechanism to choose which segment the bloc belongs to. The neurons are modeled by the simple differential equation:

$$\frac{dV}{dt} = -\frac{V}{\rho} \tag{5}$$

The neuron (i, j, k) receives input spikes with weights $\forall m \in [|1, M|], w_{k,m}$ and with probability $h_{i,j}(m)$: the relative importance of the mth grey level in the histogram of block (i, j). All K neurons of bloc (i, j) have inhibitory connections between them to implement the winner-take-all mechanism. There are also connections between contiguous blocs to smooth the segmentation: neurons coding for the same segment in direct neighbours have excitatory connections while neurons coding for different segments have inhibitory connections, so that direct neighbours tend to belong to the same segment if similar enough.

Knowing $\bar{h}_k$, the average histogram of blocs in segment k, the weights are updated at every cycle with the following rule:

$$\frac{d}{dt} w_{k,m} = \alpha \left(exp\left(w^{max} - w_{k,m}(t)\right) \bar{h}_{k,m} - 1 \right) \tag{6}$$

Implementation on GPU. This algorithm is divided in four steps:

- We first update with a pass the weights $w_{k,m}$.
- We then update in a pass $V_{i,j,k}$ taking into accounts the input spikes and the neighbour's connections.
- We calculate in a third pass $s(i, j)$: the segment to which bloc (i,j) belongs, defined by the k for which neuron (i,j,k) fires most.

– Finally we calculate the average histogram of segment k $\bar{h}_k$ on the CPU, since it is unfortunately not possible to do it efficiently on the graphic card.

Of course, implementing this is not as straightforward on the GPU as it would be on CPU. First of all we need random numbers to implement noise, used to avoid staying in local minima, and also to send the input spikes with probability $h_{i,j}(m)$. We did this by filling a giant texture with random numbers, but it is clearly not an optimal solution since the period of this pseudo-random generator is equal to the size of the random-filled texture, a number that cannot be much bigger than the size of our network.

We stored our three-dimensional variable: $V(i,j,k)$ of size (N,N,K), in a $(N \times K, N)$ texture. However we notice that we now need to have $N \times K \leq 4096$.

To avoid border effects we chose to add a null border to the texture containing V, which does not imply more calculation since we can tell the card to render only the inner part.

We are not always able to transform conditional branches into multiplications by 0 or 1 to avoid slow conditional branching. However we have observed that there exist conditional arithmetic operators in the assembly language, used if there is just a simple arithmetic operation inside the conditional branch. This enables us to make multiple tests without significant speed loss.

The last part of the algorithm is not efficiently parallelisable so we leave it on the CPU. However the transfer of the data from the GPU to the CPU and back is a costly operation. We therefore try to do that step only once every 100 steps.

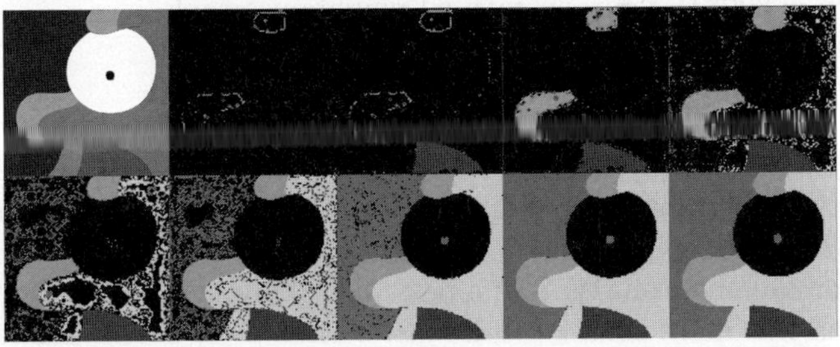

Fig. 2. The original picture followed by the different steps of its segmentation by the algorithm of Buhmann and Lange

Results. We manage to get once again a significant acceleration: 5 times faster on the 6800 Ultra and 10 times faster on the 7800 256MB. The convergence is however a little slower, due surely to the pseudo-random texture which is clearly not as good as a real pseudo-random generator, and the last part, on the CPU, not executed at each step. It is interesting to note that this last part is considered by the authors as the weakness of their algorithm in terms of biological plausibility. The results are visualised by projecting the outcome of the third pass: $s(i,j)$, see for example figure 2.

2.3 Generalisation to an Easily Modifiable Network for Use in Neuroinformatics

The implementation of the two precedent algorithms on GPU being convincing, we decided to make a more general simulator, that would be flexible enough to be adapted to different networks used in Neuroinformatics.

The main new challenge here was to provide an easy and flexible way to define the connections between neurons. The idea proposed for M connections per neuron in a $N \times N$ network was to store them in a $(N\sqrt{M/4}, N\sqrt{M/4})$ RGBA (4 colour components) texture, with a connection (two coordinates) in each colour component. This is possible since our coordinates (necessarily smaller than 4×4096 because of textures' sizes limitations) fit in 16 bits and Cg provides a function to store two 16 bits variables in a 32 bits component: `pack_2half(half2 (x,y))` Using squares of side $\sqrt{M/4}$ for each neuron's connections is motivated by the fact that textures' width and height are both limited to 4096. Very large or long rectangles are therefore not possible.

At each neuron's update we will then go through the table of connections with two nested 'for' loops. Something which is possible only in the latest generation cards. There is a limitation here in the 6800 Ultra, that each loop has to be smaller than 256, but that is not a real problem since we will face another limitation before: the size of the texture. They are limited to 4096×4096, and that number is still theoretical since in many cards such a texture would fill the entire memory on its own. But roughly, we could imagine with a 3584×3584 neighbours' texture, for example:

- $896 \times 896 = 802816$ neurons with 64 neighbours each
- $512 \times 512 = 262144$ neurons with 196 neighbours each
- $128 \times 128 = 16384$ neurons with 3136 neighbours each

However, the more connections there are, the more texture lookups are done. The observation was that for M big, the time needed increased linearly, limiting the size of M. In our experiments, we tried a 896×896 large network with 64 neighbours for each neuron. We had a pass rate of about 22 per second on a 7800 256MB, which was more than 20 times faster than a CPU-only implementation. One interesting reason for such a gain is that our model also included electrical synapses, which means that neurons adjust a little to the potential of their connections at every pass. And this is not cache-friendly, therefore favouring

Fig. 3. Detail of the simulation of an excitatory spiking neural network, with tracking of some neurons' spiking times (yellow dots) and a global spiking histogram

even more the GPU, with its faster memory. In figure 3 you can see this basic excitatory network with random but constant connections, with the red value representing V, to which we added the tracking of some neurons' spikes (the yellow pixels) and a histogram of the global spiking activity.

3 Conclusion

We have clearly seen a significant speed increase, (between 5 and 20 times faster) in all the algorithms implemented, thanks to the inherently parallel way of functioning of neural networks. But there are other practical reasons that make the GPU very interesting: it is cheap compared to clusters, more flexible than a hardware implementation, has a bright future since GPUs' speed increases faster than CPUs' speed [9] and there is always the possibility to leave parts of the algorithms on the CPU if this is more efficient.

Different limitations exist, either due to the parallel nature of the calculations or to the fact that graphic cards are still designed more towards 3-D graphics rendering than general purpose calculations. The most important ones being: the impossibility to "scatter", therefore imposing neurons to listen for a possible incoming spike at every pass, slow conditional branching, the absence of a random-generator function and the limited size of the textures. But we managed to bypass more or less all these problems in our implementations using different tricks. Some of these limitations might even disappear in future generations' card, making, in our opinion, the future of simulations of spiking neural networks on GPU very interesting.

References

1. K. Pietras GPU based multi layer perceptron as efficient method for approximation complex light models in per-vertex lighting. http://stud.ics.p.lodz.pl/~keyei/lab/atmoseng/index.html, 2005
2. F. Rosenblatt Principles of neural dynamics. Spartan Books, New York, 1962
3. M.L. Minsky, and S.A. Papert Perceptrons. MIT Press, 1969
4. L.Lapicque Recherches quantitatives sur l'excitation électrique des nerfs traitée comme une polarisation. J Physiol Pathol Gen 9:620-635, 1907
5. D. Göddeke. GPGPU–Basic Math Tutorial. Ergebnisberichte des Instituts für Angewandte Mathematik, Nummer 300, FB Mathematik, Universität Dortmund, nov 2005
6. Framebuffer Object (FBO) Class. http://www.gpgpu.org/developer/
7. S.R. Campbell, D.L. Wang, and C.Jayaprakash. Synchrony and Desynchrony in Integrate-and-Fire Oscillators. *Neural Computation 11*, pages 1595–1619, 1999.
8. J.M. Buhmann, T. Lange, and U. Ramacher. Image Segmentation by Networks of Spiking Neurons. *Neural Computation 17*, pages 1010–1031, 2005.
9. J.D. Owens, D. Luebke, N. Govindaraju, M. Harris, J. Krüger, A. E. Lefohn, and T. J. Purcell. A Survey of General-Purpose Computation on Graphics Hardware. *EuroGraphics 2005*, 2005.

SONA: An On-Chip Network for Scalable Interconnection of AMBA-Based IPs*

Eui Bong Jung[1], Han Wook Cho[1], Neungsoo Park[2], and Yong Ho Song[1]

[1] College of Information and Communications, Hanyang University, Seoul, Korea
{ebjung, hwcho, yhsong}@enc.hanyang.ac.kr
[2] Dept. of Computer Science and Engineering, Konkuk University, Seoul, Korea
neungsoo@konkuk.ac.kr

Abstract. Many recent SoCs use one or more busses to provide internal communication paths among integrated IP cores. As the number of cores in a SoC increases, however, the non-scalable communication bandwidth of bus tends to become a bottleneck to achieve high performance. In this paper, we present a scalable switch-based on-chip network, called SONA, which can be used to provide communication paths among existing AMBA-based IP cores. The network interfaces and routers for the on-chip network are modeled in register transfer level and simulated to measure the performance in latency. The simulation results indicate that the proposed on-chip network can be used to provide scalable communication infrastructure for AMBA-based IP cores with a reasonable cost.

1 Introduction

The recent improvement in semiconductor technology enables various modules with different functionality and complexity to be integrated to a single system-on-chip (SoC). A SoC often consisting of one or more processors, DSP cores, memories, I/Os and internal communication channels is used to build an embedded system such as mobile handsets, PDAs, etc. Considering that embedded systems often use battery as a power source, it is required that SoCs consume less energy for normal operation modes and nevertheless produce reasonable performance.

Traditionally one or more busses are used inside SoCs to implement communication channels among integrated components. It is because that the simple structure of this interconnection contributes to the reduction of design cost and effort. However, as a bus-based SoC integrates more and more components, the insufficiency in communication bandwidth often results in the degradation of the entire system. This problem becomes even worse when deep sub-micron technology is used for the implementation of SoCs. As the technology evolves, the relative length of global wires increases, which may make data transactions take more clock cycles and thus increase communication cost.

There have been many approaches to overcome the limitation in scalability of bus systems. One of them is to use a switch- (or router-) based network within a SoC,

* This work has been supported by a grant from Seoul R&BD Program.

V.N. Alexandrov et al. (Eds.): ICCS 2006, Part IV, LNCS 3994, pp. 244 – 251, 2006.
© Springer-Verlag Berlin Heidelberg 2006

called *on-chip network* or *network-on-chip (NoC)*. This network, an on-chip variation of high-performance system networks, effectively increases communication bandwidth and degree of operation concurrency. The communication links used in constituting on-chip networks are relatively short in length and arranged in a regular fashion, they often consume less energy for data transaction and overcome many electrical problems arisen from the use of deep sub-micron technologies. The provision of on-chip networks for SoCs effectively decouples computation from communication by the introduction of well-structured communication interfaces, which is becoming more important as the SoC density increases.

However, the change in communication layer inevitably necessitates the modification of communication interface of many existing IP (Intellectual Property) cores. In fact, the success of AMBA AHB [1] makes IP vendors develop hardware or software cores that can run in conjunction with the AMBA protocol. Considering that the reuse of IP cores plays a crucial role in reducing design cost and effort as well as preventing from taking unnecessary risk from making a new design, it is desirable to reuse the bus-based IPs in the implementation of a SoC based on a switch-based network.

This paper proposes an on-chip network, called SONA (Scalable On-chip Network for AMBA), as a scalable communication backbone which efficiently interconnects AMBA-based IPs. The network implements 2D mesh topology with a bidirectional link between a pair of switches. A network interface connects an AMBA IP to a SONA switch and converts communication protocols across the two different protocol domains. The network is modeled at register transfer level and simulated on MaxSim, a hardware/software co-simulator from ARM [2], to measure the communication performance.

2 Related Work

A variety of busses are used in SoCs to provide communication channels for IP cores within a chip. The example includes AMBA AHB from ARM [1], CoreConnect from IBM [5], MicroNetwork from Sonics [6], and Wishbone from Silicore [7]. This type of communication system provides many features for developers: simple to design and easy to develop software. However, as the amount of data to travel over bus systems increases, the insufficient bandwidth of the communication media inevitably results in long communication delay, which limits the use of bus systems only to small systems.

Switch-based networks have been long used as a communication infrastructure in the field of computer networks and parallel system networks. Such networks are brought on chip to solve the problem of insufficient communication bandwidth provided by traditional on-chip buses. Even though on-chip networks successfully inherit many useful features and techniques needed to boost communication performance, they still need to have solutions to many other constraints: buffer memories are expensive to implement, silicon budget is tight, and energy consumption needs to be kept low.

Recently there have been several attempts to design AMBA-compatible on-chip networks. However, the networks proposed in [8][9] employs a set of wide crossbar switches to simply forward AMBA signals between end nodes without the notion of packetization, resulting in limited scalability and capability far less than those needed to implement high integration in future SoCs.

3 Network Architecture

The network performance is often measured in terms of latency (i.e. the time delay for delivering information from source to destination) and throughput (i.e. the amount of information delivered in a given time window). The goals in designing SONA are to achieve high level of scalability as well as to provide high performance. In addition, less energy consumption and reduced cost in protocol conversion are also pursued in this network architecture.

3.1 Packet Structure

In AMBA, a master drives necessary bus signals when it needs to start a new transaction. In order to deliver the bus signals over a switch-based on-chip network, a network interface at the master side packages the semantics of bus signals into a packet, and another at the slave side restores the signals from the packet.

Figure 1(a) illustrates the packet structure used in SONA. Each packet consists of a header and optionally a payload. A variable-sized packet is sliced into multiple flits each being 32 bits long. The header contains a target node number, a source node number, the packet size in flit, an associated command (or transaction type), and a 32-bit address. Depending upon its transaction type, the payload may grow up to 512 flits which is the case that the burst type (HBURST) is 16 and the size (HSIZE) is 32 words.

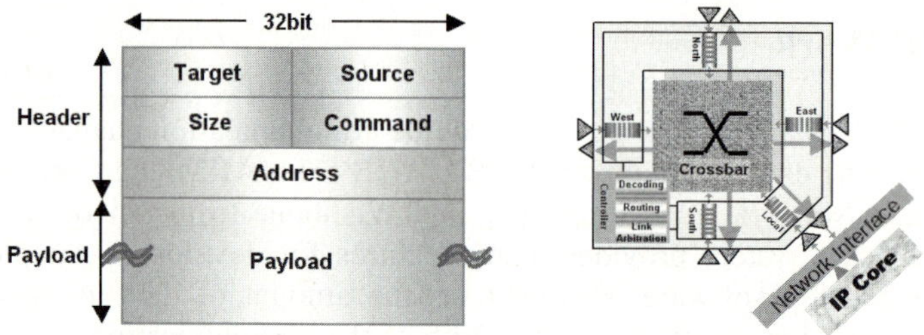

(a) (b)

Fig. 1. (a) The packet structure and (b) the router architecture used in SONA

The target and source node number is encoded in 16 bits, which allows up to 65536 nodes to be integrated in SONA. Because the maximum length of payload is 512 words, at least 9 bits are needed to record the payload size in the header. For this reason, 16 bits are reserved for packet size in the header. The command field is used to encapsulate the AMBA control signals: HTRANS (2 bits), HBURST (3 bits), HSIZE (3 bits), and HWRITE (1 bit). The address field is used to deliver 32-bit HADDR signals and the payload field is used for transmitting 32-bit HWDATA or HRDATA.

A tail flit usually carries checksum information to verify the integrity of the packet at the receiver's side. In SONA, however, no tail flits are used in packets assuming that no bit errors would occur during the transmission over communication links due to the short wire length. In order to detect the end of variable-sized packet, the receiver decodes the packet size from the header and counts the number of payload flits up to this size.

3.2 Router Architecture

Figure 1(b) shows a SONA router consisting of a central crossbar, four ports each providing a communication channel to its neighboring router, a local port through which an IP core accesses the network, and the control logic for implementing flow control and routing mechanisms. When a packet arrives at a port, it is buffered in an input queue awaiting a routing decision by the controller. Each queue in a port can hold up to 8 flits and packets are delivered in wormhole switching. The four inter-router ports are used to build a network with two-dimensional mesh topology. Each link is 32-bit wide so that a flit can move across the link in a clock cycle.

For simplicity, the on/off mechanism [10] is used for flow control over links. In order to implement this mechanism, each communication link provides a pair of control signals each for one direction. Flits can be sent over link only when the signal is set to ON. No virtual channels are implemented over each physical link. Routing deadlocks inside the network are avoid by the use of dimension-order routing [11].

3.3 Network Interface Architecture

The network interface bridging an AMBA core to the SONA network performs protocol conversion from AMBA AHB to SONA and vice versa. Depending upon the role in transaction, the AMBA core is connected to one of two network interfaces, master network interface (MNI) and slave network interface (SNI). For example, a RISC core and a memory module are connected to MNI and SNI, respectively.

Figure 2 shows the two SONA network interfaces, communicating over the network. Packets are delivered on 32 bit `flit_i/flit_o` channels. The presence of valid flits on these channels is indicated by accompanying `tx_o/rx_i` signals. Likewise, `on_off_i/on_off_o` signals are used for flow control.

Each network interface has two state machines, one for packet transmission (named MNI_Request and SNI_Resend) and the other for packet reception (named MNI_Response and SNI_Receive).

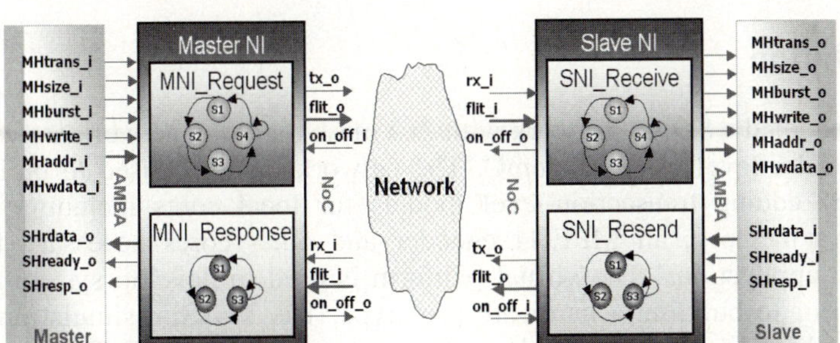

Fig. 2. Master and slave network interface for a master and a slave AMBA cores, respectively

The write operation by an AMBA bus protocol consists of three phases: arbitration phase, address phase, and data phase. During the arbitration phase, the network interface checks the buffer status of its neighboring router and reports it to the transaction-initiating local core by driving the SHREADY signal. Therefore, if there are no buffers available, the local core retries later on the reception of this signal. In the address phase, the state machine for packet transmission temporarily holds MHTRANS, MHSIZE, MHBURST, and MHWRITE into a packetization buffer and encodes them into the packet header. Likewise, when a network interface receives a packet from the network, the state machine for packet reception temporarily stores the packet into a de-packetization buffer until all the necessary signals are restored from the buffer.

In MNI, the packetization buffer is large enough to hold the AMBA signals from the local core necessary for generating a packet header. Note that the AMBA data is not stored into the buffer even when a burst write operation takes place. Instead, the MNI transmits the header flits from the buffer and payload flits from the HWDATA signal on the fly. The buffer in router becomes overflowed by the injection of long payload. There are two approaches to deal with this problem. The first is to increase the size of packetization buffer up to 512 flits to temporarily hold both the header and the data and retry the transmission later. It is not a good solution considering that memory resource within a chip is expensive. The second is to drive the SHREADY signal to low to indicate that the slave is not ready. In this case, the local core stops the generation of signals and retries the transaction later.

MNI uses a node number to identify packet destinations. It generates a destination node number by looking up a memory map which associates a node number with a memory address range. The AMBA address, MHADDR, is used for the map lookup.

SNI is responsible for delivering a request arriving from the network to the local core. It runs a de-packetization process to restore AMBA signals to the local core. For reads, it decodes a memory address from the header along with other AMBA signals such as MHTRANS, MHBURST, MHSIZE, and MHWRITE. Optionally, it generates a sequence of memory addresses needed to complete a burst type operation. When receiving a response from the core, SNI first checks if the SHREADY signal generated by the local core is set to AHB_READY. If this is the case, the data from the local memory is stored into the packetization buffer. For writes, SNI decode the payload into the MHWDATA signal. Optionally, for each word in a burst write, SNI can generate an address to drive the MHADDR signal along with MHWDATA.

4 Simulation

We have modeled a SONA network with 2x2 mesh topology network at synthesizable register transfer level using SystemC. The network model is used to build a virtual platform by adding transaction-level models for local cores including an ARM9 processor, a memory, an MPEG4 encoder and other cores necessary for mobile multimedia applications. The virtual platform is used to develop system/application software prior to building a hardware prototype. The MaxSim simulation tool from ARM is used for simulation with the traffic generated by the ARM9 transaction-level model running an application of an MPEG encoder for the 720x480 YCbCr 420 pixel format.

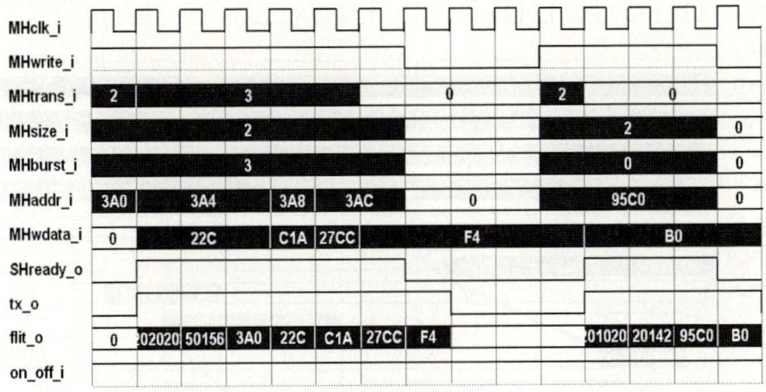

(a)

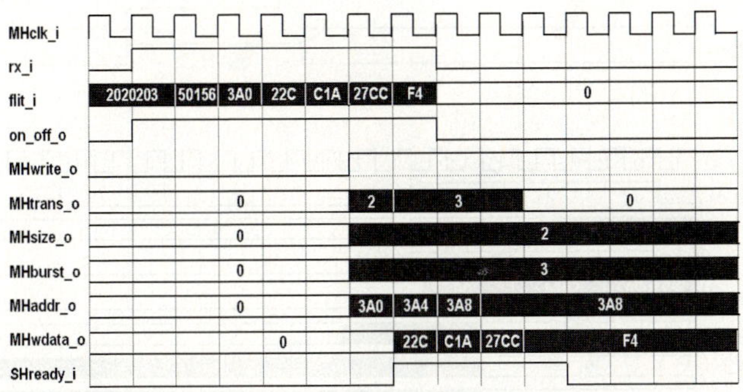

(b)

Fig. 3. The waveform indicating the operating of (a) master network interface and (b) slave network interface for a write operation

In order to evaluate the performance of SONA, we have measured only the latency for read and write operations because it is well studied that switch-based networks provide higher throughput than busses do. Figure 3(a) illustrates a burst write operation with 4 words. The first address generated at cycle 1, 3A0, is injected on `flit_o` at cycle 4, which indicates that it takes three clock cycles to inject a packet to its neighboring router at MNI upon the write request from a master core. For a burst operation, it takes an extra clock cycle for each payload flit.

Figure 3(b) shows the latency when a packet is delivered to SNI and further to the local core. The processing of a header takes three clock cycles and the payload delivery takes the same amount of delay in SNI.

Figure 4 illustrates the waveform for the signals of MNI for a read operation. It takes three clocks for MNI to inject a read request into the network as in MNI for a write (see Figure 4(a)). The transmission latency between MNI and SNI depends on the number of router hops that the packet traverses. As shown in Figure 4(b), it takes 6 clock cycles for SNI to de-packetize and deliver the request to the slave core and another 5 clock cycles to receive data from the core.

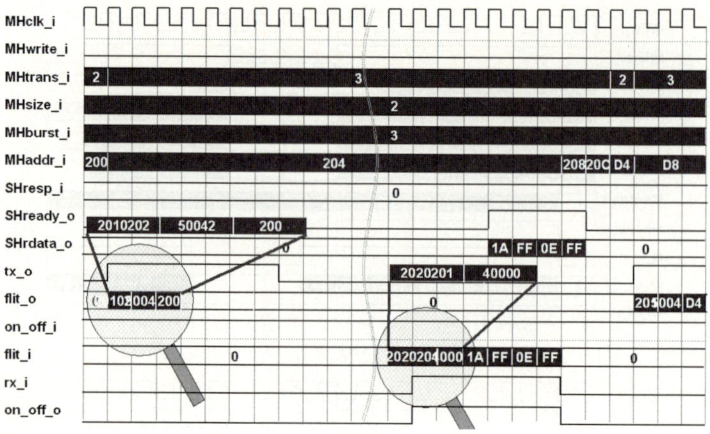

(a)

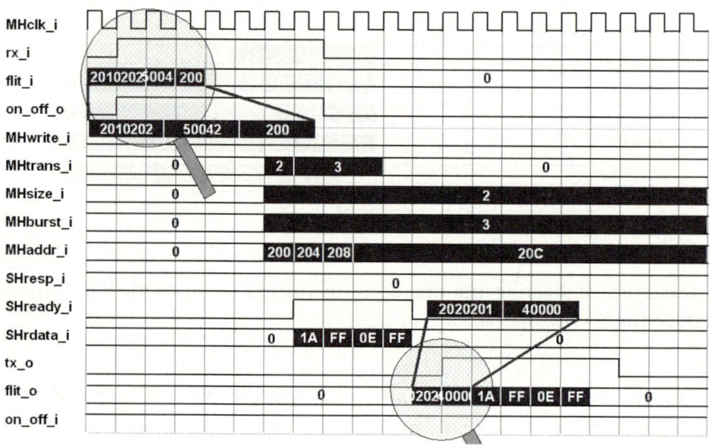

(b)

Fig. 4. The waveform indicating the operating of (a) master network interface and (b) slave network interface for a read operation

5 Conclusion

In this work, we present a scalable on-chip network for interconnecting AMBA-based IP cores. The network is modeled in SystemC to build a virtual platform for the development of software and to explore architectural space to enhance the performance. It is simulated along with other IP cores which are used to build a recent multimedia mobile SoC to observe the possibility of replacing a bandwidth-limiting on-chip AMBA with a scalable switch-based network.

Even though the use of switch-based networks brings about the increase in latency to complete a transaction, it enables IP cores to utilize increased bandwidth, making them experience less latency under high network loads. Considering that recent mobile applications requires increasing bandwidth to provide high quality multimedia service, the poor scalability of on-chip bus may become a bottleneck for achieving high performance. The increased latency can be compensated by placing the IP cores closer which make high volume of traffic. Or the network can be used to bridge multiple AMBA buses in a scalable fashion.

References

1. AMBA Bus Specification, http://www.arm.com
2. http://www.arm.com/products/DevTools/MaxSim.html
3. http://www.synopsys.com/products/logic/design_compiler.html
4. International Technology Roadmap for Semiconductors, http://public.itrs.net
5. CoreConnect Bus Architecture, http://www-03.ibm.com/chips/products/coreconnect/index.html
6. Sonics Integration Architecture, Sonics Inc., http://www.sonicsinc.com
7. W. Peterson, WISHBONE SoC Architecture Specification, Rev. B.3, Silicore Corp, 2002.
8. J. Lee, et al., SNP: A New Communication Protocol for SoC, International Conference on Communications, Circuits and Systems, Cheungdu, China, June 2004.
9. J. Jang, et al., Design of Switch Wrapper for SNA On-Chip Network, Joint Conference on Communications and Information, 2005.
10. William James Dally, Brian Towles, Principles and Practices of Interconnection Networks, Morgan Kaufmann Publishers, 2003
11. W. Dally and C. Seitz. Deadlock-free Message Routing in Multiprocessor Interconnection Networks, IEEE Transactions on Computers, 36(5):547–553, May 1987.
12. Magma Design Automation, A Complete Design Solution for Structured ASICs, white paper, http://www.magma-da.com
13. J. Liang, S. Swaminathan, R. Tessier, aSOC: A Scalable, Single-Chip Communications Architecture, Conference on Parallel Architectures and Compilation Techniques, 2000.
14. A. Radulescu, et al., An efficient on-chip network interface offering guaranteed services, shared-memory abstraction, and flexible network programming, IEEE Transactions on computer-aided design of integrated [illegible] January 2005.

Semi-automatic Creation of Adapters for Legacy Application Migration to Integration Platform Using Knowledge

Jan Pieczykolan[1,2], Bartosz Kryza[1], and Jacek Kitowski[1,2]

[1] Academic Computer Center CYFRONET-AGH,
Nawojki 11, 30-950 Cracow, Poland
[2] Institute of Computer Science, AGH University of Science and Technology,
Mickiewicza 30, 30-059 Cracow, Poland

Abstract. This paper presents a solution for semi-automatic creation of adapters – missing components between integration platforms or application migration frameworks and legacy applications – useful for collaborative environments. It introduces basic problems related to the integration process, presents a common elements of an integration platform and emphasizes a crucial role of creation of the adapter. The proposed solution is based on expert system approach; it speeds-up the whole process of development of an adapter, making the migration cost-effective, performance tuned and error prone.

1 Motivation

In the last period of time great IT expansion has been observed. Many of new applications (especially business ones) are created and deployed at enterprise organizations. But now, these applications have to coexist in a common environment with older ones – called legacy applications – which have been developed earlier often on old computer architectures with programming languages currently not supported.

Many of such legacy systems are critical – they are still running doing their job properly and efficiently. The main problem is that they often exist separated from other applications, having its own business logic, data model and data which often are the same for many of such systems (e.g. customer data).

Typical examples of such domain systems are coexisting billing and loyalty systems. Each of them has a database with similar client data. The problem appears when the company has to make a change in a particular client's data – the change must be propagated across a large number of domain systems, which often results in their inconsistency.

In response to such problems the industry has created a concept of the integration platform. The main role of this kind of software is to introduce a common data model and provide a common layer of communication between applications in a manner that enables them to invoke operations on integrated systems in a standardized way but also allows receiving notification of changes in other systems and taking suitable action to handle them.

V.N. Alexandrov et al. (Eds.): ICCS 2006, Part IV, LNCS 3994, pp. 252–259, 2006.
© Springer-Verlag Berlin Heidelberg 2006

Integration of an application still requires a creation of a bridge which will plug the application into the platform. This process needs programming effort to develop a component which will communicate with an application, providing the platform with its functionality. This component is called an adapter; the process of adapters' development can be made easier by providing intelligent tools for developers. These tools can make the development of the adapter semi-automatic by providing an expert system to select an adaptation strategy together with a knowledge base for storing a description of it. Such a solution can accelerate adapter creation making the whole process more reliable and error prone and in the effect – increasing the application ROI. Also, by using the knowledge base the best adaptation practices can be stored and reused to increase the overall performance of the final component.

This paper presents a software solution for semi-automatic creation of the adapter. It describes its basic components and shows how they use each other to provide adapter logic for communication with the application.

2 State of the Art

A state-of-the-art section is split into two parts – the former, describing the classical software created especially to help with migration of legacy code to the Grid platform and the latter, which presents sample commercial integration solutions.

2.1 Migration of Legacy Applications to the Grid

There are two existing solutions which can support migrating legacy applications to the Grid environment.

GEMLCA [1, 2] focuses on migration of software, which represents typical computational problems. The range of its usage is rather limited, because business problems in the context of an integration are more related to data and distributed application API access rather than to performing computations.

The concept presented in LGF [3] is far more universal. Mainly, LGF is dedicated to wrap a legacy system and allow transactional and concurrent invocation of its methods. But it requires the user to provide an adapter, which is responsible for communication with the legacy system. The effect of the migration process (presented in Fig. 1) is a Globus Toolkit service, which connects to the

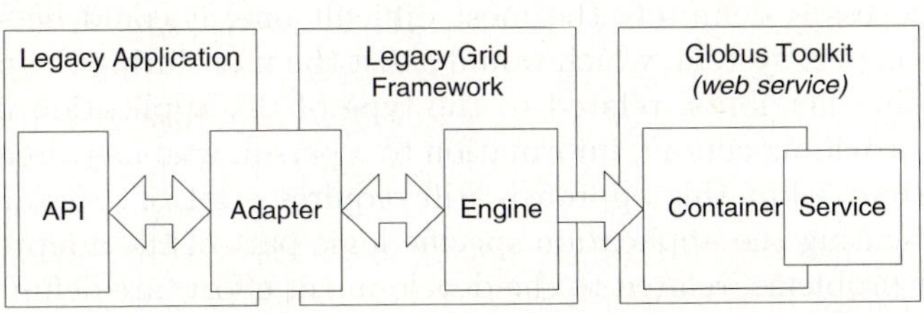

Fig. 1. Scheme of the migration process using LGF

application by Legacy Grid Framework that uses the adapter to communicate with the application.

2.2 Integration Platforms

Many commercial software products exist on the market. These are often highly complicated, sophisticated software packages, which offer complete solutions to a set of problems in the enterprise. They typically include the following components:

- Communication layer – responsible for exchanging messages between applications (e.g. IBM MQ Series, webMethods Broker, other JMS providers),
- Business process engine – that allows for easy creation of business processes using services provided by integrated applications (e.g. webMethods Modeller, Tibco StaffWare, JBoss jBPM),
- Integration server – responsible for hosting wrapped functionality of legacy applications and making it accessible on the platform.

Some products offer only partial integration functionality, for instance Tibco Rendezvous [4] offers only a middleware for connecting applications, without other services. All the above mentioned products lack functionality of easy adaptation of applications. Of course, adapters for popular products such as databases and JMS providers exist, but for tailor-made systems there is always necessity to create a dedicated adapter, that wraps its specific functionality.

3 Main Problems with Legacy Migration

Usually many problems which arise while migrating a legacy application to the Grid environment are related not only to the architecture of a particular application or a technology used, but also to development effort of the adapter, which will act as a bridge between the Grid and the application.

Specific problems of any particular application are related to its construction. A particular application can have no interface to which we can easily connect and exchange data, it is also likely to use domain-specific data structures, which have to be converted to common data structures called Canonical Data Model and the modification of the application interface may not be possible because of specific reasons, some of which had been described in [7]. This part of the migration process is definitely the most difficult one; it could be simplified by arranging an expert system, which would assist the user during the process. Such a system could offer hints, related to the type of the application interface and also help by providing contact information to a person who has already migrated a similar system – but this approach still requires a lot of programming effort while implementing the application specific logic part of the adapter.

The latter problems, related to the development effort, are definitely easier to solve. A general skeleton of the adapter can be implemented and the missing part of it, which is application specific logic, can be injected in it while the application is deployed on the integration platform. Of course, the adapter skeleton has to be dedicated for every legacy application migration framework, but it is easy to provide a set of skeletons for every framework because of a low number of them.

4 The Platform

The main responsibility of the platform is to simplify the process of developing the adapter for a particular application. As mentioned above, universal frameworks for adapting applications already exist, but they need adapters for performing communication with a particular application. The adapter, can be created on the fly, by using a semi-automatic method of creating communication logic. This can be done by providing a formal application interface description, a set of adaptation strategies and a selector for them.

4.1 Adapter

The adapter is an element which is responsible for handling communication between the application and a particular legacy application migration framework. It can be composed of two elements (cf. Fig 2):

- *Logic* – which is an application specific element; its responsibility is to handle communication with this particular application, including invocation of methods and conversion of applications' data structures to the Canonical Data Model; logic expose a well-defined interface to a skeleton,
- *Skeleton* – which is a common element; its responsibility is to provide a runtime environment for *the Logic* element.

The Logic is embedded in *the Skeleton* – it makes methods of the adapted application available to the legacy application migration framework. A method call is invoked on *the Skeleton* and it is passed to *the Logic*, which calls the legacy application method.

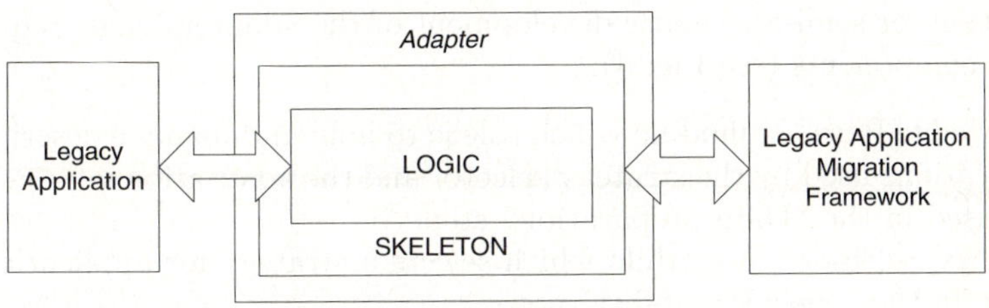

Fig. 2. Adapter block diagram

4.2 Application Adaptation Process

Development of the Adapter can be automated by providing a common Adapter Skeleton, compatible with a particular legacy application migration framework,

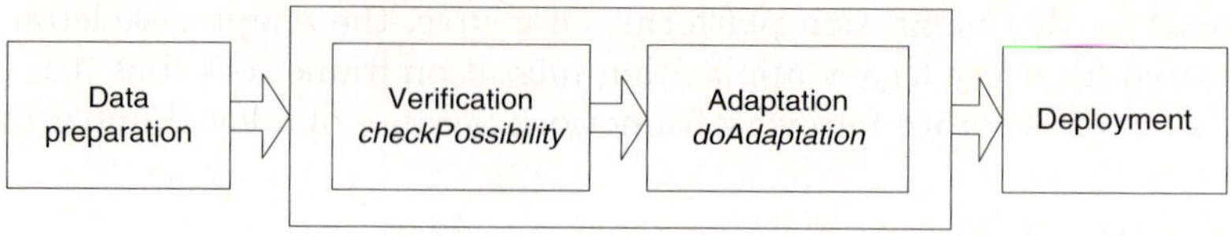

Fig. 3. The overall process of adaptation

and a mechanism for semi-automatic creation of the Adapter Logic. *The Skeleton* is delivered with the platform but *the Logic* must be generated for every application separately because it is application specific.

Development of *the Logic* can be made semi-automatic by utilizing an expert system to select a proper adaptation logic strategy together with a knowledge base to get description of this application interface. The process of adaptation consists of the following phases (see Fig. 3):

- data preparation – it consists of steps related to preparation of knowledge for a given kind of application, i.e., annotating its interfaces and defining the strategy; if already migrated, there is no need for this step,
- verification of data – the "checkPossibility" step consists of actions related to checking if there is any available strategy for a given application, if that application is described properly and if the knowledge base is ready to use,
- adaptation – the "doAdaptation" step describes all actions related to preparation of *the Skeleton* class (or classes), also, in this step the knowledge base is queried for the application description according to the strategy that has been determined in the previous step; finally *the Logic* is generated,
- deployment – the final step of adaptation, currently all required classes are prepared for use with a particular legacy application migration framework, i.e. they can be deployed as a service representing the adapted application.

4.3 Adapter Creation

The solution for semi-automatic development of the adapters' logic requires the following components (see Fig. 4):

- Knowledge Base – a module which role is to hold and manage descriptions of applications used by the strategy selector and the adaptation logic generator; it is used in the "Data preparation" step,
- Strategy Selector – a module which selects a strategy for application adaptation in the "checkPossibility" step,
- Adaptation Engine – a core part of the implementation which uses knowledge base and strategy selector to generate the adapter; it performs the mentioned "doAdaptation" step.

The above components are loosely coupled, so it is allowed to replace a concrete implementation of each component by the solution more suitable for the current needs. For example, if the knowledge base uses an ontology approach to

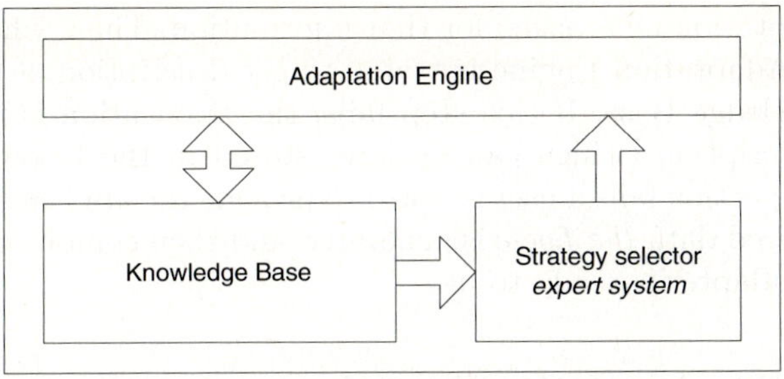

Fig. 4. Adaptation Platform architecture

represent data it can be replaced by another knowledge base implementation, using 'if-then' rules to express knowledge. The sufficient condition is that the replacing element should fulfill the interface contract with other components.

4.4 Knowledge Base

Descriptions of applications and their interfaces are stored in the knowledge base component of the system. The current implementation uses simple XML in-memory database approach to store them, but an ontological one seems to be more robust and efficient. When the application or its interface are described in this more formal way than a simple structured XML file, it is easier to reuse gathered knowledge while adapting other similar applications. This description could be an ontology individual stored in Grid Organizational Memory [5, 6].

4.5 Strategy Selector

The strategy selection process is supported by an integrated expert system. It contains descriptions of strategies with links to particular entities in the knowledge base. The adaptation strategy is a set of steps which have to be performed to create the application specific logic element of the adapter.

The current approach to strategy selection makes use of Bayesian Networks [8]. The overall process of selecting the adaptation strategy relays on the user feedback. The strategy selector asks the user a couple of questions. By the use of the Bayesian approach the system can suggest the best answer based on its percentage probability (e.g., there are more database systems than application server ones, so the system can suggest the user that it is more probably that he adapts a database system). The system can have predefined strategies (e.g. for C-language function adaptation, for Oracle PL/SQL procedure, etc.).

4.6 Adaptation Engine

The engine is a core part of the platform. Its main responsibility is to generate *the Logic* for accessing a concrete application and create an adapter by combining the provided skeleton and that logic. If the user has not provided a configuration file for the adaptation he is asked for that information. Then, when the strategy is known, the Adaptation Engine is trying to get description of the application from the Knowledge Base. If this step fails, the application is introspected to develop its description (which can be later stored in the Knowledge Base for reuse). If introspection fails a user is asked to provide the application description. Based on gathered data *the Logic* is generated and then combined with a proper Skeleton. The adapter is ready to use.

4.7 Application Adaptation Example

Many backend applications have PL/SQL interfaces. Such an interface consists of a group of database procedures which are similar to the C-language methods, they have arguments of basic or complex types and there are no return values. The only difference is that arguments can be input, output or both – input and output.

The Adaptation Engine (AE) asks the user a set of questions regarding the application to adapt. The conclusion is that PL/SQL Procedure strategy should be applied to the adaptation process. Following the selected strategy, AE requires additional information from the user about database connection parameters (hostname, port number, database name, user name and password). Based on these values it connects to this database by using JDBC [9] and performs introspection to get a list of existing procedures. The user selects the procedure to adapt and then AE gets procedures' attributes and tries to map them to a particular common data type from the Knowledge Base (KB). For instance, if the procedure name is *add_contact*, it probably will have attributes like *first name, last name, date of birth, address*, etc. KB looks through existing mappings and finds, that such attributes belong to the *ContactModel* type. AE creates an adapter for the procedure *add_contact*, which takes only one attribute of the *ContactModel* type, being able to connect to a particular database instance and invokes that procedure.

5 Conclusions and Future Work

The solutions used to migrate legacy applications to integration platforms (cf. sections 2.1 and 2.2) require an efficient adaptation layer to communicate with existing legacy applications. Such a layer can be created from scratch every time when a new application is integrated for the price of higher time-to-market period, lower efficiency, lack of serviceablity and quality. There is a need for a dynamic adaptation layer, i.e. a set of components, which can help with adaptation using semantic description of services and data. Also, there is a place for an expert system to support their selection in a semi-automatic way.

The main target of presented solution is to make an adaptation process easier and semi-automatic. Use of the expert system for adaptation strategy selection together the knowledge base to keep description of application allow the user to easily extend it with new functionality and scale to fit one's needs.

Acknowledgments

This research has been done in the framework of EU IST-2002-511385 K-Wf Grid project. Discussions with Prof. Gabriele von Voigt and Nils Jensen from the University of Hannover and AGH-UST grant are also acknowledged.

References

1. Delaitre, T., Goyeneche, A., Kacsuk, P., Kiss, T., Terstyanszky, G.Z., Winter, S.C.: GEMLCA: Grid Execution Management for Legacy Code Architecture Design, Conf. Proc. of the 30th EUROMICRO Conference, Special Session on Advances in Web Computing, Rennes, France (2004) 477-483
2. Terstyanszky, G., Delaitre, T., Goyeneche, A., Kiss, T., Sajadah, K., Winter, S.C., Kacsuk, P.: Security Mechanisms for Legacy Code Applications in GT3 Environment, Conf. Proc. of the 13th Euromicro Conference on Parallel, Distributed and Network-based Processing, Lugano, Switzerland (2005) 220-226
3. Baliś, B., Bubak, M., Węgiel, M., A Solution for Adapting Legacy Code as Web Services, in: V. Getov, T. Kielmann (Eds.), Component Models and Systems for Grid Applications, Springer (2005) 57-75.
4. Tibco: TIBCO Rendezvous(TM) Concepts, Software Release 7.0, April 2002, http://www.tibco.com.
5. Krawczyk, K., Słota, R., Majewska, M., Kryza, B., Kitowski, J.: Grid Organization Memory for Knowledge Management for Grid Environment, in:Bubak, M., Turala, M., Wiatr, K. (Eds.), Proceedings of Cracow Grid Workshop - CGW'04, Dec.13-15,2004, ACC-Cyfronet AGH, Cracow (2005) 109-115.
6. Kryza, B., Majewska, M., Słota, R., Kitowski, J.: Unifying Grid Metadata Representations through Ontologies, in Wyrzykowski, R., et al.(Eds.), Proc.Inter.Conf. PPAM2005, Sept.12-14, Poznan (2005)
7. Abramson, D.,Kommineni, J.,McGregor, J. L.,Katzfey, J.: An Atmospheric Sciences Workflow and its Implementation with Web Services, Future Generation Computer Systems, 21(1) (2005) 69-78
8. Friedman, N., Geiger, D., Goldszmidt, M.: Bayesian Network Classifiers. Machine Learning, Kluwer, 29 (1997) 131-163.
9. http://java.sun.com/products/jdbc/

A Self-configuration Mechanism for High-Availability Clusters*

Hocheol Sung[1], Sunyoung Han[1,**], Bok-Gyu Joo[2], Chee-Wei Ang[3], Wang-Cho Cheng[3], and Kim-Sing Wong[3]

[1] Department of Computer Science and Engineering, Konkuk University, Korea
{bullyboy, syhan}@cclab.konkuk.ac.kr
[2] Department of Computer and Communications, Hongik University, Korea
bkjoo@hongik.ac.kr
[3] Networking Department, Institute for Infocomm Research, Singapore
{angcw, chengwc, wongks}@i2r.a-star.edu.sg

Abstract. Most high-availability (HA) solutions currently used are based on the pre-configuration done by human administrators. If the configuration between machines participating in the HA cluster can be automated, services clustered can be provided more efficiently. For realizing this concept, the server agent and service description server (SDS) are designed and implemented in this paper. The server agent exchanges several messages with other servers and SDS. SDS is the central server that manages information needed for each machine to do "self-configuration". We also implement a web-based monitoring tool to watch the status of the overall system.

1 Introduction

Most high-availability (HA) solutions depend on human administrators to do "pre-configuration" [1][2]. For example, in the Novell clustering solution, administrators have to assign available secondary servers to a primary server. For this reason, the performance of the overall system depends on the expertise of the administrators and administrators' error can reduce overall system reliability. Moreover, in mobile network environments, doing "pre-configuration" is very difficult because servers can move to another network [3]. Another problem caused by "pre-configuration" is the quality of services clustered. The state of the secondary server can be changed during the course of service provisioning. However, in the pre-configured cluster, the state of the secondary is not reflected in configuration of cluster.

Considering this point of view, in this paper, we propose the high-availability system that is able to self-configure. Each server joins a network looks for machines that can act as its secondary server and assigns the optimal machine to its secondary server dynamically. And each server must advertise its services to other machines in the network so that they can know "who can be my secondary server in this network".

* Research based on the EDEN2 project (a collaboration between Institute for Infocomm Research, Singapore and Konkuk University, Korea).

** Corresponding author.

V.N. Alexandrov et al. (Eds.): ICCS 2006, Part IV, LNCS 3994, pp. 260 – 263, 2006.
© Springer-Verlag Berlin Heidelberg 2006

For these purpose (service discovery and advertisement), we newly introduce a central server called the Service Description Server (SDS) in this paper.

2 Operational Overview

Essentially, all participating servers including SDS need to exchange messages for self-configuration. First, when server 1 is up, it sends a registration message to SDS. Whenever receiving a registration message, SDS has to send a table update message to each server on the network. The table update message contains the list of server that can back up the primary service of the server 1 and each backup server's status information, "priority". For deciding "priority" of each server, SDS should receive a loadlevel message periodically from each server registered with SDS. After some times, server 2 is started, it will send a registration message to SDS and receive a table update message from SDS. After server 2 registered, server1 will receive another table update message from SDS. It may contain backup server information, if server 2 can be a secondary server for server 1's primary service. After getting the table update message, server1 will send an assignment message to server 2, telling server 2 to be its secondary server and listen to its keep-alive message. Similarly, when server 3 is started, server1 and server 2 will get table update message from SDS and assign secondary server for its own primary service. In this case, server 1 should choose its secondary server between server 2 and server 3 based on the status of each server. Of course, server 2 may assign server3 to its secondary server and send a keep-alive message to server 3 continuously.

In case, server 1 is down, server 2 that is secondary server will not receive a keep-alive message from server 1. After some times, server 2 will know server 1's down and send a takeover message to SDS informing SDS that it will run the service provided by server 1. As receiving a take over message, SDS will update its database and send a table update message to each server on the network. After getting the table update message form SDS, server 2 will assign another secondary server (server 3 in this scenario) for the service taken over and send a keep-alive message to server 3. After running for a while, server 1 is restarted and does registration with SDS. When server 1 receives a table update message from SDS, it will know who run its primary service and send a release service message to server 2 that took over the service from server 1 in previous scenario. When server 2 receives the release service message, it will stop the service that is the server 1's primary service and send a remove assignment message to server 3 informing server 3 of no need to listen to a keep-alive message from server 2. After getting acknowledgment from server 2, server 1 will starts the primary service and send a takeover message to SDS to update its database. After receiving the table update, server 1 will repeat the secondary assignment as mentioned earlier.

3 Experimental Results

For testing our implements, we set up test-bed in a local network as shown in Fig. 1. Although we use 5 Linux systems for our test, only 3 application servers such as the HTTP, FTP and EMAIL server participate in the self-configuration.

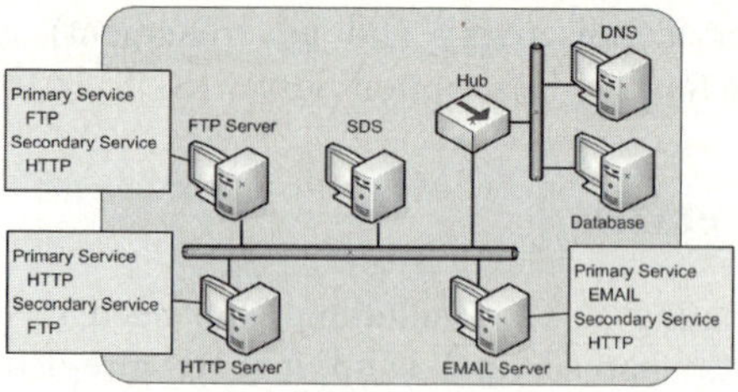

Fig. 1. Local test-bed

The database shown in Fig. 1 is for storing all the server necessary files such as configuration file, web page files and etc. Although not mentioned earlier, we also need a DNS server in local test-bed. Because most users use the FQDN to reach the server, each server has to register its name to the DNS. When the primary service is taken over by the secondary server, it should send dynamic update request to the DNS to point the service FQDN to the secondary server's IP address [4].

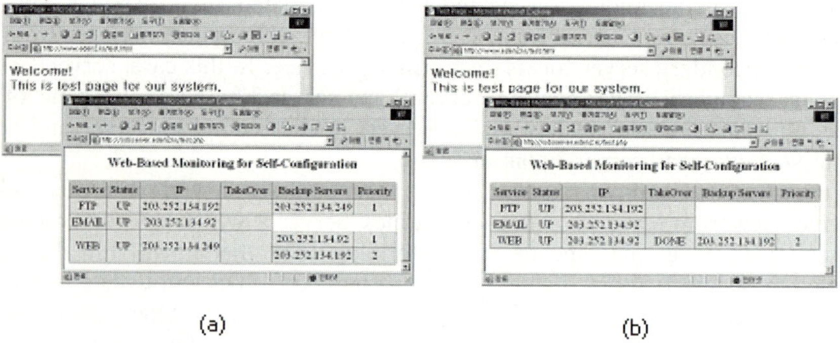

Fig. 2. (a) All application servers are normally in operation and (b) the HTTP server is down and service failover is initiated

Fig. 2 (a) shows when all application servers are registered and normally in operation. In this case, as the priority of the EMAIL server is higher than that of the FTP server (lower value is higher priority), the EMAIL server is assigned to be a secondary for HTTP service. Next, Fig. 2 (b) shows when the HTTP server is down and service failover is initiated. Because the HTTP server is down, the secondary server (the EMAIL server) for HTTP service is now providing the service and appoints another secondary server (the FTP server) for HTTP service. Finally, when the HTTP server recovers from fail and service failback is initiated, the secondary server releases the http server and revokes the assignment of another secondary server. Also, FTP server will be a secondary for HTTP server as shown in Fig 1 (a).

For evaluating the performance of the system, we simulated the actions of clusters and defined 2 types of models for this simulation, ***static clustering model*** and ***dynamic clustering model***. The static clustering model is for pre-configured clusters.

In this model, one of backup servers is assigned to a secondary server for all service time. The dynamic clustering model is for self-configured clusters. In this model, a primary server can select an optimal backup server as a secondary server at any time.

For this simulation, we configure the cluster with one primary server and 2 backup servers and run the service for 24 hours (simulation time). During the simulation, the primary server dies and recovers randomly according to its load and we calculate the response time for client's requests. From the client's point of view, the response time for service requests is the major metric for evaluating the service availability [5]. Fig.3 shows the simulation result. For static clustering model, as server's load average increase, the response time for service requests also increase. But, for dynamic clustering model, the response time for service requests doesn't increase.

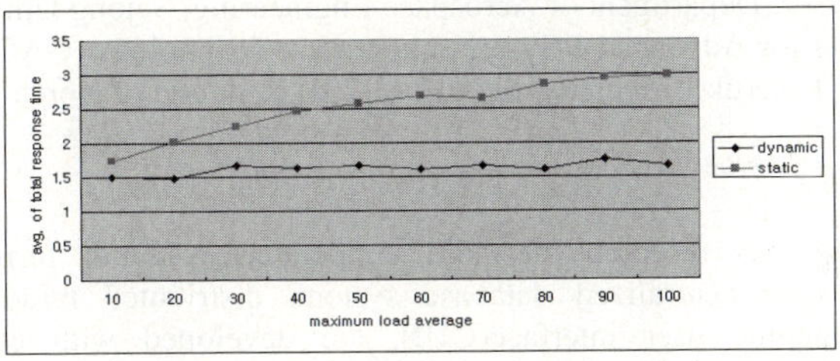

Fig. 3. Relation between each server's load average and response time for service requests

4 Conclusion

This paper presents a self-configuration concept for high-availability clustering system. Each machine participating in self-configuration can assign the secondary server for its primary service dynamically, based on the status of each backup server. As a result, when the primary server fails, its service can be automatically failover by the optimal backup server and the overall quality of service can be maintained well.

References

1. The High-Availability Linux Project, http://www.linux-ha.org
2. Novell, http://www.novell.com
3. Hocheol Sung, Sunyoung Han: Server Mobility using Domain Name System in Mobile IPv6 Networks, ICCS 2004 - LNCS3036, June 2004
4. Vixied (Ed.), P., Thomson, S., Rekhter, Y. and J. Bound: Dynamic Updates in the Domain Name System, RFC 2136, IETF, April 1997.
5. Enrique Vargas: High Availability Fundamentals, Sun BluePrints™ OnLine, November 2000.

Development of Integrated Framework for the High Temperature Furnace Design

Yu Xuan Jin[1], Jae-Woo Lee[2,*], Karp Joo Jeong[3], Jong Hwa Kim[4], and Ho-Yon Hwang[5]

[1] Graduate Research Assistant, Department of Advanced Technology Fusion
[2] Professor, Department of Aerospace Engineering
jwlee@konkuk.ac.kr
[3] Professor, Department of Internet & Multimedia Engineering
[4] Professor, Department of Industrial Engineering
[5] Professor, Department of Aerospace Engineering, Sejong University,
Center for Advanced e-System Integration Technology (CAESIT)
Konkuk University, Seoul 143-701, Republic of Korea

Abstract. In this research the high temperature vacuum furnace design framework with centralized database system, distributed middleware and integrated graphic user interface(GUI), are developed with configuration module(CAD), thermal analysis module, optimization modules, and database management system. By considering the operation under the distributed computing environment, the method of managing the analysis modules and the optimization process is proposed and the integration method of database system and CAD system under local PC environment is demonstrated by carrying out example design problem.

1 Introduction

For the product design and development of the most small and medium enterprises, the design, analysis and the production department are geographically and business wise separated. Various software programs are scattered and need integration for the efficient design and development. Therefore, dissimilar and distributed computational environment and resources must be integrated seamlessly. Because the high temperature vacuum (gas) furnace requires elaborate technology, low production cost, and fast design cycle, an integrated design framework must be developed which can integrate the database operation, the design automation, and the integrated analysis.

General process of furnace design may start with user's requirements, such as target temperature, ambient gas and uniform temperature zone size. Materials and sizes of heater and insulator are decided based on the user's requirements. Then, the chamber size and type are determined. Thereafter, the vacuum system, the temperature control system and the cooling system are consequently determined. Finally, a sample-loading type is decided. Figure 1 shows the design process of a vacuum furnace.[1]

* Corresponding author.

V.N. Alexandrov et al. (Eds.): ICCS 2006, Part IV, LNCS 3994, pp. 264 – 271, 2006.
© Springer-Verlag Berlin Heidelberg 2006

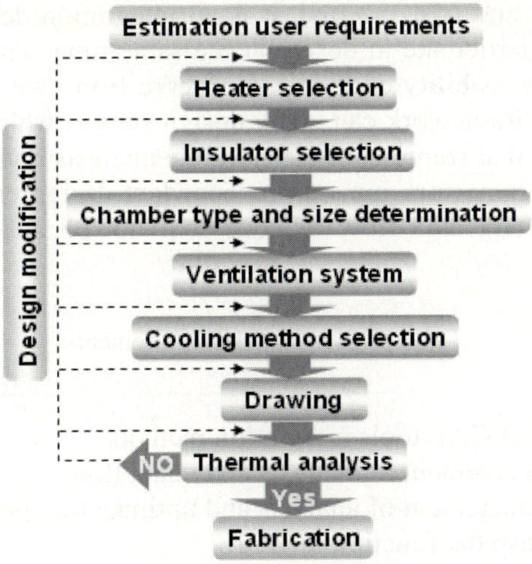

Fig. 1. Design Process

Since the design process is empirical and executed one after another, design issues that occur in the middle of the design or manufacturing can not be properly accounted. If the problems are not resolved within the design stage, whole design process must be repeated again. When the problems occur at the production stage, the development time and cost may be increased.

Existing method of design vacuum furnace is a trial and error method, which before producing vacuum furnace it is impossible to make an estimation performance. And it depends on experience. So when we produce vacuum furnace expense too much cost and it is not efficient. Consequently development of the efficient vacuum design technique with the optimization is necessary and the design framework which provides development environment is necessary. Utilizing design function user can judge the result with support data from GUI interface in the related design at the same time[2].

In this research the high temperature vacuum furnace design framework with centralized database system, distributed middleware and integrated graphic user interface(GUI), shall be developed with configuration module(CAD), thermal analysis module, optimization modules, and database management system.

2 Integration Architecture of High Temperature Vacuum Furnace Design Framework

To obtain the framework architecture under the distributed computing environment, framework's requirements and commercial framework software's architecture are to be analyzed. Table 1 shows framework requirements[3].

The framework development is summarized with two objectives. First, it is a design function which can efficiently accomplish MDO (Multidisciplinary Design

Optimization) at time and cost. Second, it is an integration design environment that the field experts can participate in design at the same time. The framework meaning of expandability is a possibility of trying to observe from two viewpoints. First, it is the expandability that framework can apply design subject which is possible. Second, it is the expandability that framework can add like analysis optimization codes which is possible. Because framework aims an independent design environment for design subject, link method and data flow of code will not be able to apply with fixed pattern.

Table 1. Framework Requirements

Basic Requirements	■ CAD/CAE tools integration method ■ Management of analysis code data flow ■ Management of analysis and optimization process ■ Grasp the function in GUI
Essential Requirements	■ Execute under the heterogeneous distributed environment ■ Centralized DBMS (Database Management System) ■ Detecting wrong data flow and process ■ Monitoring variables while process is executing ■ Delete, add, modify the design process ■ Create design variables for the complex design problem ■ Restart at wrong design point ■ The object oriented design for an expandability and standard development tool and implementation
Additional Requirements	■ Manage execution with scheduler ■ Parallel computing

The framework which considers expandability in distributed computing environment, the design plan is materializing design, analysis resources as 'component', and locating above the distributed environment system Layer which takes charge of a distributed environment. Considering this requirement, the high temperature vacuum furnace design framework centralizes distributed system(PLinda), integrated graphic user interface, configuration, database, analysis, and optimization modules. The figure 2 shows the development concept in high temperature furnace design framework.

2.1 High Temperature Vacuum Furnace Design Framework System Architecture

The high temperature vacuum furnace design framework was constructed of 6 modules. Using distributed middleware for integrating distributed resource, select Microsoft Access for managing data, and selecting Visual Basic language are easy to

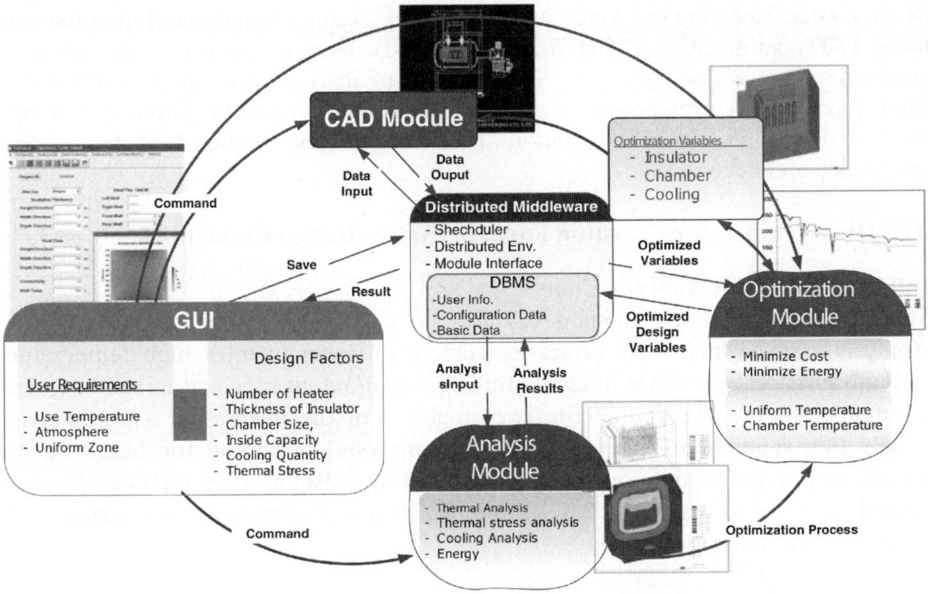

Fig. 2. Framework Development Concept[4]

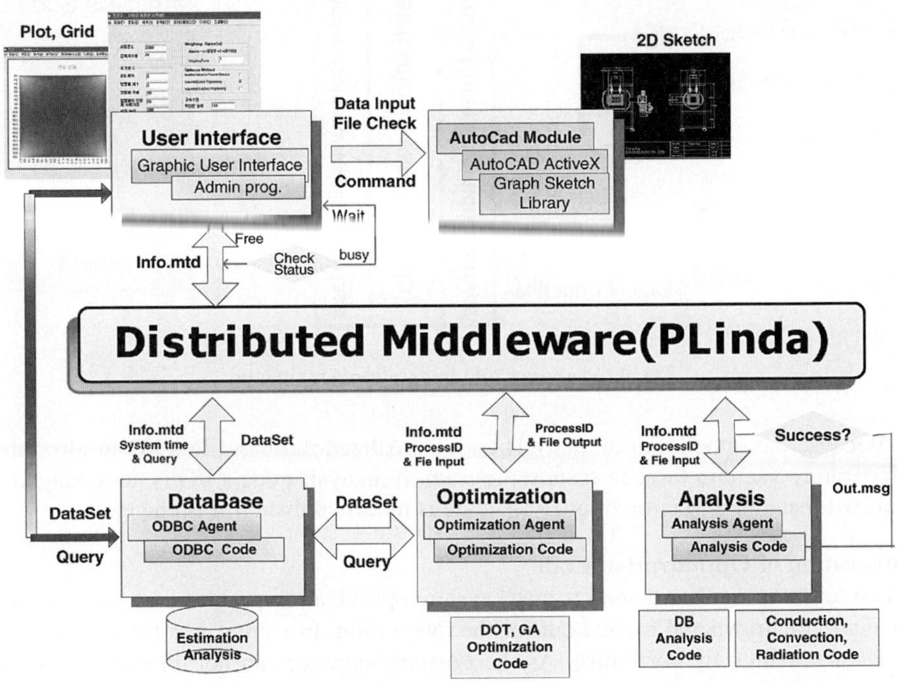

Fig. 3. System Integration Architecture

develop GUI. Selecting the AutoCAD for drawing vacuum furnace configuration, and using DOT optimization Code(Gradient Method), thermal analysis code including conduction, convection, radiation. The accuracy of analysis code can be confirmed in final report of 'Development of High Temperature Vacuum Furnace Through Integrated Design and Optimization Technology'. The figure 3 shows system integration architecture [5].

2.2 High Temperature Vacuum Furnace System Design Framework

Integration Plan of Analysis Code

So far most integration method of analysis code is developed using I/O file. Analysis codes which only offers executable program in ultra high temperature vacuum furnace system are integrated using the in/output files and mixed language is employed when the source file is available. For the integration using analysis code's I/O, standard I/O metadata file and analysis code input file based on the parameter defined by the user are created. I/O file is displayed on GUI and is saved according to the order of user's selection and data type. Figure 4 is the structure of code integration using file I/O.

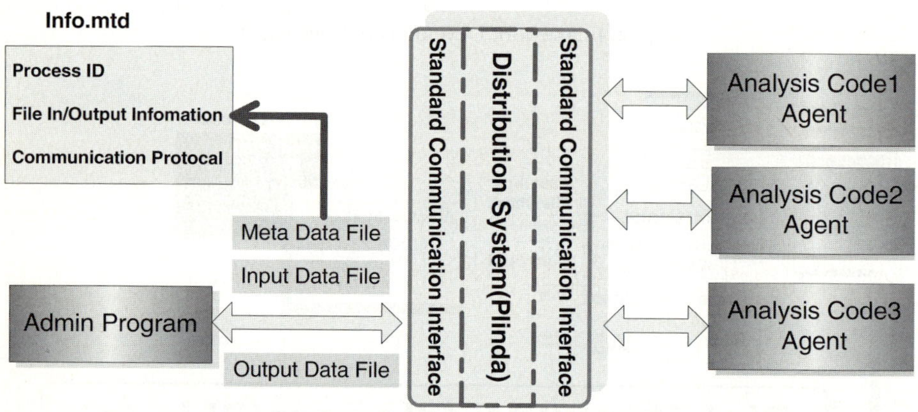

Fig. 4. Analysis Code Integration Architecture

Because the integration of analysis code is realized through file's I/O in ultra high temperature vacuum furnace system program, if analysis code's I/O is not changed, it required that a new version of analysis code is modified even if it is changed.

Integration of Optimization Code

When formulating the vacuum furnace system optimization problem, we read existing database are drawn and carried out for the integration. In many cases the optimization codes are written by Fortran[6]. As the Fortran language can not directly access the database, C++ and Fortran are used together. Figure 5 shows the integration concept of the optimization code and the analysis modules.

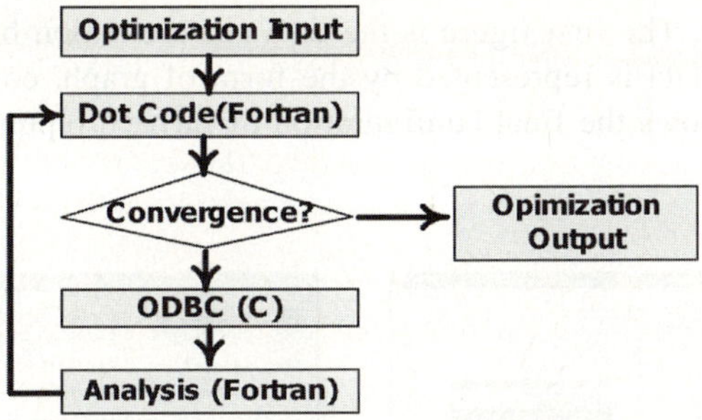

Fig. 5. Optimal Code Integration Architecture

Integration of Database

There were two kinds of methods in this system for integrating database system. First method uses file's I/O. Though analysis code is same with the integration method, input file is realized in query and data values. When user needs data for executing analysis code or optimal data, after create query and connect database, user can draw out any data which user wants. Second method doesn't directly pass middleware from user interface GUI and input data or read data about vacuum furnace system products through connecting database. Figure 3 shows the integration structure of database.

Integration of CAD

AutoCAD[7] cannot go through the distributed middleware for designing system image, but realized GUI and integration directly. Figure 6 is the integration structure of CAD and GUI. Standard vacuum furnace system design map was used and image data was drawn out in AutoCAD. When vacuum furnace design using database reaches optimal design, the fixed map make full use of AutoCAD Block function to express detailed CAD image completely using seven design parameters.

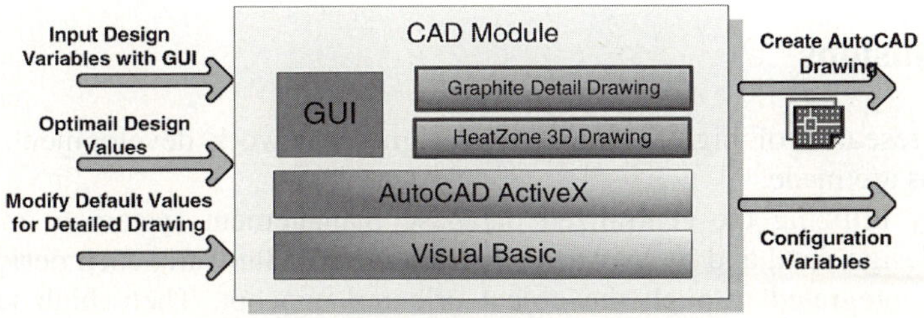

Fig. 6. AutoCAD Integration Architecture

3 Experiments

High temperature vacuum furnace design framework was developed on November 2004, which is now being used for actually furnace design and manufacturing. The following figures are the optimized result extracted by the iFuD(intelligent Furnace

Designer) system[4]. The first figure is the data representation by visualization tool-Formula One. The data is represented by the form of graph, contour and data grid. The fourth figure shows the final configuration of furnace represented by AutoCAD program.

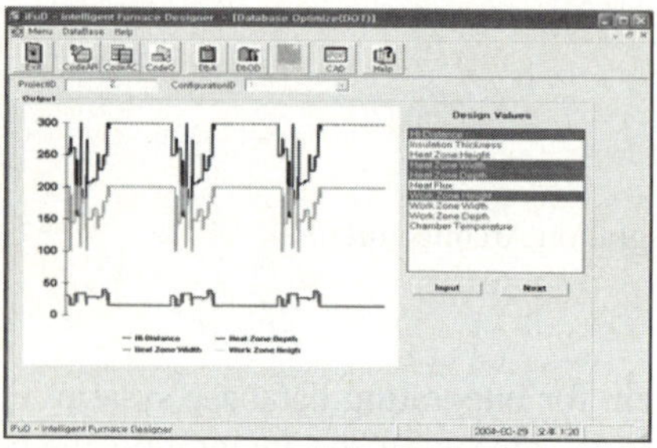

Fig. 7. Plot and Display -Graph

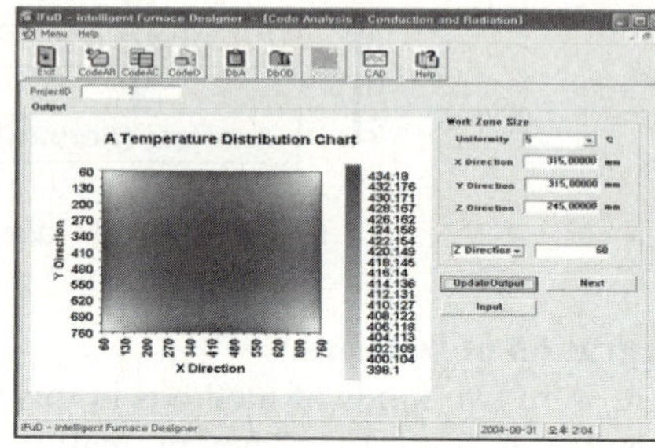

Fig. 8. Plot and Display -Contour

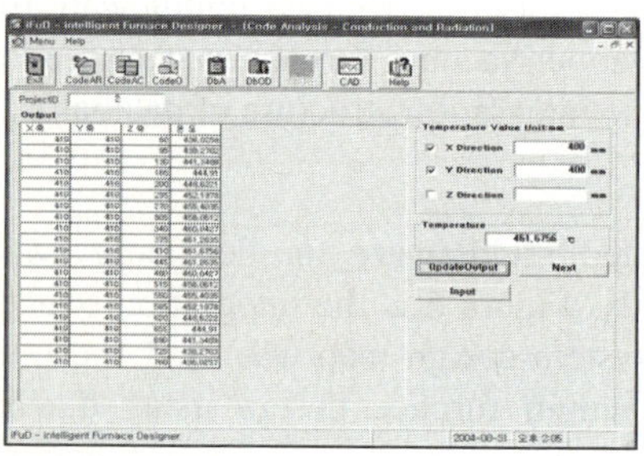

Fig. 9. Plot and Display –Data Grid

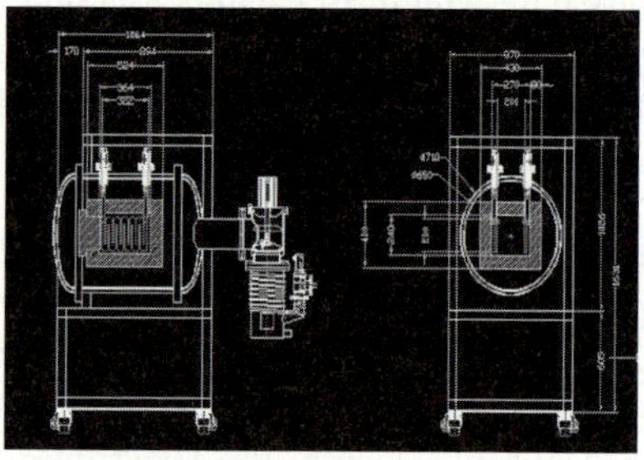

Fig. 10. Configuration for Manufacturing

4 Conclusions

From the research of high temperature design framework development, following conclusions are made.

First, by utilizing the centralized database management system to manage and integrate design data, and by using the distributive middle ware, each design analysis module is integrated through the object oriented concept. Then, high temperature design framework is developed by considering the expandability and easy modification

Second, the analysis codes are integrated using the in/output files and mixed language is employed when the source file is available.

Third, by considering the operation under the distributed computing environment, the method of managing the analysis modules and the optimization process is proposed and the integration method of database system and CAD system under local PC environment is demonstrated by carrying out example design problem.

References

1. Min-Ji Kim, Ho-Girl Jung, Jae-Woo Lee, Changjin Lee and Yung-Hwan Byun, "Optimal Design of an Energy Efficient Vacuum Furnace Using Thermal Analysis Database," *Key Engineering Materials, Vols 277-279, pp 732-740,* Jan. 2005.
2. Sang-Oh Cho, Jae-Woo Lee, and Yung-Hwan Byun, "A Study on the Integration of Analysis Modules and Optimization Process in the MDO Framework," *KSAS Journal*, Vol. 30, No. 7, Oct. 2002.
3. Salas, J. C. Townsend, "Framework Requirements for MDO Application Development, AIAA-98-4740. 1998.
4. Jae-Woo Lee, "Development of High Temperature Vacuum Furnace Through Integrated Design and Optimization Technology," Final Report, Jun. 2004.
5. Shenyi Jin, Kwangsik Kim, Karp-Joo Jeong, Jae-Woo Lee, Jong-Hwa Kim, Hoyon Hwang, Hae-Gok Suh, "MEDIC: A MDO-Enabling Distributed Computing Framework", *Lecture Notes in Artificial Intelligence,* Vol. 3613 part 1, Springer-Verlag, pp. 1092-1101, Aug. 2005.
6. DOT Users Manual, Vanderplaats Researtch & Development Inc., 1995.
7. AutoCAD User Manual(ActiveX and VBA Developer's Guide), Autodesk Inc., 2002.

A Distributed Real-Time Tele-operation System Based on the TMO Modeling*

Hanku Lee[1] and Segil Jeon[2,] **

[1] School of Internet and Multimedia Engineering, Konkuk University, Seoul, Korea
hlee@konkuk.ac.kr
[2] BioMolecular Informatics Center, Konkuk University, Seoul, Korea
Tel.: +82-11-9154-3793
sgjeon@ricl.konkuk.ac.kr

Abstract. The fast development of grid computing environments makes it possible to access geographically distributed remote instruments, experimental equipments, databases, human resources with respect to real-time in grid computing environments. With conventional programming method, it is very difficult to implement real-time models in uncontrolled distributed environments and to support well-defined interfaces from real-time systems to external systems. We propose an easy-to-use TMO-based tele-operation model with less strict real-time constraints in grid environments. Using the proposed model, we design and develop a TMO-based tele-operation system for real industrial applications used for tsunami-detecting instruments.

1 Introduction

Today we access geographically distributed remote instruments, experimental equipments, databases, human resources, high-performance computers, etc, as if accessing local resources from a long distance away. But it brings us another side: How are these instruments, devices and data well-synchronized in distributed real-time systems? With conventional programming methods it is very difficult to design and implement well-defined real-time models in uncontrolled distributed environments.

We propose a TMO-based distributed real-time tele-operation model with less strict real-time constraints in grid environments. The proposed model can be used to control tsunami-detecting instruments. For example, a remote meteorologist can control and monitor tsunami-detecting instruments and conference with local meteorologists from a long distance away using the proposed model in grid computing environments.

In the next section, we discuss related works such as TMO, Distributed Object-oriented Freeway Simulator (DOFS), Real-time CORBA. Then, we propose a TMO-based tele-operation model and mention design and implementation issues in section 3. Section 4 concludes.

---

* This paper was supported by Konkuk University in 2006.
** Corresponding author.

V.N. Alexandrov et al. (Eds.): ICCS 2006, Part IV, LNCS 3994, pp. 272 – 279, 2006.
© Springer-Verlag Berlin Heidelberg 2006

2 Related Works

The Time-Triggered Message-Triggered Object (TMO) was established in early 1990's with a concrete syntactic structure and execution semantics for economical reliable design and implementation of RT systems [1, 2, 4, 5]. TMO is a high-level real-time computing object. It is built in standard C++ and APIs called TMO Support Library (TMOSL).

TMO contains two types of methods, time-triggered methods (SpM), which are clearly separated from the conventional service methods (SvM). The SpM executions are triggered upon reaching of the RT clock at specific values determined at the design time whereas the SvM executions are triggered by service request messages from clients. Moreover, actions to be taken at real times which can be determined at the design time can appear only in SpM's. Real-time Multicast and Memory Replication Channel (RMMC) is an alternative to the remote method invocation for facilitating interactions among TMOs. Use of RMMCs tends to lead to better efficiency than the use of traditional remote method invocations does in the area of distributed multimedia applications that involve frequent delivery of the same data to more than two participants distributed among multiple nodes.

Distributed Object-oriented Freeway Simulator (DOFS) [3] is a freeway automobile traffic simulator conducting with the goal of validating the potential of the TMO structuring scheme supported by the recently implemented TMOSM. DOFS is intended to support serious studies of advanced freeway management systems by providing high-resolution high-accuracy easily expandable freeway simulation. The system can help the Driver avoiding the traffic road and supply real-time traffic information. The TMO scheme brings major improvement in the RT system design and implementation efficiency.

The Real-time CORBA (RT-CORBA) [6] is an optional set of extensions to CORBA to be used as a component of a real-time system. It is designed for applications with hard real-time requirements, such as avionics mission computing, as well as those stringent soft real-time requirements, such as telecommunication call processing. Moreover, to allow applications to control the underlying communication protocols and end-system resources, the Real-time CORBA specification defines standard interfaces that can be used to select and configure certain protocol properties.

3 The Proposed TMO-Based Tele-operation Model

3.1 Architecture

Due to the fast development of the Internet and grid computing environments, it is possible for engineers and researchers to access remote instruments and computing resources from a long distance away. However, we need to support real-time controls and the timing characteristics on these geographically distributed, grid-enabled, and real-time applications without pain during the development.

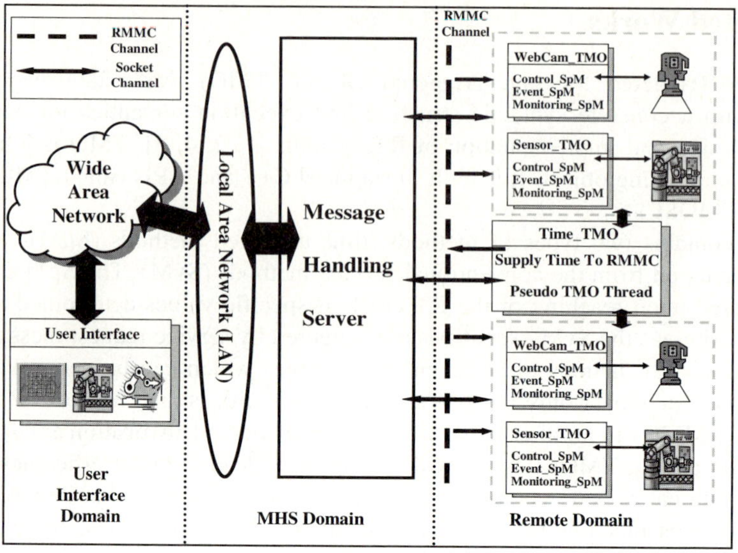

Fig. 1. The Architecture of TMO-Based Tele-Operation Model

Figure 1 depicts the architecture of the proposed TMO-based tele-operation model. One of the main issues for the proposed model is to apply the easy-to-use TMO to real-time applications that are usually hard to design and implement with conventional programming methods. The proposed model is divided to 3 domains: remote domain, message-handling-service domain, user interface domain.

The remote domain (RD) is to collect remote data and to monitor remote instruments. RD consists of the Time TMO and working TMOs. The Time TMO gives the timing characteristics to other working TMOs (e.g. WebCam_TMO and Sensor_TMO) via the Real-time Multicast and Memory Replication Channel (RMMC). The video, audio, and sensor data with the timing characteristics are transferred via the socket channel to the message-handling-service domain The time characteristics supplied by the Time TMO are more suitable to the proposed model than those supplied by the Internet or GPS time services since the Time TMO is closely located to other working TMOs and this locality avoids the network latency that makes it hard to synchronize real-time applications.

The message-handling-service domain (MHSD) is to manage message-handling servers in order to help data communication between UID and RD. MHSD provides the grid-enabled environments based on the TCP/IP-based client/server model and grid applications to handle control-messages between UID and RD to be safely and precisely transferred. MHSD should keep waking up, be started prior to other domains, and wait for control-messages. Servers in MHSD can storage a large amount of data from the remote domain and can provide the secure management of data from the remote domain to the interfaces.

Finally, the user interface domain (UID) is to provide user interfaces to check the status of the whole system, to manage incoming and outgoing control-messages between the client and remote instruments, and to handle real-time information needed for the tele-operation application for the client. This domain is implemented in MFC.

3.2 Implementation

In this section we mention several implementation issues for the remote domain in detail. Figure 2 represents the basic structure of the remote domain, called TMO-based real-time agent framework for the proposed TMO-based tele-operation system. The real-time agent framework is implemented using TMO toolkit [1]. It consists of the Time TMO and working TMOs (e.g. WebCam_TMO and Sensor_TMO). Moreover, it is basically divided into three services: control-message waiting and processing service, time service, and real-time data processing service.

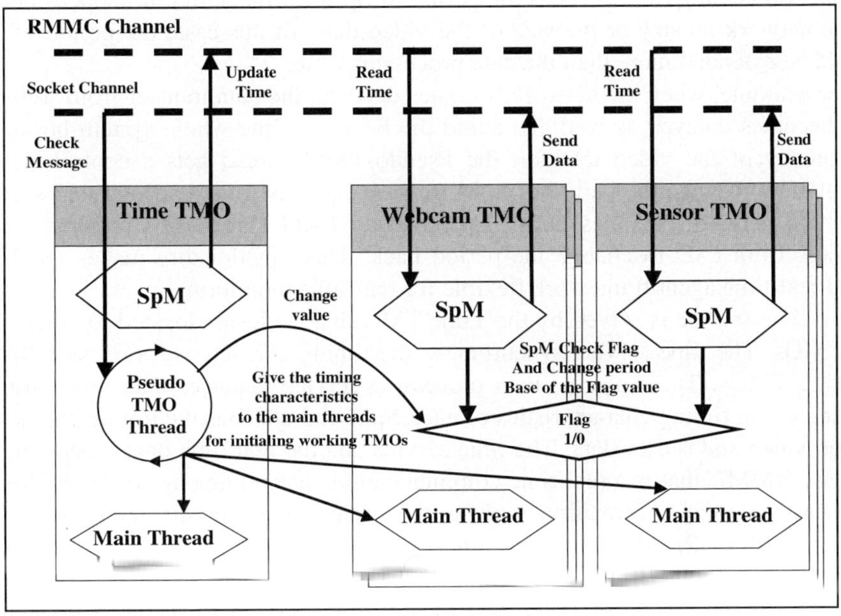

Fig. 2. The Remote Domain: TMO-Based Real-Time Agent framework

The control-message waiting and processing service waits for and processes control-messages from UID according to message types using the Pseudo-TMO-Thread. The Pseudo-TMO-Thread is located in the Time TMO and gives the timing characteristics to the main threads for initialing all working TMOs in the remote domain. But The Pseudo-TMO-Thread keeps waking up unlike SpM of TMO periodically wakes up. Major initialization steps for the agent framework are as follows:

1. The Pseudo-TMO-Thread is invoked and checks up the header information of control-messages from MHSD.
2. Each message is classified such as WebCamMessage, and is sent to its designated TMO.
3. Each TMO extracts the timing characteristics from its control-message, initializes its time, and builds up the socket channel among other TMOs.

4. The TMO middleware is activated.
5. The TMO-based real-time agent framework is activated.

The Pseudo-TMO-Thread keeps waking up and getting control-messages from UID. If a control-message is a normal command to control remote instruments, then the Pseudo-TMO-Thread does normal real-time data processing in this service. But if a control-message is a command to scale up or down the whole system in the time dimension, then the Pseudo-TMO-Thread extracts data out of the message, stops the designated SpM for a while, changes the period of the designated SpM, and restarts the designated SpM. When a working TMO needs to process video data, sometimes, the data processing time exceeds the period of SpM for the TMO. It happens because of the network latency or the size of the video data. In this case, the period of SpM should be extended more than the data processing time.

For example, when the network becomes delayed, the data transfer from web cameras becomes delayed as well. To avoid the latency of the whole system because of the latency of the video transfer, the Pseudo-TMO-Thread gets a control-message from UID to change the period of SpM for web cameras from 1 second to 3 seconds. After the network becomes normal, the Pseudo-TMO-Thread gets another control-message from UID to change the period back. This functionality makes the TMO-based real-time agent framework flexible for real-time monitoring.

The time service is served by the Time TMO that is closely located to other working TMOs. The time service synchronizes the timing characteristics of each SpM in working TMOs. The real-time data processing service manages data processing according to the timing characteristics of each SpM and attaches the timing characteristics on video and sensor data. The time service and the real-time data processing service use RMMC that is a real-time communication channel among working TMOs to broadcast common information such as the timing characteristics and memory for working TMOs. RMMC is a good alternative to the Internet or GPS time services since it avoids the network latency that makes it hard to synchronize real-time applications. SpM of the Time TMO periodically (e.g. 200 micro-seconds) updates the timing characteristics of RMMC using its own timing characteristics. Then each SpM reads the timing characteristics of RMMC, attaches it on video, audio, and sensor data, and transfer data to MHSD.

Figure 3, 4, and 5 represents a distributed, real-time tele-operation system to detect tsunami based on our proposed model. The TMO-based tele-operation system is a real-time, tele-operation, and tele-monitoring system. Using the tsunami detecting system, a meteorologist can monitor the ocean, control the instruments, and conference with local engineers from a long distance away. In detail, first, a remote meteorologist monitors the current status of a designated point of the ocean on his/her computer from a long distance away. Second, the video, audio, and sensor data from the point are collected and synchronized by the TMO-based real-time agent framework and transferred via MHSD to the remote engineer. Third, the data are updated and displayed on his/her computer. Finally, the remote meteorologist can control remote instruments by controllers on his/her computer, send control-messages such as scaling up or down the whole system in the time dimension, and in advance warn local meteorologists to prepare the natural disaster. Using the system, moreover, the remote meteorologist can chat and talk with local meteorologists.

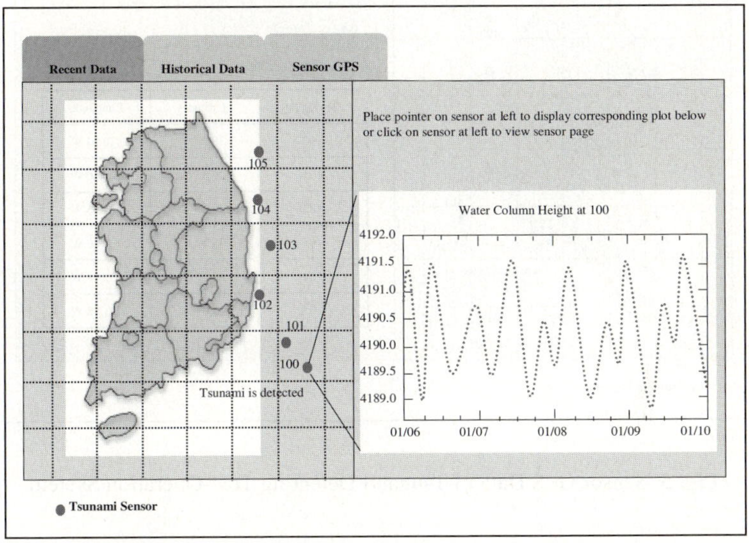

Fig. 3. Historical Data of Tsunami Detecting Tele-Operation System

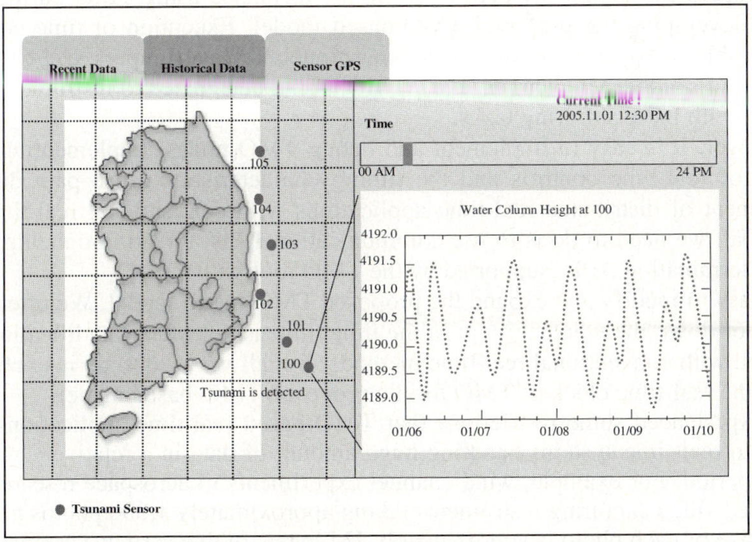

Fig. 4. Historical Data of Tsunami Detecting Tele-Operation System

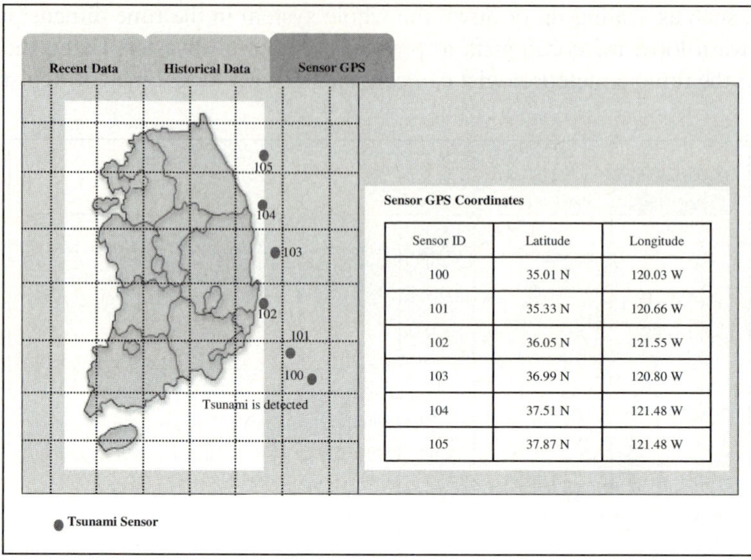

Fig. 5. Sensor GPS Data of Tsunami Detecting Tele-Operation System

3.3 Advantages and Restrictions of the Proposed Model

We experienced several advantages of adapting TMO on the proposed model during implementing a TMO-based tele-operation system.

Developers can highly predict the timing performance using TMO during designing and developing the proposed TMO-based model. Execution of time consuming unpredictable I/O operations such as video outputs, keyboard inputs, etc, can be handled by the Pseudo-TMO-Thread. Each TMO thread designated to instruments are not burdened with these suffering tasks.

Moreover, it is easy to implement and debug TMO nodes. Implementing and debugging of real-time controls and the timing characteristics cause pain during the development of distributed real-time applications with conventional real-time methods. But all we need to do is to use communication APIs, thread initializing, managing, and terminating APIs, supported by the TMO tool kit.

It is easy to modify and expand the proposed TMO-based model. We often need to scale up or down the whole system in the time dimension. Many modifications could be needed with conventional real-time methods. But all we need to do is to change the scale of the real-time clock of TMO for the proposed TMO-based model.

We experienced some restrictions that TMO-based real-time applications are not suitable to real-time systems handling huge amount of data in a relatively short SpM wakeup period. For example, wind channel experiments in aerospace researches generally need video capturing instruments taking approximately 1,000 photos per second and the size of each photo is approximately 1M bytes. In this case, we can easily scale down the period of SpM (e.g. 10 micro-seconds). But it is impossible to process this amount of video data in time with contemporary hardware and network environments.

Thus, we urge TMO-based real-time applications are suitable to systems with less strict real-time constraints such as tsunami-detecting instruments, etc, since those instruments product relatively small amount of data in the period of SpM and are not a time-critical decision model.

4 Conclusion

We proposed an easy-to-use TMO-based tele-operation model with less strict real-time constraints in grid environments. Using the proposed model, we designed and developed a TMO-based tele-operation system for real industrial applications able to be used to control and monitor tsunami-detecting instruments.

The TMO-based tele-operation model, proposed in this paper, is promising since it provides a sound TMO-based real-time agent framework, cost-effectively resolving the problems caused by conventional programming methods during the development. However, the experimental research and development with the proposed model is at an early stage. Moreover, much more research efforts are needed to develop more stable TMO-based real-time agent framework.

We will adapt the proposed model to develop a tsunami detecting system in the future research.

References

1. TMOSL_v4.0_manual_draft http://dream.eng.uci.edu/TMOdownload/
2. Kim,K.H, “APIs for Real-Time Distributed Object Programming”, IEEE Computer,June 2000,pp.72-80
3. K.H.(Kane) Kim, Juqiang Liu, Masaki Ishida and Inho Kim.: “Distributed Object-Oriented Real-Time Simulation of Ground Transportation Networks with the TMO Structuring Scheme” , Proc. COMPSAC '99 (IEEE CS Computer Software & Applications Conf.), Phoenix, AZ, Oct. 1999, pp.130-138.
4. Kim, K.H., "Real-Time Object-Oriented Distributed Software Engineering and the TMO Scheme", Int'l Jour. of Software Engineering & Knowledge Engineering, Vol. No.2, April 1999, pp.251-276.
5. Kim, K.H., "Object Structures for Real-Time Systems and Simulators", IEEE Computer, August 1997, pp.62-70.
6. Douglas Schmidt, Fred Kuhns, "An overview of the Real-time CORBA Specification", IEEE Computer special issue on Object-Oriented Real-time Distributed Computing, June 2000, pp.56-63.
7. Real-Time for Java Experts Group, “Real-time Specification for Java, Version 0.9.2,” 29 Mar. 2000, http://www.rtj.org/public.

A Sharing and Delivery Scheme for Monitoring TMO-Based Real-Time Systems*

Yoon-Seok Jeong, Tae-Wan Kim**, and Chun-Hyon Chang

Konkuk University,
Seoul 143-701, Korea
{ysjeong, twkim, chchang}@konkuk.ac.kr

Abstract. Devices and systems used in distributed environments, such as on trains or in power plants, should be able to respond to external changes in real-time. Real-time middleware such as one based on the TMO (Time-Triggered Message-Triggered Object) model is considered recently as a choice to adapt real-time properties to distributed systems. TMO middleware guarantees that systems in distributed environments operate in a reliable manner. This middleware does not have an adequate monitoring infrastructure or supporting tools used in distributed environments, however. As such, this paper proposes TSDS (TMO-based Sharing and Delivery Scheme) as a part of the monitoring infrastructure. This is configured within the TMO middleware to share and deliver monitoring data among TMO-based real-time systems. The results of experiments show that the TSDS instrumentation overhead is less than 1ms. This means that TSDS has little effect on the operation of the middleware.

1 Introduction

A real-time system aims to ensure a specific service is provided within a given period of time [1]. The performance of a real-time system depends on how accurately timing constraints are met through the use of real-time monitoring. Monitoring is essential for maintaining real-time systems in a stable manner, and many monitoring tools are being developed to support real-time systems. As well, tools that monitor the TMO (Time-Triggered Message-Triggered Object)-based real-time systems (hereinafter referred to as "TMO system") were developed [2, 8, 9]. However, existing tools that focus on single system-based monitoring cannot enable the sharing and delivery of data among systems, which is required to monitor distributed TMO systems. In addition, all or part of the monitoring function is executed at the application layer, requiring developer intervention, and eventually causing difficulty in automating monitoring.

In order to address such problems, this paper presents a TMO-based Sharing and Delivery Scheme (TSDS), which is a scheme for sharing and delivering monitoring data among TMO systems. This configures an infrastructure for monitoring within the

* This research was supported by the MIC (Ministry of Information and Communication), Korea, under the ITRC (Information Technology Research Center) support program supervised by the IITA (Institute of Information Technology Assessment).

** Corresponding author.

V.N. Alexandrov et al. (Eds.): ICCS 2006, Part IV, LNCS 3994, pp. 280 – 288, 2006.
© Springer-Verlag Berlin Heidelberg 2006

TMO middleware, an essential part of the TMO system, thereby allowing for the sharing and delivery of monitoring data.

The paper is structured as follows. Chapter 2 describes the TMO model and the problems in existing monitoring architectures. Chapter 3 provides a detailed description of the proposed TSDS. The results of the experiments on the TSDS instrumentation overhead are shown in Chapter 4. Finally, chapter 5 presents the direction for future research.

2 Backgrounds

2.1 TMO Model

The proposed scheme in this paper is modeled in such a way as to utilize the proven functions provided by the TMO model. The TMO structuring scheme was established in the early 1990's with a concrete syntactic structure and execution semantics for economical and reliable design along with an implementation of real-time systems [6, 7]. TMO is a syntactically minor and semantically powerful extension of the conventional object(s). As depicted in Fig. 1, the basic TMO structure consists of four parts.

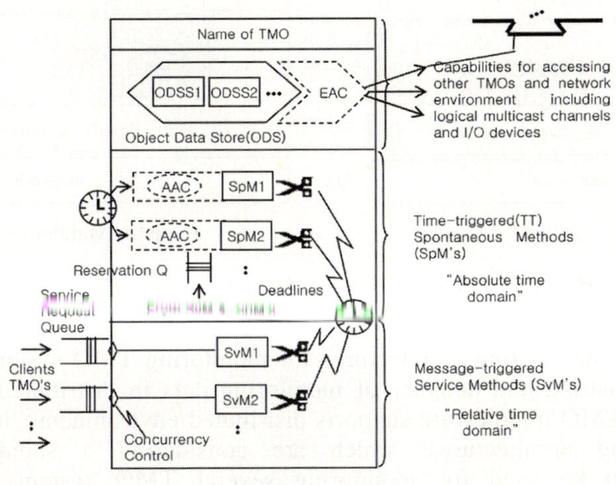

Fig. 1. The Basic Structure of TMO

- *Spontaneous Method (SpM):* A new type of method. A SpM is triggered when the real-time clock reaches specific values determined at design time. The SpM has an AAC (Autonomous Activation Condition), which is a specification of the time-windows for execution of the SpM.
- *Service Method (SvM):* A conventional service method. A SvM is triggered by service request messages from clients.
- *Object Data Store (ODS):* The basic units of storage which can be exclusively accessed by a certain TMO method at any given time or shared among TMO methods (SpMs or SvMs).
- *Environment Access Capability (EAC):* The list of entry points to remote object methods, logical communication channels, and I/O device interfaces.

2.2 Existing Monitoring Architecture

Some studies related to monitoring TMO systems have used architecture at the application layer to implement monitoring functions as shown in Fig. 1(a) [1, 8, 9]. A monitoring sensor positioned within a TMO application, an essential part of the monitor, gathers data on the corresponding object in the TMO application, and keeps them in a data store. This architecture has no direct effect on the operation of the middleware because monitoring TMO systems is conducted at the application layer. The drawback of this architecture is that a developer has to make monitoring elements such as a data store, sensors and a communication link for the sharing and delivery of monitoring data.

The other architecture used to implement monitoring functions is depicted in Fig. 2(b) [2]. A dedicated monitoring thread is added in the middleware and called up according to the middleware scheduling. While this architecture reduces developer intervention to make a store data and a sensor, the data stored in the data store can be accessed only by TMO objects in a single TMO system because the data store in this architecture is dependent on the middleware.

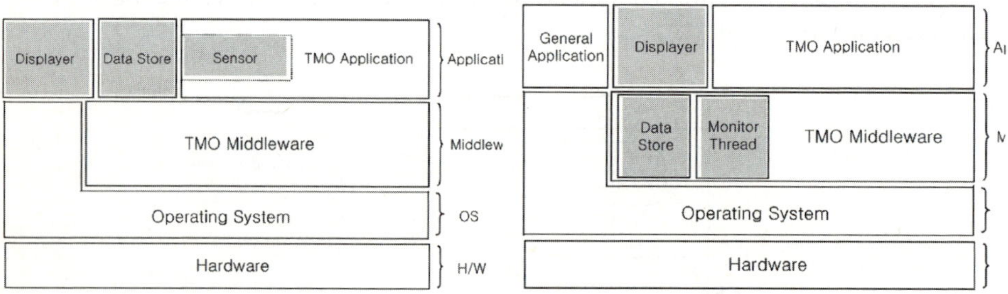

(a) at the Application Layer (b) at the Middleware Layer

Fig. 2. The Monitoring Architecture by Layer

In summary, the existing architectures for monitoring TMO systems do not have supported the sharing and delivery of monitoring data in distributed environments. Given that the TMO middleware supports distributed environments, it is difficult for both monitoring architectures, which are considering a stand-alone system environment, to be used for monitoring several TMO systems in distributed environments. Thus, a structure which supports data sharing and data delivery in distributed environments should be designed for monitoring TMO systems.

3 TMO-Based Sharing and Delivery Scheme (TSDS)

3.1 Revised TMO Model for Monitoring

The various types of middleware to realize the TMO model have been implemented in distributed environments [3, 4, 6, 10]. Each middleware has referred to the TMO model in Fig. 1 as a functional architecture. The TMO model does not include essential functions in distributed environments such as real-time monitoring, dynamic analysis, and fault-tolerance, however. As such, this paper has considered a monitoring concept from the outset of modeling in order to allow monitoring functions to be included as an infrastructure within the TMO-based middleware. Fig. 3 shows the proposed TMO model that supports monitoring functions. Some elements are added to this model in order for the monitoring concept to be applied to the TMO model.

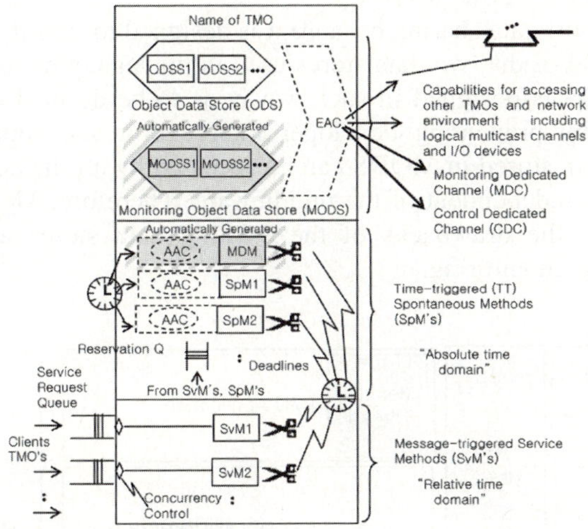

Fig. 3. The Structure of the TMO Model that Supports Monitoring Functions

- *Monitoring ODS* (MODS): Refers to the storage for sharing the results of monitoring TMO objects. It is the extended type of ODS that is assigned to each TMO object by a unit of MODS Segment (MODSS).
- *Monitoring Dedicated Method* (MDM): Refers to the dedicated SpM method for monitoring that is activated periodically in the same manner as general SpMs and transfers monitoring data in the data store.
- *Monitoring Dedicated Channel* (MDC): Refers to the dedicated channel used for the transfer of monitoring data among TMO systems.
- *Control Dedicated Channel* (CDC): Refers to the dedicated channel used for receiving control and feedback data from external systems or applications.

Our focus in this paper is the MODS, MDC, and CDC among abovementioned elements and the detailed elements in the middleware related to sharing and delivering the monitoring data. These elements are based on Linux TMO System (LTMOS) out of various types of the middleware as the basic platform for designing and implementing the sharing and delivery scheme. For more details of LTMOS, refer to [4, 10].

3.2 Monitoring Object Data Store (MODS)

The existing ODS was proposed as a storage to share data among TMO objects [6, 7]. On the other hand, the proposed MODS is designed to store monitoring data.

Fig. 4 shows the monitoring architecture proposed in this paper. Compared with ODS, MODS as a data store is positioned between the middleware and the application layer. This means that all the work related to defining and generating MODS is managed by the middleware. Thus, developers do not have to do MODS related work. This is caused by the type of data stored in the data stores. MODS keeps the formalized data such as execution time while ODS stores the various types of data that developers need. Therefore, it doesn't need additional information from the developers to define MODS.

MODS can support data sharing because it is designed to face the application layer. MODS is the hybrid of the two data stores used in the legacy monitoring schemes. In other words, MODS is designed in such a way as to be defined and created by the middleware, thereby preventing developer intervention and supporting automated monitoring. The data stored in MODS can be accessed by all the applications because it is designed to be independent of the middleware. Therefore, MODS is a data store that can overcome the drawbacks of the existing data stores and is suitable for distributed real-time monitoring.

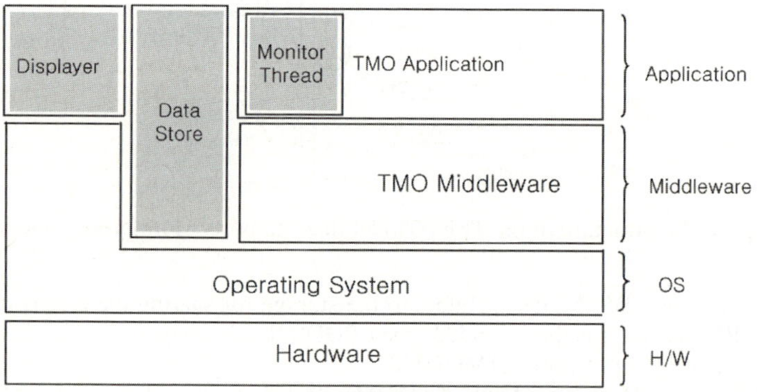

Fig. 4. The Revised Monitoring Architecture

3.3 Delivery Structure for Monitoring Data

Existing monitoring tools do not provide a special way to transfer monitoring data among TMO systems. Given that TMO systems operate in distributed environments, tools should be able to basically monitor each TMO system and deliver the gathered data to the server for analysis or other systems. As such, this paper presents a TMO-based delivery structure, which is a structure for delivery of monitoring data among TMO systems.

This structure is designed to have two separate layers that make functions of the data manipulation and data delivery independent of each other. This functional independency allows that a function for data manipulation can be used in all the TMO systems with different communication environments. The layer for data manipulation contains functions related to gathering, storing, and analyzing monitoring data. EMCB (Extended Method Control Block) as an internal data store, sensors which gather monitoring data, and MODS, which supports data sharing, are monitoring elements for data manipulation. This paper focuses only on MODS out of these monitoring elements. The layer for data delivery consists of logical and physical links that support data delivery among TMO systems. TCP/IP or Serial for physical data delivery will be selected depending on the implemented communication environments. Also, a general channel or RMMC (Real-time Multicast Memory replication Channel) as a logical link, which establishes connection among TMO systems and manages communication sessions, can be used [5].

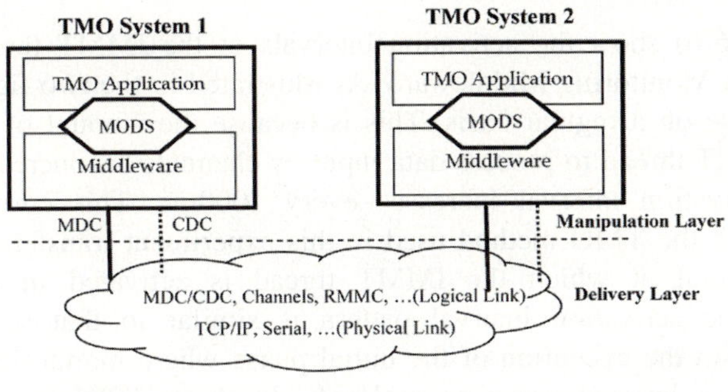

Fig. 5. A Delivery Structure for Monitoring Data

In this paper, the MDC and CDC are used as a logical link. The TMO model basically provides a logical communication link--channel--for connection among methods. Each channel is assigned a channel number, and data can be transferred via a channel that is opened with the same number. The MDC and CDC presented in this paper are designed based on such TMO model channels. Unlike RMMC and a general channel, the MDC and CDC are generated and managed by the middleware, which prevents the need for developer intervention for the delivery of monitoring data and allows automating monitoring.

(a) Monitoring Dedicated Channel (MDC): Refers to the dedicated channel for delivery of monitoring data. The middleware delivers monitoring data stored in MODS to other systems through MDC. To prevent collision with common channels, the MDC uses a specific reserved channel number.

(b) Control Dedicated Channel (CDC): Refers to the feedback channel that is designed to receive control and feedback information from external systems. Like MDC, it uses a specific reserved channel number.

4 Experiments

4.1 Purpose and Method

For the purpose of our experiment, the legacy middleware (hereinafter referred to as "Pure Middleware") and the middleware in which TSDS is applied (hereinafter referred to as "Monitoring Middleware") are experimented with under the same condition. Basically, the Monitoring Middleware is configured to share monitoring data with the other system in a periodic manner. Then, actual activation intervals were measured for OMMT (Outgoing Message Management Thread) and IMMT (Incoming Message Management Thread) which are middleware threads related to the sharing and delivery of monitoring data and should be activated at every 10ms. By conducting a comprehensive comparison between these intervals by the thread, the magnitude of loads generated by monitoring using TSDS was identified.

4.2 Experimental Results

Figs. 6(a) and 6(b) show the activation intervals of the IMMT thread in the Pure Middleware and Monitoring Middleware. As illustrated in the two figures, activation intervals increase on a regular basis. This is because the amount of time needed to occupy the IMMT thread to process data input by channels has increased. For IMMT thread, the activation interval increases every 1000ms. This corresponds to the interval at which the TMO method used in this experiment transfers data. Fig. 6(b) shows the interval at which the IMMT thread is activated in the Monitoring Middleware. The activation interval pattern is similar to that seen in the Pure Middleware. With the exception of the initial phase where monitoring functions are set, the activation interval varies at μs levels. In short, TSDS has little effect on IMMT thread, which operates within the millisecond time frame.

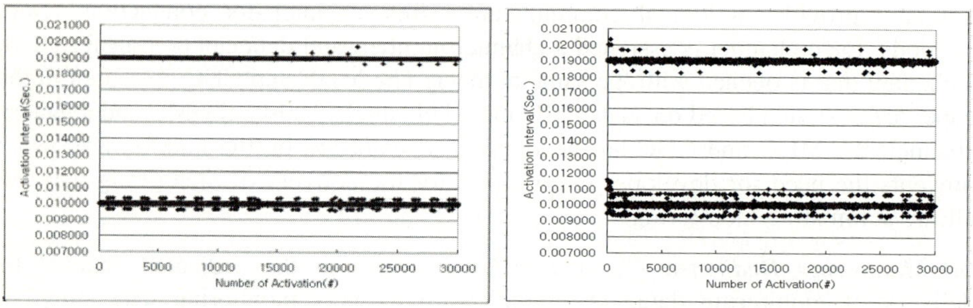

(a) Pure Middleware (b) Monitoring Middleware

Fig. 6. A Comparison of Activation Intervals of IMMT

Figs. 7(a) and 7(b) show the interval at which the OMMT thread is activated in the Pure Middleware and Monitoring Middleware. As in IMMT, the amount of time needed to occupy the OMMT thread to process data has increases. In the case of the Monitoring Middleware, as shown in Fig. 7(b), the activation interval pattern is similar to that seen in the Pure Middleware. The OMMT thread in the Monitoring Middleware varies at μs levels within 1ms. In short, TSDS has little effect on OMMT which operates within the millisecond time frame.

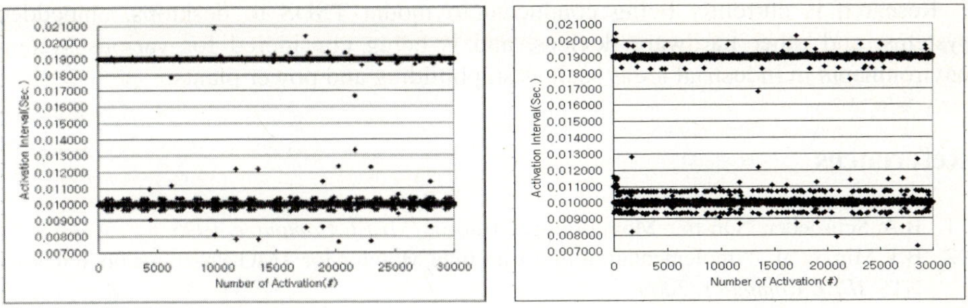

(a) Pure Middleware

(b) Monitoring Middleware

Fig. 7. A Comparison of Activation Intervals of OMMT

Table 1. Results of t-test by Thread

Thread	t-test			
	t	significance probability (Both)	An average confidence interval of 9 9%	
			Lower Limits	Upper Limits
IMMT	-.048	.962	-.00006	.00006
OMMT	.048	.962	-.00006	.00006

Finally, under the hypothesis that "the Pure Middleware and Monitoring Middleware have similarity in activation intervals," the t-test with the significance level α=.01 was conducted to identify the statistical similarity in activation interval among different threads. Table 1 shows the results of analyzing different threads. The significance probability of the IMMT and OMMT is .962. Since .962 > α=.01, hypothesis cannot be rejected, as well. This shows that the activation interval of the IMMT and OMMT threads within the two sets of middleware is statistically identical. Overall, the proposed TSDS shows little or no effect on the middleware.

5 Conclusions and Future Work

In this paper, TSDS, including both the MODS as a data store for sharing monitoring data among TMO systems and a delivery structure for monitoring data, was proposed. TSDS form a monitoring infrastructure within the middleware, supporting the sharing and delivery of monitoring data for TMO systems. This prevents developer intervention, allowing for automated monitoring functions. The overhead generated by TSDS during experiments are less than 1ms, having little or no effect on the execution of the middleware. In short, TSDS serves as an infrastructure for providing monitoring functions such as data sharing and data delivery to the middleware, making it ideally suited for monitoring TMO systems.

Research is currently being conducted to mount TSDS to desktops, embedded systems, and other hardware devices, and is being considered for various system environments in industrial areas such as shipbuilding and power plants.

References

1. B.A. Schroeder, "On-line Monitoring: A Tutorial," *IEEE Computer*, 1995
2. B.J. Min, et al., "Implementation of a Run-time Monitor for TMO Programs on Windows NT," *IEEE Computer*, 2000
3. Hyun-Jun Kim, et al., "TMO-Linux: a Linux-based Real-time Operating System Supporting Execution of TMOs," *Proceedings of the 15th IEEE International Symposium on Object-Oriented Real-Time Distributed Computing*, 2002
4. J. G Kim, and Cho, S. Y., "LTMOS: An Execution engine for TMO-Based Real-Time Distributed Objects," *Proceedings of the PDPTA*, 2000
5. Kim, K.H., "APIs for Real-Time Distributed Object Programming," *IEEE Computer*, 2000
6. Kane Kim, et al., "A Timeliness-Guaranteed Kernel Model: DREAM Kernel and Implementation Techniques," *RTCSA*, 1995
7. K.H. Kim, and Kopetz, H., "A Real-Time Object Model RTO.k and an Experimental Investigation of Its Potentials," *Proceedings of the 18th IEEE Computer Software & Applications Conference*, 1994
8. Yoon-Seok Jeong, Tae Wan Kim, Chun Hyon Chang, "Design and Implementation of a Run-time TMO Monitor on LTMOS," *Proceedings of the Embedded Systems and Applications*, 2003
9. Yoon-Seok Jeong, Tae Wan Kim, Chun Hyon Chang, "Design of an Architecture for Run-time Process Monitor," *Proc. of the 19th KIPS Spring Conference*, Vol. 10-1, 2003.
10. S.H. Park, "LTMOS(LinuxTMO System)'s Manual," HUFS, 2000

An Algorithm for the Generalized *k*-Keyword Proximity Problem and Finding Longest Repetitive Substring in a Set of Strings

Inbok Lee[1,⋆] and Sung-Ryul Kim[2,⋆⋆]

[1] King's College London, Department of Computer Science,
London WC2R 2LS, United Kingdom
inboklee@gmail.com

[2] Division of Internet & Media and CAESIT,
Konkuk University, Seoul, Republic of Korea
kimsr@konkuk.ac.kr

Abstract. The data grid may consist of huge number of documents and the number of documents which contain the keywords in the query may be very large. Therefore, we need some method to measure the relevance of the documents to the query. In this paper we propose algorithms for computing k-keyword proximity score [3] in more realistic environments. Furthermore, we show that they can be used to find longest repetitive substring with constraints in a set of strings.

1 Introduction

The data grid may consist of huge number of documents and the number of documents which contain the keywords in the query may be very large. All these documents may not be relevant to what the user wants: some may contain them in different contexts. Therefore, we need a method to measure the relevance of the documents to the query. Here we focus on the *proximity* of the keywords which means how close they appear together in the document. If they are appearing close (good proximity), it is likely that they have stronger combined meaning.

The *offset* of a word in a document is the distance (number of the words) from the start of the document. A *range* $[a, b]$ in a document represents the contiguous part of the document from a-th word to b-th word in the document. The size of the range $[a, b]$ is $b - a$.

In the data grid, we assume that documents are stored in inverted file structure. Each keyword has a list of IDs of documents which contain the keyword and a sorted list of offsets in the document. Using inverted file structure, we can easily obtain the set of documents which contain the keywords.

Kim et al. [3] defined the *generalized proximity score* and proposed $O(n \log k)$ time algorithm where k is the number of keywords and n is the number of occurrences of the keywords in the document.

⋆ This work was supported by the Post-doctoral Fellowship Program of Korea Science and Engineering Foundation (KOSEF).

⋆⋆ Contact author.

V.N. Alexandrov et al. (Eds.): ICCS 2006, Part IV, LNCS 3994, pp. 289–292, 2006.
© Springer-Verlag Berlin Heidelberg 2006

Definition 1. *Given k keywords* $w_1, w_2, \ldots, w_k$*, a set of lists* $K=\{K_1, K_2, \ldots K_k$ *where each keyword* w_i *has a sorted list of offsets* $K_i = \{o_{i1}, o_{i2}, \ldots, oij_i (1 \leq i \leq k)$*, and k positive integers* $R_1, R_2, \ldots, R_k$*, and another integer* $k' (\leq k)$*, the generalized k-keyword proximity problem is finding the smallest range that contains* k' *distinct keywords where each* w_i *of these keywords appears at least* R_i *times.*

We briefly explain Kim et al.'s algorithm. A more detailed account can be found in [3]. We need to define two terms.

Definition 2. *A* candidate range *is a range which contains at least* k' *distinct keywords where each keyword* w_i *appears at least* R_i *times in the range.*

Definition 3. *A* critical range *is a candidate range which does not contain another candidate range.*

We can easily show that the solution is the smallest critical range. Hence we need to find critical ranges and report the smallest one. First, we merge the sorted lists of offsets of k keywords into one list. This step runs in $O(n \log k)$ time. For simplicity we assume that each offsets are mapped to a number in range $[1..n]$. We store the keyword ID into a list $L[1..n]$.

The outline of the algorithm is that we first find a candidate range (*expanding* sub-step) and find a critical range from that candidate range (*shrinking* sub-step).

We maintain a range $[a, b]$. Initially $a = 1$ and $b = 0$. We also maintain k counters $c_1, c_2, \ldots, c_k$. Initially $c_1 = c_2 = \cdots = c_k = 0$. And we maintain a counter h which counts the number of c_i's $(1 \leq i \leq k)$ that are $\geq R_i$. Initially $h = 0$.

In the expanding sub-step, the range is expanded from $[a, b]$ to $[a, b+1]$. We check $L[b]$ and set $c_{L[b]} = c_{[L[b]} + 1$. We also check whether $c_{L[b]} = R_i$. If so, we update $h = h + 1$. We repeat this loop until $h = k'$. Then $[a, b]$ is a candidate range and go to the shrinking sub-step.

In the shrinking sub-step, the range is reduced from $[a, b]$ to $[a+1, b]$. We also set $c_{L[a]} = c_{L[a]} - 1$ and check whether $c_{L[a]} \leq R_{L[a]}$. If so, $h = h - 1$. And if $h < k'$, we report a critical range $[a-1, b]$. We go back to the expanding sub-step with the range $[a, b]$. These steps run in $O(n)$ time and the total time complexity is $O(n \log k)$.

2 Our Improvements

Here we can consider the following variations. First, the original problem does not specify the keyword which should be included. Some keywords may be more important than others.

Problem 1. In k-keyword proximity problem, one special keyword w_i in the query should appear in the range.

Without loss of generality, assume the keyword that must appear in the critical range is w_1. The original algorithm may report no critical range with w_1 even though the document contains w_1!

We first find a candidate for the problem and make the range narrow as much as possible. In the expanding sub-step, we move to the shrinking sub-step *only after* the current range $[a, b]$ contains w_1 at least R_1 times. We guarantee that the input to the shrinking sub-step meets the constraints of Problem 1. In the shrinking sub-step, we add one more check. If, by shrinking from $[a, b]$ to $[a+1, b]$, c_1 becomes smaller than R_1, we report $[a, b]$ as a critical range (without checking the condition $h < k'$).

Now we consider another problem when keywords in the query must be in some order (for example, "Paris Hilton" and "Hilton Paris").

Problem 2. A keyword w_i must appear before another keyword w_j.

Without loss of generality, assume that w_2 must follow w_1. It means that w_2 can appear only after w_1 appears at least R_1 times. In the expanding sub-step, we move to the shrinking sub-step *only after* the current range $[a, b]$ contains w_1 at least R_1 times before w_2 appears. We may encounter w_2 before R_1 w_1's. Then we discard the current range $[a, b]$. We restart the expanding sub-step with the range $[b + 1, b + 1]$ and initialize all the variables. In the shrinking sub-step, we do the same as we did in Problem 1.

Finally, we can consider the case where two keywords have a Boolean relation.

Problem 3. If a keyword w_i appears in the document, then also another keyword w_j must/must not appear in the document (AND/XOR relation).

Without loss of generality, assume that w_1 and w_2 forms these relations. First we consider the AND relation. In the expanding sub-step, we maintain a flag f. Initially $f = 0$. When we meet a w_1, we set $f = 1$. When we move to the shrinking sub-step in the original algorithm, we check whether the flag f is ON. If so, we postpone until we meet an occurrence of w_2. In the shrinking sub-step, we use the flag again. If $f = 0$, there is no modification at all. But if $f = 1$, each time we shrink the range we check whether it removes the occurrence of w_1. If so, we report the range $[a, b]$ as the critical range.

The procedure is almost the same when we handle the XOR relation. We use the flag again. When we meet an occurrence of w_1, we set $f = 1$. If we meet an occurrence of w_2 and $f = 1$, then we discard the current range $[a, b]$ and restart with $[b + 1, b + 1]$. The shrinking sub-step is the same as the original algorithm.

All these modification does not change the time complexity.

Theorem 1. *All the problems in Section 2 can be solved in $O(n \log k)$ time.*

3 Repetitive Longest Substring in a Set of Strings

Now we consider finding the longest substring in a set of strings.

Problem 4. Given a set of strings $\mathcal{U} = \{T_1, T_2, \ldots, T_k\}$, a set of positive integers $\mathcal{D} = \{d_1, d_2, \ldots, d_k\}$, and a positive integer $k'(\leq k)$, find the longest sting w which satisfies two conditions: (a) there is a subset $\mathcal{U}'$ of $\mathcal{U}$ such that w appears at least d_i times in each string T_i in $\mathcal{U}'$, and (b) $|\mathcal{U}'| = k'$.

We use the suffix array. The *suffix array* of a text T is a sorted array $suf[1..|T|]$ and $lcp[1..|T|]$. $suf[k] = i$ if and only if $T[i..|T|]$ is the k-th suffix of T. $lcp[k]$ is the length of the longest common prefix between each substring in the suffix array and its predecessor and $lcp(a,b) = \min_{a \le i \le b} lcp[i]$ with the following properties.

Fact 1. *$lcp(a,b) \le lcp(a',b')$ if $a \le a'$ and $b \ge b'$.*

Fact 2. *The length of the longest common prefix of $T[suf[a]..|T|]$, $T[suf[a+1]..|T|]$,..., $T[suf[b]..|T|]$ is $lcp(a+1,b)$.*

To build the suffix array for $\mathcal{U} = \{T_1, T_2, \ldots, T_k\}$, we create a new string $T' = T_1\%T_2\% \cdots T_k$ where % is a special symbol and is smaller than any other character in Σ. *suf*and *lcp*arrays can be computed in $O(|T'|)$ time by [2, 1] with one modification: % does not match itself. We use another array *ids*. $ids[j] = i$ if $T'[j]T'[j+1] \cdots \%$ was originally a suffix of T_i (we mean the first % after $T'[j]$). This computation also takes $O(|T'|)$ time.

We briefly explain the outline of [4]. Fact 1 tells that the smaller a range becomes, the longer the common prefix is. Hence, we consider IDs of strings as IDs of keywords as we did in Section 2. What we need is the smallest range that yields the longest common prefix of the suffixes (the longest common substring). We use the same algorithm in Section 2, without the merging step. Hence the time complexity is $O(n)$.

The problems in Section 2 can be transformed into these following problems except Problem 2 because we do not consider order in $\mathcal{U}$.

Problem 5. The same as Problem 4, but $\mathcal{U}'$ must contain T_i.

Problem 6. The same as Problem 4, but if $\mathcal{U}'$ contains a string T_i, it must/must not contain another string T_j (AND/XOR relation).

All these problems can be solved in $O(n)$ time with equivalent algorithm in Section 2.

References

1. T. Kasai, G. Lee, H. Arimura, S. Arikawa, and K. Park. Linear-time longest-common-prefix computation in suffix arrays and its applications. In *CPM 2001*, pages 181–192, 2001.
2. D. K. Kim, J. S. Sim, H. Park, and K. Park. Linear-time construction of suffix arrays. In *CPM 2003*, pages 186–199, 2003.
3. S.-R. Kim, I. Lee, and K. Park. A fast algorithm for the generalised k-keyword proximity problem given keyword offsets. *Information Processing Letters*, 91(3):115–120, 2004.
4. I. Lee, and Y. J. Pinzon Ardilà. Linear time algorithm for the generalised longest common repeat problem. In *SPIRE 2005*, pages 191–200, 2005.

A Grid-Based Flavonoid Informatics Portal*

HaiGuo Xu[1], Karpjoo Jeong[2,**], Seunho Jung[3], Hanku Lee[2], Segil Jeon[4], KumWon Cho[5], and Hyunmyung Kim[3]

[1] Department of Advanced Technology Fusion, Konkuk University, Seoul, Korea
haegook@ricl.konkuk.ac.kr

[2] School of Internet and Multimedia Engineering, Konkuk University, Seoul, Korea
(jeongk, hlee)@konkuk.ac.kr

[3] Department of Microbial Engineering, Konkuk University, Seoul, Korea
shjung@konkuk.ac.kr, swisdom@empal.com

[4] BioMolecular Informatics Center, Konkuk University, Seoul, Korea
sgjeon@ricl.konkuk.ac.kr

[5] Korea Institute of Science and Technology Information
ckw@kisti.re.kr

Abstract. Recently new techniques to efficiently manage biological information of biology have played an important role in the area of information technology. The flavonoids are members of a class of natural compounds that recently has been the subject of considerable scientific and therapeutic interest. This paper presents a Grid-based flavonoids web portal system. We designed relational schema, XML schema for flavonoids information and their user interfaces and proposed interoperable web service components for an efficient implementation of flavonoids web portal.

1 Introduction

The need for efficient management of biological information of biology on complex and various data is rapidly increasing in the area of information technology. Flavonoids are a class of plant pigments and found in a wide range of foods. Recent interests by the biochemical community in flavonoids information are dramatically increased [1, 2, 3]. These include antioxidative, antiallergic, anticarcinogenic effects etc. To address these needs a database and user-friendly application of the flavonoids information (i.e. flavonoid name, mass, structure, NMR, activity, related literature etc.) was developed. Today flavonoids information system deals with many different data types. But legacy database systems can't properly manage (i.e. modeling, storing, and querying) flavonoids information that has been recognized as a key component of today's flavonoids research.

Grid web portals make it possible to provide seamless access to heterogeneous information via a browser-based user interface. It provides a complete set of open, productive, and self-service tools for publishing information, building applications, and deploying and administering enterprise portal environments. These portals are typi-

* This work was supported by the Bio/Molecular Informatics Center at Konkuk University (KRF2004-F00019).

** Corresponding author. Ph: 82-2-450-3510.

V.N. Alexandrov et al. (Eds.): ICCS 2006, Part IV, LNCS 3994, pp. 293 – 300, 2006.
© Springer-Verlag Berlin Heidelberg 2006

cally built around several services including data management and user session management. These services may be built on top of Grid technologies such as Globus [4]. However, developing applications utilizing the Grid technologies remains very difficult due to the lack of high-level tools to support developers. Web service applications have recently been popularized through various domains. The development of web services protocols such as the Simple Object Access (SOAP), Web Services Description Language (WSDL) and the Universal Description, Discovery and Integration (UDDI) protocol has simplified the integration and the interaction the between other organizations [5, 6, 7]. However, existing web services protocols have not been designed for the use in science experimental domain or knowledge intensive infrastructures. The representative databases for flavonoids are the United States Department of Agriculture's (USDA) flavonoids database that provides the information of the flavonoids composition in food, and the part of the Kyoto Encyclopedia of Genes and Genomes (KEGG) pathway database that provides the information of some flavonoids related to flavonoids biosynthesis [8, 9, 10, 11]. However, any well-defined flavonoids database systems for managing of research results and related knowledge (e.g. flavonoids information, literatures etc.) are not systematically well-developed yet. An important shortcoming of existing databases is their lack of interoperability and reusability.

In this paper, we propose a data model for storing and retrieving of the flavonoids information, and design and develop a Grid-based flavonoids web portal system. We also propose web service components for efficient implementation of the portal. The main goal of the flavonoids web portal is to collaboratively work with flavonoids researchers, to provide reusable services, to provide hybrid data model for flavonoids information, and to show how effectively flavonoids information is shared and retrieved in a Grid web portal environment using web services. By adopting grid web portal technology into flavonoids research domain, we can overcome the difficulties caused by a large amount of distributed and shared flavonoids information. Moreover, compared to legacy querying models, the web portal can more effectively support complex queries and incomplete data (i.e. NMR spectroscopy data) by combing ER-model and XML-model.

2 Related Works

The flavonoids are members of a class of natural compounds that recently has been the subject of considerable scientific and therapeutic interest. These include antioxidative, antimicrobial, anticarcinogenic, and cardioprotective effects [12]. To represent these effects, flavonoids database systems started to develope. But each system only works for specific queries and data. So far only a few databases have developed. The flavonoids database developed by USDA provides the contents of flavonoids. KEGG pathway database provides the information of biosynthesis for flavonoids. KEGG mainly deals with the ligand of an enzyme involved in biosynthesis, whereas the physiological properties and chemicophysical characteristics were neglected. But these databases are inadequate for the fields of natural and medical science. Consumption of flavonoids, biologically active polyphenolic compounds found in plants, has been associated with decreased risk for some age-related and chronic diseases in humans. KEGG consists of three databases: PATHWAY for representation of higher order functions in terms of the network of interaction molecules, GENES for the collection of gene catalogs for all the completely sequenced genomes and some partial genomes, and LIGAND for the collection of chemical compounds in the cell enzyme molecules and enzymatic reactions. To the further investigation of flavonoids intake and health, the USDA published the database for the flavonoids content of selected foods in 2003.

3 The Proposed Flavonoids Grid Web Portal Model

3.1 Data Model

The proposed flavonoids data model is a hybrid model based on Entity Relationship model (i.e. ER-model) and XML-model. We have used ER-model for flavonoids information and XML-model for chemical shift table of NMR spectroscopy. This hybrid data model supports complex query and new types of flavonoids information such as NMR for flavonoids information. Flavonoids information and NMR spectroscopy are two major data types in hybrid data model. Flavonoids information represents a single flavonoids compound such as name, formula, mass, structure, related reaction, related enzyme, NMR, MS, references, etc. NMR spectroscopy, described by XML syntax, represents chemical shift table information for determining the content of a flavonoids as well as its molecular structure.

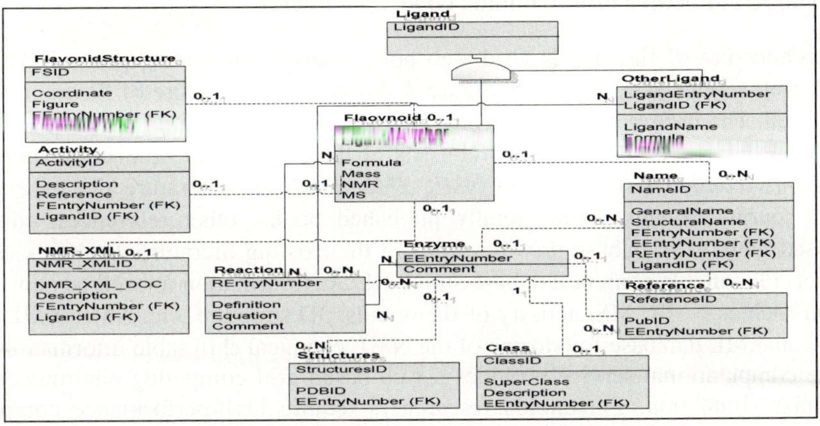

Fig. 1. E-R diagram for flavonoids information

Figure 1 represents the E-R diagram for the complete flavonoids data model. In E-R diagram, the NMR field is a pdf (i.e. portable document format) file name of the NMR structure, the MS field a pdf file name of the MassSpectroscopy structure, the Coordinate field a text file for the model in the PDB site format, and the Figure field a structure image file name, the PubID field the paperID of the PubMed site.

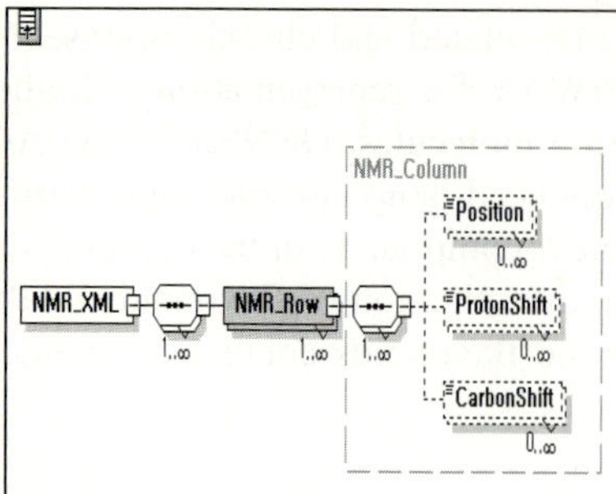


```xml
<?xml version="1.0" encoding="UTF-8"?>
<xs:schema xmlns:xs="http://www.w3.org/2001/XMLSchema" elementFormDefault="qualified"
attributeFormDefault="unqualified">
    <xs:element name="NMR_XML">
        <xs:complexType>
            <xs:sequence maxOccurs="unbounded">
                <xs:element name="NMR_Row" type="NMR_Column" maxOccurs="unbounded"/>
            </xs:sequence>
        </xs:complexType>
    </xs:element>
    <xs:complexType name="NMR_Column">
        <xs:sequence maxOccurs="unbounded">
            <xs:element name="Position" type="xs:integer" minOccurs="0" maxOccurs="unbounded"/>
            <xs:element name="ProtonShift" type="xs:string" minOccurs="0" maxOccurs="unbounded"/>
            <xs:element name="CarbonShift" type="xs:float" minOccurs="0" maxOccurs="unbounded"/>
        </xs:sequence>
    </xs:complexType>
</xs:schema>
```

a. Tree structure b. Textual version

Fig. 2. An example of a XML Schema of NMR spectroscopy

We have designed a XML schema for NMR chemical shift data. The XML model provides the extensibility for representing incomplete information. Figure 2 shows a textual version and a tree representation of this XML schema. The NMR_XML element is the root element of the XML schema that consists of zero or more NMR_Row elements. The NMR_Row element represents rows with zero or more NMR_Column elements. The NMR_Column element represents columns with zero or more Position, ProtonShift, and CarbonShift elements. The Position element, the ProtonShift element, and the CarbonShift element represent a position value, a protonshift value, and a carbonshift value, respectively, in the chemical shift table.

3.2 The Architecture of Flavonoids Grid Web Portal

The architecture of flavonoids Grid web portal mainly consists of grid middleware, portal, web services and database. Figure 4 shows the architecture of flavonoids Grid web portal. The information service provides metadata of flavonoids. The information service of the proposed web portal consists of three databases: *Literature database*, *Flavonoids content database*, and *NMR XML database*. Literature database stores related journal articles, commercially published books, other references, etc. The proposed literature database provides links of the existing literature databases via the Internet. Flavonoids content database stores KEGG information as well as new information such as NMR, MS, activity of flavonoids, 3D structure, etc. NMR XML database is an XML database for storing of the NML chemical shift table information.

The computational service provides a web-based grid computing environment for molecular simulation. Flavonoids researchers require high-performance computing resources via secure grid web portal interfaces. We have adopted a molecular simulation computational Grid System called MGrid into these computational services. MGrid (the Molecular simulation Grid system) is a very promising research technique for biochemical research areas that inevitably need high-performance environments. For example, simulations for bio-conjugates of protein, DNA, lipid, and carbohydrates definitely need HPC environments. Computational power can't solve the whole problem. It is very difficult to know appropriate simulation settings in advance. Moreover, simulation outcomes of those molecules with three-dimensional structures are difficult to validate without real experiments. These two critical problems, the requirements of computational power and the validation of simulation outcomes, have decreased the popularity of the molecular simulations. The MGrid system is designed to address these two issues and based on distributed molecular simulation architecture. Currently, the MGrid system is being ported to the test bed (http://www.mgrid.or.kr).

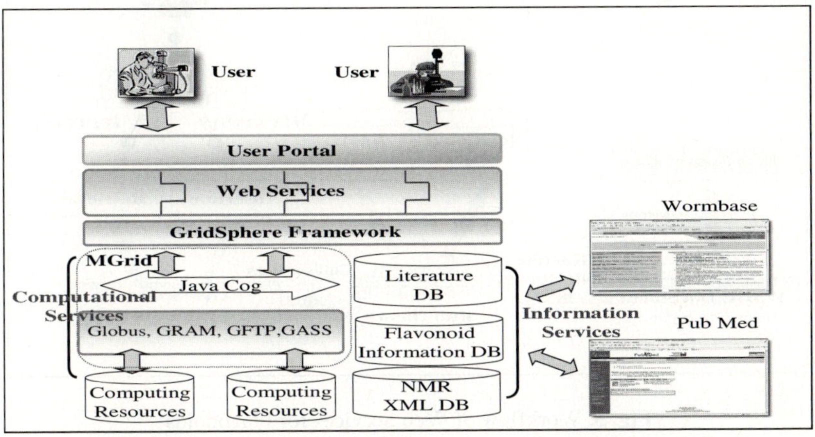

Fig. 4. Overall System Architecture

Generally, a portal is consisted of three components: portal, portlet, and portal container. The GridSphere portal framework [13] is installed and executed on the top of the Apache Tomcat that is a servlet container. The GridSphere is a standard portlet framework to build web portals. It is based on advanced web technology, web security and meta-computing technology such as PKI and Globus to provide secure and interactive services.

4 Implementation of Flavonoids Grid Web Portal Using Web Services

We developed flavonoids web services supporting flavonoids insertion and retrieval web services that are a threshold to the GridSphere portlet container. Each web service has only one interface in WSDL, so any flavonoids research groups and developers who want to use the web services can directly implement their own interface in WSDL.

The workflow of the web services is divided into 7 steps as shown in Figure 5. Step 1 uses a Java2WSDL compiler transforming a Java remote interface into a WSDL document. Step 2 uses a WSDL2Java compiler generating tie-based skeletons (i.e. you can extend a tie class and add code to your subclass that sets the target for delegation to be your implementation class). Step 3 and 4 execute insertion or retrieval portlet services through the web browser. Then, at next steps, the GridSphere portlet container will manage those portlet applications and call specific web services. Finally, data will be inserted into one of databases or retrieved from it.

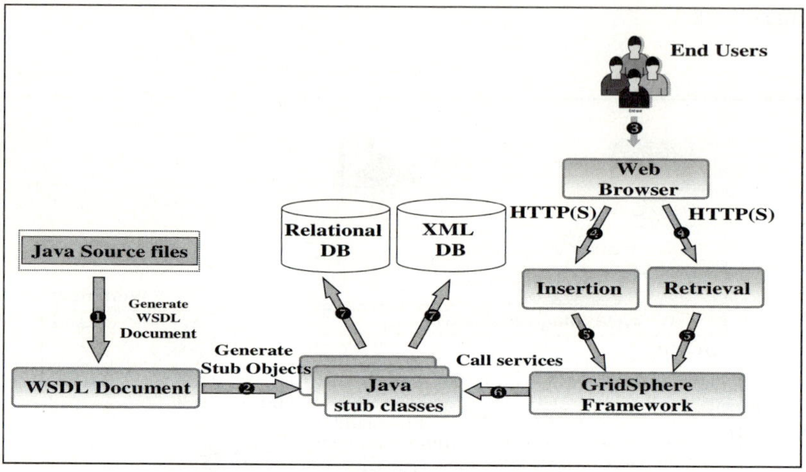

Fig. 5. Workflow of Web Services for Flavonoids

4.1 Web Services for Insertion and Retrieval of Flavonoids Information

We implemented flavonoids web services for storing and retrieving flavonoids information. The services consist of a flavonoids service, an enzyme service, a reaction service, a literature service, and a NMR data insertion web service. The web services for the insertion of flavonoids and enzyme information deal with fundamental information of flavonoids and enzyme such as mass, molecular formula, molecular name, etc. The reaction service inserts interaction information between flavonoids and ligands into ER-database. The literature service inserts related literature information into ER-database. Through the literature service, flavonoids researchers share flavonoids information.

Moreover, the flavonoids web services can more effectively support complex queries and incomplete data (i.e. NMR spectroscopy data) by combing ER-model and XML-model. For example, a user can search flavonoids information with respect to name, formula, and range of molecular weight. Also a user can send XPath query to search the NMR spectroscopy data and get a result that is presented as XSL style sheets using the NMR Retrieval web service.

4.2 User Interfaces for Grid-Based Flavonoids Web Portal

A user can easily insert or retrieve flavonoids information through the user-friendly flavonoids web portal. Figure 6 shows screen shots of the flavonoids web portal. The inserted data consists of flavonoids contents and NMR XML instances. Items of the flavonoids information are entry number, name, formula, mass, structure, reaction, enzyme, NMR(PDF format), MS(PDF format), etc. When a user clicks the 3D Structure link on the structure item (refer to c on Figure 6), the system shows the 3D structure of current flavonoids using Jmol API [14]. When a user clicks the PUBID link on the application item (refer to c on Figure 6), the system shows retrieved papers from PubMed site [15].

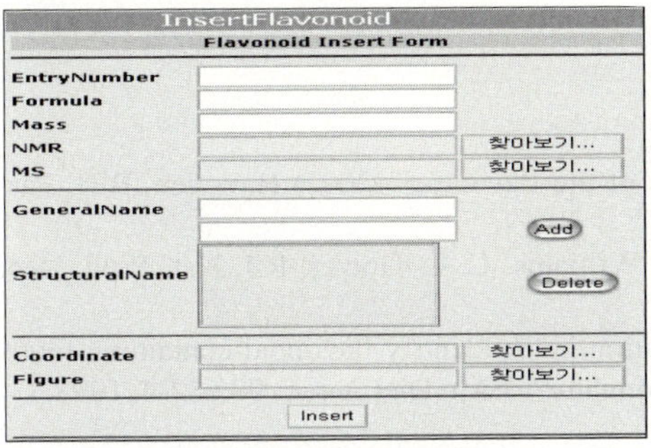

a. Insertion of flavonoids information

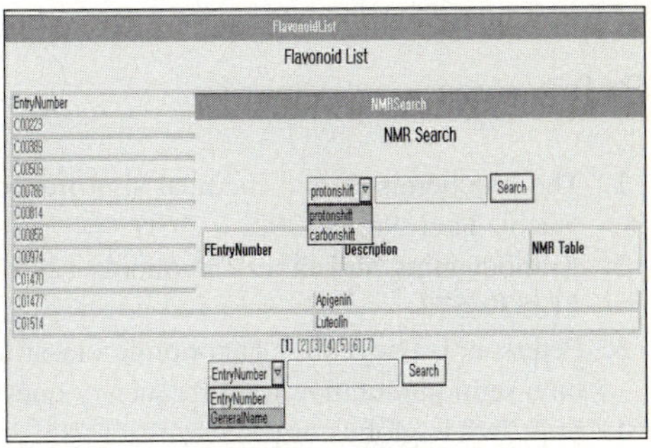

b. Retrieval using keyword and NMR values

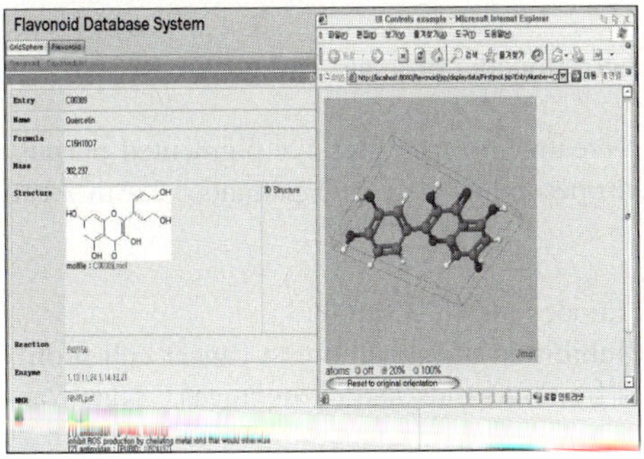

c. Results of flavonoids information retrieval

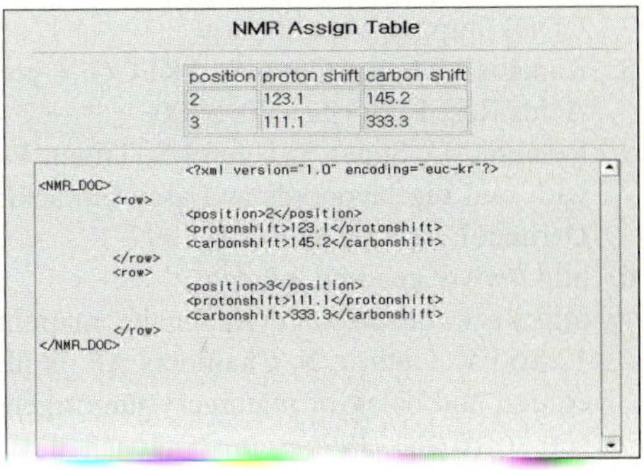

d. Results of NMR XML retrieval

Fig. 6. Insertion and Retrieval Examples

4.3 Advantages

The proposed Grid based flavonoids web portal contributes to solve the problems of current flavonoids databases by easy-implementing the flavonoids web portal and sharing flavonoids information through the web portal. It is easy for users to customize search preferences and display environments. Moreover, the GridSphere portal framework makes it easy to build and deploy portals without any modifications.

5 Conclusion

In this paper, we proposed a Grid based flavonoids web portal to easily share flavonoids information through the portal. We analyzed existing flavonoids databases such as USDA and KEGG, and designed relational schema, XML schema for flavonoids information, and their user interfaces. We explained Flavonoinformatics and showed the proposed architecture for efficient implementation of the flavonoids web portal. We implemented the web portal based on the proposed web service components.

There should be further researches on automatic collecting and processing of related literatures from other literature systems (e.g. PubMed etc.). Also, the proposed web service components should be extended to include other functions, such as update and deletion.

References

1. The biochemistry and medical significance of the flavonoids (2002) Havsteen, B.H. Parmacol. Ther. 96, 67-202.
2. Anthocyanins and other flavonoids (2004) Williams, C.A., Grayer, R.J. Nat. Prod. Rep. 21, 539-573.
3. Peterson J, Dwyer J, "Taxonomic classification helps identify flavonoid-containing foods on a semiquantitative food frequency questionnaire," J Am Diet Assoc, 98:682-4, 1998.
4. http://www.globus.org
5. SOAP, http://www.w3c.org/TR/SOAP
6. WSDL, http://www.w3c.org/TR/wsdl
7. UDDI, http://www.uddi.org/
8. Kanehisa, M. and Goto, S., "KEGG: Kyoto encyclopedia of genes and genomes", Necleic Acids Res, 1;28(1):27-30, 2000.
9. Reinhold U, Seiter S, Ugurel S, Tilgen W., "Treatment of progressive pigmented purpura with oral bioflavonoids and ascorbic acid: an open pilot study in 3 patients," J Am Acad Dermatol, 41(2 Pt 1):207-8, 1999.
10. http://www.genome.jp/kegg
11. http://www.ars.usda.gov/main/site_main.htm?modecode=12354500
12. 12.So FV, Guthrie N, Chambers AF, et al., "Inhibition of human breast cancer cell proliferation and delay of mammary tumorigenesis by flavonoids and citrus juices," Nutr Cancer, 26:167-81, 1996.
13. http://www.gridsphere.org/
14. http://jmol.sourceforge.net/
15. http://www.ncbi.nlm.nih.gov/entrez/query.fcgi

Computer Construction of Quasi Optimal Portfolio for Stochastic Models with Jumps of Financial Markets

Aleksander Janicki

Mathematical Institute, University of Wrocław,
pl. Grunwaldzki 2–4, 50–384 Wrocław, Poland
janicki@math.uni.wroc.pl

Abstract. In the paper we propose a purely computational new method of construction of a quasi–optimal portfolio for stochastic models of a financial market with jumps. Here we present the method in the framework of a Black–Scholes–Merton model of an incomplete market (see, eg. [5], [7]), considering a well known optimal investment and consumption problem with the HARA type optimization functional. Our method is based on the idea to maximize this functional, taking into account only some subsets of possible portfolio and consumption processes. We show how to reduce the main problem to the construction of a portfolio maximizing a deterministic function of a few real valued parameters but under purely stochastic constraints. It is enough to solve several times an indicated system of stochastic differential equations (SDEs) with properly chosen parametrs. This is a generalization of an approach presented in [4] in connection with a well known classical Black–Scholes model.

Results of computer experiments presented here were obtained with the use of the *SDE–Solver* software package. This is our own professional C++ application to Windows system, designed as a scientific computing tool based on Monte Carlo simulations and serving for numerical and statistical construction of solutions to a wide class of systems of SDEs, including a broad class of diffusions with jumps driven by non-Gaussian random measures (consult [1], [4], [6], [9]).

The approach to construction of approximate optimal portfolio presented here should be useful in a stock market analysis, eg. for evolution based computer methods.

1 Optimal Investment and Consumption Problem for Stochastic Models with Jumps of Financial Markets

Let us recall that the Black–Scholes–Merton model of a financial market can be understood as a spacial case of the following system of $N+1$ SDEs

$$S_0(t) = S_0(0) + \int_0^t r(s) S_0(s) ds,$$

$$S_n(t) = S_n(0) + \int_0^t \mu_n(s) S_n(s) ds + \sum_{k=1}^{N} \int_0^t \sigma_{n,k}(s) S_n(s) dB_k(s)$$

V.N. Alexandrov et al. (Eds.): ICCS 2006, Part IV, LNCS 3994, pp. 301–307, 2006.
© Springer-Verlag Berlin Heidelberg 2006

$$+\int_0^t \rho_n(s)S_n(s)d\widetilde{N}^\lambda(s),$$

for $n = 1, \ldots, N$ and $t \in (0, T]$, and where we have the money market with a price $S_0(t)$ and N stocks with prices-per-share $S_1(t), \ldots, S_N(t)$, for $t \in [0, T]$.

We assume that processes $r = r(t)$, and $\mu_n = \mu_n(t)$, for $1 \leq n \leq N$, are $\mathbf{L}^1(\Omega\times[0,T])$–integrable, and processes $\sigma_{n,k} = \sigma_{n,k}(t)$ and $\rho_n = \rho_n(t)$ are $\mathbf{L}^2(\Omega\times[0,T])$–integrable on the probability space $(\Omega, \mathcal{F}, \{\mathcal{F}_t\}, \mathrm{P})$ with a filtration $\{\mathcal{F}_t\}$, which is generated by N–dimensional Brownian motion process $(B_1(t), \ldots, B_N(t))$ and a compensated Poisson process $\widetilde{N}^\lambda(t)$.

Let the stochastic processes $\eta_0 = \eta_0(t)$ and $\eta_n = \eta_n(t)$, for $n = 1, 2, ..., N$, denote the number of shares of a bond and stocks, respectively. So, the value of the investor's holdings at time t is represented by the *wealth process*, which can be represented by

$$\mathbf{X}(t) \stackrel{\text{df}}{=} \sum_{n=0}^{N} \eta_n(t)S_n(t) = \sum_{n=0}^{N} \pi_n(t), \tag{1}$$

where

$$\pi_n(t) \stackrel{\text{df}}{=} \eta_n(t)S_n(t), \quad n = 0, 1, \ldots, N.$$

Let $\pi(t) = (\pi_1(t), \ldots, \pi_N(t))$. We say that the process

$$(\pi_0, \pi) = \{(\pi_0(t), \pi(t)) : t \in [0, T]\}$$

is the *portfolio process* or simply *portfolio* of an investor.

We define the *gains process* $\{G(t) : t \in [0, T]\}$ as a process which satisfies the following equation,

$$dG(t) = \sum_{n=0}^{N} \eta_n(t)dS_n(t) + \sum_{n=1}^{N} S_n(t)\delta_n(t)dt, \tag{2}$$

where $\delta_n = \delta_n(t)$ is the so called *divident rate payment process* of the nth stock, for all $0 < n \leq N$. By the *consumption process* we understand here any non-negative, regular enough stochastic process $c = \{c(t) : t \in [0, T]\}$. Let $x > 0$ denote the *initial wealth* (or *endowment*, i.e. an amount of money an ivestor has to his disposal at time $t = 0$), what means that we have $X(0) = x$. Let $\Gamma(t) \stackrel{\text{df}}{=} x - \int_0^t c(s)dt$. We say that the portfolio (π_0, π) is Γ*–financed*, when

$$\sum_{n=0}^{N} \pi_n(t) = \Gamma(t) + G(t), \tag{3}$$

with $G(0) = 0, \quad \Gamma(0) = x$.

Applying the general semimartingale version of Itô formula one can check that if conditions (2) and (3) are satisfied, then the wealth process

$$X \equiv X^{x,c,\pi} = \{X^{x,c,\pi}(t) : t \in [0, T]\},$$

defined by (1), can be obtained as a solution to the following SDE

$$dX(t) = (rX(t) - c(t))dt + \sum_{n=1}^{N} (\mu_n(t) + \delta_n(t) - r)\pi_n(t)dt + \\ + \sum_{n=1}^{N} \pi_n(t) \left(\sum_{m=1}^{N} \sigma_{n,m}(t)dB_m(t) + \rho_n(t)d\widetilde{N}^{\lambda}(t) \right), \quad (4)$$

with an initial condition of the form $\mathbf{X}(0) = x = \sum_{n=0}^{N} \eta_n(0)S_n(0)$.

From (2) and (3), after application of the classical Itô formula, it follows that the following – very important in our approach – equation must be satisfied

$$\sum_{n=0}^{N} S_n(t)d\eta_n(t) = \sum_{n=1}^{N} S_n(t)\delta_n(t)dt - c(t)dt. \quad (5)$$

In optimization problems utility functions can be chosen in many different ways, however the typical choice for scientific investigations is the HARA model, what means that we chose *utility function* given by

$$U^{(p)}(x) \stackrel{\text{df}}{=} x^p/p, \quad x > 0, \quad (6)$$

for $p \in (-\infty, 0)$ or $p \in (0, 1)$.

The *risk aversion coefficient* is then defined by the formula

$$R \stackrel{\text{df}}{=} -x\frac{d^2}{dx^2}U^{(p)}(x) \Big/ \frac{d}{dx}U^{(p)}(x) = 1 - p. \quad (7)$$

Here we are interested in the following optimization problem.

For a given utility function $U^{(p)} = U^{(p)}(c)$ and initial wealth $x > 0$, we look for an *optimal portfolio* $\widehat{\pi}(t) = \{\widehat{\pi}_1(t), \ldots, \widehat{\pi}_N(t)\}$ and an *optimal consumption process* $\widehat{c} = \widehat{c}(t)$, such that for the *value function* of the form

$$V_{c,\pi}(x) \stackrel{\text{df}}{=} \mathrm{E}\left[\int_0^T U^{(p)}(c(t))\mathrm{e}^{-\int_0^t \beta(s)ds}dt \right] \quad (8)$$

the following condition is satisfied

$$V_{\widehat{c},\widehat{\pi}}(x) = \sup_{(c,\pi)\in\mathcal{A}(x)} V_{c,\pi}(x). \quad (9)$$

Here the condition $(c, \pi) \in \mathcal{A}(x)$ means, that the processes $c = c(t)$ and $\pi = \pi(t)$ are subject to the stochastic constraints, what means that $c = c(t)$ is positive on $[0, T]$ and the corresponding wealth process satisfing SDE (4) is such that

$$X^{x,c,\pi}(t) \geq 0 \text{ a.s.} \quad \text{for } t \in [0, T]. \quad (10)$$

2 An Example of a Quasi–optimal Portfolio

An attempt to answer the question how to apply computational methods to solve directly and effectively optimizations problem (9) through analytical methods, e.g. making use of the Malliavin calculus (see [3], [8]) or trying to get a hedging strategy by constructing a relevant replicating portfolio (see eg. [5], etc.) is not an obvious task in our framework (see eg. [7]).

So, our aim is to describe a method of computer construction of a quasi–optimal portfolio solving approximate problem related to (9).

The method is based on the idea to maximize functional (8), taking into account only some subsets of possible portfolio processes derived from equations (4) and (5), and chosing the class of admissible positive consumption processes arbitrally, in a convenient reasonable way.

We show how to reduce the main problem to the construction of a portfolio maximizing a deterministic function of a few real valued parameters but under purely stochastic constraints.

In order to make the further exposition easier, we restrict ourselves to the one dimensional ($N = 1$) Black–Scholes–Merton model, which can be described in the following form

$$S_0(t) = S_0(0) + r \int_0^t S_0(s)ds \tag{11}$$

$$S_1(t) = S_1(0) + \mu \int_0^t S_1(s)ds + \sigma \int_0^t S_1(s)dB(s) + \rho \int_0^t S_n(s-)d\widetilde{N}^\lambda(s), \tag{12}$$

for $t \in [0, T]$, and initial conditions such that $S_0(0) > 0$, $S_1(0) > 0$.

In the model (11)–(12) all paremeters, i.e. $S_0(0)$, $S_1(0)$, r, μ, σ, and ρ are given positive real numbers. So, the processes S_0, S_1 can be described in the explicit closed form:

$$S_0(t) = S_0(0)\ \mathrm{e}^{rt}, \tag{13}$$

$$S_1(t) = S_1(0)\ \mathrm{e}^{\{(\mu-\sigma^2/2)t+\sigma B(t)+log(1+\rho)N^\lambda(t)\}}. \tag{14}$$

Our quasi–optimal portfolio is now given by $\pi = (\pi_0, \pi_1)$, where

$$\pi_0(t) = \eta_0(t)S_0(t), \quad \pi_1(t) = \eta_1(t)S_1(t), \quad t \in [0, T]. \tag{15}$$

In the example we have chosen for computer experiments presented here we reduced the class of admissible portolio process to those which are of the following form

$$\eta_1(t) = p_1 S_1(t), \quad t \in [0, T]. \tag{16}$$

We also restric ourselves to the class of consumption processes defined by

$$c(t) = c_0 S_0(t) + c_1 S_1(t), \quad t \in [0, T], \tag{17}$$

In (16) and (17) parameters p_1, c_0, c_1 are deterministic (real) variables, subject to some stochastic constraints, and which should be determined in an optimal way.

It is not difficult to notice that in such circumstances the wealth process $X(t) = X^{c_0,c_1,p_1}(t)$, defined by (4), solves the following Itô SDE

$$dX(t) = \Big(rX(t) + (\mu + \delta - r)\eta(t)S_1(t) - \big(c_0 S_0(t) + c_1 S_1(t)\big)\Big)dt \tag{18}$$

$$+\sigma\eta_1(t)S_1(t)dB(t) + \rho * \eta_1(t)S_1(t-)d\widetilde{N}^\lambda(t), \quad t \in (0, T], \quad X(0) = x.$$

Making use of the equation (5), it is also possible to check that the first component of the portfolio, i.e. the proces $\eta_0 = \eta_0(t)$ solves the following SDE

$$\begin{aligned} d\eta_0(t) &= \Big((-\mu+\delta)p_1 S_1^2(t) - \big(c_0 S_0(t) + c_1 S_1(t)\big)\Big)/S_0(t)dt - \\ &\quad -p_1\sigma S_1^2(t)/S_0(t)dB(t) - p_1\sigma S_1^2(t)/S_0(t)d\widetilde{N}^\lambda(t), \quad t \in (0,T], \quad (19) \\ \eta(0) &= (x - p_1 S_1^2(0))/S_0(0). \end{aligned}$$

In this way we arrive at the following problem.

For a given utility function $U^{(p)} = U^{(p)}(c)$ and initial wealth $x > 0$, we look for optimal values of parameters $\widehat{c_0}$, $\widehat{c_1}$, $\widehat{p_1}$, such that for the *value function* of the form

$$V_{c_0,c_1,p_1}(x) \stackrel{\rm df}{=} \mathrm{E}\left[\int_0^T U^{(p)}(c_0 S_0(t) + c_1 S_1(t))\mathrm{e}^{-\beta t}dt\right] \quad (20)$$

the following condition is satisfied

$$V_{\widehat{c_0},\widehat{c_1}\widehat{p_1}}(x) = \sup_{(c_0,c_1,p_1)\in\mathcal{A}(x)} V_{c_0,c_1,p_1}(x). \quad (21)$$

Now the condition $(c_0, c_1, p_1) \in \mathcal{A}(x)$ means, that the consumption and wealth processes, defined by (17) and (18), are such that we have

$$X^{c_0,c_1,p_1}(t) \geq 0 \text{ a.s.}, \qquad c_0 S_0(t) + c_1 S_1(t) \geq 0 \text{ a.s.} \quad \text{for} \;\; t \in [0,T]. \quad (22)$$

We see that, having to our disposal stochastic processes solving SDEs (11), (12), and (18), we are able to solve the problem (20)–(22). In order to get values of the value function (20) using the *SDE–Solver* software it is enough to solve the system of two equations

$$dY(t) = \big(c_0 r S_0(t) + c_1 \mu S_1(t)\big)\, dt + c_1\sigma_1 S_1(t)\, dB(t), \quad (23)$$

$$dZ(t) = U^{(p)}(Y(t))\mathrm{e}^{\beta t} dt, \quad (24)$$

for $t \in (0,T]$, and with initial conditions $Y(0) = c_0 S_0(0) + c_1 S_1(0)$, $Y(0) = 0$, and finally to compute

$$V_{c_0,c_1,p_1}(x) = \mathrm{E}\; Z(T). \quad (25)$$

Then, making use of formulae (19), (16), (15), and (17), one can easily construct quasi optimal portfolio and quasi optimal consumption processes.

3 Results of Computer Experiments

We solved the optimization problem desribed by formulae (11)–(22), with the following fixed values of constant parameters: $T = 1$, $r = 0.2$, $\mu = 0.15$, $\delta = 0.05$, $\sigma = 0.35$, $\rho = 0.35$, $\beta = 0$, and also with $\beta \in \{0, 0.1, 0.2, 0.4, 0.8\}$, $x = 50$, $S_{00} = 50$, $S_{10} = 50$.

The optimal solution for $\beta = 0.0$ is of the following form:

$$\widehat{c_0} = 1.0, \quad \widehat{c_1} = 0.00, \quad \widehat{p_1} = 0.0, \qquad V_{\widehat{c_0},\widehat{c_1},\widehat{c p_1}} = 16.3.$$

From a large amount of data obtained in our experiments we present here only the optimal solution for $\beta = 0$.

In Table 1 below some values of function V_{c_0,c_1,p_1} are presented with corresponding values of parameters c_0, c_1, p_1.

Table 1. Values of V_{c_0,c_1,p_1} from (20)

$\mathbf{V}_{c_0,c_1,p_1}(x)$	$\mathbf{c_0}$	$\mathbf{c_1}$	$\mathbf{p_1}$	β
16.3	**1.00**	**0.00**	**0.00**	0.00
12.6	0.30	0.30	0.00	0.00
13.8	0.80	−0.10	0.01	0.00
12.6	0.30	0.30	0.01	0.00
12.4	0.30	0.30	0.02	0.00
14.6	1.00	0.00	0.00	0.20
12.9	0.40	0.40	0.01	0.20
12.4	0.80	−0.10	0.01	0.20
11.3	0.30	0.30	0.02	0.20
13.3	1.00	0.00	0.00	0.40
11.8	0.80	−0.05	0.01	0.40
10.3	0.30	0.30	0.02	0.40
11.1	1.00	0.00	0.00	0.80
8.6	0.30	0.30	0.02	0.80

In all runs of the system of equations (11), (12), (18), (23), (24) leading to computation of values of the expression in (25) with the *SDE–Solver*, we had 1000 trajectories of the solution, which were constructed on the grid given by 1000 subintervals of lenght 0.001 of the interval $[0, 1]$. Numerical and statistical approximation methods involved are described in [1], [2], [4], [6].

Another example (simplier one, with value function depending only on two parameters) of a quasi optimal portfolio and quasi optimal consumption processes that can be generalized in the same way as in our example presented here is discussed in [1]. Instead of (16), (17), and (20), the following conditions describe the optimization problem:

$$\eta_0(t) = p_0 S_0(t), \quad c(t) = c_2 X(t),$$

$$V_{c_2,p_0}(x) \stackrel{\text{df}}{=} \mathrm{E}\left[\int_0^T U^{(p)}(c_2 X(t)) \mathrm{e}^{-\beta t} dt\right].$$

Graphical representations, visualizing trajectories and some statistical properties of quasi optimal processes $\eta_0 = \eta_0(t)$, $\eta_1 = \eta_1(t)$, $X = X(t)$, and $c = c(t)$, are included there.

4 Conclusions

We strongly insist that even such rough approximations of the optimal investment and consumption problem as presented here are of important practical interest. One can get quite useful ideas about properties of stochastic processes solving the problem, how they depend on parameters of the stochastic model of financial market, investor preferences, etc. Of course, much more work on improvement of suggested method of construction of quasi optimal portfolio has to be done. It is also quite obvious that the method can be easily extended onto much more sofisticated stochastic models of financial market. There are various questions of mathematical nature, that should be answered in the future, e.g. on the correctness and convergence of the proposed approximate method, when more and more parameters enlargig properly the sets of admissible portfolio and consumption processes, are included.

Our computer experiments indicate to some extent the direction of further development of computational methods and computer software useful in practical solution of such complicated problems as construction of optimal strategies for investors, when stochastic model of the financial market is investigated in the framework of a system of SDEs of Itô or another, more general, type. For example, our approach can be in a very simple and natural way implemented for parallel computing systems.

References

1. Janicki, A., Izydorczyk, A.: Computer Methods in Stochastic Modeling (in Polish). Wydawnictwa Naukowo-Techniczne, Warszawa, (2001)
2. Janicki, A., Izydorczyk, A., Gradalski, P.: Computer Simulation of Stochastic Models with SDE–Solver Software Package. in *Computational Science – ICCS 2003*, Lecture Notes in Computer Science vol. [illegible] ([illegible]) [illegible]–370
3. Janicki, A., Krajna, L.: Malliavin Calculus in Construction of Hedging Portfolios for the Heston Model of a Financial Market. Demonstratio Mathematica XXXIV (2001) 483–495
4. Janicki, A., Zwierz, K.: Construction of Quasi Optimal Portfolio for Stochastic Models of Financial Market, in *Computational Science – ICCS 2004*, Lecture Notes in Computer Science vol. 3039 (2004) 811–818
5. Karatzas I., Shreve, S.E.: Methods of Mathematical Finance. Springer-Verlag, Berlin, (1998)
6. Kloeden, P.E., Platen, E.: Numerical Solution of Stochastic Differential Equations, 3rd ed. Springer-Verlag, New York, (1998)
7. León, J.A., Solé, J.L., Utzet F., Vives J.: On Lévy processes, Malliavin calculus, and market models with jumps. Finance and Stochastics 6 (2002), 197–225
8. Ocone, D.L., Karatzas, I.: A generalized Clark representation formula, with application to optimal portfolios. Stochastics and Stochastics Reports 34 (1991), 187–220
9. Protter, P.: Stochastic Integration and Differential Equations – A New Approach. Springer-Verlag, New York, (2002)

A New Computational Method of Input Selection for Stock Market Forecasting with Neural Networks

Wei Huang[1,2], Shouyang Wang[2], Lean Yu[2], Yukun Bao[1], and Lin Wang[1]

[1] School of Management, Huazhong University of Science and Technology, WuHan, 430074, China
{yukunbao, wanglin}@mail.hust.edu.cn

[2] Institute of Systems Science, Academy of Mathematics and Systems Sciences, Chinese Academy of Sciences, Beijing, 100080, China
{whuang, sywang, yulean}@amss.ac.cn

Abstract. We propose a new computational method of input selection for stock market forecasting with neural networks. The method results from synthetically considering the special feature of input variables of neural networks and the special feature of stock market time series. We conduct the experiments to compare the prediction performance of the neural networks based on the different input variables by using the different input selection methods for forecasting S&P 500 and NIKKEI 225. The experiment results show that our method performs best in selecting the appropriate input variables of neural networks.

1 Introduction

The time series forecasting in stock market is characterized by data intensity, noise, non-stationary, unstructured nature, high degree of uncertainty, and hidden relationships[1]. Neural networks (NN) are particularly well suited to finding accurate solutions in an environment characterized by complex, noisy, irrelevant or partial information[2]. Some researchers have conducted work on the stock market time series forecasting by using past value or transformations of them as input variables of neural networks[3]. Neeraj et al studied the efficacy of neural networks in modeling the Bombay Stock Exchange SENSEX weekly closing values. They develop the two neural networks, which are denoted as NN1 and NN2. NN1 takes as its inputs the weekly closing value, 52-week Moving Average of the weekly closing SENSEX values, 5-week Moving Average of the same, and the 10-week Oscillator for the past 200 weeks. NN2 takes as its inputs the weekly closing value, 52-week Moving Average of the weekly closing SENSEX values, 5-week Moving Average of the same, and the 5-week volatility for the past 200 weeks[4]. Yim predicted Brazilian daily index returns. He mapped lagged returns to current returns by using the following three neural networks with backpropagation algorithm. The first neural network has nine lagged returns in the input layer (lags 1 to 9). The second one has only four neurons in the input layer (lags 1, 2, 5 and 9). The third one consists of two neurons (lags 1 and 9) in the input layer. The third neural network produced the best overall results[5]. Yu used the six input which use past prices or transformations of them. The inputs to the

V.N. Alexandrov et al. (Eds.): ICCS 2006, Part IV, LNCS 3994, pp. 308 – 315, 2006.
© Springer-Verlag Berlin Heidelberg 2006

neural networks are as follows: (1) the basis lagged six periods, (2) the RSI differential of the futures price and the index, (3) the MACD differential of the futures price and the index, (4) the change of the basis, (5) the RSI of the basis, (6) the MACD of the basis. His results for out of sample show that the neural network forecast performance is better than that of the ARIMA model[6]. Qi and Zhang use Akaike Information Criterion (AIC) and Bayesian Information Criterion (BIC) as well as several extensions to select input of neural networks for S&P 500 index. The results indicate that the information-based in-sample model selection criteria are not able to provide a reliable guide to out-of-sample performance and there is no apparent connection between in-sample model fit and out-of-sample forecasting performance[7]. Chaos analysis is a good method to analyze nonlinear dynamics in the time series. Nonlinear dynamics and chaos theory can provide information about the lag structures for the design of forecasting models using neural networks. Chaos analysis criterion (CAC) was applied to determine the lag structure for the input of neural networks based on the embedding dimensions of stock index[8, 9]. However, CAC neglect the special feature of input variables of neural networks that the input variables should not be much correlated.

Our contribution of the paper is to propose a new computational method of selecting input variables for stock market forecasting with neural networks. The computational method results from the synthetically considering the special feature of input variables of neural networks and the special feature of stock market time series. The remainder of this paper is organized as follows. Section 2 describes the new computational method. In Section 3, we conduct the experiments to compare the prediction performance of the neural networks based on the different input variables by using the different input selection methods. Finally, conclusions are given in Section 4.

2 Our Input Selection Method

In fact, neural networks for time series forecasting is a kind of nonlinear autoregressive (AR) model as follows:

$$\hat{y}_{t+n} = F\ (\ y_{t-s_1}, y_{t-s_2}, \ldots, y_{t-s_i}\ ) \tag{1}$$

where $\hat{y}_{t+n}$ is the output of the neural network, namely the predicted value when we make a prediction of n periods ahead from the present period t; y_{t-s_1}, y_{t-s_2}, …, y_{t-s_i} are the inputs of the neural network, namely the actual value at the corresponding period; s_i is the lag period from the present period t; $F(\bullet)$ is a nonlinear function determined by the neural networks. The problem is to select the appropriate input variables $(\ y_{t-s_1}, y_{t-s_2}, \ldots, y_{t-s_i}\ )$ of neural networks.

Usually, the rule of input variable selection is that the input variables should be as predictive as possible. As is well known, autocorrelation coefficient is a popular indicator to measure the correlation of time series. The autocorrelation coefficient of a series $\{y_t\}$ at lag k is estimated in the following way:

$$r_k = \frac{\sum_{t=k+1} (y_t - \bar{y})(y_{t-k} - \bar{y})}{\sum_{t=1} (y_t - \bar{y})^2} \tag{2}$$

where $\bar{y}$ is the sample mean of $\{y_t\}$. Table 1 shows the autocorrelation coefficients of daily observations of S&P 500 and NIKKEI 225. It indicates that the absolute value of autocorrelation coefficients of stock index prices become smaller when the lag period becomes longer.

Table 1. The absolute value of autocorrelation coefficients of daily observations of S&P 500 and NIKKEI 225

k	$\|r_k\|$ of S&P 500	$\|r_k\|$ of NIKKEI 225
1	0.935	0.943
2	0.877	0.872
3	0.835	0.812
4	0.790	0.764
5	0.753	0.719
6	0.702	0.660
7	0.649	0.603
8	0.598	0.535
9	0.542	0.478
10	0.487	0.433
11	0.434	0.389
12	0.368	0.345
13	0.306	0.298
14	0.229	0.265
15	0.150	0.235

The special feature of input variables of neural networks is that the input variables should not be much correlated, because the correlated input variables may degrade the prediction performance by interacting with each other as well as other elements and producing a biased effect[10]. Actually, the correlated input variables contribute the similar information for the output variable of neural networks. Therefore, the neural networks get confused and do not know to use which one. In other words, the neural networks may alternate back and forth, and over-fit.

There is a dilemma to select the input variables of neural networks. In order to let the input variables correlated to the output variable, we should choose the input variable with less lags like $\{ y_t, y_{t-1}, y_{t-2}, y_{t-3}, \ldots\}$. However, the above input variables are too correlated to each other. In order to get a trade-off of the two conflicted requirements of input variables selection, we propose a new computational method of input selection for stock market forecasting with neural networks (see Figure 1). It is a

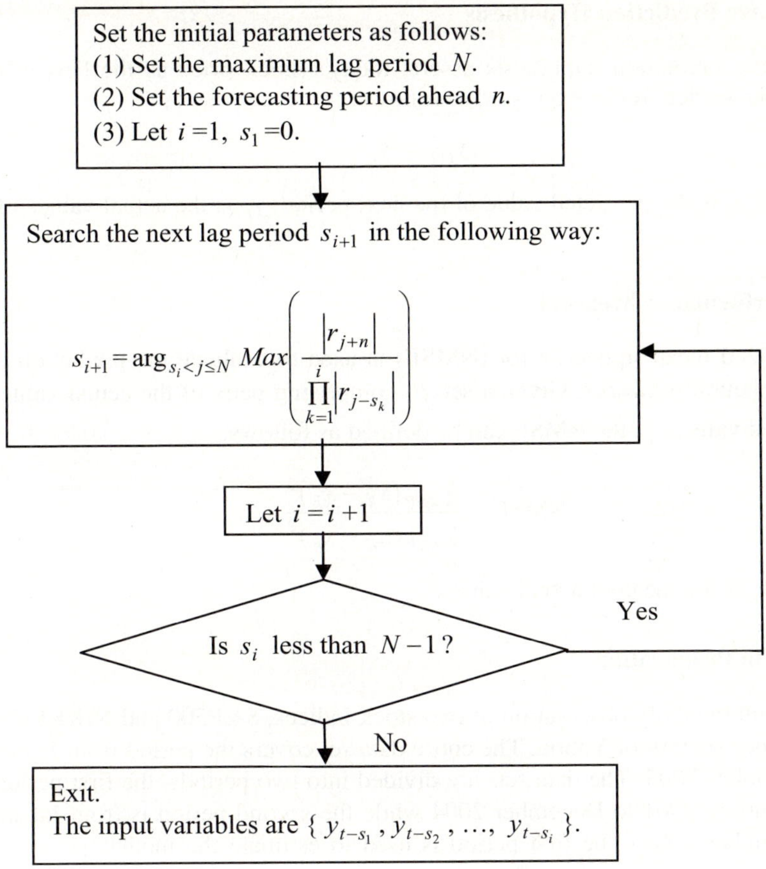

Fig. 1. Our method of input selection for stock market forecasting with neural networks

process of selecting the lagged variable which is more correlated to the predicted variable and less correlated to the already selected input variables.

3 Experiments Analysis

In order to demonstrate our method, we conduct the experiments to compare the prediction performance of the neural networks based on the different input variables by using the different input selection methods.

3.1 Neural Network Models

In order to reduce the degrees of freedom in the developed neural network models and to maintain consistency with previous research efforts, we focus on the following two popular feedforward neural network models: (1)3-layers back-propagation networks with adaptive learning rate and momentum(BPN); (2)radial basis function networks(RBFN).

3.2 Naïve Prediction Hypothesis

The naïve prediction hypothesis asserts today's stock price as the best estimate of tomorrow's price. It can expressed as follows:

$$\hat{y}_{t+1} = y_t \tag{3}$$

where $\hat{y}_{t+1}$ is the predicted value of the next period; y_t is the actual values of current period.

3.3 Performance Measure

Normalized mean squared error (NMSE) is used to evaluate the prediction performance of neural networks. Given a set P comprising pairs of the actual value x_k and predicted value $\hat{x}_k$, the NMSE can be defined as follows:

$$\text{NMSE} = \frac{\sum_{k\in P}(x_k - \hat{x}_k)^2}{\sum_{k\in P}(x_k - \overline{x}_k)^2} \tag{4}$$

where $\overline{x}_k$ is the mean of actual values.

3.4 Data Preparation

We obtain the daily observation of two stock indices, S&P500 and NIKKEI225, from the finance section of Yahoo. The entire data set covers the period from January 2001 to November 2005. The data sets are divided into two periods: the first period covers from January 2001 to December 2004 while the second period is from January 2005 to November 2005. The first period is used to estimate the models parameters. We select the appropriate size of training set by using the method in[11]. The second period is reserved for out-of-sample evaluation and comparison.

3.5 Results

Table 2 shows the prediction performances of the naïve prediction, which are used as benchmarks of prediction performance of S&P500 and NIKKEI225. In order to investigate the effects of the maximum lag period size on the prediction performance of neural networks, we let the maximum lag period $N = 8,10,12$ respectively for one forecasting period ahead, namely $n = 1$. Table 3 shows the input variables of the neural networks for forecasting S&P 500 and NIKKEI 225 by using our method, Akaike Information Criterion (AIC), Bayesian Information Criterion (BIC) and Chaos analysis criterion (CAC).

Table 2. The prediction performance of the naïve prediction

NMSE for S&P 500	NMSE for NIKKEI 225
0.2168	0.2743

Table 3. The input variables of the neural networks for forecasting S&P 500 and NIKKEI 225 by using the different input selection methods

Input selection method	Inputs variables for S&P 500	Inputs variables for NIKKEI 225
Ours ($N=8$)	$\{y_t, y_{t-3}, y_{t-7}\}$	$\{y_t, y_{t-4}, y_{t-6}, y_{t-8}\}$
Ours ($N=10$)	$\{y_t, y_{t-3}, y_{t-7}, y_{t-9}\}$	$\{y_t, y_{t-4}, y_{t-6}, y_{t-9}\}$
Ours ($N=12$)	$\{y_t, y_{t-3}, y_{t-7}, y_{t-9}, y_{t-11}\}$	$\{y_t, y_{t-4}, y_{t-6}, y_{t-9}, y_{t-11}\}$
AIC	$\{y_t, y_{t-2}, y_{t-5}, y_{t-8}\}$	$\{y_t, y_{t-1}, y_{t-6}, y_{t-8}\}$
BIC	$\{y_t, y_{t-3}\}$	$\{y_t, y_{t-3}\}$
CAC	$\{y_t, y_{t-1}, y_{t-2}, y_{t-3}, y_{t-4}\}$	$\{y_t, y_{t-1}, y_{t-2}, y_{t-3}, y_{t-4}\}$

Table 4 shows the prediction performance of 3-layers back-propagation networks with adaptive learning rate and momentum(BPN) with the different input variables determined by using the different input selection methods. Table 5 shows the prediction performance of radial basis function networks(RBFN) with the different input variables determined by using the different input selection methods. The value of NMSE by using our method is the smallest among the different input selection methods, when the initial parameter maximum lag period $N = 8,10,12$ respectively. It shows that our method performs best in selecting the appropriate input variable of the neural networks for forecasting S&P 500 and NIKKEI 225. Because our method balance the two conflicted need of input variables of neural networks: (1) the input variables should be more correlated to the output variable; (2) the input variables should be less correlated to each other. The chaos analysis criterion (CAC) performs worst in selecting the appropriate input variable of the neural networks for forecasting S&P 500 and NIKKEI 225. Compared with the naïve prediction, the neural networks perform better except when using the input variable determined by the chaos analysis criterion (CAC). Because CAC doesn't consider the special feature of input variable of neural networks, and the selected input variables are too correlated to each other. Our method doesn't require any assumptions, completely independent of particular class of model. The method makes full uses of information among sample observations even if the underlying relationships are unknown or hard to describe. Therefore, it is a very practical way to select the input variable of the neural networks when the financial time series is hard to model.

Table 4. The prediction performance of BPN for S&P 500 and NIKKEI 225 forecasting by using the different input selection methods

Input selection method	NMSE for S&P 500	NMSE for NIKKEI 225
Ours ($N=8$)	0.0907	0.0915
Ours ($N=10$)	0.0912	0.0923
Ours ($N=12$)	0.0962	0.0983
AIC	0.1254	0.1357
BIC	0.0974	0.0992
CAC	0.3256	0.3863

Table 5. The prediction performance of RBFN for S&P 500 and NIKKEI 225 forecasting by using the different input selection methods

Input selection method	NMSE for S&P 500	NMSE for NIKKEI 225
Ours ($N=8$)	0.0921	0.0929
Ours ($N=10$)	0.0932	0.0946
Ours ($N=12$)	0.0978	0.0998
AIC	0.1288	0.1396
BIC	0.1106	0.1197
CAC	0.4352	0.4879

4 Conclusions

In this paper, we propose a new computational method of input selection for stock market forecasting with neural networks. The method results from synthetically considering the special feature of input variables of neural networks and the special feature of stock market time series. The advantage of our method is data-driven in that there is no prior assumption about the time series under study. The experiment results show that our method outperforms the others in the prediction performance for stock market time series forecasting with neural networks.

Acknowledgements

This work is partially supported by National Natural Science Foundation of China (NSFC No.70221001, 70401015) and the Key Research Institute of Humanities and Social Sciences in Hubei Province-Research Center of Modern Information Management.

References

1. Hall, J. W.: Adaptive selection of US stocks with neural nets. in Trading on the edge: neural, genetic, and fuzzy systems for chaotic financial markets, G. J. Deboeck, Eds. New York: Wiley; (1994) 45-65
2. Huang, W., Lai, K.K., Nakamori, Y. & Wang, S.Y:. Forecasting foreign exchange rates with artificial neural networks: a review. International Journal of Information Technology & Decision Making, 3(2004) 145-165
3. Huang, W., Nakamori, Y. & Wang, S.Y.: Forecasting Stock Market Movement Direction with Support Vector Machine. Computers & Operations Research, 32 (2005) 2513-2522
4. Neeraj, M., Pankaj, J., Kumar, L. A. & Goutam, D.: Artificial neural network models for forecasting stock price index in Bombay Stock Exchange. Working Papers with number 2005-10-01 in Indian Institute of Management Ahmedabad, Research and Publication Department, (2005)
5. Yim, J.: A comparison of neural networks with time series models for forecasting returns on a stock market index. Lecture Notes in Computer Science, Vol. 2358, Springer-Verlag Berlin Heidelberg (2002)

6. Yu, S. W.: Forecasting and arbitrage of the Nikkei stock index futures: an application of backpropagation networks. Asia-Pacific Financial Markets, 6(1999) 341–354
7. Qi, M. & Zhang, G. P.: An investigation of model selection criteria for neural network time series forecasting. European Journal of Operational Research, 132(2001) 666-680
8. Embrechts, M., Cader, M. & Deboeck, G. J.: Nolinear dimensions of foreign exchange, stock and bond markets. in Trading on the edge: neural, genetic, and fuzzy systems for chaotic financial markets, G. J. Deboeck, Eds. New York: Wiley, (1994) 297–313
9. Oh K. J. & Kim, K.: Analyzing stock market tick data using piecewise nonlinear model. Expert Systems with Applications, 22(2002) 249-255
10. Zhang, G.P.: Neural Networks in Business Forecasting. Idea Group Inc., (2003)
11. Huang, W., Nakamori, Y., Wang, S.Y. & Zhang, H.: Select the size of training set for financial forecasting with neural networks. Lecture Notes in Computer Science, Vol. 3497, Springer-Verlag Berlin Heidelberg (2005) 879–884

Short-Term Investment Risk Measurement Using VaR and CVaR

Virgilijus Sakalauskas and Dalia Kriksciuniene

Department of Informatics, Vilnius University,
Muitines 8, 44280 Kaunas, Lithuania
{virgilijus.sakalauskas, dalia.kriksciuniene}@vukhf.lt

Abstract. The article studies the short-term investment risk in currency market. We present the econometric model for measuring the market risk using Value at Risk (*VaR*) and conditional *VaR* (*CVaR*). Our main goals are to examine the risk of hourly time intervals and propose to use seasonal decomposition for calculation of the corresponding *VaR* and *CVaR* values. The suggested method is tested using empirical data with long position EUR/USD exchange hourly rate.

1 Introduction

Trading in the stock and currency markets has many common features, yet these markets have major differences as well. Currency market has higher volatility, which causes higher risks of trade. There are many reasons which cause substantial volatility of the currency market.

- The transactions, related to the pairs of currencies exchanged, have much more trading partners, comparing to the stock trading.
- Currency exchange attracts much more instant, even unqualified traders, while stocks' trading requires at least basic minimal financial knowledge.
- The rearrangement of stock portfolio is related to quite big taxes, comparing to relatively liberate tax policy in currency trading.

The traditional way of risk estimation in the stock markets is based on periodic risk evaluations on daily basis or even by taking longer periods. This practice is in most cases based on empirical experience and is convenient for application in trading stocks. Yet even most simple analysis of currency markets indicates, that this kind of risk evaluation could be not sufficient, as during the period of 24 hours it changes several times: for particular hours it can differ even up to four times, as it is further shown in this article.

The paper aims at the estimation of the market risk for the short-term investments in currency market by suggesting the modified RiskMetrics model, based on risk evaluation according to hourly profit alterations of the financial instrument.The second part of the article describes and evaluates the theoretical settings for risk analysis in the currency markets by applying traditional models. The econometric description and substantiation of the suggested model is presented in part 3. The fourth part presents experimental verification of the method using FOREX historical data of EUR/USD hourly exchange rate fluctuations.

V.N. Alexandrov et al. (Eds.): ICCS 2006, Part IV, LNCS 3994, pp. 316–323, 2006.
© Springer-Verlag Berlin Heidelberg 2006

2 Theoretical Assumptions and Notations

One of the most widely used factors for market risk measurement is Value at Risk (*VaR*), which is extensively discussed in scientific literature starting already from the 1990. Historically, the concept of the Value-at-Risk is related to the covariance calculation method that was first adopted by the J.P.Morgan Bank as a branch standard, called RiskMetrics model [15]. The VaR measure means the biggest loss of investment R during the time interval, at the fixed rate of probability p of unfavorable event:

$$P(R > VaR) \leq p \ , \tag{1}$$

where p in most cases is selected as 0.01 or 0.05 (or 1% or 5%). The loss of investment R is understood as negative difference of the buying price P_0 and the selling price P_1 : $R = -(P_1 - P_0)$. In the article profitability is denoted as $P_1 - P_0$.

The *VaR* measurement is very popular for its relative simplicity of interpretation, as risk can be evaluated by single value- the loss rate at the occurrence of the unfavorable low-probability event. This brought the *VaR* measure acceptance almost as standard value, recognized by many researchers. Together with these advantages, application of *VaR* has disadvantages as well. One of the main drawbacks is absence of subadditivity feature: the sum of *VaR* measurements of two portfolios can be less, than the risk of value change of the joint portfolio. *VaR* measurements are also limited to estimating of the marginal loss and do not indicate to other occurrences of possible loss. For eliminating this drawback, Artzner [1] suggested the alternative measurement for market risk evaluation, which meets the subadditivity requirement. They introduced the conditional *VaR* (*CVaR*), called the expected shortfall or the expected tail loss, which indicates the most expected loss of investment, larger than indicated by *VaR*, denoted by conditional expectation:

$$CVaR = E(R \mid R \geq VaR) \tag{2}$$

The estimation of both measures of risk evaluation is based on finding adequate quantiles, according to the price distribution data of the analysed financial instrument. The calculated values of *VaR* and *CVaR* are more precise, if we have more confident information of the price distribution. Main methods of risk evaluation are based on assumption of normality of return on investment distribution. But empirical research does not confirm the normality of the real data distribution. The shape of profitability distribution has fatter tails, differences in skewness and kurtosis. The fatter tails indicate more often occurrence of extreme unpredictable values, than predicted by the assumption of normality ([3,4,6-9]). The profitability distribution is taller and narrower, than normal distribution. These empirical data indicate that by calculating *VaR* and *CVaR* with the assumption of normal distributions, we underestimate real risk value. There are several ways suggested in the research literature to reduce these deviations: substituting normal distribution with the distribution with fatter tails or to use the safety coefficient to compensate inadequacies. The theoretical background of calculating *Var* and *CVaR* risk measures on hourly basis are suggested in the next part.

3 Econometric Model for the VaR and CVaR Estimation

The presented model is based on the mathematical notation, as defined in the Risk-Metrics Technical Document [15], and is applied for risk estimation for single type of financial instrument. The input data for suggested model is collected on hourly basis, by registering the opening and closing prices of financial-instrument. Let P_{ot} be the opening price of a given financial instrument at the starting point of hour t, and P_{ct} - the closing price for the same financial instrument at the end of hour t. Then the return r_t of one hour period is defined as:

$$r_t = \ln(\frac{P_{ct}}{P_{ot}}) = \ln(P_{ct}) - \ln(P_{ot})$$

The model could be defined as adequate, if it could estimate changes of values over time and describe the distribution of return at any point of time. As stated in [11,15], the standard RiskMetrics econometric model meets this requirement only for the estimation of the investment risk for one-day returns analysis, and may give inadequate results while extending or shortening the time period. The market risk at an intraday time horizon has been quantified by Giot P. in [5]. This paper suggests alternative model, where the analysis of returns, based on continuous periods of time, is replaced by the discreet hourly-based return analysis. As the dynamics of the price of a financial instrument is best revealed by the white noise process, the modified model could be based on the following expression:

$$r_t = \mu + \sigma_t \cdot \varepsilon_t \tag{3}$$

Here μ is average alteration of return during the given period of time; σ_t - standard deviation of return, and ε_t are the independent random values, with the standard normal distribution. Consequently, the return r_t has conditional (time-dependent) normal distribution, and the equation (3) can be modified to:

$$r_t = \ln(\frac{P_{ct}}{P_{ot}}) = \mu + \sigma_t \cdot \varepsilon_t$$

According to standard RiskMetrics model $\mu = 0$, equation (3) can be simplified to:

$$r_t = \sigma_t \cdot \varepsilon_t \ .$$

Return estimations, based on the RiskMetrics model, which assumes normal distribution, slightly differ from those, observed in reality: the tails are fatter, the peak value of the return distribution is higher, and the distribution curve itself is narrower.

In most cases the inadequacies to return distribution are compensated by calculating safety factor or substituting the normal distribution by Student, Laplace, Weibul or distribution mixes [2,11-14]. In the suggested model the risk evaluation will be based on safety factor estimation from the experimental data.

By using definition (1) it is possible to calculate *VaR* as the return r_t quantile.While r_t is normal distributed with mean μ_t and standard deviation σ_t, the value $z_t = \frac{r_t - \mu_t}{\sigma_t}$ will have standard normal distribution. The value of the 5% quantile is calculated as −1.645, and the 1% quantile is 2.326. Hence:

$$P(z_t < -1.645) = P(r_t < -1.645 \cdot \sigma_t) = 0.05$$
$$P(z_t < -2.326) = P(r_t < -2.326 \cdot \sigma_t) = 0.01$$

Thus, the 5% *VaR* makes $VaR_{5\%}$=-1.645σ_t, and the 1% *VaR* makes $VaR_{1\%}$=-2.326σ_t. For the estimation of the *VaR*, $\sigma_t^{\ 2}$ must be found out.

$$\sigma_t^2 = E(r_t - E(r_t))^2 = E(r_t^2 - 2 \cdot r_t \cdot E(r_t) + E(r_t)^2 =$$
$$= E(r_t^2) - 2 \cdot E(r_t) \cdot E(r_t) + E(r_t)^2 = E(r_t^2) - E(r_t)^2$$

According to Phillipe Jorion [10], the first summand of the equation exceeds the impact of the second summand approximately for about 700 times. Therefore:

$$\sigma_t^2 = E(r_t^2) .$$

As the standard RiskMetrics model offers, the σ_t^2 is calculated by employing the method of exponential smoothing based on the past data:

$$\sigma_{t+1}^2 = \frac{\sum_{i=0}^{\infty} \lambda^i \cdot r_{t-i}^2}{\sum_{i=0}^{\infty} \lambda^i} = (1-\lambda) \cdot \sum_{i=0}^{\infty} \lambda^i \cdot r_{t-i}^2 = (1-\lambda) \cdot r_t^2 + \lambda \cdot \sigma_t^2 ,$$

where $0 < \lambda < 1$. The *CVaR* is estimated according to the definition (2). As the distribution of r_t is standard normal, for each reliability p we can apply:

$$CVaR_p = E(r_t \mid r_t \le VaR_p) = \frac{1}{p \cdot \sigma_t \cdot \sqrt{2\pi}} \int_{-\infty}^{VaR_p} x e^{-\frac{x^2}{2\sigma_t^2}} dx =$$
$$= \frac{\sigma_t}{p \cdot \sqrt{2\pi}} \cdot e^{-\frac{x^2}{2\sigma_t^2}} \Big|_{-\infty}^{VaR_p} = \frac{\sigma_t}{p \cdot \sqrt{2\pi}} \cdot e^{-\frac{VaR_p^{\ 2}}{2\sigma_t^2}} = \frac{e^{-\frac{q_p^2}{2}}}{p \cdot \sqrt{2\pi}} \cdot \sigma_t . \quad (4)$$

From the formula (4) we can calculate the values of $CVaR_{5\%} = -2{,}063 \cdot \sigma_t$, and $CVaR_{1\%} = -2{,}665 \cdot \sigma_t$. In case the return distribution is normal, the evaluations of *VaR* and *CVaR* differ only by value of constant: $CVaR_{5\%} = 1{,}254 \cdot VaR_{5\%}$ and $CVaR_{1\%} = 1{,}146 \cdot VaR_{1\%}$.

4 Experimental Verification of the Econometric Model

For the experimental verification of the suitability of the *VaR* model we will calculate 5% *VaR* and *CVaR* values for all 24 hours of the day. The long EUR/USD position data was taken from the FOREX (Foreign Exchange Market) currency market reports. The EUR/USD hourly records (total of 6782 records) of opening, closing, min and max prices have been collected during the period since 30 January 2003, 9 p.m. to 2 March 2004 9 p.m. After sorting data of selected time interval, 219 records were used for the calculation of the *VaR* values and the identification of the accuracy of estimation.

The experiment was made in the following steps:

- Verification of the hourly return data fit, under the premise of normal distribution.
- Calculation of the volatility of the trading data, collected under hourly basis.
- *VaR* and *CVaR* estimation and analysis.

To verify the data normality, the cumulative function of the observed data distribution was plotted against the theoretical cumulative distribution. The diagrams of P-P plots were slightly higher and narrower than the normal distribution and confirmed the inadequacy of standard model to return distribution, as discussed in the part 2. The calculation of hourly data volatility showed, that the trade risk increases, when the data volatility is higher. The standard deviations and the data range (max minus min value) of the observed data at the corresponding hours are shown in figure 1.

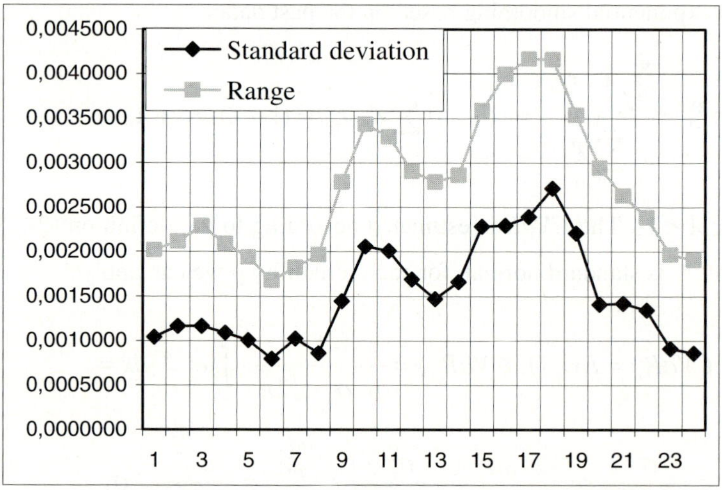

Fig. 1. Standard hourly deviations and range for 24 hours

The experimental calculations pointed out, that the highest volatility of return occured between 2 p.m. and 6 p.m., and the lowest between 10 p.m. and 8 a.m. The biggest observed difference between the highest and the lowest volatility has reached up to 400%. The difference in the volatility allows assuming, that the differences in trading risk could be similar. For calculating *VaR* using econometric model, presented in the Part 2, the standard return deviation has to be estimated by using exponential smoothing method. All calculations were made with the help of STATISTICA software, Time Series/Forecasting models. The obtained results are presented in Table 1, where $VaR_{5\%}$ and $CVaR_{5\%}$ values are estimated for each hour of the day.

Table 1. The VaR and CVaR values for 24 hours

Hours	$VaR_{5\%}$	$CVaR_{5\%}$	Hours	$VaR_{5\%}$	$CVaR_{5\%}$
00–01	–0,0014493	-0,001817422	12–13	–0,0020678	-0,002593021
01–02	–0,0016142	-0,002024207	13–14	–0,0023456	-0,002941382
02–03	–0,0016479	-0,002066467	14–15	–0,0032362	-0,004058195
03–04	–0,0015137	-0,001898180	15–16	–0,0032367	-0,004058822
04–05	–0,0014080	-0,001765632	16–17	–0,0034310	-0,004302474
05–06	–0,0011133	-0,001396078	17–18	–0,0035208	-0,004415083
06–07	–0,0014317	-0,001795352	18–19	–0,0027895	-0,003498033
07–08	–0,0012031	-0,001508687	19–20	–0,0019280	-0,002417712
08–09	–0,0020402	-0,002558411	20–21	–0,0019596	-0,002457338
09–10	–0,0028494	-0,003573148	21–22	–0,0018185	-0,002280399
10–11	–0,0027626	-0,003464300	22–23	–0,0012586	-0,001578284
11–12	–0,0023128	-0,002900251	23–24	–0,0011825	-0,001482855

Comparing the obtained $VaR_{5\%}$ with the characteristics of the hourly data volatility, revealed, that both factors of the market risk estimation possessed a very similar hourly structure, as presented in figure 2:

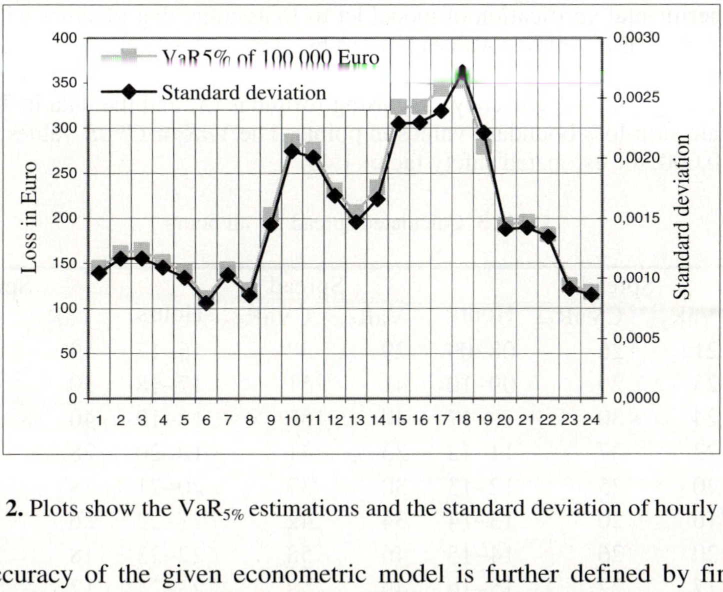

Fig. 2. Plots show the $VaR_{5\%}$ estimations and the standard deviation of hourly data

The accuracy of the given econometric model is further defined by finding out, what percent of hourly cycles exceed the estimated *VaR*. In case the achieved results do not considerably differ from 5%, it could be reasonable to rely on the potential results forecasted by the described econometric model. The theoretical and experimental values for the whole hourly cycle are presented in Table 2.

Table 2. Conformity $VaR_{5\%}$ of the theoretical to experimental values

Hours	Percent	Hours	Percent	Hours	Percent
00–01	5.47950	08–09	6.84930	16–17	5.93610
01–02	4.56621	09–10	6.39270	17–18	8.21920
02–03	7.76260	10–11	7.30590	18–19	6.39270
03–04	7.30590	11–12	7.30590	19–20	7.30590
04–05	4.56620	12–13	7.76260	20–21	7.30590
05–06	3.65300	13–14	6.84930	21–22	6.39270
06–07	5.93610	14–15	5.47950	22–23	8.21920
07–08	5.93610	15–16	2.28310	23–24	6.39270

The given percent values only slightly exceeded the anticipated 5% level (the average makes 6.31). The conformity of the model was increased by calculating the safety factor (its estimated value for the experimental data was 1.43), which was used to adjust the values of VaR and $CVaR$ in order to fit the 5% level:

$$VaR_{5\%} = -1.43 \cdot 1.645 \cdot \sigma_t \quad ; \quad CVaR_{5\%} = -1.43 \cdot 2.063 \cdot \sigma_t \tag{5}$$

According to the estimated hourly values of *VaR* or *CVaR*, the suggested model can help to set more flexible stop-loss rates. The current trading practice with the fixed stop-loss value can lead to substantial loss, where high stop-loss value increases risk to loose big part of investment, and too small value prevents from bigger gains.

The experimental verification of model let us to assume, that together with increasing risk volatility the stop-loss values have to be increased as well. The stop-loss level was evaluated by spread (difference between sell and buy prices), presented in points (1/10 000 change of base currency). By using formulas (5) and the data in Table 1 we can calculate stop-loss boundary values in points. The *VaR* or *CVaR* values are multiplied by 10,000, the estimated safety factor 1,43.

Table 3. Calculated spread for all hours

Hours	Spread		Hours	Spread		Hours	Spread	
	$VaR_{5\%}$	$CVaR_{5\%}$		$VaR_{5\%}$	$CVaR_{5\%}$		$VaR_{5\%}$	$CVaR_{5\%}$
00–01	21	26	08–09	29	37	16–17	49	62
01–02	23	29	09–10	41	51	17–18	50	63
02–03	24	30	10–11	40	50	18–19	40	50
03–04	22	27	11–12	33	41	19–20	28	35
04–05	20	25	12–13	30	37	20–21	28	35
05–06	16	20	13–14	34	42	21–22	26	33
06–07	20	26	14–15	46	58	22–23	18	23
07–08	17	22	15–16	46	58	23–24	17	21

In the Table 3 two levels of loss boundary values are presented: for more reserved trading ($VaR_{5\%}$ case) and for the player more tended to risk ($CVaR_{5\%}$ case). It can be stressed, that these coefficients are applied only for EUR/USD position in FOREX.

5 Conclusions

This paper suggests the modified RiskMetrics model of risk evaluation for the short-term investments in currency market. The method is based on calculating *VaR* and *CVaR* on hourly basis, using seasonal decomposition. The conformity of the model was increased by calculating the safety factor, which was used to adjust the values of *VaR* and *CVaR*. The experimental verification of model showed that together with increasing risk volatility the stop-loss values have to be increased as well. The main results presented in the article provide basis for further research by applying the suggested econometric model for risk evaluation of short-time investment in the currency market.

References

1. Artzner, P., Delbaen, F., Eber, J., Heath, D.: Coherent Measures of Risk. Mathematical Finance, Vol. 9(3). Backwell publishers, Malden USA (1999) 203–228
2. Benninga, S., Wiener, Z.: Value-At-Risk (Var). Mathematica in Education and Research, Vol. 7(4). Springer Verlag, New York (1998) 39-45
3. Carr, P., Geman, H., Madan, D.B., Yor, M.: The Fine Structure of Asset Returns: an Empirical Investigation. Journal of Business, No.75. The University of Chicago Press (2002) 305–332
4. Christoffersen, P., Diebold F.: How Relevant Is Volatility Forecasting For Financial Risk Management? Review Of Economics and Statistics, Vol. 82. MIT Press, Massachusetts USA (2000) 1–11
5. Giot, P.: Market risk models for intraday data. European Journal of Finance, Vol. 11(4). Routledge (Taylor & Francis), UK (2005) 309-324
6. Gopikrishnan, P., Meyer, M., Amaral, L.A.N., Stanley, H.E.: Inverse Cubic Law for the Distribution of Stock Price Variations. European Physical Journal, Springer-Verlag, Berlin Heidelberg New York (1998) 139–140
7. Danielsson, J., de Vries, C.: Value-at-Risk and Extreme Returns. Annales d'Economie et Statistique, Centre d'études de l'emploi Vol. 3 (2000) 73–85
8. Dowd, K.: Measuring Market Risk. John Wiley & Sons USA (2002)
9. Franke, J., Hardle, W., Stahl, G.: Measuring Risk in Complex Stochastic Systems. Springer-Verlag, Berlin Heidelberg New York (2000)
10. Jorion, P.: Value at Risk: the New Benchmark for Managing Financial Risk. McGraw-Hill (2000)
11. 11. Silvapulle, P., Granger, C.W.J.: Large Returns, Conditional Correlation and Portfolio Diversification: a Value-at-Risk Approach. Quantitative Finance, Vol.1(5). Routledge (Taylor & Francis), UK (2001) 542-551
12. Shapiro, S.S., Wilk, M.B., Chen, H.J.: A Comparative Study of Various Tests of Normality. Journal of the American Statistical Association, Vol 63. ASA, Boston (1968) 1343-1372
13. Rachev, S., Mittnik, S.: Stable Paretian Models in Finance. John Wiley & Sons (2000)
14. Randal, J.A., Thomson, P.J., Lally, M.T.: Non-Parametric Estimation of Historical Volatility. Quantitative Finance, Vol 4(4). Routledge (Taylor & Francis), UK (2004) 427-440
15. Riskmetrics Technical Document. 4th edn. J.P. Morgan/Reuters, New York (1996)

Computational Asset Allocation Using One-Sided and Two-Sided Variability Measures

Simone Farinelli[1], Damiano Rossello[2], and Luisa Tibiletti[3]

[1] Quantitative and Bond Research, Cantonal Bank of Zurich,
P.O. Box, CH-8010 Zurich, Switzerland
simone.farinelli@zkb.ch
[2] Department of Economics and Quantitative Methods,
University of Catania, Corso Italia, 55, 95129 Catania, Italy
rossello@unict.it
[3] Department of Statistics and Mathematics "Diego de Castro",
University of Torino, Piazza Arbarello, 8, 10122 Torino, Italy
luisa.tibiletti@unito.it

Abstract. Excluding the assumption of normality in return distributions, a general reward-risk ratio suitable to compare portfolio returns with respect to a benchmark must includes asymmetrical information on both "good" volatility (above the benchmark) and "bad" volatility (below the benchmark), with different sensitivities. Including the Farinelli-Tibiletti ratio and few other indexes recently proposed by the literature, the class of one-sided variability measures achieves the goal. We investigate the forecasting ability of eleven alternatives ratios in portfolio optimization problems. We employ data from security markets to quantify the portfolio's overperformance with respect to a given benchmark.

1 Introduction

Investment performance evaluation requires appropriate tools for ranking different risky projects. Nowadays most practitioners employ reward-risk indexes developed by academics. Nevertheless, only with normality assumption the uncertainty of future wealth is fully captured by the first two moments of the excess return's distribution. In presence of kurtosis and skewness, two-sided reward and risk measures, i.e. the average excess return and a dispersion measure considering both positive and negative deviations from the mean, do not separate information about overperformance and underperformance, causing an anomaly since investors typically distinguish between upside and downside risk. In recent years several alternatives to the Sharpe ratio has been proposed to avoid sub-optimal choices in portfolio selection problems. Some of these redefine the risk measure in the denominator, such as the Gini ratio [13], [16], the mean absolute deviation (MAD) ratio [8], the stable ratio [1], [7], the mini-max ratio [17], the Sortino-Satchell ratio [14], [15], [12], the VaR and STARR ratios, [6], [9]. The first three refer to modified versions of the dispersion (risk) measure. The Sortino-Satchell, VaR and STARR ratios have a denominator that accounts for downside risk (the mini-max ratio is a special case of STARR). Performance indexes based on one-sided variability

V.N. Alexandrov et al. (Eds.): ICCS 2006, Part IV, LNCS 3994, pp. 324–331, 2006.
© Springer-Verlag Berlin Heidelberg 2006

measures such as the Farinelli-Tibiletti ratio [3], [4], [5] and Rachev's ratios [1], [2] use instead an upper partial moment (deviations above the benchmark, for the reward), and a lower partial moment (deviations belove the benchmark, for the risk) of different orders. The higher the order, the higher the agent's inclination towards or dislike of the extreme events.[1]

The performance measures mentioned above lead to different optimal allocation. Moreover, since the joint distribution of future total returns are not known but can be only estimated, an empirical analysis is needed to exploit the ranking of alternative performance indexes. We will illustrate results from portfolio optimization problems considering five time series of total return indexes from 3 countries and employing 11 performance indexes. Historical data show high volatility affecting the joint empirical distribution of returns which is fat-tailed.

The paper is organized as follows. Section 2 reviews one-sided variability measures. Section 3 formulates the optimal portfolio problem, and contains the numerical examples and the back tests. Section 4 contains some concluding remarks.

2 One-Sided Variability Measures

Let R be a p-integrable random variable representing the total return over a fixed horizon related to a certain investment. A general performance index is given by the ratio $\Phi(R) := r(R)/\rho(R)$, for the investment opportunity having reward $r(R)$ and risk $\rho(R)$. The higher $\Phi(R)$, the more preferable the risky trade R.

Given a benchmark[2] b, the excess portfolio return writes $X = R - b$. One-sided variability measures are to be used if we want to separate information about the likelihood of a positive performance above the minimum acceptable return, $X > 0$, and the likelihood of underperforming the target return, $X < 0$. Recently, Farinelli and Tibiletti (see [3], [4], and [5]) have introduced a new reward-risk index based on the right-sided moment of p-th order[3] $\mathrm{E}[(X^+)^p]^{1/p}$ (used to value reward $r(X)$) and the left-sided moment of q-th order $\mathrm{E}[(X^-)^q]^{1/q}$ (used to value risk $\rho(X)$). The choice of p and q reflects different agent's feelings about the consequences of either overperformance or underperformance. A small value for the left order, $q < 1$, is appropriate if the agent is willing to fail the target without particular regard to the amount. When large deviations below b harm more than the small ones, then one can chooses $q > 1$. A similar reasoning applies to selecting the proper right order.

The idea behind Rachev's ratios is similar, but instead of looking at the positive or negative dispersion around the expectation, the focus is on the left and right tail of the excess portfolio return's distribution. In the Rachev's Generalized Ratio one has $r(X) := \mathrm{E}\big[(X^+)^\gamma \,|\, X \geq -VaR_{1-\alpha}\big]$ and $\rho(X) := \mathrm{E}\big[(X^-)^\delta \,|\, X \leq -VaR_\beta\big]$, where

[1] This conceptual line is similar to that of a pension fund manager's view. He synthesizes between the portfolio manager's aim (to beat the chosen benchmark, controlling upside risk) and the risk manager's aim (to control downside risk).

[2] The portfolio under consideration consists in a long position of the effective managed portfolio and a short position of the benchmark, such as a life/non-life insurance liability, a market index and a money market instrument.

[3] Here x^- denotes $-\min\{x, 0\}$ and x^+ denotes $\max\{0, x\}$.

$\alpha, \beta \in (0,1)$, the parameters $\gamma, \delta > 0$ behave similarly to p and q, while the Value-at-Risk is defined as $VaR_c := -\inf\{x \,|\, \mathrm{P}(X \leq x) > c\}$.

3 Back Test of Performance Indexes

Investors wish to maximize a performance measure finding an optimal risky portfolio regarded as the "market portfolio" in which to invest their wealth. In finite dimensional setting, let $\mathbf{r} = (r_1, \ldots, r_N)'$ be a random vector representing the total returns of N assets over a fixed horizon.[4] The optimal risky portfolio $\mathbf{w} = (w_1, \ldots, w_N)'$ solves the static stochastic optimization problem (SSO)

$$\max_{\mathbf{w} \in \mathcal{W}} \Phi(\mathbf{r}'\mathbf{w}) \tag{1}$$

$$\text{s.t. } \mathcal{W} := \left\{ \mathbf{w} \in \mathbb{R}^N \,|\, \mathbf{e}'\mathbf{w} = 0, \mathbf{l_B} \leq \mathbf{w} \leq \mathbf{u_B} \right\}$$

where $\mathbf{e}$ is a vector of ones, $\mathbf{l_B}$ and $\mathbf{u_B}$ are the vectors of lower bounds and upper bounds for portfolio weights.[5] With problem (1) in mind, we do an empirical comparison among the performance indexes in order to verify their forecasting ability when the reward measures and the risk measures used admit different definitions with respect to the traditional Sharpe ratio, particularly those ratios based on one-sided variability measures. Considering several time steps, we chose the unit time step equal to 1 month and model the re-allocation decision with a sequence of one-period SSOs.[6] In the following, we refer to $(r_{ti})_{t=-T,\ldots,0,1,\ldots,h}$ as the time series of stock index i, where T denotes the size of sample data, and h is the time lag.

3.1 Numerical Examples

To analyze the behavior of the 11 performance indexes mentioned in this paper, we propose an application to stock portfolio optimization. The primary issue being optimization, we employ the Exponential Weighted Moving Average model (EWMA) for forecasting the future total returns. One expects some of the optimized portfolio will generate an overperformance.[7]

- **Asset universe.** Five stock indexes with reinvested dividend (total return indexes) from 3 countries: S&P500 ($i = 1$), DowJones ($i = 2$), NASDAQ ($i = 3$), FTSE ($i = 4$), and NIKKEI ($i = 5$). Benchmark: T-bill as a proxy for the risk-free rate ($i = 6$). Number of assets: $N = 6$.

[4] Assume the discrete compounding convention.

[5] Given the total return $R = \mathbf{r}'\mathbf{w_p}$, the benchmark can be represented as $b = \mathbf{r}'\mathbf{w_b}$. Hence, the total excess return can be represented as $X = \mathbf{r}'\mathbf{w}$, where the difference $\mathbf{w} = \mathbf{w_p} - \mathbf{w_b}$ is termed as excess weights. Note that $\mathbf{w_p} = (0, w_2, \ldots, w_N)'$ and $\mathbf{w_b} = (1, 0, \ldots, 0)'$, hence the benchmark's return is a component of $\mathbf{r}$.

[6] This choice guarantees stochastic independence over time, since the historical returns we will use are weakly stationary. Otherwise, a stochastic dynamic programming approach were more adequate, though the pseudo-dynamic optimization approach we will use is a good starting point to estimate the solution of the former.

[7] If that is true, it is admissible to say that a performance ratio has a forecasting power.

- **Historical data.** A data set comprising monthly total returns and T-bill rates from April 2, 1984, to October 3, 2005, for a total of 258×6 observations.[8] Note that $t = 0$ corresponds to February 1, 2005 and $t = h = 8$ to October 3, 2005. The initial data set contains $T = 250$ observations and it is used for estimation and forecasting purposes.
- **Investment inequality restrictions.** Two cases considered, to test the model by imposing a set of "weak" lower and upper bounds and then "stronger" ones.
  (a) $\mathbf{l_B} = (0.1, 0.1, 0.1, 0.1, 0.1, 0)'$, $\mathbf{u_B} = (0.3, 0.3, 0.3, 0.3, 0.3, 1)'$.
  (b) $\mathbf{l_B} = (0.1, 0.02, 0.02, 0.1, 0.02, 0)'$, $\mathbf{u_B} = (0.5, 0.1, 0.1, 0.5, 0.1, 1)'$.

Assume the investors have an initial wealth $W_0 = 1$ and invest it to purchase the market portfolio $\mathbf{r}'\mathbf{w}$. On $t = 0$, the investment is rolled-forward monthly. The final value of wealth is computed recursively by the following algorithm (**A1**):[9]

```
start February 1, 2005
  for k = 1 to 9
    with each Φ(.) do
        solution w*_k of problem (1)
        sample path of cumulative wealth W_k = W_{k-1}(1 + r'_k w*_k)
  next k
stop October 3, 2005
return allocation rule: ''rank the indexes and choose always that
has performed the best during the 9 months''
```

The optimal decay factor λ^* for the returns is computed by the following algorithm (**A2**):[10]

```
for i = 1 to 5
  with time series (r_ti) do
    solution of min_{λ_i} [250^{-1} Σ_{1≤t≤250} (r_ti^2 − σ^2(λ_i))^2]^{1/2}
  return weighted decay factor λ* = Σ_{1≤j≤5} g_i λ*_i for returns
next i
```

Algorithm (**A1**) iterates algorithm (**A2**).[11] Results in table 1 show the best forecast of investment performance being with stable ratio, during the last 9 months. With the Farinelli-Tibiletti ratio the investment performance is ranked either 9th or 10th based on agent's feelings towards overperformance above the risk-free rate. If the invest-

8 In this study period all the stock indexes are very volatile, have a positive correlation, and weak stochastic dependence over time.

9 The optimization problem (1) gives raise to the optimal excess weights $\mathbf{w}^* := \mathbf{w_b} + \mathbf{w_P}^*$. Clearly, we have implemented this problem by imposing $\mathbf{w_b} = \mathbf{0}$ (no cash position).

10 The weights are given by $g_i = \theta_i^{-1} / \sum_{1 \le i \le 5} \theta_i^{-1}$, where $\theta_i = \lambda_i^* / \sum_{1 \le i \le 5} \lambda_i^*$.

11 The authors have developed an algorithm coded in MATLAB©. Moreover, to estimate the *alpha* index of return stability they used the Stable software by John P. Nolan [11]. A forthcoming version of the software will embed the quantile estimation algorithm coded by J. H. McCulloch [10]. All the values are rounded to the first fourteen decimals.

Table 1. Values of final wealth W_9

Sharpe	1.06694629842668
Minimax	1.06811617301054
MAD	1.05841393259069
Gini	1.05125992095915
Sortino-Satchell	1.01840417326684
VaR	1.07558191178067
STARR	1.07415420051962
Rachev	1.06304804004463
Generalized Rachev	1.04271604120210
Farinelli-Tibiletti	1.04342155326597 (1.03617932688779)
Stable	1.08034901565430

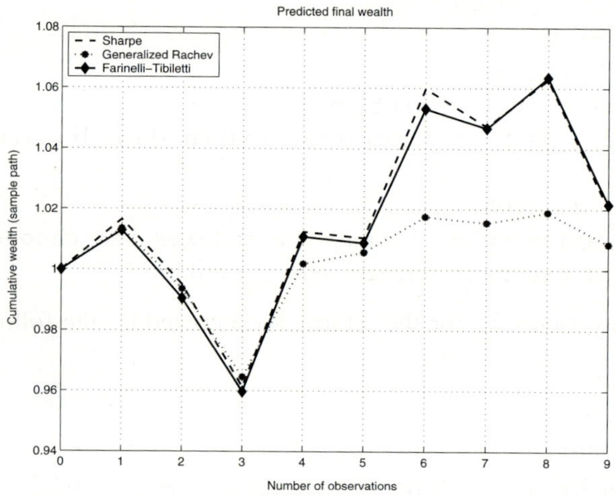

Fig. 1. Comparison among different values of final wealth computed with Sharpe ratio (*dashed line*), with Farinelli-Tibiletti ratio (*solid line*), and with Generalized Rachev ratio (*dotted line*)

ment management provides information about the history of stock prices, the investors can modify their preferences requiring alternative portfolio restrictions according to the empirical features of past returns. Case (b) corresponds to claim a greater investment in S&P500 and FSTE than in the remaining stock indexes, since these two assets have showed "better" skewness and kurtosis features during the past 258 months than the other assets as it can be seen in table 2. The values of final wealth with those new portfolio constrains considered in (**A 1**) are listed in table 3. We note the Farinelli-Tibiletti ratio with left order $p = 2$ and right order $q = 0.5$ having a superior forecasting power than the Sharpe ratio, since this parameter setting better fits the agent's willingness to a performance over the benchmark, but a lower risk.

The Sharpe ratio has a low forecasting ability, while the Sortino-Satchell ratio provides a more robust performance evaluation. Nevertheless, indexes based on one-sided

Table 2. Summary statistics

	Coef. of skewness	Coef. of kurtosis
S&P500	-1.0452	7.0024
Dow Jones	-1.2717	9.1074
NASDAQ	-1.1771	8.1169
FTSE	-0.9259	5.5730
NIKKEI	-0.3526	3.6497

Table 3. Values of final wealth W_9

Sharpe	1.02012809289746
Minimax	1.03261850863065
MAD	1.02731027560518
Gini	1.01821753891104
Sortino-Satchell	1.00079846908146
VaR	1.02847451885572
STARR	1.03561700766895
Rachev	1.00479358219820
Generalized Rachev	1.00844935672914
Farinelli-Tibiletti	1.02158214310971 (1.02018870912263)
Stable	1.02914075668800

Table 4. Values of periodic (*9 months*) wealth, $p = 2, q = 0.5$. Different results in parentheses (*rows 15th and 16th*, $p = 0.5, q = 2$)

Generalized Rachev	1.10292296179689
Generalized Rachev	0.98606688783133
Gini	1.13870280665159
MAD	1.27064360179623
Sortino-Satchell	1.16576029275820
Rachev 1	1.22929409983253
Rachev 1	1.24671887971556
Sortino-Satchell	1.14971024283881
Rachev 2	1.16075620538274
Sortino-Satchell	1.35023369205268
Generalized Rachev	0.90595224732194
Stable	1.02587651156297
Generalized Rachev	0.70932459615394
Sortino-Satchell	1.27960286499482
Rachev 2	1.19949421938987 (Farinelli-Tibiletti: 1.20454081121982)
Farinelli-Tibiletti	1.08879861877932 (Sortino-Satchell: 1.08537313315757)
Farinelli-Tibiletti	1.02346807285117

variability measures such as the Generalized Rachev ratio and the Farinelli-Tibiletti ratio have a degree of flexibility in the measurement of portfolio overperformance.

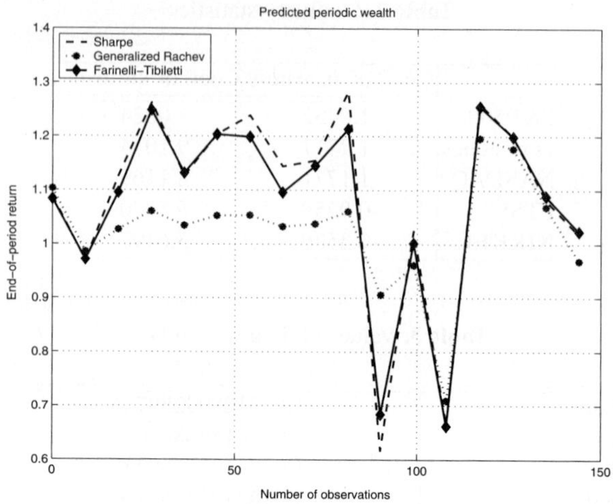

Fig. 2. Comparison among different values of periodic wealth computed with Sharpe ratio (*dashed line*), with Farinelli-Tibiletti ratio (*solid line*), and with Generalized Rachev ratio (*dotted line*)

Evaluating the portfolio's performance with respect to a benchmark additionally requires an out-of-sample analysis. We employ two moving windows: within the first (length 1 month) the final values of wealth (for each indexes) are calculated; within the second (length 9 months) the robustness of investment ranking for all the ratios involved in the optimization procedure is tested. The **robustness check algorithm** follows:

```
start December 1, 1992
  for k = 1 to 17
    with a sample of 100 data points do
        (A 1)
  next k
stop October 3, 2005
return allocation rule: ‘‘rank the indexes and choose always that
has performed the best during each period of length 9 months’’
```

Results from **robustness check algorithm** are shown in table 4 in the case (b).[12]

4 Conclusion

In this paper we solve a portfolio optimization problem comparing eleven performance indexes recently proposed by the literature, studying their behavior on wealth investment. Indexes based on one-sided variability measures are a flexible tool for modelling the agent's beliefs about either underperformance or overperformance a given benchmark. Further empirical tests will be the subject of future research. Particularly, we need

[12] We employ different parameter setting for the Rachev ratio: $\alpha = \beta = 0.05$ (Rachev 2); $\alpha = 0.01$ and $\beta = 0.5$ (Rachev 3).

to investigate the influence of the forecasting model for the expected returns, since from the mean-variance framework it is well known that optimal allocation depend in a very sensitive way on the expected returns.

References

1. Biglova, A., Huber, I., Ortobelli, S., Rachev, S.T.: Optimal Portfolio Selection and Risk Management: A Comparison Between the Stable Paretian Approach and the Gaussian One. In: Rachev, S.T. (ed.): *Handbook on Computational and Numerical Methods in Finance*. Birkhäuser, Boston (2004) 197-252
2. Biglova, A., Ortobelli, S., Rachev, S.T., Stoyanov, S.: Different Approaches to Risk Estimation in Portfolio Theory. *Journal of Portfolio Management* (2004) Fall, 103-112
3. Farinelli, S., Tibiletti, L.: Sharpe Thinking in Asset Ranking with One-Sided Measures. *European Journal of Operational Research* (2005) **5**, forthcoming
4. Farinelli, S., Tibiletti, L.: Upside and Downside Risk with a Benchmark. *Atlantic Economic Journal*, Anthology Section, (2003) **31**(4), December, 387
5. Farinelli, S., Tibiletti, L.: Sharpe Thinking with Asymmetrical Preferences. Technical Report, presented at European Bond Commission (2003) Winter Meeting, Frankfurt
6. Favre, L., Galeano, J.A.: Mean-Modified Value at Risk Optimization with Hedge Funds. *The Journal of Alternative Investment* (2002) **5**, Fall
7. Huber, I., Ortobelli, S., Rachev, S.T., Schwartz, E.: Portfolio Choice Theory with Non-Gaussian Distributed Returns. In: Rachev, S.T. (ed.): *Handbook of Heavy Tailed Distribution in Finance*. Elsevier, Amsterdam (2003) 547-594
8. Konno, H., Yamazaki, H.: Mean-Absolute Deviation Portfolio Optimization Model and its Application to Tokyo Stock Market. *Management Science* (1991) **37**, 519-531
9. Martin, D., Rachev, S.T., Siboulet, F.: Phi-Alpha Optimal Portfolios and Extreme Risk Management. *Wilmott Magazine of Finance* (2003) November, 70-83
10. McCulloch, J.H.: Simple Consistent Estimators of Stable Distribution Parameters. *Commun. Statist. Simulation* (1986) **15**, 1109-1136
11. Nolan, J.P.: Numerical Approximations of Stable Densities and Distribution Functions. *Commun. Statist. Stochastic Models* (1997) **13**, 759-774
12. Pedersen, C.S., Satchell, S.E.: On the Foundation of Performance Measures under Asymmetric Returns. *Quantitative Finance* (2003)
13. Shalit, H., Yitzhaki, S.: Mean-Gini, Portfolio Theory, and the Pricing of Risky Assets. *Journal of Finance* (1984) **39**, 1449-1468
14. Sortino, F.A., van der Meer, R.: Downside Risk. *Journal of Portfolio Management* (1991) **17**(4), 27-32
15. Sortino, F.A.: Upside-Potential Ratios Vary by Investment Style. *Pensions and Investment* (2000) **28**, 30-35
16. Yitzhaki, S.: Stochastic Dominance, Mean Variance and Gini's Mean Difference. *American Economic Review* (1982) **72**, 178-185
17. Young, M.R.: A MiniMax Portfolio Selection Rule with Linear Programming Solution. *Management Science* (1998) **44**, 673-683

Stock Trading System Based on Formalized Technical Analysis and Ranking Technique

Saulius Masteika and Rimvydas Simutis

Faculty of Humanities, Vilnius University,
Muitines 8, 44280 Kaunas, Lithuania
saulius.masteika@vukhf.lt, rimvydas.simutis@vukhf.lt

Abstract. The contribution of this paper lies in a novel application of formalized technical analysis and ranking technique by development of efficient stock trading system. The proposed system is implemented using two steps: on the first step system analyses historical data from large number of stocks and defines a quality function of each stock to specific technical trade pattern; on the second step system grades all the stocks according to the value of the defined quality function and makes suggestions to include the highest ranked stocks into the traders' portfolio. These stocks are being hold for fixed time interval in traders' portfolio, then sold and replaced with the new stocks that got the highest rank. The proposed trading system was tested using historical data records from the USA stock market (1991-2003). The trading system had given significantly higher returns when compared to the benchmark.

1 Introduction

The continuing progress of computing methods and information technologies makes significant influence on financial markets. An effective information processing is more and more important in order to succeed in the stock markets. The strategists of the investments funds use historical data and real-time information to forecast the trends of stocks' price changes and also apply these results to the decision making and formation of investment portfolios. However wide known Efficient Market Hypothesis (EMH) states that prices of stocks are just a reflection of all known information about the company. This means that having the information, no prediction of future price changes can be made. EMH also states that new information immediately forms new prices of companies' shares. If we agree with this hypothesis, we have to realize that analysis of historical information is not important for an investor as this information is already reflected in the stock price. People are not able to forecast the flow of new, important information and this information appears randomly. Therefore according to the Efficient Market Hypothesis it is impossible to get better returns in the stock market than the market benchmarks (e.g. S&P500 index) for a longer time period using the available historical information about the market. There are numerous well known research papers [1], [2], that partly confirm this hypothesis. However some published reports show [3], [4], that the efficient market hypothesis is far from the correct one. There are some papers claiming that the application of technical

V.N. Alexandrov et al. (Eds.): ICCS 2006, Part IV, LNCS 3994, pp. 332–339, 2006.
© Springer-Verlag Berlin Heidelberg 2006

analysis or nonlinear models, such as neural networks and genetic algorithms, can supply information for intelligent decision support systems, and traders using these techniques are able to beat the market for a long period of time [5], [6], [7], [8]. In this paper we reinforce these statements. The contribution of this paper lies in novel application and combination of formalized technical analysis and ranking technique for development of efficient stock trading system. The proposed system is quite simple in implementation and could attract an attention of individual stock traders and modest mutual funds. The proposed stock trading system was tested using historical data from USA stock market. The obtained results clearly contradict the statements of Efficient Market Hypothesis. The trading system based on the proposed approach has given significantly higher returns relative to the benchmark (S&P500) for a long period of time. In order to use the system in real markets it is necessary to solve problems with the risk management of trader's portfolio and to carry out additional experimental tests with the stock markets of different countries and different time periods. The paper is organized as follows: Section 2 provides the theoretical basis of the proposed system; Sections 3 is devoted for experimental investigations and evaluation of the proposed system. The main conclusions of the work are presented in Section 4.

2 Development of the Stock Trading System

The proposed stock trading system is based on formalized technical analysis and is suited for medium and short term stock traders. Over the years numerous technical indicators and patterns have been developed to describe stock performance, as well as to predict future price movements. Technical analysts and traders believe that certain stock chart patterns and shapes (e.g. moving average crosses, head and shoulder formation, range breakout, triple top/bottom formation, cup-with-a-handle formation) are signals for profitable trading opportunities [9], [10]. Many professional traders claim that they consistently make trading profits by following those signals. Short term stock traders analyze historical stock data records (price and volume) for 5-7 trading days and make decisions for buying stocks based on typical trading patterns. Normally they hold stocks for 2÷5 days. In this paper we introduce a new technical pattern, called "precursor of reverse" pattern. This pattern can be characterized as follows: after a long price decline (5-6 days) with strong trading volume, the stock price and trading volume stabilize and price prepares for rebound. The Figure 1 illustrates a typical example of the "precursor of reverse" pattern. Our extensive and rigorous historical tests have shown that this pattern has a good rebound potential. The stock trading strategy is to buy the stock as the price and volumes stabilize and hold it for 2-5 days.

2.1 Formalization of the Trading System

Classical technical analysis compares stock's price and volume history to typical chart patterns (in our case "precursor of reverse" pattern) and predicts future price behavior based on the degree of match. This technique can be formalized through trading rule of the following form: *If the technical pattern X is identified in the previous N trading days, then buy the stock and sell it on the H-th day after purchasing.* The difficult part in application of this technique is reliability of identification of the technical pattern *X*.

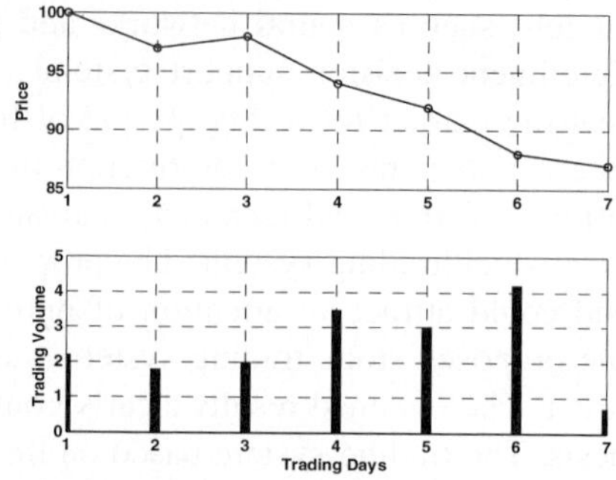

Fig. 1. A typical "precursor of reverse" pattern

The new technique that we propose here is based on the formalization of "precursor of reverse" pattern using normalized historical stock prices and trading volume data. The idea of this technique is to analyze N-days moving window for a big pool of stocks with normalized closing price and volume records and identify the technical pattern when it appears. For formalization of the "precursor of reverse" pattern we use three variables: normalized stock price P_{nN} , normalized trading volume V_{nN} at the last day of observation and normalized price change during the last day D_{nN}. During the N-days moving window the normalized stock prices P_{ni} , normalized trading volume V_{ni} and normalized price changes D_{ni} are obtained using the following equations :

$$P_{ni} = \frac{P_i}{P_1} \cdot 100 \quad , \quad V_{ni} = \frac{V_i}{\frac{\sum_{i=1}^{N} V_i}{N}} \quad , D_{ni} = P_{n(i-1)} - P_{ni} \quad i = 1...N \quad , \tag{1}$$

where P_i and V_i are stocks' closing prices and trading volumes on i-day. This equation is estimated every day for all stocks in pool. We assume that a "precursor of reverse" pattern is formed when the stock price drops approximately 10% during the seven trading days, the last day's price drops about 2% and the trading volume on last day of observation begins to decrease. In traditional technical analysis the quality of the formed pattern is evaluated using various graphical methods. Testing and application of such methods are complicated because different traders often evaluate the same trading pattern in different ways. In this paper we propose a formal equation for evaluation of quality of trading pattern Q_p. This equation is derived based on extensive interviews with experienced stock trading experts and has a following form:

$$Q_p = \frac{(P_{n1} - P_{nN})}{K_N + (P_{n1} - P_{nN})} \cdot \frac{D_{nN}}{K_D + D_{nN}} \cdot \frac{K_V}{K_V + V_{nN}} \quad . \tag{2}$$

As it can be recognized from the structure of the equation, the quality of the "precursor of reverse" pattern' Q_p is more precise when stock's price drops significantly during the observed time interval and stock's price also decrease at the last day of observation, but with decreasing trading volume. The values of parameters K_N, K_D, K_V in Equation 2 were defined using the recommendation of trading experts and are given in experimental part of this work. Having stock data records for defined moving time interval now it is straightforward to estimate the three variables of proposed pattern for every trading day and define the quality of the pattern, Q_P. This information is the backbone of the proposed trading system.

2.2 Ranking Technique

Experimental investigations of the "precursor of reverse" patterns had shown that high quality patterns are very rare. Still further, approximately only the 70% of the high quality patterns are profitable. Therefore we need to scan a lot of stocks to find potentially profitable trading opportunities. The proposed algorithm for stocks ranking consists of the following steps:

- in the first step a group of companies, whose trading patterns will be observed is selected,
- then the trading patterns - "precursor of reverse" are identified, using the moving window data,
- the quality of each trading pattern is estimated (Equation 2),
- all trading patterns are ranked following the highest pattern quality and the highest rank stocks are suggested to the trader.

2.3 Evaluation of the Trading Strategy

A portfolio of the stock trader is formed of stocks having highest rank of trading pattern. Number of stocks held in portfolio will be discussed in experimental part of this work. The efficiency of the proposed system was evaluated using three parameters: a total return on trading R_T, a total capital at the end of trading period C_T and Sharpe Ratio of the trader's portfolio S. The total return R_T was estimated using equation

$$R_T = \sum_{i=1}^{T} r_i, \quad r_i = \frac{P_i - P_{i-1}}{P_{i-1}} \cdot 100 - T_C, \tag{3}$$

where T is the number of trading transactions, P_i is stock price, T_C is transaction costs and r_i is simple return of i-th transaction. Total capital at the end of the trading period was estimated using trader's start capital and every day returns. The Sharpe Ratio is a well known measure of the risk-adjusted return of an investment [11]. Sharpe Ratio of the trader's portfolio was estimated using equation

$$S(t) = \frac{R_D(t) \cdot 250 - R_a}{\sigma_D(t) \cdot \sqrt{250}}, \tag{4}$$

where R_D and σ_D - average daily returns and standard deviation of the trader's portfolio over moving 250 days, R_a - annualized return on "risk-free" investment (we assumed it to be 3%). Mutual funds which achieve a Sharpe Ratio of over 1.0 are qualified as good. A Sharpe Ratio of a fund of over 2.0 is considered very good and Sharpe Ratios above 3.0 are outstanding.

3 Experimental Investigation

The efficiency of the proposed stock trading system was tested using historical stock market data. The data set that has been used in this study represents daily stock closing prices and trading volumes, and comes from 366 stocks on the USA stock market from October 1, 1991 till October 1, 2003. These stocks represent all the SP500 index stocks, included in the index for the whole analyzed time period.

3.1 Experimental Set-Up

We have used 7-days moving window technique and normalization technique (Eq.1) to identify the "precursor of reverse" patterns in data records. The quality of the patterns, Q_P was evaluated using Equation 2. The values of parameters here were chosen empirically using extensive interviews with stock trading experts. The values of parameters are as follows: K_N=10.0, K_D=2.0, K_V=3.5. Because the values of these parameters are crucial for the trading system more investigation must be done in the future for the optimization and adaptation of them. All trading patterns were ranked; the stocks with highest ranks were included in trader's portfolio and held for a defined time interval. The performance of the trading system was evaluated using proposed parameters, R_T, C_T and S. The detailed functioning of the proposed trading system can be described as follows:

- The stocks' prices and trading volumes were normalized and analyzed considering every trading day starting from 1991.10.01 to 2003.10.01;
- Based on 7-days moving window data (normalized price and volume) the "precursor of reverse" patterns were identified and quality of these patterns, Q_P, for every stock were estimated;
- Ranking algorithm for Q_P was applied and stocks (1-15 stocks) with the best ranks were bought and sold after defined time interval (1-5 days).

In the beginning of the testing period the start capital for trader was 1000 US$. Equal parts of this capital were invested in all selected stocks. The transaction costs of 0.5% were withdrawn from the returns per contract. This amount of transaction costs is typical for e-brokers companies (e.g. *Interactive Brokers*, USA) and reflects realistic losses that are unavoidable when changing the stocks in trader's portfolio. All income got during the stocks' trading actions were again completely reinvested and the changing of the total capital was estimated during the investment period. The structure of the proposed virtual stock trading system is presented in Figure 2. It is necessary to point out that the proposed scheme gives the possibility to apply the proposed system in real stock trading process. In this case the highest ranked stocks calculated on 7-days moving window data can be bought using market orders at the last moment before the markets are closed.

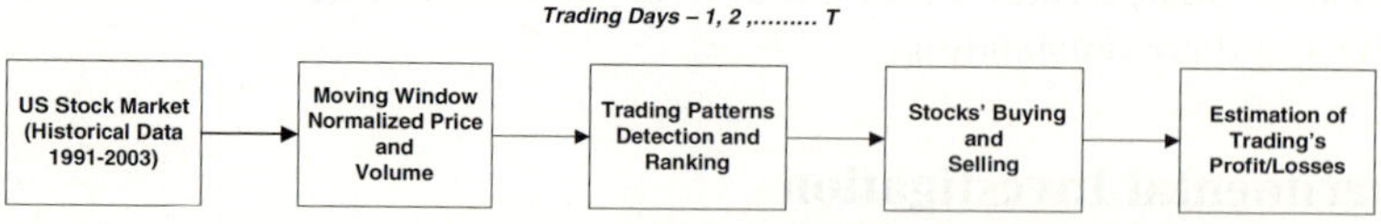

Fig. 2. The structure of the proposed stock trading system

3.2 Experimental Results

The total returns of proposed trading system are presented in Figure 3. The best results are achieved when trader buys the five best ranked stocks every day and hold them for two days. In this case the proposed trading system outperforms the benchmark significantly and the total return at the end of the trading period is 742% .The total return for the SP500 index during the same period is 112% , and total return when investing (buy and hold) in all analyzed stocks (equal part of capital in each stock) is 265%. Trading a smaller number of best ranked stocks (e.g. 1 to 4) gives even better results, but this leads to a big fluctuation in portfolio performance, therefore is not interesting for real trading process and will not be discussed here.

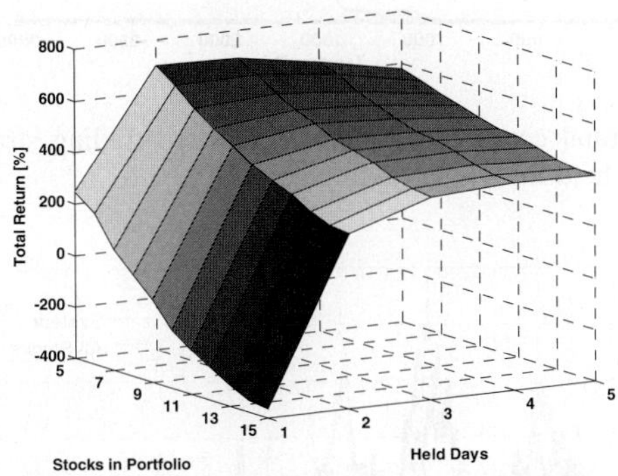

Fig. 3. The total returns depends on number of obtained stocks and how long the obtained stocks are held in trader's portfolio

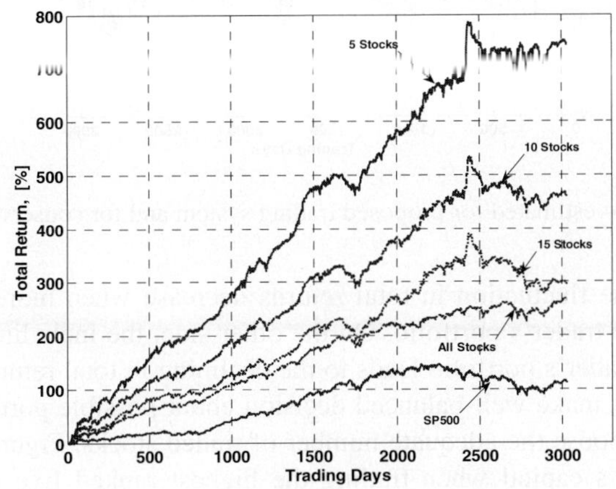

Fig. 4. The dynamic of total returns during the whole trading period for various portfolios

Figure 4 presents how the total returns of proposed trading system change during the analyzed time interval (obtained stocks are held for two days). As one can see

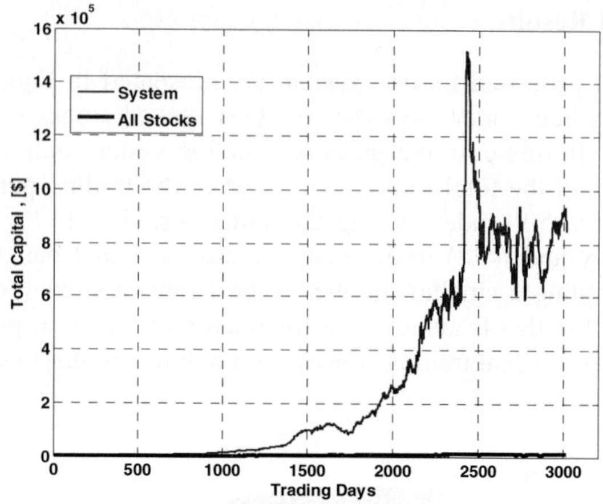

Fig. 5. The dynamic of total capital when using proposed trading strategy and conservative investment (all stocks are held for whole trading period)

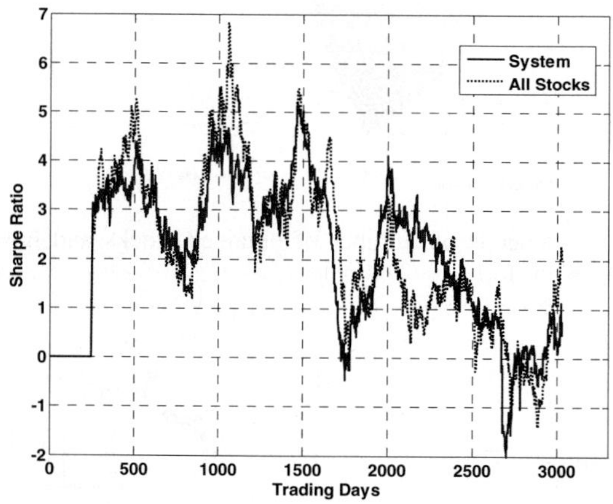

Fig. 6. Sharpe Ratio estimated for proposed trading system and for conservative investments

from the graph, the fluctuation in total returns decrease when more different stocks are included in the trader's portfolio. On the other side, the including of more different stocks in the trader's portfolio leads to the declining in total returns. Therefore the trader must always make well-balanced decision about possible portfolio fluctuations and returns, and choose the adequate number of traded stocks. Figure 5 compares the dynamic of trader's capital when trading the highest ranked five stocks and when using conservative investment strategy (start capital is invested in all analyzed stocks). The trader's capital using the proposed trading system increases dramatically from 1000 $ to 848610 $. At the same time conservative investment gives the end capital equal 13060 $. Figure 6 illustrates the Sharpe Ratio estimation for both: proposed trading system and conservative investments. As it can be recognized from the graph the Sharpe Ratio of the proposed trading system is relative high and is similar to the Ratio obtained for conservative investments. Nonetheless, the Sharpe Ratio decreases significantly for some periods of time and these periods can be very hard challenges for individual traders or mutual funds.

4 Conclusions

Preliminary experimental tests have shown that the proposed stock trading system gives very promising results and can be attractive for individual traders and modest mutual funds. Despite of these results there are two problems that will arise while using the proposed system in real stock markets: a) trading results are very sensitive to the costs of transactions: when transaction costs are over 0.8% the results of trading system's become worse than investing in benchmark; b) proposed trading system is related with high risks and the traders must be prepared for temporal decreasing of total capital even to 50 % (Fig. 5). Also more extensive tests with the system must be done (different markets, various time periods, tax regulations) to evaluate the domain of possible applications of the proposed trading system more precisely.

References

1. White, H. Economic prediction using neural networks: The case of IBM daily stock returns. In: IEEE International Conference on Neural Networks. San Diego, 1988, pp. 451-459.
2. Lowe, A.R.Webb. Time series prediction by adaptive networks: a dynamical systems perspective. In Artificial Neural Networks: Forecasting Time Series (eds. V. Rao Vemuri and Robert D. Rogers). IEEE Computer Society Press, 1994, pp. 12-19. D.
3. F. Fama. Efficient capital markets, Journal of Finance, 46(5),1991, 1575-1617.
4. D.J. [illegible], W.M. van den Bergh, H. Vaudrey. Market inefficiencies, technical trading and neural networks. In: Dunis, C. (ed.): Forecasting Financial Markets, [illegible]nomics and Quantitative Analysis, Chichester, John Wiley & Sons, 1996, pp. 245-260.
5. A.Refenes. Neural Networks in the Capital Markets, Wiley, Chichester, 1995.
6. W. Leigh, R. Purvis, J.M. Ragusa. Forecasting the NYSE composite index with technical analysis, pattern recognizer, neural network, and genetic algorithm: a case study in romantic decision support. Decision Support Systems, 32, 2002, 361-377.
7. H. Hong. J. Stein. A unified theory of underreaction, momentum trading, and overreaction in asset markets, The Journal of Finance LIV (6), 1999, 2143-2184.
8. H. Hong, T. Lim, J. Stein. Bad news travels slowly: size, analyst coverage, and the profitability of momentum strategies. The Journal of Finance LV (1), 2000, 265-295.
9. S.B. Achelis. Technical Analysis from A to Z, 2nd Editon, McGraw-Hill Professional, 2000.
10. R. Martinelli, B. Hyman. Cup-with-handle and the computerized approach. Technical Anglysis of Stocks and Commodities, 16(10), 1999, 63-66.
11. Sharpe W. F. Asset allocation: Management Style and Performance Measurement," Journal of Portfolio Management, Winter 1992, pp. 7-19.

Deriving the Dependence Structure of Portfolio Credit Derivatives Using Evolutionary Algorithms

Svenja Hager and Rainer Schöbel

Department of Corporate Finance,
Faculty of Economics and Business Administration,
Eberhard-Karls-University of Tübingen,
Mohlstraße 36, 72074 Tübingen, Germany

Abstract. The correct modeling of default dependence is essential for the valuation of portfolio credit derivatives. However, for the pricing of synthetic CDOs a one-factor Gaussian copula model with constant and equal pairwise correlations for all assets in the reference portfolio has become the standard market model. If this model were a reflection of market opinion there wouldn't be the implied correlation smile that is observed in the market. The purpose of this paper is to derive a correlation structure from observed CDO tranche spreads. The correlation structure is chosen such that all tranche spreads of the traded CDO can be reproduced. This implied correlation structure can then be used to price off-market tranches with the same underlying as the traded CDO. Using this approach we can significantly reduce the risk to misprice off-market derivatives. Due to the complexity of the optimization problem we apply Evolutionary Algorithms.

1 Introduction

Although it is still of interest to find empirical sources of correlation data, people increasingly use the market of synthetic collateralized debt obligations (CDOs) to derive information about the correlation structure of the underlying of a CDO. An observed CDO premium can be interpreted as an indicator of asset correlation. Therefore, more and more tranched products are quoted in terms of an implied correlation parameter instead of the spread or the price of a tranche. The implied correlation of a tranche is the uniform asset correlation that makes the tranche spread computed by the standard market model equal to its observed market spread. The standard market model is a Gaussian copula model that uses only one single parameter to summarize all correlations among the various borrowers' default times. But obviously a flat correlation structure is not able to reflect the heterogeneity of the underlying asset correlations since the complex relationship between the default times of different assets can't be expressed in one single number. Obviously, the standard market model doesn't reflect market opinion because the implied correlation smile emerges. Typically, mezzanine

V.N. Alexandrov et al. (Eds.): ICCS 2006, Part IV, LNCS 3994, pp. 340–347, 2006.
© Springer-Verlag Berlin Heidelberg 2006

tranches trade at lower implied correlations than equity and senior tranches on the same portfolio. This phenomenon is called implied correlation smile. Despite the questionable explanatory power of the implied correlation parameter, the implied correlation of a CDO tranche is often used to price off-market products with the same underlying as the traded CDO.

Recently, more and more researchers examine different approaches to explain and model the correlation smile. Gregory and Laurent [2] and Andersen and Sidenius [1] introduce dependence between recovery rates and defaults. In a second extension Andersen and Sidenius introduce random factor loadings to permit higher correlation in economic depressions. Both approaches are able to model a smile. Hull and White [4] discuss the effect of uncertain recoveries on the specification of the smile.

The correlation smile shows clearly that it is not appropriate to use the implied correlation of a traded CDO tranche to value non-standard tranches on the same collateral pool. To address these shortcomings we take the implied correlation approach one step further and imply a correlation matrix that reproduces all given tranche spreads of a CDO simultaneously. The dependence structure is chosen such that the resulting tranche prices are concordant with observed market prices or, respectively, such that the observed correlation smile is reproduced. Hager and Schöbel [3] showed that heterogeneous correlation structures are able to model a smile. After we derived a suitable asset correlation structure, we can use this dependency to price off-market products with the same underlying. In this study we use Evolutionary Algorithms (EAs) to derive a dependence structure from observed CDO tranche spreads. We show systematically why EAs are suitable for this kind of application. So far, there is only a limited amount of studies that connect EAs with derivative pricing. To our knowledge we are the first to apply EAs to the implied correlation problem and we are the first to derive a correlation matrix that is not necessarily flat from a set of observed tranche spreads.

2 The Optimization Problem

Suppose for the moment that we know the tranche spreads of an actively traded CDO. We assume that the CDO consists of an equity tranche, a mezzanine tranche and a senior tranche. Our goal is to derive a correlation matrix Σ that replicates the given tranche spreads of the equity, the mezzanine and the senior tranche simultaneously. Denote these target values as $\overline{s_e}$, $\overline{s_m}$ and $\overline{s_s}$. It is intuitively clear that in general there can be more than one correlation matrix that leads to the respective tranche spreads $\overline{s_e}$, $\overline{s_m}$ and $\overline{s_s}$. Hager and Schöbel [3] discuss this subject. Note that there might also be combinations of tranche spreads that can't be reproduced by any correlation matrix. However, there is no way to derive the correlation matrix Σ in closed form since even the portfolio loss distribution can't be computed in closed form for arbitrary correlation matrices.

To measure the quality of an obtained correlation matrix Σ, we first compute the appendant equity, mezzanine and senior tranche spreads $s_e(\Sigma)$, $s_m(\Sigma)$

and $s_s(\Sigma)$ and compare them with the given target values $\overline{s_e}$, $\overline{s_m}$ and $\overline{s_s}$. The goal is to find a correlation matrix such that the corresponding spreads agree. The optimization problem discussed in this study is rather complex because the search space is high dimensional and multi-modal and the objective function is non-linear, non-differentiable and discontinuous. Note that for arbitrary correlation matrices often both $s_e(\Sigma)$, $s_m(\Sigma)$, $s_s(\Sigma)$ and $\overline{s_e}$, $\overline{s_m}$, $\overline{s_s}$ are obtained via Monte-Carlo simulation. In this case we have to deal with noise. Since our optimization problem is characterized by these properties the number of applicable optimization techniques is restricted. Therefore, we choose EAs to address this challenging problem. EAs are stochastic search methods that are inspired by the Darwinian theory. They model the collective learning process within a population. The starting population is generally initialized by random. In the course of the generations the population is supposed to evolve toward successively better regions of the search space by randomized processes of selection, recombination and mutation. The generations are searched until a sufficiently good solution is found or until a termination criterion is met. Similar to other heuristic search methods, it is not guaranteed that EAs find the global optimum, but they generally find good solutions in a reasonable amount of time.

Consider the function $f(\Sigma)$ which measures the sum of the relative deviations of the obtained tranche spreads from the target spreads:

$$f(\Sigma) = \frac{|s_e(\Sigma) - \overline{s_e}|}{\overline{s_e}} + \frac{|s_m(\Sigma) - \overline{s_m}|}{\overline{s_m}} + \frac{|s_s(\Sigma) - \overline{s_s}|}{\overline{s_s}} \,.$$

In our optimization problem low values of $f(\Sigma)$ stand for high quality. In a population based optimization strategy with λ individuals, we neglect the overall performance of a certain generation and just consider the best individual in the respective generation. The objective function registers the quality of the best individual that has been generated so far. Let $h(t)$ denote the objective function at time t and let $\Sigma^{k,\tau}$ denote the kth individual in generation τ, $k \in \{1, ..., \lambda\}$, $\tau \in \{1, ..., t\}$. Consequently, the objective function, that has to be minimized, is

$$h(t) = \min_{k \in \{1,\ldots,\lambda\},\ \tau \in \{1,\ldots,t\}} \left(f(\Sigma^{k,\tau})\right) .$$

3 Pricing of Synthetic CDOs

In this study we always assume that the intensity based approach describes the default of one obligor and that the Gaussian copula model with an arbitrary correlation matrix describes the dependency between the obligors' default times. In our optimization problem all model parameters are known except for the pairwise linear correlations.

Following Laurent and Gregory [5] we consider a synthetic CDO, whose underlying consists of n reference assets, and assume that asset j has exposure $\frac{1}{n}$ and default time τ_j. $N_j(t) = 1_{\{\tau_j \leq t\}}$ denotes the default indicator process. The cumulative portfolio loss at time t is therefore $L(t) = \frac{1}{n} \sum_{j=1}^{n} N_j(t)$. A CDO is a structured product that can be divided into various tranches. The cumulative default of tranche (A, B), $0 \leq A < B \leq 1$ is the non-decreasing function

$$\omega(L(t)) = (L(t) - A)1_{[A,B]}(L(t)) + (B - A)1_{]B,1]}(L(t)) \, .$$

The tranche spread $s_{(A,B)}$ depends on the thresholds A and B. Let $B(t)$ be the discount factor for maturity t. T stands for the maturity of the CDO and let $t_1, ..., t_I$ denote the regular payment dates for the CDO margins.

The *default payments* can be written as

$$E\left[\sum_{j=1}^{n} B(\tau_j)N_j(T)(\omega(L(\tau_j)) - \omega(L(\tau_j^-)))\right] \, . \tag{1}$$

The *margin payments* are based on the outstanding nominal of the tranche. Since defaults often take place between regular payment dates we have to distinguish between the *regular payments*

$$s_{(A,B)} \sum_{i=1}^{I} B(t_i)E\left[B - A - \omega(L(t_i))\right] \tag{2}$$

and the *accrued payments.* Accrued margins are paid for the time from the last regular payment date before τ_j (this date is called $t_{k(j)-1}$) until τ_j. The *accrued payments* can be written as

$$s_{(A,B)}E\left[\sum_{j=1}^{n} B(\tau_j)N_j(T)(\tau_j - t_{k(j)-1})(\omega(L(\tau_j)) - \omega(L(\tau_j^-)))\right] \, . \tag{3}$$

A synthetic CDO can be compared with a default swap transaction because the CDO margin payments are exchanged with the default payments on the tranche. The spread $s_{(A,B)}$ is derived by setting the margin payments in (2) and (3) and the default payments in (1) equal.

4 Experimental Settings and Results

4.1 Description of the Genotype

Correlation matrices are symmetric positive semi-definite matrices whose matrix elements are in $[-1, 1]$. The diagonal of a correlation matrix always consists of ones. In the following, $\Sigma = (\Sigma_{ij})_{i,j=1,...,n}$ denotes the correlation matrix. We use Σ as phenotype and a row vector $\rho = (\rho_i)_{i=1,...,n} \in [-1, 1]^n$ as real-valued genotype. The pairwise linear correlation between asset i and asset j can be computed as $\Sigma_{ij} = \rho_i\rho_j$, $i \neq j$ and $\Sigma_{ij} = 1$, $i = j$. Using this so-called one-factor approach we can avoid Monte-Carlo simulations and provide semi-explicit expressions for CDO tranche spreads (see Laurent and Gregory [5]). Note that there are correlation matrices that can't be represented by the one-factor approach.

The initial population consists of randomly generated vectors $\rho \in [-1,1]^n$. An arbitrary vector ρ automatically leads to a symmetric, positive semi-definite matrix with elements $\Sigma_{ij} = \rho_i \rho_j$, $i \neq j$ and $\Sigma_{ij} = 1$, $i = j$.

In this study we compare several standard recombination and mutation schemes. They are carried out according to custom.

Note that recombination and mutation can breed vector elements with $|\rho_i| > 1$. To make sure that the pairwise correlations are in $[-1,1]$, define a censored vector $\rho^* = (\rho_i^*)_{i=1,\ldots,n}$ with $\rho_i^* = \min(\max(\rho_i, -1), 1)$ that replaces ρ. We maintain this modified representation of the genotype.

We consider two cases of suitable termination conditions. Naturally, reaching the optimum of the objective function with a certain precision should be used as stopping condition. Therefore, we stop our algorithm as soon as the objective function falls below a predefined value. Furthermore, we terminate the EA when the total number of function evaluations reaches a given limit.

4.2 Setup

To assess the potential of our approach, we work with simulated data. We compare the performance of a Monte-Carlo Search, a Hill-Climber ((1 + 1)-ES), an Evolution Strategy with 4 parent and 20 offspring individuals ((4, 20)-ES) and a generational Genetic Algorithm with 40 individuals (GA(40)). We mutate individuals by adding realizations of normally distributed random variables with expected value 0 and standard deviation 0.05 unless explicitly mentioned otherwise. We apply global mutation operators, i.e. every vector element is subject to mutation. In case of the (4, 20)-ES we use elite selection and 1-point crossover. Our focus is on the application of different mutation operators. We consider the 1/5-rule, global mutation without a strategy parameter and global mutation with a strategy parameter that controls the mutation step size. The mutation probability is 0.95, the crossover probability is 0.50. In case of the GA(40) we focus on the selection and crossover parameters. We use proportional selection and tournament selection with a tournament group size of 10. We use 1-point crossover and intermediate crossover and we use global mutation without a strategy parameter. The crossover probability is 0.95, the mutation probability is 0.50. In our study we just consider non-negative pairwise linear correlations and therefore non-negative genotypes for the sake of simplicity. As soon as the objective function falls below 5% we terminate the algorithm. At most 2000 function evaluations are carried out.

We consider a CDO that consists of three tranches. The respective attachment and detachment levels are 0%-5%, 5%-15%, 15%-100%. We assume that the underlying is a homogeneous pool of 10 names with equal unit nominal. The default intensity of each obligor is 1%, the recovery rate is 40%. The time to maturity is 5 years. Our goal is to find a correlation matrix that models an observed compound correlation smile. The given implied compound correlations are 0.24 for the equity tranche, 0.05 for the mezzanine tranche and 0.35 for the senior tranche. At first, we compute the spreads of the equity, the mezzanine and the senior tranche using the respective implied correlations. We get 802.2

bps, 204.4 bps and 9.0 bps. Then, we have to find a correlation matrix that reproduces all three tranche spreads simultaneously.

4.3 Performance

To make sure that we obtain reliable results, we repeat each implementation 25 times. We compute the mean value of the objective functions over the 25 runs, and we also consider the 10% and the 90% quantiles. Our focus is on the decline of the objective function in the course of the generations. To compare the performance of the different algorithms consider figure 1. It shows the objective functions for the different implementations.

At first we compare a Monte-Carlo search and a (1 + 1)-ES to analyze the search space. Generally, the Monte-Carlo search is rather inefficient especially in high dimensional search spaces. Whenever the Monte-Carlo search performs as well as a Hill-Climber or a population-based EA, the search space is probably very flat or very cragged or non-causal. However, in our case the Hill-Climbing strategy clearly outperforms the random search. Often, (1 + 1)-strategies are very efficient in simple unimodal search spaces. But if a Hill-Climber starts up the wrong hill, it has no chance to know that it has found an inferior optimal solution. Therefore, Hill-Climbing strategies frequently can't handle situations in which there are several local optima.

Then we extend the (1 + 1)-ES to a multistart (1 + 1)-ES. We obtain several different solution matrices (see figure 2 for two examples). These matrices yield tranche spreads that are sufficiently close to 802.2 bps, 204.4 bps and 9.0 bps, i.e. the sum of the percentual deviations is less than 5%. This leads to the conclusion that the search space is multimodal. A multistart (1 + 1)-ES reduces the risk of premature convergence.

We now compare different implementations of a (4, 20)-ES and a GA(40). The performance of the different ES implementations is nearly identic, the confidence intervals widely overlap. There is only a very small difference, but the global mutation strategy with one strategy parameter outperforms the other approaches. Then we compare the different GA implementations. The performance of the different GA implementations differs considerably. The GA with tournament selection combined with 1-point crossover leads to the best result. The GA with proportional selection and 1-point crossover performs moderately. The GA with proportional selection and intermediate crossover converges too fast to a dissatisfactory solution. This implementation yields the worst result. It is important to know that by means of recombination the hyperbody formed by the parents generally can't be left by the offspring individuals. Especially the intermediate crossover technique causes volume reduction since it successively narrows the search space, such that eventually the optimal solution can't be attained any more after a few generations.

We finally compare all algorithms on the basis of function evaluations needed to obtain a sufficiently good result. The (1 + 1)-ES performs slightly better than the (4, 20)-ES implementations, there is only a very small difference. However, the GA implementations cant keep up with the ES implementations.

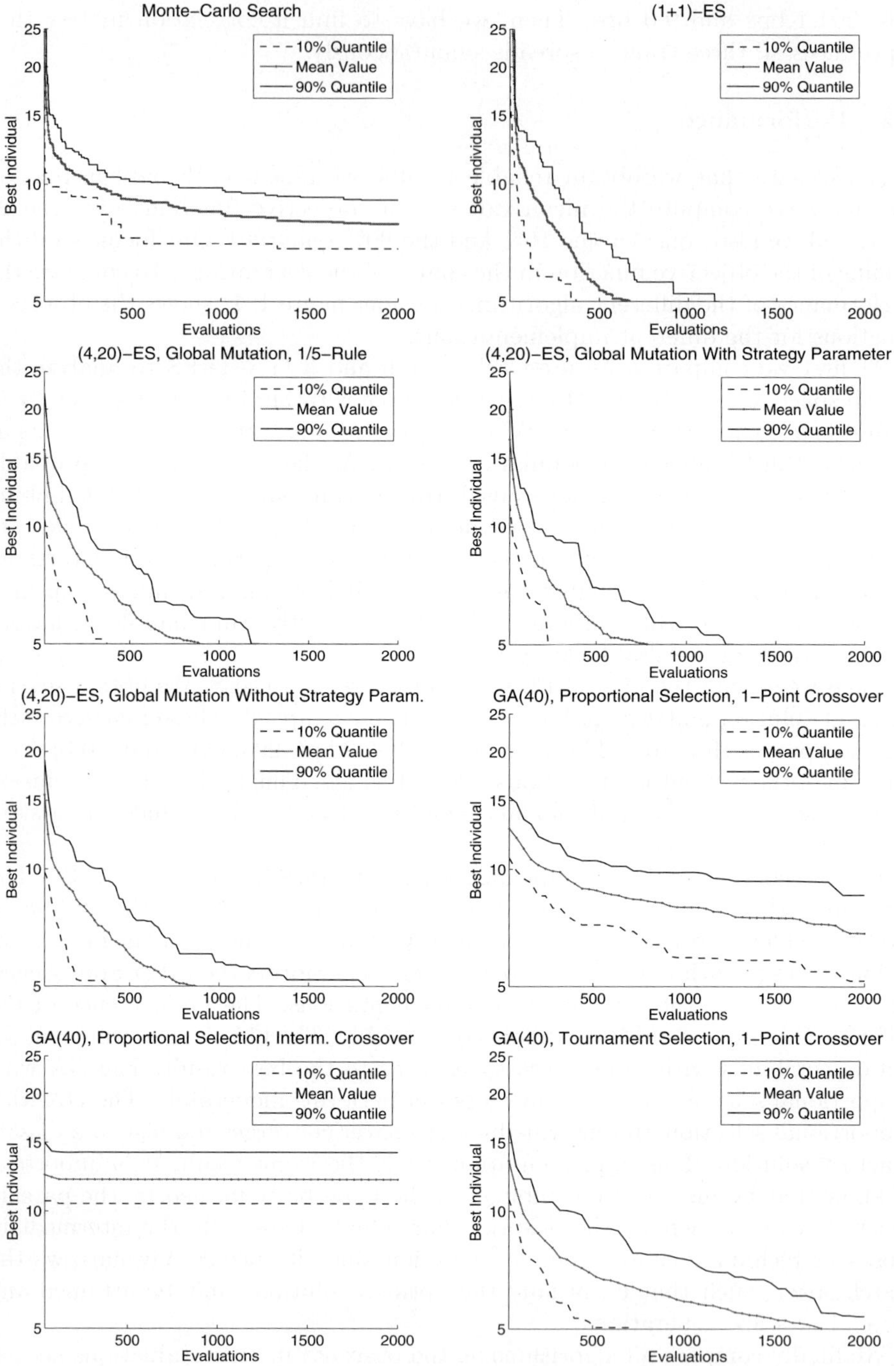

Fig. 1. Performance of different algorithms

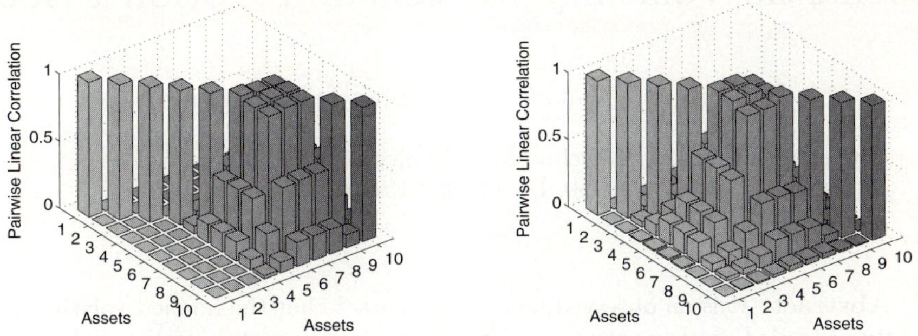

Fig. 2. Exemplary correlation matrices

5 Conclusion

In this study we used the concept of Evolutionary Algorithms to derive possible dependence structures of the underlying of a traded CDO from observed tranche spreads. These dependence structures can then be used to price off-market products with the same underlying as the CDO. Using this strategy we can reduce the pricing discrepancy that comes up when the implied correlation approach is applied. We presented several implementations of Evolutionary Algorithms and discussed their performance.

References

1. Andersen, L., Sidenius, J.: Extensions to the Gaussian copula: Random recovery and random factor loadings. The Journal of Credit Risk **1**(1) (2004) 29-70
2. Gregory, J., Laurent, J.P.: In the core of correlation. Risk, October (2004) 87-91
3. Hager, S., Schöbel R.: A note on the correlation smile. Tübinger Diskussionsbeitrag **297** (2005)
4. Hull, J., White, A.: Valuation of a CDO and an nth to default CDS without Monte-Carlo simulation. The Journal of Derivatives **12**(2) (2004) 8-23
5. Laurent, J.P., Gregory, J.: Basket default swaps, CDOs and factor copulas. The Journal of Risk **7**(4) (2005) 103-122

Stochastic Volatility Models and Option Prices

Akvilina Valaitytė and Eimutis Valakevičius

Kaunas University of Technology, Faculty of Fundamental Sciences,
Studentu st. 50, LT - 51368 Kaunas, Lithuania

Abstract. It is an observed fact in the market that the implied volatility of traded options vary from day to day. An alternative and straightforward explanation is that the instantaneous volatility of a stock is a stochastic quantity itself. The assumptions of the Black and Scholes model no longer hold. This is, therefore, one reason why Black and Scholes prices can differ from market prices of options. Having decided to make the instantaneous volatility stochastic, it is necessary to decide what sort of process it follows. The article analyzes three stochastic volatility models and considers how stochastic volatility can be incorporated into model prices of options. The investigation of stochastic volatility influence for pricing options traded in the SEB Vilnius Bank is done.

1 Introduction

The pricing of derivative instruments, such as options is a function of the movement in the price of the underlying asset over lifetime of the option. One of the main problems of financial engineering is to develop a suitable model of financial assets dynamics. The dynamics is described as a stochastic process, and pricing models describe the stochastic dynamics of asset price changes, whether this is a change in share prices, stock indices, interest rates and so on. Louis Bachelier [1] had claimed that stock prices are actually random in 1900. Comparing trajectories of random walks and stock prices, Bachielier could not find a significant difference among them. The dynamics of asset prices are reflected by uncertain movements of their values over time. Some authors [2, 3, 4, 14] state that efficient market Hypothesis (EMH) is one possible reason for the random behavior of the asset price. The EMH basically states that past history is fully reflected in present prices and markets respond immediately to new information about the asset.

The classical approach is to specify a diffusion process for the asset price, that is, a stochastic integral or stochastic differential equation where the uncertainty is driven by Wiener process. The wide spread adoption of Wiener process as a frame work for describing changes in financial asset prices is most like due to its analytic tractability.

Unfortunately, in recent years more and more attention has been given to stochastic models of financial markets which differ from traditional models. It appears that variances and covariances are not constant over time. There is now a lot of literature on time series modeling of asset prices. The reviewed literature

V.N. Alexandrov et al. (Eds.): ICCS 2006, Part IV, LNCS 3994, pp. 348–355, 2006.
© Springer-Verlag Berlin Heidelberg 2006

codes on the same path. We call the concatenated n-bit binary codes for each path by a *bit sequence* (bs). Referring Fig. 2(a) again, BS denotes a set of bit sequences derived from a single tree in $\mathbb{D}$. Similarly, SBS denotes a collection of BSs derived from $\mathbb{D}$.

4.2 Generation of DHTi

Definition 5 (SpecDirectory). *Given $\mathbb{D}$, a **SpecDirectory** is defined as a directory structure that contains each index of XML-based financial standards.*

Definition 6 (Head). *Given a SBS, a **Head**, H_d, is defined as a collection of n-bit binary codes assigned on the nodes at each depth d of every tree in $\mathbb{D}$. We assume that depth of root node is 0. We call each member in H_d by a **key**.*

At this point, note that there may exist some nodes labeled with the same names in $\mathbb{D}$. Thus, for each key, we need to correctly identify the list of trees to which the key belongs.

Definition 7 (keyTin). *A **keyTin** is defined as a single pair of (k_d, t_{id}) where k_d is a key in H_d and t_{id} is a doubly linked list of tree indexes to which k_d belongs. A $[KT]^d$ is a set of all keyTins for a depth d.*

According to some minimum support, a collection of initial $[KT]^d$ is classified into two sets.

Definition 8 (Frequent keyTin). *Given some minimum support σ and a keyTin (key, t_{id}), the key is called **frequent** if $|t_{id}| \geq \sigma \times |\mathbb{D}|$.*

Definition 9 (Frequent Set). *Given $[KT]^d$, every keyTin in $[KT]^d$ becomes a member of **frequent set** if its key is frequent. Otherwise, it belongs to **candidate set**. We denote frequent set and candidate set by $[F]^d$ and $[C]^d$, respectively.*

The initial frequent sets correspond to the frequent subtrees with only one node commonly occurring in $\mathbb{D}$. Thus, we need to further extend these frequent sets incrementally to find final maximal frequent subtrees. For this purpose, we adopt the operation `cross-reference` introduced in [8, 9]. We refer to the reader to the paper [8] for a detailed explanation of `cross-reference`.

4.3 Construction of Maximal Frequent Subtrees

To derive maximal frequent subtrees from the final frequent sets, we need to notice the following two facts: Firstly, some of the final frequent sets may be empty. An empty frequent set at depth d indicates that there does not exist any n-bit binary code satisfying the minimum support at depth d. Thus, we do not need to consider those empty frequent sets for constructing maximal frequent subtrees. Secondly, although each frequent set has a hierarchical structure, not every key in the frequent set has connected each other in tree structures. In other words, some n-bit binary codes in different frequent sets have edges between them and some have not. It is required to decide whether an edge exists between two keys being in different frequent sets. A minimum support is used to make an edge between them.

5 Conclusion

As a financial community looks for reducing costs and increasing the accessibility of markets through off-the-shelf technologies and multi-vendor standards, XML is playing a key role as the technology which suddenly made everyone want to take part in an industry consortium and work with their competitors. Due to the the generation of a huge amount of the XML-based financial data, the flood of information problem has been caused. In this paper, we have introduced the novel algorithm, FIXiT, for finding valuable information from a collection of financial data. To this end the FIXiT extracts a set of all maximal frequent subtrees from the set of heterogeneous XML-based trees. Unlike previous approaches, the FIXiT classifies each XML-based document with respect to its financial specification, represents each node of a XML-based tree as a binary code, stores them in specially devised structures, and finds all maximal frequent subtrees by expanding frequent sets incrementally. The most important beneficial effect of the FIXiT is that it not only reduces significantly the number of rounds for tree pruning, but also simplifies greatly each round by avoiding time consuming tree join operations. We are currently working to evaluate the performance and the scalability of our algorithm through extensive experiments based on both synthetic data and datasets from real applications. Some of experimental results will be presented in the workshop.

References

1. Agrawal, R., Srikant, R.: Fast algorithms for mining association rules in large databases. Proceedings of the 12th International Conference on Very Large Databases (1994) 487–499
2. Asai, T., Abe, K., Kawasoe, S., Arimura, H., Sakamoto, H., Arikawa, S.: Efficient substructure discovery from large semi-structured data. Proceedings of the 2nd SIAM International Conference on Data Mining (2002) 158–174
3. Chi, Y., Yang, Y., Muntz, R. R.: HybridTreeMiner: An efficient algorithm for mining frequent rooted trees and free trees using canonical forms. The 16th International Conference on Scientific and Statistical Database Management (2004) 11–20
4. Chi, Y., Yang, Y., Muntz, R. R.: Canonical forms for labelled trees and their applications in frequent subtree mining. Knowledge and Information Systems 8(2) (2005) 203–234
5. Inokuchi, A., Washio, T., Motoda, H.: An Apriori-based algorithm for mining frequent substructures from graph data. Proceedings of the 4th European Conference on Principles of Data Mining and Knowledge Discovery (2000) 13–23
6. Kuramochi, M., Karypis, G.: Frequent subgraph discovery. Proceedings of IEEE International Conference on Data Mining (2001) 313–320
7. Miyahara, T., Suzuki, T., Shoudai, T., Uchida, T., Takahashi, K., Ueda, H.: Discovery of frequent tag tree patterns in semistructured web documents. Proceedings of the 6th Pacific-Asia Conference of Advances in Knowledge Discovery and Data Mining (2002) 341–355
8. Paik, J., Shin, D. R., Kim, U. M.: EFoX: a Scalable Method for Extracting Frequent Subtrees. Proceedings of the 5th International Conference on Computational Science. Lecture Notes in Computer Science, Vol. 3516. Springer-Verlag, Berlin Heidelberg New York (2005) 813–817

9. Paik, J., Won, D., Fotouhi, F., Kim, U. M.: EXiT-B: A New Approch for Extracting Maximal Frequent Subtrees from XML Data. Proceedings of the 6th International Conference on Intelligent Data Engineering and Automated Learning. Lecture Notes in Computer Science, Vol. 3578. Springer-Verlag, Berlin Heidelberg New York (2005) 1–8
10. Termier, A., Rousset, M-C., Sebag, M.: TreeFinder: a First step towards XML data mining. Proceedings of IEEE International Conference on Data Mining (2002) 450–457
11. Wang, K., Liu, H.: Schema discovery for semistructured data. Proceedings of the 3rd International Conference on Knowledge Discovery and Data Mining (1997) 271–274
12. Zaki, M. J.: Efficiently mining frequent trees in a forest. Proceedings of the 8th ACM SIGKDD International Conference on Knowledge Discovery and Data mining (2002) 71–80

A Hybrid SOM-Altman Model for Bankruptcy Prediction

Egidijus Merkevicius, Gintautas Garšva, and Stasys Girdzijauskas

Department of Informatics, Kaunas Faculty of Humanities, Vilnius University
Muitinės st. 8, LT- 44280 Kaunas, Lithuania
{egidijus.merkevicius, gintautas.garsva, stasys.girdzijauskas}@vukhf.lt

Abstract. This article analyzes the problems of business bankruptcy, and the methods for bankruptcy prediction. This study proposed to join two models, one is the multi-discriminate Z-Score created by Altman, and the other is the Self-organizing maps. We proposed to generate self-organizing maps based on the financial data of public companies that are included in the NASDAQ list. These maps were used for bankruptcy prediction as well as creating classification of financial risk for Lithuanian companies. Comparing the weak results of prediction we accelerated by changing of ratios weights of the Altman Z-Score model. In this way, it can fit to conditions of the Lithuanian conjuncture. Based on the original ratio weights in Altman's Z-Score the results predicting Lithuanian bankruptcy were weak. The weights of Altman's Z-Score model were changed to fit the Lithuanian economic circumstance.

Keywords: self-organizing maps, Z-Score, bankruptcy, prediction, bankruptcy class, multivariate discriminate model, Altman.

1 Introduction

The forecasting of bankruptcy has always been a relevant task in the finance markets. Available algorithms of statistical and artificial intelligence and the combination of these methods provide more accurate and predictable results [8], [1], [7], [5]. The early history of research attempts to classify and predict bankruptcy is well documented in [4]. The historical development of statistical bankruptcy models can be divided into 3 stages: 1) univariate analysis (by Beaver in 1966); 2) multivariate (or multiple discriminate [MDA]) analysis, and 3) Logit analysis (initiated by Ohlson).

The purpose of this paper is to propose a hybrid artificial-discriminate model to be used as a predictive measure of corporate financial health (0-healthy, 1-bankrupt), based in an unsupervised artificial neural network and a multivariate discriminate model by Altman. Altman's Z-Score model was created for companies that are best characterized in a perfect market economy as evidenced by his use of USA companies financial statements. In light of this mathematical basis a second purpose to this paper has been to present a methodology for adapting Altman's Z-Score model based in the economic reality of developing countries; specifically to propose changing the weight measures in Altman's Z-Score variables.

V.N. Alexandrov et al. (Eds.): ICCS 2006, Part IV, LNCS 3994, pp. 364 – 371, 2006.
© Springer-Verlag Berlin Heidelberg 2006

Therefore, the focus is to explore the capabilities of an unsupervised learning type of artificial neural network – self-organizing map (SOM) to solve such problems as bankruptcy and financial condition. Secondly it is to describe other related work through SOM in predicting bankruptcy and financial distress. In the third part a methodology is presented using a hybrid SOM-Altman model to bankruptcy prediction, and fourthly, the results of this study are demonstrated using the proposed model. In the last section the main conclusions are presented and discussed.

2 Related Work

Artificial neural networks (ANN) are divided into supervised and unsupervised learning [8]. When working in the realm of prediction supervised ANN are normally used. The aim of this investigation is to observe the capabilities of an unsupervised ANN to predict bankruptcy classes of specifically Lithuanian companies.

The Self-organizing map (SOM) is an unsupervised learning artificial neural network that is generated without defining output values. The outcome of this process is a two-dimensional cluster map that can visually demonstrate the financial units which are scattered according to similar characteristics. In this case the bankruptcy class of data is labeled on the map and the data distribution is analyzed. A detailed description of the SOM method is presented in [11].

During the past 15 years investigations in area of SOM applications to financial analysis have been done. Doebeck described and analyzed most cases in [6]. Martin-del-Prio and Serrano-Cinca were one of the first to apply SOM in financial analysis. They generated SOM's of Spanish banks and subdivided those banks into two large groups, the configuration of banks allowed establishing root causes of the banking crisis [2].

Kiviluoto [10] made a map by means of including 1137 companies, out of which 304 companies were crashed. SOM's are said to give useful qualitative information for establishing similar input vectors. Based on Kiviluoto's study, through visual exploration one can see the distribution of important indicators (i.e. bankruptcy) on the map.

The previous authors work is based in an historical or current analysis of company and market conditions. Through this work they have been able to take past settings and predict forwards in time the outcomes of bankruptcy or crisis periods in market economy. It is proposed here that by generating SOM's to current information future segmentation of credit classes can be discerned for new or existing companies.

3 Methodology

In this section the SOM and Altman's Z-Score model are introduced as well as a hybrid SOM-Altman model. The hybrid model was created on the basis of the SOM and Altman's Z-Score with an applied new methodology to change the weights of Altman's Z-Score variables under a specific dataset test.

In the self-organizing process the output data are configured in a visualization of the topologic original data. The unsupervised learning of the SOM is based on competitive learning ("winner takes all"). A detailed description of the SOM algorithm is presented in [6], [11], [5].

Altman's Z-Score predicts whether or not a company is likely to enter into bankruptcy within one or two years. Edward Altman developed the "Altman Z-Score" by examining 85 manufacturing companies in the year 1968 [3]. Later, additional "Z-Scores" were developed for private manufacturing companies (Z-Score - Model A) and another for general/service firms (Z-Score - Model B) [4].

According to E. Altman the Z-Score bankruptcy-predictor combines several of the most significant variables in a statistically derived combination. It was originally developed on a sampling of manufacturing firms. The algorithm has been consistently reported to have a 95 % accuracy of prediction of bankruptcy up to two years prior to failure on non-manufacturing firm. Z-Score for private firms is as follows [4]:

$$Z = 0.717(X1) + 0.847(X2) + 3.107(X3) + 0.420(X4) + 0.998(X5) \tag{1}$$

where

X1 = Working capital/Total assets (captures short-term liquidity risk),

X2 = Retained earnings/Total assets (captures accumulated profitability and relative age of a firm),

X3 = Earnings before interest and taxes/Total assets (measures current profitability),

X4 = Book value of Equity/Book value of total liabilities (a debt/equity ratio captures long-term solvency risk),

X5 = Net sales/Total assets (indicates the ability of a firm to use assets to generate sales),

and

Z = Overall index.

In the original model a healthy private company has a $Z > 3$; it is non-bankrupt if $2.7 < Z < 2.99$; it is in the watch-listed zone if $1.8 < Z < 2.69$; it is unhealthy (bankrupt) if it has a $Z < 1.79$. This paper has corrected the bankruptcy classes where a healthy private company has $Z > 1.8$ and bankrupt company has a Z score of < 1.8, (e.g. we eliminated the "gray" zone).

In figure 1 is presented the algorithm of the proposed hybrid methodology for bankruptcy class prediction.

The main steps are as follows:

1 On the basis of the NASDAQ list companies financial statements, the Altman's Z-Score variables are calculated and converted to bankruptcy classes (0-healthy, 1-bankrupt).
2 Data preprocessing is executed. It consists of a normalized data set, a select map structure, a topology, a set of other options like data filter, and a set of delay etc.
3 The SOM is generated. The Inputs of SOM are the Altman's Z-Score variables and the labels are bankruptcy classes.
4 The SOM is labeled with the bankruptcy classes.
5 On the basis of the TEST list companies financial statements the Altman's Z-Score variables are calculated. Bankruptcy classes (0-healthy, 1-bankrupt) are assigned. Companies that are included in the TEST lists will be used in this study and we will be predicting their bankruptcy class and their weights Altman model weights. At this step we calculate Altman's Z-Score variables on the original weight basis.
6 The generated SOM is labeled with the bankruptcy classes of TEST companies.

7 Labeled units of the trained SOM are compared with the same units labeled with TEST bankruptcy classes.
8 Corresponded units are calculated.
9 The second part of the algorithm is created in order to increase the number of corresponding TRAIN and TEST checked labels which are located on the same SOM map unit number.

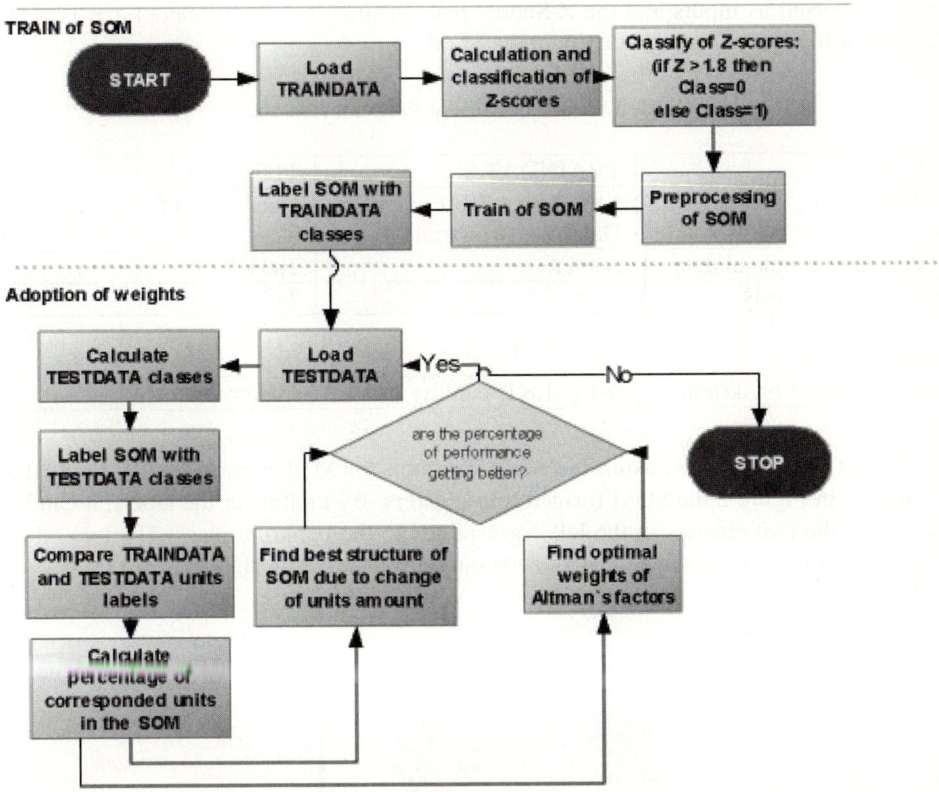

Fig. 1. Algorithm of proposed hybrid model and methodology of weights adoption

10 The attempt is made to create such a map structure within which the amount of unit numbers has the biggest corresponding label number.
11 The assessment of the influence of each of Altman's model variables to the number of corresponding labels.
12 When the performance of the prediction doesn't change the algorithm is stopped.

The results of this algorithm can be presented as follows:

1. A new SOM with a concrete prediction percentage;
2. A new multivariate discriminator model that is based on Altman's Z score but with corrected weight variables.

4 Results of Experiments

In this paper the possibilities of the use of SOM's have been studied using two real financial datasets: companies from NASDAQ list, (or TRAINDATA) loaded from EDGAR PRO Online database, and a dataset of Lithuanian company financial statements (TESTDATA) presented by one of the Lithuanian banks.

The basis for generating the SOM is TRAINDATA. The calculated bankruptcy ratios are used as inputs and the Z-Scores from Altman's Z-Score model are used as labels for the identification of units in the SOM.

Table 1. Characteristics of financial datasets

Dataset	TRAINDATA	TESTDATA
Taken from	EDGAR PRO Online Database (free trial)	Database of Lithuanian bank.
Period of financial data	2004Y	2004Y
Count of records	1108	742
Number of inputs (attributes)	5	
Risk class of bankruptcy	0-1 (>1.8 is healthy - <=1.8 is bankrupt)	

The SOM was trained using the SOM Toolbox for Matlab package [9]. From the U-matrix in Figure 2 the SOM formed four clusters. By looking at the labels, it can be seen that the two clusters in the left corresponds to the bankrupt class. The two other clusters in the right part of U-matrix correspond to the healthy class.

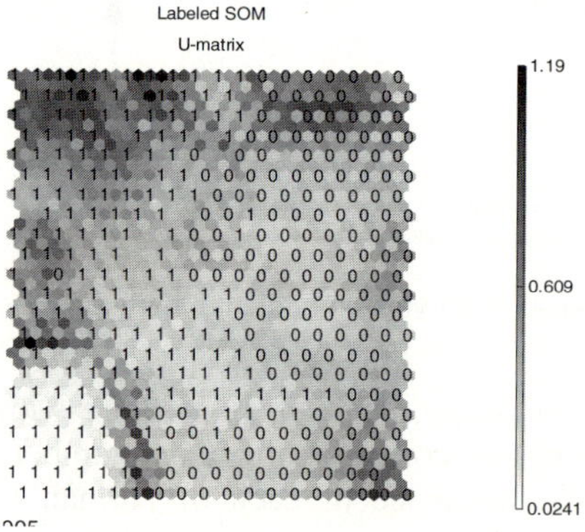

Fig. 2. U-matrix of SOM with TRAINDATA labels

The next step is to label the map with the TESTDATA labels and calculate the corresponding units between the TRAINDATA and the TESTDATA labels It is important to note that only the units that were not empty in both TRAIN and TEST cases are being compared. The ratio between corresponding units and the number of all TESTDATA labels in the SOM map defines accurateness of bankruptcy prediction which is equal to 72.678%. The U-matrix with TESTDATA labels presents at Figure 3(a).

In order to increase the accuracy of bankruptcy prediction the cycle of SOM structure change is created. By changing SOM size, the following accuracy of bankruptcy prediction results is acquired:

Table 2. Performance of bankruptcy prediction via change of SOM size

Bankruptcy prediction (x100%)	Map size		
0,69767	15	x	13
0,69811	20	x	15
0,72678	22	x	18
0,68681	25	x	20
0,68557	27	x	22
0,68182	30	x	23
0,68817	32	x	25
0,65761	33	x	27
0,66667	36	x	28
0,66486	38	x	29

From the results we can see that the bigger the size of the map, the lower the accuracy of bankruptcy prediction. The best results are acquired with the map size of 22x18.

The next investigation was concerned with the influence of each of Altman's model variables to the performance of bankruptcy prediction. In the process of the cycle the weight of each variable is being changed and the change in accuracy in bankruptcy prediction is being monitored. It was noticed that during the change of label weights the accuracy of prediction achieved its highest score and it gradually dropped afterwards. The results show that the most important influence in the performance of bankruptcy prediction is the Net Sales/Total assets variable, and secondly is the EBIT/Total assets variable. The performance result after the correction of the variable weights increased from 72.678% up to 92.352% as follows:

Table 3. Comparison of performance results via change variable weights before and after

Name	**Variable**	**Weight before**	**Weight after**
Earnings before interest and taxes/Total assets	X3	3,107	2,800
Net sales/Total assets	X5	0,998	0,400
Book value of Equity/Book value of total liabilities	X4	0,420	0,440
Working capital/Total assets	X1	0,717	0,717
Retained earnings/Total assets	X2	0,847	0,843
Performance of bankruptcy prediction (%)		**72,678%**	**92,352%**

Other variables have less influence between American and Lithuanian companies. The variables of profitability and turnover are the most important factors which have the largest weight to predict bankruptcy.

U-matrix with TESTDATA labels presented as follows (Figure 3 (b)) and can be visually compared with U-matrix before corrections of variable weights (Figure 3(a)). It can be seen that bankrupt and healthy class labels distributed comparatively to the labels in TRAINDATA (Figure 2).

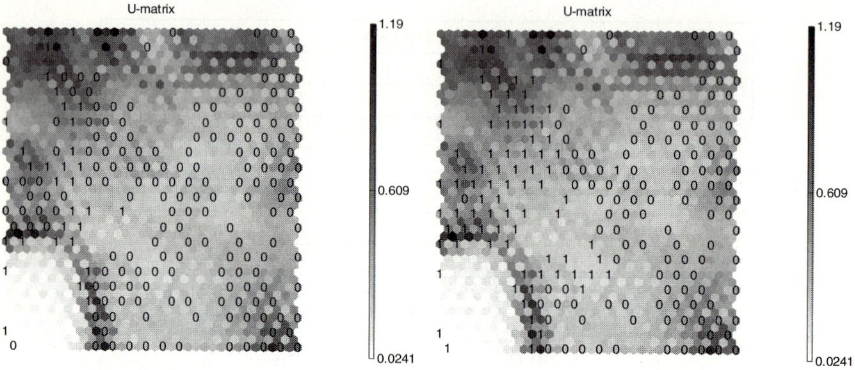

a) b)

Fig. 3. U-matrix of trained SOM with TESTDATA labels before and after correction of weights

5 Conclusions

Our experiments and the results present several conclusions:

- The presented methodology works well with real world data, hybrid SOM-Altman's bankruptcy model with presented datasets predicted with 92.352% performance.
- Methodology of presented hybrid bankruptcy model is flexible to adopt every datasets because rules and steps of methodology algorithm are universal.
- In this paper it was shown, that hybrid SOM-Altman's adopted bankruptcy model can present differences between financial statements conjuncture of two types of economic countries: west market economy and development economy.

References

1. A. Atiya. Bankruptcy prediction for credit risk using neural networks: a survey and new results. IEEE Transactions on Neural Networks, Vol. 12, No. 4, pp. 929-935, July 2001.
2. B. Martin-del-Prio, K. Serrano-Cinca, Self-Organizing Neural Network: The Financial State of Spanish Companies. In Neural Networks in Finance and Investing. Using Artificial Intelligence to Improve Real-World Performance. R.Trippi, E.Turban, Eds. Probus Publishing, 1993., 341-357

3. E. Altman. Financial Ratios, Discrimination Analysis and the Prediction of Corporate Bankruptcy. Journal of Finance, 23, September, 1968.
4. E. Altman. Predicting Financial Distress of Companies: Revisiting the Z-Score and ZETA® Models. (working paper at http://pages.stern.nyu.edu/~ealtman/Zscores.pdf) 2000
5. E. Merkevičius, R. Simutis, G.Garšva. Forecasting of credit classes with the Self-organizing maps. Information technology and control. 2004, 4(33), ISSN 1392-124X. 61-66 pp.
6. G. Deboeck. Financial Applications of Self-Organizing Maps. American Heuristics Electronic Newsletter, Jan, 1998.
7. G. Deboeck. Self-Organizing Maps Facilitate Knowledge Discovery In Finance. Financial Engineering News, December 1998.
8. J. Galindo, P. Tamayo. Credit Risk Assessment Using Statistical and Machine Learning: Basic Methodology and Risk Modeling Applications. Computational Economics. April 2000, Volumc 15.
9. J. Vesanto, J. Himberg, E. Alhoniemi, J. Parhankangas, SOM toolbox for Matlab 5, Technical report A57 (2000), Helsinki University of Technology, Finland.
10. K. Kiviluoto. Predicting bankruptcies with the self-organizing map. Neurocomputing, 21:191–201, 1998.
11. T. Kohonen. The Self-Organizing Map. Proceedings of the IEEE, 78:1464-1480.

Learning and Inference in Mixed-State Conditionally Heteroskedastic Factor Models Using Viterbi Approximation

Mohamed Saidane[1] and Christian Lavergne[2]

[1] I3M, University Montpellier II, Place Eugene Bataillon, CC - 051 34095 Montpellier, France
saidane@math.univ-montp2.fr
http://www.math.univ-montp2.fr

[2] I3M, University Montpellier II, Place Eugene Bataillon, CC - 051 34095 Montpellier, France
Christian.Lavergne@math.univ-montp2.fr

Abstract. In this paper we develop a new approach within the framework of asset pricing models that incorporates two key features of the latent volatility: co-movement among conditionally heteroskedastic financial returns and switching between different unobservable regimes. By combining conditionally heteroskedastic factor models with hidden Markov chain models (HMM), we derive a dynamical local model for segmentation and prediction of multivariate conditionally heteroskedastic financial time series. The EM algorithm that we have developed for the maximum likelihood estimation, is based on a Viterbi approximation which yields inferences about the unobservable path of the common factors, their variances and the latent variable of the state process. Extensive Monte Carlo simulations and preliminary experiments obtained with a dataset on weekly average returns of closing spot prices for eight European currencies show promising results.

1 Introduction

The factor Model, also called Index Model, is one of the basic models in finance to analyze and describe the return generation process and the risk/reward relationships of a large number of assets. It has been used extensively in finance for measuring co-movement in and forecasting financial time series. Traditionally, these issues were considered in a static framework, but recently, the emphasis has shifted toward inter-temporal asset pricing models in which agents decisions are based on the distribution of returns conditional on the available information, which is obviously changing. Several researchers have used Factor-ARCH models to provide a plausible and parsimonious parameterization of the time varying covariance structure of asset returns. Engle et al. [1] apply such structures to study the dynamic behavior of the term structure of interest rates. Diebold and Nerlove [2] use a latent factor ARCH model to describe the dynamics of exchange

V.N. Alexandrov et al. (Eds.): ICCS 2006, Part IV, LNCS 3994, pp. 372–379, 2006.
© Springer-Verlag Berlin Heidelberg 2006

rate volatility. Engle and Susmel [3] use the factor ARCH to test for common volatility in international equity markets.

A natural generalization of the different models proposed in the above literature to a multi-state model can be achieved by allowing for model transitions that are governed by a Markov chain on a set of possible models describing the different states of volatility. The originality of this work is the use of a piece-wise multivariate and linear process – which we also regard as a mixed-state dynamic linear system – for modeling the regime switches. In particular, we suppose that the observed series can be approximated using a time varying parameter model with the assumption that the evolution of these parameters is governed by a first-order hidden Markov process with m states.

2 Basic Model and Factor Structure

The model that we propose supposes that excess returns depend both on unobservable factors that are common across the multivariate time series, and on unobservable different regimes that describe the different states of volatility. In this framework, we allow a dynamic structure for the conditional variances of the underlying factors in order to investigate possible time-varying latent processes, and their implications in modeling changes in covariance matrices over time. This new specification is defined by:

$$\begin{array}{c} S_t \sim P(S_t = j/S_{t-1} = i) \\ t = 1, ..., n \quad \text{and} \quad i, j = 1, ..., m \\ \mathbf{f}_{s_t} = \mathbf{H}_{s_t}^{1/2}\mathbf{f}_t^* \qquad \text{where} \qquad \mathbf{f}_t^* \sim \mathcal{N}(\mathbf{0}, \mathbf{I}_k) \\ \mathbf{y}_t = \mathbf{X}_{s_t}\mathbf{f}_{s_t} + \varepsilon_{s_t} \ \text{ with } \ \varepsilon_{s_t} \sim \mathcal{N}(\theta_{s_t}, \mathbf{\Psi}_{s_t}) \end{array}$$

where $S_t \sim P(S_t = j/S_{t-1} = i)$ is a hidden Markov chain indicating the state or the regime at the date t, and $\mathbf{y}_t$ is a $q \times 1$ random vector of observable variables (financial returns). In an unspecified state $S_t = j$ $(j = 1, ..., m)$, θ_j are the $q \times 1$ mean vectors, $\mathbf{f}_{jt}$ the $k \times 1$ vectors of unobserved common factors, ε_{jt} the $q \times 1$ vectors of idiosyncratic noises, $\mathbf{X}_j$ the $q \times k$ factor loadings matrices, with $q \geq k$ and $rank(\mathbf{X}_j) = k$, $\mathbf{\Psi}_j$ are $q \times q$ diagonal and definite positive matrices of idiosyncratic variances, and $\mathbf{H}_{jt}$ the $k \times k$ diagonal and definite positive matrices whose elements are the variances of the common factors presumedly time varying and their parameters changes according to the regime. In particular, we suppose that the variances of the common factors follow switching GQARCH(1,1) processes, the *l-th* diagonal element of the matrix $\mathbf{H}_{jt}$ under an unspecified regime $S_t = j$ since $S_{t-1} = i$ being $h_{lt}^{(j)} = w_j^l + \gamma_j^l f_{lt-1}^{(i)} + \alpha_j^l f_{lt-1}^{(i)\,2} + \delta_j^l h_{lt-1}^{(i)}$.

3 Viterbi Approximation for Latent Structure Inference

The model developed above can be regarded as a random field with indices $i = 1, ..., q$, $t = 1, ..., n$ and $j = 1, ..., m$. Therefore, it has a switching state-space representation, with $\mathbf{f}_t$ as the continuous state variables. The measurement and transition equations are, respectively, given by:

$$\mathbf{y}_t = \theta_{s_t} + \mathbf{X}_{s_t}\mathbf{f}_{s_t} + \varepsilon_{s_t}$$
$$\mathbf{f}_{s_t} = \mathbf{0}.\mathbf{f}_{s_{t-1}} + \mathbf{f}_{s_t}$$

The task of Viterbi approximation approach is to find the best sequence of switching states S_t and common factors $\mathbf{f}_t$ that minimizes the Hamiltonian cost in equation (1) for a given observation sequence $\mathcal{Y}_{1:n}$.

$$\begin{aligned}\mathcal{H}(\mathcal{F}_{1:n}, \mathcal{S}_{1:n}, \mathcal{Y}_{1:n}) &\simeq Constant + \sum_{t=2}^{n} S_t'(-\log \mathbf{P})S_{t-1} + S_1'(-\log \pi) \\ &+ \frac{1}{2}\sum_{t=1}^{n}\sum_{j=1}^{m}\left[(\mathbf{y}_t - \mathbf{X}_j\mathbf{f}_{jt} - \theta_j)'\mathbf{\Psi}_j^{-1}(\mathbf{y}_t - \mathbf{X}_j\mathbf{f}_{jt} - \theta_j) + \log|\mathbf{\Psi}_j|\right] S_t(j) \\ &+ \frac{1}{2}\sum_{t=1}^{n}\sum_{j=1}^{m}\left[\mathbf{f}_{jt}'\mathbf{H}_{jt}^{-1}\mathbf{f}_{jt} + \log|\mathbf{H}_{jt}|\right] S_t(j) \end{aligned} \quad (1)$$

where $\mathbf{P}$ and π are, respectively, the HMM transition matrix and the vector of initial state probabilities.

Define first the "best" partial cost up to time t of the measurement sequence $\mathcal{Y}_{1:t}$ when the switch is in state j at time t:

$$J_{t,j} = \min_{\mathcal{S}_{1:t-1}, \mathcal{F}_t} \mathcal{H}\left[\mathcal{F}_{1:t}, \{\mathcal{S}_{1:t-1}, S_t = j\}, \mathcal{Y}_{1:t}\right] \quad (2)$$

Namely, this cost is the least cost over all possible sequences of switching states $\mathcal{S}_{1:t-1}$ and corresponding factor model states $\mathcal{F}_{1:t}$. In order to calculate this cost we first start by introducing some notation.

$$\begin{aligned}\mathbf{f}_{t/t-1}^{i(j)} &= \mathrm{E}\left[\mathbf{f}_t/\mathcal{Y}_{1:t-1}, S_t = j, S_{t-1} = i\right] \\ \mathbf{H}_{t/t-1}^{i(j)} &= \mathrm{E}\left[(\mathbf{f}_t - \mathbf{f}_{t/t-1}^{i(j)})(\mathbf{f}_t - \mathbf{f}_{t/t-1}^{i(j)})'/\mathcal{Y}_{1:t-1}, S_t = j, S_{t-1} = i\right] \\ \mathbf{f}_{t/t}^{i(j)} &= \mathrm{E}\left[\mathbf{f}_t/\mathcal{Y}_{1:t}, S_t = j, S_{t-1} = i\right] \\ \mathbf{H}_{t/t}^{i(j)} &= \mathrm{E}\left[(\mathbf{f}_t - \mathbf{f}_{t/t}^{j})(\mathbf{f}_t - \mathbf{f}_{t/t}^{j})'/\mathcal{Y}_{1:t}, S_t = j, S_{t-1} = i\right]\end{aligned}$$

From the theory of Kalman estimation it follows that for transition $i \rightarrow j$ the following time updates hold:

$$\mathbf{f}_{t/t-1}^{i(j)} = \mathbf{0}.\mathbf{f}_{t-1/t-1}^{i} = \mathbf{0} \qquad \forall \quad i, j = 1, ..., m \quad \text{and} \quad (3)$$

$$h_{lt/t-1}^{i(j)} = w_{lj} + \gamma_{lj} f_{lt-1/t-1}^{i} + \alpha_{lj}\left[f_{lt-1/t-1}^{i\,2} + h_{lt-1/t-1}^{i}\right] + \delta_{lj} h_{lt-1/t-2}^{i} \quad (4)$$

Given a new observation $\mathbf{y}_t$ at time t each of these predicted estimates can now be filtered using Kalman measurement update framework:

$$\mathbf{f}_{t/t}^{i(j)} = \mathbf{f}_{t/t-1}^{i(j)} + K_t(i,j)\mathbf{e}_t(i,j) \tag{5}$$

$$\mathbf{H}_{t/t}^{i(j)} = \mathbf{H}_{t/t-1}^{i(j)} - K_t(i,j)\mathbf{\Sigma}_{t/t-1}^{i(j)}K_t(i,j)' \tag{6}$$

with $\mathbf{e}_t(i,j) = \mathbf{y}_t - \theta_j - \mathbf{X}_j\mathbf{f}_{t/t-1}^{i(j)}$; $\mathbf{\Sigma}_{t/t-1}^{i(j)} = \mathbf{X}_j\mathbf{H}_{t/t-1}^{i(j)}\mathbf{X}_j' + \mathbf{\Psi}_j$ and $K_t(i,j) = \mathbf{H}_{t/t-1}^{i(j)}\mathbf{X}_j'\mathbf{\Sigma}_{t/t-1}^{i(j)\,-1}$. Each of these $i \rightarrow j$ transitions has a certain innovation cost $J_{t,t-1,i,j}$ associated with it, as defined in equation (7).

$$J_{t,t-1,i,j} = \frac{1}{2}\mathbf{e}_t(i,j)'\mathbf{\Sigma}_{t/t-1}^{i(j)\,-1}\mathbf{e}_t(i,j + \frac{1}{2}\log\left|\mathbf{\Sigma}_{t/t-1}^{i(j)}\right| - \log p_{ij} \tag{7}$$

one portion of this innovation cost reflects the continuous state transition, as indicated by the innovation terms in equation (5). The remaining cost $(-\log p_{ij})$ is due to switching from state i to state j.

Obviously, for every current switching state j there are m possible previous switching states from which the system could have originated from. To minimize the overall cost at every time step t and for every switching state j, one "best" previous state i is selected:

$$J_{t,j} = \min_i\{J_{t,t-1,i,j} + J_{t-1,i}\} \tag{8}$$

$$\delta_{t-1,j} = \arg\min_i\{J_{t,t-1,i,j} + J_{t-1,i}\} \tag{9}$$

the index of this state is kept in the state transition record $\delta_{t-1,j}$. Consequently, we now obtain a set of m best filtered continuous states and their variances at time t: $\mathbf{f}_{t/t}^{j} = \mathbf{f}_{t/t}^{\delta_{t-1,j}\,(j)}$ and $\mathbf{H}_{t/t}^{j} = \mathbf{H}_{t/t}^{\delta_{t-1,j}\,(j)}$ with $h_{lt/t-1}^{j} = h_{lt/t-1}^{\delta_{t-1,j}\,(j)}$. Once all n observations $\mathcal{Y}_{1:n}$ have been fused, the best overall cost is obtained as $J_n^* = \min_j J_{n,j}$. To decode the "best" switching state sequence, one uses the index of the best final state, $j_n^* = \arg\min_j J_{n,j}$, then traces back through $\delta_{t-1,j}$: $j_t^* = \delta_{t,j_{t+1}^*}$.

The Switching model's sufficient statistics are now simply given by $\mathrm{E}(S_t/.) = S_t(j^*)$ and $\mathrm{E}(S_tS_{t-1}'/.) = S_t(j^*)S_{t-1}(j^*)'$.[1] Given the "best" switching state sequence, the sufficient factor model statistics can be easily obtained using the Rauch-Tung-Streiber smoothing (Rosti A-V.I and Gales M.J.F [4]). For example, $\mathrm{E}(\mathbf{f}_t, S_t(j)/.) = \mathbf{f}_{t/n}^{j_t^*}$ if $j = j_t^*$ and $\mathbf{0}$ otherwise.

4 EM Algorithm

The joint likelihood of the observations sequence $\mathcal{Y}_{1:n}$, the continuous state vector sequence $\mathcal{F}_{1:n}$ and the HMM state sequence $\mathcal{S}_{1:n}$ is given by:

[1] The operator $\mathrm{E}(/.)$ denotes conditional expectation with respect to the posterior distribution, e.g. $\mathrm{E}(\mathbf{f}_t/.) = \sum_{\mathcal{S}}\int_{\mathcal{F}}\mathbf{f}_t p(\mathcal{F},\mathcal{S}/\mathcal{Y})$.

$$p(\mathcal{Y}, \mathcal{F}, \mathcal{S}) = p(S_1) \prod_{t=2}^{n} p(S_t/S_{t-1}) \prod_{t=1}^{n} p(\mathbf{f}_t/S_t, \mathcal{D}_{1:t-1}) p(\mathbf{y}_t/\mathbf{f}_t, S_t, \mathcal{D}_{1:t-1})$$

where $\mathcal{D}_{1:t-1} = \{\mathcal{Y}_{1:t-1}, \mathcal{F}_{1:t-1}, \mathcal{S}_{1:t-1}\}$, is the information set at time $t-1$, $p(S_1) = \pi_{s_1}$: the initial state probability and $p(S_t/S_{t-1}) = p_{s_{t-1}s_t}$: the transition probabilities. In the first conditional maximisation step, the model parameters i.e. π_j, p_{ij}, $\mathbf{X}_j$, θ_j and $\boldsymbol{\Psi}_j$ can be obtained by maximizing the conditional expectation of this complete log-likelihood function.

In the second step, being given the new values above and the fact that $\mathbf{y}_t/\mathcal{Y}_{1:t-1}, S_t = j, \mathcal{S}_{1:t-1} \approx \mathcal{N}\left[\theta_j, \boldsymbol{\Sigma}^j_{t/t-1}\right]$, the parameters $\phi_j = \{w_{jl}, \gamma_{jl}, \alpha_{jl}, \delta_{jl}\}$ for $j = 1, ..., m$ can be updated by maximizing the observed log-likelihood function (using a Newton-Raphson algorithm):

$$\mathcal{L}^* = c - \frac{1}{2} \sum_{t=1}^{n} \sum_{j=1}^{m} S_t(j) \left[\log |\boldsymbol{\Sigma}^j_{t/t-1}| + (\mathbf{y}_t - \theta_j)' \boldsymbol{\Sigma}^{j\ -1}_{t/t-1} (\mathbf{y}_t - \theta_j)\right]$$

5 Experimental Results

In this section, we study the performance of our proposed algorithm using synthetic and financial data.

5.1 Model Learning and Stability of the Estimates

The example used here has $q = 6$ series, $m = 3$ hidden states and only one common factor. The regime switching dates are $t_1^* = n/3 + 1$ and $t_2^* = 2n/3 + 1$. The iterations of the EM algorithm[2] stop when the relative change in the likelihood function between two subsequent iterations is smaller than a threshold value $= 10^{-4}$. In this experiment we try to estimate the parameters of the model and to study the behavior of the estimates when the size of the sequence n increases. A natural metric to measure the distance of estimators from the true parameters is the Kullback-Leibler divergence (Juang and Rabiner [5]). For each value of n, the estimation procedure was carried out a hundred times, and the distances $\widetilde{K}_n(\Theta_0, \widetilde{\Theta}_n)$ between each of the hundred estimators and the true parameter Θ_0 were evaluated on a new sequence, independent of the first hundred sequences used to obtain the estimators. Figure 1 clearly shows a general decrease in average and spread of the distances with increasing n. This imply an increasing accuracy and stability of the estimators as n increases.

5.2 Model Selection Procedure

Comparing the adequacy of different models may be done by computing a criterion for each model and comparing the criteria values. In this experiment BIC

[2] The initial parameters for the EM algorithm, were obtained by randomly perturbing the true parameter values by up to 20% of their true value.

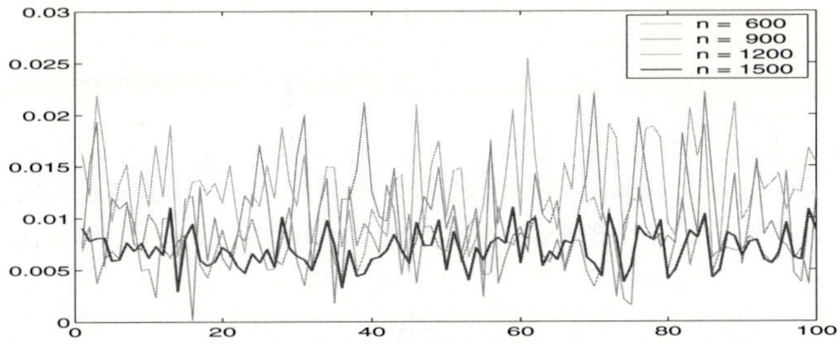

Fig. 1. Box plots of $\widetilde{K}(\Theta_0, \widetilde{\Theta}_n)$

and ICL criteria are used. The ICL criterion, is based on the maximization of the integrated complete log-likelihood function (Biernacki and Celeux [6]). Here we consider two different heteroskedastic specifications. For the two examples, $q = 6$ and $n = 900$ was used. In the first case the true model is the one used in 5.1. In the second case, we take $m = k = 2$ and $t^* = n/2 + 1$.

The steps for the model selection procedure are as follows. For each selection criterion, first, train various model configurations (obtained by varying the number of states and the number of factors). Second, use the output of EM to compute the values of the selection criterion for all configurations and select the one that yields the lowest value. In the two examples, random initialisation was used for the implementation of the learning algorithm. With this intention, we generated 100 different data experiments according to the true model. In the first case the results show that BIC and ICL chose 3 states and one factor most of the time (68%, 73%). This is the best classification, since the use of one or two states is not enough to represent the data, and choosing two factors corresponds to an overfitting. In the second case, BIC and ICL choose also the true specification most of the time (79%, 81%).

5.3 Financial Data

We have applied our model also to learn and analyze the co-movements amongst several exchange rate returns during the period where the European exchange rate mechanism has experienced a succession of crisis which reached its first culmination at the end of August 1992. What has been the impact of these changes on the nature of volatility? Has the degree of co-movement increased or decreased? Have common fluctuations become more or less volatile? Has the impact of crises on individual countries evolved over time? These questions are of interest to both policy makers and academic economists. For example, the questions of whether the common volatility has increased or declined, and whether countries have become more or less symmetric, are central to monetary and fiscal policy issues.

The time series considered here are the weekly average returns of closing spot prices relative to the US Dollar of the FRF, CHF, ITL, DEM, BEF, ESP, SEK, and GBP from 07/17/1985 to 01/22/1997 (600 observations). We trained

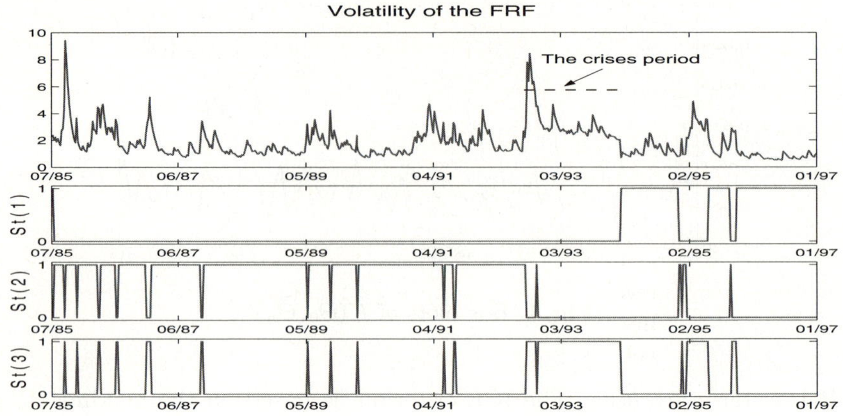

Fig. 2. Estimated Volatility of the French Franc (*Graphic 1*) and the Values of $S_t(j)$ given by the Viterbi algorithm (*Graphics 2, 3, 4*)

various model configurations (obtained also by varying the number of states and the number of factors). The key point for initialization consists in implementing a standard EM algorithm by supposing that the factors are homoskedastic[3]. Thereafter and given the output of the EM algorithm, we can use the values of $S_t(j)$ in order to obtain the optimal sequence of hidden states. At the second step, a particular simple conditionally heteroskedastic factor model is initialized for each segment. For this, one can use the empirical covariance matrices as estimates of the idiosyncratic variance matrices $\mathbf{\Psi}_j$ and the empirical means as estimates of the means θ_j. The parameters of the conditionally heteroskedastic variances are initialized by applying a GQARCH(1,1) model to each segment of data. Finally, the elements of the transition matrix $\mathbf{P}$, can be initialized by counting the number of transitions from state i to state j and dividing by the number of transitions from state i to any other state.

All the selection criteria argue that the time varying covariance structure could be modeled by two conditionally heteroskedastic common factors and three markovian regimes. For example in the cas of French Franc, figure 2 shows how the model is capable of accurately detecting abrupt changes in the time series structure and, in particular, the severe disruption by the violent storm which hit the European currency markets in September and October 1992. This segmentation shows that the third model is responsible for the high volatility segments, the second model is mainly responsible for the time period before September 1992, and the first for the lower volatility segments after 1993. This figure shows also that the average duration stay in the first regime is about 31.88 months versus 89.38 in the second and 28.73 in the third. For the two common factors, estimated α_i and δ_i are both statistically significant and their sum is slightly less than one which indicates strong GARCH effects and persistence in the volatility of exchange rates. The results show also that all the correlations between the

[3] In practice, 20 iterations of the EM algorithm are largely sufficient.

different currencies have declined just after August 1992. This is not surprising because at the end of 1992 the range of the Exchange Rate Mechanism expanded to 30%, which practically meant a return to free fluctuation.

6 Conclusion

The paper has developed a novel solution to the problem of modeling conditionally heteroskedastic financial time series subject to Markov switching within a multivariate framework. This new specification takes into account, simultaneously, the usual changing behavior of the common volatility due to common economic forces, as well as the sudden discrete shift in common and idiosyncratic volatilities that can be due to sudden abnormal events.

One of the most interesting applications of this new dynamic specification in finance is the possibility of obtaining on-line predictions of the time varying covariance matrices that is useful for dynamic asset allocation, active portfolio management and the analysis of options prices. The analysis in this paper can be also extended in several ways. First, our model can be generalized to one where one allows the idiosyncratic variances to be a stochastic function of time. Secondly, we can also think of the case where the state transition probabilities are not homogeneous in time, but depend on the previous state and the previously observed covariates levels. The study of such models would provide more flexibility in financial applications.

References

1. Engle R., Ng V.K., Rothschild M.: A Multi-Dynamic Factor Model for Stock Returns. J. of Econometrics. **52** (1992) 245–266
2. Diebold F., Nerlove M.: The Dynamics of Exchange Rate Volatility: A Multivariate Latent Factor ARCH Model. J. of Applied Econometrics. **4** (1988) 1–22
3. Engle R., Susmel R.: Common Volatility in International Equity Markets. J. of Business and Economic Statistics. **11** (1993) 369–380
4. Rosti A-V.I. Gales M.J.F.: Generalised Linear Gaussian Models. CUED/F-INFENG/TR 420, Cambridge University (2001)
5. Juang B.H., Rabiner L.: Mixture autoregressive hidden Markov models for speech signals. IEEE Trans. on ASSP. **33** (1985) 1404–1413
6. Biernacki C., Celeux G., Assessing a mixture model for clustering with the integrated completed likelihood. IEEE Trans. on PAMI. **22** (2000) 719–725

Constructing a P2P-Based High Performance Computing Platform*

Hai Jin, Fei Luo, Xiaofei Liao, Qin Zhang, and Hao Zhang

Cluster and Grid Computing Laboratory,
Huazhong University of Science and Technology, Wuhan, 430074, China
{hjin, luofeiren, xfliao, qzhang, haozhang}@hust.edu.cn

Abstract. The construction for a P2P-based high performance computing platform (P2HP) is presented to address parallel problems in this paper. P2HP utilizes idle computers in the Internet with great scalability to form an enormous computing capability for scientific supercomputing and volunteers form autonomous unstructured P2P network domains. The configuration of P2HP is easy and a programming model is provided. Its applications involve a large range of problems, and a benchmark is applied to evaluate its performance.

1 Introduction

Peer-to-Peer (P2P) fashion can construct supercomputers far beyond the power of any current computing center, which make HPC much less expensive and much more accessible than clusters [1]. Because P2P addresses failure for dynamics, *personal computers* (PCs) in the Internet are suitable to participant into P2P computing. There are countless PCs which are idle most of time. The collection of them in P2P systems has potentially enormous computing power.

Currently, most P2P-based computing systems belong to special organizations, and they are just used to solve one scientific or commercial high throughput problem. The characteristic of HPC is more instantaneous than that of *high throughput computing* (HTC), which makes them more tight constraints in HPC systems than in HTC [2].

A P2P-based HPC platform, P2HP, is depicted to assemble idle Internet resources in this paper. P2HP is constructed as a 3-layer network architecture, where volunteers form autonomous unstructured P2P domains. The computing power increases with the rise of volunteers number. It is easily configured, and the management cost is distributed into autonomous domains, which brings the advantage that P2HP fits various environments and users. A *one-sided message passing* (OMP) [3] programming model is provided for users, which makes P2HP can work on a variety of problems, especially infinite workpile applications with deadlines [4].

The rest of the paper is organized as follows. Related works are reviewed in section 2. The system architecture of P2HP is outlined in section 3, and its design issues are described in section 4. To test its high performance characteristics, sequence alignment in bioinformatics is applied as a benchmark in section 5. Finally, a conclusion is drawn in section 6.

---

* This paper is supported by National Science Foundation under grant 60433040, China CNGI project under grant CNGI-04-12-2A and CNGI-04-12-1D.

V.N. Alexandrov et al. (Eds.): ICCS 2006, Part IV, LNCS 3994, pp. 380 – 387, 2006.
© Springer-Verlag Berlin Heidelberg 2006

2 Related Works

There are many P2P-based computing systems designed to harness idle cycles of Internet-connected workstations [5], which are also known as "public-resource computing" or "global computing", such as SETI@home [6], Folding@home [7], Prediction@home [8]. Without providing a programming model, they are just used to solve one scientific or commercial HTC problem. With its OMP programming model, P2HP can solve a large range of problems.

Furthermore, a number of middleware for public-resource computing have been provided, such as BOINC [5], OmniRPC [9], and XtremWeb [10]. The clients of these systems contact directly with the server or through an agent, which is different from the communication mechanism of P2HP. With autonomous domains, the resource management of P2HP is decentralized, and most communication between entries exists in local domains.

OurGrid [11] is a solution of running applications of Bag-of-Tasks. The job scheduler is based on a provider-consumer algorithm. Different from the OMP programming model of P2HP, it provides a script-command application programming model which is limited for applications.

3 System Architecture

P2HP is constructed as a 3-layer network, as well as a pool for accessing data (Datapool) and a user interface (User), shown in Figure 1. The module User is used to manage the submission of applications, and Datapool is used to manage the storage of applications' related data.

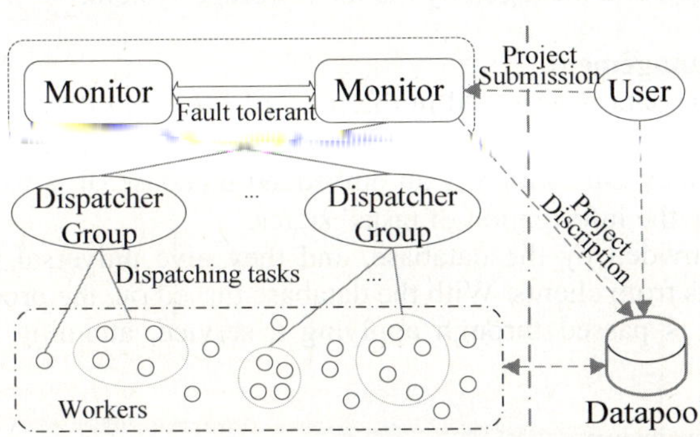

Fig. 1. Architecture of P2HP

The basic components of the 3-layer network are monitors, dispatchers and workers. The monitoring of the whole system is done by monitors, which are grouped as *Monitor Group* (MG). Dispatchers schedule tasks to the workers, and all of the dispatchers are grouped as *Dispatcher Groups* (DGs). A dispatcher can choose one or more monitors to join, and the chosen monitor is then its host monitor. Workers execute subtasks of applications and each of them is attached to a dispatcher.

The parallelism of applications in P2HP is in the application level, and applications are divided into small tasks. Each task finishes certain computing work for the application. One of them participates in the control of the tasks' execution flow, and it is defined as the main task, while the others are defined as subtasks. Programmed with OMP, applications are transferred into Datapool, and a project file (JobFile) is generated, which is submitted to a monitor in the MG. They are then performed through the scheduling process of P2HP.

4 Design Issues

4.1 Entries of 3-Layer Network

The running of the whole system is controlled by entries of the 3-layer network. Monitors inquire the information of the actions of participants and the running states of applications. Dispatchers poll the attached workers to inquire their states, which is similar to the heartbeat mechanism. The worker accepts their messages and sends back state information for its available hardware and software resources and executes dispatched tasks.

A dispatcher with a number of attached workers forms an autonomous domain, and the dispatcher is the dominator of the domain. Each domain is constructed as an unstructured P2P network. All the dominators are monitored by the monitor. The metric with autonomic domains for resource management in the 3-layer network exploits the scalability of P2HP.

4.2 Data Management

Data is accessed in the Datapool, and all the requested data is transferred by a transmission protocol and managed by the local storage system.

4.2.1 Storage Management

In the Datapool, the data is accessed in the form of files, and they are assembled in a directory. The directory management of the Datapool is combined with a lightweighted database, the Berkeley DB, which is an embedded database. It is used to record and dynamically update the information of tasks' states.

Services are provided by the database, and they give universal interfaces to the Datapool's requests from clients. With the database turned on, the process of the usage for these services is passed through applying a service, attaining information and returning its result.

4.2.2 Data Transmission

The data transmission is implemented as a *Fast Data Transfer Protocol* (FDTP). It is based on a network-connected pipe, which consists of a message channel for messages and a data channel for corresponding data files.

During the transmission, the data files are formatted as an I/O stream, and states are defined, such as BEGIN_TRANS and FINISH_ONCE for the beginning and finishing of the process, IDLE for none, SYN_RECEV and SYN_SENT for synchronization of receiving and sending a message. The transmission process is expressed by a state transition shown in Figure 2, where the real line represents the server's state transition, while the dashed one represents that of the clients.

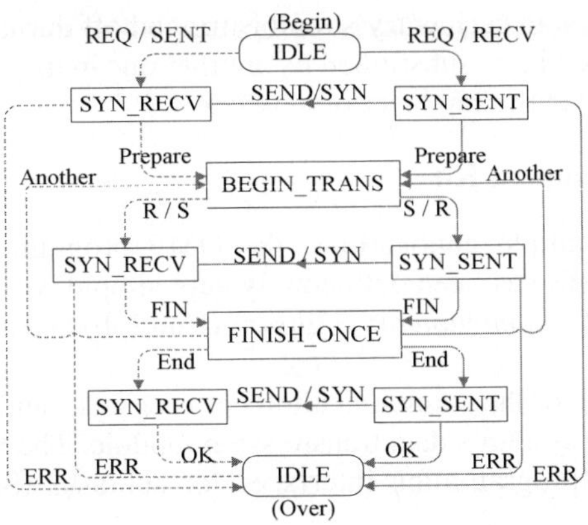

Fig. 2. State transition during data's transmission

Both clients and the server start the transmission from the IDLE state, and the transition of their states will pass through BEGIN_TRANS, EINISH_ONCE, and IDLE in the end. A file transfer starts with BEGIN_TRNAS, and ends with FINISH_ONCE. If there are other data files to be transferred, both server and client are BEGIN_TRANS again and begin to resolve the next data file. If an error occurs, the transferring fails and both of them are IDLE and prepare next transaction for next request. The procedure is iterated until all data files have been transferred, or an error occurs. Each transition will be synchronized with SYN. One of them is the sender, while the others are the receivers.

4.3 Scheduling

The running of an application is initiated by the main task, which applies subtasks from the monitor with a program ID and the number of subtasks. The Monitor selects one or several dispatchers with the lowest load and redirects the JobFile to them. Each of these dispatchers is assigned part of the project's subtasks. Then the main task requests those dispatchers to execute its subtasks.

The dispatcher schedules its workers to execute subtasks. If there are unfinished subtasks, the dispatcher distributes them to the idle workers. The worker who gets a task attains the task's program and related data from the Datapool. Then the worker performs the execution of subtasks. Finally, the worker returns the result of the subtask to the Datapool after finishing the execution. After all the subtasks have been finished, the main task gets their results from the Datapool, and further resolves them to get the final result.

4.4 Monitoring and Fault Tolerance

To ensure the availability, measures have been taken to trace P2HP's computing resources and the running states of applications' tasks. It is collaborated by monitors and dispatchers. The monitor is the supervision center of the system.

The monitor and the dispatcher cooperate as follows. In an autonomous domain, workers are traced by the dispatcher periodically. All the dispatchers are traced by the monitor. The transaction to each entry's fault is trigged off during the tracing. The load of the entry with errors will be substituted by another one in its group. This procedure is controlled by the entry in the top layer of network.

4.5 Software Development Kit

To be adaptive to multiple applications, the OMP programming model has been provided for users. OMP has been deliberately kept simple with minimum conceptual overhead. It consists of a *communication library* (ComLib) and a *software development kit* (SDK) for users.

The ComLib is used for the communication between tasks and the Datapool, and it is packaged based on the system's data transmission module. The SDK is responsible for providing application programming interfaces for subtasks (SubAPI) and the main tasks (MainAPI).

The OMP is a one-sided message passing model. As the shared remote space, the Datapool will respond to the request messages. When the client that has dispatched tasks needs the correlative data, it only locally triggers off a message like sending or receiving, and resolves this message to the Datapool. Then the requested data will be transferred by the ComLib between the Datapool and the client.

5 Experiment

To evaluate the performance of P2HP, we use a benchmark with speedup of the parallelized program in P2HP to the original sequential one in a PC. The benchmark is sequence alignment, which is a basic problem of bioinformatics.

5.1 Methods

The environment of P2HP is conducted in a network with 30 normal PCs. All computers are connected by a 100Mbps Ethernet switch. The frequencies of CPUs are from 1GHz to 2.4GHz. The Datapool is installed in a PC, where the Monitor and the Dispatcher are also configured. Other computers join the platform as workers as necessary.

CLUSTAL W and its variants are the most common used software packages for multiple alignments. First, the sequences are aligned by the sequential CLUSTAL W program in a PC. Then, as presented in [3], the sequential one is parallelized with OMP and run in P2HP. Assume that the number of sequences for multiple alignments is n, and the number of pairwise alignments in a subtask is k. Then the number of the whole pairwise alignments P_A and the number of subtasks N_S are shown in Formula 1 and 2. The phylogenetic tree is constructed according to the distance matrix of all the pairwise alignments, which is used to conduct the order of progressive alignment for multiple sequences.

$$P_A = \frac{n \times (n-1)}{2} \tag{1}$$

$$N_S = \frac{P_A}{k} = \frac{n \times (n-1)}{2k} \tag{2}$$

5.2 Evaluation

One hundred sequences are selected from NCBI (*National Center for Biotechnology Information*) which is one of the sequence databases in the world. The number of each sequence's letters is from 957 to 1534. In the CLUSTAL W program, the formation of phylogenetic tree through pairwise alignments costs about 440s.

In the parallelized execution for pairwise alignments in P2HP, it is assumed that the time for each pairwise alignment is C_P, and the communication time is C_M. The time for disposing of distance matrix is P_T, and the execution time T_P of the parallelized CLUSTAL W in P2HP is shown as Formula 3. The execution time T_S for the sequential program is as Formula 4.

The speedup V of the parallelized one in P2HP to the sequential is depicted in Formula 5. Combined with Formula 1 and 2, V is further represented in Formula 6. With the number of sequences as a constant, the speedup increases with the decreasing of k, and the maximum value is attained when k=1.

$$T_P = \frac{P_A \times (C_P + C_M)}{N_S} + P_T \tag{3}$$

$$T_S = P_A \times (C_P - E) \tag{4}$$

$$V = \frac{T_S}{T_P} = \frac{N_S \times (C_P - E)}{C_P + C_M + N_S \times P_T / P_A} \tag{5}$$

$$V = \frac{n \times (n-1)}{2k} \times \frac{C_P - E}{C_P + C_M + P_T / 2k} \tag{6}$$

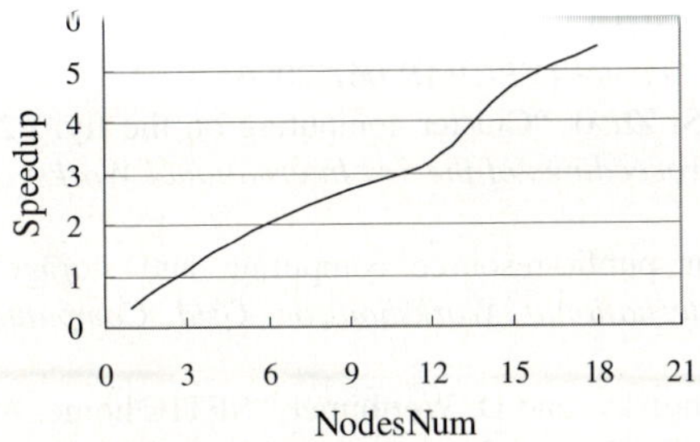

Fig. 3. Speedup with the increase of workers

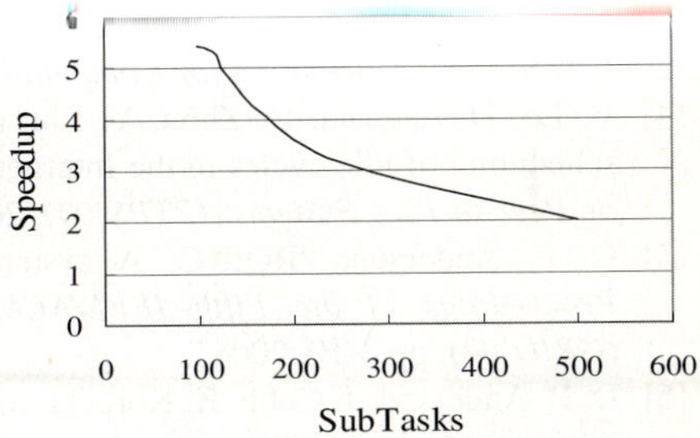

Fig. 4. Speedup with the increase of subtasks

To further verify its performance, we divide the alignments into 98 subtasks and increase the workers. The time for sequence alignment in P2HP decreases, and the speedup to the sequential CLUSTAL W increases, shown in Figure 3. If the number of workers is small, the speedup is less than 1, which means that the parallel execution of a small quantity of workers can not afford the cost of the communication.

Supposed that the number of workers is a constant as 18 and the number of subtasks is much larger than 18, the speedup to the sequential program is decreasing with the rise of the subtasks' number, as shown in Figure 4. It shows that P2HP is more adaptive to the computation intensive applications.

The experiment demonstrates that P2HP is an efficient and practical distributed HPC platform, especially to the computation intensive applications. To get the expected performance of P2HP, the number of subtasks should match that of workers in a limited environment, and applications should be divided into more subtasks if there are enough workers in the system.

6 Conclusion

P2HP is constructed as the 3-layer network that consists of monitors, dispatchers and workers, as well as a user and a Datapool. Based on unstructured P2P, P2HP is scalable and adaptive to dynamic environments. Augmenting the number of workers, its computing capability is increasing. With an easy configuration, P2HP is applicable. Moreover, the benchmark shows that it can provide scalable and enormous power for HPC applications.

References

[1] G. Bell and J. Gray, "What's next in high-performance computing", *Communications of the ACM*, 45 (2), pp.91~95, 2002.

[2] R. Raman, M. Livny, and M. Solomon, "Matchmaking: distributed resource management for high throughput computing", *Proceedings of the Seventh IEEE International Symposium on High Performance Distribute Computing (HPDC'98)*, 1998.

[3] H. Jin, F. Luo, Q. Zhang, X. Liao, H. Zhang, "OMP: a one-sided message passing programming model for P2HP", *Proceedings of the Seventh International Meeting on High Performance Computing for Computational Science (VECPAR'06)*, 2006.

[4] V. Lo, D. Zappala, D. Zhou, Y. Liu, and S. Zhao, "Cluster computing on the fly: P2P scheduling of idle cycles in the Internet", *Proceedings of the 3rd International Workshop on Peer-to-Peer Systems (IPTPS'04)*, 2004.

[5] D. P. Anderson, "BOINC: A system for public-resource computing and storage", *Proceedings of the Fifth IEEE/ACM International Workshop on Grid Computing (GRID'04)*, pp.4-10, 2004.

[6] D. P. Anderson, J. Cobb, E. Korpela, M. Lebofsky, and D. Werthimer, "SETI@home: An experiment in public-resource computing", *Communications of the ACM*, 45(11), pp.56-61, 2002.

[7] V. S. Pande, I. Baker, J. Chapman, S. P. Elmer, S. Khaliq, S. M. Larson, Y. M. Rhee, M. R. Shirts, C. D. Snow, E. J. Sorin, and B. Zagrovic, "Atomistic Protein Folding Simulations on the Submillisecond Time Scale Using Worldwide Distributed Computing", *Biopolymers*, Vol.68, pp.91-109, 2003.

[8] M. Taufer, C. An, A. Kerstens, and C. L. Brooks, "Predictor@Home: A 'protein structure prediction supercomputer' based on public-resource computing", *Proceedings of the 19th IEEE International Parallel and Distributed Processing Symposium (IPDPS'05)*, 2005.

[9] M. Sato, T. Boku, and D. Takahashi, "OmniRPC: A grid RPC system for parallel programming in cluster and grid environment", *Proceedings of 3rd International Symposium on Cluster Computing and the Grid (CCGrid'03)*, pp.206-213, 2003.

[10] G. Fedak, C. Germain, V. Neri, and F. Cappello, "XtremWeb: A generic global computing system", *Proceedings of the 1st IEEE/ACM International Symposium on Cluster Computing and the Grid (CCGrid'01)*, 2001, pp.582-587, 2001.

[11] N. Andrade, W. Cirne, F. Brasileiro, and P. Roisenberg, "OurGrid: an approach to easily assemble grids with equitable resource sharing", *Proceedings of the 9th Workshop on Job Scheduling Strategies for Parallel Processing (JSSPP'03)*, pp.61-86, 2003.

LDMA: Load Balancing Using Decentralized Decision Making Mobile Agents

M. Aramudhan[1] and V. Rhymend Uthariaraj[2]

[1] Research scholar, Dept.of. IT, MIT, Anna University, Chennai-25, Tamilnadu, India
aranagai@yahoo.co.in
[2] Professor, Dept.of. IT, MIT, Anna University, Chennai-25, Tamilnadu, India
rhymend@annauniv.edu

Abstract. This paper introduces a new load balancing algorithm, called LDMA (Load balancing using Decentralized decision making Mobile Agents), which distributes load among clustered web servers connected in a mesh topology, by a communications network and compares its performance with other load balancing algorithm: MALD (Mobile Agent based LoaD balancing). Architecture is developed for the same and all necessary attributes such as load deviation, system throughput and response time incurred as a result of the work are dealt with. In past works, a centralized decision making algorithm was used for dispatching requests to web servers in the distributed client/server environment. In the proposed approach, a decentralized decision making algorithm is used for distributing requests among web servers. The simulator is developed in C++ and its performance is evaluated. The analysis shows that LDMA is better than centralized decision making load balancing algorithms.

Keywords: Load balancing, decentralized decision making, mobile agents, clustered web servers.

1 Introduction

A distributed computer system is a collection of self-sufficient computers located at diverse or identical sites and associated by a communication network. The performance of a distributed system is enhanced to an adequate level by distributing the workload among the servers. Normally, load balancing occurs at the server side and assists to balance the load in distributed computer system. Winston [1] proved that the most excellent mechanism for achieving optimal response time is to distribute the workload equally among the servers. Incoming client requests should be evenly distributed among the servers to achieve quick response time. Traditional load balancing approaches on distributed web servers are implemented based on message passing paradigm. At present, mobile agent technology is used to implement load balancing on distributed web servers. Mobile agent is defined as a software component that can move freely from one host to another on a network and transport its state and code from home host to other host and execute various operations on the site [6]. The mobile agent based approaches have the merit of high flexibility, low network traffic and high asynchrony.

V.N. Alexandrov et al. (Eds.): ICCS 2006, Part IV, LNCS 3994, pp. 388 – 395, 2006.
© Springer-Verlag Berlin Heidelberg 2006

Distributed web servers deploy in different geographical scopes. They can be organized into cluster of web servers linked through Local Area Network (LAN), to provide high processing power and reliability. The servers are heterogeneous in terms of hardware configuration, operating systems and processing power. Generally, load balancing on Wide Area Network (WAN) is more time consuming since it involves the interaction between remote servers for gathering load information, negotiating on load reallocation and transporting the workload [2]. All approaches in this context so far has been using only centralized decision making. But, an architecture based purely on a centralized server is extremely vulnerable to congestion. In addition, it introduces a single point of failure in the Web system, as stated in [9]. Hence, we approach the problem in a totally different dimension, by introducing the concept of "decentralized decision making". LDMA uses mobile agents for this idea. In LDMA, there is no collection of load information and request transfer policy between web servers. Each server processes client requests independently and interact with others to share the workload.

2 The LDMA Framework

The overall architecture of the LDMA framework is as shown in figure 1. The LDMA framework defines two worlds, namely: client world and server world. The client world is an aggregation of all the clients in the physical world, and the server world is an aggregation of the clustered web servers, which are called replicas. The client world communicates with the server world via the dispatcher. The queue at the dispatcher has a finite buffer and a tail drop discard policy. But, unlike the other approaches, in which the dispatcher re-routes the client requests to corresponding servers (centralized decision making), the work of the LDMA dispatcher is just to broadcast client requests to all the replicas. The decision-making for load balancing among replicas take place only by the interaction of mobile agents between the replicas. The replicas are inter-connected by mesh topology. Each replica has the following two modules:

1. MASM – Mobile Agent Servicing Module
2. SND – Search aNd Destroy module.

The work of the MASM is to communicate the mobile agents with other replicas to make them decide which replica may process a request and the work of the SND module is to search for and delete (remove) a particular request from the replicas queue. The LDMA framework uses the concept of "ranked web-servers", i.e., each replica is statically assigned a rank based on which priority is given for processing a request.

2.1 LDMA Load Balancing Scheme

Initially, upon the arrival of a client request, the dispatcher broadcasts it to all the replicas, after assigning a RID (Request ID) to it. The replicas accept the request, but the request processing does not start immediately. Instead, the request is placed in its

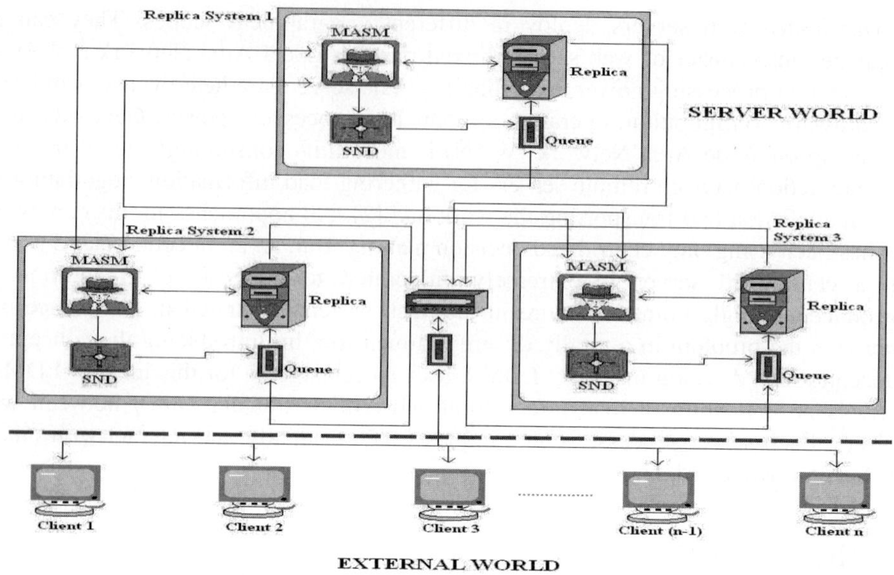

Fig. 1. The LDMA Framework

"waiting state". Each replica sends a mobile agent, with replica's rank and the RID just accepted, which we call "RID under siege". The mobile agents travel to the other replicas and check the state of the same request in the destination replicas. Then, they can return back to the source replica with either of the following messages:

1. Accepted: This case occurs when the rank of source replica is less than the rank of the destination replica, and the RID under siege is in waiting state in the destination replica. On receiving back this message, the source replica just ignores the accepted request, and chooses the next request.
2. Deleted: This case occurs,
i. when the rank of source replica is greater than the rank of the destination replica and the RID under siege is in waiting state in the destination replica, or
ii. irrelevant of the ranks, the RID under siege is in the queue in the destination replica.

The mobile agent triggers the SND module at the destination replica, which removes the RID under siege from the destination replica. On receiving back this message, the source replica starts processing the request.
3. Not Found: A mobile agent returns back to source replica, with this message, when RID under siege is not found either in its waiting state or even at the queue of the destination replica. This case occurs when the RID under siege has already been removed from the destination replica's queue, by a mobile agent from the other replica. On receiving back this message, the source replica may choose to ignore or accept the request, depending on the other mobile agent's response.

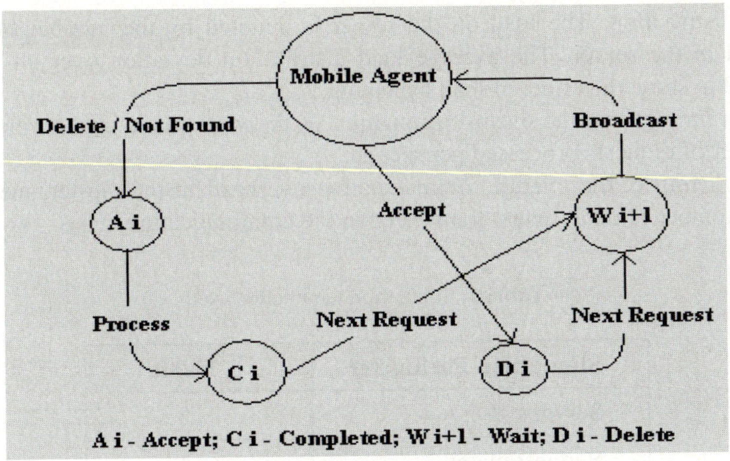

Fig. 2. LDMA Transition Diagram

To retrospect (as shown if figure 2), a replica, on accepting a request, sends mobile agents to other replicas and waits for the response. It ignores the request, if at least one of the responses is an "Accepted" message. It starts processing the request otherwise. Moreover, in case of a packet loss of a mobile agent, a replica waits for a maximum of twice the RTT (Round Trip Time) of the mobile agent. In case of no response message, the replica starts processing the request. Also, the dispatcher assigns RIDs using mod N arithmetic, i.e.,

RID = i mod N, where i = 0, 1, 2, 3…

The RIDs are in increasing order. Hence, the work of SND module at a replica is easier and it searches for RID under siege from the top (using the sequential search algorithm), till ith request in the queue, where i is "just greater than" the RID under siege. After the ith request, the RID under siege cannot be found elsewhere in the queue (since RIDs are in increasing order), except in the next cycle of RIDs.

3 LDMA Simulation Model

The software simulator was designed in C++ and implemented to model the LDMA load balancing technique in the distributed web server environment. Workload of a replica is determined by the number of request processed at each replica. To achieve best performance results a method applied needs to minimize workload difference between the replicas. For load balancing algorithm T_e and T_d are complied. T_e is the elapsed time from the start of the first client call until the entire clients call and T_d is the delay time representing the sum of all the delays associated with the client's requests. Simulation parameters governing the generation of client's events are summarized below:

Load distribution: The load on the server is denoted by the number of requests processed in the server. The average load distribution deviation over all servers is calculated to show the effect of load balancing.

System throughput: the overall throughput of the web server cluster, measured in the number of requests processed per second.

Network traffic: the overall communication overhead in the cluster, measured in the total number of data (bytes) transferred in the communication.

Table 1. Simulation parameters used

Simulation Parameter	Value
Servers	3
Request/client	1
Data rate (Transmission speed)	10 MBps
http request file size	<= 2 MB
Propagation Delay	Negligible
Mobile agent RTT	0.5 ms
Processing Delay	Negligible

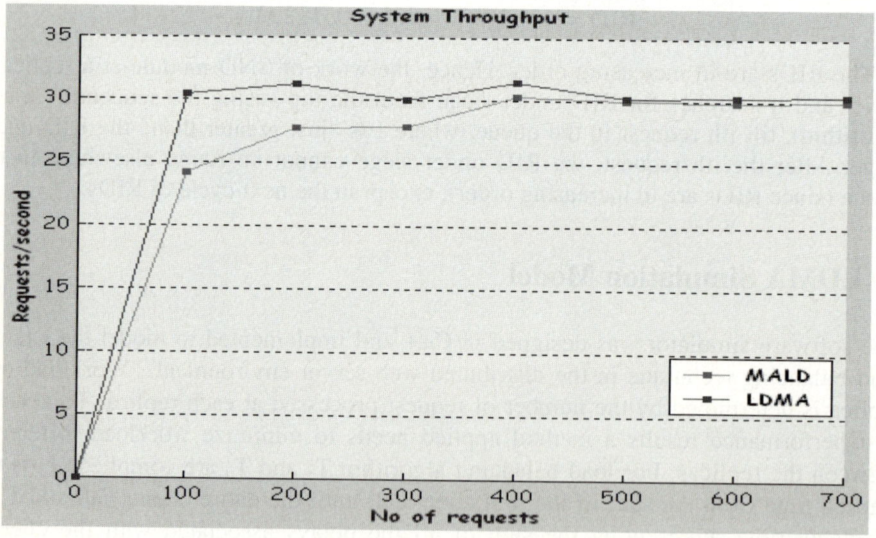

Graph 1. The LDMA system throughput

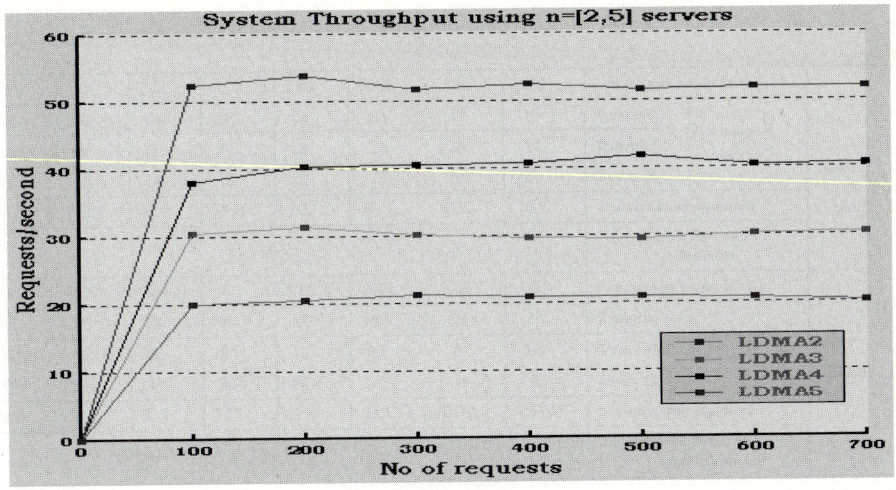

Graph 2. System throughput with web server 2,3,4,5

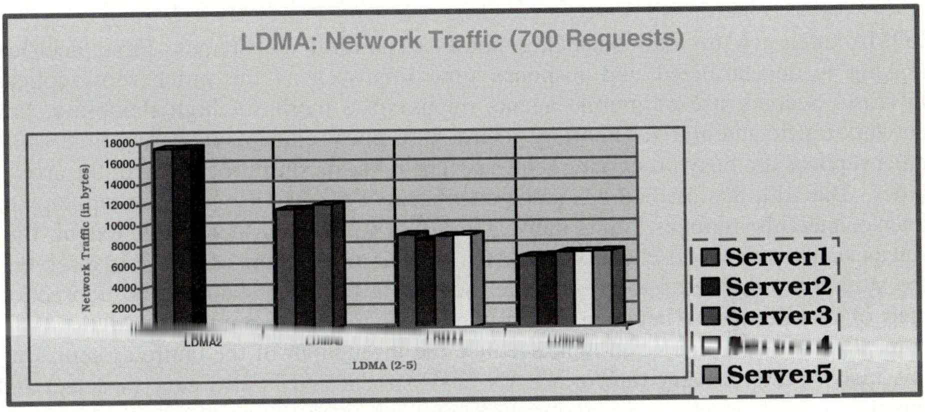

Graph 3. LDMA Network Traffic

Graph-1 shows the system throughput performance of MALD and LDMA for 700 requests. The LDMA performance is slightly better than MALD. Graph-2 shows the system throughput of LDMA having different number of servers in cluster. Graph 3 shows the network traffic of LDMA. The overhead of the packet is 50 bytes. The dispatcher broadcast the incoming requests to all servers. The communication overhead for totally N servers is as high as O (N^2). Network traffic is measured by the total number of bytes transferred in the communication. The network traffic of web server cluster having 2 to 5 servers is as shown in Graph-3. In the beginning, three server's exchanges message for processing the request takes 300 bytes communication overhead. The next two requests take 200, 100 bytes communication overhead respectively. The communication overhead of the afterward requests depends on the previous request processing time.

Table 2. Load distribution on three servers

LDMA	Total no of requests		160	202	188	207	172	143	84	115
	Requests/ server	Server1	48	71	67	67	58	45	24	38
		Server2	57	66	60	68	57	53	29	38
		Server3	55	65	61	72	57	45	31	39
	Average deviation		3.56	7.33	2.89	2.0	0.44	3.56	2.67	0.44
	Overall average deviation		2.86							
MALD	Total no of requests		160	202	188	207	172	143	83.6	115
	Requests/ server	Server1	65	63	37	74	68	42	30	33
		Server2	58	79	76	55	48	50	37	34
		Server3	37	60	75	78	56	51	16.6	48
	Average deviation		10.89	7.78	17.11	9.33	7.11	3.78	7.51	6.44
	Overall average deviation		8.74							

4 Conclusion

LDMA approach to load balancing possesses several advantages. First, decision making is decentralized and response time improves as the number of replicas increase. Second, use of mobile agents imposes the merits of high flexibility, low network traffic and high asynchrony. Third, no replica remains idle at any time while other replicas are busy processing requests. The requests start processing in the arrival order. But still, this method has some drawbacks. First, the use of mesh topology to inter-connect the replicas, limits the scalability of the system to a certain extent. But, since usually web server clusters in LAN consist of a maximum of only 7 to 8 servers, the system is considered to be scalable. Second, a failure or fault in the transaction path of mobile agents between two replicas may result in processing of the same request by many replicas and hence reduce the throughput of the entire system. But, this kind of fault is very rare in a LAN environment and is also easy to detect and repair.

References

1. W. Winston: Optimality of the Shortest Line Discipline. Journal of Applied Probability (1977) 17-28.
2. Jiannong cao, Yudong Sun, Xianbin Wang and Sajal K. Das :Scalable:Load Balancing on Distributed Web Servers Using Mobile Agents. Journal of Parallel and Distributed Computing, Vol.63, Issue 10. (2003) 996-1005.
3. Huamin Chen and Arun Iyengar :A Tiered System for Serving Differentiated Content. Journal of World Wide Web, Vol.6, Issue 4. (2003) 331-352.
4. Lang fang, Aleksander Slominski and Dennis Gannon: Web Services Security and Load Balancing in Grid Environment. Proc.of. International Conference on Grid Computing, Las Vegas, June (2005).

5. Gianfranco ciardo, Almariska and Evgenia smirni: EQUILOAD: a Load Balancing Policy for Clustered Web Servers. Proc. of Parallel and Distributed systems (2004)1420-1425.
6. Altec software business unit (2004), "Mobile Agents System for the Interconnection of Working Groups. Interconnection network vol.5 (2), (2004) 181-191.
7. Foundry Networks, White Papcr –Server Load Balancing in Today's Web-Enabled Enterprises, (2002).
8. Reinhardtriedl:Workload Modeling for Load Balancing in Distributed DB/DC Transaction Processing (1999).
9. Marco Conti, Enrico Gregori and Fabio Panzieri :Load Distribution among Replicated Web Servers: A QoS-based Approach. (1999).
10. Baruch Awerbuch, Mohammad T.Hajiaghayi, Robert D.K leinberg and Tom:Online Client-Server Load Balancing without Global Information. in proc.of the sixteenth annual ACM-SIAM symposium on Discrete Algorithms(2005) .
11. Morharchol-Balter, Bianca Schroeder, Nikhil Bansal, and Mukesh Agrawal: Size-Based Scheduling to Improve Web Performance, in ACM Transactions on Computer Systems, Vol. 21, No. 2. (2003)
12. Milan E.Soklic: Simulation of Load Balancing Algorithms: A Comparative Study. in SIGCSE Bulletin vol.34, No.4, Dec. (2002).

A Hybrid Scheme for Object Allocation in a Distributed Object-Storage System*

Fang Wang**, Shunda Zhang, Dan Feng, Hong Jiang, Lingfang Zeng, and Song Lv

Key Laboratory of Data Storage System, Ministry of Education,
School of Computer, Huazhong University of Science and Technology, Wuhan, China
wangfang@mail.hust.edu.cn, zhangshunda@163.com

Abstract. The object-based storage system, in which files are mapped onto one or more data objects stored on Object-Based Storage Devices (OSDs), has distributed storage system architecture. In such a system, the policy for object allocation is a critical aspect affecting the overall systems performance. Hashing and fragment-strip are two common techniques used for managing objects, but both have their disadvantages, and advantages, e.g. hashing achieves good workload balance and provide rather high effectiveness in allocating data, but it can not provide readily available parallelism for large file; fragment-strip takes full advantage device parallelism, simplifies the clients' operations, but this policy is not fit for small file. In this paper, we present an efficient algorithm that combines the advantages of these two approaches while avoiding their shortcomings. The key factors which can impact the performance of the whole system in the objects allocation are also be discussed.

1 Introduction

Object-based storage systems represent files as sets of objects stored on self-managed Object-Based Storage Devices (OSDs). By distributing the objects across many devices, these systems have the potential to provide high throughput, reliability, availability and scalability [1]. Much research has gone into hierarchy management, scalability, and availability of distributed file systems in projects such as AFS [3], Lustre [8], GFS [11], Coda [12] and GPFS [13], but relatively little research has been aimed at improving the efficiency of objects allocation in large scale object-based storage systems. The algorithm used for object allocation determines the performance of the system at the beginning of the communication process. It affects the workload among the devices, and it also influences the OSD-level parallelism of the object-based storage system.

The object-based storage model is emerging as architecture for distributed storage systems. Traditionally, metadata and data are managed by the same file system, on the same machine, and stored on the same device [3]. For efficiency,

* This work was supported by the National Basic Research Program of China (973 Program) under Grant No. 2004CB318201, the National Science Foundation of China under Grant No.60303032.

** Corresponding author.

V.N. Alexandrov et al. (Eds.): ICCS 2006, Part IV, LNCS 3994, pp. 396 – 403, 2006.
© Springer-Verlag Berlin Heidelberg 2006

metadata is often stored physically close to the data it describes [4]. In some modern distributed file systems, data is stored on devices that can be directly accessed through the network, while metadata is managed separately by one or more specialized metadata servers [5].

Currently, most approaches to object allocation employ one of two techniques. The first one, which we call hashing [2], allocates a file to one device by using hashing functions that map file IDs to OSD IDs. This approach converts a file to one object and sends it to only one device. The second object allocation technique, which we call fragment-strip or fragment-mapping [2], uses equal-sized fragments of each file to widely distribute the file among the OSDs.

Our object allocation scheme combines the best aspects of hashing and fragment-strip. In the algorithm, when the file is small, it is converted to a single object and directly mapped to an OSD by hashing. If the file is large, it is converted to multiple objects and each object will be distributed to an OSD.

2 Related Works

To improve the scalability of hashing, a self-adaptive hashing scheme is presented in [6]. To reduce the cost of adaptation and continue to exploit the high effectiveness of hash functions, the self-adapt hashing policy is designed to improve scalability.

In OBFS (a file system for object-based storage devices) [1], the boundary of small objects and large objects is set at 512KB. The workload characteristics of a high-performance distributed file system from Lawrence Livermore National Laboratory (LLNL) [7] were analyzed as an example of large-scale distributed file systems [1]. OBFS provides most of the files are larger than 4 KB and the majority of all files are distributed between 32 KB and 8MB. Those files that are smaller than 4 KB (a typical block size for a general-purpose file system) only account for a very small portion of the total files.

In the Panasas storage cluster [9], if a file is smaller than 64KB, it will be mirrored on the first two component objects (RAID 1). If the file is larger than 64KB, it will use additional component objects, up to full stripe worth (RAID 5). That means that 64KB is the boundary distinguishing small files from large files, and that objects are not larger than 64KB. The Lustre cluster file system [8] logical object volume management (LOVM) manages the objects as RAID.

3 Algorithm Design

3.1 The Boundary of Small and Large File

According to OBFS [1], this boundary of small and large files should be 512KB. The OBFS's conclusion was based on the analysis of LLNL [7] workload characteristics. In the LLNL [7] workload, we estimate that about 85% of all objects will be of size 512 KB and 15% of all objects will be of size smaller than 512 KB. We will refer to files that are smaller than 512KB as small objects and the rest as large objects. Small objects and large objects are treated differently by the object allocation algorithm.

3.2 The Optimal Number of Objects Mapped from One File

The relationship between the number of OSDs and parallelism is quite complex. More devices provide more transfer channels and data can be transferred in parallel. However, increasing connections can also bring down the performance, because establishing connections takes time, especially when connections are numerous. Thus, while increasing the number of devices mapping to the same file improves parallelism, it consumes extra resources of the system.

We can describe the relationship described above with the following formula:

$$\frac{T_p}{T} = \frac{n \times a + \frac{1}{n} \times b + \delta(t)}{a + b} \tag{1}$$

Tp: The time of transferring file with multiple objects in parallel.

T: The time of transferring file sequentially.

n: Number of objects mapped to a large file.

a: The sum of sender overhead, receiver overhead and the time of flight of transferring a file.

b: The time for transferring a whole file to a single device.

$\delta(t)$: Other delays of objects transmission, it is an amendment factor of the formula.

Numerator of the formula is made up of three terms, which are *n*a*, *b/n* and $\delta(t)$. The term *n*a* means connecting to n devices costs n times of the overhead connecting to a single device. And the *b/n* indicates that n devices' parallel working can make bandwidth n times wider than the single-devices situation.

Let's review the performance parameters of interconnection networks.

Depending on whether it is an SAN, LAN or WAN, the relative lengths of the time of flight and transmission may be quite different from those shown here, based on a presentation by Greg Papadopoulos of Sun Microsystems. [10]

$$\textit{Total latency} = \textit{Sender overhead} + \textit{Time of flight} + (\textit{Message size} / \textit{Bandwidth}) + \textit{Receiver overhead} \tag{2}$$

Notice that the time of flight for SANs is so short relative to overhead that it can be ignored, yet in WANs, time of flight is so long that sender and receiver overheads can be ignored. Thus, we can simplify the performance equation by combining sender overhead, receiver overhead, and time of flight into a single term called *Overhead*:

$$\textit{Total latency} \approx \textit{Overhead} + (\textit{Message size} / \textit{Bandwidth}) \tag{3}$$

In our formula:

$$a = \textit{Overhead},\ b = (\textit{Message size} / \textit{Bandwidth}) \tag{4}$$

Although the $\delta(t)$ has some relationship with *n*, their relationship is rather loose. That is to say, we can simplify $\delta(t)$ to a variable c that is irrelevant to *n*.

We can simplify our formula by replacing $\delta(t)$ with c:

$$\frac{T_p}{T} = n \times \frac{a}{a+b} + \frac{1}{n} \times \frac{b}{a+b} + \frac{c}{a+b} \tag{5}$$

The reciprocal of the above formula, T/T_p is the speedup in file transfer time as a result of parallelism. To maximize the speedup is equivalent to minimizing the above formula for T_p/T. Since we are interested in how to best map a file into objects, namely, determining an optimal n, for data transfer, although a, b are not constants, they are irrelevant to n. We now show that the sum of the two terms containing n has a minimum when $n=\sqrt{b/a}$, as follows.

Let $F(n) = n\times\frac{a}{a+b}+\frac{1}{n}\times\frac{b}{a+b}$, and solve for $\frac{dF(n)}{dn}=0$, we have $n=\sqrt{b/a}$. Since $\frac{d^2F(n)}{dn^2}>0$, $F(n)$ has a global minimum at $n=\sqrt{b/a}$. Therefore,

$$\frac{T_p}{T} \geq \frac{2\sqrt{ab}}{a+b} + \frac{c}{a+b} \tag{6}$$

The *Overhead* is the time for the processor to inject the message into the network, including both hardware and software components. For pedagogic reasons, we assume that *Overhead* is not dependent on message size. (Typically, only very large messages have larger overhead.) So we can assume that a network with a bandwidth of 1000 Mbits/second has an *Overhead* of 80 microseconds [10]. This situation is very common in today's network, and we use this representative case to estimate the parameter we need. As the factor c/(a+b) do not impact n, we can ignore it.

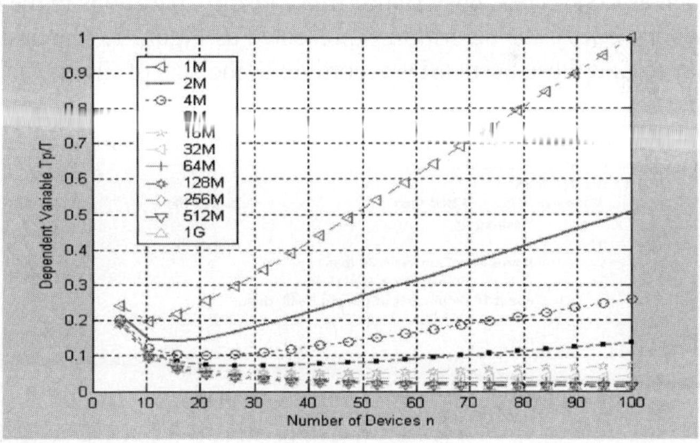

Fig. 1. The Relationship Curve of T_p/T and n

Figure 1 describes the relationship T_p/T and Number of Devices n according to formula (6). Figure 1 shows that for 1MB and 2MB files, when n >= 10 the change in T_p/T is not distinguishable. And for 4MB, 8MB and 16MB, when n >= 20 the change in T_p/T is not distinguishable. So do files above 32MB when n >= 40. We can arrive at a conclusion: n=10 (1MB-2MB), n=20 (4MB-16MB), n=40 (32MB-1GB).

3.3 How to Select OSDs for Parallel Transmission

An object-based storage system typically has hundreds of OSDs. How to select OSDs for parallel transmission of large files from numerous devices? Random choice is a good idea, which is easy to implement and can always keep workload balanced. However, it can not ensure that the fastest devices are fully utilized. We can sort the devices by some parameters such as speed, free-capacity and so on, then select the first n(10, 20 or 40) devices. This algorithm makes sure that the best-conditioned devices are used first. However, sorting hundreds of OSDs is a time consuming task for the system. Does it affect the performance to some degree? We will carry out an asymptotic time complexity analysis of the sorting algorithm to address such questions.

In a bubble sort algorithm, the time complexity of searching the first n items from a total of N items is:

$$\sum_{i=1}^{n}(N-i)=\frac{2N-(n+1)}{2}\times n \tag{7}$$

The time complexity: $T(n) = O(n^2)$ and $T(N) = O(N)$. So n impact the time complexity more noticeably than N. Because n is small (10, 20 or 40) and MDS always has high-capacity memory and high-performance CPUs, sorting algorithm will not affect the performance much. We can use $\overline{n} = (10+20+40)/3 \approx 20$ instead of n to simplify our module.

3.4 Objects Allocation Algorithm Details

The object allocation algorithm proposed here is based on sorting those devices by some parameters, such as OSD types, busy status, free capacity, partitions in the device and IP address. A pseudo-code of the algorithm is presented as Figure 2. The basic idea behind the algorithm is to find those best OSDs according to the size of files.

```
INITIALIZATION: Get number of devices (the total number of OSDs) and set dev[] empty.
Objects_allocation( the size of file )
INPUT: The size of file
      if the size of file < 512KB then                        //small file
            Hashing();
      else                                                    //large file
            if number of devices < 20 then
                  N = number of devices;
            else if 20 <= number of devices <=40 then
                  N = 20;
            else
                  N = 40;
            endif
      Sort dev[1], dev[2] ... dev[number of devices] by performance;
            if the size of file <= N * 512KB then
                  Fragment_ striping (dev[1], dev[2] ... dev[the size of file/512KB]
);
            else
                  Fragment_ striping (dev[1], dev[2] ... dev[N] );
            endif
      endif

OSDs is sorted by type, busy, freesp and partitions in function Sort.
```

Fig. 2. Objects Allocation Algorithm

4 Simulation Results

4.1 Experimental Setup

All of the experiments were executed on a PC with a 2.4 GHZ Intel Celeron CPU and 512 MB of RAM, running Red Hat Linux, kernel version 2.4.20. We used Matlab as the simulator. Matlab first generated an array of random numbers chosen from the exponential distribution of the file sizes. Then our algorithm (Section 3.4) was applied to estimate the response time of the system. We implemented the algorithm in Matlab's M file. The parameter a (Overhead) was assumed to be 80 μs and network bandwidth was assumed to be 1000 Mb/s (Section 3.2). Devices after being sorted should reduce the total response time (Section 3.3). According to the number of OSDs, the simulator considered the following situations: 16 OSDs, 32 OSDs, and 64 OSDs.

4.2 Results

Figures 3 and 4 show the simulation results with 16 OSDs, 32 OSDs and 64 OSDs. In each virtual system, we measure response times and compare among results from the hashing scheme, the fragment-strip policy and our algorithm.

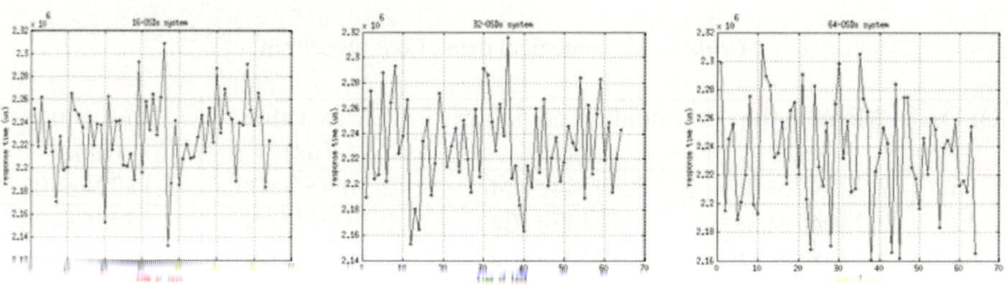

(a) (b) (c)

Fig. 3. Response time of the 16-OSDs (a), 32-OSDs (b) and 64 OSDs (c) system

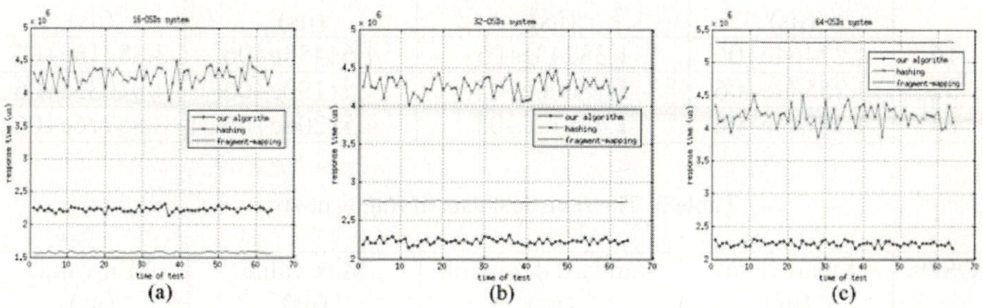

(a) (b) (c)

Fig. 4. Comparing our algorithm, hashing and fragment-strip in the 16-OSDs (a), 32-OSDs (b) and 64-OSDs (c) system

4.3 Analysis and Discussion of Experimental Results

During evaluation process, we confirmed that the file sizes generated by Matlab conforms the description of file sizes from LLNL [7] (Section 2) fairly.

Tables 1, 2 and 3 show the mean value, standard deviation, and max and min value of the three algorithms' response time in different object-based storage systems. In the 16-OSDs system, the mean value of fragment-strip's response time is the smallest of the three algorithms. However, when the number of devices increases, the fragment-strip consumes more and more response time because of the increasing the Overhead. In the 32-OSDs system, the fragment-strip's mean value of response time is larger than our algorithm's. In the 64-OSDs system, the fragment-strip's mean value of response time becomes larger than the other two. The response time of hashing is between 4 and 4.6 seconds, our algorithm's response time is between 2 and 2.3 seconds. They have not been changed much in the three different cases.

Our algorithm is faster than hashing because it makes good use of parallelism. The standard deviation of fragment-strip's response time decreases while the number of devices increases. And the standard deviation of hashing and fragment-strip is relatively steady. Our algorithm's standard deviation is smaller than hashing and larger than fragment-strip. As a whole, our algorithm is steadier than the others. And it performs best when the object-based storage system has many OSDs.

Table 1. The statistical data of our algorithm

OSDs	Mean value (μs)	Standard deviation (μs)	Max value (μs)	Min value (μs)
16	2.22947e+06	3.30344e+04	2.30907e+06	2.13224e+06
32	2.23067e+06	3.51447e+04	2.31596e+06	2.15275e+06
64	2.23583e+06	3.87005e+04	2.31128e+06	2.16086e+06

Table 2. The statistical data of hashing

OSDs	Mean value (μs)	Standard deviation (μs)	Max value (μs)	Min value (μs)
16	4.27464e+06	1.38243e+05	4.64453e+06	3.88316e+06
32	4.25589e+06	1.18006e+05	4.51130e+06	4.01033e+06
64	4.18744e+06	1.44812e+05	4.51206e+06	3.85776e+06

Table 3. The statistical data of fragment-strip

OSDs	Mean value (μs)	Standard deviation (μs)	Max value (μs)	Min value (μs)
16	1.57276e+06	8.64023e+03	1.59588e+06	1.54829e+06
32	2.75187e+06	3.68768e+03	2.75985e+06	2.74420e+06
64	5.30830e+06	2.26269e+03	5.31338e+06	5.30315e+06

5 Conclusion and Future Work

In this paper, we present a hybrid algorithm of hashing and fragment-strip, which combines the best aspects of these two algorithms while avoiding their disadvantages. Fragment-strip's good scalability is retained reasonably well, while the high efficiency of hashing makes its presence felt in our algorithm. We combine the two popular objects-allocation approaches at 512KB, the boundary of small and large file. We calculate two key factors in our algorithm, namely, the optimal number of objects mapped from one file and the scope of selecting OSDs for parallel transmission. Simulation results validate the correctness of the parameters we have calculated. Our object-allocation algorithm consumes lest time when there are numerous OSDs and its performance does not change much when the total number of devices increases.

As future work, we plan to finish the Object-Based Storage System. We will also test the algorithm in the real system instead of the simulation environment created by Matlab.

References

1. Feng Wang, Scott A. Brandt, and Ethan L. Miller, Darrell D. E. Long.: OBFS: A File System for Object-based Storage Devices. In 21st IEEE / 12th NASA Goddard Conference on Mass Storage Systems and Technologies (MSST2004), College Park, MD, April 2004.
2. Lan Xue, Yong Liu.: MDS Functionality Analysis. December 4, 2001.
3. J. H. Morris, M. Satyanarayanan, M. H. Conner, J. H. Howard, D. S. H. Rosenthal, and F. D. Smith.: Andrew: A distributed personal computing environment. Communications of the ACM, 29(3):184–201, Mar. 1986.
4. M. K. McKusick,W. N. Joy, S. J. Leffler, and R. S. Fabry.: A fast file system for UNIX. ACM Transactions on Computer Systems, 2(3):181–197, Aug. 1984.
5. G. A. Gibson and R. V. Meter.: Network attached storage architecture. Communications of the ACM, 43(11):37–45, 2000.
6. M. Spasojevic and M. Satyanarayanan.: An Empirical Study of a Wide-Area Distributed File System. ACM Transactions on Computer Systems 14(2) May 1996.
7. D. Roselli, J. Lorch, and T. Anderson.: A comparison of file system workloads. In Proceedings of the 2000 USENIX Annual Technical Conference, June 2000.
8. Peter J. Braam (with others): The Lustre Storage Architecture. Cluster File Systems, Inc. http://www.clusterfs.com August 2003.
9. J.R. Moase.: Panasas Storage Cluster & Object Storage Overview. www.panasas.com October 2004
10. John L. Hennessy, David A. Patterson.: Computer Architecture A Quantitative Approach (Third Edition).
11. S. R. Soltis, T. M. Ruwart, and M. T. O'Keefe.: The Global File System. In Proceedings of the 5th NASA Goddard Conference on Mass Storage Systems and Technologies, College Park, MD, 1996.
12. M. Satyanarayanan, J. J. Kistler, P. Kumar, M. E. Okasaki, E. H. Siegel, and D. C. Steere.: Coda: A highly available file system for a distributed workstation environment. IEEE Transactions on Computers, 39(4):447–459, 1990.
13. F. Schmuck and R. Haskin.: GPFS: A shared-disk file-system for large computing clusters. In Proceedings of the 2002 Conference on File and Storage Technologies (FAST), USENIX, Jan. 2002.

Survive Under High Churn in Structured P2P Systems: Evaluation and Strategy

Zhiyu Liu, Ruifeng Yuan, Zhenhua Li, Hongxing Li, and Guihai Chen*

State Key Laboratory of Novel Software Technology,
Nanjing University, China
gchen@nju.edu.cn

Abstract. In Peer to Peer (P2P) systems, peers can join and leave the network whenever they want. Such "freedom" causes unpredictable network environment which leads to the most complex design challenge of a p2p protocol: how to make p2p service available under churn? What is more, where is the extreme of a system's resistibility to high churn? A careful evaluation of some typical peer-to-peer networks will contribute a lot to choosing, using and designing a certain kind of protocol in special applications. In this paper we analyze the performance of Chord [1], Tapestry [2], Kelips [3], Kademlia [4] and Koorde [5], then find out the crash point [6] of each network based on the simulation experiment. Finally, we propose a novel way to help nodes survive under high churn.

Keywords: Peer-to-Peer, Fault Resilience, High Churn, Crash Point.

1 Introduction

People like to use peer-to-peer networks, because there are few restrictions. Peers can join and leave the network whenever they want. However, such "freedom" causes unpredictable network environment which leads to the most complex design challenge of P2P protocols: how to make p2p service available under churn? Almost every P2P protocol has proposed its method to deal with churn, and shows some experiment report. However, no previous work has compared those protocols in the aspect of "resilience under high churn".

Why is the problem "resilience under high churn" important? Before answering the question, let us give "churn" and "high churn" a descriptive definition.

Ordinary churn: is such a condition that nodes join the network one by one, or leave gracefully by informing their neighbors. The churn event(ie, node join and departure) happens occasionally and can be handled quickly so we can suppose that the overlay of the network is well structured before any individual churn event occurs.

High churn: is such a condition that large percent of nodes join and/or silently leave the network simultaneously and frequently.

* Corresponding author.

V.N. Alexandrov et al. (Eds.): ICCS 2006, Part IV, LNCS 3994, pp. 404–411, 2006.
© Springer-Verlag Berlin Heidelberg 2006

From above we can see that: first, high churn condition is totally different from ordinary churn: the stabilization routine which may work very well under ordinary churn could be useless under high churn. Thus, good performance under churn does not imply the same result under high churn. We have to find other ways to make sure the service is available under high churn.

Second, high churn is not just an imagination but does happen from time to time in real life. For example, a large number of maliciously controlled peers could leave the network simultaneously; the power is cut off over a wide area; temporarily hot resources like "world cup online show" may also cause a large number of peers to join and leave simultaneously.

Third, by taking a look at "resilience under high churn" we can evaluate a P2P protocol in more comprehensive way. Choosing a P2P network with better "resilience under high churn" will help us survive in extremely turbulent environment.

A lot of related works addressing to "resilience" problem have been done. Some [7, 10, 11, 14] did excellent theoretical analysis and some did carefully selected simulations [11]. So far, however, simulation of peer-to-peer systems under high churn to compare their "resilience under high churn" has not been done yet. In this paper, we first propose a measurement of "resilience under high churn" of P2P protocols: crash point, then fairly evaluate some typical P2P protocols like Chord [1], Tapestry [2], Kelips [3], Kademlia [4], Koorde [5] and so on to demonstrate their different performances of "resilience under high churn" by "crash point", and finally design a strategy to help live node survive under high churn. Here is the definition of crash point:

Crash point. If x percent of nodes' leaving simultaneously causes half (50%) randomly generated look-ups to fail, then x percent is defined as crash point.

Crash point is so defined for three reasons:

1. Both concurrent joining and leaving lead to the incorrect routing information, so without loss of generality, the percentage of concurrent leaving nodes represents the degree of churn.
2. By this definition, we can ignore the difference between two kinds of crash. One kind is that when a node becomes isolated, all look-ups from/to it will fail. The other kind is that when the whole network breaks up into some disconnected sub-nets, look-ups between nets will fail. However, LOOK-ups within a sub-net can still succeed. Then we can compare different protocols under the same criterion.
3. Successful look-up ratio is easy to record and it has certain relationship with the connectivity of the network. We discover in the simulation experiments that once half of the look-ups fail, the network could never be recovered to fully connected status.

The rest of the paper is organized as follows: section 2 introduces some related works. Section 3 discusses some important factors of a DHT structure which have impact on "crash point". Section 4 shows the experiment results.

Section 5 gives a description of the strategy DARE (Detect Automatically and Rejoin Efficiently) to help nodes survive under high churn. Finally section 6 concludes the paper and points out the future work.

2 Related Work

Liben-Nowell *et al.* [7] examine error resilience dynamics of Chord when nodes join/leave the system and derive lower bounds on the degree necessary to maintain a connected graph with high probability. Fiat *et al.* [8] build a Censorship Resistant Network that can tolerate massive adversarial node failures and random object deletions. Saia *et al.* [9] create another highly fault-resilient structure with $O(\log 3N)$ state at each node and $O(\log 3N)$ hops per message routing overhead.

Unfortunately, very few studies examine the resilience of existing graphs in comparison with each other, especially when nodes join/leave at a high rate. We are aware of only few comparison study, including that Gummadi *et al.* [11] find that ring-based graphs (such as Chord) offer more flexibility with route selection and provide the best resilience performance compared with some other DHTs routing algorithms, however they did not mention some other good DHTs like Kelips which shows very good experiment result in our simulations, Dmitri Loguinov *et al.* [14] do some theoretical analysis of the existing graphs, Jinyang *et al.* [12] compare several DHTs under churn in the aspect of look-up latency, and Simon *et al* [13] address the question of how high a rate of node dynamics can be supported by structured P2P networks, however they confine their study to hypercube only.

We began this work from 2004, and showed the resilience of ring topology of P2P overlay in [6]. Now we give more evidences to support the rationality of "crash point" and compare more topologies of structures other than ring, and finally give a strategy to help nodes survive better under high churn.

3 Analysis

One of the reasons DHTs are seen as an excellent platform for large scale distributed systems is that they are resilient in the presence of node failure. This resilience has two different aspects [11]:

Static resilience. We keep the routing table static, only delete the items of failed nodes to see whether the DHTs can route correctly without the help of stabilization routine.

Routing recovery. They repopulate the routing table with live nodes, deleting the items of failed nodes.

3.1 Static Resilience

Gummadi *et al.* [11] has concluded that flexibility is the most important factor that affects the performance of static resilience, and our simulation result is quite supportive to that conclusion. When basic routing geometry has been chosen, more flexibility means more freedom in the selection of neighbors and routes. Two cases are discussed respectively as follows:

Neighbor Selection. DHTs have a routing table comprised of neighbors. Some algorithms make purely deterministic neighbors (i.e., Koorde), others allow some freedom to choose neighbors based on other criterias in addition to the identifiers; most notably, latencies have been used to select neighbors. (i.e., Tapestry).

Route Selection. Given a set of neighbors, and a destination, the routing algorithm determines the choice of the next hop. However, when the determined next-hop is down, flexibility will describe how many other options are there for the next-hop. If there are none, or only a few, then the routing algorithm is likely to fail.

Chord, Kademlia, Klips provide both neighbor selection flexibility and routing selection flexibility. While Tapestry only provides neighbor selection flexibility, and Koorde has no flexibility in either neighbor selection or route selection. We can see their difference from simulation results present in the next section.

3.2 Routing Recovery

Three kinds of routing recovery strategies are usually used.

On demand recovery. This kind of recovery happens whenever outside environment asks the node to change. For instance, a neighbor informs you its departure, then you delete it from your neighbor list and replace it with a new neighbor.

Stabilization routine. This kind of process actively runs every certain period to eliminate the error in routing table.

Piggybacked recovery. Some protocols can use incoming messages like "look up request" to recover the routing table if necessary.

On demand recovery is useful and efficient under ordinary churn. However, it becomes useless and inefficient under high churn. On the other side, stabilization routine plays an important role under high churn. Although different stabilization policies are adopted by different DHTs, we adjust the parameters to let all the protocols run stabilization routine at the same frequency. While Kademlia, Kelips can use piggybacked recovery owing to their symmetric routing path, however Chord, Koorde, and Tapestry have no such routing recovery strategy because their routing path is asymmetric.

4 Simulation Results

In all the following simulations, we use a network of $N = 1000$ nodes, with average `round_trip_time`$(RTT) = 2s$ ($1s$ equals to 1000 simulation time units: ms) between any pair of nodes. Each node generates look-up requests at the interval exponentially distributed about the mean time of 10s, and the look-up requests are for randomly selected keys. At time 1800s, when the topology is supposed to be stable, we let a portion (from 0% to 70%, increasing 10% each time) of nodes leave the network simultaneously. Stabilization routine is trigged every 100s on average. During the simulation we record the average successful look-up ratio (SucRatio) every 30s. If a look-up does not reach the destination in 20s ($20 \approx \log_2 N * RTT$), we consider it as a failed look-up. Each simulation test case is repeated 5 times to avoid randomerror. All the simulations are run on p2psim [15].

For the rest of this section, we address two questions:

Question 1: *Does the crash point really mean something?*

We first explain why to choose 50% successful look-up ratio (SucRatio) as a sign of crash. SucRatio is easy to obtain and compute, and it has certain relationship between the degree of network connectivity. If a network crashes down, it will finally have impact on SucRatio. Let us suppose the network of size N is broken up into two equal-sized networks of size $N/2$, then, the look-ups within the subnetwork ($\frac{N}{2} \times \frac{N}{2}$) will succeed, and look-ups aimed to the other sub-network will fail. In average,

$$SucRatio = \frac{\frac{N}{2} \times \frac{N}{2} \times 2}{N \times N} \times 100\% = 50\%.$$

If SucRatio is much higher than 50%, we can expect that the network is still fully connected, but if SucRatio is much lower than 50%, the network is probably turned into pieces, thus the crash moment is not far.

Our assumption of crash sign (i.e., 50% SucRatio) is further verified by the simulation on Tapestry. Fig. 1 is the simulation result of Tapestry, it shows clearly that when 50% percent of nodes fail, the SucRatio reduces to 50% , and the network never recovers from the failure. When failed nodes are less than 40%, the network recovers very well. 50% is our best choice, however, you can change the percentage to redefine the "crash point" as long as it suits the application.

Question 2: *How do the crash points of various protocols compare?*

Here we choose five typical protocols: *Chord*, which is the representative of ring structure; *Tapestry*, which is the representative of tree structure; *Kademlia*, which implements an XOR structure; *Koorde*, which embeds a De Burijn graph into a ring structure, and *Kelips*, which claims to sacrifice memory for better lookup latency and resilience. In each protocol, every node is configured to have about 20 neighbors at the beginning, except Kelips-node has about 60 neighbors which are the least amount owing to its special join strategy.

Fig. 2 shows the successful look-up ratio (SucRatio) for varying percentage of nodes failures across different protocols. We can see that, although Kelips has much more neighbors, it does not do better than Chord. Kelips's crash point is around 70%, and Chord's crash point is behind 70% (From [11, 6] we can tell the crash point of Chord is around 80%). Tapestry has more neighbor selection flexibility than Koorde, while Koorde is based on a ring which is improved to have a better resilience than tree structure, so they have a close performance,

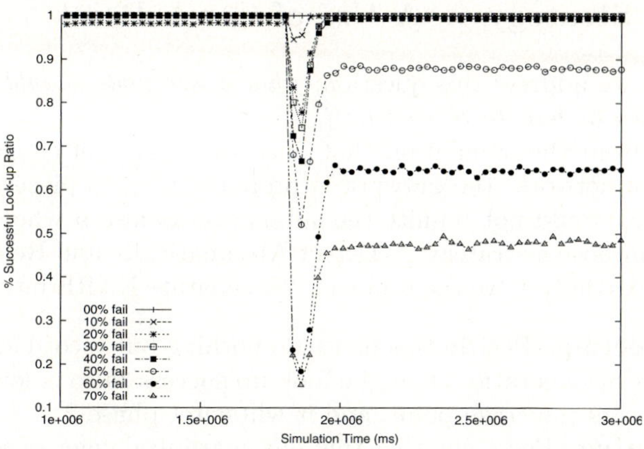

Fig. 1. Tapestry: Successful look-ups ratio for varying percentage of node failures

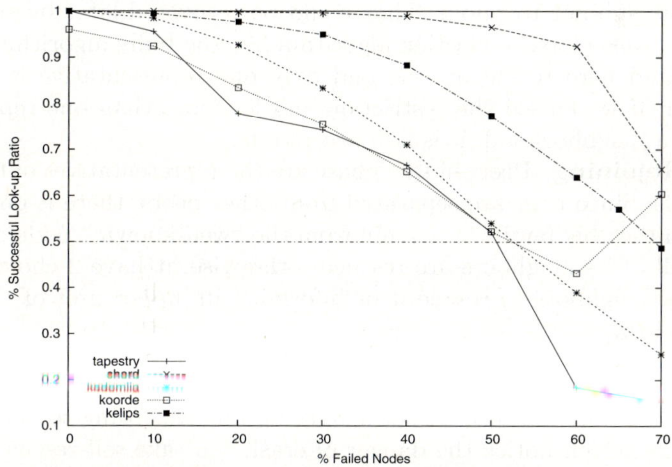

Fig. 2. Successful look-up ratio for varying percentage of node failures across different protocols

both have crash point around 50%. In comparison of Tapestry with Chord, we can conclude that flexibility in routing selection is more important than neighbor selection to improve the resilience. Kademlia is in the middle lever, however we can expect it to perform better in a larger scale network owing to its piggybacked recovery strategy. We notice that for Koorde SucRatio with 70% failure nodes is higher than that with 60%. That is because when a large percentage of nodes fail, there are much fewer look-ups generated, and some of them can quickly reach their destinations without suffering a long delivering path. However since the look-ups are very few, static method became inaccurate, that is why we stop the simulation with node failure portion at 70%.

5 Survive Strategy — A Use of Crash Point

In this section, we address this question: *what a live node should do to survive under high churn or how to rescue itself?*

We can see from the simulation that once the degree of churn exceeds the crash point of a network, the network is probably torn to pieces. The routing recovery strategy could not reunite the sub-networks into a whole all by itself. Thus we here propose a strategy — Detect Automatically and Rejoin Efficiently (DARE) — to partially solve the problem. We separate DARE into three phases:

Phase 1: Detecting. Peer in this phase do nothing but record and watch over the look-up success ratio. Once the look-up success ratio is lower than 50%, the node meets the crash point, and it will enter phase 2.

Phase 2: Electing. Peers entering this phase are in danger of isolation, they should try to rejoin from the "well-known" node(s) in order to unite again. However, aimed to reduce the overhead of rejoin, a representative in each sub-net will be elected to enter phase 3. If it succeeds, there is no need for the rest of its sub-net to rejoin. They could be recovered later by stabilization routine. Some classical election algorithm like the bully algorithm [16] could be adopted here to ensure one and only one representative is select out. However, if we loosen the restriction and let more than one representative exist in a neighborhood, it is still acceptable.

Phase 3: Rejoining. Peers in this phase are the representatives of their neighborhoods. Since they are separated from other peers, there is only one way to rejoin the "big family" — rejoin from the "well-known " node(s). If it succeeds, all of its neighbors are rescued, otherwise, it have 2 choices: waiting for other neighbors to rescue it or informing its upper user of the crash of the network.

DARE has several advantages. First, it can help nodes to discover the crash status automatically because we set a warning line according to "crash point". Second, nodes which notice the danger of crash can take self-rescue as early as possible. Third, only one node in the neighborhood will trigger the rejoin process, which can avoid a large number of concurrent joins and reduce the workload of well-known node(s).

6 Conclusion and Future Work

In this paper, we propose a measurement of "crash point" to compare the "resilience under high churn" across some typical structured P2P protocols, and present a strategy to help the alive nodes survive under high churn. It shows that Chord with ring topology has the best resilience under high churn; crash point is sensitive to flexibility in routing selection but not very sensitive to the amount of neighbors.

In the future work, we will study more about the relationship between SucRatio and the connectivity of the network. We will also develop DARE to show its performance, hoping to find a better way to discover the danger of network disconnection, and prevent the network to be separated.

Acknowledgement

The work is partly supported by China NSF grant (No. 60573131), Jiangsu Provincial NSF grant (No. BK2005208), China 973 project (No. 2002CB312002), and TRAPOYT award of China Ministry of Education. We also thank Jun Li, Wentao Zheng and all the other reviewers for their valuable suggestions.

References

1. I. Stoica, R. Morris, D. Karger, M.F. Kaashoek, and H. Balakrishnan: Chord: A Scalable Peer-to-Peer Lookup Service for Internet Applications. ACM SIGCOMM (2001)
2. B.Y. Zhao, J.Kubiatowicz, A. D. Joseph: Tapestry: a fault-tolerant wide-area application infrastructure. Computer Communication Review 32(1): 81 (2002)
3. I. Gupta, K. P. Birman, P. Linga, A. J. Demers, R. van Renesse: Kelips: Building an Efficient and Stable P2P DHT through Increased Memory and Background Overhead. IPTPS (2003)
4. P. Maymounkov and D. Mazieres: Kademlia: A peerto -peer information system based on the xor metric. In Proceedings of IPTPS '02,(2002)
5. M. F. Kaashoek and R. Karger: Koorde: A simple degreeoptpimal distributed hash table. In 2nd International workshop on P2P Systems(IPTPS03),(2003)
6. Z. Liu, G. Chen, C. Yuan, S. Lu, C.Xu: Fault Resilience of Structured P2P Systems. WISE (2004)
7. D. Liben-Nowell, H. Balakrishnan, and D. Karger: Analysis of the Evolution of Peer-to-Peer Networks. ACM PODC (2002)
8. A. Fiat and J. Saia: Censorship Resistant Peer-to-Peer Content Addressable Networks. ACM/SIAM Symposium on Discrete Algorithms (2002).
9. J. Saia, A. Fiat, S. Gribble, A.R. Karlin, and S. Saroiu: Dynamically Fault-Tolerant Content Addressable Networks. IPTPS (2002).
10. J. Aspnes, Z. Diamadi, and G. Shah: Fault-Tolerant Routing in Peer-to-Peer Systems. ACM PODC (2002)
11. K.P. Gummadi, R. Gummadi, S.D. Gribble, S. Ratnasamy, S. Shenker, and I. Stoica: The Impact of DHT Routing Geometry on Resilience and Proximity. ACM SIGCOMM (2003)
12. J. Li, J. Stribling, T. Gil, R. Morris, F. Kaashoek: Comparing the performance of distributed hash tables under churn. IPTPS (2004)
13. S.S.Lam and Huaiyu Liu: Failure Recovery for Structured P2P Networks: Protocol Design and Performance Evaluation. ACM SIGMETRICS/Performance '04 (2004)
14. D. Loguinov, A. Kumar, V. Rai, S. Ganesh: Graph-Theoretic Analysis of Structured Peer-to-Peer Systems: Routing Distances and Fault Resilience. ACM SIGCOMM (2003)
15. http://pdos.lcs.mit.edu/p2psim/howto.html
16. Garcia-Molina, H.: Elections in a Distributed Computing System. IEEE Transactions on Computers, Vol. C-31, no. 1, (January),(1982) pp. 48-59.

Analyzing Peer-to-Peer Traffic's Impact on Large Scale Networks*

Mao Yang, Yafei Dai, and Jing Tian

Peking University, Beijing, P.R. China
{ym, dyf, tj}@net.pku.edu.cn

Abstract. Currently, peer-to-peer file sharing systems are playing a dominant role in the content distribution over Internet. Therefore understanding the impact of peer-to-peer traffic on large scale networks is significant and instrumental for the design of new systems. In this paper, we focus on Maze, one of most popular P2P systems on CERNET. We perform a systematic characterization of Maze's [1] traffic impact on CERNET. We investigate the traffic volume and bandwidth on different spatial levels aggregation. According to our log-based analysis, we claim that current P2P systems have much room to improve in reducing backbone network consumption. Locality-aware content delivering mechanism can reduce the traffic on backbone network effectively. Moreover, a system with less free-rider[3] will further reduce the traffic consumption. Thus the designers of P2P system should pay more attention on incentive mechanism to reduce free-rider.

1 Introduction

Currently, peer-to-peer file sharing systems are playing a dominant role in the content distribution over Internet. The dramatically increasing traffic make the ISPs worry about the abuse of backbone bandwidth by P2P systems. In this paper, we analyze P2P traffic based on public traffic log dataset of Maze. Maze[2] has been one of the largest non-commerce P2P file sharing system over CERNET (China Education and Research Network), which is developed, deployed and operated by our academic research team. Based on its open log dataset, we can leverage Maze as a large-scale measurement platform. CERNET is an ISP which connects thousands of universities and research institutes throughout China. In CERNET, a university can be regarded as an intranet with high bandwidth. We named these intranets *zones* and the *zones* are linked by the backbone network of CERNET. There are more than 200,000 active MAZE users on CERNET every month, thus we think Maze can be an ideal platform for our measurement and analysis.

The goal of our work is to help people to understand the characteristics of P2P traffic across large scales networks and its impact on Internet backbone network. Besides, we also want to find some mechanisms that can save the backbone bandwidth utilization. The followings are of the interesting questions that we want to research and understand:

* Supported by National Grand Fundamental Research 973 program of China under Grant No.2004CB318204; This work is partially supported by an Intel Sponsored Research Project.

V.N. Alexandrov et al. (Eds.): ICCS 2006, Part IV, LNCS 3994, pp. 412 – 419, 2006.
© Springer-Verlag Berlin Heidelberg 2006

a) How do the P2P users distribute across internet on different spatial (User, IP, and zone) levels? b) What is the characterization of P2P traffic volume and traffic bandwidth? c) Do current P2P applications waste too much backbone bandwidth?

Different from previous work, we adopt a new methodology in our research. We aggregate users by zones, because zone is a more suitable level than IP-prefix or AS in analyzing the traffic impact on ISPs' backbone network. To our best knowledge, our study is the first research work base on zone level. Further more, we have most detailed logs on file transfer transactions, which enables us to perform a systematic measurement and an accurate simulation.

Based on our analysis results, we learn the following lessons:

1. The traffic volume distributions are different on different level aggregation.

2. The intra-zone bandwidth is stable and high while the inter-zone bandwidth is unstable and low. The average bandwidth decreases when the traffic is heavy.

3. The current P2P systems' content delivery model wastes much backbone bandwidth. Then we discuss potential improvement in saving backbone bandwidth of three mechanisms and claim that the P2P system is a good application model for ISPs if the systems can reduce the number of free-rider and adopt some local-aware content delivery mechanism.

The paper is structured as follows. First, we describe our research methodology in Section 2. We then discuss the host and traffic distribution in Section 3. In Section 4, we analyze the traffic impact on backbone network. Section 5 is the related works. Finally, we conclude in Section 6.

2 Methodology

Though continuous logs of Maze traffic are maintained, we perform our analysis on a log segment gathered during the period of three weeks from 09/09/05 to 09/30/05. During this period, more than 190,000 active users participated in more than 26 million file transfers. The total data traffic volume exceeded 460 Terabytes.

Table 1. Data set of Maze Traffic

Log during	9/9/2005 - 9/30/2005
# of records	26,615,75
# of unique users	190,645
# of unique IP	369,724
Total traffic volume (GBytes)	460,000
Average flows / Second	253MBytes

The data gathered for this study consists of a collection of user points during this period and the detailed traffic log. When two peers report the completion of a file transfer to the server, our log keeps only the data from the uploading peer. Each traffic log entry contains the following: *uploading peer-id*, *downloading peer-id*, *log upload time (server)*, *transfer start time (source)*, *transfer end time (source)*, *bytes transferred*, *file size*, *download peer's IP*, *upload peer's IP*, and *file md5 hash*. The bytes transferred can be different from the file size if the transfer was interrupted, or if the transfer is sourcing from multiple peers.

2.1 Map the IP Address to Locations

The 164,056 (86%) users and 152,136 IP addresses in Maze come from CERNET and this paper only analysis the users on CERNET. We aggregate IP addresses into Zones by using the WHOIS service of CERNET. As we have mentioned before, a zone refers to an intranet of a university / college or a research institution etc, and so the intra-zone transfers always have high bandwidth and do not consume any backbone bandwidth. The addresses space of CERNET currently spans 2752 zones. Most zones only own less than 32*256 IP address. There are only three zones own a whole B-Class IP addresses space.

3 Host and Traffic Distribution

3.1 Host Distribution

Figures 1 plots the cumulative distributions of users and IPs associated with Users ranked in decreasing order of number. There are 170000 unique users span on 846 different zones. We observe skews in the distributions of users after the zone aggregation. 52% users are in the top 20 zones. The same thing discovered in the users' IP addresses, 50% unique IP addresses are in the top 20 zones. The Figure 2 illustrates the number of unique IP address verses the number of unique users in each zones. The average the host density (# of user / # of IP) is 1.07. We observe that are especial large which means the users on these zones are using NATs.

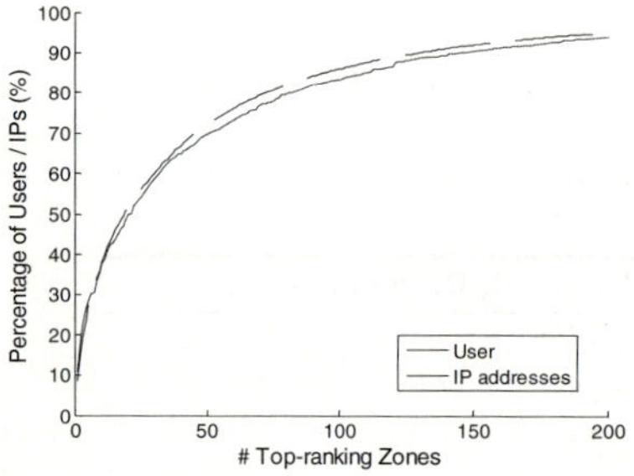

Fig. 1. The cumulative distributions of users and IPs associated with Users ranked in decreasing order of number

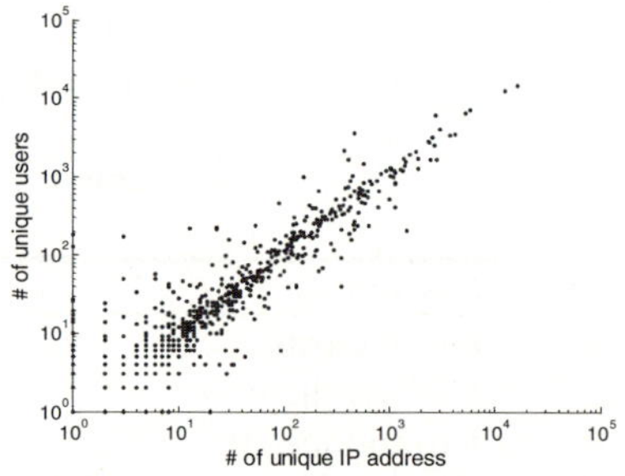

Fig. 2. The number of unique IP address vs. the number of unique users

3.2 Traffic Volume Distribution

To understand the impact of P2P systems on backbone of CERNET, we should analyze the distribution of the traffic volume. Because of the distribution of IP addresses is similar to user distribution, we analyze the traffic volume distribution only from user layer and zone layer. There is 36.20% intra-zone traffic volume and the other traffic is among zones which consumes the backbone bandwidth.

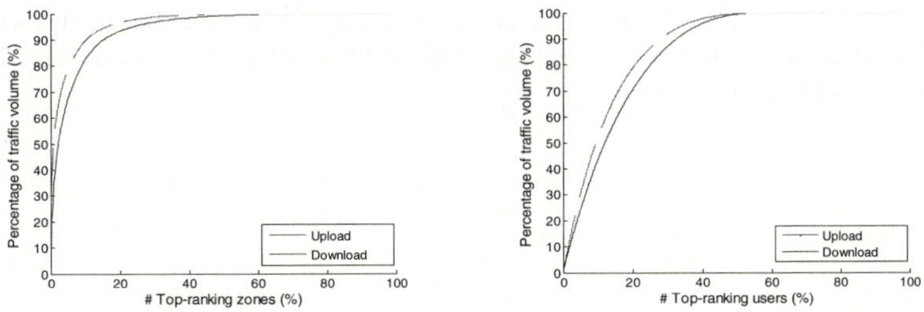

Fig. 3. The cumulative distributions of inter-zone traffic volume associated with Users / Zone level ranked in decreasing order of volume. Left: aggregate on user, Right: aggregate on zone.

Figures 3 illustrates inter-zone traffic volume aggregated on user level. Top 50% users are responsible for the whole download and upload traffic volume, and the top 20% users are responsible for over 50% traffic volume. We observe the distribution of upload volume is more skewed than the download volume, which means there are more super nodes acting as server in the system. Neither distribution of upload nor download volume follows the Zipf's law, which is not a straight line in log-log scales.

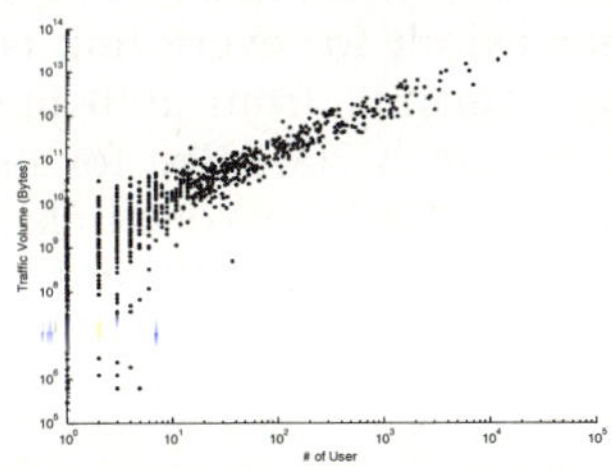

Fig. 4. Number of user vs. the total intra-zone traffic volume for each zone

What surprises us is that the distribution is quit different from user level. There are fewer zones (30%) providing the upload and more zones (60%) are responsible for download. As we investigate, this problem is cause by the limitation of IP addresses in CERNET. The users which are behind NATs or firewalls can hardly be accessed by users from other zones. Many zones in CERNET have firewalls because of the limited IP addresses space, so the users from these zones cannot not serve the user in other zones.

Figure 4 present the number of user versus the total intra-zone traffic volume for every zone. The zone which has more users induces more intra zone traffic. This is reasonable: the current P2P file sharing systems including Maze adopt a bandwidth-first mechanism to encourage user to download from proximity in a higher priority.

3.3 Bandwidth Characteristics

This subsection discusses the bandwidth characteristics of P2P systems (both on user layer and zone layer). We define the average bandwidth between a pair user as: If there were n files transferred between two peers and the file size is S_i , the transfer time is D_i. We define the average bandwidth (or transfer speed) between two peers: *AverageBandWidth* = *sum*(S_i) / *sum*(D_i). This metric help us to understanding the traffic quality between each peer.

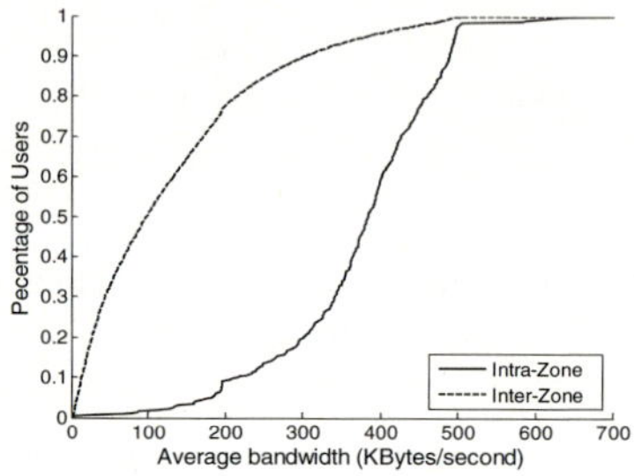

Fig. 5. The cumulative distributions of average bandwidth

Figure 5 plots the cumulative distributions of average bandwidth of intra-zone links and inter-zones links. In current P2P file sharing systems, the uploading peers always limit uploading bandwidth to the downloading peers. In Maze, the free-rider's downloading bandwidth will be limited to less than 200KBps by uploading peers, and the default max bandwidth for every link is 500KBps. Thus, the intra-zone average bandwidth ranges mainly from 200Kbps to 500kBps. We observe there is only 10% bandwidth less than 200KBps for intra-zone links, and the percentage increases to more than 70% for inter-zone links. The transfer intra-zone can provide more high speed service, and the P2P systems might enough peers exchange their data intra-zone.

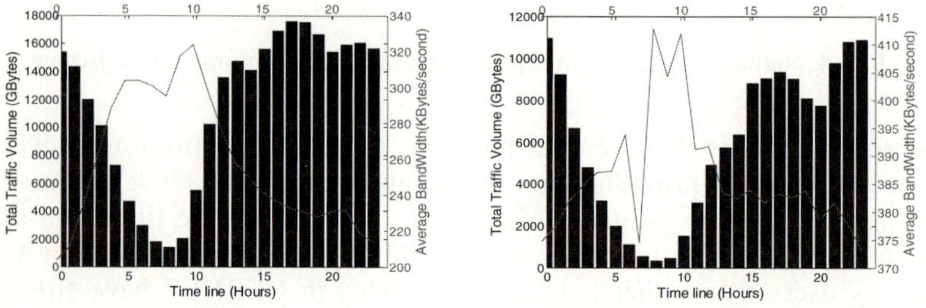

Fig. 6. The total traffic volume and average bandwidth aggregate by hours. Left: Intra-zone right: inter-zone.

We observe the traffic volume has a strong daily pattern (Figure 6.). The max flux was observed on 3 pm and 11 pm. The average bandwidth has strong correlations with the traffic volume, especial for the inter-zone traffic. When the network workload is heavy, the bandwidth for P2P decreases 10% for intra-link and 30% for inter-link, which means the bandwidth intra-zone is more stable.

4 The Traffic Impact on Backbone Network

In this section, we focus on the traffic impact on backbone network. A large scale P2P file sharing system consists of millions of users in an overlay network. Users exchange their content to each other in the overlay network. If two users delivery their content between different zones, it will consume the backbone bandwidth of ISPs.

The current P2P systems such as BitTorrent and Maze support some inner mechanisms to reduce the abuse of backbone network. The basic mechanism is bandwidth-first mechanism. It means when a peer are upon a selection of potential uploading peers, it try to select the peers who have higher bandwidth links to it. It is an approximate mechanism to implement locality-aware mechanism which lets user adapts file downloading to match the physical networks. As we know, there are more than 60% traffic is from external peers in Maze. Are there any potential in saving backbone bandwidth?

We propose some mechanisms which can reduce traffic on backbone as follow:

Origin mechanism (used by Maze): When peer Alice requests a file, the central index server will tell this peer some peers who have this file. Alice will request this file from those peers and download with bandwidth-first mechanism.

A) Locality-aware mechanism: If there are some online peer has this file in the same zone with Alice, Alice prefers downloading this file from those local peers.

B) Locality-aware on No free-rider condition: free-rider refers to the peer does not upload any file to other peer even if he has some content. If peer Bob downloaded a file, we assume he wills storage this file more than one month and can service this file to other peers whenever he is online. These assumptions will help us to understand the impact of free-rider to P2P systems.

C) Perfect proxy mechanism: We assume there are perfect proxies in every zone. When peer Bob download a file outside the zone, the proxy of this zone will cache this file forever. Alice need not download file outside of zone again if the proxy has it. This mechanism is like the traditional CDN (Content Distribution Network) solutions.

To understanding the potential of those proposed mechanisms, we conduct a trace-driven simulation. The steps of our simulation are: a) Sort the whole transfer record logs based on transfer start time. b) Parse the transfer record, when peer Alice downloads File *f* from peer Bob from time *t1* to time *t2*, we assume Peer Alice has owned file *f* in time *t2* and Bob has owned file *f* in time *t1* c) Read the transfer records one by one, when peer Alice requests a file *f* at time *t*, perform the three optimization mechanisms, calculate the hit rate (by volume) in local zone. The detail as follow:

A) If there are some other peers is in the same zone with Alice and they are uploading *f* at time *t*, Alice apt to download only from those peers. We called locality-ware mechanism.

B) If there are some other peers is in the same zone with Alice and they have owned *f* before time *t*, Alice apt to download only from those peers.

C) If the proxy of Alice's zone has owned *f* before time *t*, Alice apt to download only from the proxy.

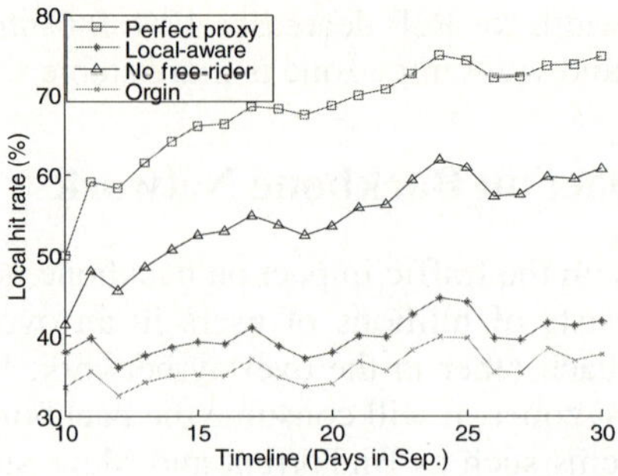

Fig. 7. The percentage hit rate in local zone in timeline (days)

Figure 7. plot the average percentage hit rate in local zone, the hit rate means the percentage of traffic volume just on local zone. The original mechanism has a stable hit rate around 36.20%, which means that more than 60% traffic occurs on of the backbone network. The locality-aware mechanism will increase the hit rate by 5% percentage. It also has a stable hit rate.

If there are no free-riders in the system, the hit rate will have a sharp increase. The average hit rate is 54.80%, and the hit rate will increase following the time from 40% to above 60%. This demonstrates if all the users maintain the download files more than 3 weeks, the hit rate will exceed 60%. Unfortunately, all of P2P file sharing systems have the free-rider problem. Many users remove their downloaded file from system and refuse to service other users. Our experiment proves that the potential improvement of P2P system.

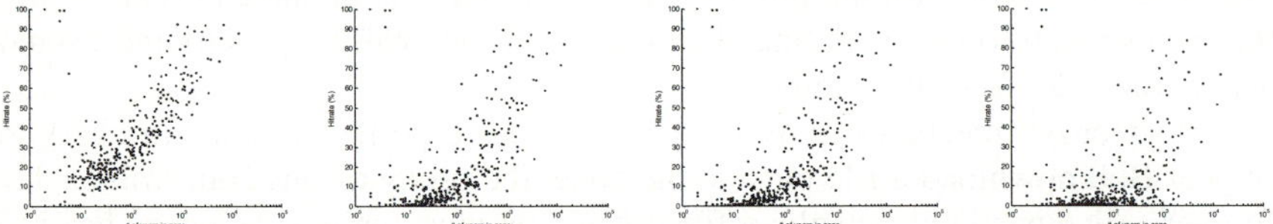

Fig. 8. The hit rate for every zone (*x*-axis is the number of user in a zone). Left to right (Original, Locality-aware, No free-rider condition, Perfect proxy).

The perfect proxy mechanism is an ideal model, and it shows the upper bound of improvement. The average hit rate is 68.19%, and the hit rate reaching 75% at the end of simulation.

Figure 8. illustrates the hit rate in different zones. We find that the hit rate is influenced by the zone size. The zones with large population have higher hit rate. Some large zones have hit rate more than 50% even in original mechanism. This demonstrates a large population zones are friendly to ISPs, they do not need too much backbone network bandwidth. The zone with low population can hardly get a high hit rate even in perfect CDN model. We also observe the high population zone has the similar hit rate in no free-rider condition with the perfect proxy. This demonstrates if there are enough users in a zone, there is high probability that somebody has your desired.

We conclude several optimizations from this simulation can be done by the designers of P2P system: a) Design a better locality-aware download mechanism. Tell peers the replica of file in his / her local zone, the bandwidth-first mechanism is not accurate. b) Design some incentive systems to reduce the free-riders in the system, which can encourage users not to remove the downloaded file away from system and encourage users to stay online longer. c) Encourage new users to join in the network, or place some super nodes in medium or small size zones. The P2P system in a large scale zone will consume less percentage backbone bandwidth than ISPs supposed.

5 Related Works

Subhabrata et al. [4] analyze the P2P system traffic in (IP, prefix, AS) level, and focus on the workload model on flew-level data. Several measurement studies have characterized the basic traffuc of P2P. Saroiu et al. [5] analyzed the behaviors of peers inside the Gnutella and Napster. Krishna et at. [6] demonstrate that KaZaA traffic did not exhibit Zipf-like behavior. Thomas et al. [7] is most similar work with us. They aggregate BitTorrent users on AS level, and demonstrated the "locality-aware" solution will reduce the bandwidth usage between ISPs.

6 Conclusion

Through the analysis of Maze's log data, we achieved a comprehensive understanding of the characterizations of P2P system's traffic volume and bandwidth. We also analyze the P2P traffic impact on the backbone network. We conclude that the current P2P systems consume too much backbone bandwidth, but the situation can be improved. And ultimately P2P system is a good solution for content delivery.

References

1. http://maze.pku.edu.cn.
2. Mao Yang, Ben Y. Zhao, et al. Deployment of a large scale peer-to-peer social network, Proceedings of the 1st Workshop on Real, Large Distributed Systems.
3. M. Yang, Z. Zhang, X. Li, Y. Dai, "An Empirical Study of Free-Riding Behavior in the Maze P2P File-Sharing System," In *Proceedings of IPTPS*, Ithaca, NY. February 2005.
4. S. Sen and J. Wang. Analyzing peer-to-peer traffic across large networks. In Proceedings of the Second SIGCOMM Internet Measurement Workshop (IMW 2002), Marseille, France, November 2002.
5. S. Saroiu, P. K. Gummadi, and S. D. Gribble. A measurement study of peer-to-peer file sharing systems. In Proceedings of Multimedia Computing and Networking (MMCN) 2002, January 2002.
6. Krishna P. Gummadi, Richard J. Dunn and et al. "Measurement, Modeling, and Analysis of a Peer-to-Peer File-Sharing Workload". Proceedings of the 19th ACM Symposium on Operating Systems Principles (SOSP-19), Bolton Landing, NY.
7. Thomas K., Pablo R., et al. Should Internet Service Providers Fear Peer-Assisted Content Distribution? Internet Measurement conference 2005.

Analyzing the Dynamics and Resource Usage of P2P File Sharing by a Spatio-temporal Model

Riikka Susitaival, Samuli Aalto, and Jorma Virtamo

Helsinki University of Technology,
P.O. Box 3000, FIN-02015 TKK, Finland
{riikka.susitaival, samuli.aalto, jorma.virtamo}@tkk.fi

Abstract. In this paper we study the population dynamics and resource usage optimization of a P2P file sharing system, where the availability of the requested file is not guaranteed. We study the system first by a deterministic fluid model and then by a more detailed Markov chain analysis that allows estimating the life time of the system. In addition, the underlying topology of the network is modelled by a simple geometry. Using the resulting spatio-temporal model we assess how much the resource usage of the network can be reduced, e.g., by selecting the nearest seed for download instead of a random one.

1 Introduction

Peer-to-peer (P2P) applications, such as file sharing, have become a significant area of Internet communication in recent years. Older examples of these applications are Gnutella, Napster and Kazaa, whereas BitTorrent is currently the most popular system. It has been widely reported that P2P related traffic forms a major part of the total traffic in the Internet. From an operator's point of view it is important that the traffic load produced by P2P applications does not encumber the underlying network too heavily. Efficient usage of the network resources would also improve the service of the individual peers by shortening average latencies.

We concentrate on BitTorrent-like P2P protocol because of its popularity but the results are applicable to other protocols as well. The idea of BitTorrent is to divide the file to be distributed into parts, named *chunks*, so that different parts can be downloaded from several peers simultaneously, where the size of the chunk is typically 256 KB, see for technical details of BitTorrent in [1]. Measurement studies [2], [3], [4], have shown that the evolution of a single file in the system can be divided into three phases. In the first *flash crowd* phase the demand for the newly released file is high. It is followed by a *steady state* and finally, the *end* means the death of the file.

A few papers have analyzed P2P file sharing systems by stochastic models so far. In paper [5], the analysis of BitTorrent-like system is divided into transient and steady state regimes. The service capacity of the transient regime is studied by a branching process and the steady state by a Markov model. Paper [6] studies

V.N. Alexandrov et al. (Eds.): ICCS 2006, Part IV, LNCS 3994, pp. 420–427, 2006.
© Springer-Verlag Berlin Heidelberg 2006

the performance of the system by a deterministic fluid model, whereas in paper [7] the network level latencies are modeled by the delay of a single class open queueing network and peer level latencies by the delay of M/G/1/K processor sharing queues. However, these models do not capture all aforementioned phases of the sharing process, namely flash crowd, steady state and especially end phase.

In this paper we study the dynamics of sharing a chunk, that is, a single piece of a file, in a P2P system. First we model the system by a deterministic fluid model and study the dynamics of the average number of downloader and seeds over time. The deterministic fluid models are, however, unable to capture all the details of the chunk sharing process such as possible unstability and extinction of the system. For this reason we construct a complete Markov chain model to obtain more information of the life cycle of chunk sharing process.

By providing the downloaders and seeds with location information we study further how the selection of the peer has an effect on the resource usage in the network. We propose a spatio-temporal model for the P2P system, in which the topology of the Internet is abstracted by a sphere, on which peers are located. Distance metric between two peers in terms of delay or bandwidth is assimilated with their geometrical distance. We consider two different peer selection policies; in the first one a random seed is selected whereas in the second the nearest one is searched. Expected values for the capacity usage for these two peer selection policies are derived and also the dynamics of the system is studied by simulations.

The paper is organized as follows: In section 2 population dynamics of the system is studied by a fluid model. Then a Markov chain model for calculating the time to extinction is constructed in section 3. In section 4 the geometric approach for modeling of chunk sharing is introduced and different peer selection policies are compared. Finally, in Section 5 we conclude our paper.

2 Deterministic Fluid Model for Chunk Sharing

In this and next sections, we analyze the population dynamics of the sharing of a single chunk of a file. We study how the number of downloaders and seeds evolves over time from the emergence of the chunk to the disappearance of it. The disappearance of a single chunk means the death of the whole file sharing process since the file is not entire anymore. The work is motivated by the model of [6] but has some differences. In paper [6] the problem of sharing of several chunks concurrently is solved by assuming that peers can forward the chunks with a constant rate. However, we find the assumption unrealistic and the model probably hides some details of the population dynamics. For this reason we consider the sharing of a single chunk at a time. In addition, among others, papers [5] and [6] assume that at least one seed stays in the system keeping the chunks available. However, measurements of BitTorrent show that the file sharing process dies sooner or later [3]. Therefore the life time of the process is also studied.

In the model, new requests for a chunk are assumed to arrive at the system with rate λ according to the Poisson process. The downloader can download the file with rate μ_d. On the other hand, the maximum upload rate of a peer for the chunk is assumed to be μ_s. After the download, the status of the downloading peer changes from a *downloader* to a *seed* and the peer can distribute the chunk forward. Note that in this context, a peer is referred to as the seed if it has the chunk in question, but not necessarily all chunks of the file. The seed leaves the system with the probability γ per time unit. Let $x(t)$ be the number of downloaders and $y(t)$ be the number of seeds at time t. In the next sections we study the evolution of the pair (x, y) both by a deterministic fluid model but by a Markov model as well.

We consider a system where a peer starts to spread a single chunk to other peers that are willing to download it. If $\mu_d x(t) < \mu_s y(t)$, the downloaders can not use all service capacity provided by the peers. On the other hand, when $\mu_d x(t) > \mu_s y(t)$ the upload capacity of seeds limits the download process. Thus the total service rate of the system is $\min\{\mu_d x(t), \mu_s y(t)\}$. First we construct a deterministic fluid model for the number of downloaders $x(t)$ and seeds $y(t)$:

$$\begin{aligned} \frac{dx(t)}{dt} &= \lambda - \min\{\mu_d x(t), \mu_s y(t)\}, \\ \frac{dy(t)}{dt} &= \min\{\mu_d x(t), \mu_s y(t)\} - \gamma y(t), \end{aligned} \tag{1}$$

where $y(0) = 1$ and $x(0) = 0$. Let $\bar{x}$ and $\bar{y}$ be possible equilibrium values of $x(t)$ and $y(t)$. If $\mu_d \bar{x} \leq \mu_s \bar{y}$, the steady state solution is $\bar{x} = \lambda/\mu_d$ and $\bar{y} = \lambda/\gamma$. From the constraint $\mu_d \bar{x} \leq \mu_s \bar{y}$ we obtain the condition for the equilibrium: $\mu_s \geq \gamma$. When $\mu_s < \gamma$ the solution of the equations (1) is $\bar{y} = 0$ and $\bar{x} \to \infty$.

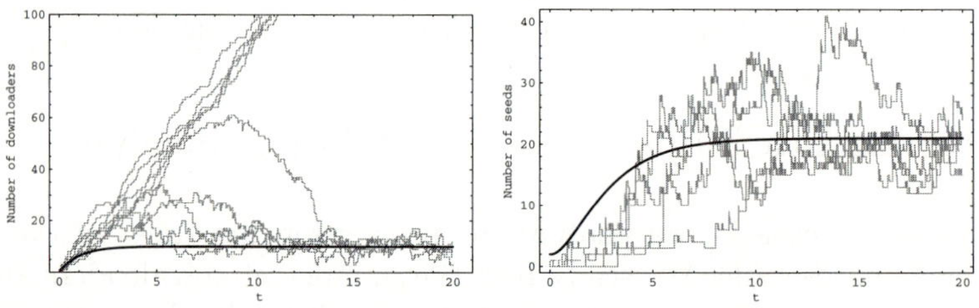

Fig. 1. The number of downloaders on the left side and the number of the seeds on the right side as a function of time (in units of $1/\mu_d$). Solid lines: fluid model (1), gray lines: simulation. $\lambda/\mu_d = 10$, $\lambda/\gamma = 20$, $\mu_s \geq \gamma$.

The evolution of the number of downloaders and seeds is depicted in Figure 1. We have fixed λ/μ_d and λ/γ to moderately small values in order to better demonstrate the dynamics of the system. The solid line corresponds to the solution of fluid model (1) and the gray lines to 10 different simulations. We can see that in the beginning the capacity of the system is not sufficient to serve chunk requests. This is seen as a dramatic increase in the number of downloaders. However, after some downloaders have changed their status to seeds, the system stabilizes. At the end time ($t = 20$) 4 simulated processes of 10 have become extinct and the number of downloaders in those processes increases without any limit.

3 Markov Chain Model for Chunk Sharing

The deterministic fluid model of the previous subsection describes the average behavior of the sharing of the chunks. However, from the simulation results we saw two effects in the population dynamics that were not captured by the fluid model. First, when the chunk became available the seeds could not serve all the downloaders, and second, if the original seed can leave the system, the death of the chunk and the whole file sharing process is irrevocable, even if $\mu_s > \gamma$. The limited life span of the file sharing process has an influence on the performance of the system and has to be analyzed by some other models. To this end, in this subsection we study the evolution of the process (x, y) in more detail by a Markov chain model with absorbtion. We construct a continuous time Markov chain process, where the state is the pair (x, y) and the transition rate matrix is Q with the elements:

$$\begin{aligned} q((x,y),(x+1,y)) &= \lambda, \\ q((x,y),(x-1,y+1)) &= \min\{\mu_d x, \mu_s y\}, \quad \text{if } x > 0, \\ q((x,y),(x,y-1)) &= \gamma y, \qquad\qquad\qquad \text{if } y > 0. \end{aligned} \tag{2}$$

The states (x, y) with $y = 0$ in the Markov chain are absorbing states. Since we are not interested in the process after entering one of the absorbing states, we combine all of them into one state 0. The mean time to absorbtion can be determined as follows: Let b_i denote the mean time to absorbtion, when the system starts from state i. Given the transition matrix Q, the mean times to absorbtion b_i are determined by a familiar Markovian recursion:

$$b_i = \frac{1}{q_i}(1 + \sum_{j \neq i} q_{i,j} b_j), \tag{3}$$

where $b_0 = 0$ and $q_i = \sum_{j \neq i} q_{ij}$. The absorbtion time starting from the initial state $(0, 1)$, i.e. the life time of the system, as a function of λ/γ is shown in the left side of Figure 2. The solid line is calculated by solving the set of linear equations (3) numerically in a truncated state space of 35×35 states. The dots are obtained from simulation of the corresponding infinite system verifying the analytical results. The figure shows that the system life time increases exponentially as a function of the expected number of the seeds λ/γ in the system.

In one limit case the absorbtion time can easily be approximated. When the mean service times $1/\mu_s$ and $1/\mu_d$ are very small, the system can be modelled as an M/M/∞-queue with arrival rate λ and departure rate γ. The mean time to absorbtion equals the average length of the busy period $E[B]$ of M/M/∞-queue:

$$E[B] = \frac{1}{\lambda}(e^{\lambda/\gamma} - 1). \tag{4}$$

The approximation and the analytical result from the Markov model are depicted on logarithmic scale on the right side of Figure 2. For $\mu_s = \mu_d = 100$ the approximation coincides with the analytical result.

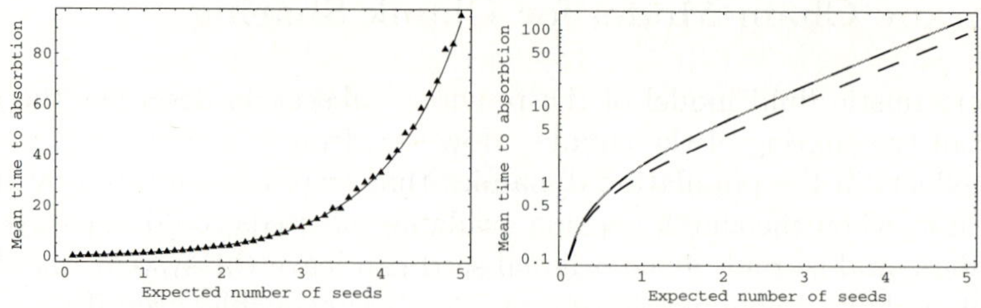

Fig. 2. Left figure: The mean time for absorbtion (in units of $1/\lambda$), solid line: analytical results, dots: simulation. Right figure: Analytical results for $\mu_s = \mu_d = 1$ (dashed line) and $\mu_s = \mu_d = 100$ (upper line) and approximation (upper line, overlapping with $\mu_s = \mu_d = 100$ line).

4 Location-Based Chunk Sharing Model

Our next objective is to study the possible reduction in network resource usage by a location-based peer selection policy, as opposed to random selection of the seed. We analyze location-aware sharing of a single chunk in a simplified setting where the underlying topology of the network is eliminated and replaced by a simple geometrical structure. By this approach we are able to estimate the capacity usage analytically.

As before, new requests for a chunk arrive in the system with rate λ according to the Poisson process. Each new request is associated with a peer i, whose location is assumed to be randomly chosen on the surface of a sphere following uniform distribution. We have chosen the spherical geometry primarily because it is symmetrical and has no artificial boundaries. It is also a natural choice if one considers a global network. Let R be the radius of the sphere and let the location of peer i be described by cylindrical coordinates z_i and ϕ_i. It is easily verified that if $z_i = -R + 2Ru$ and $\phi_i = 2\pi u'$, where u and u' are drawn from the uniform distribution $U(0,1)$, the peers are uniformly located on the sphere.

Let $D(t)$ be the *set* of downloaders and $S(t)$ be the *set* of seeds at time t. Let parameter p_i denote the selected seed j of downloader i. As a metric for distance between two peers i and j we use the shortest path between the peers on the surface of the sphere, denoted by $d_{i,j}$.

How much downloading a chunk consumes resources of the underlying network is assumed to be proportional to the distance between the peers exchanging chunks. If the peers are far apart, transferring a chunk typically needs more links than in the case of two close peers. Let $c(t)$ denote the total instantaneous capacity required for sharing chunks at time t, $c(t) = \sum_{i \in D(t), j=p_i} d_{i,j}$, i.e., $c(t)$ describes the sum of distances between the peers sharing the chunk. However, when we consider the resource usage optimization, a more interesting quantity is the average capacity usage C *per downloaded chunk* over time period $[t_0, t_{max}]$ defined as $C = \frac{1}{n}\int_{t_0}^{t_{max}} c(t)dt$, where n is the number of the chunks transferred within this period.

We consider two different peer selection policies: *Random peer selection* (RPS) policy, where the seed for download is selected randomly among all available peers, and *nearest peer selection* (NPS) policy, where the nearest possible peer in terms of the distance between the peers is selected.

4.1 Analytical Bounds for Capacity Usage

In RPS, each downloader selects one random seed. The distance to a random seed is independent of the number of seeds. Assuming, without loss of generality, that the mean download time $1/\mu_d$ of a chunk is one, the expected resource usage per a downloaded chunk is equal to the average distance between two points on a sphere (assumed to have unit area): $E[C] = E[d] = \sqrt{\pi}/4$.

In NPS, the nearest peer among $y(t) + 1$ seeds is selected for download. If N points are randomly distributed on a sphere with unit area, the expected distance to the nearest neighbor can easily be determined,

$$E[d|N = n] = \frac{\Gamma(n - \frac{1}{2})}{2\Gamma(n)} = \frac{\sqrt{\pi}}{2} \prod_{i=0}^{n-2} \frac{i + \frac{1}{2}}{i + 1}, \tag{5}$$

which is very accurately approximated by $E[d|N = n] \approx \frac{1}{2\sqrt{n-0.73}}$, with a maximum error of only 0.16% occuring at $n = 4$. At time t, N includes $y(t)$ seeds and the downloader itself, meaning that $N = y(t) + 1$. The expected resource usage for NDP policy is:

$$E[C] = \sum_{y=0}^{\infty} p\{Y = y\} E[d|N = y + 1]. \tag{6}$$

In general, the steady state distribution of $y(t)$, $p\{Y = y\}$, can be calculated from the Markov model of section 3. Due to complexity of the model, the solution cannot be expressed in a closed form. However, in a case when the service is always constrained by download rate and at least one peer stays in the system, the system of downloaders and seeds can be considered as two consecutive M/M/∞ queues, where arrival rates to the first and second queues are λ and the service rates are $x\mu_d$ and $y\gamma$, respectively. It is well known that under these assumptions the steady-state distribution of the downloaders and the seeds follows the Poisson distribution. The expected resource usage is then:

$$E[C] = \sum_{y=0}^{\infty} \frac{(\frac{\lambda}{\gamma})^y}{y!} e^{-\frac{\lambda}{\gamma}} \frac{\sqrt{\pi}}{2} \prod_{i=0}^{y-1} \frac{i + \frac{1}{2}}{i + 1}. \tag{7}$$

Note that this analytical value for capacity usage assumes that every time when the status of the seeds changes, the downloaders have to update their peers, and seek the closest peer again. This is, however, not very realistic. For this reason (7) can be viewed as a lower bound for the resource usage. Our simulations, however, suggest that this bound is not far from the resource usage of a more realistic peer selection scheme.

4.2 Simulation Results

Next we study by numerical examples how the selected policy affects the capacity usage. First, on the left side of Figure 3 we study the scenario explained in the previous subsection, where service is always constrained by download rate and at least one peer stays in the system. The capacity usage C is shown as a function of the expected number of seeds λ/γ (simulation starts at time 0, $t_0 = 1000$ and $t_{max} = 10000$). Gray triangles correspond to a simulation with RPS policy and black triangles to NPS policy. When λ/γ is small, seeds leave the system shortly after the download and the peers that want to download the chunk have to request it from the original seed. The distances from a downloader to the original seed using the two different policies are then the same. When λ/γ increases the number of seeds also increases and the selected policy has an effect on the resource usage. We can see that, e.g., for $\lambda/\gamma = 20$ the capacity usage of the policy NPS is only 23 % of the capacity usage of the policy RPS. Simulation results are very close to analytical bounds, especially when $\lambda/\gamma > 5$.

Then we consider a system in which the service capacity is constrained by both upload and download rate and the system dies if all the seeds have left the system (model introduced in Section 2). When a new downloader arrives, it seeks either a random available (RPS) or the closest available (NPS) seed for download. The simulated average capacity usage per downloaded chunk over the period from 0 to the time of extinction for random and nearest peer selection policies are shown on the right side of Figure 3. For small λ/γ the system's life time is very short, and therefore we have done $K = 1000/(\lambda/\gamma)$ simulations to ensure that also for small λ/γ we have enough simulations. When $1/\gamma \ll 1/\lambda$, after the arrival of the first downloader, most probably the system absorbs very shortly without any completed download. Considering only those simulation traces with at least one served peer distorts the realized service time of the accepted simulations close to zero. For this reason the capacity usage is also very small when $1/\gamma \ll 1/\lambda$. When $\lambda/\gamma > 1$, the realized service time is closer to the expected value $1/\mu$ and the capacity usage is also greater. Finally, when the expected number of seeds

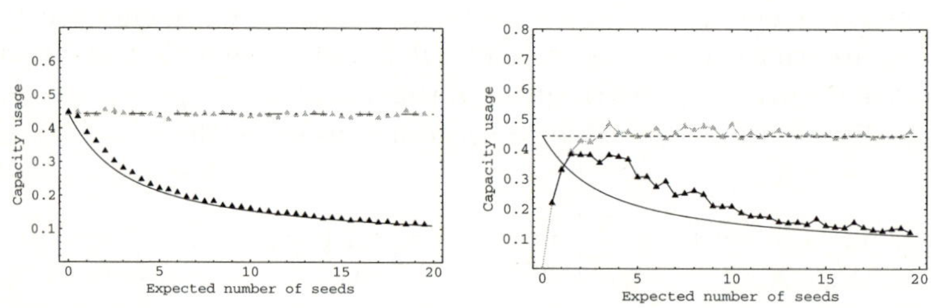

Fig. 3. Mean capacity usage as a function of λ/γ, $\mu_d = \mu_s = 1$. Gray triangles: simulation of RPS, black triangles: simulation of NPS. Dashed line: Expected resource usage for RPS policy and solid line: Expected resource usage for NPS policy. Left figure: download-constrained system. Right figure: The service capacity constrained by both upload and download rate.

λ/γ increases even more, the system most probably reaches its steady state before going to extinction. Also the capacity usage is then close to analytical bound calculated for the download-constrained system presented in the previous subsection.

5 Conclusion and Further Research Directions

In this paper we have studied the population dynamics of sharing of a single chunk in a P2P file sharing system. We have constructed a deterministic fluid model to analyze the evolution of the number of downloaders and seeds. The life time of the system is calculated by solving the absorbtion times of the corresponding Markov chain. We can see that the time to extinction increases exponentially as a function of the expected number of seeds in the system. Most important, we have proposed a spatio-temporal model to analyze the resource usage of the system. The analytical bounds for two different peer selection policies are derived. We find that by the peer selection policy where the closest peer is selected for download the resource usage of the network can be reduced to a fraction of the usage of random selection.

The arrival rate of new downloaders λ hardly remains constant over time. When a new file is released, demand for that is high adding the flash crowd effect but after some weeks or so it will be fade out. This affects the life cycle of the file sharing process. We plan to assess this issue in a future work.

In this paper we have considered the distribution of only a single chunk independently of other pieces. In the future, the model will be extended to capture the dynamics of multiple chunks as well.

References

1. B. Cohen, Incentives Build Robustness in BitTorrent, 2003, http://www.bittorrent.com/bittorrentecon.pdf.
2. M. Izal, G. Uvroy-Keller, E.W. Biersack, P.A. Felber, A.Al Hamra, and L. Garcés-Erice, Dissecting BitTorrent: Five Months in a Torrent's Lifetime, PAM, 2004.
3. J.A. Pouwelse, P. Garbacki, D.H.J. Epema, H.J. Sips, The BitTorrent P2P File-sharing system: Measurements and analysis, IPTPS, 2005.
4. L. Massoulié and M. Vojnović, Coupon replication Systems, SIGMETRICS, 2005.
5. X. Yang, G. de Veciana, Service Capacity of Peer to Peer Networks, INFOCOM 2004.
6. D. Qiu, R. Srikant, Modeling and Performance Analysis of BitTorrent-Like Peer-to-Peer Networks, SIGCOMM 2004.
7. K.K. Ramachandran, B. Sikdar, An Analytic Framework for Modeling Peer to Peer Networks, INFOCOM 2005.

Understanding the Session Durability in Peer-to-Peer Storage System*

Jing Tian, Yafei Dai, Hao Wang, and Mao Yang

Department of Computer Science, Peking University, Beijing, China
{tianjing, dyf, wanghao, ym}@net.pku.edu.cn
http://net.pku.edu.cn

Abstract. This paper emphasizes that instead of long-term availability and reliability, the short-term session durability analysis will greatly impact the design of the real large-scale Peer-to-Peer storage system. In this paper, we use a Markov chain to model the session durability, and then derive the session durability probability distribution. Subsequently, we show the difference between our analysis and the traditional *Mean Time to Failure* (MTTF) analysis, from which we conclude that the misuse of MTTF analysis will greatly mislead our understanding of the session durability. We further show the impact of session durability analysis on the real system design. To our best knowledge, this is the first time ever to discuss the effects of session durability in large-scale Peer-to-Peer storage system.

1 Introduction

Peer-to-Peer storage system is shown to be promising distributed storage architecture for its scalability. However, at the same time these systems suffer from unit failures for the existing of a large number of storage units. As a result, how to improve the availability and the reliability has become a critical and heated issue in the system design. Some approaches, such as TotalRecall[1] and OceanStore[2], have been proposed to improve the reliability as well as availability, and some analytical works have been done, for instance, [3], [4] and [5].

The availability is defined as at any given time t, the probability that data is accessible. The reliability is defined as the probability that data is not irretrievably lost at time t, and is usually measured by MTTF. From the definition, we can see that some transient failures of the storage units do not affect the reliability, but they do decrease the availability. Consequently, the reliability is a relative long-term system property. Furthermore, the availability is not a time relative property, and it only captures the probability in the long-term steady state. As a result, we argue that the short-term *session durability* property, defined as the probability that the data is always accessible before time t, is more important and practical than the long-term availability and reliability for some Peer-to-Peer storage systems. In fact the storage units are not likely to fail immediately after a data is stored. For example, consider that we store a

* This work is supported by National Grand Fundamental Research 973 program of China under Grant No.2004CB318204, National Natural Science Foundation of China under Grant No.90412008.

V.N. Alexandrov et al. (Eds.): ICCS 2006, Part IV, LNCS 3994, pp. 428 – 435, 2006.
© Springer-Verlag Berlin Heidelberg 2006

data object on a storage node n, whose lifetime and recovery time are both exponential distribution with a mean time of 10 days, then the long-term availability is only 0.5 while the probability that the data has a 24 hours continuous accessible session, is over 0.9. In contrast, if the storage node has a mean lifetime of 24 hours and a mean recovery time of 2.4 hours, the availability is more than 0.9 while the probability of a 24 hours continuous session is only about 0.37. Thus, it is clear that there is little correlation between the session durability and the availability. Consider a streaming service for hot new movies building on a Peer-to-Peer storage, for instance, we may care more about the probability that a newly added movie can be continuous accessible throughout the first 3 days rather than the long-term availability, for a transient interruption in playing will be annoying. Here, the session durability analysis can give a great help.

The session durability probability seems somewhat similar to the reliability, but it takes the transient failure into account. The transient failure makes the session durability calculation much far from the reliability calculation. As we will show in section 4, the misuse of reliability calculation can greatly mislead our understanding of the session durability in Peer-to-Peer storage system.

This paper makes the following contributions: First, we address the session durability analysis that captures the short-term data availability. Second, we present a Markov chain to model the session durability, and demonstrate how to resolve the session durability. Third, we analyze the difference between session durability calculation and reliability calculation, and conclude that MTTF calculation can not be applied in a high dynamic environment. Fourth, we show the great impact of session durability analysis on real system design.

2 Background and Motivation

Erasure Code. In the face of frequent component failures, the designers usually employ a replication or an erasure code[6] technology to achieve high data durability. Replication is a straightforward scheme which makes replicas to tolerate the failures. Erasure code provides redundancy by encoding the data into fragments and restoring data from a subset of all fragments. First of all, erasure code divides a data object into m fragments, and encodes them into n fragments, where all the fragments are in the same size and $n > m$, so the redundancy rate r is n/m. According to the erasure code theory, the original data object can be reconstructed from any m fragments out of the n fragments. In fact, we can consider the replication scheme as a subset of erasure code by making m to 1, so we use the term "*erasure code*" to refer to both redundancy strategies in the following discussions.

MTTF Calculation. In stochastic terminology, the reliability is defined as: for any target life time t, as the probability that an individual system survives for time t given that it is initially operational[7]. In a RAID system, Gibson points out that the reliability function $R(t)$, is well-approximated by an exponential function $e^{-t/MTTF}$, with the same MTTF. As a result, in stead of the reliability function, MTTF is used as the reliability metric for the convenience of computation and expression with the implicit exponential distribution assumption of system lifetime, though MTTF the mean value itself, can tell us nothing about the lifetime distribution.

3 Session Durability Model and Calculation

3.1 Exponential and Independent Unit Lifetime

Gibson [7] has shown the exponential lifetime distribution of one disk in a traditional RAID system. However, when investigating the lifetime of a storage unit in a large-scale dynamic Peer-to-Peer environment, we should take some other aspects into account besides the storage device failures, for instance, the network failures and the power problems. By analyzing the traces, Praveen et al.[8] show that the unit lifetime of PlanetLab and the web servers follows an exponential distribution. Web servers are intended to be representative of public-access machines maintained by different administrative domains, while PlanetLab potentially describes the behavior of a centrally administered distributed system. In a Peer-to-Peer storage system formed by end users, the arrival and departure behavior of an end user is unpredictable in the long run, so the behavior is best modeled by a memoryless exponential distribution[9]. Though there is no strong evidence of exponential lifetime in P2P system, some previous studies[10, 11, 12] adopt exponential lifetime in simulations or analyses. In this paper, we take the exponential lifetime assumption for a peer's lifetime.

In a large-scale Peer-to-Peer system, the storage units (servers or peers) locate in a very wide area, so there is little chance that different units fail dependently. Bhagwan et al.[13] also point out it highly unlikely to have lifetime dependency in P2P systems. In this paper, we assume that the failures are independent.

3.2 The Markov Analysis Model

By assuming all the storage units have the same independent exponential lifetime, we use a continuous-time Markov chain to model the session durability of a Peer-to-Peer storage system with erasure code. Given a system, if the erasure code's parameters are *n* and *m*, a data object will be encoded into *n* fragments and stored in *n* different storage units, called a redundancy group. Then there will be *n-m+2* states for a data object illustrated in Figure 1. State *0* is the initial state with all storage units alive, and state *k* is the state with *k* storage units failed, so obviously state *n-m+1* is the absorbing state in which we can not reconstruct the data object.

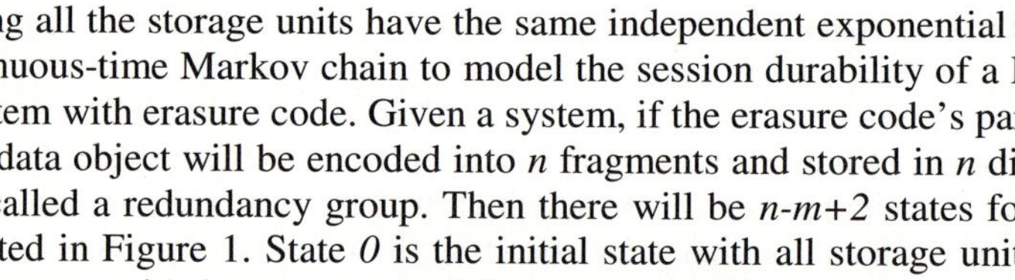

Fig. 1. Markov Model for Session Durability Analysis

The system can transit from one state to another when a failure or recovery happens. For all the storage units have the same mean lifetime t_l, the failure rate of a unit is $\lambda = 1/t_l$. In state *k*, all the *n-k* live storage units potentially fail, so the failure rate of state k is $(n-k)\times\lambda$. At the same time, the failed storage units can recover from the transient failure. By assuming that all the failed storage units have the same independent mean recovery time t_r, we derive that the recovery rate of a failed unit is u, where u is the reciprocal of t_r. Consequently, the recovery rate of state k is $k\times u$ since all the k failed units potentially recover.

3.3 Session Durability Probability Calculations

In stochastic terminology, we define the session durability probability *R(t)* as:

$$R(t) = \text{Prob}(\textit{lifetime} > t \mid \textit{initially all storage units alive})$$

In Figure 1, *R(t)* can be expressed as the probability that the system is not in the absorbing state *n-m+1* at time *t*. In this subsection, we explore and demonstrate several methods to get the *R(t)* function by resolving the Markov model. First of all, we give the transition matrix of the model

$$Q=\begin{pmatrix} -n\times\lambda & n\times\lambda & 0 & 0 & \cdots & 0 \\ \mu & -\mu-(n-1)\times\lambda & (n-1)\times\lambda & 0 & \cdots & 0 \\ 0 & 2\mu & -2\mu-(n-2)\times\lambda & (n-2)\times\lambda & & 0 \\ \vdots & \vdots & \vdots & \ddots & \ddots & \vdots \\ 0 & 0 & 0 & \cdots & -(n-m)\times\mu-m\times\lambda & m\times\lambda \\ 0 & 0 & 0 & \cdots & 0 & 0 \end{pmatrix}$$

Let $\boldsymbol{P(t)}=(p_{i,j}(t))$ be the transition matrix function, where $p_{i,j}(t)$ is the probability that the system transits from state i to state j through time t. Then we have

$$R(t) = p_{0,0}(t) + p_{0,1}(t) + \cdots + p_{0,n-m}(t) = 1 - p_{0,n-m+1}(t) \tag{1}$$

Hence, our goal is to get $p_{0,n-m+1}(t)$. According to the forward Kolmogorov equation[11], *P(t)* is determined by linear differential equations as follows

$$P'(t) = P(t)Q \tag{2}$$

Taking the Laplace transform of both sides of (2) yields

$$sP^*(s) - P(0) = P^*(s)Q$$

Under the condition *P*(0)=*I*, we have

$$P^*(s) = (p^*_{i,j}(s)) = (sI - Q)^{-1} \tag{3}$$

Consequently, we can get $p_{0,n-m+1}(t)$ by the inverse Laplace transform of $p^*_{0,n-m+1}(s)$, then we get *R(t)* from (1). Alternatively, we can directly use the unique solution[14] to (2) under the initial condition *P*(0)=*I* as follows

$$P(t) = \exp\{Qt\} = \sum_{n=0}^{\infty} \frac{Q^n t^n}{n!} \tag{4}$$

3.4 Review MTTF Calculation

The session durability is somewhat similar to the reliability in definition, so here we review the calculation of reliability for comparison. For the reliability analysis, we can still use the Markov model illustrated in Figure 1. However, there is only permanent failure but no transient failure in reliability analysis. As a result, the failure rate of a single unit is $\lambda=1/MTTF_{unit}$, where $MTTF_{unit}$ is the mean time to permanent failure. Subsequently, there is only data repair but no recovery, and the repair rate of a single unit is $u=1/MTTR_{unit}$, where $MTTR_{unit}$ is mean time to repair.

In [7], Gibson presents an iterative method to get the system MTTF, while Thomas gets the same results in[3] via the Laplace transform. The solution can be expressed as follows

$$\mathrm{MTTF} = -(1,0,\ldots,0)\cdot A^{-1}\cdot \vec{e} \tag{5}$$

A is the submatrix of transition matrix *Q* ignoring the absorbing state. By applying the exponential distribution assumption, we can identify an approximate reliability function (6) by the system MTTF. We use *exponential approximation* to refer to this approximate calculation.

$$R_{MTTF}(t) = e^{-t/MTTF} \tag{6}$$

4 Difference of Session Durability Calculation and MTTF Calculation

Because the session durability and the reliability share the similar analysis model, one may argue that we can use the traditional MTTF calculation as a substitute for the complicated session durability resolving. Unfortunately, we show in this section, that the MTTF calculation can not be applied in Peer-to-Peer environment, because the exponential function assumption does not hold in a high dynamic environment. We first give a qualitative analysis to get a preliminary understanding of the difference between two calculations, and then give a quantitative insight into the difference influenced by dynamic system parameters.

4.1 A Qualitative Analysis

There are two main conditions which the traditional exponential distribution assumption is based on: First, the failure rate is much larger than repair rate, i.e. $\lambda >> u$. Second, the number of system states is not very big, e.g. 3 in traditional RAID analysis[7]. In fact, these two conditions do not exist at all in dynamic Peer-to-Peer environment. In Peer-to-Peer environment, the unit failure caused by node leaving may be very often, and the recovery (node rejoining) usually takes a long time, so λ is comparable with u. The system designers must use more redundancy or more erasure code fragments (e.g. $n=128$ and $m=64$) to make the data more available and reliable, which enlarges the number of system states very much.

By exploring a large number of sample erasure-coded systems with randomly generalized parameters *m*, *n*, λ and *u*, we find that the patterns of the difference functions of the exact session durability calculation and the exponential approximation calculation are all alike in shape. In Figure 2, we plot a sample difference function to show the pattern. It is clear that the exponential approximation calculation first underestimates the session durability probability and then overestimates it a little bit. The underestimation is because the real system can not move too fast from the initial state to the failure state in the very beginning while it is relatively easy to fail in the two states exponential approximation. Since both the exact session durability function and the exponential approximation have the same mean session time *MST*, the integral of the difference should be *0*. Consequently, exponential approximation will definitely overestimate the session durability after the underestimation.

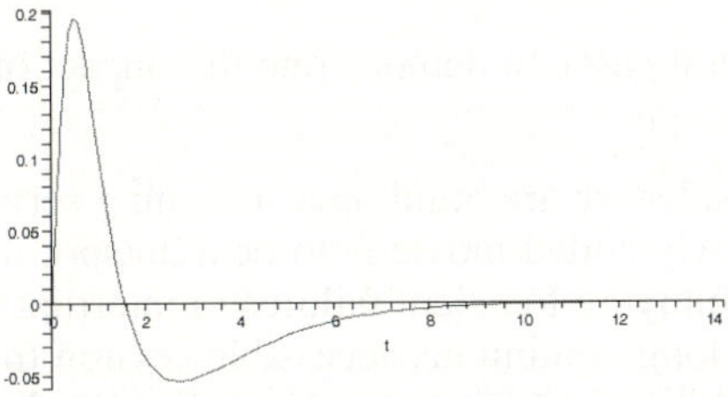

Fig. 2. The difference function of the exact session durability and the exponential approximation, *m*=4, *n*=8, λ=1, μ=1

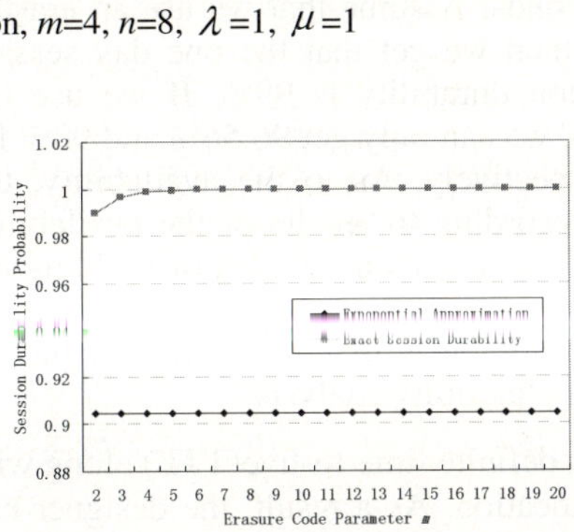

Fig. 3a. Exact session durability and the exponential approximation

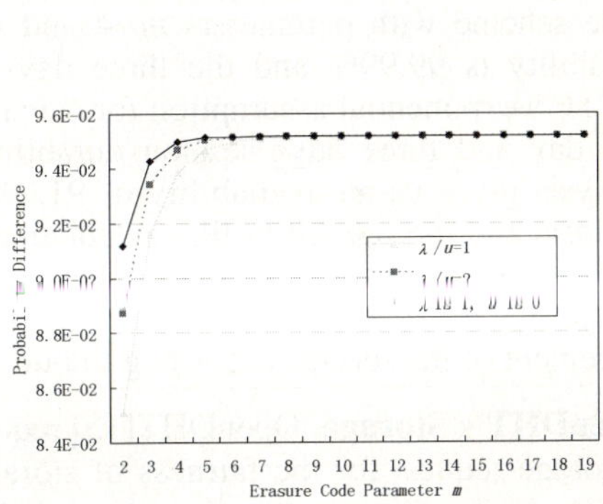

Fig. 3b. Differences between the exact session probabilities and the exponential approximations

4.2 Quantitative Insight into the Difference

To gain a deeper understanding of the difference between exact session durability and the exponential approximation influenced by the number of the states, we fix the failure rates and repair rates to investigate the difference versus the erasure code parameters *m* and *n*. By fixing the redundancy rate *r=n/m=2*, we plot the exact session durability probability as well as the exponential approximation probability at *MST/10* versus parameter *m* in Figure 3a. From the figure, we can see that the difference becomes larger when the number of states grows, and that the approximation greatly underestimates the session durability from a probability near 1 to a probability about 0.9. Furthermore, we investigate the difference trend by increase the ratio of failure rate λ to the recovery rate *u*. In Figure 3b, we plot the probability differences of three pairs of parameters λ and *u* at *MST/10*. What we find is that the differences increase when the ratio of λ to *u* increases.

We conclude the findings in this subsection that the dynamic feature of Peer-to-Peer network and the fact of using more redundancy fragments make it dangerous to use MTTF calculation as a substitute for session durability, and the misuse may greatly mislead our understanding of the session durability.

5 Impact of Session Durability on System Design

This subsection uses several cases to demonstrate the impact of our session durability analysis on real system design.

Streaming Service. Consider we are building a streaming service for new hot movies on the PlanetLab. The newly added movie is to be a hotspot in the first several days, and we do not want the annoying transient failures to interrupt the playing. Therefore, the system will require a long continuous accessible session time in first several days rather than a high availability in long-term. According to the PlanetLab trace used in[8], we find most of the nodes have a mean lifetime of 10^6 seconds, while many of them have a mean recovery time of 5×10^6 seconds. Assume that we use an erasure code scheme with parameters $m=4$ and $n=8$, then we get that the one day session durability is 99.99% and the three days session durability is 99%. If we use the MTTF's exponential assumption for calculation, we can only get 96.56% and 90% for one day and three days session durability respectively. As to the availability, the analysis gives us an availability of 91.24%. According to results of the availability analysis and the reliability like calculation, we may abandon the idea of building the service on a Peer-to-Peer environment, or use more redundancy data to enhance the durability. However, the fact is that the session durability is high enough for the requirement of the service according to our session durability analysis.

OpenDHT's Storage. OpenDHT[15] requires a definite time-to-live(TTL) along with a storage request for the fairness of storage allocation. As a result, the designer can only concentrate on how to improve session durability within the specified *TTL*, but not think about the availability and reliability. Since the availability analysis and reliability analysis give very low underestimations, the designer can use less system resource to guarantee a good enough session durability within *TTL* by using session durability analysis.

Fixed Repair Epoch for Large-Scale Distributed Storage. Large-scale distributed storage systems usually use a fixed repair epoch for the simplicity of repair mechanism. For example, [4] assumes a system with fixed repair epoch, and the system employs an erasure code with $m=32$ and $n=64$. The designer should get the knowledge about the probability that the data can survive a single epoch. Though it may not be a high dynamic system, the calculation under the exponential assumption will greatly mislead us, because there is no repair within an epoch. Assume a five years

$MTTF_{unit}$, under the exponential assumption we calculate the probabilities that a data can survive a four months epoch and a 12 months epoch respectively, and get 91.1% and 75.6%. However the real probability is greater than $1\text{-}10^{16}$ for four months, and greater than $1\text{-}10^{8}$ for 12 months.

6 Conclusions

In this paper, we first addressed the new metric, session durability, for a Peer-to-Peer storage system. Subsequently, we presented the analysis model, and demonstrated how to resolve the session durability from the model. Our experiments have shown strong evidence that MTTF calculation can not be applied in high dynamic environment, and session durability is far from the reliability analysis. We further showed the impact of session durability analysis on real system design.

References

1. R. Bhagwan, K. Tati, Y. Cheng, S. Savage, and G. M. Voelker. Total recall: System support for automated availability management. In *Proc. of the First ACM/Usenix Symposium on Networked Systems Design and Implementation (NSDI)*, 2004.
2. J. Kubiatowicz, D. Bindel, Y. Chen, S. Czerwinski, P. Eaton, D. Geels, R. Gummadi, S. Rhea, H. Weatherspoon, W. Weimer, C. Wells, and B. Zhao. OceanStore: An Architecture for Global-Scale Persistent Storage. In *ACM ASPLOS*, November 2000.
3. T. Schwarz. Generalized Reed Solomon codes for erasure correction in SDDS. In *Workshop on Distributed Data and Structures (WDAS 2002)*, Paris, France, Mar. 2002.
4. H. Weatherspoon and J. D. Kubiatowicz. Erasure Coding vs. Replication: A Quantitative Comparison, In *Proc. of IPTPS '02*, March 2002.
5. R. Rodrigues and B. Liskov. High Availability in DHTs: Erasure Coding vs. Replication. In *Proc. of IPTPS'05*, February 2005.
6. J. Plank. A tutorial on Reed-Solomon coding for fault-tolerance in raid-like systems. *Software Practice and Experience, 27(9):995-1012,* September 1997
7. G. A. Gibson. Redundant Disk Arrays: Reliable, Parallel Secondary Storage. *PhD thesis U. C. Berkeley*, April 1991.
8. P. Yalagandula, S. Nath, H. Yu, P. B. Gibbons and S. Seshan. Beyond Availability: Towards a Deeper Understanding of Machine Failure Characteristics in Large Distributed Systems. In *Proc. of the 1st Workshop on Real, Large Distributed Systems,* 2004.
9. G. Pandurangan, P. Raghavan and E. Upfal. Building low-diameter P2P networks. In *Proc. of the 42nd Annual IEEE Symposium on the Foundations of Computer Science*, Oct. 2001.
10. D. Liben-Nowell, H. Balakrishnan and D. Karger. Analysis of the evolution of Peer-to-Peer systems. In *Proc. of the 21st ACM Symposium on Principles of Distributed Computing*. Monterey, CA, USA: ACP Press, 2002
11. Y. Zhao. Decentralized Object Location and Routing: A New Networking Paradigm. *U.C. Berkeley PhD Dissertation*, August 2004
12. S. Giesecke, T. Warns and W. Hasselbring. Availability Simulation of Peer-to-Peer Architectural Styles. In *Proc. of ICSE 2005 WADS.*
13. R. Bhagwan, S. Savage and G.. Voelker. Understanding availability. In *proc. of International Workshop on Peer-to-Peer Systems (IPTPS03)*, February 2003.
14. M Kijima. Markov Processes for Stochastic Modeling. *Chapman and Hall, London*, 1997.
15. http://www.opendht.org/

Popularity-Based Content Replication in Peer-to-Peer Networks

Yohei Kawasaki, Noriko Matsumoto, and Norihiko Yoshida

Department of Information and Computer Sciences,
Saitama University, Saitama 338-8570, Japan
{yohei, noriko, yoshida}@ss.ics.saitama-u.ac.jp

Abstract. Pure peer-to-peer (P2P) networks is widely used, however they broadcast query packets, and cause excessive network traffic in particular. Addressing this issue, we propose a new strategy for content replication which prioritizes popular and attracting contents. Our strategy is based on replication adjustment, being cooperated with index caching, as well as LRU-based content replacement, and a more effective replica placement method than well-known probabilistic ones. We also present some experiment results to show that our strategy is better than other ones in regards to network traffic and storage consumption.

1 Introduction

In a P2P network, as contents are distributed to all the member nodes (peers), we must consider mechanisms to search contents. Generally, the search mechanisms for P2P networks are classified into three groups. A centralized P2P system such as Napster [1] has a central server which manages all the locations (indices) of contents. A decentralized P2P system has no central server, and is again classified into two categories. A decentralized unstructured system such as Gnutella [2] has no specific structure in network topology, and use a flooding-based query algorithm. A decentralized structured system such as Chord [3] has a well-organized structure, and has a very efficient search mechanism using a distributed hash table (DHT) on P2P network.

In centralized P2P systems, a large amount of queries cause a high load on the central server, and the server may be a single point of failure. Decentralized structured P2P systems require strictly-organized network topology which is difficult to construct in reality. Both of them can search contents efficiently, however, they have disadvantage as mentioned above. On the contrary, decentralized unstructured P2P systems are easy to construct in reality, and fault-tolerant. Therefore, they are widely used, although they have issues that the number of query packets grows exponentially, and search areas are limited to reachable nodes of query packets.

It is one of the effective improvement way for the issues of decentralized unstructured P2P systems to distribute replicas of contents in a network [4, 5]. Some researches about content replication treat all the contents equally for replication.

V.N. Alexandrov et al. (Eds.): ICCS 2006, Part IV, LNCS 3994, pp. 436–443, 2006.
© Springer-Verlag Berlin Heidelberg 2006

However, the popularity of a content (i.e. frequency of accesses to the content) is not uniform. It must be more effective to put more replicas for popular objects than unpopular ones.

This paper proposes a distributed replication mechanism for decentralized unstructured P2P networks which considers content popularity, is scalable, and is easy to implement. Section 2 summarizes related researches on content replication, and Section 3 proposes popularity-based content replication. Section 4 presents some experiment results and evaluation. Section 5 contains some concluding remarks.

2 Content Replication

Content replication in P2P network generally provides decrease of packet hops until a search succeeds, and improvement of search success rate. It also provides decrease of network load, and realizes efficient search. These effects become apparent in decentralized unstructured P2P systems in particular.

In an extreme case, allocating replicas of all contents to all nodes results in the ideal result in which reference to any content is done at no network cost. However, because a node has a limited storage resource, we cannot actually apply this extreme replication.

Thus, in the content replication, it is important to decide the number and the placements of replicas how many and where the replicas are allocated. There are some strategies already proposed to decide these.

Owner Replication. Replicas is allocated only on the requester (i.e. the node emitting a query) when the search succeeds. This strategy is simple and needs the smallest cost, but it needs enough time for replicas to spread over the network [5].

Path Replication. Replicas are allocated on all the nodes on the search path from the requester to the holder (i.e. the node having the requested content). Because one or more replications are allocated per search, a lot of storage or network resources are needed, however, a content spreads easily [5].

Path Random Replication. In this strategy, each node on the search path may or may not have a replica based on a pre-defined probability. We must decide an appropriate allocation probability [6].

Random Replication. In this strategy, replicas of the same number as of the nodes on the search path are allocated to randomly-chosen nodes in the whole network. There is no distributed algorithm presented to choose the nodes, therefore it is not certain whether the strategy can be implemented without too much cost [5].

3 Popularity-Based Content Replication

It is reported that the “Power law” can generally be applied to web content accesses [7]. There are a few contents which attract many accesses, while there are many contents which attract only a few accesses. This fact must be applied to content accesses in P2P networks as well.

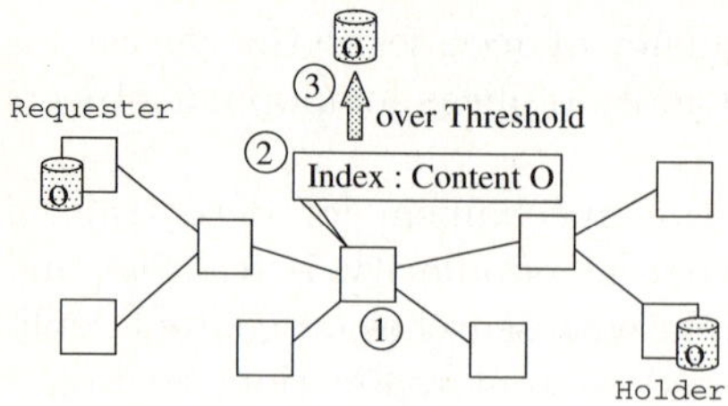

Fig. 1. Delayed content replication

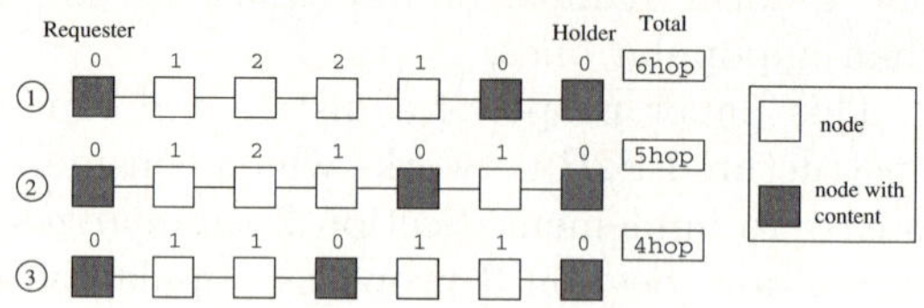

Fig. 2. Replication on search path

Allocating more replicas of more popular contents brings improvement of search success as a whole because popular contents are more often searched. It must also benefit to reduction of network traffic because the number of hops to reach a searched content becomes small. However, too many replicas causes increase of the network traffic because many replicas are found by a single query, and many "hit" packets are generated. At the same time, suppressing the number of replicas of unpopular contents brings reduction of storage consumption at every node.

There are some replica placement strategies proposed as summarized above. However, none of them considers content popularity. Therefore, these strategies result in excessive network traffic to create a content replica for low popularity contents (except for Owner Replication) and excessive storage consumption.

Square-Root Replication is one of a few exceptions, which considers content popularity [4]. The number of replicas is determined proportional to the square root of the access counts relative to other contents. This strategy is reported to exhibit, in a rough (not an exact) simulation, significant reduction of whole search traffic [5]. However, it would be necessary that any single node must have knowledge on popularities of all the contents, which is impractical.

Consequently, we construct our replication strategy based on the following design.

(1) Decision of Replica Allocation. A simple method to decide replica allocation is introduced which does not cause heavy network traffic.

Fig. 1 shows an overview of the method. A replica is allocated on the requester like Owner Replication. Another will be allocated on the node halfway between the requester and the holder, namely the most distant node from both the requester and the holder (Fig. 1-①), however it is not placed at the moment of searching immediately. As shown below, the node evaluates popularity, i.e. access rate, for that content, and decides whether placing a replica or not eventually.

Fig. 2 shows the reason why the halfway node is the best to place a replica. Sum of the number of hops is minimized if a replica is placed on the node halfway on the search path. This is more effective than probabilistic allocation.

(2) Provisional Replica Placement. A replica is not placed immediately at the halfway node. If the replica is never accessed afterward, it is just a waste

Table 1. Simulation settings

Number of nodes	2000
TTL of query packet	5
Initial max number of contents	60
Capacity of storage	100
Threshold of replication	15

Table 2. Popularities vs. contents disposition

popularity	request rate (%)	contents
(Low) 1	5	2500
2	10	1000
3	20	400
4	25	150
(High) 5	40	50

of copying cost and storage. Instead, only an index of the content, i.e. a pair of the search key and the content location, is initially placed (Fig. 1-②). The size of an index is much smaller than that of a content, therefore the cost of keeping an index is much smaller that the one of copying and storing the content.

The node has an index containing two content locations of both the holder and the requester, and it replies either of these locations in an alternate (round-robin) fashion to a query. Accesses to the content are dispersed in this way.

The index not only contributes to the efficient search, but also indicates the potential location of the replica. This cooperation of indexing and content replication is the major novelty of this research.

The node counts the number of references to each index it has, and when the number exceeds a certain threshold, a replica of the content is placed at this moment (Fig. 1-③).

(3) Replacement of Replication. Each node has limited capacity of storage, and when its storage is saturated, it decides which replicas to keep according to their popularities. When a new content or replica is added to the node whose storage is full, a replica is discarded in a LRU (Least Recently Used) manner.

4 Experiments and Evaluation

To verify the advantages of our strategy, we present some results of experiments using a simulator for virtual P2P networks. Table 1 summarizes parameter settings for the experiments.

Settings for Contents and Search. We put at each node some contents randomly up to 60 in the beginning. These initial contents will not be discarded even if the node's storage is full. We allow each node to have up to 1,000 indices, and to discard indices in a FIFO (First In First Out) manner when the index capacity overflows.
We prepare 4,100 contents as a whole, each of which has a popularity level shown in Table 2. This is to follow the Power-Law described in 3.
Our simulator selects a node randomly from 2,000 nodes per trial, and makes it search a content. We tried 100,000 trials in every experiment. The result figures below show transition at every 2,000 trials.

Settings for Network Topologies. It is reported that actual structures of P2P network topology have the nature of Power-Law [8]. In fact, a small

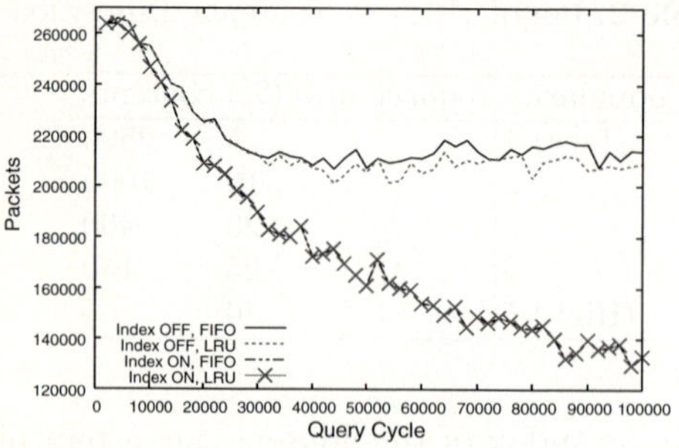

Fig. 3. Transitions of total packets

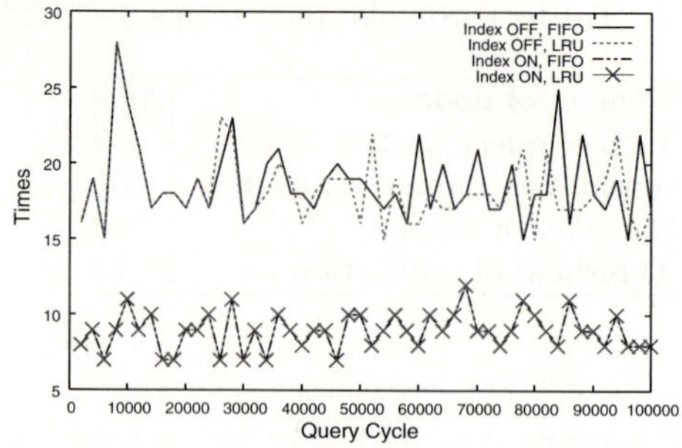

Fig. 4. Max. number of content delivery

number of nodes has many links, and most nodes has only a few links. Therefore, we follow such network topology especially in experiments described in 4.2. The Power-Law Random (PLR) topology used in the experiments is the random network in which the number of links of each node follows the Power-Law, and connections between the nodes are random.

4.1 Effect of Each Factor

Our strategy consists of replica allocation which cooperates with index allocation, and the LRU-based replacement of replicas. Hence, to clarify each effect, we compare four cases: (1) Index OFF - FIFO, (2) Index OFF - LRU, (3) Index ON - FIFO, and (4) Index ON - LRU. "Index OFF" means not allocating an index but a replica immediately. In this experiment, we simplify the network composition in which each node randomly connects with 2 to 4 neighbors.

Fig. 3 shows transitions of the total number of packets, i.e. sum of query packets and hit packets. (4) shows identical result as (3), because there is no saturation of storages in this experiment. "Index ON" brings delayed replica placement, which suppresses storage saturation. Comparing (1) and (3), or (2) and (4), we can conclude that cooperation with indexing reduces the network traffic significantly. Comparing (1) and (2), LRU-based replacement is also effective when storage saturation occurs.

Fig. 4 shows transitions of the maximum number of content deliveries. If this value is high, it means that accesses for a content are concentrated to a single node. Again, (4) shows identical result as (3). Comparing (1) and (3), or (2) and (4), accesses are dispersed by cooperation of indexing, and by round-robin handling of queries.

We also verify, although not apparent in the figure, that (3) (and (4)) shows the highest success rate of search out of the four cases, and the total number of content delivery is the highest.

4.2 Comparisons with Other Strategies

Next, we compare our strategy with other related strategies: (1) Our strategy (Proposal), (2) Owner Replication (Owner), (3) Path Replication (Path), and

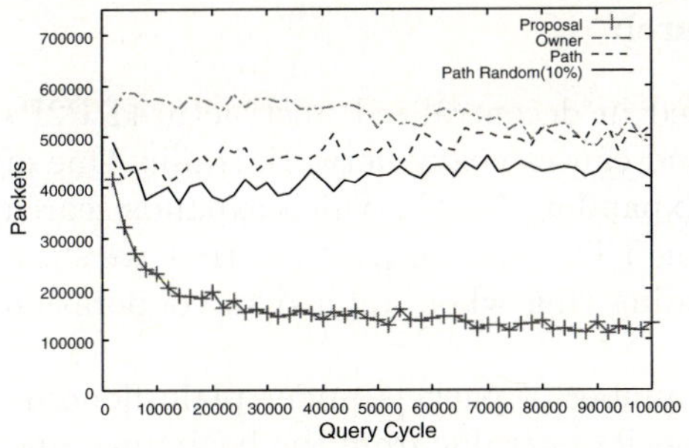

Fig. 5. Transitions of total packets

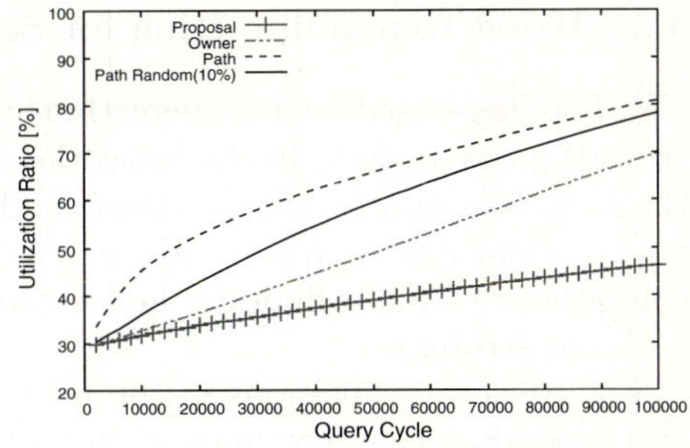

Fig. 6. Transitions of storage utilization

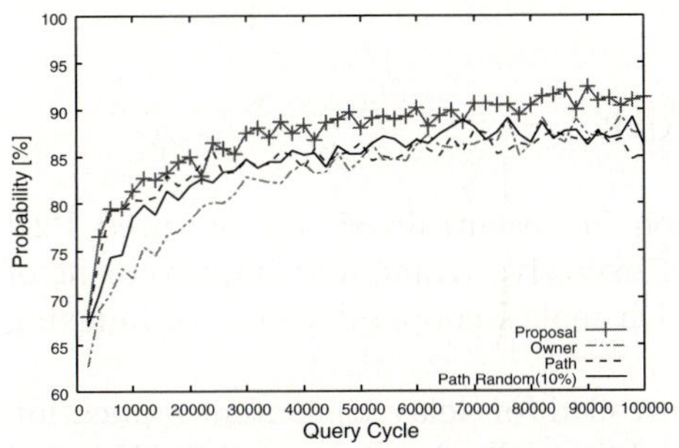

Fig. 7. Transitions of search success ratio

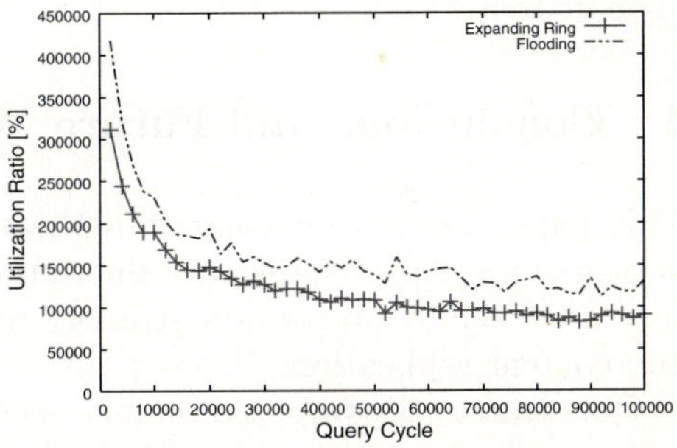

Fig. 8. Flooding vs. expanding ring

(4) Path Random Replication in which the allocation probability to each node is 10% (Path Random). The network composition we use in these experiments are the above-mentioned PLR topology.

Fig. 5 shows transitions of the total number of packets of all the strategies. Only our strategy shows the steep decline from the beginning, while the others do not.

Actually, all the strategies show declines when applied to networks of random topology. Therefore, the results shown in this figure must be affected by the characteristic of PLR topology. In the PLR topology, most packets tend to concentrate on hub nodes (i.e. nodes with many links), and our LRU-based replacement of replicas occurs more often on them. In the other strategies, search success ratio on the hubs gets lower, and query packets tend to spread much wider.

In this figure, Path Replication ("Path") shows increase of traffic. We investigated and found that many "hit" packets occupy the network which are generated by too many replicas. In fact, about 30% of total packets are "hit" packets.

Fig. 6 shows transitions of the consumption ratio of all the node storages. Our strategy shows the lowest ratio, therefore we can conclude ours achieves good performance in regard to both network traffic and storage consumption.

Fig. 7 shows transitions of search success ratio. Our strategy shows the best, that implies ours suppresses flooding of query packets.

4.3 Using Expanding Ring for Search

The flooding-based search algorithm used in decentralized unstructured P2Ps spreads query packets in the networks, and causes heavy network traffic. One of the proposals addressing this issue is "Expanding Ring", which expands search area gradually by incrementally increasing TTL (time-to-live) [5]. Here we show comparison of plain flooding and expanding ring when applied in cooperation with our strategy.

Fig. 8 shows transitions of the total number of packets under plain flooding and expanding ring. Expanding ring brings lower traffic from the beginning, and it is effective in particular at the beginning where replicas are not so many yet. However, we observe that using expanding ring, spreading of replicas becomes slow, and search success rate decreases, because searches tend to succeed within a small area.

5 Conclusions and Future Work

This paper focuses on content replication in decentralized unstructured P2P networks for the purpose of reduction of excessive traffic and improvement of search efficiency, proposes a strategy which makes cooperative use of indexing and content replication.

Our strategy is composed mainly of controlled (or delayed) content replication which cooperates with index allocation. Additionally, it implements LRU-based replica replacement, and a very simple but efficient method to decide locations of replicas. We showed advantages of our strategy by several simulated experiments. Our strategy achieved more efficient search and lower network traffic than other strategies.

Future works are as follows. First, a threshold of reference counts for switching from an index to a replica placement must be appropriately determined. Second, other P2P network topologies must be examined as well. Third, real-world experiments must be performed.

Acknowledgments

This research was supported in part by JSPS in Japan under Grants-in-Aid for Scientific Research (B) 17300012, and by MEXT in Japan under Grants-in-Aid for Exploratory Research 17650011.

References

1. Napster website, http://www.napster.com/.
2. Gnutella website, http://www.gnutella.com/.
3. I. Stoica, R. Morris, D. Karger, M. F. Kaashoek, and H. Balakrishnan, "Chord: A Scalable Peer-to-Peer Lookup Service for Internet Applications", Proc. ACM SIGCOMM 2001, 2001.

4. E. Cohen and S. Shenker, "Replication Strategies in Unstructured Peer-to-Peer Networks", Proc. ACM SIGCOMM 2002, 2002.
5. Q. Lv, P. Cao, E. Cohen, K. Li, and S. Shenker, "Search and Replication in Unstrctured Peer-to-Peer Networks", Proc. 16th ACM Int'l Conf. on Supercomputing, 2002.
6. D. Maruta, H. Yamamoto, Y. Oie, "Replication Strategies to Enable Storage Load Balancing in P2P Networks" (in Japanese), IEICE Technical Report, NS2003-319, pp.131–136, 2004.
7. S. Sato, K. Kazama, and S. Shimizu, http://www.ingrid.org/w3conf-japan/97/sato /paper.html (in Japanese).
8. L. A. Adamic, R. M. Lukose, A. R. Puniyani and B. A. Huberman, "Search in Power-Law Networks", Physical Review E, Vol.64, pp.46135–46143, 2001.

A New Method for Crude Oil Price Forecasting Based on Support Vector Machines

Wen Xie, Lean Yu, Shanying Xu, and Shouyang Wang

Institute of Systems Science, Academy of Mathematics and Systems Sciences, Chinese Academy of Sciences, Beijing 100080, China
{xiewen, yulean, xsy, sywang}@amss.ac.cn

Abstract. This paper proposes a new method for crude oil price forecasting based on support vector machine (SVM). The procedure of developing a support vector machine model for time series forecasting involves data sampling, sample preprocessing, training & learning and out-of-sample forecasting. To evaluate the forecasting ability of SVM, we compare its performance with those of ARIMA and BPNN. The experiment results show that SVM outperforms the other two methods and is a fairly good candidate for the crude oil price prediction.

1 Introduction

Oil plays an increasingly significant role in the world economy since nearly two-thirds of the World's energy consumption comes from crude oil and natural gas. Sharp oil price movements are likely to disturb aggregate economic activity, especially since Jan 2004, global oil price has been rising rapidly and brings dramatic uncertainty for the global economy. Hence, volatile oil prices are of considerable interest to many researchers and institutions. The crude oil price is basically determined by its supply and demand, but more strongly influenced by many irregular past/present/future events like weather, stock levels, GDP growth, political aspects and so on. These facts lead to a strongly fluctuating and non-linear market and the fundamental mechanism governing the complex dynamic is not understood. As Epaminondas et at. [1] reported, the oil market is the most volatile of all the markets except Nasdaq and shows strong evidence of chaos. Therefore, oil price prediction is a very important topic, albeit an extremely hard one due to its intrinsic difficulty and practical applications.

When it comes to crude oil price forecasting, most of the literatures focus only on oil price volatility analysis [1] and oil price determination within the supply and demand framework [2]. There is very limited research on oil price forecasting, including quantitative and qualitative methods. Among the quantitative methods, Huntington [3] used a sophisticated econometric model to forecast crude oil prices in the 1980s. Abramson and Finizza [4] utilized a probabilistic model for predicting oil prices and Barone-Adesi et al. [5] suggested a semi-parametric approach for forecasting oil price. Regarding the qualitative methods, Nelson et al. [6] used the Delphi method to forecast oil prices for the California Energy Commission. However, the above meth-

V.N. Alexandrov et al. (Eds.): ICCS 2006, Part IV, LNCS 3994, pp. 444 – 451, 2006.
© Springer-Verlag Berlin Heidelberg 2006

ods show poor performance and can't meet practical needs in forecasting crude oil prices. Very recently, Wang et al. [7] proposed a new integrated methodology-TEI@I methodology and showed a good performance in crude oil price forecasting with back-propagation neural network (BPNN) as the integrated technique. BPNN, a class of the most popular neural network model, can in principle model nonlinear relations but they do not lead to one global or unique solution due to differences in their initial weight set. Another drawback is that BPNN is susceptible to over-fitting problems. Consequently, it is of necessity to develop new individual methods for forecasting oil prices which can be used for further integration into other methodologies like TEI@I.

Recently, support vector machine, a novel neural network algorithm, was developed by Vapnik and his colleagues [8]. Established on the unique theory of the structural risk minimization principle to estimate a function by minimizing an upper bound of the generalization error, SVM is resistant to the over-fitting problem and can model nonlinear relations in an efficient and stable way. Furthermore, SVM is trained as a convex optimization problem resulting in a global solution that in many cases yields unique solutions. Originally, SVMs have been developed for classification tasks [9]. With the introduction of Vapnik's ε-insensitive loss function, SVMs have been extended to solve nonlinear regression and time series prediction problems, and they exhibit excellent performance [10, 11].

The goal of this paper is to propose a new method based on SVM for the task of crude oil price time series prediction. In addition, this paper examines the feasibility of applying SVM in crude oil price forecasting through the contrast with ARIMA and BPNN models. The rest of the paper is organized as follows. Section 2 describes a new SVM-based method for crude oil price prediction. To evaluate the SVM, an empirical study and its comparable results with ARIMA and BPNN are presented in section 3. Some concluding remarks are made in section 4.

2 A New SVM-Based Crude Oil Forecasting Method

In this section, a new SVM-based method for time series forecasting and its application in crude oil price prediction are presented. We first introduce a basic theory of the support vector machine model, and then present the new SVM-based method for time series forecasting.

2.1 Theory of SVM

SVMs have originally been used for classification purposes but their principles can be extended to the task of regression and time series prediction as well. In this paper, we only focus on support vector regression (SVR) for the task of time series prediction. An excellent general introduction to SVMs including support vector classification (SVC) and support vector regression (SVR) can be seen in References [8] for more details.

SVMs are linear learning machines which means that a linear function is always used to solve the regression problem. When dealing with nonlinear regression, SVMs map the data x into a high-dimensional feature space via a nonlinear mapping φ and make linear regression in this space.

$$f(x) = (\omega \cdot \varphi(x)) + b \tag{1}$$

where b is a threshold. In linear cases, $\varphi(x)$ is just x and $f(x_i)$ becomes a linear function. Thus, linear regression in a high dimensional space corresponds to nonlinear regression in the low dimensional input space. Since $\varphi(x)$ is fixed, we determine ω from the data by minimizing the sum of the empirical risk $R_{emp}[f]$ and a complexity term $\|\omega\|^2$, which enforces flatness in feature space.

$$R_{reg}[f] = R_{emp}[f] + \lambda\|\omega\|^2 = \sum_{i=1}^{l} \psi(f(x_i) - y_i) + \lambda\|\omega\|^2 \tag{2}$$

where l denotes the sample size, $\psi(\cdot)$ is a cost function and λ is a regularization constant. For the goal of regression and time series prediction, Vapnik's ε -insensitive loss function is adopted in this paper.

$$\psi(f(x) - y) = \begin{cases} |f(x) - y| - \varepsilon & \text{for } |f(x) - y| \geq \varepsilon \\ 0 & \text{otherwise} \end{cases} \tag{3}$$

For a large set of cost functions, Eq. (2) can be minimized by solving a quadratic programming problem by applying Lagrangian theory, which is uniquely solvable under Karush-Kuhn-Tucker conditions. It can be shown that we are able to rewrite the whole problem in terms of dot products in the low dimensional input space.

$$f(x) = \sum_{i=1}^{l} (\alpha_i - \alpha_i^*)(\varphi(x_i) \cdot \varphi(x)) + b = \sum_{i=1}^{l} (\alpha_i - \alpha_i^*) K(x_i, x) + b \tag{4}$$

In Eq. (4), $K(\cdot)$ is the so-called kernel function which simplifies the use of a mapping. Representing the mapping by simply using a kernel is called the kernel trick and the problem is reduced to finding kernels that identify families of regression formulas. Any symmetric kernel function $K(\cdot)$ satisfying Mercer's condition corresponds to a dot product in some feature space.

2.2 SVM-Based Forecasting

The procedure of developing a support vector machine for time series forecasting is illustrated in Fig. 1. As can be seen from Fig. 1, the flow chart can be divided into four phases. The first phase is data sampling. To develop a SVM model for a forecasting scenario, training, validating and testing data need to be collected. However, the data collected from various sources must be selected according to the corresponding criteria. The second phase is sample preprocessing. It includes two steps: data normalization, data division. In any model development process, familiarity with the available data is of the utmost importance. SVM models are no exception, and data normalization can have a significant effect on model performance. After that, data collected should be split into two sub-sets: in-sample data and out-of-sample data which are used for model development and model evaluation respectively. The third phase is SVM training and learning. This phase includes three main tasks: determination of SVM architecture, sample training and sample validation. It is the core process of SVM model. In this phase, we shall determine the time-delay τ, embedding dimension d, ε, regularization constant λ and the choice of the kernel function. The final phase is out-of-sample forecasting. When the former phases are complete, the SVM can be used as a forecaster or predictor for out-of-sample forecasting of time series.

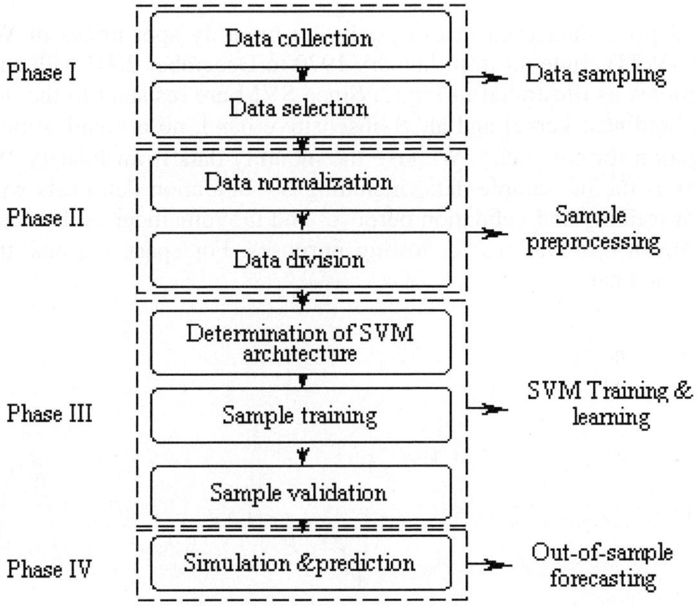

Fig. 1. A flow chart of SVM-based forecasting system

Following the above procedure, SVM-based crude oil price prediction involves four steps:

(a) Data sampling. A variety of data can be collected for this research, such as WTI, NYMEX. Data collected can be categorized into different time scales: daily, weekly and monthly. For daily data, there are various inconsistencies and missing points for the market has been closed or halted due to weekends or unexpected events. As a result, weekly data and monthly data should be adopted as alternatives.

(b) Data preprocessing. The collected oil price data may need to be transformed into certain appropriate range for the network learning by logarithm transformation, difference or other methods. Then the data should be divided into in-sample data and out-of-sample data.

(c) Training and learning. The SVM architecture and parameters are determined in this step by the training results. There are no criteria in deciding the parameters other than a trial-and-error basis. In this investigation, the RBF kernel is used because the RBF kernel tends to give good performance under general smoothness assumptions. Consequently, it is especially useful if no additional knowledge of the data is available. Finally, a satisfactory SVM-based model for oil price forecasting is reached.

(d) Future price forecasting.

3 An Empirical Study

In this section, we first describe the data, and then define some evaluation criteria for prediction purposes. Finally, the empirical results and explanations are presented.

3.1 Data

The crude oil price data used in this study are monthly spot prices of West Texas Intermediate (WTI) crude oil from January 1970 to December 2003 with a total of n = 408 observations, as illustrated in Fig. 2. Since SVM are resistant to the noise due to the use of a nonlinear kernel and an ε-insensitive band, no normalization is used in this investigation for simplicity. We take the monthly data from January 1970 to December 1999 as the in- sample data (including 60 validation data) sets with 360 observations for training and validation purposes and the remainder as the out-of-sample data sets with 48 observations for testing purposes. For space reasons, the original data are not listed here.

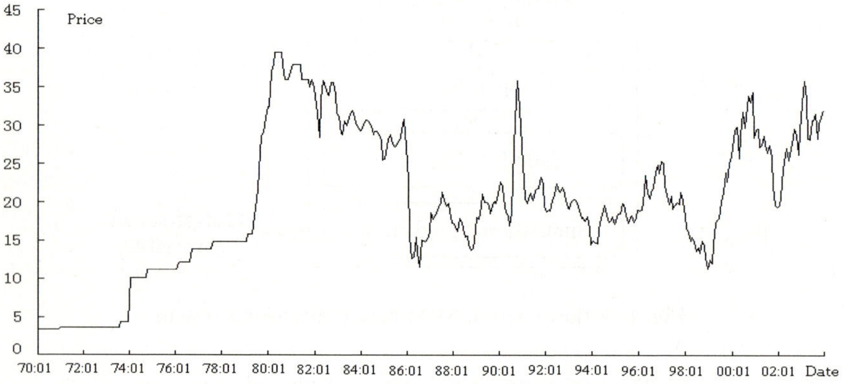

Fig. 2. The monthly oil price for the period 1970-2003

3.2 Evaluation Criteria

In order to evaluate the prediction performance, it is necessary to introduce some forecasting evaluation criteria. In this study, two main evaluation criteria, root mean square error (*RMSE*) and direction statistics (D_{stat}), are introduced. The *RMSE* is calculated as

$$RMSE = \sqrt{\frac{1}{N}\sum_{t=1}^{N}(y_t - \hat{y}_t)^2} \tag{5}$$

where y_t represents the actual value, $\hat{y}_t$ is predicted values, and N is the number of testing data sets. Apparently, the indicator RMSE describes the estimates' deviation from the real values.

In the oil price forecasting, a change in trend is more important than precision level of goodness-of-fit from the viewpoint of practical applications. Trading driven by a certain forecast with a small forecast error may not be as profitable as trading guided by an accurate prediction of the direction of movement. As a result, we introduce directional change statistics, D_{stat}. Its computational equation can be expressed as

$$D_{stat} = \frac{1}{N}\sum_{t=1}^{N} a_t \tag{6}$$

where $a_t = 1$ if $(y_{t+1} - y_t)(\hat{y}_{t+1} - y_t) \geq 0$, and $a_t = 0$ otherwise.

3.3 Results and Analysis

Each of the forecasting method described in the last section is estimated and validated by in-sample data. The model estimation selection process is then followed by an empirical evaluation which is based on the out-sample data.

The results of an augmented Dickey-Fuller (ADF) test show that the time series in level follows a unit root process. In order to utilize the ARIMA model, a first difference is necessary. Thus, ARIMA(1,1,0) is identified. The time delay τ, embedding dimension d and the prediction horizon, decided by try-and-error criteria, are respectively 4,4,1 for both BPNN and SVM. The best experiment result of each method is presented in Table 1 in which a comparison among them is performed.

Table 1. Crude oil price forecast results (Jan. 2000 - Dec. 2003)

Methods	Criteria	Full period	Sub-period I (2000)	Sub-period II (2001)	Sub-period III (2002)	Sub-period IV (2003)
ARIMA	*RMSE*	2.3392	3.0032	1.7495	1.9037	2.4868
	D_{stat} (%)	54.17	41.67	50.00	58.33	66.67
BPNN	*RMSE*	2.2746	2.9108	1.8253	1.8534	2.3843
	D_{stat} (%)	62.50	50.00	58.33	66.67	75.00
SVM	*RMSE*	2.1921	2.6490	1.8458	1.8210	2.3411
	D_{stat} (%)	70.83	83.33	50.00	58.33	91.67

Table 1 shows the detailed results of the simulated experiment via the three methods. It can be seen that the SVM method outperforms the ARIMA and BPNN models in terms of both *RMSE* and $D_{stat.}$ Focusing on the *RMSE* indicators, the values of SVM model are explicitly lower than those of ARIMA and BPNN, except in the second sub-period. From the practical application viewpoint, indicator D_{stat} is more important than indicator *RMSE*. This is because the former can reflect the trend of movements of the oil price and can help traders to hedge their risk and to make good trading decisions in advance. Concerning the SVM-based method, it can be seen in the table that although the D_{stat} value of the sub-periods II and III are some lower than BPNN, the values of D_{stat} are all above 50% and generally higher than those of the other two models, indicating that the SVM method has stronger prediction ability than the other individual models.

In addition, we observe from Table 1 that a smaller *RMSE* does not necessarily mean a higher D_{stat}. For example, for the test case of the SVM, the *RMSE* for 2001 is explicitly lower than that for 2000, while the D_{stat} for 2000 is larger than that for 2001 which implies that the D_{stat} criterion is not identical to the *RMSE* in different time series forecasting.

We can therefore conclude from the results of Table 1 and the above analysis that the proposed SVM approach performs the best among the three methods with 2.1921 and 70.83% for RMSE and D_{stat} respectively, while the ARIMA models show the worst performance.

The main reasons for the above conclusions are as follows. As Epaminondas et at. [1] reported, the crude oil market is one of the most volatile market in the world and shows strong evidence of chaos. ARIMA, typical linear models which capture time series' linear characteristics, shows insufficient to describe the nonlinear dynamics. Hence, ARIMA models perform worst among the three methods.

Both SVM and BPNN can in principle describe the nonlinear dynamics of crude oil price. Established on the unique theory of the structural risk minimization principle to estimate a function by minimizing an upper bound of the generalization error, SVM is resistant to the over-fitting problem and can model nonlinear relations in an efficient and stable way. Furthermore, SVM is trained as a convex optimization problem resulting in a global solution that in many cases yields unique solutions. Compared with the SVM's merits above, BPNN tends to suffer from over-fitting problem and does not lead to one global or unique solution owing to differences in their initial weights. Therefore, SVM generally outperforms BPNN. But, as shown in Table 1, BPNN may outperform SVM in some sub-periods. There may be two reasons: 1) the data in the sub-periods may be more suited to the BPNN's learning algorithms; 2) the chosen BPNN with the best performance outperforms SVM by chance with its initial random weight. Generally speaking, SVM outperforms ARIMA and BPNN and is more capable of oil price time series forecasting.

4 Conclusions

It has been shown in the literatures that support vector machines can perform very well on time series forecasting. The largest benefit of SVM is the fact that a global solution can be attained. In addition, due to the specific optimization procedure it is assured that over-training is avoided and the SVM solution is general.

In this paper, we propose a new method for predicting crude oil price time series based on support vector machines. There exist four phases when developing a SVM for time series forecasting: data sampling, sample preprocessing, training & learning and out-of-sample forecasting. An empirical study, in which we compare SVM's performance with those of autoregressive integrated moving average models and back-propagation neural networks, is put underway to verify the effectiveness of the SVM-based method. The results show that SVM is superior to the other individual forecasting methods in monthly oil price prediction. The prediction can be improved if irregular influences are taken into consideration in the framework of TEI@I [7], which is undoubtedly a very interesting and meaningful topic for our future study.

References

1. Panas, E., Ninni, V.: Are oil markets chaotic? A non-linear dynamic analysis. Energy Economics 22 (2000) 549-568
2. Hagen, R.: How is the international price of a particular crude determined? OPEC Review 18 (1994) 145-158
3. Huntington, H.G.: Oil price forecasting in the 1980s: what went wrong? The Energy Journal 15 (1994) 1-22
4. Abramson, B., Finizza, A.: Probabilistic forecasts from probabilistic models: a case study in the oil market. International Journal of Forecasting 11(1995) 63-72
5. Barone-Adesi, G., Bourgoin, F., Giannopoulos, K.: Don't look back. Risk August 8 (1998) 100-103
6. Nelson, Y., S. Stoner, G. Gemis, H.D. Nix: Results of Delphi VIII survey of oil price forecasts. Energy Report, California Energy Commission (1994)
7. Wang, S.Y., L.A. Yu, K.K.Lai: Crude oil price forecasting with TEI@I methodology. Journal of Systems Science and Complexity 18 (2005) 145-166
8. Vapnik, V.N.: The Nature of Statistical Learning Theory. Springer, New York (1995)
9. Burges, C.J.C.: A tutorial on support vector machines for pattern recognition. Data Mining and Knowledge Discovery 2 (1998) 121-167
10. Huang, W., Nakamori, Y., Wang, S.Y.: Forecasting stock market movement direction with support vector machine. Computers & Operations Research 32 (2005) 2513-2522
11. Muller, K.R., Smola, J.A., Scholkopf, B.: Prediction time series with support vector machines. Proceedings of International Conference on Artificial Neural Networks, Lausanne (1997) 999-1004

Credit Risk Evaluation Based on LINMAP

Tai-yong Mou[1], Zong-fang Zhou[1], and Yong Shi[2]

[1] Management School, University of Electronic Science & Technology of China, Chengdu Sichuan 610054 P.R. China
moutaiyong@163.com

[2] Chinese Academy of Science Research Center on Data Technology and Knowledge Economy, Beijing 100080 P.R. China
yshi@gucas.ac.cn

Abstract. This paper deals with customer's credit risk assessment. We put forward a method based on the Linear Programming Technique for Multidimensional Analysis of Preference (LINMAP). The method is used to develop a credit risk assessment model using a large sample of firms derived from the loan portfolio of a leading Construction Bank of China. The model gave us the method to determine the optimal credit evaluation weights and ideal point. Further we give an example for its application.

1 Introduction

Credit risk assessment is a significant area of financial management which is of major interest to practitioners, financiers and credit analysts. Considerable attention has been paid in this field from the theoretical and academic points during the last several decades[1-17]. Financial and operational researchers have tried to relate the characteristics of a firm (Financial ratios and strategic variables) to its credit risk[1-5]. According to this relationship the components of credit risk are identified, and decision models are developed to assess credit risk and the corresponding credit worthiness of firms as accurately as possible. The models include Z-score[6,7], discriminant analysis (DA)[8], logit analysis (LA) and probit analysis (PA) and more.

The main purpose of this paper is to investigate the potentials and the applicability of a new discrimination method in credit risk assessment, based on the linear programming technique for multidimensional analysis of preference (LINMAP)[9]. The LINMAP method employs a distance discrimination procedure to determine the class to which the firms under consideration belong.

2 The Linear Programming Technique Based on Data Mining

Supposing the commercial bank will provides a loan for m customers, to evaluate the customers' credit, they employ n indexes. So m firms and n indexes form a credit decision space[10-17]. In this credit decision space, every customer's credit corresponds to a point respectively. If the customer's best credit (the bank's preference) can be denoted with an ideal point in the decision space, the customer's credit can be

V.N. Alexandrov et al. (Eds.): ICCS 2006, Part IV, LNCS 3994, pp. 452 – 459, 2006.
© Springer-Verlag Berlin Heidelberg 2006

described with a weighted distance. In the credit decision space, the weighted distance between every point $(x_{i1}, x_{i2}, \cdots, x_{in})$ and ideal point $(x_1^*, x_2^*, \cdots, x_n^*)$ is

$$d_i = \left[\sum_{j=1}^{n} w_j (x_{ij} - x_j^*)^2 \right]^{\frac{1}{2}}, \ i = 1,2,\cdots,m. \tag{1}$$

Where, $w_j, (j = 1,2,\cdots,n)$ is the *i-th* index's weight. It presents the importance of *i-th* index. At the same time, the square distance can be denoted

$$S_i = d_i^2 = \sum_{j=1}^{n} w_j (x_{ij} - x_j^*)^2, \ i = 1,2,\cdots,m. \tag{2}$$

In (2), w and x^* are unknown. We can get them by using a training sample.

Using the information in the training sample, we have a sequence relation (F_k, F_l). For simplicity, we substitute (k,l) for sequence relation (F_k, F_l). The (k,l) means that *k-th* customer's credit is better than *l-th* customer's. Further, we can get a sequence relation set $Q = \{(k,l)\}$. Obviously, there are $n(n-1)/2$ elements in set *Q*, if we compare every two customer's credit.

We can calculate the weighted distance between *k-th* customer or *l-th* customer and ideal point, if we can find out w and x^*. They are

$$S_k = \sum_{j=1}^{n} w_j (x_{kj} - x_j^*)^2 \text{ and } S_l = \sum_{j=1}^{n} w_j (x_{lj} - x_j^*)^2. \tag{3}$$

In (3), x_{kj} is the value of the *k-th* customer to the *i-th* index. Similarly the value of the *l-th* customer to the *j-th* index is x_{lj}.

In set *Q*, if every sequence relation (k,l) satisfied

$$S_l \geq S_k, \tag{4}$$

the *k-th* customer's credit is not worse than the *l-th* customer's. It is consistent with the information in training sample. Otherwise, if

$$S_l < S_k, \tag{5}$$

the *k-th* customer's credit is worse than the *l-th* customer's. It is not consistent with information in training sample. Apparently, if the training sample is large enough, the weighted distance should be in accordance with the information of training sample. For measuring the degree of inconsistency between the weighted distances and the training sample information, we have to give a definition.

Definition 1. let $B = \sum_{(k,l)\in Q} (S_l - S_k)^-$, where

$$(S_l - S_k)^- = \begin{cases} 0 & , \ if \ S_l \geq S_k , \\ S_k - S_l & , \ if \ S_l < S_k , \end{cases} = \max(0;(S_k - S_l)) . \tag{6}$$

B is called inconsistent degree between the weighted distances and the training sample information.

From definition 1, we know

$$(S_l - S_k)^- = 0 \text{ if } S_l \geq S_k \text{ and } (S_l - S_k)^- > 0 \text{ if } S_l < S_k$$

The more different between S_l and S_k , the inconsistent degree is deeper. $(S_l - S_k)^- = S_k - S_l$ can measure the inconsistent degree. To finding out optimal w^* and x^* , we have to minimize *B*.

Definition 2. let $G = \sum_{(k,l)\in Q} (S_l - S_k)^+$, where

$$(S_l - S_k)^+ = \begin{cases} S_l - S_k & , \ if \ S_l \geq S_k , \\ 0 & , \ if \ S_l < S_k , \end{cases} = \max(0;(S_l - S_k)) , \tag{7}$$

calling *G* is consistent degree between the weighted distances and the training sample information. From definition 2, we know

$$(S_l - S_k)^- = 0 \text{ if } S_l < S_k ; \qquad (S_l - S_k)^- > 0 \text{ if } S_l \geq S_k .$$

The more different between S_l and S_k , the consistent degree is deeper. $(S_l - S_k)^- = S_l - S_k$ can measure the consistent degree.

We can minimize *B*, which subject to $G > B$ or $G - B = h > 0$, where *h* is a small positive real number. From definition 1 and definition 2, we have

$$(S_l - S_k)^+ - (S_l - S_k)^- = S_l - S_k . \tag{8}$$

Then

$$\begin{aligned} G - B &= \sum_{(k,l)\in Q} (S_l - S_k)^+ - \sum_{(k,l)\in Q} (S_l - S_k)^- \\ &= \sum_{(k,l)\in Q} \left[(S_l - S_k)^+ - (S_l - S_k)^- \right] \\ &= \sum_{(k,l)\in Q} (S_l - S_k) = h. \end{aligned} \tag{9}$$

So, we can get w and x^* by solving problem (10).

$$\begin{aligned} \min B = \sum_{(k,l)\in Q} \max\{0,(S_k - S_l)\} \\ s.t. \quad \sum_{(k,l)} (S_l - S_k) = h. \end{aligned} \tag{10}$$

Solving problem (10) is equivalent to solving problem (11).

$$\begin{cases} \min \sum_{(k,l)} \lambda_{kl} \\ s.t. \quad S_l - S_k + \lambda_{kl} \geq 0, \text{all}\,(k,l) \in Q \\ \sum_{(k,l)} (S_l - S_k) = h \\ \lambda_{kl} \geq 0, \quad \text{all}\,(k,l) \in Q. \end{cases} \tag{11}$$

From (3), we have

$$\begin{aligned} S_l - S_k &= \sum_{j=1}^{n} w_j (x_{lj} - x_j^*)^2 - \sum_{j=1}^{n} w_j (x_{kj} - x_j^*)^2 \\ &= \sum_{j=1}^{n} w_j (x_{lj}^2 - x_{kj}^2) - 2\sum_{j=1}^{n} w_j x_j^* (x_{lj} - x_{kj}). \end{aligned} \tag{12}$$

Because w_j and x_j^* are unknown, we institute variable v_j for $w_j x_j^*$. So, problem (11) is equivalent to problem (13)

$$\begin{cases} \min \sum_{(k,l)} \lambda_{kl} \\ s.t. \sum_{j=1}^{n} w_j (x_{lj}^2 - x_{kj}^2) - 2\sum_{j=1}^{n} v_j (x_{lj} - x_{kj}) + \lambda_{kl} \geq 0, \\ \qquad all\,(k,l) \in Q, \\ \sum_{j=1}^{n} w_j \sum_{(k,l)\in Q} (x_{lj}^2 - x_{kj}^2) - 2\sum_{j=1}^{n} v_j \sum_{(k,l)\in Q} (x_{lj} - x_{kj}) = h, \\ \qquad w_j \geq 0, j = 1,2,\cdots,n, \\ \lambda_{kl} \geq 0, all(k,l) \in Q. \end{cases} \tag{13}$$

By solving (13), we get all indexes' optimal weight w^* and ideal point in credit decision space x^*.

$$w^* = \begin{pmatrix} w_1^* \\ w_2^* \\ \vdots \\ w_n^* \end{pmatrix}, \quad v^* = \begin{pmatrix} v_1^* \\ v_2^* \\ \vdots \\ v_n^* \end{pmatrix} \text{ and } x^* = \begin{pmatrix} v_1^*/w_1^* \\ v_2^*/w_2^* \\ \vdots \\ v_n^*/w_n^* \end{pmatrix}.$$

Then we can compute the weighted distance S between every customer's credit and ideal point. The value of S is smaller, the customer's credit is better.

3 An Example for Application

To illustrate the model of the LINMAP method, consider a simple example consisting of 50 firms F_1, F_2, …, F_{50}, evaluated along five financial ratios (x_1: retained earnings/total assets, x_2: earnings before interest and taxes/total assets, x_3: net profit/total assets, x_4: net profit after taxes/total assets, x_5: liquid assets/total assets). The firms are

Table 1. Data in the training sample

Firms \ x_{ij} \ Indexes	x_1	x_2	x_3	x_4	x_5	Categories
F1	0.02	0.03	0.01	0.61	0.03	1.00
F2	0.06	0.13	0.03	0.66	0.11	1.00
F3	0.07	-0.13	-0.13	0.29	0.21	1.00
F4	0.11	0.08	0.01	0.46	0.57	1.00
F5	0.13	0.07	-0.03	0.55	0.53	1.00
F6	0.04	0.06	0.02	0.26	0.29	2.00
F7	0.05	0.32	0.01	0.24	0.17	2.00
F8	0.07	0.08	0.01	0.26	0.33	2.00
F9	0.11	0.01	0.00	0.26	0.36	2.00
F10	0.12	0.04	0.01	0.16	0.28	2.00
F11	0.20	0.08	0.00	0.66	0.79	3.00
F12	0.21	-0.08	-0.02	0.79	0.99	3.00
F13	0.23	0.00	0.00	0.77	1.00	3.00
F14	0.42	-0.04	0.00	0.81	0.99	3.00
F15	0.45	0.12	0.09	0.47	0.92	3.00
F16	0.08	0.18	0.14	0.54	0.62	4.00
F17	0.25	0.30	0.01	0.18	0.43	4.00
F18	0.32	0.17	0.11	0.52	0.84	4.00
F19	0.34	0.13	0.17	0.17	0.51	4.00
F20	0.35	0.02	0.14	0.65	1.00	4.00
F21	0.19	0.10	0.03	0.57	0.75	5.00
F22	0.40	0.09	-0.01	0.33	0.73	5.00
F23	0.42	0.37	0.10	0.50	0.92	5.00
F24	0.49	0.05	0.03	0.38	0.88	5.00
F25	0.57	0.01	0.01	0.24	0.81	5.00

Table 2. Square distance of firms in training sample

Firms	Square distance					Categories
F1～F5	0.0954	0.0966	0.1011	0.1049	0.1036	1
F6～F10	0.1050	0.1058	0.1055	0.1058	0.1058	2
F11～F15	0.1063	0.1072	0.1081	0.1058	0.1093	3
F16～F20	0.1068	0.1092	0.1086	0.1083	0.1083	4
F21～F25	0.1068	0.1083	0.1115	0.1092	0.1097	5

Table 3. Data in examination sample

Firms \ Indexes x_{ij}	x_1	x_2	x_3	x_4	x_5	Categories
F26	0.00	0.04	0.04	0.82	0.62	1
F27	0.03	0.06	0.01	0.74	0.37	1
F28	0.05	0.00	0.01	0.57	0.62	1
F29	0.07	0.18	-0.03	0.58	0.64	1
F30	0.07	0.01	0.00	0.72	0.59	1
F31	0.04	0.00	0.00	0.40	0.44	2
F32	0.04	0.10	0.01	0.50	0.34	2
F33	0.07	0.10	0.01	0.17	0.24	2
F34	0.07	0.12	0.04	0.45	0.45	2
F35	0.13	0.04	0.04	0.23	0.36	2
F36	0.09	0.08	0.00	0.70	0.79	3
F37	0.15	0.09	0.00	0.65	0.80	3
F38	0.17	-0.12	0.00	0.65	0.82	3
F39	0.41	0.09	-0.01	0.58	0.71	3
F40	0.46	-0.04	-0.01	0.79	0.99	3
F41	0.03	0.08	0.04	0.42	0.71	4
F42	0.13	0.16	0.14	0.23	0.36	4
F43	0.14	0.11	0.04	0.38	0.53	4
F44	0.17	0.12	0.18	0.24	0.41	4
F45	0.19	0.00	0.00	0.54	0.33	4
F46	0.23	0.11	0.01	0.27	0.50	5
F47	0.33	0.08	0.07	0.46	0.79	5
F48	0.37	0.19	0.01	0.62	0.99	5
F49	0.45	0.00	0.06	0.43	0.89	5
F50	0.55	0.18	0.05	0.26	0.80	5

Table 4. Square distance of firms in examination sample

Firms	Square distance					Categories
F26～F30	0.1038	0.0999	0.1051	0.1023	0.1029	1
F31～F35	0.1045	0.1042	0.1060	0.1047	0.1056	2
F36～F40	0.1070	0.1072	0.1054	0.1068	0.1057	3
F41～F45	0.1069	0.1067	0.1063	0.1065	0.1062	4
F46～F50	0.1097	0.1076	0.1099	0.1083	0.1108	5

divided into two groups (a training sample and a examination sample) and classified into five categories. Table 1 illustrates the performances of the firms according to each ratio and their classification.

From Table 1, we have a sequence relation set. So, we have a linear programming problem. Solving it, we can get the optimal weights w^* and ideal point x^*.

$$w^* = \begin{pmatrix} 0.0699 \\ 0.0100 \\ 0.0100 \\ 0.1154 \\ 0.0100 \end{pmatrix}, \qquad x^* = \begin{pmatrix} 0.8735 \\ -4.6423 \\ 0.3161 \\ 1.0913 \\ -8.9744 \end{pmatrix}.$$

Table 2 gives the square distance of firms in training sample. Then, we use an examination sample (as Table3) for model validation. Easily, we can obtain the results as table4 shows. Combining Table4 with Table3 and Table2, we can find that the model developed provided high classification accuracy. Only credit categories of F28, F33, F35, F36, F37 and F39 are different from examination sample information.

4 Conclusions

This paper focused on the identification of commercial bank customer's credit. The approach employed for developing credit assessment model is based on LINMAP approach. The method employs mathematical programming techniques to develop credit rating model. In this paper the objective was to find out a method of credit discrimination. Such a discrimination method supports credit analysts in identifying potential defaulters, thus facilitating credit-granting decisions. Of course, the applications of the LINMAP method are not only limited to credit assessment; it also involves other financial risk management fields.

References

[1] M. Doupos, K. Kosmidou, G. Baourakis, and C. Zopounidis, Credit risk assessment using a multicriteria hierarchial discrimination approach: A comparative analysis, European Journal of Operational Research, 2002(138): 392-412.

[2] West R C. A Factor-analysis Approach to Bank Condition. Journal of Banking Finance, 1985, 9:253~266.

[3] Alastair G., Corporate credit analysis, Glenlake Publishing Company Limited, 2000.

[4] Hwang ,C. L. and Yoon, K. S., Multiple attribute decision making, Springer-Verlag , Berlin,1981.

[5] Zhou Z.F., Tang X.W., Mu T.Y. and Lu Y.J. An multi-targets evaluation approaches to customers credit, Proceedings of 2003 International Conference on Management Science & Engineering, USA, Aug. 2003.

[6] Altman. E. I. and Saunders. A., Credit risk measurement: developments over the last 20 years, Journal of Banking and Finance, 1998(21): 1721-1742.

[7] Altman. E. I., Financial ratios, discriminant analysis and the prediction of corporate bankruptcy. Journal of Finance, 1968, 23: 589~609.

[8] Press S J, Wilson S. Choosing between logistic regression and discriminant analysis, J. Amer. Statist. Asoc. , 1978, 73:699~705.

[9] Srinivasan, V. and Shocker, A. D. Linear programming techniques for multimensional analysis of preference, Psychometica, 1973:337.

[10] Chen LH, Chiou TW: A fuzzy credit –rating approach for commercial loans: a Taiwan case, OMEGA.,The International Journal of Management Science, 1999(27) P.407-19.

[11] Myer P A, Pifer H. Prediction of bank failures, Journal of Finance, 1970, 25: 853~868.

[12] Barth J R, Brumbaugh R D, Sauerhaft D. Thrift institution failures: estimating the regulator's closure rule, G.G. Kaufman(Ed), Research in Financial Services, Greenwich, CT: JAI Press, 1989, 1.

[13] Caouettee J.B., Altman E.I. and Narayanan P., Managing credit risk: the next great financial challenge, John Wiley & Sons, 1998.

[14] Zhou Z.F., Tang X.W. and Shi Y., The mathematical structure on credit rating, Fur East Journal of Applied Mathematics, 2005(19).

[15] Zhou Z.F., Tang X.W., The ordering relation and dominance structure of indexes space for credit, Systems Engineering—Theory & Practice, 2004 (24): 9-15 (in Chinese).

[16] Zhou Z.F., Tang X.W. and Mou T.Y., The optimization techniques approach to customer's credit, Proceedings of ICSSSE'2003, Hongkong, 2003.10: 177-182.

[17] Zhou Z.F., Tang X.W. and Shi Y., Estimate errors on credit evaluation of commerce banks, Lecture Notes in Computer Science, 2005 (3227): 229-233.

Logic Mining for Financial Data

G. Felici[1], M.A. Galante[2], and L. Torosantucci[3]

[1] Istituto di Analisi dei Sistemi ed Informatica del CNR,
Viale Manzoni 30 - 00185 Roma, Italy
felici@iasi.cnr.it

[2] Istituto di Analisi dei Sistemi ed Informatica del CNR,
Viale Manzoni 30 - 00185 Roma, Italy
galante@iasi.cnr.it

[3] Banca Nazionale del Lavoro
luca.torosantucci@bnlmail.com

Abstract. In this paper we consider a new strategy for supporting timing decisions in stock markets. The approach uses the logic data miner *Lsquare*, based on logic optimisation techniques. We adopt a novel concept of *good* session, based on the best return expected within a given time horizon. Such definition links indirectly the buying decision with the selling decision and make it possible to exploit particular features of stock time series. The method is translated into an investment strategy and then it is compared with the standard buy & hold strategy.

1 Introduction

In this paper we investigate a new approach for the analysis and the forecasting of financial data. Such approach is based on the combination of Logic Data Mining with financial Technical Analysis. Logic Data Mining is a particular type of automated learning that derives from data a set logic rules that exploit the structure of the data, if present, and then allows to classify new records with a good level of accuracy. We use the *Lsquare* System [6], a logic data miner based on logic programming designed specifically for application to logic data that has some original features particularly useful in this context. Technical Analysis is based on a number of indicators that are computed from the values of a stock in different time intervals. It typically considers the opening, closing, minimum, and maximum values of a stock in one or more time intervals (e.g., the daily sessions) and computes from such values some indicators that suggest to buy or to sell the security under analysis. In this paper we briefly overview the basic concept of financial theory, provide a general overview of the method proposed, describe the learning algorithm used, and present computational results.

2 Financial Theory and Learning Applications

One of the most representative theories in finance is the Efficient Market Hypothesis (EMH) of Fama ([4], [5]). This hypothesis states that security prices fully, instantaneously and rationally reflect all available information about that

V.N. Alexandrov et al. (Eds.): ICCS 2006, Part IV, LNCS 3994, pp. 460–467, 2006.
© Springer-Verlag Berlin Heidelberg 2006

security. A price that fully and rationally reflects all available information implies that the market is perfectly rational and that every kind of information can be unequivocally interpreted. A price that instantaneously reflects all available information implies that the time reaction of the market is so fast that it is not possible to take advantage of new data. For these reasons, if the EMH is true, it is not possible to implement an investment strategy that constantly obtains greater returns than the market, usually represented by an index (e.g. the S&P500). The discussion on market efficiency is not only academic. Generally, investors pick the asset of a portfolio by using either *Fundamental Analysis* or *Technical Analysis* (TA). The fundamentalist states that, in general, prices do not reflect the real value of the underlying firm. This implies that the market is not rational and the stock can be either overpriced or underpriced, according to the misunderstanding of the available information. The fundamentalist also believes that every mistake will be corrected by the market. On the contrary, the follower of Technical Analysis believes that the market, on average, is rational. If a market is really efficient, both technical and fundamental analysis can not constantly ensure greater returns than the market. For this reason, according to the EMH, the best strategy for an investor should be to buy and hold a representative portfolio of the market. After the first exhaustive claim of the EMH made by Fama in 1970, a great number of papers seemed to confirm unequivocally the efficiency of the market. At the same time, a lot of research showed the unpredictability of prices and markets. In the 1980s a few papers highlight the first anomalies regarding the efficient market hypothesis, but only at the end of that decade and the beginning of the 1990s the idea of the EMH was reconsidered. In 1989 Samuelson changed his original opinion [13] and stated that efficiency is not a perfect paradigm for the market. In his work, he recalls papers which show statistically significant deviations from the random walk and the fact that it is possible to find, on long time horizons, a tendency of the price. Several papers test the effectiveness of technical analysis by means of empirical research on specific timing strategy (among them, [3] [5], [11], [12]). The work presented in this paper falls into this last category, but is characterized by the application of Data Mining techniques, that provide a large number of non parametric, non linear and non statistic models to extract knowledge from observed data. Often, the complexity of such models requires sophisticated optimisation techniques. Such tools have appeared in financial literature in recent years and seem to play a key role to uncover the hidden and local regularities that may allow to beat the market in the short to mid term, contradicting the EMH in its strong form but not necessarily in its weak or semi-strong form. Their effectiveness is often to be accounted to their capability of using models that go beyond the standard linear models of multidimensional statistic regression, as shown in the many applications described in [8], [9], [10].

3 An Overview of the Method

In this section we describe the proposed method to determine the timing to buy and to sell a stock in a profitable way. We provide the list of the TA indicators used and the definition of *good session*; we consider the problem of identifying the right training period and the problem of when is it useful to learn from data. The learning tool *Lsquare* and its role in the whole picture will be treated in Section 4.

3.1 The Learning Approach

We intend to learn from a number of previous sessions whether the current session is good to sell or to buy a given stock. To accomplish this task we must start from the definition of a number of measures of each session and a way to identify whether a session is good or not. Such data, properly processed by a learning algorithm, may identify some relation between the measures of the session and its type (good, not good), that can be used for making a decision about the right operation to make in the current session. In the following we will refer to daily sessions (technically a session may be identified by any fixed time interval). We first provide some notation. We refer to t as the current session (or day) where we want to make the decision. The *Training Period* TP is the number of days before t that we want to analyse to support our decision. We indicate with T the length in days of the training period. Each session $i \in TP$ is represented by a vector of m measures $V_i = v_i^1, v_i^2, ..., v_i^m$. Each session can belong to one of two classes: class A (*buy*) or class B (*don't buy*). Then, we apply a learning algorithm to determine a classification rule into A or B of the sessions in the TP based on their description by vectors $V_i, i \in TP$. Such rule is used to determine whether today's session, S_t, belongs to class A or class B and take appropriate action. In the next sections we will see how we determine the measures in V_i and what determines the belonging of a session to A or B.

3.2 The Indicators of Technical Analysis

Technical Analysis (TA) developed a great number of analysis instruments. Typically, four different types are considered: *Trend Instruments*, that identify the trend by means of straight lines which pass through as great a number as possible of local minimum and maximum of prices; *Moving average*, by which traders buy (sell) when the chart is over (under) its moving average; *Oscillator Indicators*, that are simple mathematical models which define critical situations of the market. The presence of a specific codified pattern (the codification is based on the analysis of the past) can describe the most probable consequence for the price in the future. Contrary to fundamental strategies, technical analysis does not use public information such as announcements of new prospects or balance sheet items; according to traders, the current price correctly reflects this information. Generally, four time series are used, which are composed of the *highest price*, the *lowest price*, the closing price and the *open price* of a significant period of time, called *session*. The length of the session depends on the time horizon of the single investor. The most common session is daily, but investors with a shorter time horizon can also use a shorter session, for example of an hour or five minutes. On the other hand, investors with a longer horizon can use a session of a week or a month. We have tested and implemented the most common TA indicators, on different time periods, and verified their effectiveness in spotting a good session when used as input data for the logic data miner. We omit for brevity a detailed description of the indicators listed below, as they are common knowledge in the financial literature: *Moving Average (MA)*, *Relative Strength Indicator (RSI)*, *Stochastic Oscillator*, *P-Momentum*, *On balance volume (OBV)*, *Price and Volume Trend (PVT)*, *Volume Accumulation Oscillator (VAO)*.

The most serious obstacle to the investigation of technical analysis is his great number of instruments which can be combined to implement a timing investment strategy. Each instrument concentrates his analysis on a particular aspect of the market; even if a timing strategy based on one or more TA indicators may be reasonable from a financial point of view, it is unlikely to ensure a better result than the market in varying conditions. Moreover, each instrument generally depends on a set of parameters and these parameters depend on the current market and on the time horizon of the investor. Typically these parameters are thresholds which permit to establish if a stock is over- or under-bought. Unfortunately, there is not a rigorous method to fix them. We have tackled these two open problems in the following way. First, we have represented each session by a large number of indicators, with different values of n, and determined experimentally what set of indicators was providing better overall performances in our learning scheme. Once such reduced set was determined, we used it to represent each session, using the selected indicators to determine the components of the vectors V_i for all sessions in the training period TP (see section 3.1). Second, we have avoided the choice of thresholds for the indicators using a particular feature of the learning method *Lsquare* that automatically maps rational values into intervals in an optimal way with respect to the learning task.

3.3 The Good Sessions and the Timing Strategy

One of the problems that is encountered when using a learning method for financial decision is what is to be considered a good (or bad) session; in other words, what are the classes (e.g., the A and B of section 3.1) that we would like to be able to recognize in the future to make good decisions. Such definition is strongly dependent on the investment strategy that will use this information. In this paper we have adopted an original definition of *good session* that is related with the time horizon in which the trader intends to operate. Let us assume that this time horizon, indicated with h, is a measure of the frequency in which we want to operate on the stock (for example, 10 days). For a given session i, let $close_i$, $open_i$, min_i, max_i, be the opening, closing, minimum, and maximum prices in the session, respectively. For session i, let the *Best Return Within h* be defined as:

$$BRW_h(i) = max_{j=i,i+h} \frac{close_j - close_i}{close_i}.$$

Such value represents the maximum increase in value that could be obtained on that stock buying it on the closing price of day i and selling it within the next h days, at closing price. For a given training period TP, consider now the distribution of $BRW_h(i), i \in TP$, and take the value λ as the splitting point that separates the highest 25% of that distribution from the rest. Now, assume one can correctly predict whether today's session t has $BRW_h(t)$ higher than λ or not. This would imply a very simple investment strategy that guarantees returns that are significantly higher than the market. Obviously such strategy cannot be applied in practice, as we cannot know in advance the value of $BRW_h(t)$; but, if we can estimate with a good level of accuracy whether it is above the current λ, we can expect high returns and sell the stock within h sessions as soon as it rises above λ. On the other hand, if the prediction turns out to be incorrect, that is, if the stock price does not rise as much as λ in h sessions, we sell the stock and tolerate that potential loss.

3.4 The Identification of the Training Period

The strategy described above is applied anew in each session t, computing new indicators and new classification rules from a TP of given length determined with a sliding time window. A relevant question to answer is what is the appropriate length of the training period; the length is obviously dependent from the time horizon h. Such parameter requires proper tuning and has been found that a value of 75 days when $h = 10$ provides good performances for the stocks analysed. Another crucial issue is whether the information contained in the training period is intersting for learning or not. In this setting, two major considerations arise. As far as the current trend of the stock is concerned, an immediate information is found in the value of λ. If in session t its training period TP has $\lambda < 0$, we know that even in the best 25% of the sessions in TP there are sessions that do not provide any gain in the value of the stock. In such case the session is skipped, no learning is applied and the stock is not purchased. When the distribution of $BRW_h(i), i \in TP$, is very flat, the information available in TP is not sufficiently varied to provide good support for differentiation of the two classes. It may also happen that all sessions in the training period are positive sessions: they provide a value of $BRW_h(i)$ sufficiently large, and thus the recognition problem is not relevant for the success of the investment. To apply this consideration is a systematic way, we adopt a filtering rule based on the *mean* μ and the *variance* σ of the distribution of $BRW_h(i), i \in TP$ and set a threshold σ^o and two thresholds μ_{low} and μ_{high} to implement the above considerations.

4 The Learning System Lsquare

The learning tool used in this application, *Lsquare*, is a particular learning method that operates on data represented by logic variables and produces rules in propositional logic that classify the records in one of two classes. The general scheme of this method is the one of automatic learning, where the records presented in a *training set* are used to infer the rules that link the class variable with some measures; these rules are then used to classify new records and predict their class. Here we intend to apply this method in each daily session, using as training data the previous T sessions, represented by a vector V_i of measures derived from a set of TA indicators. The training set, or TP, is divided into class A (good sessions, that have the value of the *Best Return Within h days* larger than λ (see Section 3.3)), and class B (the other sessions in TP). The choice of Lsquare is motivated by the fact that it uses a logic representation of the description variables, that are to all extents logic variables, and of the classification rules, that are logic formulas in Disjunctive Normal Form (DNF). Such feature enables to analyse and interpret the classification formula also from the TA point of view, exploiting the relation between the indicators and their threshold values. The learning of propositional formulas from data is tackled also by other learning methods, such as the strongly heuristic, simple and widely used decision trees (originally proposed in [2]), to the more sophisticated LAD system ([1]), to the greedy approach proposed in [14]. The *Lsquare* system and some of its additional components have been described in [6], and its detailed description is out of the scope of this paper. Here we simply mention the fact that the rules are determined using a particular problem formulation that amount to be a well know and hard combinatorial optimisation problem, the *minimum cost satisfiability problem*, or MINSAT, that is solved using a sophisticated program based on decomposition and learning techniques [15]. The DNF formulas identified are composed of conjunctive clauses that are searched for in order of coverage of the training set, and are formed by few clauses with large coverage (the interpretation of the trends present in the data) and several clauses with smaller coverage (the interpretation of the outliers in the training set). *Lsquare* is equipped with an automatic method to map continuous variables into binary ones (a complete description of this method may be found in [15]). This approach tries to solve the problem related with the choice of the thresholds in the use of TA indicators: we can in fact use the natural representation of many of them (e.g., the 0-100 value of the indicator RSI see section 3.2) and let the system choose the thresholds at which these indicators become useful to differentiate the good from the bad sessions.

5 Experiments

In this section we describe the results obtained by the application of our method to time series of stocks. In each experiment we consider an experiment period, in which each day represent a separated session. We assume to dispose of a fixed amount of capital to invest, say K, available at the beginning of the experiment period. Some necessary hypotheses were also made, that may reduce part of the risks related with the application of investment strategies in real markets: we do not allow to buy or sell a portion of the available capital, we assume to successfully buy or sell the stock at the closing price of each session, and we discount a fixed 0.02% transaction cost each time the stock is bought. Having chosen a training period TP of length T, a time horizon h, σ^o, mu_{low}, and mu_{high}, we then perform the following steps (where t is the current session and TP is the set of sessions from $t-T-1-h$ to $t-1-h$):

INVESTMENT_STRATEGY(TP, h, σ^o, mu_{low},mu_{high})

1. **Initialize**. Set $i = 0$, $d^* = -1$, $\lambda^* = -1$, goto step 1
2. **Classify** TP. For each session $i \in TP$, compute $BRW_h(i)$. From the distribution of $BRW_h(i), I \in TP$, compute λ, μ and σ. For each session $i \in TP$, if $BRW_h(i) > \lambda$ define session I to be in set A, and to be in set B otherwise; goto step 2
3. **Evaluate bad**. If ($\lambda \leq 0$) or ($\sigma < sigma^o$ and $\mu < mu_{low}$) then go to step 6, else goto step 3
4. **Evaluate good**. If ($\lambda > 0$) and ($\sigma < sigma^o$ and $\mu > mu_{high}$) then go to step 5, else goto step 6
5. **Learn and Classify**. Process the sessions in TP by *Lsquare* to learn classification rules for A and B. Apply such rules to current session t. If session t is classified in A, goto step 5, else goto step 6
6. **Buy**. If capital is available, use all of it to buy stock. set $\lambda^* = \lambda$ and $d^* = t + h$ and goto step 1
7. **Sell**. If $BRW_h(t) > \lambda^*$ or $t >= day^*$, sell all stock and goto step 1

We have considered two different stocks in different periods, making sure that the data obtained was correct and complete. The main parameters involved were initially tuned with several runs, and then used in the two experiments described below. We use a value of 10 for the time horizon h and 75 days as training period. The decision thresholds σ^o, mu_{low}, and μ_{high} were set to 1, 2.0 and 3.0 respectively. The first experiment refers to the daily sessions of the *Future Bund* (the future on the decennial german government bond), in the 200 days from November 22nd 2002 to October 10th, 2003. In this period the Bund has a positive trend with little fluctuations and our system is able to improve on Buy & Hold strategy by a significant ratio (see Figure 1, **a**)). At the end of the period, the Buy & Hold reports a gain of 1.049%, outoperformed by the 5.003% obtained by the proposed method. In the second experiment we take into account a stock with more fluctuations, in order to test the capability of the method in a different situtation. The consider case is the *Alitalia* bond from August 27th 2004 to June 6th, 2005. While Buy & Hold registers a loss of -4.639%, the proposed

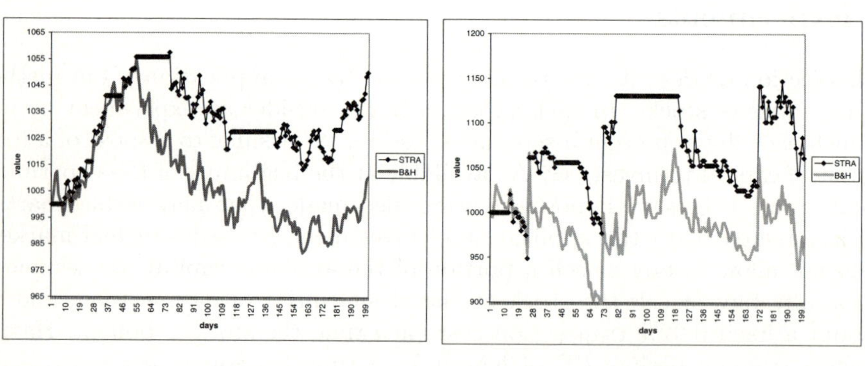

(a) **(b)**

Fig. 1. Comparison with Buy and Hold for German Bund (a) and Alitalia (b) method obtains, at the end of the period, a positive 6.148% (see Figure 1, **b**)). We also note that the proposed strategy shows better performances for each session of the experiment period.

References

1. Boros E., Ibaraki T., Makino K., 1999, Logical analysis of binary data with missing bits *Artificial Intelligence* 107, 219-263.
2. Breiman, Friedman, Olshen, Stone, 1984, *Classification & Regression Trees*,Pacific Grove, Wadsworth.
3. Elton-Gruber, 1995, *Modern Portfolio Theory and Investment Analysis*, John & Wiley & Sons Inc.
4. Fama, E. Efficient capital markets: A review of theory and empirical work, "Journal of Finance 25, 383-417, 1970
5. Fama "Efficient capital markets: II", Journal of Finance 46, 1575-1617, Dic. 1991
6. Felici G., Truemper K., 2001, "A MINSAT Approach for Learning in Logic Domains", *INFORMS Journal on Computing*, Vol.13, No.3, pp. 1-17
7. Frankfurter G., McGoun E., Anomalies in finance: what are they and what are they good for? International Review of Financial Analysis 10 (4) (2001) 407-429.
8. Leigh W., Modani N., Purvis R., Roberts T., Stock market trading rule discovery using technical charting heuristics, Expert Systems with Applicatyions, 23 (2002), 155-159
9. Leigh W., Purvis R., Ragusa J.M., Forecasting the NYSE composite index with technical analysis, pattern recognizer, neural network, and genetic algorithm: a case study in romantic decision support, Decision Support Systems 32 (2002), 361-377
10. Kovalerchuk B., Vityaev E., Data Mining in Finance: Advances in Relational and Hybrid Methods, Kluwer Academic Publishers, Norwell Massachussetts, 2000
11. Pinches G.E., The Random Walk Hypothesis and Technical Analysis, Financial Analysts Journal, March-April 1970.
12. Pring M.J., Technical Analysis Explained : The Successful Investor's Guide to Spotting Investment Trends and Turning Points, McGraw-Hill; 4 edition (February 20, 2002)
13. Samuelson, The Judgement of Economic Science on Rational Portfolio Management: Indexing, Timing, and Long Horizon Effect, Journal of Portfolio Management, n 1, 1989.
14. Triantaphyllou, E., and A.L. Soyster, On the Minimum Number of Logical Clauses Which Can be Inferred From Examples," Computers and Operations Research, Vol. 23, No. 8,1996, pp. 783-79 9
15. Truemper K. *Design of Logic-Based Intelligent Systems*, Wiley-Interscience, 2004

Mining Both Associated and Correlated Patterns

Zhongmei Zhou[1,2], Zhaohui Wu[1], Chunshan Wang[1] , and Yi Feng[1]

[1] College of Computer Science and Technology, Zhejiang University, China
[2] Department of Computer Science, Zhangzhou Teacher's College, China
{zzm, wzh, cswang, fengyi}@zju.edu.cn

Abstract. Association mining cannot find such type of patterns, "the conditional probability that a customer purchasing A is likely to also purchase B is not only greater than the given threshold, but also much greater than the probability that a customer purchases only B. In other words, the sale of A can increase the likelihood of the sale of B." Such kind of patterns are both associated and correlated. Therefore, in this paper, we combine association with correlation in the mining process to discover both associated and correlated patterns. A new interesting measure corr-confidence is proposed for rationally evaluating the correlation relationships. This measure not only has proper bounds for effectively evaluating the correlation degree of patterns, but also is suitable for mining long patterns. Our experimental results show that the mining combined association with correlation is quite a valid approach to discovering both associated and correlated patterns.

1 Introduction

Data mining is defined as the process of discovering significant and potentially useful patterns in large volume of data. Although association or correlation mining can find many interesting patterns, the following two kinds of patterns generated from only association or correlation mining are misleading or meaningless, because neither of the two cases can lead to the result that the sale of A increases the likelihood of the sale of B.

1. A and B are associated but not correlated, that is, although the conditional probability that a customer purchasing A is likely to also purchase B is greater than the given threshold, the probability that a customer purchases only B is not significantly less the conditional probability.
2. A and B are correlated but not associated, that is, although whether a customer purchases B is significantly influenced by whether she/he purchases A, the probability that a customer purchases only B is much greater than the conditional probability that a customer purchasing A is likely to also purchase B.

If pattern AB satisfies the following two conditions, then the sale of A can increase the likelihood of the sale of B.

1. The conditional probability that a customer purchasing A is likely to also purchase B is great enough.
2. The probability that a customer purchases only B is significantly less than the conditional probability that a customer purchasing A is likely to also purchase B.

V.N. Alexandrov et al. (Eds.): ICCS 2006, Part IV, LNCS 3994, pp. 468 – 475, 2006.
© Springer-Verlag Berlin Heidelberg 2006

For example, if P(B)=88% and P(B/A)=90%, then the sale of A cannot increase the likelihood of the sale of B, even though the conditional probability P(B/A)=90% is much greater than the given threshold. In this case, A and B are associated but not correlated. If P(B)=90% and P(B/A)=20%, then the sale of A cannot increase the likelihood of the sale of B, even if the purchase of B is influenced by the purchase of A. It is the case that A and B are correlated but not associated.

Patterns which satisfy the first condition are associated and patterns which satisfy the second condition are correlated, so a pattern which satisfies the two conditions is both associated and correlated. Therefore, in this paper, we combine association with correlation in the mining process to discover both associated and correlated patterns.

One difficulty is how to select a proper interestingness measure that can effectively evaluate the association degree of patterns, as there is still no universally accepted best measure for judging interesting patterns [6]. Omiccinski [5] introduced three alternative interestingness measures, called any-confidence, all-confidence and bond for mining associations. Won-young kim [9] and Young-koo lee [10] used all-confidence to discover interesting patterns although both of them defined a pattern which satisfies the given minimum all-confidence as a correlated pattern. All-confidence can be computed efficiently using its downward closure property [5], so it is employed for association mining in this paper.

Another difficulty is that there are few measures which not only have proper bounds for effectively evaluating the correlation degree of patterns but also are suitable for mining long correlated patterns. The most commonly employed method for correlation mining is that of two-dimensional contingency table analysis of categorical data using the chi-square statistic as a measure of significance. Brin et al. [2] analyzed contingency tables to generate correlation rules that identify statistical correlation in both the presence and absence of items in patterns. H. Liu et al. [3] analyzed contingency tables to discover unexpected and interesting patterns that have a low lever of support and a high level of confidence. Bing Liu et al. [1] used contingency tables for pruning and summarizing the discovered correlations etc. Although the low chi-squared value (less than the cutoff value, e.g. 3.84 at the 95% significance lever [4]) efficiently indicates that all patterns $AB, \overline{A}B, A\overline{B}, \overline{A}\overline{B}$ are independent, that is, A and B, $\overline{A}$ and B, A and $\overline{B}$, $\overline{A}$ and $\overline{B}$ are all independent. The high chi-squared value only indicates that at least one of patterns $AB, \overline{A}B, A\overline{B}, \overline{A}\overline{B}$ is dependent, so it is possible that AB is independent, i.e. A and B are independent, in spite of the high chi-squared value as showed in experimental results. Therefore, when only the presence of items is considered, in other words, when only the sale of A and B is concerned, the chi-squared value is not reasonable for measuring the dependence degree of A and B. For other commonly used measures, the measure $P(AB)\big/P(A)P(B)$ [2] does not have proper bounds. $P(AB)-P(A)P(B)\big/\sqrt{P(A)P(B)(1-P(A))(1-P(B))}$ [8] is not suitable for generating long patterns. $P(AB)-P(A)P(B)$ [7] is not rational when $P(AB)$ is compared with $P(A)P(B)$. For example, if $P(AB)=0.02$, $P(A)P(B)=0.01$, $P(A'B')=0.99$ and $P(A')P(B')=0.98$, then $P(AB)-P(A)P(B)=P(A'B')-P(A')P(B')$. The correlation degree of A and B is equal to the correlation degree of A' and B' by $P(AB)-P(A)P(B)$. However $P(AB)/P(A)P(B)=2$ and $P(A'B')/P(A')P(B')=1.01$, the correlation degree of A and B is evidently higher than the correlation degree of A' and B'. In this paper, a new interestingness measure corr-confidence is proposed for correlation mining. This measure not only has proper bounds for effectively evaluating the correlation degree of patterns, but also is suitable for mining long patterns.

The remainder of this paper is organized as follows: In section 2, some related concepts are given and an algorithm is developed for discovering both associated-correlated patterns. We report our experimental and performance results in section 3. Section 4 concludes the paper.

2 Mining Both Associated and Correlated Patterns

This section first formalizes some related concepts and then gives an algorithm for efficiently discovering all both associated and correlated patterns.

In statistical theory, $A_1, A_2, \cdots, A_n$ are **independent** if $\forall k$ and $\forall 1 \le i_1 < i_2 < \cdots < i_k \le n$,

$$P(A_{i_1} A_{i_2} \cdots A_{i_k}) = P(A_{i_1}) P(A_{i_2}) \cdots P(A_{i_k}) \tag{1}$$

In this paper, let all patterns have more than one item. A new measure corr-confidence (denoted as ρ) is given as follows using (1):

1. If a pattern has two items, such as pattern AB , then

$$\rho(AB) = P(AB) - P(A)P(B) \big/ P(AB) + P(A)P(B) \tag{2}$$

2. If a pattern has more than two items, such as pattern $X = \{i_1 i_2 \cdots i_n\}$, then

$$\rho(X) = \frac{P(i_1 i_2, \cdots, i_n) - P(i_1)P(i_2) \cdots P(i_n)}{P(i_1 i_2, \cdots, i_n) + P(i_1)P(i_2) \cdots P(i_n)}, \ (n \ge 1). \tag{3}$$

From (2) and (3), we can see that ρ has two bounds, i.e. $-1 \le \rho \le 1$.

Let η be a given minimum corr-confidence, if pattern X has two items A, B and $|\rho(AB)| > \eta$, then X is called a correlated pattern or A and B are called correlated, else A and B are called independent. If pattern X has more than two items, we define a correlated pattern and an independent pattern as follows:

Definition 1 (a correlated pattern). Pattern X is called a **correlated pattern**, if and only if there exists a pattern Y which satisfies $Y \subseteq X$ and $|\rho(Y)| > \eta$.

Definition 2 (an independent pattern). If pattern X is not a correlated pattern, then it is called **an independent pattern.**

By the definition 1, we conclude that (1) if pattern X is a correlated pattern, any super pattern of X is a correlated pattern and pattern X must have a subset which is a correlated pattern. (2) pattern X must have two subsets A and B which are correlated.

We define an associated pattern using the measure all-confidence [5].

Let $T = \{i_1, i_2, \cdots, i_m\}$ be a set of m distinct literals called *items* and D be a set of variable length transactions over T. Each transaction contains a set of items, $\{i_{j_1}, i_{j_2}, \cdots, i_{j_k}\} \subset T$. A transaction also has an associated unique identifier. Pattern X is a subset of T. Let $p(X)$ be a power set of pattern X. The interestingness measure all-confidence (denoted as α) of pattern X is defined as follows [5]:

$$\alpha = \frac{|\{d \mid d \in D \wedge X \subset d\}|}{MAX\{i \mid \forall l (l \in p(X) \wedge l \neq \phi \wedge l \neq X \wedge i = |\{d \mid d \in D \wedge l \subset d\}|)\}} \tag{4}$$

Definition 3 (an associated pattern). A pattern is called **an associated pattern,** if its all-confidence is greater than or equal to the given minimum all-confidence.

Definition 4 (an associated-correlated pattern). A pattern is called **an associated-correlated pattern** if it is not only an associated pattern but also a correlated pattern.

Let pattern X be an associated-correlated pattern, then it must have two subsets A and B which satisfy the condition that the sale of A can increase the likelihood of the sale of B.

Example 1. For the transaction database TDB in Table 1, we have $\alpha(AC) = 2/3$ and $\alpha(CE) = 2/3$, so both AC and CE have all-confidence $2/3$. We also have

$$\rho(AC) = \frac{P(AC) - P(A)P(C)}{P(AC) + P(A)P(C)} = \frac{1}{4} \quad \text{and} \quad \rho(CE) = \frac{P(CE) - P(C)P(E)}{P(CE) + P(C)P(E)} = \frac{1}{19}.$$

Let the given minimum all-confidence be 0.5 and the given minimum corr-confidence be 0.1, then both AC and CE are associated patterns. However, pattern AC is a correlated pattern and pattern CE is an independent pattern. Therefore pattern AC is an associated-correlated pattern and pattern CE is an associated but not correlated pattern. From $P(A/C) = 2/3,\ P(A) = 2/5$, we can see that the sale of C can increase the likelihood of the sale of A. Meanwhile, $P(C/A) = 1, P(C) = 3/5$, we can also see that the sale of A can also increase the likelihood of the sale of C. However $P(C/E) = 2/3$, and $P(C) = 3/5$, the sale of E cannot evidently increase the likelihood of the sale of C.

Table 1. Transaction database *TDB*

Transaction id	Items
10	A, B, C
20	C, D, E
30	A, C, D, E
40	D, E
50	B, D

We mine all frequent associated-correlated patterns in two steps. First, we discover all frequent associated patterns, and then test whether they are correlated. We use a level-wise algorithm for discovering all frequent associated-correlated patterns.

Algorithm

Input a transaction database TDB, a minimum support ξ, a minimum corr-confidence η and a minimum all-confidence λ.

Output the complete set of frequent associated-correlated patterns.

C_k : Candidate patterns of size k

L_k : Frequent associated patterns of size k

I_k : Frequent associated and independent patterns of size k

I_k^c : Frequent associated-correlated patterns of size k

L_1 = {frequent items}

$I_1 \leftarrow L_1;\ I_1^c \leftarrow \phi$

For ($k = 1; L_k \,!\neq \phi; k++$) do begin

  C_{k+1} = candidates generated from $I_k * I_k$

  C'_{k+1} =candidates generated from $I_k * I_k^c, I_k^c * I_k^c$

    For each transaction t in database do

      increment the count of all candidates in C_{k+1}, C'_{k+1} that are contained in t

  L_{k+1} = candidates in C_{k+1} with minimum support and minimum all-confidence

  L'_{k+1} = candidates in C'_{k+1} with minimum support and minimum all-confidence

  $I_{k+1}^c \leftarrow L'_{k+1}$

  For each pattern l_{k+1} in L_{k+1}

    If $|\rho(l_{k+1})| < \eta$ insert l_{k+1} into I_{k+1}

    Else insert l_{k+1} into I_{k+1}^c

Return $\cup I_k^c$

Remark: In the algorithm, the prune step is performed as follows:

Forall patterns $c \in C_{k+1}$ do

  Forall k-subsets s of c do

    If ($s \notin L_k$) or ($s \notin L'_k$) delete c from C_{k+1}

    Else if ($s \notin L_k$) then insert c into L'_{k+1}

```
Forall patterns c ∈ C'_{k+1} do
   Forall k-subsets s of c do
     If (s ∉ L'_k) then delete c from C'_{k+1}
```

3 Experiments

In this section, we report our experimental results. All experiments are performed on two kinds of datasets: 1. a dense dataset, Mushroom characteristic dataset, which consists of 8,124 transactions, each with an average length of 23 items. 2. a sparse dataset, Traditional Chinese Medicine (TCM) formula dataset, which consists of 4,643 formulas with 21689 kinds of medicine involved, each with an average length of 10 kinds of medicine. TCM formula dataset is obtained from Information Institute of China Academy of Traditional Chinese Medicine.

Table 2. Correlation confidence and chi-squared value of item pairs

Mushroom dataset	TCM formula dataset
3<->39: 0.0207035...26.5377	55<->703: 0.210326...12.0025
9<->91: 0.0112898...111.029	55<->1187: 0.250133...73.2896
34<->91: 0.0120186…7111.33	55<->3442: 0.254997...15.7207
54<->76: 0.0117676…7.36221	452<->1187: 0.246849...118.593

Table 2 shows the chi-squared value and the corr-confidence of partly item pairs in mushroom dataset and TCM formula dataset. For example, 3<->39 is an item pair. The number after ":" indicates the corr-confidence and the chi-squared value respectively. In mushroom database, item pair 34<->91 has chi-squared value 7111.33 and corr-confidence 0.0120186. Item pair 3<->39 has chi-squared value 26.5377 and corr-confidence 0.0207035. Although the chi-squared value of item pair 34<->91 is very high, the corr-confidence of item pair 34<->91 is lower than the one of item pair 3<->39. Therefore, item pair 3<->39 has a higher level dependence than item pair 34<->91 according to the corr-confidence. The chi-squared statistic simultaneously and uniformly takes into account all possible combinations of the presence and absence of the various attributes being examined as a group, so when only the presence of items in patterns is concerned, the high chi-squared value cannot infer that items in patterns are highly dependent as shown in Table 2.

Figure 1 (A) shows the runtime with limit the length of patterns and without limit the length of patterns as the minimum support ascends. Figure 1 (B) shows the runtimes with limit the length of patterns and without limit the length of patterns as the minimum all-confidence ascends with a fixed minimum support 1%. When the length of patterns produced exceeds 5, almost all frequent associated patterns are correlated patterns because of the upward closure property of correlated patterns. Therefore, we put a limit to the maximal length of patterns generated. Figure 1 indicates that if the maximum length of patterns produced does not exceed 5, the runtime decreases sharply even if the minimum support or the minimum all-confidence is low.

Table 3. Num. of patterns in mushroom data (min_sup 1%, min_len 2, max_len 5, c_conf 1%)

All_conf (%)	30	40	50	60	70	80	90
Independent	112	90	61	31	12	12	7
Associated-correlated	3678	1012	279	83	36	16	8

Table 4. Num. of patterns in TCM data (min_sup 1%, min_len2 max_len5, all_conf 10%)

Corr_conf (%)	5	10	15	20	25	30	35	40
Independent	3	7	16	31	55	76	112	160
Associated-correlated	1058	1054	1045	1030	1006	985	949	901

Table 5. Num. in mushroom data (min_sup1%, min_len 2, max_len 5, all_conf 30%)

Corr_conf (%)	1	2	3	4	5
Independent	112	324	474	541	603
Associated-correlated	3678	3466	3316	3249	3187

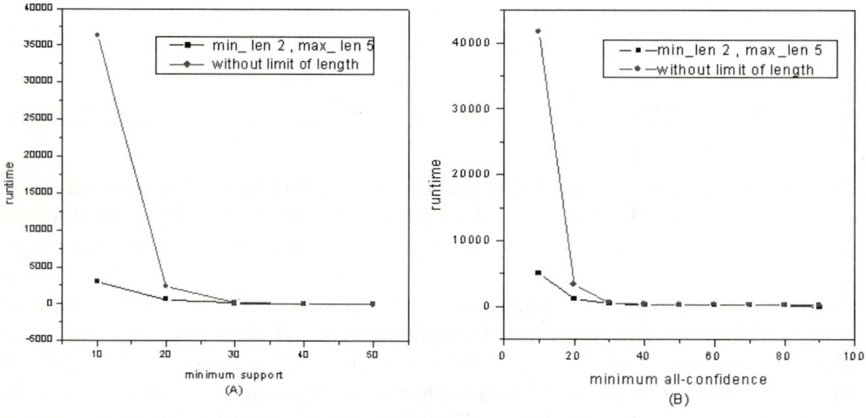

Fig. 1. The runtime of mushroom dataset

Table 3 shows the number of associated-correlated patterns and associated but not correlated patterns generated in mushroom dataset when the minimum all-confidence increases with the fixed minimum support 1%, minimum corr-confidence 1%, minimum pattern length 2 and maximum pattern length 5. From Table 3, we can see that for the minimum corr-confidence 1% and the minimum all-confidence 90%, there are seven associated but not correlated patterns and eight associated-correlated patterns in mushroom dataset. We can conclude that not all associated patterns are correlated even if the minimum all-confidence is much high. Table 4 and Table 5 show the number of associated-correlated patterns and associated but not correlated patterns generated in TCM dataset and mushroom dataset respectively as the minimum corr-confidence varies. To our surprise, when the minimum corr-confidence is 5%, there are only 0.28% associated but not correlated patterns of all associated patterns in TCM dataset, while there are 16% associated but not correlated patterns of all associated patterns in mushroom dataset.

4 Conclusions

The mining combined association with correlation can discover both associated and correlated patterns that are extraordinary useful for making business decisions. In this paper, a new interestingness measure for correlation mining is proposed, which is not only suitable for mining long correlated patterns, but also more rational and easier to control than the chi-squared test and other commonly used measures as shown in experimental results. And an algorithm is developed for efficiently discovering all frequent both associated and correlated patterns. Experimental results show that the techniques developed in this paper are feasible.

Acknowledgments

The work is funded by subprogram of China 973 project (NO. 2003CB317006), China NSF program (No. NSFC60503018) and a grant from education ministry of Fujian of China (No. JA04248).

References

1. Bing Liu, Wynne Hsu, Yiming Ma. Pruning and Summarizing the Discovered Association. In Proc. 1999 ACM SIGKDD Int. Conf. Knowledge Discovery in Databases (KDD'99), pp. 15-18.
2. S. Brin, R. Motwani, C. Silverstein. Beyond Market Basket: Generalizing Association Rules to Correlations. In Proc. 1997 ACM SIGMOD Int. Conf. Management of Data (SIGMOD'97), pp. 265-276.
3. H. Liu, H. Lu, L. Feng, F. Hussain. Efficient Search of Reliable Exceptions. In Proc. Pacific-Asia Conference on Knowledge Discovery and Data Mining (PAKDD'99), pp. 194-203.
4. F. Mills. Statistical Methods, Pitman, 1955.
5. E. Omiecinski. Alternative interesting measures for mining associations. *IEEE Trans. Knowledge and Data Engineering*, 2003(15): 57-69.
6. P.-N. Tan, V. Kumar, J. Srivastava. Selecting the Right Interestingness Measure for Association Patterns. In Proc. 2002 ACM SIGKDD Int. Conf. Knowledge Discovery in Databases (KDD'02), pp. 32-41.
7. G. Piatetsky-Shapiro. Discovery, Analysis and Presentation of Strong Rules. Knowledge Discovery in Databases, AAAI/MIT Press, 1991. pp. 229-248.
8. H. T. Reynolds. The Analysis of Cross-Classifications. The Free Press, New York, 1977.
9. W.-Y. Kim, Y.-K. Lee, J. Han. CCMine: Efficient Mining of Confidence-Closed Correlated Patterns. In Proc. 2004 Pacific-Asia Conf. Knowledge Discovery and Data Mining (PAKDD'04), pp. 569-579.
10. Y.-K. Lee, W.-Y. Kim, Y. D. Cai, J. Han. CoMine: Efficient Mining of Correlated Patterns. In Proc. 2003 Int. Conf. Data Mining (ICDM'03), pp. 581-584.

A New Multi-criteria Convex Quadratic Programming Model for Credit Analysis*

Gang Kou[1], Yi Peng[1, **], Yong Shi[1, 2], and Zhengxin Chen[1]

[1] College of Information Science & Technology,
University of Nebraska at Omaha, Omaha, NE 68182, USA
[2] Chinese Academy of Sciences Research Center on Data Technology
& Knowledge Economy, Graduate University of the Chinese Academy of Sciences,
Beijing 100080, China
{gkou, ypeng, yshi, zchen}@mail.unomaha.edu

Abstract. Mathematical programming based methods have been applied to credit risk analysis and have proven to be powerful tools. One challenging issue in mathematical programming is the computation complexity in finding optimal solutions. To overcome this difficulty, this paper proposes a Multi-criteria Convex Quadratic Programming model (MCCQP). Instead of looking for the global optimal solution, the proposed model only needs to solve a set of linear equations. We test the model using three credit risk analysis datasets and compare MCCQP results with four well-known classification methods: LDA, Decision Tree, SVMLight, and LibSVM. The experimental results indicate that the proposed MCCQP model achieves as good as or even better classification accuracies than other methods.

Keywords: Credit Analysis, Classification, Mathematical Programming, Multi-criteria Decision Making.

1 Introduction

This paper explores solving classification problem, one of the major sub-fields of data mining, through the use of mathematical programming based methods (Bradley et al 1999, Vapnik 1964 and 2000). Such methods have proven to be powerful in solving a variety of machine learning problems (Chang and Lin 2001, Fung 2003, Joachims 1999, Kou et al 2005, Mangasarian 1999, 2000 and 2005, Mitchell 1997, Zheng et al 2004). However, it is difficult to find the optimal solution of a mathematical programming problem. To overcome this difficulty, a new Multi-criteria Convex Quadratic Programming model (MCCQP) is proposed. In the proposed model, we only need to solve a set of linear equations in order to find the global optimal solution.

This paper is organized as follows: section 2 presents the MCCQP model, section 3 illustrates the numerical implementation and comparison study with several well-

* This work was supported in part by Key Project #70531040, #70472074, National Natural Science Foundation of China; 973 Project #2004CB720103, Ministry of Science and Technology, China and BHP Billion Co., Australia.

** Corresponding author.

V.N. Alexandrov et al. (Eds.): ICCS 2006, Part IV, LNCS 3994, pp. 476–484, 2006.
© Springer-Verlag Berlin Heidelberg 2006

established data mining software and reports the results, and section 4 summarizes the paper and discusses future research directions.

2 Multi-criteria Convex Quadratic Programming Model

This section introduces a new MCQP model. This model classifies observations into distinct groups via a hyperplane and based on multiple criteria. The following models represent this concept mathematically (Kou et al 2006):

Each row of a $n \times r$ matrix $\boldsymbol{A} = (\boldsymbol{A}_1, \ldots, \boldsymbol{A}_n)^T$ is an r-dimensional attribute vector $\boldsymbol{A}_i = (A_{i1}, \ldots, A_{ir}) \in \Re^r$ which corresponds to one of the records in the training dataset of a binary classification problem, $i = 1, \ldots, n$; n is the total number of records in the dataset. Two groups, G_1 and G_2, are predefined while $G_1 \cap G_2 = \Phi$ and $\boldsymbol{A}_i \in \{G_1 \cup G_2\}$. A boundary scalar b can be selected to separate G_1 and G_2. Let $\boldsymbol{X} = (x_1, \ldots, x_r)^T \in \Re^r$ be a vector of real number to be determined. Thus, we can establish the following linear inequations (Fisher 1936, Shi et al. 2001):

$$\boldsymbol{A}_i \boldsymbol{X} < b, \qquad \forall \boldsymbol{A}_i \in G_1; \tag{1}$$

$$\boldsymbol{A}_i \boldsymbol{X} \geq b, \qquad \forall \boldsymbol{A}_i \in G_2; \tag{2}$$

In the classification problem, $\boldsymbol{A}_i \boldsymbol{X}$ is the score for the ith data record. If all records are linear separable and an element $\boldsymbol{A}_i$ is correctly classified, then let β_i be the distance from $\boldsymbol{A}_i$ to b, and obviously in linear system, $\boldsymbol{A}_i\boldsymbol{X} = b - \beta_i$, $\forall \boldsymbol{A}_i \in G_1$ and $\boldsymbol{A}_i\boldsymbol{X} = b + \beta_i$, $\forall \boldsymbol{A}_i \in G_2$. However, if we consider the case where the two groups are not completely linear separable, there exist some misclassified records. When an element $\boldsymbol{A}_i$ is misclassified, let α_i be the distance from $\boldsymbol{A}_i$ to b, $\boldsymbol{A}_i\boldsymbol{X} = b + \alpha_i$, $\forall \boldsymbol{A}_i \in G_1$ and $\boldsymbol{A}_i\boldsymbol{X} = b - \alpha_i$, $\forall \boldsymbol{A}_i \in G_2$. To complete the definitions of β_i and α_i, let $\beta_i = 0$ for all misclassified elements and α_i equals to zero for all correctly classified elements. Incorporating the definitions of β_i and α_i, (1) and (2) can be reformulated as the following model:

$$\boldsymbol{A}_i\boldsymbol{X} = b - \delta + \alpha_i - \beta_i, \ \forall \boldsymbol{A}_i \in G_1$$

$$\boldsymbol{A}_i\boldsymbol{X} = b + \delta - \alpha_i + \beta_i, \ \forall \boldsymbol{A}_i \in G_2 \tag{3}$$

δ is a given scalar. $b - \delta$ and $b + \delta$ are two adjusted hyper planes for the model.

Redefine $\boldsymbol{X}$ as $\boldsymbol{X}/\delta$, b as b/δ, α_i as α_i/δ, β_i as β_i/δ, and Define a $n \times n$ diagonal matrix $\boldsymbol{Y}$ which only contains "+1" or "-1" indicates the class membership. A "-1" in row i of matrix $\boldsymbol{Y}$ indicates the corresponding record $\boldsymbol{A}_i \in G_1$ and a "+1" in row i of matrix $\boldsymbol{Y}$ indicates the corresponding record $\boldsymbol{A}_i \in G_2$. The model can be rewritten as:

$$\boldsymbol{Y}(<\boldsymbol{A}\cdot\boldsymbol{X}> - eb) = 1 + \alpha - \beta, \tag{4}$$

where $e=(1,1,\ldots,1)^{\mathrm{T}}$, $\alpha = (\alpha_1,\ldots,\alpha_n)^T$ and $\beta = (\beta_1,\ldots,\beta_n)^T$.

The proposed multi-criteria optimization problem contains three objective functions. The first mathematical function $f(\alpha) = \|\alpha\|_p^p = \sum_{i=1}^{n}(\alpha_i)^p$ (ℓ_p — norm of β_i, $1 \le q \le \infty$) describes the summation of total overlapping distance of misclassified records to b. The second function $g(\beta) = \|\beta\|_q^q = \sum_{i=1}^{n}(\beta_i)^q$ (ℓ_q — norm of β_i, $1 \le q \le \infty$) represents the aggregation of total distance of correctly separated records to b. In order to maximize the distance ($\frac{2}{\|X\|_s^s}$) between the two adjusted bounding hyper planes, the third function $h(X) = \frac{\|X\|_s^s}{2}$ should be minimized. Furthermore, to transform the generalized Multi-Criteria classification model into a single- criterion problem, weights $W_\alpha > 0$ and $W_\beta > 0$ are introduced for $f(\alpha)$ and $g(\beta)$, respectively. A single-criterion mathematical programming model can be set up:

(Model 1) *Minimize* $\frac{1}{2}\|X\|_s^s + W_\alpha \|\alpha\|_p^p - W_\beta \|\beta\|_q^q$

Subject to: $\boldsymbol{Y}(<\boldsymbol{A}\cdot\boldsymbol{X}> - \mathrm{eb}) = e - \alpha + \beta$

$$\alpha_i, \beta_i \ge 0, 1 \le i \le \mathrm{n}$$

where $\boldsymbol{Y}$ is a given $n \times n$ diagonal matrix, $e=(1,1,\ldots,1)^{\mathrm{T}}$, $\alpha = (\alpha_1,\ldots,\alpha_n)^T$, $\beta = (\beta_1,\ldots,\beta_n)^T$, $\boldsymbol{X}$ and b are unrestricted.

Please note that the introduction of β_i is one of the major differences between the proposed model and other existing Support Vectors approaches (Vapnik 1964 and 2000). It is much easier to find optimal solutions for convex quadratic programming form than other forms of nonlinear programming. To make Model 1 a convex quadratic programming form, let $s = 2$, $q = 1$ and $p = 2$. The constraints remain the same and the objective function becomes:

(**Model 2**) *Minimize* $\frac{1}{2}\| X \|_2^2 + W_\alpha \sum_{i=1}^{n} \alpha_i^2 - W_\beta \sum_{i=1}^{n} \beta_i$

Subject to: $Y\,(<A\cdot X> - eb) = e - \alpha + \beta$

Let $\eta_i = \alpha_i - \beta_i$. According to the definition, $\eta_i = \alpha_i$ for all misclassified records and $\eta_i = -\beta_i$ for all correctly separated records. The definition of η_i is one of the major differences between the proposed model and other existing approaches (Fung 2003, Gonzalez-Castano and Meyer 2000).

Add $\frac{W_b}{2} b^2$ to Model 2's objective function and the weight W_b is an arbitrary positive number.

(**Model 3**) *Minimize* $\frac{1}{2}\| X \|_2^2 + \frac{W_\alpha}{2} \sum_{i=1}^{n} \eta_i^2 + W_\beta \sum_{i=1}^{n} \eta_i + \frac{W_b}{2} b^2$

Subject to: $Y\,(<A\cdot X> - eb) = e - \eta$

where Y is a given $n \times n$ diagonal matrix, $e=(1,1,\ldots,1)^T$, $\eta = (\eta_1,\ldots,\eta_n)^T$, η, X and b are unrestricted, $1 \le i \le n$.

The Lagrange function corresponding to Model 5 is

$$L(X,b,\eta,\theta) = \frac{1}{2}\| X \|_2^2 + \frac{W_\alpha}{2}\sum_{i=1}^{n} \eta_i^2 + W_\beta \sum_{i=1}^{n} \eta_i + \frac{W_b}{2} b^2 - \theta^T (Y(<A\cdot X> - eb) - e + \eta)$$

where $\theta = (\theta_1,\ldots,\theta_n)^T$, $\eta = (\eta_1,\ldots,\eta_n)^T$, $\theta_i, \eta_i \in \Re$.

According to Wolfe Dual Theorem, $\nabla_X L(X,b,\eta,\theta) = X - A^T Y\theta = 0$, $\nabla_b L(X,b,\eta,\theta) = W_b b + e^T Y\theta = 0$, $\nabla_\eta L(X,b,\eta,\theta) = W_\alpha \eta + W_\beta e - \theta = 0$.

Introduce the above 3 equations to the constraints of Model 5, we can get:

$$Y((A\cdot A^T)Y\theta + \frac{1}{W_b} e(e^T Y\theta)) + \frac{1}{W_\alpha}(\theta - W_\beta e) = e$$

$$\Rightarrow \theta = \frac{(1+\frac{W_\beta}{W_\alpha})e}{\frac{I}{W_\alpha} + Y((A\cdot A^T) + \frac{1}{W_b} ee^T)Y} \qquad (5)$$

Proposition 1. For some $W_\alpha > 0, \theta = \frac{(1+\frac{W_\beta}{W_\alpha})e}{\frac{I}{W_\alpha} + Y((A\cdot A^T) + \frac{1}{W_b} ee^T)Y}$ exists.

Proof. Let H=Y[A $-(\frac{1}{W_b})^{\frac{1}{2}}$e], we get:

$$\theta = (1+\frac{W_\beta}{W_\alpha})(\frac{I}{W_\alpha}+HH^T)^{-1}e \quad (5')$$

$\forall H, \exists W_\alpha > 0$, when W_α is small enough, the inversion of $(\frac{I}{W_\alpha}+HH^T)$ exists.

$$\text{So } \theta = \frac{(1+\frac{W_\beta}{W_\alpha})e}{\frac{I}{W_\alpha}+Y((A\cdot A^T)+\frac{1}{W_b}ee^T)Y} \text{ exists.}$$

Algorithm 1

Input: a $n \times r$ matrix A as the training dataset, a $n \times n$ diagonal matrix Y labels the class of each record.

Output: classification accuracies for each group in the training dataset, score for every record, decision function $\text{sgn}((X^* \cdot A_i)-b^*)\{\begin{matrix} >0, \Rightarrow A_i \in G_1 \\ \leq 0, \Rightarrow A_i \in G_2 \end{matrix}$

Step 1 compute $\theta^* = (\theta_1, \ldots, \theta_n)^T$ by one of (5) or (6). W_β, W_α and W_b are chosen by standard 10-fold cross-validation.

Step 2 compute $X^* = A^T Y \theta^*$, $b^* = \frac{-1}{W_b} e^T Y \theta^*$.

Step 3 classify a incoming A_i by using decision function $\text{sgn}((X^* \cdot A_i)-b^*)\{\begin{matrix} >0, \Rightarrow A_i \in G_1 \\ \leq 0, \Rightarrow A_i \in G_2 \end{matrix}$.

END

3 Numerical Experiments in Credit Risk Analysis

The model proposed can be used in many fields, such as general bioinformatics, antibody and antigen, credit fraud detection, network security, text mining, etc. We conducted three numerical experiments to evaluate the proposed MCCQP model. All experiments are concerned about credit risk analysis. Each record in these three sets has a class label to indicate its' financial status: either Normal or Bad. Bad indicates a bankrupt credit or firm account and Normal indicates a current status account. The result of MCCQP is compared with the results of 4 widely accepted classification methods: Linear Discriminant Analysis (LDA) (SPSS 2004), Decision Tree based See5 (Quinlan 2003), SVM light (Joachims 1999) and LibSVM (Chang and Lin 2001).

The first benchmark set is a German credit card application dataset from UCI Machine Learning databases (UCI 2005). The German set contains 1000 records (700 Normal and 300 Bad) and 24 variables. The second set is an Australian credit approval dataset from See5 (Quinlan 2003). The Australian set has 383 negative cases (Normal) and 307 positive cases (Bad) with 15 attributes. The last set is a Japanese firm bankruptcy set (Kwak et al 2005). The Japanese set includes Japanese bankrupt (Bad) sample firms (37) and non-bankrupt (Normal) sample firms (111) between 1989 and 1999. Each record has 13 variables.

Credit Classification Process

Input: The Credit Card dataset A = $\{A_1, A_2, A_3, \cdots, A_n\}$, a $n \times n$ diagonal matrix $\boldsymbol{Y}$

Output: Average classification accuracies for Bad and Normal of the test set in 10-fold cross-validation; scores for all records; decision function.

Step 1 Apply several classification methods: LDA, Decision Tree, SVM, MCCQP, to A using 10-fold cross-validation. The outputs are a set of decision functions, one for each classification method.

Step 2 Compute the classification accuracies using the decision functions.

END

The following tables (Table 1, 2, and 3) summarize the averages of 10-fold cross-validation test-sets accuracies of Linear Discriminant Analysis (LDA) (SPSS 2004), Decision Tree base See5 (Quinlan 2003), SVM light (Joachims 1999), LibSVM (Chang and Lin 2001), and MCCQP for each dataset. "Type I Error" is defined as the percentage of predicted Normal records that are actually Bad records and "Type II Error" is defined as the percentage of predicted Bad records that are actually Normal records. Since Type I error indicates the potential charge-off lost of credit issuers, it is considered more costly than Type II error. In addition, a popular measurement, KS score, in credit risk analysis is calculated. The higher the KS score, the better the classification methods. The KS (Kolmogorov-Smirnov) value is defined as:

KS value = *Max* |Cumulative distribution of Bad - Cumulative distribution of Normal|.

Table 1 reports the results of the five classification methods for German set. Among the five methods, LibSVM achieves the best results for all the measurements and MCCQP achieves the second best results.

Table 1. 10-fold cross-validation result of German set

Method	Classification Accuracy			Error Rate		
	Overall	Normal	Bad	Type I	Type II	KS-Score
LDA	72.20%	72.57%	71.33%	28.32%	27.77%	43.90
See5	72.20%	84.00%	44.67%	39.71%	26.37%	28.67
SVMlight	66.50%	77.00%	42.00%	42.96%	35.38%	19.00
LibSVM	**94.00%**	**100.00%**	**80.00%**	**16.67%**	**0.00%**	**80.00**
MCCQP	73.50%	74.38%	72.00%	27.35%	26.24%	46.38

Table 2 summarizes the results for the Australian set. Among the five methods, MCCQP achieves the best overall accuracy, Normal accuracy and Type II error while LDA achieves the lowest Type I error rate and highest Bad classification accuracy and KS-score.

Table 2. 10-fold cross-validation result of Australian set

Method	Classification Accuracy			Error Rate		
	Overall	Normal	Bad	Type I	Type II	KS-Score
LDA	85.80%	80.68%	**92.18%**	**8.83%**	17.33%	**72.86**
See5	86.52%	87.99%	84.69%	14.82%	12.42%	72.68
SVMlight	44.83%	18.03%	90.65%	34.14%	47.48%	8.69
LibSVM	44.83%	86.89%	27.10%	45.62%	32.61%	13.99
MCCQP	**86.38%**	**87.00%**	85.52%	14.27%	**13.20%**	72.52

Table 3 summarizes the result for Japanese set. Among the five methods, MCCQP achieves the highest classification accuracies for overall, Normal, and Bad. In addition, MCCQP has the highest KS-score and lowest Type I and II error rates. Although See5 got the highest classification accuracy for Normal class, it's classification accuracy for Bad is only 35.14%.

Table 3. 10-fold cross-validation result of Japanese set

Method	Classification Accuracy			Error Rate		
	Overall	Normal	Bad	Type I	Type II	KS-Score
LDA	68.92%	68.47%	70.27%	30.28%	30.97%	38.74
See5	72.30%	**84.68%**	35.14%	43.37%	30.36%	19.82
SVMlight	48.15%	47.25%	52.94%	49.90%	49.91%	0.19
LibSVM	50.46%	49.45%	55.88%	47.15%	47.49%	5.33
MCCQP	**72.30%**	72.30%	**72.47%**	**27.58%**	**27.65%**	**44.77**

4 Conclusion

In this paper, a new MCCQP model for classification problem has been presented. In order to validate the model, we apply the model to three credit risk analysis datasets and compare MCCQP results with four well-known classification methods: LDA, Decision Tree, SVMLight, and LibSVM. The experimental results indicate that the proposed MCCQP model achieves as good as or even better classification accuracies than other methods.

There are still many aspects that need further investigation in this research. Theoretically, MCQP is highly efficient method in both computation time and space on large-scale problems. Since all 4 datasets used are relatively small, it will be a nature extension to apply MCQP in massive dataset. $(A_i \cdot A_j)$ in Model 3 and 4 is inner product in the vector space and it can be substituted by a kernel $K(A_i, A_j)$, which will extend the applicability of the proposed model to linear inseparable datasets. Future studies may be done on establishing a theoretical guideline for selection of kernel that is optimal in achieving a satisfactory credit analysis result.

References

Bradley, P.S., Fayyad, U.M., Mangasarian, O.L., Mathematical programming for data mining: Formulations and challenges. *INFORMS Journal on Computing,* 11, 217-238, 1999.

Chang, Chih-Chung and Lin, Chih-Jen, LIBSVM : a library for support vector machines, 2001. Software available at http://www.csie.ntu.edu.tw/~cjlin/libsvm

Fung , G. "Machine learning and data mining via mathematical programming-based support vector machines", Ph.D thesis, The University of Wisconsin - Madison. 2003

Fung, G. and Mangasarian, O. L. Multicategory Proximal Support Vector Machine Classifiers, Machine Learning 59, 2005, 77-97.

Gonzalez-Castano, F. and Meyer, R. Projection support vector machines. Technical Report 00-05, Computer Sciences Department, The University of Wisconsin – Madison, 2000

Li, J.P, Liu, J.L, Xu, W.X., Shi, Y. *Support Vector Machines Approach to Credit Assessment.* In Bubak, M., Albada, et al (Eds.), LNCS 3039, Springer-Verlag, Berlin, 892-899, 2004.

LINDO Systems Inc., *An overview of LINGO 8.0,* http://www.lindo.com/cgi/frameset.cgi?leftlingo.html;lingof.html.

Joachims, T. Making large-Scale SVM Learning Practical. Advances in Kernel Methods - Support Vector Learning, B. Schöl lkopf and C. Burges and A. Smola (ed.), MIT-Press, 1999.

Joachims, T. (2004) SVM-light: Support Vector Machine, available at: http://svmlight.joachims.org/.

Kou, G., X. Liu, Y. Peng, Y. Shi, M. Wise and W. Xu "Multiple Criteria Linear Programming to Data Mining: Models, Algorithm Designs and Software Developments" *Optimization Methods and Software* 18 (4): 453-473, Part 2 AUG 2003

Kou, G., Y. Peng, Y. Shi, M. Wise and W. Xu, "Discovering Credit Cardholders' Behavior by Multiple Criteria Linear Programming" *Annals of Operations Research* 135 (1): 261-274, JAN 2005

MATLAB. User's Guide. The MathWorks, Inc., Natick, MA 01760, 1994-2005.

Mitchell, T. M. Machine Learning. McGraw-Hill, Boston, 1997.

Murphy, P. M. and Aha, D. W. UCI repository of machine learning databases, 1992. www.ics.uci.edu/_mlearn/MLRepository.html.

Mangasarian, O. L. and Musicant, D. R. Successive overrelaxation for support vector machines. IEEE Transactions on Neural Networks, 10:1032-1037, 1999.

Mangasarian, O. L. Generalized support vector machines. In A. Smola, P. Bartlett, B. Scholkopf, and D. Schuurmans, editors, Advances in Large Margin Classifiers, pages 135-146, Cambridge, MA, 2000. MIT Press.

Mangasarian, O. L. Support Vector Machine Classification via Parameterless Robust Linear Programming, Optimization Methods and Software 20, 2005, 115-125.

Quinlan, J. See5.0. (2004) [available at:http://www.rulequest.com/see5-info.html].

Shi, Y., Peng, Y., Kou, G. and Chen, Z "Classifying Credit Card Accounts for Business Intelligence and Decision Making: A Multiple-Criteria Quadratic Programming Approach" *International Journal of Information Technology and Decision Making,* Vol. 4, No. 4 (2005) 1-19.

Vapnik, V. N. and Chervonenkis (1964), On one class of perceptrons, *Autom. And Remote Contr. 25(1).*

Vapnik, V. N. The Nature of Statistical Learning Theory. Springer, New York, second edition, 2000.

Zheng, J., Zhuang, W., Yan, N., Kou, G., Peng, H., McNally, C., Erichsen, D., Cheloha, A., Herek, S., Shi, C. and Shi, Y., "Classification of HIV-1 Mediated Neuronal Dendritic and Synaptic Damage Using Multiple Criteria Linear Programming" *Neuroinformatics* 2 (3): 303-326 Fall 2004.

Multiclass Credit Cardholders' Behaviors Classification Methods*

Gang Kou[1], Yi Peng[1,**], Yong Shi[1,2], and Zhengxin Chen[1]

[1] College of Information Science & Technology,
University of Nebraska at Omaha, Omaha, NE 68182, USA
[2] Chinese Academy of Sciences Research Center on Data Technology
& Knowledge Economy, Graduate University of the Chinese Academy of Sciences,
Beijing 100080, China
{gkou, ypeng, yshi, zchen}@mail.unomaha.edu

Abstract. In credit card portfolio management a major challenge is to classify and predict credit cardholders' behaviors in a reliable precision because cardholders' behaviors are rather dynamic in nature. Multiclass classification refers to classify data objects into more than two classes. Many real-life applications require multiclass classification. The purpose of this paper is to compare three multiclass classification approaches: decision tree, Multiple Criteria Mathematical Programming (MCMP), and Hierarchical Method for Support Vector Machines (SVM). While MCMP considers all classes at once, SVM was initially designed for binary classification. It is still an ongoing research issue to extend SVM from two-class classification to multiclass classification and many proposed approaches use hierarchical method. In this paper, we focus on one common hierarchical method – one-against-all classification. We compare the performance of See5, MCMP and SVM one-against-all approach using a real-life credit card dataset. Results show that MCMP achieves better overall accuracies than See5 and one-against-all SVM.

Keywords: multi-group classification, decision tree, See5, Multiple criteria mathematical programming (MCMP), one-against-all SVM.

1 Introduction

One of the major tasks in credit card portfolio management is to reliably predict credit cardholders' behaviors. This task has two impacts in credit management: (1) identify potential bankrupt accounts and (2) develop appropriate policies for different categories of credit card accounts. To appreciate the importance of bankrupt accounts prediction, some statistics are helpful: There are about 1.2 billion credit cards in circulation in US. The total credit card holders declared bankruptcy in 2003 are 1,625,208 which are almost twice as many as the number of 812,898 in 1993 (New Generation Research 2004). The total credit card debt at the end of the first quarter

* This work was supported in part by Key Project #70531040, #70472074, National Natural Science Foundation of China; 973 Project #2004CB720103, Ministry of Science and Technology, China and BHP Billion Co., Australia.

** Corresponding author.

V.N. Alexandrov et al. (Eds.): ICCS 2006, Part IV, LNCS 3994, pp. 485 – 492, 2006.
© Springer-Verlag Berlin Heidelberg 2006

2002 is about $660 billion (Cardweb 2004). Bankrupt accounts caused creditors millions of dollars lost each year. In response, credit card lenders have made great effort to improve traditional statistical methods and recognized that more sophisticated analytical tools are needed in this area. Development of appropriate policies for various groups of credit card accounts also has a great impact on credit card issuers' profits. From the creditor's standpoint, the desirable policies should help to keep the profitable customers and minimize the defaults. It is meaningful to conduct multiclass credit cardholders' behaviors classification because it enables card issuers to better manage credit card portfolio.

As one of the major data mining functionalities, classification has broad applications such as credit card portfolio management, medical diagnosis, and fraud detection. Based on historical information, classification builds classifiers to predict categorical class labels for unknown data. Multiclass classification refers to classify data objects into more than two classes.

Researchers have suggested various multiclass classification methods. Multiple Criteria Mathematical Programming (MCMP), decision tree, and Hierarchical Method for Support Vector Machines (SVM) are three of them. Decision tree induction is a tree structure wherein leaves represent classifications and branches represent conjunctions of features that lead to those classifications (Menzies and Hu, 2003). The decision tree software we used in this paper is See5, a Windows95/NT decision tree and rule induction product (RuleQuest 2004). Because See5 is well-known for its high classification accuracy, it is included in this study as a benchmark. MCMP and SVM are both based on mathematical programming and there is no comparison study has been conducted to date. The purpose of this paper is to compare these multiclass classification approaches. While MCMP considers all classes at once, SVM was initially designed for binary classification. It is still an ongoing research issue to extend SVM from two-class classification to multiclass classification and many proposed approaches use hierarchical approach. In this paper, we focus on one common hierarchical method – one-against-all classification. Decision tree induction is a popular classification, so we won't describe it here. For more information about decision tree, please refer to Quinlan (1993).

This paper is structured as follows. The next section discusses the formulation of multiple-group multiple criteria mathematical programming classification model. The third section describes one-against-all SVM multiclass classification method. The fourth section compares the performance of See5, MCMP, and one-against-all SVM using a real-life credit card dataset. The last section concludes the paper.

2 Multi-group Multi-criteria Mathematical Programming Model

This section introduces a MCMP model for multiclass classification. The following models represent this concept mathematically: Given an *r*-dimensional attribute vector $a = (a_1,...,a_r)$, let $A_i = (A_{i1},..., A_{ir}) \in \Re^r$ be one of the sample records, where $i = 1,..., n$; n represents the total number of records in the dataset. Suppose k groups, G_1, G_2, …, G_k, are predefined. $G_i \cap G_j = \Phi, i \neq j, 1 \leq i, j \leq k$ and

$A_i \in \{G_1 \cup G_2 \cup ... \cup G_k\}$, $i = 1,..., n$. A series of boundary scalars $b_1<b_2<...<b_{k-1}$, can be set to separate these k groups. The boundary b_j is used to separate G_j and G_{j+1}. Let $\boldsymbol{X} = (x_1,..., x_r)^T \in R^r$ be a vector of real number to be determined. Thus, we can establish the following linear inequations (Fisher 1936):

$$\boldsymbol{A}_i \boldsymbol{X} < b_1, \forall \boldsymbol{A}_i \in G_1; (1) b_{j-1} \leq \boldsymbol{A}_i \boldsymbol{X} < b_j, \forall \boldsymbol{A}_i \in G_j; (2) \boldsymbol{A}_i \boldsymbol{X} \geq b_{k-1}, \forall \boldsymbol{A}_i \in G_k; \quad (1)$$
$$2 \leq j \leq k-1,\ 1 \leq i \leq n.$$

In the classification problem, $\boldsymbol{A}_i \boldsymbol{X}$ is the score for the ith data record. If an element $\boldsymbol{A}_i \in G_j$ is misclassified into a group other than G_j, then let $\alpha_{i,j}$ be the distance from $\boldsymbol{A}_i$ to b_j, and $\boldsymbol{A}_i\boldsymbol{X} = b_j + \alpha_{i,j}$, $1 \leq j \leq k-1$ and let $\alpha_{i,j-1}$ be the distance from $\boldsymbol{A}_i \in G_j$ to b_{j-1}, and $\boldsymbol{A}_i\boldsymbol{X} = b_{j-1} - \alpha_{i,j-1}$, $2 \leq j \leq k$. Otherwise, $\alpha_{i,j}, 1 \leq j \leq k, 1 \leq i \leq n$, equals to zero. Therefore, the total overlapping of data can be represented as $\sum_{j=1}^{k} \sum_{i=1}^{n} (\alpha_{i,j})^p$. If an element $\boldsymbol{A}_i \in G_j$ is correctly classified into G_j , let $\zeta_{i,j}$ be the distance from $\boldsymbol{A}_i$ to b_j, and $\boldsymbol{A}_i\boldsymbol{X} = b_j - \zeta_{i,j}$, $1 \leq j \leq k-1$ and let $\zeta_{i,j-1}$ be the distance from $\boldsymbol{A}_i \in G_j$ to b_{j-1}, and $\boldsymbol{A}_i\boldsymbol{X} = b_{j-1} + \zeta_{i,j-1}$, $2 \leq j \leq k$. Otherwise, $\zeta_{i,j}, 1 \leq j \leq k, 1 \leq i \leq n$, equals to zero. Thus, the objective is to maximize the distance $|\zeta_{i,j}|_p$ from $\boldsymbol{A}_i$ to boundary if $\boldsymbol{A}_i \in G_1$ or G_k and is to minimize the distance $| \frac{b_j - b_{j-1}}{2} - \zeta_{i,j} |_p$ from $\boldsymbol{A}_i$ to the middle of two adjunct boundaries b_{j-1} and b_j if $\boldsymbol{A}_i \in G_j, 2 \leq j \leq k-1$. So the distances of every data to its class boundary or boundaries can be represented as $\sum_{j=1 or k} \sum_{i=1}^{n} | \zeta_{i,j} |_p$ - $\sum_{j=2}^{k-1} \sum_{i=1}^{n} | \frac{b_j - b_{j-1}}{2} - \zeta_{i,j} |_p$. As a result, the single-criterion mathematical programming model can be set up as:

(**Model 1**) *Minimize* $w_\alpha \sum_{j=1}^{k} \sum_{i=1}^{n} | \alpha_{i,j} |_p$ - w_ζ ($\sum_{j=1 or j=k} \sum_{i=1}^{n} | \zeta_{i,j} |_p$ - $\sum_{j=2}^{k-1} \sum_{i=1}^{n} | \frac{b_j - b_{j-1}}{2} - \zeta_{i,j} |_p$)

S. T.: $\boldsymbol{A}_i\boldsymbol{X} = b_j + \alpha_{i,j} - \zeta_{i,j}$, $1 \leq j \leq k-1$ (2)

$$A_iX = b_{j-1} - \alpha_{i,j-1} + \zeta_{i,j-1}, 2 \leq j \leq k \tag{3}$$

$$\zeta_{i,j} \leq b_j - b_{j-1}, 2 \leq j \leq k \text{ (a)} \quad \zeta_{i,j} \leq b_{j+1} - b_j, 1 \leq j \leq k-1 \text{ (b)}$$

where $\boldsymbol{A}_i$, i = 1, …, n are given, $\boldsymbol{X}$ and b_j are unrestricted, and $\alpha_{i,j}$, $\zeta_{i,j} \geq 0, 1 \leq i \leq n$..(a) and (b) are defined as such because the distances from any correctly classified data ($\boldsymbol{A}_i \in G_j, 2 \leq j \leq k-1$) to two adjunct boundaries b_{j-1} and b_j must be less than $b_j - b_{j-1}$. Let $p = 2$, then objective function in Model 1 can now be a quadratic objective and we have:

(Model 2) *Minimize*

$$w_\alpha \sum_{j=1}^{k}\sum_{i=1}^{n}(\alpha_{i,j})^2 - w_\zeta \left(\sum_{j=1 or j=k}\sum_{i=1}^{n}(\zeta_{i,j})^2 - \sum_{j=2}^{k-1}\sum_{i=1}^{n}[(\zeta_{i,j})^2 - (b_j - b_{j-1})\zeta_{i,j}] \right) \tag{4}$$

Subject to: (4), (5), (c) and (d)

3 SVM One-Against-All Multiclass Classification

Statistical Learning Theory was proposed by Vapnik and Chervonenkis in the 1960s. Support Vector Machine (SVM) is one of the Kernel Machine based Statistical Learning Methods that can be applied on various types of data and can detect the internal relations among the data objectives. Given a set of data, one can define the kernel matrix to construct SVM and compute an optimal hyperplane in the feature space which is induced by a kernel (Vapnik, 1995). There exist different multi-class training strategies for SVM such as one-against-all classification, one-against-one (pairwise) classification, and Error correcting output codes (ECOC).

SVM-light (Joachims 2004) is a well known software package for support vector machine *binary* classification. It is not designed to perform multiclass classification. We apply SVM-light to two-group classifications, then implement a one-against-all procedure for a four-class classification. Suppose the four groups are A, B, C and D. The four-class one-against-all procedure is: ABCD$\Rightarrow$ A|B+C+D $\Rightarrow$ A|B|C+D $\Rightarrow$ A|B|C|D. Table 1 shows the classification results and is displayed in the format of confusion matrices, which pinpoint classification accuracies. Table 2 gives an analysis of classification accuracies and false alarm rates (the percentage of misclassified records to all records which are classified to a group). The assumption of one-against-all procedure is described as following:

The classification accuracy is stable. The classification accuracy of the forecasting dataset is equal to the classification accuracy of the testing dataset as well as the classification accuracy of the training dataset. The following symbols are used in this section.

N_x Number of records in group x

N_{xyz} Number of records in group x, y and z

4 Credit Cardholders' Behaviors Classification

The model proposed can be used in many fields, such as general bioinformatics, antibody and antigen, credit fraud detection, network security, text mining, etc. This research will focus on credit card classification. The real-life credit card dataset used in this paper is come from a US bank. It contains 6000 records and 7 variables. The variables are Interest charge, Interest charge as percent of credit line, Number of months since last payment, Credit line, Average payment of revolving accounts, Last balance to payment ratio, and Average OBT revolving accounts. This dataset has been used as a classic working dataset for various data analyses to support the bank's business intelligence. We define four classes for this dataset using a label variable: The Number of Over-limits. The four classes are: Bankrupt charge-off accounts (Number of Over-Limits≥ 12), Non-bankrupt charge-off accounts (7 ≤Number of Over-Limits ≤ 11), Delinquent accounts (2 ≤ Number of Over-Limits ≤ 6), and Current accounts (0 ≤ Number of Over-Limits ≤ 2). Bankrupt charge-off accounts are accounts that have been written off by credit card issuers because of cardholders' bankrupt claims. Non-bankrupt charge-off accounts are accounts that have been written off by credit card issuers due to reasons other than bankrupt claims. The charge-off policy may vary among authorized institutions. Delinquent accounts are accounts that haven't paid the minimum balances for more than 90 days. Current accounts are accounts that have paid the minimum balances or have not balances. For decision tree method, we use See5.0. MCMP is solved by LINGO 8.0, a software tool for solving nonlinear models (LINDO Systems Inc.). SVM one-against-all is implemented using SVM-light version 6.01 (Joachims 2004), a well-known SVM software.

Table 1. A example of one-against-all 4-classes classification results

1st step	A	B+C+D
Classified as Group A	a	$N_{bcd} - bcd$
Classified as Group B,C,D	$N_a - a$	bcd
2nd step	B	C+D
Classified as Group B	b	$\frac{bcd}{N_{bcd}} \times N_{cd} - cd$
Classified as Group C,D	$\frac{bcd}{N_{bcd}} \times N_b - b$	cd
3rd step	C	D
Classified as Group C	c	$\frac{bcd}{N_{bcd}} \times \frac{cd}{\frac{bcd}{N_{bcd}} \times N_{cd}} \times N_d - d$
Classified as Group D	$\frac{bcd}{N_{bcd}} \times \frac{cd}{\frac{bcd}{N_{bcd}} \times N_{cd}} \times N_c - c$	d

The four-group classification results of See5, MCMP, and SVM-light on the credit card data are summarized in Table 3, 4, and 5, respectively. In addition, we compute Type I and II error rates. Type I error is defined as the rate of records that are misclassified as Current to records that are classified as Current. Type II error is defined as the rate of records that are actually Current but are misclassified as the other three classes (Bankrupt charge-off, Non-bankrupt charge-off, and Delinquent) to records that are classified as the other three classes. Since misclassified Current accounts contribute to huge lost in credit card business and thus creditors are more concern about Type I error than Type II error. From the confusion matrices in Table 3, 4, and 5, we observe that (1) MCMP achieves the lowest test Type I error rate: 1.65%. SVM-light has the second lowest test Type I error rate: 1.7%. See5 has the highest test Type I error rate: 2.2%; (2) Among the three classification methods, MCMP has the best test classification accuracies for Delinquent, Charge-off, and Bankrupt classes. See5 has the best test classification accuracy for Current class.

Table 2. Accuracy and False Alarm Rate analysis of 4-classes classification results

	Accuracy	False Alarm Rate
A	$\frac{a}{N_a}$	$\frac{N_{bcd}-bcd}{a+N_{bcd}-bcd}$
B	$\frac{b}{N_b}$	$\frac{\frac{bcd}{N_{bcd}}\times N_{cd}-cd+\frac{N_a-a}{bcd+N_a-a}\times b}{\frac{bcd}{N_{bcd}}\times N_{cd}-cd+b}$
C	$\frac{c}{N_c}$	$\frac{\frac{bcd}{N_{bcd}}\times\frac{cd}{\frac{bcd}{N_{bcd}}\times N_{cd}}\times N_d-d+\frac{\frac{bcd}{N_{bcd}}\times N_b-b}{cd+\frac{bcd}{N_{bcd}}\times N_b-b}\times c+\frac{N_a-a}{bcd+N_a-a}\times\frac{cd}{cd+\frac{bcd}{N_{bcd}}\times N}}{\frac{bcd}{N_{bcd}}\times\frac{cd}{\frac{bcd}{N_{bcd}}\times N_{cd}}\times N_d-d+c}$
D	$\frac{d}{N_d}$	$\frac{\frac{bcd}{N_{bcd}}\times\frac{cd}{\frac{bcd}{N_{bcd}}\times N_{cd}}\times N_c-c+\frac{\frac{bcd}{N_{bcd}}\times N_b-b}{cd+\frac{bcd}{N_{bcd}}\times N_b-b}\times d+\frac{N_a-a}{bcd+N_a-a}\times\frac{cd}{cd+\frac{bcd}{N_{bcd}}\times N}}{\frac{bcd}{N_{bcd}}\times\frac{cd}{\frac{bcd}{N_{bcd}}\times N_{cd}}\times N_c-c+d}$

Table 3. See5 Credit Card Classification Results

Evaluation on training data (280 cases):					Accuracy	Error Rate
(1)	(2)	(3)	(4)	<-classified as		
66	3	0	1	(1): Current	94.29%	Type I
6	35	27	2	(2): Delinquent	50.00%	9.59%
1	5	56	8	(3): Charge-off	80.00%	Type II
0	6	37	27	(4): Bankrupt	38.57%	1.93%
Evaluation on test data (5720 cases):						
(1)	(2)	(3)	(4)	<-classified as		
3830	609	455	87	(1): Current	76.89%	Type I
83	182	289	48	(2): Delinquent	30.23%	2.20%
3	16	83	24	(3): Charge-off	65.87%	Type II
0	1	7	3	(4): Bankrupt	27.27%	63.80%

Table 4. MCMP Credit Card Classification Results

Evaluation on training data (280 cases):					Accuracy	Error Rate
(1)	(2)	(3)	(4)	<-classified as		
50	12	8	0	(1): Current	71.43%	Type I
5	55	10	0	(2): Delinquent	78.57%	13.79%
2	5	55	8	(3): Charge-off	78.57%	Type II
1	1	5	63	(4): Bankrupt	90.00%	9.01%
Evaluation on test data (5720 cases):						
(1)	(2)	(3)	(4)	<-classified as		
3406	1012	559	4	(1): Current	68.38%	Type I
53	440	100	9	(2): Delinquent	73.09%	1.65%
4	23	92	7	(3): Charge-off	73.02%	Type II
0	0	2	9	(4): Bankrupt	81.82%	69.78%

Table 5. SVM-light Credit Card Classification Results

Evaluation on training data (280 cases):					Accuracy	Error Rate
(1)	(2)	(3)	(4)	<-classified as		
40	22	8	0	(1): Current	57.14%	Type I
1	69	0	0	(2): Delinquent	98.57%	4.76%
0	0	70	0	(3): Charge-off	100.00%	Type II
1	1	2	66	(4): Bankrupt	94.29%	12.61%
Evaluation on test data (5720 cases):						
(1)	(2)	(3)	(4)	<-classified as		
3411	1135	136	299	(1): Current	68.48%	Type I
55	199	66	282	(2): Delinquent	33.06%	1.70%
3	27	23	73	(3): Charge-off	18.25%	Type II
1	0	1	9	(4): Bankrupt	81.82%	69.78%

5 Conclusion

This is the first time that we investigate the differences among decision tree, MCMP, and one-against-all SVM for multiclass classification using a real-life credit card dataset. The results indicate that MCMP achieves better classification accuracy than See5 and one-against-all SVM. In our future research, we will focus on the theoretical differences between MCMP and one-against-all SVM. Another topic of interest is to study the subject of reducing computational cost and improving algorithm efficiency for high dimensional or massive datasets.

References

[1] Bradley, P.S., Fayyad, U.M., Mangasarian, O.L. (1999) Mathematical programming for data mining: Formulations and challenges. *INFORMS Journal on Computing,* 11, 217-238.

[2] Cardweb.com, The U.S. Payment Card Information Network, accessed April 23, 2004, [available at: http://www.cardweb.com/cardlearn/stat.html].

[3] Quinlan, J.R. (1993) C4.5: Programs for Machine Learning, Morgan Kauffman Publication.

[4] Hsu, C. W. and Lin, C. J. (2002) A comparison of methods for multi-class support vector machines, *IEEE Transactions on Neural Networks*, 13(2), 415-425.

[5] Joachims, T. (2004) SVM-light: Support Vector Machine, available at: http://svmlight.joachims.org/.

[6] Knerr, S., Personnaz, L., and Dreyfus, G. (1990), "Single-layer learning revisited: A stepwise procedure for building and training a neural network", in *Neurocomputing: Algorithms, Architectures and Applications*, J. Fogelman, Ed. New York: Springer-Verlag.

[7] Kou, G., Peng, Y., Shi, Y., Chen, Z. and Chen X. (2004b) "A Multiple-Criteria Quadratic Programming Approach to Network Intrusion Detection" in Y. Shi, et al (Eds.): CASDMKM 2004, LNAI 3327, Springer-Verlag Berlin Heidelberg, 145–153.

[8] Li, J.P, Liu, J.L, Xu, W.X., Shi, Y. *Support Vector Machines Approach to Credit Assessment.* In Bubak, M., Albada, et al (Eds.) , ICCS 2004, LNCS 3039, Springer-Verlag, Berlin, 892-899, 2004.

[9] LINDO Systems Inc., *An overview of LINGO 8.0*, http://www.lindo.com/cgi/frameset.cgi?leftlingo.html;lingof.html.

[10] New Generation Research, Inc., April 2004, [available at: http://www.bankruptcydata.com/default.asp].

[11] Menties, T. and Hu, Y. (2003) Data Mining for Very Busy People, IEEE Computer, p. 18-25.

[12] RuleQuest research (2004) [available at: http://www.rulequest.com/see5-info.html]. See 5.0. (2004) [available at:http://www.rulequest.com/see5-info.html].

[13] Stolfo, S.J., Fan, W., Lee, W., Prodromidis, A. and Chan, P.K. (2000) Cost-based Modeling and Evaluation for Data Mining With Application to Fraud and Intrusion Detection: Results from the JAM Project, *DARPA Information Survivability Conference.*

[14] Vapnik, V. N. and Chervonenkis (1964), On one class of perceptrons, *Autom. And Remote Contr. 25(1).*

[15] Vapnik, V. N. (1995), *The Nature of Statistical Learning Theory,* Springer, New York.

[16] Zhu, D., Premkumar, G., Zhang, X. and Chu, C.H. (2001) *Data Mining for Network Intrusion Detection: A comparison of Alternativest Methods*, Decision Sciences, Volume 32 No. 4, Fall 2001.

Hybridizing Exponential Smoothing and Neural Network for Financial Time Series Predication

Kin Keung Lai[1, 2], Lean Yu[2, 3], Shouyang Wang[1, 3], and Wei Huang[4]

[1] College of Business Administration, Hunan University, Changsha 410082, China
[2] Department of Management Sciences, City University of Hong Kong, Tat Chee Avenue, Kowloon, Hong Kong
{mskklai, msyulean}@cityu.edu.hk
[3] Institute of Systems Science, Academy of Mathematics and Systems Science, Chinese Academy of Sciences, Beijing 100080, China
{yulean, sywang}@amss.ac.cn
[4] School of Management, Huazhong University of Science and Technology, 1037 Luoyu Road, Wuhan 430074, China

Abstract. In this study, a hybrid synergy model integrating exponential smoothing and neural network is proposed for financial time series prediction. The proposed model attempts to incorporate the linear characteristics of an exponential smoothing model and nonlinear patterns of neural network to create a "synergetic" model via the linear programming technique. For verification, two real-world financial time series are used for testing purpose.

1 Introduction

A challenging task in financial markets such as stock market and foreign exchange market is to predict the movement direction of financial markets so as to provide valuable decision information for investors. Thus, various kinds of forecasting methods have been developed by many researchers and business practitioners. Of the various forecasting models, the exponential smoothing model has been found to be one of the effective forecasting methods. Since Brown [1] began to use simple exponential smoothing to predict inventory demand, the exponential smoothing models have been widely used in business and finance [2-3]. For example, Gardner [2] introduced exponential smoothing methods into supply chain management for predicting demand, and achieved satisfactory results. Leung et al. [3] used an adaptive exponential smoothing model to predict Nikkei 225 indices, and achieved good results.

However, the exponential smoothing method is only a class of linear model and thus it can only capture linear feature of financial time series. But financial time series are often full of nonlinearity and irregularity. Furthermore, as the smoothing constant decreases exponentially, the disadvantage of the exponential smoothing model is that it gives simplistic models that only use several previous values to forecast the future. The exponential smoothing model is, therefore, unable to find subtle nonlinear patterns in the financial time series data. Obviously, the approximation of linear models to complex real-world problems is not always sufficient. Hence, it is necessary to consider other nonlinear methods to complement the exponential smoothing model.

V.N. Alexandrov et al. (Eds.): ICCS 2006, Part IV, LNCS 3994, pp. 493–500, 2006.
© Springer-Verlag Berlin Heidelberg 2006

Recently, artificial neural network (ANN) models have shown their nonlinear modeling capability in financial time series forecasting. Since Lapedes and Farker [4] used ANN to predict the chaotic time series, the ANN models are widely used in the time series forecasting. For example, Refenes et al. [5] applied multilayer forward network models to forecast foreign exchange prices and obtained good results. Although the ANN models achieve success in financial time series forecasting, they have some disadvantages. Since the real world is highly complex, there exist some linear and nonlinear patterns in the financial time series simultaneously. It is not sufficient to use only a nonlinear model for time series because the nonlinear model might miss some linear features of time series data. Furthermore, previous studies [6-7] are shown that using ANN to model linear problems may produce mixed results.

In such situations, it is necessary to hybridize the linear model and nonlinear model for financial time series forecasting. This is because the ANN model and exponential smoothing model are complementary. On one hand, the ANN model can find subtle nonlinear features hidden in the time series data, but may miss some linear patterns when forecasting. On the other hand, the exponential smoothing model can give good results in the linear patterns of time series, but cannot capture the nonlinear patterns, which might result in inaccurate forecasts. Motivated by the previous findings, this study proposes a hybrid synergy model to financial time series prediction integrating exponential smoothing and ANN via linear programming technique.

The exponential smoothing model rather than other linear models such as ARIMA is chosen as neural network model's complement for several reasons. First of all, the major advantage of exponential smoothing methods is that they are simple, intuitive, and easily understood. These methods have been found to be quite useful for short-term forecasting of large numbers of time series. At the same time, exponential smoothing techniques have also been found to be appropriate in such applications because of their simplicity. Second, the exponential smoothing model has less technical modeling complexity than the ARIMA model and thus makes it more popular in practice. As Lilien and Kotler [8] reported, exponential smoothing models have been widely used by approximately 13% of industry. Third, Mills [9] found little difference in forecast accuracy between exponential smoothing techniques and ARIMA models. In some examples, exponential smoothing models can even obtain better results than neural network model. Foster et al. [10] once argued that the exponential smoothing is superior to neural networks in forecasting yearly data. Generally, the exponential smoothing model is regarded as an inexpensive technique that gives forecasts that is "good enough" in a wide variety of applications.

The remainder of the study is organized as follows. Sections 2 and 3 provide some basic backgrounds about the exponential smoothing and neural network methods for financial time series forecasting. In Section 4, the hybrid methodology combining the exponential smoothing and neural network model via linear programming is introduced. For verification, two experiments are performed in Section 5. Finally, Section 6 concludes the paper.

2 The Exponential Smoothing Forecasting Model

In the application of the exponential smoothing model, there are three types of models that are widely used in different time series. Simple exponential smoothing (Type I) is used when the time series has no trend. Double exponential smoothing (Type II) is an exponential smoothing method for handling a time series that displays a slowly changing linear trend. Two approaches are covered: one-parameter double exponential smoothing, which employs a single smoothing constant; and Holt-Winters' two-parameter double exponential smoothing, which employs two smoothing constants. The third is Winters' method (Type III), which is an exponential smoothing approach to predicting seasonal data. This method also contains two approaches: multiplicative Winters' method, which is appropriate for increasing seasonal variation; and additive Winters' method, which is appropriate for constant seasonal variation [1-2].

In financial time series, there is irregularity, randomicity and no trend. These features show that the simple exponential smoothing method is suitable for financial time series forecasting for the specified time period. Therefore, only the type I of the exponential smoothing model is described in detail. For the type II & III, interested readers can be referred to [1-3] for more details.

Suppose that the time series $y_1, y_2, \ldots, y_n$ is described by the model

$$y_t = \beta_0 + \varepsilon_t \tag{1}$$

where β_0 is the average of the time series and ε_t random error. Then the estimate S_t of β_0 made in time t is given by the smoothing equation

$$S_t = \alpha y_t + (1-\alpha) S_{t-1} \tag{2}$$

where α is a smoothing constant between 0 and 1 and S_{t-1} is the estimate of β_0 in t-1.

Thus a point forecast made in time t for y_{t+1} is

$$\hat{y}_{t+1} = S_t = \alpha y_t + (1-\alpha)\hat{y}_t \tag{3}$$

From Equation (2), we have

$$S_{t-1} = \alpha y_{t-1} + (1-\alpha) S_{t-2} \tag{4}$$

Substituting Equation (4) to Equation (2), then

$$S_t = \alpha y_t + (1-\alpha)(\alpha y_{t-1} + (1-\alpha) S_{t-2}) = \alpha y_t + \alpha(1-\alpha) y_{t-1} + (1-\alpha)^2 S_{t-2} \tag{5}$$

Similarly, substituting recursively for $S_{t-2}, S_{t-3}, \ldots, S_1$ and S_0, we obtain

$$S_t = \hat{y}_{t+1} = \alpha y_t + \alpha(1-\alpha) y_{t-1} + \alpha(1-\alpha)^2 y_{t-2} + \cdots + \alpha(1-\alpha)^{t-1} y_1 + (1-\alpha)^t S_0 \tag{6}$$

Here we see that S_t, the estimate made in time t of the average β_0 of the time series, can be expressed in term of observations $y_t, y_{t-1}, \ldots, y_1$ and the initial estimate S_0. The coefficients measuring the contributions of the observations $y_t, y_{t-1}, \ldots, y_1$ – that is, $\alpha, \alpha(1-\alpha), \cdots, \alpha(1-\alpha)^{t-1}$ – decrease exponentially with time. For this reason this method is referred as simple exponential smoothing.

In order to use Equation (6) to predict the time series, we need to determine the value of the smoothing constant α and the initial estimate S_0. For the smoothing constant, the ordinary least square (*OLS*) can be used to determine α; while for the initial value S_0, we can let S_0 be equal to y_1, i.e. $S_0 = y_1$, or let S_0 be equal to the simple arithmetic average of a few previous observations. For example, $S_0 = (y_1+y_2+y_3)/3$. Once determining α and S_0, exponential smoothing model can be used for prediction.

The advantage of the exponential smoothing method is that it is capable of fitting the linear patterns of the time series well and easy to use. But the financial time series is often irregular and nonlinear, it is not sufficient to use exponential smoothing for financial time series modeling.

3 The Neural Network Forecasting Model

In this study, one of the widely used ANN models, the back-propagation neural network (BPNN) [11], is used for time series forecasting. The main reason is that some studies (e.g. [11-12]) have shown that the BPNN with an identity transfer function in the output unit and logistic functions in the middle-layer units can approximate any continuous function arbitrarily well given a sufficient amount of middle-layer units [12]. Yu et al. [13] have also found that BPNN has been one popular model that can approximate various nonlinearities in the data series.

Generally, the BPNN can be trained by the historical data of a time series in order to capture the non-linear characteristics of the specific time series. The model parameters (connection weights and node biases) will be adjusted iteratively by a process of minimizing the forecasting errors. For time series forecasting, according to the previous computation process the relationship between the output (y_t) and the inputs (y_{t-1}, y_{t-2}, …, y_{t-p}) has the following mathematical representation.

$$y_t = a_0 + \sum_{j=1}^{q} a_j f(w_{0j} + \sum_{i=1}^{p} w_{ij} y_{t-i}) + e_t \tag{7}$$

where a_j ($j = 0, 1, 2, \ldots, q$) is a bias on the jth unit, and w_{ij} ($i = 0, 1, 2, \ldots, p; j = 0, 1, 2, \ldots, q$) is the connection weights between layers of the model, $f(\bullet)$ is the transfer function of the hidden layer, p is the number of input nodes and q is the number of hidden nodes. Actually, the BPNN model in (11) performs a nonlinear functional mapping from the past observation ($y_{t-1}, y_{t-2}, \ldots, y_{t-p}$). to the future value ($y_t$), i.e.,

$$y_t = \varphi(y_{t-1}, y_{t-2}, \cdots, y_{t-p}, w) + e_t \tag{8}$$

where w is a vector of all parameters and φ is a function determined by the network structure and connection weights. Thus, in some senses, the BPNN model is equivalent to a nonlinear autoregressive (NAR) model.

A major advantage of neural networks is their ability to provide flexible nonlinear mapping between inputs and outputs. They can capture the nonlinear characteristics of time series well. However, using BPNN to model linear problems may produce mixed results [6-7]. Therefore, we can conclude that the relationship between exponential smoothing and BPNN is complementary. To take full advantage of the individual strengths of two models, it is necessary to integrate the exponential smoothing and BPNN models, as mentioned earlier.

4 The Hybrid Forecasting Methodology

In real life, financial time series forecasting is far from simple due to high volatility, complexity, irregularity and noisy market environment. Furthermore, real-world time series are rarely pure linear or nonlinear. They often contain both linear and nonlinear patterns. If this is the case, there is no universal model that is suitable for all kinds of time series data. Both exponential smoothing models and BPNN models have achieved success in their own linear or nonlinear domains, but neither exponential smoothing nor BPNN can adequately model and predict time series since the linear models cannot deal with nonlinear relationships while the BPNN model alone is not able to handle both linear and nonlinear patterns equally well [6]. On the other hand, as previously mentioned, for time series forecasting the relationship between exponential smoothing and BPNN is complementary. Exponential smoothing is a class of linear models that can capture time series' linear characteristics, while BPNN models are a class of general function approximators capable of modeling nonlinearity and which can capture nonlinear patterns in time series. Hybridizing the two models may yield a robust method, and more satisfactory forecasting results may be obtained by incorporating an exponential smoothing model and a BPNN model. Therefore, we propose a hybrid methodology integrating the exponential smoothing and the BPNN for financial time series forecasting. Different from decomposition-hybrid principle described in [6], we adopt the "parliamentary" hybridization strategy [14] to create a synergetic model, as shown in Fig. 1.

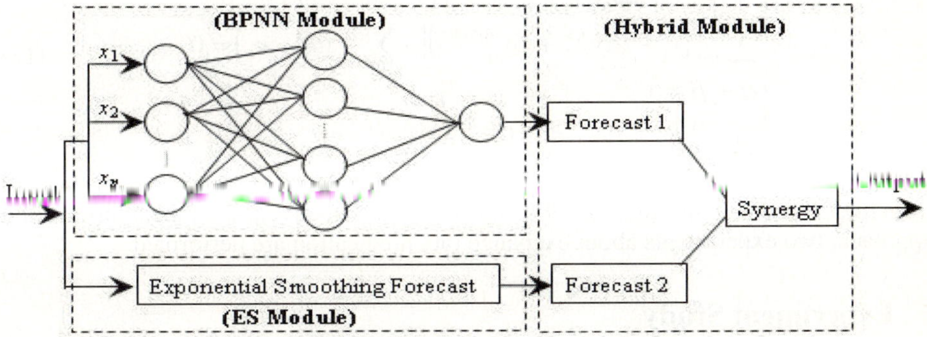

Fig. 1. The hybrid synergy forecasting model

From Fig. 1, the input is fed simultaneously into a BPNN model and an exponential smoothing forecasting model. The BPNN model generates a forecast result, while the exponential smoothing model also generates a time series forecast result. The two forecast results are entered into the hybrid forecast module and generate a synergetic forecast result as final output. In the hybridization process, the "parliamentary" hybridization strategy [14] is used, i.e.,

$$\hat{y}_t^{Hybird} = \alpha\hat{y}_t^{ES} + (1-\alpha)\hat{y}_t^{BPNN} \tag{9}$$

where $\hat{y}_t^{ES}$ is the forecast result obtained from exponential smoothing model, $\hat{y}_t^{BPNN}$ is the forecast result of the BPNN and α is the weight parameter.

Through integrating linear patterns and nonlinear patterns of financial time series, a synergetic effect will believed to be created to improve the predication performance. In Equation (9), a critical problem is how to determine the weight parameter α. Generally, the value of α can be estimated by the ordinary least square (*OLS*) method, i.e.,

$$MinQ = \sum_{t=1}^{n} (y_t - \hat{y}_t^{Hybird})^2 \tag{10}$$

However, its drawback of this approach is that the square treatment will move the fitted curve to some exceptional points and thus reducing the forecasting accuracy. One modification in this study is to minimize the sum of absolute error between estimated and the actual value, then

$$MinQ' = \sum_{t=1}^{n} \left| y_t - \hat{y}_t^{Hybird} \right| = \sum_{t=1}^{n} |e_t| \tag{11}$$

The Equation (11) can be solved by linear programming, let

$$u_t = \frac{|e_t| + e_t}{2} = \begin{cases} e_t, e_t \geq 0 \\ 0, e_t < 0 \end{cases}, \quad v_t = \frac{|e_t| - e_t}{2} = \begin{cases} 0, e_t \geq 0 \\ -e_t, e_t < 0 \end{cases} \tag{12}$$

Clearly, $|e_t| = u_t + v_t, e_t = u_t - v_t$, then the linear programming (*LP*) model can be formulated below.

$$(LP) \begin{cases} Min\ Q' = \sum_{t=1}^{n} (u_t + v_t) \\ \sum_{t=1}^{n} \left[y_t - \left(\alpha \hat{y}_t^{ES} + \beta \hat{y}_t^{BPNN} \right) \right] - \sum_{t=1}^{n} (u_t - v_t) = 0 \\ \alpha + \beta = 1 \\ \alpha \geq 0, u_t \geq 0, v_t \geq 0, t = 1, 2, \cdots, n \end{cases} \tag{13}$$

Using the simplex algorithm, an optimal hybridization parameter can be obtained from the *LP* problem in Equation (13). To verify the effectiveness of the hybridization approach, two experiments about exchange rate predication are performed.

5 Experiment Study

The data set used in this paper are daily data from 1 January 2000 till 31 December 2002 and are obtained from Pacific Exchange Rate Service (http://fx.sauder.ubc.ca/), provided by Professor Werner Antweiler, University of British Columbia, Vancouver, Canada. The daily data cover three years of observations of two major international currency exchange rates --- euros/US dollar (EUR/USD) and Japanese yen/US dollar (JPY/USD). In our empirical experiment, the data set is divided into two sample periods --- the estimation (in-sample) and the test (out-of-sample) periods. The estimation period covers observations from 1 January 2000 till 31 September 2002 and is used to estimate and refine the forecast model parameters. Meantime we take the data from 1 October 2002 to 31 December 2002 as test sets, which are used to evaluate the good or bad performance. For space, the original data are omitted, and detailed data can be obtained from the website.

In this study, the root mean square error (*RMSE*) and directional statistics (D_{stat}) [13] are used as evaluation criteria. In addition, the individual exponential and BPNN models are selected as benchmark models for comparison purposes. Finally, only one-step-ahead forecasting is considered in this study.

Based on our analysis above, we examine the forecast performances of two major currency exchange rates in terms of *RMSE* and D_{stat}. The experimental results are reported in Table 1.

Table 1. The experiment results of EUR/USD and JPY/USD

Currencies	EUR/USD			JPY/USD		
Models	ES	BPNN	Hybrid	ES	BPNN	Hybrid
RMSE	0.0047	0.0062	0.0035	0.6984	0.8226	0.6571
D_{stat} (%)	56.45	62.90	67.74	51.61	58.06	66.13

From the viewpoint of *RMSE*, the hybrid methodology is the best for both EUR/USD and JPY/USD, followed by the individual exponential smoothing and individual BPNN model. For example of EUR/USD, the *RMSE* of the individual BPNN model is 0.0062, and the individual exponential smoothing model is 0.0047, while the *RMSE* of the hybrid methodology is only 0.0035.

However, the direction prediction is more important than the accuracy prediction in the financial markets because the former can provide decision information for investors directly. Furthermore, the high *RMSE* can not lead to high D_{stat}, as the exponential smoothing and the BPNN reveals. Focusing on the D_{stat}, the hybrid method performs the best, followed by the individual BPNN model; the worst is the individual exponential smoothing. For example of JPY/USD, the D_{stat} of exponential smoothing is only 51.61%, the D_{stat} of the BPNN model is 58.06%, while that of the hybrid methodology arrives at 66.13%. The main reason is that the hybrid methodology integrating linear patterns and nonlinear patterns creates a synergetic effect and thus improves the prediction performance.

6 Conclusions

In this paper, we propose a hybrid synergy methodology incorporating exponential smoothing and neural network for financial time series forecasting and explore the forecasting capability of the proposed hybrid synergy methodology from the point of level prediction and direction prediction. Experimental results obtained reveal that the hybrid methodology performs better than the two benchmark models, implying that the proposed hybrid synergy approach can be used as an alternative solution to the financial time series forecasting.

Acknowledgements

This work is partially supported by National Natural Science Foundation of China (NSFC No. 70221001); Chinese Academy of Sciences; Key Research Institute of

Humanities and Social Sciences in Hubei Province-Research Center of Modern Information Management and Strategic Research Grant of City University of Hong Kong (SRG No. 7001677).

References

1. Brown, R.G.: Statistical Forecasting for Inventory Control. New York, McGraw-Hill, 1959
2. Gardner, E.S.: Exponential Smoothing: The State of the Art. Journal of Forecasting 4 (1985) 1-28
3. Leung, M.T., Daouk, H., Chen, A.S.: Forecasting Stock Indices: A Comparison of Classification and Level Estimation Models. International Journal of Forecasting 16 (2000) 173-190
4. Lapedes, A., Farber, R.: Nonlinear Signal Processing Using Neural Network Prediction and System Modeling. Theoretical Division, Los Alamos National Laboratory, NM Report. No. LA-UR-87-2662, 1987
5. Refenes, A.N., Azema-Barac, M., Chen, L., Karoussos, S.A.: Currency Exchange Rate Prediction and Neural Network Design Strategies. Neural Computing Applying 1 (1993) 46-58
6. Zhang, G.P.: Time Series Forecasting Using a Hybrid ARIMA and Neural Network Model. Neurocomputing 50 (2003) 159-175
7. Denton J.W.: How Good are Neural Networks for Causal Forecasting? Journal of Business Forecasting 14 (1995) 17–20
8. Lilien, G.L., Kotler, P.: Marketing Decision Making: A Model Building Approach. New York, Harper and Row Publishers (1983)
9. Mills, T.C.: Time Series Techniques for Economists. Cambridge University Press (1990)
10. Foster, B., Collopy, F., Ungar, L.: Neural Network Forecasting of Short, Noisy Time Series. Computers and Chemical Engineering 16 (1992) 293-297
11. Hornik, K., Stinchocombe, M., White, H.: Multilayer Feedforward Networks are Universal Approximators. Neural Networks 2 (1989) 359-366
12. White, H.: Connectionist Nonparametric Regression: Multilayer Feedforward Networks can Learn Arbitrary Mappings. Neural Networks 3 (1990) 535-549
13. Yu, L., Wang, S.Y., Lai, K.K.: A Novel Nonlinear Ensemble Forecasting Model Incorporating GLAR and ANN for Foreign Exchange Rates. Computers and Operations Research 32 (2005) 2523-2541
14. Wedding II, D.K., Cios, K.J.: Time Series Forecasting by Combining RBF Networks, Certainty Factors, and the Box-Jenkins Model. Neurocomputing 10 (1996) 149-168

Assessment the Operational Risk for Chinese Commercial Banks*

Lijun Gao[1,2], Jianping Li[2,**], Jianming Chen[2], and Weixuan Xu[2]

[1] Graduate University of Chinese Academy of Sciences, Beijing 100039, P.R. China
glj963217@163.com

[2] Institute of Policy & Management,
Chinese Academy of Sciences, Beijing 100080, P.R. China
{ljp, jmchen, wxu}@casipm.ac.cn

Abstract. Operational risk is one of the most important risks for Chinese commercial banks, and brings huge losses to Chinese commercial banks recent years. Using the public reported operational loss data from 1997 to 2005 of Chinese commercial banks, we simulate the operational loss distribution, find that loss frequency can be seen as Poisson distribution and the logarithm of loss is normal distribution. In accordance with the confidence level required by Basel II, aggregated loss distributions and operational Value-at-Risks (OpVaR) are calculated by Monte Carlo Simulation. Comparing with the real loss, this result is credible. We also calculate the economic capital by the $VaR_{99.9}$, and it maybe help the banks to allocate appropriate their economic capital.

1 Introduction

Operational risk has gradually become an area of risk management in global banking. The increase of sophisticated and complex banking practices have raised the needs for an effective operational risk management and measurement system both to regulators and financial industry. In the aspect of definition, the Risk Management Group (RMG) of the Basel Committee and industry representatives have agreed on a standardized definition of operational risk[1], i.e. "the risk of loss results from inadequate or failed internal processes, people and systems or from external events". This definition, which includes legal risk and excludes strategic and reputation risk, relies on the categorization of operational risks based on the underlying causes[2].

Operational risk is mainly triggered by human error, system malfunction, and operational procedure mistake or control invalidation. It may result in huge loss of financial institutes. There are some well-known cases such as the Barings' bankruptcy in 1995, the $691 million trading loss at Allfirst Financial[3], LTCM, Natwest, and Allied Irish Bank[4], and the $140 million loss at the Bank of New York due to September 11th.

* This research has been partially supported by a grant from National Natural Science Foundation of China (#70531040), the President Fund of Chinese Academy of Sciences (yzjj946) and 973 Project(#2004CB720103), Ministry of Science and Technology, China.

** Corresponding author.

V.N. Alexandrov et al. (Eds.): ICCS 2006, Part IV, LNCS 3994, pp. 501 – 508, 2006.
© Springer-Verlag Berlin Heidelberg 2006

Due to the growth of e-commerce, large-scale mergers and acquisitions, the enrichment of financial service products, and the use of more highly automated technologies, the probability of operational risk is increasing [5].

There have been several operational losses recently in Chinese commercial banks, such as the cheating case in Heilongjiang branch of Bank of China, the 4 billion loan deceived in Kaiping, etc.. The Chinese financiers and researchers have paid high attention to banks' operational risks, but there have no effective measurements to operational risk[6]. It is critical to improve the operational risk management of Chinese commercial banks.

Many of the operational risk models [7-8] such as Advanced Measurement Approaches (AMA), function correlation approach, causal models and Bayesian models, actuarial models, etc., are restricted for applying to the Chinese commercial banks by lacking of credible internal loss database. As the Monte Carlo simulation can overcome the deficiency of data by effective simulation and producing relative exact data, we use Monte Carlo Simulation to get OpVaRs and economic capital estimates for operational risk, this can help banks improve internal control and spur banks improving operational risk management.

The paper is organized as follows: In section 2, we discuss the simulation method we used; section 3 describe the original data; section 4 presents the simulation results and analysis, including the testing of the distribution of the historic operational loss data, and calculation on the operational economic capital, and the result analysis; The last past concludes the paper and introduces some future research directions.

2 The Simulation Method

Two parameters are commonly used to describe operational risk: the frequency and the loss amount. If the loss frequency and the loss amount follow certain distributions, and the statistical distribution of frequency and the loss are stable, we can use historic loss data to estimate their distribution and operational loss.

We use Monte Carlo simulation compute the frequency and the loss distributions[9]. Different OpVaRs are then simulated. The steps of the method are described as follows:

Step1. Collect historic loss data of operational events; estimate the loss frequency and the loss, Normal distribution, uniform distribution, Poisson distribution and exponential distribution are tested to find a most suitable distribution;

Step2. After generating the frequency distribution, we generate 1000 random numbers approximate to the distribution, representing the events for 1000 simulated periods.

Step3. For each period, generate required random number of loss (that is, if the simulated events for period k is X, then simulate X losses) and add them to get the aggregated loss for the period.

Step4. Run steps 1 to 3 ten times and take the average OpVaR as the final result.

3 The Data Set

We have collected operational losses of Chinese commercial banks from 1997 to 2005 by public report, such as the website of National Audit Office of the P.R.C., website of china banking regulatory commission, finance news of banks cases and others, which counts 45 loss events, including 8 major commercial banks. Table 1 is describing the number events of each year, and Fig.1 is the histogram of loss events.

Table 1. The number of operational loss events from 1997 to 2005

Year	1997	1998	1999	2000	2001	2002	2003	2004	2005
Number	5	4	1	3	4	13	5	3	7

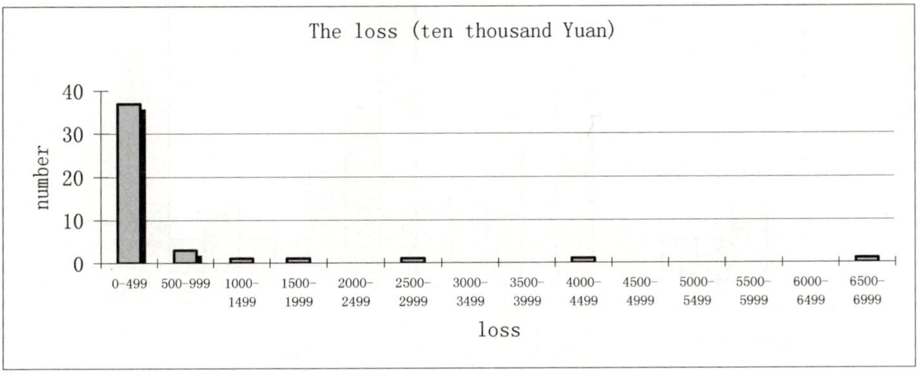

Fig. 1. Loss histogram of 1997-2005

4 Simulation Results and Analysis

4.1 Testing on the Loss Frequency Distribution and Loss Distribution

The loss varies significantly and most losses are under 5 million Yuan, so we test 11 curve regressions to simulate the loss distribution, and find that the logarithmic fits the data best. The result of curve estimations is shown in table 2, Fig. 2 is the logarithmic value of Fig.1; the software we used is SPSS 11.5.

Table 2. The loss curve estimation

Curve	Function	F value	Rsq.	Sig.
Linear	Y=20.1204+6.0E-05x	19.28	0.310	0.000
Logarithmic	Y=-23.037+5.2928lnx	850.38	0.952	0.000
Inverse	Y=25.3719-955.32/x	16.35	0.276	0.000
Quadratic	Y=17.2152+0.0002x-3.E-10x2	24.59	0.539	0.000
Cubic	Y=13.8314+.0005x-2.E-09x2+1.6E-15x3	41	0.75	0.000
Power	Y=15.4421(1.0000x)	7.29	0.145	0.01
Compound	Y=0.8556x0.3478	443.87	0.912	0.000
S-curve	Y=e3.1141-98.559/x	79.97	0.65	0.000
Logistic	Unfitted			
Growth	Y=e2.7371+2.7E-06x	7.29	0.145	0.01
Exponential	Y=15.4421e2.7E-06x	7.29	0.145	0.01

Based on table 1 and Fig. 2, we estimate the loss frequency distribution and loss distribution, test the normal distribution, uniform distribution, Poisson distribution and exponential distribution separately to find a most suitable distribution, and use the Kolmogorov-Smirnov (K-S) test to reject or accept the null hypothesis that the data originate from the selected distribution with the estimated parameters. Table 3 shows the result.

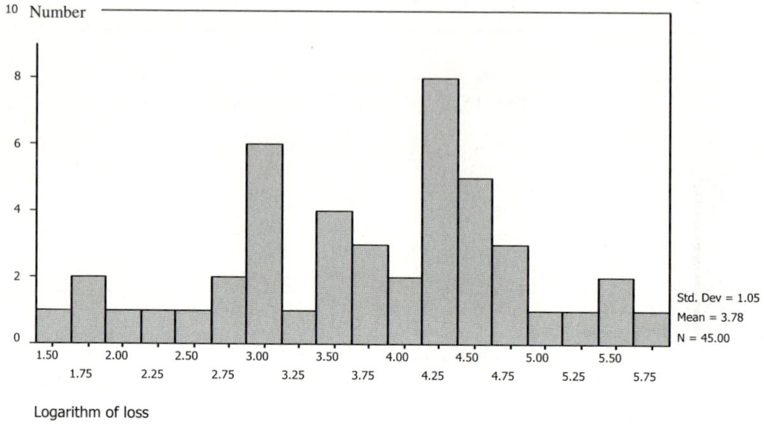

Fig. 2. Logarithm of loss histogram of 1997-2005

Table 3. Distribution test

	Loss severity distribution			Loss frequency distribution		
urve distribution	Mean/Stdev.	K-S Z	Asymp. Sig.	Mean/stdev	K-S Z	Asymp. Sig.
Normal	3.7775/1.0514	0.696	0.718	5/3.428	0.833	0.491
Uniform	1.41/5.84 (max/min)	1.192	0.117	1/13 (max/min)	1.333	0.057
Poisson	Unfitted			5	0.485	0.972
Exponential	3.7775	2.594	.000	5	1.02	0.249

The results indicate that normal distribution and Poisson distribution fit the logarithmic of loss and loss frequency better than all other distributions. As shown in table 3, the loss frequency distribution is approximately 5 events per year. Considering the historic logarithm of loss distribution, the skewness is -0.354, while the standard deviation of skewness is 0.354, then the data can be regarded as symmetry, the kurtosis is -0.231, so we can conclude that the logarithm of single loss distribution follows normal distribution.

4.2 The Losses Under Different Confidences

After we get the single loss distribution, we can calculate the OpVaRs under different confidences, table 4 shows the result.

Table 4. The OpVaRs under different confidences of single loss

VaR_{25}	VaR_{50}	VaR_{75}	VaR_{90}
1170.347	5991.009	30667.99	133344.5
VaR_{95}	VaR_{99}	$VaR_{99.9}$	$VaR_{99.99}$
321340	1673071	10634453	48741655

Then we can compute the aggregated loss in a one-year period, the steps are:

Step1: Generate the Poisson random variables $n_1, n_2, \cdots, n_{1000}$;

Step2: If $n_i = k_i$, then generate k_i normal random variables, add them up and result is the loss this period;

Step3: Run steps 1 to 2 ten times and take the average as the value;

Step4: Using the 1000 potential loss to get the operational loss.

The aggregated loss is simulated, and we get some important statistical data, the mean of the loss per year is 4659.799 million Yuan, while the standard deviation is 1246536.699.

As the new Basel Committee[10] required that the regulatory capital that covers the operational risks over a one-year period within a confidence interval of 99.9%, we compute the $VaR_{99.9}$, and other typical points of confidence level are shown as table 5:

Table 5. The aggregated operational confidence level in one year (ten thousand Yuan)

VaR_{25}	VaR_{50}	VaR_{75}	VaR_{90}
42325.3	126969.7	377572.2	996322.6
VaR_{95}	VaR_{99}	$VaR_{99.9}$	$VaR_{99.99}$
1854812	5845208	17607593	85841937

So the aggregate loss distribution as following Fig. 3:

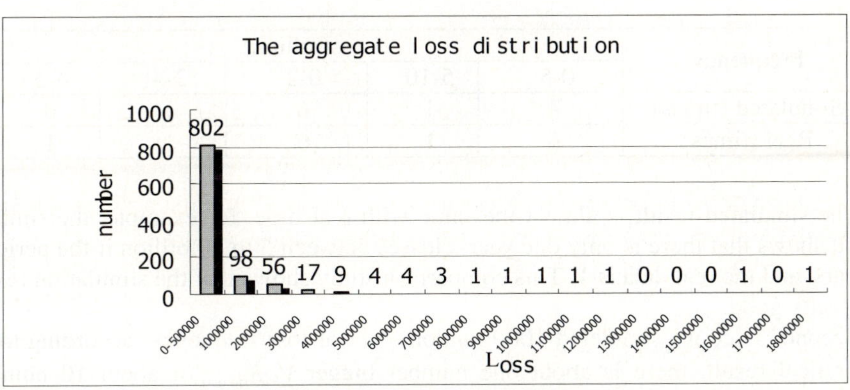

Fig. 3. The aggregate loss distribution

4.3 Result Analysis

4.3.1 There is Huge Potential Operational Loss in Bank System

Most losses are not huge and there are 802 losses under 5 billion Yuan, there are 900 events below 10 billion Yuan, but there may be very huge loss a year and the biggest loss we simulated is 176 billion, which can do harm to the survival of the banks.

4.3.2 The Loss Severity Affects the Aggregate Loss Distribution Much

Although the aggregate loss distribution is generated by the loss frequency and loss severity, the influence of the two distributions on the aggregate loss distribution is not equal, from table 4 and table 5 we can conclude that loss frequency distribution only affect the aggregate loss distribution slightly, especially for the tail distribution, and the higher frequency in one year, the less influence to aggregate loss; While the loss severity affect the aggregate loss distribution much more, which makes the curve of aggregate loss distribution is somewhat like the single loss distribution.

4.3.3 The Proposed Operational Economic Capital

Our purpose is not only to work out the aggregate loss distribution of operational loss, but also use it to help determine the approximate operational economic capital of Chinese commercial banks. Based on the results in 4.2, we can calculate the operational economic capital of Chinese commercial banks:

Economic capital = $VaR_{99.9}$-mean = 16707593-465979.9 =17141613.1 (Ten thousand)

So according to the new Basel committee's suggestion, Chinese commercial banks as a whole, should prepare 171.42 billion Yuan for operational loss.

4.3.4. The method and results are comparatively credible.

We present a simulation method and get some results to the operational risk, to verify the result. The result is tested by two ways.

The first way is to use the aggregate loss distribution to map the 1000 periods to 9 periods and then comparing them with the real loss from 1997 to 2005. Table 5 shows the detail.

Table 6. The result compared with the real data (9 years)

Frequency	Loss(billion)				
	0-5	5-10	0-2	2-4	4-5
Simulated (times)	7	1	6	1	0
Real (times)	8	1	6	1	1

The simulated result is almost the same with real time, for example, the simulate result shows that there is only one year's losses between 5 to 10 billion if the period is 9 years, and the real data is 1. This comparison study shows that the simulation results are pretty good.

Second, since we simulated 1000 random data of loss frequency, according to the theoretical result, there is about one number bigger $VaR_{99.9}$, or about 10 numbers bigger than $VaR_{99.9}$, etc.. Table 6 shows the 4 group compared results.

Table 7. The estimation of OpVaRs

Frequency	$> VaR_{99.9}$	$> VaR_{99}$	$> VaR_{95}$	$> VaR_{90}$
Theoretical	1	10	50	100
Simulated	0	11	50	101

Although there is no number bigger than the $VaR_{99.9}$, the biggest one in our simulation is 176.07135 billion, which is very close to the $VaR_{99.9}$ (176.07593 billion). The other three also accord with the theoretical result very well. It shows the simulation is reasonable.

5 Conclusions

Many operational models can't be used to measure operational risk efficiently for Chinese banks since Chinese commercial banks having little operational loss data and it is critical to quantify the operational loss. In this paper, we present a simulation method to the Chinese commercial banks as a whole, find that the logarithmic of loss follows the normal distribution while the loss frequency follows the Poisson distributions. We calculate the OpVaRs, and compare them to the real loss data per year. The results show that our method is comparatively reasonable and creditable, and the supervision capital we calculated may help the regulators to determine the operational economic capital.

Since the loss data is small, our simulation may be not exactly correct. If we have more operational loss data, we will re-calculate the certain bank's operational losses, and test more models such as the Multiple-Criteria Programming [11,12], Support Vector Machines[13] to find more suitable model. The model proposed in this paper may be also used in general bioinformatics, any antibody and any antigen. This is our further work.

References

1. Basel Committee on Banking Supervision. Operational Risk, Consultative Document. Basel, September 2001, URL: http://www.bis.org.
2. The Basel Committee on Banking Supervision Bank for International Settlements CH-4002 Basel. Third consultative paper(CP3) on the New Basel Capital Accord. Switzerland, July 30, 2003.
3. Hubner G., Peters J-P, Plunus S. Measuring operational risk in financial institutions: Contribution of credit risk modeling, March 2005.
4. Helbok G., Wagner C. Corporate financial disclosure on operational risk in the banking industry. Working Paper, September 2004.
5. Cornalba C., Giudici P. Statistical models for operational risk management. Physical A, 338, 2004, 166-172.
6. Gao, L.J, Li J.P., Chen, J.M, Wang, S.P. A New Assessment of Operational Risk of commercial bank: OpRisk+ Model, Chinese Journal of Management Science(S). 2005, Vol.13:185-188 (in Chinese).

7. Patrick F., Eric R., Jordan J. Implications of alternative operational risk modeling techniques. NBER Working Paper No. 11103, June 2004. http://papers.nber.org/papers/W11103
8. Alexander C. Statistical models of operational loss. In Operational Risk. Regulation, Analysis and Management, FT Prentice Hall Financial Times, 2003, 129-170.
9. Peters J.P., Crama Y., Hubner G. Basel II project: computation of OpVaR, Working Paper, 2003. HEC Management School, University of Liège.
10. Basel II: International Convergence of Capital Measurement and Capital Standards: a Revised Framework, Basel Committee Publications, June 2004.
11. Shi, Y., Peng, Y., Kou, G., Chen, Z . Classifying Credit Card Accounts for Business Intelligence and Decision Making: A Multiple-Criteria Quadratic Programming Approach. International Journal of Information Technology and Decision Making, Vol. 4, No. 4 (2005) 1-19.
12. Kou, G., Peng, Y., Shi, Y., M. Wise, Xu, W.X. Discovering Credit Cardholders' Behavior by Multiple Criteria Linear Programming. Annals of Operations Research 135 (1): 261-274, JAN 2005.
13. Li, J.P, Liu, J.L, Xu, W.X., Shi, Y. Support Vector Machines Approach to Credit Assessment. In Bubak, M., Albada, G.D.v., Sloot, P.M.A., Dongarra, J.J. (Eds.) , ICCS 2004, LNCS 3039, Springer-Verlag, Berlin, 892-899, 2004.

Pattern Recognition for MCNs Using Fuzzy Linear Programming

Jing He[1,3], Wuyi Yue[2], and Yong Shi[3]

[1] Institute of Intelligent Information and Communication Technology, Konan University, Kobe 658-8501 Japan
hejing@gucas.ac.cn

[2] Department of Information Science and Systems Engineering, Konan University, Kobe 658-8501 Japan
yue@konan-u.ac.jp

[3] Chinese Academy of Sciences Research Center on Data Technology, and Knowledge Economy, Beijing 100080 P.R. China
yshi@gucas.ac.cn

Abstract. This paper presents a data mining system of performance evaluation for multimedia communication networks (MCNs). Two important performance evaluation problems for the MCNs are considered in this paper. They are: (1) the optimization problem for construction of the data mining system of performance evaluation; (2) the problem of categorizing real-time data corresponding to the data mining system by means of dividing the performance data into usual and unusual categories. An algorithm is employed to identify performance data such as throughput capacity, package forwarding rate, and response time. A software named PEDM2.0 (Performance Evaluation Data Miner) is proposed to improve the accuracy and the effectiveness of the fuzzy linear programming (FLP) method compared with decision tree, neural network, and multiple criteria linear programming methods.

1 Introduction

Performance evaluation and network planning are the key tools in a reliable multimedia communication operation. Multimedia communication networks (MCNs) to support several different traffic types have become so complex that intuition alone is not sufficient to evaluate their performance. Mathematical models of performance systems range from relatively simple ones, whose solution can be obtained analytically, to very complex ones that must be simulated [1].

An important challenge for identification mining in MCNs is the identification speed that can forward the exponentially increasing volume of traffic. The data mining system can provide the new identification service that is needed by next-generation MCNs.

Research of linear programming (LP) approaches for classification problems was initiated by [3]-[5]. [6] and [7] applied the compromise solution of multiple criteria linear programming (MCLP) to deal with the same identification mining

V.N. Alexandrov et al. (Eds.): ICCS 2006, Part IV, LNCS 3994, pp. 509–516, 2006.
© Springer-Verlag Berlin Heidelberg 2006

question. [8] presented an analysis for fuzzy linear programming in classification of credit card holder behavior. In [9] the identification mining of unusual patterns for MCNs based on FLP is put forward for the first time. In this paper, we present some new research work.

The unusual pattern mining process can be described as follows: given a set of n performance evaluation data, there will be K objects that are considerably dissimilar, exceptional, or inconsistent with the remaining data.

In Section 2, we describe a data mining system for performance evaluation. The subsystems for an identification mining engine based on fuzzy linear programming are presented in Section 3. In Section 4, the results of data experiments are listed out. Finally, we conclude with a brief summary in Section 5.

2 Data Mining System for Performance Evaluation

Generally, the methods to calculate network performance include analytical, numerical and simulation methods. Nowadays, emulation is the main method for performance evaluation systems for MCNs.

A data mining system for performance evaluation can be constructed with the above three methods of analytical, numerical and simulation method. Fig. 1 shows the architecture of the data mining system for performance evaluation that we present in this paper.

These components are explained in detail as follows:

Graphical user interface
This module communicates between the users and the data mining system, allowing the user to interact with the system by specifying a performance evaluation query or task, and providing information to help focus the search.

Index system for performance evaluation
The main idea of this index system comes from [1]. The details of the index can be found from the following multi-dimensional data warehouse module.

Pattern evaluation
The process of this module is shown in Fig. 2.

An index system is acceptable if (1) it is easily understood by humans, (2) it is valid on new or test data with some degree of certainty, (3) it is potentially useful, and (4) it is novel [2].

Multi-dimensional data warehouse
Before we use our on-line analytical processing (OLAP) tools, the multi-dimensional data warehouse for performance evaluation must be constructed. The snowflake schema is a variant of the star schema, where some dimension tables are normalized thereby further splitting the data into additional tables. The snowflake schema of data warehouse is shown in Fig. 3.

Pre-computation and summarization
This module, which involves data integration and data cleaning, can be viewed as an important preprocessing step for data mining. Data from operational

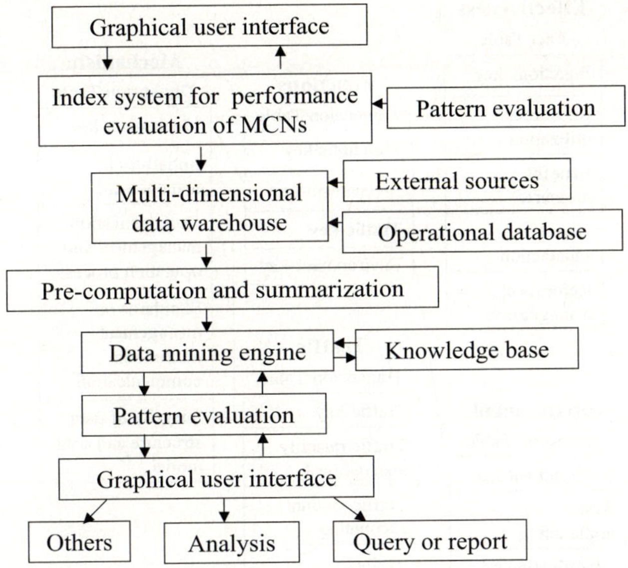

Fig. 1. Architecture of data mining system for performance evaluation

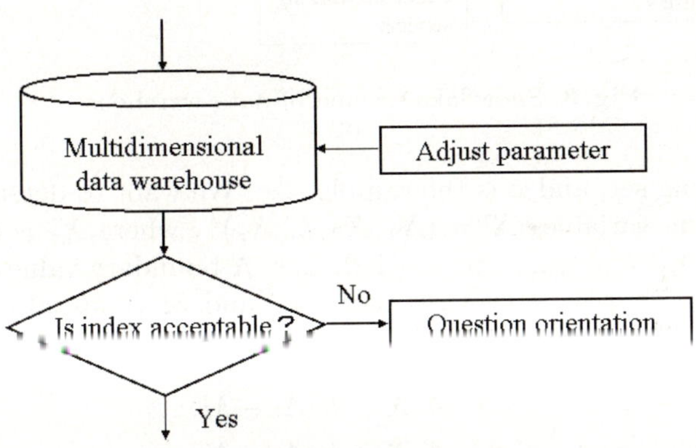

Fig. 2. Pattern evaluation process

databases and external sources (such as performance information provided by external sensors) are extracted using application program interfaces known as gateways.

3 Identification Mining Model

A basic framework of the identification mining model of unusual patterns can be presented as follows:

Given a set of r attributes about a MCN, let $\boldsymbol{A}_i = (A_{i1}, ..., A_{ir})$, $i = 1, 2, ..., n$ be training set data for the variables of every MCN, where $\boldsymbol{A}_i$ is the attributes set

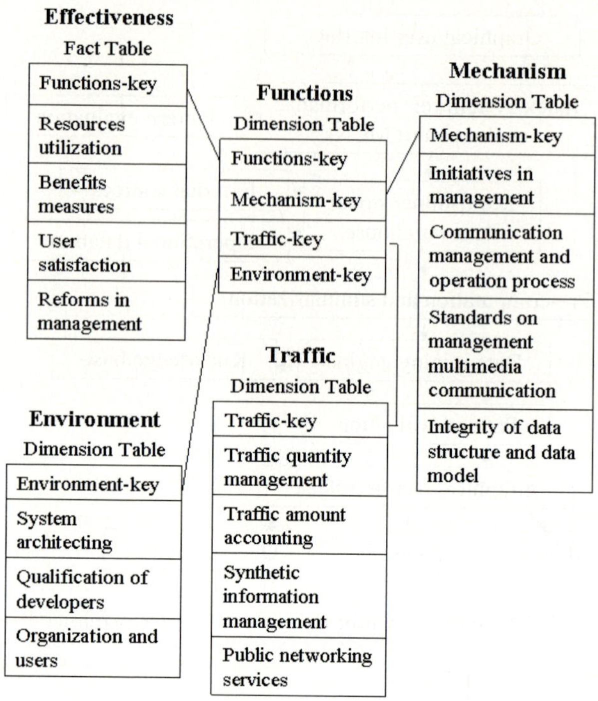

Fig. 3. Snowflake schema of data warehouse

of the ith training set, and n is the sample size. We want to determine the best coefficients of the variables $\boldsymbol{X} = (X_1, X_2, ..., X_r)^T$, where X_j is the coefficient of the variable A_{ij}, $i = 1, 2, ..., n$, $j = 1, 2, ..., r$. A boundary value b (a scalar) to separate the two classes: N (normal patterns) and M (unusual patterns), is as follows:

$$\begin{aligned} \boldsymbol{A}_i\boldsymbol{X} &\leq b, \boldsymbol{A}_i \in M, \\ \boldsymbol{A}_i\boldsymbol{X} &\geq b, \boldsymbol{A}_i \in N. \end{aligned} \tag{1}$$

To measure the separation of usual and unusual patterns, we define that α_i is the overlapping of a two-class boundary for case A_{ij} (external measurement), $i = 1, 2, ..., n$, $j = 1, 2, ..., r$. α is the maximum overlapping of a two-class boundary for case A_{ij}, $\alpha_i < \alpha$.

We define β_i to be the distance from case A_{ij} to their adjusted boundaries (internal measurement), $i = 1, 2, ..., n$, and β to be the minimum distance from case A_{ij} to their adjusted boundaries, $\beta_i > \beta$.

A model that seeks MSD (the minimal sum of the deviations of the observations from the critical value) can be written as follows:

$$\text{(M1)} \quad \text{Minimize} \sum_{i=1}^{n} \alpha_i,$$

$$
\begin{aligned}
\boldsymbol{A}_i\boldsymbol{X} &\le b+\alpha_i, & \boldsymbol{A}_i &\in M,\\
\boldsymbol{A}_i\boldsymbol{X} &\ge b-\alpha_i, & \boldsymbol{A}_i &\in N
\end{aligned} \tag{2}
$$

where $\boldsymbol{A}_i$ is given, $\boldsymbol{X}$ and b are unrestricted, and $\alpha_i \ge 0$, $i = 1, 2, ..., n$.

The alternative of the above model is to find MMD (the minimal distances of observations from the critical value are maximized). It can be written by

$$
\begin{aligned}
\text{(M2)}\quad & \text{Maximize } \sum_{i=1}^{n} \beta_i,\\
& \boldsymbol{A}_i\boldsymbol{X} \ge b-\beta_i, \quad \boldsymbol{A}_i \in M,\\
& \boldsymbol{A}_i\boldsymbol{X} \le b+\beta_i, \quad \boldsymbol{A}_i \in N
\end{aligned} \tag{3}
$$

where $\boldsymbol{A}_i$ is given, $\boldsymbol{X}$ and b are unrestricted, and $\beta_i \ge 0$, $i = 1, 2, ..., n$.

In a linear discriminate model, the misclassification of data separation can be described by two objects: MSD and MMD. Therefore, the main research aim in the identification of unusual patterns of MCNs is to seek the method that produces the higher detection accuracy.

Let y_{1L} be MSD and y_{2U} be MMD, the maximum value of $\sum_{i=1}^{n} \alpha_i$ is y_{1U} and the minimum value of $\sum_{i=1}^{n} \beta_i$ is y_{2L}. To explore this possibility, we propose a heuristic identification of the unusual pattern method by using the fuzzy linear programming for discovering the unusual patterns in MCNs as follows:

$$
\begin{aligned}
\text{(M3)}\quad & \text{Maximize } \xi,\\
& \xi \le \frac{\sum \alpha_i - y_{1L}}{y_{1U} - y_{1L}},\\
& \xi \le \frac{\sum \beta_i - y_{2L}}{y_{2U} - y_{2L}},\\
& \boldsymbol{A}_i\boldsymbol{X} = b+\alpha_i-\beta_i, \quad \boldsymbol{A}_i \in M,\\
& \boldsymbol{A}_i\boldsymbol{X} = b-\alpha_i+\beta_i, \quad \boldsymbol{A}_i \in N
\end{aligned} \tag{4}
$$

where $\boldsymbol{A}_i$, y_{1L}, y_{1U}, y_{2L}, y_{2U} are known, $\boldsymbol{X}$ and b are unrestricted, and α_i, β_i, $\xi \ge 0$, $i = 1, 2, ..., n$.

Method

(1) Create data warehouse for the performance evaluation of every MCN at every selected time spot.

(2) Generate a set of relevant attributes from the data warehouse, transform the scales of the data warehouse into the same numerical measurement, determine the two classes of usual and unusual patterns, as well as the classification threshold τ that is selected by the user, and the training set and the verifying set.

(3) Give a class boundary value b and use models (M_1), (M_2), and (M_3) to learn and compute the overall scores $\boldsymbol{A}_i\boldsymbol{X}$ $(i = 1, 2, ..., n)$ of the relevant attributes or dimensions over all observations repeatedly.

(4) If (M_1) exceeds the threshold τ, go to (7), otherwise go to (5).

(5) If (M_2) exceeds the threshold τ, go to (7), otherwise go to (6).
(6) If (M_3) exceeds the threshold τ, go to (7), otherwise go to (3) to consider to give another b.
(7) Apply the final learned scores $\boldsymbol{X}^*$ to predict the unknown data in the verifying set.
(8) Find the unusual patterns of the MCNs.

The FLP approach proposed in this paper is simpler and easier to get the meaningful results. For example, this FLP approach can get more meaningful solutions than the common classification approaches in the multiple criteria linear programming.

Real-time CNs data can be used to test our data mining system. Based on the above analysis, we have developed a software named performance evaluation data miner (PEDM2.0) [10]. This miner is an OLAP miner integrated with an OLAP whose mining is in relational databases. The development language is the C++ syntax based on Linux. This miner also combines with the algorithm of linear & non-linear programming in those softwares named Lingo9.0 and Lindo8.0 [11].

The FLP approach is not the only module in identification mining of the PEDM2.0. Statistics, decision tree, linear programming, multiple criteria linear programming, neutral networks are also used. The output results in the PEDM2.0 are the synthesis integration results based on different methods. A comparison study in terms of computational efficiency implementation will be discussed in the next section.

4 Data Experiments

Given a set of attributes, such as throughput capacity, package forwarding rate, response time, connection attempts, delay time, transfer rate and the criteria for "unusual" patterns, the purpose of pattern recognition for the MCNs is to find the better classifier through a training set and use the classifier to predict all other aspects of the performance of MCNs.

The frequently used pattern recognition in the telecommunication industry is still two-class separation technique. The key of two-class separation is to separate the "unusual" patterns called fraudulent activity from the "usual" patterns called normal activity and identify as many MCNs as possible. This is also known as the method of "detecting fraudulent list".

In this section, a real-time performance data mart with 65 derived attributes and 1000 records of a major CHINA TELECOM MCN database is first used to train the different classifiers [12]. Then, the training solution is employed to predict the performance of another 5000 records of MCNs. Finally, the classification results are compared with the decision tree, neural network and MCLP.

The results are shown in Table 1. Three known classification techniques have been used to run and test the 5000 records of the CHINA TELECOM MCN database. These results are compared with the FLP approach shown in Table 1. The software of the decision tree is the commercial version called C5.0 [13], while

Table 1. Identification rate comparisons on balanced 5000 records

Approaches	Identification rate	Time (second)
Decision Tree	79.39%	0.335
Neural Network	64.20%	0.201
MCLP	80.03%	0.936
FLP	81.74%	0.284

the software for both MCLP and FLP were developed at Chinese Academy of Science in China and Konan University in Japan [10].

Note that in Table 1 the column identification rate represents the rate of identifying the right unusual patterns in respective models as: Identification Rate = (Number of identified unusual patterns exactly) / (Number of unusual patterns) ×100%.

The identification time is calculated using different models. Because this data mining system is special for the MCNs, the FLP model is not the model with the fastest calculation speed, but it does have a higher calculation speed and higher identification rate.

The greater the identification rate is, the better the result is. As we see, the model that predicts best is the FLP with 81.74%. The second best model is the MCLP with 80.03%. The decision tree model has the third best prediction rate with 79.39% while the neural network is the worst one with 64.20%.

The shorter the identification time is, the better the result is. The fastest model on the identification time is the neural network with 0.201 seconds. The second is FLP with 0.284. Decision tree has the third fastest time with 0.335 seconds while MCLP is the slowest one with 0.936 seconds.

Both short identification time and high identification rate are important in identification mining calculations of MCNs. Therefore, if the data set is balanced, it is meaningful to implement FLP algorithm proposed in this paper. This conclusion, however, may not be true for all kinds of data sets because of the different data structure and data feature.

Many decision makers in MCNs often get a better result through an FLP approach. It has been recognized that in many decision making problems, instead of finding the exist "optimal solution" (a goal value), decision makers often approach a "satisfying solution" between upper and lower aspiration levels that can be represented by the upper and lower bounds of acceptability for objective payoffs. The model proposed in this paper also can be used in bioinformatics, antibody and antigen.

5 Conclusions

In this paper, an identification mining model of unusual patterns for MCNs has been presented. The construction flow of data mining systems of MCNs for performance evaluation, the snowflake schema of a data warehouse, and the algorithm of fuzzy linear programming were shown in detail. The data experiments proved that the fuzzy linear programming (FLP) approach we proposed in this paper has excellent accuracy and effectiveness compared with decision tree, neural network, multiple criteria linear programming methods.

Acknowledgments

This work was supported in part by GRANT-IN-AID FOR SCIENTIFIC RESEARCH (No. 16560350) and MEXT.ORC (2004-2008), Japan and in part by NSFC (No. 70472074, No. 70531040), 973 Project (No. 2004CB720103), Postdoctoral Science Foundation, China and BHP Billion Co., Australia.

References

1. Yue, W., Gu, J., Tang, X.: Performance evaluation index system for multimedia communication networks and forecasting for web-based network traffic. Journal of System Science and System Engineering **13** (1994) 44–50
2. Han, J., Kamber, M.: Data Mining: Concepts and Techniques. Morgan Kaufmann Publishers (2003)
3. Freed, N., Glover, F.: Simple but powerful goal programming models for discriminant problems. European Journal of Operation Research **7** (1981) 44–60
4. Freed, N., Glover, F.: Evaluating alternative linear programming models to solve the two-group discriminant problem. Journal of Decision Science **17** (1986) 151–162
5. Glover, F.: Improve linear programming models for discriminate analysis. Journal of Decision Science **21** (1990) 771–785
6. Kou, G., Liu, X., Peng, Y., Shi, Y., Wise, M., Xu, W.: Multiple criteria linear programming approach to data mining: models, algorithm designs and software development. Journal of Operation Methods and Software **18** (2003) 453–473
7. Kou, G., Shi, Y.: LINUX based Multiple Linear Programming Classification Program: Version 1.0. College of Information Science and Technology, University of Nebraska-Omaha (2002)
8. Shi, Y., He, J., Wang, L., Fan, W.: Computer-based algorithms for multiple criteria and multiple constraint level integer linear programming. Computers and Mathematics with Applications **49** (2005) 903–921
9. He, J., Yue W., Shi, Y.: Identification mining of unusual patterns for multimedia communication networks by using fuzzy linear programming. IEICE Technical Report **DE2005-17** (2005) 11–17
10. He, J., Shi Y.: Performance Evaluation Data Miner 2.0, CAS Research Center on Data Technology and Knowledge Economy (2005)
11. http://www.lindo.com/
12. http://www.chinatelecom.com.cn/
13. http://www.rulequest.com/see5-info.html/

Comparisons of the Different Frequencies of Input Data for Neural Networks in Foreign Exchange Rates Forecasting

Wei Huang[1,2], Lean Yu[2], Shouyang Wang[2], Yukun Bao[1], and Lin Wang[1]

[1] School of Management, Huazhong University of Science and Technology, WuHan, 430074, China
{yukunbao, wanglin}@mail.hust.edu.cn
[2] Institute of Systems Science, Academy of Mathematics and Systems Sciences, Chinese Academy of Sciences, Beijing, 100080, China
{whuang, yulean, sywang}@amss.ac.cn

Abstract. We compare the predication performance of neural networks with the different frequencies of input data, namely daily data, weekly data, monthly data. In the 1 day and 1 week ahead prediction of foreign exchange rates forecasting, the neural networks with the weekly input data performs better than the random walk models. In the 1 month ahead prediction of foreign exchange rates forecasting, only the special neural networks with weekly input data perform better than the random walk models. Because the weekly data contain the appropriate fluctuation information of foreign exchange rates, it can balance the noise of daily data and losing information of monthly data.

1 Introduction

Exchange rates are one of the most important economic indices in the international monetary markets. Because of its complicated nonlinear behavior, many researchers employ neural networks to forecast foreign exchange rates. Several design factors significantly affect the prediction performance of neural networks [1]. Although foreign exchange rates are high-frequency financial time series, fluctuating every minute, the researchers seldom employ the hourly observations for forecasting. Because there is too much noises in the observations of high-frequency, no body use the frequency that is higher than daily data. It is a common practice to predict the future value daily ahead with daily data[2, 3], to predict the future value weekly ahead with weekly data[4-6], to predict the future value monthly ahead with monthly data[7-10]. Hann and Steurer compare the prediction performances of neural network models with linear monetary models in forecasting USD/DEM by using the monthly data and weekly data. Out-of-sample results show that, for weekly data, neural networks are much better than linear models and a random walk model. However, if monthly data are used, neural networks do not show much improvement over linear models[11].

In fact, different frequencies of input data may affect the performances of neural networks, due to the volatility of currency movements. However, few researchers investigate the effect of different frequencies of input data. Our contribution of the

V.N. Alexandrov et al. (Eds.): ICCS 2006, Part IV, LNCS 3994, pp. 517 – 524, 2006.
© Springer-Verlag Berlin Heidelberg 2006

paper is to compare the prediction performance of the neural networks by using the different frequencies of input data. The remainder of this paper is organized as follows. Section 2 gives the experiment design. Section 3 discusses the experiment results. Finally, conclusions are given in Section 4.

2 Experiment Design

We employ the three frequencies of data, namely daily data { y_t^d }, weekly data { y_t^w }, monthly data { y_t^m }. The daily data is the closing price of every trading day. The weekly data is compiled from the average of daily data in that week. The monthly data is compiled from the average of daily data in that month.

2.1 Random Walk Models

The weak form of efficient market theory describes that prices always fully reflect the available information, that is, a price is determined by the previous value in the time series because all the relevant information is summarized in that previous value. An extension of this theory is the random walk (RW) model. The random walk model uses the actual value of current period to predict the future value of next period as follows:

$$\hat{y}_{t+1} = y_t \tag{1}$$

where $\hat{y}_{t+1}$ is the predicted value of the next period; y_t is the actual values of current period.

Therefore, the predicted values of next day, week, month by RW model are computed in the following way:

$$\hat{y}_{t+1}^d = y_t^d \tag{2}$$

$$\hat{y}_{t+1}^w = y_t^w \tag{3}$$

$$\hat{y}_{t+1}^m = y_t^m \tag{4}$$

2.2 Neural Network Models

In this study, we employ one of the widely used neural networks models, the three-layers back-propagation neural network (BPNN), for foreign exchange rates forecasting. The activation function used for all hidden nodes is the logistic function, while the linear function is employed in the output node. The number of input nodes is a very important factor in neural network analysis of a time series since it corresponds to the number of past lagged observations related to future values. To avoid introducing a bias in results, we choose the number of input nodes as 3, 5, 7 and 9, respectively. Because neural networks with one input nodes are too simple to capture the complex relationships between input and output, and it is rarely seen in the literatures that the number of input nodes is more than nine. Generally speaking, too many nodes in the hidden layer produce a network that memorizes the input data and lacks the ability to generalize. Parsimony is a principle for designing neural networks. Hence, the number of hidden nodes is equal to the number of input nodes.

Because there is no daily data of economic indicators such exportation, importation, we employ univariate input for the neural networks. The inputs of neural network are the past, lagged observations of exchange rates; the output is the future value. In some sense, the neural networks of univariate input are equivalent to a nonlinear autoregressive mode as follows:

$$\hat{y}_{t+1} = F(y_t, y_{t-1}, \ldots, y_{t-K}) \tag{5}$$

where $\hat{y}_{t+1}$ is the output of the neural networks, namely the predicted value of the next period; $y_t, y_{t-1}, \ldots, y_{t-K}$ are the inputs for the neural networks, namely the actual value at the period t, $t-1, \ldots, t-K$, respectively; function F is a nonlinear function determined by the neural networks; K is the max lag period, which is determined by the number of input nodes. In the study, when the number of input nodes is 3, 5, 7 and 9, K is 2, 4, 6, 8, respectively.

We employ daily data, weekly data, monthly data as inputs of the neural networks to predict the future foreign exchange rates of the next day as follows:

$$\hat{y}^d_{t+1} = F^d_d(y^d_t, y^d_{t-1}, \ldots, y^d_{t-K}) \tag{6}$$

$$\hat{y}^d_{t+1} = F^w_d(y^w_t, y^w_{t-1}, \ldots, y^w_{t-K}) \tag{7}$$

$$\hat{y}^d_{t+1} = F^m_d(y^m_t, y^m_{t-1}, \ldots, y^m_{t-K}) \tag{8}$$

where $\hat{y}^d_{t+1}$ is the output of the neural networks, namely the predicted value of the next day; function F^d_d, F^w_d, F^m_d are the nonlinear functions determined by the neural networks with the input of daily data, weekly data, monthly data, respectively.

We employ daily data, weekly data, monthly data as inputs of the neural networks to predict the future foreign exchange rates of the next week as follows:

$$\hat{y}^w_{t+1} = F^d_w(y^d_t, y^d_{t-1}, \ldots, y^d_{t-K}) \tag{9}$$

$$\hat{y}^w_{t+1} = F^w_w(y^w_t, y^w_{t-1}, \ldots, y^w_{t-K}) \tag{10}$$

$$\hat{y}^w_{t+1} = F^m_w(y^m_t, y^m_{t-1}, \ldots, y^m_{t-K}) \tag{11}$$

where $\hat{y}^w_{t+1}$ is the output of the neural networks, namely the predicted value of the next week; function F^d_w, F^w_w, F^m_w are the nonlinear functions determined by the neural networks with the input of daily data, weekly data, monthly data, respectively.

We employ daily data, weekly data, monthly data as inputs of the neural networks to predict the future foreign exchange rate of the next month as follows:

$$\hat{y}^{m}_{t+1} = F^{d}_{m}\,(y^{d}_{t}, y^{d}_{t-1}, \ldots, y^{d}_{t-K}) \tag{12}$$

$$\hat{y}^{m}_{t+1} = F^{w}_{m}\,(y^{w}_{t}, y^{w}_{t-1}, \ldots, y^{w}_{t-K}) \tag{13}$$

$$\hat{y}^{m}_{t+1} = F^{m}_{m}\,(y^{m}_{t}, y^{m}_{t-1}, \ldots, y^{m}_{t-K}) \tag{14}$$

where $\hat{y}^{m}_{t+1}$ is the output of the neural networks, namely the predicted value of the next month; function F^{d}_{m}, F^{w}_{m}, F^{m}_{m} are the nonlinear functions determined by the neural networks with the input of daily data, weekly data, monthly data, respectively.

2.3 Performance Measure

We employ root of mean squared error (RMSE) to evaluate the prediction performance of neural networks as follows:

$$\mathrm{RMSE} = \sqrt{\frac{\sum_{t} (y_t - \hat{y}_t)^2}{T}} \tag{15}$$

where y_t is the actual value; $\hat{y}_t$ is the predicted value; T is the number of the predictions

2.4 Data Preparation

From Pacific Exchange Rate Service provided by Professor Werner Antweiler, University of British Columbia, Canada, we obtain 3291 daily observations, 678 weekly data and 156 monthly data of U.S. dollar against the British Pound (GBP) and Japanese Yen (JPY) covering the period the period from Jan 1990 to Dec, 2002. First, we produce the testing sets for each neural network models by selecting 60 patterns of the latest periods from the three datasets, respectively. Then, we produce the appropriate training sets for each neural networks model from the corresponding left data in the three datasets by using the method in [12].

3 Experiments Results

Table 1 shows the prediction performances of the random walk models, which are used as benchmarks of prediction performance of foreign exchange rates for the different forecasting horizons. The prediction performance become worse as the forecasting horizon becomes longer. This pattern is consistent with the assumption of random walk model.

Table 2-5 show the 1 day ahead prediction performance of the neural networks with 3, 5, 7, 9 input nodes, respectively. In the 1 day ahead prediction of foreign exchange rates, the neural networks with weekly input data perform better than the random walk models; the neural networks with daily, monthly input data perform worse than the random walk models. Because the daily input data contain too much noise, while the monthly input data lose too much fluctuation information of foreign exchange rates at the scale of day.

Table 6-9 show the 1 week ahead prediction performance of the neural networks with 3, 5, 7, 9 input nodes, respectively. In the 1 week ahead prediction of foreign exchange rates, the neural networks with weekly input data perform better than the random walk models; the neural networks with daily, monthly input data perform worse than the random walk models. Because the daily input data can not cover the enough period which contains the behavior of foreign exchange rate at the scale of week, while the monthly input data lose some fluctuation information of foreign exchange rates at the scale of week.

Table 10-13 show the 1 month ahead prediction performance of the neural networks with 3, 5, 7, 9 input nodes, respectively. In the 1 month ahead prediction of foreign exchange rates, the neural networks with weekly input data perform little better than the random walk models when the number of the input nodes is 5 and 7; the other neural networks models perform worse than the random walk models. It indicates that the neural networks models are not suitable for the long term forecasting when the foreign exchange rates fluctuate a lot

Table 1. The prediction performance of the random walk models

Forecasting horizon	RMSE of GBP	RMSE of JPY
1 day ahead	0.0054715	0.007508
1 week ahead	0.0128701	0.01608
1 month ahead	0.0387545	0.042368

Table 2. The 1 day ahead prediction performance of the neural networks with 3 input nodes

Frequency of input data	RMSE of GBP	RMSE of JPY
daily	0.0106794	0.021370
weekly	0.0054705	0.007439
monthly	0.0381506	0.041369

Table 3. The 1 day ahead prediction performance of the neural networks with 5 input nodes

Frequency of input data	RMSE of GBP	RMSE of JPY
daily	0.0155469	0.017623
weekly	0.004491	0.00635
monthly	0.0380584	0.040172

Table 4. The 1 day ahead prediction performance of the neural networks with 7 input nodes

Frequency of input data	RMSE of GBP	RMSE of JPY
daily	0.0149991	0.017417
weekly	0.004496	0.00636
monthly	0.0380457	0.039505

Table 5. The 1 day ahead prediction performance of the neural networks with 9 input nodes

Frequency of input data	RMSE of GBP	RMSE of JPY
daily	0.0186669	0.021558
weekly	0.0054671	0.007293
monthly	0.038303	0.04163

Table 6. The 1 week ahead prediction performance of the neural networks with 3 input nodes

Frequency of input data	RMSE of GBP	RMSE of JPY
daily	0.0187851	0.022372
weekly	0.012474	0.015306
monthly	0.0346265	0.034005

Table 7. The 1 week ahead prediction performance of the neural networks with 5 input nodes

Frequency of input data	RMSE of GBP	RMSE of JPY
daily	0.0179622	0.019123
weekly	0.01154	0.01421
monthly	0.0345032	0.033285

Table 8. The 1 week ahead prediction performance of the neural networks with 7 input nodes

Frequency of input data	RMSE of GBP	RMSE of JPY
daily	0.0172154	0.019116
weekly	0.011508	0.0143
monthly	0.0336946	0.033712

Table 9. The 1 week ahead prediction performance of the neural networks with 9 input nodes

Frequency of input data	RMSE of GBP	RMSE of JPY
daily	0.018782	0.022365
weekly	0.012528	0.015432
monthly	0.035721	0.035353

Table 10. The 1 month ahead prediction performance of the neural networks with 3 input nodes

Frequency of input data	RMSE of GBP	RMSE of JPY
daily	0.0585355	0.063101
weekly	0.0402285	0.043588
monthly	0.0552075	0.059799

According to the above results, we may see that weekly data is the appropriate frequency which matches the scale of fluctuation behavior of foreign exchange rates fluctuation behavior. Weekly data balance the noise of daily data and losing information of monthly data.

We notice that the networks with 5, 7 inputs nodes perform better than the networks with 3, 9 input nodes. Because the neural networks with 5, 7 input nodes are at the appropriate level of complexity, which balances the over-fitting and under-fitting.

Table 11. The 1 month ahead prediction performance of the neural networks with 5 input nodes

Frequency of input data	RMSE of GBP	RMSE of JPY
daily	0.0574925	0.061378
weekly	0.0383246	0.041754
monthly	0.0543704	0.058285

Table 12. The 1 month ahead prediction performance of the neural networks with 7 input nodes

Frequency of input data	RMSE of GBP	RMSE of JPY
daily	0.0571355	0.060978
weekly	0.0385933	0.041929
monthly	0.0540548	0.05809

Table 13. The 1 month ahead prediction performance of the neural networks with 9 input nodes

Frequency of input data	RMSE of GBP	RMSE of JPY
daily	0.0593892	0.062386
weekly	0.0405666	0.044135
monthly	0.0558715	0.060805

4 Conclusions

In this paper, we investigate the effects of different frequencies of input data of foreign exchange rates forecasting with neural networks. The neural networks with the weekly input data performs better than those neural networks with the input of daily data and monthly data. Weekly data is the appropriate frequency which matches the scale of fluctuation behavior of foreign exchange rates fluctuation behavior.

Acknowledgements

This work is partially supported by National Natural Science Foundation of China (NSFC No. 70221001, 70401015) and the Key Research Institute of Humanities and Social Sciences in Hubei Province-Research Center of Modern Information Management.

References

1. Huang, W., Lai, K.K., Nakamori, Y. & Wang, S.Y:. Forecasting foreign exchange rates with artificial neural networks: a review. International Journal of Information Technology & Decision Making, 3(2004) 145-165
2. Walczak, S.: An empirical analysis of data requirements for financial forecasting with neural networks. Journal of Management Information System, 17(2001) 203-222

3. Yu, L.A., Wang, S.Y., Lai, K.K.: Adaptive smoothing neural networks in foreign exchange rate forecasting. Lecture Notes in Computer Science,Vol. 3516, Springer-Verlag Berlin Heidelberg (2005) 523 – 530
4. Yao, J.T. & Tan, C.L.: A case study on using neural networks to perform technical forecasting of forex. Neurocomputing, 34(2000) 79-98
5. Zhang, G.P. & Berardi, V.L.: Time series forecasting with neural network ensembles: an application for exchange rate prediction. Journal of the Operational Research Society, 52(2001) 652-664
6. Zhang, G. & Hu, M.Y.: Neural network forecasting of the British Pound/US Dollar exchange rate. Journal of Management Science, 26(1998) 495-506
7. Leung, M.T., Chen, A.S. & Dauk, H.: Forecasting exchange rate using general regression neural networks. Computer & Operations Research, 27(2000) 1093-1110
8. Lisi, F. & Schiavo, R.A.: A comparison between neural networks and chaotic models for exchange rate prediction. Computational Statistical & Data Analysis, 30(1999) 87-102
9. Qi, M. and Zhang, G.: An investigation of model selection criteria for neural network time series forecasting. European Journal of Operational Research, 132(2001) 666-680
10. Yu, L.A., Wang, S.Y., Lai, K.K.: A novel nonlinear ensemble forecasting model incorporating GLAR and ANN for foreign exchange rates. Computers and Operations Research, 32 (2005) 2523–2541
11. Hann, T.H. and Steurer, E.: Much ado about nothing? Exchange rate forecasting: Neural networks vs. linear models using monthly and weekly data. Neurocomputing, 10(1996) 323-339
12. Huang, W., Nakamori, Y., Wang, S.Y. & Zhang, H.: Select the size of training set for financial forecasting with neural networks. Lecture Notes in Computer Science, Vol. 3497, Springer-Verlag Berlin Heidelberg (2005) 879–884

Automatic Differentiation of C++ Codes for Large-Scale Scientific Computing

Roscoe A. Bartlett, David M. Gay, and Eric T. Phipps

Sandia National Laboratories**,
Albuquerque NM 87185, USA

Abstract. We discuss computing first derivatives for models based on elements, such as large-scale finite-element PDE discretizations, implemented in the C++ programming language. We use a hybrid technique of automatic differentiation (AD) and manual assembly, with local element-level derivatives computed via AD and manually summed into the global derivative. C++ templating and operator overloading work well for both forward- and reverse-mode derivative computations. We found that AD derivative computations compared favorably in time to finite differencing for a scalable finite-element discretization of a convection-diffusion problem in two dimensions.

Computing derivatives is ubiquitous in scientific computing; examples include algorithms for nonlinear equation solving, optimization, stability analysis, and implicit time integration. Computing derivatives quickly and accurately improves both the efficiency and robustness of these numerical algorithms, particularly in the presence of ill-conditioning. In this paper, we discuss computing first derivatives of element-based models implemented in ANSI/ISO C++. We use the term "element" in a broad sense to encompass any model whose computation consists of repeated evaluations of a small set of functions, each involving relatively few of the variables of the overall problem. Many classes of models fall into this category, including finite-element and finite-volume PDE discretizations and network models. We use a hybrid technique of automatic differentiation (AD) and manual assembly similar to [1, 2] to carry out the model evaluation and derivative computation one element at a time. This decomposition is discussed in more detail in Section 1, which generalizes the ideas in [2] to general element-based models and additionally describes how to compute the global adjoint.

We focus on ANSI/ISO C++ codes because much modern scientific code development is done in C++. Since no source transformation tools for C++ were available to us, we used C++ operator overloading to implement AD for computing the element-level derivatives. We assume the reader is familiar with AD and

** Sandia is a multiprogram laboratory operated by Sandia Corporation, a Lockheed Martin Company, for the United States Department of Energy under Contract DE-AC04-94AL85000.

V.N. Alexandrov et al. (Eds.): ICCS 2006, Part IV, LNCS 3994, pp. 525–532, 2006.
© Springer-Verlag Berlin Heidelberg 2006

the methods for implementing it; see [3] for a good introduction to these concepts. We used two separate AD packages: the public domain package Fad [4] for forward-mode AD and our own reverse-mode package Rad [5]. We sought to determine if AD based on operator overloading could be incorporated effectively into a large, evolving scientific application code and whether the resulting derivative calculations would be efficient enough for scientific use, particularly for reverse-mode gradient evaluations. We applied this approach to a large-scale finite-element simulation code called Charon, developed at Sandia National Laboratories for reacting fluid flows and semiconductor device simulations. Details of the implementation are presented in Section 2, along with a discussion of difficulties we encountered. To assess efficiency, in Section 3 we report flop counts and run times for Jacobian and Jacobian-transpose products and finite differences on a small convection-diffusion problem.

We believe the work presented here to be novel in a number of ways. While there have been several successful applications of automatic differentiation to Fortran-based scientific codes using source transformation, we knew of no experience with this in large C++ codes. Successfully incorporating AD by operator overloading and templating into such an application code is, we believe, both nontrivial and new, and the process we used merits discussion. While computing element derivatives was used as motivation for development of the Rad tool presented in [5], the work here represents the first measurement of the performance of Rad in a real scientific code.

1 Computing Derivatives of Element-Based Models

We are concerned with evaluating and computing derivatives of a continuously differentiable, vector valued function $f : \mathbf{R}^n \to \mathbf{R}^m$, in which m and n may be large, on the order of millions, and in which $f(x)$ is the sum

$$f(x) = \sum_{i=1}^{N} Q_i^T e_{k_i}(P_i x) \tag{1}$$

of a large number N of elements taken from a small set $\{e_k\}$ of element functions $e_k : \mathbf{R}^{n_k} \to \mathbf{R}^{m_k}$ where typically each n_k, m_k are on the order of 10 to 100. The matrices $P_i \in \mathbf{R}^{n_{k_i} \times n}$ and $Q_i \in \mathbf{R}^{m_{k_i} \times m}$ map global vectors to the local element domain and range spaces respectively. Often we seek x such that $f(x) = 0$, so we call $f(x)$ the global residual.

In some applications, such as the one we discuss in Section 3, it is convenient to deal with "interior" and "boundary" elements separately, with the boundary elements modifying or replacing some values computed by the interior elements. In effect, we compute $f(x) = (I - S^T S) f_I(x) + S^T f_B(x)$, where $(I - S^T S)$ is a projection matrix that replaces some components of the sum $f_I(x)$ of the interior elements by zeros. We suppress this extra complexity in what follows, since it is orthogonal to the other issues we discuss.

Given (1), we can clearly compute the global Jacobian $J = \partial f / \partial x$ and adjoint $\bar{J} = w^T J$ element-wise:

$$\frac{\partial f}{\partial x} = \sum_{i=1}^{N} Q_i^T J_{k_i} P_i, \qquad w^T \frac{\partial f}{\partial x} = \sum_{i=1}^{N} (Q_i w)^T J_{k_i} P_i$$

where $J_{k_i} = \partial e_{k_i} / \partial P_i x$ is the Jacobian matrix of e_{k_i}.

With these decompositions, we have translated the difficult task of computing the global Jacobian and adjoint into a series of much smaller computations on elements. In principle, any method can be used to compute these element-level derivatives: AD, symbolic differentiation, or finite differencing. This task is well suited to AD for several reasons. First, each element function e_k has only a few independent and dependent variables, often around ten and at most a few hundred, so the element Jacobians $J_i = \partial e_{k_i} / \partial P_i x$ can be treated as dense matrices, and there is no need to use sparse AD techniques. Second, each element computation is fairly simple, involving only a few operations per variable. Thus the memory burden of reverse-mode AD is reasonable and checkpointing is not generally required. Third, all parallel communication occurs during gathering of the local variables and scattering of the results to the global residuals/derivatives, which means it is not necessary to differentiate through parallel communications. Lastly, the structure of the derivative assembly closely mirrors the residual assembly, particularly when we implement AD via templating and operator overloading. This allows much of the same code for the residual evaluation to be reused for the derivative computation, as discussed next.

2 Computing Element Derivatives Via AD in C++

We turn now to some practical details of implementing AD via operator overloading in the large, element-based scientific C++ code Charon, developed at Sandia National Laboratories for simulation of reacting fluid flows and semiconductor devices. Our goals were to determine if AD based on operator overloading could be effectively incorporated into such an application code and whether the resulting derivative calculations would be efficient enough for production use.

To compute derivatives using forward AD, there are many publicly available C++ tools that in principle could be applied. We chose the Fad [4] package because of its reputation for efficiency, flexibility, and simplicity. Fad uses expression templates to eliminate much of the overhead normally associated with operator overloading. However, because the exact physics Charon is simulating is not known until run time, we were forced to use the version of Fad that uses dynamic memory allocation of the derivative array.

For reverse-mode derivative computations, we chose the Rad [5] package, which is designed precisely for element gradient computations. Rad records just enough detail during an element evaluation to permit efficient reverse accumulation of the element gradient; Rad retains scratch memory, immediately reusing it when evaluation of the next element begins.

To use these tools in Charon, we found C++ templating highly effective for computing the element functions e_k. In brief, we changed scalar floating-point types (`double` or `float`) to templated types in all C++ classes used in computing the e_k. Then by instantiating the resulting templated classes on the floating-point type, we get the original element evaluations, and by instantiating on the AD types, we compute both the element functions and their derivatives. We also templated the initialization and post-processing classes that gather and scatter to and from local variables (i.e., that compute $P_i x$ and $Q_i^T e_{k_i}$, given e_{k_i}). In addition to gathering and scattering, the AD specializations initialize the seed matrix (for Fad) and extract the element derivatives.

By providing other AD types, one could obtain many other kinds of derivatives, such as Hessian-vector products, and Taylor polynomials. This results in major savings in code development time, since only one templated residual computation needs to be written and maintained. We believe this approach is significantly more suitable to a large, evolving application code than the standard approach of copying the undifferentiated source and manually changing the type. Templating makes it impossible for the differentiated source code to become out of sync with the undifferentiated source, and forces the developer to think about how the source should be differentiated at development time.

Overall, we found our approach to be an effective way to use AD in Charon, but we did encounter some difficulties. First, interfacing the templated functions and classes for computing the e_k to the rest of the non-templated application code in a manner that easily allows new template types to be added to the application code required some significant C++ software engineering. In brief, we used container classes for storing instantiations of each templated class. This allows "glue" code to interface template and non-template code in a manner independent of the choices of AD data types.

Second, most C++ application codes use libraries written in other languages, such as Fortran. For example, Charon uses Chemkin [6] to simulate chemical reactions appearing in elements. A simple way to deal with this is to provide a templated interface class that has specializations for each AD type. These specializations extract derivative values out of the C++ classes and then compute derivatives of the Fortran source by whatever mechanism is available. In Charon, we have a forward-mode differentiated version of the Chemkin source provided by ADIFOR 2.0 [7], and this version is used by both the Fad and Rad Charon/Chemkin interface classes. We plan later to make reverse-mode differentiated Chemkin source available for the Rad specialization, provided by one or more of OpenAD [8], ADIFOR 3.0, or Tapenade [9].

Third, templating the application code classes can lengthen the time taken to compile the application significantly. Since definitions of templated functions and classes must be available at the time they are instantiated, typically when they are first referenced in a source file, the template definitions are often placed in header files along with the declarations. This results in code-bloat, and increased compile times since all of the template definitions must be recompiled in each translation unit. This additional compile time was probably the single largest hurdle to effectively incorporating AD into Charon. To cope, we split the header file for a templated class into three files, a declaration header, an implementation header, and a source file that includes both and explicitly instantiates the class on all AD types via a preprocessor macro. This drastically reduces the recompilation time of the application code, putting it on par with the original un-templated code.

Finally, passive variables gave us trouble with incorporating reverse-mode AD into Charon. Such "variables" act as constants, but are stored as AD types for flexibility. Since Charon supports multiple physics, it is hard in some parts of the code to know whether a quantity, say temperature, is a constant or an unknown being solved for. To avoid storing the temperature as a passive variable, we could provide two instantiations of the element functions, one for when temperature is an unknown (AD type) and one for when it is constant (a floating-point type). This would be necessary for any quantity that could be constant or variable, yielding a combinatorial explosion of template instantiations. To avoid this explosion, we always store potentially active variables as active. For reverse AD, this requires us to tell Rad which of these active variables are really constants (since they will not be reinitialized), so Rad can store them in memory that is not recycled at the beginning of each element evaluation. We think we can find a place in Charon where all passive variables are known, so Rad could be told before the first function evaluation to treat them as constants, but so far we have pursued more ad-hoc (and less satisfactory) approaches. Currently we use traits to mark passive variables as constants, but this requires finding all potentially passive variables, a daunting task that is unlikely to be maintainable. Another approach would be to assume a variable is constant until it is reinitialized and only to reuse memory for such non-constants. We believe this would substantially reduce Rad's efficiency, but it is an approach that would be helpful for debugging, and we are looking into it.

3 An Example Convection-Diffusion Problem

We now compare costs of alternative derivative computations in a small, two-dimensional reacting convection-diffusion problem implemented in Charon. Since we compute derivatives element-wise, the size of the AD computation is proportional to the degrees-of-freedom (DOF) per element, so we study how the costs of the Jacobian and adjoint computations scale with the DOF.

Our test problem has a two dimensional rectangular domain Ω of width 2 and height 1 containing an ideal fluid with unit density and constant but spatially varying fluid velocity $\mathbf{u}$. The fluid contains N chemical species $X_1, \dots, X_N$, with mass fractions $Y_1, \dots, Y_N$, unit molecular weights and unit diffusion coefficients. The chemical species undergo the following hypothetical chemical reactions: $2X_j \rightleftharpoons X_{j-1} + X_{j+1}$, $j = 2, \dots, N-1$, with both unit forward and reverse rate constants. For each reaction j, the rate of progress for that reaction, q_j, satisfies

$$q_j = [X_j]^2 - [X_{j-1}][X_{j+1}] = Y_j^2 - Y_{j-1}Y_{j+1}, \quad j = 2, \dots, N-1.$$

Then the production rate $\dot{\omega}_j$ of chemical species X_j is $\dot{\omega}_j = q_{j-1} - 2q_j + q_{j+1}$ for $j = 3, \dots, N-2$, with $\dot{\omega}_1 = q_2$, $\dot{\omega}_2 = -2q_2 + q_3$, and $\dot{\omega}_{N-1} = q_{N-2} - 2q_{N-1}$.

The partial differential equations governing the mass fractions of the N species are given by

$$\begin{aligned} \frac{\partial Y_j}{\partial t} + \mathbf{u} \cdot \nabla Y_j + \nabla^2 Y_j &= \dot{\omega}_j, \quad j = 1, \ldots, N-1 \\ \sum_{j=1}^{N} Y_j &= 1. \end{aligned} \tag{2}$$

Charon uses bilinear basis functions and quadrangle finite elements in a Galerkin, least-squares discretization [10]. Each element has a side length of 0.1, giving 200 total elements and four nodes per element.

Normally we would use Chemkin to compute the production rates $\dot{\omega}_j$, but to study the efficiency of the operator overloading approach, we used hand-coded C++ instead. We ignored spatial boundary conditions on the domain Ω, since they are not relevant to the computational complexity of the residual, Jacobian, and adjoint computations. To avoid complications in time integration, we made this a steady-state problem by setting $\partial Y_j/\partial t = 0$, $j = 1, \ldots, N$. While these simplifications give a highly contrived test problem that is not physically meaningful, it does have two important qualities. First, structurally it is qualitatively similar to many of the PDE problems to which Charon is applied, and second we can easily vary the number of unknowns to see how the cost of AD scales.

We computed ratios of Jacobian to residual evaluation time for the discretized form of (2), using both Fad and finite differencing to compute the element-level Jacobians. Figure 1(a) shows how these ratios vary with the degrees-of-freedom per element. We computed corresponding floating-point operation (flop) count ratios, which are shown in Figure 1(b). (Templating made getting the flop counts easy.) We used gcc 3.4.4 with `-O2` optimization on a 3.2 Ghz dual-processor (Xenon) workstation having 2 GB of RAM and a 512 KB level-1 cache, running Fedora Core 3 Linux. Note that while an individual element computation may fit entirely in cache, the entire 200 element residual evaluation does not. As expected, both the time and flop-count ratios scaled nearly linearly with the DOF per element, with slopes of about 0.27 and 1.55 respectively. While Fad Jacobian computations used roughly 50% more operations than finite differences, Fad was more than three times faster. The exact cause of this timing difference is unclear, but is likely related to improved data locality due to vectorization of the forward mode. A relative flop count slope slightly above 1.5 is not unexpected [3]. Fad recomputes each operation once for every derivative component, to give the compiler a chance to optimize temporary template objects away. We are investigating ways to cache operation results while still letting the compiler optimize temporaries away, in hopes of making Fad even more efficient.

Relative times and flop counts for an adjoint ($w^T J$) computation appear in Figures 1(c) and 1(d). The adjoint computation took between 7.5 and 9.5 times longer than the residual computation, but used only about 5.6 to 5.8 times as many operations. Compared with Fad, Rad had a larger ratio of time to flops, because of the extra memory overhead of reverse-mode AD. However, this still

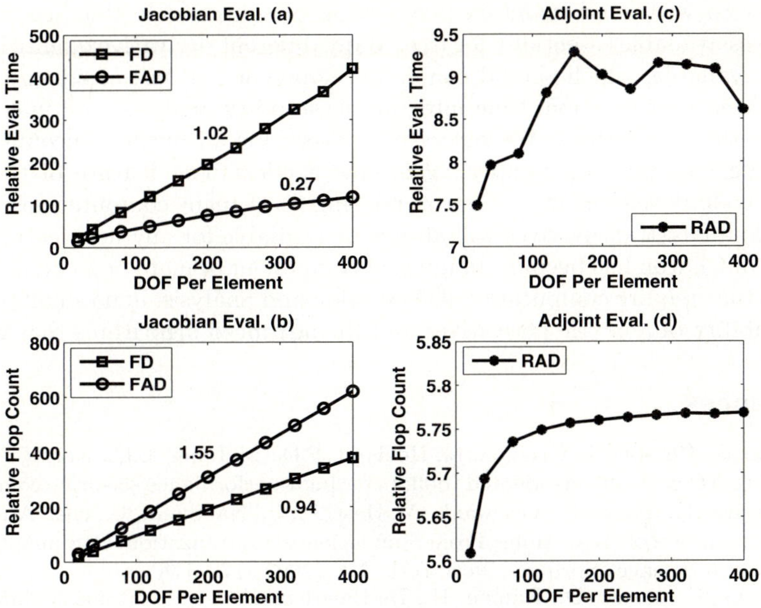

Fig. 1. Jacobian and adjoint evaluations versus degrees of freedom ($DOF = 4 \times$ number of species). (**a**) Relative Jacobian computation times. (**b**) Relative Jacobian flop counts. (**c**) Relative adjoint ($w^T J$) times. (**d**) Relative adjoint flop counts.

seems reasonably efficient: computing an adjoint with 400 DOF is ten times faster than computing the full Jacobian using Fad and multiplying by the transpose.

4 Summary and Conclusions

Our tests covered a range of 20 to 400 DOF per element, which encompasses the problem sizes normally seen in finite-element application codes. Again, since the derivatives are computed element-wise, it is this dimension that dictates the difficulty of the AD problem, not the number of elements or global number of unknowns. Thus for PDE discretizations with up to millions of unknowns, we have shown that forward-mode AD via Fad is a highly efficient method for computing the global Jacobian, more efficient than finite differencing and with better scaling to larger numbers of PDE equations. In fact Charon recently computed a transient simulation of the electric current in a finite element discretization of a bipolar junction transistor with more than 2.7 million elements on 128 processors, leveraging the Fad Jacobian computation for implicit time integration. We also found that Rad provides reverse-mode derivative computations with reasonable efficiency, which makes gradients available for use in optimization and sensitivity analysis.

We are highly encouraged by both the efficiency of forward and reverse mode AD in C++ codes, and by our experiences with implementation via templating. The Fad Jacobian computation is much faster than conventional finite differencing and provides analytic derivatives as well. Templating allows the code developer to write and maintain one version of source code that has analytic derivatives available essentially for free. Many different derivative quantities then become available, which should enable development and use of advanced nonlinear solver, optimization, time integration, stability analysis, and uncertainty quantification algorithms. We successfully overcame all hurdles encountered in templating Charon, and templating is now a permanent feature of the code. All new code development of Charon relating to element computations is templated, so analytic derivatives will always be available for any new features that are added. Charon has become an integral component of many important Sandia projects that require computational simulation and analysis, in no small part due to availability of analytic derivatives and the advanced algorithms they enable.

References

1. Abate, J., Benson, S., Grignon, L., Hovland, P.D., McInnes, L.C., Norris, B.: Integrating AD with object-oriented toolkits for high-performance scientific computing. In Corliss, G., Faure, C., Griewank, A., Hascoët, L., Naumann, U., eds.: Automatic Differentiation of Algorithms: From Simulation to Optimization. Computer and Information Science. Springer, New York, NY (2002) 173–178
2. Tijskens, E., Roose, D., Ramon, H., De Baerdemaeker, J.: Automatic differentiation for nonlinear partial differential equations: An efficient operator overloading approach. Numerical Algorithms **30** (2002) 259–301
3. Griewank, A.: Evaluating Derivatives: Principles and Techniques of Algorithmic Differentiation. Number 19 in Frontiers in Appl. Math. SIAM, Philadelphia, PA (2000)
4. Aubert, P., Di Césaré, N., Pironneau, O.: Automatic differentiation in C++ using expression templates and application to a flow control problem. Computing and Visualisation in Sciences **3** (2001) 197–208
5. Gay, D.M.: Semiautomatic differentiation for efficient gradient computations. In Bücker, H.M., Corliss, G., Hovland, P., Naumann, U., Norris, B., eds.: Automatic Differentiation: Applications, Theory, and Tools. Lecture Notes in Computational Science and Engineering. Springer (2005)
6. Kee, R.J., Rupley, F.M., Miller, J.A., Coltrin, M.E., Grcar, J.F., Meeks, E., Moffat, H.K., Lutz, A.E., Dixon-Lewis, G., Smooke, M.D., Warnatz, J., Evans, G.H., Larson, R.S., Mitchell, R.E., Petzold, L.R., Reynolds, W.C., Caracotsios, M., Stewart, W.E., Glarborg, P., Wang, C., Adigun, O., Houf, W.G., Chou, C.P., Miller, S.F., Ho, P., Young, D.J.: CHEMKIN Release 4.0, San Diego, CA. (2004)
7. Bischof, C.H., Carle, A., Khademi, P., Mauer, A.: ADIFOR 2.0: Automatic differentiation of Fortran 77 programs. IEEE Computational Science & Engineering **3**(3) (1996) 18–32
8. Utke, J.: OpenAD: Algorithm implementation user guide. Technical Memorandum ANL/MCS–TM–274, Mathematics and Computer Science Division, Argonne National Laboratory, Argonne, Ill. (2004)
9. Hascoët, L., Pascual, V.: TAPENADE 2.1 user's guide. Rapport technique 300, INRIA, Sophia-Antipolis (2004)
10. Hughes, T.J.R., Franca, L.P., Hulbert, G.M.: A new finite element formulation for computational fluid dynamics: VIII. the Galerkin/least-squares method for advective-diffusive equations. Computational Methods Applied Mechanics and Engineering **73** (1989) 173–189

A Sensitivity-Enhanced Simulation Approach for Community Climate System Model

Jong G. Kim[1], Elizabeth C. Hunke[2], and William H. Lipscomb[2]

[1] MCS Division, Argonne National Laboratory
Argonne, Illinois, U.S.A
[2] Theoretical Division, Los Alamos National Laboratory
Los Alamos, New Mexico, U.S.A
jkim@mcs.anl.gov, eclare@lanl.gov, lipscomb@lanl.gov

Abstract. A global sea-ice modeling component of the Community Climate System Model was augumented with automatic differentiation (AD) technology. The numerical experiments were run with two problem sets of different grid sizes. Rigid ice regions with high viscous properties cause computational difficulty in the propagation of AD-based derivative computation. Pre-tuning step was required to obtain successful convergence behavior. Various thermodynamic and dynamic parameters were selected for multivariate sensitivity analysis. The major parameters controlling the sea-ice thickness/volume computation were ice and snow densitives, albedo parameters, thermal conductivities, and emissivity constant. Especially, the ice and snow albedo parameters are found to have stronger effect during melting seasons. This high seasonal variability of the thermodynamic parameters underlines the importance of the multivariate sensitivity approach in global sea-ice modeling studies.

1 Introduction

The Community Climate System Model (CCSM, see www.ccsm.ucar.edu) is a fully-coupled, global climate simulation model developed by NCAR and DOE. It provides the capbility to simulate the interconnected Earth's climate systems including the atmosphere, ocean, land, and sea-ice. The parameterization schemes of the CCSM model involve a number of adjustable modeling parameters with different scales of uncertainty. This impedes new parameterization schemes since the entire model must be tuned with each new parameterization scheme. Furthermore, current sensitivity analysis and parameter tuning experiments of the CCSM model are performed by slow and labor-intensive approach: "expert judgement" of a handful of scientists. Against this background, the sensitivity-enhanced CCSM simulation approach for the global sea-ice modeling component, CICE was implemented with the AD method. This AD-based approach allows the derivative-enhanced CICE code to simultaneouly compute analytical derivatives in addition to original simulation results.

The modeling outputs of the CICE code are the global sea-ice conditions such as ice thickness, compactness, and horizontal velocity [3]. Thermodynamic, dynamic mass and energy balances coupled with transport equations are used to

V.N. Alexandrov et al. (Eds.): ICCS 2006, Part IV, LNCS 3994, pp. 533–540, 2006.
© Springer-Verlag Berlin Heidelberg 2006

derive the CICE modeling formulation. There were many early sensitivity studies to see the impact of parameter changes on simulation results. These include the works of Parkinson and Washington [6], Harder and Fischer [2], and Miller at al. [5]. From these studies, it was concluded that sea-ice modeling parameters are strongly interdependent and an objective computational scheme for tuning modeling parameters is a critical step in improving the sea-ice model development. In this paper, we used the AD source code transformation package TAPENADE [4] developed by the French National Institute for Research in Computer Science and Control (INRIA) to investigate the parameter sensitivities of the CICE model. This study was intended to help climate modelers objectively identify important modeling parameters and further use the AD-computed sensitivities as parameter tuning guide. Following a brief discussion of the model and the TAPENADE-based implementation, we discuss the numerical experiments and conclude with a brief summary.

2 Model Description

The major components of the CICE model are the thermodynamics, dynamics, and horizontal transport routines, solving the snow and ice physical status. The governing equations for each modeling component are solved on a generalized orthogonal grid by using an explicit time-step procedure. We summarize the main elements of the formulation here to identify the parameters used in the numerical sensitivity experiments. A complete description can be found in the user's manual of the CICE model [3]. Selected parameters for the sensitivity study are listed in Table 1.

2.1 Thermodynamic Parameters

The thermodynamic portion of the model determines the temperature profile and thickness changes of ice and snow based on an energy balance of radiative, turbulent, and conductive heat fluxes in each grid cell. For the energy flux from the atmosphere to the ice, the incoming shortwave flux is computed as function of α, the shortwave albedo, and i_o, the fraction of absorbed shortwave flux penetratting into the ice. Outgoing longwave radiation takes the standard blackbody form, $F_{L\uparrow} = -\epsilon\sigma\left(T_{sf}\right)^4$, where ϵ is the emissivity of snow or ice, σ is the Stefan-Boltzmann constant, and T_{sf} is the surface temperature. The minimum wind speed parameter, u_{min}, is used to maintain finite sensible and latent heat fluxes for the situation of no wind. The net absorbed shortwave flux is actually a summation over two different radiative quantities (visible and near-infrared) with two corresponding albedos. In addition to these parameters, the constants such as snow area fraction f_{snow}, penetrating fraction of visible solar readiation i_c, and bulk extinction coefficient κ_i are identified as adjustable to compute the flux of shortwave radiation. The rate of temperature change in the ice interior is computed by the conductive heat balance equation given as a function of various parameters. These include the densities and thermal conductivities of snow and ice. The thermal conductivity is the function of the conductivity of fresh ice k_o,

Table 1. Model parameters chosen for sensitivity testing

Parameter	Description	Value
ϵ	emissivity of snow and ice	0.95
u_{min}	minimum wind speed for turbulent fluxes	1 m/s
α_{iv}	visible ice albedo	0.78
α_{in}	near-IR ice albedo	0.36
α_{sv}	visible cold snow albedo	0.98
α_{sn}	near-IR snow albedo	0.70
i_c	penetrating fraction of visible solar radiation	0.7
κ_i	visible extinction coefficient in ice	1.4 m^{-1}
ρ_i	ice density	917 kg/m^3
β	T, S proportionality constant in conductivity	0.13 W/m/psu
k_o	thermal conductivity of fresh ice	2.03 W/m/deg
ρ_s	snow density	330 kg/m^3
k_s	thermal conductivity of snow	0.30 W/m/deg
S_{max}	maximum salinity, at ice base	3.2 psu
h_{mix}	ocean mixed-layer depth	20 m
D_w	drag parameter for water on ice	5.49936 kg/m^3
G^*	fractional area participating in ridging	0.15
H^*	determines mean thickness of ridged ice	25 m
C_s	fraction of shear energy contributing to ridging	0.25
C_f	ratio of ridging work to PE change in ridging	17.

an empirical constant β, the ice salinity S, and the temperature T. The salinity profile varies from $S = 0$ at the top surface ($z = 0$) to $S = S_{max}$ at the bottom surface ($z = 1$).

2.2 Dynamics and Ridging Parameters

Ice motion and deformation are determined by balancing five major stresses: wind stress from the atmosphere, water stress from the interaction between ice and ocean, Coriolis force, the stress from the tilt of the ocean surface, and the internal ice stress. A momentum balance equation is solved to obtain the ice velocity in each grid cell, using the elastic-viscous-plastic (EVP) rheology [3] to relate the internal ice stress and the rates of strain. The drag coefficient D_w is used to determine the stress between the ocean and the ice. The ice ridging scheme of the CICE code includes several tuning parameters. An empirical constant G^* is used to determine a ridging weighting function. Larger values of G^* allow thicker ice to participate in ridging, thereby increasing the ice strength. H^* determines the thickness of ridging ice. C_s is the fraction of shear dissipation energy that contributes to ridge building. Another empirical parameter, C_f, accounts for frictional energy dissipation.

3 Processing the CICE Code with TAPENADE

Given the Fortran 90 CICE code, TAPENADE successfully produced a portable Fortran 90 CICE.AD code that allows the tangent linear derivative computation of partial derivatives of ice conditions with respect to various thermodynamic and dynamic input parameters. Still under the development, TAPENADE supports most of the Fortran 90 standard. Lack of support for `count`, `present`, and dynamic memory allocation requires a preprocessing step to replace some Fortran 90 constructs with ones acceptable to TAPENADE. Also, some type mismatch problems with the Fortran `MOD` intrinsic function calls were observed in processing the CICE code. Various netCDF calls are used in the CICE code to write a restart file and history files during simulation. Working as I/O statements, these netCDF routines are not directly related to the differentiation process. We used TAPENADE's blackbox approach to bypass the netCDF routines when differentiating. To do this, we had to supply differentiation information to TAPENADE. The same approach was used to handle the MPI library routine calls in the CICE code. The number of lines of the TAPENADE-generated CICE code (50000 lines) is about double the size of the original CICE code (27000 lines). The computational time of the AD-generated tangent-linear CICE code increased proportionally with the number of independent variables.

4 Numerical Results

4.1 Valid Derivative Computation with Pre-tuning Step

In the EVP ice dynamics of the CICE model, several tuning parameters are used to maintain the stability of the EVP routine. The constraining constant Δ_{min} (named *tinyarea*) for the ice strain rate computation $P/max(\Delta, \Delta_{min})$ critically controls the convergence behavior of computed ice velocity. Figure 1 shows the ice velocity computation with different values of *tinyareas* for the EVP subcycling iterations. In the figure, the AD-generated derivatives are also plotted for each of the choices of *tunyarea*. The derivatives are computed for the ice velocity with respect to the ocean drag coefficient. Significant oscillations were observed in ice velocity computation with the smaller value of *tinyarea*, which resulted into a divergent behavior of the AD-generated derivatives. Thus, we further tuned the CICE code with increased *tinyarea* to maintain the stability of the AD-generated derivatives. Similar oscillations were reported by Bücker [1] with two-dimensional aircraft design optimization problems. They used delayed propagation approach to control the oscillation.

4.2 AD Sensitivity Experiment

Two different problem sets were used for the sensitivity experiments: 100×116 and 900×601 orthogonal grid points with 5 different ice categories and 4 ice layers in single ice category. The forcing variables of surface air temperature, specific humidity, down-welling short-wave and long-wave radiation, and geostropic winds derived from National Centers of Environmental Prediction (NCEP) reanalysis and European Center for Medium-Range Weather Forecasts (ECMWF) global analyses were used for the simulation of the year 1987 and 1996. The multi-year spin-up result of ice thickness, concentration, and velocity

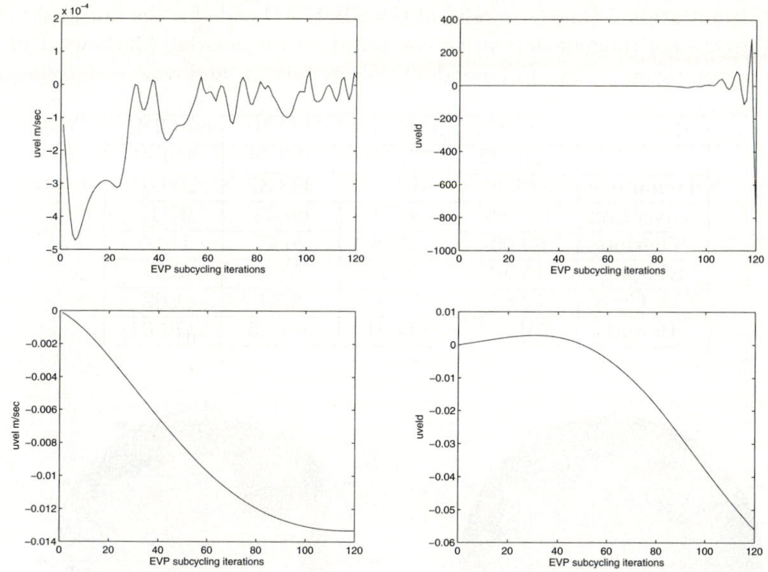

Fig. 1. Convergence behaviors of EVP subcycling step: a) u-velocity computed with *tinyarea* $= 1. \times 10^{-12}$, b) its AD derivative with *tinyarea* $= 1. \times 10^{-12}$, c) u-velocity computed with *tinyarea* $= 1. \times 10^{-6}$, and d) its AD derivative with *tinyarea* $= 1. \times 10^{-6}$

fields on January 1 was used as the initial state for the experiments. Sensitivity experiments were carried out with respect to two main dependent variables, ice thickness and hemispheric ice volume. A linux cluster system was used to run the numerical experiments. Table 2 shows the measured parallel performance of the CICE.AD code. Reasonable speed-up was achieved on up to 60 processors. Most of the computational time was spent on the ice dynamic part where the momentum equations for ice velocity are explicitly integrated.

Table 3 shows the sensitivity results of hemispheric ice volume for the 1987 ice conditions on coarse grid data. Note that the ice volume can be directly computed by the multiplication of ice thickness and area of each grid cell. The derivative numbers were nondimensionalized by using the computed ice volume and parameter values. The first column of the table lists the 12-month sum of sensitivity magnitudes for each of the 20 parameters, with larger values indicating greater sensitivity. The high sensitivity result of ice density is largely an artifact of the way density is treated in the CICE code. Increasing ice density resulted in a smaller value of the specific heat of melting and a higher temperature change. This unphysical warming reduces winter ice growth and increases summer melting. Multiple year runs are required to determine the true sensitivity of the thickness to ice density. From the table we see that parameters affecting the conductivity and radiative absorption are of paramount importance for simulating ice volume in the sea ice model; with the exception of D_w, dynamics and ridging parameters are less important than the thermodynamic parameters. If we were using ice velocity as the dependent variable, however, the dynamics

Table 2. Computational time (seconds) of the CICE.AD code for the fine grid problem set: multi-directional tangent derivative computation for one-day (January 1 of 1996) sensitivity computation of two independent parameters, ρ_i and α_{iv}

Routines	10 CPUs	20 CPUs	30 CPUs	60 CPUs
Total	1256.02	772.70	493.70	314.91
Dynamics	829.32	511.24	338.87	200.03
Advection	184.06	96.74	66.33	37.15
Thermo	93.00	42.78	26.67	11.95
Ridging	15.80	7.52	5.24	2.53
Cat Conv	12.84	5.66	3.84	1.62
Bound	701.79	444.91	291.25	173.64

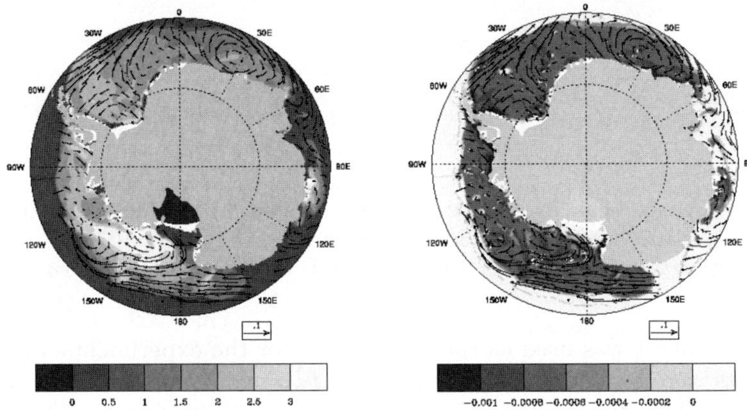

Fig. 2. Ice thickness (m) distribution of the first week of January, 1996 (*left*) and colormap of sensitivity to ice density (*right*): 900 × 601 orthogonal grid points

parameters would be more prominent. For the fine grid problem, Figure 2 shows the sensitivity distribution of ice thickness to ice density over Antarctic region. The derivative colormap shows how the ice density affects on the ice thickness computation.

5 Summary

In this study, a sensitivity-enhanced simulation approach for global sea-ice modeling was investigated through the AD method. We observed a pre-tuning step was required to obtain stable convergence behaviors of the AD-based CICE code. An important result from this study is the prominent sensitivity of ice thickness to radiative control parameters. For example, emissivity and albedo parameters have not been scrutinized in sea ice model development. This study shows that multivariate sensitivity analysis for those parameters can be easily accomplished based on the AD method. Also, the AD-based scheme for computing derivatives provides an efficient guideline to adjust those important parameters.

Table 3. Magnitudes of AD sensitivities, summed over the 12 months for northen hemisphere ice volume and its monthly sensitivity for the year of 1987

Parameter	sum	January	April	July
ϵ	0.6439×10^{-1}	0.1057×10^{-2}	0.1497×10^{-2}	0.2075×10^{-1}
u_{min}	0.4450×10^{-3}	0.4061×10^{-4}	0.1133×10^{-4}	-0.6007×10^{-5}
α_{iv}	0.1088	0.8533×10^{-4}	0.2259×10^{-2}	0.3200×10^{-1}
α_{in}	0.1057	0.2317×10^{-4}	0.7777×10^{-3}	0.3573×10^{-1}
α_{sv}	0.2050×10^{-1}	0.2577×10^{-3}	0.3578×10^{-2}	0.8714×10^{-3}
α_{sn}	0.1184×10^{-1}	0.9199×10^{-4}	0.1494×10^{-2}	0.9106×10^{-3}
i_c	0.2079×10^{-1}	-0.3489×10^{-4}	-0.3462×10^{-3}	0.7601×10^{-2}
κ_i	0.3247×10^{-2}	-0.3503×10^{-5}	0.3237×10^{-4}	0.1125×10^{-2}
ρ_i	$0.2044 \times 10^{+1}$	-0.2759	-0.1782	-0.7975×10^{-1}
β	0.1092×10^{-1}	-0.1625×10^{-2}	-0.1031×10^{-2}	-0.1974×10^{-3}
k_o	0.1796	0.2809×10^{-1}	0.1537×10^{-1}	0.1432×10^{-2}
ρ_s	0.2784×10^{-1}	0.2191×10^{-2}	0.5501×10^{-2}	0.6088×10^{-4}
k_s	0.5190×10^{-1}	0.1034×10^{-1}	0.1944×10^{-2}	0.3975×10^{-4}
S_{max}	0.1161	0.1114×10^{-1}	0.1012×10^{-1}	0.9196×10^{-2}
h_{mix}	0.4713×10^{-1}	-0.5548×10^{-2}	-0.2656×10^{-2}	0.4201×10^{-2}
D_w	0.4723×10^{-2}	-0.6713×10^{-3}	-0.4030×10^{-4}	0.2838×10^{-4}
G^*	0.2742×10^{-2}	0.2370×10^{-3}	0.1088×10^{-3}	-0.3818×10^{-3}
H^*	0.1233×10^{-2}	-0.2690×10^{-3}	-0.1638×10^{-3}	0.1209×10^{-4}
C_s	0.3252×10^{-2}	0.5369×10^{-3}	0.3284×10^{-3}	-0.3460×10^{-4}
C_f	0.3526×10^{-2}	-0.6684×10^{-3}	-0.4144×10^{-3}	0.5138×10^{-4}

Implemented by the black-box approach of TAPENADE, the parallel MPI routines of the CICE were successfully processed. Significant parallel performance was obtained for a large-size problem set. In future work, we plan to further explore the CICE model's parameter space using the best available data for both hemispheres, including satellite-derived ice concentration and ice deformation. Furthermore, we plan to tune the model using long-term (decadal) observational data.

Acknowledgement

This work was supported by the Climate Change Research Division subprogram of the Office of Biological & Environmental Research, Office of Science, U.S. Department of Energy through the Climate Change Prediction Program (CCPP), and the Scientific Discovery through Advanced Computing (SciDAC) Program under Contract W-31-109-ENG-38.

References

1. H. M. Bücker, A. Rasch, E. Slusanschi, and C. H. Bischof, *Delayed Propagation of Derivatives in a Two-dimensional Aircraft Design Optimization Problem*, Proceedings of the 17th Annual International Symposium on High Performance Computing Systems and Applications and OSCAR Symposium, Sherbrooke, Québec, Canada, May 11–14, 2003, NRC Research Press, 2003.

2. M. Harder and H. Fischer, *Sea ice dynamics in the Weddell Sea simulation with an optimized model.* J. Geophys. Res., 104: 11,151 11,162, 1999.
3. E. C. Hunke and W. H. Limpscomb, *CICE: the Los Alamos Sea Ice Model, Documentation and Software*, LA-CC-98-16, Los Alamos National Laboratory, NM, 2004.
4. L. Hascoet and V. Pascual, *TAPENADE 2.1 User's Guide, INRIA Technical Report*, `http://www-sop.inria.fr/tropics`, 2004.
5. P. A. Miller, S. W. Laxon, D. L. Feltham, and D. J. Cresswell, *Optimization of a sea ice model using basin-wide observations of Arctic sea ice thickness, extent and velocity*, J. Clim., 2005 (in press).
6. C. L. Parkinson and W. M. Washington, *A large-scale numerical model of sea ice*, J. Geophys. Res., 84:311 337, 1979.

Optimal Checkpointing for Time-Stepping Procedures in ADOL-C*

Andreas Kowarz and Andrea Walther

Institute of Scientific Computing, Technische Universität Dresden, 01062 Dresden, Germany
{Andreas.Kowarz, Andrea.Walther}@tu-dresden.de

Abstract. Using the basic reverse mode of automatic differentiation, the memory needed for the computation of discrete adjoints is proportional to the number of operations performed. This behavior is frequently not acceptable, especially for large-scale problems that involve a kind of time-stepping procedure. Therefore, we integrate a checkpointing mechanism into ADOL-C, a tool for the automatic differentiation of C and C++ programs. This checkpointing procedure is optimal for a given number of checkpoints in the sense that it yields the minimal number of recomputations. The resulting effects on the run-time behavior are illustrated by means of the derivative computation for an ODE-based optimization problem.

1 Introduction

The reverse mode of automatic differentiation (AD) provides an efficient method to compute discrete adjoint information. For example, the operation count for computing the gradient of a scalar-valued function is only a small multiple of the operation count needed to evaluate the function [3]. However, the memory needed by the reverse mode in its basic form is proportional to the operation count of the function evaluation. For real-world problems this fact may lead to an unacceptable memory requirement. For that reason, several checkpointing approaches have been developed.

If the considered function evaluation has no specific structure, one may allow the user of an AD-tool to place checkpoints somewhere during the function evaluation to reduce the overall memory requirement. This simple approach is provided for example by the AD-tool TAF [1]. As alternative, one may exploit the call graph structure of the function evaluation to place checkpoints at the entries of specific subroutines. This so-called joint reversal, see, e.g., [3], then leads to a reduction of the memory requirement. The subroutine-oriented checkpointing is used for example by the AD-tools Tapenade [8] and OpenAD [11]. As soon as one can exploit additional information about the structure of the function evaluation, an appropriate adapted checkpointing strategy can be used. This is in particular the case if a time-stepping procedure is contained in the function evaluation allowing the usage of a time-stepping oriented checkpointing. If the number of

* Partially supported by the DFG grant WA 1607/2-1.

V.N. Alexandrov et al. (Eds.): ICCS 2006, Part IV, LNCS 3994, pp. 541–549, 2006.
© Springer-Verlag Berlin Heidelberg 2006

time steps l is known a priori and the computational costs of the time steps are almost constant, one very popular checkpointing strategy is to distribute the checkpoints equidistantly over the time interval. However, it was shown in [12] that this approach is not optimal. A more advanced but still not optimal approach is the binary checkpointing used for example in [10]. However, optimal checkpointing schedules can be computed in advance to achieve an optimal, i.e. minimal, run time increase for a given number of checkpoints [2, 5]. In this paper, we present the usage of the optimal, also called binomial, checkpointing within the AD-tool ADOL-C [4] to obtain a drastic decrease in the memory requirement by taping only one instead of l time steps. For the considered numerical example the memory reduction leads even to an overall run-time reduction due to the reduced access cost for the required memory. Hence, we face for the first time the situation where checkpointing leads to a decrease in run-time despite the fact that a considerable amount of intermediate information has to be recomputed.

The structure of this paper is the following: In Sect. 2, we present the adapted implementation of ADOL-C to employ the time-stepping oriented checkpointing. This includes a description of the modified internal function representation based on nested taping. Subsequently, the derivative computation exploiting binomial checkpointing is illustrated. This section ends with a short introduction to the usage of the new checkpointing facility. Section 3 discusses possible run-time effects using as example the derivative computation for an ODE-based optimization problem. Finally, a conclusion and an outlook are given in Sect. 4.

2 Optimal Checkpointing in ADOL-C

The AD-tool ADOL-C uses operator overloading for the automatic differentiation of function evaluations $y = F(x)$ written in C or C++. For this purpose, the new class `adouble` is introduced by ADOL-C. The user has to declare the independent variables x and all quantities that directly or indirectly depend on them of type `adouble`. Other variables that do not depend on the independent variables but enter, for example, as parameters, may remain one of the passive types `double`, `float`, or `int`. During the function evaluation with `adouble` variables, ADOL-C stores for each operation the corresponding operator and the variables that are involved into a data structure called *tape*. Once, the tape, i.e., the internal representation is generated, the required derivatives are calculated on the basis of the internal function representation using the elemental differentiation rules. Due to the described approach, the derivative calculations involve a possibly substantial but always predictable amount of data that is accessed strictly sequentially. The size of the much smaller randomly accessed memory can be precalculated using information on the tape.

However, this approach may lead to the storage of a significant amount of redundant information if the function evaluation contains a time-stepping procedure. To overcome this difficulty, the next version of ADOL-C will provide a checkpointing facility based on the binomial checkpointing procedure `revolve` [5]. To exploit this additional functionality, the number l of time steps in the time-stepping part must be known before this iterative process starts. Furthermore, the user has to provide the number c of checkpoints that can be stored in an internal data structure of ADOL-C. As shown in [5], the applied checkpointing strategy is optimal if the computational costs of the time steps are identical or almost identical. Nevertheless, numerical tests presented for example in [7] show that the checkpointing procedure provided by revolve also yields surprisingly good results for varying computational costs of the time steps. Therefore, it is also feasible to use the provided checkpointing possibility in these cases.

To apply the checkpointing routine revolve inside of ADOL-C, the generation of the internal representation, i.e. the taping mechanism, as well as the derivative evaluation have to be adapted as described in the next paragraphs.

Generating the Internal Function Representation: Nested Taping

For illustrating the modified taping mechanism, we assume that the function evaluation $y = F(x)$ consists of three parts: A start up calculation G, the time-stepping procedure H, and a final calculation evaluating for example a target function J, i.e., $y = J \circ H \circ G(x)$ with

$$u = G(x), \quad v = H(u), \quad y = J(v)$$

and the evaluation of $H(u)$ is given by

$$u_1 = \tilde{H}(u), \; \ldots, \; u_i = \tilde{H}(u_{i-1}), \; \ldots, \; u_l = \tilde{H}(u_{l-1}), \; v = u_l \; .$$

The corresponding consequences for the tape generation using the "basic" taping approach as implemented in ADOL-C so far are shown in the left part of Fig. 1. As can be seen, the iterative process is completely unrolled due to

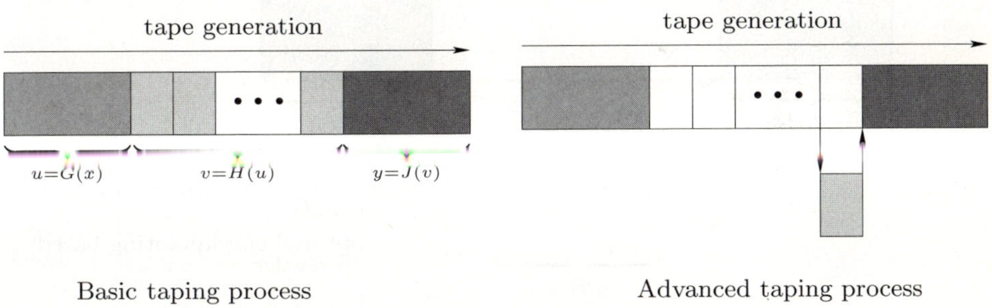

Fig. 1. Different taping approaches

the taping process. That is, the tape contains an internal representation of each time step. Hence, the overall tape comprises a serious amount of redundant information as illustrated by the light gray rectangles in Fig. 1.

To overcome the repeated storage of essentially the same information, recently we incorporated a *nested taping* mechanism into ADOL-C as illustrated on the right-hand side of Fig. 1. This new capability allows the encapsulation of the time-stepping procedure such that only the last time step $u_l = \tilde{H}(u_{l-1})$ is taped as one representative of the time steps. This is illustrated by the white rectangles where only the function is evaluated but no taping is performed. Only the last time step is taped as illustrated by the light gray rectangle. Additionally, a function pointer to the evaluation procedure $\tilde{H}$ is stored for a possibly necessary retaping during the derivative calculation as explained below. Furthermore, c checkpoints are distributed by revolve. It is important to note that the overall tape size is drastically reduced due to the advanced taping strategy. For the implementation of this nested taping we introduced a so-called "differentiating context" that enables ADOL-C to handle different internal function representations during the taping procedure and the derivative calculation. This approach allows the generation of a new tape inside the overall tape, where the coupling of the different tapes is based on the *external differentiated function* concept presented at the AD2004 conference.

Computing the Derivative Information

Applying the reverse mode of AD to the function $y = F(x)$ consists of three steps: First the discrete adjoint $\bar{v} = \bar{J}(v, \bar{y})$ of $J(v)$ is calculated, then the discrete adjoint $\bar{u} = \bar{H}(u, \bar{v})$ of $H(u)$, and finally the discrete adjoint $\bar{x} = \bar{G}(x, \bar{u})$ of G. The computation of $\bar{v}$ and $\bar{x}$ is straightforward and implemented in the current version of ADOL-C. The situation changes completely for the computation of $\bar{u}$ due to the usage of a checkpointing approach for the time-stepping procedure and the resulting nested taping. For calculating the discrete adjoint $\bar{u}$ of the time-stepping part, the routine revolve is used to steer the reverse mode differentiation as illustrated in Fig. 2. For this purpose, the nested taping in com-

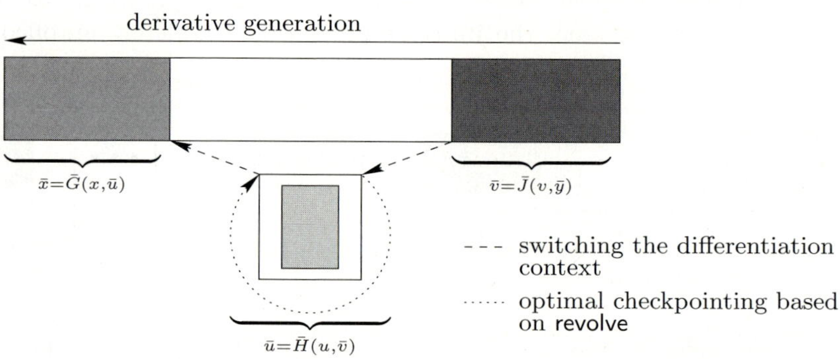

Fig. 2. Derivative computation using a checkpointing procedure

bination with the two "differentiating contexts" for the overall function F and the time-stepping part are exploited. This allows a switching from the derivative calculation for F to the more involved derivative computation for the time-stepping procedure based on a checkpointing strategy for the subfunction H. In the ideal case, the adjoint computation for H consists only of recomputations of intermediate states u_i and derivative computations based on the internal representation of one time step according to the reversal schedule provided by revolve for the given number c of checkpoints. This approach yields in the case of constant computational costs for all time steps a minimal run-time for the checkpointing approach. If the control flow changes during the time step evaluation due to a change of an adouble value, a "retaping" of the time-step function $\tilde{H}$ is automatically initiated for a corresponding update of the internal representation. Further information on the validity of tapes can be found in subsection 2.6 of the ADOL-C manual. Alternatively, an additional flag set by the user to the value 1 causes the retaping of each time step to take changes of non-adouble values into account.

The next version of ADOL-C will provide this checkpointing capability also for higher derivative computations and the vector mode although probably the most common usage will be the computation of gradient information using the scalar reverse mode of AD.

Interface of the Checkpointing Facility

Written under the objective of a lean and concise interface, the checkpointing routines of ADOL-C need only very limited information. The user must provide two routines as implementation of the time-stepping function $\tilde{H}$ with the signatures

```
int time_step_function(int n, adouble *u);
int time_step_function(int n, double *u);
```

where the function names can be chosen by the user as long as the names are unique. Furthermore, it is assumed that the result vector of one time step iteration overwrites the argument vector of the same time step. Therefore, no copy operations are required to prepare the next time step.

At first, the adouble version of the time step function has to be *registered* using the C or C++ interface, respectively.

```
C:     CpInfos *cpInfos = reg_timestep_fct(ADOLC_TimeStepFunction);
C++:   CP_Context cpc(ADOLC_TimeStepFunction);
```

Using either cpInfos or cpc and the appropriate interface the user has to provide the remaining checkpointing information:

- a pointer to the double version of the time step routine,
- the number of time steps l,
- the number of checkpoints c,
- the tape number to be used internally for the nested taping,
- the dimension of the argument vector u,
- a pointer to the argument vector storing the initial value of u and
- a pointer to the result vector storing the final value v of the time integration.

In addition a flag for enforcing the retaping of every timestep might be set. Then, the nested taping and the derivative calculation using binomial checkpointing is initiated by calling the ADOL-C function

```
C:     info = checkpointing(cpInfos);
C++:   info = cpc.checkpointing();
```

at the corresponding point of the function evaluation during the taping process. Subsequently, ADOL-C computes derivative information using the optimal checkpointing strategy provided by revolve internally, i.e., completely hidden from the user.

3 Numerical Example

The numerical example that serves to illustrate the run-time effects of the checkpointing procedure is an industrial robot as depicted in Figure 3 that has to perform a fast turn-around maneuver. We denote by $q = (q_1, q_2, q_3)$ the angular coordinates of the robot's joints, q_1 referring to the angle between the base and the two-arm system. The robot is controlled via three control functions u_1 through u_3, denoting the respective angular momentum applied to the joints (from bottom to top) by electrical motors. The control problem under consideration is to minimize the energy-related objective

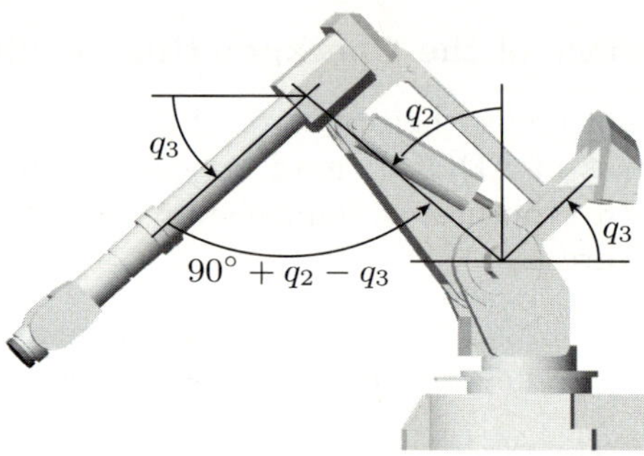

Fig. 3. Industrial robot ABB IRB 6400

$$J(q, u) = \int_0^{t_f} [u_1(t)^2 + u_2(t)^2 + u_3(t)^2] \, dt$$

where the final time t_f is given. The robot's dynamics obeys a system of three differential equations of second order:

$$M(q)\,\ddot{q} = v(q, \dot{q}) + w(q) + \tau_{\text{friction}}(\dot{q}) + \tau_{\text{reset}}(q) + u$$

where $M(q)$ is a 3×3 symmetric positive definite matrix containing moments of inertia, called a generalized mass matrix. The vector v is composed of centrifugal and Coriolis force entries, and w contains the gravitational influence. Finally, we allow for forces induced by dry friction and reset forces by means of τ_{friction} and τ_{reset}, respectively. The complete equations of motion can be found in [9]. The robot's task to perform a turn-around maneuver is expressed by means of initial and terminal conditions as well as control constraints [6]. However, for illustrating the run-time effects of the checkpointing facility integrated in ADOL-C we consider only the gradient computation of $J(q, u)$ with respect to u.

To compute an approximation of the trajectory x, we apply for the integration the standard Runge-Kutta method of order 4 resulting in about 800 lines of code. The integration and derivative computations were computed using an AMD Athlon64 3200+ (512 kB L2-cache) and 1GB main memory. The resulting averaged run-times in seconds for one gradient computation are shown in Fig. 4, where the run-time required for the derivative computation without checkpointing, i.e., the basic approach (BA), is illustrated by a dotted line. The run-time

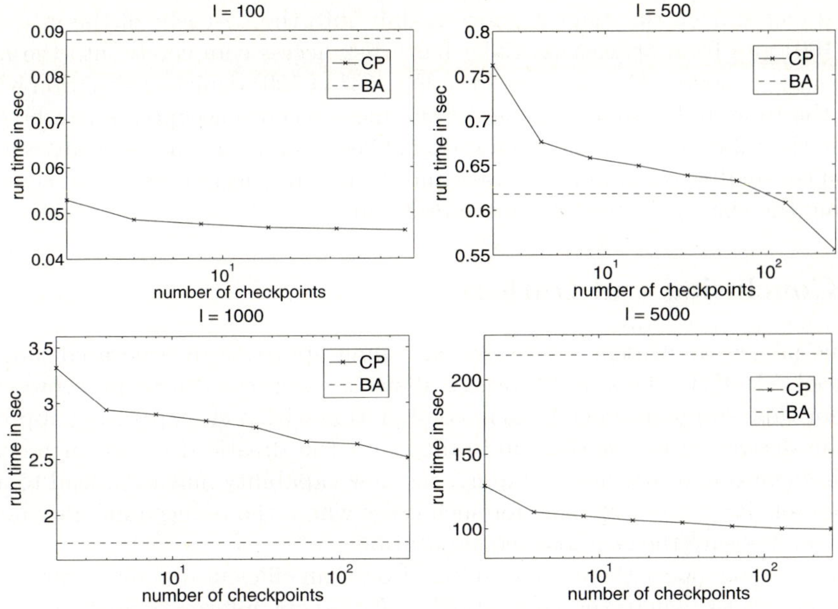

Fig. 4. Comparison of run-times for $l = 100, 500, 1000, 5000$

needed by the checkpointing approach (CP) using $c = 2, 4, 8, 16, 32, 64(, 128, 256)$ checkpoints is given by the solid line. To illustrate the corresponding savings in memory requirement, Table 1 shows the tape sizes for the basic approach as well as the tape and checkpoint sizes required by the checkpointing version. The tape size for the later varies since the number of independents is a multiple of the number of time steps l due to the distributed control u. One basic checkpointing assumption, i.e., the more checkpoints are used the less runtime the execution needs, is clearly depicted by case $l = 1000$ in Fig. 4. The smaller runtime for the basic approach completes the setting. However, the more interesting cases from this example are $l = 100$ and $l = 5000$, respectively. In these situations a smaller runtime was achieved even though checkpointing was used. These results are effected by an insight well known, i.e., computing from a level of the memory hierarchy that offers cheaper access cost may result in a significant smaller runtime. In the mentioned cases of the robot example the computation could

Table 1. Memory requirements for $l = 100, 500, 1000, 5000$

# time steps l	100	500	1000	5000
	without checkpointing			
tape size (Byte)	4.388.720	32.741.979	92.484.730	1.542.488.152
	with checkpointing			
tape size (Byte)	79.367	237.367	434.867	2.014.912
checkpoint size (Byte)	11.440	56.240	112.240	560.240

be redirected from the main memory mainly into the L2-cache of the processor ($l = 100$) and from at least partially hard disk access completely into the main memory ($l = 5000$). The last case from Fig. 4 ($l = 500$) depicts a situation where only the tape and a small number of the most recent checkpoints can be kept within the L2-cache. A well chosen ratio between l and c in this case causes a significantly smaller recomputation rate and results in a decreased overall runtime, making the checkpointing once more preferable.

4 Conclusion and Outlook

In this paper, we present a new nested taping approach incorporated into the AD-tool ADOL-C. This new strategy allows a compressed internal representation for time stepping procedures in combination with a checkpointing approach for the derivative calculation. In addition to the drastic decrease in memory requirement due to the nested taping, the new capability may even lead to a reduction of the overall run-time for such cases where the reduced memory access costs compensate the required recomputations.

We plan to employ the nested taping also for an efficient differentiation of fixpoint iterations, where the computation of gradient information can be based only on the last iteration performed during the function evaluation. Hence, also in this situation a complete unrolling and hence the storage of the full internal representation can be avoided. Therefore, also in these situations the applicability of ADOL-C will be extended.

References

1. Giering, R., Kaminski, T.: Recipes for Adjoint Code Construction. ACM Trans. Math. Software, **24** (1998) 437–474.
2. Griewank, A.: Achieving logarithmic growth of temporal and spatial complexity in reverse automatic differentiation. Optim. Methods Softw., **1** (1992) 35–54.
3. Griewank, A.: Evaluating Derivatives: Principles and Techniques of Algorithmic Differentiation. Number 19 in Frontiers in Appl. Math. SIAM, Philadelphia, 2000.
4. Griewank, A., Juedes, D., Utke. J.: ADOL-C: A package for the automatic differentiation of algorithms written in C/C++. ACM Trans. Math. Software, **22** (1996) 131–167.
5. Griewank, A., Walther, A.: Revolve: An implementation of checkpointing for the reverse or adjoint mode of computational differentiation. ACM Trans. Math. Software, **26** (2000) 19–45.
6. R. Griesse, A. Walther.: Parametric sensitivities for optimal control problems using automatic differentiation. *Optimal Control Applications and Methods*, Vol. 24, pp. 297–314 (2003).
7. Hinze, M., Sternberg, J.: A-Revolve: An adaptive memory and run-time-reduced procedure for calculating adjoints; with an application to the instationary Navier-Stokes system. Optim. Methods Softw., **20** (2005) 645–663.
8. Hascoët, L., Pascual, V.: Tapenade 2.1 user's guide. Tech. rep. 300, INRIA, 2004.
9. Knauer, M., Büskens, C.,: *Real-Time Trajectory Planning of the Industrial Robot IRB6400.* PAMM. 3, 2003, 515–516.

10. Kubota, K.: A Fortran 77 preprocessor for reverse mode automatic differentiation with recursive checkpointing. Optim. Methods Softw., **10** (1998) 319–335.
11. Naumann, U., Utke, J., Lyons, A., Fagan, M.: Control Flow Reversal for Adjoint Code Generation. Proceed. of SCAM 2004, IEEE Computer Society (2004) 55–64.
12. Walther, A., Griewank, A.: Advantages of binomial checkpointing for memory-reduced adjoint calculations. In: Feistauer, M., et al., (eds.): *Numerical mathematics and advanced applications,* Proc. ENUMATH 2003, Springer (2004) 834–843.

On the Properties of Runge-Kutta Discrete Adjoints

Adrian Sandu

Department of Computer Science,
Virginia Polytechnic Institute and State University,
Blacksburg, VA 24061
sandu@cs.vt.edu

Abstract. In this paper we analyze the consistency and stability properties of Runge-Kutta discrete adjoints. Discrete adjoints are very popular in optimization and control since they can be constructed automatically by reverse mode automatic differentiation. The consistency analysis uses the concept of elementary differentials and reveals that the discrete Runge-Kutta adjoint method has the same order of accuracy as the original, forward method. A singular perturbation analysis reveals that discrete adjoints of stiff Runge-Kutta methods are well suited for stiff problems.

Keywords: Runge-Kutta methods, discrete adjoints.

1 Introduction

Consider the ordinary differential equation (ODE)

$$y' = f(t,y) \,, \quad y\,(t_0) = y_0 \,, \quad t_0 \le t \le t_{\rm F} \,. \tag{1}$$

We will denote the Jacobian of the ODE function by $J(t,y) = \partial f(t,y)/\partial y$.

We are interested in the following optimization problem, which arises in important applications like control and data assimilation. Find the initial conditions for which a function of the system state at the final time is minimized,

$$\min_{y_0} \overline{\Psi}\,(y_0) = h\Big(y(t_{\rm F})\Big) \quad \text{subject to (1)} \,. \tag{2}$$

To apply a gradient based optimization procedure one needs to compute the derivatives of the cost function $\overline{\Psi}$ with respect to the initial conditions. It can be shown [6] that these derivatives can be obtained efficiently by solving the continuous adjoint equation

$$\lambda' = -J^T\Big(t, y(t)\Big)\,\lambda \,, \quad \lambda\,(t_{\rm F}) = \frac{\partial h}{\partial y}\Big(y(t_{\rm F})\Big) \,, \quad t_{\rm F} \ge t \ge t_0 \tag{3}$$

backwards in time from $t_{\rm F}$ to t_0 to obtain

$$\lambda(t_0) = \frac{\partial \overline{\Psi}}{\partial y_0}.$$

V.N. Alexandrov et al. (Eds.): ICCS 2006, Part IV, LNCS 3994, pp. 550–557, 2006.
© Springer-Verlag Berlin Heidelberg 2006

Note that the continuous adjoint equation (3) is formulated based on the forward solution $y(t)$.

In practice the equation (1) is solved numerically on a computer to obtain approximations of the ODE solution $y_n \approx y(t_n)$. Using a one-step numerical method (e.g., Runge-Kutta) the numerical solution is advanced in time as follows

$$y_{n+1} = \mathcal{M}_n\left(y_n\right) \;, \quad y_N = \mathcal{M}_{N-1}\Big(\mathcal{M}_{N-2}(\cdots \mathcal{M}_0(y_0))\Big) \;, \tag{4}$$

where $t_N = t_{\mathrm{F}}$ and the numerical solution at the final time is $y_N \approx y(t_{\mathrm{F}})$. The optimization problem (2) is formulated in terms of the numerical solution minimized,

$$\min_{y_0} \Psi\left(y_0\right) = h\Big(y_N\Big) \quad \text{subject to (4)} \;. \tag{5}$$

To estimate the gradient of the cost function (5) several approaches are possible. In the *continuous adjoint* approach one solves the continuous adjoint equation (3) backwards in time using any numerical discretization technique, e.g., a Runge-Kutta method. The terminal value of the adjoint variable $\lambda(t_0)$ is an approximation of the gradient of (2), and (hopefully) is also an approximation of the gradient of (5).

In the *discrete adjoint* approach the gradient of (2) is computed directly from (4) using the transposed chain rule

$$\left(\frac{d\Psi}{dy_0}\right)^T = \left(\frac{d\mathcal{M}_0}{dy_0}(y_0)\right)^T \cdots \left(\frac{d\mathcal{M}_{N-1}}{dy_{N-1}}(y_{N-1})\right)^T \left(\frac{dh}{dy_N}(y_N)\right)^T \;.$$

This calculation proceeds backwards in time, i.e. the expression is evaluated right to left as follows

$$\lambda_N = \left(\frac{dh}{dy_N}(y_N)\right)^T \cdots \quad \lambda_n = \left(\frac{d\mathcal{M}_n}{dy_n}(y_n)\right)^T \lambda_{n+1} \cdots \quad \lambda_0 = \left(\frac{d\Psi}{dy_0}\right)^T \;. \tag{6}$$

We will call λ_n discrete adjoint variables. Their evaluation requires the forward numerical solution y_0 to y_N to be available during the backward calculation.

Discrete adjoints are useful in optimization since they provide the gradients of the numerical function that is being minimized. Continuous adjoints are useful for sensitivity analysis studies. They are relatively easy to compute by applying a numerical solver of choice to the continuous equation (3), and using the forward solution $y(t)$ obtained by interpolation from a sequence of checkpoints.

The calculation of gradients by reverse automatic differentiation leads to the discrete adjoint approach. This paper is focused on analyzing some of the properties of the discrete adjoint variables λ_n and their relationship with the continuous adjoint variables $\lambda(t_n)$ when the numerical integration (both forward and backward in time) is performed by Runge-Kutta methods.

Consistency properties of discrete Runge-Kutta adjoints have been studied by Hager [5], who gives additional order conditions necessary in the context of control problems. Walther [7] has studied the effects of reverse mode automatic differentiation on explicit Runge-Kutta methods, and finds that the order of the discretization is preserved by discrete adjoints. Giles [2] has discussed Runge-Kutta adjoints in the context of steady state flows. In this paper we consider control problems where only the initial conditions are the control variables. This setting is simpler than the distributed control case considered by Hager [5] and Walther [7].

1.1 Runge-Kutta Methods

A general s-stage Runge-Kutta discretization method is defined as [3, Section II.1]

$$\begin{aligned} y_{n+1} &= y_n + h\sum_{i=1}^{s} b_i\, k_i \;, \qquad h = t_{n+1} - t_n \;, \\ k_i &= f\left(t_n + c_i h,\, Y_i\right) \;, \quad Y_i = y_n + h\sum_{j=1}^{s} a_{i,j} k_j \;, \end{aligned} \tag{7}$$

where the coefficients $a_{i,j}$, b_i and c_i are prescribed for the desired accuracy and stability properties. If $a_{i,j} \neq 0$ for $j \geq i$ the stage derivative values k_i are defined implicitly, and are obtained by solving the nonlinear system (7).

Hager [5] has shown that the discrete adjoint (6) of the Runge-Kutta method (7) is

$$\begin{aligned} \lambda_n &= \lambda_{n+1} + \sum_{j=1}^{s} \theta_j \;, \\ \theta_i &= h\, J^T(t_n + c_i h, Y_i) \left(b_i \lambda_{n+1} + \sum_{j=1}^{s} a_{j,i}\theta_j \right) \;, \quad i = 1 \cdots s \;. \end{aligned} \tag{8}$$

Hager has also shown that if all $b_i \neq 0$ then the discrete adjoint reads

$$\begin{aligned} \lambda_n &= \lambda_{n+1} + h\sum_{i=1}^{s} \overline{b}_i\, \ell_i \;, \quad \ell_i = J^T\left(t_n + c_i h,\, Y_i\right) \Lambda_i \;, \\ \Lambda_i &= \lambda_{n+1} + h\sum_{j=1}^{s} \overline{a}_{i,j}\ell_j \;, \quad \text{with} \quad \overline{b}_i = b_i \;, \quad \overline{a}_{i,j} = \frac{a_{j,i} b_j}{b_i} \;. \end{aligned} \tag{9}$$

In the continuous adjoint approach one solves the equation (3) with a Runge-Kutta method (7) with coefficients $\tilde{a}_{i,j}$, $\tilde{b}_i$, $\tilde{c}_i$ to obtain

$$\begin{aligned} \lambda_n &= \lambda_{n+1} + h\sum_{i=1}^{s} \tilde{b}_i\, \tilde{\ell}_i \;, \\ \tilde{\ell}_i &= J^T\left(t_{n+1} - \tilde{c}_i h,\, y(t_{n+1} - \tilde{c}_i h)\right) \Lambda_i \;, \quad \Lambda_i = \lambda_{n+1} + h\sum_{j=1}^{s} \tilde{a}_{i,j}\tilde{\ell}_j \;. \end{aligned} \tag{10}$$

2 Consistency of the Discrete Adjoint Method

We now regard the discrete adjoint equation (8) as a numerical method applied to the continuous adjoint equation (3) and try to assess how accurate this numerical method is. At a first glance if $b_i \neq 0$ we can use the similarity between (9) and (10) and apply the standard Runge-Kutta order conditions to the method with coefficients $\overline{a}_{i,j} = (a_{j,i} b_j)/b_i$, b_i, c_i. The difficulty consists in the fact that in the continuous adjoint (10) the transposed Jacobian is evaluated using the exact numerical solution $y(t_{n+1} - \tilde{c}_i h)$, while in the discrete adjoint (9) the Jacobian is evaluated using the stage solutions in the forward method Y_i. Therefore the accuracy with which these stage solutions are evaluated in the forward run affects the accuracy of the discrete adjoint method.

In implicit Runge-Kutta processes Y_i are the solutions of a nonlinear system of equations. The accuracy with which the nonlinear system is solved impacts further the accuracy of the discrete adjoint. In this paper we will analyze the order conditions under the assumption that the nonlinear systems are solved exactly in both the forward (7) and the discrete adjoint (8) methods. An additional complication is given by the fact that black-box application of automatic differentiation tools will result in a differentiation of the iterations needed to solve the nonlinear system in the forward method. A discussion of the behavior of the resulting (differentiated) iterations is beyond the scope of this paper.

Walther [7] found that the order of the discrete adjoints of explicit Runge-Kutta methods is the same as the order of the original method. In this section we will prove the same result in greater generality; our proof is applicable to both explicit and implicit Runge-Kutta methods.

Before we start the analysis we introduce the following "transfer functions". The discrete adjoint is obtained from the "discrete transfer function" R_D

$$\lambda_n = \left(\frac{dy_{n+1}}{dy_n} \right)^T \lambda_{n+1} = R_D \, \lambda_{n+1}$$

Similarly from the linear continuous adjoint equation (3) we derive a linear dependence between the continuous adjoint variables at different time, and call this dependence the "continuous transfer function" R_C such that $\lambda(t_n) = R_C \, \lambda(t_{n+1})$.

The analysis of the order of discrete adjoints is based on the concept of elementary differentials in the theory of order conditions explained in Hairer et al. [3, section II.2]. The numerical solution of the forward Runge-Kutta method (7) satisfies

$$\left(y^J_{n+1}\right)^{(q)}\Big|_{h=0} = \sum_{\tau \in LT_q} \gamma(\tau) \sum_j b_j \Phi_j(\tau) F^J(\tau)(y_n) \qquad (11)$$

where the superscipt J denotes the component number, the first summation is taken after all labeled trees of order q, and $F(\tau)(y_n)$ is the elementary differential associated with τ. $\Phi_j(\tau)$ is a combination of method coefficients associated with τ and $\gamma(\tau)$ is the multiplicity of tree τ.

The derivative of the solution y_{n+1} with respect to y_n satisfies

$$\left(\frac{\partial y_{n+1}^J}{\partial y_n^P}\right)^{(q)}\Bigg|_{h=0} = \frac{\partial}{\partial y_n^P}\left(y_{n+1}^J\right)^{(q)}\Bigg|_{h=0} = \sum_{\tau\in LT_q}\gamma(\tau)\sum_j b_j\Phi_j(\tau)F_P^J(\tau)(y_n) \tag{12}$$

where $F_P^J(\tau)(y_n)$ is the partial derivative of the elementary differential with respect to P-th argument, namely y_n^P. Consequently the (P,J)-th entry in the discrete transfer function has the following derivatives

$$\left(R_D^{P,J}\right)^{(q)}\Big|_{h=0} = \sum_{\tau\in LT_q}\gamma(\tau)\sum_j b_j\Phi_j(\tau)F_P^J(\tau)(y_n)\,. \tag{13}$$

The exact solution of the direct system satisfies [3, Section II.2]

$$\left(y^J\right)^{(q)}\Big|_{t=t_n} = \sum_{\tau\in LT_q} F^J(\tau)(y_n)\,. \tag{14}$$

Consider now the continuous adjoint equation. With $\Delta t = t_n - t_{n+1}$ the Taylor series of R_C about t_{n+1} is

$$\lambda(t_n) = R_C\,\lambda(t_{n+1}) = \sum_{j\geq 0}\frac{\Delta t^q}{q!}\,R_C^{(q)}\,\lambda(t_{n+1}) = \sum_{j\geq 0}\frac{\Delta t^q}{q!}\,\lambda^{(q)}(t_{n+1})$$

and therefore $R_C^{(q)}$ maps $\lambda(t_{n+1})$ to $\lambda^{(q)}(t_{n+1})$. It can be shown that the derivatives of the exact solution of the adjoint equation, taken with respect to $(-t)$ at t_n, are

$$\left(\lambda^P\right)^{(q)}\Big|_{t=t_n} = \sum_{\tau\in LT_q} F_P^J(\tau)(y_n)\lambda^J(t_n)\,. \tag{15}$$

This implies that for the continuous transfer function

$$\left(R_C^{P,J}\right)^{(q)} = \sum_{\tau\in LT_q} F_P^J(\tau)(y_n)\,. \tag{16}$$

For example taking the derivative w.r.t. $(-t)$ in (3) gives

$$\lambda^{(1)} = J^T(t, y(t))\lambda \Rightarrow \left(\lambda^P\right)^{(1)} = \sum_J f_P^J\lambda^J = \sum_J F_P^J(\tau_1)\lambda^J\,.$$

This implies that for the continuous transfer function

$$\left(R_C^{P,J}\right)^{(1)} = F_P^J(\tau_1)(y_n)\,.$$

Similarly

$$\begin{aligned}\left(\lambda^P\right)^{(2)} &= \sum_{J,K} f_{PK}^J f^K\lambda^J + \sum_J f_P^J(\lambda^J)^{(1)} = \sum_{J,K} f_{PK}^J f^K\lambda^J + \sum_{J,K} f_P^J f_J^K\lambda^K\\ &= \sum_{J,K}\left(f_{PK}^J f^K + \sum_{J,K} f_K^J f_P^K\right)\lambda^J = \sum_J F_P^J(\tau_{21})\lambda^J\,,\end{aligned}$$

and therefore for the continuous transfer function we have

$$\left(R_C^{P,J}\right)^{(2)} = F_P^J(\tau_{21})(y_n) \, .$$

Continuing this process one obtains (16).

A comparison of (13) and (16) shows that the numeric transfer function equals the continuous transfer function up to order p iff

$$\sum_j b_j \Phi_j(\tau) = \frac{1}{\gamma(\tau)} \, ,$$

for all trees τ of order $\leq p$. But these are exactly the conditions under which the forward method is of order p. Consequently, the discrete adjoint of an order p Runge-Kutta method is a discretization of order p of the continuous adjoint equation if the problem is sufficiently smooth ($y(t)$ has p+1 continuous derivatives).

3 Discrete Adjoints and Stiff Problems

The traditional linear stability analysis approach [4] considers the transfer function $R(z)$ of a Runge-Kutta method when applied to a linear scalar test problem $y' = \alpha y$. One can easily see that the transfer function of the discrete adjoint method (8) is the same as the transfer function of the forward Runge-Kutta method (7), and therefore the discrete adjoints inherits the stability properties of the original method.

To better understand the behavior of discrete Runge-Kutta adjoints on stiff systems we consider the singular perturbation model problem

$$y' = f(y, z) \, , \quad \epsilon z' = g(y, z) \, , \quad t_0 \leq t \leq t_F \qquad (17)$$

with the sub-Jacobian g_z assumed nonsingular. For this system we distinguish between the adjoint variables of the nonstiff and of the stiff components

$$\lambda(t) = \frac{\partial \Psi\big(y(t_\mathrm{F}), z(t_\mathrm{F})\big)}{\partial y(t)} \, , \quad \mu(t) = \frac{\partial \Psi\big(y(t_\mathrm{F}), z(t_\mathrm{F})\big)}{\partial z(t)} \, .$$

The adjoint variables satisfy the continuous adjoint equation

$$\begin{bmatrix} \lambda \\ \mu \end{bmatrix}' = - \begin{bmatrix} f_y^T & \epsilon^{-1} g_y^T \\ f_z^T & \epsilon^{-1} g_z^T \end{bmatrix} \begin{bmatrix} \lambda \\ \mu \end{bmatrix} . \qquad (18)$$

Consider an ϵ-expansion of the solution

$$\lambda = \sum_{i \geq 0} \epsilon^i \, \lambda^i \, , \quad \mu = \sum_{i \geq 0} \epsilon^i \, \mu^i \, , \qquad (19)$$

insert it into (18) and equate the ϵ series. The ϵ^{-1} term leads to

$$\mu^0 = 0 \,. \tag{20}$$

We equate recursively the higher order terms to obtain

$$\begin{cases} (\lambda^i)' = \left(-f_y^T + g_y^T\, g_z^{-T}\, f_z^T\right) \lambda^i + g_y^T\, g_z^{-T}\, (\mu^i)' \\ \mu^{i+1} = -g_z^{-T}\, \left((\mu^i)' + f_z^T\, \lambda^i\right) \,, \end{cases} \quad \text{for} \quad i = 0, 1, 2, \ldots \tag{21}$$

The zeroth order term evolution is given by the equation

$$(\lambda^0)' = \left(-f_y^T + g_y^T\, g_z^{-T}\, f_z^T\right) \lambda^0 \,.$$

Consider now the Runge-Kutta discrete adjoint (8) for the method applied to the singular perturbation systems (17). Denote by A, b, c the coefficient matrices of the Runge-Kutta method, by e a vector of ones, and let

$$\begin{aligned} G_Z &= \mathrm{diag}_i\big(g_z(Y_i, Z_i)\big) \,, \quad F_Z = \mathrm{diag}_i\big(f_z(Y_i, Z_i)\big) \,, \\ G_Y &= \mathrm{diag}_i\big(g_y(Y_i, Z_i)\big) \,, \quad F_Y = \mathrm{diag}_i\big(f_y(Y_i, Z_i)\big) \,, \\ S^T &= F_Y^T - G_Y^T G_Z^{-T} F_Z^T \,, \end{aligned}$$

A careful analysis based on ϵ-expansions of the discrete solution and of the stage vectors leads to the following conclusions.

If the Runge-Kutta coefficient matrix A is invertible then the adjoints of the stiff variables are integrated using

$$\mu_n^0 = R(\infty)\, \mu_{n+1}^0 \,,$$

where $R(\cdot)$ is the stability function of the RK method. We see easily that for methods with $R(\infty) = 0$ the first order term in the μ adjoint is zero, $\mu_n^0 = 0$. This is desirable considering the exact solution (20). In this case the values of μ are solved with the same accuracy as the original method, within $\mathcal{O}(\epsilon)$.

The adjoints of the non-stiff variables are integrated as

$$\begin{aligned} \lambda_n^0 = & \left(1 + h e^T \left(I - h S^T A^T\right)^{-1} S^T b\right) \lambda_{n+1}^0 \\ & + e^T \left(I - h S^T A^T\right)^{-1} G_Y^T G_Z^{-T} A^{-T} b\, \mu_{n+1}^0 \end{aligned} \tag{22}$$

Therefore the adjoint with respect to the nonstiff variable depends on both μ_{n+1}^0 and λ_{n+1}^0.

If $R(\infty) = 0$ and we are away from the initial condition then $\mu_{n+1}^0 = 0$ and we have a discretization of the reduced adjoint equation (21). The same holds if the cost function depends only on the non-stiff variables, $\Psi = \Psi(y)$. In this case the values of λ are solved with the same accuracy as the original method, within $\mathcal{O}(\epsilon)$. Note that the initialization of $\mu_N^0 \neq 0$ can introduce an $\mathcal{O}(1)$ perturbation in λ_{N-1}^0.

4 Conclusions

In this paper we analyze the consistency and stability properties of Runge-Kutta discrete adjoints. Discrete adjoints are very popular in optimization and control since they can be constructed automatically by reverse mode automatic differentiation. However, the properties of the discrete adjoints are often poorly understood.

The consistency analysis uses the concept of elementary differentials and reveals that the Runge-Kutta discrete adjoint method has the same order of accuracy as the original, forward method. The discrete adjoint also inherits the linear stability properties of the original method. A singular perturbation analysis shows that L-stable Runge-Kutta methods with an invertible coefficient matrix are well-behaved under discrete adjoint operation.

Acknowledgments

This work was supported by the National Science Foundation (NSF) through the awards NSF CAREER ACI-0413872, NSF ITR AP&IM 020519, and NSF CCF-0515170, by the National Oceanic and Atmospheric Administration (NOAA) and by the Texas Environmental Research Consortium (TERC).

References

1. M.L. Baguer and W. Romisch. Computing gradients in parametrization-discretization schemes for constrained optimal control problems. Approximation and Optimization in the Carribbean II, p. 14–34, M. Florenzano editor, Peter Lang, Frankfurt am Main, 1995.
2. M.B. Giles. On the Use of Runge-Kutta Time-Marching and Multigrid for the Solution of Steady Adjoint Equations. Technical Report NA00/10, Oxford University Computing Laboratory, 2000.
3. E. Hairer, G. Wanner, and S.P. Norsett. Solving Ordinary Differential Equations I. Nonstiff Problems. Springer Series in Computational Mathematics, 1991.
4. E. Hairer and G. Wanner. Solving Ordinary Differential Equations II. Stiff and Differential-Algebraic Problems. Springer Series in Computational Mathematics, 1996.
5. W. Hager. Runge-Kutta methods in optimal control and the transformed adjoint system. Numerische Mathematik 87(2):247–282, 2000.
6. A. Sandu, D. Daescu, and G.R. Carmichael: "Direct and Adjoint Sensitivity Analysis of Chemical Kinetic Systems with KPP: I – Theory and Software Tools", *Atmospheric Environment*, Vol. 37, p. 5083–5096, 2003.
7. A Walther. Automatic Differentiation of Explicit Runge-Kutta methods for Optimal Control. Technical University Dresden technical report WR-06-2004. To appear in Journal of Computational Optimization and Applications.

Source Transformation for MATLAB Automatic Differentiation

Rahul V. Kharche and Shaun A. Forth

Cranfield University (Shrivenham Campus), Shrivenham, Swindon SN6 8LA, UK
{R.V.Kharche, S.A.Forth}@cranfield.ac.uk

Abstract. We present MSAD, a source transformation implementation of forward mode automatic differentiation for MATLAB. MSAD specialises and inlines operations from the `fmad` and `derivvec` classes of the MAD package. The operator overloading overheads inherent in MAD are eliminated while preserving the `derivvec` class's optimised derivative combination operations. Compared to MAD, results from several test cases demonstrate significant improvement in efficiency across all problem sizes.

1 Automatic Differentiation in MATLAB

MATLAB is popular for rapid prototyping and numerical computing owing to its high-level abstraction of matrices and its rich set of function and GUI libraries. MATLAB's interpreted nature and high-level language make programming intuitive and debugging easy. Optimised BLAS and LAPACK routines for internal matrix operations facilitate good performance. MATLAB may be extended by further general purpose and application specific *toolboxes* (e.g., for optimisation, partial differential equations, control, etc.). We believe the robustness and efficiency of many MATLAB toolboxes and user's applications would benefit from an effective automatic differentiation (AD) [1] package.

Coleman and Verma's ADMAT [2] was the first significant MATLAB AD tool and implemented forward and reverse mode differentiation, with support for Jacobian compression, via operator overloading. The later ADiMat tool [3] adopted a hybrid source transformation/operator overloading implementation of forward mode AD and out-performed ADMAT on several problems. Simultaneously the `fmad` class of MAD [4], an operator overloaded implementation of forward mode AD, was also shown to outperform ADMAT. MAD's efficiency is due to appropriate data-structures and use of high-level matrix operations within its `derivvec` class which holds and propagates derivatives. Use of MATLAB's `sparse` data-type to hold and propagate sparse derivatives enables runtime sparsity exploitation – greatly enhancing performance for problems where sparsity is unknown or difficult to exploit via compression techniques.

Because there is no compilation before execution of operator-overloaded MATLAB code, performance of overloaded implementations of AD suffer due to overheads from the interpreter and the *type check* and *dispatch* mechanism of overloading. Note that MATLAB's recent just-in-time (JIT) compiler is restricted to

V.N. Alexandrov et al. (Eds.): ICCS 2006, Part IV, LNCS 3994, pp. 558–565, 2006.
© Springer-Verlag Berlin Heidelberg 2006

a subset of MATLAB's intrinsic classes and so is not applicable to the `derivvec` class. Moreover, overloaded operations typically involve substantial logic and branching dependent on the shape (scalar, vector, matrix, N-D array) or storage class (complex, sparse) used for derivatives. For example, consider the `times` operation of the `derivvec` class in Fig. 1. Here `.derivs` refers to the operand's derivative matrix, and `.shape` to the size of the operand. We see that the test on line 9 checks if operands have equal sizes and those of lines 15 and 17 test for scalar operands. Similarly, line 10 checks for `sparse` storage of derivatives. Such tests incur further run-time overheads. Other generic MATLAB overheads are described by [8] and their relevance to AD is discussed in [14].

```
function cdv = times(a,b)
if isa(b,'derivvec')
    cdv = b; mults = a;
else
    cdv = a; mults = b;
end
ssd = prod(cdv.shape); sm = size(mults); ssm = prod(sm);
mults = mults(:);
if ssd == ssm                                                    % line 9
    if issparse(cvd.derivs)                                      % line 10
        % sparse mode operations omitted for brevity
    else
        cdv.derivs = mults(:,ones(1,cdv.nderivs)).*cdv.derivs;  % line 13
    end
elseif ssd == 1                                                  % line 15
    cdv.derivs = mults*cdv.derivs; cdv.shape = sm;
elseif ssm == 1                                                  % line 17
    cdv.derivs = mults.*cdv.derivs;
end
```

Fig. 1. `derivvec - times` operation from MAD

The MSAD (MATLAB Source transformation AD) tool aims to demonstrate the benefits obtained by combining source transformation with MAD's efficient data structures. An initial, hybrid source transformation/operator overloading approach, similar to that of ADiMat, showed significant speedup compared to MAD for smaller test cases but asymptotically reached the performance of MAD as the problem size increased [6]. Section 2 of this paper describes our improved source transformation approach, which now specialises and inlines all required derivative operations. The benefits of this approach are demonstrated by the test cases of Section 3. Conclusions are presented in Section 4.

2 Source Transformation Via Specialising and Inlining

MSAD uses ANTLR-based LL(k) scanner, parser and tree parsers [5] to analyse and source transform MATLAB programs for AD [6]. Program transformation is carried out via four phases - *scanning* and *parsing* for Abstract Syntax Tree (AST) and symbol table generation, *attribute synthesis* for activity analysis [7], size and class propagation, and finally derivative *code generation*. MSAD's parser recognises the complete MATLAB (Release 14) grammar, but differentiation of code involving branches, loops, structures, cells, nested functions and programs spanning multiple files is currently not implemented. Despite these restrictions, by replacing loops with array operations, many tests cases can be differentiated.

The attribute synthesis phase propagates flags that mark a variable's activity, class, storage type and derivative storage type. Input programs are prepared by using directives to indicate the active inputs and optionally sparse storage for their derivatives. Users may optionally supply size information of input variables. For example, the directives in Fig. 2 indicate to MSAD that the size parameters (`nx, ny`) and the vortex parameter (`vornum`) are scalars. The directives also label variable `x` as an active input and that its derivatives be stored as a sparse matrix. MSAD emulates MATLAB's sparse type propagation and size computation rules for each elementary operation of the source code to deduce the storage type and size of all variables. If a variable's size and storage type cannot be determined, MSAD marks these attributes as unknown. The sizes of scalar and array constants within a program are automatically propagated.

```
function fgrad = gdgl2(nx, ny, x, vornum)
    %! size(nx) = [1, 1], size(ny) = [1, 1], size(vornum) = [1, 1]
    %! active(x), sparseDer(x)
```

Fig. 2. User directives used with gradient function of MINPACK DGL2 problem

The derivative code is generated in a final pass during which the operations from MAD's `fmad` and `derivvec` classes are specialised and inlined. Specialisation uses a variable's size, class, storage class and activity information to resolve condition checks and simplify size computations in the `fmad` and `derivvec` class operations. For variables with unknown size and storage attributes, MSAD conservatively inlines operations involving size and storage checks.

We illustrate the process of specialisation and inlining by considering the FT-BROY function of Fig. 3 [11], specifically the subexpression `(3-2*x(n)).*x(n)` of line 8. Line 3 of the program implies `n` equals the length of the vector `x`. Although this length can be determined only at run-time, `n` can safely be deduced to be a scalar. This further implies `x(n)` is a scalar, as is `3-2*x(n)`. MSAD automatically carries out this size inference during the attribute synthesis phase. During specialisation, because the operands `x(n)` and `3-2*x(n)`, held in variables `tmp_5_` and `tmp_4_` in the generated code of Fig. 4, are inferred to be scalars, the condition on line 9 from the `derivvec-times` operation in Fig. 1 is satisfied. Assuming derivatives are stored in their full form, only lines 8 and 13 from Fig. 1 need to be inserted into the generated code as seen in lines 17 to 20 of Fig. 4. Comments in Fig. 4, and the later Fig. 5, were added by hand to indicate to the reader which line computes which expression or expression's derivatives; D[a] denotes the derivatives of variable a.

```
function f = ftbroy(x)
%! active(x)
n = length(x);                                              % line 3
p = 7/3; y = zeros(n,1);
i = 2:(n-1);
y(i) = abs((3-2*x(i)) .* x(i) - x(i-1) - x(i+1) + 1).^p;    % line 6
y(n) = abs((3-2*x(n)) .* x(n) - x(n-1) + 1).^p;             % line 7
y(1) = abs((3-2*x(1)) .* x(1) - x(2) + 1).^p;               % line 8
j = 1:(n/2); z = zeros(length(j),1);
z(j) = abs(x(j) + x(j+n/2)).^p;
f = 1 + sum(y) + sum(z);
```

Fig. 3. FTBROY function

```
tmp_1_ = x(n);                                          %                  x(n)
tmp_ind_ = reshape((1:numel(x)), size(x));
tmp_ind_ = tmp_ind_(n);
d_tmp_1_ = d_x(tmp_ind_(:),:);                          %               D[x(n)]
tmp_2_ = 2 .* tmp_1_;                                   %                2*x(n)
tmp_mults_ = 2;
d_tmp_2_ = tmp_mults_(:,ones(1,res_tmp1_)).*d_tmp_1_;   %             D[2*x(n)]
tmp_3_ = 3 - tmp_2_;                                    %              3-2*x(n)
d_tmp_3_ = -d_tmp_2_;                                   %           D[3-2*x(n)]
tmp_4_ = tmp_3_;                                        %            (3-2*x(n))
d_tmp_4_ = d_tmp_3_;                                    %         D[(3-2*x(n))]
tmp_5_ = x(n);                                          %                  x(n)
tmp_ind_ = reshape((1:numel(x)), size(x));
tmp_ind_ = tmp_ind_(n);
d_tmp_5_ = d_x(tmp_ind_(:),:);                          %               D[x(n)]
tmp_6_ = tmp_4_ .* tmp_5_;                              %       (3-2*x(n)).*x(n)
tmp_mults_ = tmp_5_;                                    %               line 17
d_tmp_7_ = tmp_mults_(:,ones(1,res_tmp1_)).*d_tmp_4_;   % x(n).*D[(3-2*x(n))]
tmp_mults_ = tmp_4_;
d_tmp_8_ = tmp_mults_(:,ones(1,res_tmp1_)).*d_tmp_5_;   % (3-2*x(n)).*D[x(n)]
d_tmp_6_ = d_tmp_7_ + d_tmp_8_;                         % D[(3-2*x(n)).*x(n)]
```

Fig. 4. MSAD generated derivative code for the subexpression (3-2*x(n)).*x(n) of the TBROY function. (Comments added for clarity)

In the subexpression (3-2*x(i)).*x(i) on line 6 in Fig. 3, the size of x(i) cannot be determined since i is a vector dependent on the value of n. MSAD therefore conservatively inlines lines 7 to 19 of the derivvec-times operation. The first product of D[3-2*x(i).*x(i)], analogous to lines 17 and 18 from Fig. 4, can be seen in Fig. 5.

```
d_tmp_4= d_tmp_3                                       %      D[(3-2*x(i))]
tmp_mults_ = tmp_5_(:);                                %               x(i)
tmp_ssa_ = numel(tmp_mults_);                          %       length(x(i))
tmp_ssb_ = numel(tmp_4_);                              % length((3-2*x(i)))
if tmp_ssa_ == tmp_ssb_                                %        equal sizes
   d_tmp_7_ = tmp_mults_(:,ones(1,res_tmp1_)) .* d_tmp_4_;
elseif tmp_ssb_ == 1                                   %  (3-2*x(i)) scalar
   d_tmp_7_ = tmp_mults_ * d_tmp_4_;
elseif tmp_ssa_ == 1                                   %        x(i) scalar
   d_tmp_7_ = tmp_mults_ .* d_tmp_4_;
end
```

Fig. 5. Additional checks for vector `times` operation in `D[(3-2*x(i))].*x(i)`. (Comments added for clarity)

3 Test Results

MSAD computed derivatives were tested for correctness and performance on several optimisation, BVP and ODE problems [14]. A subset of those tests, all performed using MATLAB Release 14 on a Linux machine with a 2.8 GHz Pentium-4 processor and 512 MB of RAM, are presented here.

In Table 1 we compare use of MSAD and MAD's `fmad` class to compute derivatives by repeating the large-scale test cases from MATLAB's Optimisation Toolbox [11] performed in [4]. The test cases are: `nlsf1a`– sparse Jacobian from vector residual; `brownf`, `tbroyf` – gradient from objective function; `browng`, `tbroyg` – Hessian from hand-coded gradient. Both automatic differentiation tools may use Jacobian/Hessian compression (denoted `cmp`) [1, Chap. 7] or sparse storage (denoted `spr`) [1, Chap. 6] where appropriate. The only MSAD user directives required were those to specify the active input variables and use of sparse derivative storage. For comparison, we have included MATLAB's

Table 1. Ratio $\mathrm{CPU}(\nabla \boldsymbol{f} + \boldsymbol{f})/\mathrm{CPU}(\boldsymbol{f})$ – Jacobian/gradient (including function) to function CPU time ratio for given techniques on MATLAB Optimisation Toolbox large-scale examples. (m, n) gives the number of dependents and independents, $\hat{n}$ the maximum number of non-zero entries in a row of the Jacobian and p the number of colours for compression

Problem	$\mathrm{CPU}(\nabla \boldsymbol{f} + \boldsymbol{f})/\mathrm{CPU}(\boldsymbol{f})$ for						(m, n)	$\hat{n}$	p
	Hand-coded	sfd-(nls)	msad (cmp)	fmad (cmp)	msad (spr)	fmad (spr)			
nlsf1a(Jac)	4.4	38.3	6.9	22.5	19.4	35.1	(100,1000)	3	3
brownf(grad)	4.6	1064.9	–	–	9.3	13.7	(1,1000)	1000	–
browng(Jac)	5.2	9.5	4.2	8.4	15.3	19.6	(1000,1000)	3	3
tbroyf(grad)	3.8	810.7	–	–	8.8	15.9	(1,800)	800	–
tbroyg(Jac)	–	13.8	3.3	10.1	15.8	23.5	(800,800)	6	7

finite-difference (`sfd(nls)`) evaluation of the gradient/Jacobian/Hessian and, where available, hand-coding.

Clearly, MSAD yields significant savings compared to `fmad` in like-for-like computation of derivatives for these moderate sized problems ($n \approx 1000$). For compressed derivative computation we get savings of over 50% using `msad(cmp)` and for sparse storage gains of about 30%. Compressed AD (`msad(cmp)`, `fmad(cmp)`) out-performs compressed finite-differencing (`sfd(nls)`). For the gradient problems (`brownf`, `tbroyf`) sparse AD (`msad(spr)`, `fmad(spr)`) is several times faster than `sfd(nls)` because the functions `brownf` and `tbroy` are partially value separable [4] and the sparse derivative computation may utilise intermediate sparsity whereas finite-differencing cannot. For the `browng` problem `msad(cmp)` outperforms hand-coding due to the use of complicated expressions in the hand-coding.

Table 2 lists the total optimisation run-times with derivatives supplied using the methods of Table 1. Source transformed derivatives yield substantial savings in the total run-time compared to `fmad`'s overloading approach and run-times are comparable to those using hand-coded derivatives.

The 2-D Ginzburg-Landau unconstrained minimisation problem (GL2) [12, 13] uses an $n_x \times n_y$ mesh with 4 variables per mesh point yielding $n = 4n_x n_y$ independent variables. The objective function is again partially value separable and the gradient code is supplied. The sparse Hessian is computed as the Jacobian $\boldsymbol{Jg}$ of the gradient $\boldsymbol{g}$. Differentiated functions were generated using MSAD for full and sparse storage of derivatives; the user directives for sparse storage can be seen in Fig. 2. Table 3 gives the derivative computation ratio CPU($\boldsymbol{Jg} + \boldsymbol{g}$)/CPU($\boldsymbol{g}$) for increasing problem size. Using compression, `msad(cmp)` is nearly 80% more efficient than `fmad(cmp)` for small n. With increasing problem size, the floating point operation cost of the derivative computation of either method increases relative to its overheads and the relative advantage of source transformation decreases. However, even for n as large as 65,536 `msad(cmp)` is nearly twice as fast as overloading. With sparse derivatives (`msad(spr)`, `fmad(spr)`) we see a similar trend but smaller relative improvement due to the common overhead of manipulating MATLAB's `sparse` data structures.

The total optimisation time using MATLAB's `fminunc` solver with the different Hessian calculation techniques of Table 3 is shown in Table 4. The decrease

Table 2. Averaged CPU time for optimisation of the large-scale examples from the MATLAB Optimisation Toolbox with derivatives supplied using given techniques

Problem	Optimisation CPU time (s) for Hand-coded	sfd-(nls)	msad (cmp)	fmad (cmp)	msad (spr)	fmad (spr)
`nlsf1a`	0.16	0.36	0.17	0.31	0.20	0.35
`brownf`	0.56	–	–	–	0.7	1.25
`browng`	0.29	0.56	0.23	0.41	0.46	0.64
`tbroyf`	0.72	–	–	–	1.29	2.89
`tbroyg`	–	0.76	0.20	0.48	0.55	0.86

Table 3. Ratio CPU($\boldsymbol{Jg}+\boldsymbol{g}$)/CPU($\boldsymbol{g}$) – Hessian (including gradient) to gradient function CPU time ratio for the MINPACK 2-D Ginzburg-Landau problem using given techniques; p gives the number of colours for compression. For all problem sizes, the maximum number of non-zero entries in a row of the Jacobian is $\hat{n} = 14$.

	CPU($\boldsymbol{Jg}+\boldsymbol{g}$)/CPU($\boldsymbol{g}$) for problem size n					
Method	64	256	1024	4096	16384	65536
`msad(cmp)`	24.72	23.69	21.84	23.31	37.95	52.16
`fmad(cmp)`	115.32	105.89	89.57	72.82	72.11	90.47
`msad(spr)`	28.11	29.18	35.02	52.63	88.97	177.10
`fmad(spr)`	122.80	113.84	107.73	108.72	126.45	222.81
`#colours` p	20	23	25	24	25	25

in overall computation time obtained by using MSAD's more efficient derivative computation is seen – but this is not proportional to the decrease in derivative computation time. This is because for larger problem size the number of Newton iterations (which require a Hessian recalculation) stays fixed but the number of conjugate gradient iterations (which do not) increase [15].

Table 4. Optimisation CPU time for the MINPACK Ginzburg-Landau (GL2) problem using MATLAB's `fminunc` with derivatives supplied using given techniques

	Problem size n				
	64	256	1024	4096	16384
Method	CPU time (s) for optimisation				
`msad(cmp)`	0.74	0.59	1.34	6.95	29.62
`fmad(cmp)`	4.45	2.78	3.79	10.71	38.70
`msad(spr)`	1.23	1.14	2.87	13.05	61.93
`fmad(spr)`	4.79	3.32	5.41	17.05	73.83
`sfd`	1.41	1.29	3.60	19.91	216.30

4 Conclusion

The previous, hybrid source transformation/operator overloading implementation of MSAD [6] gave reasonable speedup over operator overloading for small problem sizes. This speedup diminished with increasing problem size. The improved implementation presented here inlines and, where possible specialises, the remaining overloaded function calls. This eliminates the type check and dispatch overhead of overloading, reduces logic and branching, and exposes a larger section of the augmented code to MATLAB's JIT acceleration. Section 3's test cases clearly demonstrate these benefits. Figure 4's code indicates the scope for further performance improvements by eliminating redundant temporaries and common subexpressions. Preliminary results obtained by implementing such improvements by hand on one test case produced a 42% speedup [14] and highlight the need for such compiler-like optimisations within a MATLAB AD-tool.

References

1. Griewank, A.: Evaluating Derivatives: Principles and Techniques of Algorithmic Differentiation. Number 19 in Frontiers in Appl. Math. SIAM, Philadelphia, Penn. (2000)
2. Coleman, T.F., Verma, A.: ADMAT: An automatic differentiation toolbox for MATLAB. Technical report, Computer Science Department, Cornell University (1998)
3. Bischof, C.H., Bücker, H.M., Lang, B., Rasch, A., Vehreschild, A.: Combining source transformation and operator overloading techniques to compute derivatives for MATLAB programs. In: Proceedings of the Second IEEE International Workshop on Source Code Analysis and Manipulation (SCAM 2002), Los Alamitos, CA, USA, IEEE Computer Society (2002) 65–72
4. Forth, S.A.: An efficient overloaded implementation of forward mode automatic differentiation in MATLAB. Accepted ACM Trans. Math Softw. (2005)
5. Parr, T., Quong R.: ANTLR: A predicated LL(k) parser generator. Software, Practice and Experience, vol. 25, p. 789, July 1995
6. Kharche, R.V.: Source transformation for automatic differentiation in MATLAB. Master's thesis, Cranfield University (Shrivenham Campus), Engineering Systems Dept., Shrivenham, Swindon SN6 8LA, UK (2004)
7. Bischof, C.H., Carle A., Khademi P., Mauer A.: ADIFOR 2.0: Automatic Differentiation of Fortran 77 Programs. IEEE Computational Science & Engineering **3**(3) (1996) 18–32
8. Menon, V., Pingali, K.: A case for source-level transformations in MATLAB. In: PLAN '99: Proceedings of the 2nd conference on Domain-specific languages, New York, NY, USA, ACM Press (1999) 53–65
9. Rose, L.D., Padua, D.: Techniques for the translation of MATLAB programs into Fortran 90. ACM Trans. Program. Lang. Syst. **21**(2) (1999) 286–323
10. Elphick, D., Leuschel, M., Cox, S.: Partial evaluation of MATLAB. In: GPCE '03: Proceedings of the second international conference on Generative programming and component engineering, New York, NY, USA, Springer-Verlag New York, Inc. (2003) 344–363
11. The MathWorks Inc. 24 Prime Park Way, Natick, MA 01760-1500: MATLAB Optimization Toolbox - User's guide. (2005)
12. Averick, B.M., Moré, J.J.: User guide for the MINPACK-2 test problem collection. Technical Memorandum ANL/MCS-TM-157, Argonne National Laboratory, Argonne, Ill. (1991) Also issued as Preprint 91-101 of the Army High Performance Computing Research Center at the University of Minnesota.
13. Lenton, K.: An efficient, validated implementation of the MINPACK-2 test problem collection in MATLAB. Master's thesis, Cranfield University (Shrivenham Campus), Engineering Systems Dept., Shrivenham, Swindon SN6 8LA, UK (2005)
14. Kharche, R., Forth, S.: Source transformation for MATLAB automatic differentiation. Applied Mathematics & Operational Research Report AMOR 2005/1, Cranfield University (Shrivenham Campus), Engineering Systems Dept., Shrivenham, Swindon, SN6 8LA, UK (2005)
15. Bouaricha, A., Moré, J.J., Wu, Z.: Newton's method for large-scale optimization. Preprint MCS-P635-0197, Argonne National Laboratory, Argonne, Illinois (1997)

The Data-Flow Equations of Checkpointing in Reverse Automatic Differentiation

Benjamin Dauvergne and Laurent Hascoët

INRIA Sophia-Antipolis, TROPICS team,
2004 Route des lucioles, BP 93, 06902 Sophia-Antipolis, France

Abstract. Checkpointing is a technique to reduce the memory consumption of adjoint programs produced by reverse Automatic Differentiation. However, checkpointing also uses a non-negligible memory space for the so-called "snapshots". We analyze the data-flow of checkpointing, yielding a precise characterization of all possible memory-optimal options for snapshots. This characterization is formally derived from the structure of checkpoints and from classical data-flow equations. In particular, we select two very different options and study their behavior on a number of real codes. Although no option is uniformly better, the so-called "lazy-snapshot" option appears preferable in general.

1 Introduction

Mathematical derivatives are a key ingredient in Scientific Computation. In particular, gradients are essential in optimization and inverse problems. The methods to compute gradients can be classified in two categories. In the first category, methods use CPU more intensively because several operations are duplicated. This can be through repeated tangent directional derivatives, or through reverse Automatic Differentation using the "Recompute-All" strategy. This is not the context of this paper. In the second category, methods spare duplicated operations through increased memory use. This encompasses hand-coded resolution of the "adjoint equations" and reverse Automatic Differentation using the "Store-All" strategy, which is the context of this work.

Being a software transformation technique, reverse AD can and must take advantage from software analysis and compiler technology [1] to minimize these efficiency problems. In this paper, we will analyze "checkpointing", an AD technique to trade repeated computation for memory consumption, with the tools of compiler data-flow analysis. Checkpointing offers a range of options that influence the resulting differentiated code. Our goal is to formalize these options and to find which ones are optimal. This study is part of a general effort to formalize all the compiler techniques useful to reverse AD, so that AD tools can make the right choices using a firmly established basis.

2 Reverse Automatic Differentiation

Automatic Differentiation by program transformation takes a program P that computes a differentiable function F, and creates a new program that com-

V.N. Alexandrov et al. (Eds.): ICCS 2006, Part IV, LNCS 3994, pp. 566–573, 2006.
© Springer-Verlag Berlin Heidelberg 2006

putes derivatives of F. Based on the chain rule of calculus, AD iserts into P new "derivative" statements, each one corresponding to one original statement of P. In particular, reverse AD creates a program $\overline{P}$ that computes gradients. In $\overline{P}$, the derivative statements corresponding to the original statements are executed in *reverse order* compared to P. The derivative statements use some of the values used by their original statement, and therefore the original statements must be executed in a preliminary "forward sweep" $\overrightarrow{P}$, which produces the original values that are used by the derivative statements forming the "backward sweep" $\overleftarrow{P}$. This is illustrated by Fig. 1, in which we have readily split P in three successive parts, U Upstream, C Center, and D Downstream. In our context, original values are made available to the backward sweep through `PUSH` and `POP` routines, using a stack that we call the "tape". Not all original values

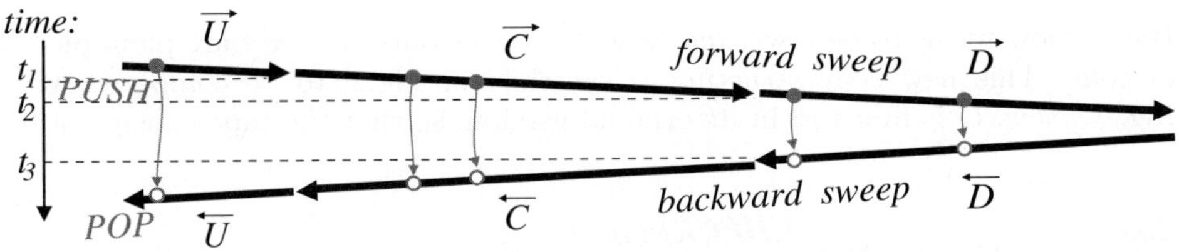

Fig. 1. Basic structure of reverse differentiated programs

are required in the backward sweep. Because of the nature of differentiation, values that are used only "linearly" are not required. The "To Be Recorded" (TBR) analysis [2, 6] finds the set of these required values denoted by *Req*. Set *Req* evolves as the forward sweep advances. For example in Fig. 1, TBR analysis of U finds the variable values required by $\overleftarrow{U}$ (i.e. actually **use**$(\overleftarrow{U})$), which must be preserved between the end of $\overrightarrow{U}$ and the beginning of $\overleftarrow{U}$. To this end, each time a required value is going to be overwritten by a statement, it is `PUSH`'ed beforehand, and it is `POP`'ped just before the derivative of this statement.

Although somewhat complex, reverse AD can be easily applied by an automatic tool, and has enormous advantages regarding the number of computation steps needed to obtain the gradient [4, chapter 3].

In [5], we studied the data-flow properties of reverse differentiated programs, in the basic case of Fig. 1, i.e. with no checkpointing. We formalized the structure of these programs and derived specialized data-flow equations for the "adjoint liveness" analysis, which finds original statements that are useless in the differentiated program, and for the TBR analysis. In this paper, we will focus on the consequences of introducing checkpointing. In this respect this paper, although stand-alone, is a continuation of [5].

3 The Equations of Checkpointing Snapshots

Checkpointing modifies the differentiated code structure to reduce the peak memory consumption. When code fragment C is "checkpointed" (notation $[C]$), the adjoint now written $\overline{[C];D}$ is formally defined by the recursive rewrite rule:

$$\boxed{Req \vdash \overline{[C];D}} = \begin{array}{l}\texttt{PUSH}(Sbk);\\ \texttt{PUSH}(Snp);\\ C;\\ \boxed{Req_D \vdash \overline{D}}\\ \texttt{POP}(Snp);\\ \boxed{Req_C \vdash \overline{C}}\\ \texttt{POP}(Sbk);\end{array} \tag{1}$$

Boxes show terms to be rewritten, whereas terms outside boxes are plain pieces of code. This new code structure is sketched in Fig.2, to be compared with Fig. 1. Now, C is first run in its original version, so that the tape consumed by

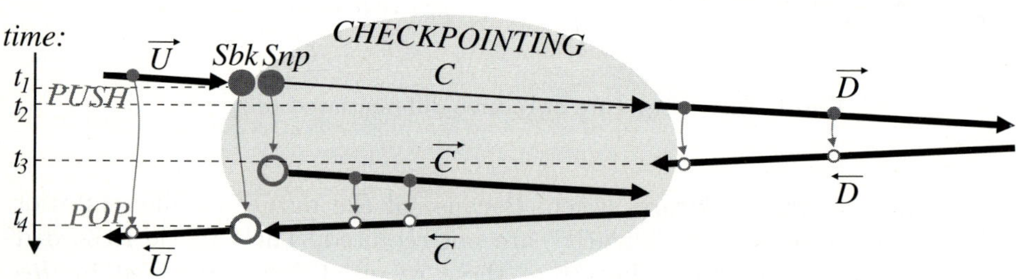

Fig. 2. Checkpointing in reverse AD

$\overline{D} \doteqdot \overrightarrow{D}; \overleftarrow{D}$ (";" denotes code sequence) is freed before execution of $\overline{C} \doteqdot \overrightarrow{C}; \overleftarrow{C}$. The peak memory consumption for $\overline{[C];D}$ is thus reduced to the maximum of the peak after $\overrightarrow{C}$ and the peak after $\overrightarrow{D}$. However, duplicate execution of C requires that "enough" variables (the "snapshot" Snp) are preserved to restore the context of execution. This also uses memory space, although less than the tape for $\overrightarrow{C}$. To not lose the benefit of checkpointing, it is therefore essential we find the smallest snapshot for a fixed C, and in further studies the placement of C that uses least memory.

This proves tricky: a larger snapshot can mean smaller tapes, and conversely. Therefore, unlike what happens with no checkpoints, there is no unique best choice for these sets. There are several "optimal" choices, among which none is better nor worse than the others. Our goal is to establish the constraints that define and link the "snapshot" and "tape" sets, and to characterize all the optimal choices. For our AD tool TAPENADE, we settled on one solution (*cf* Sect. 4) that our benchmarks indicated as a mean best choice.

Four Unknown Sets of Variables. Let's examine checkpointing definition (1) in more detail. The rewrite context *Req* is the incoming "required set" of variables imposed by U, that must be preserved across execution of $Req \vdash \overline{[C]; D}$. On the other hand, Req_D and Req_C are the sets of variables that $\vdash \overline{C}$ and $\vdash \overline{D}$ will be required to preserve, respectively. For us, Req_D and Req_C are unknowns, to be determined together with the snapshot. About the snapshot itself, due to the stack structure, there are two places where variables may be restored from the stack: before $\overrightarrow{C}$ and before $\overleftarrow{U}$. Therefore we introduce two snapshot sets: *Snp*, the "*usual snapshot*", contains variables to be restored just before $\overline{C}$, thus ensuring that their value is the same for both executions of C. *Sbk*, the "*backward snapshot*", contains variables to be restored just before $\overleftarrow{U}$. Thus, whatever happens to these variables during $Req \vdash \overline{[C]; D}$, their value is preserved for $\overleftarrow{U}$. Using *Sbk* instead of *Snp* and Req_C may improve memory traffic. In total, we have four "unknown" sets to choose: Req_D, Req_C, *Sbk* and *Snp*. Those sets must respect constraints parameterized upon *Req*, Req_D, Req_C, *Sbk*, *Snp*, and upon the fixed data-flow sets **use** (variables used) and **out** (variables partly written) of the code fragments C, D, $\overline{C}$, and $\overline{D}$. These constraints will guarantee that checkpointing preserves the semantics, i.e. the computed derivatives.

Two Necessary and Sufficient Conditions. Fig. 1 shows the differentiated program in the reference case with no checkpointing. This reference program is assumed correct. All we need to guarantee is that the result of the differentiated program, i.e. the derivatives, remain the same when checkpointing is done. This can be easily formulated in terms of data-flow sets. We observe that the order of the backward sweeps is not modified by checkpointing. Therefore the derivatives are preserved if and only if the original, non-differentiated variables that are used during the backward sweeps hold the same values. In other words, the snapshot and the tape must preserve the **use** set of $\overline{C}$ between time t_1 and t_3 i.e.

$$\mathbf{out}\begin{pmatrix} \mathtt{PUSH}(Sbk); \\ \mathtt{PUSH}(Snp); \\ C; \\ Req_D \vdash \overline{D}; \\ \mathtt{POP}(Snp); \end{pmatrix} \bigcap \mathbf{use}(Req_C \vdash \overline{C}) = \emptyset \tag{2}$$

and the **use** set of $\overleftarrow{U}$, which is *Req* by definition, between time t_1 and t_4 i.e.

$$\mathbf{out}\begin{pmatrix} \mathtt{PUSH}(Sbk); \\ \mathtt{PUSH}(Snp); \\ C; \\ Req_D \vdash \overline{D}; \\ \mathtt{POP}(Snp); \\ Req_C \vdash \overline{C}; \\ \mathtt{POP}(Sbk); \end{pmatrix} \bigcap Req = \emptyset \ . \tag{3}$$

The rest is purely mechanical. Classically, the **out** set of a code sequence is:

$$\mathbf{out}(A; B) = \mathbf{out}(A) \cup \mathbf{out}(B) \ ,$$

except in the special case of a PUSH/POP pair, which restore their argument:

$$\mathbf{out}(\texttt{PUSH}(v); A; \texttt{POP}(v)) = \mathbf{out}(A) \setminus \{v\} \ .$$

Also, the mechanism of reverse AD ensures that the variables in the required context are actually preserved, and this does not affect the variables used. Writing for short $\overline{A} \doteq \emptyset \vdash \overline{A}$, we have:

$$\begin{aligned} \mathbf{out}(\mathit{Req} \vdash \overline{A}) &= \mathbf{out}(\overline{A}) \setminus \mathit{Req} \\ \mathbf{use}(\mathit{Req} \vdash \overline{A}) &= \mathbf{use}(\overline{A}) \ . \end{aligned}$$

Also, a PUSH alone overwrites no variable. Therefore, equation (2) becomes:

$$(\mathbf{out}(C) \cup (\mathbf{out}(\overline{D}) \setminus \mathit{Req}_D)) \setminus \mathit{Snp} \bigcap \mathbf{use}(\overline{C}) = \emptyset \tag{4}$$

and equation (3) becomes:

$$((\mathbf{out}(C) \cup (\mathbf{out}(\overline{D}) \setminus \mathit{Req}_D)) \setminus \mathit{Snp} \cup (\mathbf{out}(\overline{C}) \setminus \mathit{Req}_C)) \setminus \mathit{Sbk} \bigcap \mathit{Req} = \emptyset \ . \tag{5}$$

From (4) and (5), we obtain equivalent conditions on Sbk, Snp, Req_D and Req_C:

$$\begin{aligned} \mathit{Sbk} &\supseteq ((\mathbf{out}(C) \cup (\mathbf{out}(\overline{D}) \setminus \mathit{Req}_D)) \setminus \mathit{Snp} \\ &\qquad\qquad \cup (\mathbf{out}(\overline{C}) \setminus \mathit{Req}_C)) \cap \mathit{Req} \\ \mathit{Snp} &\supseteq (\mathbf{out}(C) \cup (\mathbf{out}(\overline{D}) \setminus \mathit{Req}_D)) \cap (\mathbf{use}(\overline{C}) \cup (\mathit{Req} \setminus \mathit{Sbk})) \\ \mathit{Req}_D &\supseteq (\mathbf{out}(\overline{D}) \setminus \mathit{Snp}) \cap (\mathbf{use}(\overline{C}) \cup (\mathit{Req} \setminus \mathit{Sbk})) \\ \mathit{Req}_C &\supseteq (\mathbf{out}(\overline{C}) \setminus \mathit{Sbk}) \cap \mathit{Req} \ . \end{aligned}$$

Notice the cycles in these inequations. If we add a variable into Snp, we may be allowed to remove it from Req_D, and vice versa: as we said, there is no unique best solution. Let's look for the minimal solutions, i.e. the solutions to the equations we obtain by replacing the "$\supseteq$" sign by a simple "$=$".

Solving for the Unknown Sets. Manipulation of these equations is tedious and error-prone. Therefore, we have been using a symbolic computation system (e.g. Maple [8]). Basically, we have inlined the equation of, say, Snp into the other equations, and so on until we obtained fixed point equations with a single unknown X of the form

$$X = A \cup (X \cap B) \ ,$$

whose solutions are of the form "A plus some subset of B". The solutions are expressed in terms of the following sets:

$$\begin{aligned} \mathit{Snp}_0 &= \mathbf{out}(C) \cap (\mathbf{use}(\overline{C}) \cup (\mathit{Req} \setminus \mathbf{out}(\overline{C}))) \\ \mathit{Opt}_1 &= \mathit{Req} \cap \mathbf{out}(\overline{C}) \cap \mathbf{use}(\overline{C}) \\ \mathit{Opt}_2 &= \mathit{Req} \cap \mathbf{out}(\overline{C}) \setminus \mathbf{use}(\overline{C}) \\ \mathit{Opt}_3 &= \mathbf{out}(\overline{D}) \cap (\mathbf{use}(\overline{C}) \cup \mathit{Req}) \setminus \mathbf{out}(C) \ . \end{aligned} \tag{6}$$

For each partition of Opt_1 in two sets Opt_1^+ and Opt_1^-, and similarly for Opt_2 and Opt_3, the following is a minimal solution of our problem:

$$\begin{array}{ll} Sbk &= Opt_1^+ \cup Opt_2^+ \\ Snp &= Snp_0 \cup Opt_2^- \quad \cup Opt_3^+ \\ Req_D &= \qquad\qquad\qquad\quad Opt_3^- \\ Req_C &= Opt_1^- \cup Opt_2^- \ . \end{array} \tag{7}$$

Any quadruplet of sets *(Sbk, Snp, Req_D, Req_C)* that preserves the derivatives (compared to the no-checkpoint code) is equal or larger than one of these minimal solutions. Notice that $Opt_1 \subseteq Snp_0$, and Snp_0, Opt_2, and Opt_3 are disjoint.

4 Discussion and Experimental Results

The final decision for sets *Sbk*, *Snp*, Req_D, and Req_C depends on each particular context. No strategy is systematically best. We looked at two options.

We examined first the option that was implemented until recently in our AD tool TAPENADE [7]. We call it "*eager snapshots*". This option stores enough variables in the snapshots to reduce the sets Req_D and Req_C as much as possible, therefore reducing the number of subsequent PUSH/POP in $\overline{D}$ and $\overline{C}$. Equations (7) show that we can even make these sets empty, but experiments showed that making Req_D empty can cost too much memory space in some cases.

As always, the problem behind this is undecidability of array indexing: since we can't always tell whether two array indexes designate the same element or not, the "eager snapshot" strategy may end up storing an entire array whereas only one array element was actually concerned.

Therefore "eager snapshot" chooses Opt_1^- and Opt_2^- empty but

$$\begin{array}{ll} Opt_3^+ &= \mathbf{out}(\overline{D}) \cap (\mathbf{use}(\overline{C}) \setminus Req) \setminus \mathbf{out}(C) \\ Opt_3^- &= \mathbf{out}(\overline{D}) \cap Req \setminus \mathbf{out}(C) \end{array}$$

which gives:

$$\begin{array}{ll} Sbk &= Req \cap \mathbf{out}(\overline{C}) \\ Snp &= (\mathbf{out}(C) \cap (\mathbf{use}(\overline{C}) \cup Req \setminus \mathbf{out}(\overline{C}))) \cup \\ & \quad (\mathbf{out}(\overline{D}) \cap \mathbf{use}(\overline{C}) \setminus Req \setminus \mathbf{out}(C)) \\ Req_D &= \mathbf{out}(\overline{D}) \cap Req \setminus \mathbf{out}(C) \\ Req_C &= \emptyset \ . \end{array} \tag{8}$$

Notice that intersection between *Sbk* and *Snp* is nonempty, and requires a special stack mechanism to avoid duplicate storage space.

We examined another option that is to keep the snapshot as small as possible, therefore leaving most of the storage work to the TBR mechanism inside $\overline{D}$ and $\overline{C}$. We call it "*lazy snapshots*", and it is now the default strategy in TAPENADE. Underlying is the idea that the TBR mechanism is efficient on arrays because when an array element is overwritten by a statement, only this element is saved.

Therefore, "lazy snapshot" chooses all Opt_1^+, Opt_2^+, and Opt_3^+ empty, yielding:

$$\begin{aligned} Sbk &= \emptyset \\ Snp &= \mathbf{out}(C) \cap (Req \cup \mathbf{use}(\overline{C})) \\ Req_D &= \mathbf{out}(\overline{D}) \cap (Req \cup \mathbf{use}(\overline{C})) \setminus \mathbf{out}(C) \\ Req_C &= \mathbf{out}(\overline{C}) \cap Req \ . \end{aligned} \tag{9}$$

We ran TAPENADE on our validation application suite, for each of the two options. The results are shown in Table 1. We observe that lazy snapshots perform better in general. Actually, we could show the potential advantage of eager snapshots only on a hand-written example, where the checkpointed part C repeatedly overwrites elements of an array in *Req*, making TBR mechanism more expensive than a global snapshot of the array. On real applications, however, this case is rare and lazy snapshots work better.

Table 1. Comparison of the eager and lazy snapshot approaches on a number of small to large applications

Code	*Domain*	*Orig. time*	*Adj. time*	*Eager (8)*	*Lazy (9)*
OPA	oceanography	110 s	780 s	480 Mb	479 Mb
STICS	agronomy	1.8 s	35 s	229 Mb	229 Mb
UNS2D	CFD	2.7 s	23 s	248 Mb	185 Mb
SAIL	agronomy	5.6 s	17 s	1.6 Mb	1.5 Mb
THYC	thermodynamics	2.7 s	12 s	33.7 Mb	18.3 Mb
LIDAR	optics	4.3 s	10 s	14.6 Mb	14.6 Mb
CURVE	shape optim	0.7 s	2.7 s	1.44 Mb	0.59 Mb
SONIC	CFD	0.03 s	0.2 s	3.55 Mb	2.02 Mb
Contrived example		0.02 s	0.1 s	8.20 Mb	11.72 Mb

Whatever the option chosen, equations (7) naturally capture all interactions between successive snapshots. For example, if several successive snapshots all use an array `A`, and only the last snapshot overwrites `A`, it is well known that `A` must be saved only in the last snapshot. However, when an AD tool does not rely on a formalization of checkpointing such as the one we introduce here, it may very well happen that `A` is stored by all the snapshots.

5 Conclusion

We have formalized the checkpointing technique in the context of reverse AD by program transformation. Checkpointing relies on saving a number of variables and several options are available regarding which variables are saved and when. Using our formalization and with the help of a symbolic computation system, we found that no option is strictly better than all others and we could specify all the possible optimal options. This gives us safer and more reliable implementation in AD tools.

We selected two possible optimal options and implemented them in the AD tool TAPENADE. Experience shows that the option called "lazy snapshots" performs better on most cases.

However, we believe that for reverse AD of a given application code, the option chosen need not be identical for all checkpoints. This formal description of all the possible options allows us to look for the best option for each individual checkpoint, based on static properties at this particular code location. In this regard, we used symbolic computation again and came up with a very pleasant property: for a given checkpoint, whatever the optimal option chosen for the snapshot, the **out** set of this piece of code turns out to be always the same:

$$\mathbf{out}(\overline{C;D}) = \left(\mathbf{out}(\overline{C}) \cup ((\mathbf{out}(\overline{D}) \cup \mathbf{out}(C)) \setminus \mathbf{use}(\overline{C}))\right) \setminus \mathit{Req} \ .$$

If checkpoints are nested, this **out** set is what influences possible enclosing checkpoints. Therefore the choice of the optimal option is local to each checkpoint.

One of the current big challenges of reverse AD is to find the best possible placement of nested checkpoints. This was found [3] for one simple model case. For arbitrary programs, our formulas show that the segmentation of a code into the subsections U, C, and D has substantial impact on the memory usage, and they can help finding good such segmentations.

References

1. A. Aho, R. Sethi, and J. Ullman. *Compilers: Principles, Techniques and Tools.* Addison-Wesley, 1986.
2. C. Faure and U. Naumann. Minimizing the tape size. In G. Corliss, C. Faure, A. Griewank, L. Hascoët, and U. Naumann, editors, *Automatic Differentiation of Algorithms: From Simulation to Optimization, Computer and Information Science*, chapter 34, pages 293–298. Springer, New York, NY, 2001.
3. Andreas Griewank. Achieving logarithmic growth of temporal and spatial complexity in reverse automatic differentiation. *Optimization Methods and Software*, 1:35–54, 1992.
4. Andreas Griewank. *Evaluating Derivatives: Principles and Techniques of Algorithmic Differentiation.* Number 19 in Frontiers in Appl. Math. SIAM, Philadelphia, PA, 2000.
5. L. Hascoët and M. Araya-Polo. The adjoint data-flow analyses: Formalization, properties, and applications. In H. M. Bücker, G. Corliss, P. Hovland, U. Naumann, and B. Norris, editors, *Automatic Differentiation: Applications, Theory, and Tools*, Lecture Notes in Computational Science and Engineering. Springer, 2005.
6. L. Hascoët, U. Naumann, and V. Pascual. "to be recorded" analysis in reverse-mode automatic differentiation. *Future Generation Computer Systems*, 21(8), 2004.
7. L. Hascoët and V Pascual. Tapenade 2.1 user's guide. Technical report 0300, INRIA, 2004. http://www.inria.fr/rrrt/rt-0300.html.
8. Darren Redfern. *The Maple handbook, Maple V, release 4.* Springer, 1996.

Linearity Analysis for Automatic Differentiation*

Michelle Mills Strout[1] and Paul Hovland[2]

[1] Colorado State University, Fort Collins, CO 80523
[2] Argonne National Laboratory, Argonne, IL 60439

Abstract. Linearity analysis determines which variables depend on which other variables and whether the dependence is linear or nonlinear. One of the many applications of this analysis is determining whether a loop involves only linear loop-carried dependences and therefore the adjoint of the loop may be reversed and fused with the computation of the original function. This paper specifies the data-flow equations that compute linearity analysis. In addition, the paper describes using linearity analysis with array dependence analysis to determine whether a loop-carried dependence is linear or nonlinear.

1 Introduction

Many automatic differentiation and optimization algorithms can benefit from linearity analysis. Linearity analysis determines whether the dependence between two variables is nonexistent, linear, or nonlinear. A variable is said to be linearly dependent on another if all of the dependences along all of the dependence chains are induced by linear or affine functions (addition, subtraction, or multiplication by a constant). A variable is nonlinearly dependent on another if a nonlinear operator (multiplication, division, transcendental functions, etc.) induces any of the dependences along any of the dependence chains.

One application of linearity analysis is the optimization of derivative code generated by automatic differentiation (AD) via the reverse mode. AD is a technique for transforming a subprogram that computes some function into one that computes the function and its derivatives. AD works by combining rules for differentiating the intrinsic functions and elementary operators of a given programming language with the chain rule of differential calculus. One strategy, referred to as the forward mode, is to compute partials as the intrinsic functions are evaluated and to combine the partials as they are computed. For example, forward mode AD transforms the loop in Figure 1 into the code in Figure 2. Reverse mode AD results in less computation in the derivative code if the number of independent variables is much larger than the number of dependent variables. Figures 3 shows the adjoint code after applying the reverse mode. Notice that the temporary variable `a` must be promoted to an array to store results needed in the adjoint computation.

* This work was supported by the Mathematical, Information, and Computational Sciences Division subprogram of the Office of Advanced Scientific Computing Research, U.S. Department of Energy under Contract W-31-109-Eng-38.

V.N. Alexandrov et al. (Eds.): ICCS 2006, Part IV, LNCS 3994, pp. 574–581, 2006.
© Springer-Verlag Berlin Heidelberg 2006

```
a = 0.0
f = 0.0
for i = 1, N
  a += x[i]*x[i]
  t = sin(a)
  f += t
end
```

Fig. 1. Example loop

```
d_a = 0.0
a = 0.0
d_f = 0.0
f = 0.0
for i = 1, N
  d_a += 2*x[i]*d_x[i]
  a += x[i]*x[i]
  d_t = cos(a)*d_a
  t = sin(a)
  d_f += d_t
  f += t
end
```

Fig. 2. Example loop after forward mode AD

```
a[0] = 0.0
f = 0.0
for i = 1, N
  a[i] = a[i-1] + x[i]*x[i]
  t = sin(a[i])
  f += t
end
for i = N, 1, -1
  a_t = a_f
  a_a[i] = cos(a[i])*a_t
  a_a[i-1] = a_a[i]
  a_x[i] = 2*x[i]*a_a[i]
end
```

Fig. 3. Example loop after reverse mode automatic differentiation

```
a = 0.0
f = 0.0
for i = 1, N
  a += x[i]*x[i]
  t = sin(a)
  f += t
  a_t = a_f
  a_a = cos(a)*a_t
  a_x[i] = 2*x[i]*a_a
end
```

Fig. 4. Adjoint code after reversing the adjoint loop and fusing it with the original computation

Hascoet et al. [6] observed that the forward computation and adjoint accumulation can be fused if the original loop is parallelizable. In fact, a weaker condition suffices: the two computations can be fused whenever there are no loop-carried, nonlinear dependencies and any variables involved in linear loop-carried dependencies are scalars The example in Figure 1 includes two loop-carried dependence, and both dependences are linear; therefore, the adjoint loop may be reversed and fused with the original loop as shown in Figure 4.

Such transformations can result in significant performance improvements and storage savings, by eliminating the need to store or recompute overwritten intermediate quantities such as variable a. Data dependence analysis [2, 4, 18, 14] is used to determine whether a loop is parallelizable. Precise dependence analysis techniques can determine which variables are involved in a loop-carried dependence. In the example of Figure 1, such techniques can determine that there are loop-carried dependences involving the variable f and itself and a and itself. In

Variable	f	t	a	x
f	linear	linear	nonlinear	nonlinear
t	$\top$	$\top$	nonlinear	nonlinear
a	$\top$	$\top$	linear	nonlinear
x	$\top$	$\top$	$\top$	$\top$

Fig. 5. The variables in the first column depend on the variables in the first row in the specified way. $\top$ indicates no dependence

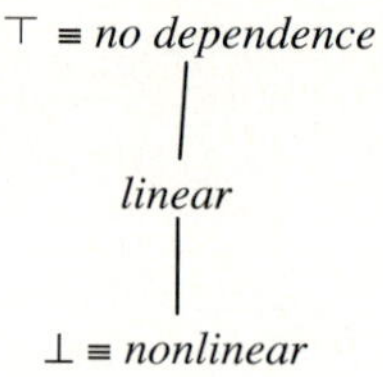

Fig. 6. Lattice for linearity analysis

this paper, we present linearity analysis as a technique for determining if the loop-carried dependence is linear or nonlinear.

The result of linearity analysis is the assignment of a dependence class to each pair of variables in the program. For the example in Figure 1, the analysis summarizes the dependences between variables as shown in Figure 5. Conservatively determining whether a loop-carried dependence is linear requires checking only whether the dependence between the variables involved in the loop-carried dependence is linear. Figure 5 indicates that `f` depends on itself linearly and the same applies to variable `a`; therefore, both of the loop-carried dependencies are linear.

2 Formulation of Linearity Analysis as a Data-Flow Analysis

Linearity analysis can be formulated as a forward data-flow analysis [10]. Data-flow analysis involves representing the subroutine to be analyzed as a control flow graph. A control flow graph contains directed edges between basic blocks indicating possible control flow in the program. Each basic block b has a set of predecessors $pred(b)$ and a set of successors $succ(b)$, and the graph contains unique entry and exit nodes.

Data-flow analysis propagates data-flow information over the control-flow graph. For linearity analysis, the data-flow information is which variables are dependent on which other variables and whether that dependence is linear or nonlinear, which we refer to as the dependence class. The analysis assigns each ordered pair of variables a value from the lattice shown in Figure 6. For example, in the statement

```
x = 3*z + y**2 + w/(v*v),
```

`x` has an linear dependence on `z`, $\langle\langle x, z\rangle, linear\rangle$, a nonlinear dependence on `y`, $\langle\langle x, y\rangle, nonlinear\rangle$, a nonlinear dependence on `w`, $\langle\langle x, w\rangle, nonlinear\rangle$, and a nonlinear dependence on `v`, $\langle\langle x, v\rangle, nonlinear\rangle$.

The set $IN(b)$ includes the dependence class assignment for each ordered variable pair that is valid at the entry of basic block b in the control-flow graph. A

Expression e	$DEPS(e)$
k k anyop k	$\{\langle v, \top\rangle \mid v \in V\}$
v	$\{\langle v, linear\rangle\} \cup \{\langle w, class\rangle \mid \langle\langle v, w\rangle, class\rangle \in IN(b)\}$
$e_1 \pm e_2$	$\{\langle v_1, (class_1 \sqcap class_2)\rangle$ $\mid v_1 = v_2$ and $\langle v_1, class_1\rangle \in DEPS(e_1)$ and $\langle v_2, class_2\rangle \in DEPS(e_2)\}$ $\cup \{\langle v, class\rangle \mid \langle v, class\rangle \in DEPS(e_1)$ and $v \notin DEPS(e_2)\}$ $\cup \{\langle v, class\rangle \mid \langle v, class\rangle \in DEPS(e_2)$ and $v \notin DEPS(e_1)\}$
$e_1 * e_2$ e_1 / e_2	$\{\langle v_1, (nonlinear \sqcap class_1 \sqcap class_2)\rangle$ $\mid v_1 = v_2$ and $\langle v_1, class_1\rangle \in DEPS(e_1)$ and $\langle v_2, class_2\rangle \in DEPS(e_2)\}$ $\cup \{\langle v, nonlinear)\rangle \mid \langle v, class\rangle \in DEPS(e_1)$ and $v \notin DEPS(e_2)\}$ $\cup \{\langle v, nonlinear)\rangle \mid \langle v, class\rangle \in DEPS(e_2)$ and $v \notin DEPS(e_1)\}$
e_1 power 1 e_1 power k	$\{\langle v, (linear \sqcap class)\rangle \mid \langle v, class\rangle \in DEPS(e_1)$ $\{\langle v, (nonlinear \sqcap class_1)\rangle \mid \langle v, class\rangle \in DEPS(e_1)\}$

Fig. 7. Definition of the *DEPS* set for each expression

transfer function $f_b(IN(b))$ calculates the $OUT(b)$ set, which includes the dependence class assignments valid upon exiting the basic block b. The dependence class for each variable pair $\langle u, v\rangle$ is initialized to $\top$, indicating no dependence, in $IN(b)$ and $OUT(b)$ for all basic blocks in the control-flow graph.

Iterative data-flow analysis visits each node in the control-flow graph computing the $IN(b)$ and $OUT(b)$ sets until the assignment of data dependence class to each ordered variable pair converges. The set $IN(b)$ is calculated by performing the pairwise meet over all the sets of data-flow facts valid upon exiting predecessor basic blocks,

$$IN(b) = \sqcap_{p \in preds(b)} OUT(p).$$

The meet operation $\sqcap$ is performed on the lattice values assigned to each variable pair. The semantics of the meet operation $\sqcap$ is defined by the lattice. For example, $linear \sqcap \top$ equals $linear$, and $linear \sqcap nonlinear$ equals $nonlinear$.

The set $OUT(b)$ is computed by applying what is referred to as a transfer function to the $IN(b)$ set. The transfer function f_b is first defined for each type of statement assuming only one statement per basic block. If there are multiple statements in block b, then f_b is the composition of all the transfer functions for the statements. To define the transfer function f_b for linearity analysis, we define the set $DEPS(e)$ as a set containing a mapping of each variable to a data dependence class. When variable v maps to dependence class $class$, $\langle v, class\rangle$, that indicates how an expression e depends upon that variable. The transfer function f_b for the assignment statement $x = e$ is then defined as

$$OUT(b) = f_b(IN(b)) = \{\langle\langle x, v\rangle, class\rangle \mid \langle v, class\rangle \in DEPS(e)\}.$$

The $DEPS(e)$ set is defined in Figure 7, where k represents a constant value, v and w represent variables in the set of all variables V, `anyop` represents any

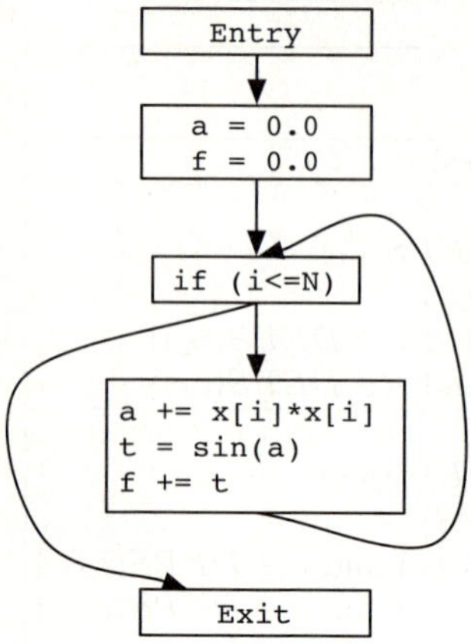

Fig. 8. Control flow graph for code in Figure 1

Statement s	$IN(s)$
`a+=x[i]*x[i]`	$\{\langle\langle v,w\rangle,\top\rangle \mid v,w \in V\}$
`t=sin(a)`	$\{\langle\langle a,a\rangle, linear\rangle, \langle\langle a,x\rangle, nonlinear\rangle\}$ $\sqcap \{\langle\langle v,w\rangle,\top\rangle \mid v,w \in V\}$
`f+=t`	$\{\langle\langle a,a\rangle, linear\rangle, \langle\langle a,x\rangle, nonlinear\rangle,$ $\langle\langle t,a\rangle, nonlinear\rangle, \langle\langle t,x\rangle, nonlinear\rangle\}$ $\sqcap \{\langle\langle v,w\rangle,\top\rangle \mid v,w \in V\}$

Expression e	$Deps(e)$
`0.0`	$\{\langle a,\top\rangle, \langle f,\top\rangle, \langle i,\top\rangle, \langle t,\top\rangle, \langle f,\top\rangle\}$
`a+x[i]*x[i]`	$\{\langle a, linear\rangle, \langle x, nonlinear\rangle\}$
`sin(a)`	$\{\langle a, nonlinear\rangle, \langle x, nonlinear\rangle\}$
`f+t`	$\{\langle f, linear\rangle, \langle t, linear\rangle,$ $\langle a, nonlinear\rangle, \langle x, nonlinear\rangle\}$

Fig. 9. Applying the linearity analysis to certain statements in the control flow graph in Figure 8

operation, and `power` represents the power operation. Notice that if a variable occurs in both subexpressions of a binary operator that $DEPS$ is computed differently than when a variable only occurs in one of the subexpressions. Figure 8 shows the control flow graph for the example program from Figure 1. Figure 9 shows the IN and $DEPS$ sets for some of the statements in the example program. Note that the IN set for a statement in a basic block is equivalent to the OUT set for the previous statement.

The worst-case complexity of linearity analysis is $O(N^4(E+V))$, where N is the number of variables, E is the number of edges in the control-flow graph, and V is the number of nodes in the control flow graph. Each pair of variables has a lattice value associated with it, and there are N^2 pairs. Each lattice value may be lowered at most twice; therefore, the graph may be visited $2 * N^2$ times. The size of the graph is $E+V$. When each node is visited, the computation may apply meet and transfer operations to each variable pair, $O(N^2)$.

2.1 Detecting Nonlinear Loop-Carried Dependences

Data dependence analysis provides information about which variable references are involved in loop-carried dependences. If a particular variable is involved in a loop-carried dependence, and the variable depends on itself nonlinearly based on the results of linearity analysis, then the loop may involve a nonlinear loop-carried dependence.

2.2 Limitations

As formulated, linearity analysis is incapable of determining that the loop shown in Figure 10 has no loop-carried, nonlinear dependences. Specifically, there is a loop-carried dependence between b and c due to `c[i] = b[i-1]`, and there is a nonlinear dependence between `b` and `c` due to `c[i] = b[i]*x[i]`. However, the nonlinear dependence is not loop carried.

One can determine whether there are nonlinear, loop-carried dependencies if the data-flow analysis is done on a use-by-use basis. This data-flow problem could be significantly more expensive than basic linearity analysis. In order to achieve higher precision at a reasonable cost, while re-using as much analysis as possible, a closer coupling between linearity analysis and data-dependence analysis may be required.

```
for i = 1 to N
  b[i] = 3*x[i]
  c[i] = b[i-1] + b[i]*x[i]
end
```

Fig. 10. Example where the current formulation of linearity analysis combined with data dependence analysis is overly conservative

3 Other Applications

Linearity analysis is also useful for a sort of "predictive slicing." In a so-called "pure" derivative computation [8], one wants to compute only the derivatives of a function and not the function itself. However, by default AD produces code that computes both the function and its derivatives, primarily because many of the intermediate function values are required to compute derivatives. However, when it can be determined that the dependent variables depend only linearly on an intermediate function value, then that intermediate value is not needed in the derivative computation. Therefore, the generated derivative code may omit the computation of these intermediates. This is equivalent to generating the derivative code, then performing a backward slice [17] from the derivative variables. Figure 11 illustrates the use of predictive slicing on the example of Figure 4. The dependent variable `f` depends nonlinearly only on `a` and `x`; therefore, `t` and `f` do not need to be computed.

```
a = 0.0
for i = 1, N
  a += x[i]*x[i]
  a_t = a_f
  a_a = cos(a)*a_t
  a_x[i] = 2*x[i]*a_a
end
```

Fig. 11. Reverse mode AD, computing only derivatives via predictive slicing

Linearity analysis can be combined with array data flow analysis to identify functions $f(x) : R^n \mapsto R$ that can be decomposed into the form: $f(x) = \sum_{i=1}^{m} F_i(x)$ where each F_i is a function of only a few elements of the vector x. This is the simplest form of partially separable function [13]. The Jacobian of $F(x) : R^n \mapsto R^m$ is sparse and this sparsity can be exploited by using compression techniques [1]. The gradient of f is the sum of the rows of this Jacobian. Thus, gradients of such functions can be computed efficiently using the forward mode.

Linearity analysis is also directly useful in numerical optimization. Optimization algorithms distinguish between linear and nonlinear constraints in order to reduce the cost of derivative evaluations (the derivatives of linear constraints are constant), to reduce the problem size via preprocessing, and to improve the performance of the optimization algorithm. Experimental results from Gould and Toint [5] indicate that preprocessing of the linear and bound constraints reduces the number of constraints by 19% and the total time to solution by 11% on average. Combined with the added savings from fewer constraint evaluations, derivative evaluations, and faster convergence, the savings can be substantial. Preliminary experiments indicate that when all constraints can be identified as linear, savings of 50% or more are possible.

4 Related Work

Karr [9] and Cousot [3] determine linear equalities and linear inequalities between variables. The focus for such techniques is to find program invariants for use with automated reasoning tools. More recent research [12, 15] discovers a subset of nonlinear relationships, polynomial relationships of bounded degree. None of these techniques distinguishes between a nonlinear dependence and a lack of dependence. Therefore, they are not suitable for the types of program optimization we have described. To-be-recorded (TBR) analysis [7] identifies the set of variables that are needed for derivative computation and thus must be recorded if overwritten. This analysis is similar to linearity analysis, but includes index variables, excludes variables that are never overwritten, and does not identify pairwise dependence. Linearity analysis can be readily extended to polynomial degree analysis. We have also extended polynomial degree analysis to a restricted form of rationality analysis. Polynomial degree analysis and rationality analysis have applications in code validation [11].

5 Conclusions and Future Work

We have presented a formal data-flow formulation for linearity analysis. Linearity analysis has several applications in automatic differentiation and numerical optimization. In addition to the applications already discussed, linearity and polynomial degree analysis have applications in code derivative-free optimization, nonlinear partial differential equations, and uncertainty quantification. We are implementing linearity and polynomial degree analysis in the OpenAnalysis framework [16] to provide compiler infrastructure-independent analysis. We are investigating ways to tightly couple linearity analysis with dependence analysis to address the limitations discussed in Section 2.2.

Acknowledgments

We would like to thank Todd Munson, Rob Kirby, and Ridgeway Scott for their suggestions, and the anonymous reviewers and Gail Pieper for their feedback.

References

1. B. M. Averick, J. J. Moré, C. H. Bischof, A. Carle, and A. Griewank. Computing large sparse Jacobian matrices using automatic differentiation. *SIAM J. Sci. Comput.*, 15(2):285–294, 1994.
2. U. Banerjee. *Dependence analysis for supercomputing.* The Kluwer international series in engineering and computer science. Parallel processing and fifth generation computing. Kluwer Academic, Boston, MA, USA, 1988.
3. P. Cousot and N. Halbwachs. Automatic discovery of linear restraints among variables of a program. In *POPL '78: Proceedings of the 5th ACM SIGACT-SIGPLAN symposium on Principles of programming languages*, pages 84–96, New York, NY, USA, 1978. ACM Press.
4. P. Feautrier. Dataflow analysis of array and scalar references. *International Journal of Parallel Programming*, 20(1), February 1991.
5. N. Gould and P. L. Toint. Preprocessing for quadratic programming. *Math. Programming*, 100(1):95–132, 2004.
6. L. Hascoët, S. Fidanova, and C. Held. Adjoining independent computations. In G. Corliss, C. Faure, A. Griewank, L. Hascoët, and U. Naumann, editors, *Automatic Differentiation of Algorithms: From Simulation to Optimization*, Computer and Information Science, chapter 35, pages 299–304. Springer, New York, NY, 2001.
7. L. Hascoët, U. Naumann, and V. Pascual. "To be recorded" analysis in reverse-mode automatic differentiation. *Future Generation Computer Systems*, 21(8), 2005.
8. T. Kaminski, R. Giering, and M. Voßbeck. Efficient sensitivities for the spin-up phase. In H. M. Bücker, G. Corliss, P. Hovland, U. Naumann, and B. Norris, editors, *Automatic Differentiation: Applications, Theory, and Tools*, Lecture Notes in Computational Science and Engineering. Springer, 2005.
9. M. Karr. Affine relationships among variables of a program. *Acta Informatica*, 6(2):133–151, 1976.
10. G. A. Kildall. A unified approach to global program optimization. In *ACM Symposium on Principles of Programming Languages*, pages 194–206, October 1973.
11. R. Kirby and R. Scott. Personal communication, 2004.
12. M. Müller-Olm and H. Seidl. Precise interprocedural analysis through linear algebra. In *POPL '04: Proceedings of the 31st ACM SIGPLAN-SIGACT symposium on Principles of programming languages*, pages 330–341, New York, NY, USA, 2004. ACM Press.
13. J. Nocedal and S. J. Wright. *Numerical Optimization.* Springer-Verlag, New York, 1999.
14. W. Pugh. Omega test: A practical algorithm for exact array dependency analysis. *Comm. of the ACM*, 35(8):102, 1992.
15. E. Rodriguez-Carbonell and D. Kapur. Automatic generation of polynomial loop invariants: Algebraic foundations. In *ISSAC '04: Proceedings of the 2004 international symposium on Symbolic and algebraic computation*, pages 266–273, New York, NY, USA, 2004. ACM Press.
16. M. M. Strout, J. Mellor-Crummey, and P. Hovland. Representation-independent program analysis. In *Proceedings of the Sixth ACM SIGPLAN-SIGSOFT Workshop on Program Analysis for Software Tools and Engineering*, 2005.
17. M. Weiser. Program slicing. *IEEE Trans. Software Eng.*, 10(4):352–357, 1984.
18. M. Wolfe and C. W. Tseng. The power test for data dependence. *IEEE Trans. Parallel Distrib. Syst.*, 3(5):591–601, 1992.

Hybrid Static/Dynamic Activity Analysis*

Barbara Kreaseck[1], Luis Ramos[1], Scott Easterday[1],
Michelle Strout[2], and Paul Hovland[3]

[1] La Sierra University, Riverside, CA
[2] Colorado State University, Fort Collins
[3] Argonne National Laboratory

Abstract. In forward mode Automatic Differentiation, the derivative program computes a function f and its derivatives, f'. Activity analysis is important for AD. Our results show that when all variables are active, the runtime checks required for dynamic activity analysis incur a significant overhead. However, when as few as half of the input variables are inactive, dynamic activity analysis enables an average speedup of 28% on a set of benchmark problems. We investigate static activity analysis combined with dynamic activity analysis as a technique for reducing the overhead of dynamic activity analysis.

1 Introduction

In forward mode Automatic Differentiation (AD), the derivative program computes a function f and its derivatives, f'. Activity analysis [5, 8, 12, 10, 7] determines which temporary variables lie along the dependence chains between inputs and outputs of the function f. When only a subset of the inputs and outputs are being studied, activity analysis can be used to identify an associated subset of local variables that are defined and used along the dependence chains from the independents (inputs of concern) to the final calculation of the dependents (outputs of concern).

Activity analysis has the potential to significantly reduce the number of calculations needed to produce the dependents from the independents. Unfortunately, static activity analysis (done at compile time) may in the presence of control flow be too conservative. On the other hand, dynamic activity analysis (done at runtime) may introduce a significant amount of overhead.

In this paper, we quantify the overhead of performing dynamic activity analysis on a number of benchmarks. Our results show that when all variables are active, the runtime checks required for dynamic activity analysis incur a significant overhead. However, when as few as half of the input variables are inactive, dynamic activity analysis enables an average speedup of 28% on a set of benchmark problems.

* This work was supported by the Mathematical, Information, and Computational Sciences Division subprogram of the Office of Computational Technology Research, U.S. Department of Energy under Contract W-31-109-Eng-38 and by the National Science Foundation under Grant No. OCE-020559.

V.N. Alexandrov et al. (Eds.): ICCS 2006, Part IV, LNCS 3994, pp. 582–590, 2006.
© Springer-Verlag Berlin Heidelberg 2006

```
void f(double x, double &y,
       double z)
{
   double a, c;

   a = z * z;
   c = x * 9;

   y = a * c;
}
```

Fig. 1. Function

```
void fprime(double x, double dx,
            double &y, double &dy,
            double z, double dz)
{ /* dx = 1, dz = 0 */
   double a, c, da, dc;

   a = z * z;
   da = dz*z + dz*z;
   c = x * 9;
   dc = 9 * dx;
   y = a * c;
   dy = da*c + dc*a;
} /* dy = ∂y/∂x */
```

Fig. 2. Derivatives

In Section 2 we provide the motivation for our studies. In Section 3 we present currently available activity analysis and our extensions. Next, we present our study of the overhead of dynamic activity analysis in Section 4. In Section 5 we present our hybrid static/dynamic analysis. Finally, we discuss future work and conclude in Section 6.

2 Motivation

We demonstrate the importance of activity analysis to AD with the following examples. In Figure 1, we show an example function `f` with an input variables `x` and `z` and an output variable `y`. AD would generate the derivative code shown in Figure 2 to calculate the derivative of `y` with respect to `x` (where we represent $\partial y/\partial x$ as just the variable `dy`). Activity analysis is applied to the original program and determines which temporary variables lie along the dependence chain between independent variables (a subset of the inputs) and dependent variables (a subset of the outputs). In the example, local variable `c` is active while local variable `a` is not. Variable `a` is inactive because it does not depend upon the value of `x`. Variable `c` is active because it depends upon the value of `x` and is used to compute the value of `y`. Activity information enables an AD tool to avoid generating the code that has been crossed out in Figure 2. In real applications, one typically uses the vector mode of AD[1] and the variables `da`, `dc`, and `dy` are arrays. Furthermore, the update `dy = a*dc + c*da;` becomes

```
for(i=0;i<nindeps;i++)
   dy[i] = a*dc[i] + c*da[i];
```

Thus, activity analysis offers the opportunity for substantial savings, especially when the number of independent variables is small.

[1] For simplicity, we restrict our discussion to the forward mode of AD. In the reverse mode, activity analysis offers substantial savings opportunities through reduced storage requirements.

```
bool flag;
double g, z;
void f2(double x, double &y)
{
    double a,b,c;
    if (flag) {
        a = g * z;
    } else {
        a = x * x;
    }
    c = x * 9;
    y = a * c;
}
```

Fig. 3. Example function, `f2`, where the control-path is not known at compile time. When `flag` is true, local variable `a` will be inactive. When `flag` is false, local variable `a` will be active.

Activity analysis can be performed within a data-flow analysis framework. Unfortunately, due to control-path uncertainty, not all variables can be statically identified as active or inactive. Consider the function `f2` in Figure 3. The control-path through `f2` is not decidable at compile time. Now, `a` will be active if `flag` is false and it will be inactive if `flag` is true. Statically we can characterize `a` as active to be conservative but this would result in more work than necessary. The amount of unnecessary work depends upon the number of independent variables that were selected when the derivative code was produced. Specifically, for `f2`, that would just be one (just `x`). For derivative code in general, that will probably not be the case.

We address the problem of activity analysis in the presence of control flow by characterizing `a` as *may active* and augmenting the derivative code to check the activity of `a` dynamically during run-time. This technique is called *dynamic analysis* or run-time analysis. Specifically we associate a boolean flag with each gradient vector (e.g., `da`) to indicate whether it is active or not.

A naive approach is to just use dynamic activity analysis on all variables. This involves the overhead of checking the active flag before every derivative computation. In the next section, we discuss current activity analysis implementations, along with our extensions. In Section 4 we will see that the overhead of dynamic activity analysis can be quite high. Thus, we describe a hybrid static/dynamic approach to activity analysis in Section 5.

3 Dynamic Activity Analysis and AD

Our work with activity analysis is based upon two AD tools: ADIC [6, 3] for C codes, and OpenAD [13] for Fortran codes. The following subsections discuss activity analysis with each tool.

Dynamic Activity Analysis in ADIC. ADIC 1.2 does not perform static activity analysis. All floating-point variables are treated as active unless they

Table 1. Characteristics of the Fortran benchmarks and dynamic overhead

Benchmark	Independent	Dependent	Size	Source	Overhead
bminsurf	x	f, fgrad	n = 400	NEOS	1.32
daerfj	x	fvec, fjac	n = 4	Minpack2	1.30
datrfj	x	fvec, fjac	n = 3	Minpack2	1.74
dchqfj	x	fvec, fjac	n = 11	Minpack2	1.68
dedffj	x	fvec, fjac	n = 5	Minpack2	1.51
dodcfg	x,lambda	f, fgrad	n = 20x20	Minpack2	1.24
dsfdfj	x,eps	fvec, fjac	n = 280	Minpack2	1.01
dsscfg	x,lambda	f, fgrad	n = 20x20	Minpack2	1.31

are specifically designated as inactive by the user. The generated derivative code will include calls to `axpy` routines, which implement the process of combining gradient vectors and local partial derivatives according to the chain rule of calculus. The `axpy` routines are implemented as C preprocessor macros. ADIC 1.2 provides a set of macros that implements dynamic activity checking. These macros augment the gradient vectors with an activity flag and check (and set) the activity flags in the `axpy` routines.

At runtime, initially the activity flags of all independent variables are set to true and of all other inputs are set to false. The gradient accumulation macros check the activity flag of each gradient vector prior to execution to affect the following:

- When a right-hand-side gradient vector is active, its values contribute to the calculation of the left-hand-side gradient vector and the activity flag of the left-hand-side gradient vector is set to active.
- When a right-hand-side gradient vector is inactive, its values do not contribute to the calculation of the left-hand-side gradient vector.
- When all right-hand-side gradient vectors are inactive, the activity flag of the left-hand-side gradient vector is set to inactive.

Static Activity Analysis in OpenAD. OpenAD does not currently support dynamic activity analysis. Instead, it uses a static *may activity analysis*. Given user-identified independent and dependent variables, the may activity analysis conservatively identifies local variables that may be active. The generated derivative code will include calls to `sax` subroutines, whose function is similar to the `axpy` routines in ADIC. These `sax` routines will only be called using gradient vectors of variables identified as may active. In Section 5, we define may activity analysis more fully.

4 Overhead of Dynamic Activity Analysis

While dynamic activity analysis can reduce the number of gradient vector operations within derivative code, it does introduce extra activity flag checking as overhead. In this section, we quantify the impact of the overhead of dynamic activity analysis.

Table 2. Characteristics of the C benchmarks

Abr	Benchmark	Independent	Dependent	Size	Source
C1	Ackley	x	f,g	n = 20	see [1, 2]
C2	Boxbetts	x	ret	n = 3	GlobOpt
C3	CamShape	par, r	obj	n = 144	ADIC
C4	GenRosenBrock	x	ret	n = 30	GlobOpt
C5	McCormic	x	ret	n = 2	GlobOpt
C6	Paviani	x	ret	n = 10	GlobOpt
C7	Plate2D	x	f, g	mx = 12	TAO
C8	Polygon	x	obj	n = 73	ADIC

4.1 Methodology

We investigated the overhead of dynamic activity analysis on two benchmark testbeds. For C codes, we generated the derivative code using ADIC 1.2, which provides two sets of `axpy` and related routines. The original set is a set of macros which do not implement the activity flag and thus provide no activity checking. The second set is a set of hand-coded macros that implement the activity flag and perform dynamic activity checking. By normalizing the pure dynamic activity analysis results to the those with no activity check, we quantify the overhead of dynamic activity analysis.

For Fortran codes, we generated the derivative code using OpenAD, which had no prior support for dynamic activity analysis. We created a tool to auto-generate `sax` subroutines that implement dynamic activity checking. Thus we created two execution units per benchmark: "No Activity Check" uses the OpenAD default routines that do not perform dynamic activity checking, while "Pure Dynamic" uses our auto-generated routines that do. Again, we normalize our Pure Dynamic results to the No Activity Check results to quantify the overhead of dynamic activity analysis.

4.2 Results

Table 1 summarizes the Fortran benchmarks used in our experiments. All are from the Minpack2 benchmark suite [11] except the bminsurf, an example problem from the TAO Toolkit [4]. The column labeled "Overhead" shows the average of four Dynamic execution times normalized against the Static execution time. Our runs represent the maximum possible overhead in that we set all inputs as independent, and all outputs as dependent prior to derivative code generation. The benchmarks display a broad range of overhead averaging 39%.

Table 2 summarizes the C benchmarks used in our experiments. Most of the problems were derived from a c++ test suite for global optimization [9]; the others are part of the ADIC test suite or TAO examples. Figure 4 displays the overhead of dynamic activity analysis for each of the C benchmarks. The arithmetic mean per analysis run is noted in parenthesis within the legend. When 100% of input variables are treated as independent variables (and are therefore

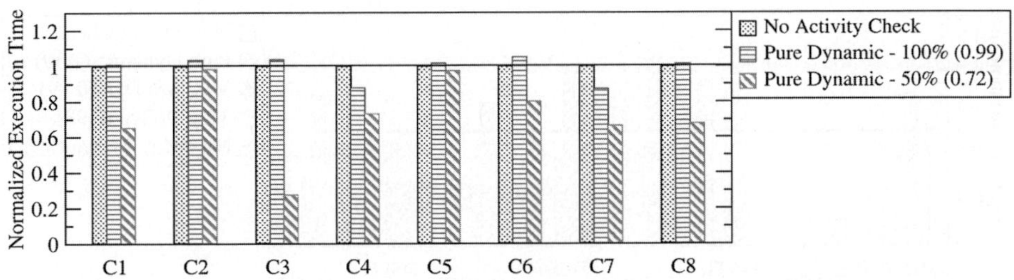

Fig. 4. Overhead of Dynamic Activity Analysis using C benchmarks and speedup when only 50% of input variables are independent. Arithmetic means are indicated within parentheses. See Table 2 for benchmark descriptions.

active), all variables are active and the cost of dynamic activity checking is pure overhead. The overhead here is significantly lower than that found with the Fortran benchmarks and can be attributed to differences in benchmarks as well as dynamic activity analysis implementation. However, in the more realistic situation where only 50% of the inputs are active, dynamic analysis pays dividends, reducing the execution time from "No Activity Check" by about 28% on average and up to 70% in the case of camshape.

5 Static/Dynamic Analysis

As we saw in Section 4, dynamic activity analysis provides full accuracy at the price of noticeable overhead. OpenAD uses a static may analysis which may incur less overhead by statically determining which local variables are provably inactive. In the generated derivative code this overestimate of the set of active variables ensures correctness but the code may be sub-optimal. We propose a hybrid static/dynamic activity analysis that uses a static forward-direction must analysis.

5.1 Must-May Static Activity Analysis

Static activity analysis is based upon the following definitions. A variable, v, is *may-vary* when there is at least one control path to a define of v where the value of v depends directly or transitively upon an independent variable. A variable, v, is *must-vary* at a point in the function when *all* control-flow paths to that point cause the value of v to depend directly or transitively upon the value of an independent variable. A variable, v, is *may-useful* when there is at least one control path from the define of v to the define of a dependent variable where the value of a dependent variable depends directly or transitively upon v.

In may activity analysis, a variable, v, is classified *may-active* when there is at least one point in the function where v is both may-vary and may-useful. OpenAD uses the OpenAnalysis [14] toolkit to implement its data-flow analysis. OpenAnalysis provides a may activity analysis. Partial derivatives for non-may-active variables never have to be calculated.

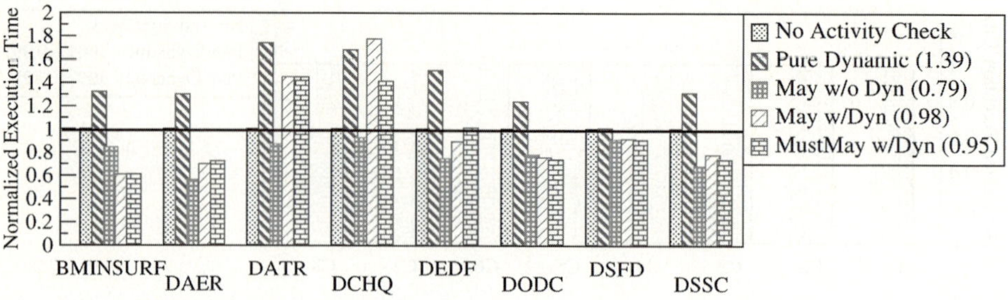

Fig. 5. Fortran results for a variety of static and/or dynamic activity analyses. *May* and *Must-May* are static analyses and *Dyn* indicates dynamic activity analysis. Arithmetic means are indicated within parentheses.

In *must-may activity analysis*, a variable, v, is *must-may-active* at a point in the function when v is both must-vary and may-useful. Should the actual execution path arrive at this point, v will be dynamically active since it is must-vary. For each memory reference that can be determined must-may-active, we can remove the activity check of dynamic activity checking. Thus, for some benchmark/independent/dependent trios that exhibit must-may-activity, our hybrid static/dynamic activity analysis may reduce the overhead of dynamic activity checking. Since we are concentrating on forward mode AD, a must-useful vs. may-useful analysis is not exploitable.

Using OpenAnalysis, we implemented the must-may activity analysis. We designed a new set of sax routines that would skip the check of the activity flag on the known must-may-active gradients. To avoid any extra checking in this regard, we re-order the arguments to the sax calls to identify by position the gradients that need to be dynamically checked and those that do not. We manually adjusted the sax calls in each benchmark's derivative code to comply with the new interface. Then we used the must-may activity results to re-order the arguments. We anticipate that this must-may activity analysis will become an option in OpenAD, generating calls under the new interface and automatically re-arranging the arguments.

5.2 Results

In Figure 5 we display the results of our hybrid static/dynamic activity analysis. All data has been normalized to the execution time of *No Activity Check*. The arithmetic mean per analysis run is noted in parentheses within the legend. The second bar represents *Pure Dynamic* activity analysis and visually shows the significant overhead (39% average) of dynamic activity checking. The third bar represents the static *May* activity analysis with no dynamic activity analysis and shows the advantage of pruning the derivative code at inactive variables with an average speedup of 21%. The fourth bar represents the hybrid combination of the static may activity analysis with dynamic activity analysis. In most benchmarks, the win from the static may analysis more than compensates for the overhead of dynamic checking. The fifth bar represents the hybrid combination of the static

Must-May activity analysis with dynamic activity analysis. Three of the benchmarks show a decrease in execution time by using must-may activity analysis rather than may activity analysis. We anticipate that as we reduce the implementation overhead of our hybrid accumulation routines and examine complex applications where more variables can be statically identified as must-active, the benefits of our hybrid strategy will become more apparent.

6 Conclusion

We have implemented a hybrid static/dynamic strategy for activity analysis. This approach offers the opportunity to use runtime information to avoid unnecessary derivative accumulation operations, as may occur with conservative static analysis, while avoiding the overhead of unneeded runtime tests, as may occur with dynamic analysis. By restricting runtime tests to variables statically identified as may active and eliminating tests for variables statically identified as must-may active, we reduce the number of runtime checks. Our experimental results indicate that this hybrid strategy can sometimes pay dividends, offering improved performance over both a conservative static strategy and a dynamic strategy. We anticipate that as we examine more complex applications and eliminate some of the implementation overhead of the hybrid strategy, the benefits of the hybrid static/dynamic strategy will be even more pronounced.

References

1. D. H. Ackley. *A connectionist machine for hillclimbing.* Kluwer Academic Publishers, Boston, 1987.
2. B. Addis and S. Leyffer. A trust-region algorithm for global optimization. Technical Report ANL/MCS-P1190-0804, Argonne National Laboratory, August 2004.
3. ADIC Webpage. `http://www-fp.mcs.anl.gov/adic/`.
4. S. J. Benson, L. C. McInnes, J. Moré, and J. Sarich. TAO user manual (revision 1.8). Technical Report ANL/MCS-TM-242, Mathematics and Computer Science Division, Argonne National Laboratory, 2005. `http://www.mcs.anl.gov/tao`.
5. C. Bischof, P. Khademi, A. Mauer, and A. Carle. Adifor 2.0: Automatic differentiation of Fortran 77 programs. *IEEE Comput. Sci. Eng.*, 3(3):18–32, 1996.
6. C. Bischof, L. Roh, and A. J. Mauer-Oats. ADIC: An extensible automatic differentiation tool for ANSI-C. *Software: Practice and Experience*, 27(12):1427–1456, December 1997.
7. C. H. Bischof, P. D. Hovland, and B. Norris. On the implementation of automatic differentiation tools. *Higher-Order and Symbolic Computation*, 2004.
8. M. Fagan and A. Carle. Activity analysis in Adifor: Algorithms and effectiveness. Technical Report TR04-21, Rice University, Dept. of Computation and Applied Mathematics, 2004.
9. Global Optimization Functions. `http://www2.imm.dtu.dk/~km/GlobOpt/testex/`.
10. L. Hascoet, U. Naumann, and V. Pascual. "to be recorded" analysis in reverse-mode automatic differentiation. *Future Generation Computer Systems*, 21(8):1401–1417, October 2005.

11. MINPACK-2 webpage.
`http://www-fp.mcs.anl.gov/otc/minpack/sectionstar3_1.html`.
12. U. Naumann. Reducing the memory requirement in reverse mode automatic differentiation by solving TBR flow equations. In *International Conference on Computational Science*, pages 1039–1048. Springer, April 2002.
13. OpenAD Webpage. `http://www-unix.mcs.anl.gov/openad/`.
14. OpenAnalysis Webpage.
`http://www-unix.mcs.anl.gov/OpenAnalysisWiki/moin.cgi`.

Automatic Sparsity Detection Implemented as a Source-to-Source Transformation

Ralf Giering and Thomas Kaminski

FastOpt, Schanzenstr. 36, 20357 Hamburg, Germany
http://www.FastOpt.com

Abstract. An implementation of Automatic Sparsity Detection (ASD) as a new source-to-source transformation is presented. Given a code for evaluation of a function, ASD generates code to evaluate the sparsity pattern of the function's Jacobian by operations on bit-vectors. Similar to Automatic Differentiation (AD), there are forward and reverse modes of ASD. As ASD code has significantly fewer required variables than AD, ASD should be operated in pure mode, i.e. without an evaluation of the underlying function included in the ASD code. In a performance comparison of ASD to AD on five small test problems, ASD is about two orders of magnitude faster than AD. Hence, for a particular class of sparse Jacobians, it is efficient to determine first the sparsity patten via ASD. In a subsequent AD step, this allows to reduce the effective dimension for the evaluation of the Jacobian by avoiding the evaluation of zero elements via a selection of seed matrices according to the sparsity pattern.

1 Introduction

Automatic Differentiation (AD) generates derivative code for evaluation of the Jacobian matrix that corresponds to a given code for evaluation of a function. Often the Jacobian is sparse, and, if this sparsity information is available, it can be exploited to compute the Jacobian more efficiently. The basic idea is that evaluation of Jacobian entries that are known to be zero is not necessary, which may allow to reduce the effective dimension for the Jacobian evaluation. Algorithms for exploiting Jacobian sparsity have been developed and demonstrated by Curtis Powell Reid (CPR) [6], Newsam and Ramsdell [13, 7], and Coleman and Verma (bi-coloring, [4]), details can be found in the respective references.

In some cases the sparsity structure is not known or changes with the input. Bischof et al. [3] describe a dynamical approach of tracking the sparsity structure (via calls to special bookkeeping routines of an extra library) during the evaluation of forward mode derivative code. Within an operator overloading framework, Geitner et.al. [7] describe how to determine the sparsity pattern by re-executing the tape, a representation of the underlying function, generated in a previous execution of the function code. The tape is built by overloading every operation to store the operands and the operation. The sparsity is represented by bit patterns and combined by logical 'or' operations.

Here we describe the source-to-source equivalent, that is, a semantical transformation of the original function code to a code that evaluates the sparsity

V.N. Alexandrov et al. (Eds.): ICCS 2006, Part IV, LNCS 3994, pp. 591–598, 2006.
© Springer-Verlag Berlin Heidelberg 2006

structure of the function's Jacobian. This transformation has been implemented in TAF (Transformation of Algorithms in Fortran, [8, 11]). TAF is an AD-tool for Fortran77-95 programs. It normalises the function code and applies a control flow analysis in order to replace old style Fortran constructs and to transform unstructured code to high-level structures. Irreducible control flow graphs are made reducible by node copying [10]. An intra-procedural data dependence analysis is applied to determine loop-carried flow, anti-flow, or output dependences. The following inter-procedural data flow analysis computes the IN and OUT sets [5] of all statements based on the given dependent and independent variables of the top-level routine. A variable is active if it depends on the independent variables and influences the dependent variables [2]. Derivative (AD) or bit-vector (ASD, see below) variables are only built for active variables, and derivative or sparsity code is only generated for active statements, i.e. statements that compute active variables. In reverse modes of AD and ASD, TAF generates recomputations for required variables by an extension of the Efficient Recomputation Algorithm (ERA [9]). ERA uses demand-driven program slicing to generate only a minimum of recomputations.

Depending on the number of independent and dependent variables, ASD is applied in forward or reverse mode to compute the Jacobian's sparsity. In the presence of a priori knowledge about the sparsity structure of a Jacobian on a block level, it is most efficient to restrict ASD to a subset of all blocks.

2 Transformation Rules

In this section we present the rules of transforming the function code into both types of ASD codes, the forward and the reverse one. To keep the notation simple, the rules are shown for computing the sparsity structure of a boolean vector times Jacobian product in forward mode and a boolean Jacobian times vector product in reverse mode. In the following f, x, y are active variables with corresponding boolean variables $\hat{f}, \hat{x}, \hat{y}$, which hold the sparsity structure that is propagated by the ASD code. In order to compute the full sparsity pattern in the above boolean product the vectors are replaced by the boolean identity matrices (true on the diagonal). In the transformation rules given below, the boolean variables are then to be replaced by boolean vectors.

For a binary operation $\circ \in \{+, -, *, /, **\}$, the assignment

$$f = x \circ y$$

is transformed by forward mode ASD to:

$$\hat{f} = \hat{x} \vee \hat{y}$$

and by reverse mode ASD to:

$$\begin{aligned} \hat{x} &= \hat{x} \vee \hat{f} \\ \hat{y} &= \hat{y} \vee \hat{f} \\ \hat{f} &= false \ , \end{aligned}$$

where $\vee$ denotes the logical 'or'. For a unary operation $\circ \in \{-,+\}$ the assignment

$$f = \circ x$$

is transformed by forward mode ASD to:

$$\hat{f} = \hat{x}$$

and by reverse mode ASD to:

$$\begin{aligned} \hat{x} &= \hat{x} \vee \hat{f} \\ \hat{f} &= false\ . \end{aligned}$$

Similar rules apply for a function invocation. For a function of one argument (e.g. sin, cos, ...), the assignment:

$$f = func(x)$$

is transformed by forward mode ASD to:

$$\hat{f} = \hat{x}$$

and by reverse mode ASD to:

$$\begin{aligned} \hat{x} &= \hat{x} \vee \hat{f} \\ \hat{f} &= false \end{aligned}$$

It is evident that in ASD, unlike AD, the transformed statements do not require any values from the original statements. Only to follow (forward mode) or to reverse (reverse mode) the control flow, values may be required (if-then-else and case constructs). They are provided in the same fashion as for AD (see [8]). Owing to the reduced number of required values, the pure forward and pure reverse modes, which compute only sparsity and do not evaluate the function itself, have a considerable advantage in efficiency. In TAF both pure modes are implemented for AD and ASD. A command line option triggers generation of the corresponding codes.

3 Implementation

The ASD implementation in TAF propagates the sparsity structure in bit-vectors. Depending on the platform used, a Fortran-90 bit-vector is an integer variable that holds 32 bits, if the kind parameter is 4 (byte), or 64 bits, if the kind parameter is 8 (byte). [1]

This way several matrix vector products are computed simultaneously. The logical operation 'or' is implemented by the IOR intrinsic function. It has two bit-vector arguments and a bit-vector result. Inside the bit-vector the value 'false' is

[1] For most efficient code the bit-vector should be as long as a word of the processor.

represented by a zero bit. A false bit-vector (all bits are false) is represented by 0 and a true one by NOT(0), where NOT is the Fortran-90 intrinsic function. For initialisation individual bits are set by IBSET and for interpretation of the results they are tested by BTEST, both of which are Fortran-90 intrinsic functions.

As an example we use the single assignment

$$f = a * x + y * sin(z)$$

which computes a new value for the variable z. It is assumed that only the variables x,y,z, and f are active.

In forward mode, ASD generates the assignment

$$sf = IOR(sx, IOR(sy, sz)) ,$$

where sf is the bit-vector corresponding to the active variable f. Other variable names are generated accordingly. In some cases the RHS expression can be simplified by applying the rules of boolean operations. Here one bit-vector corresponds to one active variable and the code computes 32 (64) matrix times vector products simultaneously. In order to compute more vectors, a bit-vector array is generated. The statements remain unchanged, since Fortran-90 elemental intrinsic functions[2] operate on scalars and arrays.

In reverse mode ASD generates the sequence of assignments

$$\begin{aligned} sx &= IOR(sx, sf) \\ sy &= IOR(sy, sf) \\ sz &= IOR(sz, sf) \\ sf &= 0 \end{aligned}$$

Similar to AD, the bit-vector to the LHS variable f is reset after the bit-vectors of all RHS variables have been updated.

4 Performance

The Minpack-2 collection [1] provides several test function codes based on real physical problems. Codes to evaluate their Jacobian, Jacobian vector products, and Jacobian sparsity are also provided. We have selected five problems of this collection to compare the performance of automatically generated ASD and AD codes:

FDC	flow in a driven cavity
FIC	flow in a channel
IER	incompressible elastic rods
SFD	swirling flow between disks
SFI	solid fuel ignition

[2] For transformational intrinsic functions such as MATMUL more complex statements must be generated.

Each function code solves a differential equation on a grid of variable size. Both the numbers of input and output variables are set equal to the number of grid points (N), i.e. the Jacobian is quadratic. The number of floating point operations in the function code and both dimensions of the Jacobian scale with N.

For each test code, TAF was first used in AD mode to generate code evaluating the full Jacobian (without any seeding) in forward and reverse modes. Next TAF's ASD mode was applied to generate forward and reverse mode codes evaluating directly the Jacobians' sparsity patterns. The codes were compiled with -O3 by the Lahey Fortran-95 compiler and run on an Athlon Linux PC for different values of N. In the code bit-vectors were represented by default integer variables (kind=4).

Fig. 1 shows the relative run-times for these cases. The runtime of ASD code is about two orders of magnitude faster than than that of AD code evaluating the full Jacobians. This is not only a consequence of the simultaneous propagation of 32 logical values by the operations on bit-vectors and of ASD's lower number of operations. An additional performance gain is achieved by the capability of TAF to generate code operating in pure forward and reverse modes. It is worth noting that the advantage of ASD over AD is almost always bigger in forward mode. Presumably this is owing to the smaller number of IOR operations required by forward model ASD as compard to reverse mode ASD, which is also illustrated by the example of section 3. Fig. 2 shows the ratio of run times for reverse and forward modes of AD, which don't change much with problem size.

The Carbon Cycle Data Assimilation System (CCDAS, http://CCDAS.org, [15, 14]) provides an example of a large-scale ASD application. CCDAS uses observed atmospheric carbon dioxide to constrain parameters in a global model of the terrestrial biosphere. A subset of the parameters are chosen to be specific to the model's plant functional types (PFTs), e.g. tundra. Uncertainties in the observations are back-propagated via the model's Hessian to parameter uncertainties. To map these parameter uncertainties onto uncertainties of target quantities diagnosed from or prognosed by the model, e.g. the land sink over a particular region, the corresponding Jacobian is needed. Now, a given region will typically only host a subset of the global model's PFTs. Consequently the sensitivity of the land sink over that region to parameters specific to PFTs not represented in that region must be zero, i.e. this Jacobian will be sparse. [12] demonstrate the use of TAF's ASD mode for the efficient computation of the sparsity structure of the CCDAS Jacobian that quantifies the sensitivity of the mean fluxes over latitudinal bands with respect the the model's parameters. While the cost of a reverse mode AD run increases by about the cost of 0.25 function evaluations per additional target quantity (from about 3.5 function evaluations for 1 target quantity to about 26 function evaluations for 96 target quantities), the cost of the ASD run remains almost constant (at about 2.5 function evaluations).

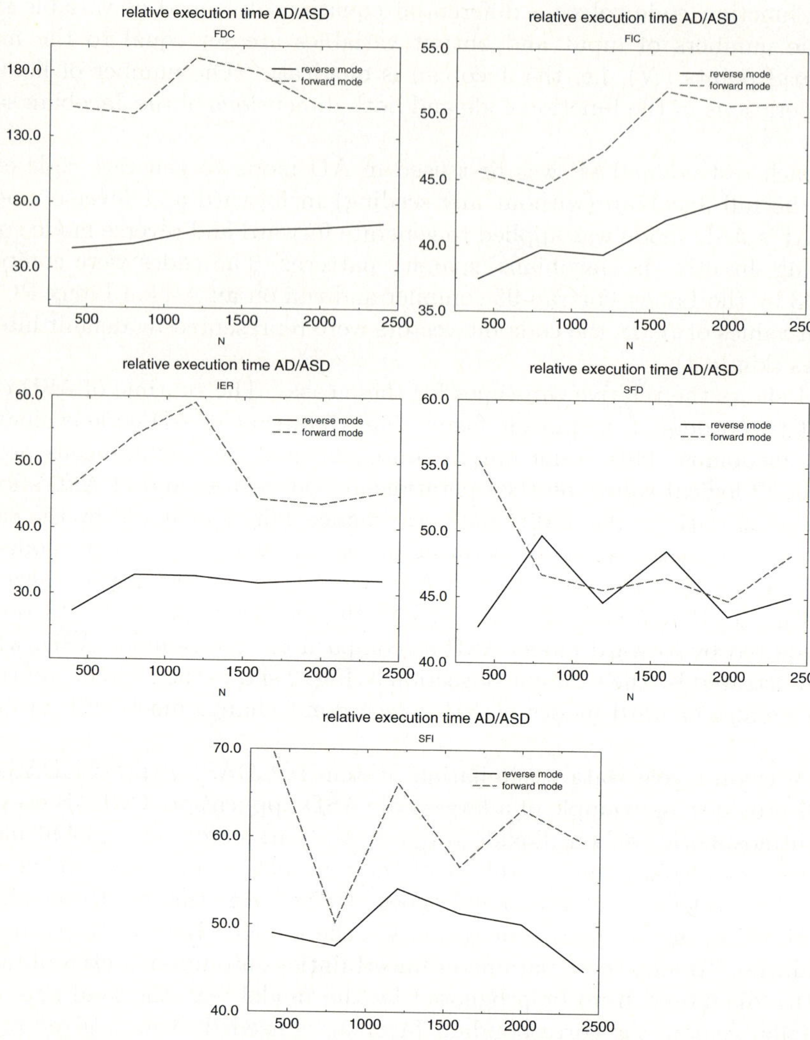

Fig. 1. Ratios CPU time(AD)/CPU time(ASD) for the five test codes over dimension of the problem. Solid lines show reverse mode ratios; dashed lines show forward mode ratios.

5 Conclusions

The rules for the new source transformation Automatic Sparsity Detection (ASD) have been presented. Given an algorithm to evaluate a function an algorithm to evaluate the sparsity pattern of the function's Jacobian is generated. As in Automatic Differentiation (AD) there are two major modes: the forward and the reverse mode. In contrast to AD code, which for its local Jacobians requires values from the function evaluation, ASD code only requires the function's control information.

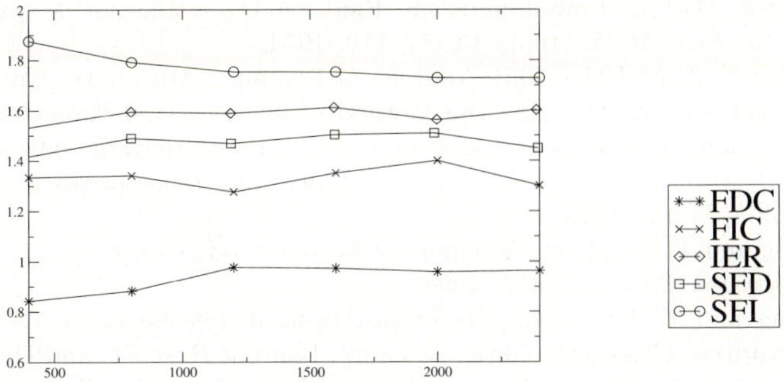

Fig. 2. Ratios CPU time(AD reverse)/CPU time(AD forward) for evaluation of the full Jacobian (without any seeding) of the five test codes over dimension of the problem

The implementation of ASD in Fortran-90 has been described and was explained by a simple example. The run-time of ASD code is about two orders of magnitude faster than that of evaluating the entire Jacobian by AD and checking for sparsity thereafter. The factor is much larger than the expected gain by computing several logical values simultaneously using bit-vectors. The reason for this is the reduced number of operations and the ability of TAF to generate pure mode ASD code, i.e. code that does not include an evaluation of the underlying function itself.

Often from prior knowledge about the function the Jacobian can be partitioned into blocks such that the sparsity structure on a block level is known but within the block is not. In these cases ASD can be restricted to the evaluation of the sparsity structure within the blocks.

References

1. Brett M. Averick, Richard G. Carter, Jorge J. Moré, and Guo-Liang Xue. The MINPACK-2 test problem collection. Preprint MCS–P153–0692, ANL/MCS–TM–150, Rev. 1, Mathematics and Computer Science Division, Argonne National Laboratory, Argonne, Ill., 1992.
2. C. Bischof, A. Carle, P. Khademi, and A. Mauer. ADIFOR 2.0: Automatic differentiation of Fortran 77 programs. *IEEE Computational Science & Engineering*, 3(3):18–32, 1996.
3. Christian H. Bischof, Peyvand M. Khademi, A. Bouaricha, and Alan Carle. Efficient computation of gradients and Jacobians by dynamic exploitation of sparsity in automatic differentiation. *Optimization Methods and Software*, 7:1–39, 1997.
4. Thomas F. Coleman and Arun Verma. Structure and efficient Jacobian calculation. In Martin Berz, Christian Bischof, George Corliss, and Andreas Griewank, editors, *Computational Differentiation: Techniques, Applications, and Tools*, pages 149–159. SIAM, Philadelphia, Penn., 1996.
5. Beatrice Creusillet and F. Irigoin. Interprocedural Array Region Analysis. Rapport CRI, A-282, Ecole des Mines de Paris, FRANCE, January 1996.

6. A. R. Curtis, M. J. D. Powell, and J. K. Reid. On the estimation of sparse Jacobian matrices. *J. Inst. Math. Appl.*, 13:117–119, 1974.
7. Uwe Geitner, Jean Utke, and Andreas Griewank. Automatic Computation of Sparse Jacobians by Applying the Method of Newsam and Ramsdell. In Martin Berz, Christian Bischof, George Corliss, and Andreas Griewank, editors, *Computational Differentiation: Techniques Applications, and Tools*, pages 161–172. SIAM, Philadelphia, Penn., 1996.
8. R. Giering and T. Kaminski. Recipes for Adjoint Code Construction. *ACM Trans. Math. Software*, 24(4):437–474, 1998.
9. R. Giering and T. Kaminski. Recomputations in reverse mode AD. In George Corliss, Andreas Griewank, Christele Fauré, Laurent Hascoet, and Uwe Naumann, editors, *Automatic Differentiation of Algorithms: From Simulation to Optimization*, chapter 33, pages 283–291. Springer Verlag, Heidelberg, 2002.
10. R. Giering and T. Kaminski. Applying TAF to generate efficient derivative code of Fortran 77-95 programs. *PAMM*, 2(1):54–57, 2003.
11. R. Giering, T. Kaminski, and T. Slawig. Applying TAF to a Navier-Stokes solver that simulates an Euler flow around an airfoil. *Future Generation Computer Systems*, 21(8):1345–1355, 2005.
12. T. Kaminski, R. Giering, M. Scholze, P. Rayner, and W. Knorr. An example of an automatic differentiation-based modelling system. In V. Kumar, L. Gavrilova, C. J. K. Tan, and P. L'Ecuyer, editors, *Computational Science – ICCSA 2003, International Conference Montreal, Canada, May 2003, Proceedings, Part II*, volume 2668 of *Lecture Notes in Computer Science*, pages 95–104, Berlin, 2003. Springer.
13. G. N. Newsam and J. D. Ramsdell. Estimation of sparse Jacobian matrices. *SIAM J. Alg. Disc. Meth.*, 4(3):404–417, 1983.
14. P. Rayner, M. Scholze, W. Knorr, T. Kaminski, R. Giering, and H. Widmann. Two decades of terrestrial Carbon fluxes from a Carbon Cycle Data Assimilation System (CCDAS). *Global Biogeochemical Cycles*, 19:doi:10.1029/2004GB002254, 2005.
15. M. Scholze. *Model studies on the response of the terrestrial carbon cycle on climate change and variability.* Examensarbeit, Max-Planck-Institut für Meteorologie, Hamburg, Germany, 2003.

Lattice Properties of Two-Dimensional Charge-Stabilized Colloidal Crystals

Pavel Dyshlovenko[1] and Yiming Li[2]

[1] Laboratory of Computer Simulations, Ulyanovsk State technical University, Ulyanovsk 432027, Russia
pavel@ulstu.ru

[2] Department of Communication Engineering, National Chiao Tung University, Hsinchu 300, Taiwan
ymli@faculty.nctu.edu.tw

Abstract. In this paper, electrostatic interaction in two-dimensional colloidal crystals obeying the non-linear Poisson-Boltzmann equation is studied numerically. We first give an overview of the recently developed approach to study of the lattice properties of colloidal crystals. The central point of the theory is determination of the force constants, which are the coefficients of the energy quadratic form of the crystal. Particular attention is given to the symmetry considerations. Some prospective topics of research are briefly discussed.

1 Introduction

Colloidal crystals are dispersions of colloidal particles arranged into a regular lattice. Besides their importance for studying structural phase transitions resembling conventional melting and freezing, they are also well-defined model systems whose macroscopic properties can be directly connected to the underlying microscopic interparticle interactions.

Theoretical analysis of lattice properties of colloidal crystals, such as normal modes of oscillations or elastic properties, is mostly based on the representation of the potential energy of the particles as a sum of pair interactions. Although the concept of pairwise interaction is adequate for many systems, such as the dipole-dipole interaction [1, 2] and purely entropic forces [3, 4], it fails for charge-stabilized colloidal crystals. In particular, it was shown that charge-stabilized colloidal crystals' elastic properties observed cannot be understood within the idea of linear superposition of pairwise interactions [5].

In this paper, we consider a two-dimensional charge-stabilized colloidal crystal obeying the general nonlinear Poisson-Boltzmann (PB) equation. Description of recently proposed approach [6, 7] is included. Our approach enables the force constant determination and successfully provides quantitative estimation of the many-particle effects. It validates the approximation of the nearest neighbor interaction in such systems. In particular, it turns out that the contribution of many-particle to the total electrostatic potential energy is significant and cannot be neglected for a broad range of particle radii and crystal lattice parameters.

V.N. Alexandrov et al. (Eds.): ICCS 2006, Part IV, LNCS 3994, pp. 599–606, 2006.
© Springer-Verlag Berlin Heidelberg 2006

This paper is organized as follows. In Section 2, we state the model formulation. In Section 3, we state the force constants determination. In Section 4, we discuss the role of symmetry. In Section 5, we present the numerical methods. Section 6 reports the results. Section 7 draws conclusions.

2 Mathematical Model Formulation

The colloidal crystal under consideration is shown in the Fig. 1. It consists of infinitely long cylindrical colloidal particles of radius R arranged in a two-dimensional hexagonal lattice with the lattice constant a. The system of the particles is immersed in symmetrical univalent electrolyte. The particles are perfectly rigid dielectric rods. The dielectric permittivity of the particles is much smaller then the one of electrolyte solution, so it is set to be zero for all numerical calculations in the paper. The particles are charged with uniform surface charge density σ which is kept constant (the so-called constant-charge or cc-model). The crystal system considered in the paper can be pertinent to the behavior of rod-like objects like DNA molecules, tobacco mosaic viruses [8] and fd viruses [9], rod-like polyelectrolytes [10] or some mesoscopic objects [11]. Throughout the present paper, length and electro-static potential are expressed in units of Debye length $\kappa^{-1} = (2nq_e^2/\epsilon kT)^{-1/2}$ and kT/q_e respectively, where n is the concentration of either of the species in the electrolyte, q_e is the absolute value of the electronic charge, ϵ is the absolute permittivity of the electrolyte, k is the Boltzmann constant, T is the absolute temperature, and the rationalised SI system of units is used to express the factors.

Electric potential ϕ in such a system obeys the non-linear PB equation [12] in the electrolyte's domain outside the particles and the Laplace equation in the interior of the particles:

$$\frac{\partial^2\phi}{\partial x^2} + \frac{\partial^2\phi}{\partial y^2} = \begin{cases} \sinh\phi, & \text{(in the electrolyte),} \\ 0, & \text{(inside the particles).} \end{cases} \tag{1}$$

Electric field at the surface of the particles meets the electrostatic boundary condition

$$E_n = \sigma, \tag{2}$$

where E_n is a normal component of the electric field in the electrolyte and dielectric permittivity of the particles is equal to zero. As usual, the tangential component of the electric field at the interface remains continuous.

There are no net forces on the particles when all the particles are located in their equilibrium positions. If one or more of the particles are displaced from the equilibrium, the non-zero net forces on them arise. The force on any particle in the system can be calculated by means of integration of the stress tensor:

$$\mathbf{F} = \oint_{\Gamma} \left[\nabla\phi \otimes \nabla\phi - \left(\frac{1}{2}|\nabla\phi|^2 + \cosh\phi - 1 \right) I \right] \cdot \mathbf{n}\, d\Gamma, \tag{3}$$

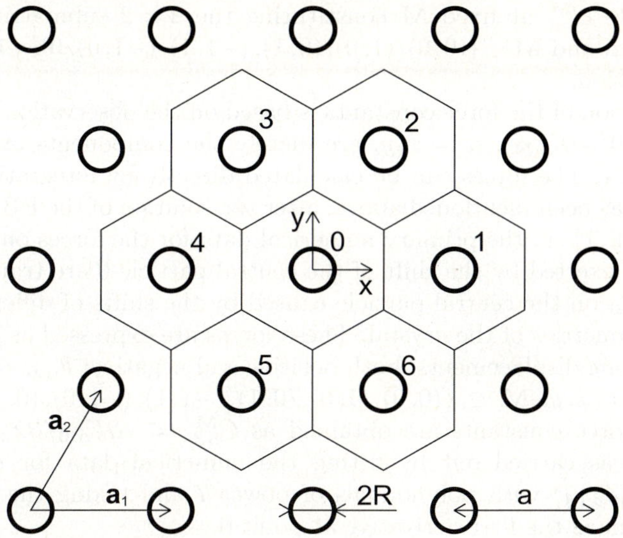

Fig. 1. Two-dimensional hexagonal colloidal crystal lattice. Particles are infinite rods perpendicular to the plane of the figure. Domain of the numerical problem comprises seven particles 0 to 6 with corresponding hexagonal Wigner-Seitz cells.

where Γ is a Wigner-Seitz cell's boundary, $\mathbf{n}$ is an outer normal vector to the element $d\Gamma$, and I is an identity matrix. Although any closed loop enclosing the particle can be employed for the integration, using of the boundary of the Wigner-Seitz cell is practical since it provides rather low errors in process of numerical implementation of the integration.

3 Force Constants Determination

Classical potential energy V of a crystal for small displacements of the particles from the equilibrium positions can be written [13] as follows:

$$V = \frac{1}{2} \sum_{\alpha,\beta,\mathbf{N},\mathbf{M}} \left(\frac{\partial^2 V}{\partial Z_{\alpha,\mathbf{N}} \partial Z_{\beta,\mathbf{N+M}}} \right) Z_{\alpha,\mathbf{N}} Z_{\beta,\mathbf{N+M}}, \tag{4}$$

where $Z_{\alpha,\mathbf{N}}$ is an α-component of the displacement Z from the equilibrium position pointed by vector $\mathbf{N}$, $\alpha = x, y$, $\beta = x, y$, $\mathbf{N}$ and $\mathbf{M}$ are vectors of the Bravais lattice. Coefficients

$$C_{\alpha\beta}^{\mathbf{M}} = \frac{\partial^2 V}{\partial Z_{\alpha,\mathbf{N}} \partial Z_{\beta,\mathbf{N+M}}} \tag{5}$$

of the quadratic form (4) are called force constants. They do not depend on $\mathbf{N}$ and can be arranged into the square matrix $C = \left\| C_{\alpha\beta}^{\mathbf{M}} \right\|$ of a quadratic form, the coefficients $C^{\mathbf{M}}_{\alpha\beta}$ at fixed $\mathbf{M}$ constituting the 2×2 submatrix $C^{\mathbf{M}}$ of the general matrix, and $\mathbf{M} \in \{(0,0),(1,0),(0,1),(-1,1),(-1,0),(0,-1),(1,-1)\}$ in our consideration.

Determination of the force constants is based on the observation that the first derivatives $(\partial V/\partial Z_{\alpha,\mathbf{N}})$, $\alpha = x, y$, are merely the components of the force on the particle $\mathbf{N}$. The forces can be calculated directly by integrating the stress tensor, as it has been mentioned above, after the solution of the PB equation has been obtained. Then, the primary numerical data for the forces on the particles of the system exerted by the shift of the central particle 0 are transformed into the forces $F_{\alpha,\mathbf{0}}$ on the central particle caused by the shifts of different particles using the symmetries of the crystal. These forces are expressed as the functions of corresponding displacements, both positive and negative: $F_{\alpha,\mathbf{0}} = F_{\alpha,\mathbf{0}}(Z_{\beta,\mathbf{M}})$, $\alpha = x, y$, $\beta = x, y$, $\mathbf{M} \in \{(0,0),(1,0),(0,1),(-1,1),(-1,0),(0,-1),(1,-1)\}$. Finally, the force constants are obtained as $C^{\mathbf{M}}_{\alpha\beta} = -\partial F_{\alpha,\mathbf{0}}/\partial Z_{\beta,\mathbf{M}}$. The differentiation was carried out by fitting the numerical data for the functions $F_{\alpha,\mathbf{0}} = F_{\alpha,\mathbf{0}}(Z_{\beta,\mathbf{M}})$ with polynomials of power 7 and taking the coefficient of the linear term as the first derivative at point 0.

4 Role of Symmetry

Taking into account the symmetry of the crystal enables significant reduction of the amount of numerical calculations. First, direct calculation of the forces on the central particle 0 would require multiple solutions of the PB equation for configurations with different particles shifted from their equilibrium positions. Translational and inversion symmetry of the crystal lattice makes it possible to reduce the calculation of the forces on the central particle 0 arising from the motion of different particles to the calculation of the forces on all the particles (seven here) due to the motion of only the central particle alone. Having the numerical solution of the PB equation for the configuration with only the central particle shifted, the forces on all the particles in the system can then be obtained by taking the integral of the stress tensor over the corresponding contours. The post-solution integration is much less expensive in the sense of computer resources required than the numerical solution itself.

Second, mirror symmetry of the problem allows the use of only a half of the problem's domain. When the particle 0 is shifted along the x-axis, the problem retains the mirror symmetry about this axis. Thus, without loss of generality, the upper half of the domain above the x-axis is required.

Finally, rotational symmetry of the crystal lattice allows further reduction of the calculations at the post-solution stage. The seven particles under consideration belong to two different orbits of the rotational subgroup of the crystal point group. The first orbit consists of only the central particle 0. The particles 1 to 6 constitute another orbit: they transform into each other when rotating about the point 0 at the angle multiple of $\pi/3$. For symmetry reasons, matrix $C^{(0,0)}$ has diagonal elements equal to each other and off-diagonal ones equal to zero. Consequently, matrix $C^{(0,0)}$ is completely determined by only one, say $C^{(0,0)}_{x,x}$, of its diagonal element. For the same reasons, matrix $C^{(1,0)}$ for the particle 1 has zero off-diagonal elements and is thus determined by only two diagonal elements $C^{(1,0)}_{x,x}$ and $C^{(1,0)}_{y,y}$. Since particles 1 to 6 belong to the same orbit of the rotational subgroup of the crystal point group the force constant matrices of these particles are not independent. If the matrix of the particle described by vector M is known, the matrices of the other particles can be obtained by the matrix transformation according to the rule of quadratic forms' matrix transformation [9]:

$$C^{\mathbf{N}} = R^{\mathrm{T}}(\phi) C^{\mathbf{M}} R(\phi), \tag{6}$$

where $\mathbf{M}$, $\mathbf{N} \in \{(1,0),(0,1),(-1,1),(-1,0),(0,-1),(1,-1)\}$, ϕ is the angle between the vectors $\mathbf{N}$ and $\mathbf{M}$, and superscript T means matrix transposition. Therefore, the complete set of 28 force constants (7 particles $\times$ 4 matrix elements, in the approximation of nearest neighbour interaction) is completely determined only by the three non-trivial independent parameters which should be obtained directly from computer experiments. The other constants are obtained then by means of symmetry transformations (6).

5 Numerical Procedures

Equations (1) and (2) are solved numerically using the method described in [14, 15]. This method combines the finite-element solution of the equation with an adaptive mesh refinement [16, 17, 18]. It is well suited for the two-dimensional problems with complicated geometry and variety of boundary conditions. The domain of the problem for numerical solution consists of the Wigner-Seitz cells of the central particle 0 and its six nearest neighbours 1 to 6. The standard von Neumann boundary conditions hold at the outer boundary of the domain.

The numerical calculations are carried out for the central particle shifted by ten equal steps along the positive direction of the x-axis so that the largest shift amounted to 10% of the separation distance between the nearest particles. The forces exerted by this shift on all the seven particles on the particles were calculated by means of numerical integration of the stress tensor, as it was described above. Since the domain of the problem is restricted to the seven particles, the interaction of the central particles with its nearest neighbours is only considered in this paper. It is shown in [6] that this is a very good approximation for a broad range of parameters a and R.

Domain of the problem with the mesh of triangular elements on it is shown in the Fig. 2. Due to the mirror symmetry about the x-axis only the upper half of the domain is required. The mesh is a Delaunay triangulation of the domain at each stage of the solution.

6 Results and Discussion

The force constants of the two-dimensional hexagonal charge-stabilized colloidal crystal for the typical set of parameters $\sigma = 2.0$, $R = 1.0$ and $a = 5.0$ are shown in Table 1. Only the three independent force constants are presented; the other

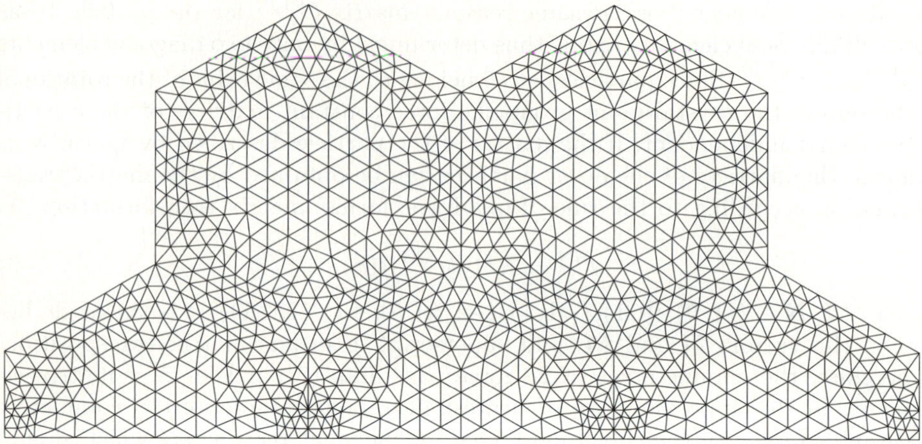

Fig. 2. Domain of the problem and irregular mesh of triangular elements on it. The round boundaries of colloidal particles are clearly observed. The mesh is obtained in the beginning of the solution after the first iteration of the numerical procedure and consists of 2160 triangular elements. The final mesh is obtained after ten iterations comprised more than 135000 elements, concentrated mostly at the outer boundaries of the particles.

ones can be obtained my means of the transformation (6). Complete set of the results and their systematic analysis are provided in [7]. Nevertheless, the example in Table 1 illustrates the main features of the data obtained. First, it turns out that the force constant $C_{y,y}^{(1,0)}$ for the system considered is definitely nonzero, while the theory of harmonic crystal based on the assumption of pairwise interaction between particles requires always this constant to be strictly equal to zero. A quantitative estimation of the contribution of the many-particle interaction into a total electrostatic interaction in a system was introduced in [6]. It was shown there that the collective electrostatic interaction in two-dimensional colloidal crystals cannot be expressed as a sum of pair interactions between the particles and that the many-particle interactions are strong enough for a broad range of charge densities σ on the particles, radii R of the particles and lattice parameters a even at very large ones when the interparticle interaction itself is weak. Another feature of the data in Table 1 is that the ratio $C_{x,x}^{(0,0)}/C_{x,x}^{(1,0)}$ is not exactly equal to -3 as it should be according to the approximation of the nearest-neighbor interaction. A quantitative measure of the validity of the nearest-neighbor interaction approximation based on this discrepancy was pro-

Table 1. Force constants of the crystal for $\sigma = 2.0$, $R = 1.0$ and $a = 5.0$

$C_{x,x}^{(0,0)}$	$C_{x,x}^{(1,0)}$	$C_{y,y}^{(1,0)}$
0.6756891	-0.2663590	0.0480621

posed in [6]. It was shown there, that this approximation remains a rather good one for a broad range of crystal parameters σ, R and a.

7 Conclusions

In this paper, we have computationally explored the electrostatic interaction in two-dimensional colloidal crystals. The non-linear Poisson-Boltzmann equation has been solved numerically with adaptive finite element method. An overview of the recently developed approach to study of the lattice properties of colloidal crystals has been reported. The central point of the theory focuses on determination of the force constants, which are the coefficients of the energy quadratic form of the crystal. Particular attention has been given to the symmetry considerations. Some prospective topics of research have briefly been discussed. The characteristics of the crystal system considered in the paper seem to be valid for broader range of the colloidal systems. In particular, the crystal with the particles obeying the constant potential model (*cp*-model) or recently proposed constant total charge model (*ctc*-model) should be investigated. Corresponding calculations are currently in progress. Different types of crystal lattices, primarily the square one, are also of interest. One more question to study is the contribution of the neighbors further then the nearest ones into the total electrostatic interaction and corresponding modification of the crystal's properties. To study this problem, further development of the program code is needed to enable sufficient number of particles in the system. There are no doubts that three-dimensional colloidal crystals possess many features of their two-dimensional counterparts. However, study of three-dimensional problems will require efforts for new program code development utilizing the power of modern libraries for partial differential equation solution and involving some kind of parallelization [19].

Acknowledgments

This work was supported in part by Taiwan National Science Council (NSC) under Contract NSC-94-2215-E-009-084 and Contract NSC-95-2752-E-009-003-PAE, by the Ministry of Economic Affairs, Taiwan under Contract 93-EC-17-A-07-S1-0011, and by the Taiwan semiconductor manufacturing company under a 2005-2006 grant. One of the author (P.D.) gratefully acknowledges the financial support from the Mianowski Fund of Foundation for Polish Science during his visit to the Institute of Catalysis and Surface Chemistry (Cracow) where a part of the present work was carried out.

References

1. Keim, P., Maret, G., Herz, U., von Grünberg, H.H.: Harmonic lattice behavior of two-dimensional colloidal crystals. Phys. Rev. Lett. **92** (2004) 215504
2. Hay, M.B., Workman, R.K., Manne, S.: Two-dimensional condensed phases from particles with tunable interactions. Phys. Rev. E **67** (2003) 012401

3. Cheng, Z., Zhu, J., Russel, W.B., Chaikin, P.M.: Phonons in an entropic crystal. Phys. Rev. Lett. **85**(7) (2000) 1460–1463
4. Penciu, R.S., Kafesaki, M., Fytas, G., Economou, E.N., Steffen, W., Hollingsworth, A., Russel, W.B.: Phonons in colloidal crystals. Europhys. Lett. **58**(5) (2002) 699–704
5. Weiss, J.A., Larsen, A.E., Grier, D.G.: Interactions, dynamics, and elasticity in charge-stabilized colloidal crystals. J. Chem. Phys. **109**(19) (1998) 8659–8666
6. Dyshlovenko, P.E.: Evidence of many-particle interactions in two-dimensional charge-stabilized colloidal crystals. Phys. Rev. Lett. **95** (2005) 038302
7. Dyshlovenko, P.E. (the paper in preparation)
8. Adams, M., Fraden, S. Biophys. J. **74** (1998) 669
9. Purdy, K.R., Dogic, Z., Fraden, S., Rühm, A., Lurio, L., Mochrie, S.G.J.: Measuring the nematic order of suspensions of colloidal fd virus by x-ray diffraction and optical birefringence. Phys. Rev. E **67** (2003) 031708
10. Guilleaume, B., Blaul, J., Ballauff, M., Wittemann, M., Rehahn, M., Goerigk, G.: The distribution of counterions around synthetic rod-like polyelectrolytes in solution. Eur. Phys. J. E **8** (2002) 299–309
11. de A. A. Soler-Illia, G.J., Sanchez, C., Lebeau, B., Patarin, J.: Chemical strategies to design textured materials: from microporous and mesoporous oxides to nanonetworks and hierarchical structures. Chem. Rev. **102** (2002) 4093–4138
12. Israelachvili, J.N.: Chap. 12. In: Intermolecular and Surface Forces. Academic Press (1991)
13. Feynman, R.P.: Chap. 1. In: Statistical Mechanics. W. A. Benjamin, Inc., Massachusetts (1972)
14. Dyshlovenko, P.E.: Adaptive mesh enrichment for the poissonboltzmann equation. J. Comp. Phys. **172** (2001) 198–208
15. Dyshlovenko, P.E.: Adaptive numerical method for poissonboltzmann equation and its application. Comp. Phys. Commun. **147** (2002) 335–338
16. Li, Y., Sze, S.M., Chao, T.S.: A Practical Implementation of Parallel Dynamic Load Balancing for Adaptive Computing in VLSI Device Simulation. Comp. Phys. Commun. **147** (2002) 335–338
17. Li, Y., Chao, T.S., Sze, S.M.: A Domain Partition Approach to Parallel Adaptive Simulation of Dynamic Threshold Voltage MOSFET. Eng. Comput. **18** (2002) 124–137
18. Li, Y., Yu, S.M.: A Parallel Adaptive Finite Volume Method for Nanoscale Double-gate MOSFETs Simulation. J. Comput. Appl. Math. **175** (2005) 87–99
19. Li, Y.: A Parallel Monotone Iterative Method for the Numerical Solution of Multi-dimensional Semiconductor Poisson Equation. Comp. Phys. Commun. **153** (2003) 359–372

Self-consistent 2D Compact Model for Nanoscale Double Gate MOSFETs

S. Kolberg[1], T.A. Fjeldly[2], and B. Iñiguez[3]

[1,2] UniK – University Graduate Center and Norwegian University of Science and Technology, N-2027 Kjeller, Norway
{kolberg,torfj}@unik.no
[3] Universitat Rovira i Virgili (URV), Tarragona, E-43001, Spain
benjamin.iniguez@urv.net

Abstract. 2D modeling results of the electrostatics and the drain current in nanoscale DG MOSFETs are presented. The modeling of the 2D capacitive coupling within the device is based on the conformal mapping technique. In moderate above-threshold conditions, we obtain self-consistent results, which are in excellent agreement with numerical simulations.

1 Introduction

In nanoscale double gate (DG) MOSFETs, the electron barrier topology is a critical factor when determining conduction paths and currents in the device. A prerequisite to obtaining a precise description of such devices is to include two-dimensional (2D) effects in the models, based on a self-consistent solution of the 2D field pattern in the device. In such an approach, short-channel effects and scaling properties will be intrinsic to the model which, accordingly, will require only a minimal parameter set of clear physical origin.

The basic modeling problem is to obtain an analytical or semi analytical solution of a 2D Poisson's equation where the four contacts (source, drain and the two gates) and the dielectric gaps define the boundary conditions. According to the superposition principle, Poisson's equation can be separated into a 2D Laplace equation for the capacitive coupling and the remainder involves the potential distribution established by the body charges [1 - 10]. The latter can usually be treated by simplifying considerations.

The Laplace problem for the DG MOSFET can be solved in different ways. One possibility is to perform a full Fourier expansion of the potential or by using a low-order truncation [1-4]. A corresponding procedure by means of expansion in Bessel functions has been used for cylindrical surrounding gate transistors [5]. An alternative approach is to apply the conformal mapping technique [6], which was first used for classical, long-channel MOSFETs [7]. Later, the technique was enhanced and applied to sub-100 nm devices [8] and to the subthreshold regime of undoped, nanoscale DG MOSFETs [9,10]. Here, we present new modeling results based on this technique applied to a wider range of operation.

V.N. Alexandrov et al. (Eds.): ICCS 2006, Part IV, LNCS 3994, pp. 607 – 614, 2006.
© Springer-Verlag Berlin Heidelberg 2006

In Section 2, we discuss the conformal mapping as applied to the Laplaces problem in DG MOSFETs. In Section 3, we present a classical analysis of the self-consistent electrostatics of the device at gate voltages where the concentration of free electrons is significant.

The specific device considered is assumed to have a gate length of L =25 nm, a silicon thickness of t_{Si} = 12 nm, a p-type body doping of $N_a = 10^{15}$ cm-3, an aluminum metal gate, and a high-k gate insulator (nitrated Si-oxide) with a relative permittivity of 7 and a thickness of t_{ox} = 1.6 nm [11]. In such a small device volume, the depletion charge can be neglected. Because of the small dimensions of this device, the drain current will have the character of both drift-diffusion and ballistic transport. However, we only consider the drift-diffusion formalism since it allows us to make comparisons with numerical calculations using the Atlas simulator from Silvaco.

2 Conformal Mapping and Capacitive Coupling

The method of conformal mapping is applied since the solution of the Laplace equation is more easily derived in the new plane into which the device is mapped. This solution is then mapped back to the normal plane using a mapping function for the coordinates between the two planes.

2.1 Conformal Mapping

The device body described in the normal (x, y) -plane, as depicted in Fig. 1 (with 2D equipotential lines indicated), is mapped into the upper half of a complex (u, iv) -plane.

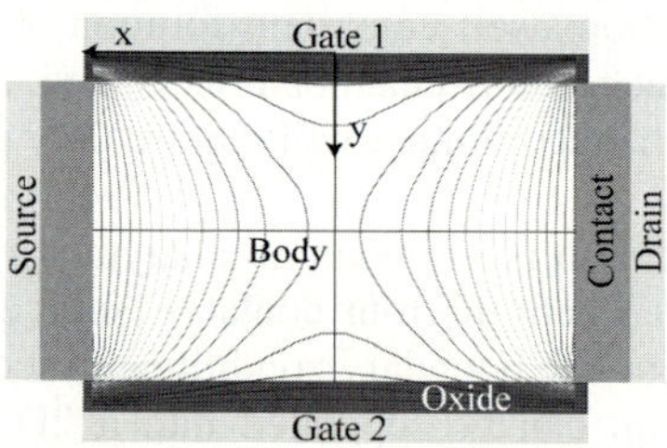

Fig. 1. Schematic view of the DG MOSFET device structure indicating the 2D equipotential lines associated with capacitive coupling between the contacts

The boundary of the body is mapped into the real u-axis [9], as shown in Fig. 2. The iv-axis represents the gate-to-gate symmetry line through the body center. To simplify the discussion, we replace the insulator of thickness t_{ox} by an electrostatically equivalent silicon layer of thickness $t'_{ox}=t_{ox}\varepsilon_s/\varepsilon_{ox}$, where ε_s and ε_{ox} are the permittivities of the silicon and the insulator, respectively. Laplace's equation is then considered for this extended body whose boundary is defined by the inner surfaces of the gate electrodes, the source, the drain, and the insulator gaps in the four corners between the contacts. In the strongly doped source and drain contacts, the depletion widths will be

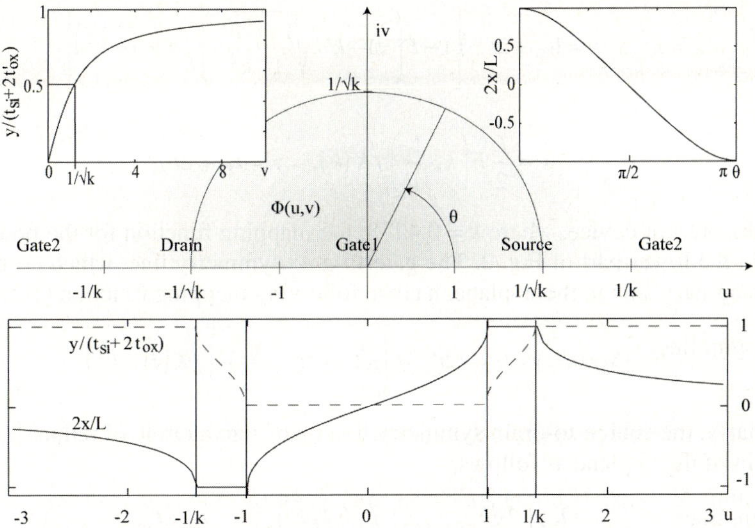

Fig. 2. The body of the DG MOSFET mapped into the upper half of the (u, iv)-plane. The insets show the mapping functions for the u-axis (lower), the iv-axis (upper left) and the circle with radius $1/\sqrt{k}$. These represent the boundary, the gate-to-gate symmetry line, and the source-to-drain symmetry line, respectively.

small compared to the body dimensions and can be neglected, although the potential drops within this region should be counted.

The mapping of the boundary is defined by the following Schwartz-Christoffel transformation [5,9,10]:

$$z = x + iy = \frac{L}{2}\frac{F(k,w)}{K(k)} \quad \text{where} \quad F(k,u) = \int_0^w \frac{dw'}{\sqrt{\left(1-w'^2\right)\left(1-k^2w'^2\right)}} \tag{1}$$

Here, $F(k,u + iv)$ is the elliptic integral and $K(k) \equiv F(k,1)$ is the complete elliptic integral, both of the first kind. The modulus k is a constant between 0 and 1 determined by the geometric ratio $L/(t_{si} + 2t'_{ox})$. For real arguments in the standard range $0 \le u \le 1$, ample approximate expressions, series expansions, and iteration routines exist for $F(k,u)$. $F(k, u + iv)$ can also be expressed in terms of the standard elliptic integral for some other values of the argument, both real and complex. In addition, routines exist for calculating the values for general complex arguments w.

Note that in Fig. 2, $u = 0$ corresponding to $x = 0$ defines the middle point on the upper gate contact (Gate 1). The four corners of the body map to $u = \pm 1$ and $u = \pm 1/k$. The middle point in the lower gate contact (Gate 2) is at $u = \pm\infty$ or $v = \infty$.

For the boundary, the mapping functions are given by the following expressions in terms of the standard elliptic integral (note that $F(k,-u) = - F(k,u)$).

Gate1: $$x = \frac{L}{2}\frac{F(k,u)}{K(k)}, \quad y = 0 \tag{2}$$

S and D:
$$x = \pm\frac{L}{2}, \quad y = (t_{Si} + 2t'_{ox})\left[1 - F\left(\sqrt{1-k^2}, \sqrt{\frac{1-k^2u^2}{1-k^2}}\right)\Big/K\left(\sqrt{1-k^2}\right)\right] \tag{3}$$

Gate2:
$$x = \frac{L}{2}F\left(k, \frac{1}{ku}\right)\Big/K(k), \quad y = t_{Si} + 2t'_{ox} \tag{4}$$

For the present device, where $k = 0.4278$, the mapping function for the boundary is shown in the lower part of Fig. 2. The gate-to-gate symmetry line, which corresponds to the imaginary axis in the w-plane, has the following mapping function [10]:

G1-G2 sym. line:
$$x = 0, \quad y = (t_{Si} + 2t'_{ox})F\left(\sqrt{1-k^2}, \frac{v}{\sqrt{1+v^2}}\right)\Big/K\left(\sqrt{1-k^2}\right) \tag{5}$$

Similarly, the source-to-drain symmetry line maps into a circle of radius $1/\sqrt{k}$ about the origin of the w-plane as follows,

S–D sym. line:
$$x = \frac{L}{2}F\left(\frac{2\sqrt{k}}{1+k}, \cos(\theta)\right)\Big/K\left(\frac{2\sqrt{k}}{1+k}\right), \quad y = \frac{t_{Si} + t'_{ox}}{2} \tag{6}$$

where θ is the angle measured anticlockwise from the positive u-axis. The mapping functions for these two symmetry lines are shown in the upper left and right insets of Fig. 2, respectively.

Once the potential distribution has been calculated in the (u, iv)-plane, it can be mapped back into the (x, iy)-plane using the above expressions.

2.2 Capacitive Coupling

The potential distribution throughout the body can generally be expressed as [6]

$$\varphi(u,v) = \frac{v}{\pi}\int_{-\infty}^{\infty}\frac{\varphi(u')}{(u-u')^2 + v^2}du' \tag{7}$$

where $\varphi(u')$ is the electrostatic potential along the boundary, i.e. for all values of u', and the integral runs over the entire boundary. The major contributions to this integral come from the four equipotential contacts and minor terms come from the insulator at the four corners. In the limit of zero insulator thickness, Eq. (7) results in the following analytical expression for the potential distribution in the w-plane [10],

$$\varphi(u,v) = \frac{1}{\pi}\left\{\begin{aligned}&(V_{GS2} - V_{FB})\left[\pi - \tan^{-1}\left(\frac{1-ku}{kv}\right) - \tan^{-1}\left(\frac{1+ku}{kv}\right)\right] + (V_{GS1} - V_{FB})\times \\ &\left[\tan^{-1}\left(\frac{1-u}{v}\right) + \tan^{-1}\left(\frac{1+u}{v}\right)\right] + V_{bi}\left[\tan^{-1}\left(\frac{1-ku}{kv}\right) - \tan^{-1}\left(\frac{1-u}{v}\right)\right] \\ &+ (V_{bi} + V_{DS})\left[\tan^{-1}\left(\frac{1+ku}{kv}\right) - \tan^{-1}\left(\frac{1+u}{v}\right)\right]\end{aligned}\right\} \tag{8}$$

In our calculations, the effects of a finite insulator thickness are included. V_{GS1} and V_{GS2} are the potentials of Gate 1 and Gate 2, respectively, referred to the source contact, V_{FB} is the flat-band voltage for the gates and the silicon body, V_{bi} is

takes place in these units. The activation of a hidden unit (neuron j) is a function f_j of the weighted inputs plus a bias, as given in equation (1).

$$x_{pj} = f_j\left(\sum_i w_{ji}\, x_{pi} + \theta_j \right) = f_j(y_{pj}). \tag{1}$$

Where w_{ji} is the weight of input i to neuron j, x_{pi} is input i, that is, output i from the previous layer, for input pattern p and θ_j is the threshold value. The output of the hidden units is distributed over the next layer of $x_{h,2}$ hidden units until the last layer of hidden units, of which the outputs are fed into a layer of x_o output units [1].

3 Development of Neural Network Transistor Models

To find the most appropriate channel parameters for different voltage and current values, MOSFET has to be simulated in a large region of input voltages. It is essential for a successful approximation. The variable parameters for simulations are W, L channel parameters and V_{GS}, V_{DS} node voltages. Adding the output value I_D of the simulations, the data files are created in HSPICE environment to train the neural network in MATLAB 6.0 program. Fig. 1 shows the n- and p-channel MOSFET circuit connections to obtain the training and test data for single transistors.

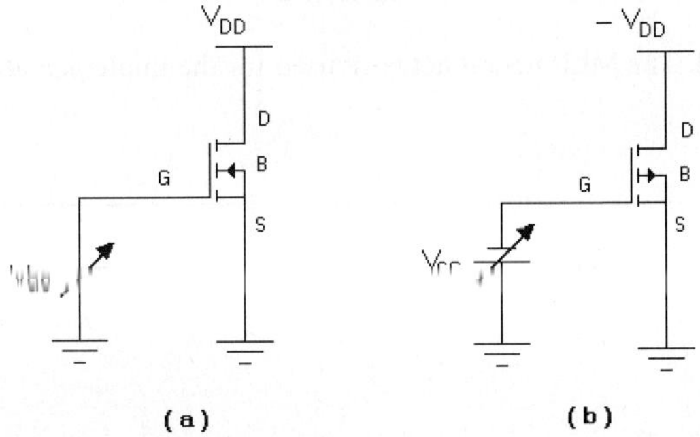

Fig. 1. MOSFET circuits for producing training data: a) n-channel, b) p-channel

The MOSFETs were modeled between 1 V to 3.3 V range. The neural network structure shown in Fig.2 was used to model different operation regions. The MLP network consists of three inputs, three hidden layers and two output neurons. Activation functions of hidden units were tangent hyperbolic sigmoid and output was purelinear. The model has three separate training datasets between 1V and 3.3V input voltage range since the MOSFET cannot be accurately modeled in a wide input gate-source voltage range.

Gate-source potential V_{GS}, drain-source potential V_{DS} and drain current I_D were applied to the inputs of MLPs. The outputs of the MLPs were channel width W and effective channel length L. Both channel length and width were varied between 0.18μ to 7μ. In this range, training data were obtained with different step sizes for each channel parameter. Once again using different step sizes for gate-source and drain-source voltages between 1V to 3.3V depending on the interval, the training is occurred. The test data were obtained randomly in the same range and they were different from the training data. Over 200K data points were obtained from HSPICE simulations. The simulation of the neural network was performed in the MATLAB 6.0 program. 100 randomly chosen test data were applied for testing.

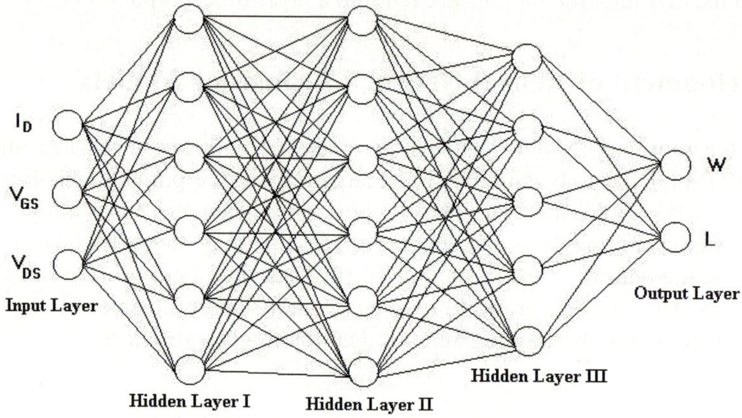

Fig. 2. The MLP neural network used for the implementation

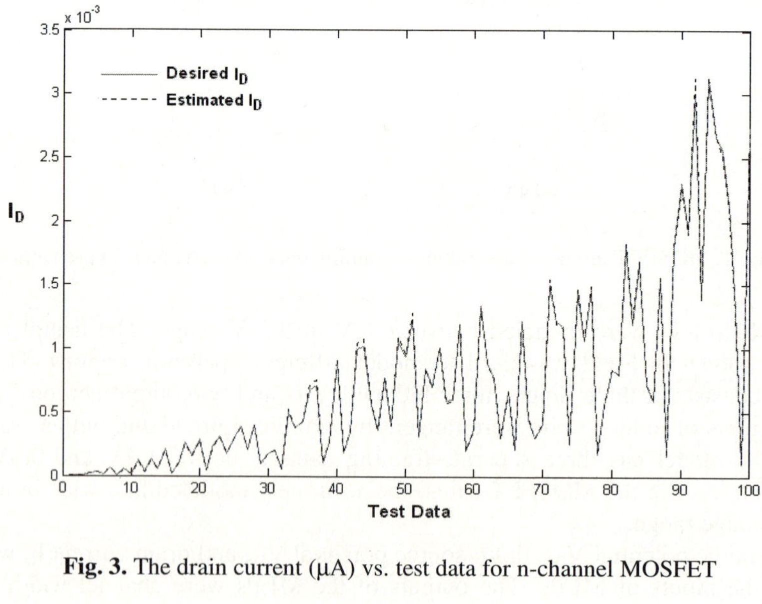

Fig. 3. The drain current (μA) vs. test data for n-channel MOSFET

After the gate-source, drain-source voltages and required drain currents were applied to the neural networks, the estimated aspect ratios were simulated in HSPICE to check the validity of drain currents at the same input voltages.

The neural network size in Fig. 2 was obtained by trial and error. Different network architectures were trained and tested. Finally, architecture shown in Fig. 2 gave the best overall performans.

The figures illustrate that the estimated and required drain currents are very close to each other with a maximum error of 8.3%. Since the channel parameters might increase with the half of the resolution steps, error reached the given value. However, the neural network output channel parameters were more accurate with respect to this error, estimated channel parameters were applied with the suitable resolution values. This proves the success of the neural network estimation. Performances of the test data for n-channel and p-channel MOSFETs were shown in Fig.3 and 4. For all figures the vertical axis is the amplitude of the drain current and the horizontal one represents the test data. The sign showing current direction is not considered. The dotted black lines in all figures represent the drain current with estimated aspect ratio and the solid grey lines show the desired drain currents.

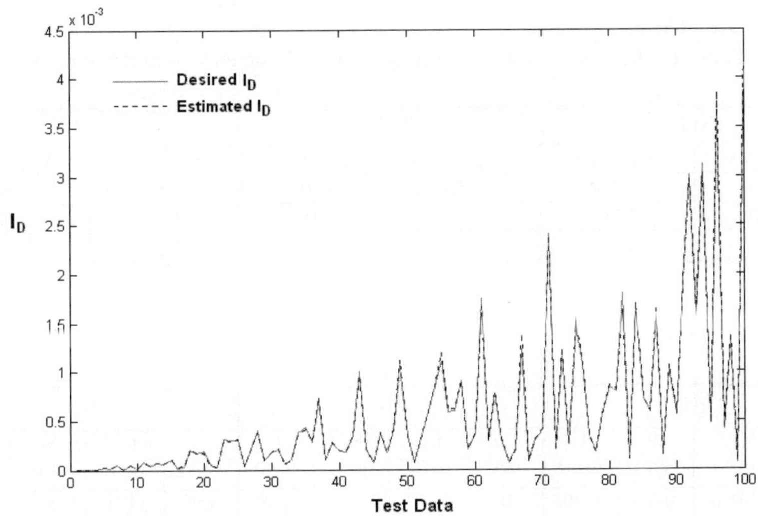

Fig. 4. The drain current (μA) vs. test data for p-channel MOSFET

4 The Implementation Circuits

The developed neural network was applied to some main building blocks of analog integrated circuit design. These are the basic current mirror circuit and a differential amplifier which are very essential for most analog circuits. Each transistor in the circuit blocks assumed as a single transistor and designer decided the gate-source, drain-source voltages and drain current. This flexibility of design is supported by the neural network.

4.1 The Basic Current Mirror Circuit

The circuit in Fig. 5 is designed using the developed neural network. The node voltages for each numbered node are shown in Table 1. The required current, estimated current, estimated channel length (L) and width (W) values for the current mirror circuit are given in Table 2.

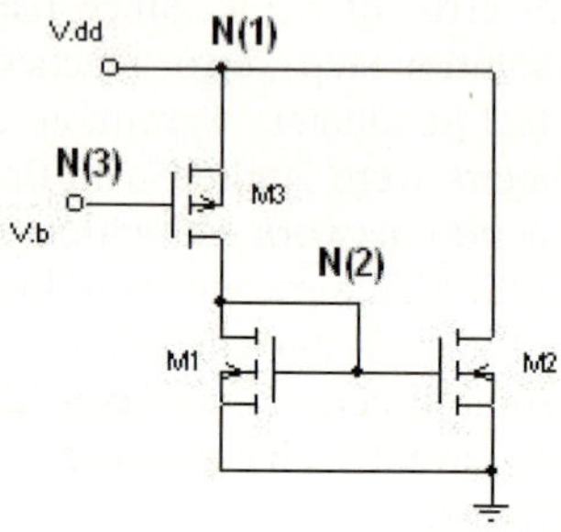

Fig. 5. The Basic Current Mirror Circuit

Table 1. The voltage values of the nodes in te basic current mirror circuit

N(1)		N(2)		N(3)	
V_{des} (V)	V_{sim} (V)	V_{des} (V)	V_{sim} (V)	V_{des} (V)	V_{sim} (V)
3.30	3.30	1.50	1.49	1.50	1.50
3.30	3.30	1.00	1.01	2.00	2.00
3.30	3.30	1.30	1.29	2.20	2.20

Table 2. The desired and simulation current values, estimated channel width and length

M1				M2				M3			
I_{des} (μA)	I_{sim} (μA)	W (μm)	L (μm)	I_{des} (μA)	I_{sim} (μA)	W (μm)	L (μm)	I_{des} (μA)	I_{sim} (μA)	W (μm)	L (μm)
150	148.0	0.69	0.49	300	294.3	1.38	0.49	150	148.0	1.68	0.52
100	102.6	1.37	0.50	30	31.8	0.38	0.54	100	102.6	2.49	0.51
50	50.4	0.33	0.54	900	895.3	2.40	0.22	50	50.4	1.98	0.52

4.2 The Basic Differential Amplifier Circuit

The circuit in Fig.6 is designed using the developed neural network. The node voltages for each numbered node are shown in Table 3. The required current, estimated current, estimated channel length and width values for the differential amplifier circuit are shown in Table 4.

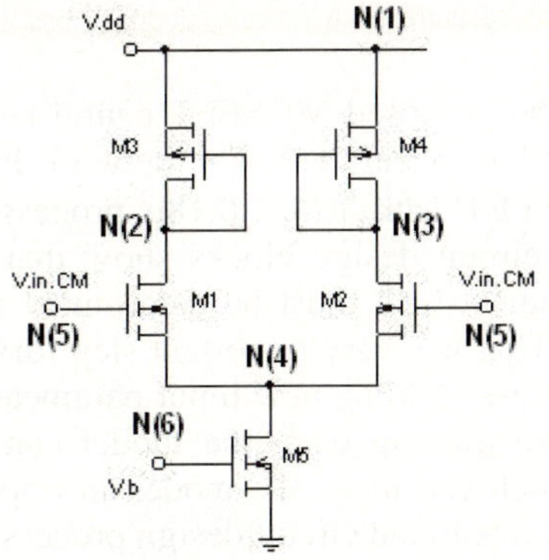

Fig. 6. The Basic Differential Amplifier Circuit

Table 3. The voltage values of the nodes in differential amplifier circuit

N(1)		N(2)		N(3)	
V_{des} (V)	V_{sim} (V)	V_{des} (V)	V_{sim} (V)	V_{des} (V)	V_{sim} (V)
3.30	3.30	2.30	2.31	2.30	2.31
3.30	3.30	2.00	1.93	2.00	1.93
3.30	3.30	2.00	2.02	2.00	2.02
N(4)		**N(5)**		**N(6)**	
V_{des} (V)	V_{sim} (V)	V_{des} (V)	V_{sim} (V)	V_{des} (V)	V_{sim} (V)
0.70	0.70	1.70	1.70	1.00	1.00
0.80	0.78	1.80	1.80	1.30	1.30
0.75	0.75	1.60	1.60	0.90	0.90

Table 4. The desired and simulation current values, estimated channel width and length

M1				M2				M3			
I_{des} (μA)	I_{sim} (μA)	W (μm)	L (μm)	I_{des} (μA)	I_{sim} (μA)	W (μm)	L (μm)	I_{des} (μA)	I_{sim} (μA)	W (μm)	L (μm)
140	139	1.94	0.49	140	139	1.94	0.49	140	139	4.96	0.34
375	402	2.53	0.20	375	402	2.53	0.20	375	402	5.11	0.31
25	24.1	0.59	0.52	25	24.1	0.59	0.52	25	24.1	0.61	0.53

M4				M5			
I_{des} (μA)	I_{sim} (μA)	W (μm)	L (μm)	I_{des} (μA)	I_{sim} (μA)	W (μm)	L (μm)
140	139	4.96	0.34	280	278	2.33	0.25
375	402	5.11	0.31	750	805	2.69	0.18
25	24.1	0.61	0.53	50	48.3	0.96	0.50

5 Conclusion

The test results prove that the proposed MOSFET neural network model can decide channel width and length values accurately. The network has a very close function approximation for the MOSFET with TSMC 0.18µ process technology parameters. The applications of analog circuit design blocks show that the model can find the appropriate channel parameters which must be determined by the designer with his experience and knowledge. That is a very important step forward for complex analog and digital VLSI design process. Adding new input parameters to the neural network structure and obtaining more training data, the model can produce more accurate results in a wider range, which can make the model an important tool for designers during the analog and digital integrated circuit design process.

Acknowledgements

This research has been supported by Yildiz Technical University Scientific Projects Coordination Department. Project Number: 24-04-03-02.

References

1. Avci, M., Babac, M. Y., Yildirim, T.,: Neural Network Based MOSFET Channel Length and Width Decision Method for Analogue Integrated Circuits, International Journal of Electronics, Vol. 92, No. 5, May 2005, 281-293
2. Avci, M., Babac, M. Y., Yildirim, T.,: Neural Network Based Transistor Modeling and Aspect Ratio Estimation for YITAL 1.5□ process, ELECO 2003, International Conference Proceedings, Bursa, Turkey, 2003, pp. 54-57
3. Gunes, F., Gurgen, F., Torpi, H.: Signal-Noise Neural Network Model for Active Microwave Devices, Circuits Devices and Systems, Vol. 143. IEE Proceedings (1996) 1-8
4. Gunes, F., Torpi, H., Gurgen, F.: Multidimensional Signal-Noise Neural Network Model, Circuits Devices and Systems, Vol. 145. IEE Proceedings (1998) 111-117
5. Yildirim, T., Torpi, H., Özyilmaz, L.: Modelling of Active Microwave Transistors Using Artificial Neural Networks, Proceedings of IJCNN'99 Int. Joint Conf. on Neural Networks, Vol. 6. IEEE publication, Washington (1999) 3988–3991
6. Gunes, F., Torpi, H., Çetiner, B.A.: Neural Network Modeling of Active Devices for Use in MMIC Design, Artificial Intelligence in Engineering, Vol. 13. Elsevier (1999) 385–392
7. Kothapalli, G. M.,: Artificial Neural Networks as Aid in Circuit Design, Microelectronics Journal, 26, 1995, 569-578
8. Langenheine, J., Folling, S., Meier, K., Schemmel, J.: Towards a Silicon Primordial Soup: A Fast Approach to Hardware Evaluation with a VLSI Transistor Array, ICES 2000, LNCS 1801, 2000, 123-132
9. www.mosis.org : for TSMC 0.18µ technology parameters. (2003)
10. Hush, D., R., Horne, B., G.: Progress in Supervised Neural Networks, IEEE Signal Processing Magazine, January (1993), 8-39
11. Geva, S., Sitte, J.: A Constructive Method for Multivariate Function Approximation by Multilayer Perceptrons, IEEE Transactions on Neural Networks, Vol. 3 (4) (1992) 623-624

Vlasov-Maxwell Simulations in Singular Geometries

Franck Assous[1] and Patrick Ciarlet Jr.[2]

[1] Bar-Ilan University, 52900 Ramat-Gan, Israel
and College of Judea&Samaria, Ariel, Israel
franckassous@netscape.net

[2] ENSTA,32 bvd Victor, 75739, Paris Cedex 15
ciarlet@ensta.fr

Abstract. This paper is devoted to the solution of the time-dependent Vlasov-Maxwell equations in singular geometries, i.e. when the boundary includes reentrant corners or edges. Indeed, computing the electromagnetic fields in this case is a challenge *per se*, as these geometrical singularities generate very strong solutions in their neighborhood. Moreover, they have also an influence over the solution of the Vlasov equation, through the coupling. We propose here a method to solve this problem, illustrated by numerical examples.

1 Introduction

The numerical simulation of charged particles beams or plasma physics phenomena requires methods of solution for the time-dependent coupled Vlasov-Maxwell system. Within this framework, we developed a numerical method (see [1]), with continuous approximations of the electromagnetic field, which is recommended in order to reduce spurious oscillations. In addition, the time stepping numerical scheme, which is explicit by construction, can be solved very efficiently. Finally, the conditions on the divergence of the fields are considered as constraints, and are dualized, using a Lagrange multiplier, which yields a saddle-point variational formulation.

In practical examples, the boundary of the computational domain includes reentrant corners and/or edges (called *geometrical singularities*) that generate strong fields. Hence, they require a careful computation of the electromagnetic field in their neighborhood.

We developed a method (see [2]), the so-called *Singular Complement Method* (*SCM* hereafter), which consists in splitting the space of solutions into a two-term sum. The first subspace is made of regular fields, and coincides with the whole space of solutions, provided that the domain is either convex or regular. So, one can compute the regular part of the solution with the help of an *ad hoc* – classical – method [1]. The second one is called the subspace of *singular fields*, and is computed with the help of specifically designed methods: they originate from relations between the electromagnetic singularities and the singularities of the Laplace operator.

V.N. Alexandrov et al. (Eds.): ICCS 2006, Part IV, LNCS 3994, pp. 623–630, 2006.
© Springer-Verlag Berlin Heidelberg 2006

As a first attempt, the *Singular Complement Method* was constructed in a divergence-free framework (cf. [2]). When the divergence of the electric field no longer vanishes, $\operatorname{div}\mathcal{E} = f(t)$, with $f \neq 0$, a simple solution is to substract a gradient, to reach the divergence-free field

$$\tilde{\mathcal{E}} = \mathcal{E} - \operatorname{grad}\phi.$$

Unfortunately, to determine ϕ, one has to solve the time-dependent (*via* the data) problem

$$-\Delta\phi = f(t)\,,$$

which slows down drastically the numerical implementation.

To alleviate this drawback, we studied in detail different splittings of the electromagnetic space, which could be used for the SCM (cf. [3]). In addition to the divergence-free splittings, we propose new splittings, direct and possibly orthogonal, with curl-free singular fields, or with singular fields with L^2 divergence, etc.

As an important application – actually, the origin of this study – we present the computation of strong electromagnetic fields, *via* the numerical solution to the coupled, non linear, Vlasov-Maxwell system of equations. In Section 2, we recall Vlasov-Maxwell equations, and describe the *SCM* well-suited to this framework. Section 3 is devoted to the numerical algorithms. Numerical experiments of the coupled Vlasov-Maxwell are presented in Section 4.

2 Vlasov-Maxwell Equations

Let Ω be a bounded, open, polyhedral subset of $\mathbb{R}^3$, and Γ its boundary. We denote by $\mathbf{n}$ the unit outward normal to Γ. The Vlasov equation models the transport of charged particles, under the influence of an electromagnetic field. The Vlasov equation reads

$$\frac{\partial f}{\partial t} + \mathbf{v}\cdot\nabla_{\mathbf{x}} f + \frac{\mathbf{F}}{m}\cdot\nabla_{\mathbf{v}} f = 0\,. \tag{1}$$

Above, the unknown f is the distribution function of the particles, $f(\mathbf{x},\mathbf{v},t)$, $\mathbf{v}$ stands for the velocities of the particles, m is the mass of a particle, and

$$\mathbf{F} = e(\mathcal{E} + \mathbf{v}\times\mathcal{B}) \tag{2}$$

is the the well-known Lorentz force. The electric and magnetic fields $\mathcal{E}$ and $\mathcal{B}$ are solution of the Maxwell equations in vacuum

$$\frac{\partial\mathcal{E}}{\partial t} - c^2\,\mathbf{curl}\,\mathcal{B} = -\frac{1}{\varepsilon_0}\mathcal{J}, \tag{3}$$

$$\frac{\partial\mathcal{B}}{\partial t} + \mathbf{curl}\,\mathcal{E} = 0, \tag{4}$$

$$\operatorname{div}\mathcal{E} = \frac{\rho}{\varepsilon_0}, \tag{5}$$

$$\operatorname{div}\mathcal{B} = 0, \tag{6}$$

where c, ε_0 and μ_0 are respectively the light velocity, the electric and magnetic constants ($\varepsilon_0\mu_0 c^2 = 1$). The charge and current densities ρ and $\mathcal{J}$ have to verify the charge conservation equation

$$\frac{\partial \rho}{\partial t} + \operatorname{div} \mathcal{J} = 0. \tag{7}$$

Let us assume for simplicity that the boundary Γ is a perfect conductor, so we have $\mathcal{E} \times \mathbf{n} = 0$ and $\mathcal{B} \cdot \mathbf{n} = 0$ on Γ. These equations are supplemented with appropriate initial conditions. Remark that the coupling occurs on the one hand, by the right-hand sides of Maxwell equations, ρ and $\mathcal{J}$, which are computed from the solution to the Vlasov equation $f(\mathbf{x}, \mathbf{v}, t)$, thanks to the relations

$$\rho = \int_{\mathbf{v}} f \, d\mathbf{v}, \qquad \mathcal{J} = \int_{\mathbf{v}} f \, \mathbf{v} \, d\mathbf{v}\,; \tag{8}$$

on the other hand, the electromagnetic field $(\mathcal{E}, \mathcal{B})$ determines the forces that act on the particles *via* the Lorentz force $\mathbf{F}$.

Geometrical singularities have no effect *per se* on the regularity of the solution to the Vlasov equation. Therefore, we resort to a particle method [4]: it consists in approximating the distribution function $f(\mathbf{x}, \mathbf{v}, t)$ by a linear combination of Dirac masses in the phase space $(\mathbf{x}, \mathbf{v})$

$$f(\mathbf{x}, \mathbf{v}, t) \simeq \sum_k w_k \, \delta(\mathbf{x} - \mathbf{x}_k(t))\delta(\mathbf{v} - \mathbf{v}_k(t)), \tag{9}$$

where each term of the sum can be identified with a macro-particle, characterized by its weight w_k, its position $\mathbf{x}_k$ and its velocity $\mathbf{v}_k$. This distribution function is a solution of the Vlasov equation (1) if and only if $(\mathbf{x}_k, \mathbf{v}_k)$ is a solution of the differential system:

$$\frac{d\mathbf{x}_k}{dt} = \mathbf{v}_k \,, \tag{10}$$

$$\frac{d\mathbf{v}_k}{dt} = \mathbf{F}(\mathbf{x}_k, \mathbf{v}_k) \,, \tag{11}$$

which describes the time evolution of a macro-particle k, submitted to the electromagnetic force $\mathbf{F}$.

The system (10-11) can be numerically solved by an explicit time discretization algorithm. We used a leapfrog scheme which is well-adapted in this case. Given a constant time step Δt, the particles positions are defined at time $t_n = n\Delta t$ and the particle velocities are computed at time $t_{n+1/2} = (n + 1/2)\Delta t$. We refer the reader to [5] for more details. It is a classical approach in Particle In Cell (PIC) approach.

Even if geometrical singularities have no effect on the regularity of the solution to the Vlasov equation, they have an influence over f, through the *coupling*, i.e through the force $\mathbf{F}$. Hence, the electromagnetic field must be computed accurately. To this purpose, the *SCM* was introduced in [2], first for divergence-free problems.

Let us present here a way to generalize this approach to Vlasov-Maxwell problems. As we are interested in non divergence-free solutions, we will only present the electric formulation. Details on the magnetic counterpart can be found in [6]. Let us recall the definitions of the following spaces

$$\begin{aligned}\mathbf{H}(\mathbf{curl}\,,\Omega) &= \{\mathbf{u} \in \mathbf{L}^2(\Omega), \mathbf{curl}\,\mathbf{u} \in \mathbf{L}^2(\Omega)\}\\ \mathbf{H}(\mathrm{div}\,,\Omega) &= \{\mathbf{u} \in \mathbf{L}^2(\Omega), \mathrm{div}\,\mathbf{u} \in L^2(\Omega)\}\\ \mathbf{H}^1(\Omega) &= \{\mathbf{u} \in \mathbf{L}^2(\Omega), \mathbf{grad}\,\mathbf{u} \in \mathbf{L}^2(\Omega)\}\,.\end{aligned}$$

We define the space of electric fields $\mathcal{E}$, called $\mathbf{X}$,

$$\mathbf{X} = \{\mathbf{x} \in \mathbf{H}(\mathbf{curl}\,,\Omega) \cap \mathbf{H}(\mathrm{div}\,,\Omega)\ :\ \mathbf{x} \times \mathbf{n}_{|\Gamma} = 0\}\,.$$

When the domain is convex, the spaces $\mathbf{X}$ is regular. That is not the case anymore in a singular domain (see for instance [7]). Hence, one introduces the regular subspace for electric fields (indexed with $_R$)

$$\mathbf{X}_R = \mathbf{X} \cap \mathbf{H}^1(\Omega)\,, \tag{12}$$

which is actually closed in $\mathbf{X}$ (cf. [6]), so that one is able to consider its orthogonal, and then define a two-part, direct, and orthogonal sum of the space. The orthogonal subspace is called singular subspacs (indexed with $_S$). One can write

$$\mathbf{X} = \mathbf{X}_R \overset{\perp_{\mathbf{X}}}{\oplus} \mathbf{X}_S \tag{13}$$

Thus, one can split an element $\mathbf{u}$ of the space $\mathbf{X}$ into an orthogonal sum of a *regular* part and of a *singular* part: $\mathbf{u} = \mathbf{u}_R + \mathbf{u}_S$. We have now to characterize the singular electric fields. Following [6], elements $\mathbf{x}_S \in \mathbf{X}_S$ satisfy

$$\boldsymbol{\Delta}\mathbf{x}_S = 0 \quad \text{in } \Omega\,, \tag{14}$$

$$\mathbf{x}_S \times \mathbf{n}_{|\Gamma} = 0\,. \tag{15}$$

3 Numerical Algorithms

The numerical method consists in first computing numerically the basis of the singular subspace. Then we solve the problem by coupling a classical method (to compute the regular part of the solution) to the linear system, which allows to compute the singular part of the solution.

To compute $\mathbf{x}_S \in \mathbf{X}_S$, we introduce its divergence- and curl-free parts $\mathbf{v}_S$ and $\mathbf{l}_S$, which verify the following Helmholtz decomposition

$$\mathbf{x}_S = \mathbf{v}_S + \mathbf{l}_S\,. \tag{16}$$

We also introduce s_N and s_P, the non-vanishing (singular) solutions of

$$\Delta s_N = 0, \qquad \Delta s_P = 0 \qquad \text{in } \Omega, \tag{17}$$

respectively with Neumann and Dirichlet homogeneous boundary condition. Assume that s_N and s_P are known, then one can use the singular mappings (see [6]), to find ϕ_S and ψ_S such that

$$-\Delta\phi_S = s_N, \qquad -\Delta\psi_S = s_P \qquad \text{in } \Omega, \tag{18}$$

still with Neumann and Dirichlet homogeneous boundary condition. Finally, one obtains the singular basis functions $\mathbf{x}_S$ of $\mathbf{X}_S$ by the aid of the relations

$$\mathbf{v}_S = \mathbf{curl}\,\phi_S, \qquad \mathbf{l}_S = \mathbf{grad}\,\psi_S\,, \tag{19}$$

together with relation (16). Hence, the keypoint is to compute s_N or s_P. To this purpose, there exists several methods ([2], [8], etc.). We recall here briefly the method we choose, the *Principal Part Method* (see [9] for details). Consider, for simplicity reasons, a domain with one singularity. The *Principal Part Method* consists in splitting s_N in a regular part $\tilde{s}_N$ (which belongs to $H^1(\Omega)$) and a known singular part s_N^P

$$s_N = s_N^P + \tilde{s}_N\,.$$

Above, s_N^P belongs to $L^2(\Omega)$ but not to $H^1(\Omega)$, and verifies $\Delta s_N^P = 0$. One thus computes, for instance by a P^1 finite element method, $\tilde{s}_N$ by solving

$$\Delta\tilde{s}_N = 0 \text{ in } \Omega, \qquad \frac{\partial\tilde{s}_N}{\partial\boldsymbol{\nu}} = -\frac{\partial s_N^P}{\partial\boldsymbol{\nu}} \text{ on } \Gamma, \tag{20}$$

avoiding thus mesh rafinement techniques.

One still uses singular mappings [6] to find the singular field $\phi_S \in H^1(\Omega) \cap L_0^2(\Omega)$ such that

$$-\Delta\phi_S = s_N \text{ in } \Omega, \qquad \frac{\partial\phi_S}{\partial\boldsymbol{\nu}} = 0 \text{ on } \Gamma$$

Again, one can split this field in a regular part $\tilde{\phi}_S$ (which belongs to $H^2(\Omega)$) and a singular part ϕ_S^P

$$\phi_S = \tilde{\phi}_S + C_\phi \phi_S^P\,,$$

where C_ϕ is a constant to be determined. The principal parts ϕ_S^P is harmonic and does not belong to $H^2(\Omega)$. Its analytic expression is known. The regular part $\tilde{\phi}_S$ can be computed easily, by solving a standard variational formulation. The singular function s_P and the singular field ψ_S are computed in the same way.

Then, using a final time the singular mappings, one can compute the singular electromagnetic basis functions. To determine the basis $\mathbf{v}_S$ (resp. $\mathbf{l}_S$), one simply takes the curl of ϕ_S (resp. the gradient of ψ_S)

$$\mathbf{v}_S = \mathbf{curl}\,\tilde{\phi}_S + C_\phi \mathbf{curl}\,\phi_S^P\,, \tag{21}$$
$$\mathbf{l}_S = \nabla\tilde{\psi}_S + C_\psi \nabla\psi_S^P\,, \tag{22}$$

and $\mathbf{x}_S$ is easily obtained with (16). The singular constant C_ϕ can be computed by using a formula derived from integration by parts (cf. [9]). As an example, one gets

$$C_\phi = \frac{\|s_N\|^2}{\displaystyle\int_\Gamma s_N \frac{\partial \phi_S^P}{\partial \boldsymbol{\nu}} d\Gamma}.$$

Now, we have to modify a classical method by handling the decomposition (13). We first write Ampère and Faraday's laws as two second-order in time equations. To enforce the divergence constraints on the electromagnetic field, we introduce two Lagrange multipliers (say p for the electric field), to dualize Coulomb's and absence of free magnetic monopole's laws. In this way, one builds a mixed variational formulation (*VF*) of Maxwell equations. For the electric field (the case of the magnetic field is handled similarly), this formulation reads
Find $(\mathcal{E}, p) \in \mathbf{X} \times L^2(\Omega)$ *such that*

$$\frac{d^2}{dt^2}\int_\Omega \mathcal{E}\cdot\mathbf{F}\,d\mathbf{x} + c^2\int_\Omega \mathbf{curl}\,\mathcal{E}\cdot\mathbf{curl}\,\mathbf{F}\,d\mathbf{x} + c^2\int_\Omega \operatorname{div}\mathcal{E}\operatorname{div}\mathbf{F}\,d\mathbf{x} + \int_\Omega p\operatorname{div}\mathcal{E}\,d\mathbf{x}$$
$$= -\frac{1}{\varepsilon_0}\frac{d}{dt}\int_\Omega \mathcal{J}\cdot\mathbf{F}\,d\mathbf{x} + \frac{c^2}{\varepsilon_0}\int_\Omega \rho\operatorname{div}\mathbf{F}\,d\mathbf{x}, \qquad \forall\mathbf{F}\in\mathbf{X}, \tag{23}$$

$$\int_\Omega \operatorname{div}\mathcal{E}\,q\,d\mathbf{x} = \frac{1}{\varepsilon_0}\int_\Omega \rho\,q\,d\mathbf{x} \qquad \forall q\in L^2(\Omega). \tag{24}$$

To include the *SCM* in this formulation, the electric field $\mathcal{E}$ is split like $\mathcal{E}(t) = \mathcal{E}_R(t) + \mathcal{E}_S(t)$. The same splitting is used for the test functions of the variational formulation, which is discretized in time, with the help of the leap-frog scheme. From a practical point of view, we choose the Taylor-Hood, P_2-iso-P_1 Finite Element. In addition to being well-suited for discretizing saddle-point problems, it allows to build diagonal mass matrices, when suitable quadrature formulas are used. Thus, the solution to the linear system, which involves the mass matrix, is straightforward [1].

Then, one can write down the discrete singular part as a finite sum. Assuming again that there is one singularity, and let $(\mathbf{x}_S)$ be a given basis of the discrete singular space. One has $\mathcal{E}_S(t) = \kappa(t)\mathbf{x}_S$, where κ is continous time-dependent function. This results in a fully discretized VF:

$$\mathbb{M}_\Omega \mathbf{E}_R^{n+1} + \mathbb{M}_{RS}\boldsymbol{\kappa}^{n+1} + \mathbb{L}_\Omega \mathbf{p}^{n+1} = \mathbf{F}^n\,, \tag{25}$$
$$\mathbb{M}_{RS}^T \mathbf{E}_R^{n+1} + \mathbb{M}_S\boldsymbol{\kappa}^{n+1} + \mathbb{L}_S \mathbf{p}^{n+1} = \mathbf{G}^n\,, \tag{26}$$
$$\mathbb{L}_\Omega^T \mathbf{E}_R^{n+1} + \mathbb{L}_S^T\boldsymbol{\kappa}^{n+1} = \mathbf{H}^n\,. \tag{27}$$

Above $\mathbb{M}_\Omega$ denotes the usual mass matrix, and $\mathbb{L}_\Omega$ corresponds to the divergence term involving $\mathbf{x}_R^h$ and the discrete Lagrange multiplier $p_h(t)$. Then, $\mathbb{M}_{RS}$ is a rectangular matrix, which is obtained by taking $\mathbf{L}^2$ scalar products between regular and singular basis functions, $\mathbb{M}_S$ is the "singular" mass matrix, and finally, $\mathbb{L}_S$ corresponds to the divergence term involving $\mathbf{x}_S$ and $p_h(t)$. To solve this system, one first removes the unknown $\boldsymbol{\kappa}^{n+1}$, so that the unknowns $(\mathbf{E}_R^{n+1}, \mathbf{p}^{n+1})$

can be obtained with the help of a Uzawa-type algorithm. Finally, one concludes the time-stepping scheme by computing $\boldsymbol{\kappa}^{n+1}$ with the help of (26).
For error estimates for the singular functions s_N, s_P and ϕ_S, ψ_S, we refer the reader to [10] . As far as the electromagnetic sigular fields are concerned, error estimates are provided in [3]. Let us briefly recall that, using the electric field splitting

$$\mathcal{E}(t) = \mathcal{E}_R(t) + \mathcal{E}_S(t) = \mathcal{E}_R(t) + \kappa(t)\mathbf{x}_S\,,$$

one can prove, for a two-dimensional domain [11]

$$|\kappa - \kappa_h| \leq Ch^{2\alpha}, ||\mathcal{E} - \mathcal{E}_h||_{\mathbf{X}} \leq Ch^{2\alpha-1-\varepsilon}, ||\mathcal{E} - \mathcal{E}_h||^2_{\mathbf{L}} \leq Ch^{4\alpha-2-\varepsilon},$$

where π/α denotes here the reentrant angle of the non-convex domain Ω.

4 Numerical Experiments

We consider an L-shaped domain Ω, and assume its boundary is entirely perfectly conducting. Initial conditions are uniformly set to zero. A bunch of particles – electrons – is emitted from the top-most part of the boundary, with an initial velocity equal to $\mathbf{v} = v_y\boldsymbol{e}_y$, with $v_y = -2.10^8\,ms^{-1}$. The electromagnetic field is therefore a *self-consistent* field. Particles are absorbed at the down-most part of the boundary.

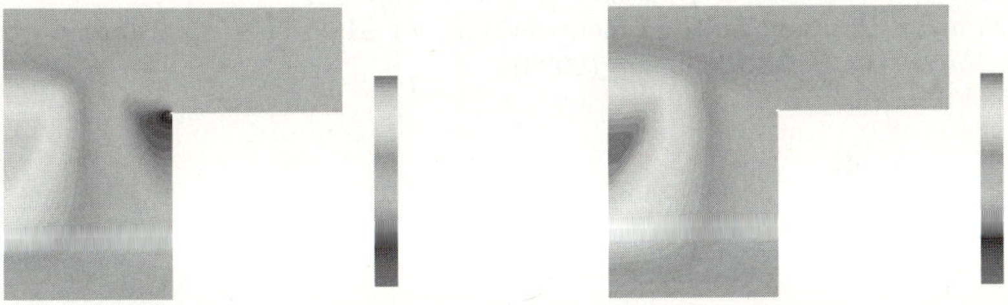

Fig. 1. electric field with (left) and without (right) SCM

As shown Figure 1 the electric field – $\mathcal{E}_x$ component – obtained after 500 time-steps is very different, when it is computed with and without the *SCM*. Recall that the electromagnetic field is the result of the motion of the charged particles only. Therefore, differences are entirely due to the coupling between the Vlasov and Maxwell equations.

5 Conclusion

In this paper, we proposed a new scheme for the time-dependent coupled Vlasov-Maxwell system of equations, in domains with singular geometries. The numerical experiments we have shown and their performance, make the study of such physical configurations possible.

References

1. F. Assous, P. Degond, E. Heintzé, P. A. Raviart, J. Segré, On a finite element method for solving the three-dimensional Maxwell equations, *J. Comput. Phys.*, **109**, 222-237, 1993.
2. F. Assous, P. Ciarlet, Jr., J. Segré, Numerical solution to the time-dependent Maxwell equations in two-dimensional singular domain: The Singular Complement Method, *J. Comput. Phys.*, **161**, 218-249, 2000.
3. F. Assous, P. Ciarlet, Jr., E. Garcia, J. Segré, Time-dependent Maxwell's equations with charges in singular geometries, submitted to *Comput. Methods Appl. Mech. Engrg.*.
4. P.A. Raviart, *An Analysis of Particle Methods*, Springer-Verlag, Berlin, 1985.
5. F. Assous, P. Degond, J. Segré, A particle-tracking method for 3D electromagnetic PIC codes on unstructured meshes, *Comput. Phys. Com.*, **72**, 105-114, 1992.
6. E. Garcia, *Résolution des équations de Maxwell avec charges dans des domaines non convexes*, PhD Thesis, University Paris 6, France, 2002. (in French)
7. P. Grisvard, *Singularities in boundary value problems*, **22**, *RMA* Masson, Paris, 1992.
8. C. Hazard, S. Lohrengel A singular field method for Maxwell's equations: numerical aspects for 2D magnetostatics, *SIAM J. Appl. Math.*, **40**, 1021-1040, 2002.
9. F. Assous, P. Ciarlet, Jr., S. Labrunie, J. Segré, Numerical solution to the time-dependent Maxwell equations in axisymmetric singular domain: The Singular Complement Method, *J. Comput. Phys.*, **191**, 147-176, 2003.
10. P. Ciarlet, Jr.,J. He, The singular complement method for 2D scalar problems, *C. R. Acad. Sci. Paris, Ser. I*, **336**, 809-814 (2005).
11. E. Jamelot, A nodal finite element method for Maxwell's equations, *C. R. Acad. Sci. Paris, Ser. I*, **339**, 353-358 (2003).

Fast Rigorous Analysis of Rectangular Waveguides by Optimized 2D-TLM

Ayhan Akbal and Hasan H. Balik

University of Firat, Department of Electrical and Electronics Engineering, Elazig, Turkey
ayhanakbal@gmail.com, hasanbalik@gmail.com

Abstract. In this paper, The optimized 2D-TLM as been introduced and applied to rectangular waveguides which is widely used. Results obtained by using optimized 2D-TLM were compared with analytic results and shown to be accurate.

1 Introduction

Rectangular waveguide is one of the earliest type of the transmission lines and still commonly used in many current applications. A lot of components such as isolators, detectors, attenuators, couplers and slotted lines are available to use for various standard waveguide bands between 1 GHz to above 220 GHz [1]. At the operating frequencies where these waveguides commonly used, the assumptions which are valid only low frequencies can not be applied to gain accurate results. Therefore full-wave analysis techniques must be required. Some of these full-wave numerical techniques solve the problem in time domain [2-4] whereas others solve in frequency domain [4-7]. Although full-wave numerical technique gives accurate results, it requires more time and computer resources for solutions. The demands of the design engineer require a technique which is accurate, yet retains the interactive design capabilities of the simpler techniques.

In this contribution, time and frequency domain analysis of rectangular waveguide has been accurately analyzed by enhanced 2-D TLM method and shown to be accurate to find mode cut-off frequency.

2 Review to Rectangular Waveguides

The rectangular waveguide shown in Fig. 1 supports both TM and TE modes; therefore it is not possible to define unique voltage by only using TEM waves at the recent operating frequencies. The waves cannot propagate trough the rectangular waveguide if the operating frequency is below then some certain frequency. This frequency is called cut-off frequency. The mode frequency must be higher then this cut-off frequency. If mode frequency is lower then cut-off frequency, propagating waves decay rapidly in the direction of waveguides axes. When the operating frequency is higher then cut-off frequency, waves have two modes. These are TE and TM modes respectively. The cut-off frequency has been only determined by geometry of the wave guides.

V.N. Alexandrov et al. (Eds.): ICCS 2006, Part IV, LNCS 3994, pp. 631 – 637, 2006.
© Springer-Verlag Berlin Heidelberg 2006

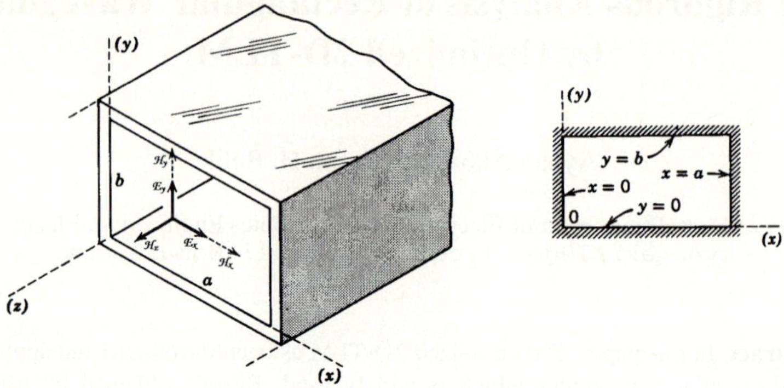

Fig. 1. Rectangular Waveguides

Mode cut-off frequency can be analytically calculated by;

$$f_{c,mn} = \frac{1}{2\pi\sqrt{\mu\varepsilon}}\sqrt{\left(\frac{m\pi}{a}\right)^2 + \left(\frac{n\pi}{b}\right)^2} \quad (1)$$

where m and n are mode degrees respectively.

3 Optimized 2D-TLM Method

TLM was first introduced by P. N. John in 1970. This technique is based on the field theory – the circuit theory similarities. Transmission line modeling divides the structure into unit cells and structure model is carried out by solving each cell

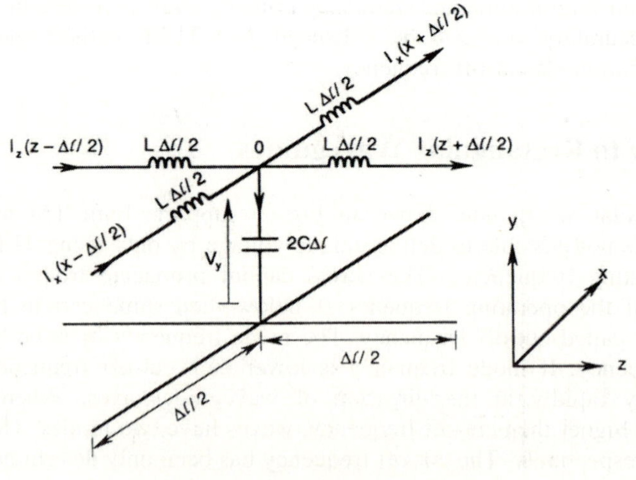

Fig. 2. Two Dimensional TLM Cell

separately. Current and voltage are set to be independent variables. The correlation between input and output voltage is found by applying Kirchoff current and voltage laws onto Fig. 2 which is circuit model of the cell analyzed.

TLM method like FDTD is interested in Maxwell equations. Given microwave structure which is rectangular waveguide in this application has been divided into cells. Each cell has been traded as electrical circuit and therefore electrical circuit solution has been applied onto the every cell repeatedly. The main advantage of TLM against MoM or SDM, TDM does not require any pre-calculation. As a result of this advantage, any optimization can be applied to any microwave circuits without refinements. Another reason to choose TLM for this contribution is that TLM method is very easy to adapt on the computer programming.

2D-TLM Equation is given by;

$$\frac{\partial^2 \Phi}{\partial u^2} + \frac{\partial^2 \Phi}{\partial v^2} = \mu\varepsilon \frac{\partial^2 \Phi}{\partial t^2} \tag{2}$$

For 2D applications $\frac{\partial}{\partial y} = 0$ and $E_x=E_y=H_y=0$. Therefore;

$$\frac{\partial H_x}{\partial z} - \frac{\partial H_z}{\partial x} = +\varepsilon \frac{\partial E_y}{\partial t} \tag{3}$$

$$\frac{\partial^2 E_y}{\partial x^2} + \frac{\partial^2 E_y}{\partial z^2} = \mu\varepsilon \frac{\partial^2 E_y}{\partial t^2} \tag{4}$$

Both equation (3) and equation (4) are very similar. If above equations are rewritten as voltage and current,

$$E_y \equiv V_y \; , \quad H_z \equiv I_x \; , \quad H_x \equiv -I_z \; , \quad \mu = L \; , \quad \varepsilon = 2C \tag{5}$$

$$\mu_r = \varepsilon_r = 1 \text{ and } 1/\sqrt{LC} = 1/\sqrt{\mu_0 \varepsilon_0} = c \tag{6}$$

can be easily found. c is free space light speed in Equation (6).

4 Computer Simulation and Numerical Results

First analyzed mode and then maximum frequency of the interest must be determined. Because this process is necessary to specify time step and cell size of TLM simulation. To avoid numerical dispersion, the ratio of minimum wavelength and cell must be chosen carefully. Rectangular waveguide analyzed by TLM has divided into $N_x x N_y$ number of cells so that Δx and Δy are cell sizes in x –y axes respectively. As a source, Gauss pulse of which durations have been chosen according to maximum operating frequency used. Gauss pulse is applied at one point, and the calculated field's components of observation points are saved. The frequency response of rectangular waveguides has been derived from the time response.

4.1 TM Mode Analysis Results

The analyzed rectangular waveguide's dimensions are given 90mm in width and 45mm in height respectively. Chosen parameters used throughout the computer simulation of TM Mode by optimized TLM technique are given in Table 1

Table 1. 2D-TLM Parameters for TM Modes

f_{max} (maximum frequency)	10 GHz
Δx (cell size in x-axes)	1.125mm
Δy (cell size in y-axes)	1.125mm
N_x (number of cell in x-axes)	80
N_y (number of cell in y-axes)	40
Δt (time step)	2.76 pico second
T (simulation duration)	10000Δt
Δf (frequency resolution)	36.23 MHz

Table 2 compares optimized results and analytical results. It is demonstrated that optimized TLM algorithm presented here has good agreement with analytical results and error is less then 0.2%. Time and frequency response of TM modes are given in Figure 3.

Table 2. 2D-TLM Simulation Results and Analytic Result TM Modes Frequencies

	Analytical Results (GHz)	**TLM Results (GHz)**	**Error**
TM_{11}	**3.7268**	**3,6978**	**0,0290**
TM_{21}	**4.7140**	**4,6994**	**0,0146**
TM_{31}	**6.0093**	**5,9869**	**0,0224**
TM_{12}	**6.8718**	**6,8703**	**0,0015**
TM_{22} TM_{41}	**7.4536**	**7,4155**	**0,0381**
TM_{32}	**8.3333**	**8,3150**	**0,0183**
TM_{51}	**8.9753**	**8,9279**	**0,0474**
TM_{42}	**9.4281**	**9,4059**	**0,0222**

4.2 TE Mode Analysis Results

Chosen parameters used throughout the computer simulation of TE Mode by optimized TLM technique are given in Table 3.

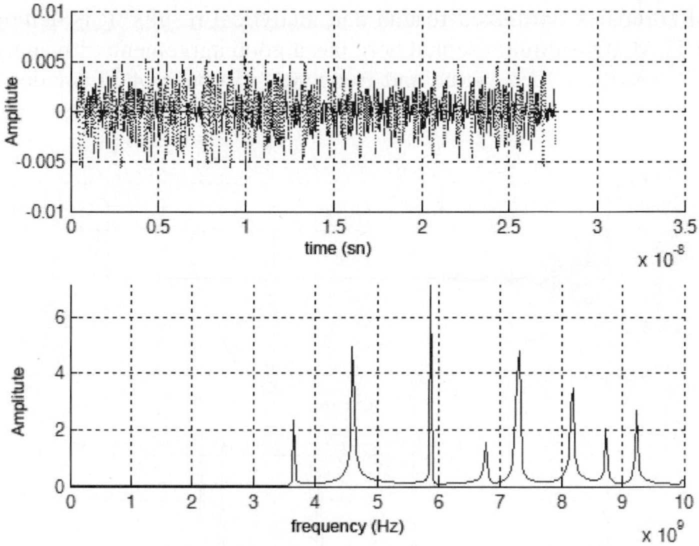

Fig. 3. TM Mode Time and Frequencies Response

Table 3. 2D-TLM Parameters for TE Modes

f_{max} (maximum frequency)	10 GHz
Δx (cell size in x-axes)	2.25mm
Δy (cell size in y-axes)	2.25mm
N_x (number of cell in x-axes)	40
N_y (number of cell in y-axes)	20
Δt (time step)	5.46 pico second
T (simulation duration)	10000Δt
Δf (frequency resolution)	18.31 MHz

Table 4. 2D-TLM Simulation Results and Analytic Result TE Modes Frequencies

	Analytical Results (GHz)	TLM Results (GHz)	Error
TE_{10}	1.6667	1.6476	0,0191
TE_{01} ve TE_{20}	3.3333	3.2962	0,0371
TE_{11}	3.7268	3.6980	0,0288
TE_{21}	4.7140	4.9001	-0,1861
TE_{30}	5.0000	4.9866	0,0134
TE_{31}	6.0093	5.9670	0,0423
TE_{40} ve TE_{02}	6.6667	6.5540	0,1127
TE_{12}	6.8718	6.8130	0,0588
TE_{41} ve TE_{22}	7.4536	7.3594	0,0942
TE_{50} ve TE_{32}	8.3333	8.2039	0,1294
TE_{51}	8.9753	8.8240	0,1513
TE_{42}	9.4281	9.4002	0,0279

Table 4 compares optimized results and analytical results. It is demonstrated that optimized TLM algorithm presented here has a good agreement with analytical results and error is less then 0.2%. Time and frequency response of TE modes are given in Figure 4.

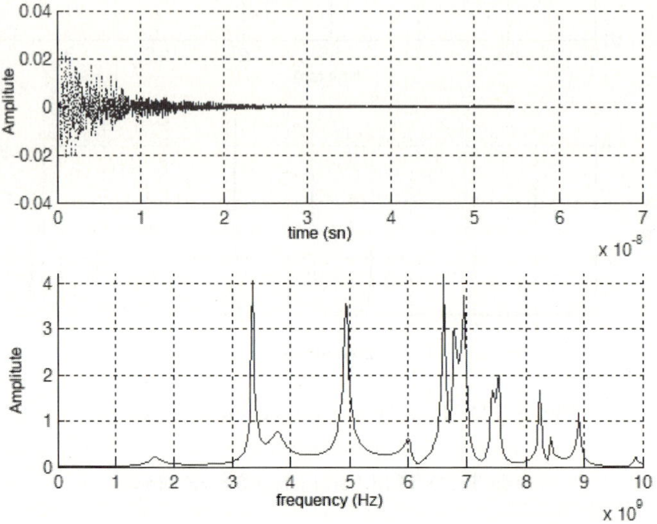

Fig. 4. TE Mode Time and Frequencies Response

5 Conclusion

In this paper, optimized 2D-TLM method has been introduced to analyze widely used rectangular waveguides. It is found and demonstrated in this contribution that the results are in very good agreements to analytical results.

References

[1] D.M. Pozar, Microwave Engineering, USA, 1998.

[2] K. L.Tsakmakidis, C. Hermann, A. Klaedtke, C. Jamois, O. Hess, "Systematic Modal Analysis of 3-D Dielectric Waveguides Using Conventional and High Accuracy Nonstandard FDTD Algorithms", IEEE Photonics Technology Letters, 17(12), pp 2598 – 2600, Dec.2005.

[3] J. Hesselbarth and R. Vahldieck, "Resonance frequencies calculated efficiently with the frequency-domain TLM method," Microwave and Wireless Components Letters, vol. 13, pp. 190 – 192, May 2003.

[4] Kreczkowski and M. Mrozowski, "Efficient multimode mixed time-frequency domain analysis and optimization of waveguide structures," Microwaves, Radar and Wireless Communications, Vol. 3, pp. 803 – 806, May 2004.

[5] Hasan H. Balik and C. J. Railton , New Compensation Functions for Efficient Excitation of Open Planar Circuits in SDM , IEEE Trans. On Microwave Theory and Technique , 47 , 106-108 , Jan 1999

[6] I.A. Eshrah, A.A. Kishk, A.B. Yakovlev and A.W. Glisson, "Spectral Analysis of Left-Handed Rectangular Waveguides With Dielectric-Filled Corrugations," Antennas and Propagation, vol. 53, pp. 3673 – 3683, Nov. 2005.

[7] Hasan H. Balik , Final Remedy of the Excitation in the Analysis of Open Planar Circuits , International Journal for Engineering Modelling , Vol. 16, No. 3-4 , 99-103 , 2003.

A New Approach to Spectral Domain Method: Functional Programming

Hasan H. Balik[1], Bahadir Sevinc[2], and Ayhan Akbal[3]

[1] Dept. of Electrical and Electronics Engineering, University of Firat, Elazig, Turkey
Tel.: +90 424 241 00 99
hasanbalik@gmail.com
http://www.hasanbalik.com
[2] Dept. of Informatics, University of Firat, Elazig, Turkey
bahadirsevinc@firat.edu.tr
[3] Dept. of Electrical and Electronics Engineering, University of Firat, Elazig, Turkey
ayhanakbal@gmail.com

Abstract. The Spectral Domain Method is powerful technique to analyze planar microwave circuits. But available conventional programming languages used in the literature does not give the enough speed to use the Spectral Domain Method to develop package analysis program. Functional approach to Spectral domain Method gives a high level of programming and a variety of features which help to build elegant yet powerfully and general libraries of functions.

1 Introduction

For the last two decades, open microstrip structures have received special attention from the electromagnetic community because of their potential applications in the design of new devices and components. Meanwhile, high–speed computer has influenced the computation of electromagnetic problem to the point that most practical computations of the fields can be solved numerically on the computer. The reason why, most of the analysis of the devices and components can be achieved numerically but is almost impossible to be solved analytically. A lot of efforts have still been done on improving numerical techniques because complexity of the problems always overstretch the speed of the processors. Moreover the operating frequency raised up for more available bandwidth, full–wave techniques which require more computer power and resources must be used.

A number of numerical full–wave techniques are reported in the literature for the analysis of microstrip antennas [1], resonators [2] and circuits [3]. All techniques reported in the literature have been either written by conventional programming languages such as pascal and C or developed by using commercial analysis tools such as Matlab. In the author knowledge none of the papers can explored the idea of the way of programming such as functional or logic.

This contribution presents a functional programming approach to Spectral Domain Method which is one of the full–wave numerical technique and widely used for the analysis of the microwave and millimeter wave devices and components. With this approach, Spectral Domain Method have gained a high level of

V.N. Alexandrov et al. (Eds.): ICCS 2006, Part IV, LNCS 3994, pp. 638–644, 2006.
© Springer-Verlag Berlin Heidelberg 2006

programming giving its user a variety of features which help to build elegant yet powerfully and general libraries of functions. Numerical results have also given and compared with publish date to show the accuracy of the re–written Spectral Domain Method by Haskel which is widely used functional programming language instead of conventional language such as pascal used in [3].

2 Functional Approach to SDM

2.1 Introduction

In this paper, five fundamental modules have been rewritten by using functional approach instead of conventional programming language such as pascal to show applicability of the approach. All of the modules are described in the sections below.

2.2 Input Functions

In this module the input parameters are taken and passed to other modules. Operating frequency, substrate layer parameters, k_x, k_z which are Fourier transform variables in x and z directions respectively, n which is number of layers, l_x and l_z which are dimensions of rooftop function are used as input values. The re–written module becomes as follows:

```
type Ind = Double
type D = Double
type M = Double
type E = Double
type Layer = [(Ind,D,M,E)]
type OneLayer = (Ind,D,M,E)
type Kx = Complex Double
type Kz = Complex Double
type Lx = Double
w :: Double
w  = 2*3.1456*saveF
saveLayer :: IO -  Layer
saveKx :: IO -   Kx
saveKz :: IO -  Kz
saveLx :: IO -  Lx
saveN :: IO -  Double
saveF :: IO- Double
nx = saveKx/sqrt(saveKx*saveKx+saveKz*saveKz)
nz = saveKz/sqrt(saveKx*saveKx+saveKz*saveKz)
```

2.3 Empedans Functions

In this module the Green Function in the spectral domain has been calculated by using functional approach. Mathematical formulation of the Green function can be found in the literature such as [4, Chapter 4].

```
findlayer :: Ind -> OneLayer
findlayer indx = head [(ind,d,m,e) | (ind,d,m,e)
    <- saveLayer , ind==indx ]
layparD :: OneLayer -> D
layparD (ind,d,m,e) = d
layparM ::  OneLayer -> M
layparM (ind,d,m,e) = m
layparE ::  OneLayer -> E
layparE (ind,d,m,e) = e
ztm :: Double -> Complex Double
ztm i  = (gama i)/(w*(layparE(findlayer i)):+0)
gama :: Double -> Complex Double
gama i = sqrt((saveKz*saveKz)+(saveKx*saveKx)-
   ((w*w*(layparM (findlayer i))*(layparE(findlayer i))):+0))
zte :: Double -> Complex Double
zte i  = (w*(layparM(findlayer i)):+0) /gama i
zelist :: Double -> (Complex Double,Double)
zelist n = ((zeN n) ,(n-1) )
zeN :: Double -> Complex Double
zeN n = ztm n / atanh((gama n)*((layparD (findlayer n)):+0))
zhN :: Double -> Complex Double
zhN n = zte n / atanh((gama n)*((layparD (findlayer n)):+0))
zhlist :: Double -> (Complex Double,Double)
zhlist n = ((zhN n) ,(n-1) )
coth :: Double -> Complex Double
coth i = atanh((gama i)*((layparD (findlayer i)):+0))
ze2 :: (Complex Double,Double) -> (Complex Double,Double)
ze2 (n,2)  = (n,2)
ze2 (n,s) =
      ze2
        (
          (
           (ztm s * (n*(coth s ))+ztm s)/
           (ztm s *(coth s)+n)
          )
            ,(s-1)
        )
zh2 :: (Complex Double,Double) -> (Complex Double,Double)
zh2 (n,2)  = (n,2)
zh2 (n,s) =
      zh2
        (
          (
           (zte s * (n*(coth s ))+zte s)/
           (zte s *(coth s)+n)
```

```
              )
                 ,(s-1)
           )
  ze1 = ztm 1
  zh1 = zte 1
  ze :: Double -> Complex Double
  ze n =
         1/
      (
         ((1:+0)/ze1)
         +
         ((1:+0)/fst((ze2 (zelist n))))
      )
  zh :: Double -> Complex Double
  zh n =
         1/
      (
         ((1:+0)/zh1)
         +
         ((1:+0)/fst((zh2 (zhlist n))))
      )
  gzz n = nz*nz*(ze n) + nx*nx* (zh n)
  gzx n = nx*nz*(-(ze n)+(zh n))
  gxz n = gzx n
  gxx n = nx*nx*(ze n)+nz*nz*(zh n)
```

2.4 Current Functions

In this module current basis functions functions which are rooftop functions [4, Chapter3] are calculated by functional approach.

```
  jz :: Double ->Complex Double
  jz n = (2/saveKx)*sin(saveKx*(saveLx:+0))*
      exp(saveKx*(n*(saveLx):+0))
  jx :: Double -> Complex Double
  jx n = (2/(saveKx*saveKx))* (1-cos(saveKx*
     (saveLx:+10)))*exp(saveKx*(n*(saveLx):+0))
  makeNpar :: Double -> Double -> [Double]
  makeNpar 0 n = []
  makeNpar n a =(-((n-1)-a)):(makeNpar (n-1) a)
  jzn :: Double -> Double -> [Complex Double]
  jzn n a = map jz (makeNpar n a)
  jxn :: Double -> Double -> [Complex Double]
  jxn n a = map jx (makeNpar n a)
  mux :: [Complex Double] -> [(Complex Double)]
  mux xs = concat (map (fun xs) xs)
```

```
fun :: [Complex Double] ->Complex Double-> [(Complex Double)]
fun as a = [(a*b)| b<-as]
```

2.5 Integral Functions

This model is used to calculate the each element of the impedance matrix

```
makeMat :: Integer -> Integer-> Double ->
[(Complex Double)] -> [(Complex Double)]
makeMat a c k [] = []
makeMat a c k (n:ns)
     | (a>0 && a<=(1*c))  = (n*(gzz k)):(makeMat (a+1) c k ns)
  | (a>(1*c) && a<=(2*c))  = (n*(gzx k)):(makeMat (a+1) c k ns)
  | (a>(2*c) && a<=(3*c))  = (n*(gzz k)):(makeMat (a+1) c k ns)
  | (a>(3*c) && a<=(4*c))  = (n*(gzx k)):(makeMat (a+1) c k ns)
  | (a>(4*c) && a<=(5*c))  = (n*(gxz k)):(makeMat (a+1) c k ns)
  | (a>(5*c) && a<=(6*c))  = (n*(gxx k)):(makeMat (a+1) c k ns)
  | (a>(6*c) && a<=(7*c))  = (n*(gxz k)):(makeMat (a+1) c k ns)
  | (a>(7*c) && a<=(8*c))  = (n*(gxx k)):(makeMat (a+1) c k ns)
```

3 Numerical Results

In these sections below includes several analyzed example microwave structures to show accuracy of the re–written program by using Spectral Domain Method in Haskell which is one of the functional programming language. Total program code have been optimized 55% compared to pascal code. As a result runtime has been reduced 50% compared to pascal code run on the computer. The computer has intel P4 2.4 GHz with 1 GB RD RAM. The operating system is Redhat Linux 8.0.

3.1 Simple Low–Pass Filter

Measurement results are available for the microstrip low–pass filter [5] shown in figure 1. The dimensions and parameters of the dielectric substrate are given in figure 1.

The S–parameter results are plotted in figures 2 where it can be seen that the calculated results and measurements are in very good agreement.

3.2 Edge–Coupled Filter

In order to further prove the accuracy of the re-written program, the analysis of the microstrip edge–coupled filter shown in fig. 6 in [6] is considered. The measurements performed by Shibata *et al* [7] for this filter.

As seen in fig. 3, there is a clear agreement between the newly written program and measured data.

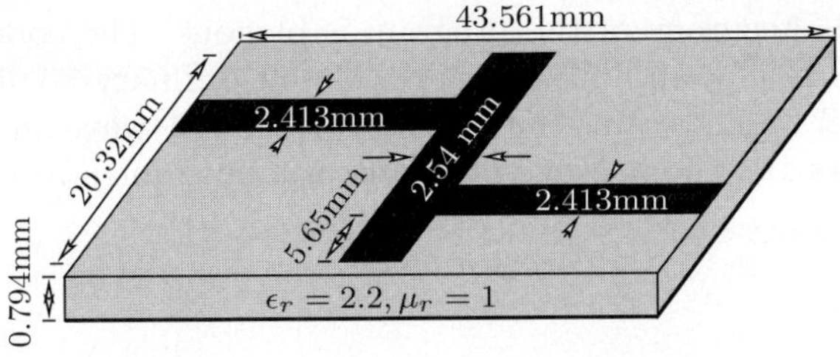

Fig. 1. Low–pass filter detail

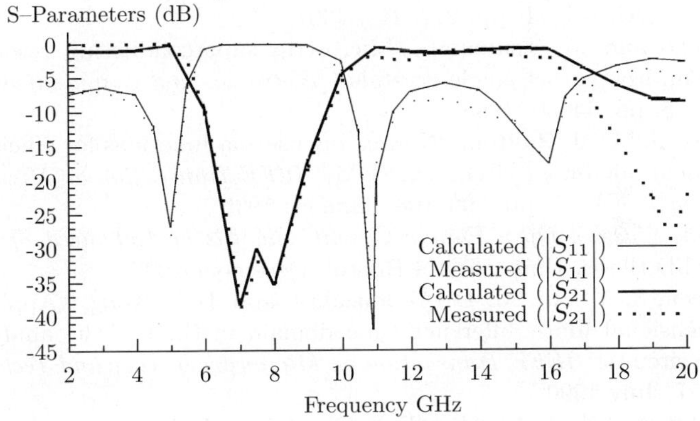

Fig. 2. Plot of S–parameters' magnitude for the low–pass filter

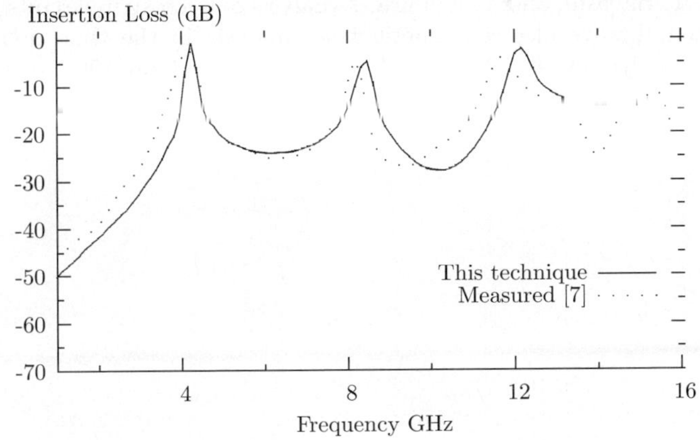

Fig. 3. Magnitude of S–parameters for the edge–coupled filter

4 Conclusion

We have shown that realistically complex microstrip circuits can be rigorously analyzed by re–written program which uses functional approach to Spectral

Domain Method. Accuracy of the program is obvious. The code size and run-time reduction are 55% and 50% respectively on ordinary computers. By this approach a model which retains the accuracy of the full–wave analysis technique as well as the speed of the package programs has been introduced.

References

1. A. Gharsallah, A. Mami, R. Douma, A. Gharbi, and H. Baudrand, “Analysis of microstrip antenna with fractal multilayer substrate using iterative method,” *INTERNATIONAL JOURNAL OF RF AND MICROWAVE COMPUTER-AIDED ENGINEERING*, vol. 11, pp. 212–218, 2001.
2. T. Fukusako and M. Tsutsumi, “Microstrip superconducting resonators loaded with yttrium iron garnet single crystals,” *Eletronics and Communication in Japan*, vol. 81, no. 5, pp. 44–50, 1998.
3. H. H. Balik and C. J. Railton, “New compensation functions for efficient excitation of open planar microwave circuits in SDM,” *IEEE Transaction on Microwave Theory and Technique*, vol. 47, pp. 106–108, January 1999.
4. H. H. Balik, *Passive Open Planar Circuit Analysis by Enhanced Spectral Domain Method.* PhD thesis, University of Bristol, December 1997.
5. D. M. Sheen, S. M. Ali, M. D. Abdouzahra, and J. A. Kong, “Application of the three–dimensional finite-difference time–domain method of the analysis of planar microstrip circuits,” *IEEE Transaction on Microwave Theory and Technique*, vol. 38, pp. 849–857, July 1990.
6. C. J. Railton and S. A. Meade, “Fast rigorous analysis of shielded planar filters,” *IEEE Transaction on Microwave Theory and Technique*, vol. 40, pp. 978–985, May 1992.
7. T. Shibata, T. Hayashi, and T. Kimura, “Analysis of microstrip circuits using three–dimensional full–wave electromagnetic field analysis in the time domain,” *IEEE Transaction on Microwave Theory and Technique*, vol. 36, pp. 1064–1070, June 1988.

Optimized Design of Interconnected Bus on Chip for Low Power*

Donghai Li, Guangsheng Ma, and Gang Feng

College of Computer Science & Technology, Haerbin Engineering University, Haerbin, Heilongjiang 150001, China
ldh12151@tom.com

Abstract. In this paper, we firstly propose an on-chip bus power consumption model, which includes the self transition power dissipated on the signal lines and the coupled transition power dissipated between every two signal lines. And then a new heuristic algorithm is proposed to determine a physical order of signal lines in bus. Experimental results show an average power saving 26.85%.

1 Introduction

With the advent of portable and high density micro-electronic devices such as the laptop personal computer and wireless communication equipment, power consumption of very large scale integrated (VLSI) circuits has become a critical concern [1]. Further ultra-deep submicron (UDSM) VLSI and system-on-chip have resulted in a considerable portion of power dissipated on buses, especially in communication and multi- media applications, a large fraction of power is consumed during accessing memory and data transfer. So we must consider low power optimization on the interconnected buses.

Because of the UDSM, the major sources of power consumption in buses are the self transition activities and the coupled activities of the lines of each bus [1]. The traditional power model $P = \alpha C_L V^2 / T$ is no longer valid [2,3]. The main research in this paper is to minimize the self transition activities and the coupled transition activities of on-chip buses.

There are many researches who have addressed the problem of minimizing power on buses. Reference [4] proposed a bus-binding method for minimizing transitions activities by integrating the scheduling results, but this approach only minimizes the self transition activities. Reference [5] proposed a method to determine a relative placement order of bus line to reduce the coupled transition activities, but it only minimize the coupled transition activities of adjacent bit-lines and neglect the coupled transition activities of nonadjacent signal lines. Reference [6,7] proposed an approach of combing wire swapping and spacing to minimize the coupled transition activities, but it neglects the self transition activities. Reference [8,9] proposed different encoding scheme to minimize the coupled transition activity, which need additional encode and decode circuits and thus increases the hardware overhead. Reference [1] proposed

* This work is supported by NFS No.60273081& HRBEU Foundation No.F0488.

V.N. Alexandrov et al. (Eds.): ICCS 2006, Part IV, LNCS 3994, pp. 645 – 652, 2006.
© Springer-Verlag Berlin Heidelberg 2006

a high-level interconnection synthesis algorithm which bind the data transfer to buses and determine the physical order of signal lines in each bus, the algorithm minimizes the self transition activities and the coupled transition activities on buses, but the coupled transition activities between the nonadjacent signal lines are neglected.

In this paper, we propose a on-chip buses power consumption model which involves both the self transition activities and the coupled transition activities between every two signal lines, and then a new heuristic algorithm is proposed to determine the physical order of signal lines of bus to minimize the weight of the transition activities. There is no additional hardware overhead during the whole optimization process.

2 Power Model

In the UDSM VLSI, the dynamic consumption on interconnect buses include not only the self transition activities on single line, but also the coupled transition activities on adjacent and nonadjacent two single lines as shown in figure 1 [2,3].

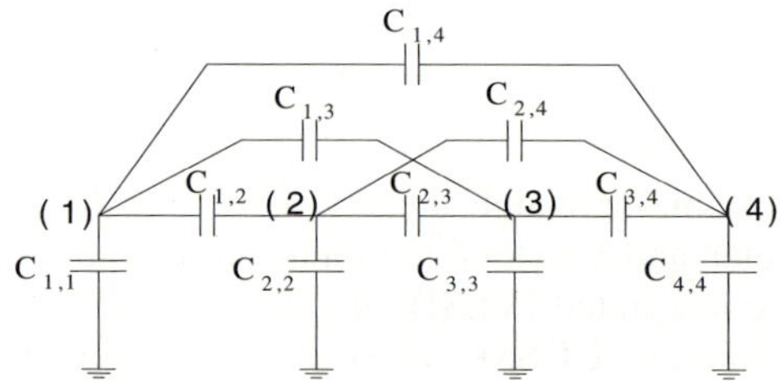

Fig. 1. Self and Coupled capacitances

2.1 Self Transition Power

As shown in figure 2, there are three types of transitions on single line. In type 1, the signal transits from low to high. In type 2, the signal transits from high to low. In type 3, no signal transits on the line. Type 1 transition will generate power consumption, which is a valid transition. We can conclude the following theorem.

Theorem 1: At t clock step, the valid self transition number of the signal line i on bus K can be expressed as follow

$$X_{K,i}^{t} = \chi_{K,i}^{t}\left(\chi_{K,i}^{t} - \chi_{K,i}^{t-1}\right). \tag{1}$$

Where $\chi_{K,i}^{t} \in \{0,1\}$ is the value of signal i on bus K at clock t, $X_{K,i}^{t} \in R$.

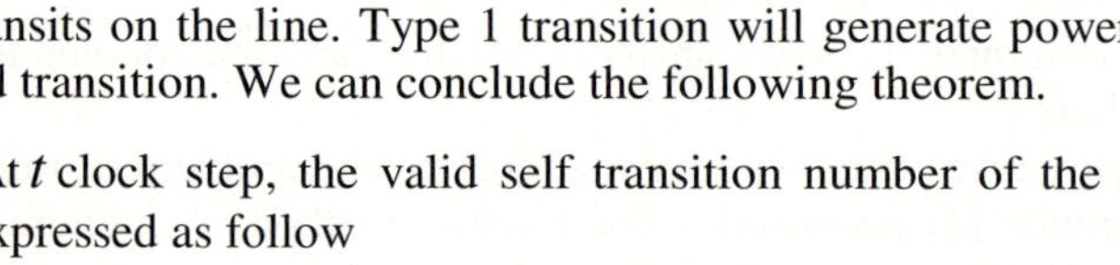

Fig. 2. Self transition type

From (1), it can be concluded that the self transition number of the signal line i on the bus K in T clock step is

$$X_{K,i}^{T} = \sum_{t=1}^{T} X_{K,i}^{t} \; . \tag{2}$$

From (2), it can be concluded that the dynamic self transition power consumption of the signal line i on bus K in T clock step is

$$P_{S,K,i}^{T} = X_{K,i}^{T}\left(C_S + C_L\right)V_{dd}^{2}/T \; . \tag{3}$$

Where C_S and C_L are self capacitance on signal line i, V_{dd} is the voltage.

2.2 Coupled Transition Power

As shown in figure 1, in UDSM VLSI every two signal lines can generate coupled capacitance and then generate coupled transition power. As shown in figure 3, there are five transition types between two signal lines. In type 1, no signal transitions occur on either line. In type 2, both signals make transitions to the same states. In type 3, exactly one of the two signals makes a transition and finally the two signals have the same states. In type 4, exactly one of the two signals makes a transition and finally the two signals have different states. In type 5, one signal transits from low to high and another from high to low. Among these five types, type 1,2,3 will not generate dynamic charge on couple capacitance and thus no dynamic power consumption. Type 4 and 5 generate dynamic consumption and also the charge generated by type 5 is two times as much as type 4. Type 4 transition is a valid transition. We can conclude following theorem.

Theorem 2: At t clock step, the valid coupled transition number for every two signal lines on bus K can be expressed as follow

$$Y_{K,i,j}^{t} = \left(\chi_{K,i}^{t} - \chi_{K,j}^{t}\right)\left(\left(\chi_{K,i}^{t} - \chi_{K,i}^{t-1}\right) - \left(\chi_{K,j}^{t} - \chi_{K,j}^{t-1}\right)\right) \tag{4}$$

where $\chi_{K,i}^{t}, \chi_{K,j}^{t} \in \{0,1\}$ are the value of signal i and signal j on bus K at t clock step, $Y_{K,i,j}^{t} \in R$.

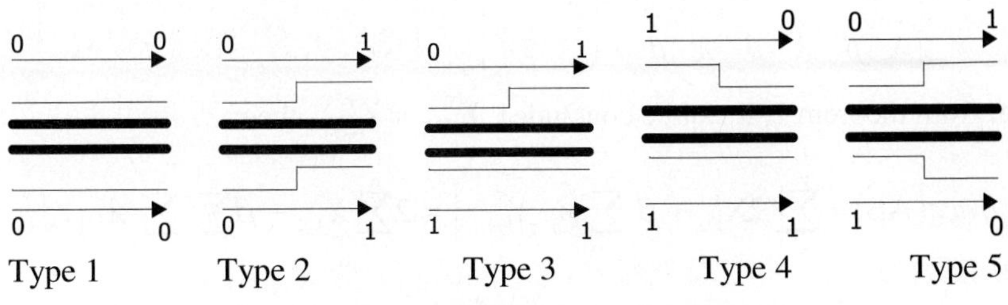

Fig. 3. Coupled transition type

From (4), it can be concluded that the coupled transition number between signal line i and j on the bus K in T clock step is

$$Y^{T}_{K,i,j} = \sum_{t=1}^{T} Y^{t}_{K,i,j} \quad . \tag{5}$$

From (5), it can be concluded that the dynamic coupled transition power consumption between the signal line i and the signal j on bus K in T clock step is

$$P^{T}_{C,K,i,j} = Y^{T}_{K,i,j} C_{i,j} V^{2}_{dd} / T \,. \tag{6}$$

Where $C_{i,j}$ is the coupling capacitance between the signal i and j on bus K .

2.3 Dynamic Power Consumption Model

From (3) and (6), the dynamic power consumption on bus K in T clock step is

$$P^{T}_{K} = \sum_{i=1}^{n} P^{T}_{S,K,i} + \sum_{i=1}^{n-1}\sum_{j=i+1}^{n} P^{T}_{C,K,i,j} = \left((C_S + C_L) \sum_{i=1}^{n} X^{T}_{K,i} + \sum_{i=1}^{n-1}\sum_{j=i+1}^{n} Y^{T}_{K,i,j} C_{i,j} \right) V^{2}_{dd} \Big/ T$$

We define the coupling capacitance between the adjacent signal lines as C_C and $\lambda_{|i-j|} = C_{i,j} / C_C$; $\beta = C_C / (C_S + C_L)$ then

$$P^{T}_{K} = (C_S + C_L) \left(\sum_{i=1}^{n} X^{T}_{K,i} + \beta \sum_{i=1}^{n-1}\sum_{j=i+1}^{n} \lambda_{|i-j|} Y^{T}_{K,i,j} \right) V^{2}_{dd} \Big/ T \,\cdot \tag{7}$$

We define the transition weighted of self and coupled transition activities on bus K in T clock steps is Z^{T}_{K} , then

$$Z^{T}_{K} = \sum_{i=1}^{n} X^{T}_{K,i} + \beta \sum_{i=1}^{n-1}\sum_{j=i+1}^{n} \lambda_{|i-j|} Y^{T}_{K,i,j} \,\cdot \tag{8}$$

Theorem 3: $Z^{T}_{K} = \frac{1}{2} \mathrm{trace}(AB)$. (9)

Where A and B is the matrix as follow, B is the matrix of transition

$$A = \begin{bmatrix} 2 & \lambda_1\beta & \lambda_2\beta & \cdots & \lambda_{n-1}\beta \\ \lambda_1\beta & 2 & \lambda_1\beta & \cdots & \lambda_{n-2}\beta \\ \lambda_2\beta & \lambda_1\beta & 2 & \cdots & \lambda_{n-3}\beta \\ \vdots & \vdots & \vdots & \ddots & \vdots \\ \lambda_{n-1}\beta & \lambda_{n-2}\beta & \lambda_{n-3}\beta & \cdots & 2 \end{bmatrix} \quad B = \begin{bmatrix} X^{T}_{K,1} & Y^{T}_{K,1,2} & Y^{T}_{K,1,3} & \cdots & Y^{T}_{K,1,n} \\ Y^{T}_{K,2,1} & X^{T}_{K,2} & Y^{T}_{K,2,3} & \cdots & Y^{T}_{T,2,n} \\ Y^{T}_{K,3,1} & Y^{T}_{K,3,2} & X^{T}_{K,3} & \cdots & Y^{T}_{K,3,n} \\ \vdots & \vdots & \vdots & \ddots & \vdots \\ Y^{K}_{K,n,1} & Y^{T}_{K,n,2} & Y^{K}_{K,n,3} & \cdots & X^{T}_{K,n} \end{bmatrix}$$

Prove: from theorem 2, it can be concluded $Y^{T}_{K,i,j} = Y^{T}_{K,j,i}$, then

$$trace(AB) = \sum_{i=1}^{n} \left(2X^{T}_{K,i} + \beta \sum_{j=1, j\neq i}^{n} \lambda_{|i-j|} Y^{T}_{K,i,j} \right) = 2\sum_{i=1}^{n} X^{t}_{k,i} + \beta \sum_{i=1}^{n}\sum_{j=1, j\neq i}^{n} \lambda_{|i-j|} Y^{T}_{K,i,j}$$

$$= 2\sum_{i=1}^{n} X^{T}_{K,i} + 2\beta \sum_{i=1}^{n-1}\sum_{j=i+1}^{n} \lambda_{|i-j|} Y^{T}_{K,i,j} = 2Z^{T}_{K}$$

So the conclusion is right. □

Give a matrix of n ranks $\tilde{A}$

$$\left|\tilde{A}\right|_{i,j} = \begin{cases} 1 & if \quad i = j \\ \lambda_{j-i}\beta & if \quad j \succ i \\ 0 & if \quad j \prec i \end{cases}$$

It can be concluded

$$Z_K^T = \mathrm{trace}\left(\tilde{A}B\right) . \tag{10}$$

3 Optimized for Low Power

In section 2, from (7) and (8), it can be concluded that minimizing P_K^T is equivalent to minimizing Z_K^T ; and from (10), the bus transition weight can be obtained, if the physical order of signal lines on the bus can be adjusted, minimum Z_K^T can be obtained. A heuristic algorithm is proposed to determine a physical order of signal lines in bus. The input to the algorithm is an initial order set$\Psi$of signal lines (the elements in Ψcorrespond to each signal line in the bus) and the coupled-transition weight of the adjacent signal lines, The output is the adjusted physical order set ψ of signal lines given by the algorithm is shown in algorithm 1.

The algorithm is divided into two parts, the first part is decomposing as shown in algorithm 2, the elements in$\Psi_{m,n}$ are classified into three parts, for the signal lines which have smaller coupled- transition weight than others (hence larger $p(v_i)$), put the elements into the middle of the order, that is in set $\Psi_{m,3n-1}$; for the signal lines which have bigger coupled-transition weight with others, put the corresponding elements into the two sides of the order, that is in $\Psi_{m,3n-2}$ and$\Psi_{m,3n}$ respectively, until the number of elements in leaf nodes sets of tree T is no more than 3 ; the second part is arranging as shown in algorithm 3, from the deepest leaf node, incorporate other leaf nodes to their father node at the same time make the expense smallest, and delete the leaf nodes, until there is only one root node. We can obtain a sorted set whose order is the result.

Algorithm 1. Heuristic algorithm of signal lines ordering

```
Procedure Signal lines ordering( )
  begin
      m:=0;n:=1;t:=m;
       m,n= ;
      Build a tree T and the root is    m,n;
      decomposing ( m,n);
      Arrange element (T);
      M:=0;n:=1;
      Return ψ;
  end.
```

Algorithm 2. Heuristic algorithm to construct tree T

```
Procedure Decomposing ( m,n)
 begin
  Construct a graph G, the vertex corresponding to the
  element in m,n and the weight of the edge
  corresponding the coupled transition weight;
  m:=m+1;
  set L:=sorted the edge of G ,the smallest weight first;
  L:=(e1,e2,…en);
  for k:=1 to k:=⌈|L|⌉/2
    begin
     for two vertexes  of  ek
       begin
        p(vi):= p(vi)+1/|L|;
        p(vj):= p(vj)+1/|L|;
       end
     end;
   for every vertex of G
    begin
     if  P(Vi)>P    then    { P is a threshold }
       begin
        delete the vertex Vi  from G;
        m,3n-1:= m,3n-1∪{Vi};
      end;
     P(Vi):=0;
    end;
   for the left part of G, find the edge with maximum
     weight in turns and delete the edge until G is
     divided into tow subgraphs G1 and G2;
   Add the vertexes of G1 to set  m,3n-2;
   Add the vertexes of G1 to set  m,3n;
   Add  m,3n-1,  m,3n-2,  m,3n to tree T as the middle,
    left, right  leaf node of   m-1,n  respectively;
  for every  m,i
    begin
      if | m,i|>3  then
        begin
         Decomposing ( m,j);
         t:=max(t, m);
      end
    else  m:=m-1;
   end;
  return T ;
end.
```

Algorithm 3. Heuristic algorithm to find the final order

```
Procedure Arrange element (T)
  begin
   for k:=t-1 downto 0
    begin
     for every  t-1,i
      begin
         t-1,i:=MIN(Cost(Ψt,3i-2∪ t,3i-1∪ t,3i),
```

```
            Cost( t,3i-2∪ t,3i-1∪Reverse( t,3i)),
            Cost( t,3i-2∪Reverse( t,3i-1)∪ t,3i),
            Cost(Reverse( t,3i-2)∪ t,3i-1∪ t,3i),
         Cost(Reverse( t,3i-2)∪Reverse( t,3i-1)∪ t,3i),
         Cost(Reverse( t,3i-2)∪ t,3i-1∪Reverse( t,3i)),
         Cost( t,3i-2∪Reverse( t,3i-1)∪Reverse( t,3i)),
         Cost(Reverse( t,3i-2)∪Reverse( t,3i-1)∪
             Reverse( t,3i)));
         delete the leaf nodes of  t-1,i;
      end;
    end;
end.
```

4 Experimental Results

The algorithm in section 3 is implemented in C++ and is executed on a Pentium IV computer with clock speed of 1.7GHz. During the experiment, we test some Benchmark Circuits. First using some random data to simulate, then obtain the profile of these data, which is input to our algorithm and obtain the results of the experiment.

Among the Benchmark Circuits, DIFFEQ is to solve a particular differential equation; GCD is to compute the greatest-common-divisor of two numbers; KALMAN is an implementation of the Kalman filter.

We use this algorithm to test every Benchmark Circuits two times, and each time $\beta = 3, 4$ and $\lambda_{|i-j|} = 1/|i-j|$. The experiment result is in table 1, the third column is the result of simulation without using our algorithm and the fourth column is the result of executing our algorithm. From table 1, we can see that a reduction in on-chip power consumption of an average of 26.85% can be had by utilizing the heuristic algorithm.

Table 1. Comparisons of the transition weight

	β	Simulate result	Heuristic algorithm	Reduction (%)
DIFFEQ	3	602.5	413.4	31.39
	4	758.6	531.8	29.9026
GCD	3	1127.8	816.4	27.62
	4	1389.4	1065.7	23.30
KALMAN	3	3948.7	2861.6	27.53
	4	4683.4	3824.1	21.37
Average				26.85

5 Conclusion

In this paper, we propose a on-chip bus power consumption model, which considers not only the self-transition activities but also the coupled transition activities between every two signal lines in each bus. We minimize the transition activities by adjusting the physical order of signal lines in bus. The experimental results show that average 26.85% of on-chip bus power consumption can be saved without additional hardware overhead.

References

1. Chun-Gi Lyuh,Taewhan ,Ki-wookkin. Coupling-aware high-level interconnect synthesis. IEEE Trans. on Computer-aided Design 2004; 23 (1): 157-164
2. Paul P.Sotiriadis, Ananhta P.Chandrakasan. A bus energy model for deep submicron technology, IEEE Trans. on VLSI syst 2002; 10 (6): 341-349.
3. Paul P.Sotiriadis, Ananhta P.Chandrakasan. Power estimation and power optimal communication in deep sub-micron buses: analytical models and statistical measures. Journal of Circuits,Systems and Computers 2002; 11 (6): 637-658.
4. C.Lyuh, T.kim. High-Level synthesis for low-power based on network flow method. IEEE Trans. on VLSI Syst. 2003; 11 (6): 364-375.
5. Y.shin, T.Sakura. Coupling-Driven bus for low-power application-specific systems In Proc. of DAC. 2001; 750-753
6. Luca Macchiarulo, Enrico Macii, Massimo Ponicno. Wire placement for crosstalk energy minimization in address buses. In Proc. of DATE. 2002; 158-162.
7. Enrico Macci, Massimo Poncino, Sabino Salerno. Combining wire swapping and spacing for low-power deep-submicron buses. In Proc. of VLSI 2003; 198-202.
8. T.lv, J.henkel, H.Lekatsas, W.wolf. An adaptive dictionary encoding scheme for SOC data buses.In Proc. of DATE 2002; 1059-1064.
9. Srinivas R.sndhara, Naresh R shanbhay. Coding for system-on-chip networks: A unified framework. In Proc. of DAC. 2004; 103-106.

A Conservative Approach to SystemC Parallelization

B. Chopard[1], P. Combes[1], and J. Zory[2]

[1] University of Geneva - Department of Computer Science, Switzerland
[2] STMicroelectronics - AST, Geneva, Switzerland

Abstract. SystemC has become a very popular language for the modeling of System-On-Chip (SoC) devices. However, due to the ever increasing complexity of SoC designs, the ever longer simulation times affect SoC exploration potential and time-to-market. We investigate the use of parallel computing to exploit the inherent concurrent execution of the hardware components, and thus to speed up the simulation of complex SoC's. A parallel SystemC prototype based on the open source OSCI kernel is introduced and preliminary results are discussed.

1 Introduction

The design of modern Systems-on-Chips (SoC) becomes more and more demanding as the complexity and the functionality of new circuits and applications increase. The ability to develop and test these systems with completely virtual platforms and reasonably fast simulations is a key enabler for tomorrow's technology. The SystemC language was developed by the Open SystemC Initiative (OSCI) to enable system level design. It has quickly become a very popular modeling solution, for engineers can represent the functionality, communications, software and hardware components at multiple levels of abstraction with a single common language.

This paper explores the use of parallel techniques to speed up the simulation of SoC executable specifications. Section 2 describes a parallel SystemC kernel prototype built from the public domain OSCI simulator. A performance analysis is then derived from a straightforward pipeline application in Section 3. Finally, experimental results of the parallel simulation of a complex telecom application are presented in Section 4.

2 A Parallel SystemC Kernel

SystemC [1] exhibits two features that motivated our work. First of all, SystemC is most often used to describe the behavior of a complete system where several hardware and/or software components concurrently perform some tasks. The purpose of this work is to exploit this inherent concurrency to implement parallel simulations.

The second important feature of SystemC is its C++ open source library. The SystemC paradigm combines the flexibility, portability and ease-of-use of object-oriented C++ programming with dedicated concepts and constructs for SoC modeling practices [2]. Our choices in developing a parallel SystemC framework are driven by the willingness to keep the modeling style as open as possible.

V.N. Alexandrov et al. (Eds.): ICCS 2006, Part IV, LNCS 3994, pp. 653–660, 2006.
© Springer-Verlag Berlin Heidelberg 2006

2.1 The OSCI SystemC Kernel

A typical SystemC application is characterized by both a structural part and a behavioral part. The structure will basically describe the way (hardware and software) components are connected to each other ; this translates into various modules connected via channels. Hierarchical structures are supported as well: modules may instantiate other (sub-)modules. The behavioral part of the system is captured in SystemC with the notion of process. Each module may contain one or several processes. The execution of a given process is driven by events (such as a value change on the channels connected to the module) that might awaken or even restart the process.

Once the user has fully described both aspects of the application, it is the task of the SystemC kernel to simulate the whole system, given certain input stimuli. The kernel has to schedule, on a sequential processor, the multiple processes of the system in response to those events generated by the application.

To make sure the concurrent processes use coherent inputs/outputs, the scheduler runs all the processes that are ready to be executed first, and only then it updates their changes on the channels. Both evaluate and update phases constitute a δ-cycle: multiple δ-cycles can occur at the same simulation time, which is very useful for modeling fully-distributed, time-synchronized computation (as in Rgeister Transfer Level) [1].

In the OSCI open source kernel, the scheduler relies on a stack of "runnable processes" (cf. Figure 1) that is emptied during the evaluate phase, and then filled again in with those processes that are sensitive to any event triggered during the update phase (such as a "value-changed" event on a channel). If the stack is empty at the end of the update phase, then the simulation time can be updated to the time of the next event to be triggered, and if there are none, the simulation stops.

2.2 Towards a Parallel SystemC Kernel

There exists some distributed SystemC platforms ([3, 4]) which aim at coordinating various simulators over geographically distant sites. They allow Intellectual Property (IP) vendors to expose their products for testing by potential clients, while hiding their behavior details. Hence, the primary goal is not performance, but usability, whereas we look for fast simulations at the scale of cheap clusters of PC's.

In [5], Savoui et al. encapsulate SystemC processes into POSIX threads to transparently benefit from shared memory SMP architectures. This is integrated in the last version of the OSCI kernel, but complex designs would require costly huge computers.

To our knowledge no true parallel implementation of SystemC exists so far. However much work has been done in the last two decades about parallel VHDL or Verilog simulation kernels, two famous Hardware Description Languages. It is related to the wider field of Parallel Discrete Event Simulations (PDES). Indeed, VHDL or Verilog applications, as well as SystemC applications, are conceptually Discrete Event Systems (DES) [6, 7]: in short, upon the occurrence of some events, processes are awaken and some computation produce in turn new events. There is a large body of literature about PDES, their implementation [8, 9, 10] and their utility for parallel Verilog or VHDL simulation [11, 12, 13], but they have not yet been introduced for SystemC.

The so-called conservative approach strongly enforces causality constraints and thus requires a lot of synchronizations between the interacting processes located on different computing nodes. The so-called optimistic approach is more permissive as it allows processes not to synchronize as often as they strictly should, using forecast mechanisms. When a causality error is detected, a rollback procedure is used.

Optimistic PDES are rather complex and require knowledge of the underlying concepts for the tuning of many application dependent parameters [14]. This constraint is a major drawback for SoC designers, and thus these PDES are not widespread in their community [15]. Furthermore, as VHDL/Verilog applications are mostly low-level abstraction designs, rollbacks can be cost-effective, for little work is to be undone, but with large data sets and coarse grain processes such as in functional-level SystemC models, these mechanisms would consume tremendous amounts of memory, whereas the reduced synchronization/computation ratio makes the overhead for conservative synchronizations acceptable. For these reasons we consider here a conservative approach when parallelizing SystemC. If the CPU time between successive synchronizations is long enough and a fair load balancing among the processors can be achieved, we may expect a reasonable speedup from this approach.

In order to have an open source kernel, to preserve compatibility with SystemC libraries and semantics as well as to keep SystemC users still feel familiar with a parallel kernel, we choose to implement the conservative algorithms directly in the OSCI open source reference kernel.

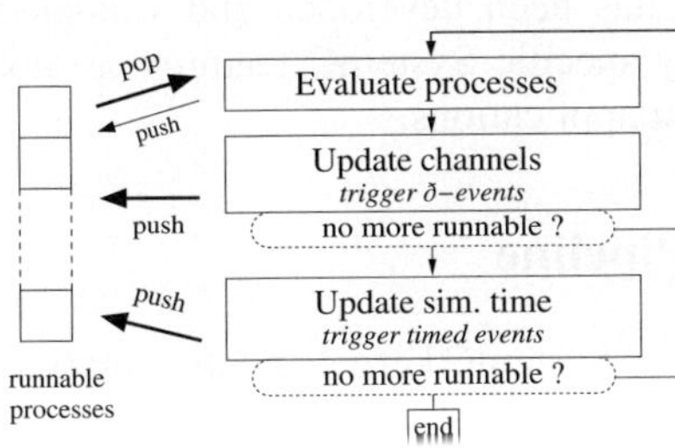

Fig. 1. SystemC kernel scheduler

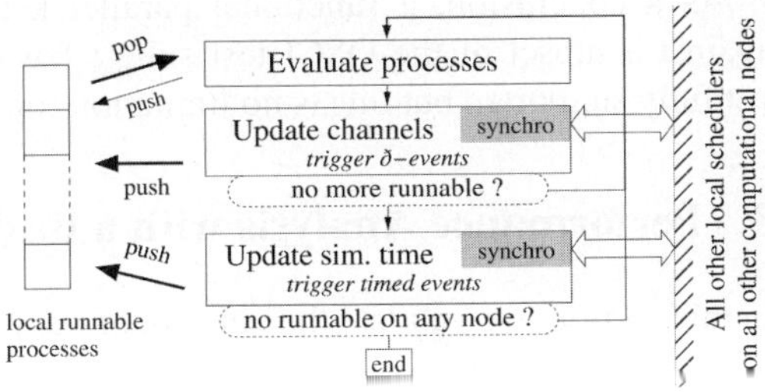

Fig. 2. Scheduler local to one node

2.3 Parallelization of the OSCI Kernel

Given the goals discussed in section 2.2, our parallel kernel has to face two major issues: respect for the informal SystemC semantics and minimum changes in the API (ie. the language itself), while still offering the flexibility to specify the process partitioning.

Our approach is to have a copy of the OSCI SystemC scheduler running on every processing node, each simulating a subset of the application modules. The SystemC semantics specify that, within a δ-cycle, the execution order of the processes is not pre-defined[1] [1], but which processes are to be run depends on the previous δ-cycle. Thus, within a δ-cycle, it is possible to execute the processes in parallel, but all local schedulers must synchronize at its end. The conservative approach discussed in section 2.2 leads us to implement a strong synchronization of both channels and time (Figure 2).

[1] But it is the same for two executions of the same simulator, so that some dependency bugs can be hidden; this is still true on every node of the parallel simulator, but not at the global scale.

The synchronization of one channel only involves the nodes that it connects: if its value has changed, the nodes trigger the associated event so that the sensitive processes are pushed back in the local runnable processes stacks, like in a serial kernel.

The second synchronization deals with simulation time. Our implementation defines a master node which collects the next timed events from every other node, computes the next simulation time and then sends it to all nodes for local update. If this time is the current simulation time, then it means that the current δ-cycle loop has not finished yet on some node: all other nodes must run another δ-cycle, even if no process is to be run locally. The end of the simulation is reached (.i.e. the local schedulers can stop) when the simulation time has reached its maximum value.

With regard to the changes in the language, they are limited to the addition of a new kind of sc_module: sc_node_module. From the user point of view, a "node module" looks like a classical purely hierarchical module, except that it cannot have other node modules as submodules. Internally, it gathers all the modules that are to be managed by one processing node; the partitioning of the application is thus the responsibility of the designer. The channels connecting node modules are internally duplicated, and their bindings to the node modules ports fully define how to synchronize them. It has been a major reverse-engineering and development work to build, from this very simple way of partitioning, all the information necessary to perform the required synchronizations, but this would be too technical to be reported here in details.

As a conclusion, a functional parallel kernel has been developed and validated against a subset of the OSCI test suite ; few very specific SystemC features are not yet fully supported but this is no limitation for most applications.

3 Performance Analysis with a Basic Pipeline

In this section, we validate the concept of our parallel SystemC kernel and its prototype implementation against a regular pipeline test-application.

3.1 Description of the Application

Our test-application is a pipeline loop made of $N \times P$ stages distributed in N modules and P submodules. In the parallel simulator, N is also the number of computational nodes. Every (sub)module has one input and one output port, interconnected as shown Figure 3. The channels used are sc_signal⟨char[L]⟩'s: they hold two versions of an L-long array, the current one and the one for the next update - see 2.1. Every submodule on its turn defines a process which is sensitive to its input signal; each time the process is awaken (actually, every δ-cycle), the behavior is to "increment" and copy data from input to output and to perform n_{flop} floating point divisions to emulate CPU load.

3.2 Performance Model

The model below uses the terminology defined in Figure 3. It only considers CPU processing time and communications over the network. δ-cycle synchronizations and possible overheads introduced by the sequential and parallel SystemC kernels are on-purpose not addressed in this "reference" model. Potential discrepancies between this

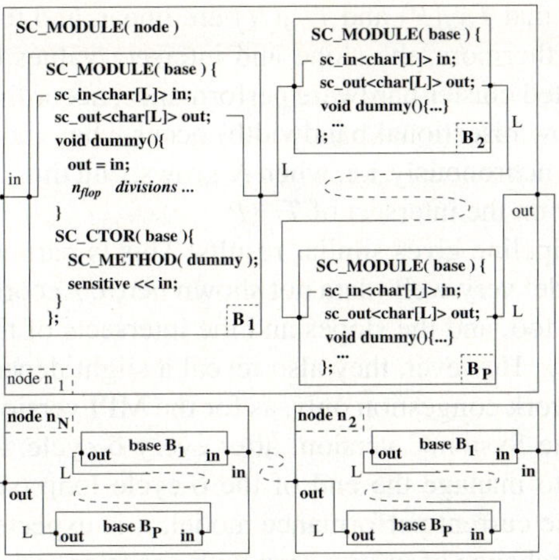

N : Number of modules / computing nodes
P : Number of submodules
L : Size of data transferred on the signals at every δ-cycle
n_{flop} : Number of dummy floating point divisions per process and δ-cycle
n_δ : Number of δ-cycles executed for the simulation
α : CPU time to execute one floating point division
β : Transmission time for one byte over the network
γ : Time to access one byte in memory
t_{st} : Time for initializing a communication

Fig. 3. Structure and parameters of the application

model and the simulation results would demonstrate the impact of SystemC scheduling in the overall performance.

The total execution time can be written as : $T = t_{init} + n_\delta . t_\delta + t_{finish}$. If n_δ is large enough, then t_{init} and t_{finish} can be neglected. Thus they are ignored by our timers for the experiments. Assuming that the scheduler overhead is null, one has

$$t_{\delta_{seq}} = N(t_{evaluate} + t_{update_{seq}}) \qquad t_{\delta_{//}} = t_{evaluate} + t_{update_{//}}$$

Here $t_{evaluate}$ represents the computation time for the n_{flop} divisions in all P submodules, added to the time of the memory accesses for the data copies and increments:

$$t_{evaluate} = P\left(\frac{n_{flop}}{\alpha} + \frac{2L}{\gamma} + \frac{2L}{\gamma}\right)$$

The intra-signals updates consist of $(P-1)$ copies. The inter-signals updates still consist of copies, but in the parallel version, there are also network communications:

$$t_{update_{seq}} = (P-1)\frac{2L}{\gamma} + \frac{2L}{\gamma} \qquad t_{update_{//}} = (P-1)\frac{2L}{\gamma} + \frac{2L}{\gamma} + \left(t_{st} + \frac{L}{\beta}\right)$$

When all terms are gathered, the performance model is:

$$T_{//} = n_\delta\left[\frac{6}{\gamma}LP + \frac{L}{\beta} + \frac{n_{flop}}{\alpha}P + t_{st}\right] \quad (1) \qquad T_{seq} = n_\delta NP\left(\frac{6}{\gamma}L + \frac{n_{flop}}{\alpha}\right) \quad (2)$$

3.3 Model Validation

All our experiments run on a cluster of up to 52 1.5GHz Pentium IV mono-processor nodes, with 500MB RAM, and connected through a Fast-Ethernet network.

A pure MPI/C++ version of the pipeline test-application was first developed to serve as a reference. This version simply avoids all the possible SystemC scheduler overheads and hence almost perfectly matches the performance model described above. For space reasons and to put a stress on the SystemC version, no figure is given here to illustrate this, but our experiments clearly highlight that $T_{seq}(P)$ and $T_{seq}(N)$ are linear, and that $T_{//}(P)$, $T_{seq}(L)$ and $T_{//}(L)$ are affine. Furthermore, the slope and intersect values of the lines approximate quite well the expected cluster hardware performance, according to the model. Yet, network congestions (low bisectional bandwidth) occur when more and more communications are requested synchronously, i.e. when N grows, and thus an additional dependency upon N may appear on the intersect of $T_{//}(P)$.

Globally, the SystemC version of the pipeline gives similar results. This is particularly true for T_{seq} that still matches the model very well (data not shown here). According to Figure 4, $T_{//}(L)$ is obviously affine too, and the slopes and the intersects of the lines figure out the influence of P and n_{flop}. However, they also reveal a slight dependency on N. It cannot be explained by network congestion only, as for the MPI version, because the impact is more visible : in the SystemC version, after every δ-cycle, an additional synchronization is performed, to manage the end of the δ-cycle loop (see section 2.3). Although it was ignored in the current performance model, it is expected to grow at least as $\log N$ as it requires an exchange of information with a master node.

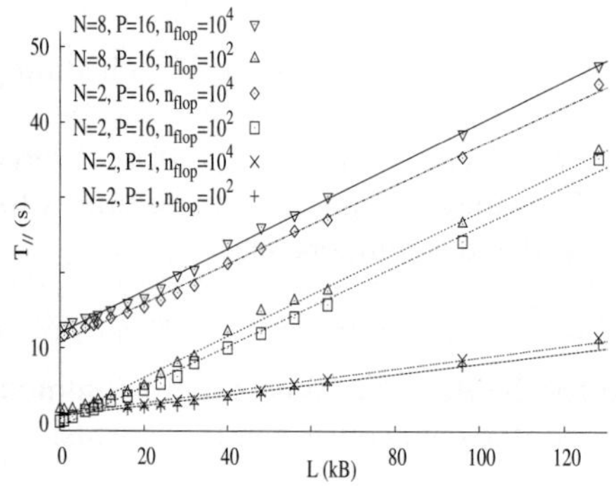

Fig. 4. $T_{//} = n_\delta \left[\left(\frac{6P}{\gamma} + \frac{1}{\beta} \right) L + P \frac{n_{flop}}{\alpha} + t_{st} \right]$

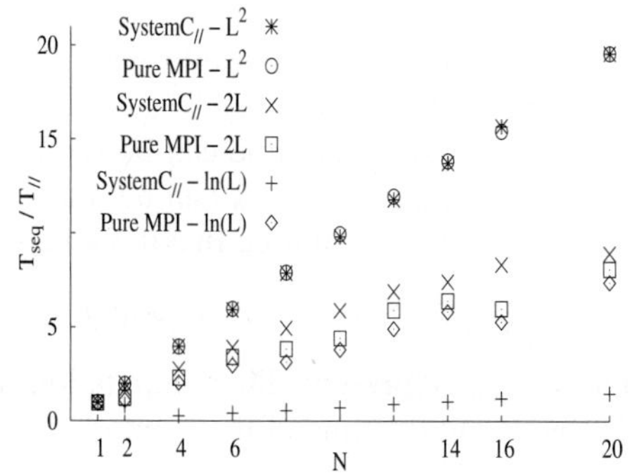

Fig. 5. Speedups (with L=1000)

3.4 Performance

Figure 5 compares the speedups of the reference (namely "pure MPI") and the SystemC versions, with realistic values for n_{flop}: indeed, the amount of computation is often dependent on the amount of data. We have tested three kinds of such dependencies, following the complexities of usual algorithms: $\ln(L)$, $2L$ and L^2.

To explain the shape of the curves, let us consider the efficiency $E = T_{seq}/T_{//}$. Equations 1 and 2 of the performance model lead to the relation :

$$P\left(\frac{6\beta}{\gamma} + \frac{\beta}{\alpha}\frac{n_{flop}}{L}\right) = \frac{E}{1-E}\left(\frac{1}{\beta} + \frac{t_{st}\beta}{L}\right)$$

With our cluster, $\frac{6\beta}{\gamma} \approx \frac{\beta}{\alpha}$, and since our tests run with $L = 1000 \gg 1$, one has $\frac{t_{st}\beta}{L} \ll 1$. Hence, when $n_{flop} = O(\log(L))$ or $O(L)$, $E \approx \frac{1}{1+\frac{1}{P}} < 1$: the curve is "attracted" by the

N axis, for P cannot grow too much. Yet, when $n_{flop} = O(L^2)$, $E \approx \frac{1}{1+\frac{1}{LP}} \approx 1$. This not only explains the general shape of the curves but also why the $\ln(L)$ and $2L$ results are so close for the pure MPI version. For the parallel SystemC version however, the "time synchronizations" (see section 2.3), ignored in our model, become significant if n_{flop} is very small ($\ln(1000) \approx 7$).

4 Case Study: Beyond 3G Modem

The previous section demonstrates that the achievable speedup of a distributed SystemC application heavily depends on the number of processors and the computation to communication time ratio. Here, we further explore those aspects with a real complex application, developed in the frame of the IST MATRICE project. This project involved many academic and industrial partners, from the field of cellular telecommunication Research and Development. It aimed at investigating "Multi Carrier-Code Division Multiple Access" techniques for the broadband component of Beyond 3G systems.

The platform is based on a pipeline, composed of three major stages, made of multiple SystemC modules, which model the major physical layer components of a wireless transmitter, receiver and channel. Please see [16] for a full description of this application. It is only worth noting here that it implements more than 26000 lines of C++ style code, with 27 channels and 26 processes.

A first parallel version of this application is almost immediately available. Indeed, the canonical partition is to have every of the three major stages running on its own computing node. Thus, the work is limited to replacing the SC_MODULE keyword by SC_NODE_MODULE (see section 2.3) in the declarations of these three modules. However, as it could be already anticipated from the pipeline performance model, a good speedup can only be achieved when CPU loads are well balanced over the computational nodes. Some quick investigations revealed that one of the three stages already accounts for 47% of the total serial simulation time. Thus, with the 3-node canonical partition, the speedups are intrinsically limited to 2.1. We reached a speedup of 1.92.

To overcome this heterogeneity, a 4-node partition has been implemented, with an effort of gathering the submodules according to their needs for CPU time. Nonetheless, 28% of the total serial time is still spent simulating one unsplittable submodule, which limits our absolute speedup to 3.6. We reached a speedup of 3.07. This new partitioning requires a bit more work because the major hierarchical modules have to be split, and the whole structure must be reorganized. But someone familiar with the application can do this within twenty minutes, and even less with a GUI.

Further investigations about the discrepancies between the theoritically possible speedups and the real ones reveal that this application require many useless channel updates. This first means that many sychronizations of empty data are performed and then, indirectly, that the processes do not have a regular load all along the simulation. Thus, to further improve speedups, we may introduce disbalances in our regular pipeline test-application (see 3), so that we can study quantitatively their impacts on simulation time and investigate a dynamic load balancing at the granularity of a few δ-cycles rather than the static approach based on the global CPU load.

5 Conclusion

We demonstrate that it is possible to develop a parallel SystemC kernel with a user-friendly interface. An ideally balanced application shows that, despite the high level of synchronization required, speedups comparable to the number of computing nodes can be achieved. A performance model extracts figures about the CPU granularity with respect to synchronization needs in order to reach acceptable efficiency levels.

Even with a real-life coarse-grain application, where the CPU load is not equally distributed in space (over nodes) nor time (along the simulation), we showed that speedups close to their theoritical maximum can be achieved at very low development costs.

Our parallel SystemC kernel has still to be improved before being packaged for open source distribution. In particular the adaptation of existing optimistic PDES algorithms to avoid synchronization that are rarely required will be investigated.

References

1. OSCI `http://www.systemc.org/`: SystemC 2.0.1 Documentation: User's Guide, Functional Specifications, Language Reference Manual. (2002)
2. Martinelli, P., Wellig, A., Zory, J. In: IEEE International Workshop on Rapid System Prototyping. (2004) 193 – 200
3. Aboulhamid, E.M., et al.: eSYS.net (2004) `http://www.esys-net.org/`.
4. Meftali, S., et al.: SOAP based distributed simulation environment for System-on-Chip (SoC) design. In: Forum on Specification and Design Languages. (2005) To appear.
5. Savoiu, N., et al. In: Design, Automation and Test in Europe. (2002) 875–881
6. Ziegler, et al.: Theory of Modeling and Simulation - 2nd Edition. Academic Press (2000)
7. Skold, S., Ayani, R. Technical Report TRITA-IT R 94-19, Dept of Teleinformatics, Royal Institute of Technology, Stockholm (1992)
8. Misra, J. In: Computing Surveys. Volume 18. (1986) 39–65
9. Fujimoto, R.M. In: Communications of the ACM. Volume 33. (1990) 30–53
10. Ferscha, A.: Parallel and Distributed Simulation of Discrete-Event Simulations. In: Handbook of Parallel and Distributed Computing. McGraw-Hill (1995)
11. Naroska, E. In: Design, Automation and Test in Europe. (1998) 159–165
12. University of Cincinnati: SAVANT proj. (1999) `http://www.ececs.uc.edu/~paw/savant/`.
13. Cadwell, B., Browy, C. `http://www.avery-design.com/web/avery_hdlcon02.pdf` (2005)
14. Low, Y.H., et al. In: SIMULATION. Volume 72. (1999) 170–186
15. Fujimoto, R.M. ORSA Journal on Computing **5** (1993) 213–230
16. IST: MATRICE proj (2004) `http://www.ist-matrice.org/`.

Modular Divider for Elliptic Curve Cryptographic Hardware Based on Programmable CA*

Jun-Cheol Jeon, Kee-Won Kim, Jai-Boo Oh, and Kee-Young Yoo**

Department of Computer Engineeing, Kyungpook National University,
Daegu, 702-701 Korea
{jcjeon33, nirvana, jboh}@infosec.knu.ac.kr,
yook@knu.ac.kr

Abstract. This study presents an efficient division architecture using irreducible trinomial in GF(2^n), based on programmable cellular automata (PCA). The most expensive arithmetic operation in elliptic curve cryptosystems (ECC) is division, which is performed by multiplying the inverse of a multiplicand. The proposed architecture is highly regular, expandable, and has reduced latency. The proposed architecture can be efficiently used in the hardware design of crypto-coprocessors.

1 Introduction

Finite field GF(2^n) arithmetic operations have recently been applied to a variety of fields, including cryptography and error-correcting codes [1]. A number of modern public key cryptography systems and schemes, for example, Diffie-Hellman key pre-distribution, ElGamal cryptosystem, and ECC, require division and inversion operations [2].

The main operation of ECC is the inverse/division operation, which can be regarded as a special case of exponentiation [3]. Since division, however, is quite time consuming, efficient algorithms are required for practical applications. Division operations can generally be classified into two approaches: a fast architecture design or a novel algorithm development. This current study focuses on the former approach.

Cellular automata have been used in evolutionary computations for over a decade. They have been used in a variety of applications, such as parallel processing and number theory. CA architecture has been used in the design of arithmetic computations that Zhang [4] proposed architecture with programmable cellular automata, Choudhury [5] designed an LSB multiplier based on CA, and Jeon [6] proposed simple and efficient architecture based on periodic boundary CA.

This paper proposes efficient hardware architecture for division based on PCA. We focused on the architecture in ECC, which uses restricted irreducible polynomials, especially, trinomials. The structure has a time complexity of $n(n\text{-}1)(T_{\_AND}+ T_{\_XOR} + T_{\_MUX})$ and a hardware complexity of (nAND+(n+2)XOR+ nMUX+ 4nREGISTER). In addition, our architecture can easily be expanded for other public key cryptosystems with additional (n-2)XOR gates. Our architecture focuses on both area and time complexities.

* This work was supported by the Brain Korea 21 Project in 2006.
** Corresponding author.

V.N. Alexandrov et al. (Eds.): ICCS 2006, Part IV, LNCS 3994, pp. 661 – 668, 2006.
© Springer-Verlag Berlin Heidelberg 2006

The rest of this paper is organized as follows: The theoretical background, including finite fields, ECC, and CA, is described in Section 2. Section 3 presents the proposed division architecture based on PCA, and we present our discussion, together with a comparison of the performances between the proposed architecture and previous research, in Section 4. Finally, the conclusion is presented in Section 5.

2 Preliminary

In this section, we discuss the mathematical background in the finite field and ECC, and the characteristics and properties of cellular automata.

2.1 Finite Fields

A finite field or Galois Field (GF), which is a set of finite elements, can be defined by commutative law, associative law, distributive law and it contains facilitates for addition, subtraction, multiplication, and division. A number of architectures have already been developed to construct low complexity bit-serial and bit-parallel structures by using various irreducible polynomials to reduce the complexity. Since a polynomial basis operation has regularity and simplicity, the ability to design and expand it into high-order finite fields, with a polynomial basis, is easier to realize than with other basis operations [7].

A finite field can be viewed as a vector space of dimensions n over GF(2^n). That is, there exists a set of n elements $\{1, \alpha, \ldots, \alpha^{n-2}, \alpha^{n-1}\}$ in GF(2^n) such that each $A \in$ GF(2^n) can be written uniquely in the form $A = \sum A_i \alpha^i$, where $A_i \in \{0,1\}$. This section provides one of the most common bases of GF(2^n) over GF(2), which are polynomial bases [7, 8]. Let $f(x) = x^n + \sum_{i=0}^{n-1} f_i x^i$, where $f_i \in \{0,1\}$, for $i = 0, 1, \ldots, n-1$, be an irreducible polynomial of degree n over GF(2). For each irreducible polynomial, there exists a polynomial basis representation. In such a representation, each element of GF(2^n) corresponds to a binary polynomial of less than n. That is, for $A \in$ GF(2^n) there exist n numbers $A_i \in \{0,1\}$ such that $A = A_{n-1}\alpha^{n-1} + \ldots + A_1\alpha + A_0$. In many applications, such as cryptography and digital communication applications, the polynomial basis is still the most widely employed criterion [9-11]. In the following, we confine our attention to computations that use the polynomial basis.

2.2 Elliptic Curve Cryptosystem

In ECC, computing kP is the most important arithmetic operation, where k is an integer and P is a point on the elliptic curve. This operation can be computed by the addition of two points k times. ECC can be done with at least two types of arithmetic, each of which gives different definitions of multiplication [12]. Two types of arithmetic are, namely, $\mathbf{Z}_p$ arithmetic (modular arithmetic with a large prime p as the modulus) and GF(2^n) arithmetic, which can be done with shifts and exclusive-ors. This can be thought of as the modular arithmetic of polynomials with coefficients mod 2.

We focused on GF(2^n) arithmetic operation. Let GF(2^n) be a finite field by definition. Then, the set of all solutions for equation E: $y^2 + xy = x^3 + a_2x^2 + a_6$, where a_2, a_6 $\in$ GF(2^n), $a_6 \neq 0$, together with special point called the point at infinity $\boldsymbol{O}$, is a non-supersingular curve over GF(2^n). Let $P_1 = (x_1, y_1)$ and $P_2 = (x_2, y_2)$ be points in E(GF(2^n)) given in affine coordinates [13]. Assume that P_1, $P_2 \neq \boldsymbol{O}$, and $P_1 \neq -P_2$. The sum $P_3 = (x_3, y_3) = P_1 + P_2$ is computed as follows; if $P_1 \neq P_2$ then $\lambda = (y_1 + y_2)/(x_1 + x_2)$, $x_3 = \lambda^2 + \lambda + x_1 + x_2 + a_2$, $y_3 = (x_1 + x_3)\lambda + x_3 + y_1$, and if $P_1 = P_2$ (called point doubling), then $\lambda = y_1 / x_1 + x_1$, $x_3 = \lambda^2 + \lambda + a_2$, $y_3 = (x_1 + x_3)\lambda + x_3 + y_1$.

In either case, the computation requires one division, one squaring, and one multiplication. Squaring can be substituted by multiplication. From the point addition formula, it should be noted that no computation, except for addition, is performed at the same time due to data dependency. Therefore, sharing hardware between division and multiplication is more desirable than the separated implementation of division and multiplication [3, 8]

The additive inverse and multiplicative inverses in GF(2^n) can be calculated efficiently using the extended Euclidean algorithm. Division and subtraction are defined in terms of additive and multiplicative inverses: A-B is A+(-B) in GF(2^n) and A/B is $A\cdot(B^{-1})$ in GF(2^n). Here, the characteristic 2 finite fields GF(2^n) used should have $n \in \{113, 131, 163, 193, 233, 239, 283, 409, 571\}$ [3]. This restriction is designed to facilitate interoperability while enabling implementers to deploy efficient implementations that are capable of meeting common security requirements [13].

The rule that is used to pick acceptable reduction polynomials is the following: if a degree n binary irreducible trinomial, $f(x) = x^n + x^k + 1$, for $n > k \geq 1$ exists, then, the irreducible trinomial with the smallest possible k should be used. These polynomials enable the efficient calculation of field operations.

2.3 Programmable Cellular Automata

A CA is a collection of simple cells arranged in a regular fashion. CAs can be characterized based on four properties: a cellular geometry, a neighborhood specification, number of states per cell, and a rule to compute to a successor state. The next state of a CA depends on the current state and rules [14]. Only 2-state and 3-neighborhood CAs are considered in this paper. Table 1 shows all possible states and rules. Each mapping is called a 'rule' of the CA.

The next state transition for the ith cell can be represented as a function of the present states of the ith, (i+1)th, and (i-1)th cells for a 3-neighberhood CA: $Q_i(t+1) = f(Q_{i-1}(t), Q_i(t), Q_{i+1}(t))$, where '$f$' represents the combinational logic function as a CA rule, which is implemented by a combinational logic circuit (CL), and $Q(t+1)$ denotes the next state for cell $Q(t)$.

Table 1 specifies the three particular sets of transition from a neighborhood configuration to the next state. CA can also be classified as linear or non-linear. If the neighborhood is only dependent on an XOR operation, the CA is linear, whereas if it is dependent on another operation, the CA is non-linear. If the neighborhood is only dependent on an EXOR or EXNOR operation, then the CA can also be referred to as an additive CA.

Table 1. State transition and functions according to different rules

Rules	Logical functions	State transition							
		111	110	101	100	011	010	001	000
90	$Q_{i-1}(t) \oplus Q_{i+1}(t)$	0	1	0	1	1	0	1	0
150	$Q_{i-1}(t) \oplus Q_i(t) \oplus Q_{i+1}(t)$	1	0	0	1	0	1	1	0
240	$Q_{i-1}(t)$	1	1	1	1	0	0	0	0

Furthermore, if the same rule applies to all cells in a CA, the CA is called a uniform or regular CA, whereas if different rules apply to different cells, it is called a hybrid CA. A CA can be divided into three patterns based on the boundary conditions: a null boundary CA, intermediate boundary CA, and periodic boundary CA (PBCA). For the remainder of this paper, a CA will be regarded as a PBCA, unless otherwise mentioned. A PBCA regards the leftmost and rightmost cells as neighbors.

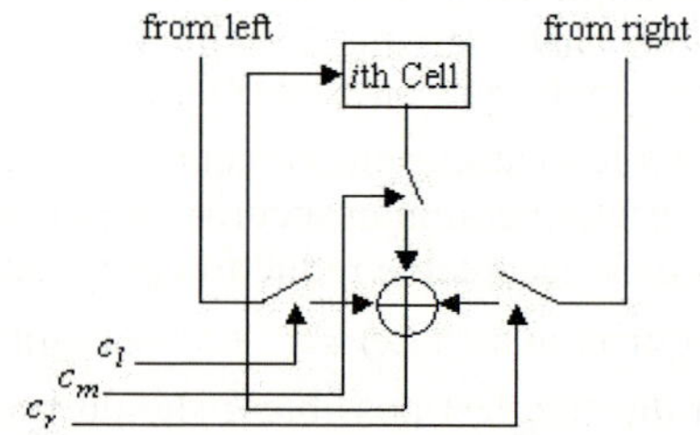

Fig. 1. A 3-neighborhood linear PCA cell structure

Table 2. The PCA corresponding rules according to the control signals

Control signals			Corresponding Rules	
C_l	C_m	C_r		
0	0	1	Q_r	170
0	1	0	Q_m	202
0	1	1	$Q_m \oplus Q_r$	102
1	0	0	Q_l	240
1	0	1	$Q_l \oplus Q_r$	90
1	1	0	$Q_l \oplus Q_m$	60
1	1	1	$Q_l \oplus Q_m \oplus Q_r$	150

A programmable CA (PCA) is a CA whose CL is not fixed for each cell, but is controlled by a number of signals such that different functions (rules) can be generated. Fig. 1 shows a 3-neighberhood linear PCA cell structure. A combination of control signals decides the rule of each PCA cell. The value of the control signals and corresponding rules of the PCA are presented in Table 2. If the '90' rule presented in Table 2 is used for the renewal of the next cell's value, the control signals of the C_l

and C_r values are '1' and C_m has a value of '0'. Also, if the '150' rule is used, the C_l, C_m, and C_r, all possess '1'values. We use the '170' rule which only depends on the right cell.

3 Division Architecture Based on PCA

This section presents $A(x)/B(x)$ architecture based on PCA. Finite field division in GF(2^n) can be performed using multiplication and inverse; that is, $A(x)/B(x) = A(x)B(x)^{-1}$. The division can be implemented efficiently by repeatedly applying $A(x)B(x)^2$ multiplications. Let us suppose that $A(x)$ and $B(x)$ are the elements on GF(2^n). Then, the two polynomials $A(x)$, $B(x)$ are as follows:

$$A(x) = A_{n-1}x^{n-1} + \ldots + A_1x^1 + A_0,\ B(x) = B_{n-1}x^{n-1} + \ldots + B_1x^1 + B_0$$

From the above equation, we have

$$B(x)^2 = B_{n-1}x^{2n-2} + B_{n-2}x^{2n-4} + \ldots + B_1x^2 + B_0$$

Then, $A(x)B(x)^2 \bmod T(x)$ can be induced from the two equations. The definitive algorithm for implementation is as follows:

$$\{\ldots[A(x)B_{n-1}x^2 \bmod T(x) + A(x)B_{n-2}]\, x^2 \bmod T(x) + \ldots + A(x)B_1\}\, x^2 \bmod T(x) + A(x)B_0$$

[Algorithm 1] $A(x)B(x)^2$ multiplication algorithm

Input : $A(x)$, $B(x)$, $T(x)$
Output : $A(x)B(x)^2 \bmod T(x)$

Step 1 : $M(x) = 0$
Step 2 : for $i = n$-1 to 0
Step 3 : $M(x) = M(x)\, x^2 \bmod T(x) + A(x)B_i$

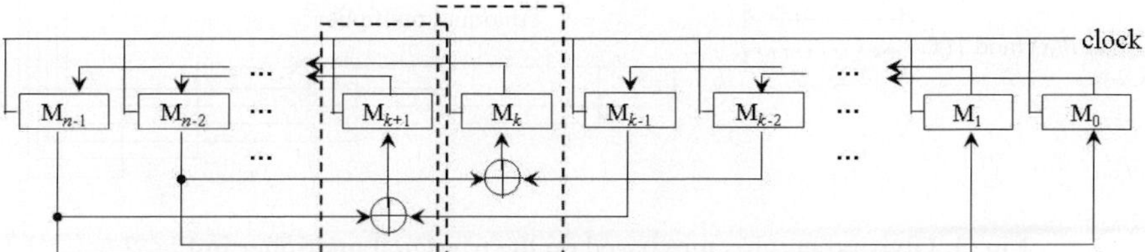

Fig. 2. Multiplication architecture using irreducible trinomials in Step 3 of Algorithm 1

Let $M(x)x^2 \bmod T(x)$ be $M_{n-1}x^{n-1} + \ldots + M_1x^1 + M_0$, where $T(x) = x^n + x^k + 1$. Then the following equation holds: $M(x)x^2 \bmod T(x) = M_{n-3}x^{n-1} + \ldots (M_{n-1} \oplus M_{k-1})x^{k+1} + (M_{n-2} \oplus M_{k-2})x^k + \ldots + M_{n-1}x^1 + M_{n-2}$. The equation is illustrated based on two PCA cell structures where C_l, $C_r = 1$ and C_m=0.

Here, the $A(x)B(x)^2$ operation can be used as an efficient method in division algorithms as follows [7]:

[Algorithm 2] $A(x)/B(x)$ Division algorithm

Input : $A(x)$, $B(x)$, $T(x)$
Output : $D(x) = A(x)/B(x) \bmod T(x)$

Step 1 : $D(x) = B(x)$
Step 2 : for $i = n$-2 to 1
Step 3 : $D(x) = B(x)D(x)^2 \bmod T(x)$
Step 4 : $D(x) = A(x)D(x)^2 \bmod T(x)$

The result is $D(x) = A(x)B(x)^{-1}$ and when $A(x) = 1$, the algorithm realizes the inverse operation $B(x)^{-1}$. In this case, the $A(x)B(x)^2$ operation can be used to compute the operations in Step 3 and 4. Fig. 3 shows the proposed architecture for division. Each initial value is such that cellular automata have all zeros, and that the B register and Shift register have $B(x)$ values.

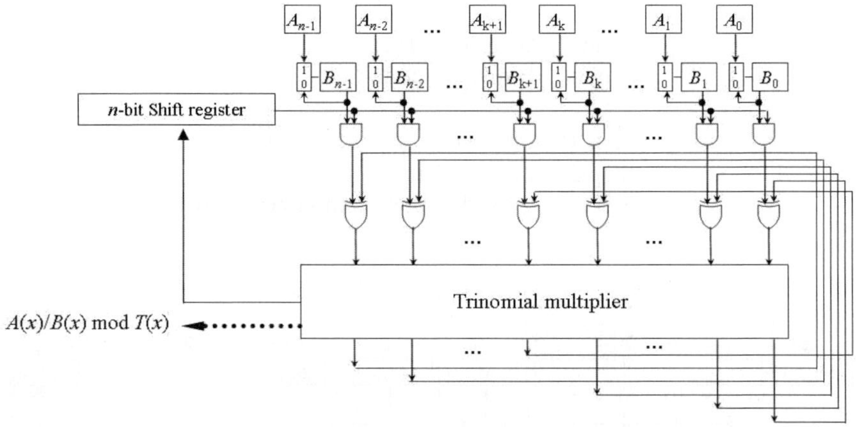

Fig. 3. Division architecture based on the trinomial multiplier in Fig. 2

After the $A(x)B(x)^2$ operation in Fig.2, the computed values transfer to the Shift register. $B(x)^{-1}$ is computed after the above mentioned process n-2 times. After the process, the system chooses $A(x)$ instead of $B(x)$ in the upper registers by Muxes for the final resultant values. It is possible to perform $A(x)/B(x) \bmod T(x)$ in $n(n-1)$ clock cycles by using n AND gates, n+2 XOR gates, n Muxes, and $4n$ bits registers, plus extra equipment such as control signals for the transfer of results in cellular automata to the shift register, and in order to change the values in the B register, immediately after deriving the values of $B(x)^{-1}$.

Moreover, our architecture can be easily applied to other public key cryptosystems by using general irreducible polynomials. In Fig. 3, by using additional n-2 XOR gates, the proposed architecture can perform a general division operation. Although the architecture is used as a general divider, it has the same latency as of that in Fig. 3 because of the parallel property.

4 Comparison and Analysis

A comparison of the proposed division architecture, with existing structures was performed, focusing on time and hardware complexity issues. As such, Wang's [10] and Kim's [11] division architecture was chosen.

Wang proposed parallel-in parallel-out division architecture, which has a latency of $n(2n\text{-}1.5)$ and a critical path of ($T_{\_AND}$+ $3T_{\_XOR}$) over $GF(2^n)$. Kim proposed a serial-in serial-out divider, which has a latency of $2n(n\text{-}1)$ and critical path of a ($2T_{\_AND}$+$3T_{\_XOR}$ + $T_{\_MUX}$). However, our serial-in parallel-out architecture has only a latency of $n(n\text{-}1)$ and a critical path of ($T_{\_AND}$+$T_{\_XOR}$ +$T_{\_MUX}$) over $GF(2^n)$.

We usually try to find the design that will best satisfy a given set of requirements when we implement the arithmetic unit design. We consider construction simplicity, which is defined by the number of transistors that are needed for its construction and the time that is needed for the signal change to be propagated through the gates [15]. Table 3 shows a comparison of area-time products.

Table 3. A comparison of area-time products among dividers

Circuits	Area-time product
Proposed architecture	$128n^3(n\text{-}1)(30n\text{-}22.5)$
Wang et al. [15]	$92n(105n^2\text{-}176n\text{+}69)(n\text{-}1)$
Kim et al.[16]	$336n(n\text{-}1)(4n\text{+}1)$

As in Table 3, our architecture has less complexity than the other architectures. In particular, the proposed architecture has $O(n^3)$ area-time complexity whereas the other architecture have $O(n^4)$ and $O(n^5)$ complexity, respectively. We have shown that our architecture has less complexity than the serial or parallel architectures with regard to area and time. Our architecture only focuses on ECC, which is restricted by using irreducible trinomials. Our architecture, however, can be easily applied to other public cryptosystems with additional n-2 XOR gates, while existing systolic architectures including those of Wang's and Kim's, hardly reduce the level of complexity, although they apply irreducible trinomials for ECC. Moreover, our architecture does not influence latency after it has been applied to a general divider.

5 Conclusion

This paper has presented efficient hardware architecture in order to compute the $A(x)/B(x)$ modulo irreducible trinomials, which are restricted in the Certicom Standard for ECC. We have proposed a simple hardware architecture that is the most expensive arithmetic operation scheme, such as inverse and division in ECC over GF(2^n). The proposed architecture includes the characteristics of both PCA and irreducible trinomials, and it has minimized both time and hardware complexity. Moreover, our architecture can be easily applied to a general division architecture with no additional latency needed. Therefore, we have shown that our architecture has outstanding advantages, as compared to typical structures.

References

1. T. R. N. Rao and E. Fujiwara, Error-Control Coding for Computer Systems, Englewood Cliffs, NJ: Prentice-Hall (1989)
2. W. Drescher, K. Bachmann, and G. Fettweis, "VLSI Architecture for Non Sequential Inversion over GF(2^m) using the Euclidean Algorithm," The International Conference on Signal Processing Applications and Technology, Vol. 2. (1997) 1815-1819
3. A.J.Menezes, Elliptic Curve Public Key Cryptosystems, Boston, MA: Kluwer Academic Publishers (1993)
4. C. N. Zhang, M. Y. Deng, and R. Mason, "A VLSI Programmable Cellular Automata Array for Multiplication in GF(2^n)," PDPTA '99 International Conference (1999)
5. P. Pal. Choudhury and R. Barua, "Cellular Automata Based VLSI Architecture for Computing Multiplication and Inverses in GF(2^m)," IEEE 7th International Conference on VLSI Design (1994) 279-282
6. Jun-Cheol Jeon and Kee-Young Yoo, "An Evolutionary Approach to the Design of Cellular Automata Architecture for Multiplication in Elliptic Curve Cryptography over Finite Fields," Lecture Notes in Artificial Intelligence PRICAI 2004: Trends in Artificial Intelligence (LNAI 3157), Springer-Verlag, Vol. 3157. (2004) 241-250
7. A. J. Menezs, Applications of Finite Fields, Boston, MA: Kluwer Academic Publishers (1993)
8. IEEE P1363, Standard Specifications for Public Key Cryptography (2000)
9. S. W. Wei, "VLSI architecture of divider for finite field GF(2^m)", IEEE International Symposium on Circuit and Systems, Vol. 2. (1998) 482-485
10. C. L. Wang and J. H. Guo, "New Systolic Arrays for C+ AB^2, inversion, and division in GF(2^m)", IEEE Trans. on Computer, Vol. 49, No. 10. (2000) 1120-1125
11. N. Y. Kim and K. Y. Yoo, "Systolic architecture for inversion/division using AB^2 circuits in GF(2^m)", Integration, the VLSI Journal, Vol. 35. (2003) 11-24
12. C. Kaufman, R. Perlman, and M. Speciner, Network Security private communication in a public world, New Jersey: Prentice Hall (2002)
13. SEC 1: Elliptic Curve Cryptography version 1.0, Certicom Reserch (2000)
14. O. Lafe, Cellular Automata Transforms: Theory and Applications in Multimedia Compression, Encryption, and Modeling, Kluwer Academic Publishers (2000).
15. D. D. Gajski, Principles of digital design : Prentice-Hall International Inc. (1997)

A General Data Grid: Framework and Implementation

Wu Zhang, Jian Mei, and Jiang Xie

Department of Computer Science and Technology, Shanghai University,
Shanghai, 200072, China
zhang@mail.shu.edu.cn, meijian_2003@yahoo.com.cn

Abstract. Today, data grids have become an important emerging platform for managing and processing a very large amount of data distributed across multiple grid nodes and stored in relational databases. However, there are still obstacles for potential grid users to be involved into the trend and the data grid application development is far from the data grids. While the traditional established data grid architectures are not particularly suitable for the some grid service, such as data replica services, user authentication services, and the Optimal Path selection services. Our project aims to exploit a novel architecture named General Data Grid, which integrates the metadata services, data replica services, java message services, and the Optimal Path selection on Data Grid environment. At the end of this paper, we describe the key implement on the GDGrid and present a simple example application concerning finding the optimal route. Our experiment of GDGrid shows the algorithm of the route selection (Heart-Beat algorithm) is effectively, improve the performance greatly, and afford fault-tolerance management to the great extent.

1 Introduction and Relate Work

With the emergence of the massive data collections of terabyte scale and the high-performance computers, how to utilize these huge data sets and the high-performance computing capacity becomes a great challenge.

Under the circumstances, "Grid" is appearing as are an approach for building dynamically constructed problem-solving environments using geographically and organizationally dispersed high-performance computing and data-handling resources [1]. Thus today can be used as effective tools for distributed computing and data processing in many domains such as astronomy, geography, and earthquake, etc. In all the Grid applications, Data Grid plays an important role in the data handling. It is a dynamic logical namespace that enables coordinated sharing of heterogeneous distributed storage resources and digital entities based on local and global policies across administrative domains in a virtual enterprise[2].

In the Architecture, European Data Grid Project [3] is the development of a new environment to support globally distributed scientific exploration involving terabytes datasets. And the GriPhyN Project [4] provides a new degree of transparency in how data-handling and processing capabilities are integrated to deliver data products to end-users or applications. These two projects are strongly related to grid services architecture, but they do not address more sophisticated planners, which can take into

V.N. Alexandrov et al. (Eds.): ICCS 2006, Part IV, LNCS 3994, pp. 669 – 676, 2006.
© Springer-Verlag Berlin Heidelberg 2006

account performance and reliability as well as provide feedback to the user, so that the user can decide whether to go ahead with a request. In the data replicas management, a simulation framework— OptorSim was introduced in [5] where data replication was combined with job scheduling. They uses a prediction function based on spatial and time locality regardless of the overall data access cost on the Data Grid.

In this paper we propose a novel framework for access control, message services, data replicas services and the selection of the OP (Optimal Path) in Data Grid environment. Our key contributions include the following aspects: (1) JMS (Java Message Service) is employed on subscribe/cancel service, replicas service, and Heart-Beat service. Through the combine of the GridFTP and JMS, we can ensure the replicas' consistency and validity. (2) In the access control management, we introduce the mechanism of inspection certificates. This module is responsible for verifying the certificates when the users visit the specified access binaries. (3) In the selection of the OP (Optimal Path), we employ the Heart-Beat algorithm. By means of this algorithm, we can find the temporal OP with a little expense.

The rest of the paper is organized as follows. In Section 2, we describe the related work. We describe the general data grid framework in detail in Section 3. In Section 4, we show the key implementation status of the general data grid and Section 5 concludes the paper and discusses future work.

2 The General Data Grid Framework

The GDGrid (General Data Grid) is built on the OGSA-DAI [6]. We present a framework of GDGrid as Fig.1 shows. The GDGrid is composed of three Layers, which are the SRL (Service Requesting Layer), the SML (Service Managing Layer) and the SPL (Service Providing Layer) [7].

2.1 SRL (Service Requesting Layer)

In the SRL, the SC (Service Client) can visit the SWP (Service Web Portal) of the GDGrid. Through the SWP, the SC can register the own data services into the GDGrid and become a SP (Service Provider). The SC can attain the service provided by the GDGrid without the protected data resources in the SWP. In case that the SC wants to get the services of the protected data resources, it must log on the AM (Access Management) module and get the relevant certificates of the protected data resources. In the GDGrid, once the SC would like to become a SP, it must download the Middleware of the SPL (Service Providing Layer) and perform other operations what we will discuss in the following sections. In this layer, the SC will interact with the SM (Scheduling Management) module. The SC advances a request, and then the SM will offer the temporal optimal route according to the heart-beat algorithm to the SC.

2.2 SML (Service Managing Layer)

The Service Managing Layer is the core of the GDGrid. Its main goals are to furnish the data grid interface to the SCs, integrate the heterogeneous data, offer a secure environment and find the Optimal Path of the request execution. Moreover, this layer

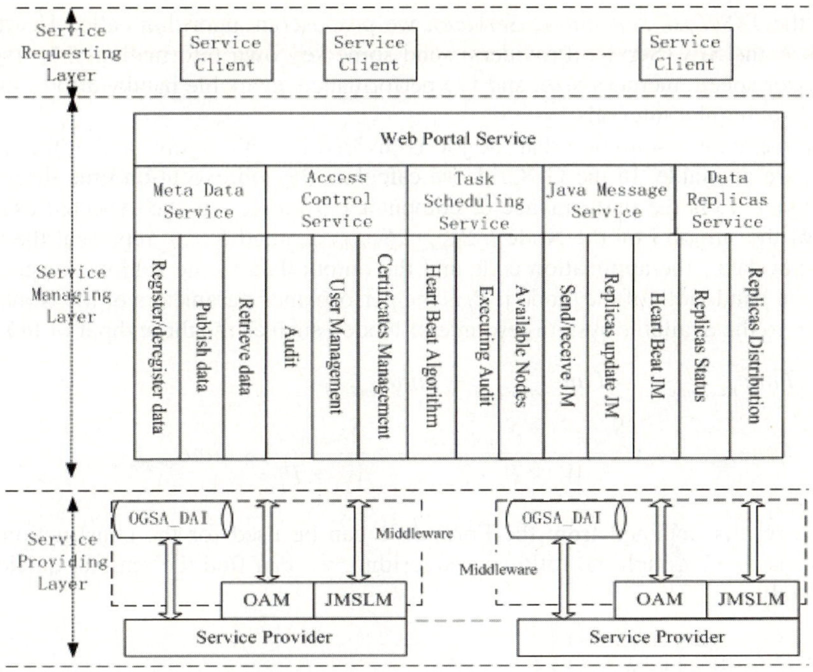

Fig. 1. The GDGrid Architecture

coordinates the application execution by attempting to fulfill the application requirements and the available data resources. On the whole, this layer consists of six main services.

The *WPS (Web Portal Service)* builds the relationship between the SRL and the SPL. In the GDGrid, the WPS is responsible for maintaining a description of the GDGrid. In the other words, we can think it as an interface of the GDGrid. To be the SC, it can log in the Data Grid, obtain the GDGrid's services freely and register its services to be a SP. To be the SP, it can register or deregister its services and the descriptions, change its service, obtain the GDGrid's services, request a certificate of its own, and authorize its own certificate replicas to the other dependable SP, etc.

The *MDS (Meta Data Service)* offer all the metadata describing features of data sources, and maintain the catalogue of the original data and the catalogue of replicas data. The *ACS (Access Control Service)* focuses are on managing the certificates, empowering the users to utilize the data resources and presenting the monitoring of the logged users. In the monitoring management, we make use of a web based monitoring tool (Map Center [8]) which provides access to status information.

The *JMS (Java Message Service)* offer three kinds of service. One service aims to send the real-time messages to any other nodes, and the second service is used to customize the some services, such as getting the latest public information, the state of the original data and so on. The last service is the Heart-Beat message service.

The *DRS (Data Replicas Service)* is used to coordinate data resources replication across the GDGrid from one node to another. There are including three parts: *Replicas Catalog*, *Replicas States* and *Replicas Descriptions*.

In the *TSS(Task Scheduling Service)*, we propose an algorithm called Heart-Beat, which is the SPs (Service Providers) send some key own information [11] (such as processor speed, memory size, and I/O performance, available bandwidth) to the TSS module at regular intervals.

We make an assumption that the job is divided into *N* subjobs, and at that time *M* nodes are available. In the GDGrid, we calculate the job execution time through the Formula 1. P_j is the performance of computer j. $Time_{Exec\_ij}$ is the expected execution time of the subjob i on the Node j. S_{Input_j}, $S_{Application\_j}$ and S_{output_j} represent the size of the input data , the application code and the output data of the subjob j respectively. W_j is the bandwidth of the Node j. P_j is the performance parameters of the Node j. and T_j refer to the available system resource of Node j such as the throughput of I/O.

$$Time_{Response\_j} = Time_{arrived\_j} - Time_{send\_j}$$

$$Time_{Exec\_ij} = \left(\frac{S_{Input\_i} + S_{Application\_i}}{W_j \times P_j \times T_j} + \frac{S_{Output\_i}}{W_j \times P_j}\right) \times Time_{Response\_j} \tag{1}$$

The results obtained from the Formula 1 can be used for the construction of an M×N matrix. Through the following algorithm, we can find the optimal quickly and accurately.

```
for(i=1,i<=n,i++){
 for(j=1,j<=m,j++){
   Node=Min() /*select the node of the minimize
                 executing time*/
   }
 if(SearchNodes(Node)==False) //check this Node
       AddNodes(Node) //add this Node to the path
     else
       NodeA=Compare(Node) /**Compare the total
        runtime in this Node to the value in
        column i, then return the node of the
        minimize executing time **/
     AddNodes(NodeA) //add this Node to the path
}
return Path    //return the optimal path
```

2.3 SPL (Service Providing Layer)

The SRL is basically used for organizing the SPs (Service Providers) orderly, utilizing the SPs' resource efficiently and administering the certificates securely. In the SRL, the OGSA-DAI are applied to integrate various heterogeneous data resources seamlessly and to build the relationship between the SPs and the SML.

As for a SP, JMSLM (Java Message Service Local Module) is used to receive the input JM issued by other SPs and to send the output JM to the destination. In this GDGrid, the input JM can be considered a manner which the other SPs notify the current SP. We assume that a current SP holds a replica of a remote SP's data resource. Once the remote SP's data resource is changed, the remote SP will send a notice of the changes to the JMS of the SML through the JMSLM, the JMS will distribute the notice to all the nodes which possess the replicas, and finally the JMSLM will proceed with the notice in the SP. As for the remote SP the notice is an output JM, while as for the current SP the notice is an input JM. In addition, the Heart-Beat is considered as an output JM for a SP and send through the JMSLM in the same way.

In the SP, there is another module named OAM (Own Authentication Module) which is adopted with the purpose of recording the own and other certificates information.

3 The General Data Grid Key Implementation

3.1 Java Message Service

Java message service is implemented based on the JMS of the SML and the JMSLM of the SPL. It includes three major message services: the subscribe/cancel service, the replicas service, and the Heart-Beat service.

During the course of designing the JMS, the advanced EJB technology and the design patterns are employed to the GDGrid. If the SCs have interest in some SPs, they can subscribe the latest information of some SPs by the Subscribe Message service. In case that the SPs accept the subscribe request, the SPs will send to the real-time latest information to the SCs timely. If the SCs have no interest in the SPs any more, they can cancel the order by the Cancel Message service.

The Replicas Message service associating with the GridFTP makes it possible that the data replicas can keep the consistency and validity. Once a SP's data resource is modified, the SP will send a notice of the changes to the Replicas Message service, it will distribute the notice to all the nodes possessing the replicas. If a node possessing a replica is closed when the original data are changed, the Replicas Message service will put the notice to the massage cache area. Once the node is available, the Replicas Message service will resend the notice to the node. In the management of the massage cache area, GDGrid provides the service mechanism called Least Recently Used (LRU). That is the node which is accessible in the first time owns the headmost services and receives the message about itself.

As for the Heart-Beat message service, it plays a vital role to the selection of the optimal path. Through the service, the GDGrid can attain the correct real-time information about the usable nodes.

3.2 Authorization Service

About the authorization service, we can explain through the Fig.2.

Step 1: The user A sends a request for owning an own unique certificate in order to protect some protected data resources.

Step 2: The ACS module assigns a unique certificate (For example CTF001, X509) for the protected data resources of user A. Once other users want to visit those data resources, they should own the CTF001.

Step 3: User B sends a request for visiting the protected data resources of user A.

Step 4: The ACS module transmits the request to the User A. The request consists of the information of the demander (User B) username, IP address, the position etc.

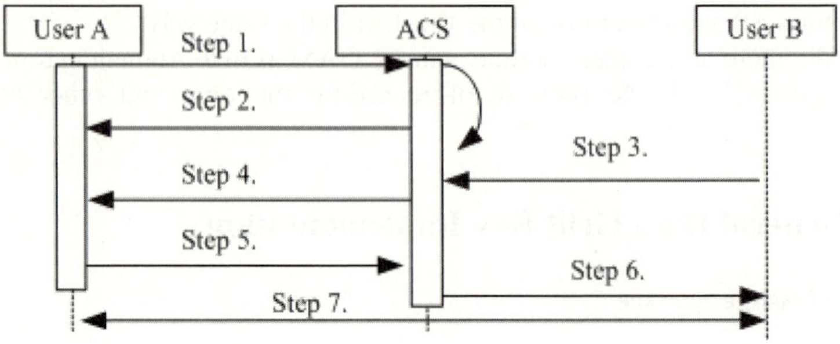

Fig. 2. Procedure In Authorization

Step 5: The User A confirms the request and returns a message (yes or no). If the message is no, the ACS module will refuse the request which is send by the demander. In the Fig.2, the User A sends the message which is yes.

Step 6: The ACS module receives the message form User B, checks the certificates management sub module, and sends a replica of the User A's certificate to the User B.

Step 7: Once the User B owns the replica of the User A's certificate, he can use the User A data resources directly without through the ACS module certification.

By this means, we not only save the authentication time and the bandwidth, but also make utmost use of the data resources in the safe situation.

3.3 The Optimal Path Discovery

Experiments are taken to analyze the process of finding the optimal path. The job is divided into four subjobs and there are six accessible and available nodes at the experiments. The parameters used in the experiments are list as follows: *<CPU, Memory, Bandwidth, Available_Resouce, Response_Time>*, { Node_1<1G, 256M, 1.85M, 80%, 50S>, Node_2<2G, 512M, 1.7M, 100%, 45S>,Node_3<2G, 512M, 1.6M, 80%, 45S >,Node_4<2.5G, 512M, 17M, 70%, 45S >,Node_5<2G 2G, 1G, 1.8M, 55%, 45S >,Node_6<2.6G, 512M, 1.74M, 60%, 50S>}.

Table 1. The forcase of executing subjob. the Zero means the node has the subjob's input data.

Nodes	Forecast Subjob1			Forecast Subjob2			Forecast Subjob3			Forecast Subjob4		
	S_in	S_ap	S_ou	S_in	S_ap	S_ou	S_in	S_ap	S_ou	S_in	S_ap	S_ou
Node_1	30M	10M	12M	25M	15M	10M	0M	8M	8M	12M	10M	10M
Node_2	30M	10M	12M	25M	15M	10M	15M	8M	8M	12M	10M	10M
Node_3	0M	10M	12M	25M	15M	10M	15M	8M	8M	0M	10M	10M
Node_4	30M	10M	12M	25M	15M	10M	15M	8M	8M	12M	10M	10M
Node_5	30M	10M	12M	0M	15M	10M	15M	8M	8M	12M	10M	10M
Node_6	30M	10M	12M	25M	15M	10M	15M	8M	8M	12M	10M	10M

According to calculating these parameters through the Formula 1, the 6×4 matrix is shown as follows:

	Subjob1	Subjob2	Subjob3	Subjob4
Node_1	1675.7	1621.6	486.5	1013.5
Node_2	611.8	588.2	**364.7**	376.5
Node_3	**344.5**	843.7	516.8	316.4
Node_4	650.8	631.9	384.5	**389.9**
Node_5	528.5	**233.0**	311.4	312.5
Node_6	753.5	734.4	443.8	447.0

	Subjob1	Subjob2	Subjob3	Subjob4
Node_1	1675.7	1621.6	486.5	1013.5
Node_2	**258.8**	294.1	364.7	235.3
Node_3	344.5	404.3	516.8	**316.4**
Node_4	650.8	631.9	384.5	389.9
Node_5	528.5	**233.0**	311.4	312.5
Node_6	753.5	734.4	**204.3**	447.0

Fig. 3. The Execution Time. Left matrix is without data replicas, and the right is with replicas.

When choosing the optimal path, the SM will use the Heart-Beat algorithm. In this experiment without data replicas, the optimal path is *Node_3→Node_5→Node_2→Node_4*. Also another experiment is done during the period of executing this job, we shut down the *Node_2*. The SM will process with the adaptive executing path choice. That is in this experiment the subjob_3 executed on the *Node_2*. When the *Node_2* is inaccessible, the Heart-Beat algorithm will select the optimal node to execute the subjob_3 expect the *Node_2*. So the optimal path is changed to be *Node_3→Node_5→Node_1→Node_4* in this experiment. Through this experiment, it proves that the Heart-Beat algorithm is effectively, improve the performance greatly, and afford fault-tolerance management to some extent.

For the purpose of finding the data replicas' effect on the scheduling, we make some data replicas on the different nodes (in this experiment, the time of the replicas' transfer is not considered.): employ the subjob1's data on the Node_2 and Node_3; the subjob2's data on Node_2, Node_3 and Node_5; the subjob3's data on Node_1 and Node_6; the subjob4's data on Node_2 and Node_3. By using the Formula 1, the execution time with data replicas 6×4 matrix is gotten, and the optimal path is: *Node_2→Node_5→Node_6→Node_3*.

From the two 6×4 matrixes, we can figure out the probable total runtime. One with no data replica is 389.9s, and the other with some data replicas is 316.4s. Obviously, the data replicas must short the runtime of the task.

4 Conclusion and Future Work

The GDGrid facilitates the usability of the distributed heterogeneous data resources on the Grid. It integrates Metadata services, Data Replica services, Java Message services, Replicas management services and route selection within its framework. In this paper, we have described the three level of the GDGrid in detail and introduce the functions of different sub-modules briefly. We present the key implementations and do some experiments that validate the efficiency of our Heart-Beat algorithms.

Additional future work of this project will be to integrate and classify data resources. In addition, to extend the performance of data transfers, we intend to investigate protocols based on Quality of Service concerning the transformation of large quantities of data. Further, since in our GDGrid the dynamic transfer of the data replicas is considered in the optimal path selection, we plan to add this consideration into the Heart-Beat algorithm and improve the algorithm in the efficiency and speed.

References

1. I.Foster and C.Kesselman (Eds): The Grid: Blueprint for a New Computing Infrastructure. http://www.mkp.com/books_catalog/1-55860-475-8.asp, Morgan Kanfmann, Los Altos, CA,1988.
2. Reagan.Moore, Arcot rajasekar, and Michal Wan, MEMBER, IEEE: Data Grids, Digital Libraries, and Persistent Archives: An Integrated Approach to Sharing, Publishing, and Archiving Data. Proceedings of the IEEE, VOL. 93, NO. 3, 578-588, March 2005.
3. Segal, B.; Robertson, L.; Gagliardi, F.; Carminati, F.: Grid computing: the European Data Grid Project. Nuclear Science Symposium Conference Record, 2000 IEEE Volume 1, 15-20 Oct. 2000 Page(s):2/1 vol.1
4. Deelman, E., Kesselman, C., Mehta, G., Meshkat, L.: GriPhyN and LIGO, building a virtual data Grid for gravitational wave scientists. High Performance Distributed Computing, 2002. HPDC-11 2002. Proceedings. 11th IEEE International Symposium on Page(s):225 - 234
5. William H. Bell, David G. Cameron, Luigi Capozza, A. Paul Millar, Kurt Stockinger, Floriano Zini: Simulation of Dynamic Grid Replication Strategies in OptorSim. Proc. Of the 3rd Int'l IEEE workshop on Grid Computing (Grid 2002), Baltimore, USA.
6. Open Grid Services Architecture – Data Access and Integration Project: http://www.ogsadai.org.uk.
7. Nong Xiao, Dongsheng Li, Wei Fu, Bin Huang, Xicheng Lu: GridDaen: A Data Grid Engine. Second International Workshop,GCC2003, Shanghai, China, Page(s):519-528
8. Map Center Home Page: Http://mapcenter.in2p3.fr
9. G. Aloisio, M. Cafaro, I. Epicoco: Early experiences with the GridFTP protocol using the GRB-GSIFTP library. Future Generation Computer Systems, Volume 18, Number 8 (2002), pp. 1053-1059, Special Issue on Grid Computing: Towards a New Computing Infrastructure, North-Holland.
10. Ann Chervenak, Ian Foster, Carl Kesselman, Charles Salisbury and Steven Tuecke: The Data Grid: Towards an Architecture for the distributed Management and Analysis of Large Scientific Datasets. Journal of Network and Computer Applications, 23:187-200, 2001.
11. Sang-Min Park, Jai-Hoon Kim: Chameleon: A Resource Scheduler in A Data Grid Environment*. Cluster Computing and the Grid, 2003. Proceedings. CCGrid 2003.3rd IEEE/ACM International Symposium on12-15 May 2003 Page(s):258 - 265.

Path Following by SVD

Luca Dieci[1], Maria Grazia Gasparo[2], and Alessandra Papini[3]

[1] School of Mathematics, Georgia Institute of Technology,
Atlanta GA 30332, USA
dieci@math.gatech.edu

[2] Università di Firenze, Dip. Energetica "S. Stecco", via C. Lombroso 6/17,
I-50134 Firenze, Italia
mariagrazia.gasparo@unifi.it

[3] Università di Firenze, Dip. Energetica "S. Stecco", via C. Lombroso 6/17,
I-50134 Firenze, Italia
alessandra.papini@unifi.it

Abstract. In this paper, we propose a path-following method for computing a curve of equilibria of a dynamical system, based upon the smooth Singular Value Decomposition (SVD) of the Jacobian matrix. Our method is capable of detecting fold points, and continuing past folds. It is also able to detect branch points and to switch branches at such points. Algorithmic details and examples are given.

Subject Classifications: 65F15, 65F99.

1 Introduction

One of the most important and recurring problems in applications is that of finding solutions of overdetermined nonlinear systems, $f(u) = 0$, where f is a C^k map, $k \geq 1$, from (part of) $R^{n+1} \to R^n$. Standard occurrences of this situation are homotopy techniques for solving nonlinear systems, see [15], and also techniques to find equilibria or periodic solutions of parameter dependent dynamical systems, see [11]. In such case, the system is usually written as

$$f(x, \alpha) = 0, \quad f : R^n \times R \to R^n. \tag{1}$$

As it is well understood, assuming that 0 is a regular value for f, the solution set of $f(u) = 0$, is a C^k curve (e.g., see [10]); we stress that, for (1), this does not mean that the curve can be globally parametrized in α. Computation of this regular curve of equilibria is the task of path-following algorithms, or continuation methods. The successful continuation methods are of *predictor-corrector* type: From knowledge of a point on the curve, they seek a new point on the curve at a certain (arc-length) distance from the present one, by iteratively *correcting* an initial *prediction* of this new point. The standard prediction stage is carried out by a tangent (or Euler) approximation, and the standard correction is performed by Newton's method or one of its variants. We refer to [1] for an excellent overview of continuation techniques. Now, while continuing the curve of

V.N. Alexandrov et al. (Eds.): ICCS 2006, Part IV, LNCS 3994, pp. 677–684, 2006.
© Springer-Verlag Berlin Heidelberg 2006

equilibria, a robust algorithm must also be able to adaptively choose the steps to be taken, and to detect *bifurcation* values (e.g., points where two solution curves intersect, or points where –for (1)– the curve fails to be parametrizable in α). Predictor-corrector algorithms are well understood and reliable implementations exist; see [9, 15]. In practical terms, the largest cost of prediction-correction methods is given by the need for frequent factorizations of the Jacobian matrices involved.

One aspect which has not been satisfactorily resolved in previous works on predictor-corrector algorithms is that the dynamical point of view, inherent in the continuation context, gets lost at the linear algebra level: the linear algebra is done in a *static* way, namely canned linear algebra software is used to factor the Jacobians. Our approach, which we present in this paper, is to view the Jacobians as smooth functions and to **use** the **underlying smoothness** of the factors in their decompositions to device better continuation strategies; at the same time, we will monitor variation of the factors to determine how to choose the continuation steps. Motivated by their relevance in detecting bifurcation phenomena, and by the provable smoothness in situations of practical interest (see [6] for the C^k case, and [4] for the analytic case), our particular emphasis here is on SVD decompositions. We must point out right away that a smooth SVD will ordinarily require a singular value to cross zero when the Jacobian becomes singular (one obtains a *signed* SVD), and two singular values will exchange ordering when they coalesce (one obtains an *unordered* SVD).

2 Path Following Methods Via SVD

In this work, we focus on computing curves of equilibria of (1): $f(x, \alpha) = 0$. By using arc-length parametrization, the problem is rewritten as

$$f(x(s), \alpha(s)) = 0, \;\; \dot{x}(s)^T \dot{x}(s) + \dot{\alpha}(s)^2 = 1,$$

where $(\dot{x}(s), \dot{\alpha}(s))^T$ is the tangent vector and satisfies

$$f_x \dot{x} + f_\alpha \dot{\alpha} = 0. \tag{2}$$

A point (x, α) on the curve is a regular point if $f_x(x, \alpha)$ is invertible, is a fold (turning) point if $f_x(x, \alpha)$ is singular and $f_\alpha(x, \alpha) \notin \text{range}(f_x(x, \alpha))$, is a branch point if $f_x(x, \alpha)$ is singular and the enlarged matrix $M = \begin{bmatrix} f_x & f_\alpha \\ \dot{x}^T & \dot{\alpha} \end{bmatrix}$ has rank equal to n. At a regular point there is a unique tangent; this also occurs at a fold point with $\dot{\alpha} = 0$; at a branch point instead, there are two tangents and two branches of equilibria crossing at the point. A branch point may occur with $f_\alpha \in \text{range}(f_x)$ and $\text{rank}(f_x) = n - 1$, and also with $\text{rank}(f_x) = n - 2$.

The pseudo arc-length path following approach (see e.g. [10], [13]) can be sketched as follows. Given a point $(x_0, \alpha_0) = (x(s_0), \alpha(s_0))$ on the path (i.e. $f(x_0, \alpha_0) = 0$), a tangent $(\dot{x}_0, \dot{\alpha}_0)$ and $s_1 = s_0 + h$ for some steplength h, we seek $(x_1, \alpha_1) = (x(s_1), \alpha(s_1))$ by solving

$$F(x,\alpha) \equiv \begin{bmatrix} f(x,\alpha) \\ \dot{x}_0^T(x-x_0)+\dot{\alpha}_0(\alpha-\alpha_0)-h \end{bmatrix} = 0. \tag{3}$$

To solve (3), we can use an iterative method starting from the tangent predictor

$$(x_1^{(0)},\alpha_1^{(0)}) = (x_0,\alpha_0) + h(\dot{x}_0,\dot{\alpha}_0).$$

Typically, the stationary Newton method is used as corrector: for $k=0,1,\ldots$

$$\begin{bmatrix} f_x(x_1^{(0)},\alpha_1^{(0)}) & f_\alpha(x_1^{(0)},\alpha_1^{(0)}) \\ \dot{x}_0^T & \dot{\alpha}_0 \end{bmatrix} \begin{bmatrix} \Delta x \\ \Delta\alpha \end{bmatrix} = -F(x_1^{(k)},\alpha_1^{(k)}) \tag{4}$$

$$(x_1^{(k+1)},\alpha_1^{(k+1)}) = (x_1^{(k)}+\Delta x, \alpha_1^{(k)}+\Delta\alpha)\,.$$

The idea to use SVD for computing bifurcations of vector fields appeared in [5] several years ago. Afterwards, as far as we could determine, it has not been considered any more. Why? Surely, one reason is that existing techniques typically work just fine. But, we believe that a more cogent reason is that by using the standard linear algebra SVD, whereby singular values are kept always positive and ordered, it is nearly impossible to locate exactly a fold or a branch point: One would need to step exactly at such points! Simply monitoring small singular values, as done in [5], is at best inefficient and potentially misleading. The answer to this impasse is beautifully provided by smoothness: If the Jacobian f_x becomes singular, a singular value will go through 0, and monitoring the signs of the singular values will suffice. As we will make clear below, our smooth SVD method provides not only a theoretically sound, but also a computationally interesting, way to compute curves of equilibria. As a matter of fact, we know of no other technique which allows **at once** to compute the points on the curve, provide tangents to form the predictor, detects folds and branch points, and also allows easily to switch branches at branch points.

Motivated by the work [7], where we studied several methods to compute smooth curves of SVD, we have written a `Matlab` code implementing a SVD-based path following method for tracking smooth path of equilibria. Our code computes a smooth SVD of f_x at points on the curve and uses this SVD to locate folds and generic branch points (switching branches and continuing new branches). Moreover the code avoids to compute f_x and f_α at the predictor and solves (3) by a Newton-type method, where $f_x(x_0,\alpha_0)$ and $f_\alpha(x_0,\alpha_0)$ are used instead of $f_x(x_1^{(0)},\alpha_1^{(0)})$ and $f_\alpha(x_1^{(0)},\alpha_1^{(0)})$ (cfr. (4)). From now on, the symbols f_x and f_α will be used to denote $f_x(x_0,\alpha_0)$ and $f_\alpha(x_0,\alpha_0)$ respectively.

Notice that if $f_x = U\Sigma V^T$, then the Newton-type system

$$\begin{bmatrix} f_x & f_\alpha \\ \dot{x}_0^T & \dot{\alpha}_0 \end{bmatrix} \begin{bmatrix} \Delta x \\ \Delta\alpha \end{bmatrix} = \begin{bmatrix} c \\ d \end{bmatrix}$$

becomes simply

$$\begin{bmatrix} \Sigma & U^T f_\alpha \\ \dot{x}_0^T V & \dot{\alpha}_0 \end{bmatrix} \begin{bmatrix} V^T\Delta x \\ \Delta\alpha \end{bmatrix} = \begin{bmatrix} U^T c \\ d \end{bmatrix}$$

which has a very simple structure and can be solved with a computational cost of $O(n)$ flops, by ad hoc substitution techniques tailored for the two cases Σ invertible and Σ singular. So, taking into account the matrix-vector products, each iteration costs $O(n^2)$ flops.

Analogously, the SVD of f_x is easily exploited in the computation of the tangents. Indeed, problem (2) becomes

$$\Sigma V^T \dot{x} + U^T f_\alpha \dot{\alpha} = 0.$$

If Σ is invertible, we set $\dot{\alpha} = \pm 1$ (the sign is chosen so that we do not trace back the same piece of the curve), $\dot{x} = -\dot{\alpha} V \Sigma^{-1} U^T f_\alpha$ and then we normalize. If we are at a fold point (Σ singular and $f_\alpha \notin \text{range}(f_x)$), the solution is $(\dot{x}, \dot{\alpha}) = (Ve_n, 0)$. Finally, at a branch point ($\text{rank}[f_x \ , \ f_\alpha] = n-1$, but $\text{rank}(M) = n$) we have two tangents in the kernel of $[f_x \ , \ f_\alpha]$, and we proceed as follows: (1) We approximate the tangent, call it u, to the branch that we are following, by normalizing the secant approximation obtained using the last two points on the curve; (2) we seek a point on the other branch by looking for solutions lying on a hyperplane parallel to the tangent line we have, and at a distance h from it (this we can do using a vector v in the null space of $[f_x \ , \ f_\alpha]$ orthogonal to u).

3 Continuation of the SVD

In [7], we proposed several techniques to compute a smooth block-SVD for a smooth matrix-valued function $A(s) : [a, b] \to R^{n \times n}$. In the context of path following for equilibria, we have $A(s) = f_x(x(s), \alpha(s))$ and want to smoothly compute its complete signed SVD. The theory developed in [7] guarantees that one of the algorithms there studied, namely Algorithm BSVD, can be successfully used to this scope as long as the singular values remain distinct. We refer to [7] for details. Here we only recall that this method is essentially based on the solution of suitable algebraic Riccati equations at each continuation step. These equations are solved by Newton's method, for which we can construct a second order approximation as initial guess. Convergence is ensured as long as the continuation steplength h is sufficiently small.

If two singular values coalesce within a continuation step, the Newton iteration for the Riccati equations may either converge to a wrong solution or fail to converge. Now, having two singular values coalescing is a non-generic occurrence for one-parameter problems (see [6]), and thus one should not witness it in practice; naturally, even less likely (and thus not very interesting) is the case of three or more singular values coalescing at once, a case which we can thus safely disregard. However, special symmetries in the problem make the occurrence of two singular values coalescing (and crossing) possible, and we want methods robust enough to handle this case. Futhermore, singular values nearing each other can cause numerical difficulties. To overcome this critical situation, in the present paper, we implemented a novel technique, which we call the **accordion** strategy. The idea is to adaptively switch from a complete SVD to a block SVD in order to bypass coalescing (or just close) singular values. In practice, at each step we monitor the singular values to detect if we are nearing a possible crossing. If this is the case, we group into a 2×2 block the two singular values which seem to coalesce and apply Algorithm BSVD to smoothly continue a block-SVD with one block of dimension 2 and the others of dimension 1. This structure is possibly maintained over several continuation steps, until the crossing has been safely passed, and we can split the 2×2 block and proceed with a complete SVD. Of course, as the continuation proceeds, the 2×2 block must be further decomposed to get always a smooth complete SVD. To do this, we solve a simple Procrustes problems (see [12] for analogous refining methods in the context of computing analytic path of SVD). In the context of continuation methods for large bifurcation problems, strategies where one dynamically chooses the block sizes have also been considered in [2, 3].

The *accordion* strategy works very well in practice so long as coupled with a reliable criterion to select the steplength at each continuation step. In the continuation context, the stepsize control is generally convergence-dependent through the number of performed iterations ([8], [9],[10]). In our code, instead, we implemented a technique based on the distance between the predictors and the converged values. This is similar in spirit to the stepsize control in codes for solving initial value problems of differential equations.

Finally, the code allows to locate fold and branch points accurately. Given an interval $[s_0, s_1]$ such that $\sigma_n(s_0) \cdot \sigma_n(s_1) < 0$, we use the secant method to find s^* s.t. $\sigma_n(s^*) = 0$ and compute the point $(x(s^*), \alpha(s^*))$ on the equilibria curve. Obviously, each secant iteration involves a continuation step from s_0 to the current iterate.

4 Algorithmic Details

We now discuss in more details the three key algorithmic choices we have adopted in our code.

1. *When and how to group/split singular values.* Given s_0, assume we have the decomposition $A(s_0) = U(s_0)\Sigma(s_0)V(s_0)^T$ with $\sigma_1(s_0) > \sigma_2(s_0) > \ldots > \sigma_n(s_0) \geq 0$. We predict the singular values at $s_1 = s_0 + h$ by [6]

$$\sigma_i^{(\mathrm{pred})} = \sigma_i(s_0) + (U^T(s_0)(A(s_1) - A(s_0))V(s_0))_{ii} = \sigma_i(s_0) + h\dot{\sigma}_i(s_0) + O(h^2)$$

and declare a possible crossing if

$$\text{either} \quad \frac{\sigma_{i+1}^{(\mathrm{pred})} - \sigma_i^{(\mathrm{pred})}}{\sigma_{i+1}(s_0) - \sigma_i(s_0)} < 0 \quad \text{or} \quad \frac{\sigma_{i+1}^{(\mathrm{pred})} - \sigma_i^{(\mathrm{pred})}}{\sigma_i^{(\mathrm{pred})} + 1} \leq \eta$$

for some small $\eta > 0$ ($\eta = 10^{-4}$ is the default value). In this case, we group σ_i, σ_{i+1} into a 2×2 block.

Once the singular values $\sigma_{i+1}(s_1)$ and $\sigma_i(s_1)$ have been computed, the 2×2 block is unrolled if

$$\frac{\sigma_{i+1}(s_1) - \sigma_i(s_1)}{\sigma_i(s_1) + 1} > \eta.$$

2. *Refinement of a* 2×2 *block to get a complete SVD.* Given a block C, we get a standard decomposition $U_C^T C V_C = \mathrm{diag}(\sigma_1, \sigma_2)$ with $\mathrm{sign}(\sigma_i) = \mathrm{sign}(\sigma_i^{(\mathrm{pred})})$ and then possibly modify it to recover smoothness in the complete SVD. To do this, we solve a Procrustes problem to minimize the distance (in Frobenius norm) between U_C, V_C and suitable reference orthogonal factors. If $\sigma_1 \neq \sigma_2$, U_C and V_C are correct up to permutation and/or changes of signs of their columns. If $\sigma_1 = \sigma_2$, then U_C and V_C may have to be changed into $U_C Q$ and $V_C Q$ for a suitable orthogonal matrix Q.

3. *Stepsize control.* Denote by $x^{(\mathrm{pred})}$, $\alpha^{(\mathrm{pred})}$, $\Sigma^{(\mathrm{pred})}$ the predictors used for $x(s_1)$, $\alpha(s_1)$ and $\Sigma(s_1)$, respectively. Our steplength choice tries to perform a mixed absolute/relative error control, monitoring the errors $\|x^{(\mathrm{pred})} - x(s_1)\|$, $\|\alpha^{(\mathrm{pred})} - \alpha(s_1)\|$, $\|\Sigma^{(\mathrm{pred})} - \Sigma(s_1)\|$, in such a way that the iterative procedure involved in a continuation step will converge in a few iterations. Moreover, starting from the second continuation step, we try to control also $\|U^{(\mathrm{pred})} - U(s_1)\|$ and $\|V^{(\mathrm{pred})} - V(s_1)\|$, where $U^{(\mathrm{pred})} = U(s_0) + h\dot{U}(s_0)$, $V^{(\mathrm{pred})} = V(s_0) + h\dot{V}(s_0)$, and the derivatives are approximated by differences using the orthogonal factors obtained at the last two steps. At the end of a continuation step of size h, we compute the weighted norm $\rho_\sigma = \sqrt{\frac{1}{n}\sum_{i=1}^{n}(\frac{\sigma_i^{(\mathrm{pred})} - \sigma_i}{\epsilon_r|\sigma_i| + \epsilon_a})^2}$, where ϵ_r and ϵ_a are relative and absolute error tolerances, and analogously we compute ρ_x, ρ_α, ρ_U, ρ_V. Then, we set $\rho = \max(\rho_\sigma, \rho_x, \rho_\alpha, \rho_U, \rho_V)$ and $h_{\mathrm{new}} = \frac{h}{\sqrt{\rho}}$. If $\rho \leq 1.5$, h_{new} is used to proceed the continuation; otherwise the step just completed is rejected and retried with the new steplength.

A continuation step may also fail because of lack of convergence either when solving a Riccati equation or when computing a new point on the curve. In both cases, the steplength is halved and the step retried.

5 Some Numerical Results

We refer on some experiments on two standard test problems. The results have been obtained with the following data:

- Initial steplength: 10^{-3}; minimum steplength: 10^{-8};
- Tolerances for solving algebraic Riccati equations: 10^{-8};
- Tolerances for computing fold and branch points: 10^{-12};
- Tolerances for computing points on the curve: 10^{-14}.

In the tables we give:

`Nsteps`: required number of continuation steps to complete the path;
`hmax`: maximum used steplength;
$\texttt{Nits}_1$: average number of iterations required to solve a Riccati equation (this is needed to update the SVDs);
$\texttt{Nits}_2$: average number of iterations per continuation step required to compute the point on the curve. The numbers $\texttt{Nits}_1$ and $\texttt{Nits}_2$ take into account all performed iterations, on both successful and unsuccessful steps.

Example 1. This is known as the aircraft problem [14]. The dimension is $n = 5$. The curve has to be followed from $(x_0, \alpha_0) = (0, 0)$ until α is out of $[-1, 1]$. Starting with $\dot{\alpha}_0 = 1$, two folds are found at $\alpha = 0.18608332702306$ and $\alpha = -0.50703056119994$. The figure on the right shows the first component of the solution vs. α. Table 1 shows the results obtained with several choices of the parameters ϵ_r and ϵ_a for the steplength control. Starting with $\dot{\alpha}_0 = -1$, two folds are detected at $\alpha = -0.18690833269959$ and $\alpha = 0.51015853464868$; the performance is identical to that in Table 1.

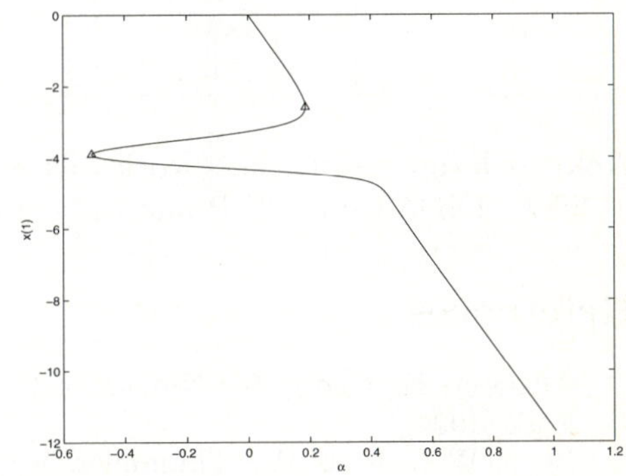

Table 1. Example 1

$\epsilon_a (= \epsilon_r)$	Nsteps	Nits_1	Nits_2	hmax
10^{-4}	694	2	4	0.24
10^{-3}	226	3	6	0.70
10^{-2}	82	4	11	1.72

Example 2. This is a test problem from the AUTO manual [9]:

$$f(x, \alpha) = \begin{bmatrix} x_1(1 - x_1) - 3x_1x_2 \\ -\frac{1}{4}x_2 + 3x_1x_2 - 3x_2x_3 - \alpha(1 - e^{-5x_2}) \\ -\frac{1}{2}x_3 + 3x_2x_3 \end{bmatrix}.$$

The continuation process starts from $(x_0, \alpha_0) = (1, 0, 0, 0)$ and stops when α is out of $[0, 0.6]$. The code detects two branch points, at $\alpha = 0.55$ and $\alpha = 0.36846953170021$, and a fold at $\alpha = 0.56459590997250$. In the figure on the right, we show the first solution component vs. α. At each branch point we started new runs to follow the branches. In Table 2 we give a summary of performance of the code relatively to completion of the three paths in the figure.

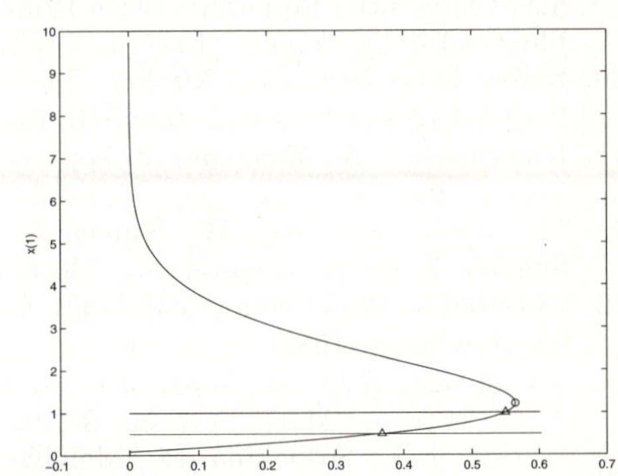

In both examples, $3 \div 5$ secant iterations were needed to locate either a fold or a branch point.

Table 2. Example 2

$\epsilon_a(=\epsilon_r)$	Nsteps	Nits_1	Nits_2	hmax
10^{-4}	792	2	2	0.76
10^{-3}	281	2	3	2.25
10^{-2}	121	3	3	5.27

Acknowledgement. This work was supported in part under NSF-DMS Grant 0139895, INDAM-GNCS Rome-Italy, and MIUR Rome-Italy.

References

1. Allgower, E., Georg, K.: Numerical Continuation Methods, Springer-Verlag, New York (1990)
2. Beyn, W.J., Kleß, W., Thümmler, V.: Continuation of low-dimensional invariant subspaces in dynamical systems of large dimension. In: Fiedler, B. (ed.): Ergodic Theory, Analysis and Efficient Simulation of Dynamical Systems. Springer (2001) 47–72
3. Bindel, D., Demmel, J., Friedman, M.: Continuation of Invariant Subspaces for Large Bifurcation Problems. In: Proceedings of the SIAM Conference on Applied Linear Algebra. The College of William and Mary, Williamsburg, VA (2003) http://www.siam.org/meetings/la03/proceedings.
4. Bunse-Gerstner, A., Byers, R., Mehrmann, V., Nichols, N. K.: Numerical Computation of an Analytic Singular Value Decomposition by a Matrix Valued Function, Numer. Math. **60** (1991) 1–40
5. Chow, S., Shen, Y.: Bifurcations via Singular Value Decompositions. Appl. Math. Comput. **28** (1988) 231–245
6. Dieci, L., Eirola, T.,: On Smooth Decomposition of Matrices. SIAM J. Matrix. Anal. Appl. **20** (1999) 800–819
7. Dieci, L., Gasparo, M.G., Papini, A.: Continuation of Singular Value Decompositions. Mediterr. j. math. **2** (2005) 179–203
8. Dhooge, A., Govaerts, W., Kuznetsov, Y.A.: A MATLAB Package for Numerical Bifurcation Analysis of ODE. ACM Trans. on Math. Software **29** (2003) 141–164
9. Doedel, E.J., et al.: AUTO97: Continuation and Bifurcation Software for Ordinary Differential Equations (with HomCont), User's Guide. Concordia University, Montreal, P.Q, Canada (1997) http://indy.cs.concordia.ca
10. Keller, H.B.: Numerical Methods in Bifurcation Problems. Springer Verlag, Tata Institute of Fundamental Research, Bombay (1987)
11. Kuznetsov, Y.A.: Elements of Applied Bifurcation Theory. Springer-Verlag, New York (1995)
12. Mehrmann, V., Rath., W.: Numerical Methods for the Computation of Analytic Singular Value Decompositions. Electron. Trans. Numer. Anal. **1** (1993) 72–88
13. Rheinboldt, W.C.: Numerical Analysis of Parametrized Nonlinear Equations. Wiley, New York (1986)
14. Rheinboldt, W.C., Burkardt, J.V.: A locally parametrizied continuation process. ACM Trans. on Math. Software **9** (1983) 215–235
15. Watson, L.T., Sosonkina, M., Melville, R.C., Morgan, A.P., Walker, H.F.: Algorithm 777: HOMPACK 90: a suite of Fortran 90 codes for globally convergent homotopy algorithms. ACM Trans. Math. Software **23** (1997) 514–549

Comparing Leja and Krylov Approximations of Large Scale Matrix Exponentials*

L. Bergamaschi[1], M. Caliari[2], A. Martínez[2], and M. Vianello[2]

[1] Dept. of Math. Methods and Models, University of Padova
berga@dmsa.unipd.it

[2] Dept. of Pure and Appl. Math., University of Padova
{mcaliari, acalomar, marcov}@math.unipd.it

Abstract. We have implemented a numerical code (ReLPM, Real Leja Points Method) for polynomial interpolation of the matrix exponential propagators $\exp(\Delta t A)\mathbf{v}$ and $\varphi(\Delta t A)\mathbf{v}$, $\varphi(z) = (\exp(z) - 1)/z$. The ReLPM code is tested and compared with Krylov-based routines, on large scale sparse matrices arising from the spatial discretization of 2D and 3D advection-diffusion equations.

1 Introduction

The systematic study and application of the so-called "exponential integrators" began about two decades ago, but has received a strong impulse in recent years; see, e.g., [5, 7, 8] and references therein. A building-block of exponential integrators is the efficient evaluation of the underlying matrix exponential functions, like $\exp(\Delta t A)\mathbf{v}$ and $\varphi(\Delta t A)\mathbf{v}$, $\varphi(z) = (\exp(z) - 1)/z$ (here $A \in \mathbf{R}^{n \times n}$, $\mathbf{v} \in \mathbf{R}^n$, and $\Delta t > 0$ is a time step). To this respect, most authors regard Krylov-like (cf. e.g. [7, 14]) as the methods of choice. Nevertheless, an alternative class of polynomial methods has been developed since the beginning (cf., e.g., [6, 15, 16]), which are based on direct interpolation or approximation of the exponential functions on the spectrum (or the field of values) of the relevant matrix. Despite of a preprocessing stage needed to get an estimate of some marginal eigenvalues, the latter are competitive with Krylov-like methods in several instances, namely on large scale, sparse and in general nonsymmetric matrices, arising from the spatial discretization of parabolic PDEs; see, e.g., [4, 10, 11].

Among others, the ReLPM (Real Leja Points Method), proposed in [4] and applied to advection-diffusion models in [2], has shown very attractive computational features. It rests on Newton interpolation of the exponential functions at a sequence of Leja points on the real focal interval of a family of confocal ellipses in the complex plane. The use of Leja points is suggested by the fact that they guarantee maximal (and thus superlinear) convergence of the interpolant on every ellipse of the confocal family, and thus superlinear convergence

* Work supported by the MIUR PRIN 2003 project "Dynamical systems on matrix manifolds: numerical methods and applications" (co-ordinator L. Lopez, University of Bari), by the ex-60% funds of the University of Padova, and by the GNCS-INdAM.

V.N. Alexandrov et al. (Eds.): ICCS 2006, Part IV, LNCS 3994, pp. 685–692, 2006.
© Springer-Verlag Berlin Heidelberg 2006

of the corresponding matrix polynomials to the matrix exponential functions. This feature is shared also by other set of interpolation points, like e.g. standard Chebyshev points, but differently from the latter, at the same time Leja points allow to increase the interpolation degree just by adding new nodes of the same sequence; see [1, 4] for the scalar and matrix features of interpolation at Leja points. A key step in the approximation procedure is given by estimating cheaply a real focal interval, say $[a, b]$, such that the "minimal" ellipse of the confocal family which contains the spectrum (or the field of values) of the matrix is not too "large" (the underlying theoretical notion is that of "capacity" of a compact complex set). The numerical experience with matrices arising from stable spatial discretizations of parabolic equations (which are the main target of the ReLPM code) has shown that good results can be obtained at a very low cost, simply by intersecting the Gershgorin's circles of the matrix with the real axis. Indeed, it is worth stressing that the ReLPM method works well with "stiff" matrices, whose spectrum (or whose field of values) has a projection on the real axis which is nonpositive and much larger than the projection on the imaginary axis; cf. [4].

We give now two formulas, which are the basis for practical implementation of the ReLPM method. The kernel of the ReLPM code is given by interpolation of $\varphi(h\lambda)$, for suitable $h \leq \Delta t$, at Leja points of the real focal interval $[a, b] = [c - 2\gamma, c + 2\gamma]$. Observe that once $\varphi(hA)\mathbf{v}$ is computed, then $\exp(hA)\mathbf{v} = h\varphi(hA)\mathbf{v} + \mathbf{v}$. In practice, it is numerically convenient to interpolate the function $\varphi(h(c + \gamma\xi))$ at Leja points $\{\xi_s\}$ of the reference interval $[-2, 2]$ (since it has capacity equal to 1, cf. [15]). Then, given the corresponding divided differences $\{d_i\}$ for such a function, the matrix Newton polynomial of degree m is

$$p_m(A) = \sum_{i=0}^{m} d_i \Omega_i \approx \varphi(hA) \; , \quad \Omega_i = \prod_{s=0}^{i-1} ((A - cI)/\gamma - \xi_s I) \; . \qquad (1)$$

In general, it is not feasible to interpolate with the original time step Δt, which has to be fractionized. This happens, for example, when the expected degree for convergence is too large. The ReLPM code subdivides dynamically Δt into smaller substeps $h = h_k$, and recovers the required vector $\varphi(\Delta t A)\mathbf{v}$ according to the time marching scheme

$$\mathbf{y}_{k+1} = \mathbf{y}_k + h_k \varphi(h_k A)(A\mathbf{y}_k + \mathbf{v}) \; , \quad k = 0, 1, \ldots, k^* \; ; \; \mathbf{y}_0 = \mathbf{0} \; , \qquad (2)$$

where $\sum h_k = \Delta t$. Here we use the fact that $\Delta t \varphi(\Delta t A)\mathbf{v}$ is the solution at $t = \Delta t$ of the differential system $\dot{\mathbf{y}}(t) = A\mathbf{y}(t) + \mathbf{v}$, $\mathbf{y}(0) = \mathbf{0}$.

2 The ReLPM Code

In this section we present the pseudo-codes of the three subroutines which compose the ReLPM code. They are displayed in Tables 1–3, and accompanied by a detailed documentation.

Table 1. SUBROUTINE RELPM

1. INPUT: A, $\mathbf{v}$, Δt, *exptype*, *tol*
2. CONSTANTS: $M = 124$, $(\xi_0, \ldots, \xi_M)$ array of $M + 1$ Leja points in $[-2, 2]$
3. $k := 0$, $\rho := \Delta t$, $\mathbf{p} := \mathbf{0}$, $\mathbf{w} := \mathbf{v}$
4. a, b:="extrema of the real points in the Gersghorin's circles of A"
5. $c := (a + b)/2$, $\gamma := (b - a)/4$, $\nu := 3\gamma$, $h := \min\{\Delta t, M/\nu\}$, $oldh := 0$
6. REPEAT
    7. IF $h \neq oldh$ THEN
        8. CALL DIVDIFF$(h, c, \gamma, M, (\xi_0, \ldots, \xi_M), (d_0, \ldots, d_M))$
        9. $oldh := h$
    10. ENDIF
    11. CALL INTERP$(A, \mathbf{w}, h, tol, c, \gamma, M, (\xi_0, \ldots, \xi_M), (d_0, \ldots, d_M), \mathbf{q}, err, m)$
    12. IF $m > M$ THEN $h := h/2$
    13. ELSE
        14. $\rho := \rho - h$, $\mathbf{p} := \mathbf{p} + h\mathbf{q}$
        15. IF $\rho > 0$ THEN
            16. $\mathbf{w} := A\mathbf{p}$, $\mathbf{w} := \mathbf{w} + \mathbf{v}$, $k := k + 1$, $\sigma := h\gamma/m$
            17. IF $\sigma > 1$ THEN $h := \min\{\sigma h, M/\gamma, \rho\}$
            18. ELSE $h := \min\{h, \rho\}$
            19. ENDIF
        20. ENDIF
    21. ENDIF
22. UNTIL $\rho = 0$
23. $k^* := k$, IF $exptype = 0$ THEN $\mathbf{w} := A\mathbf{p}$, $\mathbf{p} := \mathbf{w} + \mathbf{v}$ ELSE $\mathbf{p} := \mathbf{p}/\Delta t$ ENDIF
24. OUTPUT: the vector $\mathbf{p}$ such that $\mathbf{p} \approx \exp(\Delta t A)\mathbf{v}$ ($exptype = 0$), or $\mathbf{p} \approx \varphi(\Delta t A)\mathbf{v}$, $\varphi(z) = (e^z - 1)/z$ ($exptype = 1$); the total number of substeps k^*

Comments to Table 1. This is the main subroutine. It accepts a matrix $A \in \mathbf{R}^{n \times n}$, a vector $\mathbf{v} \in \mathbf{R}^n$, a time step $\Delta t > 0$, and the type of exponential function (exp or φ). The output is a vector which approximates the corresponding function of the matrix $\Delta t A$, applied to the vector $\mathbf{v}$. The underlying method is the time-marching scheme (2), with a dynamical managing of the variable substeps $h = h_k$, and Newton interpolation as in (1) of $\varphi(hA)$ at real Leja points related to spectral estimates for A.

1.) *exptype*: type of exponential function (0 for exp, 1 for φ); *tol*: the relative error tolerated for the result $\exp(\Delta t A)\mathbf{v}$ or $\varphi(\Delta t A)\mathbf{v}$.
2.) M: maximum interpolation degree allowed "a priori" in the Newton interpolation of $\varphi(hA)$, $h \leq \Delta t$; $(\xi_0, \ldots, \xi_M)$ is an array of Leja interpolation points, like e.g. the Fast Leja points in [1]. It is worth stressing that the default $M = 124$ is tuned on spatial discretization matrices of linear advection-diffusion models, in order to guarantee that all the divided differences computed by the subroutine DIVDIFF are accurate (see [3]).
3.) Initializations: ρ is the portion of Δt still to be covered; $\mathbf{p} = \mathbf{y}_0$ and $\mathbf{w} = A\mathbf{y}_0 + \mathbf{v}$, cf. (2).

4. - 5.) Approximation of the real focal interval of a "minimal" ellipse which contains the numerical range of the underlying matrix, and the associated parameters: c is the center and γ the capacity (length/4) of the interval. Hereafter, h and *oldh* are the current and the previous (sub)steps. For any given substep h, theoretical estimates show that superlinear convergence of the matrix interpolation polynomial should start at a degree between $h\gamma$ and $2h\gamma$, cf. [10, 4], provided that the capacity of the minimal ellipse above is relatively close to γ. Convergence at reasonable tolerances of the matrix interpolation polynomial is expected for a degree lower than $h\nu = 3h\gamma$ (the factor 3 is an "empirical" choice, based on numerical experience). Hence, the input step Δt is possibly reduced in such a way that $[h\nu] \leq M$ (here $[\cdot]$ denotes the integer part).
6. - 22.) Main loop: implements the time marching scheme (2) with dynamical managing of the substeps $h = h_k$.
7. - 10.) When the current and the previous substeps are different the divided differences $(d_0, \ldots, d_M)$ are (re)computed.
11.) Computes $\mathbf{q} = p_m(A)\mathbf{w} \approx \varphi(hA)\mathbf{w}$, where $\mathbf{w} = A\mathbf{p} + \mathbf{v}$, $\mathbf{p} \approx \mathbf{y}_k$.
12. - 21.) If the interpolation process has not converged, the substep is halved and control returns to point 6, otherwise the current substep has been successful.
14.) The remaining portion ρ of Δt and $\mathbf{p}$ are updated: now $\mathbf{p} \approx \mathbf{y}_{k+1}$.
15. - 20.) If Δt has not been completed, prepares the next substep.
16.) Computes $\mathbf{w} \approx A\mathbf{y}_{k+1} + \mathbf{v}$; σ is a parameter used to detect fast convergence, i.e. convergence degree smaller than $h\gamma$.
17. - 19.) When the actual convergence degree m is smaller than $h\gamma$ (fast convergence), the next substep h is increased but in such a way that $h\gamma \leq M$ (since convergence is expected again at a degree lower than $h\gamma$).
23.) Now $\mathbf{p} \approx \mathbf{y}_{k^*} = \Delta t \varphi(\Delta t A)\mathbf{v}$: computes the right type of matrix exponential function according to *exptype*.

Table 2. SUBROUTINE DIVDIFF

1. INPUT: h, c, γ, M, $(\xi_0, \ldots, \xi_M)$
2. "computes $(d_0, \ldots, d_M)$, the divided differences of $\varphi(h(c + \gamma\xi))$, $\xi \in [-2, 2]$, at the Leja points $(\xi_0, \ldots, \xi_M)$, by the accurate matrix algorithm in [3]"
3. OUTPUT: $(d_0, \ldots, d_M)$

Comments to Table 2. This subroutine accepts a time step h, two parameters related to the spectral features of an external matrix A, a maximum interpolation degree M and a corresponding array of Leja interpolation points. It returns the divided differences for the Newton polynomial interpolation of $\varphi(hA)$ up to degree M. The subroutine is thought to work in double precision.

1.) h: time step; c and γ: see point 4 in the comments to the subroutine RELPM; M and $(\xi_0, \ldots, \xi_M)$: maximum interpolation degree and corresponding Leja interpolation points, see the comment to point 2 of the subroutine RELPM.

2.) Computes the $M+1$ divided differences in double precision as the first column of $\varphi(h(c+\gamma\,\Xi_M))$, where Ξ_M is the $(M+1)\times(M+1)$ bidiagonal matrix with the Leja points $(\xi_0,\ldots,\xi_M)$ on the main diagonal and $(1,\ldots,1)$ on the diagonal immediately below. The matrix $\varphi(h(c+\gamma\,\Xi_M))$ is approximated via 16-term Taylor expansions by the scheme proposed in [3], on the basis of [9]. Differently from the standard divided differences table, this algorithm computes accurately the divided differences even when their size goes below machine precision, provided that M is not too large for the given h, c and γ (to avoid the underflow of some intermediate quantities in the computation); see the comment to point 2 of the subroutine ReLPM. A more sophisticated implementation could discard the possible tail of divided differences that are not sufficiently accurate (see [3]).
3.) Given $(d_0,\ldots,d_M)$, the Newton interpolation polynomial of degree $m\le M$ for $\varphi(h\lambda)$ is $p_m(\lambda)=\sum_{i=0}^{m} d_i \prod_{s=0}^{i-1}((\lambda-c)/\gamma-\xi_s)$, $\lambda\in[c-2\gamma,c+2\gamma]$.

Table 3. Subroutine INTERP

1. INPUT: A, $\mathbf{w}$, h, tol, c, γ, M, $(\xi_0,\ldots,\xi_M)$, $(d_0,\ldots,d_M)$
2. CONSTANTS: $\ell=5$
3. $\mathbf{u}:=\mathbf{w}$, $\mathbf{q}:=d_0\mathbf{w}$, $e_0:=\|\mathbf{q}\|_2$, $\beta:=\|\mathbf{w}\|_2$, $m:=0$
4. REPEAT
    5. $\mathbf{z}:=(A\mathbf{u})/\gamma$, $\mathbf{u}:=\mathbf{z}-(c/\gamma+\xi_m)\mathbf{u}$
    6. $m:=m+1$, $e_m:=|d_m|\,\|\mathbf{u}\|_2$
    7. $\mathbf{q}:=\mathbf{q}+d_m\mathbf{u}$
    8. IF $m\ge\ell-1$ THEN $err:=(e_m+\ldots+e_{m-\ell+1})/\ell$ ENDIF
9. UNTIL $err\le\beta\,tol$ or $m\ge M$
10. IF $err>\beta\,tol$ THEN $m:=M+1$ ENDIF
11. OUTPUT: the vector $\mathbf{q}=p_m(A)\mathbf{w}$, the estimated error err, and the interpolation degree m, such that $\|\mathbf{q}-\varphi(hA)\,\mathbf{w}\|_2\approx err$, with $err\le\|\mathbf{w}\|_2\,tol$ when $m\le M$, whereas a convergence failure occurred when $m>M$.

Comments to Table 3. This subroutine tries to compute an approximation $\mathbf{q}=p_m(A)\mathbf{w}\approx\varphi(hA)\mathbf{w}$, cf. (1), up to an error (relative to $\|\mathbf{w}\|_2$) less than a given tolerance, where $A\in\mathbf{R}^{n\times n}$, $\mathbf{w}\in\mathbf{R}^n$ and $\Delta t\ge h>0$.
1.) $(d_0,\ldots,d_M)$ are the divided differences for the Newton interpolation up to degree M of the function $\varphi(h(c+\gamma\xi))$ at the Leja points $(\xi_0,\ldots,\xi_M)\subset[-2,2]$.
2.) ℓ: number of consecutive error estimates to be averaged in order to filter error oscillations (the default $\ell=5$ has shown a good numerical behavior).
3.) Initializations: $\mathbf{u}=\Omega_0\mathbf{w}$, $\mathbf{q}=p_0(A)\mathbf{w}$ (cf. (1)).
4. - 9.) Main loop: Newton interpolation of the matrix operator $\varphi(hA)\mathbf{w}$.
5.) Computes the vector $((A-cI)/\gamma-\xi_m I)\mathbf{u}=A\mathbf{u}/\gamma-(c/\gamma+\xi_m)\mathbf{u}=\Omega_{m+1}\mathbf{w}$, avoiding to shift and scale the matrix.
6. - 7.) Computes e_m, the norm of the new term in the Newton interpolation, and the vector $\mathbf{q}=p_m(A)\mathbf{w}$.

8.) Averages the last ℓ values of e_i to get the actual interpolation error estimate.
9.) Exits the loop as soon as the estimated error is below the relative tolerance, or the maximum interpolation degree M has been reached.
10.) Sets the output degree m to a value $> M$ in case that convergence has not been attained.

3 Numerical Tests and Comparisons

In this section we present some numerical examples, where we have tested the ReLPM code on large scale matrices arising from spatial discretizations of 2D and 3D advection-diffusion equations. The model problem is

$$\frac{\partial u}{\partial t} = \mathrm{div}(D\nabla u) - \langle \overrightarrow{\vartheta}, \nabla u \rangle \ , \quad x \in \Omega \ , \ t > 0 \ , \tag{3}$$

with initial condition $u(x,t) = u_0(x)$, $x \in \Omega$, and mixed Dirichlet and Neumann boundary conditions on $\Gamma_\mathcal{D} \cup \Gamma_\mathcal{N} = \partial\Omega$, $\Omega \subset \mathbf{R}^d$, $d = 2, 3$. In (3), D is a $d \times d$ diffusion matrix, and $\overrightarrow{\vartheta} \in \mathbf{R}^d$ a constant velocity field. Finite Difference (FD) or Finite Elements (FE) discretizations produce a system of ODEs like $\dot{\mathbf{y}}(t) = A\mathbf{y}(t) + \mathbf{b}$, $\mathbf{y}(0) = \mathbf{u}_0$, where A is large, sparse and nonsymmetric. In the FE case it is obtained cheaply by left-multiplying the stiffness matrix with the inverse of a diagonal mass matrix, via the mass-lumping technique (cf. [2]).

In all the examples we have computed $\varphi(\Delta tA)\mathbf{v}$ with $\mathbf{v} = (1, \ldots, 1)^t$ (corresponding to $u_0 \equiv 1$), for two values $(\Delta t)_1$ and $(\Delta t)_2$ of the time step Δt, depending on the specific matrix. The tests have been performed in double precision by a Fortran version of the ReLPM code, on an IBM Power5 processor with 1.8Gb of RAM. We also give the comparisons with the PHIPRO Fortran code by Y. Saad (again in double precision), which is based on Krylov subspace approximations (cf. [12]). The results are collected in Table 4. We report, for both methods, the number of substeps (steps), the number of total iterations (i.e., of matrix-vector products), and the CPU time. In addition, for PHIPRO we show also the chosen dimension for the Krylov subspace. This is a delicate choice, for which a simple criterion does not seem to be available; in any case, the default $m = 60$ is not suitable in the examples. The tolerances for both methods have been tuned in order to have an error, relative to the "exact" result of the exponential operator, of about 10^{-6} in the 2-norm (which is compatible with the underlying spatial discretization error). It is worth observing that even using smaller tolerances the performances of both methods would not change significantly, since superlinear convergence has already been reached.

Example 1 (FE-2D). We have taken a 2D equation like (3), with $\Omega = (0,1)^2$, $D = I$, $\overrightarrow{\vartheta} = (60, 60)$, and homogeneous Dirichlet boundary conditions. We have adopted a standard Galerkin FE discretization with linear basis functions, on a uniform mesh with $n = 490\,000$ nodes and $977\,202$ triangular elements, which produces a matrix with $3\,424\,402$ nonzeros (average nonzeros per row ≈ 7). Here $(\Delta t)_1 = 0.001$ and $(\Delta t)_2 = 0.01$.

Table 4. Comparing RELPM and PHIPRO on the advection-diffusion discretization matrices in Examples 1–4 (the CPU times are in seconds)

Δt	Code	FE-2D			FE-3D			FD-2D			FD-3D		
	PHIPRO	steps	iter	CPU	steps	iter	CPU	steps	iter	CPU	steps	iter	CPU
	$m=$ 10	126	1386	57.1	81	891	59.8	54	594	44.2	35	385	349.3
	$m=$ 20	40	840	45.5	28	588	50.2	21	441	43.1	†	†	†
$(\Delta t)_1$	$m=$ 25	29	754	43.2	21	546	45.2	16	416	45.6	†	†	†
	$m=$ 30	23	713	45.4	17	527	48.1	13	403	55.4	†	†	†
	$m=$ 50	13	663	58.4	10	510	58.9	8	408	67.5	†	†	†
	RELPM	17	857	27.0	12	585	32.0	5	392	20.6	3	234	133.0
	PHIPRO	steps	iter	CPU	steps	iter	CPU	steps	iter	CPU	steps	iter	CPU
	$m=$ 10	868	9548	414.3	428	4708	302.1	431	4741	375.4	158	1738	1593.2
	$m=$ 20	287	6027	318.8	152	3192	242.1	166	3486	323.4	†	†	†
$(\Delta t)_2$	$m=$ 25	190	4940	280.9	117	3042	250.4	126	3276	353.2	†	†	†
	$m=$ 30	149	4619	306.6	94	2914	268.0	103	3193	415.6	†	†	†
	$m=$ 50	73	3723	343.3	53	2703	301.0	58	2958	516.6	†	†	†
	RELPM	131	7720	227.3	74	4335	231.7	49	3617	186.3	16	1094	633.4

Example 2 (FE-3D). Here we have a 3D advection-dispersion equation, where D is the hydrodynamic dispersion tensor, $D_{ij} = \alpha_T |\overrightarrow{\vartheta}| \delta_{ij} + (\alpha_L - \alpha_T)\vartheta_i\vartheta_j / |\overrightarrow{\vartheta}|$, $1 \le i,j \le d$ (α_L and α_T being the longitudinal and transverse dispersivity, respectively). The domain is $\Omega = [0,1] \times [0.0.5] \times [0,1]$, discretized by a regular mesh of $n = 161 \times 81 \times 41 = 524\,681$ nodes and $3\,072\,000$ tetrahedral elements. The boundary conditions are homogeneous Dirichlet on $\Gamma_{\mathcal{D}} = \{0\} \times [0.2, 0.3] \times [0,1]$, whereas homogeneous Neumann conditions are imposed on $\Gamma_{\mathcal{N}} = \partial\Omega \setminus \Gamma_{\mathcal{D}}$. The velocity is $\overrightarrow{\vartheta} = (1,0,0)$, the transmissivity coefficients are piecewise constant and vary by an order of magnitude depending on the elevation of the domain, $\alpha_L(x_3) = \alpha_T(x_3) \in \{0.0025, 0.025\}$. The resulting FE matrix has $7\,837\,641$ nonzeros (average per row ≈ 14). Here $(\Delta t)_1 = 1$ and $(\Delta t)_2 = 10$.

Example 3 (FD-2D). Again a 2D model, with $\Omega = (0,10)^2$, $D = I$, $\overrightarrow{\vartheta} = (100,100)$, and homogeneous Dirichlet boundary conditions. We have adopted a second order central FD discretization on a uniform grid with stepsize 0.01 ($n = 1\,002\,001$ nodes), generating a pentadiagonal matrix with $5\,006\,001$ nonzeros. Here $(\Delta t)_1 = 0.01$ and $(\Delta t)_2 = 0.1$.

Example 4 (FD-3D). Here a 3D equation, with $\Omega = (0,1)^3$, $D = I$, $\overrightarrow{\vartheta} = (200,200,200)$, and homogeneous Dirichlet boundary conditions. Here we have adopted a second order central FD discretization on a uniform grid with stepsize 0.005 ($n = 8\,120\,601$ nodes), generating an eptadiagonal matrix with $56\,601\,801$ nonzeros. Here $(\Delta t)_1 = 10^{-3}$ and $(\Delta t)_2 = 5.2 \times 10^{-3}$.

Comments on the results. First, notice that RELPM performs better than PHIPRO in all the tests, even with an optimal choice of the Krylov subspace dimension (underlined CPU times), in spite of a smaller number of total Krylov iterations. Indeed, it is worth stressing that RELPM computes only 1 matrix-vector product, 2 daxpys, 1 vector scaling and 1 scalar product per iteration inside INTERP. Moreover, it allocates only the matrix and 6 vectors, whereas PHIPRO has to allocate, besides the matrix, all the m Krylov subspace generators and 4 vectors (neglecting a matrix and some vectors of dimension m). In addition, when m increases, PHIPRO decreases the total number of iterations, but is penalized by the long-term recurrence in the orthogonalization process. For small m, it is penalized by a larger number of substeps. The difference in storage requirements produces remarkable consequences in the last example. There, the matrix is extremely large, and PHIPRO can work only with $m \leq 10$ due to the memory limitations (1.8Gb), becoming about 2.5 times slower than ReLPM.

References

1. Baglama, J., Calvetti, D., Reichel, L.: Fast Leja points. Electron. Trans. Numer. Anal. **7** (1998) 124–140
2. Bergamaschi, L., Caliari, M., Vianello, M.: The ReLPM exponential integrator for FE discretizations of advection-diffusion equations. Springer LNCS vol. 3039 - part IV, 2004 (Proc. of ICCS 2004) 434–442
3. Caliari, M.: Accurate evaluation of divided differences for polynomial interpolation of exponential propagators. Draft (2005)
4. Caliari, M., Vianello, M., Bergamaschi, L.: Interpolating discrete advection-diffusion propagators at Leja sequences. J. Comput. Appl. Math. **172** (2004) 79–99
5. Cox, S. M., Matthews, P. C.: Exponential time differencing for stiff systems. J. Comput. Phys. **176** (2002) 430–455
6. Druskin, V. L., Knizhnerman, L. A.: Two polynomial methods for calculating functions of symmetric matrices. U.S.S.R. Comput. Math. and Math. Phys. **29** (1989) 112–121
7. Hochbruck, M., Lubich, C., Selhofer, H.: Exponential integrators for large systems of differential equations. SIAM J. Sci. Comput. **19** (1998) 1552–1574
8. Hochbruck, M., Ostermann, A.: Exponential Runge-Kutta methods for parabolic problems. Appl. Numer. Math. **53** (2005) 323–339
9. McCurdy, A., Ng, C., Parlett, B. N.: Accurate Computation of Divided Differences of the Exponential Function. Math. Comp. **43** (1984) 501–528
10. Moret, I., Novati, P.: The computation of functions of matrices by truncated Faber series. Numer. Funct. Anal. Optim. **22** (2001) 697–719
11. Novati, P.: A polynomial method based on Feiér points for the computation of functions of unsymmetric matrices. Appl. Numer. Math. **44** (2003) 201–224
12. Saad, Y.: SPARSKIT: a basic tool kit for sparse matrix computations. Dept. of Computer Science and Engineering, University of Minnesota. Version 2 (2005)
13. Schaefer, M. J.: A polynomial based iterative method for linear parabolic equations. J. Comput. Appl. Math. **29** (1990) 35–50
14. Sidje, R. B.: Expokit. A software package for computing matrix exponentials. ACM Trans. Math. Software **24** (1998) 130–156
15. Tal-Ezer, H.: Polynomial approximation of functions of matrices and applications. J. Sci. Comput. **4** (1989) 25–60

Combined Method for Nonlinear Systems of Equations*

Peng Jiang[1], Geng Yang[2], and Chunming Rong[3]

[1,2] School of Computer Science and Technology, P.O.Box 43,
Nanjing University of Posts and Telecommunications, 210003, Nanjing, China
alice20006@hotmail.com,
angg@njupt.edu.cn
[3] Department of Electrical and Computer Engineering, University of Stavanger, Norway
chunming.rong@uis.no

Abstract. This paper proposes Block Broyden Method combined with SSOR preconditioning technique for solving nonlinear systems of equations. The implementation process is described in detail and the time complexity is analyzed. This method has faster solving speed and better performance than the unpreconditioned one, which is shown by several numerical tests arising from the Bratu problem. Therefore, it can be used in the large-scale problems arising from scientific and engineering computing.

1 Introduction

The Block Broyden Method[1] is an effective iterative algorithm for solving nonlinear equations. It parallelizes very well and gives good CPU time savings. However, the iterative matrix in the Block Broyden Algorithm is a block diagonal matrix so that partial relevant information among the nodes is lost, therefore affecting the convergence speed of the algorithm to some extent. Hence, seeking for proper preconditioning methods[2, 3] is one of the effective ways to solve this problem. Some preconditioners are proposed and analyzed in paper [4, 5].

In this article, we combine SSOR preconditioner with the Block Broyden Method. We organize our article as follows. Section 2 introduces SSOR preconditioner. Section 3 gives general remarks on SSOR Preconditioner based on Block Broyden Algorithm, and it also shows the implementation details and analyzes the time complexity. Some numerical results and interpretation of these results are included in Section 4. Section 5 contains the summary remarks.

2 SSOR Preconditioner

We suppose there are large sparse linear systems of the form as (1)

$$Ax = b\,. \tag{1}$$

* This work was supported in part by the Natural Science Foundation of Jiangsu Province under Grant No 05KJD520144 and the Foundation of the QingLan Project （KZ0040704006）.

V.N. Alexandrov et al. (Eds.): ICCS 2006, Part IV, LNCS 3994, pp. 693 – 699, 2006.
© Springer-Verlag Berlin Heidelberg 2006

where $A=[a_{i,j}]$ is an $n\times n$ matrix and b a given right-hand-side vector. If the original, symmetric, matrix is decomposed as $A=D+L+L^T$ in its diagonal, lower and upper triangular part, the SSOR matrix is defined as

$$M=(D+L)D^{-1}(D+L)^T. \tag{2}$$

Then we can calculate the inverse of the preconditioner and get the transformed linear system $M^{-1}Ax=M^{-1}b$ which has the same solution as (1).

3 SSOR Preconditioner Based on Broyden Algorithm

3.1 General Remarks

In the following discussion, we are concerned with the problem of solving the large system of nonlinear equations as (3):

$$F(x)=0. \tag{3}$$

where $F(x)=(f_1,\cdots.f_n)^T$ is a nonlinear operator from R^n to R^n, and $x^*\in R^n$ is an exact solution. Suppose that the components of x and F are divided into q blocks:

$$F=\begin{pmatrix}F_1\\ \vdots\\ F_q\end{pmatrix}\qquad x=\begin{pmatrix}x_1\\ \vdots\\ x_q\end{pmatrix}$$

We consider generalization of the Block Broyden algorithm with Preconditioning method (BBP) as follows:

Algorithm 3.1.1. BBP Method

1. Let x^0 be an initial guess of x^*, and B^0 an initial block diagonal approximation of $J(x^0)$. Calculate $r^0=F(x^0)$.
2. For k = 0, 1…until convergence:
   - 2.1 Solve $B^k s^k=-r^k$:
     - 2.1.1 Calculate SSOR preconditioner according to (2) and get the preconditioned equation.
     - 2.1.2 Solve the preconditioned equation by Jacobi method.
   - 2.2 Update the solution $x^{k+1}=x^k+s^k$.
   - 2.3 Calculate $r^{k+1}=F(x^{k+1})$. If r^{k+1} is small enough, stop.
   - 2.4 Calculate $(s^k)^T s^k$ and update B^{k+1} by

$$B_i^{k+1} = B_i^k + \frac{r_i^{k+1}(s_i^k)^T}{(s^k)^T s^k} \,. \tag{4}$$

Then set $k = k+1$, and go to step 2.

3.2 Implementation

The first step in Algorithm 3.1.1 is the preparation stage. The second step is an iterative process. We update the solution x^{k+1} according to step 2.2, and calculate the residual according to step 2.3. If the residual is smaller than a given value or the number of iterations is larger than an upper bound value, the iteration process is stopped. Otherwise, the inner product of s^k is calculated and B^{k+1} is updated to repeat the iteration. Fig.1 describes the computing flow.

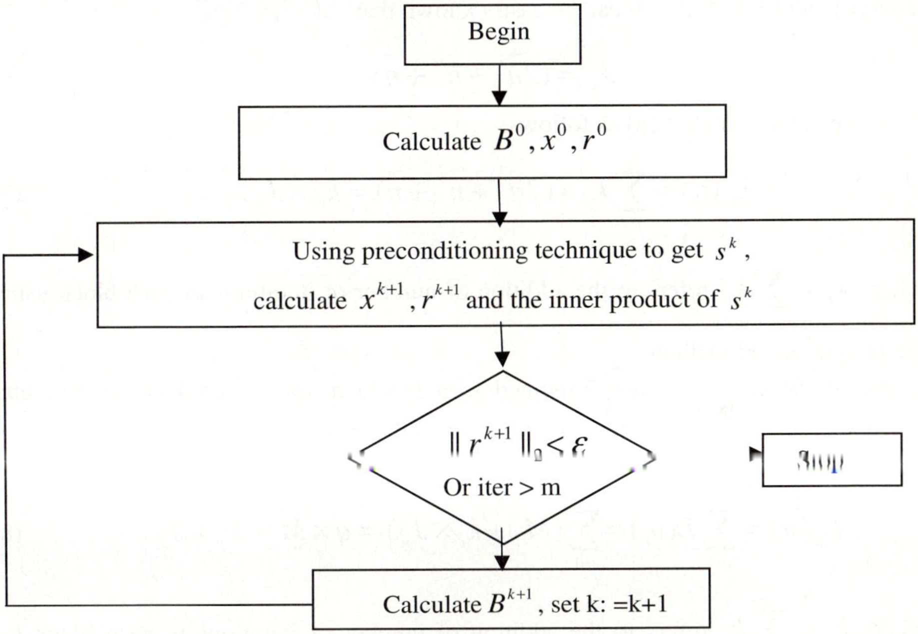

Fig. 1. The computing flow of BBP method

3.3 Time Complexity

In Algorithm 3.1.1, let $L(n_i)$ be the complexity of solving each block linear equations $B_i^k s_i^k = -r_i^k$, then the complexity of step 2.1 is $L(n) = \sum_{i=1}^{q} L(n_i)$. Step 2.2 needs n additions and the complexity is $n = \sum_{i=1}^{q} n_i$. The residual is calculated in step 2.3, and let R(n) be the corresponding calculative cost. The complexity of step 2.4 is $\sum_{i=1}^{q}(3n_i - 1 + 2n_i^2)$. Therefore the total complexity of Algorithm 3.1.1 is:

$$U = \sum_{i=1}^{q}(L(n_i) + n_i + 3n_i - 1 + 2n_i^2) + R(n) = \sum_{i=1}^{q}(4n_i - 1 + 2n_i^2) + L(n) + R(n). \quad (5)$$

We can know that from (5) that the total complexity U differs in $L(n)$, which means the complexity of solving q block linear equations $B_i^k s_i^k = -r_i^k$. We will calculate $L(n)$ for unpreconditioned method and SSOR preconditioning method respectively.

For unpreconditioned method, we simply use Jacobi method to solve each block linear system. So $L(n_i)$ is the complexity of jacobi, which is (number of steps)× (cost per step) $= k_i \times J_n$. It can be easily known that

$$J_n = (2\bar{n}^3 + \bar{n}^2 + \bar{n}). \quad (6)$$

So we can calculate L(n) as follows:

$$L_n(n) = \sum_{i=1}^{q} k_i \times (2\bar{n}^3 + \bar{n}^2 + \bar{n}) = k_n \times J_n. \quad (7)$$

where $k_n = \sum_{i=1}^{q} k_i$ refers to the addition of number of iterations in each block using unpreconditioned method.

For SSOR preconditioner combined with Jacobi method, $L(n_i)$ is the computational cost of construct the preconditioner plus the complexity of Jacobi, then we can get L(n) as follows:

$$L_p(n) = \sum_{i=1}^{q} L(n_i) = \sum_{i=1}^{q}(M_i + k_i \times J_n) = q \times \overline{M} + k_p \times J_n. \quad (8)$$

where $k_p = \sum_{i=1}^{q} k_i$ refers to the addition of number of iterations in each block for SSOR method, and $\overline{M}$ means the complexity of constructing SSOR Preconditioner. $\overline{M}$ can be easily known as follows:

$$\overline{M} = \frac{\bar{n}^4}{3} + 5\bar{n}^3 + \frac{\bar{n}^2}{6} + \frac{\bar{n}}{2}. \quad (9)$$

Now the performance of each method can be compared exactly according to (7) and (8).

4 Numerical Experiments

In the numerical experiments, we compare the convergence speed and efficiency of SSOR preconditioning method with the unpreconditioned one for solving a Bratu problem in computational physics. The programming language is C++, using double float variables to calculate the problem.

The nonlinear partial differential equation can be written as

$$\begin{cases} -\Delta u + u_x + \lambda e^u = f, \\ u|_{\partial\Omega} = 0 \end{cases} (x, y) \in \Omega = [0,1]\times[0,1] . \tag{10}$$

on the unit square Ω of R^2 with zero Dirichlet boundary conditions. It is known as the Bratu problem and has been used as a test problem by Yang in [1] and Jiang in [4]. The function f is chosen such that the solution of the discretized problem is unity. Using a linear element P, a standard finite element approximation gives a large system of nonlinear equations of size N^2. For$\lambda \geq 0$, the system has a unique solution. In the following tests, we suppose f=e, λ=1 and N=50, 80 and 110, giving three grids, M1, M2 and M3, with 2500, 6400 and 12100 unknowns, respectively.

Let r be the nonlinear discretized function of dimension $n = N^2$ obtained from (10), so the residual vector norm is $\| r^k \|_2$, where k is the number of nonlinear iterations. We also set an upper bound value m=5000. The stopping criterion used in the following tests is either $\| r^k \|_2 < 10^{-5}$ or m>5000. From boundary condition we can get the initial solution $x^0 = 0$ and the initial block jacobi matrix B^0.

When the grid is M1, we set the block number as q1=800. When the grid is M2, we set q2=1600. When the grid is M3, we set q3=1000 and q4=2000 respectively. Table 1, 2, 3, and 4 show the number of nonlinear iterations, which is denoted by "k" and the sum of numbers of iterations for solving each linear block during the i-th nonlinear iteration, which is denoted by "k[i]".

Table 1. Comparison of the total number of iterations in M1, q1

	SSOR	No Preconditioner
k	983	1289
k[145]	1467	5618
k[440]	880	3173

Table 2. Comparison of the total number of iterations in M2, q2

	SSOR	No Preconditioner
k	3526	4003
k[675]	2941	7597
k[1100]	2356	7032

Table 3. Comparison of the total number of iterations in M3, q3

	SSOR	No Preconditioner
k	3987	4479
k[790]	1369	5839
k[2420]	528	5010

Table 4. Comparison of the total number of iterations in M3, q4

	SSOR	No Preconditioner
k	4013	4499
k[1200]	2417	8919
k[2400]	1100	7661

From Table 1 to Table 4, the following observations can be made: The number of nonlinear iterations is almost the same, but the number of iterations in each block is much smaller for SSOR method. This is because we do not apply any preconditioning technique to solve the nonlinear system (3), but combine it with Block Broyden Algorithm to solve each linear block, as described in step 2.1 of Algorithm 3.1.1.

However, the performance of each method should not be determined by the numbers of iterations, but be judged according to formulas given in Section 3.3. Data shown in Table 4 is used as an example. During the 2400–th iteration, the following can be known:

$$k_p = 1100,\ k_n = 7661,\ q_4 = 2000,\ \bar{n} = M_3 / q_4 = 12100/2000 = 6$$

Thus we can calculate $\overline{M}$ according to (9):

$$\overline{M} = \frac{\bar{n}^4}{3} + 5\bar{n}^3 + \frac{\bar{n}^2}{6} + \frac{\bar{n}}{2} = \frac{6^4}{3} + 5\times 6^3 + \frac{6^2}{6} + \frac{6}{2} = 1521$$

J_n is calculated according to (6):

$$J_n = 2\bar{n}^3 + \bar{n}^2 + \bar{n} = 2\times 6^3 + 6^2 + 6 = 474$$

The value of $L_n(n)$ for the unpreconditioned method can be calculated according to (7):

$$L_n(n) = k_n \times J_n = 7661\times 474 = 3631314$$

The value of $L_p(n)$ for SSOR method can be calculated according to (8):

$$L_p(n) = q\times \overline{M} + k_p \times J_n = 2000\times 1521 + 1100\times 474 = 3563400$$

We find that $L_p(n) < L_n(n)$, which can be verified by other experimental data, so the performance of SSOR method is much better than the unpreconditioned one.

5 Conclusions

We have proposed Block Broyden Method combined with SSOR preconditioner for solving nonlinear systems arising from scientific and engineering computing. This method has faster solving speed and better performance than the unpreconditioned one. One of the reasons is that a proper preconditioner is used to transform the system, so the spectral properties of the broyden matrix is improved and quick convergence speed is gained. On the other hand, as the iterative matrix is block diagonal, the algorithm only needs to store the diagonal matrix, thus largely reduces memory space.

In future work, we expect to develop more effective preconditioning methods combined with Block Broyden Algorithm so that the preconditioned system can converge quickly and meanwhile can be constructed as easily as possible.

References

1. Yang G, Dutto L, Fortin M: Inexact block Jacobi Broyden methods for solving nonlinear systems of equations. SIAM J on Scientific Computing (1997) 1367-1392
2. Y. Saad, H. A. van der Vorst: Iterative solution of linear systems in the 20th century. J Comput Appl Math (2000) 1-33
3. Von Hagen. J, Wiesbeck.W: Physics-based preconditioner for iterative algorithms in MoM-problems. IEEE Trans on Antennas and Propagation (2002) 1315-1316
4. Peng Jiang, Geng Yang: Performance Analysis of Preconditioners based on Broyden Method. Applied Mathematics and Computation (Accepted for publication)
5. Peng Jiang, Geng Yang: A Preconditioning Method based on Broyden Algorithm. Journal of Nanjing University of Posts and Telecommunications (in Chinese) (Accepted for Publication)

A General Family of Two Step Runge-Kutta-Nyström Methods for $y'' = f(x, y)$ Based on Algebraic Polynomials

Beatrice Paternoster

Dipartimento di Matematica e Informatica
Universitá di Salerno, Italy
beapat@unisa.it

Abstract. We consider the new family of two step Runge–Kutta–Nyström methods for the numerical integration of $y'' = f(x, y)$, which provide approximation for the solution and its first derivative at the step point, and depend on the stage values at two consecutive step points. We derive the conditions to obtain methods within this family, which integrate algebraic polynomials exactly, describe a constructive technique and analyze the order of the resulting method.

1 Introduction

We are concerned with the analysis of a family of two step methods for the numerical integration of second order Ordinary Differential Equations, in which the first derivative does not appear explicitly,

$$y''(t) = f(t, y(t)), \quad y(t_0) = y_0, \quad y'(t_0) = y'_0, \qquad y(t), f(t, y) \in R^n, \qquad (1)$$

having a periodic or an oscillatory solution. The initial value problem (1) often arises in applications of molecular dynamics, orbital mechanics, seismology. When the response time is extremely important, for example in simulation processes, there is the need of obtaining an accurate solution in a reasonable time frame, and therefore there is a great demand of efficient methods for the direct integration of problem (1).

Many methods with constant coefficients have already been derived for second order ODEs (1) with periodic or oscillatory solutions: see for example [5, 8, 11, 12, 14, 15] for an extensive bibliography.

In this paper we consider the following generalization of the two step Runge–Kutta–Nyström (TSRKN) methods introduced in [14], by introducing the stage values at two consecutive step points, in order to increase the order of the methods, as already done in [9] for first order ODEs:

V.N. Alexandrov et al. (Eds.): ICCS 2006, Part IV, LNCS 3994, pp. 700–707, 2006.
© Springer-Verlag Berlin Heidelberg 2006

$$
\begin{aligned}
Y_n^j &= u_{j1}y_{n-1} + u_{j2}y_n + h\,(\overline{u}_{j1}y'_{n-1} + \overline{u}_{j2}y'_n) + \\
&\quad + h^2 \textstyle\sum_{s=1}^{m} (a_{js} f(x_{n-1}+c_s h, Y_{n-1}^s) + b_{js} f(x_n + c_s h, Y_n^j)), \\
y_{n+1} &= \theta_1 y_{n-1} + \theta_2 y_n + h(\eta_1 y'_{n-1} + \eta_2 y'_n) + \\
&\quad h^2 \textstyle\sum_{j=1}^{m} (v_j f(x_{n-1}+c_j h, Y_{n-1}^j) + w_j f(x_n + c_j h, Y_n^j)), \\
hy'_{n+1} &= \theta'_1 y_{n-1} + \theta'_2 y_n + h(\eta'_1 y'_{n-1} + \eta'_2 y'_n) + \\
&\quad h^2 \textstyle\sum_{j=1}^{m} (v'_j f(x_{n-1}+c_j h, Y_{n-1}^j) + w'_j f(x_n + c_j h, Y_n^j)),
\end{aligned}
\tag{2}
$$

c_j, θ_1, θ_2, u_{j1}, u_{j2}, $\overline{u}_{j1}$, $\overline{u}_{j2}$, v'_j, w'_j, v_j, w_j, a_{js}, b_{js}, $j, s, = 1, \ldots, m$ are the coefficients of the methods, which can be represented by the following array:

$$
\begin{array}{c|cc|c|c}
\mathbf{c} & \mathbf{u} & \bar{\mathbf{u}} & \mathbf{A} & \mathbf{B} \\
\hline
 & \theta & \eta & \mathbf{v^T} & \mathbf{w^T} \\
 & \theta' & \eta' & \mathbf{v'^T} & \mathbf{w'^T}
\end{array}
$$

In [14] the TSRKN method was derived as an indirect method from the two step Runge–Kutta methods introduced in [9]. The reason of interest in methods TSRKN (1.2) lies in the fact that, advancing from x_i to x_{i+1} we only have to compute Y_i, because Y_{i-1} has alreay been evaluated in the previous step. Therefore the computational cost of the method depends on the matrix A, while the vectors $\mathbf{v}$ and $\bar{\mathbf{v}}$ add extra degrees of freedom.

Our aim is to analyze two step implicit methods of type (2) which integrate algebraic polynomials exactly. The main motivation for the development of implicit methods (2), as those considered in the present paper, is their property of having a high stage order which make them suitable for stiff systems, also because their implicitness. Collocation–based methods also belong to this class.

In Section 2 we extend Albrecht's approach [1, 2] to the family (2), with the aim to derive the conditions for TSRKN methods to integrate algebraic polynomials exactly. In Section 3 we consider the collocation–based methods of type (2).

2 TSRKN Methods Based on Algebraic Polynomials

Let us consider the TSRKN methods (2). It is known that the method (2) is zero–stable if [14]

$$
-1 < \theta \leq 1 \tag{3}
$$

We treat formulas (2) by extending Albrecht's technique [1, 2] to the numerical method in concern, as we already have done in [11] for Runge–Kutta–Nyström methods, and in [13] for two step Runge–Kutta methods. According to this approach, we regard the two step Runge–Kutta–Nyström method (2) as a composite linear multistep scheme, but on a non–uniform grid.

We associate a linear difference operator with each stage, in the following way:

$$\begin{aligned} &\mathcal{L}_j[z(x);h] = z(x+c_jh) - u_{j,1}z(x-h) - u_{j,2}z(x) - \\ &h\left(\overline{u}_{j,1}z'(x-h) + \quad \overline{u}_{j,2}z'(x)\right) - \\ &h^2\textstyle\sum_{s=1}^{m}(a_{js}z''(x+(c_s-1)h) + b_{js}z''(x+c_sh), \end{aligned} \tag{4}$$

for $j = 1, \ldots, m$, is associated with the internal stage Y_n^j of (2).

$$\begin{aligned} &\bar{\mathcal{L}}[z(x);h] = z(x+h) - \theta_1 z(x-h) - \theta_2 z(x) - \\ &h(\eta_1 z'(x-h) + \eta_2 z'(x)) - \\ &h^2\textstyle\sum_{j=1}^{m}(v_j z''(x+(c_j-1)h) + w_j z''(x+c_jh)), \end{aligned} \tag{5}$$

is associated with the stage y_{n+1} in (2). Finally

$$\begin{aligned} &\bar{\mathcal{L}}'[z(x);h] = hz'(x+h) - \theta_1' z(x-h) - \theta_2' z(x) - \\ &h(\eta_1' z'(x-h) + \eta_2' z'(x)) - \\ &h^2\textstyle\sum_{j=1}^{m}(v_j' z''(x+(c_j-1)h) + w_j' z''(x+c_jh)) \end{aligned} \tag{6}$$

is associated with the final stage y'_{n+1} in (2).
Obviously the following relation holds:

$$\mathcal{L}_j[1;h] = 0, \quad j = 1, \ldots, m,$$

and

$$\bar{\mathcal{L}}[1;h] = \bar{\mathcal{L}}'[1;h] = 0$$

from which the parameters of the method have to satisfy the following relations:

$$u_{j,1} + u_{j,2} = 1, \quad j = 1, \ldots, m \tag{7}$$

$$\theta_1 + \theta_2 = 1, \qquad \theta_1' + \theta_2' = 0. \tag{8}$$

In order to have the *consistency* of the internal stages, the following relation hold:

$$\mathcal{L}_j[x;h] = 0, \quad j = 1, \ldots, m,$$

which holds if

$$c_j + u_{j1} = \overline{u}_{j,1} + \overline{u}_{j,2} \qquad (9).$$

(9) implies that $y(x_i + c_j h) - Y_i^j = O(h)$ for $h \to 0$. In the same way the final stages are consistent if $\bar{\mathcal{L}}[x;h] = \bar{\mathcal{L}}'[x;h] = \bar{\mathcal{L}}'[x^2;h] = 0$, that is

$$1+\theta_1 = \eta_1 + \eta_2, \qquad 1+\theta_1' = \eta_1' + \eta_2',$$
$$\textstyle\sum_{j=1}^m (v_j' + w_j') = \frac{2-\theta_1' + 2\eta_1'}{2}. \tag{10}$$

If (4) is identically equal to zero when $z(x) = x^p$, i.e. if $\mathcal{L}_j[x^p;h] = 0$, then it results:

$$\textstyle\sum_{s=1}^m (a_{js}(c_s-1)^{p-2} + b_{js}c_s^{p-2}) =$$
$$= \frac{c_j^p - (-1)^p u_{j,1} - (-1)^{p-1} p\, \overline{u}_{j,1}}{p(p-1)}. \tag{11}$$

Moreover, if (5) results equal to zero when $z(x) = x^p$, i.e. $\bar{\mathcal{L}}[x^p;h] = 0$, then

$$\textstyle\sum_{j=1}^m (v_j(c_j-1)^{p-2} + w_j c_j^{p-2}) =$$
$$= \frac{1-(-1)^p\theta_1 - (-1)^{p-1} p\, \eta_1}{p(p-1)}. \tag{12}$$

Finally, if we annihilate (6) on the function $z(x) = x^{p+1}$, then from $\bar{\mathcal{L}}'[x^{p+1};h] = 0$, it follows that

$$\textstyle\sum_{j=1}^m (v_j'(c_j-1)^{p-1} + w_j' c_j^{p-1}) =$$
$$= \frac{(p+1) - (-1)^{p+1}\theta_1' - (-1)^p (p+1)\, \eta_1'}{p(p+1)}. \tag{13}$$

We can now give the following definitions:

Definition 1. *An m stage TSRKN method (2) is said to satisfy the simplifying conditions $AB_2(p)$, $VW_2(p)$ and $V'W_2'(p)$ if its parameters satisfy respectively:*

Condition $AB_2(p)$:

$$\textstyle\sum_{s=1}^m (a_{js}(c_s-1)^{k-2} + b_{js}c_s^{k-2}) =$$
$$= \frac{c_j^h - (-1)^k u_{j,1} - (-1)^{k-1} k\, \overline{u}_{j,1}}{k(k-1)}$$

for $k = 1, \ldots, p,\ j = 1, \ldots, m$.

Condition $VW_2(p)$:

$$\sum_{j=1}^m (v_j(c_j-1)^{k-2} + w_j c_j^{k-2}) = \frac{1-(-1)^k\theta_1 - (-1)^{k-1} k\, \eta_1}{k(k-1)}$$

for $k = 1, \ldots, p$.

Condition $V'W_2'(p)$:

$$\sum_{j=1}^{m}(v_j'(c_j-1)^{k-1}+w_j'c_j^{k-1}) =$$

$$=\frac{(k+1)-(-1)^{k+1}\theta_1'-(-1)^k(k+1)\,\eta_1'}{k(k+1)}.$$

for $k=1,\ldots,p$.

$AB_2(p)$, $VW_2(p)$ and $V'W_2'(p)$ allow the reduction of order conditions of trees in the theory of two step RKN methods, which is under development by the authors of this paper; moreover they also mean that all the quadrature formulas represented by the TSRKN method have order at least p, similarly as it happens in the theory of Runge–Kutta methods [3].

As far as the order is concerned, we follow the classical definition of convergence of order p, related with the local truncation error. As a consequence, the conditions which we are going to formulate, are given *for exact starting values* y_1, y_1', Y_0^j, $j=1,\ldots,m$, as already done, for instance, in [6].

Definition 2. *An m–stage TSRKN method (2) has order p if for sufficiently smooth problems (1), and for exact starting values y_1, y_1', Y_0^j, $j=1,\ldots,m$,*

$$y(x_1+h)-y_1=O(h^{p+1}),\ hy'(x_1+h)-hy_1'=O(h^{p+2}) \tag{14}$$

By using Albrecht's theory [1, 2], it is easy to prove the following theorem:

Theorem 1. If $AB_2(p)$, $VW_2(p)$ and $VW_2'(p)$ hold, then for exact starting values the m–stage TSRKN method (2) has order of *convergence* p.

Proof. $AB_2(p)$, $VW_2(p)$ and $V'W_2'(p)$ imply that all the stages of the method have order p or, in Albrecht's terminology, that each stage in (4)–(6) has order of *consistency* p, so that the method has order of *consistency* p. In this case the method converges with order at least p.

It is worth mentioning that the conditions $AB_2(p)$, $VW_2(p)$ and $V'W_2'(p)$ are only sufficient conditions for the TSRKN method to have order p, but not necessary. If all the stages have order of *consistency* p, then all the stages are exact on any linear combination of the power set $\{1,x,x^2,\ldots,x^p\}$, and this implies that the TSRKN method results exact when the solutions of the system of ODEs (1) are algebraic polynomials. Moreover the simplifying conditions $AB_2(p)$, $VW_2(p)$ and $V'W_2'(p)$ are a constructive help for the derivation of new numerical methods within the family of TSRKN methods.

3 Collocation Methods

Let us generalize the Definition 3.2 of [7], in order to derive collocation methods for (1):

Definition 3. Consider m real numbers $c_1, \ldots, c_m \in [0,1]$, the solution values y_n, y_{n-1} and the derivative values y'_n, y'_{n-1}. The *collocation polynomial* $P(x)$ of degree $2m+3$ is then defined by:

$$P(x_{n-1}) = y_{n-1}, \quad P(x_n) = y_n \tag{14}$$

$$P'(x_{n-1}) = y'_{n-1}, \quad P'(x_n) = y'_n \tag{15}$$

$$P''(x_{n-1} + c_i h) = f(x_{n-1} + c_i h, P(x_{n-1} + c_i h)), \tag{16}$$

$$P''(x_n + c_i h) = f(x_n + c_i h, P(x_n + c_i h)). \tag{17}$$

Then the numerical solution of (1) is given by

$$y_{n+1} = P(x_{n+1}), \quad y'_{n+1} = P'(x_{n+1}) \tag{18}$$

(14)–(18) constitute a Hermite interpolation problem with incomplete data, because the function values at $x_n + c_i h$ are missing. Following [7], to compute the *collocation polynomial* $P(x)$, we introduce the dimensionless coordinate $t = (x - x_n)/h$, $x = x_n + th$, with nodes $t_1 = -1, t_2 = 0$, and define the following polynomials, which constitute a generalized Lagrange basis:

– $\phi_i(t)$, $i = 1, 2$, of degree $2m+3$, defined by

$$\phi_i(t_j) = \delta_{ij}, \qquad \phi'_i(t_j) = 0, \quad i,j = 1,2, \tag{19}$$

$$\phi''_i(c_j - 1) = 0, \; \phi''_i(c_j) = 0, \; i = 1,2, \;\; j = 1, \ldots, m, \tag{20}$$

– $\psi_i(t)$, $i = 1, 2$, of degree $2m+3$, defined by

$$\psi_i(t_j) = 0, \qquad \psi'(t_j) = \delta_{ij}, \quad i,j = 1,2, \tag{21}$$

$$\psi''_i(c_j - 1) = 0, \quad \psi''_i(c_j) = 0, \;\; i = 1,2, \;\; j = 1, \ldots, m, \tag{22}$$

– $\chi_{i,n-1}(t)$ and $\chi_{i,n}(t)$, $i = 1, \ldots, m$, of degree $2m+3$, defined by

$$\chi_{i,n-1}(t_j) = 0, \quad \chi_{i,n}(t_j) = 0, \quad i = 1, \ldots, m, \; j = 1,2 \tag{23}$$

$$\chi'_{i,n-1}(t_j) = 0, \quad \chi'_{i,n}(t_j) = 0, \quad i = 1, \ldots, m, \; j = 1,2 \tag{24}$$

$$\chi''_{i,n-1}(c_j - 1) = \delta_{ij}, \quad \chi''_{i,n-1}(c_j) = 0, \quad i,j = 1, \ldots, m, \tag{25}$$

$$\chi''_{i,n}(c_j - 1) = 0, \quad \chi''_{i,n}(c_j) = \delta_{ij}, \quad i,j = 1, \ldots, m. \tag{26}$$

δ_{ij} denotes the Kronecker tensor. Then the expression of the collocation polynomial $P(x)$ in terms of these polynomials is given by:

$$P(x_n + th) = \phi_1(t)\, y_{n-1} + \phi_2(t)\, y_n + h(\psi_1(t)\, y'_{n-1} +$$

$$\psi_2(t)\, y'_n) + h^2 \sum_{j=1}^{m} (\chi_{j,n-1}(t) \quad P''(x_{n-1} + c_j h) \quad +$$

$$+ \quad \chi_{j,n}(t) \; P''(x_n + c_j h)).$$

After constructing $\phi_i(t)$, $\psi_i(t)$, $\chi_{i,n-1}(t)$ and $\chi_{i,n}(t)$, by putting $t = c_i$, by writing $P(x_n + c_i h) = Y_n^i$ and by inserting the collocation conditions (14)–(17), we obtain the expression of the two step Runge–Kutta–Nyström (TSRKN) collocation method of type (2), where the parameters of the method are given by the following relations:

$$\theta_i = \phi_i(1), \quad u_{j,i} = \phi_i(c_j), \quad i = 1,2, \; j = 1, \ldots, m$$

$$\eta_i = \psi_i(1), \quad \overline{u}_{j,i} = \psi_i(c_j), \quad i = 1,2, \; j = 1, \ldots, m$$

$$v_j = \chi_{j,n-1}(1), \quad a_{js} = \chi_{j,n-1}(c_s), \quad j, s, = 1, \ldots, m,$$

$$w_j = \chi_{j,n}(1), \quad b_{js} = \chi_{j,n}(c_s), \quad j, s, = 1, \ldots, m$$

$$\theta'_i = \phi'_i(1), \quad \eta'_i = \psi'_i(1), \quad i = 1, 2,$$

$$v'_j = \chi'_{j,n-1}(1), \quad w'_j = \chi'_{j,n}(1), \quad j = 1, \ldots, m,$$

and $\phi_i(t)$, $\psi_i(t)$, $\chi_{i,n-1}(t)$ and $\chi_{i,n}(t)$ are the polynomials defined by conditions (19)–(26).

4 Conclusions

We have considered the family of TSRKN methods for $y'' = f(x, y)$ which integrate algebraic polynomials exactly. Following the procedure showed in this paper, that is annihilating the linear difference operators (4)–(6) on the set of power functions, and solving the arising systems $AB_2(p + 1)$, $VW_2(p + 1)$ and $V'W'_2(p)$, it is possible to derive TSRKN methods for ODEs (1) of order of convergence p. If $p = 2m + 2$, then the method is of collocation type.

Following the procedure showed in this paper, that is annihilating the linear difference operators (4)–(6) on different basis of functions, it is possible to derive TSRKN methods for ODEs having solutions with an already known behaviour. For example, it is worth considering TSRKN methods for ODEs (1) having periodic or oscillatory solution, for which the dominant frequency ω is known in advance; in this case a proper set of functions is the basis $\{1, \cos\omega x, \sin\omega x, \cos 2\omega x, \sin 2\omega x, \ldots\}$ for trigonometric polynomials, as already considered in [11, 13] for Runge–Kutta–Nyström and two step Runge-Kutta methods. The technique used in this paper can also be applied for the construction of collocation methods within family (2).

References

1. P. Albrecht, Elements of a general theory of composite integration methods, *Appl. Math. Comp.* **31** (1989), 1–17.
2. P. Albrecht, A new theoretical approach to RK methods, *SIAM J. Numer. Anal.* **24**(2) (1987), 391–406.
3. J. C. Butcher, The Numerical Analysis of Ordinary Differential Equations: Runge–Kutta and General Linear Methods, Wiley, Chichester (1987).
4. M. Cafaro and B. Paternoster, Analysis of stability of rational approximations through computer algebra, *Computer Algebra in Scientific Computing CASC-99* (Munich 1999), V.G.Ganzha, E.W.Mayr, E.V.Vorozhtsov Eds., pp. 25-36, Springer Verlag, Berlin (1999).
5. J.P.Coleman and L.Gr.Ixaru, P–stability and exponential–fitting methods for $y'' = f(x, y)$, *IMA J. Numer. Anal.* **16** (1996) 179–199.
6. Coleman J.P., Order conditions for a class of two–step methods for $y'' = f(x, y)$, IMA J. Num. Anal., vol.23 (2003) p.197-220.
7. Hairer E., Wanner G., Solving Ordinary Differential Equations II: Stiff and Differential–Algebraic Problems, Springer, Berlin, 1991.
8. L. Gr. Ixaru, B. Paternoster, A conditionally P-stable fourth order exponential-fitting method for y”=f(x,y), *J. Comput. Appl. Math.* **106** (1999), 87–98.
9. Z. Jackiewicz, R. Renaut and A. Feldstein, Two–step Runge–Kutta methods, *SIAM J. Numer. Anal.* **28**(4) (1991), 1165–1182.
10. J. D. Lambert, Numerical methods for ordinary differential systems: The initial value problem, Wiley, Chichester (1991).
11. B. Paternoster, Runge–Kutta(–Nyström) methods for ODEs with periodic solutions based on trigonometric polynomials, *Appl. Numer. Math.* (**28**) 2–4 (1998), 401–412.
12. B. Paternoster, A phase–fitted collocation–based Runge–Kutta–Nyström method, *Appl. Numer. Math.* **35**(4) (2000), 239–355.
13. B. Paternoster, General Two-Step Runge-Kutta methods based on algebraic and trigonometric polynomials, *Int. J. Appl. Math.* **6**(4) (2001), 347–362.
14. B. Paternoster, Two step Runge-Kutta-Nyström methods for $y'' = f(x, y)$ and P–stability, *Computational Science – ICCS 2002, Lecture Notes in Computer Science 2331, Part III*, P. M. A. Sloot, C. J. K. Tan, J. J. Dongarra, A. G. Hoekstra Eds. (2002), 459–466, Springer Verlag, Amsterdam.
15. L.R.Petzold, L.O.Jay, J.Yen, Numerical solution of highly oscillatory ordinary differential equations, *Acta Numerica*, 437–483, Cambridge University Press (1997).

Schur Decomposition Methods for the Computation of Rational Matrix Functions

T. Politi[1] and M. Popolizio[2]

[1] Dipartimento di Matematica, Politecnico di Bari,
Via Amendola 126/B, I-70126 Bari (Italy)
politi@poliba.it

[2] Dipartimento di Matematica, Università degli Studi di Bari,
Via Orabona 4, I-70125 Bari (Italy)
popolizio@dm.uniba.it

Abstract. In this work we consider the problem to compute the vector $\boldsymbol{y} = \Phi_{m,n}(A)\boldsymbol{x}$ where $\Phi_{m,n}(z)$ is a rational function, $\boldsymbol{x}$ is a vector and A is a matrix of order N, usually nonsymmetric. The problem arises when we need to compute the matrix function $f(A)$, being $f(z)$ a complex analytic function and $\Phi_{m,n}(z)$ a rational approximation of f. Hence $\Phi_{m,n}(A)$ is a approximation for $f(A)$ cheaper to compute. We consider the problem to compute first the Schur decomposition of A then the matrix rational function exploting the partial fractions expansion. In this case it is necessary to solve a sequence of linear systems with the shifted coefficient matrix $(A - z_j I)\boldsymbol{y} = \boldsymbol{b}$.

1 Introduction

Matrix functions arise in different fields of Mathematical Sciences and Engineering, in particular in connection with the solution of ordinary differential systems, with applications in control theory (see [1]), nuclear magnetic resonance, Lie group methods for geometric integration (see [8]), and in the numerical solution of stiff differential equations (see [11], and references therein). Usually the matrix function can be defined via power series or as the solution of nonlinear systems. Several methods have been proposed in past to solve one of the following problems:

- Given a complex matrix $A \in \mathbb{C}^{n\times n}$ compute $f(A)$, where f is an analytic function on a region of the complex plane;
- Given a matrix A with a given geometric structure compute the matrix function $f(A)$, having a special structure too;
- Given a matrix A and a vector $\boldsymbol{x}$ compute $f(A)\boldsymbol{x}$.

The second problem has a great interest in very important cases when $f(z)$ is the exponential function. In fact it is well known that the exponential map is a function defined from a Lie Algebra to its related Lie Group, i.e. if A is skew-symmetric (i.e. $A^T = -A$) then $Q = exp(A)$ is orthogonal. In these cases it is very important in the numerical approximation to preserve the property.

V.N. Alexandrov et al. (Eds.): ICCS 2006, Part IV, LNCS 3994, pp. 708–715, 2006.
© Springer-Verlag Berlin Heidelberg 2006

Recently this problem has been considered using approaches based on the Krylov subspaces technique but also exploiting the Schur decomposition of Hamiltonian skew-symmetric matrices ([5, 9]).

In past many algorithms have been proposed for the approximation of matrix $f(A)$ (see [7]) and most of these are based on the rational approximation of f,

$$\Phi_{m,n}(z) = \frac{R_m(z)}{Q_n(z)},$$

being $R_m(z)$ and $Q_n(z)$ polynomials of degree m and n, respectively. Hence

$$f(A) \simeq \Phi_{m,n}(A) = R_m(A)\left[Q_n(A)\right]^{-1}.$$

A classical way to obtain the function $\Phi_{m,n}(z)$ is to use the Padé approximation (usually with $m = n$) together with a pre-processing technique for matrix A (usually the scaling and squaring technique, see [6]).

In this work we decribe briefly some numerical methods based on the Schur decomposition of the matrix, in order to compute the rational function approximation of the exponential matrix function, then we consider the use of the partial fraction expansion. In Section 2 we describe the numerical methods based on the Schur decomposition, while in Section 3 we describe the algorithms to compute rational matrix functions based on the partial fractions expansion technique. Finally in Section 4 some numerical issues are shown.

2 Numerical Methods Based on the Schur Decomposition

A general approach to compute $f(A)$, with $A \in \mathbb{C}^{n\times n}$ is to employ similarity transformation

$$A = SDS^{-1} \tag{1}$$

where $f(D)$ is easily computable. In fact

$$f(A) = Sf(D)S^{-1}. \tag{2}$$

If A is diagonalizable then we could take D diagonal hence the computation of $f(D)$ is trivial since it is a diagonal matrix. For stability problems it is a good choice to take well conditioned matrix S in (1). The better way to take a well conditioned matrix is S to be orthogonal, so that the decomposition (1) becomes the Schur decomposition of A:

$$A = QTQ^*,$$

where T is upper triangular or block upper triangular. The computation of the Schur factors is achieved with perfect backward stability by the QR algorithm (see [6]). Once the decomposition has been performed the problem is the computation of $f(T)$. If T is upper triangular also

$$F = f(T)$$

is upper triangular, so that Parlett (see [10]) proposed the following recurrence relation, which comes equating the (i, j) elements $(i < j)$ in the commutativity relation $FT = TF$:

$$f_{ij} = t_{ij}\frac{f_{ii} - f_{jj}}{t_{ij} - t_{ji}} + \sum_{k=i+1}^{j-1} \frac{f_{ik}t_{kj} - t_{ik}f_{kj}}{t_{ii} - t_{jj}} \tag{3}$$

The problem with this approach is that the recurrence breaks in when $t_{ii} = t_{jj}$ for some $i \neq j$, that is when T has repeated eigenvalues, and it can give inaccurate results in floating point arithmetic when T has close eigenvalues. Parlett observed that if $T = (T_{ij})$ is block upper triangular then $F = (F_{ij})$ has the same block structure and, for $i < j$,

$$T_{ii}F_{ij} - F_{ij}T_{jj} = F_{ii}T_{ij} - T_{ij}F_{jj} + \sum_{k=i+1}^{j-1} (F_{ik}T_{kj} - T_{ik}F_{kj}) \, . \tag{4}$$

The recurrence can be used to compute F a superdiagonal at a time, provided we can evaluate the blocks $F_{ii} = f(T_{ii})$ and solve the Sylvester equation (4) for the F_{ij}. For the equation (4) to be nonsingular we need that T_{ii} and T_{jj} have no common eigenvalue. For the Sylvester equations to be well conditioned a necessary condition is that the eigenvalues of T_{ii} and T_{jj} are well separated. An alternative approach is first to compute

$$A = XDX^{-1}$$

where X is well conditioned and D is block diagonal. Then

$$f(A) = Xf(D)X^{-1}$$

and the problem reduces to compute $f(D)$. The usual way to compute a block diagonalization is first to compute the Schur form and then to eliminate off-diagonal blocks by solving Sylvester equations (see [6]). In order to gurantee a well conditioned matrix X a bound must be imposed on the condition of the individual transformations, this bound will be a parameter of the algorithm (see [4]).

Computing $f(D)$ reduces to compute $f(D_{ii})$ for each diagonal block D_{ii}. The D_{ii} are triangular but, unlike the Schur method, no particular eigenvalue distribution is guaranteed, because of limitations on the condition of the transformations; therefore $f(D_{ii})$ is still a nontrivial calculation.

3 Partial Fractions Expansions

A different approach in the computation of the matrix function $f(A)$ is to use a rational approximation, $\Phi_{m,n}(z)$:

$$\Phi_{m,n}(z) = \frac{R_m(z)}{Q_n(z)} \tag{5}$$

where $R_m(z)$ and $Q_n(z)$ are polynomials of degree m and n respectively. Usually function $\Phi_{m,n}(z)$ could be the (m,n) Padé approximation of the exponential function or the Chebyshev approximation but also another approaches are possible. The problem becomes the computation of the vector $\Phi_{m,n}(A)\boldsymbol{y}$. Supposing all the roots of polynomial $Q_n(z)$ to be distinct and $n \geq m$, the rational function $\Phi_{m,n}(z)$ can be represented using the partial fractions expansion:

$$\frac{R_m(z)}{Q_n(z)} = \alpha_0 + \sum_{j=1}^{n} \frac{\alpha_j}{z - z_j} \tag{6}$$

where

$$\alpha_0 = \lim_{z\to\infty} \frac{R_m(z)}{Q_n(z)}$$

and

$$\alpha_j = \frac{R_m(z)}{Q'_n(z)} \tag{7}$$

where z_j are the roots of $Q_n(z)$. Hence

$$\Phi_{m,n}(A) = R_m(A)\left[Q_n(z)\right]^{-1}$$

and, from (6):

$$\Phi_{m,n}(A)\boldsymbol{x} = \alpha_0\boldsymbol{x} + \sum_{j=1}^{n} \alpha_j (A - z_j I)^{-1}\boldsymbol{x}.$$

The vector $\boldsymbol{y}$ can be computed using the following algorithm:

1. Compute

$$\alpha_0 = \lim_{z\to\infty} \frac{R_m(z)}{Q_n(z)}, \qquad \alpha_j = \frac{R_m(z)}{Q'_n(z)}$$

   for $j = 1, \ldots, n$.
2. For $j = 1, \ldots, n$, compute the vector $\boldsymbol{x}_j$ solving

$$(A - z_j I)\boldsymbol{x}_j = \boldsymbol{b}$$

3. Set

$$\boldsymbol{y} = \alpha_0 \boldsymbol{b} + \sum_{j=1}^{n} \alpha_j \boldsymbol{x}_j.$$

The main problem is the solution of the linear systems at step 2. A good choice for this solution is to use Krylov-subspace iterations, since they are invariant under the shifts z_j. Hence the work required to solve n linear systems is the same to solve one system. Another advantage of this technique is that it can easily implemented on a parallel architecture since the linear systems to be solved in step 2 are independent (see [2]). Moreover there is a second level parallelism since also each system could be solved using a parallel algorithm. As last remark we consider that, when the rational function has poles and residuals as complex conjugate pairs, it is possible to exploit the following property (see [12]):

$$\alpha_j(A - z_jI)^{-1}\boldsymbol{x}_j + \overline{\alpha}_j(A - \overline{z}_jI)^{-1}\boldsymbol{x}_j = 2\Re\left(\alpha_j(A - z_jI)^{-1}\boldsymbol{x}_j\right). \qquad (8)$$

In [3] it is shown that the algorithm based on partial fractions representation can be very sensitive to perturbations. In fact the coefficients (7) are

$$\alpha_j = \frac{R_m(z_j)}{a_0 \displaystyle\prod_{k=1,k\neq j}^{n} (z_j - z_k)}.$$

and it follows that the presence of closed poles z_k may cause some coefficient α_j to be very large. In this case the error on the computation of vector $\boldsymbol{x}_j$ is amplified. In ([3]) these difficulties are remedied using an incomplete partial fraction representation of the rational function. The representation consists in writing the function as

$$\frac{R_m(z)}{Q_n(z)} = \prod_{k=1}^{t} \frac{r_l(z)}{q_l(z)} \qquad (9)$$

and using a partial fraction representation for each of the factors $r_l(z)/q_l(z)$. The functions $r_l(z)$ and $q_l(z)$ are polynomials such that the degree of $q_l(z)$ is not greater than the degree of $p_l(z)$ for each l. The polynomials $r_l(z)$ and $q_l(z)$ are chosen so that the partial fraction coefficient α_{lj} of the representation

$$\frac{r_l(z)}{q_l(z)} = \alpha_{0l} + \sum_{j=1}^{n_l} \frac{\alpha_{jl}}{z - z_{jl}}$$

where n_l is the degree of $q_l(z)$, and $j = 1, \ldots, n_l$, and $1 \leq l \leq t$, are not so large and the number of factors t is small. The incomplete partial representation algorithm can be described through the following steps:

1. Given the sets z_{jl} and α_{jl} with $j = 1, \ldots, n_l$, and $l = 1, \ldots, t$;
2. Put $\boldsymbol{x} = \boldsymbol{b}$;
3. For $l = 1, \ldots, t$:
   3.1 Solve the systems
   $$(A - z_{jl}I)\boldsymbol{x}_j = \boldsymbol{x}, \qquad j = 1, \ldots, n_l$$
   3.2 Compute the vector
   $$\boldsymbol{x} = \alpha_{0l}\boldsymbol{x} + \sum_{j=1}^{n_l} \alpha_{jl}\boldsymbol{x}_j.$$

The methods described previously could be used together computing first the Schur decomposition of matrix A:

$$A = QTQ^*$$

and then using the partial fractions expansions of the rational approximation of function $f(z)$ applied to matrix T. Hence given the vector $\boldsymbol{x}$ we have the approximation

$$f(A)\boldsymbol{x} \simeq Q\Phi_{m,n}(T)\boldsymbol{y}$$

where $\boldsymbol{y} = Q^*\boldsymbol{x}$.

4 Numerical Examples

In this section we show an example of the application of the algorithm described in the previous section to the computation of the exponential matrix applied to a real vector. Given the unsymmetric matrix $A \in \mathbb{R}^{n\times n}$ and the vector $\boldsymbol{x} \in \mathbb{R}^n$ our aim is to approximate the vector $e^A\boldsymbol{x}$. We recall that the method is based on the following steps:

1. Computation of the real Schur decomposition of the real matrix $A = QTQ^T$;
2. Computation of the Chebyshev rational approximation for the exponential function $C_{n,n}(z)$;
3. Computation of the poles z_j and the residuals α_j of rational function $C_{n,n}(z)$;
4. Solve the linear systems $(T - z_jI)\boldsymbol{x}_j = \boldsymbol{y}$, where $\boldsymbol{y} = Q^T\boldsymbol{x}$;
5. Compute the vector
$$\boldsymbol{w} = \alpha_0\boldsymbol{y} + \sum_{j=1}^{n} \alpha_j\boldsymbol{x}_j;$$
6. Compute the vector $\boldsymbol{z} = Q\boldsymbol{w}$.

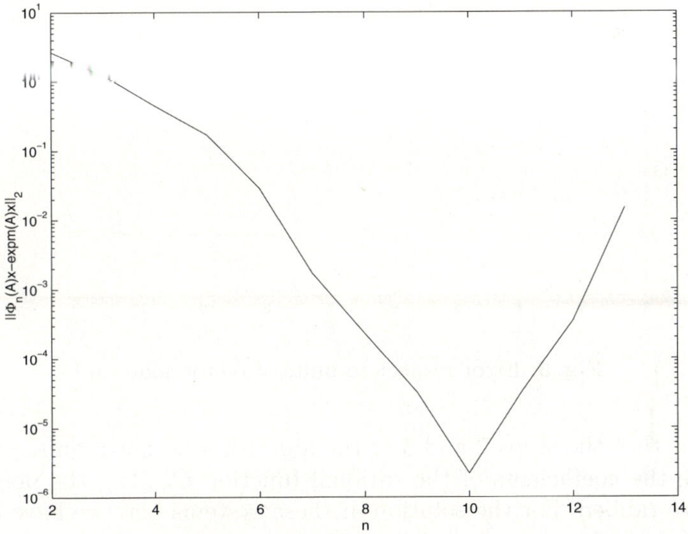

Fig. 1. Estimate of the error

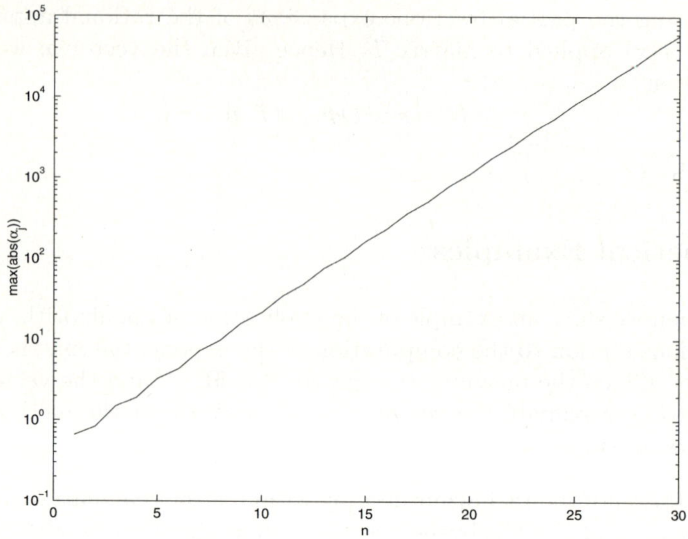

Fig. 2. Maximum Residuals of the Partial Fractions Expansion for Chebyshev Approximations of degree n

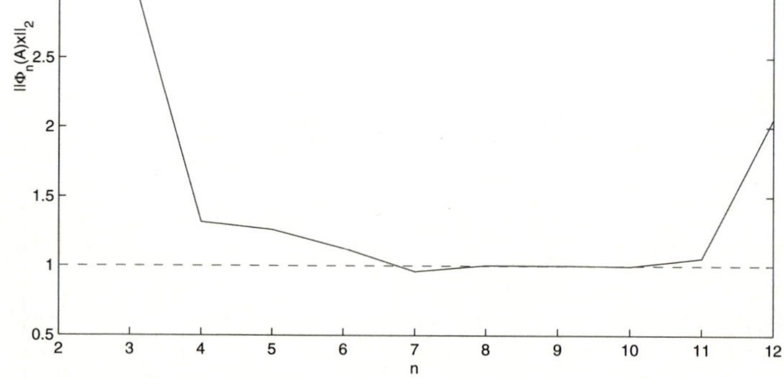

Fig. 3. Error respect to unitary vector solution

It is obvious that the steps 2 and 3 of the algorithm do not require any computation since the coefficients of the rational function $C_{n,n}(z)$, the poles and the residuals are tabbed. For the solution of these systems here we have considered just direct methods neglecting their shifted structure, in future we shall consider the application of iterative methods based on Krylov subspaces. A further remark is related to the use of Chebyshev approximation in step 2: we recall that using the property (8) the number of linear systems can be reduced. In Figure 3 we show the difference between the vector computed by MatLab function `expm` times a random vector $\boldsymbol{x}$ taking a random matrix $A \in \mathbb{R}^{20\times 20}$ and the vector computed by the algorithm described in the present section. We observe that taking the result computed by MatLab routine as correct the error decreases when the degree of the Chebyshev approximation is growing until $n = 10$, then the error increases. The reason of this behaviour is explained in Figure 3 where we show that the maximum of the residuals $|\alpha_j|$ increases with n, and as observed in the Section 3 the error in the solution of the linear systems is amplified. Finally we have considered a random skew-symmetric matrix A and a unitary 2-norm vector $\boldsymbol{x}$. In this case also the vector $e^A\boldsymbol{x}$ has unitary 2-norm. We observe that the behaviour of the error is exactly the same observed in the first example.

References

1. Åstrom K.J. , Wittenmark B.: Computer-Controlled Systems: Theory and Design. Prentice-Hall, Englewoods Ciffs, NJ, (1997)
2. Baldwin C., Freund R.W., Gallopoulos E.: A Parallel Iterative Method for Exponential Propagation. Proceedings of the Seventh SIAM Conference on Parallel Processing for Scientific Computing. D.H. Bailey et al. ed. SIAM (1995) 534–539
3. Calvetti D., Gallopoulos E., Reichel L.: Incomplete Partial Fractions for Parallel Evaluation of Rational Matrix Functions. J. Comp. Appl. Math. **59** (1995) 349–380
4. Davies P.J., Higham N.J.: A Schur-Parlett Algorithm for Computing Matrix Functions. SIAM J. Matr. Anal. Appl. **25** (2) (2003) 464–485
5. Del Buono N., Lopez L., Politi T.: Computation of functions of Hamiltonian and skew-symmetric matrices. Preprint (2006)
6. Golub G.H., Van Loan C.F.: Matrix Computation. The John Hopkins Univ. Press, Baltimore, (1996)
7. Higham N.J.: Functions of Matrices. MIMS EPrint 2005.21 The University of Manchester
8. Iserles A., Munthe-Kaas H., Nørsett S., Zanna A.: Lie-Group Methods. Acta Numerica **9** (2000) 215–365
9. Lopez L., Simoncini V.: Analysis of projection methods for rational function approximation to the matrix exponential. SIAM J. Numer. Anal. (to appear)
10. Parlett B.N.: A Recurrence among the Elements of Functions of Triangular Matrices. Lin. Alg. Appl. **14** (1976) 117–121
11. Saad Y.: Analysis of some Krylov subspace approximation to the matrix exponential operator, SIAM J. Numer. Anal. **29** (1) (1992) 209–228
12. Schmelzer T.: Rational approximations in scientific computing. Computing Laboratory, Oxford University, U.K. (2005)

Piecewise Constant Perturbation Methods for the Multichannel Schrödinger Equation

Veerle Ledoux*, Marnix Van Daele, and Guido Vanden Berghe

Vakgroep Toegepaste Wiskunde en Informatica, Ghent University,
Krijgslaan 281-S9, B-9000 Gent, Belgium
{Veerle.Ledoux, Marnix.VanDaele, Guido.VandenBerghe}@UGent.be

Abstract. The CPM$\{P, N\}$ methods form a class of methods specially devised for the propagation of the solution of the one-dimensional Schrödinger equation. Using these CPM$\{P, N\}$ methods in a shooting procedure, eigenvalues of the boundary value problem are obtained to very high precision. Some recent advances allowed the generalization of the CPM$\{P, N\}$ methods to systems of coupled Schrödinger equations. Also for these generalised CPM$\{P, N\}$ methods a shooting procedure can be formulated, solving the multichannel bound state problem.

1 Introduction

There are many problems in quantum chemistry, theoretical physics, atomic and molecular physics, and physical chemistry that can be transformed into the solution of coupled differential equations of Schrödinger type. Such a system of coupled equations may be written in matrix notation

$$\frac{d^2\mathbf{y}(x)}{dx^2} = [\mathbf{V}(x) - E\mathbf{I}]\,\mathbf{y}(x)\ . \tag{1}$$

If there are n channels, $\mathbf{y}(x)$ is a column vector of order n, $\mathbf{I}$ is the $n \times n$ unity matrix. The potential energy $n \times n$ matrix $\mathbf{V}(x)$ will be assumed throughout the following to be symmetric, which is often the case in molecular scattering and bound state applications.

There are various approaches to the solution of the coupled equations (1) (see a.o. [1, 2, 5, 14, 15]). In the more early work approximate schemes were used which attempt to reduce the coupled equations to a set of one-dimensional problems (e.g. in [14]). A more modern approach is to propagate the solutions numerically, without reducing them to one-dimensional form. A large number of numerical methods have been suggested for carrying out the propagation. However when bound state boundary conditions are applied, acceptable solutions of the coupled equations exist only when E is an eigenvalue of the Hamiltonian and additional techniques are needed to locate the eigenvalues. Early methods for doing this were developed by Gordon [4] and Johnson [10].

* Research Assistant of the Fund for Scientific Research - Flanders (Belgium) (F.W.O.-Vlaanderen).

V.N. Alexandrov et al. (Eds.): ICCS 2006, Part IV, LNCS 3994, pp. 716–723, 2006.
© Springer-Verlag Berlin Heidelberg 2006

Recently it has been shown in [12] that higher order Piecewise Constant Perturbation (CP) methods can be constructed for the propagation of the solution of Eq. (1). Here we will show how such a CP method can be used in a shooting procedure to solve the associated eigenvalue problem, that is to estimate the values of E for which a solution of Eq. (1) exists satisfying some boundary conditions in the endpoints of the integration interval.

2 The CP Algorithm for the Multichannel Case

The Piecewise Perturbation methods (PPM) have been successfully applied for the propagation of the solution of a one-dimensional Schrödinger problem (see [6, 7]). The PPM idea is to replace (piecewisely) the given equation by another differential equation, called the reference differential equation, which can be solved exactly. The deviation of the solution of the reference equation from the solution of the original equation is further estimated by means of the perturbation theory.

The CP methods form a subclass of the PPM where the potential function V of the reference equation is a piecewise constant. In [7] and [11] some higher order CP versions, the so-called CPM$\{P,N\}$ methods were found to be well suited for the solution of the one-dimensional Schrödinger problem. More recently these CPM$\{P,N\}$ formulae were generalised to the coupled channel case (see [12]). In this section the main elements of the CPM$\{P,N\}$ algorithm are recapitulated briefly. In the following description bold type indicates a column vector or matrix.

Let $[X, X+h]$ be the current one step interval of the partition of the integration interval. The multichannel Schrödinger problem

$$\mathbf{y}'' = (\mathbf{V}(x) - E\mathbf{I})\,\mathbf{y} \tag{2}$$

is considered on this interval. The algorithm of the CP method links the values of the solution at the two ends of the interval in the following two ways, to be used for forwards and backwards propagation,

$$\begin{bmatrix}\mathbf{y}(X+h)\\ \mathbf{y}'(X+h)\end{bmatrix} = \mathbf{T}^f(h)\begin{bmatrix}\mathbf{y}(X)\\ \mathbf{y}'(X)\end{bmatrix}, \quad \begin{bmatrix}\mathbf{y}(X)\\ \mathbf{y}'(X)\end{bmatrix} = \mathbf{T}^b(h)\begin{bmatrix}\mathbf{y}(X+h)\\ \mathbf{y}'(X+h)\end{bmatrix}.$$

To construct $\mathbf{T}^f$ and $\mathbf{T}^b$ we use two particular solutions of the local problem

$$\mathbf{y}'' = (\mathbf{V}(X+\delta) - E\mathbf{I})\,\mathbf{y}, \quad \delta \in [0,h]\,. \tag{3}$$

Specifically, if $\mathbf{u}(\delta)$ and $\mathbf{v}(\delta)$ are the $n \times n$ solutions corresponding to the initial conditions $\mathbf{y}(0) = \mathbf{I}$, $\mathbf{y}'(0) = \mathbf{0}$ and $\mathbf{y}(0) = \mathbf{0}$, $\mathbf{y}'(0) = \mathbf{I}$, respectively ($\mathbf{0}$ is the n by n zero matrix) then $\mathbf{T}^f$ and $\mathbf{T}^b$ have the form

$$\mathbf{T}^f(\delta) = \begin{bmatrix}\mathbf{u}(\delta) & \mathbf{v}(\delta)\\ \mathbf{u}'(\delta) & \mathbf{v}'(\delta)\end{bmatrix}, \quad \mathbf{T}^b(\delta) = \begin{bmatrix}\mathbf{v}'(\delta) & -\mathbf{v}(\delta)\\ -\mathbf{u}'(\delta) & \mathbf{u}(\delta)\end{bmatrix}, \tag{4}$$

To determine **u** and **v** the potential matrix is approximated by a truncated series over the shifted Legendre polynomials $P_n^*(\delta/h)$. The used parametrization is

$$\mathbf{V}(X+\delta) = \sum_{m=0}^{N} \mathbf{V}_m h^m P_m^*(\delta/h) \tag{5}$$

where the matrix weights are calculated by quadrature ($\bar{\mathbf{V}}_m = \mathbf{V}_m h^{m+2}, m = 1, 2, ...$),

$$\mathbf{V}_0 = \frac{1}{h}\int_0^h \mathbf{V}(X+\delta)d\delta,$$
$$\bar{\mathbf{V}}_m = (2m+1)h\int_0^h \mathbf{V}(X+\delta)P_m^*(\delta/h)d\delta, \; m = 1, 2, 3, \ldots . \tag{6}$$

The symmetric matrix $\mathbf{V}_0$ is then diagonalized and let **D** be the diagonalization matrix. In the **D** representation Eq. (3) becomes

$$\mathbf{y}^{\mathbf{D}\prime\prime} = \left(\sum_{m=0}^{N} \mathbf{V}_m^{\mathbf{D}} h^m P_m^*(\delta/h) - E\mathbf{I}\right)\mathbf{y}^{\mathbf{D}}, \quad \delta \in [0, h] \tag{7}$$

and this is solved for $\mathbf{u}^{\mathbf{D}}$ and $\mathbf{v}^{\mathbf{D}}$; the initial conditions are the same as in the original representation. The perturbation procedure is used, in which the diagonal matrix $\mathbf{V}_0^{\mathbf{D}}$ is the reference potential and

$$\Delta\mathbf{V} = \sum_{m=1}^{N} \mathbf{V}_m^{\mathbf{D}} h^m P_m^*(\delta/h) \tag{8}$$

is the perturbation.

First, the matrices of functions $\mathbf{u}^{\mathbf{D}}(\delta)$ and $\mathbf{v}^{\mathbf{D}}(\delta)$, denoted generically as $\mathbf{p}(\delta)$, are written as a perturbation series:

$$\mathbf{p}(\delta) = \mathbf{p}_0(\delta) + \mathbf{p}_1(\delta) + \mathbf{p}_2(\delta) + \mathbf{p}_3(\delta) + \ldots \tag{9}$$

where the zeroth order term $\mathbf{p}_0(\delta)$ is the solution of

$$\mathbf{p}_0'' = \left(\mathbf{V}_0^{\mathbf{D}} - E\right)\mathbf{p}_0 \tag{10}$$

with $\mathbf{p}_0(0) = \mathbf{I}, \mathbf{p}_0'(0) = \mathbf{0}$ for $\mathbf{u}_0$ and $\mathbf{p}_0(0) = \mathbf{0}, \mathbf{p}_0'(0) = \mathbf{I}$ for $\mathbf{v}_0$. Let the functions $\xi(Z)$ and $\eta_0(Z)$ be defined as follows:

$$\xi(Z) = \begin{cases} \cos(|Z|^{1/2}) & \text{if } Z \leq 0\,, \\ \cosh(Z^{1/2}) & \text{if } Z > 0\,, \end{cases} \tag{11}$$

and

$$\eta_0(Z) = \begin{cases} \sin(|Z|^{1/2})/|Z|^{1/2} & \text{if } Z < 0\,, \\ 1 & \text{if } Z = 0\,, \\ \sinh(Z^{1/2})/Z^{1/2} & \text{if } Z > 0\,, \end{cases} \tag{12}$$

while $\eta_s(Z)$ with $s > 0$ are further generated by recurrence :

$$\eta_1(Z) = [\xi(Z) - \eta_0(Z)]/Z,$$
$$\eta_s(Z) = [\eta_{s-2}(Z) - (2s-1)\eta_{s-1}(Z)]/Z\,, \quad s = 2, 3, \ldots\,. \tag{13}$$

The zeroth order propagators $\mathbf{u}_0(\delta)$ and $\mathbf{u}_0(\delta)$ are then diagonal matrices, defined as follows:

$$\mathbf{u}_0 = \mathbf{v}_0' = \xi(\mathbf{Z}) \tag{14}$$

$$\delta\mathbf{u}_0' = \mathbf{Z}(\delta)\eta_0(\mathbf{Z}) \tag{15}$$

$$\mathbf{v}_0 = \delta\eta_0(\mathbf{Z}) \tag{16}$$

where

$$\mathbf{Z}(\delta) = (\mathbf{V^D}_0 - E\mathbf{I})\delta^2 \tag{17}$$

and $\xi(\mathbf{Z})$, $\eta_m(\mathbf{Z})$ two $n \times n$ diagonal matrices of functions with $\xi(Z_i)$, resp. $\eta_m(Z_i)$ as i^{th} diagonal element (with $Z_i(\delta) = (V_{0_{ii}}^D - E)\delta^2$).

The $n \times n$ 'correction' matrix of functions $\mathbf{p}_q$ is the solution of the system

$$\mathbf{p}_q'' = (\mathbf{V}_0^\mathbf{D} - EI)\mathbf{p}_q + \Delta\mathbf{V}(\delta)\mathbf{p}_{q-1}, \quad \mathbf{p}_q(0) = \mathbf{p}_q'(0) = 0\,. \tag{18}$$

The following iteration procedure exists to construct the corrections.

Correction $\mathbf{p}_{q-1}$ ($\mathbf{p} = \mathbf{u^D}, \mathbf{v^D}$) is assumed to be known and of such a form that the product $\Delta\mathbf{V}(\delta)\mathbf{p}_{q-1}$ reads

$$\Delta\mathbf{V}(\delta)\mathbf{p}_{q-1}(\delta) = \mathbf{Q}(\delta)\xi(\mathbf{Z}) + \sum_{m=0}^{+\infty} \delta^{2m+1}\mathbf{R}_m(\delta)\eta_m(\mathbf{Z})\,. \tag{19}$$

Then $\mathbf{p}_q(\delta)$ and $\mathbf{p'}_q(\delta)$ are of the form

$$\mathbf{p}_q(\delta) = \sum_{m=0}^{+\infty} \delta^{2m+1}\mathbf{C}_m(\delta)\eta_\mathbf{m}(\mathbf{Z})\,, \tag{20}$$

$$\mathbf{p'}_q(\delta) = \mathbf{C}_0(\delta)\xi(\mathbf{Z}) + \sum_{m=0}^{+\infty} \delta^{2m+1}\left(\frac{d\mathbf{C}_m(\delta)}{d\delta} + \delta\mathbf{C}_{m+1}(\delta)\right)\eta_m(\mathbf{Z})\,, \tag{21}$$

where all $\mathbf{C}_m$ matrices are given by quadrature:

$$\mathbf{C}_0(\delta) = \frac{1}{2}\int_0^\delta \mathbf{Q}(\delta_1)d\delta_1$$

$$\mathbf{C}_m(\delta) = \frac{1}{2}\delta^{-m}\int_0^\delta \delta_1^{m-1}\left(\mathbf{R}_{m-1}(\delta_1) - \frac{d^2\mathbf{C}_{m-1}(\delta_1)}{d\delta_1^2} - [\mathbf{C}_{m-1}(\delta_1), \mathbf{V}_0^\mathbf{D}]\right)d\delta_1$$

where $[\mathbf{C}_{m-1}, \mathbf{V}_0^\mathbf{D}]$ is the commutator of the matrices $\mathbf{C}_{m-1}$ and $\mathbf{V}_0^\mathbf{D}$.

To calculate successive corrections for $\mathbf{u}^\mathbf{D}$, the starting functions in $\Delta\mathbf{V}(\delta)\mathbf{p}_0(\delta)$ are $\mathbf{Q}(\delta) = \Delta\mathbf{V}$ and $\mathbf{R}_0(\delta) = \mathbf{R}_1(\delta) = ... = \mathbf{0}$. For $\mathbf{v}^\mathbf{D}$ the starting functions are $\mathbf{Q}(\delta) = \mathbf{0}$, $\mathbf{R}_0(\delta) = \Delta\mathbf{V}(\delta)$, $\mathbf{R}_1(\delta) = \mathbf{R}_2(\delta) = ... = 0$. We have thus all ingredients necessary to evaluate the perturbative corrections.

Once the values at h of the $\mathbf{u}^\mathbf{D}$, $\mathbf{v}^\mathbf{D}$ matrices and of their derivatives have been evaluated, they are reconverted to the original representation to obtain the desired $\mathbf{T}^f$ and $\mathbf{T}^b$.

This theory was used to construct some CP versions which are identified as CPM$\{P,N\}$ methods. Such a CPM$\{P,N\}$ algorithm is of maximum order P at low energies and of order N in the asymptotic regime. As shown in [12] a CPM$\{P,N\}$ method can be used to efficiently propagate the solution of Eq. (2) (forwards and backwards).

3 The Boundary Value Problem

We now consider the eigenvalue problem arising when boundary conditions are introduced at both ends of the integration interval. Values of E have to be found for which a solution of (1) exists satisfying the boundary conditions.

Before proceeding to the multichannel case, it is instructive to consider the procedure used for the single-channel (one-dimensional) Schrödinger equation. In the one-dimensional case $y(x)$ and $V(x)$ are both scalar quantities

$$y''(x) = [V(x) - E]y(x), \quad x \in (a \geq -\infty, b \leq +\infty) . \tag{22}$$

In this simple case, the eigenvalue problem has been solved since the early work of Cooley [3] based on the widely used shooting method. Such a shooting method proceeds as follows: A trial value E_{trial} of E is chosen and the following steps are used: (i) we start at the beginpoint of the integration interval a and propagate the solution (e.g. using a CP method) towards the endpoint of the integration interval b. We stop at an arbitrary point x_{match}. (ii) We then start at b and we propagate backwards until x_{match} is reached. (iii) At x_{match} the left-hand and right-hand solutions are matched. The two solutions are arbitrarily normalised, so their values can always made to agree at the matching point by renormalising them. However, the criterion for E_{trial} to be an eigenvalue is that the derivatives y' should match, as well as the values. The matching condition is thus

$$\frac{y_L'}{y_L} = \frac{y_R'}{y_R} , \tag{23}$$

or equivalently

$$\det \begin{pmatrix} y_L & y_R \\ y_L' & y_R' \end{pmatrix} = 0 , \tag{24}$$

where the subscripts L and R indicate outwards and inwards solutions originating at a and b respectively. The matching function $y_L y_R' - y_R y_L'$ is thus a function of the energy that is zero when E_{trial} is an eigenvalue. If E_{trial} is not found to be an eigenvalue, steps (i)-(iii) are repeated with an adjusted value of E_{trial}. It is possible to obtain a new E_{trial} value simply by using one of the standard numerical procedures for finding a zero of a function. For the CP methods a Newton-Raphson iteration procedure can be formulated, because the CP algorithm allows a direct evaluation of the first derivatives of the solution with respect to the energy E (see [7]). The software packages SLCPM12 [8] and MATSLISE [13] use the CP methods in combination with this Newton-Raphson process to obtain very accurate eigenvalue estimations for the one-dimensional Schrödinger problem.

For a system of n coupled equations, a procedure can be used which is largely inspired from the method outlined above for the one-dimensional problem. In the multichannel case, the desired wavefunction is a column vector $\mathbf{y}(x)$. In order to start propagating a solution to the coupled equations, it is necessary to know not only the initial values of the elements of $\mathbf{y}(x)$ at a and b (which are given by the boundary conditions) but also their derivatives. In the single-channel case, the initial derivatives are arbitrary, because their effects are cancelled by renormalising the function after the matching point was reached. However, in the multichannel case the situation is more complicated: here the ratios of the initial derivatives in the different channels are significant. In early methods, schemes were devised for converging upon suitable initial derivatives as well as the eigenvalue. However, Gordon [4] has devised a method that avoids the problem of searching for the correct values of the initial boundary derivatives. Instead of propagating a single wavefunction vector, a complete set of n vectors is propagated, spanning the space of all possible initial derivatives. So the wavefunction becomes an $n \times n$ matrix $\mathbf{Y}(x)$ instead of a column vector $\mathbf{y}(x)$. Since the columns of $\mathbf{Y}(x)$ span the space of all possible initial derivatives, any wavefunction that satisfies the boundary conditions can be expressed as a linear combination of them. The true wavefunction vector $\mathbf{y}(x)$ can thus be expressed as

$$\mathbf{y}(x) = \mathbf{Y}_L(x)\mathbf{c}_L, \quad x \le x_{\mathrm{match}}, \tag{25}$$

$$\mathbf{y}(x) = \mathbf{Y}_R(x)\mathbf{c}_R, \quad x \ge x_{\mathrm{match}}, \tag{26}$$

where $\mathbf{Y}_L$ and $\mathbf{Y}_R$ are the wavefunction matrices propagated from a and b and $\mathbf{c}_L$ and $\mathbf{c}_R$ are x-independent column vectors that must be found. For an acceptable wavefunction, both $\mathbf{y}$ and its derivative must match at $x = x_{\mathrm{match}}$,

$$\mathbf{y}(x_{\mathrm{match}}) = \mathbf{Y}_L(x_{\mathrm{match}})\mathbf{c}_L = \mathbf{Y}_R(x_{\mathrm{match}})\mathbf{c}_R, \tag{27}$$

$$\mathbf{y}'(x_{\mathrm{match}}) = \mathbf{Y}'_L(x_{\mathrm{match}})\mathbf{c}_L = \mathbf{Y}'_R(x_{\mathrm{match}})\mathbf{c}_R\,. \tag{28}$$

These two equations can be combined into the single equation

$$\begin{bmatrix} \mathbf{Y}_L & \mathbf{Y}_R \\ \mathbf{Y}'_L & \mathbf{Y}'_R \end{bmatrix} \begin{bmatrix} \mathbf{c}_L \\ -\mathbf{c}_R \end{bmatrix} = \mathbf{0}\,, \tag{29}$$

where the matrix on the left-hand side is evaluated at $x = x_{\mathrm{match}}$. It is a matrix of order $2n \times 2n$ and is a function of the energy at which the wavefunctions are calculated. A non-trivial solution of Eq. (29) exists only if the determinant of the matrix on the left is zero, and this is only true if the energy used is an eigenvalue of the coupled equations. It is then straightforward to find the vectors $\mathbf{c}_L$ and $\mathbf{c}_R$.

As for the one-dimensional case it is possible to construct a Newton-Raphson process to localise the eigenvalues by using the derivatives of the wavefunction with respect to E. As shown in [9] and [12] these derivatives can be propagated by the CP methods simultaneously with the wavefunction itself at a relatively low extra cost.

The procedure described above requires that the wavefunction matrix and its derivative be propagated explicitly. However there is one well known difficulty in the theory of close-coupled equations. The propagation of the wavefunction into the so-called classically forbidden region (where $V(x) > E$) is numerically unstable. It is due to the fact that the exponentially growing component y_j of the wave function in the most strongly closed ($V_{jj}(x) > E$) channel soon dominates the entire wave function matrix and destroys the required linear independence of the solutions. One approach to overcoming the difficulty is to apply certain stabilizing transformations during propagation (see [4]). In [9] a stabilizing transformation based on LU decomposition is described for the propagation by CP methods. After some propagation steps this regularization procedure can be applied to re-establish the linear independence of the columns.

Another way to avoid the difficulty is to use a so-called "invariant imbedding" method, in which the propagated quantity is not the wave function matrix $\mathbf{Y}(x)$ but rather its logarithmic derivative $\mathbf{Y}'(x)\mathbf{Y}(x)^{-1}$ (see e.g. [10]). For the CP methods we can use the knowledge of the components of the matrix $\mathbf{T}^f$ and $\mathbf{T}^b$ for the propagation of the log derivative of the solution $\mathbf{\Psi} = \mathbf{Y}'\mathbf{Y}^{-1}$:

$$\begin{aligned} \mathbf{\Psi}(X+h) &= [\mathbf{u}'(h) + \mathbf{v}'(h)\mathbf{\Psi}(X)][\mathbf{u}(h) + \mathbf{v}(h)\mathbf{\Psi}(X)]^{-1} \\ \mathbf{\Psi}(X) &= [-\mathbf{u}'(h) + \mathbf{u}(h)\mathbf{\Psi}(X+h)][\mathbf{v}'(h) - \mathbf{v}(h)\mathbf{\Psi}(X+h)]^{-1} . \end{aligned} \tag{30}$$

The matching condition can then also be expressed in terms of $\mathbf{\Psi}(x)$.

4 Example

As a test problem, we consider a multichannel Schrödinger problem of which the exact eigenvalues are known

$$\mathbf{y}'' = \begin{bmatrix} x^2 - E & -1 \\ -1 & x^2 - E \end{bmatrix} \mathbf{y}, \quad \mathbf{y}(0) = \mathbf{y}(10) = \mathbf{0} . \tag{31}$$

The CPM{10,8} method was used as the propagation method and at x_{match} a Newton-Raphson process was applied. Doing all calculations in standard precision (16 significant figures), the results listed in Table 1 were generated. The table shows the first four eigenvalue estimations for different values of the input tolerance *tol*. The second column contains the number of meshpoints m in the corresponding partition. For more details on the generation of the partition, we refer to [12].

Knowing that the first four exact eigenvalues are 2, 4, 6 and 8, it is clear that we were able to obtain very precise eigenvalue approximations. Also other (more complicated) testcases show that the multichannel CPM$\{P, N\}$ methods have the power to estimate eigenvalues accurately. Moreover, the CPM$\{P, N\}$ methods are very efficient to use in a shooting process: the relatively time consuming

Table 1. Approximations of the first four Eigenvalues of Problem (31)

tol	m	$n=0$	$n=1$	$n=2$	$n=3$
10^{-6}	23	1.99999995	3.999999956	5.99999991	7.99999991
10^{-8}	38	1.9999999995	3.9999999995	5.9999999992	7.9999999992
10^{-10}	62	1.999999999998	3.999999999998	5.999999999998	7.999999999998
10^{-12}	102	1.99999999999998	3.99999999999998	5.99999999999999	7.99999999999999

task of generating the partition can be performed only once (at the very beginning of the run) and can be completely separated from the repeatedly asked but fast executable task of propagating the solution at various values of E.

References

1. Allison, A.C.: The numerical solution of coupled differential equations arising from the Schrödinger equation. J. Comput. Phys. **6** (1970) 378–391
2. Allison, A.C.: The numerical solution of the equations of molecular-scattering. Adv. At. Mol. Phys. **25** (1988) 323–341
3. Cooley, J.W.: An Improved Eigenvalue Corrector Formula for Solving the Schrödinger Equation for Central Fields. Math. Comput. **15** (1961) 363-374
4. Gordon, R.G.: New Method for Constructing Wavefunctions for Bound States and Scattering. J. Chem. Phys. **51** (1969) 14–25
5. Hutson, J.M.: Coupled channel methods for solving the bound-state Schrödinger equation. Comput. Phys. Commun. **84** (1994) 1–18.
6. Ixaru, L.Gr.: Numerical Methods for Differential Equations and Applications. Reidel, Dordrecht-Boston-Lancaster (1984)
7. Ixaru, L.Gr., De Meyer, H., Vanden Berghe, G.: CP methods for the Schrödinger equation, revisited. J. Comput. Appl. Math. **88** (1997) 289–314
8. Ixaru, L.Gr., De Meyer, H., Vanden Berghe, G.: SLCPM12 - A program for solving regular Sturm-Liouville problems. Comp. Phys. Comm. **118** (1999) 259–277

[illegible] - A package for the solution of the coupled channel Schrödinger equation. Comput. Phys. Commun. **147** (2002) 834–852

10. Johnson, B.R.: Renormalized Numerov method applied to calculating bound-states of coupled-channel Schrödinger equation. J. Chem. Phys **69** (1978) 4678–4688
11. Ledoux, V., Van Daele, M., Vanden Berghe, G.: CP methods of higher order for Sturm-Liouville and Schrödinger equations. Comput. Phys. Commun. **162** (2004) 151–165
12. Ledoux, V., Van Daele, M., Vanden Berghe, G.: CPM$\{P, N\}$ methods extended for the solution of coupled channel Schrödinger equations. Comput. Phys. Commun. **174** (2006) 357–370
13. Ledoux, V., Van Daele, M., Vanden Berghe, G.: MATSLISE: A MATLAB package for the Numerical Solution of Sturm-Liouville and Schrödinger equations. ACM Trans. Math. Softw. **31** (2005)
14. Levine, R.D.: Adiabatic approximation for nonreactive subexcitation molecular collisions. J. Chem. Phys. **49** (1968) 51
15. Rykaczewski, K., Batchelder, J.C., Bingham, C.R. et al.: Proton emitters ^{140}Ho and ^{141}Ho: Probing the structure of unbound Nilsson orbitals. Phys. Rev. C. **60** (1999) 011301

State Dependent Symplecticity of Symmetric Methods*

Felice Iavernaro and Brigida Pace

Dipartimento di Matematica, Università di Bari, Italy
felix@dm.uniba.it, pace@dm.uniba.it

Abstract. Despite symmetric one-step methods applied to Hamiltonian dynamical systems fail in general to be symplectic, we show that symmetry implies, however, a relation which is close to symplecticity and that we called *state dependent symplecticity*. We introduce such definition for general maps and analyze it from an analytical viewpoint in one simpler case. Some numerical tests are instead reported as a support of this feature in relation with the good long time behaviour of the solutions generated by symmetric methods.

Keywords: Hamiltonian and Poisson systems, symplecticity, symmetric methods.

Subject Classification: 65P10, 65L05, 37M15.

1 Introduction

In this paper we link the property of symmetry of a one step numerical integrator applied to the Hamiltonian system

$$\dot{y} = J\nabla H(y), \quad J = \begin{pmatrix} 0 & -I \\ I & 0 \end{pmatrix}, \quad y = (p,q)^T, \quad H(p,q) : \mathbb{R}^m \times \mathbb{R}^m \longrightarrow \mathbb{R}, \quad (1)$$

with I the identity matrix, to a relation (satisfied by the map representing the method itself) called *state dependent symplecticity* (sd-symplecticity) which is close to the standard symplecticity property of symplectic integrators (we refer to [5] for the general theory on Hamiltonian problems). Although we do not report exhaustive theoretical results, we give some insights on how such feature in turn relies on the good stability properties shared by symmetric methods when applied to particular but important Hamiltonian systems in a neighborhood of an equilibrium point. In the following we assume that

$$y_1 = \phi_h(y_0) \quad (2)$$

is a one-step method of order p applied to the problem (1) with stepsize h (we assume regularity of the transformation ϕ_h).

* This work was supported by COFIN-PRIN 2004 (project "Metodi numerici e software matematico per le applicazioni").

V.N. Alexandrov et al. (Eds.): ICCS 2006, Part IV, LNCS 3994, pp. 724–731, 2006.
© Springer-Verlag Berlin Heidelberg 2006

Definition 1. *The one-step method* (2) *is called sd-symplectic if, when applied to the problem* (1)*, its Jacobian matrix satisfies*

$$\left(\frac{\partial y_1}{\partial y_0}\right)^T \widehat{J}(y_1, -h) \left(\frac{\partial y_1}{\partial y_0}\right) = \widehat{J}(y_0, h), \tag{3}$$

where $\widehat{J}$ *is a skew-symmetric nonsingular matrix for all* $h \leq h_0$.

Although for our purposes we have attached property (3) to a numerical method, we may extend its applicability to any parametric transformation in the form (2), where h stands for the parameter. Furthermore, one easily realizes that the matrix $\widehat{J}(y, h)$ approximates J up to the order of the method: $\widehat{J}(y, h) = J + O(h^p)$.

The dependence of relation (3) on the stepsize h is both implicit (since y_1 depends on h) and explicit. In the particular cases where the explicit dependence is missing or where $\widehat{J}(\cdot, \gamma)$ is an even function of γ, we may recast (3) as

$$\left(\frac{\partial y_1}{\partial y_0}\right)^T \widetilde{J}(y_1) \left(\frac{\partial y_1}{\partial y_0}\right) = \widetilde{J}(y_0), \tag{4}$$

where $\widetilde{J}(y) = \widehat{J}(y, \pm h)$, and we are led back to the definition of a Poisson map with respect to the bracket

$$\{F, G\} = \nabla F(y)^T \, \widetilde{J}(y) \, \nabla G(y), \qquad \text{with} \quad F, G : \mathbb{R}^{2m} \to \mathbb{R}.$$

Poisson systems generalize Hamiltonian systems in that they are defined by substituting to the matrix J in (1) any skew-symmetric matrix $\widetilde{J}(y)$ satisfying the Jacobi identity $\{\{F, G\}, H\} + \{\{G, H\}, F\} + \{\{H, F\}, G\} = 0$. The flow of a Poisson system

$$\dot{z} = \widetilde{J}(z) \nabla H(z) \tag{5}$$

is a Poisson map, and this justifies the study of Poisson integrators, that is numerical methods satisfying (4) when applied to (5).

Thus Definition 1 weakens the properties of a Poisson map. In general, neither the matrix $\widehat{J}(y, h)$ satisfies the Jacobi identity nor can it be stated that a sd-symplectic method is a Poisson integrator for a given set of almost Poisson problems. On the other hand, from a geometric viewpoint it seems like that property (3) may still imply, under suitable assumptions on the structure of the Hamiltonian and on the dynamics of the solution, an almost preservation of volumes of any bounded regions in the phase space, under the iterations of the method. This circumstance has already been detected and analysed in the simpler case of problems with one degree of freedom [4], where Definition 1 simplifies as $\mu(y_1, -h)(\frac{\partial y_1}{\partial y_0})^T J (\frac{\partial y_1}{\partial y_0}) = \mu(y_0, h) J$ with μ a scalar function. In this paper we consider the case of higher dimensional problems. For the special case of the trapezoidal method the computation simplifies remarkably and therefore we will use such simpler method to retrieve some theoretical results, while we will provide numerical evidence that other symmetric methods do exhibit similar behaviours.

2 Sd-Symplecticity of Symmetric Runge-Kutta Methods

To see that any symmetric consistent RK method is sd-symplectic, we start the computation by considering the trapezoidal formula, which is the simplest symmetric non symplectic one step method. The application of the trapezoidal method to the Hamiltonian problem (1) defines the mapping

$$y_1 = y_0 + \frac{h}{2} J(\nabla H(y_0) + \nabla H(y_1)).$$

By differentiating y_1 with respect to y_0, we get the variational equation

$$\left(I - \frac{h}{2} J\, \nabla^2 H(y_1)\right) \frac{\partial y_1}{\partial y_0} = \left(I + \frac{h}{2} J\, \nabla^2 H(y_0)\right),$$

where $\nabla^2 H(y)$ is the Hessian matrix of $H(y)$. Setting $A^{\pm}(y) = I \pm \frac{h}{2} J\, \nabla^2 H(y)$ yields

$$(\frac{\partial y_1}{\partial y_0})^T \left(A^-(y_1)\right)^T J \left(A^-(y_1)\right) \frac{\partial y_1}{\partial y_0} = \left(A^+(y_0)\right)^T J \left(A^+(y_0)\right).$$

A direct computation shows that $(A^-(y))^T\, J\, A^-(y)$ and $(A^+(y))^T\, J\, A^+(y)$ define the same skew-symmetric matrix

$$\widehat{J}(y,h) = J + \frac{h^2}{4} \nabla^2 H(y)\, J\, \nabla^2 H(y). \tag{6}$$

For s-stage symmetric RK methods, we follow a similar approach (see [4] for further details). In this case the mapping (2) reads

$$y_1 = y_0 + hJ(b^T \otimes I)\nabla H(K), \tag{7}$$

where $K = [K_1^T, \ldots, K_s^T]^T$ is the block vector of the internal stages

$$K = e \otimes y_0 + h(A \otimes J)\nabla H(K),$$

and $\nabla H(K) \equiv [\nabla^T H(K_1), \ldots, \nabla^T H(K_s)]^T$. Due to symmetry, we can split the term $(b^T \otimes I)\nabla H(K)$ of (7) in two (symmetric) terms depending uniquely on y_0 and y_1 respectively:

$$y_1 = y_0 + \frac{h}{2} J(b^T \otimes I) \left[\nabla H(K^+(y_0)) + \nabla H(K^-(y_1))\right], \tag{8}$$

where

$$K^{\pm}(y) = e \otimes y \pm h(A \otimes J)\nabla H(K^{\pm}(y)). \tag{9}$$

Differentiation of (8) with respect to y_0 yields a variational equation that looks similar to the one obtained for the trapezoidal method:

$$\left(I - \frac{h}{2} J\, F_{-h}(y_1)\right) \frac{\partial y_1}{\partial y_0} = \left(I + \frac{h}{2} J\, F_h(y_0)\right), \tag{10}$$

with

$$F_{\pm h}(y) \equiv (b^T \otimes I)\nabla^2 H(K^{\pm}(y))\frac{\partial K^{\pm}(y)}{\partial y}.$$

By defining

$$\widehat{J}(y,\gamma) = (I + \frac{\gamma}{2}F_\gamma(y))^T\, J\,(I + \frac{\gamma}{2}F_\gamma(y)), \tag{11}$$

and exploiting symmetry, we finally arrive at (3).

2.1 The Trapezoidal Method as a Simple Example

The presence of the internal stages in a RK-method is responsible of the loss of symmetry of the matrix $\widehat{J}(y,\gamma)$ with respect to the second argument γ; in fact, in general, $F_\gamma(y) \neq F_{-\gamma}(y)$. Looking at (6) one realizes that for the trapezoidal method sd-simplecticity reduces to the standard preservation of a Poisson bracket structure (hereafter, to simplify the notation, we set again $\widetilde{J}(y) \equiv \widehat{J}(y,\pm h)$). This comes not as a surprise since the trapezoidal method is conjugate to the midpoint implicit method which is symplectic [3]. Therefore the simpler condition (6) is well understood in terms of measure preserving properties[1]. In the phase space $\mathbb{R}^{2m}$ consider a 2-dimensional sub-manifold $\mathcal{M}$ obtained as the image of a compact set $K \subset \mathbb{R}^2$ through a continuously differentiable function $\psi : (s,t) \in K \mapsto (p,q) \in \mathcal{M}$, that is $\psi(K) = \mathcal{M}$. The scalar quantity

$$\widehat{\Omega}(\mathcal{M}) = \iint_K \left(\frac{\partial\psi}{\partial s}(s,t)\right)^T \widetilde{J}(\psi(s,t)) \left(\frac{\partial\psi}{\partial t}(s,t)\right) ds\, dt. \tag{12}$$

is the sum of the *scaled* oriented areas of the projections of $\mathcal{M}$ onto the orthogonal planes (p_i,q_i), $i = 1,\dots,m$. The term “scaled areas” means that $\widetilde{J}$ acts as a weight function. Therefore it turns out that the trapezoidal method preserves the quantity $\widehat{\Omega}(\mathcal{M})$, i.e.,

$$\widehat{\Omega}(\phi_h(\mathcal{M})) = \widehat{\Omega}(\mathcal{M}), \tag{13}$$

with $\widetilde{J}(\cdot) = \widehat{J}(\cdot,\pm h)$ defined in (6). For $h \to 0$ we get $\widetilde{J} \to J$ and (13) reduces to the classical geometrical interpretation of symplecticity.

From (3), for the trapezoidal method we get

$$\left(\frac{\partial y_n}{\partial y_0}\right)^T \widetilde{J}(y_n) \left(\frac{\partial y_n}{\partial y_0}\right) = \left(\frac{\partial y_n}{\partial y_{n-1}}\frac{\partial y_{n-1}}{\partial y_0}\right)^T \widetilde{J}(y_n) \left(\frac{\partial y_n}{\partial y_{n-1}}\frac{\partial y_{n-1}}{\partial y_0}\right)$$
$$= \left(\frac{\partial y_{n-1}}{\partial y_0}\right)^T \widetilde{J}(y_{n-1}) \left(\frac{\partial y_{n-1}}{\partial y_0}\right),$$

and an induction process allows us to link the generic state vector y_n to the initial one:

$$\left(\frac{\partial y_n}{\partial y_0}\right)^T \widetilde{J}(y_n) \left(\frac{\partial y_n}{\partial y_0}\right) = \widetilde{J}(y_0). \tag{14}$$

[1] As guide lines for the following description, we adopt the same argument and notations exploited in [3] to describe the geometric interpretation of symplecticity.

A symplectic transformation is volume preserving. Analogously a Poisson transformation preserves a non-Euclidean measure (*scaled volume*). To find out the expression of the scaled volume preserved by the trapezoidal method, we consider the determinants of both sides of (14)

$$\det\left(\frac{\partial y_n}{\partial y_0}\right) = \left(\frac{\det\left(\widetilde{J}(y_0)\right)}{\det\left(\widetilde{J}(y_n)\right)}\right)^{\frac{1}{2}}, \tag{15}$$

from which we obtain the following scaled volume conservation property:

$$\text{scaled-Vol}(S_n) = \text{scaled-Vol}(S_0), \tag{16}$$

where the scaled volume of a $2d$-dimensional region $S \subset \Omega$ defined by the trapezoidal method is

$$\text{scaled-Vol}(S) \equiv \int_S (\det(\widetilde{J}(y))^{\frac{1}{2}} dy = \int_S \sqrt{\det(J + \frac{h^2}{4}\nabla^2 H(y) J \nabla^2 H(y))} dy. \tag{17}$$

In particular, for separable Hamiltonian systems $H(p,q) = T(p) + U(q)$, (17) reduces to

$$\text{scaled-Vol}(S) = \int_S \det\left(I + \frac{h^2}{4} U''(q) T''(p)\right) dp\, dq.$$

The Taylor expansion of (15) yields

$$\det\left(\frac{\partial y_n}{\partial y_0}\right) = 1 + O(h^2),$$

where the $O(h^2)$ term may be assumed independent of n if the Hessian matrix $\nabla^2 H(y_n)$ remains bounded. This is true for systems (like the nonlinear pendulum) where the entries of the Hessian matrix are bounded functions or, more importantly, when the solution y_n itself lies in a compact set of the phase space[2]. In such a case we obtain

$$\int_{S_n} dy_n = \int_{S_0} dy_0 + O(h^2), \tag{18}$$

where the $O(h^2)$ term is independent of the integration time, which states a nearby-preservation property of the volumes.

Due to the appearance of the internal stages, it is not possible to retrieve a relation so easy as (14) for higher order RK methods. Consequently, the remainder in the analogue expression of (18) for a symmetric RK method, is expected

[2] This is a standard assumption when investigating the long time behaviour of the solutions of Hamiltonian systems (see for example [3], Theorem 8.1, page 312).

to depend on the time t_n, but this is not always true. From (10) we obtain the analogue of (15) for symmetric RK methods:

$$\det\left(\frac{\partial y_n}{\partial y_0}\right) = \frac{\prod_{i=0}^{n-1} \det\left(I + \frac{h}{2} J F_h(y_i)\right)}{\prod_{i=0}^{n-1} \det\left(I - \frac{h}{2} J F_{-h}(y_i)\right)}. \tag{19}$$

It turns out that the left hand side remains indeed bounded for the dynamics of many interesting Hamiltonian system. Such circumstance has been analysed for two-dimensional problems in [4], and is related to a global (rather than local) character of the solution. In the next section we report a few examples to give numerical evidence that the same may occur in higher dimensional systems.

3 Numerical Results

Hereafter we list four problems used for our tests together with a brief description (for further details see Chap. I of [3] and reference therein). All of them have separable Hamiltonian function in the form $H(p,q) = 1/2p^T p - U(q)$ (they come indeed from the application of Newton's second law), with the potential U satisfying the symmetry relation $U(-q) = U(q)$. These conditions seem to be right ingredients that makes the determinants in (19) $O(h^p)$-bounded, independently of the time t_n.[3]

- TEST 1: *Two-body Problem.* The dynamics of two bodies attracted by their gravitational forces lies in a plane and it is identified by the (normalized) Hamiltonian function

$$H(p_1,p_2,q_1,q_2) = \frac{1}{2}(p_1^2 + p_2^2) - \frac{1}{(q_1^2 + q_2^2)^{\frac{1}{2}}},$$

  where $p = (p_1, p_2)^T$ and $q = (q_1, q_2)^T$ are the velocity and position vectors of one body in a coordinate system centred at the second body.
- TEST 2: *Perturbed two-body Problem.* The same as the two body-problem with the addition of a perturbation term that accounts for non-Newtonian interactions:

$$H(p_1,p_2,q_1,q_2) = \frac{1}{2}(p_1^2 + p_2^2) - \frac{1}{(q_1^2 + q_2^2)^{\frac{1}{2}}} - \frac{\mu}{(q_1^2 + q_2^2)^{\frac{3}{2}}},$$

  where $|\mu|$ is a small real number (here set equal to 10^{-2}).
- TEST 3: *Fermi-Pasta-Ulam Problem.* In the presented form, this problem describes the interaction of $2m$ mass points linked with alternating soft nonlinear and stiff linear springs, in a one-dimensional lattice with fixed end points ($q_0 = q_{2m+1} = 0$). The Hamiltonian function is

$$H(p,q) = \frac{1}{2}\sum_{i=1}^{m}(p_{2i-1}^2 + p_{2i}^2) + \frac{\omega^2}{4}\sum_{i=1}^{m}(q_{2i} + q_{2i-1})^2 + \sum_{i=0}^{m}(q_{2i+1} + q_{2i})^4.$$

  We chose $m = 3$ (6 degrees of freedom) and $\omega = 50$.

[3] Some counterexamples of reversible Hamiltonian systems with $U(-q) \neq U(q)$ for which symmetric non symplectic RK-methods are not appropriate can be found in [1].

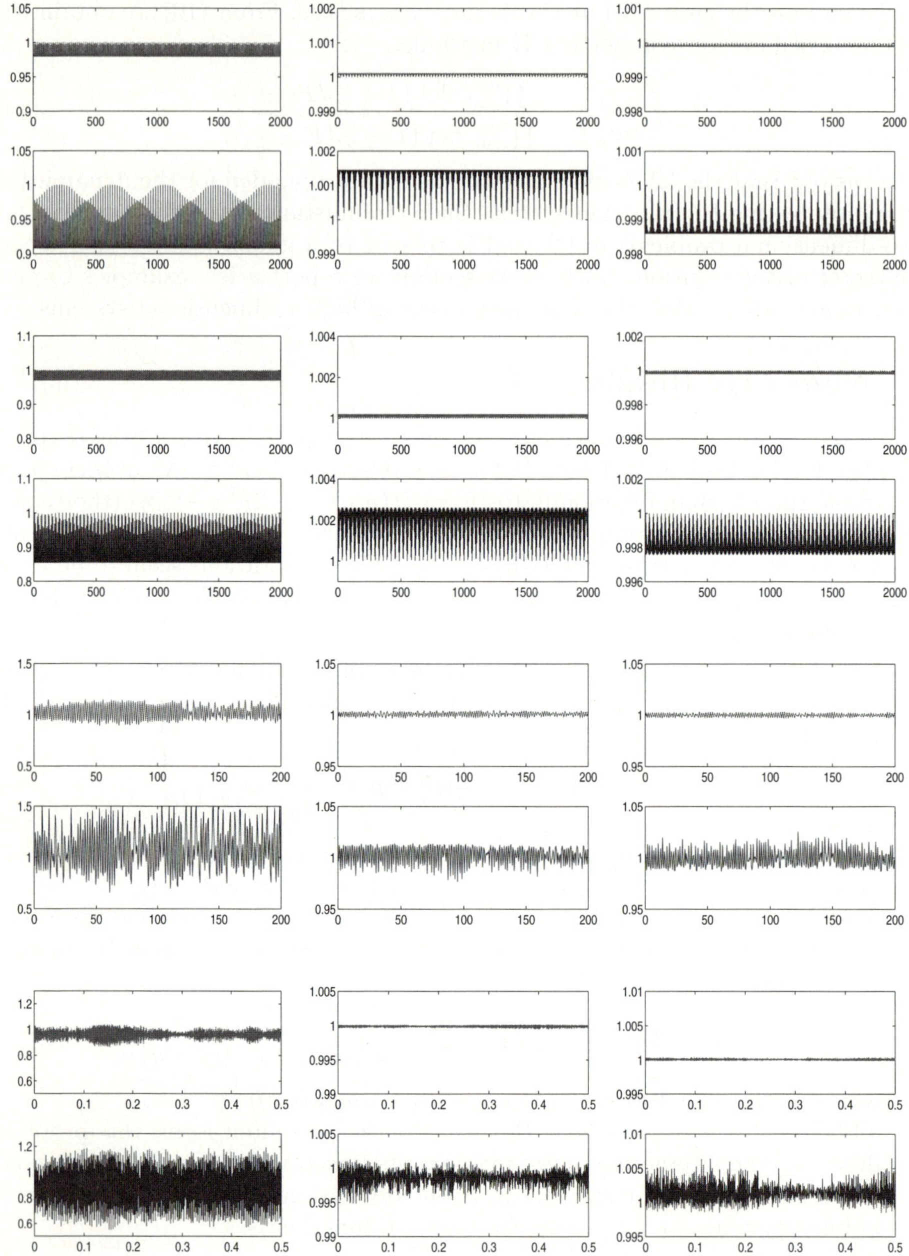

Fig. 1. Nearby volume preservation of the trapezoidal method (first column), Lobatto IIIA of order 4 (second column) and Lobatto IIIB of order 4 (right column). The i-th row displays the results for TEST i, for $i = 1, \ldots 4$. Each figure consists of two pictures reporting the quantity defined in (19) of the methods applied with a given stepsize h (upper plot) and $\bar{h} = 2h$ (lower plot). The list of the stepsizes h is: TEST 1-2: $h = 0.1$; TEST 3: $h = 0.25$; TEST 4: $h = 5 \cdot 10^{-5}$.

- TEST 4: *Molecular dynamics.* Neutral atoms and molecules are subject to two distinct forces, one attracting and the other repelling, in the limit of large distance and short distance. This may be accounted for by considering pair-potentials like the Lennard-Jones potential (also known as the 6-12 potential) which, for the atoms i and j at a distance r, reads

$$V_{ij}(r) = 4\varepsilon_{ij}\left(\left(\frac{\sigma_{ij}}{r}\right)^{12} - \left(\frac{\sigma_{ij}}{r}\right)^{6}\right).$$

The resulting system, simulating the dynamics of a network of N particles, has Hamiltonian

$$H(p,q) = \frac{1}{2}\sum_{i=1}^{N}\frac{1}{m_i}p_i^T p_i + \sum_{i=1}^{N}\sum_{j=i+1}^{N} V_{ij}(||q_i - q_j||).$$

In our experiment we have considered $N = 7$ argon atoms ($m_i = m = 66.34 \cdot 10^{-27}\mathtt{Kg}$) lying on a plane, with six equilibrium points located at the vertices of a regular hexagon and the remaining one at its centre, and

$$\varepsilon_{ij} = \varepsilon \simeq 1.6540 \cdot 10^{-21}\mathtt{J}, \qquad \sigma_{ij} = \sigma = 0.341 \cdot 10^{-9}\mathtt{m}$$

As initial conditions, we chose null velocities and positions slightly far away from the equilibria.

As numerical integrators we have used the LobattoIIIA and LobattoIIIB methods of order 4 and, for comparison purposes, the trapezoidal method. The related results have been displayed in the central, right and left columns of Figure 1 respectively. They report the quantity $\det(\partial y_n/\partial y_0)$ defined in (19) in correspondence of two different stepsizes in order to better infer its independence of the time integration interval. We have avoided to plot the residual $||(\partial y_n/\partial y_0)^T J(\partial y_n/\partial y_0) - J||$ since in general it fails to remain bounded even for the trapezoidal method and therefore it does not make sense

References

1. E. Faou, E. Hairer and T.-L. Pham, *Energy conservation with non-symplectic methods: examples and counter-examples*, BIT Numerical Mathematics, **44** (2004), 699–709.
2. E. Hairer and C. Lubich, *Symmetric multistep methods over long times*, Numer. Math., **97** (2004), 699-723.
3. E. Hairer, C. Lubich, and G. Wanner, *Geometric numerical integration. Structure-preserving algorithms for ordinary differential equations.*, Springer Series in Computational Mathematics, **31**, Springer-Verlag, Berlin, 2002.
4. F. Iavernaro and D. Trigiante, *State dependent symplecticity and area preserving numerical methods*, (submitted).
5. K. R. Meyer, G. R. Hall, *Introduction to Hamiltonian dynamical systems and the N-body problem*, Applied Mathematical Sciences 90, Springer-Verlag, New York, 1992.

On the Solution of Skew-Symmetric Shifted Linear Systems

T. Politi[1] and A. Pugliese[2]

[1] Dipartimento di Matematica, Politecnico di Bari,
Via Amendola 126/B, I-70126 Bari (Italy)
politi@poliba.it

[2] School of Mathematics, Georgia Institute of Technology,
Atlanta, GA 30332 U.S.A.
pugliese@math.gatech.edu

Abstract. In this paper we consider the problem of solving a sequence of linear systems with coefficient matrix $A_\alpha = I + \alpha A$ (or $A_\alpha = \alpha I + A$), where α is a real paramater and A is skew-symmetric matrix. We propose to solve this problem exploiting the structure of the Schur decomposition of the skew-symmetric matrix and computing the Singular Value Decomposition of a bidiagonal matrix of halved size.

1 Introduction

In this paper we consider the solution of a sequence of linear systems of the form

$$A_\alpha \boldsymbol{x}_\alpha = \boldsymbol{b}_\alpha \tag{1}$$

where

$$A_\alpha = I + \alpha A \tag{2}$$

or

$$A_\alpha = \alpha I + A$$

with I identity matrix of order n, α positive real parameter belonging to $]0, \alpha_{\max}]$ and A is a real skew-symmetric matrix of order n. The question is how to solve efficiently the linear systems for subsequent values of the shift α. The goal is to obtain a solution procedure that is cheaper, in terms of total solution costs, trying to save much computations as possible. In [2, 3] the authors consider the solution of a sequence of linear systems (1)-(2) having A symmetric and positive definite. In this case they try to obtain an efficient preconditioning matrix to save computational operations when the parameter α changes, observing that using the same preconditioner for different values of α brings to a very slow convergence. Sequences of linear systems of the form (1)-(2) arise in different fields of applied mathematics. If we consider the numerical solution of a Kortweg-de Vries partial differential equation using a particular discretization (see [4]) a differential system

$$y'(t) = A(y)y(t)$$

V.N. Alexandrov et al. (Eds.): ICCS 2006, Part IV, LNCS 3994, pp. 732–739, 2006.
© Springer-Verlag Berlin Heidelberg 2006

with initial condition $y(0) = y_0$ has to be solved. Applying numerical schemes of the integration, at each step, a sequence of linear systems (1)-(2) has to be solved.
The paper is organized as follows: in Section 2 we derive the structure for a decomposition of matrix (2), in Section 3 we analyze the computational cost of the proposed method, while in Section 4 possible applications are described.
The following theorem is a general result that will be useful later.

Theorem 1. *Let A be the following $n \times n$, n even, square matrix*

$$A = \begin{bmatrix} B & C \\ -C & B \end{bmatrix}$$

with $C, B \in \mathbb{R}^{m \times m}$ symmetric matrices and $m = n/2$. If A is nonsingular then A^{-1} retains the same structure of A, i.e.

$$A^{-1} = \begin{bmatrix} X & Y \\ -Y & X \end{bmatrix}$$

with $X, Y \in \mathbb{R}^{m \times m}$ symmetric matrices.

2 The Decomposition of the Matrix $I + \alpha A$

The factorization technique that we describe in this section has the same of starting point of the one shown in [5] in order to compute the exponential of the skew-symmetric matrix A times a unitary norm vector $\boldsymbol{v}$, and, in general, to compute analytic matrix functions. First of all we consider the problem to find the Schur decomposition of the skew-symmetric matrix A. Hence we consider the real skew symmetric matrix $A \in \mathbb{R}^{n \times n}$, a vector $\boldsymbol{q}_1 \in \mathbb{R}^n$, such that $\|\boldsymbol{q}_1\|_2 = 1$, the Krylov matrix $K(A, \boldsymbol{q}_1, m) = [\boldsymbol{q}_1 \; A\boldsymbol{q}_1 \; A^2\boldsymbol{q}_1 \dots \; A^{m-1}\boldsymbol{q}_1] \in \mathbb{R}^{n \times m}$ and the Krylov subspace $\mathcal{K}_m = span\{\boldsymbol{q}_1, A\boldsymbol{q}_1, \dots, A^{m-1}\boldsymbol{q}_1\}$. The following result states the conditions for the existence of a Hessenberg form of A (see [6]).

Theorem 2. *Suppose $Q = [\boldsymbol{q}_1 \; \boldsymbol{q}_2 \dots \; \boldsymbol{q}_n] \in \mathbb{R}^{n \times n}$ is an orthogonal matrix. Then $Q^\top A Q = H$ is an unreduced Hessenberg matrix if and only if $R = Q^\top K(A, \boldsymbol{q}_1, n)$ is nonsingular and upper triangular.*

Thus, when $K(A, \boldsymbol{q}_1, n)$ is of full rank n, from the QR factorization of $K(A, \boldsymbol{q}_1, n)$, it follows that an unreduced Hessenberg form H of A exists. The Hessenberg decomposition, $A = QHQ^\top$, is essentially unique when the first column $\boldsymbol{q}_1$ of Q is fixed and its unreduced Hessenberg form H is skew-symmetric and possesses the following tridiagonal structure

$$H = \begin{bmatrix} 0 & -h_2 & 0 & \dots & 0 \\ h_2 & 0 & -h_3 & \ddots & \vdots \\ 0 & \ddots & \ddots & \ddots & 0 \\ \vdots & \ddots & h_{n-1} & 0 & -h_n \\ 0 & \dots & 0 & h_n & 0 \end{bmatrix}. \qquad (3)$$

For skew-symmetric matrices the reduction to the above tridiagonal form can be performed using the following Lanczos tridiagonalization process:
Let $\boldsymbol{q}_1$ be a vector of $\mathbb{R}^n$ with $\|\boldsymbol{q}_1\| = 1$ and set $h_1 = 0$ and $\boldsymbol{q}_0 = 0$.

$$
\begin{array}{l}
\texttt{for } j = 1 : n \\
\qquad \boldsymbol{w}_j = A\boldsymbol{q}_j + h_j \boldsymbol{q}_{j-1} \\
\qquad h_{j+1} = \|\boldsymbol{w}_j\| \\
\qquad \boldsymbol{q}_{j+1} = \boldsymbol{w}_j / h_{j+1} \\
\texttt{end}
\end{array}
$$

We have to notice that, in exact arithmetic, the above algorithm is equivalent to the Arnoldi process applied to skew symmetric matrices. It provides an orthogonal matrix $Q = [\boldsymbol{q}_1 \ \boldsymbol{q}_2 \ \ldots \ \boldsymbol{q}_n] \in \mathbb{R}^{n\times n}$ such that $Q^\top AQ = H$ where H is the tridiagonal matrix (3). Moreover, it allows one to take full advantage of the possible sparsity of A due to the matrix-vector products involved. However, in floating-point arithmetic the vectors $\boldsymbol{q}_j$ could progressively lose their orthogonality, in this case, the application of a re-orthogonalization procedure is required [6, 7]. We suppose n to be even, but the case of n odd may be approached in a similar way. Let us consider the permutation matrix

$$P = [\boldsymbol{e}_1, \boldsymbol{e}_3, \ldots, \boldsymbol{e}_{n-1}, \boldsymbol{e}_2, \boldsymbol{e}_4, \ldots, \boldsymbol{e}_n],$$

where $\boldsymbol{e}_i$ is the i–th vector of the canonical basis of $\mathbb{R}^n$. Then, if H is as in (3) we have

$$P^\top HP = \begin{bmatrix} 0 & -B \\ B^\top & 0 \end{bmatrix}, \tag{4}$$

where B is the lower bidiagonal square matrix of size $m = n/2$ given by:

$$B = \begin{bmatrix} h_2 & 0 & \ldots & \ldots & 0 \\ -h_3 & h_4 & \ldots & & 0 \\ 0 & -h_5 & \ddots & \ddots & \vdots \\ \vdots & \ddots & \ddots & h_{n-2} & 0 \\ 0 & \ldots & 0 & -h_{n-1} & h_n \end{bmatrix}. \tag{5}$$

Since all diagonal and subdiagonal entries of B are non zero, the m singular values of B are distinct and different from zero.
Let us consider the singular value decomposition of B

$$B = U\Sigma V^\top,$$

with $\Sigma = \mathrm{diag}(\sigma_1, \sigma_2, \ldots, \sigma_m)$ and $\sigma_1 > \sigma_2 > \ldots \sigma_m > 0$, and the orthogonal $n \times n$ matrix

$$W = \begin{bmatrix} U & 0 \\ 0 & V \end{bmatrix}.$$

Hence

$$W^\top P^\top HPW = \begin{bmatrix} U^\top & 0 \\ 0 & V^\top \end{bmatrix} \begin{bmatrix} 0 & -B \\ B^\top & 0 \end{bmatrix} \begin{bmatrix} U & 0 \\ 0 & V \end{bmatrix} = \begin{bmatrix} 0 & -\Sigma \\ \Sigma & 0 \end{bmatrix}.$$

The matrix $I + \alpha A$ can be decomposed as

$$I + \alpha A = QPW(I + \alpha\widehat{\Sigma})W^\top P^\top Q^\top \tag{6}$$

where

$$\widehat{\Sigma} = \begin{bmatrix} O & -\Sigma \\ \Sigma & O \end{bmatrix}.$$

If we have to solve the sequence of linear systems (1)-(2) we could exploit the factorization (6) computing vector $\boldsymbol{x}_\alpha$, for a given value α through the following steps:

1. Compute $\boldsymbol{y}_\alpha = Q^\top \boldsymbol{b}_\alpha$;
2. Compute $\boldsymbol{c}_\alpha = W^\top P \boldsymbol{y}_\alpha$;
3. Solve the linear system

$$(I + \alpha\widehat{\Sigma})\boldsymbol{u}_\alpha = \boldsymbol{c}_\alpha \qquad \Leftrightarrow \qquad \begin{bmatrix} I_m & -\alpha\Sigma \\ \alpha\Sigma & I_m \end{bmatrix} \boldsymbol{u}_\alpha = \boldsymbol{c}_\alpha \tag{7}$$

4. Compute $\boldsymbol{x}_\alpha = QPW\boldsymbol{u}_\alpha$.

The solution of (7) is very simple, in fact exploiting Theorem 1 it is easy to compute the inverse of the matrix $I + \alpha\widehat{\Sigma}$. In fact:

$$\begin{bmatrix} I_m & -\alpha\Sigma \\ \alpha\Sigma & I_m \end{bmatrix}^{-1} = \begin{bmatrix} X & Y \\ -Y & X \end{bmatrix}$$

where

$$X = (I_m + \alpha^2\Sigma^2)^{-1}$$

$$Y = \alpha\Sigma(I_m + \alpha^2\Sigma^2)^{-1}$$

and, explicitly

$$X = \mathrm{diag}\left((1 + \alpha\sigma_1^2)^{-1}, (1 + \alpha\sigma_2^2)^{-1}, \ldots, (1 + \alpha\sigma_m^2)^{-1}\right)$$

$$Y = \mathrm{diag}\left(\alpha\sigma_1(1 + \alpha\sigma_1^2)^{-1}, \alpha\sigma_2(1 + \alpha\sigma_2^2)^{-1}, \ldots, \alpha\sigma_m(1 + \alpha\sigma_m^2)^{-1}\right).$$

Decomposing the vectors $\boldsymbol{u}_\alpha$ and $\boldsymbol{c}_\alpha$ as

$$\boldsymbol{u}_\alpha = \begin{bmatrix} \boldsymbol{u}_1 \\ \boldsymbol{u}_2 \end{bmatrix}, \qquad \boldsymbol{c}_\alpha = \begin{bmatrix} \boldsymbol{c}_1 \\ \boldsymbol{c}_2 \end{bmatrix}$$

with $\boldsymbol{u}_1, \boldsymbol{u}_2, \boldsymbol{c}_1, \boldsymbol{c}_2 \in \mathbb{R}^m$, the solution of system (7) is

$$\boldsymbol{u}_1 = (I_m + \alpha^2\Sigma^2)^{-1}(\boldsymbol{c}_1 + \alpha\Sigma\boldsymbol{c}_2)$$

$$\boldsymbol{u}_2 = (I_m + \alpha^2\Sigma^2)^{-1}(\boldsymbol{c}_2 - \alpha\Sigma\boldsymbol{c}_1).$$

3 Some Remarks on the Computational Cost

In this section we consider the computational cost of the algorithm described in the previous section for the solution of a single linear system and the cost to solve a different linear system changing just the vector $\boldsymbol{b}_\alpha$ and the parameter α. We suppose that matrix A has a dense structure, that is the worst case from the computational point of view. It is obvious that this process cannot be used when we have to solve a small number of linear systems since it involves the computation of the Schur decomposition of a matrix of order n and of all the factor of the singular value decomposition of a bidiagonal matrix of order $n/2$. It is well-known (see [6]) that the computational cost of the Schur decomposition of A is

$$2n^3 + 3n^2 + 7n$$

plus $n-1$ square roots, while the cost of the SVD decomposition of a bidiagonal matrix of order m is

$$12m^2 + 30m + 2m \text{ square roots,}$$

and, in this particular case is $3n^2 + 15n$ plus n square roots. In the following we shall neglect the number of square roots. In Table 1 we recall the number of operation required by the decompositions we need to compute just once. In Table 2 we recall the computational costs required by the operations that we need to perform at each step (i.e. when the value of parameter α changes).

Table 1. Computational costs for the different steps of the algorithm

Step	Flops
Schur Decomposition	$2n^3 + 3n^2 + 7n$
SVD Decomposition	$3n^2 + 15n$
Total	$2n^3 + 6n^2 + 22n$

We observe that the lower cost of the solution of the linear system is due to the special structure of the coefficient matrix. If we have to solve K_α linear systems (with K_α different values of α) the global computational cost is

$$N_1 = 2n^3+6n^2+22n+K_\alpha\left(5n^2+4n\right) = 2n^3+(6+5K_\alpha)n^2+2(11+2K_\alpha)n. \quad (8)$$

Now we compare the computational with a similar algorithm, which starts considering the same Schur decomposition of A but solves, for a fixed value of α a tridiagonal non-symmetric system. In detail we consider the decomposition

$$I + \alpha A = Q(I + \alpha H)Q^\top$$

Table 2. Computational costs for each step the algorithm (α fixed)

Step	Flops
Step 1 (Matrix-vector product)	$2n^2$
Step 2 (Matrix-vector product)	$n^2/2$
Step 3 (Solution of the linear system)	$4n$
Step 2 (Matrix-vector products)	$5n^2/2$
Total	$5n^2 + 4n$

hence the single linear system becomes

$$Q(I + \alpha H)Q^\top \boldsymbol{x}_\alpha = \boldsymbol{b}_\alpha.$$

Once this decomposition has been computed the following steps need to be carried on:

1. Compute $\boldsymbol{y}_\alpha = Q^\top \boldsymbol{b}_\alpha$;
2. Compute the LU decomposition of the tridiagonal matrix $I + \alpha H$;
3. Solve the linear system $(I + \alpha H)\boldsymbol{u}_\alpha = \boldsymbol{y}_\alpha$;
4. Compute $\boldsymbol{x}_\alpha = Q\boldsymbol{u}_\alpha$.

Also for this algorithm we recall in Table 3 the computational costs for the single steps.

Table 3. Computational cost for the second algorithm

Step	Flops
Schur decomposition of A	$2n^3 + 3n^2 + 7n$
Step 1 (Matrix-vector product)	$2n^2$
Step 2-3 (LU decomposition + linear solver)	$11n$
Step 4 (Matrix-vector products)	$2n^2$

The total computational cost for K_α linear systems is

$$N_2 = 2n^3 + 3n^2 + 7n + K_\alpha \left(4n^2 + 11n\right) = 2n^3 + (3 + 4K_\alpha)n^2 + (7 + 11K_\alpha)n. \quad (9)$$

Both costs have the same coefficient for n^3 hence it is important to analyze the coefficients for n^2 and n. Comparing (8) and (9) and in particulat their dependence on the number K_α of systems that must be solved we observe that the computational cost N_1 of the first algorithm, for large K_α, will be greater than N_2. Just for small values of n, i.e. when the linear coefficient for n, N_1 is smaller than N_2. In order to underline this phenomenon in Figure 1 we have sketched the number of flops required by the two algorithms (solid line for the

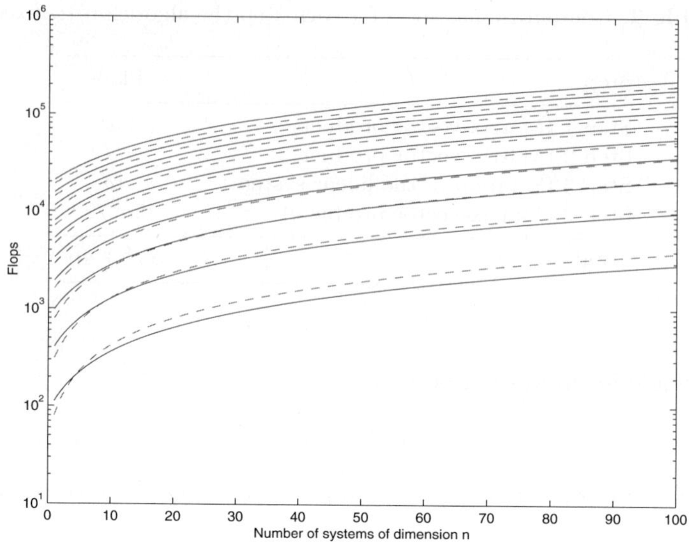

Fig. 1. Computational costs for considered algorithms

first, dashed line for the second), taking different value for $n = 2, 4, \ldots, 20$, and K_α.

4 Applications

In this section we describe just an efficient application of the algorithm described in Section 2. Considering the Kortvieg-de Vrijes partial differential equation

$$u_t = -uu_x - \delta^2 u_{xxx},$$

with periodic boundary conditions $u(0,t) = u(L,t)$, where L is the period and δ is a small parameter. As shown in [4], appropriate methods of space discretization lead to solve a set of ODEs of the form

$$y' = A(y)y, \qquad y(0) = y_0,$$

evolving on the sphere of radius $\|y_0\|$, where $y(t) = (u_0(t), u_1(t), \ldots, u_{N-1}(t))^\top$, $u_i(t) \approx u(i\Delta x, t)$ for $i = 0, 1, \ldots, N-1$, and where $\Delta x = \frac{2}{N}$ is the spatial step of $[0, 2]$. For instance if we consider the space discretization method in [8] we have

$$A(y) = -\frac{1}{6\Delta x} g(y) - \frac{\delta^2}{2\Delta x^3} P \tag{10}$$

where both $g(y)$ and P are two $N \times N$ skew symmetric matrices given by:

$$[g(y)]_{i,j} = \begin{cases} u_{i-1} + u_i & \text{if } j = i+1 \\ -(u_0 + u_{N-1}) & \text{if } i = 1, j = N \\ -(u_{j-1} + u_j) & \text{if } i = j+1 \\ u_0 + u_{N-1} & \text{if } i = N, j = 1 \\ 0 & \text{otherwise.} \end{cases}$$

and

$$P = \begin{bmatrix} 0 & -2 & 1 & 0 & \dots & 0 & -1 & 2 \\ 2 & 0 & -2 & 1 & \ddots & & 0 & -1 \\ -1 & 2 & 0 & -2 & \ddots & \ddots & & 0 \\ 0 & -1 & 2 & 0 & \ddots & \ddots & \ddots & \vdots \\ \vdots & \ddots & \ddots & \ddots & \ddots & \ddots & 1 & 0 \\ 0 & & \ddots & \ddots & 2 & 0 & -2 & 1 \\ 1 & 0 & & \ddots & -1 & 2 & 0 & -2 \\ -2 & 1 & 0 & \dots & 0 & -1 & 2 & 0 \end{bmatrix}.$$

Applying the Backward Euler Method a system like (1) must be solved at each step, especially the paramter δ is time dependent and a splitting technique for (10) is used using an explicit method for $g(y)$ and an implicit one for P.

References

1. Bai Z., Golub G., Ng M.K.: Hermitian and skew-Hermitian splitting methods for non-Hermitian positive definite linear systems. SIAM J. Matr. Anal. **24** (3) (2003) 603–626
2. Benzi M., Bertaccini D.: Approximate inverse preconditioning for shifted linear systems. BIT **43** (2003) 231–244
3. Bertaccini D.: Efficient solvers for sequences of complex symmetric linear systems. ETNA **18** (2004) 49–64
4. Chen J.B., Munthe-Kaas H., Qin M.Z.: Square-conservative schemes for a class of evolution equations using Lie group methods. SIAM J. Num. Anal. **39** (6) (2002) 2164–2178
5. Del Buono N., Lopez L., Peluso R.: Computation of the exponential of large sparse [illegible]
6. Golub G.H., Van Loan C.F.: Matrix Computation. The John Hopkins Univ. Press, Baltimore, (1996)
7. Saad Y.: Analysis of some Krylov subspace approximation to the matrix exponential operator, SIAM J. Numer. Anal. **29** (1) (1992) 209–228
8. Zabusky N.J., Kruskal M.D.: Interaction of solitons in a collisionless plasma and the recurrence of initial states. Phys. Rev. Lett. **15** (1965) 240–243

Search Based Software Engineering

Mark Harman

King's College London, Strand, London, WC2R 2LS

Abstract. This paper was written to accompany the author's keynote talk for the Workshop on Computational Science in Software Engineering held in conjunction with International Conference in Computational Science 2006 in Reading, UK. The paper explains how software engineering activities can be viewed as a search for solutions that balance many competing constraints to achieve an optimal or near optimal result.

The aim of Search Based Software Engineering (SBSE) research is to move software engineering problems from human-based search to machine-based search, using a variety of techniques from the metaheuristic search, operations research and evolutionary computation paradigms. As a result, human effort moves up the abstraction chain to focus on guiding the automated search, rather than performing it. The paper briefly describes the search based approach, providing pointers to the literature.

1 Introduction

Software engineers often face problems associated with the balancing of competing constraints, trade-offs between concerns and requirement imprecision. Perfect solutions are often either impossible or impractical and the nature of the problems often makes the definition of analytical algorithms problematic.

Like other engineering disciplines, Software Engineering is typically concerned with near optimal solutions or those which fall within a specified acceptable tolerance. It is precisely these factors that make robust metaheuristic search-based optimization techniques readily applicable [33].

The growing international Search Based Software Engineering community has shown that search-based solutions using metaheuristic search techniques can be applied to software engineering problems right through the development life-cycle. For example, work has shown the applicability of search-based approaches to the 'next release' problem (requirements engineering) [5], project cost estimation [2, 10, 13, 14, 45], testing [7, 9, 16, 23, 24, 25, 58, 59], automated remodularisation (software maintenance) [29, 51], transformation [17, 18, 19, 6, 31] and studies of software evolution [8].

In exploring these applications, a range of search-based techniques have been deployed, from local search (for example [45, 51]) to genetic algorithms (for example [7, 29, 58, 59]) and genetic programming (for example [8, 14, 13, 16]). Techniques are also being developed to support search-based software testing by transforming software to assess [24] and improve [30] its evolutionary testability.

V.N. Alexandrov et al. (Eds.): ICCS 2006, Part IV, LNCS 3994, pp. 740–747, 2006.
© Springer-Verlag Berlin Heidelberg 2006

2 What Is Search Based Software Engineering?

Search Based Software Engineering, as its name implies, treats software engineering problems as search problems, and seeks to use search techniques in order to solve the problems. Key to the approach is the re-formulation of a software engineering problem as a search problem [11, 27]. The term Search Based Software Engineering was coined in 2001 [33], since which time there has been a rapidly developing community working on this area with its own conferences and journal special issues. However, there was significant work on the application of search techniques to problems in software testing [20, 41, 42, 52, 55, 56, 60] and restructuring [15, 48] before the term 'Search Based Software Engineering' was coined to encompass the wider application of search to software engineering as a whole.

The search techniques used are a set of generic algorithms taken from the fields of metaheuristic search, operations research and evolutionary computation. These algorithms are concerned with searching for optimal or near optimal solutions to a problem within a large (possibly) multi-modal search space [21, 22, 40, 57].

For such problems, it is often infeasible to apply a precise analytic algorithm that produces the 'best' solution to the problem, yet it is possible to determine which is the better of two candidate solutions. Search techniques have been applied successfully to a number of engineering problems ranging from load balancing in the process industries (pressing of sugar pulp), through electromagnetic system design, to aircraft control and aerodynamics [61]. Search Based Software Engineering simply represents the application of these search algorithms to software engineering problems and the investigation of the implications of this novel application area.

Harman and Clark [27] identify four important properties in order for the Search Based Software Engineering approach to be successful:

1. **Large search space**
   If the fitness function is only used to distinguish a *few* individuals from one another, then the value of the fitness function for each individual can be computed and the search space explored exhaustively. There would be no need to use a search-based technique to sample the search space. Of course, most search spaces are *very* large. That is, most fitness functions apply to large (conceptually infinite) search spaces, such as the space of all expressible programs in some language or the space of all expressible designs in some design notation.
2. **Low computational complexity**
   Search based algorithms sample a portion of a very large search space. The portion sampled is typically non-trivial, requiring many thousands (possibly hundreds of thousands) of fitness evaluations. Therefore the computational complexity of the fitness function has a critical impact on the overall complexity of the search process. Fortunately, most fitness functions are relatively cheap to compute, since they can be constructed in terms of the structural or syntactic properties of the programs, designs and systems which they assess and computed in time linear in the size of the program design or system.

3. **Approximate continuity**
   It is not necessary for a function to be continuous to be useful as a fitness function, but too much discontinuity can mislead a search, because all search–based optimisation approaches rely upon the guidance given by the fitness function. Continuity ensures that this guidance is perfect; the less continuous is the fitness function, the less guidance it gives.
4. **Absence of known optimal solutions**
   If there is a known optimal solution to a problem, then clearly there is no need to use a search-based approach to seek optimal (or near optimal) solutions.

Fortunately, these four problem characteristics are very prevalent in software engineering, where problems typically do involve a large search space (such as the number of possible designs, test cases or system configurations that may exist). Also, in many situations, there is no known optimal solution to the problem. The properties of 'low computational complexity' and 'approximate continuity' may not be present in all cases. However, even in cases where they are absent, it may be possible to transform the problem into one that is more amenable to Search Based Software Engineering [6, 31].

Interest in Search Based Software Engineering has grown rapidly in the past five years. For example, the work on search based testing is now sufficiently developed to merit its own survey paper [50], while there has been a healthy and growing Search Based Software Engineering track of the Genetic and Evolutionary Computation Conference GECCO, since 2002 and special issues and workshops on Search Based Software Engineering [26, 34].

3 Search Based Software Engineering Can Yield Fresh Insight

It has been widely observed that search techniques are good at producing unexpected answers. This happens because the techniques are not hindered by implicit human assumptions. One example is the discovery of a patented digital filter using a novel evolutionary approach [54]. Another example is the discovery of patented antenna designs [46] which are available commercially. The human formalises their (explicit) assumptions as a fitness function. Many of these are already available in the form of software metrics [27]. The machine uses this fitness function to guide the search. Should the search produce unexpected results then this reveals some **implicit** assumptions and/or challenges the human's intuition about the problem.

Unlike human–based search, automated search techniques carry with them no bias. They automatically scour the search space for the solutions that best fit the (stated) human assumptions in the fitness function. This is one of the central strengths of the approach. Software engineering is often hampered by poor human intuition and the presence of unstated and implicit assumptions. Automated search techniques will effectively work in tandem with the human, in an iterative process of refinement, leading to better fitness functions and, thereby, to better encapsulation of human assumptions and intuition.

Insight can also come from visualization of the landscape [39, 43, 44, 53]. That is, to use the fitness function values as a measure of height (or vertical co-ordinate), in a landscape where each individual in the search space potentially occupies some location within the horizontal co-ordinates.

Harman and Clark [27] describe other ways in which the SBSE approach can provide insight in the field of software metrics research, by providing a way to understand software metrics as fitness functions and to consider the effect of the metrics in terms of the optimizations that they produce when used as fitness functions.

4 Conclusion

Software engineering is essentially a search for a solution that balances many competing constraints to achieve an optimal or near optimal result. Currently, this search process is a highly labour–intensive human activity. It cannot scale to meet the demands of the new and emerging software engineering paradigms. Search Based Software Engineering addresses this problem head on, moving software engineering problems from human–based search to machine–based search. As a result, human effort will move up the abstraction chain, to focus on **guiding** the automated search, rather than **performing** the search itself.

Acknowledgements

This keynote arose as a result of recent work [1,3,4,6,11,12,17,18,19,27,28,29,30, 31,32,33,35,47,49] undertaken by the author with many other colleagues in the growing Search Based Software Engineering community. The work is currently funded by a large EPSRC project, SEBASE, for which the other partners are John Clark (University of York) and Xin Yao (University of Birmingham) and industrialists from DaimlerChrysler Berlin, Motorola UK and IBM UK. This keynote draws on ideas from the SEBASE project and from other keynotes and tutorials prepared by the author in collaboration with Joachim Wegener at DaimlerChrysler [38, 37, 36].

References

1. Konstantinos Adamopoulos, Mark Harman, and Robert Mark Hierons. Mutation testing using genetic algorithms: A co-evolution approach. In *Genetic and Evolutionary Computation Conference (GECCO 2004), LNCS 3103*, pages 1338–1349, Seattle, Washington, USA, June 2004. Springer.
2. Jesús Aguilar-Ruiz, Isabel Ramos, José C. Riquelme, and Miguel Toro. An evolutionary approach to estimating software development projects. *Information and Software Technology*, 43(14):875–882, December 2001.

3. Giulio Antoniol, Massimiliano Di Penta, and Mark Harman. A robust search–based approach to project management in the presence of abandonment, rework, error and uncertainty. In 10th *International Software Metrics Symposium (METRICS 2004)*, pages 172–183, Chicago, Illinois, USA, September 2004. IEEE Computer Society Press, Los Alamitos, California, USA.
4. Giulio Antoniol, Massimiliano Di Penta, and Mark Harman. Search-based techniques applied to optimization of project planning for a massive maintenance project. In 21st *IEEE International Conference on Software Maintenance (ICSM 2005)*, pages 240–249, Budapest, Hungary, 2005. IEEE Computer Society Press, Los Alamitos, California, USA.
5. A.J. Bagnall, V.J. Rayward-Smith, and I.M. Whittley. The next release problem. *Information and Software Technology*, 43(14):883–890, December 2001.
6. André Baresel, David Wendell Binkley, Mark Harman, and Bogdan Korel. Evolutionary testing in the presence of loop–assigned flags: A testability transformation approach. In *International Symposium on Software Testing and Analysis (ISSTA 2004)*, pages 108–118, Omni Parker House Hotel, Boston, Massachusetts, July 2004. Appears in Software Engineering Notes, Volume 29, Number 4.
7. André Baresel, Harmen Sthamer, and Michael Schmidt. Fitness function design to improve evolutionary structural testing. In *GECCO 2002: Proceedings of the Genetic and Evolutionary Computation Conference*, pages 1329–1336, New York, 9-13 July 2002. Morgan Kaufmann Publishers.
8. Terry Van Belle and David H. Ackley. Code factoring and the evolution of evolvability. In *GECCO 2002: Proceedings of the Genetic and Evolutionary Computation Conference*, pages 1383–1390, New York, 9-13 July 2002. Morgan Kaufmann Publishers.
9. Leonardo Bottaci. Instrumenting programs with flag variables for test data search by genetic algorithms. In *GECCO 2002: Proceedings of the Genetic and Evolutionary Computation Conference*, pages 1337–1342, New York, 9-13 July 2002. Morgan Kaufmann Publishers.
10. Colin J. Burgess and Martin Lefley. Can genetic programming improve software effort estimation? A comparative evaluation. *Information and Software Technology*, 43(14):863–873, December 2001.
11. John Clark, José Javier Dolado, Mark Harman, Robert Mark Hierons, Bryan Jones, Mary Lumkin, Brian Mitchell, Spiros Mancoridis, Kearton Rees, Marc Roper, and Martin Shepperd. Reformulating software engineering as a search problem. *IEE Proceedings — Software*, 150(3):161–175, 2003.
12. Karnig Derderian, Qiang Quo, Mark Harman, and Robert Hierons. Computing unique input/output sequences using genetic algorithms. In *3rd International Workshop on Formal Approaches to Testing of Software (FATES 2003)*, pages 164–177, Montréal, Canada, 2003. LNCS 2931.
13. Jose J. Dolado. On the problem of the software cost function. *Information and Software Technology*, 43(1):61–72, 1 January 2001.
14. José Javier Dolado. A validation of the component-based method for software size estimation. *IEEE Transactions on Software Engineering*, 26(10):1006–1021, 2000.
15. D. Doval, S. Mancoridis, and B. S. Mitchell. Automatic clustering of software systems using a genetic algorithm. In *International Conference on Software Tools and Engineering Practice (STEP'99)*, Pittsburgh, PA, 30 August - 2 September 1999.

16. Maria Cláudia Figueiredo Pereira Emer and Silva Regina Vergilio. GPTesT: A testing tool based on genetic programming. In *GECCO 2002: Proceedings of the Genetic and Evolutionary Computation Conference*, pages 1343–1350, New York, 9-13 July 2002. Morgan Kaufmann Publishers.
17. Deji Fatiregun, Mark Harman, and Rob Hierons. Evolving transformation sequences using genetic algorithms. In 4th *International Workshop on Source Code Analysis and Manipulation (SCAM 04)*, pages 65–74, Chicago, Illinois, USA, September 2004. IEEE Computer Society Press, Los Alamitos, California, USA.
18. Deji Fatiregun, Mark Harman, and Rob Hierons. Search-based amorphous slicing. In 12th *International Working Conference on Reverse Engineering (WCRE 05)*, Carnegie Mellon University, Pittsburgh, Pennsylvania, USA, November 2005. To appear.
19. Deji Fatiregun, Mark Harman, and Robert Hierons. Search based transformations. In *Genetic and Evolutionary Computation – GECCO-2003*, volume 2724 of *LNCS*, pages 2511–2512, Chicago, 12-16 July 2003. Springer-Verlag.
20. Roger Ferguson and Bogdan Korel. The chaining approach for software test data generation. *ACM Transactions on Software Engineering and Methodology*, 5(1):63–86, January 1996.
21. F. Glover. Tabu search: A tutorial. *Interfaces*, 20:74–94, 1990.
22. David E. Goldberg. *Genetic Algorithms in Search, Optimization & Machine Learning*. Addison-Wesley, Reading, MA, 1989.
23. Hans G. Groß, Bryan F Jones, and David E Eyres. Structural performance measure of evolutionary testing applied to worst-case timing of real-time systems. *IEE Proceedings Software*, (2):25–30, 2000.
24. Hans-Gerhard Groß. A prediction system for evolutionary testability applied to dynamic execution time. *Information and Software Technology*, 43(14):855–862, December 2001.
25. Hans-Gerhard Groß and Nikolas Mayer. Evolutionary testing in component-based real-time system construction. In *GECCO 2002: Proceedings of the Genetic and Evolutionary Computation Conference*, page 1393, New York, 9-13 July 2002. Morgan Kaufmann Publishers.
26. Walter Gutjahr and Mark Harman. Focussed issue on search based software engineering. *Journal Computers and Operations Research*, 2006. To appear.
27. Mark Harman and John Clark. Metrics are fitness functions too. In 10th *International Software Metrics Symposium (METRICS 2004)*, pages 58–69, Chicago, Illinois, USA, September 2004. IEEE Computer Society Press, Los Alamitos, California, USA.
28. Mark Harman, Chris Fox, Robert Mark Hierons, Lin Hu, Sebastian Danicic, and Joachim Wegener. Vada: A transformation-based system for variable dependence analysis. In *IEEE International Workshop on Source Code Analysis and Manipulation (SCAM 2002)*, pages 55–64, Montreal, Canada, October 2002. IEEE Computer Society Press, Los Alamitos, California, USA. Voted best paper by attendees.
29. Mark Harman, Robert Hierons, and Mark Proctor. A new representation and crossover operator for search-based optimization of software modularization. In *GECCO 2002: Proceedings of the Genetic and Evolutionary Computation Conference*, pages 1351–1358, New York, 9-13 July 2002. Morgan Kaufmann Publishers.
30. Mark Harman, Lin Hu, Robert Hierons, André Baresel, and Harmen Sthamer. Improving evolutionary testing by flag removal ('best at GECCO' award winner). In *GECCO 2002: Proceedings of the Genetic and Evolutionary Computation Conference*, pages 1359–1366, New York, 9-13 July 2002. Morgan Kaufmann Publishers.

31. Mark Harman, Lin Hu, Robert Mark Hierons, Joachim Wegener, Harmen Sthamer, André Baresel, and Marc Roper. Testability transformation. *IEEE Transactions on Software Engineering*, 30(1):3–16, January 2004.
32. Mark Harman and Bryan Jones. SEMINAL: Software engineering using metaheuristic innovative algorithms. In 23rd *International Confernece on Software Engineering (ICSE 2001)*, pages 762–763, Toronto, Canada, May 2001. IEEE Computer Society Press, Los Alamitos, California, USA.
33. Mark Harman and Bryan F. Jones. Search based software engineering. *Information and Software Technology*, 43(14):833–839, December 2001.
34. Mark Harman and Bryan F. Jones. The seminal workshop: Reformulating software engineering as a metaheuristic search problem. *Software Engineering Notes*, 26(6):62–66, November 2001.
35. Mark Harman, Stephen Swift, and Kiarash Mahdavi. An empirical study of the robustness of two module clustering fitness functions. In *Genetic and Evolutionary Computation Conference (GECCO 2005)*, Washington DC, USA, June 2005. Association for Computer Machinery. to appear.
36. Mark Harman and Joachim Wegener. Evolutionary testing. In *Genetic and Evolutionary Computation (GECCO)*, Chicago, July 2003.
37. Mark Harman and Joachim Wegener. Getting results with search–based software engineering. In 26th *IEEE International Conference and Software Engineering (ICSE 2004)*, Edinburgh, UK, 2004. IEEE Computer Society Press, Los Alamitos, California, USA. To Appear.
38. Mark Harman and Joachim Wegener. Search based testing. In 6th *Metaheuristics International Conference (MIC 2005)*, Vienna, Austria, August 2005. To appear.
39. E. Hart and P. Ross. GAVEL - a new tool for genetic algorithm visualization. *IEEE-EC*, 5:335–348, August 2001.
40. John H. Holland. *Adaption in Natural and Artificial Systems.* MIT Press, Ann Arbor, 1975.
41. B.F. Jones, H.-H. Sthamer, and D.E. Eyres. Automatic structural testing using genetic algorithms. *The Software Engineering Journal*, 11:299–306, 1996.
42. Bryan F. Jones, David E. Eyres, and Harmen H. Sthamer. A strategy for using genetic algorithms to automate branch and fault-based testing. *The Computer Journal*, 41(2):98–107, 1998.
43. Yong-Hyuk Kim and Byung-Ro Moon. Visualization of the fitness landscape, A steady-state genetic search, and schema traces. In *GECCO 2002: Proceedings of the Genetic and Evolutionary Computation Conference*, page 686, New York, 9-13 July 2002. Morgan Kaufmann Publishers.
44. Yong-Hyuk Kim and Byung-Ro Moon. New usage of sammon's mapping for genetic visualization. In *Genetic and Evolutionary Computation – GECCO-2003*, volume 2723 of *LNCS*, pages 1136–1147, Chicago, 12-16 July 2003. Springer-Verlag.
45. Colin Kirsopp, Martin Shepperd, and John Hart. Search heuristics, case-based reasoning and software project effort prediction. In *GECCO 2002: Proceedings of the Genetic and Evolutionary Computation Conference*, pages 1367–1374, New York, 9-13 July 2002. Morgan Kaufmann Publishers.
46. D. S. Linden. Innovative antenna design using genetic algorithms. In D. W. Corne and P. J. Bentley, editors, *Creative Evolutionary Systems*, chapter 20. Elsevier, Amsterdam, The Netherland, 2002.
47. Kiarash Mahdavi, Mark Harman, and Robert Mark Hierons. A multiple hill climbing approach to software module clustering. In *IEEE International Conference on Software Maintenance (ICSM 2003)*, pages 315–324, Amsterdam, Netherlands, September 2003. IEEE Computer Society Press, Los Alamitos, California, USA.

48. Spiros Mancoridis, Brian S. Mitchell, C. Rorres, Yih-Farn Chen, and Emden R. Gansner. Using automatic clustering to produce high-level system organizations of source code. In *International Workshop on Program Comprehension (IWPC'98)*, pages 45–53, Ischia, Italy, 1998. IEEE Computer Society Press, Los Alamitos, California, USA.
49. Phil McMinn, David Binkley, and Mark Harman. Testability transformation for efficient automated test data search in the presence of nesting. In *UK Software Testing Workshop (UK Test 2005)*, Sheffield, UK, September 2005.
50. Philip McMinn. Search-based software test data generation: A survey. *Software Testing, Verification and Reliability*, 14(2):105–156, June 2004.
51. Brian S. Mitchell and Spiros Mancoridis. Using heuristic search techniques to extract design abstractions from source code. In *GECCO 2002: Proceedings of the Genetic and Evolutionary Computation Conference*, pages 1375–1382, New York, 9-13 July 2002. Morgan Kaufmann Publishers.
52. F. Mueller and J. Wegener. A comparison of static analysis and evolutionary testing for the verification of timing constraints. In *4th IEEE Real-Time Technology and Applications Symposium (RTAS '98)*, pages 144–154, Washington - Brussels - Tokyo, June 1998. IEEE.
53. Hartmut Pohlheim. Visualization of evolutionary algorithms - set of standard techniques and multidimensional visualization. In Wolfgang Banzhaf, Jason Daida, Agoston E. Eiben, Max H. Garzon, Vasant Honavar, Mark Jakiela, and Robert E. Smith, editors, *Proceedings of the Genetic and Evolutionary Computation Conference*, volume 1, pages 533–540, Orlando, Florida, USA, 13-17 July 1999. Morgan Kaufmann.
54. T. Schnier, X. Yao, and P. Liu. Digital filter design using multiple pareto fronts. *Soft Computing*, 8(5):332–343, April 2004.
55. N. Tracey, J. Clark, and K. Mander. Automated program flaw finding using simulated annealing. In *International Symposium on Software Testing and Analysis*, pages 73–81. ACM/SIGSOFT, March 1998.
56. Nigel Tracey, John Clark, Keith Mander, and John McDermid. Automated test-data generation for exception conditions. *Software Practice and Experience*, 30(1):61–79, 2000.
57. P. J. M. van Laarhoven and E. H. L. Aarts. *Simulated Annealing: Theory and Practice.* Kluwer Academic Publishers, Dordrecht, the Netherlands, 1987.
58. Joachim Wegener, André Baresel, and Harmen Sthamer. Evolutionary test environment for automatic structural testing. *Information and Software Technology Special Issue on Software Engineering using Metaheuristic Innovative Algorithms*, 43(14):841–854, 2001.
59. Joachim Wegener and F. Mueller. A comparison of static analysis and evolutionary testing for the verification of timing constraints. *Real-Time Systems*, 21(3):241–268, 2001.
60. Joachim Wegener, Harmen Sthamer, Bryan F. Jones, and David E. Eyres. Testing real-time systems using genetic algorithms. *Software Quality*, 6:127–135, 1997.
61. G Winter, J Periaux, M Galan, and P Cuesta. *Genetic Algorithms in Engineering and Computer Science.* Wiley, 1995.

Modular Monadic Slicing in the Presence of Pointers*

Zhongqiang Wu[1], Yingzhou Zhang[2], and Baowen Xu[1]

[1] Dep. of Computer Science and Engineering, Southeast Univ., Nanjing 210096, China
[2] College of Computer, Nanjing Univ. of Posts and Telecommunications, Nanjing 210003, China
bwxu@seu.edu.cn, zhangyz@njupt.edu.cn

Abstract. Program slicing is a family of program decomposition techniques. For traditional slicing methods lack modularity and flexibility, we have proposed a new formal method for program slicing—modular monadic slicing. This paper presents an approach to extend the modular monadic slicing for handling pointers. With the inspiration of forward program slicing, our approach obtains the point-to information through the data-flow iteration. By combining the forward monad slicing with data-flow iteration, our method computes the point-to information and slices the program in the same phase. So our approach has the same precision as traditional data-flow iteration methods, but needs less space. In addition, our approach also inherits the properties of language independence and reusability from the original monadic slicing method.

1 Introduction

Program slicing is a family of program decomposition techniques proposed by M. Weiser [1], which is widely applied to program debugging, program integration, reverse engineering and so on [2]. For the traditional slicing methods lack modularity, we have proposed a novel formal slicing method, called modular monadic slicing [4]. This paper extends the original monadic slicing method to handle the pointers. With the inspiration of forward program slicing [3], our approach obtains the point-to information through the data-flow iteration that can make the point-to information and slicing be computed in the same phase. As a result, instead of recording point-to information for every statement, we only need to record the information for several current statements. So our method saves space without losing precision.

The rest of the paper is organized as follows: In Section 2, we briefly introduce the concepts of monad and monad transformers. We discuss original modular monadic slicing algorithm in Section 3. The extended slicing algorithm is presented in more detail in section 4 and 5, including the chief difficulty in dealing with pointers, the extended algorithm, and complexity analysis. A case study is shown in sections 6. In Section 7, we conclude this paper.

---

* This work was supported in part by the NSFC (60425206, 60373066, 90412003), National Grand Fundamental Research 973 Program of China (2002CB312000), Advanced Armament Research Project (51406020105JB8103), and Natural Science Research Plan for Jiang Su High School (05KJD520151).

V.N. Alexandrov et al. (Eds.): ICCS 2006, Part IV, LNCS 3994, pp. 748 – 756, 2006.
© Springer-Verlag Berlin Heidelberg 2006

2 Preliminaries

A monad is a triple (m, $return_m$, $bind_m$) which must satisfy three laws: Left unit, Right unit and Associativity [7], where m is a type constructor; $return_m$ and $bind_m$ are two primitive operators. A monad transformer consists of a type constructor t and an associated function $lift_t$, where t maps any given monad (m, $return_m$, $bind_m$) to a new monad ($t\ m$, $return_{t\ m}$, $bind_{t\ m}$). In [4], we have presented the slice monad transformer SliceT as shown below (where L denotes a set of labels of expressions that were required to compute the current expression), to uniformly describe the slicing computation.

> Abstract Syntax:
> S :: = ide := l.e | S_1; S_2 | skip | **read** ide | **write** l.e
> | **if** l.e **then** S_1 **else** S_2 **endif** | **while** l.e **do** S **endwhile**
>
> **Fig. 1.** Abstract syntax of language W

type SliceT $L\ m\ a = L \rightarrow m\ (a, L)$
$return_{SliceT\ L\ m}\ x\ = \lambda L.\ return_m\ (x, L)$
$m\ \text{‘}bind_{SliceT\ L}\ m\text{’}\ f\ = \lambda L.\ \{(a, L') \leftarrow m\ L\ ; f\ a\ L'\}_m$
$lift_{SliceT\ L}\ m\ = \lambda L.\ \{a \leftarrow m\ ;\ return_m\ (a, L)\}_m$
updateSlice f $= \lambda$L. return_m (f L, L)

In modular monadic semantics, the monad definition is a combination of some monad transformers that are applied to a base monad. In this paper, we use the identity monad Id as the base monad. We apply some monad transformers, such as EnvT, StateT [5], to the monad Id, and obtain the resulting monad ComptM: ComptM ≡ (StateT · EnvT) Id. With the ComptM, the semantics functions are shown below:

$\boldsymbol{C}$ :: Cmd → ComptM (), $\boldsymbol{E}$ :: Exp → ComptM Value

For the convenience of discussion, in [4], we consider an example language **W**, and define *Syn*(s, *L*) where s is a **W**-program analyzed. The language **W** contains assignment, branch statement, loop block and I/O statement, its abstract syntax is provided in Figure 1.

3 Modular Monadic Static Slicing Algorithm

In [4], the data structure of slices was shown as follows:

type Var = String **type** Labels = [Int] **type** Slices = [(Var, Labels)]
getSli :: ComptM Slices, *setSli* :: Slices → ComptM Slices
lkpSli :: Var → Slices → ComptM Labels
updSli :: (Var, ComptM Labels) → Slices → ComptM ()
mrgSli :: Slices → Slices → ComptM Slices

The main idea of static slicing algorithm can be briefly stated as follows: we firstly apply the slice transformer SliceT to semantic building blocks, which makes the resulting semantic description include program slice semantic feature. According to the semantic description, we then compute static slices of each statement in sequence.

Finally we will obtain the static slices of all single variables in the program. The concrete steps are shown in Figure 2.

In the algorithm, the underlying monad ComptM in semantic building blocks is: ComptM ≡ (SliceT · StateT · EnvT) Id. Meanwhile, if a computation of a labeled expression l.e is included during applying $bind_{\text{SliceT}Lm}$, the intermediate set L' should be reified as following:

$$L' = \{l\} \cup L \cup \bigcup_{r \in Refs(l.e)} lkpSli\ (r, getSli\ )$$

where *Refs*(l.e) denotes the set of variables occurred in expression l.e. The relation above reflects when and how to change the set *L*.

Since we finally obtain the slices of all variables after the last statement is analyzed, the program slice of each variable, on the average, costs $O(n + m)$, where *m and n* refer to the number of labeled expressions in the program and the number of all labeled expressions appeared (perhaps repeatedly) in the sequence of analyzing the program, respectively. The total space cost is $O(v \times m)$, which is unrelated to *n*, where *v* refers to the number of single variables. For *n* is no less than *m*, the time complexity is $O(n)$. When considering special loop [2], the worst time complexity can reach $O(m^2)$.

Input: Slicing criterion <p, v>
Output: Static slice

1. Initialize the set *L* and the table Slices.
2. Add semantic feature of program slicing into semantic building blocks in a modular way, through slice transformer SliceT.
3. Compute static slices of each statement in sequence basing on the semantic description in Step 2, obtaining the final Slices.
4. Returning the final static slicing result according to Slices and *Syn*(s, *L*).

Fig. 2. Static slicing algorithm

4 Static Analysis in the Presence of Pointers

In the presence of pointers, program analysis has three problems. Firstly, with the pointers, the unbounded data structures such as linked list can be built, which need to be represented in a finite way. Secondly, we should give a new definition of data dependence to fit the new circumstance. Finally, for a variable may point to a set of locations in the presence of pointers, point-to analysis is needed. The solution to the first problem is helpful to the new data dependence definition and point-to analysis, which is a base. The second problem provides a goal by giving a new definition; and the solution to the third problem helps to implement the goal.

Many researchers have proposed solutions to the first problem [9, 10]. Generally, these methods use some approximation to deal with the unbounded data structures. A sort of rough approximation is to consider the heap as a whole. To be more precise, it needs shape analysis. Chase [10] distinguished the variables in the heap by the pointer variables, and obtained the Storage Shape Graph (SSG) for every statement. This paper uses a method whose precision is between the two methods mentioned above. It considers all the heap space allocated in the same statement as an array, and implements the array as a whole.

The second problem is new definition of data dependence. Because the solution to the first problem gives a finite way to represent the variables in the program, in this case we can define the data dependence in terms of potential definitions and uses of abstract memory locations [9]. In our extended slicing algorithm, this new definition is implemented by point-to analysis, redefining the *Refs*(l.e) and considering the reference through the left-value expression of assignments.

The third problem is point-to analysis. According to whether using control-flow information, there are two classes of point-to analysis—flow-sensitivity and flow-insensitivity. Flow-sensitive analysis considers the control-flow information. It is lower in efficiency and needs more space, but has high precision. However, flow-insensitive analysis suggests that the statements in the program can be executed in any order. It can only obtain the point-to information for the whole program (or for a certain area). For example, [6] obtains some point-to sets of every variable for every function. So the flow-insensitive analysis isn't very precise, but has high efficiency. Especially, with the use of Union-find, its efficiency is nearly linear [8]. With the inspiration of forward program slicing, our approach obtains the point-to information by the data-flow iteration which is a flow-sensitive analysis. After integrating with original slicing algorithm, the point-to relation and slicing are computed in the same phase. Then instead of recording point-to information for every statement, we only need to record the information for some current statements. So our method needs less space compared with the traditional flow-sensitive analysis.

5 Modular Monadic Slicing Algorithm in the Presence of Pointers

5.1 Data Structure

With the pointers, we may need to update the slices of some variables at the same time, so we extend the operator *updSli* of the abstract datatype Slices as follows:

updSli :: [Var] → Labels → Slices → ComptM ()

We also need to design the abstract datatype PT for point-to analysis:

type Var = String **type** PT = [(Var, [Var])]
getPT :: ComptM PT, *setPT* :: PT → ComptM PT
lkpPT :: Var → PT → ComptM [Var], *mrgPT* :: PT → PT → ComptM PT
updPT :: [Var] → [Var] → PT → ComptM ()

The abstract datatype PT is a table of pairs of a single variable and its associated point-to set (a set of variables). It has five operators *getPT*, *setPT*, *lkpPT*, *updPT* and *mrgPT*, which return and set the current table of point-to sets, lookup a point-to set corresponding to a variable in a given table of point-to sets, update some point-to sets corresponding to a list of variables in a given table of point-to sets, and merge two table of point-to sets into one table, respectively.

5.2 The Operations of the Expressions

In the premise of only considering simple dereference (Multi-dereference can be divided into some simple dereferences), the assignments in language **W** can be divided into eight classes, as shown below (the syntax is similar with C language):

(1) x := l.e′ (2) x := l.y (3) x := l.*y (4) x := l.&y
(5) *x := l.e′ (6) *x := l.y (7) *x := l.*y (8) *x := l.&y

And the fourth kind x := l.&y can be used to represent the statements that allocate heap memory (y is considered as an array to address as a whole).

Table 1. The operations of the two classes of left-value expressions

Class of expressions	Variables to update	Updating method	Reference
X	x	Strong update	ϕ
*x	*lkpPT*(x, *getPT*)	Weak update	{x}

Table 2. The operations of the four classes of right-value expressions

Expression	Point-to set	Reference
e′	ϕ	*Refs*(l.e′)
y	*lkpPT*(y, *getPT*)	{y}
*y	$\bigcup_{v \in lkpPT(y, getPT)} lkpPT(v, getPT)$	{y}∪ *lkpPT*(y, *getPT*)
&y	{y}	ϕ

Among the eight classes of assignments, there are two classes of left-value expressions (x and *x) and four classes of right-value expressions (e′, y, *y, &y). The right-value expression e′ represents pure value computation (this paper doesn't consider the pointer arithmetic). For example, we suggest that the expression 'x + 2' won't return an address. And the expression y either returns a value or returns an address. To deal with the two situations in the same way, we associate every variable with a point-to set which initially is null [8]. For the assignments can be divided into left-, and right-value expressions to consider respectively, the operations of the six classes of expressions can be seen as atoms.

Before discussing these atomic operations, we need to redefine *Refs*(l.e). The variables appeared in the expression e can be divided into three classes. The first class includes the reference variables. The second class includes the dereference variables. The third class includes the variables whose address is obtained. *Refs*(l.e) includes the first two classes of variables and the variables that may be referred after derefer the variables in the second class. The formal definition is shown below:

Refs(l.e) = {x | x is in the first class}∪{y | y is in the second class}
∪ {z | z ∈ *lkpPT*(y, *getPT*), y is in the second class}

The extended algorithm also needs to update the information of a variable. The process that uses new information to cover original information is called **strong update**. And the process that adds new information to the origin is called **weak update**.

Now, we show the six atomic operations. The operation for the left-value expression mainly relates to deciding which variable's information need to be updated and which variable is referred by the analyzed expression. As presented in table 1, the solution to the left-value expression x is: strongly updating the slice and point-to set of variable x because we only update one variable; this left-value expression doesn't refer any variables. As presented in table 1, the solution to the left-value expression *x is: weakly updating the information of variables in *lkpPT*(x, *getPT*) because we update a list of variables; this left-value expression refers to variable x.

The operation for the right-value expression mainly relates to deciding which variable is referred by the analyzed expression and obtaining the point-to set. The variables that is referred by the analyzed expression can be obtained by *Refs*(l.e), so we focus on the way to obtain the point-to set. As presented in table 2, the expression e′ doesn't give a point-to set; the expression y gives the point-to set—*lkpPT*(y, *getPT*), which is the point-to set of variable y; the expression *y gives the point-to set—$\bigcup_{v \in lkpPT(y, getPT)} lkpPT(v, getPT)$, which is the union of the point-to sets of the variables in *lkpPT*(y, *getPT*); the expression &y gives a simple point-to set {y}.

5.3 The Solution to the Statements

With the combination of the operations of left- and right- value expression, we can address the assignments. And we pick four classes of operations to show below:

$$(1)\ [\![x := l.{*}y]\!] = \lambda L.\{L' \leftarrow \{l\} \cup L \cup \bigcup_{r \in Refs(l.{*}y)} lkpSli(r, getSli);\ updSli([x], L', getSli);$$
$$V' \leftarrow \bigcup_{v \in lkpPT(y, getPT)} lkpPT(v, getPT);\ updPT([x], V', getPT)\}$$

$$(2)\ [\![x := l.\&y]\!] = \lambda L.\{L' \leftarrow \{l\} \cup L \cup \bigcup_{r \in Refs(l.y)} lkpSli(r, getSli);\ V' \leftarrow [y];$$
$$updSli([x], L', getSli); updPT([x], V', getPT)\}$$

$$(3)\ [\![{*}x := l.e']\!] = \lambda L.\{L' \leftarrow \{l\} \cup L \cup \bigcup_{r \in Refs(l.e') \cup \{x\}} lkpSli(r, getSli);\ T \leftarrow getSli;$$
$$updSli(lkpPT(x, getPT), L', getSli); T' \leftarrow getSli;\ mrgSli(T, T')\}$$

$$(4)\ [\![{*}x := l.y]\!] = \lambda L.\{L' \leftarrow \{l\} \cup L \cup \bigcup_{r \in Refs(l.y) \cup \{x\}} lkpSli(r, getSli);\ T \leftarrow getSli;$$
$$updSli(lkpPT(x, getPT), L', getSli); T' \leftarrow getSli;$$
$$mrgSli(T, T'); V' \leftarrow lkpPT(y, getPT);\ P \leftarrow getPT;$$
$$updPT(lkpPT(x, getPT), V', getPT); P' \leftarrow getPT;\ mrgPT(P, P')\}$$

Because of the similarity of forward slicing and dataflow iteration, we can simply add the point-to analysis to original implementation of the branch statement and loop block as shown below:

$$[\![\textbf{if } \text{l.e}' \textbf{ then } S_1 \textbf{ else } S_2 \textbf{ endif}]\!] = \lambda L.\{L' \leftarrow \{\text{l}\} \cup L \cup \bigcup_{\text{r} \in \mathit{Refs}(\text{l.e}')} \mathit{lkpSli}\ (\text{r}, \mathit{getSli})\ ;$$

$$\text{T} \leftarrow \mathit{getSli};\ \text{P} \leftarrow \mathit{getPT};\ [\![S_1]\!] L'; \text{T1} \leftarrow \mathit{getSli}; \text{P1} \leftarrow \mathit{getPT};\ \mathit{setSli}(\text{T});$$

$$\mathit{setPT}(\text{P}); [\![S_2]\!] L';\ \text{T2} \leftarrow \mathit{getSli}; \text{P2} \leftarrow \mathit{getPT};\ \mathit{mrgSli}(\text{T1},\text{T2}); \mathit{mrgPT}(\text{P1}, \text{P2})\}$$

$$[\![\textbf{while } \text{l.e}' \textbf{ do } S \textbf{ endwhile}]\!] = \textbf{\textit{Fix}}\ (\lambda f.\ \lambda L.\{L' \leftarrow \{\text{l}\} \cup L \cup \bigcup_{\text{r} \in \mathit{Refs}(\text{l.e}')} \mathit{lkpSli}\ (\text{r}, \mathit{getSli})\ ;$$

$$\text{T} \leftarrow \mathit{getSli};\ \text{P} \leftarrow \mathit{getPT}; fL'\{ [\![S]\!] L'; \text{T}' \leftarrow \mathit{getSli}; \text{P}' \leftarrow \mathit{getPT};$$

$$\mathit{mrgSli}(\text{T},\text{T}');\ \mathit{mrgPT}(\text{P}, \text{P}')\}\ \})$$

5.4 The Complexity of the Extended Algorithm

Based on the slicing algorithm shown in section 3, the extended algorithm augments the point-to analysis. For the cost of point-to analysis is less than the cost of slicing, we can only consider the cost of slicing. The extended algorithm may need to update the slices of all variables, so the operation of one statement may cost $O(v)$, where v is the number of variables. Furthermore, our extended algorithm will cost $O(n \times v)$, where n refer to the number of all labeled expressions appeared (perhaps repeatedly) in the sequence of analyzing the program. Since we finally obtain the slices of all variables, the program slice of each variable, on the average, costs $O(n)$, which is the same as the worst complexity of the algorithm in [4]. So our extended algorithm doesn't add complexity.

About the space complexity, we pay attention to the constructions *Refs*(l.e), Slices, L', L and point-to sets. We can use space $O(v \times m)$ to store *Refs*(l.e) and Slices, where m is the number of labeled expressions. The label set L' and L will cost $O(m)$. And the point-to sets may cost $O(v \times v)$. For v isn't larger than m, our extended algorithm cost $O(v \times m)$, which is the same as the original algorithm.

```
1     flag := 1;
2     if flag < 5 then
3        s = &a
      else
4           s = &b
      endif;
5        c = 1;
6        while flag < 5 do
7           t = &c;
8           *s = *t;
9           flag = flag + 1
      endwhile;
10    write a
```

Fig. 3. A sample program

6 A Case Study

To clearly explain our extended algorithm, we analyze a concrete program (shown in figure 3). Each expression is uniquely labeled (marked in source program). For example the second expression is flag < 5. We suppose S(v) and P(v) represent the slice and point-to set of variable v, respectively. Initially, the slices and point-to sets of all the variables are null. Below, we only give the variant part of the point-to set and slice. After the first expression is executed, the slice of variable flag is: S(flag) = {1}.

The second to fourth expressions combine a branch statement, it assignment the variable s. After executing this statement, the point-to set and slice of variable s is:

S(s) = {1, 2, 3, 4} P(s) = {a, b}

Similarly, after executing the fifth expression, the slice of variable c is: S(c) = {5}.

The next four expressions combine a loop block, which needs iteration to obtain the slices and point-to sets. After the first loop, the variant part is:

S(flag)={1, 6, 9} S(t)={7} S(a)={1, 2, 3, 4, 5, 6, 7}
S(b) = {1, 2, 3, 4, 5, 6, 7} P(t) = {c}

At this time, the computation hasn't been convergent. After the second loop, the information of variable a and b is:

S(a) = {1, 2, 3, 4, 5, 6, 7, 9} S(b) = {1, 2, 3, 4, 5, 6, 7, 9}

The slices and point-to sets hasn't been convergent yet. After the third loop, the iteration is convergent, so the computing is end.

The tenth expression has no influence on the information, so the final result is:

S(flag) = {1, 6, 9}, S(s) = {1, 2, 3, 4}, S(c) = {5}, S(t) = {7},
S(a) = {1, 2, 3, 4, 5, 6, 7, 9}, S(b) = {1, 2, 3, 4, 5, 6, 7, 9};
P(flag) = ϕ, P(s) = {a, b}, P(c) = ϕ, P(t) = {c}, P(a) = ϕ, P(b) = ϕ

From the above information, we can get the final result of slices by *Syn*(s, *L*) [4]. For example, the final result of slice criterion <10, a > is {1, 2, 3, 4, 5, 6, 7, 9, 10}, where 10 is included because of the rule of **write** statement in *Syn*(s, *L*) [4].

7 Conclusions

For the traditional slicing methods lack modularity and flexibility, we have proposed a novel formal slicing algorithm, called Modular monadic slicing algorithm [4]. To addressing the pointers, this paper combines the point-to analysis with the original monadic slicing algorithm. The feature of this algorithm is: our approach uses the data-flow iteration which is a precise flow-sensitive method to obtain the point-to information, but needs less space compared with the traditional data-flow iteration. In addition, our approach also inherits the properties of language independence and reusability from the original monadic slicing method.

References

1. Weiser, M.: Program slicing, IEEE Transactions on Software Engineering, 1984, 16(5), pp. 498-509
2. Binkley, D., Gallagher, K. B.: Program slicing, Advances in Computers, 1996, 43, pp. 1-50
3. Song, Y., Huynh, D.: Forward dynamic object-oriented program slicing, Application-Specific Systems and Software Engineering and Technology (ASSET '99). IEEE CS Press, 1999, pp. 230-237
4. Yingzhou. Zhang, Baowen. Xu, etc: Modular Monadic Program Slicing, COMPSAC 2004, Hong Kong, China, 2004, pp. 66-71
5. Liang, S., Hudak, P., Jones, M.: Monad transformers and modular interpreters, The 22nd ACM SIGPLAN-SIGACT Symposium on Principles of Programming Languages, POPL'95, ACM Press, New York, 1995, pp. 333-343

6. Michael Burke, Paul Carini, Jong-Deok Choi and Michael Hind: Flow-insensitive interprocedural alias analysis in the presence of pointers, In Proceedings of the Seventh International Workshop on Languages and Compilers for Parallel Computing, Aug 1994
7. Wadler, P.: Comprehending monads, ACM Conference on Lisp and Functional Programming, 1990, pp. 61-78
8. Bjarne Steensgaard: Points-to analysis in almost linear time, In 23rd Annual ACM SIGACT-SIGPLAN POPL, Jan 1996, pp. 32-41
9. Susan Horwitz, Phil Pfeiffer and Thomas Reps: Dependence analysis for pointer variables, In Proceedings of the SIGPLAN Conference on Programming Language Design and Implementation, Jun 1989, pp. 28-40
10. David R. Chase, Mark Wegman and F. kenneth Zadek: Analysis of pointers and structures, Proceedings of the SIGPLAN '90 Conference on Program Language Design and Implementation, White Plains, NY, Jun 1990, pp. 296-310

Modified Adaptive Resonance Theory Network for Mixed Data Based on Distance Hierarchy

Chung-Chian Hsu[1], Yan-Ping Huang[1,2], and Chieh-Ming Hsiao[1]

[1] Department of Information Management, National Yunlin University of Science and Technology, 123, Sec. 3, University Road, Douliu, Yunlin 640, Taiwan, R.O.C
hsucc@mis.yuntech.edu.tw, g9120817@yuntech.edu.tw

[2] Department of Management Information System, Chin Min Institute of Technology, 110, Hsueh-Fu Road, Tou-Fen, Miao-Li 351, Taiwan, R.O.C
sunny@chinmin.edu.tw

Abstract. Clustering of data is a fundamental data analysis step that has been widely studied across in data mining. Adaptive resonance theory network (ART) is an important algorithm in Clustering. ART is also very popular in the unsupervised neural network. Type I adaptive resonance theory network (ART1) deals with the binary numerical data, whereas type II adaptive resonance theory network (ART2) deals with the general numerical data. Several information systems collect the mixing type attitudes, which included numeric attributes and categorical attributes. However, ART1 and ART2 do not deal with mixed data. If the categorical data attributes are transferred to the binary data format, the binary data do not reflect the similar degree. It influences the clustering quality. Therefore, this paper proposes a modified adaptive resonance theory network (M-ART) and the conceptual hierarchy tree to solve similar degrees of mixed data. This paper utilizes artificial simulation materials and collects a piece of actual data about the family income to do experiments. The results show that the M-ART algorithm can process the mixed data and has a great effect on clustering.

Keywords: adaptive resonance theory network (ART), distance hierarchy, clustering algorithm, data mining, software engineering.

1 Introduction

Clustering is the unsupervised classification of patterns into groups. It is an important data analyzing technique, which organizes a collection of patterns into clusters based on similarity [1-3]. Clustering is useful in several exploratory pattern-analysis, grouping, decision-making, and machine-learning situations. This includes data mining, document retrieval, image segmentation, and pattern classification. Clustering methods have been successfully applied in many fields including image processing [1], pattern recognition [4], biology, psychiatry, psychology, archaeology, geology, geography, marketing and information retrieval [5,6], software engineering [7-10]. Intuitively, patterns with a valid cluster are more similar to each other than they are to a pattern belonging to a different cluster.

V.N. Alexandrov et al. (Eds.): ICCS 2006, Part IV, LNCS 3994, pp. 757 – 764, 2006.
© Springer-Verlag Berlin Heidelberg 2006

The majority of the software clustering approaches presented in the literature attempt to discover clusters by analyzing the dependencies between software artifacts, such as functions or source files [7-13]. Software engineering principles, such as information hiding or high-cohesion and low-coupling are commonly employed to help determine the boundaries between clusters.

Most of clustering algorithms consider either categorical data or numeric data. However, many mixed datasets including categorical and numeric values existed nowadays. A common practice to clustering mixed dataset is to transform categorical values into numeric values and then proceed to use a numeric clustering algorithm. Another approach is to compare the categorical values directly, in which two distinct values result in distance 1 while identical values result in distance 0. Nevertheless, these two methods do not take into account the similarity information embedded between categorical values. Consequently, the clustering results do not faithfully reveal the similarity structure of the dataset. This article is based on distance hierarchy [2-3] to propose a new incremental clustering algorithm for mixed datasets, in which the similarity information embedded between categorical attribute is considered during clustering. In the setting, each attribute of the data is associated with a distance hierarchy, which is an extension of the concept hierarchy [21] with link weights representing the distance between concepts. The distance between two mixed data patterns is then calculated according to distance hierarchies.

The rest of this article is organized as follows. Section 2 reviews clustering algorithms and discusses the shortcomings of the conventional approaches to clustering mixed data. Section 3 presents distance hierarchy for categorical data and proposes the incremental clustering algorithm based on distance hierarchies. In Section 4, experimental results on synthetic and real datasets are presented. Conclusions are given in Section 5.

2 Related Work

Adaptive resonance theory neural networks model real-time prediction, search, learning, and recognition. ART networks function as models of human cognitive information processing [15-19]. A central feature of all ART systems is a pattern matching process that compares an external input with the internal memory of an active code. ART1 deals with the binary numerical data and ART2 deals with the general numerical data [18]. However, these two methods do not deal with mixed data attributes.

About clustering mixed data attributes, there are two approaches for mixed data. One is resorted to a pre-process, which transferred the data to the same type, either all numeric or all categorical. For transferring continuous data to categorical data, some metric function is employed. The function is based on simple matching in which two distinct values result in distance 1, with identical values of distance 0 [20]. The other is to use a metric function, which can handle mixed data [21]. Overlap metric is for nominal attributes and normalized Euclidean distance is for continuous attributes.

Among problems with simple matching and binary encoding, a common approach for handling categorical data is simple matching, in which comparing two identical categorical values result in distance 0, while two distinct values result in distance 1

[21, 22]. In this case, the distance between patterns of Gary and John in the previous example becomes d(Gary, John) =1, which is the same as d(John, Tom) = d(Gary, Tom) = 1. Obviously, the simple matching approach disregards the similarity information embedded in categorical values.

Another typical approach to handle categorical attributes is to employ binary encoding that transforms each categorical attribute to a set of binary attributes and a categorical value is then encoded to a set of binary values. As a result, the new relation contains all numeric data, and the clustering is therefore conducted on the new dataset. For example, as the domain of the categorical attribute: Favorite_Drink. The set of it is {Coke, Pepsi, Mocca}. Favorite_Drink is transformed to three binary attributes: Coke, Pepsi and Mocca in the new relation. The value Coke of Favorite_Drink in a pattern is transformed to a set of three binary values in the new relation, i.e. {Coke=1, Pepsi=0, Mocca=0}. The Manhattan distance of patterns Gary and John is d_M(Gary, John) =2, which is the same as d_M(Gary, Tom) and d_M(John, Tom), according to the new relation. Traditional clustering algorithm transfers Favorite_Drink categorical attributes into a binary numerical attribute type as shown in figure 1.

ID	Favorite Drink	Amt.
Gary	Coke	70
John	Pepsi	70
Tom	Coffee	70

ID	Coke	Pepsi	Coffee	Amt.
Gary	1	0	0	70
John	0	1	0	70
Tom	0	0	1	70

Fig. 1. Traditional clustering algorithm transfers Favorite_Drink categorical attributes into binary numerical attribute type

The ART network is a popular incremental clustering algorithm [1]. It has several variants [23, 24], in which ART1 handles only the binary data and ART2 can handle only the arbitrary continuous data. K-prototype [25] is a recent clustering algorithm for mixed data. It transfers categorical data attributes to the binary data format, however, the binary data do not reflect the similar degree. It influences the clustering quality. Therefore, this paper proposes a modified adaptive resonance theory network algorithm and the conceptual hierarchy tree to solve the similar degree of mixed data.

3 Clustering Hybrid Data Based on Distance Hierarchy

The distance hierarchy tree is a concept hierarchy structure. It is also a better mechanism to facilitate the representation and computation of the distance between categorical values. A concept hierarchy consists of two parts: a node set and a link set [2, 3, 26, 27]. According to binary encoding approach, it does not reflect the similar degree. However, it influences the clustering quality. Maintenance was difficult when the domain of a categorical attribute changes, because the transformed relation schema also needs to be changed. The transformed binary attributes cannot preserve the semantics of the original attribute. Because of the drawbacks resulting from the binary-encoding approach, this paper uses distance hierarchy to solve the similar degree of mixed data.

This paper extends the distance hierarchy structure with link weights. Each link has a weight representing a distance. Link weights are assigned by domain experts. There are several assignment alternatives. The simplest way is to assign all links as a uniform constant weight. Another alternative is to assign heavier weights to the links closer to the root and lighter weights to the links away from the root. For simplicity, unless stated explicitly, each link weight is set to 1 in this article. The distance of two concepts at the leaf nodes is the total weight between those two nodes.

A point X in a distance hierarchy consists of two parts, an anchor and a positive real-value offset, denoted as X(N, d), that is, anchor(X) =N and offset(X) =d. The anchor is a leaf node and the offset represents the distance from the root of the hierarchy to the point. A point X is an ancestor of Y if X is in the path from Y to the root of the hierarchy. If neither one of the two points is an ancestor of the other point, then the least common ancestor, denoted as LCA(X, Y), is the deepest node that is an ancestor of X as well as Y.

A special distance hierarchy calls numeric distance hierarchy for a numeric attribute, say $\mathbf{x}_i$, is a degenerate one, which consists of only two nodes, a root MIN and a leaf MAX, and has the link weight w being the domain range of $\mathbf{x}_i$, i.e. w = (max_i-min_i). A point p in such a distance hierarchy has the value (MAX, d_p) where the anchor is always the MAX and the offset d_p is the distance from the point to the root MIN.

About measuring distance, the distance between two data points can be measured as follows: Let x = [x_1 x_2 … x_n] and y = [y_1 y_2 … y_n]. The distance between a training pattern **x** and an M-ART neuron **y** is measured as the square root of the sum of the square differences between each-paired components of **x** and **y**. Specifically, **x** and **y** represent a training data and a map neuron, respectively, with n-dimension, and C is a set of n distance hierarchies, then the distance between **x** and **y** can be expressed as

$$d(\mathbf{x},\mathbf{y}) = \|\mathbf{x}-\mathbf{y}\| = (\sum_{i=1,n} w_i (x_i - y_i)^2)^{1/2} = (\sum_{i=1,n} w_i (h(x_i) - h(y_i))^2)^{1/2} \qquad (1)$$

Where $h(x_i)$ and $h(y_i)$ are the mapping of x_i and y_i to their associated distance hierarchy h_i and w_i , the attribute weight, is a user specified parameter allowing the domain expert to give different weights. For a numeric attribute I, $h(x_i)$ - $h(y_i)$ is equal to x_i-y_i, since $h(x_i)$ - $h(y_i)$ = (MIN, $d_{h(xi)}$) - (MIN, $d_{h(yi)}$) = (MIN, x_i-min_i)- (MIN, y_i-min_i) = (x_i-y_i).

This paper proposes the distance hierarchy tree structure to overcome the expression for similar degree. This distance hierarchy tree algorithm combines the adaptive resonance theory network algorithm and it can be effective with mixed data in data clustering. This section presents distance hierarchy for categorical data and it proposes the incremental clustering algorithm based on distance hierarchies.

The categorical utility function [28] attempts to maximize the probability that the two objects in the same cluster have attribute values in common and the probability that the objects from different clusters have different attributes. The categorical utility of a set of clusters can be calculated as

$$CU = \sum_k (\frac{|C_k|}{|D|} \sum_i \sum_j [P(A_i = V_{ij} \mid C_k)^2 - P(A_i = V_{ij})^2] \qquad (2)$$

Here, $P(A_i{=}V_{ij}|C_k)$ is the conditional probability that the attribute i has the values V_{ij} given the cluster C_k, and $P(A_i{=}V_{ij})$ is the overall probability of the attribute i having the values V_{ij} in the entire set. The function aims to measure if the clustering improves the likelihood of similar values falling in the same cluster. Obviously, the higher the CU values, the better the clustering result [29].

For numeric attributes, the standard deviation represents the dispersion of values. Variance (σ^2) can be used for evaluating the quality of clustering numeric data. Several cluster validity indices, such as Davies-Bouldin (DB) Index and Calinski Harabasz (CH) Index [27, 30], have been published; however, they are only suitable for the numeric data. Hence, in order to evaluate the effectiveness of clustering mixed data, this paper uses CV index [31], which combined the category utility (CU) function with variance. The CV is defined as in Equation (3), where the CU and variance are the validity index for categorical and numeric data, respectively. The higher the CV values, the better the clustering result.

$$CV = \frac{CU}{1 + \text{Variance } (\sigma^2)} \tag{3}$$

4 Experiments

This paper develops a prototype system with Borland C++ Builder 6. A series of experiments have been performed in order to verify the method. A mixed synthetic dataset and a UCI dataset have also been designed to show the capability of the M-ART in reasonably expressing and faithfully preserving the distance between the categorical data. It also reports the experimental results of artificial and actual data.

These experiments use a real Adult dataset from the UCI repository with 48,842 records of 15 attributes, including eight categorical attributes, six numerical attributes, and one class attribute.

This experiment uses 7 attributes, which include three categorical attributes, such as Relationship, Marital_status, and Education; and four numeric attributes, Capital_gain, Capital_loss, Age, and Hours_per_week. The concept hierarchies are constructed in figure 2.

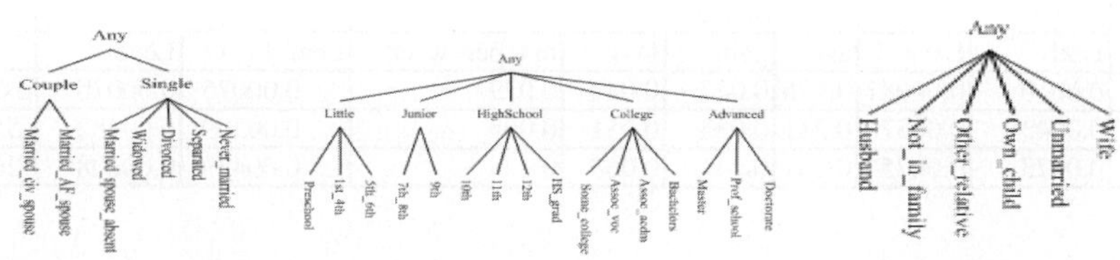

(a) marital-status (b) education (c) relationship

Fig. 2. Concept hierarchies for (a) Marital-status (b) Education and (c) Relationship attributes of the Adult dataset

The M-ART parameters are established as follows: the initial warning value is 0.55 and it increases progressively 0.05 until 0.75. The initial learning rate is 0.9. The stop condition t occurs when the momentum of the output layer is lower than 0.000015.

This paper collects the dataset with different methods. These methods divide adult datasets into 5, 6, 7 and 8 groups. Concerning the CU values for categorical attributes, the higher the CU values, the better the clustering result. The CU value of clustering M-ART method is the highest, K-prototype method is second, and traditional ART2 is the lowest. The symbol " ***" means that it does not find the suitable parameter to divide into group with the datasets. The parameter of ART2 reaches 7, it is unable to divide seven groups all the time. The problem occurs because there are too many parameters in ART2.

This paper normalizes the variance between 0 and 1 for numeric results. The normalized variance is useful in CV index. Table 1 shows the CV values of the clustering results by M-ART, ART2 and k-prototypes on level 1 and the leaf level in individual concept hierarchies with cluster numbers 5, 6, 7 and 8. The higher the value of CV values, the better the clustering result. The CV value in M-ART method is the highest, K-prototype method is second, and traditional ART2 is the lowest.

Table 1. The CV values for Adult dataset with 5, 6, 7, 8 clusters by M-ART, ART2 and K-Prototypes

M-ART									
Cluster	CU		Variance				CV		Increased
No.	Leaf	Level 1	age	gain	loss	hrs_per_week	Leaf_Level	Level 1	
5	1.069	1.16	0.127	0.022	0.038	0.071	0.85	0.59	31.08%
6	1.113	1.2	0.163	0.023	0.044	0.085	0.85	0.60	29.05%
7	1.115	1.21	0.182	0.011	0.044	0.100	0.83	0.61	27.05%
8	1.177	1.31	0.218	0.014	0.052	0.115	0.84	0.65	23.00%

K-Prototype									
Cluster	CU		Variance				CV		Increased
No.	Leaf	Level 1	age	gain	loss	hrs_per_week	Leaf_Level	Level 1	
5	0.859	0.834	0.114	0.033	0.046	0.073	0.68	0.46	32.85%
6	1.002	0.977	0.424	0.72	1.454	2.937	0.15	0.16	-1.72%
7	1.039	0.919	0.515	0.893	1.801	3.635	0.13	0.12	7.01%
8	1.088	1.087	0.610	1.068	2.154	4.342	0.12	0.13	-5.56%

ART2									
Cluster	CU		Variance				CV		Increased
No.	Leaf	Level 1	age	gain	loss	hrs_per_week	Leaf_Level	Level 1	
5	0.0010	0.00081	0.176	0.027	0.042	0.079	0.00075	0.00070	6.64%
6	0.0043	0.00570	0.211	0.033	0.051	0.095	0.00309	0.00482	-55.87%
8	0.0073	0.00757	0.281	0.044	0.068	0.128	0.00479	0.00608	-26.71%

5 Conclusions and Future Work

Most traditional clustering algorithms can only handle either categorical or numeric value. Although some research results have been published for handling mixed data, they still cannot reasonably express the similarities among categorical data.

This paper proposes the distance hierarchy tree structure to overcome the expression for similar degree. This distance hierarchy tree algorithm combines the adaptive resonance theory network algorithm and it can be effective with mixed data in data clustering. It presents a MART algorithm, which can handle mixed dataset directly. The experimental results on synthetic data sets show that the proposed approach can better reveal the similarity structure among data, particularly when categorical attributes are involved and have different degrees of similarity, in which the traditional clustering approaches do not perform well. The experimental results on the real dataset have better performances than other algorithms.

MART is a clustering algorithm for any field with mixed data. The future work will try to use this method in finding out the pattern rules from software engineering databases

References

1. Jain, A., and Dubes, R.: Algorithms for clustering Data. Prentice-Hall, Englewood Cliffs. NJ. (1988).
2. Hsu, C. C.: Generalizing Self-Organizing Map for Categorical Data. IEEE Transactions on Neural Networks. (2006, In press).
3. Hsu, C. C. and Wang, S. H.: An Integrated Framework for Visualized and Exploratory Pattern Discovery in Mixed Data. IEEE Transactions on Knowledge and Data Engineering. vol. 18, no. 2. (2006) 161-173.
4. Anderberg, M.: Cluster Analysis for Applications. Academic Press, New York. (1973).
5. Rasmussen, E.: Clustering Algorithms. Information Retrieval: Data Structures & Algorithms. William B. Frakes and Ricardo Baeza-Yates (Eds.). Prentice Hall. (1992).
6. Salton, G. and Buckley, C.: Automatic Text Structuring and Retrieval-experiments in Automatic Encyclopedia Searching. Proceedings of the Fourteenth International ACM SIGIR Conference on Research and Development in Information Retrieval. (1991) 21-30.
7. Kadamuddi, D. and Tsai, J. P.: Clustering algorithm for parallelizing software systems in multiprocessors environment. Software Engineering, IEEE Transactions, vol. 26, Issue 4, (2000) 340-361.
8. Tian, J.: Better reliability assessment and prediction through data clustering. Software Engineering, IEEE Transactions, vol. 28, Issue 10, (2002) 997-1007.
9. Chen K., Zhang W., Zhao H., Mei H.: An approach to constructing feature models based on requirements clustering. Requirements Engineering, 2005. Proceedings. 13th IEEE International Conference. (2005) 31-40.
10. Andritsos, P. and Tzerpos, V.: Information-theoretic software clustering. Software Engineering, IEEE Transactions, vol. 31, Issue 2. (2005) 150-165.
11. Hutchens, D. H. and Basili, V.R.: System Structure Analysis: Clustering with Data Bindings. Software Engineering, IEEE Transactions, vol. 11, no. 8. (1985) 749-757.
12. Schwanke, R. W.: An Intelligent Tool for Re-Engineering Software Modularity. Proc. 13th Int'l Conf. Software Engineering. (1991) 83-92.
13. Mancoridis, S., Mitchell, B., Chen, Y. and Gansner, E.: Bunch: A Clustering Tool for the Recovery and Maintenance of Software System Structures. Proc. Int'l Conf. Software Maintenance. (1999).
14. Can, F.: Incremental clustering for dynamic information processing. ACM Transaction for Information Systems, vol. 11. (1993) 143-164.
15. Carpenter, G. A.: Distributed learning, recognition, and prediction by ART and ARTMAP neural networks. Neural Networks, vol. 10, no. 8. (1997) 1473–1494.
16. Carpenter, G. A., and Grossberg, S.: Normal and amnesic learning, recognition, and memory by a neural model of cortico-hippocampal interactions. Trends in Neuroscience, vol. 16, no. 4. (1993) 131–137.

17. Grossberg, S.: How does a brain build a cognitive code? Psychological Review, vol. 87. (1980) 1–51.
18. Grossberg, S.: The link between brain, learning, attention, and consciousness. Consciousness and Cognition, vol. 8. (1999) 1–44.
19. Grossberg, S.: How does the cerebral cortex work? Development, learning, attention, and 3D vision by laminar circuits of visual cortex. Behavioral and Cognitive Neuroscience Reviews, vol. 2, no. 1. (2003) 47–76.
20. Guha, S., Rastogi, R. and Shim, K.: ROCK: A robust clustering algorithm for categorical attributes. Proceedings of the IEEE Conference on Data Engineering. (1999) 512-521.
21. Wilson, D.R. and Martinez, T. R.: Improved heterogeneous distance functions. Journal of Artificial Intelligence Research, vol. 6. (1997) 1-34.
22. Ester, M., Kriegel, H. P., Sander, J., Wimmer, M. and Xu, X.: Incremental clustering for mining in a data warehousing environment. Proceedings of the 24th Intl. Conf. on Very Large Data Bases (VLDB). (1998) 323-333.
23. Carpenter, G. A. and Grossberg, S.: ART 2 : Self-organization of stable category recognition codes for analog input patterns. Applied Optics : Special Issue on Neural Networks, vol. 26. (1987) 4919-4930.
24. Carpenter, G., Grossberg, A., S. and Rosen, D. B.: Fuzzy ART: fast stable learning and categorization of analog patterns by an adaptive resonance system. Neural Networks, vol. 4. (1991) 759-771.
25. Huang, Z.: Extensions to the k-means algorithm for clustering large data sets with categorical values. Data Mining and Knowledge Discovery, vol. 2, no. 3. (1998) 283-304.
26. Dash, M. and Choi, K., Scheuermann, P. and Liu, H.: Feature selection for clustering - a filter solution, IEEE International Conference on Data Mining. (2002) 115 – 122.
27. Maulik, U. and Bandyopadhyay, S.: Performance evaluation of some clustering algorithms and validity indices. IEEE Transactions on Pattern Analysis and Machine Intelligence. (2002) 1650-1654.
28. Gluck, M. A. and Corter, J. E.: Information, uncertainty, and the utility of categories. Proceedings of the Seventh Annual Conference of the Cognitive Science Society. (1985).
29. Barbara, D., Couto, J. and Li, Y.: COOLCAT: an entropy-based algorithm for categorical clustering. Proceedings of the eleventh international conference on Information and knowledge management. (2002) 582-589.
30. Halkidi, M., Batistakis, Y. and Vazirgiannis, M.: On clustering validation techniques. Journal of Intelligent Information Systems, vol. 17. (2001) 107-145.
31. Hsu, C. C. and Chen, Y. C.: Mining of Mixed Data with Application to Catalog Marketing. Expert Systems with Applications. (2006, In press).

Checking for Deadlock, Double-Free and Other Abuses in the Linux Kernel Source Code

Peter T. Breuer and Simon Pickin

Universidad Carlos III de Madrid, Leganes, Madrid, 28911 Spain
ptb@inv.it.uc3m.es, spickin@it.uc3m.es

Abstract. The analysis described in this article detects about two real and uncorrected deadlock situations per thousand C source files or million lines of code in the Linux kernel source, and three accesses to freed memory, at a few seconds per file. In distinction to model-checking techniques, the analysis applies a configurable "3-phase" Hoare-style logic to an abstract interpretation of C code to obtain its results.

1 Introduction

The pairing of formal methods practitioners and the Linux kernel has sometimes seemed more than unlikely. On the one hand kernel contributors have preferred to write elegant code rather than elegant specifications; "the code is the commentary" has been one of the mantras of the Linux kernel development, meaning that the code should be so clear in its own right that it serves as its own specification. That puts what is usually the first question of formal methods practitioners, "what should the code do" out of bounds. And on the other hand, formal methods practitioners have not been able to find a way into the six million lines of ever-changing C code that comprises the Linux kernel source.

This article describes the application of *post-hoc* formal logical analysis to the Linux kernel. The technology detects real coding errors that have been missed by thousands of eyes over the years. The analyser itself is written in C (thus making it easy to compile and distribute in an open source environment) and is itself licensed under an open source license. The analysis is configurable, which means that it is possible to re-program and extend it without rewriting its source.

By way of orientation, note that static analysis is in general difficult to apply to C code, because of pointer arithmetic and aliasing, but some notable efforts to that end have been made. David Wagner and collaborators in particular have been active in the area (see for example [7], where Linux user space and kernel space memory pointers are given different types, so that their use can be distinguished, and [8], where C strings are abstracted to a minimal and maximal length pair and operations on them abstracted to produce linear constraints on these numbers). That research group often uses model-checking to look for violations in possible program traces of an assertion such as "`chroot` is always followed by `chdir` before any other file operations". In contrast, the approach in this article assigns a (customisable) approximation semantics to C programs, via a (customisable) program logic for C. A more lightweight technique still is

V.N. Alexandrov et al. (Eds.): ICCS 2006, Part IV, LNCS 3994, pp. 765–772, 2006.
© Springer-Verlag Berlin Heidelberg 2006

that exemplified by Jeffrey Foster's work with CQual [5, 6], which extends the type system of C in a customisable manner. In particular, CQual has been used to detect "spinlock under spinlock", a variant of the analysis described here. The technology described in this article was first described in prototype in [3], being an application of the generic "three-phase" program logic first described in [1] and developed in [2]. The tool itself now works at industrial scales, treating millions of lines of code in a few hours when run on a very modest PC.

In the analysis here, *abstract interpretation* [4] forms a fundamental part, causing a simplification in the symbolic logic description of state that is propagated by the analysis; for example, "don't know" is a valid abstract literal, thus a program variable which may take any of the values 1, 2, or 3 may be described as having the value "don't know" in the abstraction, leading to a state s described by one atomic proposition, not a disjunct of three.

To be exact, the analysis constructs two abstract approximations, a state s and a predicate p describing the real state x such that

$$x \in s \cap p \tag{1}$$

The approximated state s assigns a range of integer values to each variable.

Predicates are restricted to the class of disjunctions of conjuncts of simple ordering relations $\mathtt{x} \leq \mathtt{k}$, and there is a simple decision procedure for implication.

In [3] we focussed on checking for a particular problem in SMP systems – "sleep under spinlock". A function that can sleep (i.e., that can be scheduled out of the CPU) ought never to be called from a thread that holds a "spinlock", the SMP locking mechanism of choice in the Linux kernel. Trying to take a locked spinlock on one CPU provokes a busy wait ("spin") that occupies the CPU completely until the spinlock is released on another CPU. If the thread that has locked the spinlock is scheduled out while the lock is held, then the only thread that likely has code to release the spinlock is not running. If by chance that thread is rescheduled in to the CPU before any other thread tries for the spinlock then all is well. But if another thread tries for the spinlock first, it will spin uselessly, keeping out of that CPU the thread that would have released the spinlock. If yet another thread tries for the spinlock, then on a 2-CPU SMP system, the machine is dead, with both CPUs spinning waiting for a lock that will never be released. Such vulnerabilities are denial of service vulnerabilities that any user can exploit to take down a system. 2-CPU machines are also common – any Pentium 4 of 3.2GHz or above has a dual "hyper-threading" core. So, calling a function that may sleep while holding the lock on a spinlock is a serious matter. Detecting it is one application of the abstract logic that may be applied by the analyser.

2 Example Run

About 1000 (1055) of the 6294 C source files in the Linux 2.6.3 kernel were checked for spinlock problems in a 24-hour period by the analyser running on a 550MHz (dual) SMP PC with 128MB ram. About forty more files failed to

files checked:	1055	
alarms raised:	18	(5/1055 files)
false positives:	16/18	
real errors:	2/18	(2/1055 files)
time taken:	~24h	
LOC:	~700K	(unexpanded)

```
1 instances of sleep under spinlock
      in sound/isa/sb/sb16_csp.c
1 instances of sleep under spinlock
      in sound/oss/sequencer.c
6 instances of sleep under spinlock
      in net/bluetooth/rfcomm/tty.c
7 instances of sleep under spinlock
      in net/irda/irlmp.c
3 instances of sleep under spinlock
      in net/irda/irttp.c
```

Fig. 1. Testing for sleep under spinlock in the 2.6.3 Linux kernel

File & function	Code fragment
`sb/sb16_csp.c:`	619 `spin_lock_irqsave(&p->chip->reg_lock, flags);`
`snd_sb_csp_load`	... ...
	632 `unsigned char *kbuf, *_kbuf;`
	633 `_kbuf = kbuf = kmalloc (size, GFP_KERNEL);`
`oss/sequencer.c:`	1219 `spin_lock_irqsave(&lock,flags);`
`midi_outc`	1220 `while (n && !midi_devs[dev]->outputc(dev, data)) {`
	1221 `interruptible_sleep_on_timeout(&seq_sleeper,HZ/25);`
	1222 `n--;`
	1223 `}`
	1224 `spin_unlock_irqrestore(&lock,flags);`

Fig. 2. Sleep under spinlock instances in kernel 2.6.3

parse at that time for various reasons (in one case, because of a real code error, in others because of the presence of gnu C extensions that the analyser could not cope with at that time, such as `__attribute__` declarations in unexpected positions, case statement patterns matching a range instead of just a single number, array initialisations using "`{ [1,3,4] = x }`" notation, enumeration and typedef declarations inside code blocks, and so on). Five files out of that selection showed up as suspicious under the analysis, as listed in Fig. 1.

Although the flagged constructs are indeed calls of the kernel memory allocation function `kmalloc` (which may sleep) under spinlock, the arguments to the call sometimes render it harmless, i.e. cause it not to sleep after all. The `kmalloc` function will not sleep with `GFP_ATOMIC` as second argument, and such is the case in several instances, but not in the two instances shown in Fig. 2.

3 Analytic Program Logic

The C code analyser is based on a compositional program logic called NRBG (for "normal", "return", "break", "goto", reflecting its four principal components). The four components, N, R, B, G, represent different kinds of control flows: a "normal" flow and several "exceptional" flows.

Program fragments are thought of as having three phases of execution: *initial*, *during*, and *final*. The initial phase is represented by a condition p that holds as the program fragment is entered. The only access to the internals of the during phase is via an exceptional exit (R, B, G; return, break, goto) from the fragment. The final phase is represented by a condition q that holds as the program fragment terminates normally (N).

The N part of the logic represents the way control flow "falls off the end" of one fragment and into another. I.e., if p is the condition that holds before program $a;b$ runs, and q is the condition that holds after, then

$$p\ N(a;b)\ q \quad = \quad p\ N(a)\ r\ \wedge\ r\ N(b)\ q \tag{2}$$

To exit normally with q, the program must flow normally through a, hitting an intermediate condition r, then enter fragment b and exit it normally.

The R part of the logic represents the way code flows out of the parts of a routine through a "return" path. Thus, if r is the intermediate condition that is attained after normal termination of a, then:

$$p\ R(a;b)\ q \quad = \quad p\ R(a)\ q\ \vee\ r\ R(b)\ q \tag{3}$$

That is, one may either return from program fragment a, or else terminate a normally, enter fragment b and return from b.

The logic of break is (in the case of sequence) equal to that of return:

$$p\ B(a;b)\ q \quad = \quad p\ B(a)\ q\ \vee\ r\ B(b)\ q \tag{4}$$

where again r is the condition attained after normal termination of a.

Where break and return logic differ is in the treatment of loops. First of all, one may only return from a forever `while` loop by returning from its body:

$$p\ R(\texttt{while(1)}\,a)\ q \quad = \quad p\ R(a)\ q \tag{5}$$

On the other hand, (counter-intuitively at first reading) there is no way of leaving a forever `while` loop via a break exit, because a break in the body of the loop causes a normal exit from the loop itself, not a break exit:

$$p\ B(\texttt{while(1)}\,a)\ F \tag{6}$$

The normal exit from a forever loop is by break from its body:

$$p\ N(\texttt{while(1)}\,a)\ q \quad = \quad p\ B(a)\ q \tag{7}$$

To represent the loop as cycling possibly more than once, one would write for the R component, for example:

$$p\ R(\texttt{while(1)}\,a)\ q \quad = \quad p\ R(a)\ q\ \vee\ r\ R(\texttt{while(1)}a)\ q \tag{8}$$

where r is the intermediate condition that is attained after normal termination of a. However, in practice it suffices to check that $r \to p$ holds, because then (8) reduces to (5). If $r \to p$ does not hold, p is *relaxed* to $p' \geq p$ for which it does.

Typically the precondition p is the claim that the spinlock count ρ is below or equal to n, for some n: $\rho \leq n$. In that case the logical components $i = \mathrm{N}, \mathrm{R}, \mathrm{B}$ have for each precondition p a strongest postcondition $p\ \mathrm{SP}_N(a)$, $p\ \mathrm{SP}_R(a)$, $p\ \mathrm{SP}_B(a)$, compatible with the program fragment a in question. For example, in the case of the logic component N:

$$p\ N(a)\ q \quad \leftrightarrow \quad p\ \mathrm{SP}_N(a) \leq q \tag{9}$$

Each logic component X can be written as a function rather than a relation by identifying it with a postcondition generator no stronger than SP_X. For example:

$$(\rho \leq n)\ N \begin{pmatrix} \texttt{spin\_lock(\&}x\texttt{)} \\ \texttt{spin\_unlock(\&}x\texttt{)} \end{pmatrix} \quad = \quad \begin{pmatrix} \rho \leq n+1 \\ \rho \leq n-1 \end{pmatrix} \tag{10}$$

Or in the general case, the action on precondition p is to substitute ρ by $\rho \pm 1$ in p, giving $p[\rho - 1/\rho]$ (for `spin_lock`) and $p[\rho + 1/\rho]$ (for `spin_unlock`) respectively:

$$p\ N \begin{pmatrix} \texttt{spin\_lock(\&}x\texttt{)} \\ \texttt{spin\_unlock(\&}x\texttt{)} \end{pmatrix} \quad = \quad \begin{pmatrix} p[\rho - 1/\rho] \\ p[\rho + 1/\rho] \end{pmatrix} \tag{11}$$

The functional action on sequences of statements is then described as follows:

$$p\ N(a;b) = (p\ N(a))\ N(b) \tag{12}$$
$$p\ R(a;b) = p\ R(a)\ \vee\ (p\ N(a))\ R(b) \tag{13}$$
$$p\ B(a;b) = p\ B(a)\ \vee\ (p\ N(a))\ B(b) \tag{14}$$

The G component of the logic is responsible for the proper treatment of `goto` statements. To allow this, the logic – each of the components N, R, B and G – works within an additional *context*, e. A context e is a set of labelled conditions, each of which are generated at a `goto` x and are discharged/will take effect at a corresponding labelled statement x`:` .... The G component manages this context, first storing the current precondition p as the pair (x, p) (written $x{:}p$) in the context e at the point where the `goto` x is encountered:

$$p\ G_e(\texttt{goto}\ x) = \{x{:}p\} \uplus e \tag{15}$$

The $\{x{:}p\}$ in the equation is the singleton set $\{(x, p)\}$, where x is some label (e.g. the "`foo`" in "`foo: a = 1;`") and p is a logical condition like "$\rho \leq 1$".

In the simplest case, the operator $\uplus$ is set theoretic disjunction. But if an element $x{:}q$ is already present in the context e, signifying that there has already been one `goto` x statement encountered, then there are now two possible ways to reach the targeted label, so the union of the two conditions p and q is taken and $x{:}q$ is replaced by $x{:}(p \cup q)$ in e.

Augmenting the logic of sequence (12-14) to take account of context gives:

$$p\ N_e(a;b) = (p\ N_e(a))\ N_{pG_e(a)}(b) \tag{16}$$
$$p\ R_e(a;b) = p\ R_e(a)\ \vee\ (p\ N_e(a))\ R_{pG_e(a)}(b) \tag{17}$$
$$p\ B_e(a;b) = p\ B_e(a)\ \vee\ (p\ N_e(a))\ B_{pG_e(a)}(b) \tag{18}$$

The N, R, B semantics of a `goto` statement are vacuous, signifying one cannot exit from a `goto` in a normal way, nor on a break path, nor on a return path.

$$p\ N_e(\texttt{goto}\ x) = p\ R_e(\texttt{goto}\ x) = p\ B_e(\texttt{goto}\ x) = F \tag{19}$$

The only significant effect of a `goto` is to load the context for the logic with an extra exit condition. The extra condition will be discharged into the normal component of the logic only when the label corresponding to the `goto` is found (e_x is the condition labeled with x in environment e, if any):

$$\begin{array}{llll} p\ N_{\{x:q\}\cup e}(x{:}) & = p \vee q & p\ R_e(x{:}) & = F \\ p\ B_e(x{:}) & = F & p\ G_e(x{:}) & = e - \{x{:}e_x\} \end{array} \tag{20}$$

This mechanism allows the program analysis to pretend that there is a "short-cut" from the site of the `goto` to the label, and one can get there either via the short-cut or by traversing the rest of the program. If label `foo` has already been encountered, then we have to check at `goto foo` that the current program condition is an invariant for the loop back to `foo:`, or raise an alarm.

The equations given can be refined by introducing temporal logic (CTL). Consider the return logic of sequence, for example. If $\mathbf{EF}p$ is the statement that there is at least one trace leading to condition p at the current flow point, then:

$$\frac{pR(a)\mathbf{EF}r_1 \qquad pN(a)\mathbf{EF}q \qquad qR(b)\mathbf{EF}r_2}{pR(a;b)(\mathbf{EF}r_1 \wedge \mathbf{EF}r_2)} \tag{21}$$

The deduction that $\mathbf{EF}r_1 \wedge \mathbf{EF}r_2$ holds is stronger than $r_1 \vee r_2$, which is what would be deduced in the absence of CTL.

The above is a *may* semantics, because it expresses the possible existence of a trace. A *must* semantics can be phrased via the the operator $\mathbf{AF}p$, which expresses that all traces leading to the current point give rise to condition p here:

$$\frac{pR(a)\mathbf{AF}r_1 \qquad pN(a)\mathbf{AF}q \qquad qR(b)\mathbf{AF}r_2}{pR(a;b)(\mathbf{AF}(r_1 \vee r_2))} \tag{22}$$

In general, the deduction systems prove $pX\mathbf{AF}q_1$, pXq_2, $pX\mathbf{EF}q_3$ with $q_1 \leq q_2$ and $q_2 \leq q_3$, which brackets the analysis results between forced and possible.

4 Configuring the Analysis

The static analyser allows the program logic set out in the last section to be specified in detail by the user. The motive was originally to make sure that the logic was implemented in a bug-free way – writing the logic directly in C made for too low-level an implementation for what is a very high-level set of concepts. Instead, a compiler into C for specifications of the program logic was written and incorporated into the analysis tool.

The *logic compiler* understands specifications of the format

```
ctx precontext, precondition :: name(arguments) =
    postconditions with ctx postcontext ;
```

Table 1. Defining the single precondition/triple postcondition NRB logic of C

ctx e, p::for(stmt)	= (n∨b, r, F) with ctx f where ctx e, p::stmt = (n,r,b) with ctx f;
ctx e, p::empty()	= (p, F, F) with ctx e;
ctx e, p::unlock(label l)	= (p[n+1/n], F, F) with ctx e;
ctx e, p::lock(label l)	= (p[n-1/n], F, F) with ctx e;
ctx e, p::assembler()	= (p, F, F) with ctx e;
ctx e, p::function()	= (p, F, F) with ctx e;
ctx e, p::sleep(label l)	= (p, F, F) with ctx e;
ctx e, p::sequence(s_1, s_2)	= (n_2, r_1∨r_2, b_1∨b_2) with ctx g where ctx f, n_1::s_2 = (n_2,r_2,b_2) with ctx g and ctx e, p::s_1 = (n_1,r_1,b_1) with ctx f;
ctx e, p::switch(stmt)	= (n∨b, r, F) with ctx f where ctx e, p::stmt = (n,r,b) with ctx f
ctx e, p::if(s_1, s_2)	= (n_1∨n_2, r_1∨r_2, b_1∨b_2) with ctx f_1∨f_2 where ctx e, p::s_1 = (n_1,r_1,b_1) with ctx f_1 and ctx e, p::s_2 = (n_2,r_2,b_2) with ctx f_2;
ctx e, p::while(stmt)	= (n∨b, r, F) with ctx f where ctx e, p::stmt = (n,r,b) with ctx f;
ctx e, p::do(stmt)	= (n∨b, r, F) with ctx f where ctx e, p::stmt = (n,r,b) with ctx f;
ctx e, p::goto(label l)	= (F, F, F) with ctx e∨{l::p};
ctx e, p::continue()	= (F, F, p) with ctx e;
ctx e, p::break()	= (F, F, p) with ctx e;
ctx e, p::return()	= (F, p, F) with ctx e;
ctx e, p::labeled(label l)	= (p∨e.l, F, F) with ctx e\\l;

Legend			
assembler	– *gcc inline assembly code;*	if	– *conditional statement;*
sleep	– *calls to C functions which can sleep;*	switch	– *case statement;*
function	– *calls to other C functions;*	while	– *while loop;*
sequence	– *two statements in sequence;*	do	– *do while loop;*
		labeled	– *labelled statements*

where the *precondition* is an input, the entry condition for a code fragment, and *postconditions* is an output, a tuple consisting of the N, R, B exit conditions according to the logic. The *precontext* is the prevailing `goto` context. The *postcontext* is the output `goto` context, consisting of a set of labelled conditions.

For example, the specification of the empty statement logic is:

```
ctx e, p::empty() = (p, F, F) with ctx e;
```

signifying that the empty statement preserves the entry condition `p` on normal exit (`p`), and cannot exit via return (`F`) or break (`F`). The context (`e`) is unaltered. The full set of logic specifications is given in Table 1. To translate back into the logic presentation in Section 3, consider that

$$\texttt{ctx } e,\ p\ ::\ k = (n, r, b) \texttt{ with ctx } e';$$

means

$$p\ N_e(k) = n \qquad p\ R_e(k) = r$$
$$p\ B_e(k) = b \qquad p\ G_e(k) = e'$$

when written out in the longer format.

5 Software

The source code of the software described here is available for download from `ftp://oboe.it.uc3m.es/pub/Programs/c-1.2.13.tgz` under the conditions of the Gnu Public Licence, version 2.

6 Summary

A C source static analyser for the Linux kernel has been created, capable of dealing with the millions of lines of code in the kernel on a reasonable timescale, at a few seconds per file. It is based on a "three-phase" logic of imperative programming, as described in this article.

References

1. P.T. Breuer, N. Martínez Madrid, L. Sánchez, A. Marín, C. Delgado Kloos: A formal method for specification and refinement of real-time systems. In *Proc. 8'th EuroMicro Workshop on Real Time Systems*, pages 34–42. IEEE Press, July 1996. L'aquilla, Italy.
2. P.T. Breuer, C. Delgado Kloos, N. Martínez Madrid, A. López Marin, L. Sánchez: A Refinement Calculus for the Synthesis of Verified Digital or Analog Hardware Descriptions in VHDL. *ACM Transactions on Programming Languages and Systems (TOPLAS)* 19(4):586–616, July 1997
3. Peter T. Breuer, Marisol Garciá Valls: Static Deadlock Detection in the Linux Kernel, pages 52-64 In *Reliable Software Technologies - Ada-Europe 2004, 9th Ada-Europe International Conference on Reliable Software Technologies, Palma de Mallorca, Spain, June 14-18, 2004*, Eds. Albert Llamosí and Alfred Strohmeier, ISBN 3-540-22011-9, Springer LNCS 3063, 2004.
4. P. Cousot, R. Cousot: Abstract interpretation: A unified lattice model for static analysis of programs by construction or approximation of fixpoints. In *Proc. 4th ACM Symp. on the Principles of Programming Languages*, pages 238–252, 1977.
5. Jeffrey S. Foster, Manuel Fähndrich, Alexander Aiken: A Theory of Type Qualifiers. In *Proc. ACM SIGPLAN Conference on Programming Language Design and Implementation (PLDI'99)*. Atlanta, Georgia. May 1999.
6. Jeffrey S. Foster, Tachio Terauchi, Alex Aiken: Flow-Sensitive Type Qualifiers. In *Proc. ACM SIGPLAN Conference on Programming Language Design and Implementation (PLDI'02)*, pages 1-12. Berlin, Germany. June 2002.
7. Rob Johnson, David Wagner: Finding User/Kernel Pointer Bugs With Type Inference. In *Proc. 13th USENIX Security Symposium, 2004* August 9-13, 2004, San Diego, CA, USA.
8. David Wagner, Jeffrey S. Foster, Eric A. Brewer, Alexander Aiken: A First Step Towards Automated Detection of Buffer Overrun Vulnerabilities. In *Proc. Network and Distributed System Security (NDSS) Symposium 2000*, February 2-4 2000, San Diego, CA, USA.

Generating Test Data for Specification-Based Tests Via Quasirandom Sequences

Hongmei Chi[1,2], Edward L. Jones[2], Deidre W. Evans[2], and Martin Brown[2]

[1] School of Computational Science,
Florida State University, Tallahassee, FL 32306-4120
chi@csit.fsu.edu

[2] Department of Computer and Information Sciences,
Florida A& M University, Tallahassee, FL 32307-5100

Abstract. This paper presents work on generation of specification-driven test data, by introducing techniques based on a subset of quasirandom sequences (completely uniformly distributed sequences) to generate test data. This approach is novel in software testing. This enhanced uniformity of quasirandom sequences leads to faster generation of test data covering all possibilities. We demonstrate by examples that well-distributed sequences can be a viable alternative to pseudorandom numbers in generating test data. In this paper, we present a method that can generate test data from a decision table specification more effectively via quasirandom numbers. Analysis of a simple problem in this paper shows that quasirandom sequences achieve better data than pseudorandom numbers, and have the potential to converge faster and so reduce the computational burden. Functional test coverage, an objective criteria, evaluates the quality of a test set to ensure that all specified behaviors will be exercised by the test data.

Keywords: automatic test case generation, specification-driven test, functional test coverage, quasirandom numbers, well-distributed sequences.

1 Introduction

Software testing [14] is a costly process that is critical for accessing system behavior. The two common strategies are black-box testing, driven by specification of software [22], and white box testing, driven by the software structure [16]. Specification-based testing, whose inputs are derived from a specification, is black-box testing. Specification-based testing of software is to increase the effectiveness of software testing [13]. A formal software specification is one of the most useful documents to have when testing software, since it is a concise and precise description of functionality. Specification-based testing focuses on obtaining test data from specification [19]. Generating test data to cover all specification is a challenge for a complex system [6, 21].

We are developing an approach to deriving test data from quasirandom sequences [20] instead of pseudorandom sequences. Quasirandom sequences are

V.N. Alexandrov et al. (Eds.): ICCS 2006, Part IV, LNCS 3994, pp. 773–780, 2006.
© Springer-Verlag Berlin Heidelberg 2006

constructed to minimize the *discrepancy*, a measure of the deviation from uniformity and therefore quasirandom sequences are more uniformly distributed than pseudorandom sequences. In the past, pseudorandom number generators, such as linear congruential generators [9] have been used in the implementation of random testing. Recently, it has been recognized that the convergence rate of Monte Carlo approaches based on pseudorandom numbers is slow and that an important improvement of the convergence rate can be achieved by using quasi-Monte Carlo methods [12, 15]. This observation is the motivation for the investigation described in this paper.

We will explore the use of completely uniformly distributed sequences in generating test data. The organization of this paper is following. An overview of completely uniformly sequences (quasirandom sequences) is given in § 2. An overview for specification-based tests and test data generation is presented in § 3. In § 4, we analyze a simple specification-based test problem using completely uniformly sequences and numerical results are shown. A discussion of results and conclusion are presented in § 5.

2 Quasirandom Sequences

Pseudorandom numbers are constructed to mimic the behavior of truly random numbers, whereas highly uniform numbers, called quasirandom numbers, are constructed to be as evenly distributed as is mathematically possible. Pseudorandom numbers are scrutinized via batteries of statistical tests that check for statistical independence in a variety of ways, and are also checked for uniformity of distribution, but not with excessively stringent requirements. Thus, one can think of computational random numbers as either those that possess considerable independence, such as pseudorandom numbers; or those that possess considerable uniformity, such as quasirandom numbers [15].

From Fig. 1, we can see the difference between pseudorandom and quasirandom sequences. Pseudorandom numbers are only a subsitute for true random numbers and tend to show clustering effects; while quasirandom numbers tends to more uniformly distributed. There are many applications that do not really require randomness, but instead require numbers that uniformly cover the sample space. Quasirandom sequences are more suitable for such applications. In particular, fewer quasi-random samples are needed to achieve a similar level of accuracy as obtained by using pseudo-random sequences [11, 18].

The original construction of quasirandom sequences is related to the Weyl sequence [10] and the van der Corput sequence. Weyl sequence is based on irrational numbers while the van der Corput sequence is a one-dimension quasirandom sequence based on digital inversion. This digital inversion method is the central idea behind the construction of current quasirandom sequences, such as Halton, Faure and Sobĺ sequences. Niederreiter [15] extended this method to arbitrary bases and dimensions. Weyl sequences is used this paper for numerical experiments in § 4. The definition of Weyl sequence is as follows:

Definition 1. *If θ is an irrational number, then the Weyl sequence $n * \theta$ (mod 1), n=1,2,3,...., is uniformly distributed.*

Here (mod 1) is operation of keeping the fraction part of any number, for example 2.345 (mod 1) = 0.345. The Weyl sequence is easy to implement and well-distributed. For different θ, the different dimensions of the Weyl sequence can be generated.

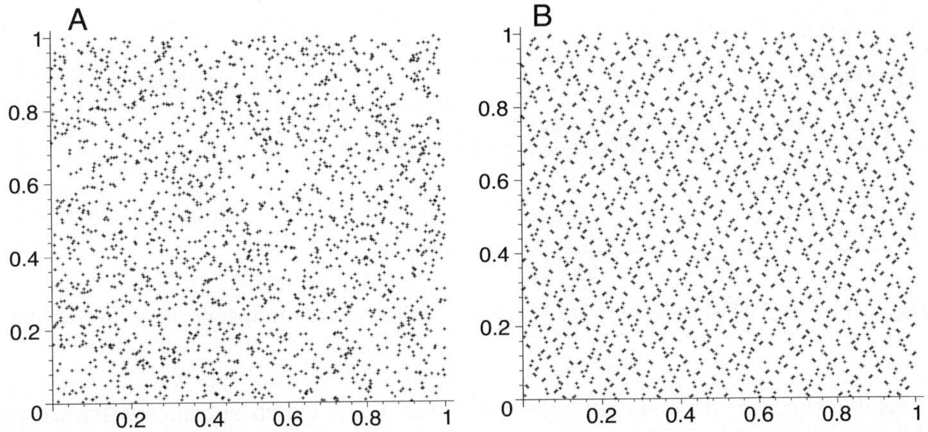

Fig. 1. Comparison of pseudorandom numbers and quasirandom numbers in two dimensions. A: 2000 pseudorandom numbers (linear congruential generator); B: 2000 quasirandom numbers (Weyl sequence).

3 Specification-Based Tests

Although a formal software specification is one of the most useful document to have when testing software, most of software specifications are stated informally in practice and that leaves a lot of ambiguities. Additional specification notations are needed to clarify these statements. A decision table is a rule-based specification in which responses are specified in terms of combinations of conditions met by input data. The decision table is a specification technique that can be used as the basis for test case design [3] [5] [8]. In this section we show by an example how a decision table can provide the basis for defining specification-based tests. We also show that how quasirandom sequences produce the test data based on the decision table.

One measure for test case effectiveness is defined as functional coverage, which measures the thoroughness of testing based on the specification. This is a ratio of the number of rules triggered by the set of test data to the number of rules in the decision table.

Definition 2. *Functional coverage* $= \frac{\#rules-satisfied}{\#rules-in-the-decision-table}$.

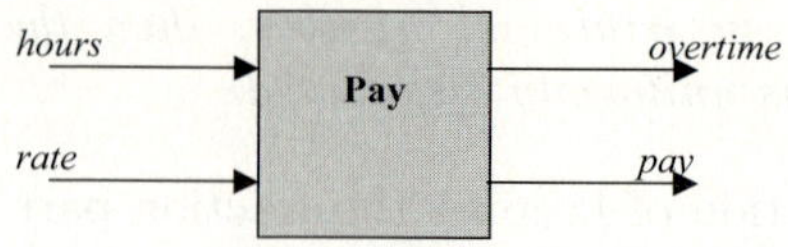

Fig. 2. Black-box schematic specification for Pay

```
CONDITIONS                                          | DECISION RULES
---------------------------------------------------  --------------
 hours>40                                           |  N Y N Y
 rate<10                                            |  Y Y N N
---------------------------------------------------  --------------
 ACTIONS                                            | ACTION RULES
---------------------------------------------------  --------------
 pay = hours * rate; pay = pay;                     |  X - - -
 pay = 1.5 * rate * (hours - 40) + 40 * rate;       |  - X - -
 pay = 40 * rate;                                   |  - - X X
 overtime  = 0;                                     |  X - X X
 overtime = 1.5 * rate * (hours - 40);              |  - X - -
---------------------------------------------------  --------------
```

Fig. 3. Payroll Decision Table (DT1) based on the Narrative Specification in Table 1

Consider the narrative specification in Table 1 [7], which specifies software to compute the weekly pay of employees. The first step in transforming this narrative specification is to identify the stimuli and responses. From the specification, we can deduce that the necessary stimuli (input data) are the hours worked and the hourly salary rate. The responses are the amount of the pay and the resulting overtime paid as shown in Figure 2. According to the specification, the software must determine whether an employee is hourly (rate $\leq$10) or salaried, and whether the employee has exceeded 40 hours of work (hours $\geq$ 40). Figure 3 is a summary of all rules and actions for Pay Specification in Table 1.

Table 1. Payroll Specification

Calculate employee pay, including overtime paid at 1.5 times the hourly rate of hourly employees for time in excess of 40 hours. Salaried employees are not paid overtime, nor do they lose pay when they work less than the normal work week of 40 hours. Hourly employees earn less than 10 per hour.

When testing complicated software with a static specification, it is difficult to determine manually whether each rule has been covered and if there are anomalies in the decision table specification [4]. Jones [7] [8] has developed a tool that uses test data to identify specification anomalies, while using the specification to determine adequacy of the test data. We use quasirandom sequences to provide the test data, and functional coverage as the criterion for measuring the test data. The procedure is simple: according Figure 2, we generate two-dimension

Table 2. Payroll Specification (Extended from Table 1)

Calculate employee pay, including overtime paid at x times the hourly rate. Salaried employees are paid overtime only if they work more than 50 hours, but they do not lose pay when they work less than the normal work week of 40 hours. Hourly employees earn less than \$30 per hour. Employees who work more than 50 hours receive 1.5 times the hourly rate for each overtime hour. Employees who work more than 60 hours are paid 1.5 times the hourly rate for hours up to 60, and 1.6 times the hourly rate for each hour after 60. Employees who work more than 70 hours are paid 1.5 times the hourly rate for hours up to 60, 1.6 times the hourly rate for hours between 60 and 70, and 1.7 times the hourly rate for each hour after 70. Those who work more than 80 hours receive 1.5 times the hourly rate for hours up to 60, 1.6 times the hourly rate for hours between 60 and 70, 1.7 times the hourly rate for hours between 70 and 80, plus a bonus of \$100.

Conditions	**Decision Rules**							
hours > 40	N	N	Y	Y				
hours > 50					Y			
hours > 60						Y		
hours > 70							Y	
hours > 80								Y
rate >= 30	N	Y	N	Y	Y	Y	Y	Y
Actions	**Action Rules**							
regular_pay = hours * rate	X							
regular_pay = 40 * rate		X	X	X	X	X	X	X
over_pay = 0	X	X						
over_pay = 1.5 * rate * (hours – 40)			X		X			
over_pay = 1.5 * rate * (20) + 1.6 * rate * (hours – 60)						X		
over_pay = 1.5 * rate * (20) + 1.6 * rate * (10) + 1.7 * rate * (hours – 70)							X	
over_pay = 1.5 * rate * (20) + 1.6 * rate * (10) + 1.7 * rate * (10) + 100								X

Fig. 4. Payroll Decision Table 2 (DT2) based on the Narrative Specification in Table2

Table 3. Test Results for Decision Tables (#test data for full funtional coverage)

Generator	Decision Table DT1		Decision Table DT 2	
	#rules	#test data pairs	#rules	#test data pairs
PRNG	4	6	8	29
QRNG	4	5	8	11

Table 4. Test data Generated by a QRNG for Decision Tables DT 1 and DT 2

	Test Data for DT 1					Test Data DT 2										
# DT rules	4					8										
# test pairs	5					11										
hours	48	58	25	29	70	88	76	43	110	36	62	67	51	15	78	1
rate	15	13	8	19	6	16	40	32	52	40	56	40	51	31	38	13
rule	2	2	3	4	1	3	5	7	4	2	6	6	7	2	5	1
Uncovered rules	none					none										

Table 5. Test data Generated by a PRNG for DecisionTables DT 1 and DT 2

	Test Data for DT 1						Test Data for DT 2										
# DT rules	4						8										
#Test pairs	6						11										
hours	4	53	54	56	34	37	6	79	81	84	52	56	14	103	67	26	100
rate	6	5	11	3	1	13	19	15	33	10	3	41	20	55	22	3	41
rule	3	1	2	1	3	4	1	3	4	3	3	7	1	4	3	1	4
Uncovered rules	none						2, 5, 6, 8										

test data sets (hour, rate), and check functional coverage to see how many decision table rules are satisified (covered) by one or more test data pairs. The measure of interest for comparing pseudo-random and quasirandom generation of data sets is the number of test data needed to reach functional coverage of 1. The numerical results are shown in Section 4.

4 Numerical Experiments

We need a more complicated decision table for generating test data. Therefore, we extend the decision table (in Table 1 and Figure 3) and make more rules. The new specification narrative and decision table are shown in Table 2 and Figure 4, respectively.

In order to compare the effectiveness of quasirandom numbers, we use quasirandom sequences and pseudorandom sequences to produce the test data. The numerical results are shown in Table 2. The pseudorandom number generator(PRNG) we used in this paper is one of linear congruential generators (LCGs) in Numerical Recipe in C [17]. This LCG is defined as following:

Definition 3. *The LCG determined by* $x_n = ax_{n-1} \pmod{m}$ *with* $a = 16807$ *and* $m = 2^{31} - 1$ *has a period of* $2^{31} - 2$.

The quasirandom number generator is Weyl sequences and we used the same implementation in [2]. The Weyl sequence we used in this paper with $\theta = 2$. The results in Table 3 show that quasirandom number generator(QRNG) significantly converges faster, i.e., covers all rules with fewer test data. The quasirandom test data are presented in Table 4, the pseudorandom test data in Table 5. The faster convergence is more marked for the large decision table.

5 Conclusions and Future Work

A new scheme for generating test data via quasirandom sequences is proposed. The advantage of this scheme is that we can provide test data based on a specification automatically. This scheme is an alternative to generate test data manually or from pseudorandom numbers. Our numerical results, though preliminary , are promising. Should our observations about faster convergence (full coverage with fewer test data) hold, quasirandom test generation may offer economical advantages over pseudo-random testing. A broader question is whether quasirandom testing is superior to pseudo-random testing, in terms of efficiency and effectiveness. Addressing this question may require a replication of past studies such as in Abdurazik [1].

In the future, we will extend the study given in this paper to support the test-driven specification paradigm of Jones [7, 8] when applied to more complex problems requiring large, complex decision table specifications. Ongoing work includes the development of a library of quasirandom generation routines to support specification-based test generation.

One of the limitations of this scheme may occur when input is extremely distributed instead of uniformly distributed. On the other hand, because many of the accepted software testing practices are based on partitioning and sampling, the impact of non-uniform distributions may be negligible.

References

1. A. Abdurazik, P. Ammann, W. Ding, and J. Offutt. Evaluation of three specification-based testing criteria. *Sixth IEEE International Conference on Complex Computer Systems (ICECCS'00)*, pages 179–187, 2000.
2. P. Beerli, H. Chi, and E. Jones. Quasi-monte carlo method in population genetics parameter estimation. *Mathematics and Computers in Simulation*, In press, 2006.
3. R. V. Binder. *Testing Object-oriented systems: models, patterns and tools.* Addison-Wesley, Reading, Massachusetts, 1999.
4. K. H. Chang, S. Liao, and R. Chapman. Test scenario geneneration based on formal specification and usage. *International Journal of Software Engineering and Knowledge Engineering*, 10(2):1–17, 2000.
5. N. Glora, H. Pu, and W. O. Rom. Evaluation of process tools in systems analysis. *Information and Technology*, 37:1191–1126, 1995.
6. J. B. Goodenough and S. L. Gerhart. Toward a theory of test data selection. In *Proceedings of the international conference on Reliable software*, pages 493–510, 1975.
7. E. L. Jones. Automated support for test-driven specification. In *Proceedings of the 9th IASTED International Conference on Software Engineering and Applications*, pages 218–223, 2005.
8. E. L. Jones. Test-driven specification: Paradigm and automation. In *44th ACM Southeast Conference, March 10-12*, 2006.

9. D. E. Knuth. *The Art of Computer Programming, vol. 2: Seminumerical Algorithms.* Cambridge University Press, New York, 1997.
10. L. Kuipers and H. Niederreiter. *Uniform Distribution of Sequences.* John Wiley and Sons, New York, 1974.
11. W. L. Loh. On the asymptotic distribution of scrambled net quadrature. *Annals of Statistics*, 31:1282–1324, 2003.
12. W.J. Morokoff and R.E. Caflish. Quasirandom sequences and their discrepancy. *SIAM Journal on Scientific Computing*, 15:1251–1279, 1994.
13. H. Muccini, A. Bertolino, and P. Inverardi. Using software architecture for code testing. *IEEE Trans. on Software Engineering*, 30(3):160–171, 2004.
14. G. Myers, C. Sandler, T. Badgett, and T. Thomas. *The Art of Software Testing.* John Wiley and Sons, New Jersey, 2004.
15. H. Niederreiter. *Random Number Generations and Quasi-Monte Carlo Methods.* SIAM, Philadelphia, 1992.
16. S. Ntafos. A comparison of some structural testing strategies. *IEEE Trans. on Software Engineering*, 14(6):868—874, 1988.
17. W. H. Press, S. A. Teukolsky, W. T. Vetterling, and B.P. Flannery. *Numerical Recipes in C.* Addison-Wesley, Reading, Massachusetts, 1992.
18. J. Spanier and E. Maize. Quasirandom methods for estimating integrals using relatively small sampling. *SIAM Review*, **36**:18–44, 1994.
19. P. Stocks and D. Carrington. Test template framework: A specification-based testing case study. In *Proceedings of Int. Symp. Software Testing and Analysis*, pages 11–18, 1993.
20. S. Tezuka. *Uniform Random Numbers, Theory and Practice.* Kluwer Academic Publishers, IBM Japan, 1995.
21. P. Variyam. Specification-driven test generation for analog circuits. *IEEE Trans. on Computer-aided Design*, 19(10):1189—1201, 2000.
22. G. Wimmel, H. Ltzbeyer, A. Pretschner, and O. Slotosch. Specification based test sequence generation with propositional logic. *Software Testing, Verification and Reliability*, 10(4):229–248, 2000.

Support Vector Machines for Regression and Applications to Software Quality Prediction

Xin Jin[1], Zhaodong Liu[1], Rongfang Bie[1,*], Guoxing Zhao[2,3],and Jixin Ma[3]

[1] College of Information Science and Technology,
xinjin796@126.com, liuzd661@163.com, rfbie@bnu.edu.cn
Beijing Normal University, Beijing 100875, P.R. China
[2] School of Mathematical Sciences, Beijing Normal University, Beijing 100875, P.R. China
[3] School of Computing and Mathematical Science,
The University of Greenwich, London SE18 6PF, U.K
G.Zhao@gre.ac.uk,
j.ma@gre.ac.uk

Abstract. Software metrics are the key tool in software quality management. In this paper, we propose to use support vector machines for regression applied to software metrics to predict software quality. In experiments we compare this method with other regression techniques such as Multivariate Linear Regression, Conjunctive Rule and Locally Weighted Regression. Results on benchmark dataset MIS, using mean absolute error, and correlation coefficient as regression performance measures, indicate that support vector machines regression is a promising technique for software quality prediction. In addition, our investigation of PCA based metrics extraction shows that using the first few Principal Components (PC) we can still get relatively good performance.

1 Introduction

Software quality management, which is an important aspect of software project development, is an ongoing comparison of the actual quality of a product with its expected quality [16]. Software metrics are the key tool in software quality management. Many researchers have sought to analyze the connection between software metrics and code quality [8][13][14][15][17][23]. The methods they used fall into four mainly categories: association analysis (association rules), clustering analysis (k-means, fuzzy c-means), classification analysis (decision trees, layered neural networks, Holographic networks, logistic regression, genetic granular classification [12]) and prediction analysis (linear regression).

In this paper we propose to use support vector machines for regression to predict software quality. Support vector machine technique has attracted many researchers in optimization and machine learning areas [19,22]. In the case of regression, the objective is to choose a hyperplane with small norm while simultaneously minimizing the sum of the distances from the data points to the hyperplane.

* Corresponding author.

V.N. Alexandrov et al. (Eds.): ICCS 2006, Part IV, LNCS 3994, pp. 781 – 788, 2006.
© Springer-Verlag Berlin Heidelberg 2006

The remainder of this paper is organized as follows. In Section 2, we describe the software metrics and benchmark dataset we used. Section 3 presents the support vector regression method. Section 4 describes three comparison algorithms. Section 5 introduces PCA. Section 6 presents the performance measures and the experiment results. Conclusions are covered in Section 7.

2 Software Metrics

We investigated the twelve software metrics, as shown in Table 1, which are used in the famous benchmark dataset MIS [15,12]. Simple counting metrics such as the number of lines of source code or Halstead's number of operators and operands describe how many "things" there are in a program. More complex metrics such as McCabe's cyclomatic complexity or Bandwidth attempt to describe the "complexity" of a program, by measuring the number of decisions in a module or the average level of nesting in the module, respectively.

Table 1. Description of the MIS dataset with a detailed characterization of the software metrics [18]

Metrics	*Detailed Description*
LOC	Number of lines of code including comments, declarations and the main body of the code
CL	Number of lines of code, excluding comments
TChar	Number of characters
TComm	Number of comments
MChar	Number of comment characters
DChar	Number of code characters
N	Halstead's Program Length N = N1+N2, N1 is the total number of operators, N2 is the total number of operands
N'	Halstead's Estimate of Program Length $N' = n_1 \log_1 n_1 + n_2 \log_2 n_2$, n_1 is the number of unique operators, n_2 is the number of unique operands
NF	Jensen's Estimate of Program Length Metric $\log_1 n_1! + \log_2 n_2!$
V(G)	McCabe's Cyclomatic Complexity Metric, where V(G) = e−n+2, and e represents the number of edges in a control graph of n nodes
BW	Belady's Bandwidth measure $BW = (\Sigma_i i L_i)/n$, L_i represents the number of nodes at level "i" in a nested control flow graph of n nodes. This measure indicates the average level of nesting or width of the control flow graph representation of the program
Changes	Number of changes

In this study, MIS is represented by a subset of the whole MIS data with 390 modules written in Pascal and FORTRAN. These modules consist of approximately 40,000 lines of code. Our goal is to develop a prediction model of software quality in which the number of modifications (changes) is projected on a basis of the values of the 11 software metrics that is used to characterize a software module. We cast the problem in the setting of regression, the explanatory variables are the first eleven software metrics and the dependent variable is the number of changes. Software modules, which have no changes, could be deemed to be fault-free, while software modules with the number of changes being too big, for example, over 10, can be sought as potentially highly faulty modules.

3 Support Vector Machine Regression (SVR)

Specifically, the ε-insensitive support vector regression will be used for predicting software quality. In the ε-insensitive support vector regression, our goal is to find a function $f(x)$ that has an ε-deviation from the actually obtained target y_i for all training data and at the same time is as flat as possible [10]. Suppose $f(x)$ takes the following form:

$$f(x) = wx + b \qquad w \in X, b \in \Re. \tag{1}$$

Then, we have to solve the following problem:

$$\begin{aligned} &\min \frac{1}{2}\|w\|^2 \\ \text{Subject to} \quad & y_i - wx_i - b \le \varepsilon \\ & wx_i + b - y_i \le \varepsilon \end{aligned} \tag{2}$$

In the case where the constraints are infeasible, we introduce slack variables. This case is called soft margin formulation, and is described by the following problem.

$$\begin{aligned} &\min \frac{1}{2}\|w\|^2 + C\sum_{i=1}^{l}(\xi_i + \xi_i^*) \\ \text{Subject to} \quad & y_i - wx_i - b \le \varepsilon + \xi_i \qquad \xi_{i,}\xi_i^* \ge 0 \\ & wx_i + b - y_i \le \varepsilon + \xi_i^* \qquad C \succ 0 \end{aligned} \tag{3}$$

where C determines the trade-off between the flatness of the $f(x)$ and the amount up to which deviations larger than ε are tolerated. This is called ε-insensitive loss function:

$$|\xi|_\varepsilon = \begin{cases} 0 & if|\xi| \le \varepsilon \\ |\xi| - \varepsilon & if|\xi| \succ \varepsilon \end{cases} \tag{4}$$

By constructing the Lagrangian function, we formulate the dual problem as

$$Max \ -\frac{1}{2}\sum_{i=1}^{l}\sum_{j=1}^{l}(\lambda_i - \lambda_i^*)(\lambda_j - \lambda_j^*)x_i x_j - \varepsilon\sum_{i=1}^{l}(\lambda_i + \lambda_i^*) + \sum_{i=1}^{l} y_i(\lambda_i - \lambda_i^*) \tag{5}$$

$$\text{Subject to} \qquad \sum(\lambda_i - \lambda_i^*) = 0 \qquad \lambda_i, \lambda_i^* \in (0, C)$$

At the optimal solution, we obtain

$$w^* = \sum_{i=1}^{l}(\lambda_i - \lambda_i^*)x_i \tag{6}$$

$$f(x) = \sum_{i=1}^{l}(\lambda_i - \lambda_i^*)x_i x + b^* \tag{7}$$

We compute the optimal value of b from the complementary slackness conditions:

$$b^* = y_i - w^* x_i - \varepsilon \quad \lambda_i \in (0, C)$$

$$\text{and } b^* = y_i - w^* x_i + \varepsilon \quad \lambda_i^* \in (0, C) \tag{8}$$

In some case, we need to map input space into feature space and try to find a hyperplane in the feature space by using the trick of kernel functions:

$$Max \ -\frac{1}{2}\sum_{i=1}^{l}\sum_{j=1}^{l}(\lambda_i - \lambda_i^*)(\lambda_j - \lambda_j^*)K(x_i, x_j) - \varepsilon\sum_{i=1}^{l}(\lambda_i + \lambda_i^*) + \sum_{i=1}^{l} y_i(\lambda_i - \lambda_i^*) \tag{9}$$

At the optimal solution, we obtain

$$w^* = \sum_{i=1}^{l}(\lambda_i - \lambda_i^*)K(x_i), \tag{10}$$

$$f(x) = \sum_{i=1}^{l}(\lambda_i - \lambda_i^*)K(x_i, x) + b \tag{11}$$

where K(.,.) is a kernel function. Any symmetric positive semi-definite function, which satisfies Mercer's conditions, can be used as a kernel function in the SVMs context [10, 20]. In this paper, we use the linear kernel [11].

4 Regressors for Comparison

4.1 Multivariate Linear Regression

Multivariate Linear Regression (MLR) finds a set of basis vectors wx_i and corresponding regressors β_i in order to minimize the mean square error of the vector y. The basis vectors are described by the matrix $Cxx^{-1}Cxy$. A low-rank approximation to this problem can be defined by minimizing

$$\varepsilon^2 = E\left[\left\| y - \sum_{i=1}^{M} \beta_i w_{x_i}^T x w_{y_i} \right\|^2\right] \tag{12}$$

where M=dim(y), $N<M$ and the orthogonal basis wy_i span the subspace of y which gives the smallest mean square error given the rank N.

4.2 Conjunctive Rule

Conjunctive Rule (CR) consists of antecedents "AND"ed together and the consequent (prediction value) for the regression. This learner selects an antecedent by computing the Information Gain of each antecendent and prunes the generated rule using Reduced Error Prunning (REP) or simple pre-pruning based on the number of antecedents [1]. The Information is the weighted average of the mean-squared errors of both the data covered and not covered by the rule. In pruning, the weighted average of the mean-squared errors on the pruning data is used.

4.3 Locally Weighted Regression

Locally Weighted Regression (LWR) is a memory-based method that performs a regression around a point of interest using only training data that are local to that point [3,4]. We consider here a form of locally weighted regression that is a variant of the LOESS model [5]. The LOESS model performs a linear regression on points in the data set, weighted by a kernel centered at *x*. The kernel shape is a design parameter for which we use the Linear.

5 Principal Component Analysis

Principal Component Analysis (PCA) is a famous multivariate data analysis method that is useful in linear feature extraction [6,7]. The PCA finds a linear transformation y=Wx such that the retained variance is maximized. Each row vector of W corresponds to the normalized orthogonal eigenvector of the data covariance matrix.

One simple approach to PCA is to use singular value decomposition (SVD). Let us denote the data covariance matrix by $R_x(0) = E\{x(t)x^T(t)\}$. Then the SVD of $R_x(0)$ gives $R_x(0) = UDU^T$, where $U = [U_s, U_n]$ is the eigenvector matrix (i.e. modal matrix) and D is the diagonal matrix whose diagonal elements correspond to the eigenvalues of $R_x(0)$ (in descending order). Then the PCA transformation from *m*-dimensional data to *n*-dimensional subspace is given by choosing the first *n* column vectors, i.e., *n* principal component vector y is given by $y = U_s^T x$.

6 Experiment Results

10-fold cross-validation on the benchmark MIS dataset, available on [9], is used for estimating prediction performance.

6.1 Performance Measures

We use the following performance measures:

Mean Absolute Error (*MAE*): MAE provides a measure of how close a prediction model is to the actual data.

$$\mathrm{MAE} = \frac{1}{n}\sum_{i=1}^{n} | a_i - p_i | \tag{13}$$

where a_i and p_i is the actual and predicted value for the *i*th test case. MAE ranges from 0 to infinity, with 0 corresponding to the ideal. the smaller the MAE the better.

Correlation Coefficient (*CC*): CC is a measure of how well trends in the predicted values follow trends in past actual values [21]. It measures how well the predicted values from a forecast model "fit" with the real-life data. A perfect fit gives a CC of 1.0.

$$\mathrm{CC} = \frac{\sum_{i=1}^{n}(p_i - \bar{p})(a_i - \bar{a})}{\sum_{i=1}^{n}(p_i - \bar{p})^2 \sum_{i=1}^{n}(a_i - \bar{a})^2} \tag{14}$$

The higher the CC the better.

6.2 Results

Fig.1 shows MAE of different regressors on the original MIS metrics and the PCA extracted data. On the original metrics, SVR with the linear kernel (SVR_L) get the best performance by achieving a minimum of 3.98 MAE. As shown in Fig.1, using the first few Principal Components (PC) we can still get relatively good performance. For CRule, using PCs is even better than using the original metrics.

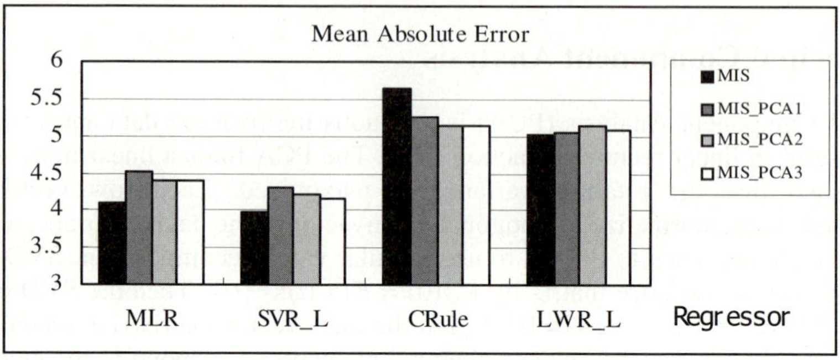

Fig. 1. MAE of different regressors on the original MIS metrics and the PCA extracted data. MLR= Multivariate Linear Regression, SVR_L=Support Vector Regression with Linear kernel, CRule=Conjunctive Rule, LWR_L=Locally Weighted Regression with Linear kernel. MIS_PCA1 means using the first Principal Component, MIS_PCA2 means using the first two Principal Components, etc.

Fig.2 shows CC of different regressors on the original MIS metrics and the PCA extracted data. SVR with Linear kernel get the best performance by achieving a maximum of 0.77 CC. For CRule, using PCs is better than using the original metrics.

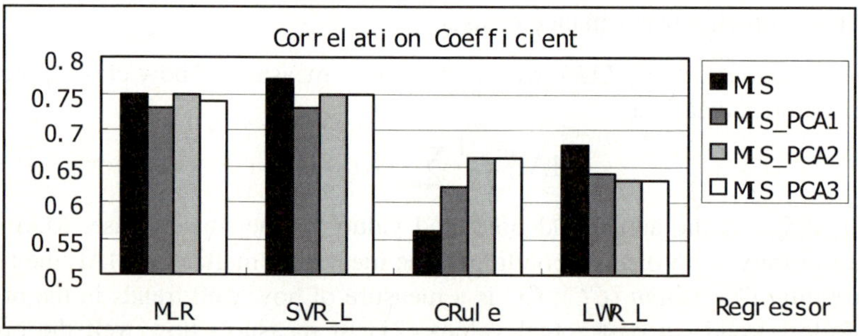

Fig. 2. CC of different regressors on the original MIS metrics and PCA extracted data. MLR= Multivariate Linear Regression, SVR_L=Support Vector Regression with Linear kernel, CRule=Conjunctive Rule, LWR_L=Locally Weighted Regression with Linear kernel. MIS_PCA1 means using the first Principal Component, MIS_PCA2 means using the first two Principal Components, etc.

7 Conclusions

In this paper we propose to use support vector machines for regression applied to software metrics to predict software quality. Comparison is done with three other techniques: Multivariate Linear Regression, Conjunctive Rule and Locally Weighted Regression. Results on benchmark dataset MIS, using MAE and CC as performance measures, indicate that support vector machines regression is a promising technique for software quality prediction. The investigation of PCA based metrics extraction show that using the first few Principal Components we can still get relatively good performance.

Acknowledgments

This work was supported by the National Science Foundation of China under the Grant No. 10001006 and No. 60273015.

References

1. I.Witten, E.Frank: Data Mining –Practical Machine Learning Tools and Techniques with Java Implementation. Morgan Kaufmann (2000)
2. Friedman, J.H.: Stochastic Gradient Boosting. Technical Report, Stanford University (1999)
3. Atkeson, C., A. Moore, S. Schaal: Locally Weighted Learning. AI Reviews (1996)
4. Cohn, D.A., Ghahramani, Z., Jordan, M.I.: Active Learning with Statistical Models. Journal of Artificial Intelligence Research, Vol. 4, pp. 129-145 (1996)
5. Cleveland, W., Devlin, S., Grosse, E.: Regression by Local Fitting. Journal of Econometrics, 37, pp. 87-114 (1988)
6. Zeng, X.Y., Chen, Y.W., et al: A New Texture Feature based on PCA Maps and Its Application to Image Retrieval. IEICE Trans. Inf. and Syst., E86-D 929-936 (2003)
7. Diamantaras, K.I. and Kung, S.Y.: Principal Component Neural Networks: Theory and Applications. John Wiley & Sons, INC (1996)
8. D. Garmus, D. Herron: Measuring The Software Process, Prentice Hall, Upper Saddle River, NJ (1996)
9. MIS: http://www.win.tue.nl/~jromijn/2IW30/2IW30_statistics/LTU/DATA/CH12 (2006)
10. Theodore B. Trafalis, Huseyin Ince: Support Vector Machine for Regression and Applications to Financial Forecasting, ijcnn, p.6348, IEEE-INNS-ENNS International Joint Conference on Neural Networks (IJCNN'00) Volume 6 (2000)
11. A.J. Smola and B. Scholkopf, A Tutorial on Support Vector Regression, NEUROCOLT2 Technical Report Series, NC2-TR-1998-030 (1998)
12. Witold Pedrycz, Giancarlo Succi: Genetic Granular Classifiers in Modeling Software Quality. Journal of Systems and Software 76(3): 277-285 (2005)
13. W. Pedrycz, G. Succi, M.G. Chun, Association Analysis of Software Measures, Int. J of Software Engineering and Knowledge Engineering, 12(3): 291-316 (2002)
14. K.H. Muller, D.J. Paulish, Software Metrics, IEEE Press/Chapman & Hall, London, 1993.
15. J. C. Munson, T. M. Khoshgoftaar: Software Metrics for Reliability Assessment, in Handbook of Software Reliability and System Reliability, McGraw-Hill, Hightstown, NJ, 1996.
16. 16.Scott Dick, Aleksandra Meeks, Mark Last, Horst Bunke, Abraham Kandel: Data Mining in Software Metrics Databases. Fuzzy Sets and Systems 145(1): 81-110 (2004)
17. W. Pedrycz, G. Succi, P. Musilek, X. Bai: Using Self-Organizing Maps to Analyze Object Oriented Software Measures. J. of Systems and Software, 59, 65-82 (2001)

18. P. K. Simpson: Fuzzy Min-Max Neural Networks. Part 1: Classification, IEEE Trans. Neural Networks, Vol. 3, pp. 776-786 (1992)
19. M.S. Bazaraa, H.D. Sherali, and C.M. Shetty: Nonlinear Programming: Theory and Algorithms, John Wiley &Sons Inc., New York (1993)
20. C. Cortes and V. Vapnik, Support Vector Networks, Machine Learning, 20, 273-297 (1995)
21. Correlation Coefficient: http://www.neatideas.com/cc.htm (2006)
22. T. Joachims, Making Large-Scale SVM Learning Practical, Technical Report, LS-8-24, Computer Science Department, University of Dortmund (1998)
23. R. Subramanyan and M.S. Krishnan: Empirical Analysis of CK Metrics for Object-Oriented Design Complexity: Implications for Software Defects, IEEE Trans. Software Eng., Vol. 29, pp. 297-310, Apr (2003)

Segmentation of Software Engineering Datasets Using the M5 Algorithm

D. Rodríguez[1], J.J. Cuadrado[2], M.A. Sicilia[2], and R. Ruiz[3]

[1] The University of Reading,
Reading, RG6 6AY, UK
d.rodriguezgarcia@rdg.ac.uk
[2] The University of Alcalá,
28805 - Alcalá de Henares (Madrid), Spain
{jjcc, msicilia}@uah.es
[3] The University of Seville,
41012 - Avda Reina Mercedes s/n., Sevilla, Spain
rruiz@rdg.ac.uk

Abstract. This paper reports an empirical study that uses clustering techniques to derive segmented models from software engineering repositories, focusing on the improvement of the accuracy of estimates. In particular, we used two datasets obtained from the International Software Benchmarking Standards Group (ISBSG) repository and created clusters using the M5 algorithm. Each cluster is associated with a linear model. We then compare the accuracy of the estimates so generated with the classical multivariate linear regression and least median squares. Results show that there is an improvement in the accuracy of the results when using clustering. Furthermore, these techniques can help us to understand the datasets better; such techniques provide some advantages to project managers while keeping the estimation process within reasonable complexity.

Keywords: Effort estimation, Data mining, Tress, M5.

1 Introduction

Effort estimation is one of first and more important activities that project managers face at the beginning of each project. Accurate estimates are an integral part of all process improvement activities. The current state-of-practice for effort and cost estimates entails the use one of the many public (e.g., COCOMO [2]) or private (e.g., PRICE S [13]) parametric models. Both public or commercial models use classical regression as their foundation for deriving equations from historical project databases. During the last decade, several organizations such as the International Software Benchmarking Standards Group (ISBSG) [10] have started to collect project management data from a variety of organizations. In this way, companies without historical datasets could use these generic databases for estimation or companies already collecting data could compare themselves

V.N. Alexandrov et al. (Eds.): ICCS 2006, Part IV, LNCS 3994, pp. 789–796, 2006.
© Springer-Verlag Berlin Heidelberg 2006

with other industries, i.e., benchmarking. One problem faced by project managers using such databases is the heterogeneity of the projects (e.g., the latest release of the ISBSG has more than 50 attributes and 3000 instances). Specifically, heterocestacity (non-uniform variance) is known to be a problem affecting datasets that combine data from heterogeneous sources. For example, a straightforward application of least square regression using the 709 projects, which can be obtained from the Reality estimation tool accompanying the ISBSG repository, results in measures of median relative error of 280% and less than 23% of the estimates are within the 75% of actual values. This leads to poor estimates in the software engineering. In this paper, we describe an empirical study on the appropriateness of model trees, i.e., a decision tree with a linear regression model for each subset, with relation to predictive quality. Concretely, we derived several datasets from the ISBSG repository and also compare the outputs of using M5 with multivariate Linear Regression (LR) and Least Median Squares (LMS) .

This paper is organized as follows. Section 2 and 3 describe respectively the methods of analysis and evaluation techniques used in this paper. Section 4 describes the datasets used for this analysis and the results of applying the methods of analysis, followed by a discussion (Section 5). Finally, Section 6 concludes the paper and outlines our future work.

2 Related Work

2.1 Regression Models

Regression techniques [16] are a kind of algorithmic techniques which looks for an equational model to fit a set of observed data values.

Linear Regression (LR) is the classical linear regression model. It is assumed that there is a linear relationship between a dependant variable (e.g., effort) with a set of or independent variables, i.e., attributes (e.g. *size in function points, team size, development platform,* etc.). The aim is to adjust the data to a model so that $y = \beta_0 + \beta_1 x_1 + \beta_2 x_2 + ... + + \beta_k x_k + e$.

The linear least square method finds the line that minimises the sum of squared errors. A problem with this method is that it assumes a normal distribution and cannot cope well with outliers. Regression analysis can be made more robust for outliers using the least median squares method. Least median square regression analysis is suitable for small data-sets, and it is sensitive to abnormal observations and errors. Least Median Square (LMS) is a robust regression technique that includes outlier detection.

2.2 Rules, Decision Trees and Model Trees

A *rule-based system* [9] consists of a library of rules of the form: *if* (assertion) *then* action. Such rules are used to elicit information or to take appropriate actions when specific knowledge becomes available. These rules reflect a way to reason about the relationships within the domain. Their main advantage is their simplicity. However, they are mainly appropriate for deterministic problems, which is not usually the case in software engineering [1]. To overcome this problem, rules can also contain a certainty measure in the premises and/or in the conclusions.

Decision trees [3] are used for predicting or explaining outputs from observations. In such a tree, each node is a leaf indicating a class or an internal decision node that specifies some test to be carried out. If the output values conform to intervals, then the decision trees are called regression trees, whereas if they do correspond to a nominal or ordinal scale they are called classification trees. There are many tree-building algorithms such as C4.5 (Quinlan, 1993) which determine which attributes best classifies the remaining data, and then the tree is constructed iteratively. The main advantage of decision trees is their immediate conversion to rules that can be easily interpreted by decision-makers. For numeric prediction in data mining, it is common to use regression trees or model trees [3]. Both techniques build a decision tree structure where each leaf is responsible for a particular local regression of the input space, in our case the software engineering dataset. The difference between them is that while a regression tree generates constant output values for subsets of input data (zero-order models), model trees generate linear (first-order) models for each subset.

The *M5 algorithm* builds trees whose leaves are associated to multivariate linear models and the nodes of the tree are chosen over the attribute that maximizes the expected error reduction as a function of the standard deviation of output parameter. In this paper, we have applied a version of M5, called M5P -Prime- implemented in the WEKA toolkit [21] to the ISBSG dataset [10]. Weka's M5 algorithm actually builds a decision tree which divides the attribute space in an orthohedric clusters, with the border parallel to the axis. An advantage of model trees is that they can be easily converted into rules; each branch of the tree has a condition as follows: *attribute* $\leq$ *value* or *attribute* $>$ *value*.

We will now explain how the M5 algorithm works using the *Reality* dataset, which is a subset of the ISBSG repository with 709 projects used by the Reality estimation tool accompanying the ISBSG repository and further explained in the next section. For example, Weka's M5 algorithm created only a decision node to calculate the *NormalisedWorkEffort*, and therefore, two linear models, LM1 and LM2:

```
UnadjustedFunctionPoints <= 343:
LM1 (510/53.022\%)
NormalisedWorkEffort =
 90.5723 * DevelopmentPlatform=MF,MR
 + 63.5148 * LanguageType=ApG,3GL,2GL
 + 628.9547 * LanguageType=3GL,2GL
 + 184.9949 * ProjectElapsedTime
 + 10.9211 * UnadjustedFunctionPoints
 - 545.8004
```

```
UnadjustedFunctionPoints > 343:
LM2(199/318.225\%) NormalisedWorkEffort =
  10189.7332 * DevelopmentPlatform=MF,MR
  - 5681.5476 * DevelopmentPlatform=MR
  + 155.8191 * LanguageType=ApG,3GL,2GL
  + 5965.379 * LanguageType=3GL,2GL
  + 551.4804 * ProjectElapsedTime
  + 4.3129 * UnadjustedFunctionPoints
  - 8118.3275
```

The two branches generated are interpreted as follows: if *UnadjustedFunctionPoints*, i.e. size of the system, is less than 343 then we apply LM1 otherwise LM2. In this case, both LM1 and LM2 are lineal models where the *Normalised-*

WorkEffort is the dependent variable; M5 however can assign to the dependent variable either a constant or a linear equation (in the majority of the cases).

The categorical data of the linear regression function obtained by Weka is calculated by substituting the value for the appropriate value wherever it occurs. For example, if we had an instance with *DevelopmentPlatform* equals to MF, *LanguageType* equals to *ApG* and *UnadjustedFunctionPoints* less than 343.

For evaluating each categorical expression, if the value of the category on the left hand side is equal to any of the categories on the right hand side of the equation, then we substitute the entire equation with value 1; otherwise with the value 0. Following the example we obtain:

```
LM num: 1 NormalisedWorkEffort =
    90.5723 * 1, MR
    + 63.5148 * 1
    + 628.9547 * 0
    + 184.9949 * ProjectElapsedTime
    + 10.9211 * UnadjustedFunctionPoints
    - 545.8004
```

Further information provided by the M5 algorithm within brackets is as interpreted as follows. For the LM1 branch, there are 510 instances and the error in that leave is 53.022%; for LM2, there were 199 projects and the error was 318.225%.

3 Evaluation of the Techniques

We have created 2 subsets of the ISGBN repository. One of them, called the Reality dataset, is included in the repository as part of the ISBSG Reality Checker tool (used for effort estimation). Datasets are divided into training and test, such that approximately 2/3 of the instances are used for training and the other 1/3 of the instances are used for testing, i.e. evaluation. This is a common practice in data mining; other other techniques such as cross validation can be more accurate, they are however more complex. In statistics as well as in data mining, with linear regression models used in this work, the goodness of fit of a model is usually measured by the correlation and by the *mean squared error*. In the software Engineering domain however, it is common to compare the goodness of each technique using the Mean Magnitude of Relative Error (MMRE) and Pred(%), proposed by Conte et al. [4]:

- MMRE is calculated as $\frac{1}{n} \cdot \sum_{i=1}^{n} \left| \frac{e_i - a_i}{e_i} \right|$, where n is the sample size , e_i is the estimated value for the *i-th* element and a_i is the actual value.
- Prediction at Level l –$Pred(l\%)$– is defined as the number of cases whose estimations are under the $l\%$, divided by the total number of cases. For example, $Pred(25) = 0.75$ means that 75% of cases estimates are within the inside 25% of its actual value.

In Software Engineering, standard criteria for a model to acceptable are roughly $Pred(25) \geq 0.75$ and $MMRE \leq 0.25$ [6].

4 Approach and Results

The International Software Benchmarking Standards Group (ISBSG), a non-profit organization, maintains a software project management repository from a variety of organizations. The ISBSG checks the validity and provides benchmarking information to companies submitting data to the repository. Furthermore, it seems that the data is collected form large and successful organizations. In general, such organizations have mature processes and well established data collection procedures. In this work, we have used release 8, which contains 2028 projects and more than 55 attributes per project.

Before using the dataset, there are a number of issues to be taken into consideration. An important attribute is the quality rating given by the ISBSG can be from A (where the submission satisfies all criteria for seemingly sound data) to D (where the data has some fundamental shortcomings). According to ISBSG only projects classified as A or B should be used for statistical analysis. Also, many attributes in ISGSB are categorical attributes or multi-class attributes that need to be pre-processed for this work (e.g. the project scope attribute which indicates what task were included in the project work effort –planning , specification, design, development and testing– were grouped). Another problem of some attributes is the large number of missing instances. As a result, we had to do some pre-processing. We selected some attributes and instances manually. There are quite a large number of variables in the original dataset that we did not consider relevant or they had too many missing values to be considered in the data mining process. From the original database, we only considered the IFPUG estimation technique and those that can be considered very close variations of IFPUG such as NESMA [12] or Dreger [5] .

In our study, we have selected *NormalisedWorkEffort* or *SummaryWorkEffort* as dependent variable provided by the ISBGN dataset . The normalized work effort is an estimate of the effort for the whole software life-cycle even if the project did not cover all the phases in the software development life-cycle. Summary work effort is the actual effort even if the project did not carry out the whole life-cycle. Both values are the same for projects covering the whole life-cycle or projects where it is not known if they covered the whole life-cycle. For each dataset, we divided the dataset into a training and test using Weka utilities to creates stratified cross-validation folds [7]. This means that per default class distributions are approximately retained within each fold. The training dataset contains approximately 2/3 of instances and the remaining 1/3 of instances was used for validation.

We compared the outputs using algorithms provided by Weka's M5 models, Multivariate Linear Regression (MLR), or just Linear Regression (LR) for simplicity, and Least Median Squares (LMS) for each dataset.

DS1. The Reality dataset if composed of 709 instances and 6 attributes (*DevelopmentType, DevelopmentPlatform, LanguageType, ProjectElapsedTime, NormalisedWorkEffort, UnadjustedFunctionPoints*). The dependant variable that we used for this dataset is the *NormalisedWorkEffort.* Table 1 compares the goodness of the results for the reality dataset.

Table 1. DS1. Reality dataset results

	M5	*LeastMedSq*	*LR*
Correlation coefficient	0.36	0.06	0.37
Mean absolute error	4829.08	5817.38	5244.29
Root mean squared error	15715.46	18583.45	15612.01
Relative absolute error	74.18%	89.36%	80.56%
Root relative squared error	93.54%	110.6143%	92.92%
MMRE	1.99	1.44	2.62
Pred(25)	0.20	0.17	0.13
Pred(30)	0.24	0.22	0.16

Table 2. DS2. Dataset results

	M5	*LeastMedSq*	*LR*
Correlation coefficient	0.87	0.8108	0.78
Mean absolute error	1191.08	3497.23	3340.26
Root mean squared error	7978.20	12437.35	9482.81
Relative absolute error	20.32%	59.67 %	56.99%
Root relative squared error	54.07%	84.30%	64.27%
MMRE	0.39	0.76	1.69
Pred(25)	0.70	0.29	0.23
Pred(30)	0.73	0.36	0.28

DS2. The dataset DS2 is composed of 1390 instances and 15 attributes (*FP, VAF, MaxTeamSize, DevType, DevPlatf, LangType, DBMUsed, MethodUsed, ProjElapTime, ProjInactiveTime, PackageCustomisation, RatioWEProNonPro, TotalDefectsDelivered, NormWorkEff, NormPDR*). The dependant variable for this dataset is the *NormalisedWorkEffort*.

5 Discussion

The M5 models provide some advantages from both quantitative and qualitative point of view when compared with simpler regression methods. From a quatitative point of view, we can conclude that the goodness, i.e., estimation accuracy, of M5 is better than the classical linear regression or least mean squares (cf. Tables 1 and 2). The M5 algorithm greatly improved estimates to levels that could be considered as acceptable by the software engineering community. Also, when compared with other techniques, the M5 algorithm is able to handle both continuous and categorical variables.

In addition to greater accuracy, the M5 model provides other advantages from a qualitative point of view. Firstly, each subset is clearly defined in the sense that new instances are easily assigned to a local model. The second benefit is that decision trees are easily understandable by users in general and by project managers in particular as we can read them as rules. Each branch of the tree has a condition as follows: *attribute* $\leq$ *value* or *attribute* $>$ *value*. Such conditions are frequently used by experts in all sciences. Also, this provides a clear indication of which variables are most important for prediction (as conditionals in the branches of the tree). The leaves of the tree allow project managers to gain further knowledge into the characteristics of the dataset.

6 Conclusions

This paper presented the results of applying the M5 algorithm as a estimation technique to 2 datasets generated from the ISGSB repository. The M5 algorithm assigns to each of the generated local model a linear regression formula. We also applied the classical linear regression and least median squares and M5 for effort estimation. When we compared the goodness of such methods, results show that there is an improvement in the accuracy of the results when using such local models. Furthermore, M5 can help us to understand the datasets better; the tree (rules) generated with M5 provides project managers with a better understanding of what attributes are more important in a particular dataset. Finally, we believe that M5 in particular and clustering techniques in general, provide some advantages to project managers while keeping the estimation process within reasonable complexity.

Further work will consist of using data mining techniques for characterizing not only the instances but also the attributes (in this work, the attributes were selected manually using expert knowledge). More needs to be done also understanding and comparing different clustering techniques to create segmented models and its usefulness for project managers.

Acknowledgments

The research was supported by the University of Reading and the Spanish Research Agency (CICYT TIN 2004-06689-C03 – The INGESOFT Project).

References

1. Aguilar–Ruiz J.S., Riquelme J.C., Ramos I. and Toro M.: An evolutionary approach to estimating software development projects. *Information and Software Technology*, 14(43):875–882, 2001
2. Boehm, B. 1981, Software Engineering Economics, Prentice-Hall, 1981
3. Breiman, L., Friedman, J., Olshen, R. and Stone, C. J.,*Classification and Regression Trees*, Chapman and Hall, New York, 1984.
4. Conte SD, Dunsmore HE and Shen V. 1986. *Software Engineering Metrics and Models*, Benjamin/Cummings.
5. Dreger, J. 1989, Function Point Analysis, Englewood Cliffs, NJ, Prentice Hall,
6. Dolado, J.J.: On the problem of the software cost function. *Information and Software Technology, 43:61–72, 2001.*
7. *Fayyad, U.M. and Irani, K.B.: Multi-interval discretisation of continuous valued attributes for classification learning. In* Proceedings of the Thirteenth International Joint Conference on Artificial Intelligence. Morgan Kaufmann, 1993.

8. Finnie, G.R., Wittig, G.E., and Desharnais, J.-M.: A comparison of software effort estimation techniques: using function points with neural networks, case-based reasoning and regression models. *Journal of Systems and Software*, 39(3):281–289, 2000.
9. Holte, R.C.: Very simple classification rules perform well on most commonly used datasets. *Machine learning*, 11:63–91, 1993.
10. ISBSG, 2004. International Software Benchmarking Standards Group (ISBSG), Web site: http://www.isbsg.org/
11. Mitchell, T.: *Machine Learning.* McGraw Hill, 1997.
12. NESMA, 1996. NESMA FPA. Counting Practice Manual Version 2.0.
13. PRICE, 2005. Price S. Web Site: http://www.pricesystems.com/
14. Quinlan JR. 1992. Learning with continuous class. *Proc. of the 5th Australian Joint Conference on Artificial Intelligence*: 343-348, World Scientific.
15. Quinlan, J.R.: *C4.5: Programs for machine learning.* Morgan Kaufmann, San Mateo, California, 1993.
16. Rousseeuw, P. J. and Annick M. L., *Robust Regression and Outlier Detection*, New York: John Wiley & Sons, 1987.
17. Shepperd, M. and Schofield, C.: Estimating software project effort using analogies. *IEEE Transactions on Software Engineering*, 23(12):736–743, 2000.
18. Srinivasan, K. and Fisher, D.: Machine Learning Approaches to Estimating Software Development Effort. *IEEE Transactions on Software Engineering*, 21(2): 126–137, 1995.
19. Walkerden, F. and Jeffery, R.: An empirical study of analogy-based software effort estimation. *Empirical Software Engineering*, 42:135–158, 1999.
20. Wang, Y. and Witten, I.H.: Induction of model trees for predicting continuous classes. *Proceedings of the poster papers of the European Conference on Machine Learning.* University of Economics, Faculty of Informatics and Statistics, Prague.
21. Witten I. and Frank E. 1999. *Data Mining Practical: Machine Learning Tools and techniques with Java implementations.* Morgan Kaufmann

A Web User Interface of the Security Requirement Management Database Based on ISO/IEC 15408

Daisuke Horie, Shoichi Morimoto, and Jingde Cheng

Department of Information and Computer Sciences,
Saitama University, Saitama, 338-8570, Japan
{horie, morimo, cheng}@aise.ics.saitama-u.ac.jp

Abstract. In order to support design and development of secure information systems, we have proposed a security requirement management database based on the international standard ISO/IEC 15408. Design and development of secure information systems concern issues of information security engineering as well as software engineering. Our security requirement management database will be useful in practices only if we can provide its users with a highly usable user interface. This paper presents the design and development of a web user interface of our security requirement management database. We analyze and define usability requirements that the database should satisfy, present design and implementation of the web user interface, and show some examples for evaluating the interface from the viewpoint of usability engineering.

1 Introduction

Nowadays, security is an important factor in information system development. There is rising demand for security in information systems. Database technologies have successfully been applied to software engineering [9,6,3]. Similarly, we can apply database technologies to information security engineering. Therefore, we have proposed a concept for a database to support design and development of secure information systems, named "ISEDS (Information Security Engineering Database System) [7]." ISEDS can collect and manage the past knowledge and experience for secure system design.

However, we had not sufficiently discussed how to use ISEDS yet. In order to achieve the purpose of ISEDS, it is necessary to clarify usage of ISEDS and usability requirements for ISEDS from the viewpoint of usability engineering. Moreover, it is desirable to implement a useful and convenient user interface which satisfies the requirements. In this paper, we analyze and define the requirements and present design and implementation of a web user interface. Users of ISEDS can easily and effectively use it for ISEDS anytime and anywhere by web browsers.

V.N. Alexandrov et al. (Eds.): ICCS 2006, Part IV, LNCS 3994, pp. 797–804, 2006.
© Springer-Verlag Berlin Heidelberg 2006

2 The Outline of ISEDS

We developed ISEDS based on the international standard ISO/IEC 15408, because ISO/IEC 15408 defines common criteria that should be applied to validate a secure information system [5]. ISEDS can store the knowledge or experience in security specifications for secure system design according to the common criteria. The data can be a guideline to design and development of secure information systems.

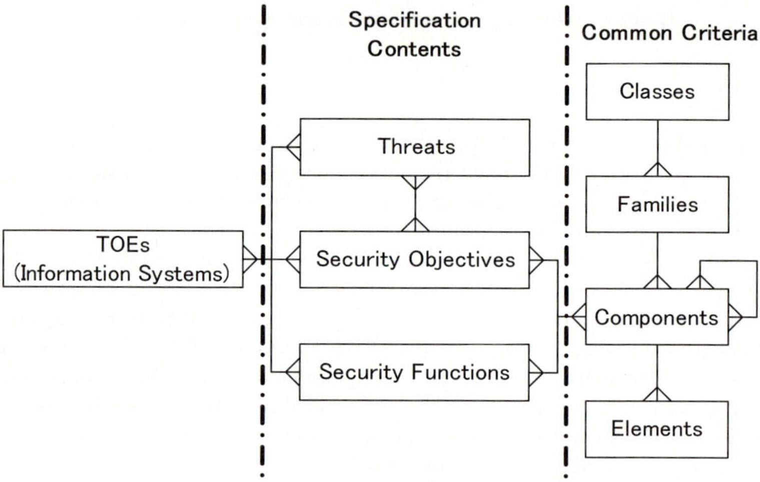

Fig. 1. The entity-relationship diagram of ISEDS

ISEDS has the structure as shown in Fig. 1. It is the entity-relationship diagram in relational notation [2]. Below, an italic word denotes a schema or its attribute in Fig. 1.

Applicants who apply to obtain the evaluation of ISO/IEC 15408 have to make and submit the specification document called 'security target' or 'protection profile' to the evaluation organization. A security target, ST for short, must describe the range of the target system which is evaluated (target of evaluation, TOE for short). A protection profile, PP for short, is a template for an ST. The schema *TOE* has attributes that are written to an ST or a PP, i.e., *TOE Name*, *TOE Abstract*, *Author*, *Small Classification*, *Middle Classification*, and *Large Classification.* The classifications denote TOE kinds. So far, *Large Classification* is only 'IT products.' *Middle Classification* is 'software,' 'hardware,' 'middleware,' and so on. *Small Classification* shows the concrete TOE kind, e.g., 'Database,' 'Firewall,' 'IC card,' 'OS,' 'Copier,' and so on. In addition to these attributes, we defined *Document Type* and *Document File*. *Document Type* is a flag for distinguishing an ST or a PP. *Document File* is an attribute for storing the binary file of the ST or PP.

An ST or a PP must describe threats assumed in its category. Next, the documents must describe security objectives which oppose these threats. Thirdly, the documents must describe security criteria required for achievement of these objectives. In particular, the required security criteria must be quoted from the security functional requirements which are defined in ISO/IEC 15408 Part 2. For the implementation of security functional requirements, the documents must also describe that they are actually implemented as what functions in systems. These functions are called a TOE security function, TSF for short. *TOE* relates to one or more *Threats*, *Security Objectives*, and *Security Functions*. The schemata *Threat*, *Security Objective*, and *Security Function* have their ID numbers, abbreviation names, formal names, texts of the definition on the documents, and foreign keys to *TOE*. ISEDS can store such data of the documents.

Security functional requirements of Part 2 have a hierarchical structure which is composed by classes, families, components, and elements. Each element is an indivisible security requirement, and describes a security functional requirement exactly and directly. A component is a group of elements, and may be mutually dependent on the other components. A family is a group of components, and a class is a group of families. The schemata of security functional requirements have the attributes defined in ISO/IEC 15408 Part 2. ISEDS can also store such data of the criteria.

3 The Web User Interface of ISEDS

ISEDS can achieve its purpose only if we can provide its users with a usable user interface. Thus, we examine requirements to ISEDS from the viewpoint of software engineering, information security engineering, and usability engineering. In compliance with the requirements we have implemented the prototype of the user interface.

3.1 Requirements for ISEDS

We herein estimate and enumerate requirements for ISEDS.

1. From the viewpoint of software engineering, ISEDS must structurally collect and manage data of knowledge or experience for secure system design.
Requirement 1-1: Users must be able to structurally refer to the data using ISEDS.
Requirement 1-2: Users must be able to update the data using ISEDS.

2. From the viewpoint of information security engineering, ISEDS must provide useful information for secure system design and development.
Requirement 2-1: Users must be able to know what systems comply with ISO/IEC 15408 using ISEDS.

Users may refer to what TOEs exist. That is, the users may want to search whether or not the TOEs in firewall category exist. Moreover, users may refer to what threats are assumed in a category and what security objectives, security functions, or criteria of ISO/IEC 15408 are required in a category. For example, the users may refer to what threats are assumed in the TOEs of firewall category.

Requirement 2-2: Users must be able to know what security criteria are in ISO/IEC 15408 using ISEDS.

Users may refer to what criteria exist. The users may refer to what criteria about user data protection are. Moreover, users may refer to what categories, threats, or objectives require a criterion. The users may refer what systems require the criterion FTA_TSE.1. In addition, users may refer to hierarchical structure of ISO/IEC 15408. The users may refer to which criteria are included in the class FTA.

Requirement 2-3: Users must be able to know what threats are in the past systems using ISEDS.

Users may refer to what threats exist. The users may want to search whether or not the threats about spoofing exist. Moreover, users may refer to what categories assume a threat and what security objectives, security functions, or criteria can resist a threat. The user may refer to what security functions can resist spoofing.

Requirement 2-4: Users must be able to know what security countermeasures are in the past systems using ISEDS.

Users may refer to what security objectives exist. The users may want to search whether or not security objectives about concealment of IP addresses exist. Moreover, users may refer to what categories or threats require an objective and what security criteria are required for achievement of an objective. The users may refer to what security criteria are required for concealment of IP addresses.

At least, ISEDS should satisfy the above requirements. In order to provide useful functions satisfying the requirements, it is necessary to consider usability engineering. ISO 9241-11 defines that usability is extent to which a product can be used by specified users to achieve specified goals with effectiveness, efficiency and satisfaction in a specified context of use [4]. The usability consists of four elements, i.e., effectiveness, efficiency, satisfaction, and context of use. The effectiveness is accuracy and completeness with which users achieve specified goals. The efficiency shows resources expended in relation to the accuracy and completeness with which users achieve goals. The satisfaction is freedom from discomfort, and positive attitudes towards the use of the product. The context of use shows users, tasks, equipment (hardware, software and materials), and the physical and social environments in which a product is used. ISEDS's interface should also satisfy these requirements for usability.

3.2 Functions Provided by the Web User Interface

We have implemented a prototype user interface satisfying the above requirements. The user interface provides the following functions.

Function 1: Certified System's Specifications Search; the function to refer to all or specific data of certified system specifications. This function satisfies the requirements 1-1, 2-1, 2-3, and 2-4.
Function 2: Common Criteria Search; the function to refer to all or specific security criteria in ISO/IEC 15408. This function satisfies the requirements 1-1 and 2-2.
Function 3: Free Word Search; the function to freely retrieve data in ISEDS. This function satisfies the efficiency and satisfaction.
Function 4: Update Function; the function to update the data in ISEDS. This function satisfies the requirement 1-2.

The effectiveness is satisfied by all functions. Moreover, we implemented the user interface using web technologies so as to satisfy the context of use, i.e., the requirement for 'anytime and anywhere.' The user interface is implemented with PHP, because it can dynamically generate web pages and has high affinity with PostgreSQL which implements ISEDS. The user interface generates SQL sentences based on user's requests and sends the SQL sentences to ISEDS. Next, the interface receives data from ISEDS and legibly displays the results. Users who are not well informed about SQL can easily and ubiquitously use ISEDS with web-browsers. Thus, the user interface satisfies the context of use.

3.3 Examples

We show some examples using the user interface.

Certified System's Specifications Search	Display TOE list Display Threat list Display Security Objective list Display Security Function list
Common Criteria Search	Display Element list Display Component list Display Family list Display Class list
Free Word Search	
Update Function	

Fig. 2. The top page of the user interface

Example 1: If users require retrieval such as the requirements 2-1, 2-3 or 2-4; Click the required link from Certified System's Specifications Search menus in Fig. 2. All of the data corresponding to the selected menu will be displayed as Fig. 3. Moreover, one can narrow the range to input a specific keyword into the text box, select the radio button of the attribute including the specific keyword, and push the button 'search more.' For example, click the link 'TOE List.' All data of TOEs will be displayed. Input a keyword 'firewall' into the text box,

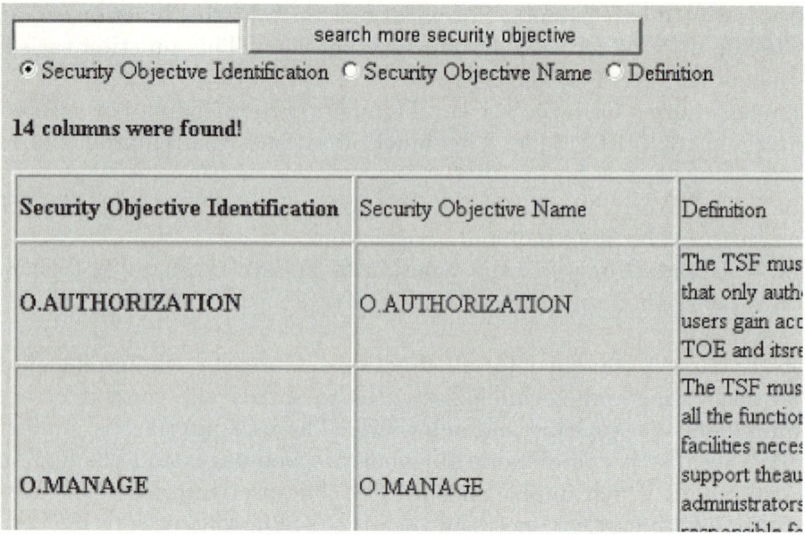

Fig. 3. The result display in the user interface

select the radio button 'Small Classification,' and push the button 'search more TOEs.' Then, data of documents in the firewall category will be displayed.

Example 2: If users require such as the requirement 2-2;
Click the required link from Common Criteria Search menus in Fig. 2. All of the data corresponding to the selected menu will be displayed as Fig. 3. Moreover, one can narrow the range as well as the example 1. For example, click the link 'Element List.' All data of elements displayed. Input a keyword 'recording' into the text box, select the radio button 'Definition,' and push the button 'search more elements.' Then, data of elements including the keyword 'recording' in the attribute 'Definition' will be displayed.

Example 3: If users require further retrieval;
Click the link 'Free Word Search' in Fig. 2. Users can freely choose the items to display and to narrow retrieval range. For example, click the link 'Free Word Search,' select the radio button 'Small Classification' of 'TOE' for narrowing the range, check the check boxes 'Threat name' and 'Definition' of 'Threat' for display, input a keyword 'firewall' into the text box, and push the button 'Search.' Then, the threat names and definitions of threats which are assumed in the firewall category will be displayed.

Example 4: If users want to update the data;
Click the link of 'Update Function' in Fig. 2. If the users have no authority to update, they must get their account and password. Moreover, the users need to make a security specification document file for updating data according to STs or PPs. After that, click the link 'Update Function,' input the user name and password, push the button 'Reference,' select the document file, and push the

button 'Upload.' The administrator will extract data from the uploaded file and update the data.

4 Discussion

We implemented the prototype user interface which has four functions. Function 1 supplies to easily and effectively retrieve the required data of design in secure system's specifications. Function 2 supplies to easily and effectively retrieve the required data of security criteria in ISO/IEC 15408. Function 3 supplies to freely retrieve the relationship of secure system design data. Function 4 supplies to update the data in ISEDS and to store secure system design data which are newly defined by users. These functions satisfy the requirements that we defined for ISEDS.

Furthermore, we evaluate the user interface from the viewpoint of usability engineering. There are two methods in usability evaluation, i.e., quantitative evaluation and qualitative evaluation. The quantitative evaluation is a questionnaire and the qualitative evaluation is heuristic evaluation or user tests [8]. Since it is difficult to perform questionnaires and user tests now, we evaluate the user interface with heuristic evaluation. The user interface satisfies five out of ten usability heuristics. ISEDS and its interface were implemented in a server which has suitable performance. Thus, the interface satisfies 'match between system and the real world,' one of the heuristic principles. Function 3 satisfies 'user control and freedom.' We unified the usage and display in every function. Therefore, the interface satisfies 'consistency and standards' and 'recognition rather than recall.' Since we prepared some explanations of the usage in the web pages, it also satisfies 'help and documentation.'

However, the interface does not satisfies the other heuristic principles 'visibility of system status,' 'error prevention,' 'flexibility and efficiency of use,' 'aesthetic and minimalist design,' and 'help users recognize, diagnose, and recover from errors.' For example, users may refer to what the other threats that are included in categories assuming a certain threat are. The users may also refer to what security functions are implemented in categories which require a certain criterion. The user interface cannot generate SQL sentences for such operations at once now. Users must have many intricate steps to execute such operations. We need to carefully discuss further improvement of the user interface for the efficiency and satisfaction.

5 Concluding Remarks

We have presented a web user interface of ISEDS. We designed, implemented, and evaluated the web user interface according to the requirements which are analyzed and defined from the viewpoint of software engineering, information security engineering, and usability engineering. Users can easily and effectively refer to secure system design data in ISEDS with web browsers. Therefore, ISEDS and its user interface can support design and development of secure information systems.

Because the user interface has been evaluated only with qualitative method, we should also quantitatively evaluate it. Moreover, the structure of STs, PPs, security requirements, and the security functional requirements can easily, exactly and rigorously be expressed in XML. Therefore, we are improving ISEDS as a native XML database into which users can directly store the XML documents and are developing its web service. Moreover, ISEDS is deficient in some attributes of ISO/IEC 15408, STs, and PPs now, e.g., creation dates of STs or PPs, evaluation assurance levels, and security assurance requirements defined in ISO/IEC 15408 Part 3. We are discussing whether or not these attributes are necessary.

References

1. Advanced Information Systems Engineering Laboratory, Saitama University.: ISEDS: Information Security Engineering Database System. `http://www.aise.ics.saitama-u.ac.jp/`
2. Chen, P.: The Entity-Relationship Model - Toward a Unified View of Data. ACM Transactions on Database Systems (TODS), Volume 1, Issue 1, pp. 9-36 (1976)
3. International Software Benchmarking Standard Group.: Empirical Databases of Metrics Collected from Software Projects. http://www.isbsg.org/
4. ISO 9241-11 standard.: Ergonomic Requirements for Office Work with Visual Display Terminals – Part 11: Guidance on Usability (1998)
5. ISO/IEC 15408 standard.: Information Technology - Security Techniques - Evaluation Criteria for IT Security (1999)
6. Jiao, J. and Tseng, M.: A Requirement Management Database System for Product Definition. Journal of Integrated Manufacturing Systems, Vol. 10, No. 3, pp. 146-154 (1999)
7. Morimoto, S., Horie, D., and Cheng, J.: A Security Requirement Management Database Based on ISO/IEC 15408, in Computational Science and its Applications - ICCSA 2006, International Conference, Glasgow, UK, May 8-11, 2006, Proceedings. Lecture Notes in Computer Science, Springer-Verlag, May (2006)
8. Nielsen, J. and Molich, R.: Heuristic Evaluation of User Interfaces. Proceedings of the SIGCHI conference on Human factors in computing systems: Empowering people, pp. 249-256, Seattle, WA, April (1990)
9. Software Engineering Institute.: Software Engineering Information Repository. http://seir.sei.cmu.edu/large

Domain Requirements Elicitation and Analysis - An Ontology-Based Approach*

Yuqin Lee and Wenyun Zhao

Software Engineering Lab, Computer Science and Technology Department,
Fudan University, Shanghai, China
li_yuqin@yahoo.com.cn, wyzhao@fudan.edu.cn

Abstract. Domain requirements are fundamental for software reuse and are the product of domain analysis. This paper presents an approach to elicit and analyze domain requirements based on ontology. Using subjective decomposition method, problem domain is decomposed into several sub problem domains. The top-down refinement method is used to refine each sub problem domain into primitive requirements, which are specified using ontology definition. Abstract stakeholders are used instead of real ones when decomposing problem domain and ontology is used to represent domain primitive requirements. Not only domain commonality, variability and qualities are presented, but also reasoning logic is used to detect and handle incompleteness and inconsistency of domain requirements. In addition, a case of 'spot and futures trading' e-business is used to illustrate the approach.

1 Introduction and Related Work

Requirements engineering (RE) is the branch of software engineering concerned with the real world goals for, functions of, and constraints on software systems. [1] It is generally accepted that the ultimate quality of the delivered software depends on the requirements upon which the system has been built. [3][4] Studies performed at many companies have measured and assigned costs to correct defects in software discovered at various phases of the lifecycle. Generally, the later in the software lifecycle a defect is discovered the more expensive it is to rectify. [5] [6].To achieve large scale reuse of software, a domain is usually mentioned. [2]

Among those domain requirements elicitation and specification approaches, there are mainly three types. One is the traditional method. Domain requirements are presented in natural language. The main drawback of this method is that there are gaps between domain users and requirements engineers. This will be the main source of generating inconsistent and ambiguous requirements. The second method is the scenario-based method. Domain users can take part in elicitation conveniently, and the requirements are the true reflection of the real system. But its disadvantage is the problem how to guarantee a set of scenarios that are the complete requirements of the

* This paper was supported by the National High Technology Development 863 Program of China under Grant No.2005AA113120.

V.N. Alexandrov et al. (Eds.): ICCS 2006, Part IV, LNCS 3994, pp. 805 – 813, 2006.
© Springer-Verlag Berlin Heidelberg 2006

system.[7] The overlap of different scenarios is the source of requirements inconsistency. The third solution is knowledge-based methods.[7][8][17] The methods have been researched hotly, but there are few sound achievements yet. Among the third methods, feature-based and ontology-based methods are usually researched. However, the feature-based method[17] concentrates on commonality and variability analysis having a shortcoming in reasoning logic, while ontology-based method[7][8] focusing on representing domain concepts and relations of concepts lack of distinguishing concepts to common or variation sets. Our approach combines the advantages of two kinds of knowledge-based methods, not only representing commonality, variability and qualities clearly, but having reasoning logic which can be used in requirements analysis also.

In this paper, an approach is suggested to systematically develop domain requirements based on ontology. Using subjective decomposition method, domain problem is decomposed into several sub problem domains. Subjectivities are the viewpoints of abstract stakeholders of domain. Top-down refinement method is used to refine each sub problem domain into primitive requirements, which are specified using ontology definition. The advantages of our approach are using abstract stakeholders other than practical ones when decomposing domain, getting commonality, variability and quality features from each sub problem domain, using ontology in domain elementary requirements representation, and using reasoning logic based on ontology relations to analyze completeness and consistency easily.

The paper is organized as follows: Section 2 gives out ontology definition. Section 3 discusses domain requirements metamodel. Section 4 provides requirements elicitation and refinement methods for ontology-based domain requirements. Section 5 discusses domain requirements analysis, focusing on completeness and consistency. Section 6 illustrates our approach by an example of spot and futures e-business domain model. Section 7 draws a conclusion and some suggestions for future work.

2 Ontology Definition

The term ontology was taken from philosophy. According to Webster's Dictionary, an ontology is a branch of metaphysics relating to the nature and relations of being. In knowledge engineering area, ontology was first defined by Neches.[8][9] Another ontology definition is newly defined by Gruber:" Ontology is a formal, explicit specification of a shared conceptualization." [10]

We give a definition of ontology that a quadruple consists of the core elements of ontology, i.e. concepts, relations, hyperspace and theorems.

O={C,R,H,T}. C(Concepts) and R(Relations) are two sets which don't intersect with each other. H stands for hyperspace consisting of concepts and relations. T is the set of ontology theorems.

R consists of the relations as follows:

Is-a: describes the equivalence relation;

Is part of: describes the part and whole relation.

Mutually exclusive: describes the relation of two concepts which can't be in a system contemporary;

Association: describes two concepts having relations with each other; such as pre condition and post condition, etc.

Based on this ontology definition, we have the following definitions or theorems:

Definition 1: functional (one by one mapping) relation. If c_1, $c_2 \in C$, and $r \in R$, $r(c_1, c_2)$. If one c_1 corresponds to only one c_2, then r is functional. Vice versa, if one c_2 corresponding to only one c_1, then r is inverse functional.

Definition 2: symmetric relation. $\exists c_1, c_2 \in C$, if $\exists r \in R$, $r(c_1,c_2)$, then $r(c_2,c_1) \in R$. We call r is symmetric.

Definition 3: Transitive relation. If $c_1,c_2,c_3 \in C$, $\exists r \in R$, $r(c_1, c_2)$ and $r(c_2,c_3)$, then $r(c_1,c_3)$. We call r is transitive.

Definition 4: Homomorphic mapping. Let $O_1= \{C_1, R_1, H_1, T_1\}$, and $O_2= \{C_2, R_2, H_2, T_2\}$ be different ontology, define a homomorphic mapping M from O_1 to O_2 as:

M= {M(C1), M(R1), M(H1), M(T1)}, satisfies the following rules:

1. $M(C_1)$is an one by one mapping. For each $c_1 \in C_1$,then $M(c_1) \in C_2$; if $c_1, c_2 \in C_1$, and $c1 \neq c2$, then $M(c_1) \neq M(c_2)$;
2. $M(R_1)$ is a one by one mapping as $M(C_1)$;
3. $M(H_1)$ is a one by one mapping too;
4. T_1 is a subset of T_2.

Definition 5: Ancestor relation. Let $O_1=\{C_1, R_1, H_1, T_1\}$, and $O_2=\{C_2, R_2, H_2, T_2\}$, $O_1 \neq O_2$; If there is a homomorphic mapping M from O_1 to O_2, we call O_1 ancestor of O_2.

Definition 6: Inheritance relation. Let $O_1=\{C_1, R_1, H_1, T_1\}$, and $O_2=\{C_2, R_2, H_2, T_2\}$, $O_1 \neq O_2$; If O_1 is ancestor of O_2, we say O_2 has an inheritance relation with O_1.

Definition 7: Nesting. Ontology can be nested. This feature will be used in multi-viewpoints domain analysis.

3 Domain Requirements Metamodel

The metamodel (see fig 1) for a domain requirement is adopted from [2] and be improved.

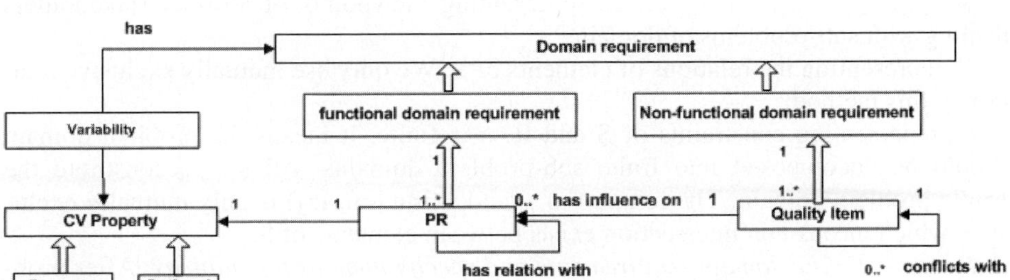

Fig. 1. Metamodel for Domain Requirement

The metamodel has a domain requirement as the central model element. Domain requirements are categorized into functional and nonfunctional requirements. Functional requirements can be refined into PRs (primitive requirements). The common and optional property is relative. The common or optional property of PR is analyzed regarding in how many systems it exists. Optional property is realized as variability of requirements. There are relations between PRs. Nonfunctional requirements have relations with and can influence functional requirements. We use the ontology definition above to represent domain requirements including not only functional and nonfunctional requirements, but also common and variant points in functional requirements. In this method, we categorize common requirements into several sets according to subjects. The variants are realized as configuration or value of dimension in PR level. Nonfunctional requirements are classified as a separate set.

4 Domain Requirements Elicitation and Refinement

Based on ontology description above, domain requirements are elicited and refined. The process consists of three phases. The result of this process is to get PR specifications and relations of PRs.

Phase 1: Define domain terms. To develop precise and consistent domain requirements, the basic terms used in the domain should be defined. [2] The principle of defining terms in a domain is to guarantee them within the domain scope. Domain terms constitute a terminology dictionary, which evolves along with domain decomposition. Domain terms based on ontology contains domain concepts and relations of concepts.

Phase 2: Decompose domain requirements. To decrease complexity, a problem domain is decomposed into several sub-problem domains. Subjective decomposition used in this paper is based on modeling different subjective perspectives of different stakeholders on a system or a domain. [16] In the method illustrated in [16], subjectivity is accounted for in DEMRAL by considering different stakeholders and annotating features by their stakeholders. We don't use different viewpoints of stakeholders directly, but subjectivity is abstracted from a group of stakeholders who have similar needs for the systems.

Domain is specified using the following representation: *D=(S, R, C)*.

S= $\{s_1, s_2,..., s_n\}$, s_1, s_2 .... s_n, representing viewpoints of abstract stakeholders dealing with sub-problems of domain;

R: representing the relations of elements of S. We only use mutually exclusive relation in this method;

C: representing constraints of S and R. n is finite. It means the problem domain should be decomposed into finite sub-problem domains. All $s_1...s_n$ constitute the whole problem domain. The relation of s_i and $s_j(i<n,j<n, i\neq j)$ is only mutually exclusive, which means non intersection exists between elements of S.

Phase 3: Refine domain requirements and specify them using ontology. After problem domain is decomposed into several sub-problem domains, the requirements of each sub-problem domain are refined into functional and nonfunctional requirements. After this all functional requirements are refined into PRs.

Top-down method [14] is used to refine functional requirements in a hierarchy structure, which we call function tree. At each level, each function is decomposed into a number of functions at the next level, until a level is reached at which the functions are regarded as elementary functions [15]. The elementary functions are called PRs (primitive requirements). Different levels are related using whole-part association (WPA[17]).

Specify each sub problem domain as $S_i = (C_i, R_i, H_i, T_i)$. According to the definition above, domain requirements can then be: Requirements ={C, R, H, T}

$C= \cup\{C_i\}, i=1..n$; $C_i= \{\cup\{PR_{ij}\}\} \cup Q_i$, $j=1..r_i$. Q_i represents quality features of sub problem domain i;

$R= \cup\{R_i\}$; $T= \cup\{T_i\}$. *H* is the hyperspace consisting of C and R.

According to the definition above, we get a triple table DR= (sc, re, dc). DR stands for domain requirements; sc and dc stand for source and destination concepts; re stands for relation of sc and dc. One row represents relation of a couple of concepts. To analyze requirements conveniently, a row represents a direct relation of sc and dc.

5 Domain Requirements Analysis

Based on the triple table DR= (sc, re, dc), we can analyze completeness and consistency of requirements.

- Detection and Handling of Requirements Incompleteness

From [18], the REQ represents complete requirements of a domain, if the following formula is satisfied.

$$Domain(REQ) \supseteq Domain\ (NAT) \tag{1}$$

Domain(REQ) stands for requirements of domain, *Domain (NAT)* stands for needs of stakeholders.

Completeness of requirements is relative. We discuss completeness in the point of view whether concepts and relations set are complete.

Definition 1. If c_2 has direct or transitive association with c_1, and $c_1 \in C$, $c_2 \notin C$, then we call concepts is incomplete. The solution is to add c_2 to C.

Definition 2. $\exists c_1, c_2 \in C$, and c_2 has direct association with c_1, i.e. $r(c_1, c_2)$, but $r \notin R$, so we call the relation incomplete. The solution is to add r to R.

Definition 3. Completeness is detected by inherent relation. If two requirements have inherent relation in domain knowledge base, the related requirement should be included in requirements set when the source requirement is selected. The analysis process of completeness is iterative, and it will end until the result concepts and relations set are stable.

- Detection of Requirements Inconsistency

Subject overlap is the main source of inconsistency. We assume that different sub problem domains don't intersect in this paper, so that inconsistency doesn't appear.

6 Case Study

We take spot and futures e-business domain as an example. The domain provides an e-business platform to support multi-to-multi real time transactions in B2B area.

- Requirements Refinement and Specification of Spot and Futures E-business

Phase 1: define domain terms. By analyzing spot and futures e-business domain, the commonalities are that there are order, match, settlement and delivery phases. The variations exist in each phase. Many different areas use the B2B trading method, so contract which is the base unit for transaction has different features for different commodities. Several match modes are supported, such as: forward, special performance, auction, etc. We get a terminology dictionary including all these terms.

Phase 2: Decompose e-business domain requirements. The abstract stakeholders are roles responsible of trading, settlement and delivery. The problem domain can be decomposed into three sub problem domains: trade, settlement and delivery.

$D=(S,R,C)$

S={trade, settlement, delivery}.

R: mutually exclusive relation between elements of S, complying with the whole transaction flow.

C: constraint of S and R. S has three elements, and is thus finite. The elements of S deal with different phases of transaction flow, so they don't intersect with each other.

Phase 3: Refine e-business domain requirements and specify them by ontology. After using the refinement method, the requirements tree of spot and futures transaction domain is in figure 2.

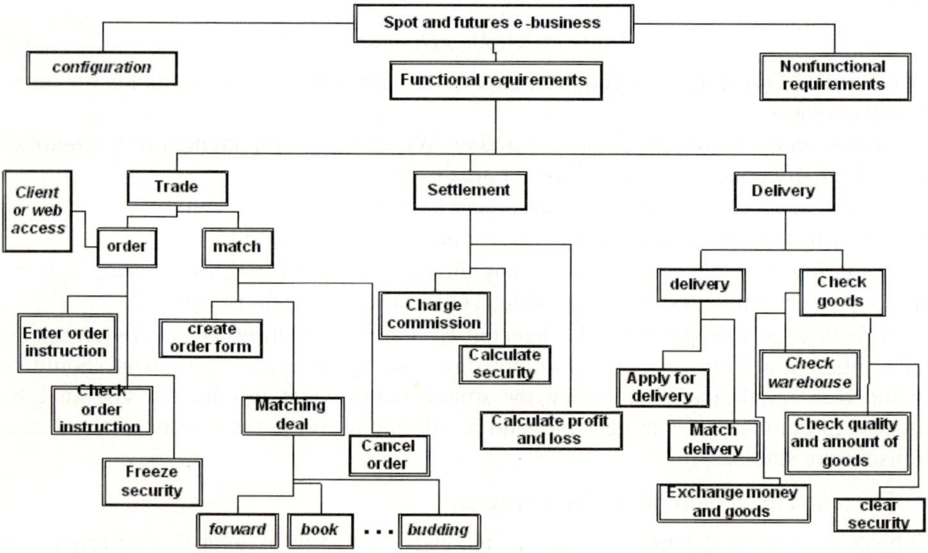

Fig. 2. Requirements Tree of Spot and Futures Transaction

In figure 2, the italic represents variant points. The leaves are primitive requirements. We only specify these leaves and their relations using ontology.

The relations of PRs are {is-a, is part of, is precondition of, is post condition of, transition}. The theorems are corresponding to those defined in the ontology.

Requirements of the case are represented as follows:

$C= \cup\{C_1,C_2,C_3, Q\}$.

C_1= {order, match}

△*order*={client access, web access}

⊙*order*={enter order instruction, check order instruction, freeze security}

match={ create order form, ●match deal, cancel order}

●*match deal*= {forward, special field, book, talk, list, auction, bidding}

C_2={charge commission, calculate security, calculate profit and loss}

C_3={apply for delivery, match delivery, ○check special field , exchange money and goods, check quality and amount of goods, clear security}

○*check special field*= optional { check warehouse in special field}

Q= {order transaction should complete in 0.1 seconds, etc}.

Where,

Q--quality features i.e. nonfunctional requirements of the domain

△-- mutually exclusive relation

⊙--whole and part relation

●--alternative option, every element is optional ,but at least one is selected

○—optional

- Requirements Analysis of Spot and Futures E-business Domain

The domain is decomposed by using subject-orient method according to different business phases, so time guarantees sub problem domains don't overlap. This method decreases inconsistency of requirements from different sub problem domains. We focus on completeness analysis.

After requirements refinement, we get triple table DR= (sc, re, dc) representing domain requirements, where sc and dc are nodes of requirements tree, and re is branch of the tree representing direct relation between primitive requirements and implied relations between brother nodes.

Take sub problem settlement as example, we get the table as follows.

Table 1. Requirements table of sub problem settlement

• Sc	• re	• dc
charge commission	is part of	settlement
calculate security	is part of	settlement
calculate profit and loss	is part of	settlement
charge commission	is precondition	calculate security
calculate security	is precondition	calculate profit and loss

After analysis according to rules above, we can get a stable requirements table.

7 Conclusion and Future Work

The domain requirements stand for requirements for families of similar systems in one area, but it's difficult to get domain requirements from domain users because there are gaps between domain users and software developers. In this approach, we use ontology to represent domain requirements so that domain users can take part in requirements elicitation easily. Using subjective decomposition method, domain problem is decomposed into several sub problem domains. Subjectivities are the viewpoints of abstract stakeholders of domain. Top-down refinement method is used to refine each sub problem domain into primitive requirements, which are specified using ontology definition. The requirements specification is precise and easy to analyze. A case of spot and futures e-business is used to illustrate our approach. The advantages of our approach are using abstract stakeholders other than practical ones when decomposing domain, and getting commonality, variability and quality features form each sub problem domain, and using ontology in domain elementary requirements representation, and using reasoning logic based on ontology relations to analyze completeness and consistency easily.

In this paper, mutual exclusion is the predicted relation. In reality, subjects may crosscut many aspects of a domain so that intersection often exists. We will improve the approach to deal with more complex situations, and to develop map rules from the domain model into reusable architecture and components.

References

1. Axel van L.,Handling obstacles in goal-oriented requirements engineering, IEEE Transactions on software engineering, vol.26, NO.10, pp978-1005, Oct 2000.
2. M.Moon, An approach to developing domain requirements as a core asset based on commonality and variability analysis in a product line, IEEE Transactions on software engineering, vol.31, NO.7, pp 551-569, Jul 2005
3. Brooks, F.P., No Silver Bullet: Essence and Accidents of Software Engineering, IEEE Computer, 20, 4 (April 1987), 10-19.
4. Hrones, J.,Defining Global Requirements with Distributed QFD. Digital Technical Journal 5, 4, 36-46.
5. W. James, Effectiveness of Elicitation techniques in distributed requirements engineering, Proceedings of the IEEE Joint International Conference on RE, 2002.
6. Davis, Alan M. Software Requirements: Objects, Functions, and States. Englewood Cliffs, NJ: Prentice-Hall, 1993.
7. Zhi, Jin. Ontology-based requirements elicitation automatically. Chinese J. Computers. Vol.23,No.5, May 2000. pp486-492.
8. R.Q.Lu, Ontology-based requirements analysis Journal of Software. 2000,11(8). 1009~1017.
9. NechesR, Enabling technology for knowledge sharing.AIMagazine,1991,12(3):36~56.
10. GruberTR. A translation approach to portable ontology specifications. Knowledge Acquisition,1993,5(3):199~220.
11. Gruninger,M; Lee,J.; Introduction to the ontology application and design section, guest editors-communications of the ACM, Feb,2002, Vol 45, No,2 , pp39-41.
12. G.Booch, The Unified Modeling Language Reference Manual. Addison-Wesley, 1999
13. M.Jackson. Problem Frames: Analyzing and Structuring Software Development Problems. Addison-Wesley, 2001
14. D.Knuth. The Art of Computer Programming. Addison-Wesley, 1973

15. Xuefeng.Z, Inconsistency Measurement of Software Requirements Specifications an Ontology-Based Approach. Proceedings of the 10th IEEE ICECCS. 2005
16. Krzysztof Czarnecki, Feature Modeling, July, 1998, pp1-31.
17. Z. wei, M.Hong, A feature-oriented domain model and its modeling process. JOS. 2003. Vol. 14, No. 8. pp1345-1356.
18. Konstantin K., David L.,P., On Documenting the Requirements for Computer Programs Based on Models of Physical Phenomena. models3 august, pp1-14.
19. James.C, Commonality and Variability in Software Engineering. IEEE software 1998 November. pp37-45

Integrative Computational Frameworks for Multiscale Digital Human Modeling and Simulation*

Richard C. Ward[1], Line C. Pouchard[2], and James J. Nutaro[1]

[1] Computational Sciences and Engineering Division, Oak Ridge National Laboratory, P.O. Box 2008, Bethel Valley Road, Oak Ridge, Tennessee, USA
{wardrc1, nutarojj}@ornl.gov

[2] Computer Sciences and Mathematics Division, Oak Ridge National Laboratory, P.O. Box 2008, Bethel Valley Road, Oak Ridge, Tennessee, USA
pouchardlc@ornl.gov

Abstract. Integrated digital human modeling has seen increased interest over the last decade. We describe two efforts to develop computational frameworks for digital human modeling and describe the progress toward understanding the requirements for implementation. Both projects addressed data repository, computational environment, and visualization of results. But neither environment was a true problem-solving environment in that integration of computations with visualization capabilities was limited or absent. We detail the development of the computational environments for each effort and then provide proposals for improving the integration of the various components of a future "Digital Human" computational environment.

1 Introduction

Integrated digital human modeling has seen increased interest over the last decade. Here we report on two efforts; the first led by Oak Ridge National Laboratory (ORNL) and a more recent larger effort in which ORNL played a role.

* Portions of this research were supported by a grant from the Defense Advanced Research Projects Agency, executed by the U.S. Army Medical Research and Materiel Command/TATRC Cooperative Agreement, Contract W81XWH-04-2-0012. Portions of this research were sponsored by the Laboratory Directed Research and Development Program of Oak Ridge National Laboratory, managed by UT-Battelle, LLC, for the U. S. Department of Energy under Contract No. DE-AC05-00OR22725.

The submitted manuscript has been authored by the U.S. Department of Energy, Office of Science of the Oak Ridge National Laboratory, managed for the U.S. DOE by UT-Battelle, LLC, under contract No. DE-AC05-00OR22725. Accordingly, the U.S. Government retains a non-exclusive, royalty-free license to publish or reproduce the published form of this contribution, or allow others to do so, for U.S. Government purpose.

Approved for Public Release, Distribution Unlimited.

V.N. Alexandrov et al. (Eds.): ICCS 2006, Part IV, LNCS 3994, pp. 814–821, 2006.
© Springer-Verlag Berlin Heidelberg 2006

The first effort, Virtual Human (VH), culminated in the development of a Web-based, Java client/server environment within which some simple cardiovascular circuit models and various other compartment models were incorporated. The second effort, the Defense Advance Research Projects Agency (DARPA)-funded Virtual Soldier (VS) Project, was a much more substantial effort to address human modeling and predict the consequences of a soldier being wounded on the battlefield. A more comprehensive project, this second effort addressed the development of an integrated data repository, computational environment, and visualization of results and predictions in a concept referred to as the holographic medical electronic representation (Holomer) [1].

Both these projects contributed significantly to our understanding of how a computational framework can be built to address virtual human simulation. However, the degree of integration of various components of the environment left much to be desired. Neither environment fully integrated computations with other elements of the environment; continued work must be done to bring the design of digital human modeling environments in line with present thinking in Service-Oriented Architecture (SOA) (http://www.service-architecture.com/) for flexibility of integrating data, computational, and visualization components.

Finally, we describe some concepts that might facilitate full integration and provide an end-to-end capability for data acquisition, model computation, and display of results and predictions.

2 Virtual Human

The concept of a VH was developed during 1996-2000. The VH was a "concept...to combine models and data to build a comprehensive computational capability for simulating the function as well as the structure of the human body and allow trauma simulations as well as many other applications" [2]. In the fall of 1999 a workshop was held in Rockville, Maryland under ORNL auspices. The general interest in the concept and enthusiasm for developing a VH, as well as discussions in the Architecture breakout group, laid a foundation for subsequent design and development of a computational framework.

2.1 VH Computational Environment

ORNL then initiated an effort to develop a prototype VH computational environment. We chose to develop a Web-delivered environment using Java Remote Method Invocation (RMI) with the human anatomical geometry defined using Virtual Reality Modeling Language (VRML) [9] [8]. The software underlying the computational environment was derived from a generic client-server simulation framework [3]. The environment allowed for integration of models written in different programming languages integrated using Java Native Interface (JNI). Remote steering of the computation was also incorporated into the VH computational environment. The user interface is displayed in Fig. 1, showing windows with the anatomical geometry, a circuit model for the left side of the heart, and

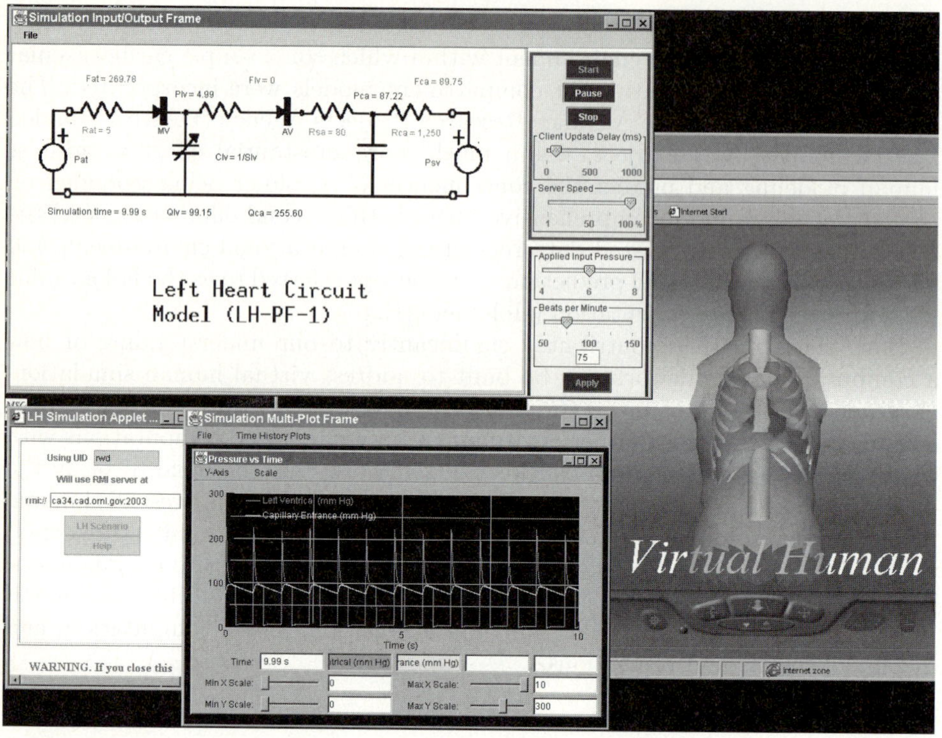

Fig. 1. Virtual Human User Interface

the physiological response of that model. The anatomical geometry used here was a portion of the National Library of Medicine Visible Human male data converted to VRML format.

2.2 PhysioML

The most significant development, however, was to create PhysioML, a physiological Extensible Markup Language (XML) to support model description [9]. PhysioML contains tags to describe the model parameters and variables and associated units. Initial conditions are also defined and the user can select particular variables which can be altered during the computation (computational steering). Finally, PhysioML has tags to describe the user interface including placement of parameters and variables on the interface and graphical representation of the results of the simulation.

Thus, PhysioML incorporates the parameters for the physiological model equations, control of variables for steering the computation, and tags for display of the results in the user interface. While other XML languages such as SBML (http://sbml.org/) and cellML (http://www.cellml.org/) incorporated the first concept, PhysioML is unique in providing a capability to control the interface display and computational steering. At the moment there is no means to incorporate the model description (functions) in the XML format, although eventually

Table 1. Main XML tags used in PhysioML

Display		**User Interface Definition**
	panel	defines a window panel
	image	URL for screen image
	label	screen display
Model		**Model Definition**
	variable	define variable (name, initial value)
	transfer	transfer matrix
	box	define a compartment
	boxstuff	image displayed in compartment
	boxtrigger	threshold for compartment

it is envisioned that this can be accomplished using MathML. For details on PhysioML and examples see http://www.ornl.gov/~rwd/VH/xmlfiles.html.

3 Virtual Soldier

The purpose of the DARPA VS Project, started in 2004, was to use physiological models and data to predict the location of a wound to the heart (left ventricle or right ventricle) caused by a fragment. The specific examples modeled included small fragments wounds to myocardial zones 7 and 12 of the left ventricle with the medical consequences being either tamponade or exsanguination. While different software was used to implement the computational framework for VS and more sophisticated models were utilized, many of the concepts for the VS, especially for the user interface, extrapolate concepts originally developed for the VH.

3.1 The VS Holomer Concept

An important concept developed in the VS Project was that of the Holomer. The Holomer incoporated both the computational framework, including the data and properties (molecular, biochemical, cellular, physiologic, organ, tissue and whole body), computational models, and the display environment. The focus of Phase I was on the heart, so the anatomy considered in the VS Holomer was restricted to the heart and surrounding major vessels.

3.2 Computational Framework for Phase I

For Phase I two types of computational modeling were conducted: 1) high-level integrative physiological (HIP) models (circuit models) and 2) three-dimensional finite element (FE) models, including electrophysiology and mechanical motion. The HIP models were optimized to the physiological characteristics and results were then passed, via file transfer, to the FE models.

The results of both types of computations were integrated using a visualization environment based on SCIRun [7]. In addition, visualization using the

SCIRun environment was linked to the VS Knowledge Base (VSKB), an ontology containing definitions for anatomical terms (the Foundational Model of Anatomy) and physiology. The project also developed an XML format for describing the fragment wound. The integration of output using SCIRun was what was generally referred to as the Holomer.

ORNL and its partner, the Center for Information Technology at the University of South Carolina, developed two components of the VS computational framework. First, we developed middleware to support the project including Web services for the data repository, a client to connect to the Web service for the VSKB, and services and associated client API for the HIP model computations. The University of Washington developed a Web service for the Foundational Model of Anatomy [6]. These middleware components, various Web services for data repository, computations, and ontologies, provide a good infrastructure for a future comprehensive computational framework. Unlike the Phase I framework, this infrastructure would facilitate launching computations from the environment.

The second component of the work conducted at ORNL involved the development of a "HotBox" within the SCIRun visualization environment. The "HotBox" facilitated interaction between the VSKB ontologies (specifically the anatomical ontology) and the geometric anatomical models and between the anatomy and associated physiology (see Fig. 2). The problem was to display

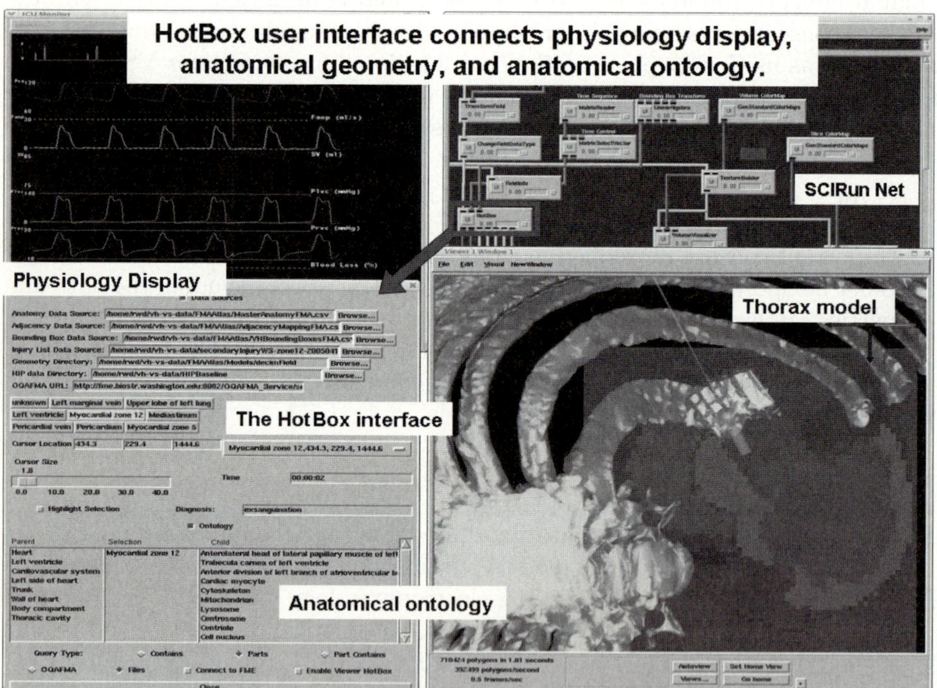

Fig. 2. Virtual Soldier Holomer User Interface

this information in such a way as to capture the three-dimensional (3D) nature of the human body and to correlate that with extensive information about both the anatomy and the physiology of the wounded soldier [5]. The VS "HotBox" succeeded in providing this connectivity.

4 Future

In the future we should begin to see true integrative environments for multiscale human modeling and simulation. A major challenge is to integrate two very different modeling approaches, one based on discrete information (e.g., stoichiometric biochemical reactions) and one based on continuous, time-dependent simulation (e.g., differential equation-based systemic organ models). Two concepts are suggested as potential ways to bridge these different modeling approaches.

1. **Layering of information.** This involves the use of autonomous agents to enable knowledge discovery in support of modeling and simulations. This knowledge discovery process can mine what is known about relevant anatomic, metabolic, or physiological information that impact simulations.
2. **Discrete-event simulation (DEVS).** Discrete event simulation is being used to model continuous simulations [4]. Since DEVS by definition can also incorporate discrete reaction kinetics, it has the potential to provide the necessary bridge to bring about integration between these approaches. A conceptualization of our future vision is contained in Fig. 3.

In the concept vision, the ontological information layers (obtained both by existing libraries of information and supported by intelligent agent searches) support a computational layer based on a combination of DEVS and continuous

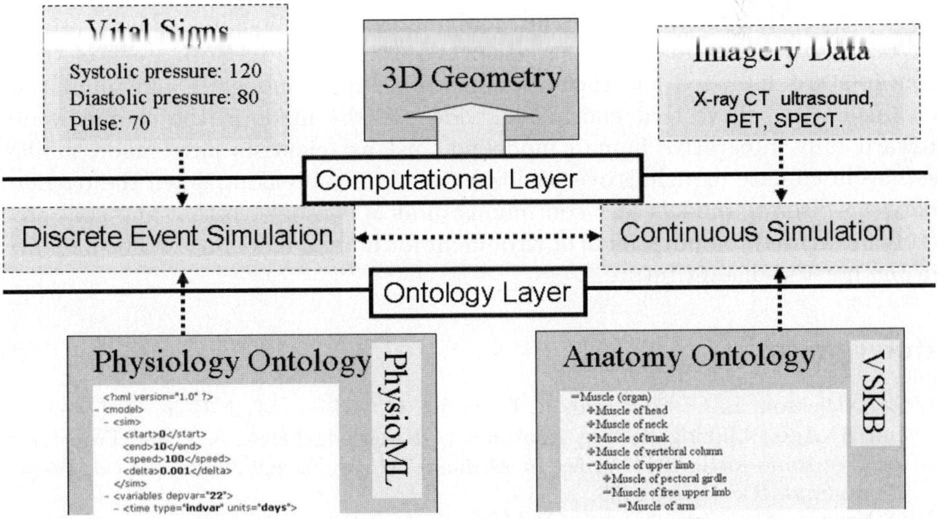

Fig. 3. Future vision of the Virtual Human or Digital Human Infrastructure

simulation approaches, which in turn supports the data and user interface layers. This concept uses capabilities and characteristics developed in the VH and VS projects.

DEVS has a performance advantage over continuous simulation which can be understood intuitively as follows. The time advance function determines the frequency with which state updates are calculated at the spatial grid points. The time advanced at each grid point is inversely proportional to the magnitude of the time derivative at that point, and so "*regions with slow change will have large time advances relative to regions that are changing quickly. This causes the simulation algorithm to focus effort on the changing portion of the solution, with significantly less work being devoted to portions that are changing slowly*" [4].

Our suggestion is that DEVS can be used to implement multiscale computations when there is *loose coupling* between "fast" states and "slow" states of the model system. DEVS can incorporate both discrete models (such as stoichiometric chemical reactions) and continuous models (organ systems) to provide a comprehensive computational framework for the Digital Human.

5 Summary

We have presented a brief historical account of development of integrative human modeling, using the VH Project and the DARPA VS Projects as examples. The lessons learned from these examples and the successes attained have been discussed. In addition, we have outlined the difficulties faced by these projects in attaining a truly integrated human modeling and simulation environment. Finally, we have addressed what we believe to be the future in this effort, the push to attain fully-integrated, multiscale modeling, incorporating both discrete metabolic reactions and continuous modeling based on differential equation models.

We have proposed two different capabilities which might overcome some of these difficulties and make possible fully integrative modeling. These are: 1) **layering of information** and 2) **discrete-event simulation.** Each of these concepts are discussed and their usefulness to human modeling and simulation outlined. We believe that enormous strides will be made in the coming years toward fully-integrative human modeling, just as scientists have made similar strides in climate modeling over the last decade. By breaking down the barriers between discrete models and continuous models, we believe that the goal of a truly integrated computational environment for human modeling will be a reality in the not-too-distant future.

References

1. S. P. Dickson, L. C. Pouchard, R. C. Ward, G. Atkins, M. J. Cole, B. Lorensen, and A. Ade. Linking Human Anatomy to Knowledgebases: A Visual Front End for Electronic Medical Records. In *Medicine Meets Virtual Reality-13 Conference Proceedings*. IOS Press, 2005.
2. C. Krause. The Virtual Human Project: An Idea Whose Time Has Come? *ORNL Review*, 33(1), 2000.

3. C. S. Lindsey, J. S. Tolliver, and T. Lindblad. *JavaTech, an Introduction to Scientific and Technical Computing with Java.* Cambridge University Press, Cambridge, UK, 2005.
4. J. Nutaro. Discrete event simulation of continuous systems. To appear in *Handbook of Dynamic Systems Modeling*, 2005.
5. L. C. Pouchard and S. P. Dickson. *Ontology-Based Three-Dimensional Modeling for Human Anatomy*. Technical Report ORNL/TM-2004/139, Oak Ridge National Labortory, 2004.
6. C. Rosse and J. L. V. Mejino. Ontology for bioinformatics: The foundational model of anatomy. *Journal of Biomedical Informatics*, 36:478–500, 2003.
7. Scientific Computing and Imaging Institute (SCI). *SCIRun: A Scientific Computing Problem Solving Environment*, 2002.
8. R. C. Ward, K. L. Kruse, G. O. Allgood, L. M. Hively, K. N. Fischer, N. B. Munro, and C. E. Easterly. Virtual Human Project. In *Proceedings of the SPIE: Visualization of Temporal and Spatial Data for Civilian and Defense Applications*, pages 158–167, 2001.
9. R. C. Ward, D. J. Strickler, J. S. Tolliver, and C. E. Easterly. A Java User Interface for the Virtual Human. In *Proceedings of the Joint BMES/EMBS Conference*, page 1211, Atlanta, GA, 1999.

Multi-scale Modeling of Trauma Injury

Celina Imielinska[1], Andrzej Przekwas[2], and X.G. Tan[2]

[1] Dept. of Biomedical Informatics and Dept. of Computer Science,
Columbia University,
701 W. 168th Str. HHSC 201, New York, NY 10032
ci42@columbia.edu
[2] CFD Research Corp., 215 Wynn Drive, Huntsville, AL 35505
{ajp, xgt}@cfdrc.com

Abstract. We develop a multi-scale high fidelity biomechanical and physiologically-based modeling tools for trauma (ballistic/impact and blast) injury to brain, lung and spinal cord for resuscitation, treatment planning and design of personnel protection. Several approaches have been used to study blast and ballistic/impact injuries. Dummy containing pressure sensors and synthetic phantoms of human organs have been used to study bomb blast and car crashes. Large animals like pigs also have been equipped with pressure sensors exposed to blast waves. But these methods do not anatomically and physiologically biofidelic to humans, do not provide full optimization of body protection design and require animal sacrifice. Anatomy and medical image based high-fidelity computational modeling can be used to analyze injury mechanisms and to optimize the design of body protection. This paper presents novel approach of coupled computational fluid dynamics (CFD) and computational structures dynamics (CSD) to simulate fluid (air, cerebrospinal fluid) solid (cranium, brain tissue) interaction during ballistic/blast impact. We propose a trauma injury simulation pipeline concept staring from anatomy and medical image based high fidelity 3D geometric modeling, extraction of tissue morphology, generation of computational grids, multiscale biomechanical and physiological simulations, and data visualization.

1 Introduction

Primary blast injury (PBI) results from an interaction of pressure wave with the human body, and gas filled organs, ear, lungs, and gastrointestinal track. A typical explosion creates a shock wave traveling at three to five times the speed of sound. Temperature across the shock wave can increase over 1000 degrees Celsius near the explosion and pressure can increase abruptly to more than 20 atmospheres, Fig. 1. Closed spaces such as rooms and street canyons cause reflections and shock wave interference patterns that can amplify pressure changes. A large percentage of fatal blast injuries result from PBIs, in which the shock wave directly damages the lungs through violent, localized pressure changes. Because a shock wave travels faster in liquids and solids than air, organs like the lungs, intestines, and inner ear are damaged by shear stresses when a shock wave reaches tissue-gas interfaces. The lungs are most

V.N. Alexandrov et al. (Eds.): ICCS 2006, Part IV, LNCS 3994, pp. 822 – 830, 2006.
© Springer-Verlag Berlin Heidelberg 2006

vulnerable to PBIs because they contain large surface areas of fragile alveoli where oxygen is exchanged for carbon dioxide. Differences in wave propagation mechanics at the interface between air and blood in alveolar membranes cause large deformations of lung tissue that collapse alveolar sacs, tear alveolar membranes and rupture blood vessels. The extent of lung injury is a decisive parameter for mortality in victims surviving an explosion. Because the symptoms of primary blast injury are often delayed, victims may not receive timely treatment.

Fig. 1. Blast wave dynamics – injury potential

We are seeking to mathematically model blast wave interaction with human body and ballistic/impact injury to the vital human organs. This modeling effort requires a combination of diverse disciplines including gas and human body dynamics, tissue biomechanics, and the pathophysiology of organ damage.

2 Material and Methods

We demonstrate a 3D modeling pipeline starting with acquisition of data from various imaging modalities that can provide the most detailed anatomy, to high fidelity FEM modeling. This high fidelity simulation combining right mathematical models describing body dynamics, tissue biomechanics, and injury pathophysiology requires detailed geometry and material properties of the organs (torso/lung, head/brain). Such information has to be extracted from anatomical data and multimodal medical imaging. Computational models have to be calibrated and validated against available experimental models such as physical surrogates and animal "models". The computational model will be invaluable in better understanding of injury mechanisms, in better planning of experimental study, and in optimized design of personnel protection armor.

2.1 Design of Imaging System for Organ Injury

We define a 3D imaging pipeline that provides image acquisition, processing, segmentation, and modeling of 3D anatomy of human anatomical structures. The images come from CT, MR patient scans and also from CT and color cryosections of the Visible Human datasets. Methods for function and physiology are applied to depict, assess and classify the full extent of brain, spine, and lung injury under trauma. The structure are later used for FEM impact and biomechanics simulations.

2.2 Modeling Head Anatomy

We use two sets of data depicting head anatomy:

(a) the Visible Human Male color cryosection data segmented, Fig. 2, and reconstructed/rendered with 3D Vesalius Visualizer [1], Fig. 3.

(b) Visible Human Male CT, Fig. 4, segmented with hybrid segmentation method [2,3], Fig. 5., and reconstructed with t-shells [4], Fig. 6.

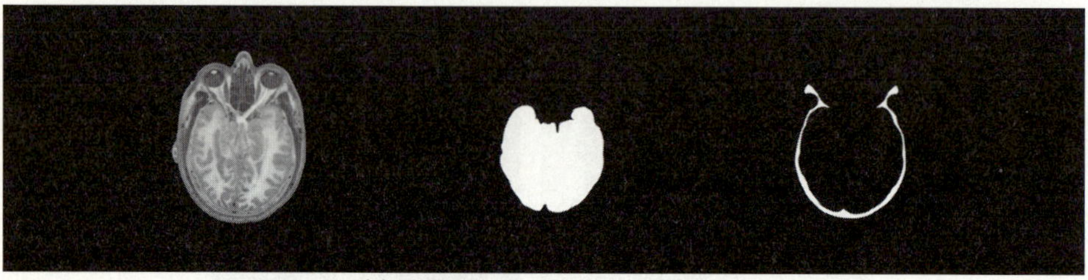

(a) (b) (c)

Fig. 2. (a) Visible Human data with (a) brain and (b) skull regions segmented

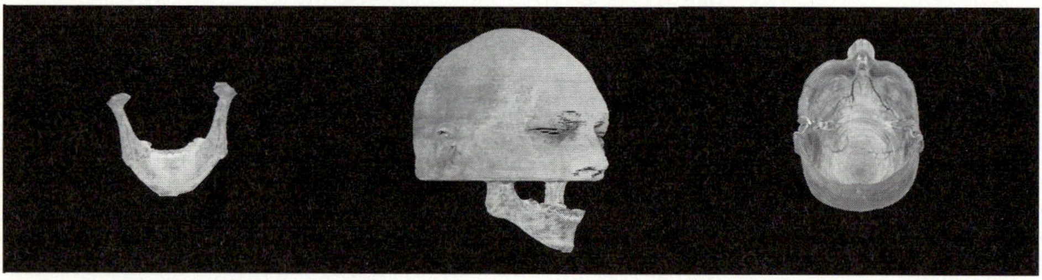

Fig. 3. Selected 3D models from the Visible Human Make data visualizes with 3D Vesalius Visualizer [1]

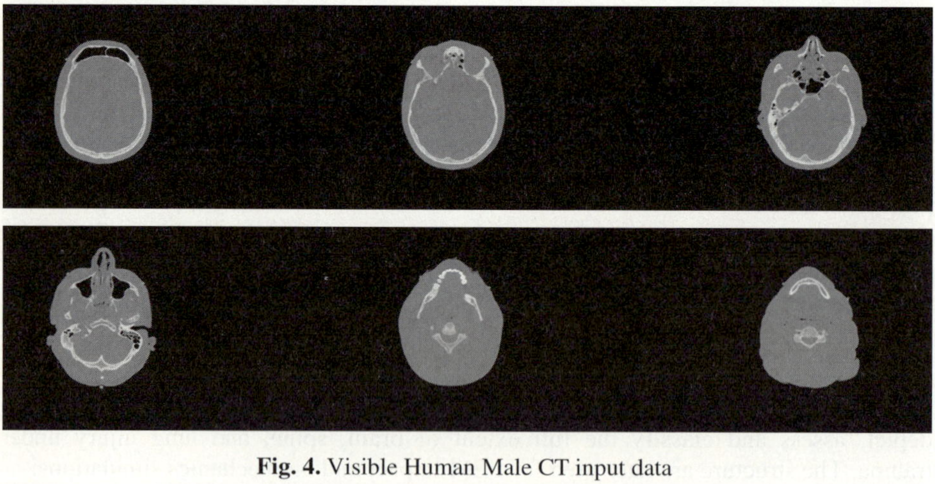

Fig. 4. Visible Human Male CT input data

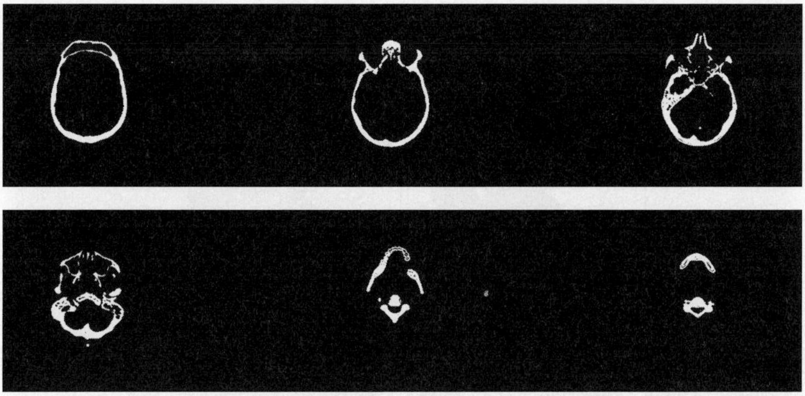

Fig. 5. Visible Human Male CT segmented data

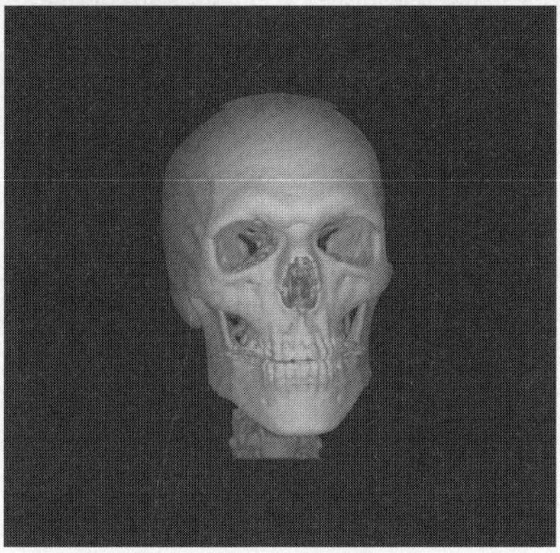

Fig. 6. Visible Human Male skull reconstructed with t-shells. [3].

2.3 Geometry Preserving Automated Coarsening of Triangulated Surfaces

High resolution medical images and image processing software generate large number of surface triangles. This is desired for the image visualization but is not practical for FEM biomechanics modeling. It is not uncommon to see millions of triangles on the cranium and brain surfaces. A FEM tetrahedral volume mesh generated from such number of surface mesh would have hundreds of millions elements. We have developed a novel algorithm and a software module for automated coarsening of triangulated surfaces while preserving the geometrical fidelity. We have tested it on automated coarsening of the surface triangle meshes for a human head depicted in Fig. 7. Figures 6a-c present the original human head mesh with 950,000 triangles and two levels of coarsened grids. The algorithm provides very powerful capability of generating coarse tessellation which can be used directly for tetrahedral meshing or as a surface geometry for generation of NURB surfaces and further high quality structured FEM grids. In the future, we plan to implement this module into a data processing software framework for semi-automated generation of high quality FEM grids for human body, with emphasis on human head, neck, and spine.

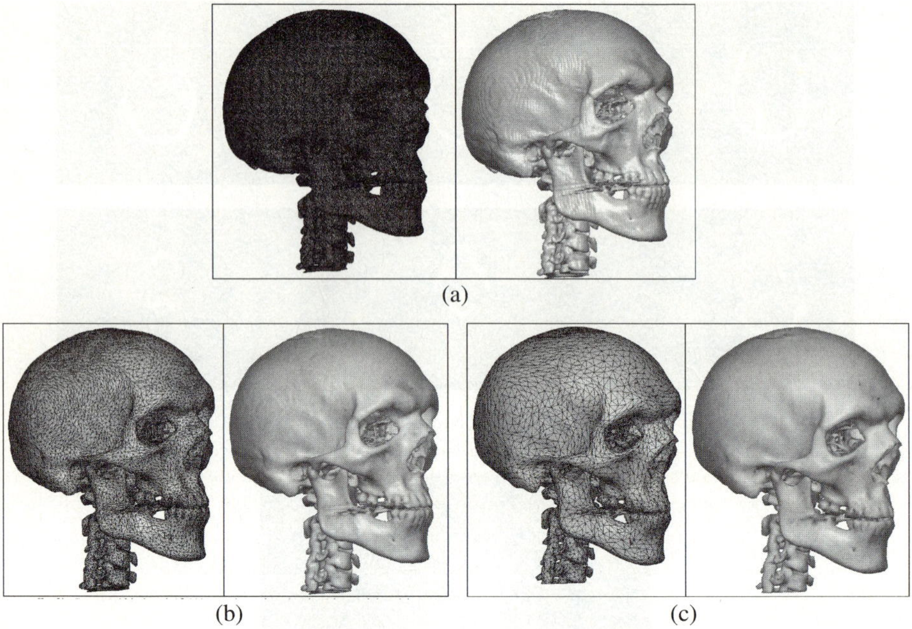

Fig. 7. (a) original model with 950,000 triangles (too fine to see) and rendered solid model, (b) coarsened mesh with 95,000 triangles and rendered resultant solid model, (c) coarsened mesh with 37,500 triangles and rendered resultant model

2.4 Adaptation of FEM Model for Ballistic Injury

CFDRC is developing software tools FEM-Bi, for multiscale modeling of blast injuries to thorax, lungs, brain, and spine coupled to physiological trauma injury model CoBi. We are adapting FEM-Bi for modeling ballistic impact of head protective gear materials in proximity to animal/human head and for modeling head injury. The FEM model can be validated on existing experimental data of mechanical behavior of material used in the protective gear and protection padding.

2.5 Explicit Dynamics Systems

In the current helmeted head model, larger number of finite elements (100K or more) is needed to properly resolve the complex geometry and stress/strain localization. As a result, the solution time for the *implicit FEM solution* for large grid counts (>100K) will increase dramatically and become prohibitively expensive. To speed up the computations and take less computer memory, we developed an *explicit solver for the FEM equations*. For large-scale high-speed event, it is advantageous to use the conditional-stable explicit solver because:

1) A small time step is required to track the high-frequency portion of the response anyway, making it impossible to take advantage of the large time steps permitted by implicit methods;

2) Time integration of the discrete momentum equations does not require the solution of any equations due to the use of diagonal mass matrix. The construction and solving of linear system in the implicit method in each iteration of time step is very time-consuming, especially for the large finite element model;
3) Robustness for the nonlinear problem compared to the Newton iterations in nonlinear implicit solver, in which the convergence failure can often occur due to 'rough' events such as contact-impact or the large time step.

2.6 Explicit Solution of Frictional Contact Problem

The development includes the virtual work and contact searching, which appears to be highly desirable for the general robustness of explicit finite element techniques. Since the explicit formulation has no convergence problem caused by the nonlinear iterations, the resulting frictional contact model is suitable for applications displaying significant non-linear behavior. An important aspect of contact problems is the method used to convert the associated variational inequality to an equality suitable for finite element solution. In the explicit analysis we still adopt the penalty method since it offers distinct advantages on the integration algorithm for the frictional tractions, compared to the Lagrange multipliers method. At beginning of each time step, the estimate for an appropriate choice of the penalty parameter is based on the material properties and geometric size of the shell element. The increasing of the default penalty parameter becomes necessary if the results show visible penetration, and thus does not fulfill the constraint equation in a correct way. On the other hand, the larger penalty parameter may decrease the explicit time step size.

The most important practical aspect of computing the contact force vector is acquisition of the projection for the finite element node, i.e., the contact searching. Briefly stated, the algorithm consists of the following three phases: 1) Identification of the closest target node for the FE node; 2) Among all target elements sharing the closest node, identify the element containing the projection; 3) Calculate the coordinates of the projected point. In the current implementation, we calculate the closest target node once at the beginning of each global time step, which is used for the fluid domain in the fluid-structure interaction (FSI) simulation. During one global time step the local search for the projection is conducted for each FE explicit time step.

2.7 Parametric Simulations

We will use in the future the medical imaging data and resulting 3D models to set-up parametric model of the head, spine and other anatomical regions. We will perform ballistic impact FEM simulations and analyze the dynamics of protective gear in contact with underlying anatomy, the material-structure impact loads and potential injury to the anatomical structure(s).

3 Results

In the following we show examples to verify the correctness of the new developed explicit finite element solver, and the newly developed truss model used in the gear protected head under impact.

3.1 Test on Explicit Finite Element Solver

We have implemented the explicit solver of the finite element model for the FSI applications, in addition to the existing implicit solver. Here are some preliminary results to show the effectiveness of the developed explicit solver.

The key features of the current explicit solver implementation include:

1) Optimal explicit time step size to maintain numerical stability and accuracy,
2) Global/local search algorithm for the contact problem,
3) Full-integrated element to avoid the hourglassing modes,
4) Numerical dissipation for the central difference time stepping.

The new explicit formulation for the structures calculations was then tried out on a large model with 44,791 eight-node solid element (155,394 dofs) as shown in Fig. 8. We made the FE mesh from the geometry data provided by [5]. The FE takes into account the cerebrum, cerebellum, falx and tentorium, CSF, dura, 3 layered cortical and trabecular skull bone, scalp, and facial bone. Table 1 shows the CPU time and computer memory needed for the test case, in which the head is moving forward with an initial velocity. The stable time step size for the explicit algorithm is calculated internally and adjusted.

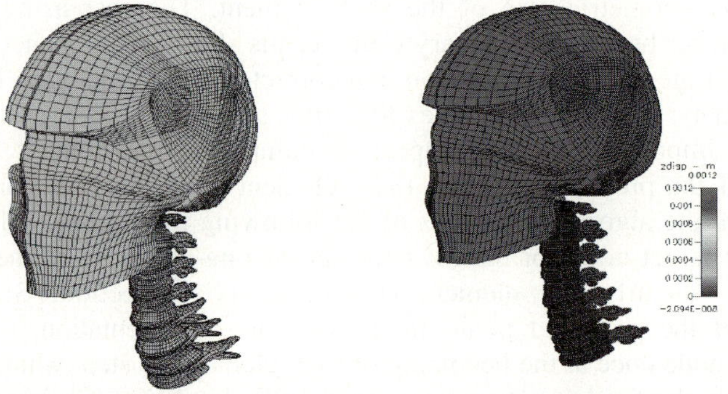

Fig. 8. The head-spine FE model and explicit solution for the head-spine model at t=0.12ms

Table 1. CPU time and Memory requirements of explicit FE solvers for one time step

Model Size (no. of dofs)	CPU (sec)	Memory (MB)
155,394	16.1	150

3.2 2 D Head-Spine Model under Blunt Impact

To test the computational method a 2D multi-body test problem comprised of a helmet-head-cervical spine under rear impact has been setup. A rigid projectile is impacting the outer surface of the helmet with initial velocity of 30m/s. The helmet, protection pad, head and C0-T1 spine are approximately modeled with 2D solid elements. The neck muscle and the ligament between vertebrae are modeled with the viscoelastic truss element. The bottom surface of T1 vertebrate is fixed. In Fig. 9, as the projectile is decelerated and rebounded during the impact on the helmet, both the helmet and the protection pad are deformed and rotated along the skull, and part of the absorbed energy by the helmet is transferred to the entire head in the form of rapid, typically angular acceleration of the head. Severe injuries of the skull, brain and cervical spine can be identified under this impact based on certain head injury criteria (HIC) such as the critical strain/strain rate at the tissue level.

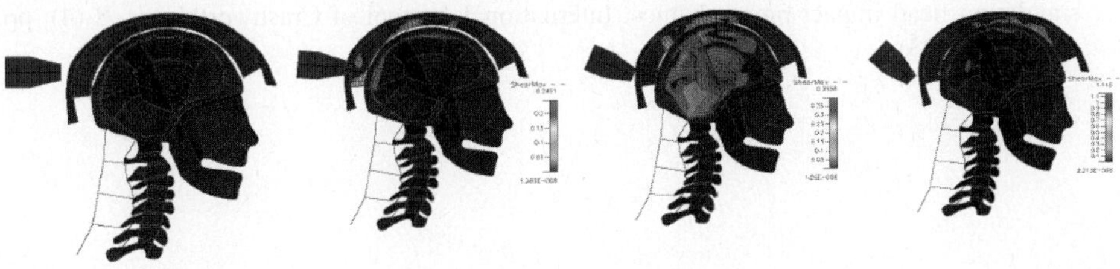

Fig. 9. Solid-Truss multi-body system under impact at time 1) 0ms, 2) 0.5 ms, 3) 1.2 ms, 4) 1.8 ms

It is noted that since we model the cerebrospinal fluid between the skull and the brain as the solid elements, the motion of the brain may be more constrained than in the real situation. We will report in the near future how the brain deforms surrounded by CSF modeled with the fluid elements.

4 Discussions

We demonstrated a simulation pipeline for modeling trauma to head and spine. We plan to develop robust, generic advanced modeling tools and experimental procedures for modeling trauma under ballistic and explosion blast injury. We would like to develop streamlined procedure to generate good quality FEM (volumetric) meshes for human organs from medical images. Databases are needed with examples of parametric head/neck geometries, and tissue specific material properties. The overall goal is to develop of tightly coupled FEM-CFD software tool for modeling a wide range of trauma cases.

Acknowledgements

We would like to thank Autodesk for generous donation of Maya Software, and Dr. Udupa from University of Pennsylvania for letting us use 3DVIEWNIX to model head anatomy.

References

1. Imielinska, C., Molholt, P., "Incorporating 3D Virtual Anatomy into the Medical Curriculum", *CACM,* pp.49-54, Feb. 2005.
2. Imielinska C., Udupa, J.K., Metaxas, D., Jin, Y., Angelini, E., Chen, T., Zhuge, Y., "Hybrid Segmentation Methods", book chapter in "Insight into Images: Principles and Practice for Segmentation, Registration, and Image Analysis", edited by T. Yoo, A.K. Peters, March 2004.

3. Bathe, K.-J., "Finite Element Procedures", Prentice Hall, New Jersey, 1996.G.J.
4. Grevera, G.J., Udupa, J.K., Odhner, D., "An Order of Magnitude Faster Isosurface Rendering in Software on a PC than Using Dedicated, General Purpose Rendering Hardware", *IEEE Trans. on Visualization and Computer Graphics,* 6 (4), pp. 335-345, 2000.
5. Horgan, T.J., Gilchrist, M.D., The creation of three-dimensional finite element models for simulating head impact biomechanics. International Journal of Crashworthiness, 8 (4), pp. 353-366 (2003).

Investigation of the Biomechanic Function of Cruciate Ligaments Using Kinematics and Geometries from a Living Subject During Step Up/Down Motor Task

Luigi Bertozzi[1], Rita Stagni[2], Silvia Fantozzi[2], and Angelo Cappello[2]

[1] Department of Electronics, Informatics and Computer Science,
University of Bologna Via Venezia 52, 47023 Cesena, Italy
lbertozzi@deis.unibo.it

[2] Department of Electronics, Informatics and Computer Science,
University of Bologna Viale Risorgimento 2, 40136 Bologna, Italy
{rstagni, sfantozzi, acappello}@deis.unibo.it

Abstract. The modeling approach is the only possible way to estimate the biomechanic function of the different anatomical sub-structures of the knee joint in physiological conditions. Subject-specific geometry and kinematic data were the foundations of the 3D quasi-static model adopted for the present work. A previously validated cruciate ligaments model was implemented taking the anatomical twist of the fibers into account. The anatomical load components, developed by the modeled ligaments, were estimated during step up/down motor tasks. The anterior cruciate ligament never developed force, along every directions. The posterior cruciate ligament developed increasing forces with the increasing of the flexion angle until at about 70° of flexion. Bigger repeatability in the force curves was obtained in extension with respect to the flexion movement. In conclusion the proposed model was effective in evaluating loads in the anterior and posterior cruciate ligament during the execution of daily living activities.

1 Introduction

In the human knee joint, the harmonious interaction among all its different anatomical sub-units provides the well known mobility and stability characteristics. The knowledge of the biomechanic function of the knee passive structures, like the cruciate ligaments, is of fundamental importance and of great clinical interest for the development of new effective rehabilitative and surgical procedures. This interest is demonstrated by over 8 million of injury related visits for knee symptoms by physicians and in emergency rooms, 381000 total knee replacements and 12000 other repair of cruciate ligaments performed in the USA in 2002 as reported by the American Association of Orthopedic Surgeons (AAOS) [1].

During its normal function, the knee lets the shank move with respect to the thigh, maintaining the stability under external articular load and torque. This is the result of several contributions: inter-segmental contact loads, ligaments tensioning, muscle forces, inertia of body segments. Thus, if we want to quantify the contribution of each anatomical structure, the only possible way is a modeling approach.

V.N. Alexandrov et al. (Eds.): ICCS 2006, Part IV, LNCS 3994, pp. 831 – 838, 2006.
© Springer-Verlag Berlin Heidelberg 2006

The problem of the knee modeling has been tackled from different points of view and at different levels of complexity. Many two-dimensional models on sagittal plane were proposed in literature and several of these were based on a four–bar linkage modeling approach [2-6]. These models allowed to investigate the function of the knee ligaments only in the sagittal plane in different loading conditions [7;8]. Three-dimensional mathematical and finite elements models were also developed [9-12]. These can include sub-models of anatomical articular surfaces and contact forces, of articular deformations, of different passive structures, like ligaments, capsule and menisci, and of active structures like muscles. Nevertheless, these complex models were unusable in physiological context for their computational weight. The logical evolution of this approach could be the evaluation of a 3D model during the execution of a motor task characteristic of daily living activity [13;14]. In this context, even if the model is designed properly for a specific application, its potential can be nullified by the errors resulting from the anatomical, geometrical and mechanical parameters definition. In the cited papers [2;3;7-9;11-13], these errors were due to disagreement in the origins of parameters and inputs, which were often obtained from different and non-homogeneous sources.

Thus, in this work special attention was paid to the input data and parameter, in particular to the geometry influencing mechanics. Subject specific geometries and kinematic data are the foundations of the 3D quasi-static model adopted. The cruciate ligament models took the twisting of the fibers into account and the reference length of each fiber was estimated from the subject-specific passive flexion kinematics.

The aim of this study was the evaluation of the biomechanic role, in terms of forces, of the anterior and the posterior cruciate ligaments during a step up/down motor task.

2 Material and Methods

2.1 Subject and Experimental Acquisitions

The selected subject (male, height 168 cm, weight 62 kg, and age 30 years) underwent a high resolution nuclear magnetic resonance (NMR) scan of his right knee with a 1.5T Gemsow scanner (*GE Medical Systems, Milwaukee, Wisconsin*) [14], as reported in Table 1. The subject performed 2 repetitions of step up/down motor tasks while acquired by means of fluoroscopy (*SBS 1600, Philips Medical System Nederland B.V.*) at 10 images per second. The knee under analysis was kept inside the fluoroscopic field of view during the execution of the selected task. Moreover, for the

Table 1. The NMR scanning procedure parameters

Scanning sequence	Spin Echo (T1 weighted)
Number of slices	54
Pixel spacing	0.037x0.037 (cm·cm)
Scanned region length (across the knee)	15.9 (cm)
Slice thickness	2.5 (mm)
Slice spacing	3 (mm)

detection of the subject specific fibers reference length [2;3], passive flexion was performed with the help of a qualified operator and acquired by means of the same fluoroscopic set-up.

2.2 Knee Geometrical Model

A 3D tiled surface geometrical representation of the distal femur, the proximal tibia, and the insertion areas of the anterior and posterior cruciate ligaments was generated from the NMR dataset using the software Amira (*Indeed - Visual Concepts GmbH, Berlin, Germany*). For each NMR slice, the outer contour of the structures of interest was detected and outlined with an entirely manual 2D segmentation technique. The resulting stacks of contours were interpolated to generate polygonal surfaces of each structure [14], as shown in Fig. 1(*a-b*).

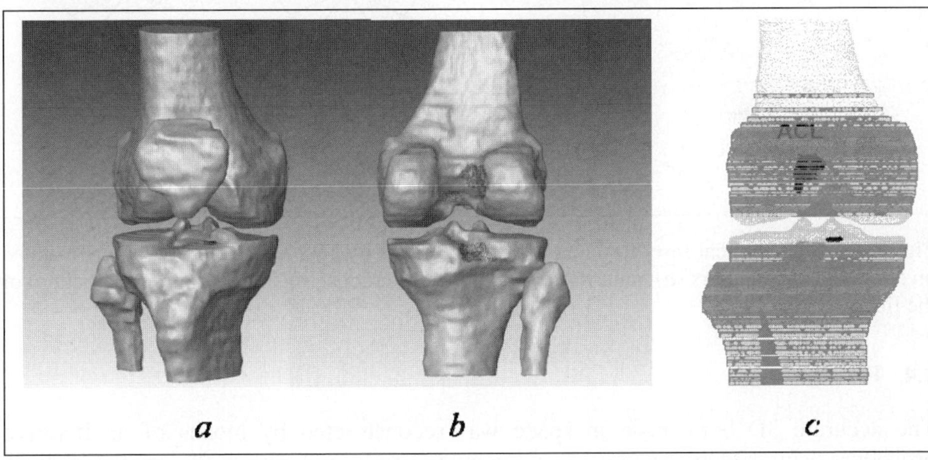

Fig. 1. Anterior and posterior view of the reconstructed bony geometries, (*a*) and (*b*) respectively. Anterior view of the bony geometries and the ligament insertion areas (dotted regions) on the femur and the tibia (*c*).

2.3 Ligament Geometrical Model

The anatomical insertion areas of both cruciate ligaments were described by a set of points. These were also calculated by means of the software Amira as prints of the ligament geometrical volume on the 3D bony surface, see Fig. 1(*c*). The inertia tensor was calculated from each cloud of points, and its principal axes and planes were calculated. The anatomical points were then projected on the first principal plane. A quadratic equation for each planar insertion area was estimated to fit the contour line of the projected anatomical points, and in each case an ellipse was obtained. The planar insertion points were then selected uniformly mapping 25 points on these elliptical areas. The 25 points were distributed: 1 in the centre of the ellipse, 12 uniformly distributed on the contour of the evaluated ellipse and 12 uniformly distributed along the contour of an ellipse having the same centre and semi-axes half of the previous one. The 25 planar insertion points selected on each elliptical area were then fitted on the 3D anatomical insertion area using the "thin plate splines" (TPS) method [15] as shown in Fig. 2(*a*).

The joining method between the femoral and tibial insertion points took the anatomical twist of the ligament fibers into account. The anterior cruciate ligament had the order of the fibers on the tibial insertion area rotated by 90° laterally with respect to the femoral insertion area coherently with the anatomical external twist of the ligament, see Fig. 2(*b*). The posterior cruciate ligament had the order of the fibers on the tibial insertion area rotated by 90° medially with respect to the femoral insertion area [16], see Fig. 2(*c*).

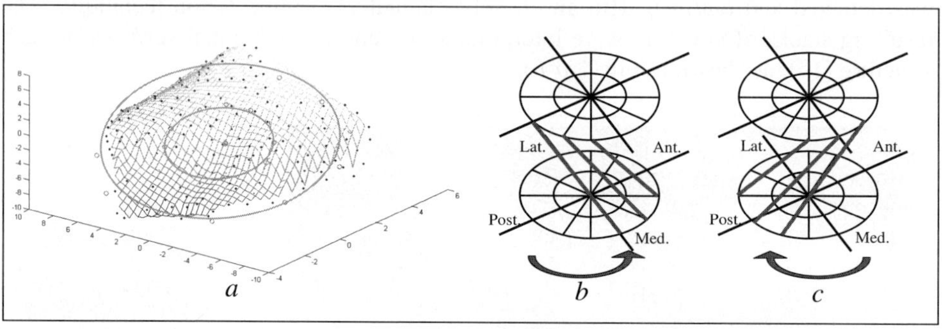

Fig. 2. A 3D anatomical insertion area with the 25 fitted by TPS fiber insertion points mapped on two elliptical contours (*a*); anterior (*b*) and posterior cruciate ligament (*c*) ordering pattern of the fibers

2.4 Kinematics

The accurate 3D bone pose in space was reconstructed by means of an iterative procedure using a technique based on tangency condition between the projection lines and the surface of the geometrical model. The accuracy of this technique was assessed to be 1 degree for rotations and 1 mm for translations [17]. The passive flexion and step up/down kinematics were then reconstructed.

2.5 Mechanical Ligament Properties

The two cruciate ligaments, in both models, were modeled with 25 different linear-elastic elements. The elastic modulus E of each ligament was the same and it was considered constant from literature equal to 175 MPa [3]. The reference length l_{0j} of each fiber j was defined according to Goodfellow's hypothesis [18], as the maximal length reached by each fiber during passive flexion. In a previous study, the authors validated this technique using the drawer test along the anterior and the posterior directions [19]. From the NMR dataset the total insertion area was known. The relative cross-sectional area A_j for each fiber was calculated proportionally to the distance of each modeled insertion point from its adjacent ones after the TPS deformation on the anatomical surface.

The stiffness coefficient K_j was calculated for each fiber j with the equation (1) where E, l_{0j} and A_j were the variables mentioned above.

$$K_j = \frac{E \cdot A_j}{l_{0j}} \tag{1}$$

The force expressed from each fiber was shown in equation (2) where ΔL_j was the difference between instant length l_j and the reference length l_{0j} of the fiber.

$$F_j = -K_j \cdot \Delta L_j \tag{2}$$

The total ligament force was the vectorial sum of all fiber forces of the ligament. Obviously, the force expressed by each fiber was imposed to be zero if the distance between its two insertions was smaller than the reference length.

2.6 Simulation and Post-processing Tools

The mechanical system, composed from the bony geometries and the ligaments geometrical model including its mechanical properties, was implemented and animated with the acquired experimentally kinematics in ADAMS/View 2005 (*MSC.Software Corporation 2 MacArthur Place Santa Ana,CA 92707 USA*). This simulator of mechanical systems allowed to estimate each variable in the model, in particular, for each relative position between the femur and the tibia. The three components of the forces, anterior-posterior (A/P), proximal-distal (P/D) and medial-lateral (M/L) NMR projections, and the magnitude for each fiber were calculated and exported for both cruciate ligaments. Post-processing elaborations were computed with Matlab 7 (*The MathWorks, Inc, MA 01760-2098*). The three components were set to zero when the magnitude of the force of each fiber was positive (compression), see equation (2). All these forces were transposed to the anatomical tibial reference system.

3 Results

The global qualitative behavior of the posterior cruciate ligament was very similar among the three anatomical directions, in particular considering the extension movements. In the A/P and in the M/L components similar and bigger forces were always reached than those reached in the P/D direction, see Fig. 3. The maximum forces reached were three times bigger in the extension movements and over five times bigger in the flexion movements.

In extension movements the mechanical contribution of the posterior cruciate ligament was very limited from the full extension to about 30°-40° of flexion. Then a rapid and quite linear increasing of its contribution was recorded until reaching the maximum force until about 70° of flexion. At this angle of flexion a little decreasing of the forces was observed, in particular in the P/D direction, see in Fig. 3(*b*).

The behavior of the two flexion movements was different with respect to each other, in particular along the P/D direction. Like during the extension movements, also in this case, very little forces were expressed from the full extension to a variable angle at about 20°-40° of flexion. From this point, both two curves showed an increasing of their contribution until reaching bigger maximum forces with respect to those calculated in the extension movements. An isolated difference between the two curves was clearly evident in the P/D direction, where one of these began to decrease at about 45° of flexion, see Fig. 3(*e*).

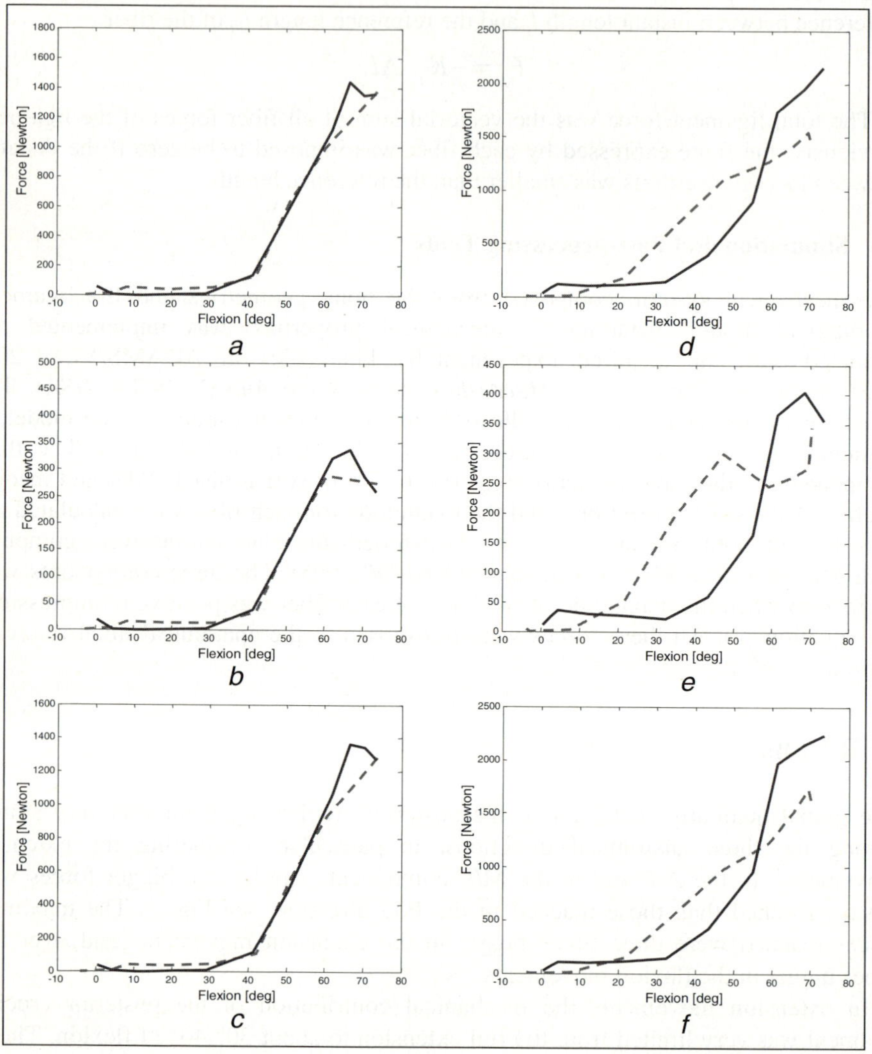

Fig. 3. Posterior cruciate ligament component forces on the anatomical tibial reference system versus the knee flexion angle (2 motor task repetitions: solid and dashed line): A/P, P/D and M/L component forces of the three extension movements (*a*, *b* and *c*, respectively), and A/P, P/D and M/L component forces of the two flexion movements (*d*, *e* and *f*, respectively)

Regarding the anterior cruciate ligament the results obtained, during the simulations of both flexion and extension movements, were always equal to zero along every anatomical directions.

4 Discussion

Anterior and posterior cruciate ligament models were implemented using geometrical parameters and kinematics data from a single living subject. Regarding the posterior cruciate, the greater repeatability obtained in the extension (step up movement) was probably due to a major activity of the muscles for controlling the movement. These had the goal to perform the movement against the gravity force and their concentric contractions were more controlled by the nervous system. On the other hand the flexion (step down movement) was an according to the gravity movement. Thus the eccentric contraction of the muscles was less controlled and a minor repeatability was obtained. The total inactivity of the anterior cruciate ligament was probably due to the typology of the movement that tends to slack the anterior cruciate and to stretch the posterior cruciate ligament. Indeed the biggest anatomical force components, regarding the posterior cruciate ligament, were obtained along the A/P and the M/L directions, where bigger contributions to the joint stabilization function were necessary.

Although the linearity of the mechanical characteristic assumed for each ligament fiber, the produced results were in agreement with physiology. In conclusion the proposed model, including all experimental acquisitions and data elaborations, was effective in evaluating subject-specific cruciate ligament loads during the execution of daily living activities.

References

1. AAOS. Internet site address: http://www.aaos.org
2. Zavatsky, A.B., O'Connor, J.J.: A model of human knee ligaments in the sagittal plane. Part 1: Response to passive flexion. Proc Inst.Mech.Eng [H.] 206 (1992) 125-134
3. Zavatsky, A.B., O'Connor, J.J.: A model of human knee ligaments in the sagittal plane. Part 2: Fibre recruitment under load. Proc Inst.Mech.Eng [H.] 206 (1992) 135-145
4. Gill, H.S., O'Connor, J.J.: Biarticulating two-dimensional computer model of the human patellofemoral joint. Clin Biomech 11 (1996) 81-89
5. Lu, T.W., O'Connor, J.J.: Lines of action and moment arms of the major force-bearing structures crossing the human knee joint: comparison between theory and experiment. J Anat 189 (Pt 3) (1996) 575-585
6. Zavatsky, A.B., Wright H.J.: Injury initiation and progression in the anterior cruciate ligament. Clin. Biomech. 16 (2001) 47-53
7. Zavatsky A.B., O'Connor J.J.: Ligament forces at the knee during isometric quadriceps contractions. Proc. Inst. Mech. Eng [H.] 207 (1993) 7-18
8. Shelburne K.B., Pandy M.G.: A musculoskeletal model of the knee for evaluating ligament forces during isometric contractions. J. Biomech. 30 (1997) 163-176
9. Wismans J., Veldpaus F., Janssen J., Huson A., Struben P.: A three-dimensional mathematical model of the knee-joint. J. Biomech. 13 (1980) 677-685
10. Blankevoort L., Kuiper J.H., Huiskes R., Grootenboer H.J.: Articular contact in a three-dimensional model of the knee. J. Biomech. 24 (1991) 1019-1031
11. Mommersteeg T.J., Huiskes R., Blankevoort L., Kooloos J.G., Kauer J.M.: An inverse dynamics modeling approach to determine the restraining function of human knee ligament bundles. J. Biomech. 30 (1997) 139-146

12. Moglo K.E., Shirazi-Adl A.: Cruciate coupling and screw-home mechanism in passive knee joint during extension—flexion. J. Biomech. 38 (2005) 1075-1083
13. Piazza S.J., Delp S.L.: Three-dimensional dynamic simulation of total knee replacement motion during a step-up task. J. Biomech. Eng 123 (2001) 599-606
14. Stagni R., Fantozzi S., Davinelli M., Lannocca M.: Comparison of knee cruciate ligaments models using kinematics from a living subject during chair rising-sitting. Lecture Note in Computer Science Vol. 3036 Springer-Verlag (2004)1073-1080
15. Bookstein F.L.: Principal Warps: Thin-Plate Splines and the Decomposition of Deformations. IEEE Trans. Pattern Anal. Mach. Intell. 11 (1989) 567-585
16. Mommersteeg T.J., Kooloos J.G., Blankevoort L., Kauer J.M., Huiskes R., Roeling F.Q.: The fibre bundle anatomy of human cruciate ligaments. J. Anat. 187 Pt. 2 (1995) 461-471
17. Zuffi S., Leardini A., Catani F., Fantozzi S., Cappello A.: A model-based method for the reconstruction of total knee replacement kinematics. IEEE Trans. Med. Imaging 18 (1999) 981-991
18. Goodfellow J., O'Connor J.: The mechanics of the knee and prosthesis design. J. Bone Joint Surg. Br. 60-B (1978) 358-369
19. Bertozzi L., Stagni R., Fantozzi S., Cappello A.: 3D subject-specific model of the human knee from in-vivo measurements: validation on the knee drawer test. In submission to the Med. Eng. Phys.

Optimization Technique and FE Simulation for Lag Screw Placement in Anterior Column of the Acetabulum

Ruo-feng Tong, Sheng-hui Liao, and Jin-xiang Dong

State Key Laboratory of CAD and CG,
Department of Computer Science and Engineering,
Zhejiang University, China
liaoshenhui@zju.edu.cn

Abstract. This paper presents an optimization technique for determining the lag screw placement in the anterior column of the acetabulum, and investigates new method for generating accurate finite-element (FE) model for biomechanics analysis. For prepare once measure, an accurate hemi-pelvis model is reconstructed from the volume-of-interest extracted from computed-tomography (CT) data, and the initial position of the lag screw is determined by traditional manual like method. Then, an objective function, for improving the placement of lag screw, is build by adaptive sampling the weighted distance of screw to the acetabulum boundary according to surgical requirement, and the two end points of the lag screw are modified iteratively to reduce the objective value. 30 hemi-pelvis models are tested by the optimization technique, and the statistical measure data are provided according to new anatomic reference landmarks for clinical use. In the second part, FE method is employed to evaluate the optimization result. To generate accurate and high quality FE model, a semi-automatic FE preprocessor specifically adapted to the pelvis anatomy is developed. The produced volume mesh has a very regular mesh structure and achieves a smooth change of element size transition. The final simulation stress distribution pattern justifies the placement of the lag screw in the anterior column of the acetabulum.

1 Introduction

Several studies have shown that open reduction and internal fixation of displaced acetabular fractures improve functional and clinical results [1]. Lag screw fixation along the long axis of the anterior column has been recommended for the treatment of transverse and T type fractures. However, proper placement of a lag screw in the anterior column is challenging because of its unique anatomy, relatively small cross sectional area, and the risk of violation of the hip joint [2].

V.N. Alexandrov et al. (Eds.): ICCS 2006, Part IV, LNCS 3994, pp. 839–846, 2006.
© Springer-Verlag Berlin Heidelberg 2006

2 Technology for Determine the Placement of Lag Screw

2.1 Previous Work

There are several approaches proposed in the literature to determine the placement of the lag screw in anterior column of the acetabulum in recent years. Mears and Rubash [3] suggested that a starting point for lag screw fixation of the anterior column should be chosen 2.5 cm above the roof of the acetabulum, and the screw should be directed parallel to the iliopectineal line. Letournel and Judet [4] advocated an entry point 3 to 4 cm above the acetabular roof and recommended that the direction of screw placement be controlled visually and by palpation of the iliopectineal eminence with a finger. The study of Ebraheim et al [5] showed that the entry point for lag screw placement along the functional axis of the anterior column in a sagittal plane can be localized intraoperatively by palpating the bony landmarks and measuring a mean distance of 42 mm posteriorly from the anterior interspinous notch, and 46 mm from the superior acetabular rim. And the inclination of the screw placement from this starting point should be 90.6 ± 5.0 degree in the sagittal plane and 29.0 ± 4.4 degree in a transverse plane.

All of the previous approaches were based on clinic experience or measured directly on embalmed pelvis. One of the shortcomings of these methods is the embalmed pelvises are difficult to get, often with poor bone quality, and always have the virus risk. Second, because of the irregular bony structure, manual work is tend to introduce measure error. Third, physical methods usually need to dissect the models, and can measure once only. Fourth, some geometric information, such as some angle, is hard to measure by simple instrument on the physical model.

2.2 Investigate Optimal Lag Screw Placement

This paper selects the CT volume as the source data because 3D CT imaging of pelvis patients is routine at present, and can provide accurate data information which is needed for the experimentation.

For prepare once measure, an accurate hemi-pelvis model is reconstructed from the volume-of-interest extracted from CT volume, and the initial position of the lag screw is determined by manual method like paper [5]. First, selecting the initial cut as the inferior bony acetabular margin, the model is sectioned at 1 cm intervals perpendicular to the anterior surface of the anterior column. A line designated AB, is drawn perpendicular to the pelvis brim on the caudal surface of the cross section 1 and 3 cm superior to the initial cut connecting the (A) lateral bony acetabular margin and the (B) pelvic brim, and the midpoint for this line is determined. A second perpendicular line, CD, is drawn on the midpoint of line AB. The midpoint of line CD is determined to yield the midpoint (X) of the two cross sections of the anterior column, which are on the functional axis of the anterior column of the acetabulum. Then, the projection points of the functional axis on the outer table of the iliac wing and the pubic bone are calculated. Such as Fig.1 shows.

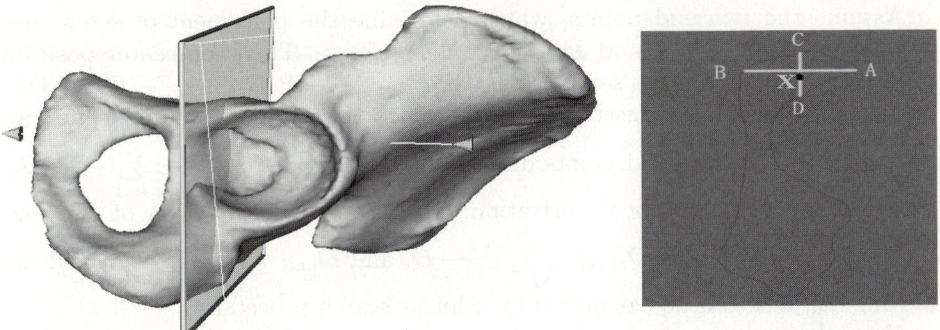

(a) (b)

Fig. 1. The initial placement of the lag screw

While, as the configuration of the entire anterior column is irregular and has a curved true axis, the functional axis determined by only two cross sections can not guarantee for the globe optimal placement. This paper designs an objective function to improve the placement of lag screw in the anterior column of the acetabulum.

Our consideration is based on the clinical requirement: to avoid the violation of the acetabulum and cortical penetration, the distance from the lag screw to the boundary of the anterior column of the acetabulum should as large as possible.

Start from the initial placement of the screw, the algorithm adaptively samples the weighted distance of screw to the acetabulum boundary. Assume the current number of sample cross sections, which are perpendicular to the lag screw, is N. For each cross section $i, i = 1, 2 \ldots N$, the algorithm calculates the shortest vector from the center of the screw $\boldsymbol{P}_i$ on the section plane to the acetabulum boundary:

$$\boldsymbol{SV}_i = \min\left[(\boldsymbol{V}_i)_\alpha\right], 0 \leq \alpha \leq 350 \tag{1}$$

where $(\boldsymbol{V}_i)_\alpha$ is an function of $\boldsymbol{P}_i$ and searches in the extracted volume-of-interest directly rather than done a significant number of intersection calculations with the discrete surface model. Define the weight value as inverse proportion to the length of the shortest vector: $\frac{c}{\|\boldsymbol{SV}_i\|}$, where c is a constant. Then the hint moving vector of section i is:

$$\boldsymbol{D}_i = \frac{c}{\|\boldsymbol{SV}_i\|}\left(-\frac{\boldsymbol{SV}_i}{\|\boldsymbol{SV}_i\|}\right) = -c\frac{\boldsymbol{SV}_i}{\|\boldsymbol{SV}_i\|^2} \tag{2}$$

The globe objective function can be constructed as the sum of these N hint moving vectors' norm:

$$Obj = \sum_{i=1}^{N} \|D_i\| \tag{3}$$

It is clear that when Obj reduces to the minimal value, the lag screw get the optimal placement.

Assume the two end points, which determine the placement of screw, are $\boldsymbol{P}_{start} =< X_1, Y_1, Z_1 >$ and $\boldsymbol{P}_{end} =< X_2, Y_2, Z_2 >$. Then the center position of the screw on each cross section plane is $\boldsymbol{P}_i = \frac{r_2}{r_1+r_2}\boldsymbol{P}_{start} + \frac{r_1}{r_1+r_2}\boldsymbol{P}_{end}$. This proportion gives us a suggest that the influence of each hint moving vector can be brought to the two end points directly, $\sum_{i=1}^{N} \boldsymbol{D}_i = \sum_{i=1}^{N} \frac{r_2}{r_1+r_2}\boldsymbol{D}_i + \sum_{i=1}^{N} \frac{r_1}{r_1+r_2}\boldsymbol{D}_i$, rather than do a numerical derivation. So, the moving direction of two end points can be defined as $\boldsymbol{D}_{start} = \sum_{i=1}^{N} \frac{r_2}{r_1+r_2}\boldsymbol{D}_i$ and $\boldsymbol{D}_{end} = \sum_{i=1}^{N} \frac{r_1}{r_1+r_2}\boldsymbol{D}_i$, and the moving distance can be calculated by a linear search procedure.

Now, the two end points of the lag screw are modified iteratively to reduce the objective value until ΔObj reaches a threshold. During the iteration, the number of sample cross sections N can be adaptively increased until the CT spacing resolution is reached, this is helpful to prevent local minimal result.

2.3 Statistical Measure Data

For convenient clinical use, we investigate new anatomic reference landmarks, especially for the inclination of the functional axis, such as Fig.2 shows. In all, 30 hemi-pelvis models are tested by the optimization technique, and the statistical measure data are listed as follows:

(1) The mean distance FG, between the anterior interspinous notch (F) and the intersection of the perpendicular line from the projection point (P) to the curve line EF connecting the apex of the sciatic notch (E) with the anterior interspinous noth (F), is 39.48 ± standard deviation of 2.12 mm.

(2) The mean distance EG, between apex of the sciatic notch (E) and the intersection of the perpendicular line from the projection point (P) to the line EF, is 41.38 ± 2.23 mm.

(3) The mean distance PG, between the projection point (P) and its perpendicular intersection on the curve line EF, is 16.33 ± 3.31 mm.

(4) The mean distance GH, between the perpendicular intersection point on curve line EF (G) and the superior rim of acetabulum (H), is 25.65 ± 0.82 mm.

(5) The mean length of the screw for anterior column fixation is 104.24 ± 4.63 mm.

(6) For the inclination of the functional axis of the anterior column, we use the angle $\angle SPA$, between the axis and the line PA connecting the projection point (P) with the posterior inferior iliac spine (A), which is 41.52 ± 0.92 degree. And the angle $\angle SPQ$, between the axis and the line PQ connecting the projection point (P) with the point (Q) on the radial line GP as we found the point G, P and Q are almost on a common plane, which is 35.35 ± 1.12 degree.

3 Finite Element Simulation

To evaluate the optimization placement result, the FE method is used for predicting the biomechanical behavior of the pelvis with the lag screw in the anterior column of the acetabulum. However, creating an accurate FE pelvis model is not

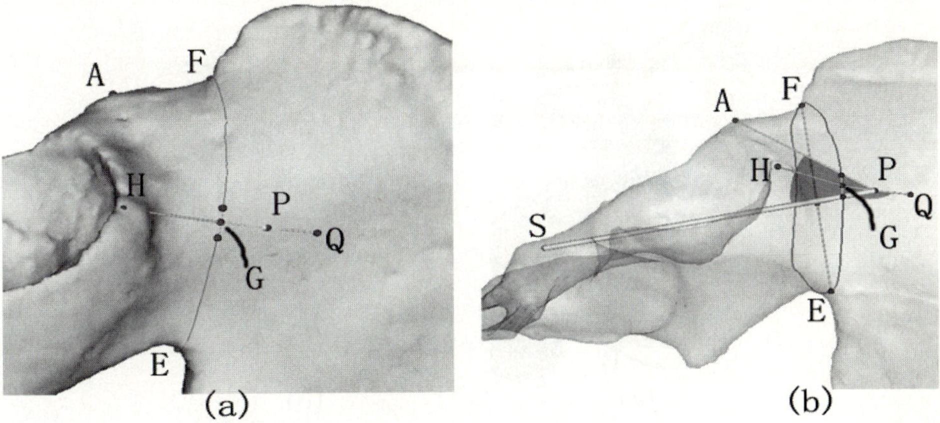

Fig. 2. The distance and angle measurements

an easy work, which has lots of complex anatomical features, such as acetabulum, foramen obturaturm, and so on.

For general automatic volume mesh generation, tetrahedral is by far the most common form. Most current techniques can fit into one of three main categories: the octree method [6], Delaunay triangulation [7] [8] [9] [10] and the advancing front approach [11]. While, general volume meshing methods do not work well to create a satisfactory FE pelvis model efficiently, as they did not consider the patient-specific shape configuration.

There are many FE modelling work specially for biomechanical analysis proposed in the literature in recent years, most of them extracted geometric information from embalmed physics model or CT data at a serial of parallel cross section, and the generated meshes align inherently with the orthogonal plane, which does not properly account for the preferential orientation feature of the physical model.

This paper presents a semi-automatic FE preprocessor to extract the crucial geometry of the highly irregular bony structure of the hemi-pelvis, to input into downstream process for the development of FE model. The mesh developed herein has to be well-represented in terms of original geometry as well as achieving a high quality of mesh as poor representation and quality of finite element would lead to inaccurate results.

3.1 Semi-automatic FE Preprocessor to Generate Volume Mesh

The core consideration of our algorithm is to use curve cut surface, which properly account for the preferential orientation feature of the physical model, to extract the crucial geometry. This is accomplished by our semi-automatic FE preprocessor.

To control the placement of these curve cut surfaces, the preprocessor first creates pairs of control spline by setting several anatomical landmarks interactively, such as these dark blue splines in Fig.3 shows. The shape of the curve cut

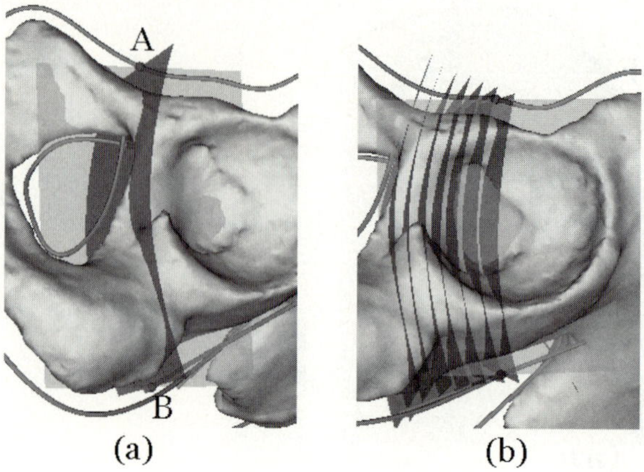

Fig. 3. The curve cut Surfaces

surface is determined as follows: for each selected pair points A and B on the control splines, a direction plane is determined by the axis AB and the average direction of the two control splines at the point A and B, such as the green plane in Fig.3(a) shows; then, the initial cut surface, with AB as its axis, is generated perpendicular to the direction plane; the axis AB, which is a B-spline, can be bent arbitrarily on the direction plane, and result to a free bend curve cut surface, such as the red curve cut surface shown in Fig.3(a).

To create the whole model, some key curve cut surfaces are first determined manually, then the in-between curve cut surfaces are generated automatically by linear interpolation, such as Fig.3(b) shows.

After all of the curve cut surfaces are determined, the sample points on the cut contours are produced by an adaptive scheme. The main factors of the sample distance on the cut contour are: (1) the distance of current point to the adjacent contour; (2) the curvature of local surface; (3) the anatomical features

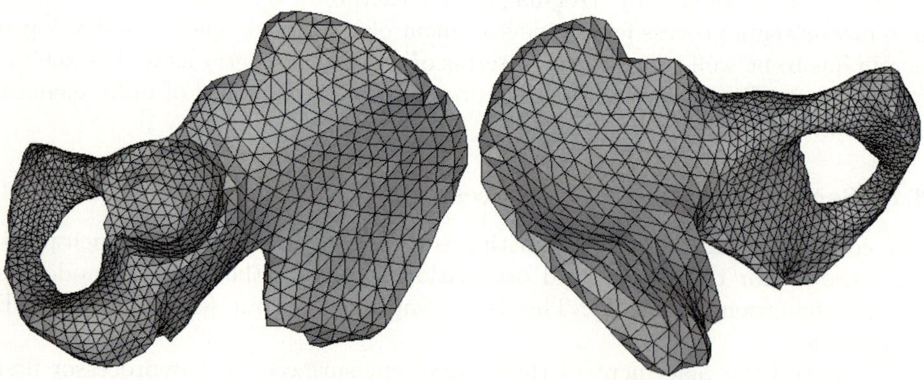

Fig. 4. The result hemi-pelvis FE model

of the physical model. Then the surface mesh is created based on these extracted geometric information.

And as the curve cut surfaces are expandable, the FE preprocessor creates additional insertion points by 2D triangulation, then mapped back to the curve cut surfaces. Finally, the volume mesh is created using the surface mesh together with these insertion points by 3D Delaunay triangulation. And the bony material properties are assigned corresponding to the Hounsfield value of CT. The detailed discussion is ignored because of space limitation.

Fig.4 shows the result, we can see that the finite element model developed from this process has a highly regular mesh structure and achieves a smooth change of element size transition, as well as good representation of its original geometry, which would give a better prediction of its biomechanical response in vivo and in vitro situations.

3.2 FE Biomechanics Analysis

Finally, the pelvis model combined with the lag screw in the anterior column of the acetabulum is created, such as Fig.5(a) shows. Axis direction loading is added on the screw nailhead, and the simulation stress distribution has a uniform pattern, which justifies the placement of the lag screw in the anterior column of the acetabulum, such as Fig.5(b) shows.

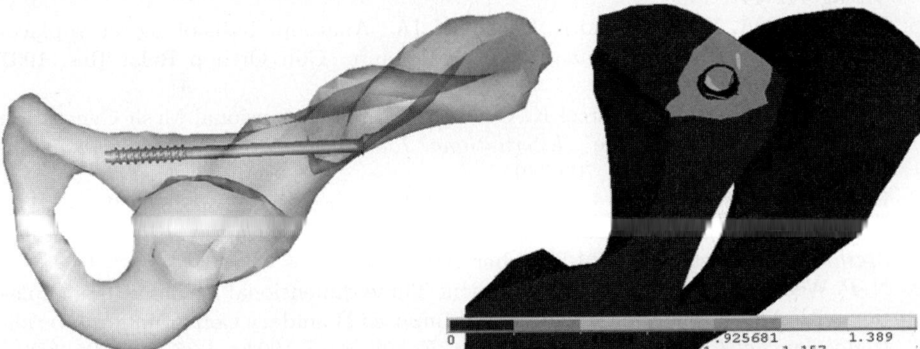

(a) (b)

Fig. 5. The combined model and simulation stress distribution pattern

4 Conclusion

We have presented an automatic optimization technique for improving the lag screw placement in the anterior column of the acetabulum. The objective function is constructed directly based on the clinical requirement, and the adaptive iteration scheme is bound to find the globe minimal value. In addition, a particular convenient FE preprocessor which take advantages of the curve cut surface, is developed. And the result model has a mesh structure of high quality and preserves the original geometry feature, which is of the essence in the study of accurate biomechanics Analysis. The method here can be applied easily to create other finite element volume meshes. We intend to do some advanced biomechanical study in the near future and improve robustness of the algorithm.

Acknowledgments

This project was supported by the Natural Science Foundation (No.M603129) of Zhejiang Province, China.

References

1. Schopfer A, Willett K. Powell J, Tile M: Cerclage wiring in internal fixation of acetabular fractures. J Orthop Trauma 7:236-241, 1993.
2. Anglen JO, DiPasquale T: The reliability of detecting screw penetration of the acetabulum by intraoperative auscultation. J Orthop Trauma 8:404-408, 1994.
3. Mears DC, Rubash HE: Techiques of Internal Fixation. In Mears DC, Rubash HE (eds). Pelvic and Acetabular Fractures. Thorofare, NJ, Slack 299-318, 1986.
4. Letournel E, Judet R: Operative Treatment of Specific Type of Fractures. In Letournel E, Judet R (eds). Fractures of the Acetabulum. Ed 2. Berlin, Springer-Verlag 442-447, 1993.
5. Ebraheim NA, Xu R, Biyani A, Benedetti JA.: Anatomic basis of lag screw placement in the anterior column of the acetabulum. Clin Orthop Relat Res. 1997 Jun;(339):200-5.
6. Mark S. Shephard and Marcel K. Georges, "Three-Dimensional Mesh Generation by Finite Octree Technique". *International Journal for Numerical Methods in Engineering*, 1991,vol 32, pp. 709-749.
7. H. Borouchaki, F. Hecht, E. Saltel and P. L. George. "Reasonably Efficient Delaunay Based Mesh Generator in 3 Dimensions", *Proceedings 4th International Meshing Roundtable*, pp.3-14, October 1995.
8. N. P. Weatherill and O. Hassan. "Efficient Three-dimensional Delaunay Triangulation with Automatic Point Creation and Imposed Boundary Constraints". *International Journal for Numerical Methods in Engineering*, 1994,vol 37, pp.2005-2039.
9. S. Rebay. "Efficient Unstructured Mesh Generation by Means of Delaunay Triangulation and Bowyer-Watson Algorithm", *Journal Of Computational Physics*, 1993,vol. 106, pp.125-138.
10. David L. Marcum and Nigel P. Weatherill. "Unstructured Grid Generation Using Iterative Point Insertion and Local Reconnection", *AIAA Journal*, September, 1995,vol 33, no.9, pp.1619-1625.
11. R. Lohner. "Progress in Grid Generation via the Advancing Front Technique", *Engineering with Computers*,1996, vol 12, pp.186-210.

Model of Mechanical Interaction of Mesenchyme and Epithelium in Living Tissues

Jiří Kroc**

Department of Mechanics, University of West Bohemia in Pilsen
Univerzitni 22, 306 14 Pilsen, Czech Republic
kroc@c-mail.cz
http://www.c-mail.cz/kroc

Abstract. Developmental biology describes how tissues, organs, and bodies are made from living cells. There exists a large body of biological data about developmental processes but there is still not ultimate understanding of how the whole orchestra of all involved processes is working. It is the place where mathematical modelling could help to create biologically relevant models of morphological development. The morphological development could be mathematically decomposed into three distinct but mutually interconnected parts, namely to mechanical response of tissues, signalling by chemicals, and switching of cells into different types by a gene regulatory network. This paper is focussed to the part dealing with mechanical interaction of growing mesenchyme and epithelium within a living tissue modelled by a set of nodes interconnected by deformable bars as in tensegrity models.

1 Introduction

Developmental biology is describing—simply said—how the whole body of a living creature could be created from a single cell. It is a very vital field that produce a lot of genetic, signalling and morphological data—but current knowledge does not cover the whole field so far. New, clever experimental approaches bring a constant flux of data. The most important point is that there is not unique understanding of how those single parts are working together. From a certain distance and using a kind of metaphor, we could say that we roughly know how every single player perform its part but we do not understand how the whole orchestra is working. This is the point where mathematical models take their part. They enable us to build adequate models of developmental biology using biologically relevant input data where outputs could be directly compared to another biologically observed data.

From the mathematical point of view, the whole problem could be easily decomposed into three distinct parts. Namely, evolution of morphology—i.e., topology in the mathematical sense—by growth, signalling by specific chemicals

** This work was supported by the Czech Ministry of Education, Youth and Sports under grant number MSM 4977751303.

V.N. Alexandrov et al. (Eds.): ICCS 2006, Part IV, LNCS 3994, pp. 847–854, 2006.
© Springer-Verlag Berlin Heidelberg 2006

produced by different types of cells, and actions of gene regulatory network which tells every single cell if it has to undergo—in the specific morphological and chemical context—a cell type change or not.

The whole problem becomes quite complicated not only due to size of growing morphology but, as well, due to size of gene regulatory network itself. Therefore, there is a good reason to start with a much simpler case that will capture the essence of this type of modelling. The natural beginning is to model growth of a tooth. From biological point of view, it is well known that most of—if not all—interactions among all cells in developing tooth are encapsulated inside of growing tooth egg and its subsequent stages.

A growing tooth egg enter several developmental stages going from the egg, across cap, bell and to the final shape. It is well known that the crucial role in development of mammalian tooth—having crown with a relatively complex structure—is played by epithelial and mesenchyme growth which compete one with the other in speed of growth. Mesenchyme is encapsulated in epithelium, i.e. epithelium grows in two-dimensions and mesenchyme in three-dimensions. This leads to the situation that some parts of epithelium—which generates mechanical force—are imposed to higher strains. Places which encounter larger strains are transformed into so called knots—via gene regulatory network—which influence speed of growth of epithelium in surrounding and naturally leads to creation of crown by invagination of epithelium into mesenchyme.

Mathematical model of morphological development of one tooth will be built—and published in a series of papers—using complex systems where cellular automaton is employed as the mathematical tool expressing the complexity of the model through its parts [1, 2, 3], i.e. growth by cell division, signalling to neighbouring cells by chemicals and transformation of cells into different cell types due to gene regulatory network. Concept of cellular automata enables a very detailed—spatially and in time—definition of behaviour of every part of the simulated topology. Some models of morphological development are already known but the problem is that they do not fully reflect biologically observed behaviour, e.g. some non-local computations of mechanical interactions are used what is not in coherence with biological observations. Those models typically use simplified gene regulatory networks, they do not work with correct mechanical interactions of cells, and use some other simplifications which might lead to improper biological outputs.

The general idea of the CA-model of mechanical behaviour of mesenchyme presented here comes from the tensegrity models [4, 5, 6, 7, 8] where the structure is composed from a set of two generic types of elements, one is under compression load (bars) and the other one under tensile load (strings). Spatial combination of those two types of elements leads to light and stable structures which are able to sustain large loads compared to classical structures. In the model, we work only with bars interconnecting nodes—originally located in the centre of each cell—where one node belongs exactly to one cell. We employ knowledge achieved in structural design [9, 10, 11] regarded to elastic properties of structures to model mesenchyme but we know that behaviour of epithelium is richer then is used there. Gene regulatory network and diffusion of chemicals will be involved in the model later.

2 Model

The model of the mesenchymal and pseudo-epithelial interaction is built step by step. Firstly, one-dimensional model of mesenchyme and pseudo-epithelium is defined, studied, and carefully tested on tensile and compressible examples of metallic materials. Then a two-dimensional model is proposed with special attention to mesenchyme where mechanical properties of epithelium are simplified. It is fascinating how mechanical force could be created in living cells and tissues [12, 13, 14, 15]. It is known that cells are working with small local actions leading to large global shape and force changes. We use *ad hoc* mechanisms leading to a very similar mechanical effects as in living tissues.

The model works with a network of nodes interconnected by bars. Each cell manage one node, see Figure 1. Bars are deformable by tension and compression. Strain is defined as

$$\epsilon(t) = \frac{L(t) - L_0}{L_0}, \tag{1}$$

where L_0 is the initial length and $L(t)$ is its actual value at given time t. Use of this equation allows working with relative values used in the Hook's law.

The Hook's law represents linear dependence of force/stress on strain and is defined as

$$\sigma(t) = \frac{F(t)}{A} = E \cdot \epsilon, \tag{2}$$

where $\sigma(t)$ is stress [N/M^2], $F(t)$ is force [N], and A is the cross-section of bars [m^2]. This equation is elastic—i.e., linear—for the whole range of strain ϵ what is rather physically unrealistic because it allows to compress material to physically impossible strains.

Therefore, force F is composed from two distinct parts. One is defined by Hooks law for tensile and compressible deformation which is taken from Equation 2, and the other one represents incompressibility of material. In our case, we expect that material could not be compressed below $\epsilon = -0.8$ of its original length L_0.

$$\begin{aligned} F &= E \cdot A \cdot \epsilon, & \epsilon &\in (-0.8, +\infty), \\ \epsilon &= -0.08, & F &< E \cdot A \cdot (-0.8). \end{aligned} \tag{3}$$

In words, there is linear dependence of force F on strain ϵ above $\epsilon = -0.8$, and the constant value of $\epsilon = -0.8$ is taken for force below the value of $E \cdot A \cdot (-0.8)$ due to incompressibility.

The reason why we use—as the first approximation—such dependence of force F on strain ϵ is strictly defined by one well known physical constrain. It is known that compressibility of solids and liquids has a limit. The value of compression from the principle could not go bellow $\epsilon = -1$; it is even very difficult to approach values close to it. It stems from the law of mass conservation because mass could not be compressed to negative volumes.

To elucidate what happen if we take linear dependence of stress/force on deformation the following sequence of values of L which is computed from L_0 by multiplication it with a factor of 2, 1.1, 0.9, 0.5, and 0.01 is inserted into Equation 1 and gives ϵ equal to 1, 0.1, -0.1, -0.5, and -0.99, respectively. There is an obvious limit in compression equal to $\epsilon = -1$.

If force F depends on strain ϵ linearly in the case of compression then we could not apply force lower then $(-E \cdot A)$. Values bellow are not physically relevant. Therefore, we have to use some kind of nonlinear dependence of force on deformation in region of compression as is done, for example, in Equation 3.

The following set of three equations 4, 5, and 6 represents local equilibria for three types of cells used in the two-dimensional model, i.e. for top, bulk, and bottom—similarly for the left and right—cells

$$\sum\nolimits_{(k,l)\in((-1,0),(0,-1),(1,0))} F^{(k,l)} + F^{ext}_{surf} = 0 \tag{4}$$

$$\sum\nolimits_{(k,l)\in((-1,0),(0,-1),(1,0),(0,1))} F^{(k,l)} + F^{ext}_{bulk} = 0 \tag{5}$$

$$\sum\nolimits_{(k,l)\in((1,0),(0,1),(-1,0))} F^{(k,l)} + F^{fix} = 0 \tag{6}$$

where F^{ext} represents an externally applied force to the cell in our case and F^{fix} represents fixing force applied at the bottom, left, and right border cells. The last equation 6 represents the situation where cells are fixed during the whole simulation to the initially defined locations. Force F^{fix} balance the other forces, and therefore, no movement of those cells occurs. It could be simply done by keeping of all cell coordinates constant through the whole simulation.

In one-dimensional case—where a column of cells/bars is taken and the only allowed deformation works vertically—,no equilibrium is computed at the bottom cell. The position is simply kept constant, i.e. the coordinates x and y are constant.

Vertical displacement of the top cell y_N in the one-dimensional case is computed according to Equations 1, 2 and 4 (where (k,l) = {(0,-1)}) and gives the following formula for tensile deformation

$$y_N(t+1) = \frac{F^{ext}}{E \cdot A} \cdot L_0 + y_N(t) + L_0 \tag{7}$$

Vertical displacement of bulk cells in the one-dimensional case is computed from Equations 1, 2, and 5 (where (k,l) = {(0,-1),(0,1)})—and after certain rearrangement—by mere averaging of positions

$$y_N(t+1) = \frac{1}{2}(y_{N+1}(t) + y_{N-1}(t)) \tag{8}$$

it could be done due to use of the linear Hooks law in compression. Please, note which simulation steps are taken on the left and right sides of the equation. Vertical displacement of all cells in the case of compression works with the same equations as in the case of tensile deformation.

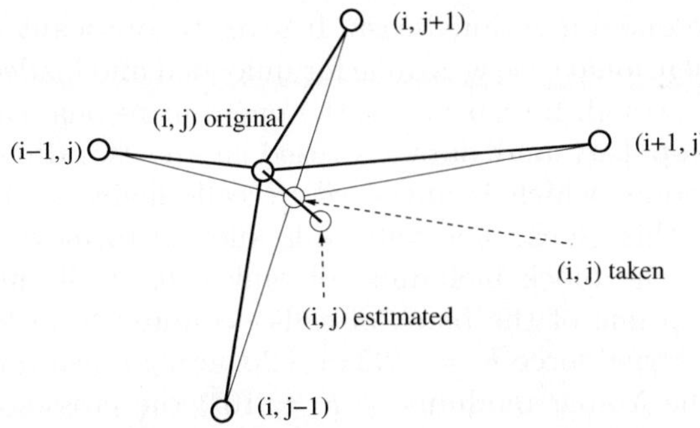

Fig. 1. The figure depicting the main idea of the used algorithm. The original position of the point managed by the updated cell (i, j) is moved to the taken value according to values provided by four neighbouring cells which lays on the line connecting the original position with the estimated one.

Displacements in the two-dimensional case behaves in a much more complicated way then in the one-dimensional one. The main idea of the algorithm is explained in the Figure 1. Whereas in the one-dimensional case the new position of the central cell could be simply computed by averaging positions of upper and lower cells, in the two-dimensional case we have to use an iteration method to find an estimated value.

In Figure 1, original position of the point managed by the updated cell (i, j) is expected to be moved to the estimated position—given by solution of the equilibrium equations—which is the optimal solution of the problem regarded to data given by the neighbouring cells $(i-1, j)$,$(i, j-1)$,$(i+1, j)$, and $(i, j+1)$. If the point is moved to this estimated position then local change could be too fast and the updating algorithm looses its stability. Such instability could be removed when we take some value laying on the line interconnecting original and estimated value—called taken value. The distance from original point to the taken one is a predefined fraction p from the distance of original and estimated points. Situation at the top cells is similar to the situation explained for the bulk cells except the fact that only three neighbouring cells are presented there, i.e. $(i-1, j)$,$(i, j-1)$, and $(i+1, j)$.

Initially, the position of the cell under consideration is estimated using equations 4, 5 and 6 representing local equilibrium at the node. The algorithm used to find new position of nodes is working with halving of intervals in two dimensions. Then deformation limits are tested and the value called estimated achieved, i.e. compression could not go bellow $\epsilon = -0.8$. Finally, p% shift of the old position towards the estimated position of the cell under consideration is taken where p is typically equal to 10% or 20%.

3 Results and Discussions

Vertical displacement of the top cell y_i was tested for several different external forces F^{ext} computed according to Equation 7 for tensile deformation and for compression in the one-dimensional case. It gives theoretically expected values. Testing of two-dimensional case was done for unloaded and loaded cases. Analysis of results is not as straightforward as in the one-dimensional case.

The following topology and data are used in the two-dimensional case—a block of 10×10 nodes which is anchored by cells laying at the left, bottom, and right edges of this block. The only cells allowed to move are those laying at the top edge of the block including all bulk cells. Cells number 5, 6, and 7 located at the top line of the block of cells—counted from left to right—are subjected to the external force $F = -2.5$ or -20 acting downwards, see Figure 2. Bars are having the Young modulus of $E = 10^5$; the cross-section of them is $A = 0.01 \cdot 0.01 = 10^{-4}$, the incompressibility threshold is set to 0.2, and the initial distance of nodes $L_0 = 0.1$.

Snapshots depicting evolution of topology of a block of 10×10 nodes subjected to the force $F = -2.5$ or -20 which acts downwards at 5th, 6th, and 7th cell at the top block of cells could be seen in Figures 2 and 3. In the figures, nodes are depicted without bars. Several important observations were made upon this sequence. Firstly, symmetry breaking is present in the model because cells are moving to the left in the horizontal direction what is the most profound effect at the top cells. Secondly, as expected, the largest deformation is present at the place where external forces are applied. Thirdly, applied force squeeze the block and some top cells without applied external force are moved upwards for $F = -20$, e.g. third and fourth top nodes from the left. All those observations are coherent with experiments.

Model allows use of volume forces and it is possible to change the Young modulus locally from a place to a place but those properties of the model are not studied in this contribution. It is reserved for the future use of the model and for better description of behaviour of living tissues.

Solutions of the local equilibria equations 4, 5, and 6 have to be found by an iterative formula because it is the well known fact that there does not exist, in general, an analytical solution of a set of two or more nonlinear equations. It is found that presented algorithm fails when large simulation steps are used,

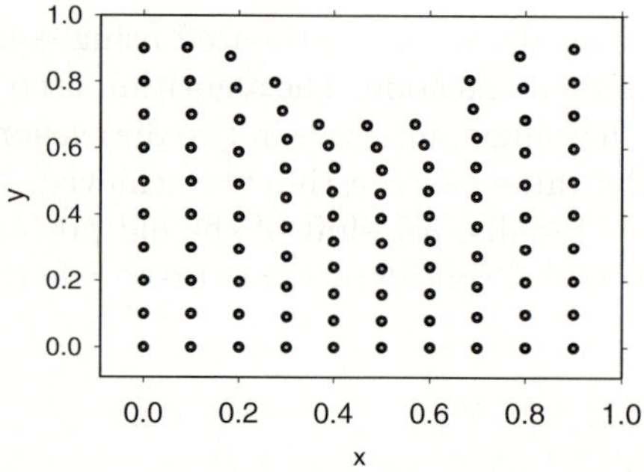

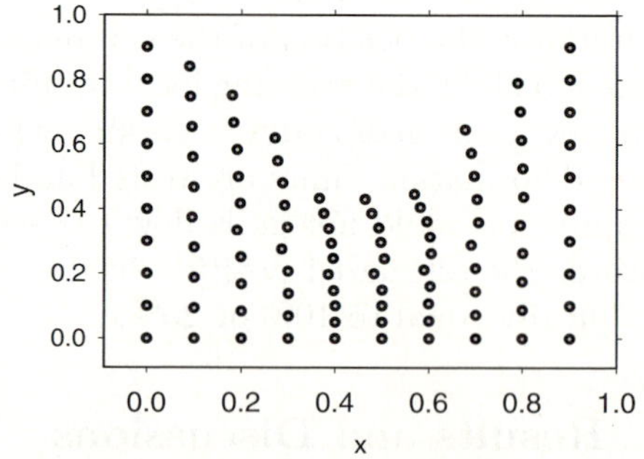

Fig. 2. Two configurations taken at simulation steps 100 and 300—depicting evolution of topology of a block of 10×10 nodes stressed by force the $F = -20$ N

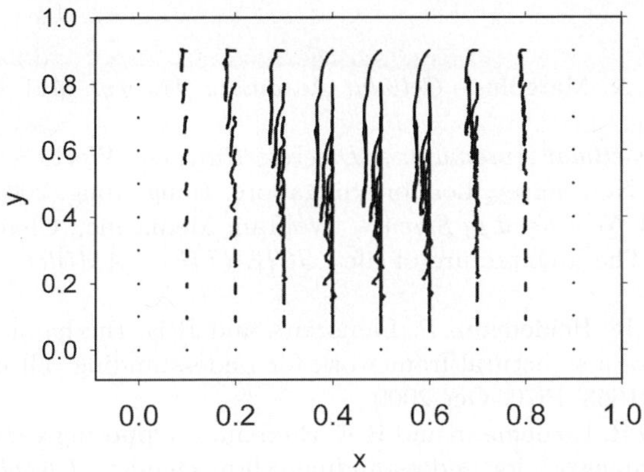

Fig. 3. A cumulative plot of all positions of all nodes for all simulated steps from 0 to 300 for the force $F = -20$ N is shown here. Trajectories belonging to different nodes could be easily distinguished.

i.e. when the value of p is approaching or equal to one. The reason is simple. Algorithm could exchange positions of neighbouring nodes for large steps, what is not physically acceptable. It automatically leads to instability of simulation and mixing of nodes. The proposed CA-model is working with physically relevant values as models using finite element method (FEM) do.

The topology used in this contribution represents the first, testing step towards a new model of tooth growth. The model describing behaviour of mesenchyme is defined here, i.e. mesenchyme is pressed by force from the top by a force that pseudo-epithelium generates, but—in general—mesenchyme could be pressed from any direction. In the future model, mechanical influence of epithelium would be taken into the game together with a gene regulatory network, and signalling chemicals produced according to this network which switch cells from one type into another one.

4 Conclusions

The first part—i.e. mechanical behaviour of mesenchyme and pseudo-epithelium—of the model describing morphological development of tooth is proposed and tested in this contribution. In this model, mechanical influence of epithelium is mimicked by use of external force acting at the top side of square block of cellular automata cells. Hence, the CA-model is prepared to take into account mechanical influence of epithelium generating mechanical pressure to mesenchyme. It is shown that a tissue composed from living cells could be simulated by use of a tensegrity like structure which is composed from a set of nodes mutually interconnected by deformable bars.

References

1. T. Toffoli and N. Margolus. *Cellular Automata Theory.* MIT Press, Cambridge, 1987.
2. A. Ilachinski. *Cellular Automata: A Discrete Universe.* World Scientific Publishing Co. Pte. Ltd., New Jersey, London, Singapore, Hong Kong, 2001.
3. S. Wolfram. *A New Kind of Science.* Wolfram Media Inc., Champaign, 2002.
4. D.E. Ingber. The architecture of life. *SCIENTIFIC AMERICAN*, 278(1):48–57, Jan 1998.
5. D.E. Ingber S.R. Heidemann P. Lamoreux and R.E. Buxbaum. Opposing views on tensegrity as a structural framework for understanding cell mechanics. *J Appl Physiol*, 89(4):1663–1670, Oct 2000.
6. P. Lamoreux S.R. Heidemann and R.E. Buxbaum. Opposing views on tensegrity as a structural framework for understanding cell mechanics. *J Appl Physiol*, 89:1670–1674, Oct 2000.
7. D.E. Ingber. Opposing views on tensegrity as a structural framework for understanding cell mechanics - rebuttals. *J Appl Physiol*, 89:1674–1677, Oct 2000.
8. D.E. Ingber S.R. Heidemann P. Lamoreux and R.E. Buxbaum. Opposing views on tensegrity as a structural framework for understanding cell mechanics - rebuttals. *J Appl Physiol*, 89:1677–1678, Oct 2000.
9. P. Hajela and B. Kim. On the use of energy minimization for ca based analysis in elasticity. *Struct Multidisc Optim*, 23:24–33, Dec 2001.
10. E. Kita and T. Toyoda. Structural design using cellular automata. *Struct Multidisc Optim*, 19:64–73, Mar 2000.
11. Z. Gurdal and B. Tatting. Cellular automata for design of truss structures with linear and nonlinear response. In *8th AIAA/USAF/NASA/ISSMO Symposium on Multidisciplinary Analysis and Optimization*, pages 1–11, Long Beach, CA, Sep 2000. American Institute for Aeronautics and Astronautics.
12. Pilot F. and Lecuit T. Compartmentalized morphogenesis in epithelia: From cell to tissue shape. *DEVELOPMENTAL DYNAMICS*, 232(3):685–694, Mar 2005.
13. Hay E.D. The mesenchymal cell, its role in the embryo, and the remarkable signaling mechanisms that create it. *DEVELOPMENTAL DYNAMICS*, 233(3):706–720, Jul 2005.
14. Ball EMA and Risbridger GP. Activins as regulators of branching morphogenesis. *DEVELOPMENTAL BIOLOGY*, 238(1):1–12, Oct 2001.
15. S.R. Heidemann and D. Wirtz. Towards a regional approach to cell mechanics. *TRENDS in Cell Biology*, 14(4):160–166, Apr 2004.

Three-Dimensional Virtual Anatomic Fit Study for an Implantable Pediatric Ventricular Assist Device

Arielle Drummond[1], Timothy Bachman[2], and James Antaki[1]

[1] Department of Biomedical Engineering,
Carnegie Mellon University,
700 Technology Drive, Pittsburgh PA 15219
{adrummon, antaki}@andrew.cmu.edu
[2] Department of Bioengineering, University of Pittsburgh,
749 Benedum Hall, Pittsburgh, PA 15213
tbachman@engr.pitt.edu

Abstract. An innovative pediatric ventricular assist device (PVAD) is being developed to treat young patients (2.5kg-15kg) with severe heart failure that otherwise have very few options due to their small size. To optimize the design of the PVAD for the target patient population, three-dimensional anatomical compatibility studies must be conducted. The aim of this project was to evaluate the utility of three dimensional reconstructions to obviate fit studies in human subjects. Serial CT scans of the thorax of one child were obtained as part of routine treatment. The images were enhanced by adjusting the contrast of the images and segmented semi-automatically prior to 3-D reconstruction. The results were visualized as surface renderings of the rib cage and heart. This data was then amended with solid models of the implantable hardware, including the PVAD and cannulae. Manipulation of the relative orientation of the components revealed surgical challenges that may be anticipated and motivated design modifications to improve the anatomic compatibility. Unique challenges associated with these data sets include the availability of pediatric CT images and difficulty of segmentation due to the small scale of the anatomic features as compared to the resolution of the images.

1 Introduction

The limited options to treat ventricular failure in children with congenital heart disease has motivated the development of a highly reliable and biocompatible ventricular assist device (VAD) for chronic support (up to six months) for children 2.5kg to 15kg. An innovative pediatric VAD (PVAD) is being developed by a consortium from the University of Pittsburgh, Carnegie Mellon University, World Heart Corporation (Oakland, CA) and LaunchPoint Technologies (Goleta, CA) based on a miniature axial flow pump with magnetic levitation. Design requirements to assure biocompatibility include minimization of blood trauma and anatomic compatibility of the implanted components, including the pump, cannulae sets, and drive line (See Figure 1).

V.N. Alexandrov et al. (Eds.): ICCS 2006, Part IV, LNCS 3994, pp. 855 – 861, 2006.
© Springer-Verlag Berlin Heidelberg 2006

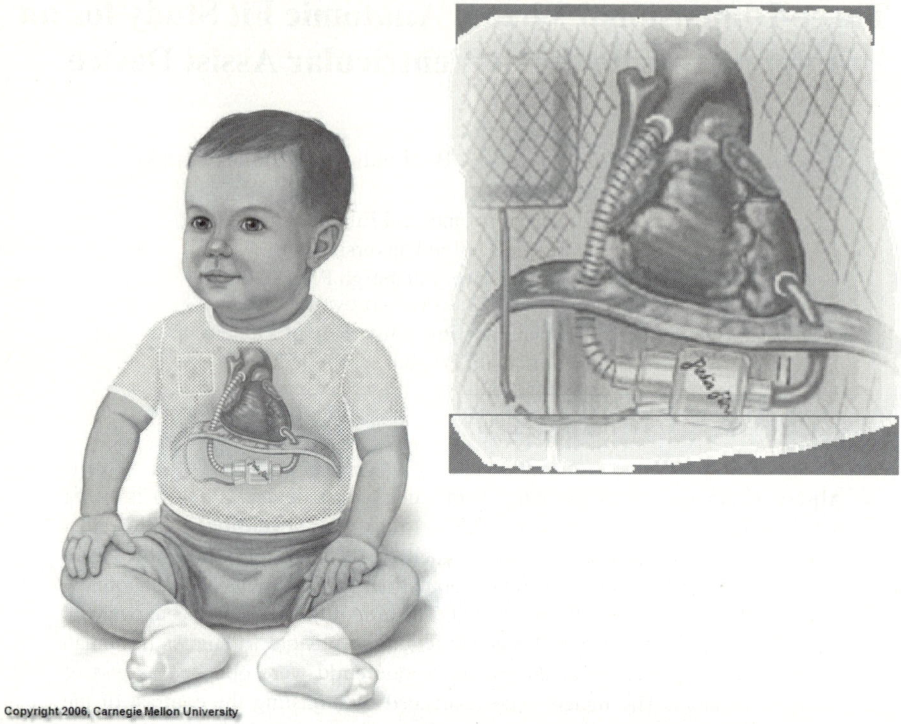

Fig. 1. Pediatric patient with an implanted pediatric ventricular assist device

Various decisions must be made concerning the placement of the VAD in addition to the decision on cannula length, diameter and angle of insertion. Advanced computer simulations are being conducted to assure compatibility with the blood; however the typical method for determining anatomic fit is usually through physical experiments on cadavers or living subjects.[1] While cadaver studies allow direct anatomical dimensions to be taken; there are several disadvantages of such an approach. The major disadvantages include the limited availability of subjects and the inability to rapidly modify the implanted hardware. Additional complications due to tissue fixation, lung deflation and myocardial compression prevent accurate representation of the anatomy of living subjects.[1] Intra-operative studies addresses some of these limitations but are extremely constrained in time, subject availability, and extent of manipulations that are permissible.

The purpose of this project therefore is to employ a geometric computer model to facilitate both the design and the surgical strategy to provide the optimal fit for as wide a range of patient sizes and diagnosis as possible. The anatomic data for young patients is rather sparse in the literature. Therefore an additional goal of this study is to develop a database of thoracic and abdominal measurements, similar to that developed by Mussivand et al.[2] Additional advantages of 3D reconstruction include the ability to obtain anatomical measurements from live subjects, minimal invasiveness, and the ability to generate a CAD solid model manipulated for fit and cannulae design. However, there are also few challenges to using this method, which are addressed herein.

2 Methods

2.1 Acquisition of Image Data

Retrospective images were acquired from a patient (male, 8 years of age) who underwent a high resolution gated CT scan of the chest with a 64 detector GE CT scanner (GE Medical Systems, Waukesha, WI) at the Children's Hospital of Pittsburgh. Image slice thickness was limited to 2mm due to radiation dosage concerns in pediatric patients. The images were re-sliced by interpolation to 1.25mm slick thickness using conventional post-processing software.

2.2 Three-Dimensional Reconstruction

The serial CT data were imported as DICOM files into visualization software, Mimics (Materialise, Ann Arbor, MI), to generate a 3-D surface model of the chest cavity. The raw images were manipulated by pre-processing enhancement, followed by segmentation, reconstruction and visualization, illustrated in Figure 2.

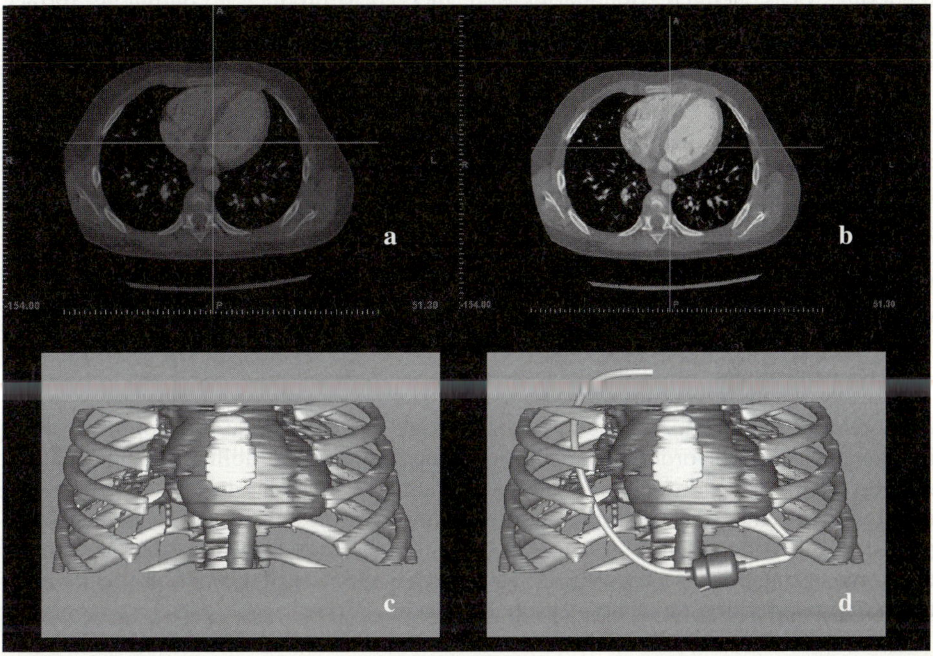

Fig. 2. 3-D reconstruction process, (a) Images are imported into Mimics and contrast-enhanced (b) Segmentation of bone by thresholding, (c) Reconstruction of ribs, heart, vessels, and bronchi (d) Introduction of solid models of implanted hardware: pediatric VAD and cannulae

2.3 Enhancement

Images acquired from the CT scan were enhanced to differentiate anatomical structures from each other. The DICOM files, a standard medical image file format, produced by the CT scanner were imported to the data visualization software, Mimics. A project file of the images was created to view the images in the software. During automatic segmentation, the surrounding structures pixels are often classified as soft tissue leading to undesirable result, therefore the images were enhanced by the manual contrast adjustment tool, through a histogram equalization method, to differentiate surrounding tissue from the heart muscle.

2.4 Segmentation

The cardiac boundaries were identified using a combination of methods, specifically: a thresholding tool was used to select the desired organ for reconstruction. This tool selects the organ based on the density of the pixels; generally the tool selects surrounding tissues as well. Additional editing of the boundaries was made to a number of image slices to eliminate outlier data and patch dropout regions.

2.5 Reconstruction

Following the segmentation of each image slice in the data series a 3D surface rendering was generated. Boundaries were merged to produce a continuous surface rendering. Finally, a surface contour tool was used to smooth extraneous structures from the reconstructed geometry.

2.6 Visualization

The final step to 3D reconstruction is visualization of the 3D image. There are various ways to manipulate the image. SolidWorks (Concord, MA), a CAD modeling software was used to visualize reconstructed anatomy. The Mimics software has the ability to export files in IGES and STL format to this modeling software. In SolidWorks, various visualization tools can be used to adjust the section view, rotation, lighting, and coloring.

2.7 Measurements

Anatomic measurements of pediatric patients are needed for the development of the implantable VAD in order to develop a pump that has the ability to fit in patient ranging from 2.5kg to 15kg. Anatomic landmarks relevant to the insertion and positioning of the PVAD were recorded within Mimics. Landmarks included the position of the ribcage in reference to left ventricular apex as well as the diameter of the aorta. Using Mimics as well as SolidWorks, the described measurements were taken to prove feasibility of the measurement tool.

Additional critical measurements can be recorded using Mimics to develop a database of anthropometric data of pediatric patients in the target patient population, specifically:[1, 2]

1. Volume ratio of the device with respect to vital organs: liver, heart, and lung.
2. Distances between organs:
    a. Distance between the left ventricular apex and diaphragm.
    b. Distance between the aortic arch and ventricular apex.
    c. Distance between left ventricular apex and chest wall.
    d. Diameter of descending aorta.

3 Results

A three-dimensional model of and 8 year old male subject was constructed as shown in Figure 3. The modeled anatomy includes the chest wall (ribs, sternum and vertebrae), heart and descending aorta. Measurements of the cardiac anatomy yielded an estimated inner diameter of the aorta of 13.7mm. This will affect the choice of cannula used for the outlet of the VAD. The distance between the left ventricular apex to the chest was measured at 12.1mm. This affects the ability to access the heart by the inflow cannula.

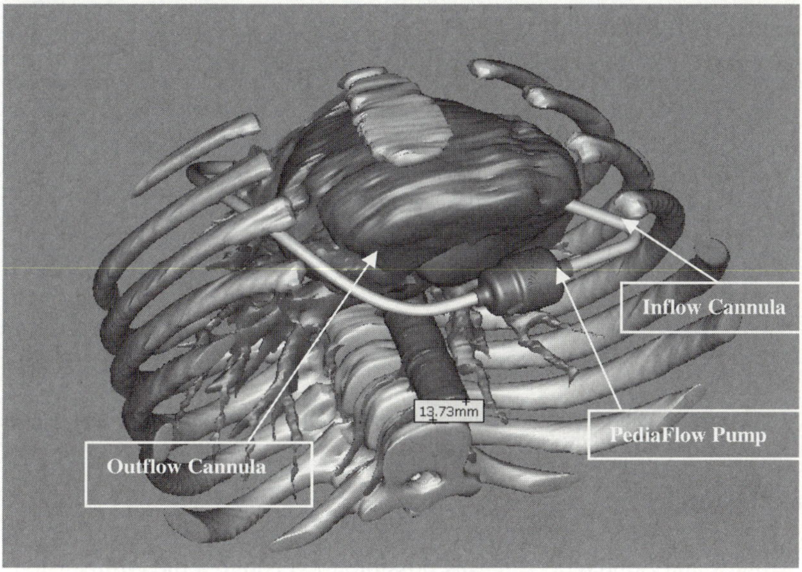

Fig. 3. 3-D reconstruction of the chest, heart, PVAD (gold) and cannula (white). Measurement of internal diameter of the descending aorta was 13.7mm.

Various locations were considered for locating the implanted pump. The x, y, z locations as well as the "patch", "yaw", & "roll" angles were varied to achieve quantitatively and compromise between:

- Length of cannula, hence pressure drop (minimize)
- Compression/ abrasion of organs (minimize)
- Target sites conveniently assessable.
- Kinks or sharp bends that might obstruct or disturb flow.

Figure 3 depicts one preferred configuration with the pump positioned below the diaphragmatic margin. Due to the relative size of the current VAD with respect to chest cavity, the preferred location of the VAD will be in the abdomen with cannula penetrating the diaphragm. Hence both abdominal and thoracic surgery will be required.

Cannulae constructed in SolidWorks were modified to connect the VAD from the LV apex to the ascending aorta. The outflow cannula was designed to be 70 mm in length and diameter of 5mm.

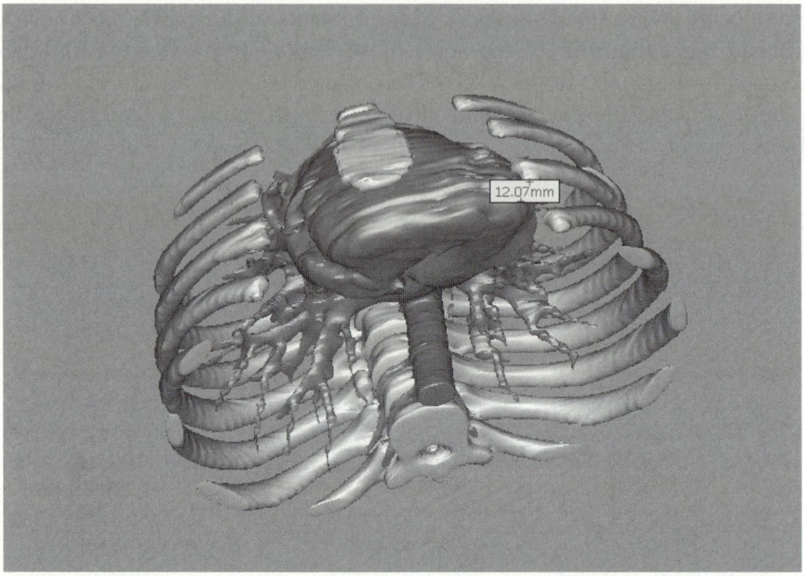

Fig. 4. Anterior measurement of distance between the left ventricular apex and the ribcage

4 Discussion

Using Mimics and SolidWorks, it was possible to generate a three-dimensional solid assembly to simulate the implantation of a Pediatric VAD in the chest of a young patient. These models can be easily measured and manipulated for the anatomical fit studies as described by Warriner et al.[5] The models have proven helpful for development of the cannulae, specifically to determine the proper insertion angle into the LV chamber and anastomosis location at the aorta.[6] These results will enable future computational fluid dynamic simulations to optimize the geometry according to flow requirements.

However there are still several challenges to pursuit of 3D reconstruction of pediatric patients. Foremost, the spatial resolution required for accurate representation necessitates CT data series with very fine slice thickness, ideally less than 1mm. Additional cardiac gating is needed to obtain temporal accuracy and avoid geometric distortion throughout the cardiac cycle. Obtaining these data however would require radiation dosage exceeding levels that would typically be indicated.

Secondly, segmentation, a perennial problem, is exacerbated in these very small patients due to limited resolution and contrast. Manually editing is almost always necessary, and it is unlikely that the process will be fully automated.

Future work will entail collection of additional data as well as utilization of the current data set for subsequent simulation studies. Future image collection will include a prospective study of brain dead patients which will enable slice thickness down to 0.63mm. By collecting data over a statistical sampling of patients, it will be possible to develop an anthropometric database of thoracic and abdominal anatomy that may be made available for a variety of purposes. Validation of accuracy for the generation of 3D models by a human expert will be accomplished by obtaining cadaveric measurements shortly after the brain dead patients are withdrawn from ventilator support. The cadaveric measurements will be compared to the measurements acquired from the 3D models. Ongoing research with the current data set includes flow visualization experiments based on transparent replicas of the fluid geometry. Combined with computational fluid dynamics simulation, these studies will collectively be used to optimize the biocompatibility efficacy, and overall safety of the pediatric ventricular assist device.

References

1. Zhang, B.M., T. Tatsumi, E. Taenaka, Y. Uyama, C. Takano, H. Takamiya, M., Three-Dimensional Thoracic Modeling for an Anatomical Compatibility Study of Implantable Total Artificial Heart. Artificial Heart, 1999. 23(3): p. 229-234.
2 Mussivand, T., et al., Critical anatomic dimensions for intrathoracic circulatory assist devices. Artif Organs, 1992. 16(3): p. 281-5.
3. Fujimoto, L.K., et al., Human thoracic anatomy based on computed tomography for development of a totally implantable left ventricular assist system. Artif Organs, 1984. 8(4): p. 436-44.
4. Fujimoto, L.K., et al., Anatomical considerations in the design of a long-term implantable human left ventricle assist system. Artif Organs, 1985. 9(4): p. 361-74.
5. Warriner, R.K., et al., Virtual anatomical three-dimensional fit trial for intra-thoracically implanted medical devices. Asaio J, 2004. 50(4): p. 354-9.
6. May-Newman, K.D., Hillen, B.K, Sironda, C. S, Dembitsky, W., Effect of LVAD outflow conduit insertion angle on flow through the native aorta. Journal of Medical Engineering & Technology, 2004. 28(3): p. 105-109.

Soft Computing Based Range Facial Recognition Using Eigenface

Yeung-Hak Lee[1], Chang-Wook Han[1], and Tae-Sun Kim[2]

[1] School of Electrical Engineering and Computer Science, Yeungnam University, 214-1 Dae-dong, Gyongsan, Gyongbuk, 712-749 South Korea
{annaturu, cwhan}@yumail.ac.kr

[2] Department of Digital Electronic Engineering, Kyungwoon University, 55 Induk-ri, Sandong-myun, Kumi, Kyungbuk, 730-852 South Korea
tskim@ikw.ac.kr

Abstract. The depth information in the face represents personal features in detail. In particular, the surface curvatures extracted from the face contain the most important personal facial information. These surface curvature and eigenface, which reduce the data dimensions with less degradation of original information, are collaborated into the proposed 3D face recognition algorithm. The principal components represent the local facial characteristics without loss for the information. Recognition for the eigenface referred from the maximum and minimum curvatures is performed. The normalized facial images are also considered to enhance the recognition rate. To classify the faces, the cascade architectures of fuzzy neural networks, which can guarantee a high recognition rate as well as parsimonious knowledge base, are considered. Experimental results on a 46 person data set of 3D images demonstrate the effectiveness of the proposed method.

1 Introduction

Today's computer environments are changing because of the development of intelligent interface and multimedia. To recognize the user automatically, people have researched various recognition methods using biometric information – fingerprint, face, iris, voice, vein, etc [1]. In a biometric identification system, the face recognition is a challenging area of research, next to fingerprinting, because it is a no-touch style. For visible spectrum imaging, there have been many studies reported in literature [2]. But the method has been found to be limited in their application. It is influenced by lighting illuminance and encounters difficulties when the face is angled away from the camera. These factors cause low recognition. To solve these problems a computer company has developed a 3D face recognition system [2][3]. To obtain a 3D face, this method uses stereo matching, laser scanner, etc. Stereo matching extracts 3D information from the disparity of 2 pictures which are taken by 2 cameras. Even though it can extract 3D information from near and far away, it has many difficulties in practical use because of its low precision. 3D laser scanners extract more accurate depth information about the face, and because it uses a filter and a laser, it has the advantage of not being influence by the lighting illuminance when it is angled away from the cam-

V.N. Alexandrov et al. (Eds.): ICCS 2006, Part IV, LNCS 3994, pp. 862 – 869, 2006.
© Springer-Verlag Berlin Heidelberg 2006

era. A laser scanner can measure the distance, therefore, a 3D face image can be reduced by a scaling effect that is caused by the distance between the face and the camera [4][5].

Broadly speaking the two ways to establish recognition employs the face feature based approach and the area based approach [5]-[8]. A feature based approach uses feature vectors which are extracted from within the image as a recognition parameter. An area based approach extracts a special area from the face and recognizes it using the relationship and minimum sum of squared difference. Face recognition research usually uses 2 dimensional images. Recently, the 3D system becomes cheaper, smaller and faster to process than it used to be. Thus the use of 3D face image is now being more readily researched [3][9]-[12]. Many researchers have used 3D face recognition using differential geometry tools for the computation of curvature [9]. Hiromi et al. [10] treated 3D shape recognition problem of rigid free-form surfaces. Each face in the input images and model database is represented as an Extended Gaussian Image (EGI), constructed by mapping principal curvatures and their directions. Gordon [11] presented a study of face recognition based on depth and curvature features. To find face specific descriptors, he used the curvatures of the face. Comparison of the two faces was made based on the relationship between the spacing of the features. Lee and Milios [13] extracted the convex regions of the face by segmenting the range of the images based on the sign of the mean and Gaussian curvature at each point. For each of these convex regions, the Extended Gaussian Image (EGI) was extracted and then used to match the facial features of the two face images.

One of the most successful techniques of face recognition as statistical method is principal component analysis (PCA), and specifically eigenfaces [14][15]. In this paper, we introduce novel face recognition for eigenfaces using the curvature that well presenting personal characteristics and reducing dimensional spaces. Moreover, the normalized facial images are considered to improve the recognition rate.

Neural networks (NNs) have been successfully applied to face recognition problems [16]. However, the complexity of the NNs increases exponentially with the parameter values, i.e. input number, output number, hidden neuron number, etc., and becomes unmanageable [17]. To overcome this curse of dimensionality, the cascade architectures of fuzzy neural networks (CAFNNs), constructed by the memetic algorithms (hybrid genetic algorithms) [18], are applied to this problem.

2 Face Normalization

The nose is protruded shape and located in the middle of the face. So it can be used as the reference point, firstly we tried to find the nose tip using the iterative selection method, after extraction of the face from the 3D face image [20]. Usually, face recognition systems suffer drastic losses in performance when the face is not correctly oriented. The normalization process proposed here is a sequential procedure that aims to put the face shapes in a standard spatial position. The processing sequence is panning, rotation and tilting [21].

3 Surface Curvatures

For each data point on the facial surface, the principal, Gaussian and mean curvatures are calculated and the signs of those (positive, negative and zero) are used to determine the surface type at every point. The z(x, y) image represents a surface where the individual Z-values are surface depth information. Here, x and y is the two spatial coordinates. We now closely follow the formalism introduced by Peet and Sahota [19], and specify any point on the surface by its position vector:

$$R(x,y) = xi + yj + z(x,y)k \tag{1}$$

The first fundamental form of the surface is the expression for the element of arc length of curves on the surface which pass through the point under consideration. It is given by:

$$I = ds^2 = dR \cdot dR = Edx^2 + 2Fdxdy + Gdy^2 \tag{2}$$

where

$$E = 1 + \left(\frac{\partial z}{\partial x}\right)^2, \quad F = \frac{\partial z}{\partial x}\frac{\partial z}{\partial y}, \quad G = 1 + \left(\frac{\partial z}{\partial y}\right)^2 \tag{3}$$

The second fundamental form arises from the curvature of these curves at the point of interest and in the given direction:

$$II = edx^2 + 2fdxdy + gdy^2 \tag{4}$$

where

$$e = \frac{\partial^2 z}{\partial x^2}\Delta, \quad f = \frac{\partial^2 z}{\partial x \partial y}\Delta, \quad g = \frac{\partial^2 z}{\partial y^2}\Delta \tag{5}$$

and

$$\Delta = (EG - F^2)^{-1/2} \tag{6}$$

Casting the above expression into matrix form with;

$$V = \begin{pmatrix} dx \\ dy \end{pmatrix}, \quad A = \begin{pmatrix} E & F \\ F & G \end{pmatrix}, \quad B = \begin{pmatrix} e & f \\ f & g \end{pmatrix} \tag{7}$$

the two fundamental forms become:

$$I = V'AV \quad I = V'BV \tag{8}$$

Then the curvature of the surface in the direction defined by V is given by:

$$k = \frac{V'BV}{V'AV} \tag{9}$$

Extreme values of *k* are given by the solution to the eigenvalue problem:

$$(B - kA)V = 0 \tag{10}$$

or

$$\begin{vmatrix} e - kE & f - kF \\ f - kF & g - kG \end{vmatrix} = 0 \tag{11}$$

which gives the following expressions for k_1 and k_2, the minimum and maximum curvatures, respectively:

$$\begin{aligned} k_1 = \{ & gE - 2Ff + Ge - [(gE + Ge - 2Ff)^2 \\ & - 4(eg - f^2)(EG - F^2)]^{1/2} \} / 2(EG - F^2) \end{aligned} \tag{12}$$

$$\begin{aligned} k_2 = \{ & gE - 2Ff + Ge + [(gE + Ge - 2Ff)^2 \\ & - 4(eg - f^2)(EG - F^2)]^{1/2} \} / 2(EG - F^2) \end{aligned} \tag{13}$$

Here we have ignored the directional information related to k_1 and k_2, and chosen k_2 to be the larger of the two. For the present work, however, this has not been done. The two quantities, k_1 and k_2, are invariant under rigid motions of the surface. This is a desirable property for us since the cell nuclei have no predefined orientation on the slide (the x – y plane).

The Gaussian curvature *K* and the mean curvature M is defined by

$$K = k_1 k_2 \ , \quad M = (k_1 k_2)/2 \tag{14}$$

which gives k_1 and k_2, the minimum and maximum curvatures, respectively. It turns out that the principal curvatures, k_1 and k_2, and Gaussian are best suited to the detailed characterization for the facial surface, as illustrated in Fig. 1. For the simple facet model of second order polynomial of the form, i.e. a 3 by 3 window implementation in our range images, the local region around the surface is approximated by a quadric

$$z(x, y) = a_{00} + a_{10}x + a_{01}y + a_{01}y + a_{20}x^2 + a_{02}y^2 + a_{11}xy \tag{15}$$

and the practical calculation of principal and Gaussian curvatures is extremely simple.

4 Eigenface

4.1 Computing Eigenfaces [14]

Consider face images of size N by N, extracted contour line value. These images can be thought as a vector of dimension N^2, or a point in N^2 – dimensional space. A set of images, therefore, corresponds to a set of points in this high dimensional space. Since facial images are similar in structure, these points will not be randomly distributed, and therefore can be described by a lower dimensional subspace. Principal component analysis gives the basis vectors for this subspace. Each basis vector is of length N^2, and is the eigenvector of covariance matrix corresponding to the original face images.

Fig. 1. Six possible surface type according to the sign of principal curvatures for the face surface; (a) concave (pit), (b) convex (peak), (c) convex saddle, (d) concave saddle, (e) minimal surface, (f) plane

Let $\Gamma_1, \Gamma_2, \ldots, \Gamma_M$ be the training set of face images. The average face is defined by

$$\Psi = \frac{1}{M}\sum_{n=1}^{M}\Gamma_n \tag{16}$$

Each face differs from the average face by the vector $\Phi_i = \Gamma_n - \Psi$. The covariance matrix

$$C = \frac{1}{M}\sum_{n=1}^{M}\Phi_n\Phi_n^T \tag{17}$$

has a dimension of N^2 x N^2. Determining the eigenvectors of C for typical size of N is intractable task. Once the eigenfaces are created, identification becomes a pattern recognition task. Fortunately, we determine the eigenvectors by solving an M by M matrix instead.

4.2 Identification

The eigenfaces span an M-dimensional subspace of the original N^2 image space. The M significant eigenvectors are chosen as those with the largest corresponding eigenvalues. A test face image Γ is projected into face space by the following operation: $\omega_n = u_n^T(\Gamma - \Psi)$, for n=1, …, M, where u_n is the eigenvectors for C. The weights ω_n from a vector $\Omega^T = [\omega_1\ \omega_2 \ldots \omega_{M'}]$ which describes the contribution of each eigenface in representing the input face image. This vector can then be used to fit the test image to a predefined face class. A simple technique is to use the Euclidian distance $\varepsilon_n = \|\Omega - \Omega_n\|$, where Ω_n describes the *n*th face class. In this paper, we used the cascade architectures of fuzzy neural networks to compare with the distance as described next chapter.

5 Cascade Architectures of Fuzzy Neural Networks (CAFNNs)

As originally introduced in [17], the structure of the CAFNNs is the cascade combination of the logic processors (LPs) which consist of fuzzy neurons. The sequence of relevant input subset and the connections were optimized by memetic algorithms in [18] to construct parsimonious knowledge base, but accurate one. As illustrated in [18], the memetic algorithms are more effective than the optimization scenario in

[17]. Therefore, the optimization scenario in [18] will be considered in this approach. For more details about the CAFNNs and its optimization, please refer to [17][18].

To apply the CAFNNs to classification problems, the output (class) should be fuzzified as binary. For example, if we assume that there are 5 classes (5 persons) in the data sets, the number of output crisp set should be 5 that are distributed uniformly. If the person belongs to the 2nd-class, the Boolean output can be discretized as "0 1 0 0 0". In this classification problem, the winner-take-all method is used to decide the class of the testing data set. This means that the testing data are classified as the class which has the biggest membership degree.

6 Experimental Results

In this study, we used a 3D laser scanner made by a 4D culture to obtain a 3D face image. First, a laser line beam was used to strip the face for 3 seconds, thereby obtaining a laser profile image, that is, 180 pieces and no glasses. The obtained image size was extracted by using the extraction algorithm of line of center, which is 640 by 480. Next, calibration was performed in order to process the height value, resampling and interpolation. Finally, the 3D face images for this experiment were extracted, at 320 by 320. A database is used to compare the different strategies and is composed of 92 images (two images of 46 persons). Of the two pictures available, the second photos were taken at a time interval of 30 minutes.

Table 1. The comparison of the recognition rate (%)

		Best1	Best5	Best10	Best15
k_1	CAFNN(normalized)	64.5	78.4	89.6	95.9
	CAFNN	56.2	73.9	85.2	90.5
	k-NN	42.9	57.1	66.7	66.7
k_2	CAFNN (normalized)	68.1	86.4	90.1	96.3
	CAFNN	63.7	80.3	85.8	90.1
	k-NN	61.9	78.5	83.3	88.1

From these 3D face images, finding the nose tip point, using contour line threshold values (for which the fiducial point is nose tip), we extract images around the nose area. To perform recognition experiments for extracted area we first need to create two sets of images, i.e. training and testing. For each of the two views, 46 normal-expression images were used as the training set. Training images were used to generate an orthogonal basis, as described in section 3, into which each 3D image in training data set is projected in section 4. Testing images are a set of 3D images extracted local area we wish to identify.

Once the data sets have been extracted with the aid of eigenface, the development procedure of the CAFNNs should be followed for the face recognition. The used parameter values are the same as [18]. Since a genetic algorithm is a stochastic optimization method, ten times independent simulations were performed to compare the results with the conventional classification methods, as described in Table 1 and Fig. 2. In Table 1 and Fig. 2, the results of the CAFNN are averaged over ten times independent simulations, and subsequently compared with the results of the conventional method (k-nearest neighborhood: k-NN). Also, the normalized facial images were considered to generate the curvature-based data set. As can be seen from Table 1 and Fig. 2, the recognition rate is improved by using normalized facial images.

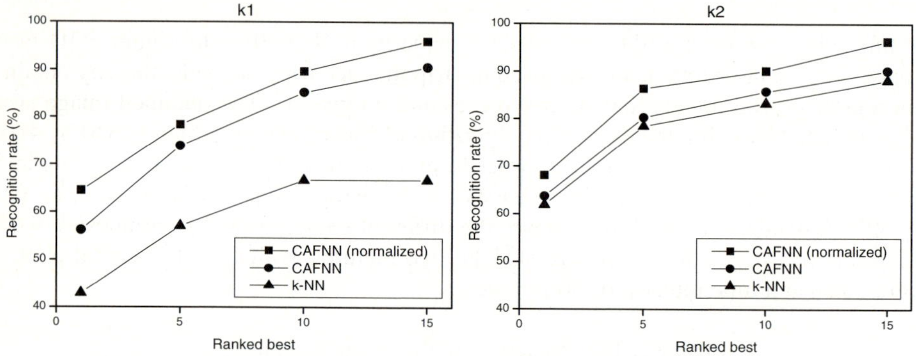

Fig. 2. The recognition results using eigenfaces for each area: (a) k_1, (b) k_2

7 Conclusions

The surface curvatures extracted from the face contain the most important personal facial information. We have introduced, in this paper, a new practical implementation of a person verification system using the local shape of 3D face images based on eigenfaces and CAFNNs. The underlying motivations for our approach originate from the observation that the curvature of face has different characteristic for each person. We found the exact nose tip point by using an iterative selection method. The low-dimensional eigenfaces represented were robust for the local area of the face. The normalized facial images were also considered to improve the recognition rate. To classify the faces, the CAFNNs were used. The CAFNNs have reduced the dimensionality problem by selecting the most relevant input subspaces too. Experimental results on a group of face images (92 images) demonstrated that our approach produces excellent recognition results for the local eigenfaces.

From the experimental results, we proved that the process of face recognition may use low dimension, less parameters, calculations and less same person images (used only two) than earlier suggested. We consider that there are many future experiments that could be done to extend this study.

References

1. Jain, L. C., Halici, U., Hayashi, I., Lee, S. B.: Intelligent Biometric Techniques in Fingerprint and Face Recognition. CRC Press (1999)
2. 4D Culture. http://www.4dculture.com
3. Cyberware. http://www.cyberware.com
4. Chellapa, R., et al.: Human and Machine Recognition of Faces: A Survey. UMCP CS-TR-3399 (1994)
5. Hallinan, P. L., Gordon, G. G., Yuille, A. L., Giblin, P., Mumford, D.: Two and Three Dimensional Pattern of the Face. A K Peters Ltd. (1999)
6. Grob, M.: Visual Computing. Springer Verlag (1994)
7. Nikolaidis, A., Pitas, I.: Facial Feature Extraction and Pose Determination. Pattern Recognition 33 (2000) 1783-1791
8. Moghaddam, B., Jebara, T., Pentland, A.: Bayesian Face Recognition. Pattern Recognition 33 (2000) 1771-1782
9. Chua, C. S., Han, F., Ho, Y. K.: 3D Human Face Recognition using Point Signature. Proc. of the 4th ICAFGR (2000)
10. Tanaka, H. T., Ikeda, M., Chiaki, H.: Curvature-based Face Surface Recognition using Spherial Correlation. Proc. of the 3rd IEEE Int. Conf. on Automatic Face and Gesture Recognition (1998) 372-377
11. Gordon, G. G.: Face Recognition based on Depth and Curvature Feature. Proc. of the IEEE Computer Society Conf. on Computer Vision and Pattern Recognition (1992) 808-810
12. Chellapa, R., Wilson, C. L., Sirohey, S.: Human and Machine Recognition of Faces: A survey. Proceedings of the IEEE 83(5) (1995) 705-740
13. Lee, J. C., Milios, E.: Matching Range Image of Human Faces. Proc. of the 3rd Int. Conf. on Computer Vision (1990) 722-726
14. Turk, M., Pentland, A.: Eigenfaces for Recognition. Journal of Cognitive Neuroscience 3(1) (1991) 71-86
15. Hesher, C., Srivastava, A., Erlebacher, G.: Principal Component Analysis of Range Images for Facial Recognition. Proc. of CISST (2002)
16. Zhao, Z. Q., Huang, D. S., Sun, B. Y.: Human Face Recognition based on Multi-features using Neural Networks Committee. Pattern Recognition Letters 25 (2004) 1351-1358
17. Pedrycz, W., Reformat, M., Han, C. W.: Cascade Architectures of Fuzzy Neural Networks. Fuzzy Optimization and Decision Making, 3 (2004) 5-37
18. Han, C. W., Pedrycz, W.: A New Genetic Optimization Method and Its Applications. submitted to International Journal of Approximate Reasoning
19. Peet, F. G., Sahota, T. S.: Surface Curvature as a Measure of Image Texture. IEEE Trans. PAMI 7(6) (1985) 734-738
20. Lee, Y., Park, G., Shim, J., Yi, T.: Face Recognition from 3D Face Profile using Hausdorff Distance. Proc. of PRIA-6-2002 (2002)
21. Lee, Y.: 3D Face Recognition using Longitudinal Section and Transection. Proc. of DICTA-2003 (2003)

A Privacy Algorithm for 3D Human Body Scans

Joseph Laws and Yang Cai

Carnegie Mellon University,
Visual Intelligence Studio, Cylab, CIC 2218,
4720 Forbes Avenue, Pittsburgh, PA 15213, USA
ycai@cmu.edu

Abstract. In this paper, we explore a privacy algorithm that detects human private parts in a 3D scan dataset. The intrinsic human proportions are applied to reduce the search space by an order of magnitude. A feature shape template is constructed to match the model data points using Radial Basis Functions in a non-linear regression. The feature is then detected using the relative measurements of the height and area factors. The method is tested on 100 datasets from CAESER database.

1 Introduction

The rapidly growing market of three-dimensional holographic imaging systems has created significant interest in possible security applications. Current devices operate using a millimeter wave transceiver to reflect the signal off the human body and any objects carried on it. Unlike current metal detectors, the system can also detect non-metal threats or contraband, including plastic, liquids, drugs and ceramic weapons hidden under clothing. The technology has also been used to create body measurements for custom-fit clothing. The holographic imager creates a high-resolution 3-D scan of a body, allowing shops to provide tailored measurements to designers or provide recommendations on best-fit clothing. These high resolution scanned images reveal human body details and have raised privacy concerns. Airport and transport officials in several countries are refusing to run a test trial with the scanners until a more suitable way to conceal certain parts of human body is found.

The scanner creates a three-dimensional point cloud (voxels) around the human body. Since the millimeter wave signal can not penetrate the skin, a three-dimensional human surface can be found. Furthermore, since the typical pose of a subject is standing, we can segment the 3-D dataset into 2-D contours, which reduces the amount of data processing significantly. The goal of this study is to develop a method that can efficiently find and conceal the private parts of a human.

2 Relevant Studies

From a computer vision point of view, detecting features from 3D body scan data is nontrivial because human bodies are flexible and diversified. Function fitting has been used for extracting special landmarks, e.g. ankle joints, from 3D body scan data

V.N. Alexandrov et al. (Eds.): ICCS 2006, Part IV, LNCS 3994, pp. 870 – 877, 2006.
© Springer-Verlag Berlin Heidelberg 2006

[1,2], similar to the method for extracting special points on terrain [5]. Curvature calculation is also introduced from other fields such as the sequence dependent DNA curvature [3]. These curvature calculation use methods such as chain code [7] circle fit, ratio of end to end distance to contour length, ratio of moments of inertia, and cumulative and successive bending angles. Curvature values are calculated from the data by fitting a quadratic surface over a square window and calculating directional derivatives of this surface. Sensitivity to the data noise is a major problem in both function fitting and curvature calculation methods because typical 3D scan data contain loud noises. Template matching appears to be a promising method because it is invariant to the coordinate system [1,2]. However, how to define a template and where to match the template is challenging.

In summary, there are two major obstacles in this study: robustness and speed. Many machine learning algorithms are coordinate-dependent and limited by the training data space. Some algorithms only work within small bounding boxes that do not warrant an acceptable performance. For example, if a feature detection algorithm takes one hour to process, then it is not useful for a security screening system. In this paper, we present a fast and robust algorithm for privacy protection.

3 Intrinsic Proportions

Our first step is to reduce the search space of the 3D body scans. We start by dividing the 3D data points into 2D slices. The points are 'snapped' to the nearest planes enabling us to convert a 3D problem to 2D. In this study, we assume that body features can be detected from the contours on the cutting planes. Examining each slice from top to bottom is rather an expensive process. Here we present a novel approach to reduce the search space by making use of intrinsic proportions. It is a relative measurement that uses an object in the scene to measure other objects [15].

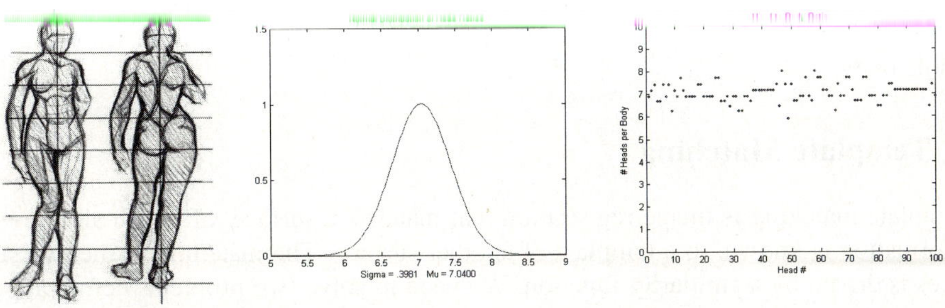

Fig. 1. Body height measured by head example (left), normal distribution of heads per body (middle), spread of actual # of heads per body size for all models (right)

Intrinsic proportion measurements have been used in architecture and art for thousands years. Roman architect Vitruvius said that the proportions of a building should correspond to those of a person, and laid down what he considered to be the relative measurements of an ideal human. Similarly in art, the proportions of the human body in a statue or painting have a direct effect on the creation of the human figure. Artists use analogous measurements that are invariant to coordinate systems. For example, using the head to measure the height and width of a human body, and using an eye to measure the height and width of a face. Figure 1 shows a sample of the vertical proportion in a typical art book and the actual distribution of head to body proportions calculated from our data set. Based on our observations from 100 3D scan data sets of adults from 16 to 65 years old, including subjects from North America, Europe and Asia, we found that the length of one and a half head units from the bottom of the head is enough to cover the chest area. In addition, the chest width is about three heads wide. Figure 2 shows an output from the intrinsic proportion calculation based on the sample from CARSER database.

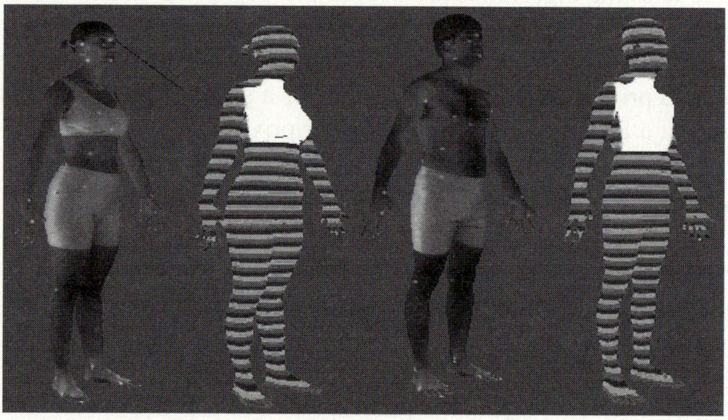

Fig. 2. Detected chest in white

We found that intrinsic proportion method can reduce the search space by an order of magnitude. In addition, it reduces the risk of the local optima while searching the whole body.

4 Template Matching

Template matching is image registration that matches a surface, of which all relevant information is known, to a template of another surface. The matching of the two surfaces is driven by a similarity function. We need to solve two problems before applying template matching on the regions of interest. First, a suitable template had to be created. Second, a similarity function had to be selected so that a minimization algorithm can align the template onto the region of interest. For each plane of the scan data, the back of the body contour can be removed. By assign the X-axis between the two points with the longest distance, we can obtain the front part of the body contour. We then use three radial basis functions to configure the template for female breast pattern.

$$Y = \sum_{i=1}^{3} a_i * \exp(-(n - s_i)^2) \quad (1)$$

where, $a = a_1 = a_2$, $b = a_3$, $s = s_1 = s_2$, and $s_3 = 0$.

We use non-linear regression on the variables *a, b,* and *s* to match the template with the scan data. Figure 3 shows the matching results for the female and male samples.

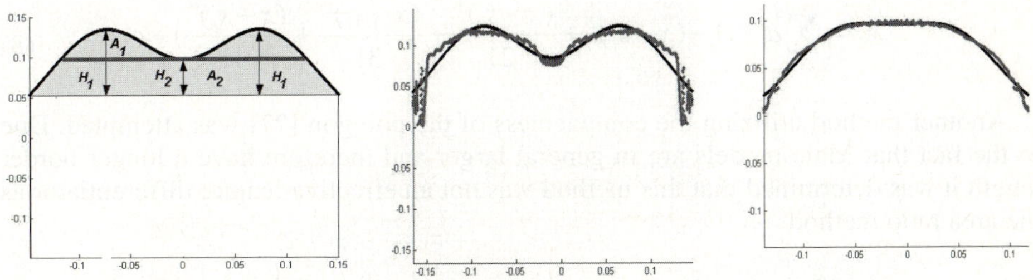

Fig. 3. Variable definitions for the breast template (left), matching results for the female sample (middle) and male sample (right). The solid black curves are the template contours. The red points are the 3D scan data.

5 Coordinate Invariant Measurements

Most shape descriptions depend on coordinate systems and viewpoint, meaning that the algorithm can only work within the training data space. Our shape invariant measurements are aimed to compute the shape properties from the ratio, rather than absolute values.

Template matching not only filters out noises but also describes the characteristics of a shape. We define the following invariant similarity functions to the coordinate system: height ratio and area ratio. The height ratio is defined as

$$H_r = \frac{H_1}{H_2} = \frac{Y_{mid/2}}{Y_{mid}} \quad (2)$$

The area ratio is defined as the ratio of the area of curvature feature (A_1) to the total area (A_2) of the model by the following formula:

$$A_r = \frac{A_2}{A_1} \quad (3)$$

where,

$$A_2 = \int_l \sum_{i=1}^{3} a_i * \exp(-(x - s_i)^2) dx \quad (4)$$

$$A_1 = \int_l (\sum_{i=1}^{3} a_i * \exp(-(x - s_i)^2) - c) dx \quad (5)$$

$$c = \sum_{i=1}^{3} a_i * \exp(-(mid - s_i)^2) \tag{6}$$

We used the Taylor series to find an appropriate approximation of the areas.

$$A_1 = \sum_{i=1}^{3} a_i * (1 - (x - s_i)^2 + \frac{(x - s_i)^4}{2!} + \frac{(x - s_i)^6}{3!} + \frac{(x - s_i)^8}{4!}) \tag{7}$$

$$A_2 = \sum_{i=1}^{3} a_i * (1 - (x - s_i)^2 + \frac{(x - s_i)^4}{2!} + \frac{(x - s_i)^6}{3!} + \frac{(x - s_i)^8}{4!}) - c \tag{8}$$

Another method utilizing the compactness of the polygon [27] was attempted. Due to the fact that Male models are in general larger and therefore have a longer border length it was determined that this method was not an effective feature differentiator as the area ratio method.

5 Results

We tested our algorithm with the subset of CAESER database that contains 50 males and 50 females ages 16-65, where 50 of them are North American with a black mix, 24 are Asian, and 26 are from the European survey from Italy and the Netherlands. We try to find the breast features from known female and male scan data samples. Figure 4 shows the test results. From the plot, we can see the distinguishable two groups for male (without the curvature feature) and female group (with the curvature feature). There is a 'dilemma' zone where some over-weight males do have the curvature features. However, the over-lapped zone is small, less than 8% of the total 100 samples.

Fig. 4. Chest feature detected

After the area and height factors have been calculated we determine the feature area, as seen in Figure 5. Once we find the feature area, we reduce the polygon resolution so that the area is blurred. Figure 6 shows the results of the blurring effects.

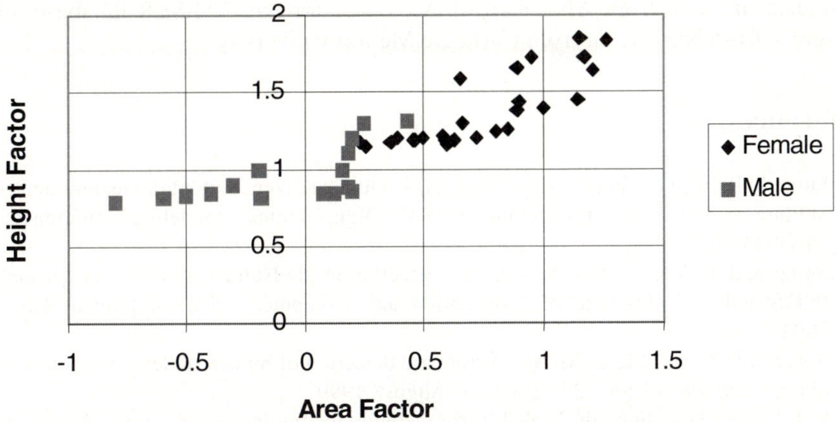

Fig. 5. Classification test results with male and female samples

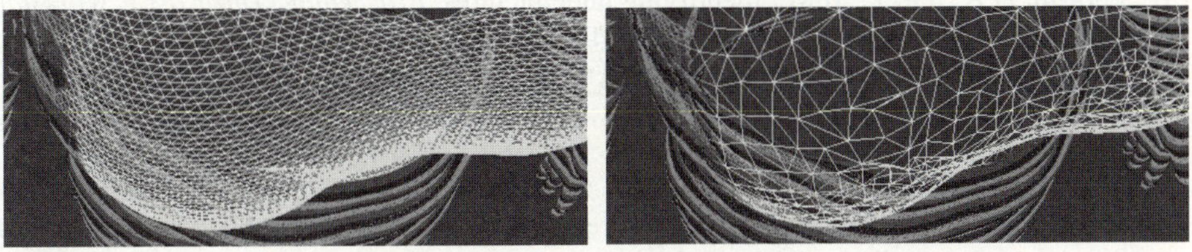

Fig. 6. Blurred surface rendering

6 Conclusions

In this paper, we explored the algorithm to recognize the body feature areas and blur them to protect a subject's privacy. Human intrinsic proportions are used to drastically reduce the search space and reduce the chance of detecting local optima. The feature template is defined with Radial Basis Functions whose parameters are determined by non-linear regression. Feature factors of the height and area are then used to recognize the curvature feature. The relative measurements are coordinate invariant. With non-linear regression method, the template matching is effective and convergent within a given error range. We tested 100 body scans from the CAESER database and found the algorithm can classify the male and female bodies based on the curvature features at the successful rate of over 90%.

Our future work includes the development of more robust coordinate invariant method to detect the predefined body features, and to fine tune the algorithms both for protecting privacy and detecting concealed weapons. Ultimately, we will work with the real field data to fine tune the algorithms.

Acknowledgement

We would like to thank for Cylab, Carnegie Mellon University for the support on security research and we are in debt to ARO for the research grant. We would also appreciate the help from Alva Karl of Air Force for the CAESER database and the assistance from Karl X. Fu from Carnegie Mellon University.

References

1. Suikerbuik, R., H. Tangelder, H. Daanen, A. Oudenhuijzen, Automatic feature detection in 3D human body scans, Proceedings of SAE Digital Human Modeling Conference, 2004, 04-DHM-52
2. Suikerbuik C.A.M. Automatic Feature Detection in 3D Human Body Scans. Master thesis INF/SCR-02-23, Institute of Information and Computer Sciences. Utrecht University, 2002
3. Forsyth, D.A. and Fleck, M. M., Automatic detection of human nudes, *International Journal of Computer Vision* , 32 , 1, 63-77, August, 1999
4. S. Ioffe and D.A. Forsyth, Probabilistic methods for finding people, *International Journal of Computer Vision* , Volume 43, Issue 1, pp45-68, June 2001
5. Forsyth, D.A. and Fleck, M.M., Body Plans, Proc. CVPR-97, 678-83, 1997.
6. Forsyth, D.A.; Fleck, M.M., Identifying nude pictures, *Proceeding. Third IEEE Workshop on Applications of Computer Vision.* 103-108, 1996.
7. M.M. Fleck, D.A. Forsyth and C. Bregler, Finding naked people, *Proc. European Conf. on Computer Vision* , Edited by: Buxton, B.; Cipolla, R. Berlin, Germany: Springer-Verlag, 1996. p. 593-602
8. http://www.dace.co.uk/proportion.htm
9. Jones, P.R.M. and Rioux, M. 1997. Three-dimensional surface anthropometry: applications to the human body. Optics and Lasers in Engineering, 28, 89-117.
10. Robinette, K.M., Blackwell, S., Daanen, H.A.M., Fleming, S., Boehmer, M., Brill, T., Hoeferlin, D., Burnsides, D. 2002. Civilian American and European Surface
11. Anthropometry Resource (CAESAR), Final Report, Volume I: Summary, AFRL-HE-WP-TR-2002-0169, United States Air Force Research Laboratory, Human Effectiveness Directorate, Crew System Interface Division, 2255 H Street, Wright-Patterson AFB OH 45433-7022 and SAE International, 400 Commonwealth Dr., Warrendale, PA 15096.
12. Gordon G. Face recognition based on depth and curvature features. Proceedings of the IEEE Computer Society Conference on Computer Vision and Pattern Recognition (Champaign Illinois), pages 108-110, 1992.
13. Calladine C.R. Gaussian curvature and shell structures. The Mathematics of Surfaces, Oxford University Press, pages 179-196, 1985.
14. Li, P., Corner, B.D., Paquette, S. Evaluation of a surface curvature based landmark extraction method for three dimensional head scans. International Ergonomics Conference, Seoul, 2003.
15. Ratner, P. 3-D human modeling and animation, John Wiley & Sons, Inc. 2003
16. Liu, X. W. Kim, and B. Drerup, 3D Characterization and Localization of Anatomical Landmarks of the Foot, Proceeding (417), Biomedical Engineering , Acta Press, 2004, http://www.actapress.com/PaperInfo.aspx?PaperID=16382
17. Bansal, M. Analysis of curvature in genomic DNA. http://www.ibab.ac.in/bansal.htm
18. Coleman, R., M. Burr, D. Souvaine, A. Cheng, An intuitive approach to measuring protein surface curvature, Proteins: structure, function and bioinformatics, vol. 61, no.4, pp 1068-1074
19. Goldgof, D.B., T.S.Huang, H.Lee, A Curvature-Based Approach to Terrain Recognition, November 1989 (Vol. 11, No. 11) pp. 1213-1217

20. Besl, P.J. and R. C. Jain, "Three-dimensional object recognition," ACM Comput. Surveys, vol. 17, no. 1, pp. 75-145, Mar. 1985.
21. Brady, M., J. Ponce, A. Yuille, and H. Asada, "Describing surfaces," *Comput. Vision, Graphics, Image Processing*, vol. 32, pp. 1-28, 1985.
22. Fan, T.G., G. Medioni, and R. Nevatia, "Description of surfaces from range data using curvature properties," in *Proc. CVPR*, May 1986.
23. Chen, H.H. and T. S. Huang, "Maximal matching of two three-dimensional point sets," in *Proc. ICPR*, Oct. 1986.
24. Haralick, R.M., S.R. Sternberg, and X. Zhuang, "Image analysis using mathematical morphology," IEEE Trans. Pattern Anal. Machine Intell., vol. PAMI-9, no. 4, pp. 532-550, 1987.
25. Goldgof, D.B., T. S. Huang, and H. Lee, "Feature extraction and terrain matching," in Proc. IEEE Comput. Soc. Conf. Comput. Vision Pattern Recognition, Ann Arbor, MI, May 1988.
26. Goldgof, D.B., T. S. Huang, and H. Lee, "Curvature based approach to terrain recognition," Coord. Sci. Lab., Univ. Illinois, Urbana-Champaign, Tech. Note ISP-910, Apr. 1989.
27. Sonka, M. et al. Image processing, analysis and machine vision, PWS Publishing, 1999

The Study of the Detection and Tracking of Moving Pedestrian Using Monocular-Vision*

Chang Hao-li, Shi Zhong-ke, and Fu Qing-hua

College of Automation, Northwestern Polytechnical University, Xi'an 710072, China
Chl8201@sohu.com, {zkeshi, fuqinghua}@nwpu.edu.cn

Abstract. To ensure the safety and efficiency of the pedestrian traffic, this paper presents a real-time system for moving pedestrian detection and tracking in sequences of images of outdoor scenes acquired by a stationary camera. The self-adaptive background subtraction method and the dynamic multi-threshold method were adopted here for background subtraction and image segmentation. During the process of tracking, a new method based on gray model GM(1,1) was proposed to predict the motion of pedestrians. And then a template for tracking pedestrian continuously was presented by fusing several characters of targets. Experimental results of two real urban traffic scenes demonstrate the efficiency of this method, then the application of this method is discussed in real transportation system.

1 Introduction

Pedestrian is one of the most principle participants in transportation system, ensuring the safety and smoothness of the pedestrian traffic is an important goals to construct the city transportation system. But in present research, only vehicles are considered[1][3], while pedestrians are mostly neglected. When designing the traffic control scheme, the delay of vehicles and the length of vehicles' queue are the most important targets to evaluate the performance of the system. The obstacles of intelligent vehicles are thought mostly vehicles too. But in real traffic scenes, the influence of pedestrians can not be neglected, especially in the traffic environment of China which is a typical mixed traffic. The information of pedestrians can not only be used to realize a traffic control application such as a pedestrian control scheme at intersection, but also can be used to design the safety system and navigation system of Intelligent Vehicles. Pedestrian detection is the most common approach to obtain the pedestrian information in traffic scenes. It is the foundation to analysis and comprehend pedestrian behavior.

Tracking moving pedestrians in video streams has been an active area of research in computer vision. In recent years, many researchers have begun to do research in this area. Stereovision is the most common method to detect pedestrians[4][5]. It is efficient in middle and short distance, but is too computationally expensive and the system is complex. Neural network[4][5] and the step rhythm[6][7] are used to recognize pedestrians in some systems. This method is complex and not credible. To detect the

* Supported by the national natural science foundation of China (Grant NO. 60134010).

V.N. Alexandrov et al. (Eds.): ICCS 2006, Part IV, LNCS 3994, pp. 878 – 885, 2006.
© Springer-Verlag Berlin Heidelberg 2006

pedestrian the legs must be detected exactly, so this method will be disabled when the legs can not be detected for some reason. So an approach to detect moving pedestrians in congested traffic scenes based on monocular vision was presented. It includes two course of processing: segmentation and tracking. The self-adaptive background subtraction method and the dynamic multi-threshold method are adopted here for background subtraction and image segmentation. A new method based on gray model GM(1,1) was proposed to predict the motion of pedestrians, which will improve the accuracy. A simple criterion based on multi-features is used for classification and template matching, guided by motion prediction for tracking. Tested using real traffic video sequences, the system are able to track multiple isolated pedestrians robustly.

2 Moving Pedestrian Extraction

Reliable tracking requires that the pedestrians can be segmented out reliably. To be useful, the segmentation method needs to accurately separate pedestrians from the background, be fast enough to operate in real time, be insensitive to lighting and weather conditions, and require a minimal amount of initialization. This can be done by using either models describing the appearances of the targets or a model describing the appearance of the background.

At the images level, we perform background subtraction and thresholding to produce difference images. On the assumption that the camera is still, the background difference method is comparatively efficient and simply to detect moving objects. The moving object can be segmented out according to the following equation.

$$dif\,(i, j) = I_k(i, j) - B(i, j)\;. \tag{1}$$

$I_k(i, j)$, $B(i, j)$ represent the current image and current background respectively.

2.1 Adaptive Background Segmentation

A self-adaptive background subtraction method is used for segmentation. This method is much simpler and more robust to update the background. In addition, this method is also insensitive to lighting conditions and has the further advantage of not requiring initialization with a background image.

Take the first frame as the initial background. For each frame of video sequence, we take the difference between the current image and the previous image giving the difference image BW_k:

$$BW_k = \begin{cases} 1 & \text{if abs } (\mathrm{I_k} - \mathrm{I_{k-1}}) \geq T \\ 0 & \text{if abs } (\mathrm{I_k} - \mathrm{I_{k-1}}) < T \end{cases}. \tag{2}$$

Here, I_k, I_{k-1} represent the two continuous images of the image serial, the value of T is 10% of the peak value.

We update the background by taking a weighted average of the current background and the current frame of the video sequence. However, the current image also contains foreground objects. Therefore, before we do the update we need to classify the pixels as foreground and background and then use only the background pixels from the current image to modify the current background. The binary object mask is used to distinguish the foreground pixels from the background pixels.

$$B_k = \begin{cases} B_{k-1} , & BW_k = 1 \\ (1-\alpha)I_k + \alpha B_{k-1} , & BW_k = 0 \end{cases} . \tag{3}$$

B_k , B_{k-1} and I_k represent the new background, the instantaneous background and the current image respectively. The weight assigned to the current and instantaneous background affect the update speed. We want the update speed to be fast enough so that changes in illumination are captured quickly, but slow enough so that momentary changes do not persist for an unduly long amount of time. The weight α has been empirically determined to be 0.1. We have found that this gives the best tradeoff in terms of update speed and insensitivity to momentary changes.

2.2 Select the Two Thresholds Dynamically

After subtracting the current image from the current background, the resultant difference image has to be thresholded to get the binary object mask. The resulting connected regions are then grouped into pedestrians and tracked. Since the object mask itself is used to update the current background, a poorly set threshold would result in poor segmentation. So the choice of the threshold is critical. The pedestrians' appearances change dynamically, so a static threshold cannot be used to compute the object mask. Therefore we need a way to update the threshold as the current background changes. The difference image is used to update the threshold.

The reason using two different thresholds is for detecting the nature of occlusion. In our images, a major portion of the image consists of the background. Therefore the difference image would consist of a large number of pixels having high values. And the histogram contains mainly two parts: the noise and the statistical characters of the motion object. We use this observation to decide the thresholds. On assumption that the noise parameters obey the Gaussian model, which can be described as follows:

$$f = \begin{cases} f_l = (\sqrt{2\pi}\sigma_l)^{-1} e^{-(x-\mu)^2/2\sigma_l^2} & x \le \mu \\ f_r = (\sqrt{2\pi}\sigma_r)^{-1} e^{-(x-\mu)^2/2\sigma_r^2} & x > \mu \end{cases} . \tag{4}$$

Among these, μ represents the mean value, σ_l, σ_r represents the parameters of the two parts of noise at the both sides of the mean value.

The histogram of the difference image will be filtered using the middle-value filter. Then find the two segmentation thresholds at the filtered histogram. The histogram of the difference image will have high values for low pixel intensities and low values for the higher pixel intensities. To set the left threshold, we look for the first dip in the histogram that occurs to the left of mean value, starting from the pixel value $x = \mu - 3\sigma_l$ corresponding to the histogram. The corresponding pixel value is used as the left threshold T_l. To set the right threshold T_r, the method is similar as looking for the left threshold. We need to look for the first dip in the histogram that occurs to the right of the mean value, and start from $x = \mu + 3\sigma_r$.

2.3 Segmentation

Moving entities are then extracted as follows:

$$BW(i, j) = \begin{cases} 1, & dif(i, j) \le T_l \text{ or } dif(i, j) \ge T_r \\ 0, & others \end{cases} . \tag{5}$$

In order to eliminate noise from being classified as foreground, a threshold is used so that any blob with area smaller than the threshold is deleted from the foreground. Several measures were taken to further reduce the effect of noise. A single step of erosion followed by a step of dilation is performed on the resulting image and small clusters are totally removed. And also, the background image is updated using a very slow recursive function to capture slow changes in the background.

3 Pedestrian Tracking

The purpose of tracking pedestrians is to obtain the tracks of the pedestrians. The key is to detect and track pedestrians continuously. In every frame, a relation between the blobs in the current frame is sought with those in the previous frame. Then the pedestrian in the detection region can be tracked. Achieving robust tracking in outdoor scenes is a hard problem owing to the uncontrollable nature of the environment. Furthermore, tracking in the context of an intersection should be able to handle non free-flowing traffic and arbitrary camera views.

3.1 Detection and Recognition

The individual regions are computed using a connected components extraction method. The various attributes of the blob such as centroid, area, and elongation that can be computed during the connected component extraction.

Although the pedestrians are much different, the shapes of them are similar and the inverse proportion of the height and the width comply with certain criterion. We model the pedestrian as a rectangular patch with a certain dynamic behavior. The area of the pedestrian is smaller and the dispersibility of the shape is larger comparing with vehicle. There we can define the dispersibility as follows:

$$dis = P^2/A . \tag{6}$$

Here the equation (6) represents the shape dispersibility; the l, w, P and A stand for the height, width, perimeter and the area. According to the two characters, we can distinguish pedestrians from vehicles. We found that this simple model adequately resembles the pedestrian's shape and motion dynamics for the purpose of tracking.

3.2 Motion Prediction

To improve the efficiency of the approach, we need to have an estimation of where to place the box corner which represents the detected pedestrian with respect to the distance of one bounding box corner and the centroid of the pedestrian are obtained.

Hence, the parameters estimated are the distance of a corner point from the centroid, the length and the height of the pedestrian.

The motion of the pedestrian is random and difficult to depict. Therefore, we choose the grey model GM(1,1) as the motion model to predict the motion of the pedestrian. The future state of the pedestrian is predicted by processing the records using the GM(1,1). The definition of the GM(1,1) is as follows:

Supposed that there is a date serial $X^0 = \{x_1^0, x_2^0, \cdots, x_n^0\}$, we can get a new date serial $X^1 = \{x_1^1, x_2^1, \cdots, x_n^1\}$ by accumulating all these dates. In this equation, $X_k^1 = \sum_{i-1}^{k} x_k^0$, so the differential equation of the GM(1,1) is as follows:

$$\frac{dX^1}{dt} + aX^1 = b \cdot \tag{7}$$

The equation $\hat{a} = (a,b)^T$ represents the predicting parameter serial. The following result will be obtained:

$$\hat{a} = (BB^T)^{-1} BX^0 \cdot \tag{8}$$

$$B = \begin{bmatrix} -\frac{1}{2}(x_1^1 + x_2^1) & -\frac{1}{2}(x_2^1 + x_3^1) & \cdots & -\frac{1}{2}(x_{n-1}^1 + x_n^1) \\ 1 & 1 & \cdots & 1 \end{bmatrix} \cdot \tag{9}$$

Using these estimated parameters, request the difference equation, and the following prediction model can be obtained:

$$\hat{x}_{k+1}^0 = \left(\frac{b}{1+0.5a} - \frac{a}{1+0.5a} x_1^0\right) e^{-a(k-1)} \cdot \tag{10}$$

Using the model, the previous three positions can be used to predict pedestrians' next position. The new data will be used to update the record. The process is quickly enough and the date needed is few. So the pedestrians' up-to-the-minute movement law can be tracked and the position conditions can be predicted exactly.

3.3 Matching

Tracking is then performed by finding associations between the pedestrians in the current image with those in the previous frames based on the proximity of the blobs. The pedestrian in the current image inherit the timestamp, label and other attributes such as velocity from a related pedestrian. So, the parameters such as condition, shape, area and statistical characteristics are computed for each pedestrian.

Because of the association of the pedestrian motion in temporal and spatial, the position can be considered as the most important feature. First search the area of around the prediction position, if detect nothing then the tracking is failed. Otherwise we can continue judge according to other features.

Although blobs obtained from difference images can be sufficient to decide the location of pedestrians in many cases, the intensity information may be useful to resolve certain ambiguities. The use of such information in the form of statistical distribution of intensities may add to the robustness of the current system and is worth pursuing. We use the following two statistical characteristics: the mean gray value and the coefficient of the consecutive frames which are computed out for matching.

Then the template of the detected objects can be build according to these dates and used to track moving pedestrians. During the process of tracking, the matching model is build up for each object, so we can judge whether the object appears around the detecting region and the template are matching or not. If succeed, then track the object and the template is updated by the object's new characters. Otherwise, the template remains, and the next frame will be processed. A blob is considered at the pedestrian level only if it is tracked successfully for a certain period of time. This eliminates blobs that appear then disappear momentarily, such as most blobs that appear due to tree motion back and forth.

4 Results and Application

We test the proposed method using the two video sequences acquired by a stationary camera. One is that there is only single people in the traffic scene, the other is there are many pedestrians. The size of the image is 320×240, the image collection rate is 0.1fra/s. The velocity of process is 11fra/s, so it is quickly enough for the system.

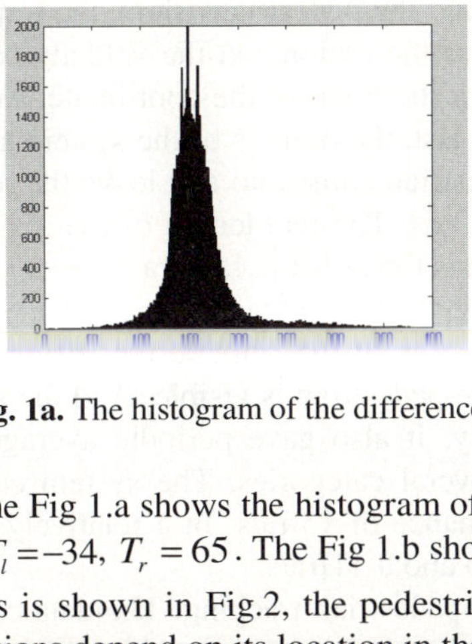

Fig. 1a. The histogram of the difference image

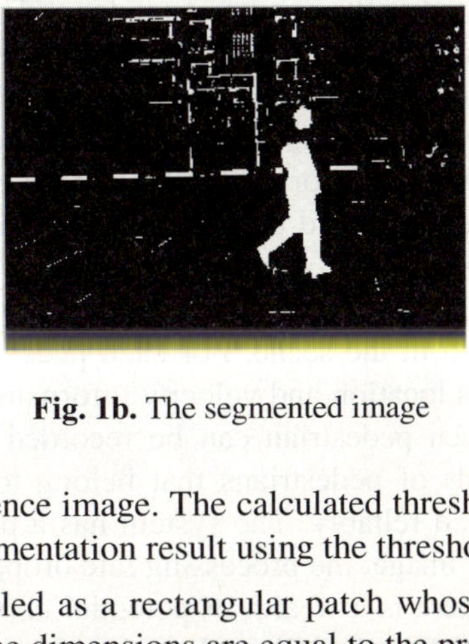

Fig. 1b. The segmented image

The Fig 1.a shows the histogram of the difference image. The calculated thresholds are $T_l = -34$, $T_r = 65$. The Fig 1.b shows the segmentation result using the thresholds.

As is shown in Fig.2, the pedestrian is modeled as a rectangular patch whose dimensions depend on its location in the image. The dimensions are equal to the projection of the dimensions of an average size pedestrian at the corresponding location in the scene. As the following figures shown, pedestrians close to each other and several pedestrians appearing at the same time can be detected exactly.

The results of the GM(1,1) position estimation for a pedestrian are shown in Fig.3. The position estimations of the GM(1,1) are presented against the actual measurements in Fig 3.a. The two tracks are similar and the result showed that this approach can track the pedestrian perfectly. Fig 3.b is the real image showing the detection result, the thinner line rectangle represents the predicted position of the pedestrian, and the wider one is the real position.

Fig. 2. The result of pedestrian detection

Fig. 4. The result of pedestrian tracking

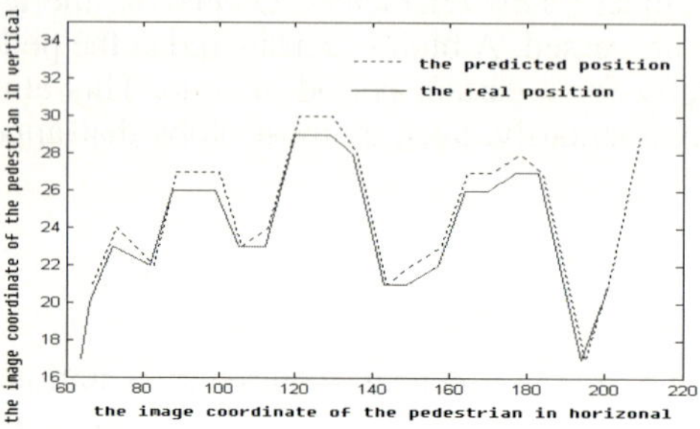

Fig. 3a. The result of tracks comparing

Fig. 3b. The result of position comparing

Fig 4 is the results of tracking. The lines behind the pedestrians show the trajectories of the pedestrians. And the time used to pass the region and the velocity can be calculated out given the frame rate. By analyzing the trend of the coordinate, we can know the moving direction of the pedestrian. At last, the outputs of the system are all these results. For example, in the Fig.4, the pedestrian come into and leave the region at the 231th and the 367th frame. The time is 13.6s. The real length of the region is 15.5m, and the velocity is 1.14m/s. The direction is towards the camera.

In most cases, pedestrians were tracked correctly throughout the period they appeared in the scene. For each pedestrian in the view of the camera, the system produces location and velocity information when the pedestrian is visible. And the speed of each pedestrian can be recorded successfully. It also gave periodic averages of speeds of pedestrians that belong to one of several categories. The system can be applied reliably. The system has a peak performance of 15fra/s. In a relatively cluttered image, the processing rate dropped down to about 11fra/s.

There is a wealth of potential applications of pedestrian tracking. The data can be used in a scheduling algorithm to control walk signals at an intersection. And it can detect and track not only humans in front of or around vehicles but it can also be employed to track several diverse traffic objects of interest. One should note that the reliable detection and tracking of traffic objects is important in several vehicular applications. In this paper, we are mainly interested in applications related to traffic control with the goal of increasing both safety and efficiency of existing roadways. For example, information about pedestrians crossing the streets would allow for automatic control of traffic lights at an intersection. Pedestrian tracking also allows the use of a warning system, which can warn drivers and workers at a work zone from possible collision risks. The proposed approach also has a large number of potential applications, such as security monitoring, event recognition, pedestrian counting, traffic control, and traffic-flow pattern identification applications emphasize tracking on a coarser level.

5 Conclusions

This paper presents a real-time system for moving pedestrian detection and tracking in images sequences of outdoor scenes acquired by a stationary camera. This approach was tested using two traffic video images, and the potential application is discussed. This approach can detect and track pedestrians exactly and robustly in most cases. The system outputs the spatio-temporal coordinates of each pedestrian during the period when the pedestrian is in the scene.

To improve the precision and stability of the system, the approach should be ameliorated in many aspects. For example, how to track several pedestrians in the clustered traffic scenes; and also how to track pedestrians at night; these are all problems should be solved in the next research.

References

1. Chu J W, Ji L S, Guo L, et al, C: Study on Method of Detecting Preceding Vehicle Based on Monocular Camera. IEEE Intelligent Vehicles, Italy(2004), June 14-17:750–755
2. Surendra G, Osama M, Robert F, et al, J: Detection and Classification of Vehicles. IEEE Transactions on Intelligent Transportation Systems Vol.3(1) 37–47(2002)
3. Osama M, Nikolaos P P, J: A novel method for tracking and counting pedestrians in real-time using a single camera. IEEE Transactions on Vehicles Technology(2001)
4. Wohle J C, Anlauf J, Pörtner T, et al. A time delay neural network algorithm for real-time pedestrian recognition[C], IEEE Intelligent Vehicles(1998) 247–252
5. Zhao L, Thorpe C,C: Stereo and neural network based pedestrian detection, ITSC(1999)289–303
6. Pai C J, Tyan H R, Liang Y M, J: Pedestrian detection and tracking at crossroads. IEEE (2003)101–104
7. Cristovbal C, Johann E, Thomas K, J: Walking pedestrian recognition, IEEE Transactions on Intelligent Transportation Systems Vol.3(1) 155–163(2000)
8. Alan J L, Hironobu F, Raju S P,J: Moving target classification and tracking from real-time video, IEEE(1998) 8-14

An Implementation of Real Time-Sentential KSSL Recognition System Based on the Post Wearable PC

Jung-Hyun Kim, Yong-Wan Roh, and Kwang-Seok Hong

School of Information and Communication Engineering, Sungkyunkwan University, 300, Chunchun-dong, Jangan-gu, Suwon, KyungKi-do, 440-746, Korea
{kjh0328, elec1004}@skku.edu, kshong@skku.ac.kr
http://hci.skku.ac.kr

Abstract. Korean Standard Sign Language (hereinafter, "KSSL") is a complex visual-spatial language that is used by the deaf community in the South Korea. Wire communications net and desktop PC-based a traditional study on sign language linguistics with small vocabulary (words) have several restrictions (e.g. limitation of the motion, conditionality in the space) and general problems (e.g. inaccuracy in measuring, necessity of complex computation algorithm) according to using of vision technologies with image capture and video processing system as input module of sign language signals. Consequently, in this paper we propose and implement ubiquitous-oriented wearable PC-based sentential KSSL recognizer that improve efficiency of KSSL input module according to wireless sensor network, recognizes and represents continuous KSSL with flexibility in real time, and analyze and notify definite intention of user more efficiently through correct measurement of KSSL gestures using wireless haptic devices. The experimental result shows an average recognition rate of 93.7% for continuous 44 KSSL sentences.

1 Introduction

Human beings usually interact with each other either by natural language channel such as speech and writing, or by body language channel such as hand gesture, head gesture, facial expression, lip motion and so on. Thus, the study on the perception model of body language, the information fusion of body language channel and natural language channel is very important for the improvement of computer's human language understanding and the increase of human-computer interaction applicability. As a part of natural language understanding, sign language recognition is very important: on one hand, it is one of the main methods of human-computer interaction in VR (Virtual Reality); on the other hand, it is an auxiliary tool for a deaf-mute to communicate with ordinary people through computer [1]. The related studies about recognition and representation of sign language are progressing actively by numerous researchers in the many countries. Especially, "standard sign language translation system" developed jointly by the professors of the KIST (The Korea Institute of Science and Technology) and the Samsung Electronics Co., Ltd. in the South Korea is sign language translation system that recognizes and represents small vocabulary-based sign language. And, the Hitachi Ltd. in the Japan announced "Japanese - sign language translation technology" that a sign language animation is created automatically

V.N. Alexandrov et al. (Eds.): ICCS 2006, Part IV, LNCS 3994, pp. 886 – 893, 2006.
© Springer-Verlag Berlin Heidelberg 2006

if input the Japanese sentence [2]. Also, Wu jiangqin et al. implemented 26 word-level sign language recognition system using neural network and HMM hybrid method in the China [1]. However, usually, wire communications net and desktop PC-based these traditional studies on sign language recognition and representation with small vocabulary(words and morpheme) have not only several restrictions such as limitation of representation, conditionality on space, complexity between transmission mediums and limitation of motion but also some problems such as variation of recognition performance, uncertainty of measurement and necessity of complex computation algorithms according to using of image capture system or video processing system with changing of background's colors and illumination condition as acquisition of sign language signals.

Accordingly, we propose and implement real time-sentential KSSL recognizer using blue-tooth module, wireless haptic devices and fuzzy logic based on the post wearable PC platform(embedded i.MX21 board) for embedded-ubiquitous computing that guarantee mobility of portable terminal. The advantages of our approach are as follows: 1) it improves efficiency of KSSL input module according to wireless sensor network and contributes to user's the convenience. That is, it recognizes and represents continuous sign language of users with flexibility in real time, and 2) because the ability of communication and representation of sentential sign language recognizer are very superior more than word (morpheme)-level sign language recognizer, it is possible more effective and free interchange of ideas and information between the hearing-impaired and hearing person.

The composition of this paper are 1) regulation of components the KSSL in Section 2, 2) KSSL input module using wireless haptic devices, and training and recognition models using RDBMS in Section 3, 3) fuzzy max-min module for the KSSL recognition in Section 4, and 4) experiments and results in Section 5. Finally, this study is summarized in Section 6 together with an outline of challenges and future directions.

2 The Regulation of the KSSL

The phonemic systems of oral languages are primarily sequential: that is, the majority of phonemes are produced in a sequence one after another, although many languages also have non-sequential aspects such as tone. As a consequence, traditional phonemic writing systems are also sequential, with at best diacritics for non-sequential aspects such as stress and tone [3]. In this paper, to implement sentential KSSL recognition system in real time, this study selected 25 basic KSSL gestures according to the "Korean Standard Sign Language Tutor(hereinafter, “KSSLT”)[4]" and a point of reference of 'sign language morpheme'. And 23 hand gestures necessities of KSSL gestures are classified as hand shapes, pitch and roll degree. Consequently, we constructed 44 sentential KSSL recognition models according to associability and presentation of hand gestures and basic KSSL gestures. The example of basic KSSL gestures and hand gestures for “the date (numbers-day-month)” in sentential KSSL recognition models are shown in Table 1.

Table 1. The example of sign language about "the date (finger numbers-day-month)"

Object Language	Description of the KSSL
Finger-Number	1~30(or 31) : Signing of numbers
Month	Represent signing of numbers that correspond to "1(one)" with a left hand (indication of month), and draw crescent with thumb, index finger of a right hand
Day	Spread out thumb, index fingers of a both hands and rises on in front of chest.
Necessary Hand Gestures	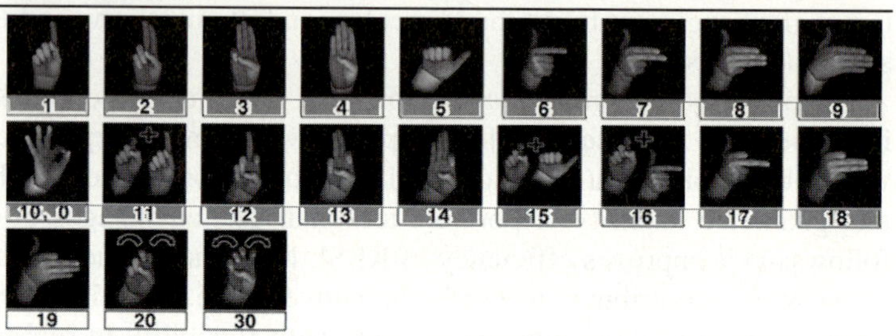

3 KSSL Input Module and Recognition Models Using RDBMS

Virtual reality sensors provide a powerful technology for human-computer interaction (HCI) and have been applied to many fields such as medical and on-line game service. Their particularly useful feature is that the user may use the technology easily and routinely by using ready-to-wear articles of clothing e.g. headsets or data gloves. Also the VR sensors could be simply plug into a computer system and allow the user uninhibited control and interaction with both the local computer system [5].

To implement real time-sentential KSSL recognizer, we used 5DT company's wireless 2 data gloves to acquire structural information of hands and fastrak® for motion tracking which are one of popular input devices in the haptic application field, and blue-tooth module for wireless sensor network as KSSL input module. The data glove is basic gesture recognition equipment that can capture the degree of finger stooping using fiber-optic flex sensor and acquires data through this. The structural motion information of each finger in data glove are captured by f1=thumb, f2=index, f3=middle, f4=ring and f5=little in regular sequence. Each flexure value has a decimal range of 0 to 255, with a low value indicating an inflexed finger, and a high value indicating a flexed finger. Also, the fastrak® is electromagnetic motion tracking system, a 3D digitizer and a quad receiver motion tracker. And it provides dynamic, real-time measurements of six degrees of freedom; position (X, Y, and Z Cartesian coordinates) and orientation (azimuth, elevation, and roll) [6] [7].

The captured continuous KSSL data of various users is transmitted to embedded i.MX21 board and database server for training and recognition models through blue-tooth module. The KSSL data that transmitted to server is used to as a training data for KSSL recognition models by analytic function of RDBMS SQL. And, KSSL data that transmitted to embedded i.MX21 board is used as input variables of fuzzy recognition module for significant KSSL recognition. The architecture and composition of KSSL input module is shown in Fig. 1.

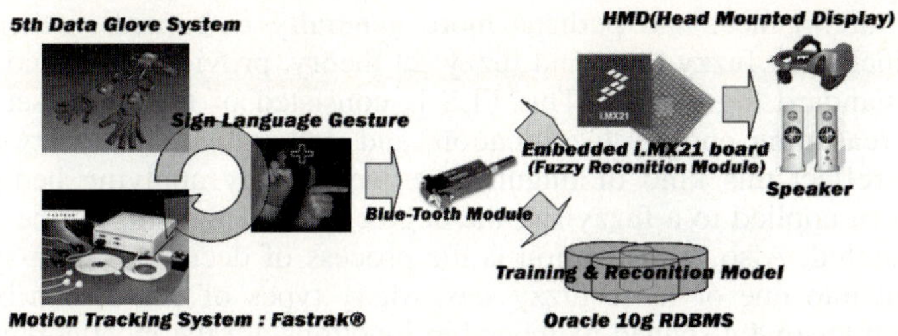

Fig. 1. The architecture and composition of KSSL input module

The RDBMS is the main stream database management system that maintains data records and indices in tables and their relationships may be created and maintained across and among the data and tables [8][9]. The RDBMS is used to classify and segment KSSL data that are transmitted from KSSL input module to database server by valid gesture record set and invalid record set (that is, invalid record set is status transition record set) and to analyze valid record set efficiently. A rule to segment valid gesture record set and invalid record-set is shown in Fig. 2.

1. Difference Between Preceding Average(preceding 3 and 1) and Current Row Value
2. Decide validity : Preceding Average - Current Value = 52-17 = 35

Average Between 3 Preceding and 1 Preceding from Current Row : 52

1. Difference between Preceding Average and Current Row Value > 5
2. 'X'(Invalidity Record) check

SerialNO	F1THUMB	F1INDEX	F1MIDDLE	F1RING	F1LITTLE	F1X-POSI.	F1Y-POSI.	F1Z-POSI.	F1_Px	F1_Py	GESTURE DATA...	VALIDITY
22	73	86	255	255	255	50.90	-14.53	-4.74	55.42	-5.71		X
23	51	48	255	255	255	52.05	-11.97	-2.13	53.29	-4.95		X
24	32	15	255	255	255	53.99	-9.50	0.01	50.02	-4.75		X
25	17	0	255	255	255	51.03	-10.50	-1.05	51.11	-4.83	Omission...	X
26	7	0	255	255	255	52.01	-9.85	-1.02	49.85	-5.23		
27	0	0	255	254	255	52.78	-8.77	-1.75	51.85	-6.02		
28	0	0	253	254	255	49.87	-10.18	-2.12	48.97	-5.27		
[illegible]	[illegible]	[illegible]	[illegible]	[illegible]	[illegible]	[illegible]	[illegible]	[illegible]	[illegible]	[illegible]		
30	0	0	255	255	255	51.09	-8.75	-1.75	47.93	-4.94		

Fig. 2. The segmentation rule and record set. If the difference between preceding average (preceding 3 and 1) and current row value is over 5, the current value is regarded as transition KSSL record. Also, if one of KSSL data values of data gloves and motion tracker is over 5, current KSSL data is also regarded as transition KSSL record. According to the logic in source code, even though a record set is, based on the above process, decided as valid, the record set is regarded as a status transition KSSL record set.

4 Fuzzy Max-Min Composition for KSSL Recognition

The fuzzy logic is a powerful problem-solving methodology with a myriad of applications in embedded control and information processing, and is a paradigm for an alternative design methodology which can be applied in developing both linear and nonlinear systems for embedded control and has been found to be very suitable for embedded control applications [6][10]. The Fuzzy Logic System (hereinafter, "FLS") will be demonstrated that the use of fuzzy systems makes a viable addition to the field of artificial intelligence, and perhaps more generally to formal mathematics as a whole, and including fuzzy logic and fuzzy set theory, provide a rich and meaningful addition to standard logic [10]. This FLS is consisted of the fuzzy set, fuzzy rule base, fuzzy reasoning engine, fuzzification, and defuzzification. Fuzzy sets can be modified to reflect this kind of linguistic refinement by applying hedges. Once a hedge has been applied to a fuzzy set, the degrees of membership of the members of the set are altered. Also, fuzzification is the process of decomposing a system input and/or output into one or more fuzzy sets. Many types of curves can be used, but triangular or trapezoidal shaped membership functions are the most common because they are easier to represent in embedded- controllers [6]. The fuzzy max-min CRI for fuzzy relations that is proposed in this paper is defined in Fig. 3.

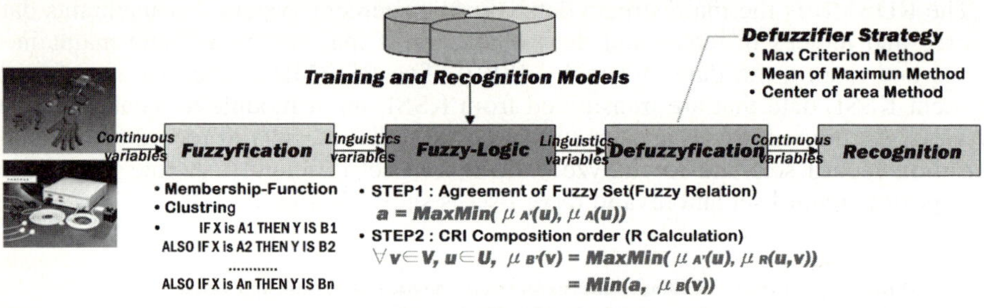

Fig. 3. Fuzzy Max-Min CRI (Direct Method)

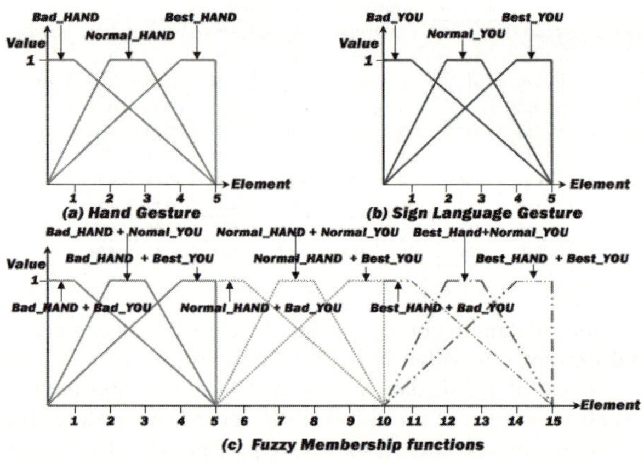

Fig. 4. The fuzzy membership functions. Fig. 4 describes the formula (1) pictorially (because fuzzy input variables are very various, we represent membership functions partially: "YOU" in KSSLT). If compare easily, sign language "YOU" in KSSLT consist of hand gesture that correspond to "paper" in "the game of paper, stone and scissors" and hand motion that correspond to "point at a person with one's finger". We prescribe gesture behavior such as the shape of one's hand (fingers) and the direction of back of the hand by 'Hand Gesture', and classified by three types of "Bad_HAND, Normal_HAND and Best_HAND" according to accuracy in "(a) Hand Gesture". Also, we prescribe hand's position and gesture in spatial dimensions by "Sign Language Gesture", and classified by three types of "Bad_YOU, Normal_YOU and Best_YOU" according to accuracy in "(b) Sign Language Gesture" in spatial dimensions.

The training and recognition model using the RDBMS is used as an independent variables for comparison and reasoning with input variables(KSSL data) in fuzzy logic (fuzzy max-min composition), and recognizes user's dynamic KSSL through efficient and rational fuzzy reasoning process. Therefore, we decide to take the characteristics of KSSL input data as the average value over repeated experiment result values, where the repetition number is controlled by several parameters. Input scale factors transform the real inputs into normalized values, and output scale factors transform the normalized outputs into real values. The proposed membership function of fuzzy set is defined as in the following formula (1) and the fuzzy membership functions are shown in Fig. 4.

$$\mu_{tz} = \begin{bmatrix} \frac{1}{(s-p)}(x-s)+1, & p<x\le s \; ; \textit{Slopes-up} \\ 1 \quad , & s<x\le t \; ; \textit{Horizontality} \\ -\frac{1}{(q-t)}(x-t)+1, & t<x\le q \; ; \textit{Slopes-down} \end{bmatrix} \tag{1}$$

5 Experiments and Results

Our experiment environment is consisted of blue-tooth module, wireless haptic devices and i.MX21 board based on embedded LINUX operating system for embedded-ubiquitous computing. That is, data gloves transmits 14 kinds of hand's structural motion data (10 fingers gesture data, 4 pitch & roll data) and motion tracker transmits 12 kinds of hand's spatial motion data with both hands to embedded i.MX21 board via blue-tooth module. The proposed whole file size of sentential KSSL recognizer is 283 Kbytes (including images and composite sounds for visual and auditive representation) and it can process and calculate 200 samples per seconds on i.MX21 board.

Overall process of sentential KSSL recognizer consists of three major steps. In the first step, while the user inputs prescribed the KSSL to i.MX21 board using 2 data gloves, motion tracker and blue-tooth module, the KSSL input module captures user's sign language data. And, in the second step, the KSSL recognizer changes characteristics of inputted KSSL data by parameters for fuzzy recognition module. In the last step, it calculates and produces fuzzy value for user's dynamic KSSL through a fuzzy reasoning and composition, and we decide to give a weight to each parameter. KSSL recognizer decides user's dynamic KSSL according to degree of the produced fuzzy value. The process (flow-chart) of automatic real time-sentential KSSL recognizer is shown in Fig. 5.

The experimental set-up is as follows. The distance between KSSL input module and embedded i.MX21 board for KSSL recognition processing is about radius 10M's ellipse form. The reagents take KSSL motion moving 2 wireless data gloves and 2 receivers of the motion tracker to prescribed position. For every 20 reagents, we repeat this action 15 times. Experimental results, Fig. 6 shows an average recognition rate of 93.7% with fuzzy recognition module for 44 KSSL recognition models. Also, the user interface for visual representation using HMD (Head Mounted Display) of the KSSL recognition is shown in Fig. 7.

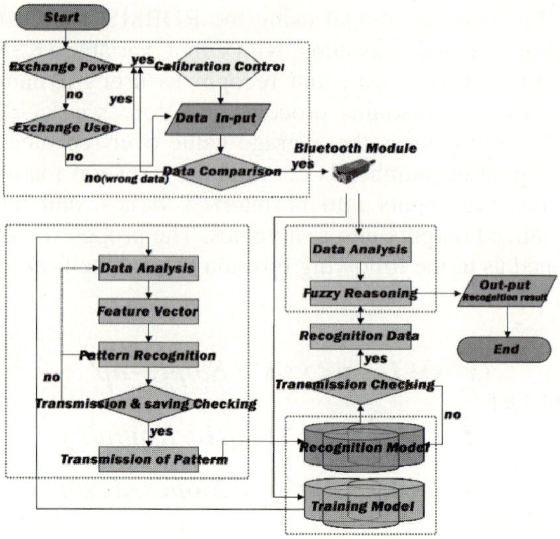

Fig. 5. The flow-chart of the fuzzy sign language recognition system

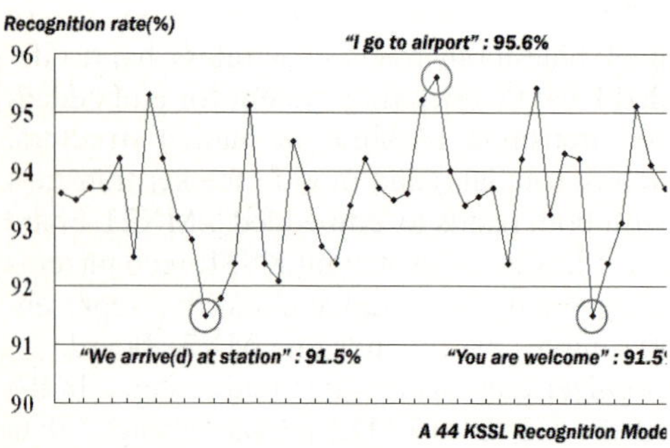

Fig. 6. The average recognition rate

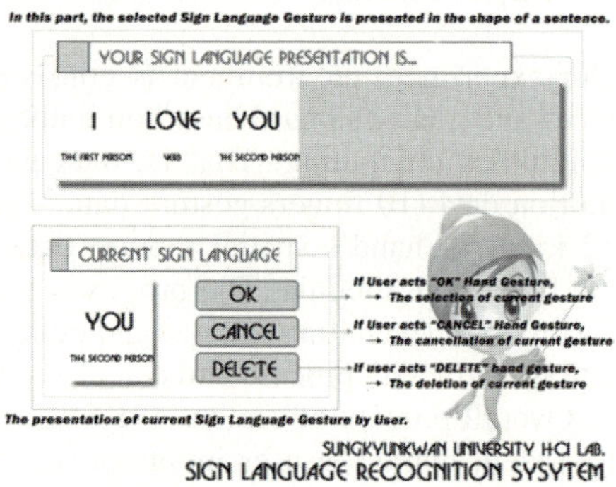

Fig. 7. The GUI for visual representation

6 Conclusions

Wearable computing is an active topic of study, with areas of study including user interface design, augmented reality, pattern recognition, using of wearable for specific applications or disabilities. Depending on the application, the primary input to a wearable might be a chording keyboard, gesture, speech recognition or even just passive sensors (context awareness). Output might be presented via speech, audio tones, a head-mounted display or haptic output.

As preliminary study for recognition and representation of KSSL, our researchers implemented hand gesture recognition system that recognize 19 hand gestures according to shape and stoop degree of hand. Accordingly, with this preliminary study in this paper, we implemented real time-sentential KSSL recognition system that analyzes user's intention between hearing-impaired and hearing person more efficiently and recognizes and represents 44 continuous KSSL sentences with flexibility more accurately based on the post wearable PC platform. Also, we clarify that this study is fundamental study for implementation of multi modal recognition system to take the place of traditional uni-modal recognition system in sign language recognition.

Furthermore, we would like to develop advanced multi-modal HCI technology through integrating of sign language recognition system and other haptics such as smell, taste, hearing and sight.

Acknowledgement

This research was supported by the MIC (Ministry of Information and Communication), Korea, under the ITRC (Information Technology Research Center) support program supervised by the IITA (Institute of Information Technology Assessment)" (IITA-2005- C1090-0501-0019).

References

1. Wu jiangqin et al.: A Simple Sign Language Recognition System Based on Data Glove. ICSP98. IEEE International Conference Proceedings (1998) 1257-1260
2. H.-Y.Jang. et al.: A Study on Hand-Signal Recognition System in 3-Dimensional Space. Journal of IEEK, Vol. 2004-41CI-3-11. IEEK (2004) 103-114
3. S.-G.Kim.: Standardization of Signed Korean. Journal of KSSE, Vol. 9. KSSE (1992)
4. S.-G.Kim.: Korean Standard Sign Language Tutor. 1st edn., Osung Publishing Company, Seoul (2000)
5. Y SU et al.: Three-Dimensional Motion System ("Data-Gloves"): Application for Parkinson's Disease and Essential Tremor. Virtual and Intelligent Measurement Systems. IEEE International Conference Proceedings (2001) 28-33
6. J.-H.Kim. et al.: Hand Gesture Recognition System using Fuzzy Algorithm and RDBMS for Post PC. FSKD2005. Lecture Notes in Artificial Intelligence, Vol. 3614. Springer-Verlag, Berlin Heidelberg New York (2005) 170-175
7. 5DT Data Glove 5 Manual and FASTRAK® Data Sheet.: http://www.5dt.com
8. Relational DataBase Management System.: http://www.auditmypc.com/acronym/RDBMS.asp
9. Oracle 10g DW Guide.: http://www.oracle.com
10. C.H.Chen, Fuzzy Logic and Neural Network Handbook. 1st edn., McGraw-Hill, New York (1992)

Patient Modeling Using Mind Mapping Representation as a part of Nursing Care Plan*

Hye-Young Ahn[1], Eunja Yeon[2], Eunmi Ham[2], and Woojin Paik[3,**]

[1] Dept. of Nursing, Eulji University,
143-5 Yongdoo-dong, Jung-gu, Daejeon, 301-832, Korea
ahanaya@eulji.ac.kr
[2] Dept. of Nursing Science
[3] Dept of Computer Science, Konkuk University,
322 Danwol-Dong, Chungju-Si, Chungcheongbuk-Do, 380-701, Korea
{eunice, hem2003, wjpaik}@kku.ac.kr

Abstract. Nursing care plan reports are one of the most important documents in the application of nursing processes. In this paper, we describe how a text discourse analysis and an information extraction system can be used to convert a traditional nursing care plan into a mind mapping representation. Mind mapping is a process to allow the nurses to focus on the patients rather than on a disease process. Mind mapping encourages the nurses to maintain a holistic view of the patient. A mind mapping representation refers to a visual picture of a patient at the center with various nursing care related information visually linked to the patient's form. Our goal is to develop visually browsable models of the patients to aid in the nursing process education and also help the nurses focus on the patients in the actual care settings.

1 Motivation: Information Extraction from Nursing Care Plan

According to North American Nursing Diagnosis Association (NANDA), a nursing diagnosis is defined as a critical judgment about individual, family, or community responses to actual or potential health problems or life processes. The goal of a nursing diagnosis is to identify health problems of the patient and his/her family. A diagnosis provides a direction for the following nursing care [1]. The nursing care plan as the repositories of knowledge involved in the overall nursing process, which leads up to the nursing diagnosis [2].

A complementary approach to the traditional tabular and narrative based nursing care plan is referred as mind mapping [3]. A mind map is designed to move away from the linear nature of the traditional nursing plan. A mind map is a graphical representation of the connection between concepts and ides that are related to a patient. From the patient representation placed in the middle, associated information, such as nursing diagnosis, expected outcome, suggested interventions, observation summaries, and the evaluation by the nurses, are grouped the linked to the patient. A mind

* This paper was supported by Konkuk University in 2005.
** Corresponding author.

V.N. Alexandrov et al. (Eds.): ICCS 2006, Part IV, LNCS 3994, pp. 894 – 901, 2006.
© Springer-Verlag Berlin Heidelberg 2006

map is for the nurses to develop a whole patient picture that is comprised of a variety of information, which are related to the identified patient problems. A mind map is used to reflect how nurses in practices truly think.

Our text preprocessing program first identifies and extracts the nursing diagnoses, outcomes, and interventions, which are usually in a table. This process is fairly straightforward as each component is placed under clear descriptive heading. However, extracting observed information summaries about the patients and the evaluations by the nurses is difficult. The main focus of the research in this paper is about the extracting the summary and evaluation information in the narrative portion of the nursing care plans. Our text discourse analysis program classifies each clause in the narrative portions of the care plan according to eight categories. When the classification is completed, the semantic relation extraction program identifies the summary and evaluation pairs, which are linked by a causal relation. Then, the causally linked summary and evaluation pairs will become a part of the mind map visual representation. The mind map will be used as a teaching aid for the novice nurses to learn how to evaluate and represent the patient conditions.

2 Classifying Clauses in Nursing Care Plan Narratives According to a Nursing Process Text Discourse Model

A text discourse model specifies the necessary classes of knowledge to be identified in order to develop the skeletal conceptual structure for a class of entities. Based on a discourse linguistics observation [4], writers are often influenced by the schema of a particular text type if they produce texts of that type repeatedly. This means that the writers consider both specific content they wish to convey and usual structure for that type of text on the basis of the purpose it is intended to serve.

The existence of such predictable structures in texts is consistent with findings in cognitive psychology which suggest that human cognitive processes are facilitated by the ability to 'chunk' the vast amount of information encountered in daily life into larger units of organized data [5]. Based on schema theories, humans recode individual units of perception into increasingly larger units, which will eventually reach at the level of a schema. It has also been argued that humans possess schema for a wide range of concepts, events, and situations [6]. In discourse linguistics, this schema theory was extended to suggest that schema exist for text-types that participate regularly in the shared communication of a particular community of users.

As the text structure of a particular text type is discovered, the text's discernible and predictable superstructure is also revealed. Superstructure is defined as the text-level syntactic organization of semantic content. It can be also referred to as the global schematic structure or the recognizable template that is filled with different meaning in each particular example of that text type [7]. Some of the text types for which schemas or models have been developed with varying degree of details are: newspaper articles [7, 11], arguments [8], editorials [9], abstracts [10], and legal texts [12].

To develop a text schema for the narratives in the nursing care plan, we analyzed randomly selected fifteen nursing care plans as a training data set. The care plans were a part of a course assignment for the 'Psychiatric Nursing' course, which was offered at the Department of Nursing Science, Konkuk University in Chungju, Korea. The course was for Juniors who were majoring in the nursing science. Each student developed a detailed case study report of one patient while the student was working as a student intern at a psychiatric warden for four weeks. The nursing care plan was one section of the case report, which was submitted at the end of the internship period. The case reports were from the course offered in the Fall 2004 semester. All case reports were mainly written in Korean with English translations for a number of important concepts. We also used another set of fifteen randomly selected nursing care plans as a testing data set.

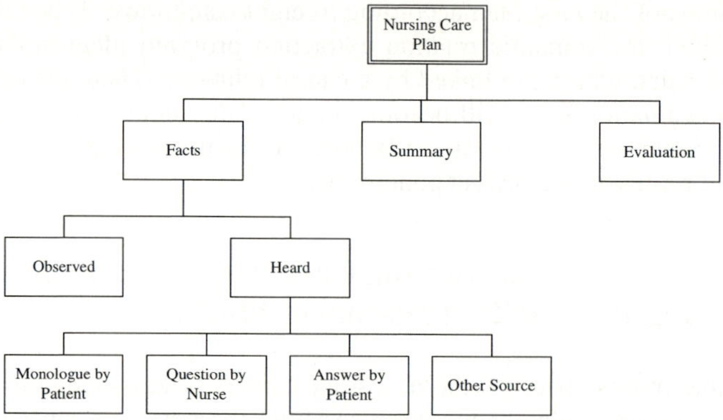

Fig. 1. Nursing Care Plan Text Schema

The nursing care plan text schema is shown in the Figure 1. It is based on the qualitative content analysis of the training data set. At the most general level, there are three major categories. They are 'Facts', 'Summary', and 'Evaluation'. 'Facts' refer to the factual information about the patients. 'Facts' major category is further divided into 'Observed' and 'Heard' categories according to how the information was collected. 'Observed' is for the information directly observed or measured by the nurses. 'Heard' is what the nurses heard about the patient. 'Heard' category is further divided into four sub-categories. Four sub-categories are: 'Monologue by Patient', 'Question by Nurse', 'Answer by Patient', and 'Other Sources'. 'Monologue by Patient' is for the information, which is based on what the patient said about him/herself without external inquiry. 'Question by Nurse' is somewhat problematic sub-category as it does not independently convey information about the patient. Nurses ask questions either to learn a particular aspect about the patient's condition or to encourage the patient to continue talk about him/herself. Thus, most of the clauses classified under this category do not convey substantive information. 'Answer by Patient' sub-category is for the information provided by the patient in response to a question by the attending nurses. Finally, 'Other Sources' sub-category refers to the patient information provided by the family members, friends, other nurses, or the physicians.

We decided to code each clause with the discourse categories at the most specific level. The following shows two paragraphs in two different nursing care plans. These plans were used as the training data set. Although, all qualitative content analysis and computational processing is directly applied to the original Korean text, the example paragraphs are manually translated into English to help non-Korean to understand the meaning of the examples. '<*category name*>' and '</category name>' are used to show the beginning and end of a clause, which is coded as a particular discourse category.

Example 1. The example is from a paragraph explaining the insight of a patient HS in the mental state assessment section.

<monologueByPatient> HS said "I hope my nervousness goes away soon'. </monologueByPatient> <evaluation> Based on the statement, HS seems to understand about his condition well. </evaluation> <otherSource> As HS voluntarily admitted to the hospital last time, </otherSource> <evaluation> it is believed that HS is positively thinking about the treatments that HS is receiving. </evaluation>

Example 2. The example is from a paragraph describing the life when the patient JK was in teens in the developmental history part.

<questionByNurse> Nurse asked 'Have you ever had a girl friend?' </questionByNurse> <answerByPatient> JK said 'No, I never had a girl friend.' </answerByPatient> <questionByNurse> Nurse asked 'Do you have many friends?' </questionByNurse> <answerByPatient> JK said 'No, I do not have many friends.' </answerByPatient> <summary> As the patient stated that he never had a relationship with a female and also did not have many male friends, </summary> <evaluation> it is likely that JK have been having a difficult interpersonal relationship since he was a teenager. </evaluation>

2.1 Extracting Text Classification Features

While coding the training data, we developed both defining features and properties for each category. The defining features convey the role and purpose of that category within the nursing care plan text schema while the properties provide suggestive clues for the recognition of that category. The manual coding suggested to us that we were relying on five types of linguistic information during our coding. The data, which would provide these evidence sources, were then analyzed statistically and translated into computationally recognizable text characteristics. The five sources of evidences are described in the following.

Lexical Evidences: This source of evidence is a set of one, two, three word phrases for each component. The set of lexical evidences for each component was chosen based on observed frequencies and distributions. Only the words or phrases with sufficient occurrences and statistically skewed observed frequency of occurrences in a particular component were used.

After all coded training data are processed by the lexical evidence extraction module, the frequency distribution of each piece of lexical evidence is further processed to generate the probability information. For example, '있는 듯 하다 (English translation: it seems to be)' occurred three times in the clause coded as 'Facts', six times in the 'Summary' clauses, and 15 times in 'Evaluation' in the training data set. This tells us that the clause should be coded as 'Facts' three out of 24 times if the sentence includes '있는 듯 하다' as an example of three word lexical evidences. In the same manner, '있는 듯 하다' is indicated six out of 24 times as 'Summary', and 15 out of 24 times as 'Evaluation'. After all one, two, and three-word lexical evidence based on probability calculation is completed, a series of normalization processes is applied to weed out unreliable lexical evidences and also to remove other influences such as likelihood of component occurring. As each clause is processed, the lexical evidences for the words and phrases in the clause are combined using the Dempster-Shafer Theory of Evidence [13]. At the end of all processing, each clause is assigned with four numbers ranging from zero to one. One number represents the clause's probability of having been correctly coded as 'Facts'. Other numbers are the probability figures for 'Summary', and 'Evaluation'.

Syntactic Evidences: We utilize two types of syntactic evidences: 1) typical sentence length as measured in the average number of words per clause for each category and 2) individual part-of-speech distribution based on the output of the part-of-speech tagging. This evidence helps to recognize those categories, which tend to have a disproportionate number of their words be of a particular part of speech. For example, 'Observed' category clauses tend to have comparatively small number of adjectives.

Tense Evidences: Some components tend to contain verbs of a particular tense more than verbs of other tenses. For example, 'Fact' clauses are almost always in the past or present perfect tense. The tense evidence is a byproduct of part-of-speech tagging.

Document Structure Evidences: There are two types of document structure evidences. Firstly, nursing care plans have section headings, which are closely aligned with three major categories. For example, 'Health related Information' section rarely had clauses, which are categorizes as 'Summary' or 'Evaluation'. Thus, we utilized the heading information as another evidence source. Secondly, we included the relative position of each clause with respect to the source paragraph as another evidence source.

Order of Component Evidences: This source of evidence replies on the tendency of categories to occur in a particular, relative order. We calculated the frequency with which each category followed every other category and the frequency with which each category preceded every other category. The results are stored in two seven-by-seven matrices.

2.2 Text Classification

To assign a basic category label to each clause, each clause in the training data set is categorized according to the predetermined nursing care plan text schema. The first text classification task involves manually coding all clauses in a set of training documents in preparation for feeding the automatic system. Each clause is classified as "in" or "out" of the individual categories as outlined by the category definitions. The next step is to take these manually classified clauses and process them through the trainable text classification system. During this process, it builds a vector of lexical evidences, syntactic evidences, tense evidences, and document structure evidences. Multi-level Natural Language Processing outputs are the basis for these textual data feature representations.

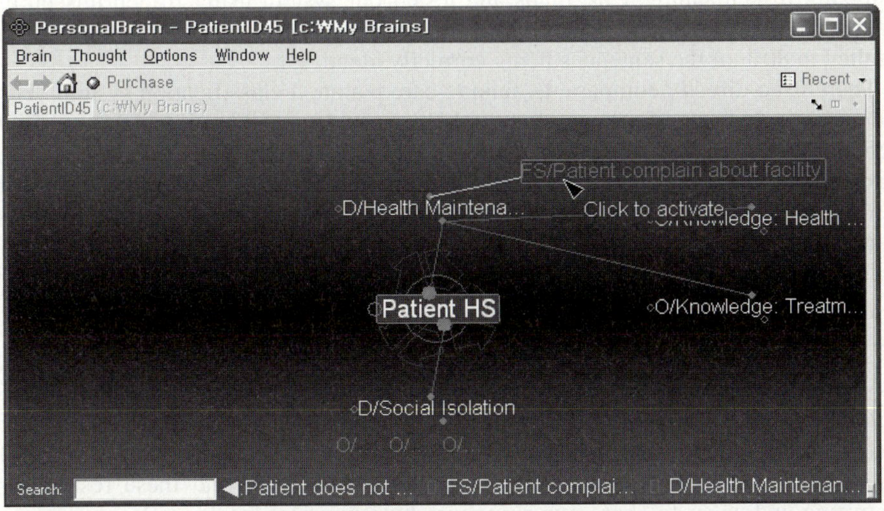

Fig. 2. Mind Map for Patient HS

3 Making a Mind Map Visual Representation

Our goal is to make a mind map of a patient along with all related factual and evaluative information about the patient. The Figure 2 shows a mind map representation about a patient, who is referred as HS.

For this research, we used PersonalBrain, a commercial off the shelf software from The Brain Technologies Corp (http://www.thebrain.com/). PersonalBrain is an information visualization system, which enables the users to link information into a network of logical associations. PersonalBrain uses 'Thoughts' to refer 'concepts' and 'Links' to refer 'relations'. The Figure 2 shows 'Health Maintenance, Altered' and 'Social Isolation' as the nursing diagnoses for the patient HS. Each nursing diagnosis is preceded by 'D'. Nursing Outcome is preceded by 'O' and Intervention is preceded by 'I'. PersonalBrain truncates the labels of each node in the graph if the label is too long. The full label is revealed when the user moves the cursor on top of the node. The user can make any node a central node by double clicking on it. The mind map represented in PersonalBrain is dynamically repositioned to minimize the overloading of the visual representation in small screen space. At the top of the mind map, there is a factual information summary observed by the nurse. "Patient complain about facility' is preceded by 'FS', which stands for 'factual summary'. 'E' is for the evaluative statement by the nurses.

4 Implementation and Evaluation

The computational modeling of instantiating a discourse-level model of the nursing care plan texts and converting the traditional nursing care plan to the mind map visual representation is an ongoing effort. We developed a prototype system by manually analyzing fifteen sample care plans and tested our system using fifteen unseen care plans. The first run and evaluation of the correctly categorizing four basic categories resulted in 80% of the sentences being correctly identified.

There is no directly comparable nursing care plan text classification system. However, a news text classification system, which assigned sentences into one out of fourteen categories, performed at 72% correct rate in the fully automatic mode and 80% correct rate with various manual heuristic adjustments [11]. It should be noted that our text classifier did not utilize the second iteration of incorporating the sentences, with certainty membership value close to zero, as a part of new training data set. We believe the addition of this process will improve the correctness of our system.

5 Summary

Although we are clearly in the early stages of developing a patient modeling system based on the mind map representation scheme through the use of nursing care plan discourse modeling and information extraction system, we find these results quite promising and eager to share our premature but empirical results and experiences in creating an operational system with other researchers.

We expect the resulting mind map to aid both nursing students and practitioners finding the nursing process examples of making proper nursing diagnosis. They can review what others have done to learn from the sound decisions and the mistakes.

There are many tasks that we have yet to finish. The major missing task is the usability evaluation of the resulting mind map. We also need to increase the size of test data set to improve the reliability of the evaluation results.

We have applied the nursing care plan discourse model to the actual care plans by coding a small set of sample texts. This effort in conjunction other automatic discourse modeling works, we argue that it is possible to extract a particular section of the texts by utilizing a text type specific discourse model.

References

1. Sparks. S.M. and Taylor, C.M., Nursing Diagnosis Reference Manual 5th Edition, Springhouse, Springhouse, Pennsylvania, 2000.
2. Doenges, M., and Moorehead, M.F., Application of Nursing Process and Nursing Diagnosis: An Interactive Text for Diagnostic Reasoning 4th Edition, F.A. Davis Co., Philadelphia, Pennsylvania, 2003.
3. Mueller, A., Johnston, M. & Bligh, D.: Joining mind mapping and care planning to enhance student critical thinking and achieve holistic nursing care. Nursing Diagnosis 13(1):24-27 (2002).
4. Jones, L.B.: Pragmatic aspects of English text structure. Summer Institute of Linguistics, Arlington, TX (1983).

5. Rumelhart, D.: Understanding and summarizing brief stories. In D. LaBerge and S.J. Samuels (Editors) Basic processes in reading: Perception and comprehension. Hillsdale, NJ: Lawrence Earlbaum Associates: 265-303. (1977)
6. Rumelhart, D.: Schemata: the building blocks of cognition. In R. Spiro, B. Bruce, & W. Brewer (Editors) Theoretical issues in reading comprehension: Perspectives from cognitive psychology, linguistics, artificial intelligence and education. Hillsdale, NJ: Lawrence Earlbaum Associates: 33-8. (1980)
7. van Dijk, T.A.: News analysis: Case studies of international and national news in the press. Hillsdale, NJ: Lawrence Earlbaum Associates. (1988)
8. Cohen, R.: Analyzing the structure of argumentative discourse. Computational Linguistics, 13. 11-24. (1987)
9. Alvarado, S.J.: Understanding editorial text: A computer model of argument comprehension. Boston, MA: Kluwer Academic Publishers. (1990)
10. Liddy, E.D.: The discourse-level structure of empirical abstracts: An exploratory study. Information Processing and Management 27.1, Tarry Town, NY, Pergamon Press: 55-81. (1991)
11. Liddy, E.D., McVearry, K.A., Paik, W., Yu, E., and McKenna, M.: Development, Implementation and Testing of a Discourse Model for Newspaper Texts. Proceedings of a Human Language Technology Workshop. Plainsboro, NJ, Morgan Kaufmann Publishers: 159-164. (1993)
12. Paik, W. and Lee, J.: Extracting Legal Propositions from Appellate Decisions with Text Discourse Analysis Methods. Lecture Notes in Computer Science (LNCS) Vol. 3292, Springer-Verlag: 621-633. (2004)
13. Pearl, J.: Probabilistic reasoning in intelligent systems: networks of plausible inference, Morgan Kaufmann Publishers, San Francisco, CA. (1988)

A Technique for Code Generation of USN Applications Based on Nano-Qplus*

Kwangyong Lee[1], Woojin Lee[2], Juil Kim[2], and Kiwon Chong[2]

[1] Ubiquitous Computing Middleware Team, ETRI, Daejeon, Korea
kylee@etri.re.kr
[2] Department of Computing, Soongsil University, Seoul, Korea
{bluewj, sespop}@empal.com, chong@ssu.ac.kr

Abstract. A technique for automatic code generation of USN applications based on Nano-Qplus is proposed in this paper. Nano-Qplus is a sensor network platform developed by ETRI. Programs of nodes such as sensors, routers, sinks and actuators in a sensor network are automatically generated through the technique. Developers can implement USN applications from models of sensor networks. The execution code is generated by setting attribute values of each node according to the model through the script proposed in this paper. Through the technique of this paper, developers can easily implement USN applications even if they do not know the details of low-level information. The development effort of USN applications also will be decreased because execution codes are automatically generated. Furthermore, developers can perform early test through rapid code generation, so the verified code is generated by correcting errors in the early development stage.

1 Introduction

Ubiquitous sensor network (USN) is a wireless network which consists of a lot of lightweight, low-powered sensors. A lot of sensors which are connected to a network sense geographical and environmental changes of the field. They transmit the sensing data to a base station and the data is transmitted to users through sensor network server. Collection of information in USN is performed through this process. Through USN, things can recognize other things and sense environmental changes, so users can get the information from the things and use the information anytime, anywhere. The sensor networks can be used for various application areas such as home, health, and robot.

However, it is difficult to construct USN applications. Resources of nodes in a sensor network are limited and wireless communication between nodes is unreliable. Nodes should also perform low-power operations. Developers should consider these facts, so it is very difficult to construct USN applications. Therefore, it is need to make developers can simply design USN applications by abstracting the details of low-level communication, data sharing, and collective operations.

Accordingly, a technique for automatic code generation from a simple design of USN application is proposed in this paper. Programs of nodes such as sensors,

* This work was supported by the Soongsil University Research Fund.

V.N. Alexandrov et al. (Eds.): ICCS 2006, Part IV, LNCS 3994, pp. 902–909, 2006.
© Springer-Verlag Berlin Heidelberg 2006

routers, sinks and actuators in a sensor network are automatically generated by setting attribute values of a script proposed in this paper. Therefore, developers can easily develop USN applications even if they do not know the details of low-level communication, data sharing, and collective operations. The technique of this paper brings focus to USN application on a sensor network platform known as Nano-Qplus [1, 2].

2 A Technique for Code Generation of USN Applications

A programming model to construct USN applications based on Nano-Qplus is presented in this section. It is compared to the existing programming models for USN applications. Moreover, the script for the design of an application and the algorithm for automatic code generation of the application are presented.

2.1 Concepts of the USN Programming

Figure 1 presents the concept of USN programming described in existing works [3, 4, 5, 6, 7, 8]. A modeling is done and a simple program based on the model is written using the high level language or the simple script. Then the code is automatically generated according to the program. It is important that the program is written using the high level language or the script. The high level language or the script helps users to construct applications, even though they do not know the details of low-level information of USN. A specification-level language, a script language, or APIs were proposed in order to abstract the low-level information in the related works. However, users should learn the proposed language, the script language or APIs in order to develop USN applications using these techniques.

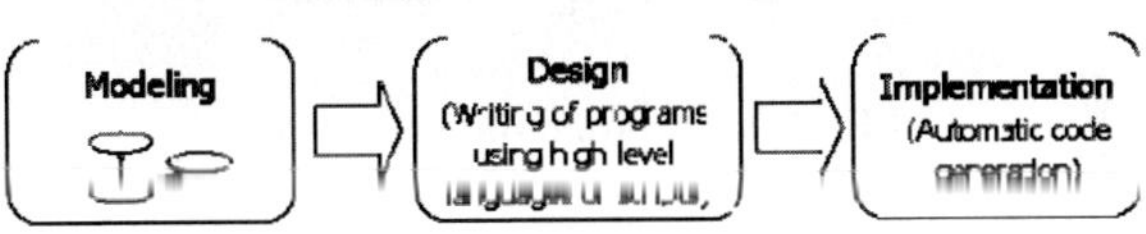

Fig. 1. The concept of USN programming in the existing works

A technique to complement the existing techniques for construction of USN applications is proposed in this paper.

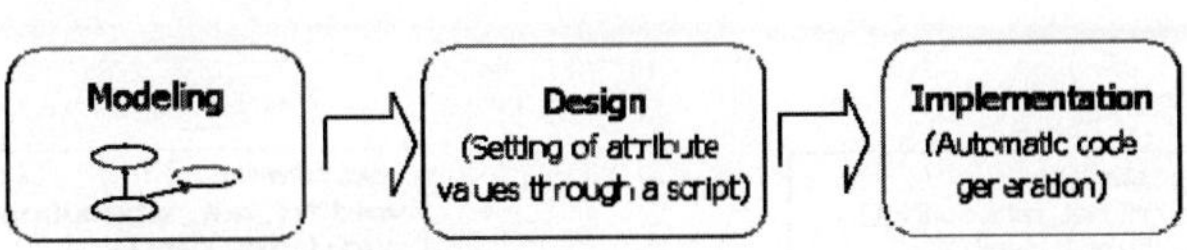

Fig. 2. The concept of USN programming in this paper

Figure 2 presents the concept of USN programming proposed in this paper. Developers can implement USN applications by automatically generating execution code of each node in the sensor networks after they do modeling the sensor networks and set attribute values of each node according to the model. The execution code is automatically generated by setting attribute values of each node through a script. Therefore, users can construct USN applications without learning a language or APIs.

2.2 The USN Programming Model

Figure 3 shows the USN programming model proposed in this paper. USN model is designed, and attribute values for sensor nodes, router nodes, sink nodes, and actuator nodes in the model are set through scripts. Modules and code templates which are provided by Nano-Qplus are selected according to the attribute values of scripts, so C codes for each node are automatically generated.

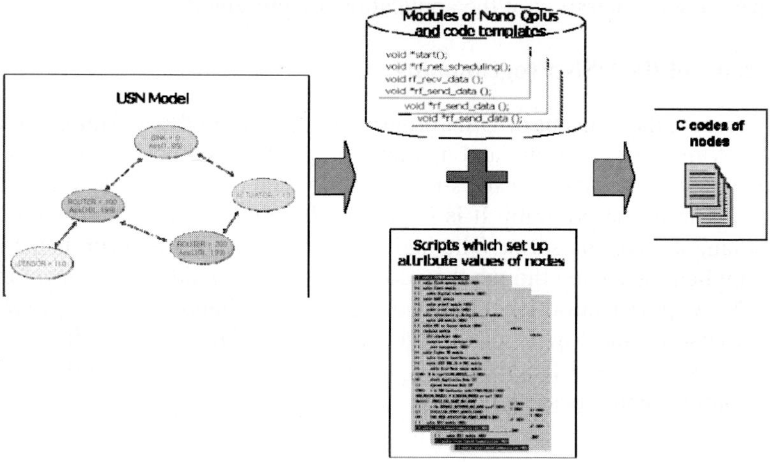

Fig. 3. USN programming model

Nodes in a USN communicate with each other through Store&Forward method. Templates to generate programs of nodes according to that method are presented in this paper. Figure 4, 5, 6 and 7 show templates which contain modules provided by Nano-Qplus in order to generate programs for sensor nodes, router nodes, sink nodes, and actuator nodes.

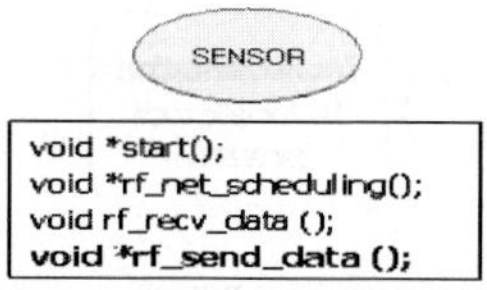

Fig. 4. Modules of a sensor node

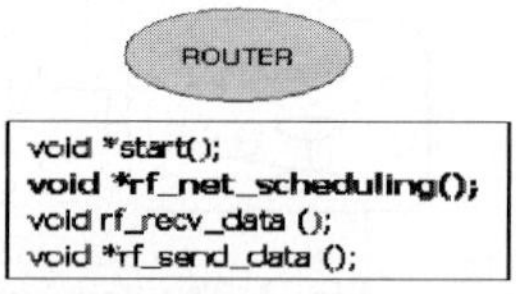

Fig. 5. Modules of a router node

Four types of nodes contain the same modules generally, but the contents of the modules are dependent upon the node type. For example, main role of a sensor node is to send sensing data to other nodes, so the contents of rf_send_data() module are dependent upon the script setting for the sensor node. Main role of a router node is to send data received from a node to other nodes, so the contents of rf_net_scheduling() module are dependent upon the script setting for the router node. Moreover, main role of a sink node and an actuator node is to process data, so the contents of rf_recv_data() module are depedent upon the script setting for the sink node and the actuator node.

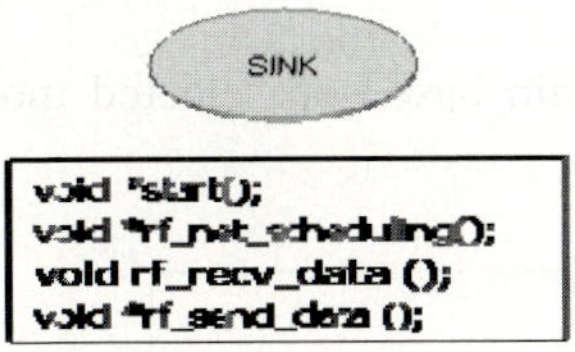

Fig. 6. Modules of a sink node

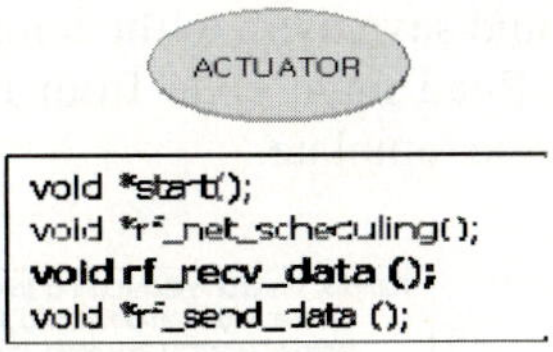

Fig. 7. Modules of a actuator node

2.3 A Script for Generation of USN Applications

To generate the application from the designed model, the script which can easily configure the attributes of each node in USN is proposed. Figure 8 shows the script.

The script has been written as a typical script language [9] which is used for configuring environment in Linux, so most of users can easily use it. If users configure the attributes of each node using the script, a program code for each node is automatically generated based on the script.

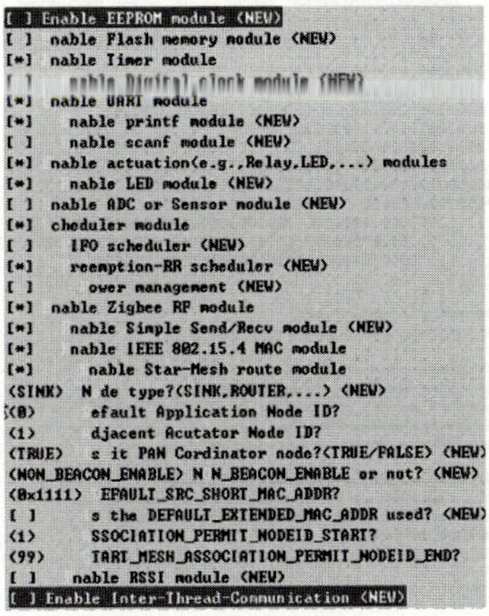

Fig. 8. A script for automatic code generation

The script showed in figure 8 has been written for generating the application on a sensor network platform known as Nano-Qplus.

2.4 An Algorithm for Generation of USN Applications

The following is the process for generating source code to control each node based on the script.

Step 1 – Read Config_Info(.config) file in order to get the attribute values of a node.
Step 2 – Parse Config_info(.config) file and find out selected modules. Then read headers, data and function codes from HashTable_Module according to the selected modules and save them to the template.
Step 3 – Read main code from HashTable_Main based on selected modules and save it to the template.

```
Templet Transformation(Config_Info config_info) {
    templet = getTemplet(config_Info.NODETYPE);
    templet.setAttribute(config_Info.Attribute);

    HashTable_Module hashTable_Module = new HashTable_Module();
    HashTable_Main     hashTable_Main = new HashTable_Main();

//  Set the NODETYPE of HashTable_Module
    hashTable_Module.setType(config_Info.NODETYPE);
// Construct code according to the config_Info of each node
    Iterator iterator = Parser.getIterator(config_Info);
    while( iterator.hasNext() ) {
            templet.addHeader( hashTable_Module.getHeader( iterator.ModuleName ) );
            templet.addData( hashTable_Module.getData( iterator.ModuleName ) );
            templet.addFunction( hashTable_Module.getFunction( iterator.ModuleName ) );
            templet.addMain ( hashTable_Main.getMain( iterator.ModuleName ) );
            iterator.next();
    }
```

Fig. 9. An algorithm for generating USN application

```
public class HashTable_Module {
        private Hashtable moduleTable;
        private String[] moduleKey = {"Zigbee_Simple", "Zigbee_MAC","Zigbee_MAC_StarMesh", "Scheduler_FIFO",
                        "Scheduler_PreemptionRR", "Sensor_LIGHT","Sensor_GAS","Sensor_Temperature"};
        String nodeType = "";
        private final String HEADER = "H";
        private final String DATA = "D";
        private final String FUNCTION = "F";
        public void setType(String nodeType) {
                TempletHashTable templetTable = new TempletHashTable();
                moduleTable = new Hashtable();
                int size = moduleKey.length;
                String templetModule = "";
                this.nodeType = nodeType;
                for(int i = 0; i < size; i++) {
                        templetModule = (String)templetTable.get(moduleKey[i]);
                        templetModule = templetModule.replaceFirst("&1", nodeType);
                        moduleTable.put(moduleKey[i], templetModule);
                }
        }
        public String getHeader(String moduleName) {
                return getReplaceModule(moduleName, HEADER);
        }
        public String getData(String moduleName) {
                return getReplaceModule(moduleName, DATA);
        }
        public String getFunction(String moduleName) {
                return getReplaceModule(moduleName, FUNCTION);
        }
        public String getReplaceModule(String moduleName, String value) {
                String templetModule = (String)moduleTable.get(moduleName);
                templetModule = templetModule.replaceFirst("&2", value);
                return templetModule;
        }
}
```

Fig. 10. HashTable_Module class

Figure 9 presents the algorithm for generating source code of each node. Headers, data and function codes are generated by calling the functions of HastTable_Module class according to the type of the target node.

The HashTable_Module class used in the algorithm is presented in figure 10. The class includes getHeader(), getData() and getFunction() to generate the program for each node. These functions use the key and the value corresponding to the key defined in a hash table through the getReplaceModule() function in order to get source codes. HashTable_Module class generates the proper source code using a hash table dynamically because the codes of headers, data and functions are dependent upon the type of the node.

Table 1. Structure of hash table

Key	Value
Zigbee_Simple	"&1_Zig_Simple_&2"
Zigbee_MAC	"&1_Zig_MAC_&2"
Zigbee_MAC_StarMesh	"&1_Zig_StarMesh_&2"
Scheduler_FIFO	"&1_Sche_FIFO_&2"
Scheduler_PreemptionRR	"&1_Sche_PreemptionRR_&2"
Sensor_LIGHT	"&1_Sensor_LIGHT_&2"
Sensor_GAS	"&1_Sensor_GAS_&2"
Sensor_Temperature	"&1_Sensor_Temperature_&2"

It is necessary to define many hash tables according to the number of node type if a static hash table which has the keys and values for one node type is used. Moreover, the codes to control the hash tables must be written additionally according to the number of hash tables. It is necessary to define a new hash table and to write the code to control the hash table if a new type of the node is added.

However, there is no need to define a new hash table when a new type of the node is added if a dynamic hash table is used. No additional codes are needed such as control codes for hash tables. The structure of dynamic hash table used in the HashTable_Module class is presented in table 1. The key of the hash table is the name of a module provided by Nano-Qplus. Strings such as "&1" and "&2" in the key value are dynamically replaced according to the type of module and node. When the type of each node is determined, the string "&1" is replaced with the type, and "&2" is replaced with "H" (means Header), "D" (means Data) or "F" (means Function) based on the type of required module. Following is an example.

i.e.) If the Zigbee_Simple module for radio frequency communication of a node is selected, "&1_Zig_ Simple_&2" value is selected from the hash table. Then, the value is replaced as follows by calling functions of HashTable_Module class.
setType("SINK"); → "SINK_Zig_Simple_&2"
getHeader("Zig_Simple") → "SINK_Zig_Simple_H"
getFunction("Zig_Simple") → "SINK_Zig_Simple_F"
"SINK_Zig_Simple_H" is the name of a file which contains header codes of the ZigBee_Simple module for a sink node, and "Sink_Zig_Simple_F" is the name of a file which contains function codes of the ZigBee_Simple module for a sink node.

3 Case Study with Gas Monitoring System

An USN application for Gas Monitoring System such as figure 11 has been developed using the proposed technique in this paper.

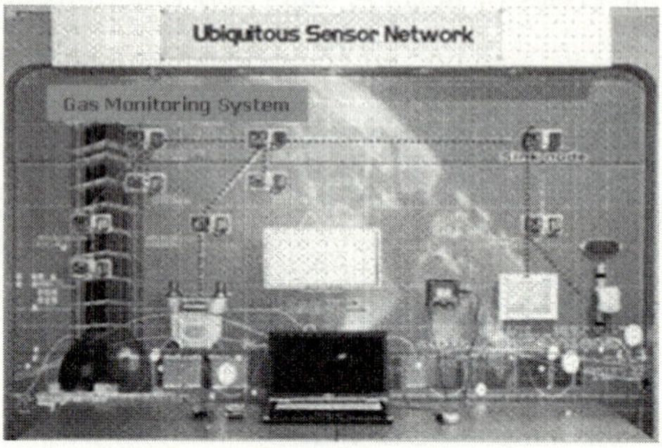

Fig. 11. Gas Monitoring System

A USN model was designed for the Gas Monitoring System. In the model, sensor nodes sense gas data and transmit the data to router nodes. The router nodes receive the data and transmit it to the sink node. The sink node receives the data, computes it and transmits it to the actuator node. The actuator node performs an action according to the threshold value.

The system of figure 11 was developed after the application was automatically generated using a script based on designed model. Result that applies, sensing data was forwarded from sensor node to router node and router node sent forwarded data to sink node. An action command according to a gas value was forwarded from sink node to router node and router node sent forwarded the action command to actuator.

The figure 12 is an example to set the attribute values of the script in order to generate automatically the program of the sink node. The USN application was automatically generated by setting the attribute values of each node through the script.

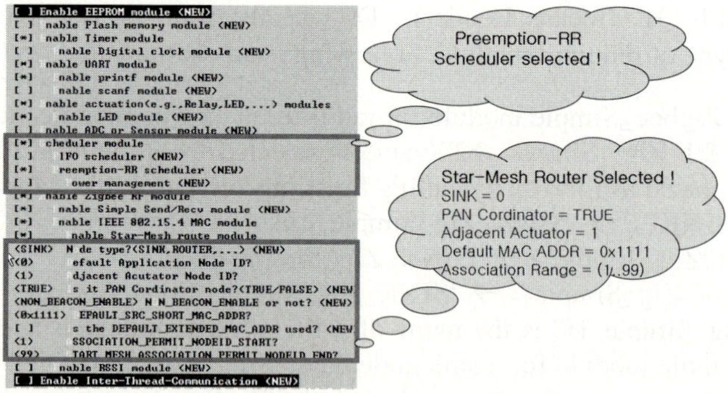

Fig. 12. An example of attribute values setting using the script

4 Conclusion

The technique for automatic code generation of USN applications based on Nano-Qplus is proposed in this paper. Developers can implement USN applications by automatic generation of execution code of each node in the sensor networks after they make models of the sensor networks and set attribute values of each node according to the model using the script. The script for automatic code generation of each node is proposed in this paper. The templates and an algorithm for automatic code generation are also presented. Through the technique of this paper, developers will easily implement USN applications even if they do not know the details of low-level communication, data sharing, and collective operations. The development effort of USN applications also will be decreased because execution codes are automatically generated. Furthermore, developers can perform early test through rapid code generation, so the verified code is generated by correcting errors in the early development stage.

References

1. Kwangyong Lee et al., "A Design of Sensor Network System based on Scalable & Reconfigurable Nano-OS Platform," IT-SoC2004, October 2004.
2. ETRI Embedded S/W Research Division, "Nano-Qplus," http://qplus.or.kr/
3. E. Cheong, J. Liebman, J. Liu, and F. Zhao, "Tinygals: a programming model for event-driven embedded systems," SAC, 2003.
4. M. Welsh and G. Mainland, "Programming sensor networks using abstract regions," NSDI, 2004.
5. R. Newton and M. Welsh, "Region streams: Functional macroprogramming for sensor networks," DMSN, 2004.
6. A. Boulis, C. Han, and M. B. Srivastava, "Design and implementation of a framework for efficient and programmable sensor networks," MobiSys, 2003.
7. B. Greenstein, E. Kohler, and D. Estrin, "A sensor network application construction kit (SNACK)," SenSys, 2004.
8. Ramakrishna Gummadi, Omprakash Gnawali, and Ramesh Govindan, "Macro-programming Wireless Sensor Networks Using Kairos," LNCS 3560, pp. 126–140, 2005.
9. Neil Matthew, Richard Stones, "Beginning Linux Programming 3rd Edition," WROX PRESS, 2003.

A Study on the Indoor Real-Time Tracking System to Reduce the Interference Problem*

Hyung Su Lee[1,2], Byunghun Song[2], and Hee Yong Youn[1,**]

[1] School of Information and Communication Engineering, Sungkyunkwan University, 300 Chunchun Jangan Suwon, Kyunggido 440-746 South Korea
hslee@keti.re.kr, youn@ece.skku.ac.kr
[2] Korea Electronics Technology Institute, 68 Yatap Dong Pundang Sungnam, Kyunggido, 463-816, South Korea
bhsong@keti.re.kr

Abstract. The real-time tracking system is an essential component in the development of low cost sensor networks to be used in pervasive and ubiquitous computing. In this paper we address the interference problem of the sensor network platform that uses ultrasonic for location tracking. We also present a novel scheme reducing the error rate caused by interference, which is particularly suited for supporting context-aware computing. It is achieved by considering the speed variance of the mobile node and thereby correcting the interference errors. Performance evaluation using an actually implemented platform, Pharos, reveals that the proposed scheme outperforms the three existing schemes. It also identifies that error rate decreases as the number of packets transmitted for distance estimation increases, while it increases as the speed of the mobile node increases.

Keywords: Indoor tracking, interference problem, location-aware, sensor network, ubiquitous.

1 Introduction

The wireless sensor network is an emerging technology that may greatly aid humans by providing ubiquitous sensing, computing and communication capabilities through which people can more closely interact with the environment wherever they go. To be context-aware, one of the central issues in the sensor network is location- aware and tracking, whose goal is to monitor the path of a moving object. Ubiquitous and pervasive environment presents opportunities for a rich set of location-aware applications such as car navigation and intelligent robots, interactive virtual games, habit monitoring, logistics service, tracking, etc [1,2]. Typical indoor applications require better accuracy than what the current outdoor location systems provide. Outdoor location technologies such as GPS [3] have poor indoor performance because of the harsh

---

* This research was supported by the Ubiquitous Autonomic Computing and Network Project, 21st Century Frontier R&D Program in Korea and the Brain Korea 21 Project in 2005.

** Corresponding author.

V.N. Alexandrov et al. (Eds.): ICCS 2006, Part IV, LNCS 3994, pp. 910–917, 2006.
© Springer-Verlag Berlin Heidelberg 2006

nature of indoor environments. Indoor environments often contain substantial amounts of metal and other reflective materials that affect the propagation of radio frequency signals in non-trivial ways, causing severe multi-path effects, dead-spots, noise, and interference [4,5,6].

Cricket is an indoor location system for pervasive and sensor-based computing environments such as those envisioned by MIT's Project Oxygen [7]. Cricket has been widely adopted as the indoor tracking platform. It provides fine-grained location information such as space identifiers, position coordinates, and orientation to the applications running on handhelds, laptops, and sensor nodes. However, Cricket has serious interference problem when the cricket-based node moves.

In this paper we present a novel scheme reducing the interference problem of the location tracking system that is particularly suited for supporting context-aware computing. It is achieved by considering the speed variance of the mobile node and thereby correcting the interference errors. The proposed location platform is called Pharos and the new algorithm for interference problem is called IAA (Interference Avoidance Algorithm). Performance evaluation using the actually implemented platform, Pharos, reveals that the proposed scheme outperforms the three existing schemes in terms of the error rate in the location estimation. It also identifies that the error rate decreases as the number of packets transmitted for distance estimation increases, while it increases as the speed of the mobile nodes increases.

The rest of the paper is organized as follows. We first present the related works in Section 2. In Section 3 we describe the proposed new algorithm and system. Section 4 illustrates the performance evaluation of the proposed scheme. Our conclusions and future work are presented in Section 5.

2 The Related Works

2.1 Localization Method with Cricket

The researchers distinguish two different indoor location architectures. The active mobile architecture has an active transmitter on each mobile device, which periodically broadcasts a message on a wireless channel. The receivers deployed in the infrastructure like the ceilings and walls listen to such broadcasts and estimate the distance to the mobile device on each broadcast they hear. This architecture has a limited scalability in the ubiquitous environment. In contrast, the passive mobile architecture exchanges the locations of the transmitter and receiver. Here, the beacons deployed at known positions in the infrastructure periodically transmit their location or identity on a wireless channel, and the passive receivers on mobile devices listen to each beacon. Each mobile device estimates its distance to every beacon it hears and uses the set of distances to estimate its position [6,7]. There exist merit and demerit in the two methods above. Generally, mobile nodes are tightly constrained in terms of transmission power, on-board energy, processing capacity and storage. Therefore, Cricket employed the passive mobile architecture.

Cricket decided to use a combination of RF and ultrasound hardware to enable a listener to determine the distance to the beacons, from which the closest beacon can be more unambiguously inferred. Figure 1 shows how a cricket achieves this by measuring the one-way propagation time of the ultrasonic signals emitted by a beacon, taking advantage of the fact that the speed of sound in air (about 1.13 ft/ms at room temperature) is much smaller than the speed of light (RF) in air. On each transmission, a beacon concurrently sends information about the space over RF, together with an ultrasonic pulse. When a listener hears the RF signal, it uses the first few bits as training information and then turns on its ultrasonic receiver. It then listens to the ultrasonic pulse, which will usually arrive a short time later. The listener uses the time difference between the receipt of the first bit of RF information and the ultrasonic signal to determine the distance to the beacon. Of course, the value of the estimated distance is not as important as the decision on which the closest beacon is. The use of time-of-flight of signals for measuring the distance is not a new concept.

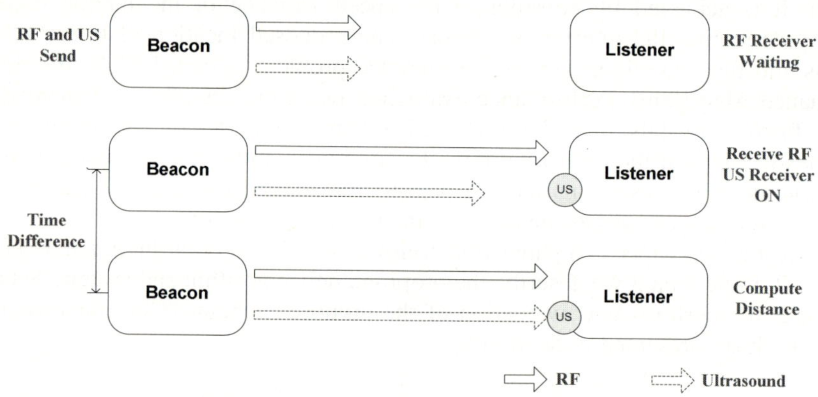

Fig. 1. Measuring the location of a Cricket

2.2 Interference Problems

While Cricket has an attractive property that its decentralized beacon network is easy to configure and manage, it comes in the absence of explicit coordination. There is no explicit scheduling or coordination between the transmissions of different beacons that may be in close proximity, and the listeners do not transmit any information to avoid compromising privacy. This lack of coordination can cause RF transmissions from different beacons to collide, and may cause a listener to wrongly correlate the RF data of one beacon with the ultrasonic signal of another, yielding false results. Furthermore, ultrasonic reception suffers from severe multi-path effects caused by reflections from the walls and other objects, and these are orders of magnitude longer in time than RF multi-path because of the relatively long propagation time for sound waves in air. In fact, this is one of the reasons why it is hard to modulate data on the ultrasonic signal, which makes it a pure pulse. Thus, the listener's task is to gather various RF and ultra-sound (US) samples, deduce and correlate the {RF,US} pairs that were sent concurrently by different beacons, and choose the space identifier sent from the pair with the smallest distance [7].

To better understand the effects of interference and multi-path (due to reflected signals) on distance estimation, we characterize different RF and ultrasonic signals that a listener can hear in Figure 2. Consider the RF and ultrasonic signals sent by a beacon, A, and an interfering beacon, I. The listener potentially hears the followings. We only need to consider the cases when a US pulse arrives while some RF signal is being received. The reception of the first ultrasonic signal US-A, US-RA, US-I, or US-RI while RF-A is being received will cause the listener to calculate the distance to A using the time interval between the detection of RF-A and the particular ultrasonic signal. This is because the listener, after receiving the RF signal from a beacon, waits for the first occurrence of an ultrasonic pulse to determine the distance. All subsequent ultrasonic receptions during the reception of the RF message are ignored. Of course, if the direct signal US-A is the first one to be received, the listener correctly estimates the distance to A. However, wrong correlation of any other ultrasonic signal with RF-A could be problematic. RF-A is the RF signal from A, US-A is the direct ultrasonic signal from A, US-RA is the reflected ultrasonic signal from A, RF-I is the RF signal from I, US-I is the direct ultrasonic signal from I, and US-RI is the reflected ultrasonic signal from I, respectively.

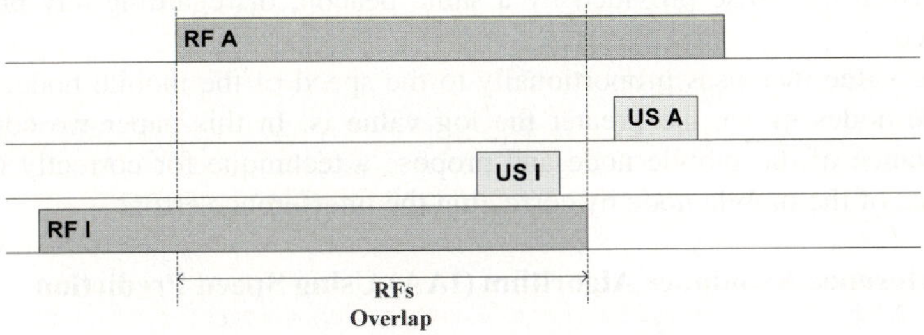

Fig. 2. The effect of interference problem

2.3 The Existing Solution for Interference Problem

Priyantha et al. [8] developed and compared three simple algorithms determining the closest beacon and overcoming the interference problem discussed above. These algorithms are useful with fixed or slowly moving nodes.

- **Majority Algorithm**
  This is the simplest algorithm, which pays no attention to estimating the distance but simply picks the beacon of the highest frequency in the received data set. This algorithm does not use ultrasonic signals for determining the closest beacon, and as we find from our experiments, its performance is not good. We consider this algorithm primarily for the comparison with other algorithms.
- **MinMean Algorithm**
  Here, the listener calculates the mean distance from each unique beacon for the set of data points. Then, it selects the beacon with the minimum mean as the closest one. The advantage of this algorithm is that it can be computed with small information as a new sample updates the mean in a straightforward way.

The problem with this algorithm is, however, that it is not immune to the multi-path effects that cause the distance estimates to display modal behavior; where computing a statistical metric like mean (or median) is not reflective of any actual beacon position.

- **MinMode Algorithm**
  Since the distance estimates often show a significant modal behavior due to reflections, this approach of highest-likelihood estimate computes the per-beacon statistical modes over the past *n* samples (or time window). For each beacon, the listener then picks the distance corresponding to the mode of the distribution, and uses the beacon that has the minimum distance value among the modes.

3 The Proposed System

The aforementioned interference avoidance algorithms show good performance for only slowly moving node. Especially, referring to the result above, the MinMode algorithm performs the best. However, since the nodes have relatively high mobility in real-life, a different log is chosen for the mobile node to correctly distinguish the real data from the noise provided by a same beacon, disregarding any presence of interference.

The log value increases proportionally to the speed of the mobile node; the faster the mobile nodes move, the greater the log value is. In this paper we consider the speed variance of the mobile node and propose a technique for correctly measuring the distance of the mobile node by correcting the interference errors.

3.1 Interference Avoidance Algorithm (IAA) Using Speed Prediction

Figure 3 shows a mobile node moving from $A(X_A, Y_A)$ to $B(X_B, Y_B)$ during a time interval $\Delta T(T_B - T_A)$. The beacons are placed apart with a constant distance M. If D_A and D_B are the distance measured at time T_A and T_B, respectively, the speed prediction ΔV_{real} can be derived with the following equation.

$$\Delta V_{real} = \frac{B - A}{\Delta T} \tag{1}$$

The mobile node is considered to be out of range if ΔT does not satisfy the following equation.

$$\Delta T \leq \frac{M}{V_{real}} (\, if \;\Delta T \times V_{real} \geq M \,, then\; out\; of\; range) \tag{2}$$

The speed of the mobile node moving from A to B can be predicted using the following equation.

$$\Delta V_{pred} = \alpha \Delta V_{last} + (1 - \alpha)\Delta V_{history} \;\; (0 \leq \alpha \leq 1) \tag{3}$$

The predictive value D_{pred} of D_B can be derived by applying the reversed trilateration to Eq. (3). Eventually, any value exceeding D_{pred} is regarded as an error caused by interference and thus excluded from the calculation to obtain correct D_B value.

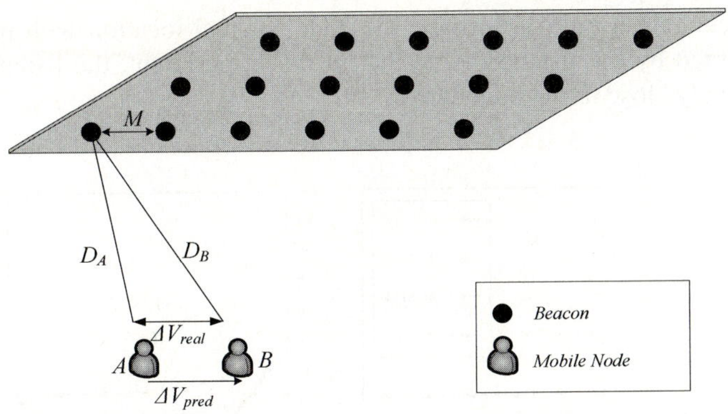

Fig. 3. The relationship between the velocity and distance measurement

3.2 Hardware Architecture

Figure 4 shows the Pharos system developed for real-time tracking of localization in indoor environments. The Pharos is compatible with MIT's Cricket. The Pharos contains a TI MSP430 processor with 8 kilobytes of flash memory for program and 512 bytes of SRAM for data. It uses a 2.4GHz RF wireless transceiver and three LEDs for output indication. A low power 40 KHz ultrasonic transceiver is used for distance measurement.

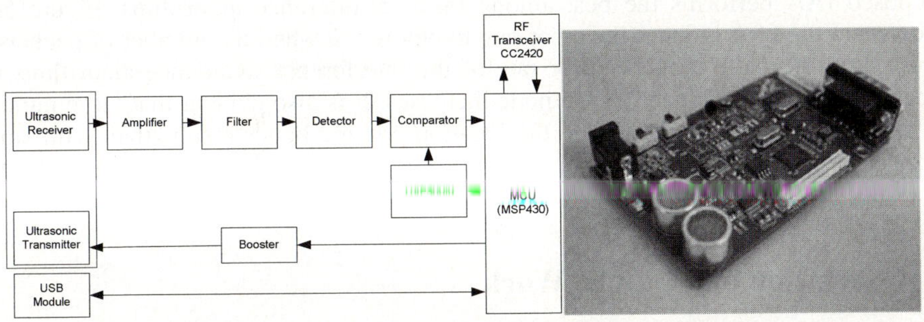

Fig. 4. The Pharos real-time tracking platform and hardware block diagram

4 Performance Evaluation

The experiment aims at determining the system performance when the listener is mobile. For a mobile listener, being able to obtain accurate location information in short time is important. The listener crosses the boundary at two speeds each time (vt = 1, 3). Each time the listener crosses a boundary, a transition event and a timestamp are logged. Once crossing the boundary, the listener remains standstill for a short period of time to determine how long it takes to stabilize to a correct value, and then the experiment is repeated for the next boundary. When analyzing the data, the logged transition events are used to determine the node's actual location with respect to the location reported by the listener. Note that in this experiment, the listener is always located relatively close to the boundaries.

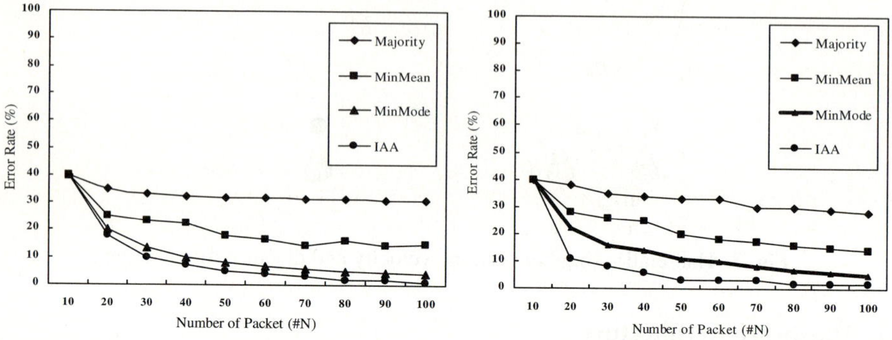

(a) vt = 1m/sec (b) vt = 3m/sec

Fig. 5. The error rates for a mobile Pharos listener

Figure 5 shows the location error rate at the listener for the four algorithms compared. The error rate is calculated over the number of packets during which the listener moves around a location with a velocity (vt). Referring to the result above the proposed IAA performs the best among the four inference algorithms. Figure 5(a) shows that the IAA reduces the error rate to below 5% when the number of packets is larger than 60. Note that the error rate of the interference avoidance algorithms increases if the moving speed of the node increases. It is also evident that large number of packets transmitted to estimate the location provide better results than with small number of packets, which is expected since a larger transmission counts provides the algorithm with more information to work on.

5 Conclusion and Future Work

We have studied how to overcome the problems in the existing methods measuring the distance in indoor ubiquitous environment. Through actual experiment, we found that real-time tracking is difficult because of interference between the sensor nodes. In this paper we have presented a new IAA (Interference Avoidance Algorithm) for reducing the error in the location identification due to interference within the infrastructure based sensor network by considering the velocity of the moving node. The proposed IAA calculates the distance using the velocity predicted based on the previously measured velocity. The calculated distance corrects the error induced by interference. We also proposed a formula for velocity prediction and implemented a real platform. The experiment results show that the proposed IAA can reduce the error to below 5%, and it is always better than the existing interference avoidance algorithms. As the speed of the moving object grows, the results show that the error rate slightly increases as expected.

The proposed Pharos is intended for use indoors or in urban areas where outdoor systems like the Global Positioning System (GPS) do not work well. It can provide distance ranging and positioning of the precision of between 1cm and 3cm. Our future work directions include extending the prototype and using it for further applications. We also plan to develop a more efficient inference algorithm. We hope that the findings in this paper will be helpful for design and implementation of real-time tracking system in ubiquitous computing environment.

References

[1] P. Bahl and V. Padmanabhan, "RADAR: An In-Building RF-Based User Location and Tracking System," in Proc. of the IEEE INFOCOM'00, Vol.2, pp. 755-784, March 2000.

[2] J. Hightower, G. Boriello and R. Want, "SpotON: An indoor 3D Location Sensing Technology Based on RF Signal Strength," University of Washington CSE Report 2000-02-02, Feb. 2000.

[3] J. Liu, P. Cheung, L. Guibas, and F. Zhao, "A Dual-Space Approach to Tracking and Sensor Management in Wireless Sensor Networks," in Proceedings of ACM WSNA'02, pp. 131-139, Sept. 2002.

[4] J. Cadman, "Deploying Commercial Location-Aware Systems," in Proceedings of 2003 Workshop on Location-Aware Computing, pp. 4-6, Jan. 2003.

[5] D. Ganesan, B. Krishnamachari, A. Woo, D. Culler, D. Estrin and S. Wicker, "Complex Behavior at Scale: An Experimental Study of Low-Power Wireless Sensor Networks," Technical Report UCLA/CSD-TR 02-0013, Jan. 2002

[6] A. Mainwaring, J. Polastre, R. Szewczyk, D. Culler, and J. Anderson, " Wireless Sensor Networks for Habitat Monitoring," in Proc. of ACM WSNA'02, pp. 88-97, Sept. 2002.

[7] N. Priyantha, and H. Balakrishnan, "The Cricket Indoor Location System: Experience and Status.," in Proc. of the 2003 Workshop on Location-Aware Computing, pp. 7-9, Jan. 2003.

[8] N. Priyantha, A. Chakraborty, and H. Balakrishnan, "The Cricket Location-Support System," in Proc. of 6th Annual ACM International Conference on Mobile Computing and Networking (MOBICOM), pp. 32-43, Aug. 2000.

A Task Generation Method for the Development of Embedded Software

Zhigang Gao, Zhaohui Wu, and Hong Li

College of Computer Science, Zhejiang University,
Hangzhou, Zhejiang, P.R. China, 310027
{gaozhigang, wzh, lihong}@zju.edu.cn

Abstract. The development of embedded software has the tendency towards higher levels of abstraction. This development paradigm shift makes the synthesis of embedded software a key issue in the development of embedded software. In this paper, we present a new method of generating real-time tasks, which uses Component Sequence Diagram to organize tasks and assign their priorities. Moreover, we use simulated annealing algorithm to explore design space and iterate the process of task generation until an optimization implementation is obtained. Experimental evaluation shows this method can yield correct implementation and has better time performance.

1 Introduction

Synthesis of embedded software refers to the process from function specification to code implementation on a specific platform. It is a key problem in the design of embedded system. Non-functional requirements are implemented in task generation phase when synthesizing embedded software. Current research works often ignore or deal with non-functional properties roughly [3, 5]. In this paper, we focus our attention on the problem of task generation on single processor and propose a novel task generation method. Gu et al. [6] proposed a method of synthesizing real-time implementation from component-based software models. Our work is an extension of the work of Gu et al., which combines simulated annealing (SA) algorithm with Component Sequence Diagram (CSD). We use SA to explore design space and CSD to organize tasks, to assign their priorities, and to prune the search space of simulated annealing algorithm. The process of task generation is iterated until an optimization implementation is obtained.

The rest of this paper is organized as follows. Section 2 presents the computational model. Section 3 describes the process of task generation. Section 4 gives the experimental results and the conclusions of this paper.

2 Computational Model

The software model before task generation is called the structural model, and the software model after task generation is called the runtime model.

V.N. Alexandrov et al. (Eds.): ICCS 2006, Part IV, LNCS 3994, pp. 918 – 921, 2006.
© Springer-Verlag Berlin Heidelberg 2006

Structural Model. We use the structural model proposed in [1]. A transaction is defined as a series of related components' actions triggered by an event. An example of structural model is shown in Fig. 1.

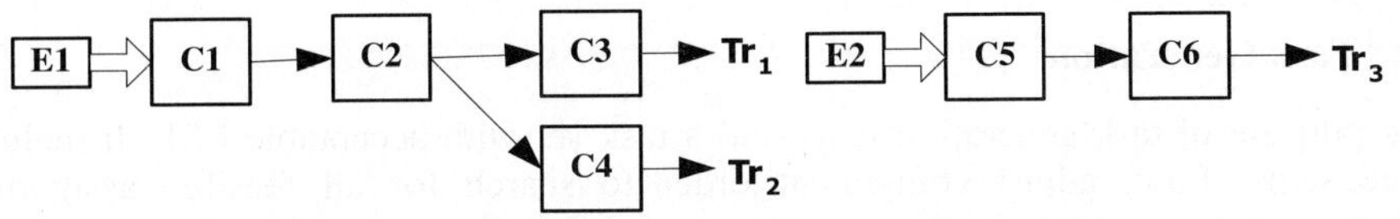

Fig. 1. Structural model consisting of three transactions

Runtime Model. The runtime model is based on task sequences. It is a sequence of related tasks communicating with each other through events.

3 The Implementation of Task Generation

The problem of assigning priorities to tasks for obtaining a schedulable system is an NP-hard problem [7]. Similarly, the problem of task generation is an NP-hard problem, too. It usually resorts to heuristic algorithms. Unfortunately, heuristic algorithms would result in high computation overhead because of large search space. Here we use a diagram, CSD, to describe assignment sequence of components. It has no scale and only denotes the constraints of time sequence (CTS). Since the assignment of components on CSD denotes a kind of execution orders of components according to the constraints of time sequence, we can organize components into tasks and assign their priorities according to their places on CSD. The process of task generation is shown in Fig. 2. It includes two phases: initialization and task generation. Task generation is performed iteratively until an acceptable Critical Scaling Factor (CSF) [2] is obtained or no schedulable task set can be found.

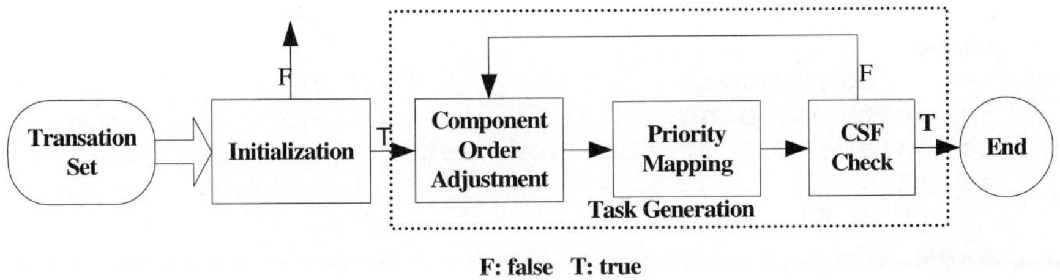

Fig. 2. The process of task generation

3.1 Initialization

The purpose of initialization is to abstract the constraints of time sequence between components and to obtain a feasible component arrangement scheme that accords with their CTS. It includes three steps: First, generate the transaction set. Only the transactions within the least common multiple (LCM) of all periods of transactions in the structural model need to be generated. We introduce the notion of virtual transaction. A transaction with the period less than LCM is divided into multiple virtual transactions. Second, abstract the constraints in transactions and virtual transactions. Third, assign all the components of transactions and virtual transactions obtained in the first step on CSD, while satisfying the constraints obtained in the second step.

3.2 Task Generation

The purpose of task generation is to find a task set with acceptable CSF. It includes three steps: First, adjust component order to search for all feasible assignment schemes of components in the search space of SA. Second, merge the components that belong to the same transaction and are adjacent into a task and assign their priorities. Third, check Whether CSF is acceptable. The worst-case local response time of a task can be obtained using the formula presented in [4]. A task sequence's worst-case local response time (WCRT) is the sum of the worst-case local response time of all its tasks. After all task sequence's WCRT is computed, we can obtain the CSF of a system according to the deadlines of all task sequences.

The complete process of task generation is shown in **Algorithm TG.**

```
Algorithm TG (TS)
 /* TS is the set of transactions. TE is the temperature
of SA. E and E^1 are the energy of components assignment on
CSD. CSF_exp is the expected CSF. T_min is the threshold of
temperature of SA */
  TE = 10*max{D_Tr_i / ∑C_ik | Tr_i ∈ RM} ;
  Initialize components assignment on CSD;
  E = 0; CSF = CSF_old = 0;
  Priority mapping;
  Calculate CSF of the task set;
  E^1 = -CSF;
  do {
      Select two components C_i and C_j randomly on CSD;
      if (C_i and C_j are exchangeable)
          Exchange their positions on CSD;
      else
            continue;
      Priority mapping;
      Calculate CSF of the task set;
      if (E < E^1)
          E^1 = E;
      else {
            if ( e^((E-E^1)/TE) < 0.5 )
                  E = E^1;
            else
                  Return to the components assignment
            before exchanging;
            }
        if (CSF > CSF_old ) TE = TE*0.95;
  } while (CSF < CSF_exp and T > T_min)
  if (T <= T_min) return false;
  return true;
```

4 Experiments and Conclusions

In order to evaluate the effect of our algorithm, we compared our method, named SA+CSD, with the method only using SA. The results are shown in Fig. 3.

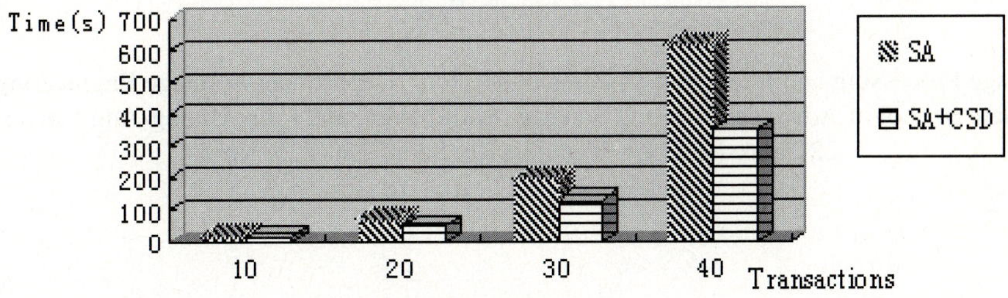

Fig. 3. The result of experiments

In this paper, we present a new method of task generation. It uses CSD to generate tasks and assign their priorities. By using of CTS, the speed of task generation is improved greatly. Our future work will focus on the influence of other resource constraints, such as memory, energy, on task generation.

References

1. Wang, S., Shin, K.G.: An Architecture for Embedded Software Integration Using Reusable Components. CASES. (2000) 110-118
2. Vestal, S.: Fixed-Priority Sensitivity Analysis for Linear Compute Time Models. IEEE Trans. Software Eng., Vol. 20. (1994) 308-317
3. Kodase, S., Wang, S., Shin, K.G.: Transforming Structural Model to Runtime Model of Embedded Software with Real-Time Constraints. DATE. (2003) 20170-20175
4. Tindell, K., Clark, J.: Holistic Schedulability Analysis for Distributed Hard Real-Time Systems. Microprocess & Microprogram, Vol. 40. (1994) 117-134
5. Gomaa, H.: Designing Concurrent Distributed, and Real-Time Applications with UML. Addison–Wesley. (2000)
6. Gu, Z., Shin, K.G.: Synthesis of Real-Time Implementations from Component-Based Software Models. Proc. IEEE Real-Time Systems Symposium (RTSS 2005). Miami, FL
7. Burns, A.: Scheduling Hard Real-Time Systems: A Review. Software Engineering Journal, Vol. 6, No. 3. (1991) 116-128

Active Shape Model-Based Object Tracking in Panoramic Video

Daehee Kim, Vivek Maik, Dongeun Lee, Jeongho Shin, and Joonki Paik

Image Processing and Intelligent Systems Laboratory, Department of Image Engineering, Graduate School of Advanced Imaging Science, Multimedia, and Film, Chung-Ang University, 221 Huksuk-Dong, Tongjak-Ku, Seoul 156-756, Korea
http://ipis.cau.ac.kr

Abstract. Active Shape Model (ASM) paradigm is a popular method for image segmentation where a priori information about the shape of the object of interest is available. The effectiveness of the method is contingent upon a correct correspondence between model points and the features extracted from the image. Extensive application of these models soon revealed one of their limitations when, for a given model point, no obvious salient point can be found in the image. The primary cause of such limitation is due to weak edges and presence of abrupt noise which is the case with low light surveillance video images. In this paper we propose a fusion-based panoramic tracking algorithm of in low light images using multiple sensors. The proposed algorithm uses an IR and CCD sensor for image capture. The proposed tracking system consists of three steps: (i) pyramid based fusion algorithm, (ii) reconstruction of panoramic image, and (iii) active shape model (ASM)-based tracking algorithm. The experimental results show that the proposed tracking system can robustly extract and track objects on panoramic images in real-time.

1 Introduction

Video surveillance system is useful for monitoring large and complex environments such as large building, airport, etc. Motion detection and object tracking play an important role in video-based surveillance system, some of which are, i) adaptive background generation and background subtraction [1, 2], ii) selective pixel integration [3], and iii) region based tracking [4]. In this paper we target our algorithm for low light visual images where above mentioned method fails to perform. Using fusion-based tracking we can integrate information from CCD and IR sensors and perform effective target tracking. For integrating and processing multi-sensor image we used panoramic imaging. The main advantage by doing so is they improve field of view to incorporate wide image observation region. Panoramic algorithm is carried out using direct linear transformation (DLT) and the singular value decomposition (SVD) algorithm. Panoramic algorithm is merged with the pyramid bared fusion to over come heavy computational overhead and provide better visual quality.

V.N. Alexandrov et al. (Eds.): ICCS 2006, Part IV, LNCS 3994, pp. 922 – 929, 2006.
© Springer-Verlag Berlin Heidelberg 2006

2 Pyramid Based Fusion Algorithm

2.1 Pyramid Construction for Image Fusion

The pyramid representation can be used both for assessing the salience of the source image features, and for the reconstruction of the final image result. The following definitions for the pyramid are used. The fusion method described within this paper use a Laplacian pyramid representation. Laplacian pyramids are constructed for each image using the filter subtracts decimates (FSD*)* method. Thus the k'th level of the FSD Laplacian pyramid, k L, is constructed from the corresponding Gaussian pyramid level *k* based on the relationship.

$$L_k = G_k - wG_k = G_k(1-w), \tag{1}$$

Where *w* represents a standard binomial Gaussian filter, usually of 5x 5 spatial pixels extent. When constructing the FSD Laplacian, due to the decimation process and the fact that w is not an ideal filter, a reconstruction of the original image based on the FSD Laplacian pyramid incurs some loss of information.

2.2 Feature Saliency Computation

The feature saliency computation process, labeled sigma, expresses a family of functions that operate on the pyramids of both images yielding saliency pyramids. In practice, these functions can operate on the individual pixels or on a local region of pixels within the given pyramid level. The saliency function captures the importance of what is to be fused. When combining images having different focus, for instance, a desirable saliency measure would provide a quantitative measure that increases when features are in better focus. Various such measures, including image variance, image gradients, have been employed and validated for related applications such as auto focusing. The saliency function only selects the frequencies in the focused image that will be attenuated due to defocusing. Since defocusing is a low pass filtering process, its effects on the image are more pronounced and detectable if the image has strong high frequency content. One way to high pass filter an image is to determine its Laplacian or second derivative in our case.

$$\nabla^2 L_K = \frac{\partial^2 L_k}{\partial x^2} + \frac{\partial^2 L_k}{\partial y^2}, \tag{2}$$

In order to accommodate for possible variations in the size of texture elements, we computer the partial derivative by using a variable spacing between the pixels used to compute the derivatives. Hence a discrete approximation to the modified Laplacian is given by,

$$ML(i,j) = \left|2I(i,j) - I(i-1,j) - I(i+1,j)\right| + \left|2I(i,j) - I(i,j-1) - I(i,j+1)\right|, \tag{3}$$

Finally, the focus measure at a point (i, j) is computed as the sum of modified Laplacian values, in a small window around (i, j), that are greater than a threshold value.

$$F(i,j)=\sum_{x=i-N}^{i+N}\sum_{y=j-N}^{j+N} M_k(x,y)\, for M_k(x,y)\geq T_1 \quad (4)$$

The parameter determines the window size used to compute the focus measure. In contrast to auto focusing methods, we typically use a small window of size, i.e. $N=1$. The above equation can be referred to as sum modified Laplacian (SML).

3 Reconstruction of Panoramic Image Construction Algorithm

Most feature-based correspondence algorithms use DLT and SVD for 3D-transformation and interpolation. We begin with a simple linear algorithm for determining H given a set of four 2D to 2D point correspondences, $x_i \leftrightarrow x_i'$. The transformation is given by the equation $x_i' = Hx_i$. Note that this is an equation involving homogeneous vectors; thus the 2-vectors x_i' and Hx_i are not equal, they have the same direction but may differ in magnitude by a non-zero scale factor. The equation may be expressed in terms of the vector cross product as $x_i' \times Hx_i = 0$. This form will enable a simple linear solution for H to be derived.

The SVD is one of the most useful matrix decompositions, particularly for numerical computations. Given a square matrix A , the SVD is a factorization of A as $A = UDV^{\mathrm{T}}$, where U and V are orthogonal matrices, and D is a diagonal matrix with non-negative entries. Note that it is conventional to write V^{T} instead of V in this decomposition. The decomposition may be carried out in such a way that this is always done. Thus a circumlocutory phrase such as "the column of V corresponding to the smallest singular value" is replaced by "the last column of V ."

4 Active Shape Tracker

The original ASM was first proposed by Cootes [2], [3]. Detection, analysis, and tracking a human body in a video sequence is a major application area for the ASM because the shape of the human body has unique combination of head, torso, and legs, which can be modeled with small number of parameters. In this section we briefly revisit ASM theory including three steps: (a) shape variation modeling, (b) model fitting, and (c) local structure modeling.

4.1 Shape Variation Modeling

Given a frame of input video, initial landmark points should be assigned on the contour of the object either manually or automatically. Good landmark points should be at or close to the desired boundary of each object. A particular shape X is represented by a set of n landmark points which approximate its outline as

$$X = [x_1, x_2, \cdots, x_n, y_1, \cdots, y_n]^T. \tag{5}$$

Different sets of such landmark points make a training set. A shape in the training set is normalized in scale, and aligned with respect to a common frame, as shown in Fig.1. Although each aligned shape is in the $2n$-dimensional space, we can model the shape with a reduced number of modes using the principal component analysis (PCA) analysis. The main modes of the template model, X , are then described by the eigen-vectors φ of the covariance matrix C, with the largest eigen values [9].

4.2 Model Fitting

We can find the best shape and pose parameters to match a shape in the model coordinate frame, x, to a new shape in the image coordinate frame, y, by minimizing the following error function

$$E = (y - Mx)^T W^T (y - Mx), \tag{6}$$

where M represents the geometric transformation of scaling (s), translation (t), and rotation (θ). For instance if we apply the transformation to a single point, denoted by $[p,q]^T$, we have

$$M\begin{bmatrix} p \\ q \end{bmatrix} = s\begin{bmatrix} \cos\theta & \sin\theta \\ -\sin\theta & \cos\theta \end{bmatrix}\begin{bmatrix} p \\ q \end{bmatrix} + \begin{bmatrix} t_x \\ t_y \end{bmatrix}. \tag{7}$$

After a set of pose parameters, $\{\theta, t, s\}$, is obtained, the projection of y on to the model coordinate frame is given as

$$x_p = M^{-1} y. \tag{8}$$

Finally, the parameters are updated as

$$b = \phi^T (x_p - \bar{x}). \tag{9}$$

4.3 Local Structure Modeling

In order to interpret a given shape in the input image based on ASM, we must find a set of parameters that best match the model to the input shape. If we assume that the shape model represents boundaries and strong edges of the object, a profile across each landmark point has an edge like local structure. Let g_i, i=1..., n, be the normalized derivative of a local profile of length K across the i-th landmark point, $\bar{g}$ and S_g the corresponding mean and covariance, respectively. The nearest profile can be obtained by minimizing the following Mahalanobis distance between the sample and mean of the model as

$$f(g_{i,m}) = (g_{i,m} - \overline{g})^T S_g^T (g_{i,m} - \overline{g}), \quad (10)$$

where $g_{i,m}$ represents the shifted version of g_i by m samples along the normal direction of the corresponding boundary. Figures 1 and 2 gives a schematic representation of the ASM. Fig.2. shows the result of PCA and local structure model fitting.

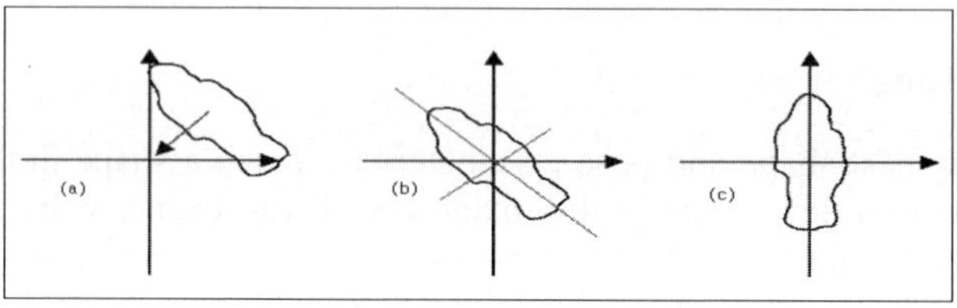

Fig. 1. Shape alignment: (a) Move centroid to origin, (b) find major axes of the shape,(c) rotation for alignment

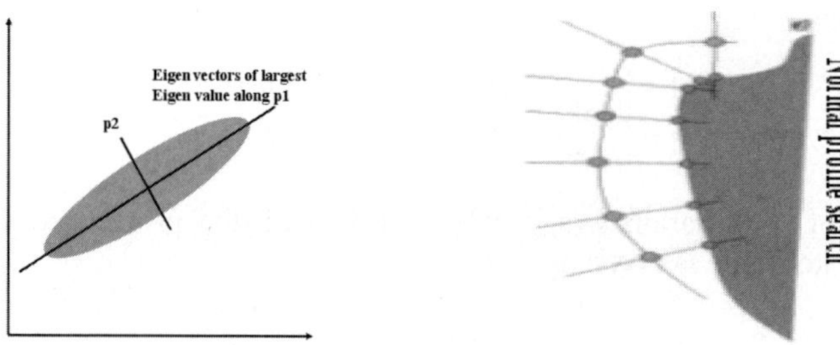

(a) (b)

Fig. 2. (a) PCA analysis and (b) Local structure model fitting

5 Experimental Results

In this section we present the results of the proposed algorithm. Fig. 6 gives some examples of pyramid banned fusion algorithm. As can be seen the CCD sensor fails to provide complete object information in both canes due to low illumination environment. But fusion with IR image in Fig 3(f) results in much more distinguishable object for tracking. In Fig 4 we provide results of panorama view. The panoramic reconstruction is carried out on fused image to provide wide view for object tracking. The object tracking results after panorama reconstruction is carried out using ASM.

Experimental results show tracking in the dark no problem. This makes the best use of an advantage of CCD-camera and IR-camera.

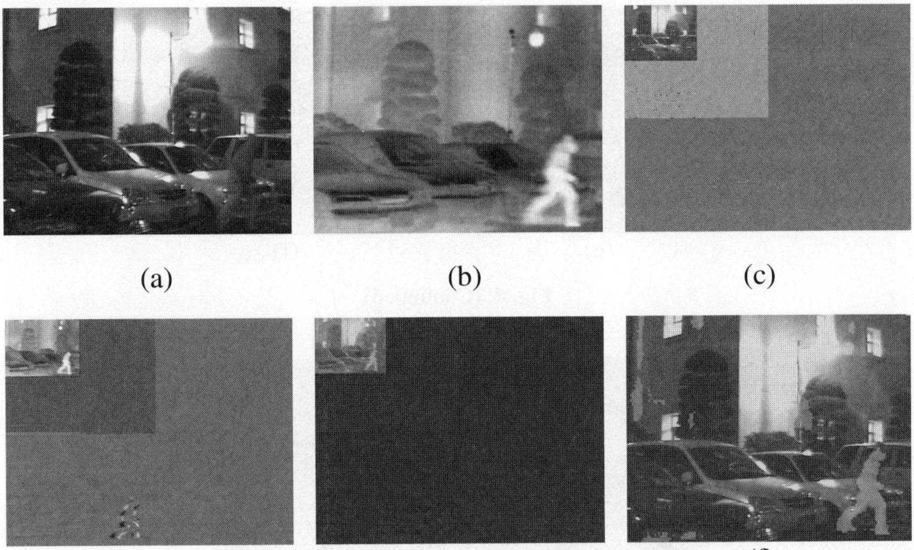

(a) (b) (c)

(d) (e) (f)

Fig. 3. Results of multi-dimensional fusion of IR and CCD camera sequences. (a) input ccd image, (b) input IR image, (c) daubechies wavelet representation of CCD, (d) daubechies wavelet representation of IR, (e) fusion result in wavelet domain, (f) reconstruction of fused wavelet.

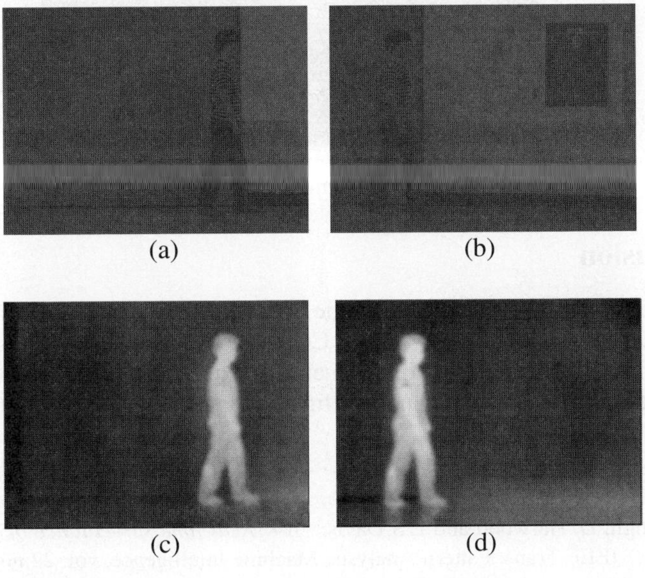

(a) (b)

(c) (d)

Fig. 4. Results of multi-dimensional fusion of IR and CCD camera sequences. (a), (b) input CCD image, (c),(d) input IR image, (e), (f) reconstruction of fusion

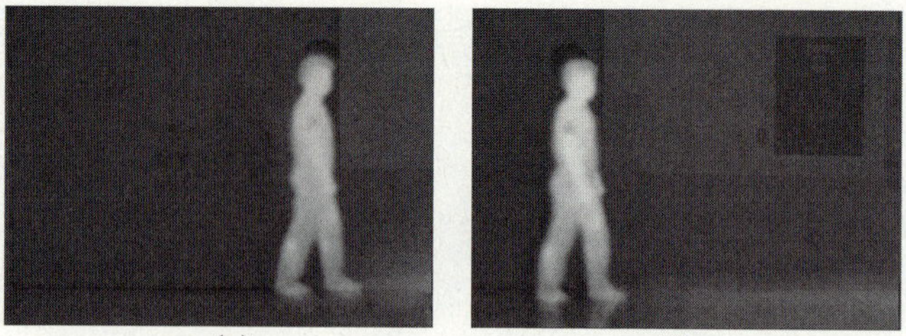

(e) (f)

Fig. 4. (Continued)

Fig. 5. Generated panoramic image

Fig. 6. ASM tracking on panoramic image

6 Conclusion

In this paper we proposed an automatic fusion-based panoramic tracking system. Pyramid fusion was extended to IR and CCD sensors. Active shape tracker was used to carry about object tracking. Experimental results prove the effectiveness of the proposed algorithm in real-time and low light environment.

References

[1] I. Haritaoglu D. Harwood and L.S Davis, "*W4: Real-time surveillance of people and their activities,*" IEEE Trans. Pattern Analysis, Machine Intelligence, vol. 22 no. 8, pp. 809-830, August 2000.

[2] A. Koschan, S. K. Kang, J. K. Paik, B. R. Abidi, and M. A. Abidi, "*Video object tracking based on extended active shape models with color information,*" Proc. 1st European Conf. Color in Graphics, Imaging, Vision, pp. 126-131, University of Poitiers, France, April, 2002.

[3] P. Chang and J. Krumm, "*Object recognition with color occurrence histograms*," IEEE Conf. on Computer Vision and Pattern Recognition, Fort Collins, CO, June, 1999.

[4] T. Horprasert, D. Harwood, and L.S. Davis, "*A robust background subtraction and shadow detection*," Proc. ACCV'2000, Taipie, Taiwan, January 2000.

[5] C. Chen, W. Hsieh, J. Chen, "*Panoramic appearance-based recognition of video contents using matching graphs*," IEEE Transactions on Systems, Man, and Cybernetics-PART B: Cybernetics, vol.34, no. 1, February 2004.

[6] S. Kim, J. Kang, J. Shin, S. Lee, J. Paik, S. Kang, B. Abidi, and M. Abidi, "*Optical flow-based tracking of deformable object using a non-prior training active feature model*," PCM 2004, LNCS, vol. 3333, pp. 69-78, December 2004.

[7] Z. Zhu, A. Hanson, E. Riseman, "*Generalized parallel-perspective stereo mosaics from airborne video*," IEEE Transactions on Pattern Analysis and Machine Intelligence, vol. 26, no. 2, pp. 226-237, February 2004.

[8] R. Patil, P. Rybski, T.Kanade, and M. Veloso, "*People detection and tracking in high resolution panoramic video mosaic*," Proceedings of 2004 IEEE/RSJ International Conference on Intelligent Robots and Systems, pp. 1323-1328, September 2004.

[9] C. R. Wren, A. Azerbayejani, T. Darrell, and A. P. Pentland, "*Pfinder: Real-time tracking of the humand body,*" IEEE Trans. Pattern Anal. Machine Intell., vol. 19, pp. 780-785, July 1997.

Interworking of Self-organizing Hierarchical Ad Hoc Networks and the Internet*

Hyukjoon Lee[1], Seung Hyong Rhee[1], Dipankar Raychaudhuri[2], and Wade Trappe[2]

[1] Kwangwoon University, 447-1 Wolgye-Dong, Nowon-Gu, Seoul 139-701, Korea
{hlee, shrhee}@daisy.kw.ac.kr
[2] WINLAB, Rutgers University, 73 Brett Road, Piscataway, NJ 08854, USA
{ray, trappe}@winlab.rutgers.edu

Abstract. Self-organizing hierarchical ad hoc network (SOHAN) is a new network architecture that has been proposed to increase the scalability property of flat ad hoc networks. This paper describes how SOHAN interoperates efficiently with the Internet based on IPv6. Procedures for the autoconfiguration of a globally-routable address, routing and gateway discovery are presented. The amount of control overheads is reduced by taking advantage of cross-layer interaction and limited-scope broadcast (LSBC) techniques. Simulation results display SOHAN with the proposed interworking procedures outperforms conventional flat ad hoc networks in interworking with the Internet in terms of throughput capacity and scalability.

Keywords: Ad hoc network, IPv6, Mobile IP, Interworking, Cross-layer interaction.

1 Introduction

Ad hoc networks have been studied extensively by the research community during the past several years motivated by their high potential for rapid deployment and cost benefits. The focus of most research works has been on stand-alone “flat” networks, in which no hierarchical relationship between nodes is assumed and every node contributes to multi-hop communication. One of the fundamental problems of the flat wireless networks is that they do not scale well. Gupta and Kumar describe that the throughput of a wireless network is bounded above and decreases as $O\left(1/\sqrt{n}\right)$ as n becomes large [1]. This motivates the investigation of a new wireless network architecture based on a hierarchical structure.

SOHAN (Self-Organizing Hierarchical Ad hoc Network) is a novel ad hoc network architecture proposed by Ganu *et al.* that consists of three tiers, i.e. access points (AP’s), forwarding nodes (FN’s) and mobile nodes (MN’s) [2]. In this new architecture, the AP’s are connected together by high-speed wired links. Using high-speed wired links not only increases the system capacity, but also provides a convenient framework for interconnecting to a wired network. This paper describes how SOHAN interoperates efficiently with the Internet based on IPv6.

---

* This work was supported by Grant No. R01-2001-00349 from the Korea Science & Engineering Foundation and Research Grant of Kwangwoon University in 2004.

V.N. Alexandrov et al. (Eds.): ICCS 2006, Part IV, LNCS 3994, pp. 930 – 937, 2006.
© Springer-Verlag Berlin Heidelberg 2006

The interworking of ad hoc networks with the Internet presents interesting challenges. For example, the global routing of the Internet cannot be directly applied, since each type of networks uses a different address architecture (i.e., flat vs. hierarchical) and routing protocols (i.e., host-specific vs. prefix matching). There exist several research works published in the literature that propose different approaches to interconnect the ad hoc networks and the Internet [3-8]. Most of them use Mobile IP to provide the mobile nodes (MN's) with globally-routable addresses, i.e., care-of addresses (CoA's) [3-9]. In this approach, a foreign agent (FA) acts as the interworking gateway. One of the main concerns of this approach is that it requires flooding-based operations. For example, duplicate address detection (DAD) in address autoconfiguration, gateway discovery and route discovery flood messages over the entire ad hoc networks. Flooding could decrease the throughput capacity of the network significantly.

The interworking methods for SOHAN proposed in this work takes advantage of the hierarchical structure and routing mechanism such that efficient interworking is achieved without interworking gateways. Based on the expectation that IPv6 would become the unifying packet transport protocol in both the core and access networks of the future communication systems, we focus on the interworking with IPv6.

The rest of this paper is organized as follows. In section 2, we introduce SOHAN architecture. In section 3, we present the detailed discussion on how the interworking can be achieved. In section 4, we show some simulation results. Finally, in section 5, we conclude our discussion.

2 Self-organizing Hierarchical Ad Hoc Networks

2.1 Architecture

SOHAN is a novel self-organizing hierarchical ad hoc network architecture designed to provide significant improvements in systems capacity and performance relative to conventional float ad hoc networks. It consists of three tiers of radio nodes: low-power mobile node (MN) at the lowest tier, high-power forwarding nodes (FN) at the mid-tier, and wired access points (AP) at the highest tier (Fig. 1).

The MN operates on a single radio (e.g. 802.15.4) and, instead of directly connecting to other MN's, it connects to an AP or FN of the best link quality. As a user device, the MN does not forward packets for other nodes. The bandwidth and energy-constrained MN's do not communicate with an Internet host, but merely send/receive their data to/from the directly-connected AP's or FN's of much higher bandwidth and energy.

The FN can have a direct radio connection with all three types of nodes. The main function of FN is to forward packets for other nodes by multi-hop ad hoc routing. It can be equipped with a single or two radio interfaces, one for FN-MN traffic and the other for the intra-FN and FN-AP traffic. The FN can be either fixed or mobile. The AP has both a radio interface (e.g. 802.11) and a wired interface to the wired links. The AP's can be configured as an access router (AR). Multiple FN's can be directly connected to the AP in ad hoc mode. The transmission speed of the wired links that connect the AP's is assumed orders of magnitude faster than that of the wireless links.

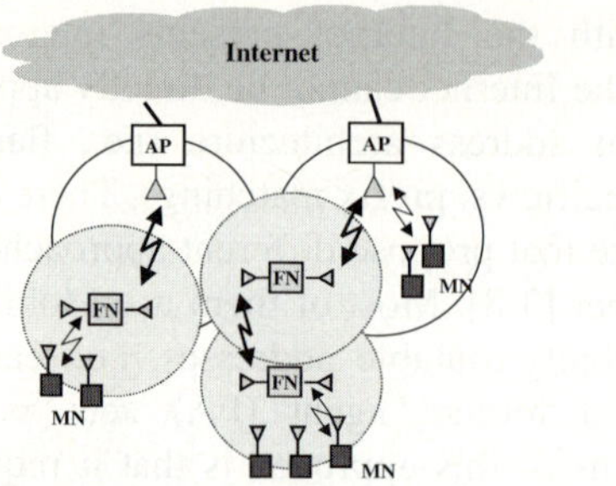

Fig. 1. Self-Organizing Hierarchical Ad hoc Network (SOHAN) architecture

Msg Type	Node Type	Node ID	Channel	Cost to AP	Beacon No.	BSSID	AP IPv6 Addr.

Fig. 2. Message format of SOHAN beacon and association

2.2 Topology Discovery

The AP's and FN's use a self-organizing topology discovery protocol based on beacons. The beacon is an application-level message based on the 802.11 MAC beacons and used by the AP's and FN's to identify their one-hop neighbors. It carries information about the quality of physical links to other nodes within a radio range, which is used as the basis of determining the optimal logical network topology.

Upon bootstrapping, all nodes enter a self-organizing phase by transmitting the beacons on their predetermined channels and repeat this phase periodically. Based on the beacons received, the FN's and MN's update their neighbor table and transmit the association messages to the best one-hop parent. The beacon and association messages share the same augmented message format (Fig. 2).

2.3 Routing

The routing protocol proposed in [2] is based on the topology discovery protocol. The neighbor table generated by the topology discovery protocol contains information about the next hop node (i.e., parent node) for each node to reach an AP. The neighbor tables are periodically exchanged between neighboring devices such that information about multi-hop paths can be incrementally built up in a similar way to the distance-vector routing protocol. The main purpose of this protocol is to enable data to flow from the MN's towards the AP's. Hence, entries for the MN's are excluded from the neighbor table exchange. A packet sent across the boundaries of two subtrees would be routed via the wired links. This routing strategy is based on the assumption that the wired links connecting the AP's provide much higher bandwidth than the wireless links. It may be sub-optimal since an ad-hoc path may exist that goes through the FN's in less number of hops.

More recently, a new L2.5 routing protocol based on AODV with appropriate modifications has been proposed [10]. This routing protocol, operating with the MAC addresses, sets up the initial routing table based on the neighbor table and adds an entry as a new route is found on-demand. The routes from the AP to the FN/MN's are found by flooding RREQ (Route Request) messages or by reverse route setup when data packets are transmitted by the FN's towards the AP.

3 Interworking of SOHAN and the Internet

The topology discovery and routing protocols discussed in the previous section can make the SOHAN appear as a set of Ethernet-like wireless LAN segments. Therefore, IPv6 operations used in interworking, such as router discovery and address autoconfiguration, can be directly applied in SOHAN (Fig. 3). In what follows, we assume that every AP is configured as an IPv6 router in order to simply our discussion. All aspect of the interworking functionalities discussed in this paper can easily be extended with minor modifications when multiple AP's exist in a subnet.

3.1 Address Autoconfiguration

An FN/MN in SOHAN must be configured with a globally-routable IP address to communicate with an Internet host. IPv6 stateless address autoconfiguration allows a globally-routable IP address for the FN/MN to be constructed from the prefix of an AR when the FN/MN joins or boots up in an ad hoc network [11]. If the FN/MN does not have its own IPv6 home address, the FN/MN should temporarily configure an *initia*l address simply by forming a link-local address, or using the IPv6_MANET initial prefix [8]. The uniqueness of these tentative addresses can be verified using the strong DAD described in [12, 13].

DAD operation based on NDP, when applied to an ad hoc network, must be performed in a multi-hop fashion over the entire network (i.e., flooding), if a single interworking gateway is used. However, flooding causes scalability problem when there are a large number of nodes in the network. Moreover, it is meaningless to flood the DAD packets to the nodes with possibly different prefixes (i.e., the nodes associated with different AR's) as in case of SOHAN. Therefore, the flooding in SOHAN is confined within the subset of nodes that are associated with the same AP. This so-called limited-scope broadcast (LSBC) uses the information about each node's association with AP's stored in the neighbor table of the MAC layer. The node can determine whether it is associated with the same AP as the source of flooding by checking the Basic Service Set ID (BSSID).

The *globally routable* IP address is formed by appending to the prefix of the AP the interface identifier of the FN/MN. We modify the format of a beacon message specified in [2] to include the IPv6 address of the AP associated with each FN (Fig. 2). The FN/MN that is multi-hop away from an AP receives the address of the AP associated with its upstream neighbor. Thus, the periodic broadcast of router advertisement is not used. Therefore, a considerable amount of valuable radio resource can be saved.

3.2 Routing

Address autoconfiguration discussed above logically maps a part of the network corresponding to a subtree with an AP as its root in the network topology to an IP subnet (Fig. 4). Since the ad hoc routing is performed below the network layer, the subtree appears as a single-hop wireless LAN segment to the IPv6. The interworking between the two routing protocols becomes a cross-layer interaction problem.

Since the FN/MN's use MAC addresses as their identifier for ad hoc routing at L2.5, the IP addresses must be translated to the corresponding MAC addresses before packets are routed. IPv6 address resolution operation based on requires broadcast on a single-link [13]. Hence, the LSBC is again used to perform the address resolution. In a usage scenario where the FN's and MN's are highly mobile, frequent update of neighbor caches and route tables is necessary. LSBC is expected to reduce the overhead of flooding significantly.

When an IP packet originated from an Internet host and destined to an FN/MN is received by the AP that announces the route for the FN/MS, the IP layer of the AP hands over the packet to the ad hoc routing layer for multi-hop forwarding since all FN/MN's appear as if they were a single hop away from the AP. Before the route table is searched, address resolution is performed. If the destination node exists within the subnet, the MAC address is returned within address resolution reply message by unicast. Using this MAC address, the AP searches the route table for the entry of the destination node. A new route table entry is added when an outbound data packet is received by the AP and when a router solicitation message arrives at the AP. Hence, the AP does not flood RREQ messages for destination nodes within its subnet. If an entry is a stale one, the AP would receive a RRER message. In this case, the AP should flood a RREQ message to find an alternate path. Note that there is no tunneling or examination of routing headers involved in AP's forwarding a packet to a destination node in SOHAN.

In order to deliver a packet to an Internet host, the FN/MN's use the default route to forward the packet through the ad hoc network to the AP. The FN/MN's determine its destination is located outside the subnet by examining the destinations prefix. Packet forwarding between two nodes that are associated with different AP's must go through the wired links. Once the FN determines to use the default route to the AP, normal ad hoc routing proceeds in the ad hoc routing layer, which is transparent to the upper layer. The AP then forwards the packet towards the destination through the Internet as usual.

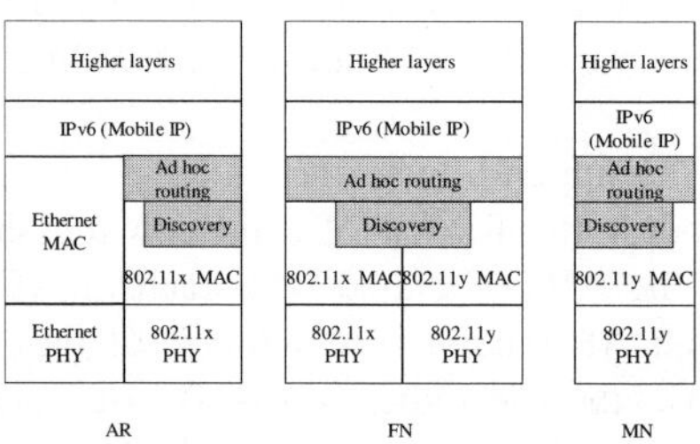

Fig. 3. Protocol stack for interworking of SOHAN and Internet

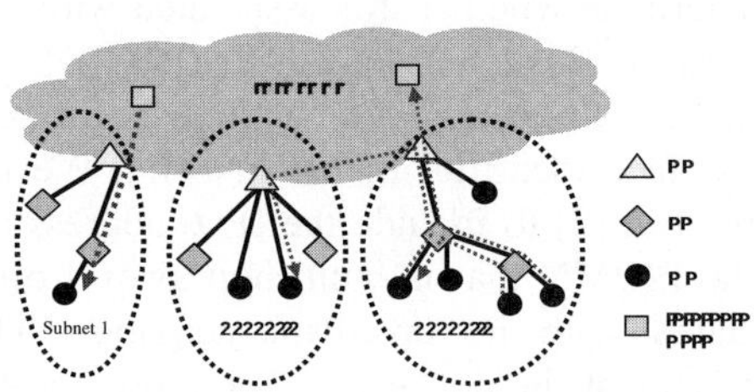

Fig. 4. Hierarchical structure of SOHAN

4 Simulation

4.1 Simulation Environments

We used ns-2 simulator with Monarch extensions to evaluate the performance of the proposed methods for interworking between SOHAN and the Internet. For the three tiers of SOHAN nodes, we implemented the topology discovery and routing protocols based on 802.11 MAC and AODV with appropriate modifications to operate directly on top of 802.11 MAC layer with MAC addresses. Procedures for LSBC and reverse path setup were implemented and added to the AODV module. Both IPv6 and mobile IPv6 modules were also added with the implementation of address autoconfiguration, routing, and gateway discovery. The main simulation parameters are summarized in Table 1. Five different combinations of AP's, FN's and MN's were used to study the performance with respect to the total number nodes. The number of AP's, FN's and MN's were chosen in accordance with the result by Liu *et al.* which states that the throughput capacity increases linearly with the number of AP's if it grows faster than the square root of the number of nodes [14]. All AP's are configured as routers and connected to a backbone router in a star topology. This backbone router is directly connected to another router to which all the Internet hosts used in the simulation are connected via a LAN. All of the wired links between the AP's and Internet hosts are given enough bandwidth (100 Mbps) that congestion does not occur in the wired section of end-to-end path between the MN and Internet host. Every communication session is established between a MN and an Internet host. That is, no traffic flows exist between two MN's or two Internet hosts.

In order to compare the performance of SOHAN in interworking with the Internet against that of flat ad hoc networks, we ran a series of simulations with the same parameters using the interworking procedures proposed by Wakikawa et al. for flat ad hoc network [8]. The flat ad hoc network consists of the AP's and MN's only. This implies all MN's are capable of forwarding packets among themselves. In each case of the simulation, the number of MN's is made equal to the number of FN's plus the number of MN's.

Table 1. Simulation Parameters

Simulation area	$1000\ m \times 1000\ m$				
Number of AP's	4	6	8	10	12
Number of FN's	6	12	22	34	48
Number of MN's	10	24	42	66	96
Number of Internet hosts	10				
Number of pkts/sec generated	4 pkts/sec				
Packet size	512 bytes				
Number of communication pairs	20				
Mobility model	Random waypoint model				
Max speed	0 m/s, 20 m/s				
Pause time	30 sec				
Simulation time	1000 sec				

4.2 Simulation Results

We measure the system throughput, average end-to-end delay, and normalized overhead while increasing the number of nodes in order to compare the performance of the two networks with respect to scalability. The number of bytes is measured for MAC frames, instead of IP packets, since routing for SOHAN takes place in L2.5.

Fig. 5(a) shows the curves for normalized control overheads. One can observe that the control overheads incurred by the flat ad hoc network increase at an exponential rate whereas they increase at a near linear rate. This indicates that flooding of control messages for gateway and route discovery in the flat ad hoc network can reduce the performance significantly. On the other hand, the amount of flood packets in SOHAN is better controlled thanks to LSBC.

Fig. 5(b) clearly indicates that SOHAN has better scalability properties than the flat network in terms of throughput. Notice that SOHAN produces lower throughput than the flat network when the number of nodes is 16. This is because SOHAN has only 10 AP/FN's and some MN's lose connections. Since all nodes are capable of forwarding in flat networks, the loss of connection is less likely to happen. Finally, Fig. 5(c) also clearly illustrates the improved scalability property achieved by SOHAN compared to that of the flat network.

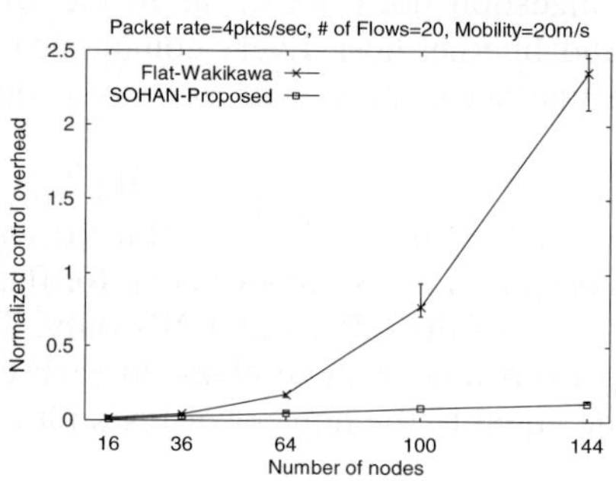

(a) Routing overhead vs. Number of nodes

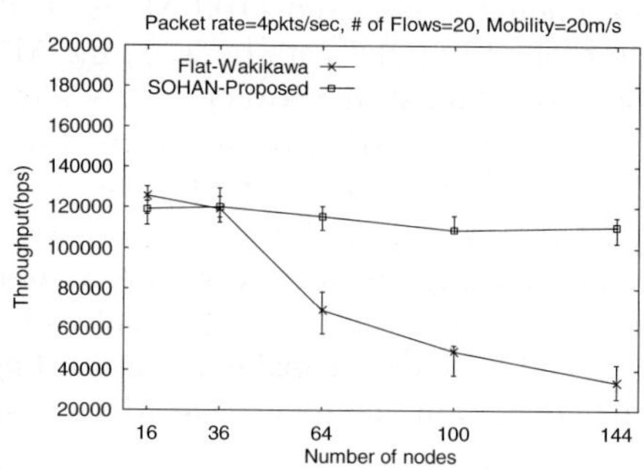

(b) Throughput vs. Number of nodes

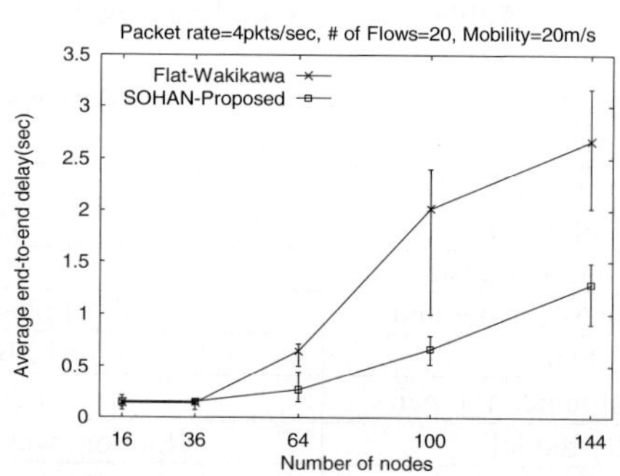

(c) Average end-to-end delay vs. Number of nodes

Fig. 5. Simulation Results

5 Conclusions

We presented a method of interworking with the Internet for self-organizing ad hoc networks based on IPv6. The most distinctive feature of our method is that, at the network layer ad hoc nodes perform normal IPv6 operations while interworking gateways act as a normal IPv6 access router. That is, all interworking functionalities along with ad hoc routing and topology discovery are hidden in the sub-IP (extended MAC) layer. This allows the interworking to be scalable as well as to exploit the hierarchical structure of SOHAN. As a result, a significant improvement in performance can be achieved in terms of control overhead, throughput and delay compared to an interworking procedure proposed for a flat ad hoc network.

References

1. Gupta, P., Kumar, P.: The Capacity of Wireless Networks. IEEE Transactions on Information Theory, Mar 2000, Vol. IT-46(2), 388-404
2. Ganu, S., Raju, L., Anepu, B., Seskar, I., Raychaudhuri, D.: Architecture and Prototyping of an 802.11-based Self-Organizing Hierarchical Ad-Hoc Wireless Network (SOHAN). *submitted to* MobiHoc (2004)
3. Lei, H., Perkins, C.: Ad Hoc Networking with Mobile IP. Proc. of the 2nd European Personal Mobile Communications Conference. (Oct, 1997) 197-202
4. Broch, J., Maltz, D., Johnson, D.: Supporting Hierarchy and Heterogeneous Interfaces in Multi-hop Wireless Ad hoc Networks. Workshop on Mobile Computing. (1999) 370-375
5. Jonsson, U., Alriksson, F., Larsson, T., Johansson, P., Maguire Jr. G.: MIPMANET - Mobile IP for Mobile Ad Hoc Networks. MobiHOC'00. (Aug, 2000) 75-85
6. Sun, Y., Belding-Royer, E., Perkins, C.: Internet connectivity for ad hoc mobile networks. International Journal of Wireless Information Networks special issue on Mobile Ad hoc Networks. (2002) Vol. 9(2), 75-88
7. Xi, J., Bettstetter, C.: Wireless Multihop Internet Access : Gateway Discovery, Routing and Addressing. Proc. of the Int. Conf. on 3G and Beyond 3G Wireless. (May, 2002) 109-114
8. Wakikawa, R., Malinen, J., Perkins, C., Nilsson, A., Tuominen, A.: Global Connectivity for IPv6 Mobile Ad Hoc Networks. draft-wakikawa-manet-globalv6-04.txt. IETF Internet Draft. (Jul, 2005)(work in progress)
9. Johnson, D., Perkins, C., Arkko, J.: Mobility Support in IPv6. IETF RFC 3775 (Jun, 2004)
10. Yang, S.: A Joint MAC Discovery-Routing Protocol for Self-Organizing Hierarchical Ad Hoc Networks. Ph.D. Thesis (2004)
11. Thomson, S., Narten, T.: IPv6 Stateless Address Autoconfiguration. IETF RFC 2462. (Dec, 1998)
12. Jeong, J., Park, J., Kim, H., Kim, D.: Ad Hoc IP Address Autoconfiguration. draft-jeong-adhoc-ip-addr-autoconf-02.txt. IETF Internet Draft. (Feb, 2004)(work in progress)
13. Narten, T., Nordmark, E., Simpson, W.: Neighbor Discovery for IPv6. IETF RFC 2461. (Dec, 1998)
14. Liu, B., Liu, Z., Towsley, D.: On the Capacity of Hybrid Wireless Networks. IEEE INFOCOM '03. (Apr, 2003) Vol. 2, 1543-1552

A Dependable Communication Network for e-Textiles

Nenggan Zheng, Zhaohui Wu, Lei Chen, Yanmiao Zhou, and Qijia Wang

College of Computer Science and Technology,
Zhejiang University, 310027, Hangzhou, P.R. China
{zng, wzh, leisteven, yanmiaozhou}@zju.edu.cn,
gotowqj@163.com

Abstract. Due to high frequent wear and tear or other faults in use, it is important to implement a fault-tolerant communicating network for e-textiles that can be easily woven into a fabric. In this paper, we introduce token buses to connect the nodes, instead of the single rings in the original e-textile token grid. The topology of the new network is described. And we also discuss the media access control protocols and the reliable operations. Simulation results show that the new e-textile communication network can improve the ability of e-textile applications to tolerate faults and provide the communicating services of less delay.

1 Introduction

Electronic textiles (e-textiles), also called smart fabrics, are emerging new computing substrates, which combine the advantages and abilities of electronic modules and textiles into one [1]. People in this research field wear off-the-shelf electrical components such as microprocessors, sensors and conductive strands into traditional clothing materials. Potential applications for e-textiles include medical monitoring, military uniforms and ambient computing devices [2]. And several prototypes based on e-textiles are presented in the papers and websites available [1, 3-5].

When the fabrics are tailored as a wearable garment or when the applications are in use, tear and wear are highly frequent, which potentially introduce link or node failures into the communication network. The failures will lead to the destruction of a local part in the communication network and even the collapse of the whole distributed system. Consequently, it is important to implement a fault-tolerant communicating network for e-textiles that can be easily woven into fabrics. The physical layout of the communication network should be run in one of two perpendicular directions, which is determined by the weaving process. Z. Nakad et al. modify the Token Grid Network (as shown in Fig. 1(a)) of [6] and propose the e-textile token grid that has an added "transverse" dimension between two token grids [7].

With the added "transverse" dimension, the e-textile token grid has the capacity to support large numbers of nodes, thus not limited by the width direction of the fabric size. By interconnecting the nodes in different grids, the added dimension can decrease the number of the nodes in a ring and directly reduce the delay time to wait for the tokens. Accounting for the demand of routing around the dormant or disabled nodes, as shown in Fig. 1, communication protocols implement the "Wrong Routing" technology [8].

V.N. Alexandrov et al. (Eds.): ICCS 2006, Part IV, LNCS 3994, pp. 938–945, 2006.
© Springer-Verlag Berlin Heidelberg 2006

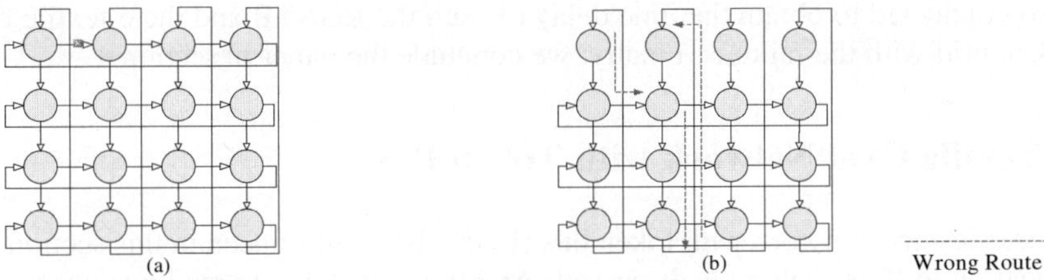

Fig. 1. (a) A link fault on the top Row ring (The red square in the figure represents a link fault.) (b) The communication on the top Row ring breaks down. "Wrong Routing" technology is used to route the data packet between the nodes on this Row ring with a link fault.

The e-textile token grid provides a communicating scheme with considerable fault-tolerant ability for e-textiles in dynamic and harsh environments. However, Z. Nakad's e-textile token grid network can not tolerate faults simultaneously present on every ring. The e-textile token grid network offers the fault-tolerant operations for the communication services, but the distributed system still breaks down in the presence of the simultaneous faults in every ring (as illustrated in Fig. 2). The reason for this result lies in the fact that the nodes are interconnected by single rings. A link failure is sufficient to block off the data traffic on a single ring. And a node fault will stop the communications on the two perpendicular rings converging on this node, regarded as two output link faults of the node. For the case (simultaneous faults on every row or column rings) is potentially high frequent in the manufacturing process or in use, the communication network for e-textiles should have the ability to recover from the corrupt state. In this paper, we aim to introduce a token bus to connect the nodes, instead of the single ring in the original token grid and construct *the e-textile grid network with token bus* (EGNTB). With the token bus and its fault-tolerant media access control protocol, faults can only affect the nodes with errors and have result in less delay time.

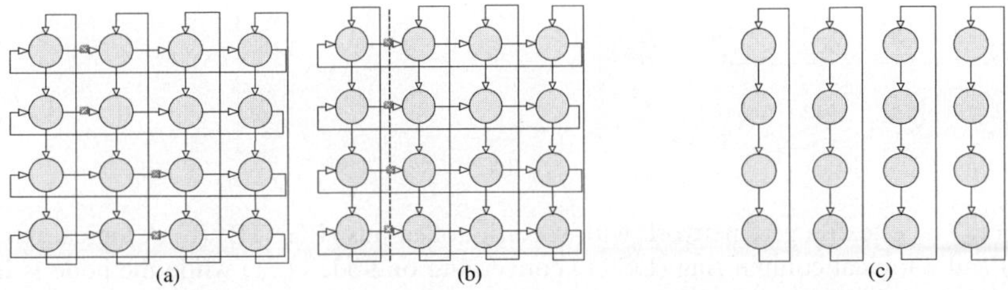

Fig. 2. Serious faults resulting in the destructions of the communication on the whole token grid. (c) is the result of (a) or (b). Note that the faults are indicated as red squares in the figure.

The remainder of this paper is organized as follows: In section 2, the topology of the e-textile token grid with specific token bus is illustrated and the operations of the networks in presence of tears are also discussed. Next, section 3 evaluates the throughput of the new network in the case of uniform load. Simulation experiments are also conducted to obtain the time delay of both the EGNTB and the e-textile original token grid with the failures. Finally, we conclude the paper in section 4.

2 e-Textile Grid Network with Token Bus

The e-textiles grid network with token bus (EGNTB) is introduced in this section. The modification to the topology is discussed. And the fault-tolerant media control access protocols are introduced in two subsections. The basic operations are described firstly in subsection 2.1, and then we will discuss the operations on the new network with the serious faults.

2.1 Topology and Basic Protocols of the EGNTB

The EGNTB (e-textile grid network with token bus) is a two-dimensional network structure arranged in M rows and N columns. With token buses in row and column, each node is connected to a row bus and a column bus. Fig. 3(a) depicts an example of four columns and four rows. The address of every node is denoted as (RowID, ColumnID). Each token bus is referred by its respective row or column number. For instance, the top row token bus is named as row-token-bus 1 (RTB1) while the left-most column token bus is identified as column-token-bus 1 (CTB1). For each node is connected onto two token buses, two pairs of transmitters and receivers are necessary to implement the EGNTB. Thanks to the bus topology of the new network, the "transverse" dimension can be easily implemented by connecting the corresponding row/column token bus of the grids in two perpendicular directions. Additional hardware interfaces are saved.

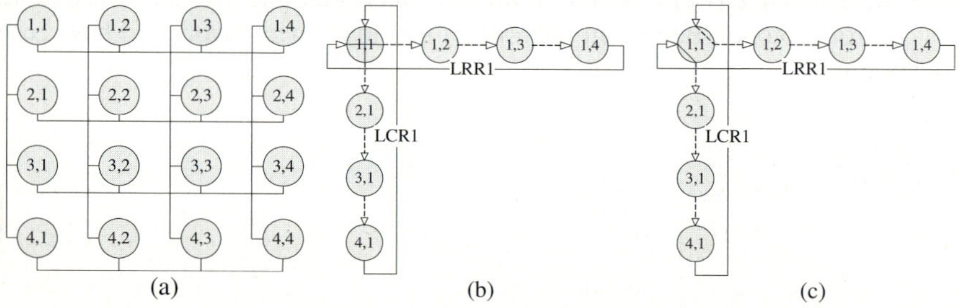

Fig. 3. (a) The e-textile grid network with specific token bus (EGNTB) (b) A logical row ring (LRR1) and a logical column ring (LCR1) converging on node (1, 1) while the node is in DR configuration (c) Node (1, 1) in SR configuration

The *specific* MAC protocol controls the tokens that are passed from a node to its active subsequent one on the buses. Different from IEEE 802.4 standard for LAN, the nodes on a row/column token bus of the EGNTB form a logical token ring without token competition. As shown in Fig. 3(b), two logical token rings converging on Node (1, 1) which are named as logical-row-ring1 (LRR1) and logical-column-ring1 (LCR1) respectively. To simplify the description of the network, we assume that all of the logical row rings have the directions pointing to the right and all of the logical column rings have the directions pointing to the bottom of the figure. Each node has the same chance to grasp the token and keep it for sending data packets on the logical token ring. The basic protocols assign every node has the same period that can be adapted to the real-life requirements of the applications. As an advantage of the token rings, except for the usual few bits of station latency, other bits of communicating buffer are not required in the node of the EGNTB.

The token circulated on every logical ring to keep the information of the network and control the access chance to the communication channel. Node set on the logical ring is preserved in the nodes by a bitmap of variable length. The bits can be mapped to the ID table for routing the packets. Dormant or disabled nodes are considered by the inactive bits that are encapsulated in the token. Because the EGNTB implement a logical topology of the original token rings, the basic operation protocol of EGNTB are similar to the original e-textile token grid networks in [6] and [7]. The basic token grid operation protocol is described formally in [6]. The nodes in the network have two configurations: the single-ring connection (SR) and the double-ring (DR) connection (shown in Fig. 3(c) and (b)). For a node with the DR configuration, the row ring and the column ring converging at this node are separate. While a node is in the merged configuration, the two rings are merged into a new token ring. If an active node wants to send a packet to another node on the same row or the same column, the node should wait for the corresponding ring token and seize it when the token arrives. And while the destination of the data packet on the different row and column, that is, in the case of the source node (R_1, C_1) and the destination node (R_2, C_2), the operation protocol will make the node (R_1, C_2) or the node (R_2, C_1) is in the SR configuration to connect the source node with the destination one into a ring. For example, if node (4, 1) sends a packet to node (1, 2), the operation protocol can require node (1, 1) or node (4, 2) be in the SR configuration to form a new merged ring and carry the data packet. The protocol also supports the inter communication between grids by supporting the address of the different grid IDs and buffers of appropriate sizes.

2.2 Fault Tolerant Operations

As discussed in the section 1, there are still two potential cases of the faults that can disable the whole network, though the existing e-textiles networks provide considerable robust operations for applications. By reason of the high frequent abrasions or for the need of manufacturing e-textile applications, the simultaneous link failures on the row rings or the column rings often happen. Firstly, a tear across the width of a fabric will sever the token grid into two parts as shown in Fig. 2(b). Both the parts will only own the column rings and the full connectivity in every part is lost. The remaining nodes can not communicate with any node on different columns or rings. Secondly, the power-efficient characteristics of e-textiles require the power consuming nodes to enter into a sleep state to conserve power energy. A node in the failure state or in the dormant state is equivalent to two link failures. That is, a node failure or dormant is treated as the link failures of its two communication outputs. In the case of every ring with a node failure or dormant, the communication of original token grid network will be completely broken out. For the cases discussed above, the existing token grid network for e-textiles needs to enhance its fault-tolerant ability to achieve more robust operations.

The difference between Z. Nakad's e-textile token grid network and EGNTB lies in that each node in latter is connected by logical token rings based on the specific MAC protocol. The new network benefits from the token bus and the protocol, thus owning the inherent high reliability. This is the point of how we enhance the fault-tolerant ability of the EGNTB and make it more suitable for e-textiles applications. The error detection algorithm with delay-time counters is used for checking the link or node faults [4] [8]. When a fault is found by the time counters in the nodes, the protocol update the inactive bits in the token. Both the node fault and the bus fault are detected and treated as two link failures on the logical token ring. For a node fault, as shown in Fig. 4(a), the fault is regarded as its input link failure and output link failure in Fig. 4(b). While in the presence of the fault on the token bus, the fault will partition the node set on the bus into two subsets, which is illustrated in Fig. 5(a) and (b). Note that the faults are indicated as red squares in the figure.

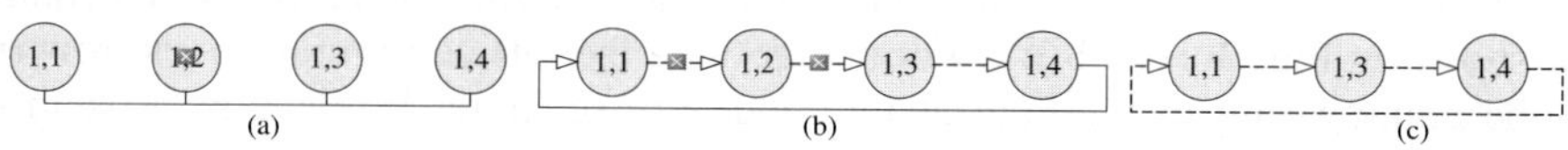

Fig. 4. Node fault: (a) Node (1, 2) in failure state (b) Two link faults equivalent on the logical token ring (c) The logical ring after fault-tolerant operations is performed

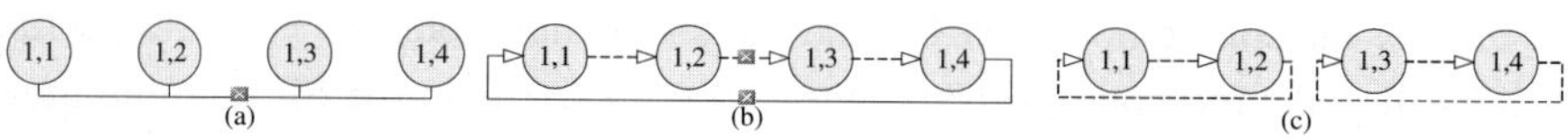

Fig. 5. A fault on the token bus: (a) A fault on the bus between node (1, 2) and node (1, 3) (b) Two link faults equivalent on the logical token ring which will partition the former ring into two sub rings (c) The result of fault-tolerant operations

When a node faults are checked, the inactive bits in the token are updated to reflect the dynamic variation on the bus and the failure node is eliminated from the logical ring. The number of the nodes on the logical token ring decreases by one, as shown in Fig. 4 (c). The drop in number of the nodes, however, leads to less time delay in transferring the data packet.

In the case of a fault on the token bus, the fault cuts the bus into two parts. Network information including inactive bits and the node set of the logical ring is updated by an error-broadcasting token. The node set of the logical ring is divided into two subsets by the fault. And for those nodes in the other subset, the associated inactive bits are set up to present that the node is not connected onto the logical ring of the current subset. Consequently, the former logical ring is partitioned into two logical token rings without any faults. The result is illustrated in Fig. 5(c).

With the token bus and its specific MAC protocols, the faults introduced in the e-textiles only affect the nodes in failure state or the local piece of the bus. When the simultaneous faults are presented on the every ring as in Fig. 2(a) and (b), local full connectivity is still preserved. No remaining node will be abrupt by the faults as in the original token grid network.

3 Network Performance

In this section, the performance of the EGNTB is described. We evaluate the approximate performance of maximum throughput and conduct simulation experiments to obtain the delay time with several faults.

Given that there is no fault on the network, the EGNTB is an original token grid network logically. Thus, the maximum capacity of the token grid network that is proposed by T.D. Todd in [10] can be used as the approximate throughput of the EGNTB. Let τ represent the node-to-node latency, t_{token} denote the token transmission time and T is the transmission time of a data packet. R presents the number of the rows. The approximate throughput of the square EGNTB with uniform load can be calculated as the following equation [10]:

$$C = \frac{2R(R+1)}{2R+1} \bullet \frac{1}{1 + t_{token}/T + \tau/T} . \quad (1)$$

And to test the network performance with faults, we also conduct simulation experiments by using the physical layer model of the EGNTB on the Matlab. An EGNTB network of 16 nodes (4 Rows, 4 Columns) is studied on different fault occasions to obtain communication costs. An original e-textile token grid of 16 nodes is also tested under the same conditions and the results are used for comparison. Node (1, 1) is selected to send a data packet to every other node in the network. A data packet transmission comes to an end at the instant when the Node (1, 1) receives a reply from the destination node. Then a new transmission follows in the same way. Due to the virtual communicating bus we implement in the Matlab model, the time cost is recorded as a factor of a time unit. The time unit is defined as the delay that a token moves from a node to its direct successor on the token ring without any faults. The cost for transmitting fifteen data packets (fifteen destination node in the grid) is recorded.

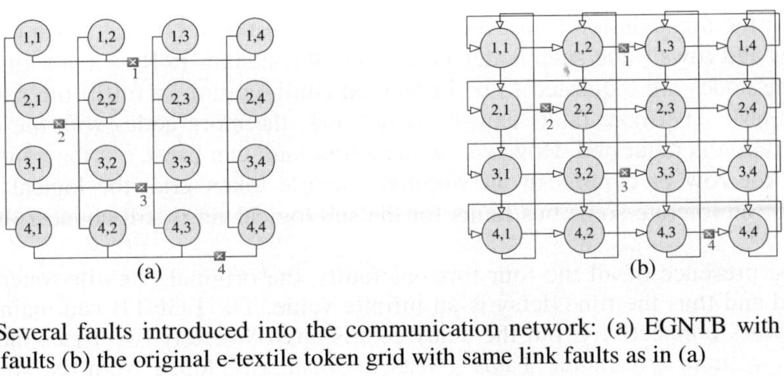

Fig. 6. Several faults introduced into the communication network: (a) EGNTB with detailed cases of faults (b) the original e-textile token grid with same link faults as in (a)

As shown in Fig. 6 (b) and (a), four cases of the link/bus faults are introduced into the communication networks. To conserve space, we only discuss the experiments on the link/bus faults.

Table 1. Simulation results

Case	EGNBT	Original e-textile Token Grid
No Faults	252	252
One Fault of case 2/3/4	252	252
The Fault of case 1	340	384
One Fault of case 2/3/4 and the fault 1 (2 faults)	386	450
Two faults of case 2/3/4 and the fault 1 (3 faults)	430	514
All four faults	476	Infinite value

It is noted that single fault of case 2 or 3 or 4 has no effect on the total cost of the delay time. The fault is tolerated by the grid topology of the network.

In the case of fault 1, when a data packet is transmitted to a destination node on the Row 1, the original token grid with link fault 1 has to bypass the disconnected ring by a "Wrong Route". For example, if the destination of the data packet is node (1, 2), node (1, 1) routes the packet by two merge configurations at Node (2, 1) and Node (4, 2) respectively to turn around the link with fault 1. For the EGNTB with the fault 1, additional delay is observed in the two transmissions which have the destination nodes are node (1, 3) and node (1, 4) respectively. The RTB 1 is cut into two pieces and thus additional merge configuration is requested to route the data packet. Local connection of the two logical sub rings on row 1 is preserved. Meanwhile, the data packet transmission from node (1, 1) to node (1, 2) has less delay on the sub logical ring than that on the original proper logical ring. Because of the factors, the additional delay cost is less than that of the original token grid.

With the link fault 1, the addition of the fault 2, 3 or 4 to the networks has the same effect on the transmission from Node (1, 1) to the destination nodes not on the same columns and rows (nine such nodes in the Fig. 3(a) neither in Row 1 nor in Column 1). Some nodes are requested to be in merged configuration to fulfill the data transmission task. The more faults exit in the network, the more nodes with the merging configuration is requested. However, while a link fault can break out the communication on the row or column in the original e-textile token grid, the logical rings in EGNBT can tolerate some bus faults for the sub logical rings still connect the nodes locally.

In the presence of all the four link/bus faults, the original e-textile token grid is disabled and thus the time delay is an infinite value. The EGNTB can maintain the full network connectivity, but the delay cost is approximately two times that of the network without any faults. Table 1 lists the results obtained from the simulation experiments. These results show the EGNTB can improve the ability of e-textile applications to tolerate faults. On the occasions of same faults discussed above, the new network can provide the communication services of less delay time. Furthermore, the feature of a graceful variation of performance with the number of faults is also inherited.

4 Conclusions

The harsh environment of e-textile applications requires a fault-tolerant communication scheme to reduce the time cost of faults or low power operations. In this paper, we introduce token buses to connect the nodes in e-textiles, instead of the single rings. Based on the specific protocol, the EGNTB implements the operations logically similar to the original token grid network. Thanks to the token bus, the new network inherits the high reliability of the bus topology and has the same approximate throughput with uniform load.

Simulation results show the new communication network proposed in this paper can enhance the ability of e-textile applications to tolerate faults. In the presence of same faults, the new network can provide the communication services of less delay time. And even when a fabric is torn into several parts, the full-connectivity in the local parts is also preserved in every fragments of the fabric. Furthermore, the EGNTB inherits the feature that the network performance degrades gracefully as the number of faults increases.

References

1. D. Marculescu, R. Marculescu, N. H. Zamora, P. Stanley-Marbell, P. K. Khosla, S. PARK, S. Jayaraman, S. Jung, C. Lauterbach, W. Weber, T. Kirsein, D. Cottet, J. Grzyb, G. TrÖster, M. Jones, T. Martin, and Z. Nakad, "Electronic Textiles: A Platform for Pervasive Computing", Proceedings of the IEEE, VOL. 91, NO. 12, 1995-2018, December 2003.
2. M. Jones, T. Martin, Z. Nakad, R. Shenoy, T. Sheikh, D. Lehn, and J. Edmison, "Analyzing the Use of E-textiles to Improve Application Performance", IEEE Vehicular Technology Conference 2003, Symposium on Wireless Ad hoc, Sensor, and Wearable Networks (VTC 2003)(extended abstract), October 2003.
3. Tanwir Sheikh, Modeling of Power Consumption and Fault Tolerance for Electronic Textiles, Bradley Department of Electrical and Computing Engineering, Virginia Tech, September 2003.
4. Zahi Nakad, Architecture for e-Textiles. PhD thesis, Bradley Department of Electrical and Computing Engineering, Virginia Tech, 2003.
5. The Georgia Tech wearable motherboard: The intelligent garment for the 21st century (1998). [Online]. Available:http://www.smartshirt.gatech.edu.
6. T. D. Todd, "The Token Grid: Multidimensional Media Access for Local and Metropolitan Networks", Proceedings of the eleventh Annual Joint Conference of the IEEE Computer and Communications Societies, pp. 2415-2424, 1992.
7. Z. Nakad, M. Jones, and T. Martin, "Communications in Electronic Textile Systems", Proceedings of the 2003 International Conference on Communications in Computing, pp. 37-43, June 2003.
8. Z. Nakad, Mark Jones, and Thomas Martin, "Fault-Tolerant Networks for Electronic Textiles", Proceedings of the 2004 International Conference on Communications in Computing, Las Vegas, June 2004.
9. F. E. Ross, "An Overview of FDDI: The Fiber Distributed Data Interface", IEEE J. Select. Areas Commun., vol. 7, pp. 1043-1051, Sept. 1989.
10. T. D. Todd, "The Token Grid Network", IEEE/ACM Transactions on Networking, vol. 2, No. 3, pp. 279-287, June, 1994.

EAR-RT: Energy Aware Routing with Real-Time Guarantee for Wireless Sensor Networks*

Junyoung Heo[1], Sangho Yi[1], Geunyoung Park[1], Yookun Cho[1], and Jiman Hong[2,]**

[1] Seoul National University
{jyheo, shyi, gypark, cho}@ssrnet.snu.ac.kr
[2] Kwangwoon University
gman@daisy.kw.ac.kr

Abstract. Most energy aware routing algorithms focus on the increasing the lifetime and long-term connectivity of the wireless sensor networks. But energy efficiency sacrifices the communication delay between source and sink node. Therefore, many researchers have mentioned of energy and delay trade-offs. But the delay was not the main concern. In this paper we propose EAR-RT, an real-time guaranteed routing protocol for wireless sensor networks without harming energy awareness. Simulation results show that our real-time routing algorithm provides real-time guaranteed delivery while network is stable.

1 Introduction

Wireless sensor networks typically consist of hundreds or thousands of sensor nodes deployed in a geographical region to sense events. Wireless sensor networks provide a high-level description of the events being sensed. They are used in many applications such as environmental control, civil engineering, automatic manufacturing, habitant monitoring and so on. They can be used even in harsh environment [1, 2]. Therefore, their developing entails significant technical challenges due to the many environmental constraints.

There are many research issues in the area of sensor networks because of existing a lot of constraints of sensor nodes. The energy constraints is the most important one of them. Sensor nodes are supplied with power generally by built-in battery. In addition, transmission of a packet in wireless sensor networks is expensive. Therefore, routing, which is highly correlated with energy consumption, is a critical factor determining the performance of network.

Most previous researches on routing [3, 4, 5] focused on the algorithm design and performance evaluation in terms of the packet transmission overhead

* The present research was conducted by the Research Grant of Kwangwoon University in 2006, and was supported in part by the Brain Korea 21 Project.

** Corresponding author.

V.N. Alexandrov et al. (Eds.): ICCS 2006, Part IV, LNCS 3994, pp. 946–953, 2006.
© Springer-Verlag Berlin Heidelberg 2006

and loss rate. On the other hand, some routing algorithms were proposed in [1, 2, 6, 7, 8] in order to improve scalability of routing algorithms for large sensor networks.

In our proposed algorithm, the energy drain rate and the residual energy of each sensor node are used as the routing information, updated per almost every communication with energy awareness. To obtain such information, our algorithm takes advantage of the characteristic of wireless network, called overhearing. When two nodes communicate with each other, the neighboring nodes can overhear the packet being transmitted. Overhearing enables each sensor node to obtain the energy information of neighboring nodes without additional overhead. Then, they update their routing tables using this information. Therefore, our algorithm can provide longer connectivity of sensor networks through efficient use of energy among the nodes in the network. In addition, our algorithm considers real-time communication. Real-time packets must be arrived at sink node before deadline. But existing energy-aware routing delayed packets to save energy. So, these routing algorithms are not suitable for real-time packet transmission.

The rest of the paper is organized as follows. In Section 2 we present related works. Section 3 describes a new energy aware routing algorithm with real-time property. Section 4 presents and evaluates the performance of the proposed algorithm against the prior algorithm. Finally, some conclusions are given in Section 5.

2 Related Works

There are a lot of research results on sensor network routing for energy awareness and real-time communications.

In [3], Ganesan et al. proposed the use of braided multi-paths instead of completely disjoint multi-paths in order to keep the cost of maintaining the multi-paths low. The costs of such alternate paths are also comparable to the primary path because they tend to be much closer to the primary path.

In [4], Chang and Tassiulas proposed an algorithm for maximizing the lifetime of a network by selecting a path whose nodes have the largest residual energy. In this way, the nodes in the primary path retain their energy resources, and thus avoid having to continuously rely on the same route. This contributes to ensuring longer life of a network.

In [7], Chang and Tassiulas also applied this combined metric concept for direct routing. Their algorithm is proposed to maximize the lifetime of a network when data rate is known. The main idea is to avoid using the nodes with low energy and to choose the shortest path.

In [5], Li et al. proposed an algorithm in which the constraint of the residual energy of a route is relaxed slightly to select a more energy efficient route. It is generally known that a route with the largest residual energy for routing a packet entails high energy consumption. Therefore, there is a trade-off between minimizing the total consumed energy and the residual energy of the network. In that paper, groups of the sensors in geographic proximity were clustered together into a zone and each zone was treated as a separate entity and allowed to determine how it will route a packet across.

Lu et al. proposed real-time communication architecture for sensor networks [9]. They proposed velocity monotonic scheduling for packet scheduling on a node. They did not consider end-to-end delay between source and sink.

He et al. proposed end-to-end real-time communication protocol, SPEED[10]. SPEED maintains a desired delivery speed through a combination of feedback control and non-deterministic geographic forwarding. By doing so, it achieved real-time communication. However it did not consider energy awareness.

3 EAR-RT: Energy Aware Routing with Real-Time Guarantee

We had proposed energy aware routing with dynamic probability scaling (EAR-DPS) in our previous work[11]. This work is based on the energy aware routing (EAR) proposed by Shah and Rabaey[1]. EAR uses maintenance phase to update residual energy of sensor nodes. But EAR-DPS uses overheard packets without maintenance phase to update residual energy. Our proposed algorithm to support real-time communication is based on EAR-DPS.

EAR and EAR-DPS find multiple routes, if any, from source to destination nodes. Each route is assigned a probability of being selected to transmit a packet, based on residual energy and the energy for communications at the nodes along the route. Then, based on these probabilities, one of the candidate routes is chosen in order to transmit a packet. The probability is proportional to the energy level at each node, so the routes with higher energy are more likely to be selected. EAR protects any route from being selected all the time, preventing the energy depletion [1, 11]. However, the end-to-end delay of communication may increase. So, EAR and EAR-DPS cannot be used for real-time communication. EAR-RT resolves this problem by considering the deadline of packets.

The operation of EAR-RT consists of two phases: setup phase and data communication phase. Basically EAR and EAR-DP use similar process. In setup phase, the sink node initiates a route request and a routing table is built up by finding all the paths from a source to the sink and their energy cost and time delay. In data communication phase, data packets are sent from the source to the sink. Each intermediate node forwards the packet to a neighboring node, which is chosen randomly among neighboring nodes that can deliver the packet in time. This probability is inversely proportional to the energy cost of neighboring nodes.

Before describing our EAR-RT algorithm, we make a list of some common notations used in this paper in Table 1.

3.1 Setup Phase

The sink node initiates the setup phase by flooding a route request message. A route request message has energy cost($Cost_i$), time delay($Time_i$) and residual energy(R_i) of a node. $Cost_{sink}$, energy cost of the sink node is 0. $Time_{sink}$ is also 0.

Table 1. List of notations used in this paper

$C_{i,j}$	Expected energy cost to send a packet from node i to the sink node via node j.
$Cost_i$	Expected energy cost to send a packet from node i to the sink node.
$E_{i,j}$	Energy cost to send a packet from node i to node j directly.
$T_{i,j}$	Expected time delay to send a packet from node i to the sink node via node j.
$Time_i$	Expected time to send a packet from node i to the sink node.
$H_{i,j}$	Time to send a packet from node i to node j directly. It includes transmission delay, queuing delay and so on.
$P_{i,j}$	Probability to select node j to forward a packet at node i.

When a node receives a route request message from neighboring nodes, it adds neighboring nodes only which are closer to sink node than itself into routing table. Then it calculates its energy cost and time delay. Then it forwards a route request message with its energy cost and time delay.

If node i receives a route request message from node j which is closer to sink node than node i, then

$$
\begin{aligned}
Cost_i &= \sum_{j \in RT} P_{i,j} C_{i,j} \\
Time_i &= \sum_{j \in RT} P_{i,j} T_{i,j} \\
C_{i,j} &= Cost_j + E_{i,j} \\
T_{i,j} &= Time_j + H_{i,j} \\
P_{i,j} &= \frac{1/C_{i,j}}{\sum_{k \in RT} 1/C_{k,i}} \\
E_{i,j} &= (d_{ij})^{\alpha} / R_j^{\beta},
\end{aligned}
$$

where RT is routing table, d_{ij} is the distance between node i and node j, R_j is the residual energy at node j normalized to its initial energy, α and β are weighting factors[1, 11].

The R_i, residual energy decreases as time goes on. So the residual energy should be informed to neighbor nodes frequently. EAR-DPS and EAR-RT make use of overhearing packets to update the residual energy. In the wireless networks, when two nodes communicate with each other, the neighboring nodes of a sender can hear the packet being transmitted.

3.2 Data Communication Phase

After setup phase is completed, each sensor node sends data packets that are collected to the sink node. We consider only a real-time packet with deadline. The deadline may be predefined or determined by nodes at every transmission. The laxity is embedded in a packet and re-calculated at every node on a path to the sink node.

The path to the sink node is built up while transmission of a packet. Every node - including source node - selects the next node according to the following rules:

1. Select nodes in routing table which can deliver a packet within deadline.

$$RT' = \{j | T_{i,j} \leq laxity\_of\_a\_packet\}$$

2. Calculate probability based on the RT'

$$P'_{i,j} = \frac{1/C_{i,j}}{\sum_{k \in RT'} 1/C_{k,i}}$$

3. Select the next node randomly by the probability, P'.
4. When a node i selects node j as the next hop, node i estimates the energy needed to transmit a data packet to node j and its residual energy after transmitting.
5. Node i forwards the data packet including its residual energy R_i, current cost $Cost_i$, $Time_i$ and new laxity that decreases by $H_{i,j}$.
6. Every node k that overhear the packet and its routing table has the node i recomputes $C_{k,i}$ based on R_i and $Cost_i$ in the packet. Also $T_{k,i}$ is recomputed based on $Time_i$. Then it recomputes the probability P, $Cost_k$ and $Time_k$ based on new $C_{k,i}$ and $T_{i,i}$. At this moment, the residual energy of node i is reflected in its neighbors' routing table.
7. This process continues until the data packet reaches the sink node.

4 Experimentation

We ran a simulation to evaluate our EAR-RT. In this simulation, the area of the sensor networks was assumed to be 100m×100m. The number of the nodes in the network was assumed to be 100 nodes – one node is sink node while the others are sensor nodes. The sink node was located at the center of the field and the sensor nodes are placed randomly in the field. All of the sensor nodes sent data to the sink node at fixed interval. The length of interval between transmissions at each node was all identical. This interval can be considered as a virtual time unit. That is, all of the sensor nodes send their data to the sink node just once a single time unit. Every packet from sensor nodes has a deadline. The deadline is randomly chosen between 5 and 8 hops.

Every node was given an identical amount, 0.3J, of initial energy. Energy for transmission was assumed 20nJ/bit + 1pJ/bit/m^3. Energy for reception was assumed 30nJ/bit[1, 11]. The packet length was assumed 256 bits. Energy metric function with α=1 and β=50 was used. In this simulation, the following assumptions were made:

- Every node knows its position and distance between itself and other nodes.
- Every node has identical maximum radio range, 20m.

Figure 1 shows the residual energy of each node at time 600 of a network using EAR-RT and EAR-DPS. Each rectangle indicates sensor node and number in a

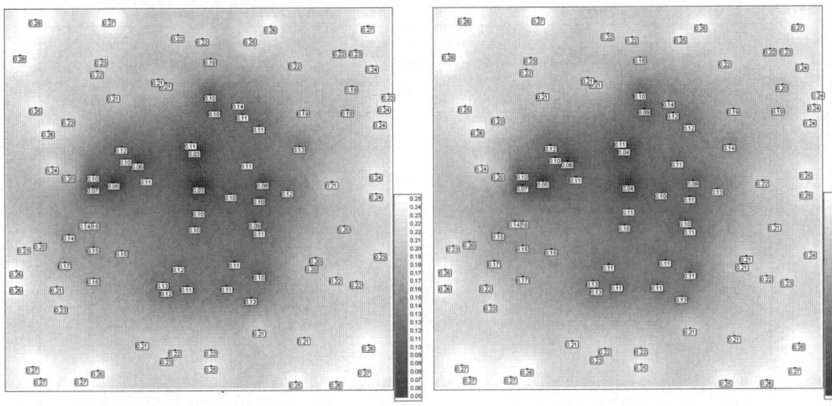

Fig. 1. (a) Residual energy of EAR-RT (b) Residual energy of EAR-DPS

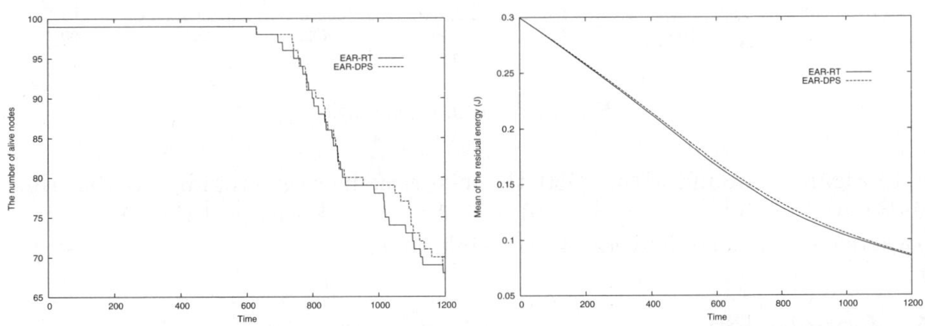

Fig. 2. (a) The number of active nodes during communication phase (b) Mean of residual energy of each node

rectangle indicates residual energy of a node. Nodes in dark area have smaller energy than those in light area. This figure shows that both algorithms affect the residual energy of nodes alike.

Figure 2(a) shows the number of active nodes in the network. Figure 2(b) shows the average of residual energy of each node. These figures tell us that the energy awareness of EAR-RT is not inferior to EAR-DPS. In case of EAR-RT, the number of alive nodes and average residual energy are slightly less than EAR-DPS. However EAR-RT improved the real-time property by sacrificing its energy awareness a little.

Figure 3 shows the deadline miss ratio of EAR-RT and EAR-DPS. EAR-RT delivered more packets in time compared to EAR-DPS. While the network is stable, EAR-RT delivered most packets in time. But some packets of EAR-DPS cannot meet the deadline because considering only energy. From these results, our EAR-RT is suitable for real-time communications while keeping energy awareness in sensor networks.

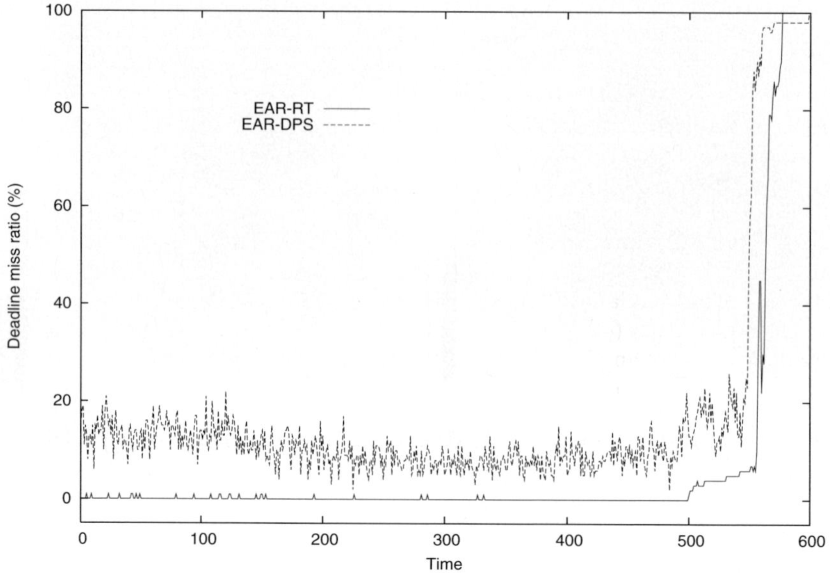

Fig. 3. Deadline miss ratio

In Figure 3, around of time 550, the miss ratio increases rapidly. At that time, nodes around sink node were dead and the network was partitioned. So, most packets could not be delivered to the sink node.

5 Conclusion

We proposed EAR-RT, an energy aware routing algorithm with real-time guarantee for high survivability and real-time communication of wireless sensor networks. Our algorithm provide real-time communication without harming energy awareness of existing energy aware routing algorithm, EAR-DPS. Simulation results showed that our algorithm reduced deadline miss ratio compared to existing algorithm. Consequently our algorithm can provide longer connectivity of sensor networks and ensure efficient use of energy among the nodes in the sensor network with guaranteeing real-time properties.

References

1. Shah, R., Rabaey, J.: Energy aware routing for low energy ad hoc sensor networks. In: IEEE Wireless Communications and Networking Conference (WCNC), Orlando, FL. (2002)
2. Heinzelman, W.R., Chandrakasan, A., Balakrishnan, H.: Energy-efficient communication protocol for wireless microsensor networks. In: HICSS '00: Proceedings of the 33rd Hawaii International Conference on System Sciences-Volume 8, Washington, DC, USA, IEEE Computer Society (2000) 8020

3. Ganesan, D., Govindan, R., Shenker, S., Estrin, D.: Highly-resilient, energy-efficient multipath routing in wireless sensor networks. SIGMOBILE Mob. Comput. Commun. Rev. **5** (2001) 11–25
4. Chang, J.H., Tassiulas, L.: Maximum lifetime routing in wireless sensor networks. IEEE/ACM Trans. Netw. **12** (2004) 609–619
5. Li, Q., Aslam, J., Rus, D.: Hierarchical power-aware routing in sensor networks. In: DIMACS Workshop on Pervasive Networking. (2001)
6. Schurgers, C., Srivastava, M.B.: Energy efficient routing in wireless sensor networks. In: IEEE Military Communications Conference (MILCOM). (2001) 357–361
7. Chang, J.H., Tassiulas, L.: Energy conserving routing in wireless ad-hoc networks. In: INFOCOM (1). (2000) 22–31
8. Braginsky, D., Estrin, D.: Rumor routing algorthim for sensor networks. In: WSNA '02: Proceedings of the 1st ACM international workshop on Wireless sensor networks and applications, New York, NY, USA, ACM Press (2002) 22–31
9. Lu, C., Blum, B.M., Abdelzaher, T.F., Stankovic, J.A., He, T.: Rap: A real-time communication architecture for large-scale wireless sensor networks. Technical report, Charlottesville, VA, USA (2002)
10. He, T., Stankovic, J.A., Lu, C., Abdelzaher, T.: Speed: A stateless protocol for real-time communication in sensor networks. In: ICDCS '03: Proceedings of the 23rd International Conference on Distributed Computing Systems, Washington, DC, USA, IEEE Computer Society (2003) 46
11. Park, G., Yi, S., Heo, J., Choi, W.C., Jeon, G., Cho, Y., Shim, C.: Energy aware routing with dynamic probability scaling. Lecture Notes in Computer Science **3642** (2005) 662–670

A Design of Energy-Efficient Receivers for Cluster-Head Nodes in Wireless Sensor Networks*

Hyungkeun Lee[1] and Hwa-sung Kim[2]

[1] Department of Computer Engineering and
[2] Department of Electronic Communications Engineering
Kwangwoon University
Seoul, Korea
{hklee, hwkim}@daisy.kw.ac.kr

Abstract. In wireless sensor networks, embedded sensor nodes equipped with sensing, computation, and communication resources are generally constrained in energy supply. An efficient way to save energy of the network is to partition the network into distinct clusters with specific nodes called cluster-head. Since, however, higher level of energy consumption at cluster-head nodes might cause more damage to the network, energy-saving in cluster-head nodes is critical. In this paper, we propose an energy-efficient receiver for cluster-head nodes in two-tiered wireless sensor networks. The receiver performs multiuser detection and channel decoding jointly in a CDMA system, where information about channel codes is utilized for multiuser detection. This receiver exhibits improved performance of the sensor networks in terms of overhead, delay and power, and its excellent performance is shown via simulation.

1 Introduction

Recent advances in MEMS (micro electro-mechanical system) technology and embedded software have resulted in cheap and small devices with sensing, computing and wireless communication capabilities. A network of these devices could be utilized for information gathering and distributed sensing in many civil, military and industrial applications. The use of wireless medium for communication provides the network operating to convey collected information to a sink node without any fixed infrastructure. Wireless sensor networks pose many new challenges primarily because the sensor nodes are resource constrained, where sensor nodes are powered by small batteries that cannot be replaced [1]. Under this hard energy constraint, sensor nodes can only transmit a finite number of bits in their lifetime. Consequently, reducing the energy consumption per bit for end-to-end transmissions becomes an important design consideration for such networks. Since all layers of the protocol stack contribute to the energy per bit consumed in its end-to-end transmission, energy minimization requires a joint design of the underlying hardware where the energy is actually expended. Another efficient way to save energy of the network is to partition the network into

---

* This work is supported by the ubiquitous Autonomic Computing and Network Project, the Ministry of Information and Communication (MIC) 21st Century Frontier R&D Program in Korea, and the Research Grant of Kwangwoon University in 2005.

V.N. Alexandrov et al. (Eds.): ICCS 2006, Part IV, LNCS 3994, pp. 954 – 961, 2006.
© Springer-Verlag Berlin Heidelberg 2006

distinct clusters with specific nodes called cluster-head, which is the multi-tiered wireless sensor network. However, higher level of energy consumption at cluster-head nodes might cause more damage to the network since such cluster-head nodes also limit accessibility of other sensor nodes. Therefore, the energy-saving mechanism in cluster-head nodes is critical issue and determines the performance and lifetime of such networks [2].

We only consider the simplified case where interference is eliminated by using direct sequence code division multiple access (DS-CDMA) schemes. DS-CDMA is a popular choice as a multiple access scheme within a cluster of wireless sensor networks, where nodes are assigned different spreading codes. The transmitter of each sensor node sends its data by modulation with its own sequence. The conventional DS-CDMA receiver consists of a bank of matched filters to obtain nodes' data from the received signal. However, in multiuser detection (MUD) receivers, interference from the other signals is removed by subtracting it from the desired signal. This is possible because the correlation properties between the signals are known at the cluster-head node. MUD provides a means of reducing the effect of multiple access interference (MAI), and hence increases the system capacity [3]. The error correction capability of channel codes plays an important role in achieving performance improvement in noisy and interference-limited environments. Therefore, the transmitted data from sensor nodes to the cluster-head node are usually protected against errors due to MAI and noise by employing channel codes in DS-CDMA systems, as depicted in Figure 1.

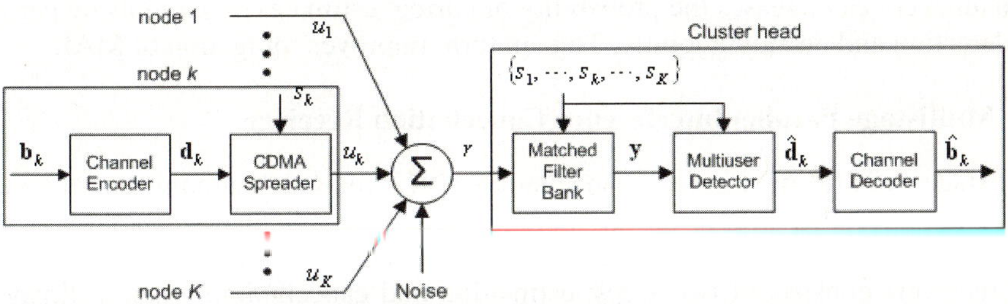

Fig. 1. *K*-node Cluster System with MUD and Channel Codes

New design approaches for MUD that include the consideration of channel codes were proposed in [4], where MUD provides soft outputs including reliability measurements to the channel decoder. Joint detection-decoding (JDD) receivers have been proposed recently to improve the performance of receivers [5]. These JDD receivers repeat multiuser detection and error correction in an iterative manner. Therefore, the complexity of receiver increases linearly with the number of iterations and the complexity of each stage depends on the frame size of the channel codes. In [6], a JDD receiver was proposed employing multistage parallel interference canceller (PIC), where each stage utilizes convolutional codes for estimation and cancellation. Multistage PIC is known to be a simple and effective technique for mitigating MAI in DS-CDMA systems. However, its performance can be significantly degraded due to incorrect interference decisions that are subtracted from the received signal. It minimizes the effect of incorrect cancellation with the aid of convolutional codes, and also decreases the computation complexity. Cluster-head nodes with such receivers have interference and noise reduction, so that the nodes can operate with less overhead and less latency. The remainder of this paper is organized as follows. Section 2 presents the concept of JDD receivers employing PIC and convolutional codes, and the analysis of the JDD receiver performance. The application of the receivers in wireless sensor networks is evaluated by simulation in Section 3. Finally, concluding remarks are given in Section 4.

2 Joint Detection-Decoding Receiver

MUD receivers jointly estimate the transmitted signals of all nodes in the system. A JDD receiver performs multiuser detection and channel decoding jointly, where information about channel codes is utilized during multiuser detection. Let us assume a DS-CDMA system employing channel codes, where a sequence of information bits $\mathbf{b}_k$ is encoded into the sequence of channel symbols $\mathbf{d}_k$ and the sequence of channel symbols is transmitted via the multiuser channel resulting in the matched filter bank output $\mathbf{y}_k$, for a node *k*. The outputs $\mathbf{y}_k$ include MAI due to the other *k*-1 nodes and are correlated over a frame of channel codes. Therefore, the symbol estimates $\hat{\mathbf{d}}_k$ from the matched filter bank output $\mathbf{y}_k$ can be jointly detected over nodes and decoded in time to obtain the information estimates $\hat{\mathbf{b}}_k$. The error correction capability of channel codes increases the probability of correct estimates of symbols by performing detection and decoding jointly. This, in turn, improves mitigation of MAI.

2.1 Multistage Parallel Interference Cancellation Receiver

Multistage parallel interference cancellation (PIC) receivers attempt to remove the MAI through signal processing at multiple stages for all nodes. They have much lower complexity than linear MUD receivers. As depicted in Figure 2, each stage of PIC receivers consists of two steps: estimation and cancellation, where estimates of all nodes' signals are generated and MAI for all nodes are constructed and canceled from the received signal based on the estimates. As the accuracy of the estimates improves through the multiple stages, the performance of receiver also improves [5].

The estimation step computes the estimates of symbols $\hat{d}_k(l \mid l-1)$ from the *a posteriori* signal of the previous stage $y_k(l-1)$ for a node *k*, where *l* is the number of stage. The estimates $\hat{d}_k(l \mid l-1)$ are treated as *a priori* information for the next stage. The cancellation step tries to remove MAI present in the node's signal $y_k(l)$ using the correlation matrix $\mathbf{R}$,

$$y_k(l) = y_k(0) - \sum_{i \neq k} \rho_{ik} \hat{d}_i(l \mid l-1) \tag{1}$$

The final decision is made based on the *a posteriori* signal of the last stage, $y_k(L)$. In a conventional multistage PIC receiver, the estimation step makes a hard decision on the *a posteriori* signal of the previous stage $y_k(l-1)$ to decide the estimate of symbols $\hat{d}_k(l \mid l-1)$. However, this estimation step may lead to an incorrect decision that may cause an incorrect cancellation of interference. Such incorrect cancellation is propagated to the subsequent stages of the receiver. Therefore, the performance of multistage PIC receivers improves as the probability of correct decision on the estimates of symbols $\hat{d}_k(l \mid l-1)$ increases at each estimation step. In order to improve the performance of such receivers, JDD receivers that utilize channel codes at estimation steps are described in the next subsections.

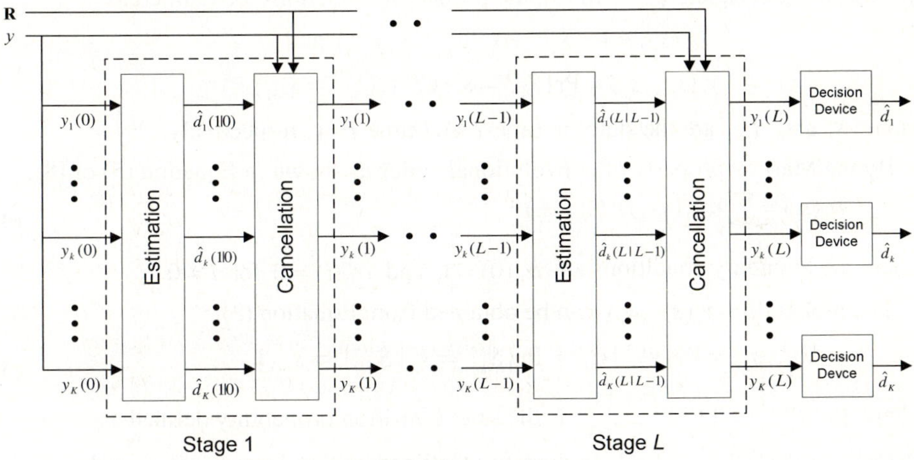

Fig. 2. Multi-stage PIC Receiver

Performance evaluation of the JDD receivers is shown in [6]. These results show the obvious performance improvement of JDD receivers, where it is shown that the systems employing JDD schemes increase the user capacity remarkably or decrease the bit error rate.

2.2 Joint Detection-Decoding Receiver with Partial Decoding Information

The JDD receiver employs a generalized version of the decoding method in that it minimizes the probability of error in partial words consisting of one or more channel symbols. In the trellis diagram of the Viterbi algorithm, a path consists of states and branches representing state transitions. Each entire codeword $\mathbf{d}_k = \left[\mathbf{d}_k^{(1)}, \cdots, \mathbf{d}_k^{(N+l)}\right] = \left[d_k^{(1)}, \cdots, d_k^{(n(N+l))}\right]$ can be represented by a unique state sequence $\mathbf{S}_k = \left[S_k^{(0)}, \cdots, S_k^{(t)}, \cdots, S_k^{(N+l+1)}\right]$, where $S_k^{(0)} = 0$ and vice versa. The partial codeword $\mathbf{d}_k^{(t)} = \left[d_k^{((n-1)t+1)}, \cdots, d_k^{(nt)}\right]$ corresponds to each state transition (branch) from the state $S_k^{(t-1)}$ at time $t-1$ to the state $S_k^{(t)}$ at time t, where $t = 1, \cdots, N+l$. Therefore, the entire codeword of channel symbols $\mathbf{d}_k$ and the partial codeword of channel symbols $\mathbf{d}_k^{(t)}$ correspond to the path and branches of convolutional codes, respectively. The partial codeword $\mathbf{d}_k^{(t)}$ also contains partial decoding information about the transmitted symbols. It can reduce the complexity of the receiver at the expense of suboptimum decoding at estimation steps.

The estimation steps in the proposed JDD receiver employ $\mathbf{y}_k^{(t)}$, i.e., the *a posteriori* signal corresponding to the partial codeword of channel symbols $\mathbf{d}_k^{(t)}$. Then they obtain the estimates of the partial codeword $\hat{\mathbf{d}}_k^{(t)}$ based on the signal $\mathbf{y}_k^{(t)}$ and the probabilities of starting states $S_k^{(t-1)}$, and calculate the probabilities of ending states $S_k^{(t)}$ for the next signal $\mathbf{y}_k^{(t+1)}$. In [8], the probability functions are defined as

$$\alpha_t(s_t) = \Pr\{S_k^{(t)} = s_t; \mathbf{y}_k^{(1)}; \cdots; \mathbf{y}_k^{(t)}\} \text{ and}$$
$$\gamma_t(s_{t-1}, s_t) = \Pr\{S_k^{(t)} = s_t; \mathbf{y}_k^{(t)} \mid S_k^{(t-1)} = s_{t-1}\}, \tag{3}$$

where s_t and s_{t-1} are the states at time t and time $t-1$, respectively.

By the Markov property of convolutional codes as shown in Equation (5) of [8],

$$\alpha_t(s_t) = \sum_{s_{t-1}} \alpha_{t-1}(s_{t-1}) \gamma_t(s_{t-1}, s_t), \tag{4}$$

where the boundary conditions are $\alpha_0(0) = 1$, and $\alpha_0(i) = 0$ for $i \neq 0$.

The probability $\gamma_t(s_{t-1}, s_t)$ can be obtained from Equation (3),

$$\gamma_t(s_{t-1}, s_t) = \Pr\{\mathbf{y}_k^{(t)} \mid \mathbf{d}_k^{(t)}\} \cdot \Pr\{S_k^{(t)} = s_t \mid S_k^{(t-1)} = s_{t-1}\}, \tag{5}$$

where $\Pr\{S_k^{(t)} = s_t \mid S_k^{(t-1)} = s_{t-1}\}$ is the state transition probability defined as $\Pr\{S_k^{(t)} = s_t \mid S_k^{(t-1)} = s_{t-1}\} = 1$, if there is a trellis transition from s_{t-1} to s_t, and $\Pr\{S_k^{(t)} = s_t \mid S_k^{(t-1)} = s_{t-1}\} = 0$, otherwise. From Equation (4), the MAP decoding problem on a state transition (branch) becomes the procedure to maximize the probability of the next state based on $\alpha_{t-1}(s_{t-1})$ and $\gamma_t(s_{t-1}, s_t)$,

$$\hat{S}_k^{(t)} = \arg\max_{s_t \in \mathrm{S}} [\log \Pr(\alpha_t(s_t))] \text{ for } t = 1, \cdots, N + l, \tag{6}$$

where S is the set of states, and s_t and s_{t+1} at time t and time t+1, respectively.

Finally, the estimates of partial code $\hat{\mathbf{d}}_k^{(t)}$ for $t = 1, \cdots, N + l$ can be obtained as

$$\hat{\mathbf{d}}_k^{(t)} = \arg\max_{s_t, s_{t-1} \in \mathrm{S}} [\log \Pr(\alpha_{t-1}(s_{t-1}) \gamma(s_{t-1}, s_t))] \text{ for } t = 1, \cdots, N + l. \tag{7}$$

Once the estimates of partial codeword $\hat{\mathbf{d}}_k^{(t)}$ are obtained, they are fed to the cancellation step in the same stage to eliminate MAI. In the cancellation step, MAI is constructed using the correlation matrix $\mathbf{R}$ and partial codewords $\hat{\mathbf{d}}^{(t)}$ of nodes. Then the MAI is subtracted from the output of matched filter $\mathbf{y}_k^{(t)}$. These new output signals form the input to the estimation step in the next stage.

3 JDD Receivers for Cluster-Head Nodes in Sensor Networks

Two-tiered wireless sensor networks, where the network is partitioned into distinct clusters with specific nodes called cluster-head, usually uses CDMA technology for a multiple access scheme as described in [9]. In such networks, the higher level of energy consumption at cluster-head nodes might cause more damage to the network since such cluster-head nodes also limit accessibility of other sensor nodes. Therefore, the energy-saving mechanism in cluster-head nodes is critical issue and determines

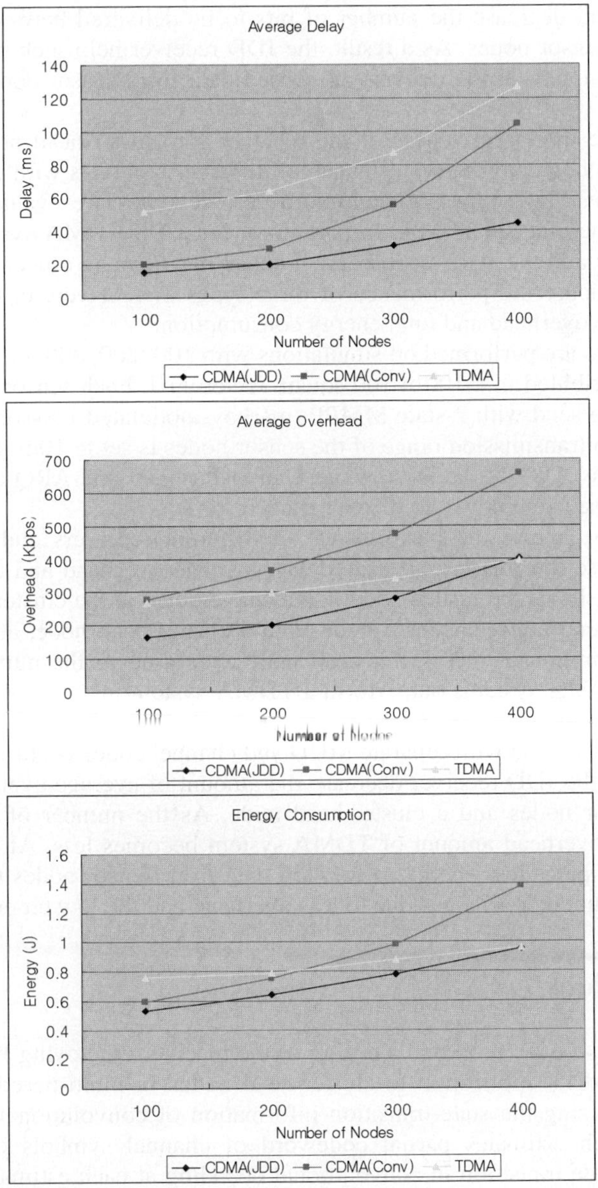

Fig. 3. Simulation Results of Receivers in Wireless Sensor Networks

the performance and lifetime of networks. The receiver of the cluster-head node within a dense cluster, however, suffers from MAI due to non-orthogonal PN codes. MAI in addition to noise causes the bit errors and results in frame retransmissions or information losses from sensor nodes. In the receiver of cluster-head nodes, MUD can be employed to eliminate MAI, while channel codes are utilized to cope with bit errors. Therefore, JDD receivers are expected to improve the performance of cluster-head nodes and decrease the number of bits to be delivered between a cluster-head node and its sensor nodes. As a result, the JDD receiver help such networks save the energy consumption in the delivery of sensed data from sensor nodes to the cluster-head node.

To evaluate the effectiveness of the receiver for cluster-head nodes, we perform simulations to compare the performance of different receivers with different multiple access schemes. Two types of multiple access schemes are evaluated; one is DS-CDMA and the other is TDMA. In particular, for a DS-CDMA system, the conventional receiver and the JDD receiver are compared. The main objective of the simulation is to compare the performance of three types of receivers in terms of average delay, average overhead and total energy consumption.

Experiments are performed on simulations with 100, 200, 300, or 400 sensor nodes randomly distributed in a 100 × 100 square meter area. Each sensor node is assumed to have data to send with 2-state MMPP (markov-modulated Poisson process) traffic. The maximum transmission range of the sensor nodes is set to 10m. It is assumed that the channel has AWGN (additive white Gaussian noise) and ARQ (automatic repeat request) scheme is employed to recover frame errors.

In order to capture the performance of different receivers and multiple access schemes we use the metrics, average delay, average overhead and energy consumption. Figure 3 shows the results about the average delay at the cluster-head nodes, the average overhead between sensor nodes and a cluster-head node, and the amount of energy consumption at the cluster-head node according to the number of nodes. It shows that CDMA systems outperform a TDMA system in terms of delay in cluster-head nodes. Furthermore, the CDMA system utilizing the JDD receiver has less average delay than the one with separate MUD and channel codes at cluster-head nodes. It also indicates the JDD receiver decrease the amount of average overhead remarkably between sensor nodes and a cluster-head node. As the number of nodes increases, however, the overhead amount of TDMA system becomes less. At last, the receiver with JDD dissipates less energy to forward data from sensor nodes to parent cluster-head nodes than other schemes due to less overhead and the low bit error rate.

4 Conclusions

In this paper, we have proposed a design of the receiver employing PIC and convolution codes, and its application to cluster-head nodes in multi-tiered wireless sensor networks. Utilizing the state-transition information of convolutional codes, our proposed approach estimates partial codeword of channel symbols corresponding to branches of state transition in convolutional decoding at each estimation step in PIC. We have evaluated the performance of the receiver and the performance of the wireless sensor network when the receiver is used for cluster-head nodes. It is shown that the receiver outperforms the existing receiver when CDMA is used for a multiple access scheme. Furthermore, the receiver experiences much less delay than a TDMA scheme. In particular, the proposed JDD receiver is quite attractive when a number of users (nodes) exist within a cluster in terms of delay and energy consumption.

References

1. I.F. Akyildiz, W. Su, Y. Sankarasubramaniam, and E. Cayirci, "A Survey on Sensor Networks," IEEE Communication Magazine, August, 2002.
2. A. J. Goldsmith and S. B. Wicker, "Design challenges for energy-constrained Ad Hoc wireless networks," IEEE Wireless Communications Magazine, pp. 8-27 Aug. 2002.
3. S. Verdu, Multiuser Detection, Cambridge University Press, 1998.
4. Y. Shama, B. R. Vojcic and B. Vucetic, "Suboptimum Soft-output Detection Algorithm for Coded Multiuser Systems," IEEE Transactions on Communications, vol. 48 no. 10, pp. 1622-1625, October 2000.
5. M. C. Reed, C. B. Schlegel, P. D. Alexander and J. A. Asenstorfer, " Iterative Multiuser Detection for CDMA with FEC: Near-Single-User Performance," IEEE Transactions on Communications, vol. 46 no. 12, pp. 1693-1699, December 1998.
6. H.Lee and P.K.Varshney, "A joint detection-decoding receiver with reduced complexity," Proc. of the IEEE Vehicular Technology Conference, pp. 2533-2537, April 2003.
7. S. Lin and D. J. Costello, Error Control Coding: Fundamentals and Applications, Prentice Hall, Englewood Cliffs, NJ, 1983.
8. L. R. Bahl, J. Cocke, F. Jelinek and J. Raviv, "Optimal Decoding of Linear Codes for Minimizing Symbol Error Rate," IEEE Transactions on Information Theory, IT-20, no. 2, pp. 284-287, March 1974.
9. W. Heinzelman, A. Chandrakasan, and H. Balakrishnan, "An Application-Specific Protocol Architecture for Wireless Microsensor Networks," IEEE Transactions on Wireless Communications, vol. 1, no. 4, pp. 660-670, October 2002.
10. B. Deb, S. Bhatnagar and B. Nath, "A Topology Discovery Algorithm for Sensor Networks with Applications to Network Management," Technical Report DCS-TR-441, Rutgers University May, 2001.

An Error Control Scheme for Multicast Video Streaming on the Last Hop Wireless LANs*

Junghoon Lee[1], Mikyung Kang[1], Gyungleen Park[1], Hanil Kim[2], Choelmin Kim[2], and Seongbaeg Kim[2]

[1] Dept. of Computer Science and Statistics,
[2] Dept. of Computer Education,
Cheju National University, 690-756, Jeju Do, Republic of Korea
{jhlee, mkkang, glpark, hikim, cmkim, sbkim}@cheju.ac.kr

Abstract. This paper proposes and analyzes the performance of an error control scheme for multicast video streaming over IEEE 802.11 WLAN. The proposed scheme makes all packets include the field indicating the number of packets of the message to which they belong, the receiver nodes report errors in a best-effort manner through contention period, and finally access point retransmits the requested packets through the overallocated slot that is unavoidably brought by QoS guarantee. Simulation results show that the proposed scheme can not only efficiently utilize the network bandwidth by reusing the wasted bandwidth for error control but also reduce the deadline miss ratio by 23 % compared with fixed length-based error control scheme without affecting other streams.

1 Introduction

The great success of IEEE 802.11 technology for WLANs (Wireless Local Area Networks) is creating new opportunities for the deployment of advanced multimedia services such as video conferencing, video multicast, and so on[1]. Audio/video streaming will be a critical part of the wireless communication, as multicast streaming offers an efficient paradigm for transmitting video from a sender to a group of receivers using mobile devices such as PDA, telematics, and so on, with lower network and end-system costs. Unlike the general data traffic, video traffic is delay-sensitive and somewhat tolerant to packet loss through the use of error concealment technique at the video decoder. However, the use of WLANs for the transport of video accompanies some problems resulted from the strict delay constraints of the video traffic and the inherent unpredictability of the wireless link[2].

The streaming service on WLAN requires a certain type of QoS (Quality of Service) guarantee and run-time error control to provide an acceptable video quality. QoS guarantee means a reservation of resources that can meet the delay

* This research was supported by the MIC, Korea, under the ITRC support program supervised by the IITA.

V.N. Alexandrov et al. (Eds.): ICCS 2006, Part IV, LNCS 3994, pp. 962–969, 2006.
© Springer-Verlag Berlin Heidelberg 2006

constraint of video stream[3]. The guarantee mechanism inevitably depends on the underlying MAC (Medium Access Control) layer and IEEE 802.11 WLAN standard consists of a basic DCF (Distributed Coordination Function) and an optional PCF (Point Coordination Function)[4]. Due to packet collisions intrinsic to CSMA/CA (Carrier Sense Multiple Access with Collision Avoidance), DCF is not appropriate for video streaming. The QoS guarantee cannot be provided without developing a deterministic access schedule on top of collision-free PCF[5]. However, DCF cannot be totally excluded from WLAN operation because the standard demands that the DCF should be at least long enough to transmit one maximum data unit, to prevent starvation of stations that are not allowed to send during the PCF.

Meanwhile, lost packets constitute one of the main causes of video quality degradation, while the wireless channel error is characterized as bursty and location dependent[6]. Due to the delay constraints, the number of retransmission that can be used is limited and usually small. In case of multiple clients, each client will have different channel conditions, processing capabilities, and only limited feedback channel capabilities. The error control scheme for multicast video is not aiming at recovering all lost packets but recovering as many packets as possible. Most importantly, the error control procedure should not affect the other guaranteed traffics in WLAN.

The real-time guarantee inevitably generates overallocation during PCF as the time constraints of message stream are usually described with the maximum value of message size. Retransmission via this overbooked bandwidth does not interfere the transmission of other guaranteed messages. In the other hand, if we let error report containing retransmission request be delivered via the DCF period in a best-effort manner, the entire error control procedure can be carried out without any influence to other real-time messages. Finally, though the variable message size makes it hard to decide when to report the error list, the receivers can determine the completion of message transmission by counting *Beacon* frame AP (Access Point) generates periodically in WLAN. Based on the requirements described previously, this paper will propose and analyze the performance of an error control scheme for multicast video stream on IEEE 802.11 WLAN. We focus on the video streaming scenario in the last mile network, namely, between AP and the mobile devices.

The rest of this paper is organized as follows: After reviewing some related works in Section 2, Section 3 describes the background of this paper including network, error, and message models. Section 4 proposes an error control scheme and Section 5 shows the result of performance measurement. Finally, Section 6 summarizes and concludes this paper.

2 Related Works

Error control in wireless network has been intensively studied for both unicast and multicast. Most of the approaches use ARQ (Automatic Retransmission reQuest), FEC (Forward Error Correction), or a combination of both[7]. Pure

ARQ-based schemes are less appropriate for the multicast case due to ACK explosions and the requirement to retransmit different packets to the respective users. For significant packet loss rates, each user will require frequent packet replacement, and different receivers are most likely to require different packets. To respond to requests by multiple users, we may have to resend a significant fraction of the original data even for small loss rates. For a packet network with dynamic bandwidth, a different class of the source coding technique called progressive coding is better suited. However, this method has the problem in deciding optimal coding parameter on wireless channel.

As a hybrid ARQ mechanism, Majumdar et al. have proposed a method that combines the reliability and fixed delay advantage of forward error control coding with bandwidth-conserving channel-adaptive properties of ARQ protocol[2]. Masala has also proposed a multicast scheme that aims at globally optimizing the parameters involved in a real-time video transmission, ranging from video encoding and packetization to the 802.11 MAC interface parameters[3]. These schemes are mainly built on top of DCF, so they didn't consider the effect of QoS reservation.

Lu et al. have proposed a timestamp-based content-aware adaptive retry mechanism where MAC dynamically determines whether to send or discard a packet by its retransmission deadline, which is assigned to each packet according to its temporal relationship and error propagation characteristics with respect to other video packets within the same group of pictures[7]. However, their scheme is too complex to be exploited in the WLAN standard, as it crosses the protocol layer boundaries.

3 Background

3.1 Network Model

We consider a wireless-cum-wired network scenario as shown in Fig. 1[8]. A fixed node is connected with an AP through a wired link which is overprovisioned so that no packets are dropped at its end. This model enables us to concentrate on the behavior of inside WLAN part as it assumes ideal environment of no packet loss or jitter for outside WLAN part. Each cell is assumed to consist of an AP and multiple MSs (Mobile Stations), while each flow is either uplink (from MS

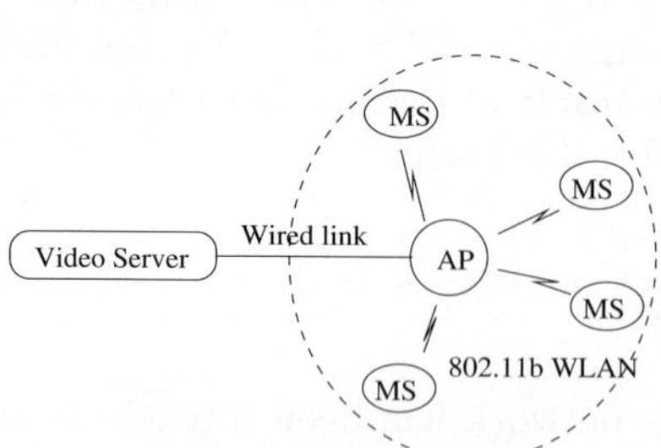

Fig. 1. Network model

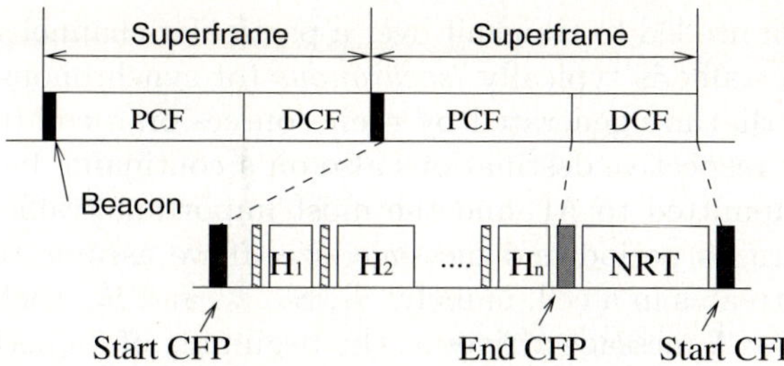

Fig. 2. Time axis of wireless LAN

to AP) or downlink (from AP to MS). Most flows are downlink, and the stream flow arrives periodically from the remote server outside the cell via reliable wired link. The multicast on WLAN prevents the automatic ACK transmission from the receiver.

The IEEE 802.11 was developed as a MAC standard for WLAN[4]. The standard divides the time axis into the series of superframes, and each of them consists of CFP (Contention Free Period) and CP (Contention Period), as shown in Fig. 2. In CFP, AP polls each stream once a polling round, providing a predictable access to each delay sensitive streams. For simplicity, this paper assumes that the polling round coincides with a superframe, but this assumption can be easily generalized. Appropriate time amount is allocated to each multicast stream, and the length of time interval is calculated according to the QoS requirement. The multicast stream is also an instance of message stream that flows from AP to WLAN. On the contrary, during CP any station can send non-real-time message after contending the shared medium via CSMA/CA.

3.2 Error Model

WLANs may experience location-dependent channel errors, now that some MSs can correctly communicate with the AP, while at the same time others may suffer packet drops due to errors on the channel. Therefore, given N MSs, there are also N independent error models. An 802.11 radio channel is modeled as a Gilbert channel[8], where two states, *good* and *bad*, linked with a Marcov chain, represent the state of channel during an 802.11 slot time. A MAC protocol data unit is received correctly if the channel is in state *good* for the whole duration of transmission, otherwise, it is received in error. We denote the transition probability from state *good* to state *bad* by p and the probability from state *bad* to state *good* by q. The pair of p and q representing a range of channel conditions, has been obtained by using the trace-based channel estimation. After all, the average error probability, denoted by ϵ, and the average length of a burst of errors are derived as $\frac{p}{p+q}$ and $\frac{1}{q}$, respectively.

3.3 Message Model

The video stream traffic occupies the network as a form of streaming-specific packets. For example, H.264 standard decouples the coding aspect from the bit stream adaptation needed to transmit over a particular channel such as WLAN. Such multimedia traffic is typically *isochronous* (or synchronous), consisting of message streams that are generated by their sources on a continuing basis and delivered to their respective destinations also on a continuing basis[9]. The QoS requirement is submitted to AP and the most important traffic characteristics of each stream are its period and message size. If we assume that there are n multicast video streams in a cell, namely, S_1, S_2, ..., and S_n, each stream can be modeled as follows: A message arrives at the beginning of its period and it must be transmitted by the end of period. The deadline is soft in that some delayed packets are permissible and transmission jitter is absorbed by a playback buffer in the video player. The period of S_i, is denoted as P_i, and the maximum length of a message as C_i.

4 Error Control Scheme

4.1 Bandwidth Allocation

By allocation, we mean the procedure of determining capacity vector, $\{H_i\}$, for the given superframe time, F, and message stream set described as $\{S_i(P_i, C_i)\}$. Fig. 2 also illustrates that the slot size is not fixed, namely, $H_i \neq H_j$ for different i and j. At this figure, a message of size up to C_i is generated and buffered at regular intervals of P_i, and then transmitted by H_i every time the node is polled. This subsection briefly describes the allocation scheme of Lee's work[5].

Let δ denote the total overhead of a superframe including polling latency, IFS, exchange of beacon frame, and the like, while D_{max} the maximum length of a non-real-time data packet. In addition, P_{min} denotes the smallest element of set $\{P_i\}$. Then the requirement for the superframe time, F, can be summarized as in Ineq. (1). Within this range, the scheme can select F and slightly modify P_i's such that they are harmonic[9].

$$\sum H_i + \delta + 2 \cdot D_{max} \leq F \leq P_{min} \tag{1}$$

In addition, the least bound of H_i that can meet the time constraint of S_i is calculated as in Eq. (2).

$$\begin{aligned} H_i &= \frac{C_i}{(\lfloor \frac{P_i}{F} \rfloor - 1)} \qquad if(P_i - \lfloor \frac{P_i}{F} \rfloor \cdot F) \leq D_{max} \\ H_i &= \frac{C_i}{\lfloor \frac{P_i}{F} \rfloor} \qquad Otherwise \end{aligned} \tag{2}$$

The allocation vector calculated by Eq. (2) is a feasible schedule if the vector meets Ineq. (1). Finally, we can determine the length of CFP (T_{CFP}) and that of CP (T_{CP}) as follows:

$$T_{CFP} = \sum H_i + \delta, \quad T_{CP} = F - T_{CFP} \geq D_{max} \tag{3}$$

This scheme is based on the analysis of the worst case available time. Since such an allocation usually considers the upper bound of message size, especially in multimedia applications, unused bandwidth degrades the network throughput. For detailed description of bandwidth allocation, refer to [5].

4.2 Error Report

A message is divided into packets of size H_i, hence, for a message to be correctly assembled at a receiver, the receiver should receive all packets successfully. If a packet experiences an error during transmission, additional explicit messages should be sent to request the retransmission of that packet. Per-packet response to the receiver increases overhead, but if the receiver reports the list of damaged packets after it estimates the end of delivery of one message, this waste can be eliminated. It is desirable to send error report via the contention period in order not to interfere the transmission of other stream. If the size of every message is fixed and known in advance, the receiver can easily decide the end of one message transmission. However, the size varies message by message in multimedia streams. Hence, we make all packets include a field specifying the number of packets of the message to which they belong so that the receiver can decide when to report error packet list as long as it receives at least one packet.

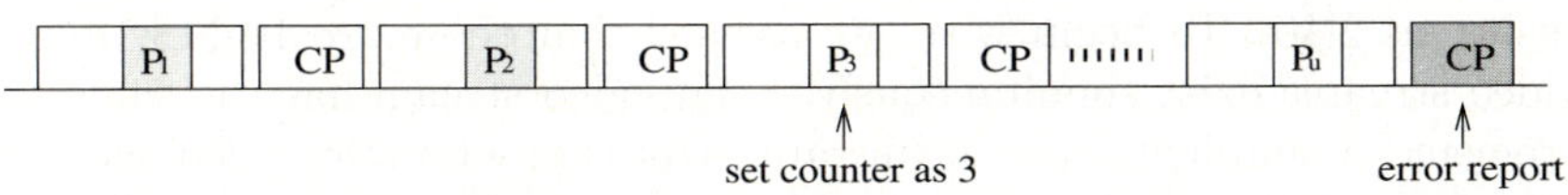

Fig. 3. An example of error reporting

To construct the error report message, the receiver initializes the error packet list when it receives a packet arrives. If the sequence number of this packet is not 1 but k, the receiver appends the numbers of 1 through $(k-1)$ to the list. From then, the receiver appends each number of the erroneously received packets. As the receiver also hears *Beacon* and a stream sends a packet per a polling round, it knows the end of transmission even if the last message is omitted. Fig. 3 shows the example, where a message is transmitted with u packets and each of them has an information that specifies u as well as its sequence number and message identifier. At first, a packet numbered as 3 (not 1) arrives. Then the receiver initializes and adds 1 and 2 to error list as well as sets counter to 3. The counter increases each time the receiver hears *Beacon* frame. When the counter reaches u, receiver sends error report back to the sender via contention period.

4.3 Retransmission

AP receives error report containing error packet list from the receivers of multicast via CP each time it completes a message transmission until it begins a new transmission. AP builds a retry list sorted by the frequency of appearance in the set of error reports. A new error list arrival reorders the list. Then AP resends the packet according to the order in the retry list when it meets an extra slot for the corresponding video stream. The retransmission scheme can be extended to consider the priority of packet given by the coding scheme performed at the upper layer.

5 Performance Analysis

This section begins with the description of analytic model for the recovered error for the proposed scheme. The average extra bandwidth for a message, R_i, can be calculated as shown in Eq. (4).

$$R_i = (\frac{H_i}{F} - \frac{\bar{C}_i}{P_i}) \cdot P_i \tag{4}$$

Then the average number of extra access time, w, can be approximated as R_i/H_i. If there are w extra frames for a message composed of k frames, the probability of successful transmission of a message, $T(k, w)$, is calculated recursively. Namely,

$$T(k,w) = \sum_{i=0}^{m} {}_kC_i \cdot \epsilon^i (1-\epsilon)^{k-i} \cdot T(i, w-i), T(0,w) = 1, T(k,0) = (1-\epsilon)^k \tag{5}$$

where m is the smaller of k and w while ϵ denotes the frame error rate.

Now we will show the performance measurement results obtained via simulation using ns-2[10]. To begin with, we assume that there are 10 MSs in a cell, 3-5 video streams exist simultaneously. Each video stream has the same traffic requirement for simplicity such as bit rate, error characteristic, video packet size, and the number of receivers. Fig. 4 and Fig. 5 plots the average deadline meet ratio observed at each receiver according to BER (Bit Error Rate) and peak-average ratio. In these figures, the number of messages not packets is counted. The curve denoted as fixed length is for the case where the receiver only knows the maximum length of message. Fig. 4 exhibits that the proposed scheme can reduce the deadline miss ratio by 23 % compared with fixed length scheme and by 48 % compared to no error control case, respectively, when BER is around 10^{-6}. In Fig. 5, when the peak-average ratio of the message is 1 (all message sizes are equal), both proposed and fixed length schemes show same performance as expected. However, the proposed scheme improves the deadline meet ratio due to the efficient error report mechanism when the ratio increases over 1.0. Other results are omitted due to space limitation. After all, the simulation results demonstrate that overbooked bandwidth can be efficiently used for error recovery of multicast video stream.

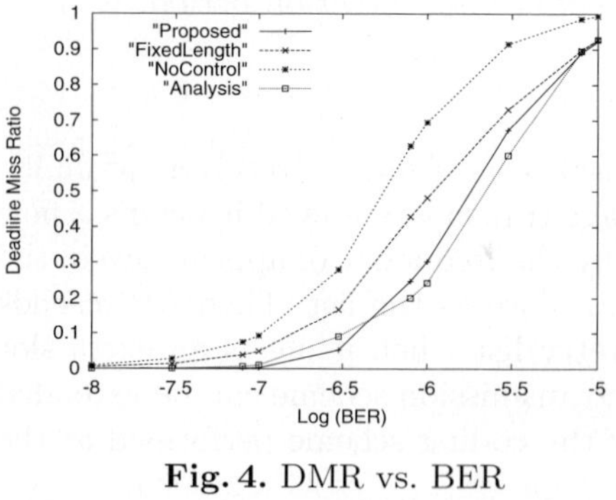

Fig. 4. DMR vs. BER

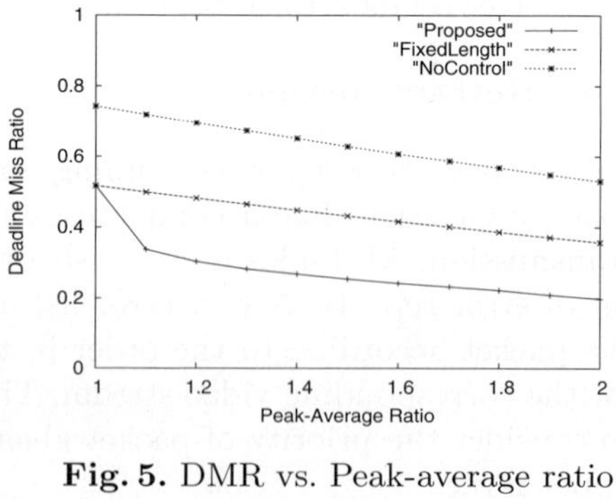

Fig. 5. DMR vs. Peak-average ratio

6 Conclusion

In this paper, we have proposed and analyzed the performance of an error control scheme for multicast video streaming over IEEE 802.11 WLAN. The proposed error control scheme makes the receiver node report errors in a best-effort manner via CP, and also makes AP retransmit the requested packets through the overallocated slot that is unavoidably brought by QoS guarantee in CFP. Therefore, this scheme is able to eliminate the interference to the guaranteed stream transmission, while the message size contained in each frame enables the timely report of error list as long as a packet of message arrives at the receiver. A continuous trace of each video stream channel shows that the proposed scheme can not only efficiently utilize the network bandwidth, but also reduce the deadline miss ratio by 23 % compared with fixed length scheme and by 48 % compared with no error control case, respectively, for the given simulation parameters. As a future work, we are planning to develop a video streaming on dual wireless links focusing on the guarantee scheme and transmission schedule that can efficiently cope with network errors based on the temporarily redundant channel.

References

1. Mao, S., Lin, S., Wang, Y., Panwar, S. S., Li, Y.: Multipath video transport over wireless ad hoc networks. IEEE Wireless Communications (2005)
2. Majumdar, A., Sachs, D., Kozintsev, I., Ramchandran, K.: Multicast and unicast real-time video streaming over wireless LANs. IEEE Trans. Circuit and Systems for Video Technology, 12 (2002) 524-534
3. Masala, E., Chiasserini, C., Meo, M., De Martin, J.: Real-time transmission of H.264 video over 802.11-based wireless ad hoc networks. Proceedings of Workshop on DSP in Mobile and Vehicular Systems (2003)
4. IEEE 802.11-1999: Part 11: Wireless LAN Medium Access Control (MAC) and Physical Layer (PHY) Specifications (1999) also available at *http://standards.ieee.org/getieee802*
5. Lee, J., Kang, M., Jin, Y., Kim, H., Kim, J.: An efficient bandwidth management scheme for a hard real-time fuzzy control system based on the wireless LAN. Lecture Notes in Artificial Intelligence, Vol. 3642. Springer-Verlag, Berlin Heidelberg New York (2005) 644-659
6. Shah, S., Chen, K., Nahrstedt, K.: Dynamic Bandwidth Management for Single-hop Ad Hoc Wireless Networks. ACM/Kluwer Mobile Networks and Applications (MONET) Journal. 10 (2005) 199-217
7. Lu, A., Chen, T., Steenkiste, P.: Video Streaming over 802.11 WLAN with Context-aware Adaptive Retry. IEEE International Conference on Multimedia and Expo (2005)
8. Bottigliengo, M., Casetti, C., Chiaserini, C., Meo, M.: Short term fairness for TCP flows in 802.11b WLANs. Proc. IEEE INFOCOM (2004)
9. Carley, T., Ba, M., Barua, R., Stewart, D.: Contention-free periodic message scheduler medium access control in wireless sensor/actuator networks. IEEE Real-Time Systems Symposium (2003)
10. Fall, K., Varadhan, K.: Ns notes and documentation. Technical Report, VINT project, UC-Berkeley and LBNL (1997)

Design of a Fast Handoff Scheme for Real-Time Media Application on the IEEE 802.11 Wireless LAN*

Mikyung Kang[1], Junghoon Lee[1,**], Jiman Hong[2], and Jinhwan Kim[3]

[1] Dept. of Computer Science and Statistics, Cheju National University,
[2] School of Computer Science and Engineering,
Kwangwoon University
[3] Dept. of Multimedia Engineering, Hansung University,
690-756, Jeju Do, Republic of Korea
{mkkang, jhlee}@cheju.ac.kr, gman@daisy.kw.ac.kr, kimjh@hansung.ac.kr

Abstract. This paper proposes and analyzes a fast handoff scheme that exploits the overallocated bandwidth inevitably generated to guarantee the QoS requirement of real-time multimedia stream on the IEEE 802.11 wireless LAN. By using the reserved but not used network time and making the priority of the probe frame higher than any other data frames, AP and station can exchange RTS/CTS to negotiate when to send probe message, making AP immediately respond to the probe request with probe response message in the CFP. The result of simulation that focuses on the effect of the amount of overalloation and average number of simultaneous requests, shows that the proposed scheme reduces the AP scan time maximally by 16 % for the given experiment parameters.

1 Introduction

IEEE 802.11 based WLANs(Wireless Local Area Networks) have seen immense growth in the last few years. Because of the mobility-enabling nature of wireless networks, there is an opportunity for many promising multimedia and peer-to-peer applications such as VoIP, 802.11 phones and mobile video conferencing[1]. However, mobile clients suffer from quality degradation resulted from frequent handoff since each cell may cover just a small area, i.e., rooms or sections of a highway. Frequent handoffs and disconnections incur disruptions and instability of the connection between mobile host and server, even in the middle of an application session. To the worse, according to the handoff procedure defined in WLAN standard, the network connection as perceived by the application may be affected by the jittery and unpredictable handoff latencies. Such problem is particularly serious for the fast moving device such as telematics.

* This research was supported by the MIC, Korea, under the ITRC support program supervised by the IITA.

** Corresponding author.

V.N. Alexandrov et al. (Eds.): ICCS 2006, Part IV, LNCS 3994, pp. 970–978, 2006.
© Springer-Verlag Berlin Heidelberg 2006

Wireless LAN STA (Station) is the most basic component of the wireless network[2] and it means any device that contains the functionality of the 802.11 protocol. BSS (Basic Service Set) is the basic building block of an 802.11 WLAN, and each BSS consists of any number of stations. As a SSID (Service Set IDentifier) is a unique identifier that distinguishes one WLAN from the others, all APs (Access Points) and STAs attempting to join a specific WLAN must have the same SSID. A WLAN handoff is performed at the MAC layer when a mobile STA moves beyond the radio range of the current AP and enters another BSS[3]. During the handoff, management frames are exchanged between the STA and the AP. The handoff procedure essentially requires the transfer of STA information such as authentication, authorization, and accounting information, from the old AP to the new AP.

Previous research results show that the probe phase overwhelms the total handoff latency while the variation in the probe-wait time also causes the large variations in the overall handoff latency. For the client, the service is ceased during the handoff. Because the STA must scan the channel to which an AP may belong for the maximum scanning period, and it must repeat iteratively for all channels, the probe time occupies the biggest part of the handoff latency. Thus any handoff scheme built upon the techniques/heuristics that either cache or deduce AP information without having to actually perform a complete active scan definitely should cope with the dominating cost of the scan process.

To solve such a problem by reducing AP scanning delay of handoff latency at MAC layer, this paper proposes and analyzes a fast handoff scheme that exploits the overallocated bandwidth necessarily accompanied in providing QoS guarantee to real-time multimedia stream. Using the reserved but not used network time, AP and STA can exchange RTS (Request To Send)/CTS (Clear To Send) to negotiate when to send a probe message, making AP immediately respond to the *probe request* with matching *probe response* message in the next CFP. In addition, by making the priority of the probe frame higher than any other data frames, collision between *probe response* messages and ordinary data frames can be eliminated, and dynamic adjustment of the channel scan time further improves AP scanning time. With the reduced AP scan time, a seamless handoff process is performed, minimizing the deadline miss ratio.

This paper is organized as follows: After issuing the problem in Section 1, Section 2 introduces the related works. With the description on the handoff procedure in IEEE 802.11 WLANs in Section 3, Section 4 proposes the fast handoff scheme. Section 5 shows and discusses the performance measurement results and then Section 6 finally concludes this paper with a brief summarization and the description of future works.

2 Related Works

A lot of works have been already carried out to reduce the handoff latency for the roaming client station. However, existing handoff schemes are not suitable for meeting requirements of real-time multimedia application due to its long and occasionally unbounded delay. The sequence of messages being exchanged during the handoff process can be categorized into three groups, namely, probe, authentication, and association. Accordingly, existing works are also classified by the delay element to reduce.

First, the researches to improve the probe delay are as follows: Kim et al. proposed a selective scanning algorithm using the neighbor graphs[4]. This approach forces changes in the network infrastructure and use of IAPP (Inter Access Point Protocol) though it narrows the search space with neighbor graphs. Moreover, this scheme does not consider the time amount required by the client to process the received *probe responses*. Shin et al. proposed a new handoff procedure which reduces the MAC layer handoff latency, in most cases, to a level where VoIP communication becomes seamless using both selective scanning algorithm and caching mechanism[2]. It needs just an insignificant modification in the client-side wireless card driver such as channel mask and improved cache dimensioning. According to the analysis result by Jain[5] and Mishra[3], there are remarkable variations in handoff latencies with change in SSID and channel of APs, and probe delay is the major malicious factor to the total handoff performance.

Second, to improve the authentication delay, Pack et al. proposed a fast Inter-AP handoff scheme using the predictive authentication method based on IEEE 802.1x model [6][7]. The IEEE 802.1x authentication delay is reduced by using the FHR (Frequent Handoff Region) selection algorithm that makes the candidate APs selected by the predictive algorithm, perform the pre-authentication, directly taking into account traffic patterns and user characteristics, which are collected and managed in the centralized system.

Third, to improve the association delay, Mishra et al. focused on reducing the reassociation delay[8]. The reassociation delay is reduced by using a caching mechanism on the AP side. This caching mechanism is based on the IAPP protocol in order to exchange the client context information between neighboring APs. The cache in the AP is built by observing the information contained in an IAPP *Move-Notify* message or in the *reassociation request* sent to the AP by the client. By exchanging the client information with the old AP, the new AP prevents the client from sending its context information, resulting in the reduction of the reassociation delay.

3 Backgrounds

3.1 IEEE 802.11 WLAN

The wireless LAN operates on both CP (Contention Period) and CFP (Contention Free Period) phases alternatively in BSS as shown in Fig. 1(a). Each superframe consists of CFP and CP, which are mapped to PCF (Point Coordination Function) and DCF (Distributed Coordination Function), respectively[9]. Though PCF is optional, QoS guarantee cannot be provided without PCF. Recently, the PCF-enabled schemes are increasingly being applied to the WLAN showing a reasonable throughput. Moreover, previous researches based on the

DCF can not avoid both the collision between the probe messages and normal data frames, and the probe delay according to the backoff time of DCF.

Fig. 1(a) shows the operation of polling procedure as well as the allocation of capacity vector. The DCF exploits collision-based CSMA/CA (Carrier Sense Multiple Access with Collision Avoidance) protocol for non-real-time messages, and RTS/CTS clearing technique to further reduce the possibility of collisions as shown in Fig. 1(b).

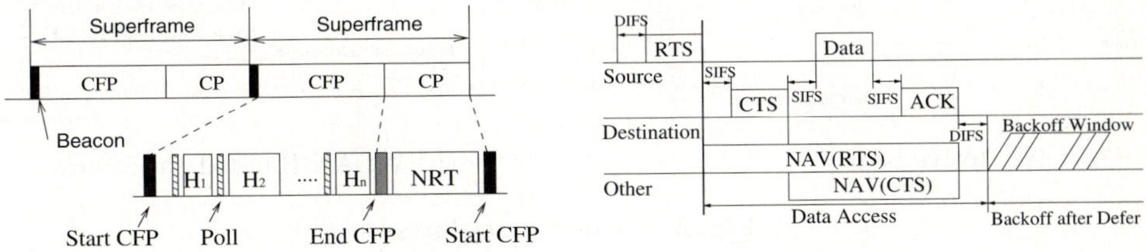

(a) Time axis of wireless LAN (b) RTS/CTS frame exchange

Fig. 1. IEEE 802.11 WLAN operations

PC (Point Coordinator) node, typically AP, sequentially polls each station during CFP. AP maintains the polling list ordered by a polling sequence. The PC attempts to initiate CFP by broadcasting a *Beacon* at regular intervals derived from a network parameter of *CFPRate*. The polled node transmits its message for a predefined time interval, and it always responds to a poll immediately whether it has a pending message or not. Only after the medium is idle the coordinator will get the priority due to the shorter IFS (InterFrameSpace).

3.2 The Handoff Procedure in WLAN

The handoff process can be divided into two logical steps of discovery and reassociation[3]. The discovery process involves handoff initiation and scanning phases. As signal strength and signal-to-noise ratio from a station's current AP get weaker, STA loses connectivity and initiates a handoff. Then the client is not able to communicate with its current AP, so the client needs to find the other APs available. This scan function is performed at a MAC layer, and the station can create the available AP list ordered by the received signal strength.

For the scan phase, STA can perform scan operation either in passive or active mode. In passive scan mode, using the information obtained from beacon frames, STA listens to each channel of the physical medium to try and to locate an AP. In the active mode (the wireless NICs do by default), as shown in Fig. 2(a), STA broadcasts additional probe packets on each channel and receives responses from APs. Thus the STA actively probes for the APs, and the actual number of messages varies from 3 to 11. Fig. 2(b) shows the sequence of messages typically observed during a handoff process. The handoff process starts with the first *probe request* and ends with a *reassociation response* from the new AP.

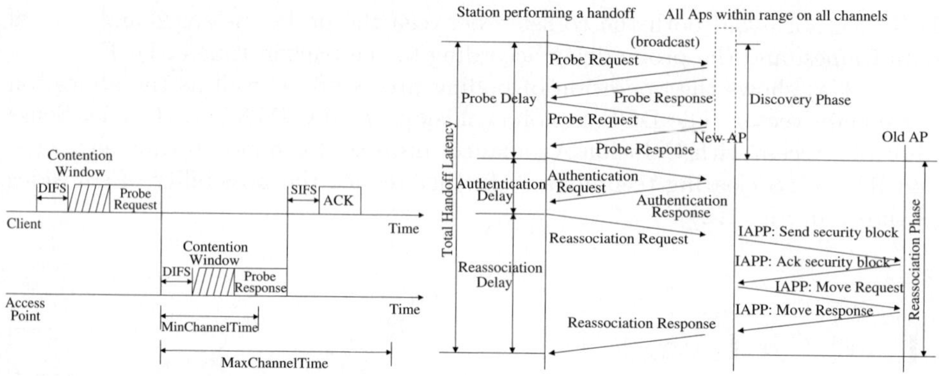

(a) Active Scan

(b) WLAN Handoff Procedure

Fig. 2. Handoff procedures

The probe function follows the IEEE 802.11 MAC active scan function and the standard specifies a scanning procedure as follows[2][3]:

1. Using CSMA/CA, acquire the access right to the medium.
2. Transmit a *probe request* containing the broadcast address as destination, SSID, and broadcast BSSID (Basic SSID).
3. Start a *ProbeTimer*.
4. If medium is not busy before the *ProbeTimer* reaches *MinChannelTime*, scan the next channel. Otherwise, process all received *probe responses*.
5. Move to next channel and repeat the above steps.

After all channels have been scanned, informations received from *probe response* are scrutinized by STA to select a new AP. Once the STA decides to join a specific AP, authentication messages are exchanged between the STA and the selected AP, and after a successful authentication, the STA sends a reassociation request and expects a reassociation response back from the AP.

4 Proposed Scheme

4.1 AP Channel Management

Real-time guarantee is provided based on the worst case available transmission time, so a stream can meet extra slots in some periods. Moreover, as C_i is usually just the upper bound of message size in multimedia applications, some period, P_i, has message to send less than C_i. As a result, a node possibly has no pending message when it receives a poll. Though there have been plenty of bandwidth allocation schemes for the real-time message stream or sensor data stream, we exploit Lee's scheme, as it completely conforms to WLAN standard[10].

To begin with, let δ denote the total overhead of a superframe including polling latency, IFS and the like, while D_{max} the maximum length of a data

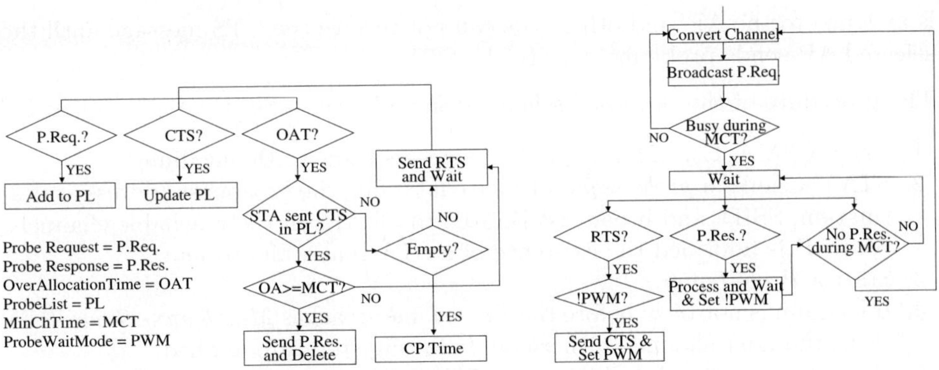

(a) AP Process (b) STA Process

Fig. 3. AP and STA processes

packet. For each superframe, at least a time amount as large as D_{max}, should be reserved for a data packet so as to keep compatibility with WLAN standard. The capacity vector, H_i, is large enough to exchange RTS/CTS. If P_{min} is the smallest element of set $\{P_i\}$, the requirement for the superframe time, F, can be summarized as follows:

$$\begin{aligned} H_i &= \frac{C_i}{(\lfloor \frac{P_i}{F} \rfloor - 1)} && if(P_i - \lfloor \tfrac{P_i}{F} \rfloor \cdot F) \leq D_{max} \\ H_i &= \frac{C_i}{\lfloor \frac{P_i}{F} \rfloor} && Otherwise \end{aligned} \qquad (1)$$

By this, we can determine the length CFP period (T_{CFP}) and that of CP (T_{CP}) as follows:

$$T_{CFP} = \sum H_i + \delta, \quad T_{CP} = F - T_{CFP} \geq D_{max} \qquad (2)$$

For detailed description, refer to [10].

4.2 AP Scanning Procedure

In general, it takes more time to perform a passive scan than active scan in collecting the necessary information. Thus current WLAN equipments use active scan mode in order to reduce handoff delay. In the standard active scanning procedure, though the probe phase of one channel can be terminated before the *ProbeTimer* reaches *MaxChannelTime*, the STA has to wait during the *MaxChannelTime*. In addition, the *probe response* messages and other ordinary data frames contend for the shared channel, so they can collide with one another. To minimize the collision using active scan mode, we will assign the higher priority to the *probe response* message, and also provide variable scan time.

As shown in Fig. 3, the proposed AP scanning scheme is performed through the unused slots. AP can not only send RTS message to the STA which waits for *probe response* message, but also STA can respond with CTS message to the appropriate AP which will transmit the *probe response* message. Once the RTS/CTS messages are exchanged, the priority to send *probe response* message is assigned to the AP, and other APs can not receive the CTS message until the selected AP sends probe message to the STA.

The procedure of the proposed scheme using adaptive scan time is as follows:

1. Using CSMA/CA, STA acquires the access right to the medium.
2. STA transmits a *probe request* frame containing the broadcast address as destination, SSID, and broadcast BSSID, to all APs in the reachable channels. And AP is informed the existence of a STA that waits to join.
3. Start a *ProbeTimer*.
4. If medium is not busy before the *ProbeTimer* reaches *MinChannelTime*, STA scans the next channel. Otherwise, following steps are applied.
5. Using the unused slot, AP sends RTS to STA specifying the ID submitted in step 2.
6. STA responses with CTS if it still wants to receive the *probe response*.
7. AP sends *probe response* to STA if overallocated slot time amount is large enough to send the probe message. Otherwise, *probe response* is postponed until AP meets such a slot.
8. If the number of RTS messages is equal to the number of *probe responses*, and if SSID of being transmitted packet is equal to that of already received *probe response* during the *MinChannelTime*, that is, after sensing that the *probe response* doesn't arrive at STA any more, STA scans the next channel.

5 Performance Measurements

This section will show performance measurement results performed via simulation. The experiments are based on some assumptions to concentrate on the major performance parameters, namely, the amount of overallocation and the number of pending handoff request. Every stream has equal period and communication time, while each time variable is aligned to F and total number of streams is set to 5 for simplicity. However, these assumption can be easily generalized into the more realistic circumstances. Finally, we compared the handoff time of

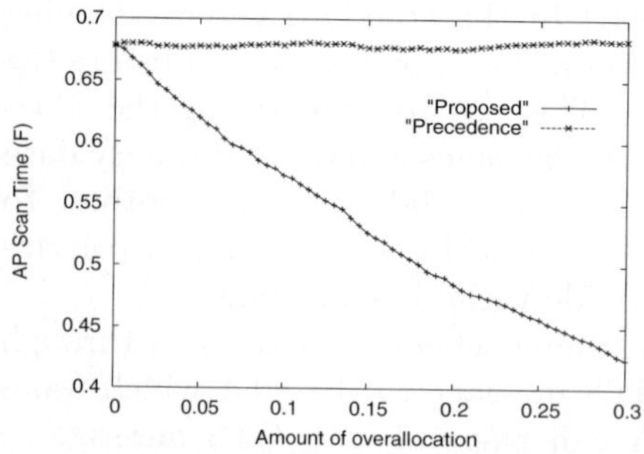

Fig. 4. Scan time vs. overallocation

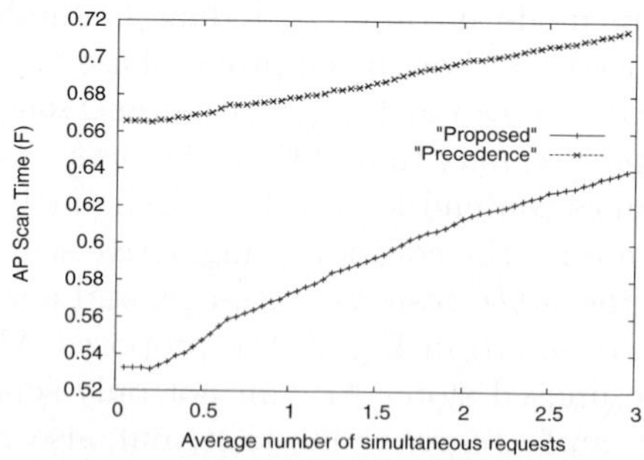

Fig. 5. Scan time vs. # of requests

the proposed scheme with that of a scheme which just gives the precedence to the packets relevant to handoff procedure.

Fig. 4 shows the effect of overallocated bandwidth to the AP scan time. The probability of unused slot due to overallocation can be estimated as $(\frac{H_i}{F} - \frac{C_i}{P_i})$. The efficient usage of overallocated bandwidth can speed up the handoff time by 16 % when the overallocation value is 0.1. Fig. 5 also plots the AP scan time according to the number of simultaneous requests. The performance gap gets narrow when more handoff requests are submitted to the network as the proposed scheme can expect the improvement only if CFP has an overallocation larger than the handoff procedure. The total AP scan time can be calculated by the sum of time needed to scan a used channel and time to scan on empty channel.

6 Conclusion and Future Work

This paper proposes and analyzes a fast handoff scheme that exploits the overallocated bandwidth inevitably generated to guarantee the QoS requirement of real-time multimedia stream on the IEEE 802.11 Wireless LANs. Using the reserved but not used network time, AP and STA can exchange RTS/CTS to negotiate when to send probe message, making AP immediately respond to the probe request with probe response message in the next CFP. In addition, by making the priority of the probe frame higher than any other data frames, collision of probe response messages and ordinary data frames can be minimized. Simulation results show that the proposed scheme improves the AP scan time according to the amount of overalloation and average number of simultaneous requests.

The channel with the best signal is not necessarily the best channel to connect to because it could be much more congested than a channel with a lower signal strength. Because of this, a heuristic which considers bit rate information together with signal strength can achieve optimal performance. As a future work, we consider an error control mechanism as well as a heuristic algorithm that signal strength can achieve optimal performance.

References

1. Mao, S., Lin, S., Wang, Y., Panwar, S. S., Li, Y.: Multipath video transport over wireless ad hoc networks. IEEE Wireless Communications (2005)
2. Shin, S., Rawat, A., Schulzrinne, H.: Reducing MAC Layer Handoff Latency in IEEE 802.11 Wireless LANs. ACM MobiWac'04 (2004)
3. Mishra, A., Shin, M., Arbaugh, W.: An Empirical Analysis of the IEEE 802.11 MAC Layer Handoff Process. ACM Computer Communications Review, Vol. 33, No.2 (2003) 93-102
4. Kim, H. Park, S. Park, C, Kim, J., Ko, S.: Selective channel scanning for fast handoff in wireless lan using neighbor graph. International Technical Conference of Circuits/Systems, Computer and Communications (2004)
5. Jain, A.: Handoff Delay for 802.11b Wireless LANs, Project Report (2003)

6. Pack, S., Choi, Y.: Fast Inter-AP Handoff using Predictive-Authentication Scheme in a Public Wireless LAN. IEEE Networks (2002)
7. Pack, S., Choi, Y.: Pre-Authenticated Fast Handoff in a Public Wireless LAN based on IEEE 802.1x Model. IFIP TC6 Personal Wireless Communications (2002)
8. Mishra, A., Shin, M., Arbaugh, W.: Context Caching using Neighbor Graphs for Fast Handoffs in a Wireless Network. IEEE INFOCOM (2004)
9. IEEE Part 11: Wireless LAN Medium Access Control (MAC) and Physical Layer (PHY) Specifications. IEEE Standard 802.11 (1999)
10. Lee, J., Kang, M., Jin, Y., Kim, H., Kim, J.: An efficient bandwidth management scheme for a hard real-time fuzzy control system based on the wireless LAN. Lecture Notes in Artificial Intelligence, Vol. 3642. Springer-Verlag, Berlin Heidelberg New York (2005) 644-659

Accuracy Enhancement by Selective Use of Branch History in Embedded Processor

Jong Wook Kwak[1], Seong Tae Jhang[2], and Chu Shik Jhon[1]

[1] Department of Electrical Engineering and Computer Science,
Seoul National University, Shilim-dong, Kwanak-gu, Seoul, Korea
{leoniss, csjhon}@panda.snu.ac.kr

[2] Department of Computer Science, The University of Suwon,
Suwon, Gyeonggi-do, Korea
stjhang@suwon.ac.kr

Abstract. The branch prediction accuracy is one of essential parts of performance improvement in embedded processors as well as modern microarchitectures. Until now, the length of branch history has been statically fixed for all branch instructions, and the history length is usually selected in accordance with the size of prediction table. In this paper, we propose a dynamic per-branch history length adjustment policy, which can dynamically change the history length for each branch instruction. The proposed solution tracks data dependencies of branch instructions and identifies strongly correlated branches in branch history. Compared to the previous *bimodal* style predictors and the fixed history length predictors in embedded processors, our method provides better history length for each branch instruction, resulting in substantial improvement in prediction accuracy.

Keywords: Branch Prediction, Branch History, History Length Adjustment, Data Dependency, *gshare* Predictor

1 Introduction

To achieve higher performance, recent microarchitectures have made use of deeper pipeline, dynamic scheduling and multi-issue superscalar processor technologies. In the field of embedded processors, such technology trends are expected to be realized in a near future as well. Consequently, accurate branch predictor will be one of essential parts of modern embedded processors, because the penalty for a branch miss-prediction increases as the number of pipeline stages and the number of instructions issued per cycle increase [1]. Although there have been many proposals of complex hybrid branch predictors in modern microarchitectures, they can not directly be implemented in embedded processors, due to the strict hardware resource constraints in their environments. Instead, most branch predictors in embedded processors have used the small-scale *bimodal* style predictors[2]. In *bimodal* predictors, the only information in branch prediction is the address of branch instruction(PC), without Global Branch History(GBH).

V.N. Alexandrov et al. (Eds.): ICCS 2006, Part IV, LNCS 3994, pp. 979–986, 2006.
© Springer-Verlag Berlin Heidelberg 2006

However, as shown in many previous works, branch instructions have a special feature in their executions, called *correlation*, and the prediction accuracy is expected to increase if the predictor additionally utilizes the correlation features of branch instructions[3][4]. In this paper, we propose a new novel algorithm and hardware structure, which can be implemented in embedded processors with providing the correlation features, as the alternative of *bimodal* predictors in embedded processors. Our policy is much more profitable to high-performance embedded systems. The rest of this paper is organized as follows. Section 2 describes the backgrounds and the related works about this paper. Section 3 proposes dynamic per-branch history length adjustment and explains its distinctive algorithm and required hardware structure. In section 4, we simulate our proposal and discuss the simulation results. Finally, section 5 concludes this paper.

2 Backgrounds and Related Works

Various vectors of information can be used as an input of branch predictors. The branch address was firstly proposed to index the Pattern History Table (PHT). This style of predictors is often called as the *bimodal* predictor and it is still widely used in current microarchitectures, because the bimodal distribution of branch execution is one of main characteristics of most branch instructions[5]. Further, in the field of embedded environments, *bimodal* predictor is an essential branch predictor. For examples, *XScale* Processor implements the *bimodal* predictor with 2 bit saturation counter in Branch Target Buffer(BTB)[2]. In the family of ARM processors, they implement simple *bimodal* style branch predictors as well, due to the strict hardware resource constraints[6]. However, it has been shown that some branches, especially conditional branches, are strongly correlated to the outcomes of a few past branch instructions. The 2-level adaptive branch predictor and its variations utilize this idea. This mechanism uses two levels of branch history to make the predictions; the history of the last k branches encountered and the branch behavior for the last s occurrences of the specific pattern of these k branches[3].

Meanwhile, the length of branch history is statically fixed for all branch instructions, and the history length is usually selected in accordance with the size of PHT. However, as shown in previous works, different branch instructions require different length histories to achieve high prediction accuracies [7][8][9]. In previous works, [7] and [8] require *prior-profiling* to change the history length. Although [9] provides dynamic history length fitting method, it also requires many *intervals* which consist of a fixed number of consecutive branch instructions, to profile and change the history length, thus it is semi-dynamic. On the whole, these intervals induce the overall decrease of IPC. However, our policy adjusts the history length in program execution time, that is *fully-dynamic*. Moreover, our policy is *per-branch* method and provides optimal history length for each branch instruction, which is not the case of previous works.

3 History Length Adjustment in Branch Prediction

In this section, we propose *Dynamic Per-Branch History Length Adjustment* (called DpBHLA Policy), and additionally we show its algorithm and required hardware structure. The proposed solution tracks data dependencies of branch instructions and identifies strongly correlated branches(called *key branch*) in branch history, based on *operand register* in branch instruction. By identifying the key branch, the DpBHLA policy selectively uses the information of key branch in GBH, resulting in different history length for each branch instruction.

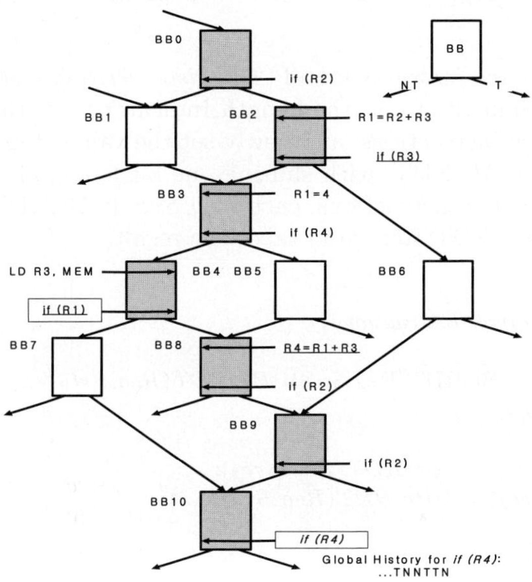

Fig. 1. Example Scenario

Figure 1 shows the example scenario to illustrate the DpBHLA mechanism. Figure 1 is a kind of Control Flow Gragh(CFG), composed of Basic Blocks (BBs). Each basic block consists of one branch instruction and a few other instructions which do not change the execution path. *if (R4)* in BB10 is the branch instruction in which we want to predict the execution path. The GBH for *if (R4)* in BB10 is currently (...TNNTTN). To identify the key branch, we introduce Branch Register Dependency Table (Br_RDT), which stores the key branch information. Figure 2 shows its structure. In Figure 2, the number of entries in Br_RDT is the same as the number of architectural registers and the width of each entry is the same as the history length n, according to the 2^n PHT size. To select the optimal history length for each branch instruction, each entry in Br_RDT is handled by the following algorithm, shown in Figure 3.

The algorithm in Figure 3 mainly controls each entry of Br_RDT, based on the instruction format. The important instruction formats in our algorithm are

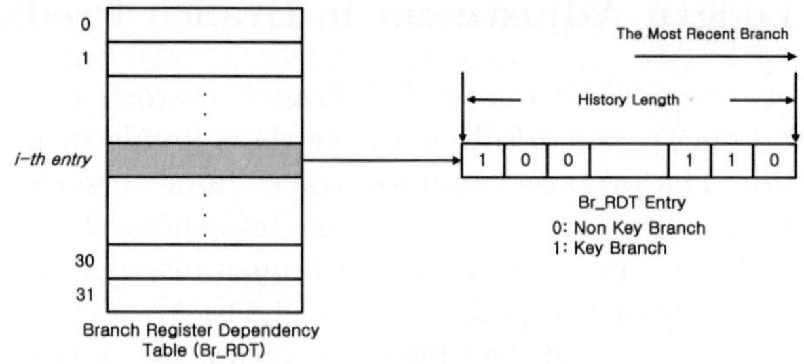

Fig. 2. Branch Register Dependency Table

(1) *conditional branch instruction* and (2) *register writing instruction.* In case of conditional branch instructions, the Length_Indicator takes the entries of source operands in branch instructions, and newly set the value of logic 1 into the LSB field of Reg_Src_i in Br_RDT, with shifting one bit left in all Br_RDT entry. In case of register writing instructions, each Reg_Src_i in Br_RDT is *bit-wise-or*ed, and the entry of Br_RDT(Reg_dest) takes the result.

if (Conditional Branch Instruction){

Length_Indicator $=$ Br_RDT(Reg_Src$_1$)$(|Br\_RDT(Reg\_Src_2)|...)$;
$AllBr\_RDTentry << 1$;

$Br\_RDT(Reg\_Src_1) = 1|Br\_RDT(Reg\_Src_1)$;
$(Br\_RDT(Reg\_Src_2) = 1|Br\_RDT(Reg\_Src_2); ...)$
}

else if (Register Writing Instruction){

$Br\_RDT(Reg\_dest) = Br\_RDT(Reg\_Src_1)(|Br\_RDT(Reg\_Src_2)|...)$;
}
(...) depends on the number of source operands in branch instruction

Fig. 3. History Length Adjustment Algorithm

Figure 4 illustrates the example contents for selecting the history length of *if (R4)* in BB10. In Figure 4, the entry of R4 in BB10 is (...0000100) and the Length_Indicator uses the entry of R4 in Br_RDT at the end of the algorithm. The entry of Br_RDT indicates whether *i-th* field of each Br_RDT entry is BB which contains the key branch or not. Therefore, this vector describes that the BB which has the data dependency with R4 in BB10 is 3 ahead BB (i.e., third field) from BB10, and it additionally indicates that the strongly correlated branch (i.e., key branch) with *if (R4)* in BB10 exists in *3 ahead BB* (i.e., BB4). Actually, R4 is affected by the value of R1 and R3 in BB8

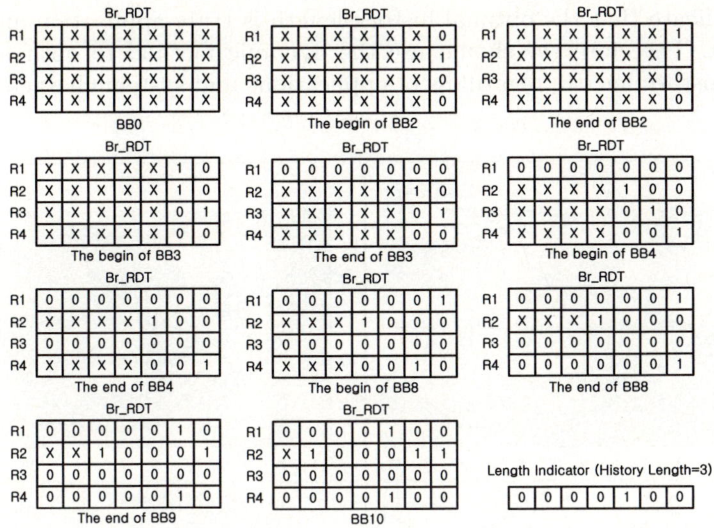

Fig. 4. The Example Contents of Br_RDT

(*R4=R1+R3*), and the value of R1 and R3 is used by BB4 (*if(R1)*) and BB2 (*if(R3)*) respectively. By the *or*-ing function of the algorithm ($Br\_RDT(R4) = Br\_RDT(R1) \| Br\_RDT(R3)$), R4 in Br_RDT has the value of (...0000001) at the end of BB8. In this way, R4 has (...0000100) at the end of the algorithm. This result indicates that the key branch for *if (R4)* in BB10 is *if (R1)* in BB4. Consequently, when predicting *if (R4)* in BB10, we use history length 3 (TTN, i.e., BB4, BB8 and BB9), instead of predefined static history length n for the 2^n PHT size(...TNNTTN). As shown in this example, by identifying the key branch, the different history length can be used for each branch instruction.

4 Performance Evaluation

In this section, we evaluate the performance of our proposal. We use an event-driven simulator, *SimpleScalar*[10], for our simulation. As benchmark programs, we use *SPEC 2000* application suits[11].

At first, Figure 5 shows *miss-prediction rate(%)* vs. history length, for each application. In Figure 5, we vary the number of PHT entries from 512 to 4K, and we change the history length from 1 to n for each 2^n PHT size. Although the results are strongly application dependent, the selected four simulation results show the representative result patterns, compared to other applications. As shown in Figure 5, the prediction accuracy is significantly affected by the history length. For example, the difference between the best accuracy and the worst accuracy is 2.1%, in case of *181.mcf* in 512 PHT. Furthermore, the best history length is also dependent on the PHT size, as shown in *181.mcf*: (1) history length 6 in 512 PHT vs. (2) history length 8 in 4K PHT. Overall, these results indicate that the optimal history length is truly application and system dependent. Therefore, we should use the dynamic method to optimally select the best history length, regardless of applications and system characteristics.

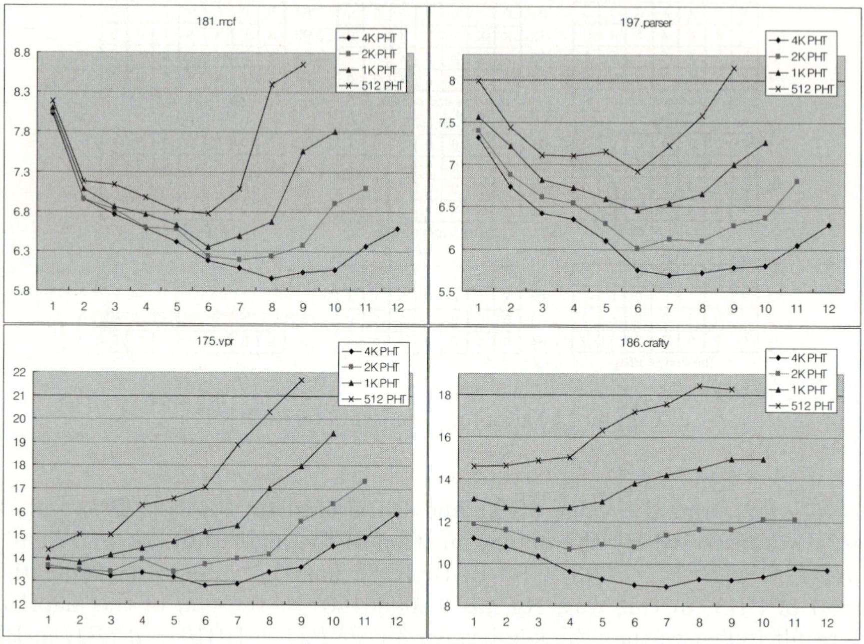

Fig. 5. Branch Miss-Prediction Rate vs. History Length

On the other hand, Figure 6 shows the *branch prediction accuracy(%)* vs. PHT size. Generally, it has been known that the simple *bimodal* predictor is superior to the 2-level predictor or the *gshare* predictor in case of small PHT size, because a short length of the GBH can not provide enough information to distinguish each branch instruction. Our simulation shows this phenomenon as well, in case of simulation results in 512 PHT entries, as shown in average result. Due to this reason, the *bimodal* predictor is usually adopted by the small-scale embedded processors. Besides, as the PHT size increases, the 2-level predictor and the *gshare* predictor, on the contrary, provide better prediction accuracy than the *bimodal* predictor.

Figure 7 shows the *miss-prediction rate(%)* of DpBHLA policy in 512, 1K, and 2K PHT, respectively. In our simulation, Fixed Length depicts the result of conventional fixed and static history length policy, and Length Adjustment is our proposal. As shown in Figure 7, DpBHLA policy outperforms Fixed Length policy, 2.1% in average, and the improvement is up to 4.7% in case of *175.vpr*. Moreover, in small-scale system environment(such as 512 PHT), DpBHLA significantly outperforms Fixed Length policy, and it even outperforms the *bimodal*

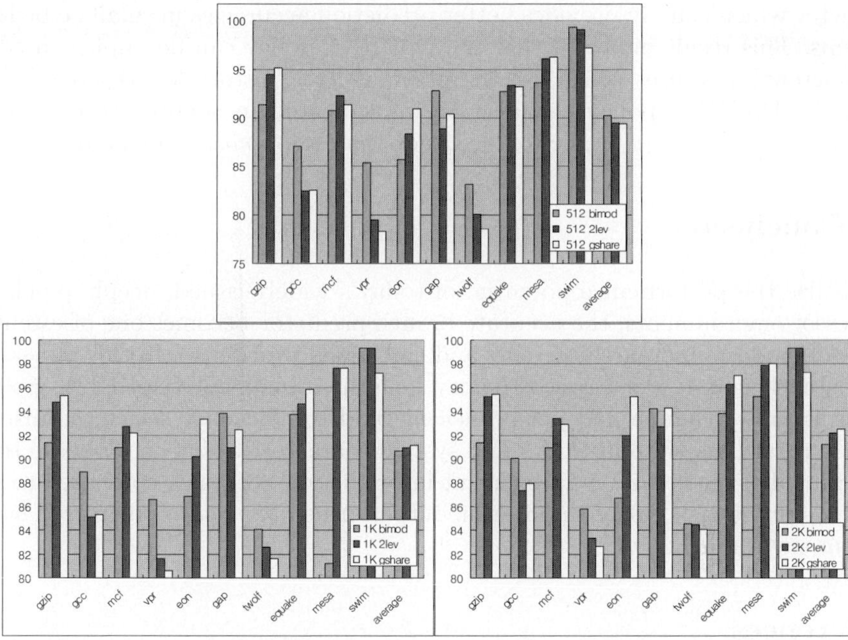

Fig. 6. Branch Prediction Accuracy vs. PHT Size

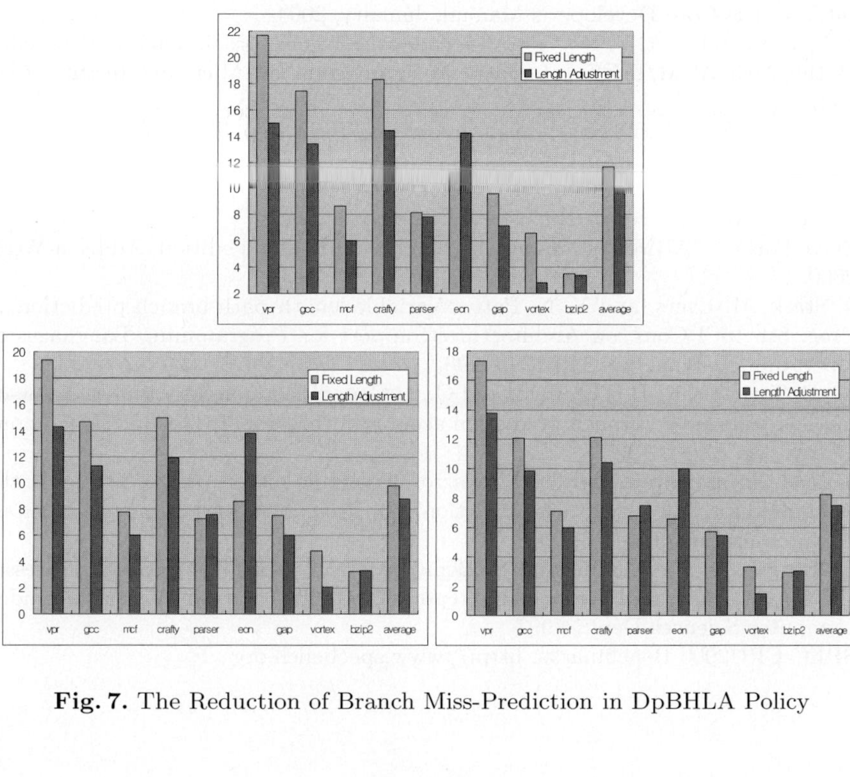

Fig. 7. The Reduction of Branch Miss-Prediction in DpBHLA Policy

predictor which usually provides better prediction accuracies in small embedded systems. This result indicates that the DpBHLA policy can be implemented in the microarchitectures which use the previous fixed history length policy. Further, the DpBHLA policy can also be implemented in small-scale embedded processors, with better prediction accuracy than the *bimodal* predictors.

5 Conclusion

To realize the performance potential of today's widely-issued, deeply-pipelined processor environments, the accurate branch predictor becomes one of essential parts of modern microarchitectures and embedded processors. In this paper, we presented Dynamic Per-Branch History Length Adjustment (DpBHLA) policy, which is fully-dynamic and per-branch method to efficiently select the history length, regardless of applications and system characteristics. The proposed solution provided up to 4.7% improvement in prediction accuracy, compared to the fixed history length policy, and it even outperformed *bimodal* predictors in small system environments.

References

1. J. L. Hennessy and D.A Patterson, "Computer Architecture : A Quantitative Approach", Third Edition, Morgan Kaufmann Publishers, Inc, 2001.
2. Intel XScale Core Developer's Manual, January, 2004
3. Yeh, T. Y. and Patt, Y. N., "Two-level adaptive branch prediction", In Proceedings of the 24th ACM/IEEE International Symposium on Microarchitecture, 51-61, 1991
4. McFarling, S., "Combining branch predictors. Tech. Rep. TN-36m", Digital Western Research Lab., June, 1993
5. Keith Diefendorff, "K7 Challenges Intel, New AMD Processor Could Beat Intel's Katmai", Microprocessor Report, Vol. 12, No. 14, 1998
6. Steve Furber, "ARM System-on-Chip Architecture", 2nd edition, Addison-Wesley, 2000
7. J. Stark, M. Evers, and Y. N. Patt, "Variable length path branch prediction", In Proc. 8th Int'l Conf. on Architectural Support for Programming Languages and Operating Systems, pp. 170-179, 1998.
8. M.-D. Tarlescu, K. B. Theobald, and G. R. Gao, "Elastic history buffer: A low-cost method to improve branch prediction accuracy", In Proc. Int'l Conf. on Computer Design, pp. 82-87, 1997.
9. T.Juan, S. Sanjeevan, and J. J. Navarro, "Dynamic history length fitting: A third level of adaptivity for branch prediction", In Proc. 25th Int'l Symp. on Computer Architecture, pp. 155-166, 1998.
10. D. Burger, T. M. Austin, and S. Bennett, "Evaluating future micro-processors: the SimpleScalar tool set", Tech. Report TR-1308, Univ. of Wisconsin-Madison Computer Sciences Dept., 1997
11. SPEC CPU2000 Benchmarks, http://www.specbench.org

A Novel Method of Adaptive Repetitive Control for Optical Disk Drivers

Kyungbae Chang and Gwitae Park

ISRL, Korea University,
1, 5ga Anam-dong Sungbuk-Gu 136-713 Seoul, South Korea
{lslove, gtpark}@korea.ac.kr
http://control.korea.ac.kr

Abstract. In optical disk drives, which support various speeds, it is not avoidable to have a varying periodic disturbance. However, it is still possible to control optical disk drives by using a controller which repeatedly control the drive to change sampling frequency to follow the change of reference period. This paper introduces an adaptive repetitive control method to attenuate the periodic disturbances. The proposed adaptive repetitive control is built with two parts, the repetitive controller and the frequency multiplier. The repetitive controller uses a varying sampler operating at a variable sampling rate maintained at fixed multiple times of the disturbance frequencies and the frequency multiplier generates the varying sampling frequencies based on the disturbance frequency.

1 Introduction

In this paper, an adaptive repetitive control that can accommodate the varying period disturbance is proposed. Fig. 1 shows the block diagram of the proposed controller, consisting of an actuator, a feedback compensator, an optical sensor, a frequency multiplier, and a repetitive controller. In the diagram, e represents a tracking error, r is a disturbance, and FG stands for the pulse generating signal which has an integer numbers in each period. The pulse generating signal is generated by a spindle motor. The frequency multiplier counts the number of periods of FG. This frequency multiplier produces the fixed number of cycles for each varying period of the FG signal. The repetitive controller uses a generated sampling frequency per period. In the system, there is no need of adjusting sampling time, because the generated sampling period is always synchronized with the period of the disturbance. The number of the sampling point has the same value in case that the period is not considered. sw1 and sw2 are the track search functions controlled by firmware. The details of the frequency multiplier, the repetitive controller, and the search operation are introduced in the following sections.

1.1 Frequency Multiplier

Most optical disk driver systems have FG signals, generated by the spindle motor. The FG signal has the integer numbers per each rotation period. The rotation period is relative to the disturbance and the eccentricity period. In this paper, FG signals are used to estimate the period of the disturbance signal. To generate multiple clocks by

V.N. Alexandrov et al. (Eds.): ICCS 2006, Part IV, LNCS 3994, pp. 987 – 990, 2006.
© Springer-Verlag Berlin Heidelberg 2006

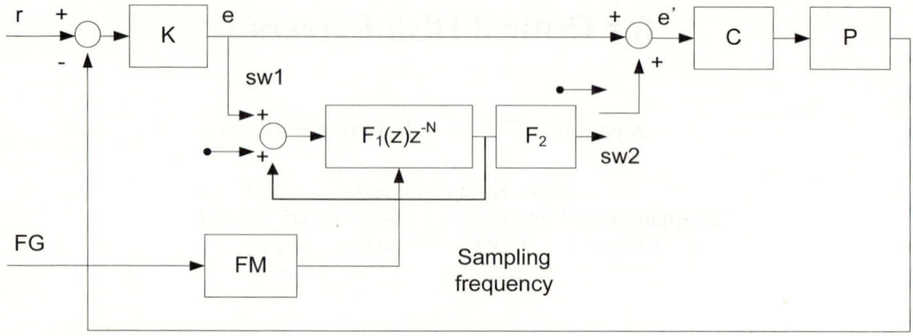

Fig. 1. Adaptive repetitive control system

using an FG signal, the period of the FG signal is estimated followed by dividing the FG signal by integer number. In a digital system, a simple way of measuring the duration of a signal is to count the number of pulses from a source at much higher frequency than the measured signal. The higher frequency is used, the better accuracy of the measurement can be achieved.

Fig. 2 illustrates the timing diagram of the frequency multiplier. The sampling clock of this block is the fixed clock of the existing digital controller. Since the fixed sampling frequency of the existing digital controller is an even higher frequency than the FG signal, it is possible to generate multiple frequencies. This block counts the duration between the rising edges, falling edges, or the edges of the FG signal with the fixed frequency. The counted value is M, and M is divided by L which is the defined number of each period. Then the matched counter counts until the match counter reaches the value of M/L, and this block generates the pulse. This pulse is the varying sampling clock with the varying sampling frequency. So the N samples of the repetitive controller are expressed as;

$$N=L\times N_{FG}$$

N_{FG} is the number of FG signals in each rotation period. L is the defined data, the number per the FG signal. The N samples always have the same number per each period. In this case, the maximum error period T_e of the generated sampling frequency is expressed as;

$$T_e=T_s\times(L-1)$$

If the sampling period T_s is small enough, the change of the controller's sampling times is small. For example, in case of T_e=39.2uSec for T_s=5.6uSec, L=8, and 6.25 mSec of the disturbance period, the change deviation is 0.63%. So the effect of the error period is not noticeable.

1.2 Adaptive Repetitive Controller

The previous period M of the FG signal is used instead of the current period M+Δ. When the current speed is changed into another speed, there is the difference between

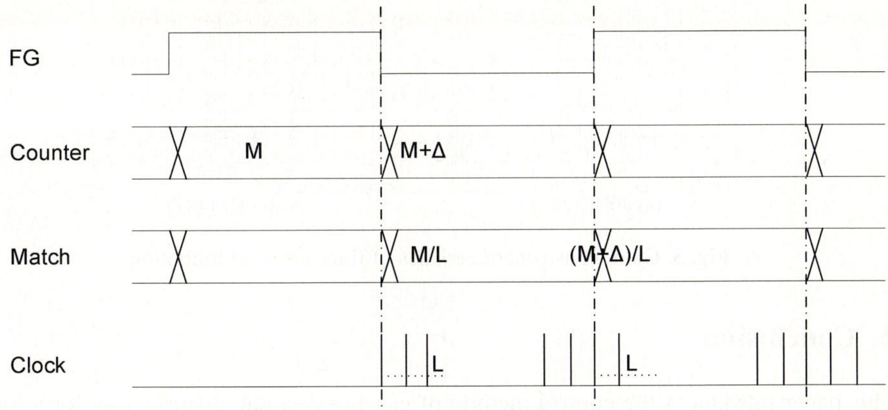

Fig. 2. Timing diagram of the Frequency Multiplier (FM)

the above two periods. Assume that we have a previously recorded waveform e(i) in one revolution with N equally spaced samples per revolution, N=0 to N-1. Then e(i) can be decomposed into sinusoidal components. The varying period disturbance can be expressed as the following form:

$$e(i) = m\sin(2\pi f\frac{i}{N} + \phi)$$

Where, m and ϕ are unknown amplitude and phase. According to the standard of optical disks, m is usually less than 280um. In CAV mode, when the speed is changed from 24x to 48x, the average change of the FG period is 1.25%. This indicates that the difference between the previous period and the current one is 1.25%. The frequency multiplier generates the sampling frequency per each pulse. The change deviation δ between the edges is characterized by

$$\delta = \frac{T_{FG} \times 0.0125}{T_{ecc}} \quad (T_{FG} = T_{ecc} / N_{FG})$$

T_{ecc} is the period of the eccentric disturbance. For example, δ =0.14% for T_{ecc}=1.25mSec and NFG =9. Therefore, in practice, the system stability is possible to be maintained without any change to the repetitive controller's parameters in all cases.

For the repetitive control, it is very important to keep the control period synchronized to the signal period. The frequency multiplier generates the same number of frequencies which has the pulse per each period and is synchronized to the periodic disturbance. Then the repetitive controller is executed at the generated sampling frequency from the frequency multiplier. Hence there is no need to keep the control period synchronized to the periodic disturbance.

In this paper, F1 is chosen as a band pass filter with linear phase characteristic in center frequency. The filter F1 is a 3rd order filter. However, the band pass filter has some of the phase in other frequencies. The sampling frequency of the filter F1 uses the varying sampling frequency from the frequency multiplier. In digital filter, in case of the coefficient having the same value, the phase characteristic is in proportion to the sampling frequency. The characteristic is shown as following Fig. 3.

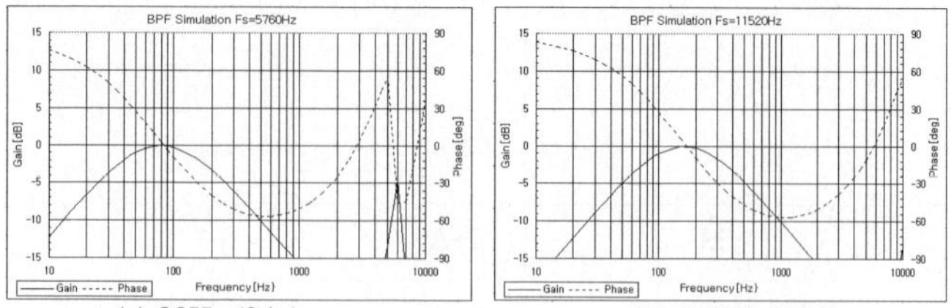

(a) 80Hz (24x) (b) 160Hz (48x)

Fig. 3. Gain, Phase characteristics of filter for band limitation

2 Conclusion

This paper introduces the control method of varying periodic disturbances for a track-following servo system of an optical disk driver. An adaptive repetitive control is proposed and implemented on the real optical disk driver system. The frequency multiplier is used to generate the adaptive sampling frequency and to exactly synchronize the disturbance and the control period. The result of experiments verifies the effectiveness of the proposed adaptive repetitive controller.

A discrete system is built to track a variable periodic signal with varying sampling time. The proposed adaptive repetitive control is useful enough to be applied to the real world optical disk drive products.

References

1. Z.Cao and Gerard F. Ledwich. “Adaptive repetitive control to track variable periodic signals with fixed sampling rate”, IEEE ASME Transactions on Mechatronics. Vol.7, No.3 Sep, 2002, pp378-384.
2. G.M Dotsch and Henk T. Smakman, "Adaptive Repetitive Control of a Compact Disc Mechanism", the 34th Conference on Decision & Control, IEEE 1995, pp1720-1725.
3. T.Y.Doh, J.R.Ryoo, M.J.Chung “Repetitive controller design for track-following servo system of Optical Disk Driver”, IEEE AMC2002, 2002.
4. Gerard Ledwich, A. Bolton, “Tracking periodic inputs using sampled compensators” Proc. Inst. Elect. Eng. Pt. D, vol. 138 no. 3, 1991.
5. Tadashi Inoue, "Practical repetitive control system design", 29th Conference on Decision and Control, 1990, pp1673-1678 .
6. C.Cosner, G.Anwar, M.Tomizuka “Plug In Repetitive Control for Industrial Robotic Manipluators”, IEEE, 1990, pp1970-1975.
7. M.Tomizuka,T.Tsao, and K. Chew, “Discrete-Time Doman Analysis and Synthesis of Repetitive Controller”, Proc America of Control Conference, 1988, pp860-866.

A Real Time Radio Link Monitoring Using CSI*

Hyukjun Oh and Jiman Hong**

Kwangwoon University, Seoul, Korea
{hj_oh, gman}@kw.ac.kr

Abstract. In this paper, a real time radio link monitoring scheme for wireless communication applications is proposed. It can be used as a part of the radio resource management for the efficient radio link control in wireless communication systems. The proposed method is based on the use of the channel state information transmitted or generated by mobile terminals and base stations. In contrast to the existing radio link monitoring schemes, the proposed method can provide very fast responses to the channel variations so that it is appropriate for the real time operations. In addition, the proposed scheme is computationally efficient to implement.

1 Introduction

Efficient utilization and allcation of the spectrum for cellular communications is certainly one of the major challenges in cellular system design [1]. In radio transmission subsystems, techniques such as deployment of time and space diversity systems, use of low noise filter and efficient equalizers, and deploy of efficient modulation schemes can be used to suppress interference and extract the desired signal [1]. Co-channel interference caused by frequency reuse is the most restraining factor on the overall system capacity in wireless networks, and the main idea behind channel assignment strategies is to make use of radio propagation path loss characteristics in order to maximize the carrier-to-interference ratio (CIR) and hence increase the radio spectrum reuse efficiency. Some parameters that can be used for indicating such radio propagation channel conditions is often called channel state information. For example, signal-to-noise ratio (SNR), carrier-to-interference ratio (CIR), received signal strength indicator (RSSI), and much more.

The purpose of radio resource management (RRM) can be summarized as follows [2], [3]:

- Ensure planned coverage for each service.
- Ensure required connection quality.
- Ensure planned (low) blocking.
- Optimize the system usage in run time.

* The present research has been conducted by the Research Grant of Kwangwoon University and Seoul Metropolitan Government in 2006.

** Corresponding author.

V.N. Alexandrov et al. (Eds.): ICCS 2006, Part IV, LNCS 3994, pp. 991–998, 2006.
© Springer-Verlag Berlin Heidelberg 2006

To achieve the above goals, RRM has several dedicated functions. RRM functions can be itemized to six elements: admission control, load control, packet scheduler, resource manager, handover control, and power control [3]. Any of them is very closely tied with the radio link status or condition [3]-[5]. Especially, connection based RRM functions are critically dependent on the radio link condition. Therefore, the radio link monitoring functionality is one of the most critical elements in the wireless communication system combined with RRM. In addition, the response time of radio link monitoring function determines the performance of overall RRM operations. The real time operationality of the radio link monitor or even faster response time in the real time radio link monitoring function is the key design factor.

The overall performance of RRM is much more improved as RRM response time is shortened further. As mentioned above, the basic foundation for good RRM is knowledge about the current propagation channel condition because the best RRM is supposed to react to the time-varying propagation channel condition most properly. RRM gets the required propagation channel information from the radio link monitoring function. This leads to emphasizing of the radio link monitoring functionality and the real time operation of the radio link monitoring with very fast response time is critical to improve the overall system performance. Recently, it has been noted that the real time radio link monitoring functionality plays an important role in several radio resource management related works in real time wireless communication systems recently for this reason [5], [6].

Existing radio link monitoring schemes are mostly dealt by higher layers [3]-[5]. In results, the response time for radio link status change of previous radio link monitoring schemes are slow and they are appropriate for non-real time or limited real time RRM functions. Hence, there is a big room for possible performance improvement of the overall system by adding the fast real-time radio link monitor. Existing previous schemes are utilizing a long-term observation of the propagation channel condition and they take reactions to the cases where there is a noticeable big change in the propagation channel condition such as radio link failure [3]-[5]. In other words, they cannot react to the current instantaneous channel condition properly. It is possible to extend the current existing schemes to deal with the instantaneous radio link state, but it is quit complex and it is even not implementable in some real-time communication systems.

In this paper, a real time radio link monitoring scheme for wireless communication applications is proposed. It can be used as a part of the radio resource management for efficient radio link control in wireless communication systems. The proposed method is based on the use of the channel state information transmitted or generated by mobile terminals and base stations. In contrast to the existing radio link monitoring schemes, the proposed method can provide very fast responses so that it is appropriate for the real time operations. In addition, the proposed scheme is computationally efficient to implement and it is appropriate for the real time operations. The proposed scheme can provide very efficient real time radio link monitoring functionality in the wireless communication systems combined with the radio resource management.

This paper is structured as follows: Section 2 describes our system model with the applications of wireless communications used for our work. Section 3 addresses the proposed radio link monitoring algorithm appropriate for the fast real-time wireless communications using the channel state information.

2 System Model

In this section, the target system model is addressed. A usual RRM system with radio link monitoring function is shown in Fig. 1 [3], which comprises several system parameters that can be used or controlled by joint combination of RRM and radio link monitor. The input of the real-time RRM system consists of two tables and a set of traffic parameters such as the number of carriers, the number of data users, the multi-slot classes, and the traffic loads of voice and data. All of these are controlled by RRM given radio link monitor results. The control outputs of the system are the radio resource related parameters such a delay calculated as Time to First Bit (TTFB), average throughput, bandwidth utilization, number of retransmissions, modulation and coding schemes, and resource allocations [3].

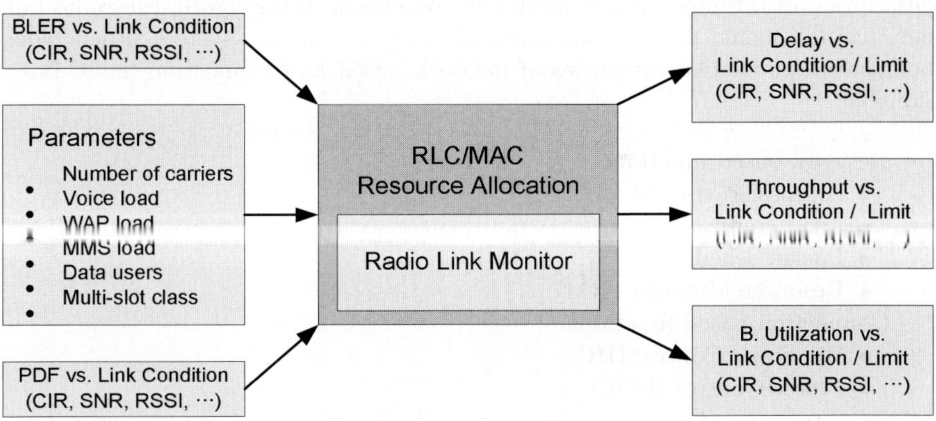

Fig. 1. A system model

So far, such important system tables have been built based on non-real time radio link monitor results obtained through long-term observable radio link parameters. It causes the performance degradation when the radio link is fast time-variant because it can reflect only long-term variations of the propagation channel. During the short-term variation of the radio link, the existing radio link scheme cannot update its radio link state fast enough because it doesn't have detailed state of the radio link or it doesn't support real-time processing [3]-[5]. In results, the design of real-time radio link monitor that is able to detect fast time-varying channel condition is indispensable. The importance of such a radio link monitoring capability gets more important as wireless communication systems gets more evolved to support very high speed mobile over 250km/h.

Another thing that we must point out is what parameters or what indicators are available and appropriate for this purpose of the real-time radio link monitoring that can reflect fast varying radio link conditions. As mentioned above, several parameters are being utilized by RRM and radio link monitor. None of them is appropriate for the short-term radio link state monitoring. Most wireless communication systems have the parameter of channel state information (CSI) for the reliable transmissions. CSI's are available in the form of the signals from the base stations or in the form of direct estimations by the end-user equipment. In general, CSI is used in physical layer to react to the fast-varying propagation channel condition. It is quite complex in the sense of multiple detailed levels that it is represented. Many variants or forms of system parameters can be regarded as CSI. Any form of CSI's can be used for this purpose. The most promising and preferable one is instantaneous or short-term signal to interference ratio (SIR) or some other indicators directly related to the instantaneous SIR values. In this paper, we consider the use of one candidate of the TPC bit that is directly generated from short-term SIR values. The details are addressed in Section 3.

The overall RRM function combined with radio link monitor provide several RRM functionalities as shown in Fig. 2. It shows an example of function-level description of targeted system model [3]. As shown in the figure, the radio link monitoring scheme plays a fundamental basic role in the system. Each block belongs to one of two categories of network based and connection based functions [3]:

- Network based functions
  - Admission control (AC).
  - Load control (LC).
  - Packet scheduler (PS).
  - Resource Manager (RM).
- Connection based functions.
  - Handover Control (HC).
  - Power Control (PC).

Because our interest is focused on the system with the application of wireless communications, the best target system would be cellular phone. Mainly, it consists of three parts: application related components like several multimedia peripherals and application processors, modem related components including baseband processors, and RF chains [6]. It is assumed that the dedicated application or modem processor is equipped with an operating system and it serves as a brain for performing the proposed radio link scheme in this paper. A real time operating system (RTOS) would be appropriate for this purpose, but we do not limit the form of possible operating system in our study, because the operating system is not of our interest in this paper. It simply provides a room for central control of the RRM and radio link monitoring scheme. The embedded

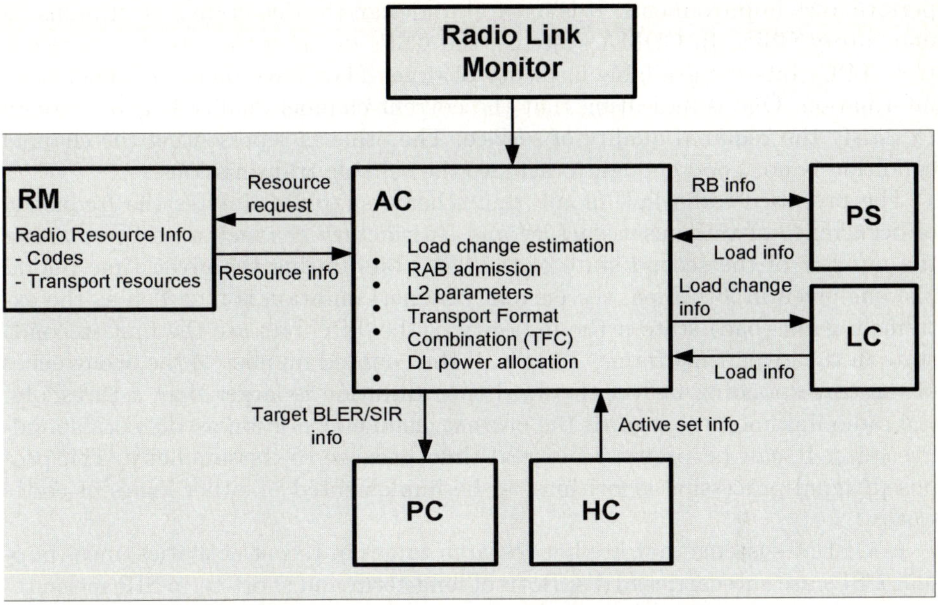

Fig. 2. An example of function-level description of the overall system

processors currently used in designs take two possible shapes: microprocessors or digital signal processors (DSP's). They can be used separately as a single component, or they can be incorporated in a larger silicon chip in the form of embedded cores along with program/data memory and other dedicated logics. In our system model, any form and any kind of dedicated processors are allowed. The proposed scheme is simple enough to support real time processing even with SW.

3 Proposed Real Time Computationally Efficient Radio Link Monitoring Scheme

A computationally efficient signal processing algorithm is proposed to monitor the propagation radio link condition. The algorithm should be simple to implement so that it can provide the fast response in the real-time processing and it does not add another noticeable power dissipation to the overall system. The proposed algorithm is based on CSI that is available in most wireless communication systems for the reliable transmissions [6]. In this section, we consider the CDMA and OFDM cellular systems that are the most widely accepted third generation standards for the cellular communications in the world. However, the proposed scheme is not limited to those systems. It can be generalized easily to other systems.

In CDMA systems, several different CSI's are estimated or transmitted by the end-user terminal. Some of them are mandatory and some of them are for performance improvements. It is natural to design the detection algorithm using mandatory CSI's. In CDMA systems, the CSI's of the transmitted power control (TPC) bits are available in plenty of time. They have binary channel state information. One is indicating that the current channel quality is good enough to satisfy the required quality of service. The other is representing the channel condition is not good enough to achieve the reliable transmissions.

The proposed radio link monitoring scheme simply estimates the frequency of occurrence of this unsatisfactory and satisfactory channel state by counting the number of the second state of the TPC bits during the given time period like one given in [6]. Then, the current channel condition is regarded as the extremely good (bad) state if the majority of the TPC bits are the first (second) state in the given time frame. That is, if the counted number of the occurrences of the corresponding CSI in the fixed time duration is larger than a threshold, the radio link monitor regards the current channel condition as the corresponding state. It can be processed in real time because of its simplicity. The proposed signal processing algorithm can be implemented in either logics or codes easily.

In OFDM systems that are hot research topics in recent activities, most popular CSI's are sub-carrier SIR's. Both of long-term and short-term SIR measurements are possible. The short-term SIR values are good real time indicators for the instantaneous radio link quality. However, it is a burden work to process short-term sub-carrier SIR values with faster response because they are quiet large amount of information to process in real-time when the number of subcarriers is large. A complex dedicated hardware would be required to process them in real time for radio link monitoring.

In this paper, we propose a novel scheme to utilize such burden CSI's efficiently for radio link monitoring. The proposed method is very simple to implement and make use of them possible for real-time processing even with SW codes.

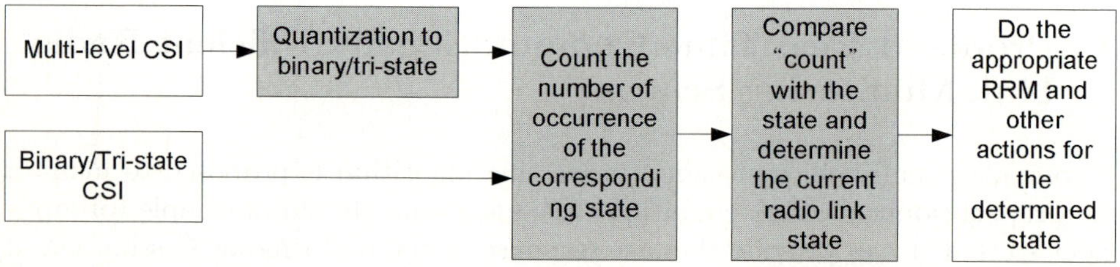

Fig. 3. The proposed real time radio link monitoring method including the quantization

The proposed method is to quantize such burden CSI information to binary or tri-states. In fact, one can try even larger number of states but it make the radio link monitoring system fail to meet our purpose of real time processing with simple codes due to the increased complexity. The proposed real time radio link monitoring scheme based on the quantization is described in Fig. 3 and 4. In any case, it can be processed in real time with fast response indicators of the radio link monitor. It requires only a couple of counters and a simple quantizer to implement. In Fig. 4, an example of tri-state based radio link monitoring is given.

Using our proposed method, any form of CSI's can be utilized for achieving the goal of the desirable radio link monitoring. For instances, ACK/NACK, several event-indicating flags, CRC check flag, hybrid ARQ state information, and TPC bits are appropriate for the binary state based radio link monitoring without quantization. On the other hand, SIR, RSSI, CIR, noise variance, estimated channel responses, and energy metric of decoders are useful for the real-time radio link monitoring with quantization into binary or tri-state information.

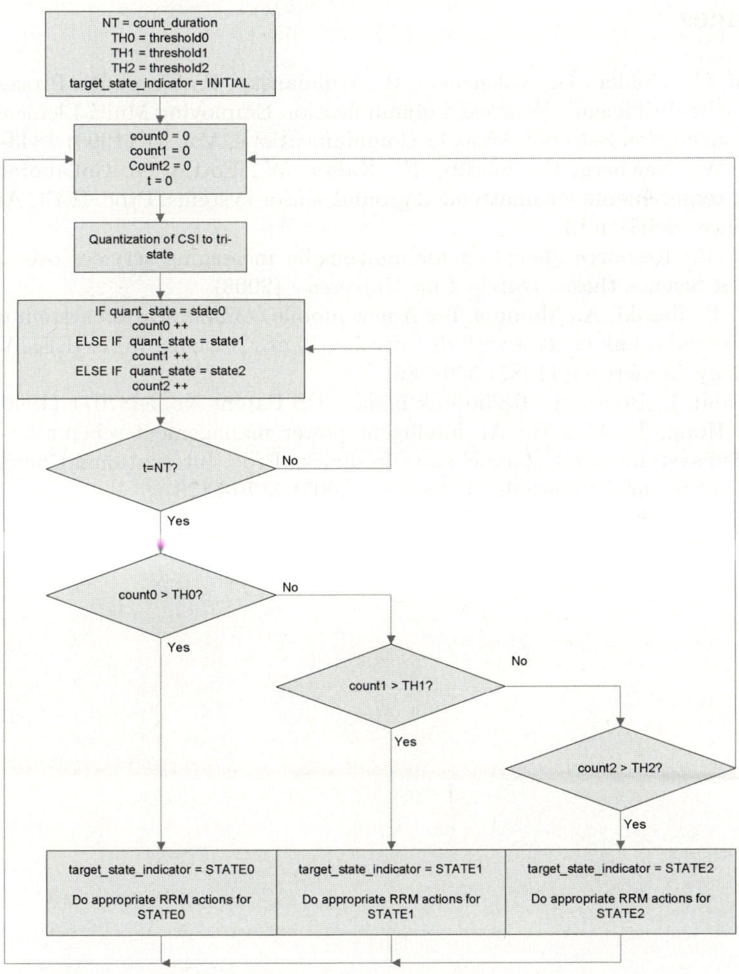

Fig. 4. An example of the proposed radio link monitoring scheme based on tri-state radio link condition

4 Conclusions

In this paper, a possible way to improve the performance of the radio resource management using the real-time radio link monitoring function based on the channel state information available in usual wireless communication system has been proposed. The proposed scheme is very simple to implement and to monitor the propagation channel condition effectively so that it can provide the real-time operation even with SW codes. A signal processing algorithm to monitor the radio link channel condition is also proposed using the available channel state information combined with the quantization methodology.

References

1. Foschini, G., Golden, G., Valenzuela, R., Wolniansky, P.: Simplified Processing for High Spectral Efficiency Wireless Communication Employing Multi-Element Arrays IEEE Journals on Selected Areas in Communications, Vol. 17 (1999) 1843-1852
2. Merrill, W., Newberg, F., Sohrabi, K., Kaiser, W., Pottie, G.: Collaborative networking requirements for unattended ground sensor systems. Proc. IEEE Aerospace Conference (2003) 1-13
3. Nowicki, E.: Resource allocation for multimedia messaging services over EGPRS. Master of Science thesis, Dublin City University (2003)
4. Hattori, T., Sasaki, A., Momma, K.: A new mobile communication system using autonomous radio link control with decentralized base stations. Proc. IEEE Vehicular Technology Conference (1987) 579–586
5. Nordstrand, I., Bodin, S.: Radio link failure. US Patent No. 5487071 (1996)
6. Oh, H., Hong, J., Ahn, H.: An intelligent power management scheme for wireless embedded systems using channel state feedbacks Proc. International Conference on Fuzzy Systems and Knowledge Discovery (2005) 1170-1173

Adaptive Encoding of Multimedia Streams on MPSoC

Julien Bernard[1], Jean-Louis Roch[1], Serge De Paoli[2], and Miguel Santana[2]

[1] INRIA/MOAIS
Laboratoire Informatique et Distribution
51, avenue Jean Kuntzmann
38330 Montbonnot Saint Martin, France
[2] STMicroelectronics
850, rue Jean-Monnet
F-38926 Crolles Cedex, France

Abstract. This paper describes a dynamic scheduling technique based on work-stealing that is proved to be efficient on SMP and clusters. We apply this technique in the MPSoC field, using a simulation in SystemC. We experiment on a MPEG-4 encoding application and we demonstrate that the work-stealing scheduling is more efficient than a static placement scheduling in terms of time and use of resources.

Keywords: MPSoC, scheduling, work-stealing.

1 Introduction

In order to improve the performance of current embedded systems, Multiprocessor System-on-Chip (MPSoC) offers many advantages, especially in terms of flexibility and low cost.

Applications require more and more intensive computations, especially multimedia applications such as video encoding. The system should be able to exploit the resources as much as possible in order to save power and time. This challenge may be addressed by a technique based on parallel computing coupled with performant scheduling.

In this paper, we present a dynamic scheduling technique based on work-stealing. It is proved to be efficient in the SMP and cluster area and we make a proof-of-concept adaptation for an MPSoC platform based on SystemC. We use an MPEG-4 encoding algorithm to compare the work-stealing scheduling with a static placement scheduling.

This paper is organized as follows. Section 2 gives an overview of different scheduling techniques used on MPSoC and explains the principles of work-stealing. Section 3 presents the key points in the implementation of work-stealing and our choices for the implementation on top of SystemC. Section 4 describes various MPEG-4 implementations including our implementation using work-stealing. Section 5 exposes the results we obtained with our implementation on our platform, compared to a static sheduling. Finally, section 6 concludes the paper.

V.N. Alexandrov et al. (Eds.): ICCS 2006, Part IV, LNCS 3994, pp. 999–1006, 2006.
© Springer-Verlag Berlin Heidelberg 2006

2 Related Work and Scheduling on MPSoC

In this section, we discuss about the state of the art related to scheduling, and in particular scheduling on MPSoC.

2.1 Scheduling by Mapping and Pipelining

Mapping and pipelining are two static scheduling methods to improve the performances of MPSoC.

Mapping consists in sharing data by making a static placement on the available resources. The way the placement is done is closely linked to the application. Moreover, the time of computation highly depends on the input data. So the performance are irregular and impredictible.

Pipelining consists in sharing computation by cascading several processors, each one making a part of the whole function. By nature, the number of processors is limited and fixed by the application. And the global computation rate is limited by the rate of the slowest processor.

So, all these approaches are neither scalable nor efficient. First, the number of processors is fixed and depends on the application itself. And more generally, the program and the hardware are closely linked. Second, the program does not adapt to the input data. This results in poor performances for the worst-case data.

2.2 Dynamic Work-Stealing

Scheduling constraints such as architecture independence and input data independence are close to the ones considered for fine grain multithreaded computations [15]. To schedule such computations, many works focus on *work-stealing*, from both a theoretical point of view [1] [8] and a practical point of view [9] [3] [10].

A work-stealing scheduling is based on a classical greedy scheme. It consists in mapping to an idle processor a task that is ready to be executed. Following [1], we note T_∞ the execution time of an algorithm on an infinite number of processors and T_1 the sequential time of this algorithm. Then, neglecting the cost of the interpretation, R.L. Graham [13] proved that the time T_p required for execution on p processors verifies:

$$T_p \leq \frac{T_1}{p} + T_\infty \tag{1}$$

This time appears asymptotically optimal in the case of very parallel applications where $T_\infty \ll T_1$. However, realizing this scheduling also has a cost that must be taken into account. It is *a priori* bounded by the number n of tasks. Since $n > T_1/T_\infty$, this overhead can be very important for a fine-grained algorithm.

Work-stealing schedulers try to minimize this overhead by generating parallelism only when required, i.e. when a processor becomes idle. Efficient work-stealing is based on the *work-first principle* [3]: move the cost of parallelism to the critical path. Indeed, the number of idle tops is bounded by T_∞ on each processor; thus, if the processors are able to easily find some tasks that are ready to be executed, the scheduling overhead, bounded by $O(p.T_\infty)$, will be negligible for algorithms that have a high level of parallelism.

Initially developped for SMP architectures [3], the principle has been extended to processors with different speeds [8] and then to distributed architecture [7], SMP clusters and heterogeneous grids [2].

2.3 Conclusion

So, on the one hand, usual scheduling techniques on MPSoC such as mapping and pipelining seems to have many drawbacks, especially in terms of adaptability. On the other hand, a proven efficient scheduling technique based on work-stealing exists for distributed systems. Our approach is to import work-stealing in the MPSoC field.

3 Implementation of Work-Stealing on MPSoC

This section describes our technical choices for the implementation of work-stealing on MPSoC.

In our experiments, we use a platform to simulate a MPSoC. It is implemented over SystemC, using a Transaction Level Modeling. It is composed of severals nodes linked together by a component called network. A node has a processor, a ROM, a RAM, an interrupt controller and a timer. The network is simply a shared memory in which we added extra-functionnalities. Figure 1 shows this platform.

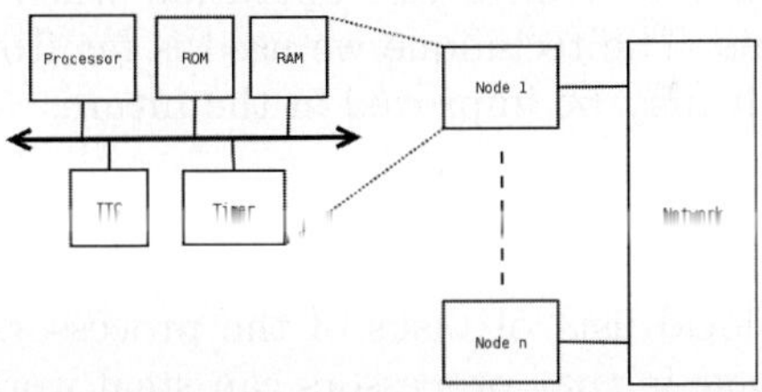

Fig. 1. The MPSoC platform of our experiments

To implement the schedule while reducing contention, work-stealing is often based on a randomized distributed algorithm. A task is locally managed on the processor that creates it and the default sequential (depth-first) execution is optimized. Each processor then locally handles its own list of tasks. When it is required (synchronization between some tasks on different processors, or idle time of one processor), a processor can access to a part of the list owned by another processor, in mutual exclusion, to steal a task (theft operation).

3.1 Choice of the Victim

First, when a processor becomes idle, it steals the oldest ready task on a chosen processor. Here, two ways are possible. A deterministic approach [2]: the victim processor is chosen cyclicly (round-robin). A probabilistic approach: the victim processor is chosen randomly [3]. Then, for parallel computations on p identical processors, it is proved that $T_p < \frac{T_1}{p} + O(p).T_\infty$ with high probability.

To choose a victim processor, we applied the deterministic approach for two main reasons. First it is simpler to implement, there's no need for a pseudo-random number generator, a simple counter is enough. Moreover, a processor can easily determine the end of the computation: if none of the other processors has work left then, it's finished.

3.2 Mutual Exclusion of the Stacks

The synchronization between processors being rare ($O(p.T_\infty)$ for parallel computations), most exclusive accesses are local. Then, an arithmetic lock based on an atomic instruction (such as CompareAndSwap) may be used to implement mutual exclusion. The synchronization may even be implemented basically through very light counters if both process access to distinct parts of the list, such as the head and the tail typically (THE protocol [14] [3]).

Working with SystemC, we could not use an architecture-dependent instruction like CompareAndSwap. In fact, we implemented a very trivial method that may be improved a lot: we added a hardware lock in the network component. It is a special address in memory that, when read, activates a lock. There is one lock for each processor attached to the network. Each one is used to protect each processor's stack.

More generally speaking, embedded processors may have or not such an atomic instruction. It is necessary to have this operation when dealing with efficient distributed applications. The technique we used is far from efficient due to the contention it implies: it may be improved in the future.

3.3 Local Stack Management

We implemented the local lists of tasks of the processors in the shared memory. The main advantage is that processors can steal work themselves to other processors. So, a processor is never interrupted and is always in activity until there is no more work anywhere.

The memory is shared between all the processors. Each list of task is simply organized as a stack of equal size chunks. This is possible because we know the application and then, we can make some optimizations. More precisely, a chunk consists in 32 bytes that stores structure information (mainly to reproduce the calling stack) and the parameters of the functions.

We make this choice because we want to keep things simple ; having only a single level of memory to manage is easier in this first attempt. We don't want to make a fully functional portable system but rather a proof of concept.

In the future, it will be possible to implement the local stacks in the local memories of the processors. It will probably be more efficient as an access in a local memory is faster than an access in a shared memory.

4 Case sudy: MPEG-4 Encoder

In this section, we will first quickly analyze an existing parallel implementations and then, explain how we did an implementation using work-stealing. We assume that the reader has a knowledge of the MPEG-4 standard.

4.1 Analysis of an Existing Implementation

An approach for a parallel implementation is to use fine grain parallelism. This method is fully described in [5]. The idea is to search for the data dependencies in the algorithm at a very fine grain level. Then, this model is transformed to add a proper mapping and scheduling. Finally, with all the added meta-data, the compiler is able to generate a parallel code for an SMP machine.

This approach of parallelism is very application-dependent. Moreover, it does not allow to adapt the computations to the actual picture. It is close to the static scheduling problems we talked about previously (see 2.1).

4.2 Adaptive Parallel MPEG-4 Encoding

Our approach is to use the work-stealing scheduler that we implemented (see 3).

We first make some simplifications to the MPEG-4 encoder, based on the data dependencies analysis: we only consider the encoding of one frame, as each frame has to be processed after the previous one. Then, we also consider to use a motion compensation algorithm that only depends on the previous picture and not the current one. This allows to compute all the macroblocks in parallel.

Then, we introduce a little overhead in the algorithm to improve the performance of work-stealing. We make a recursive cut of the image i.e. we make a function that simply cut the image in several parts and apply itself recursively on each part until there is only one macroblock in the part. Then, the normal function effectively treats the macroblock.

This overhead allows to create tasks of different weight at each step of the recursion so that big task will be stolen first. As a consequence, the number of steals decreases as each processor computes big tasks before idling.

This configuration given, we can make some theoretical analysis relative to our application.

The total work T_1 is the sum of the computation of all the macroblocks (which is in fact the sequential work T_s) plus the overhead implied by the recursive cut and by the work-stealing mechanisms.

T_∞ is, in our simple case, the largest computation time among the computation times of all the macroblocks. It should not be too far from the average computation time of one macroblock. In fact, to be in the condition of the greedy scheduling theorem [4] (see 2.2), the average parallelism T_1/T_∞ should be close to the number p of processors.

In our case, we make our tests with CIF pictures. That represents roughly 400 macroblocks[1]. And we consider that p is not higher than 20. So, that allows the

[1] In fact, 22*18=396 macroblocks.

worse macroblock to be computed 20 times slower than an average macroblock, which should be largely enough.

5 Experiments on an MPSoC Platform and Results

This section presents the results we obtained in our experiments.

For our experiments, we used usual test sequences in CIF format: coastguard, football, foreman, news and stefan. We encoded them with the algorithm presented in 4.2.

We made a first series of experiments on each sequence. We encoded 30 consecutive frames of each sequence. Then, we calculated the average T_∞ and the average *average parallelism* (T_1/T_∞) on the overall frames. Finally, we chose a picture whose caracteristics were close to this average for the second series of experiments.

The reason for doing this is that we wanted to make our experiments on real pictures and not on average unreal pictures, as the work-stealing scheduler takes advantage of the non-uniformity of input data. So we adopted this compromise to have an average real picture for our test.

In the second series of experiments, we compare our work-stealing scheduling with a static placement scheduling. The static placement scheduling consisted in sharing the macroblocks in strips of 18 macroblocks (the height of the picture). Each processor receives the same number of strips with a maximum difference of 1.

We tested both on the same frame (the one chosen previously) of each sequence and we calculated the parallel efficiency ($T_1/(p * T_p)$).

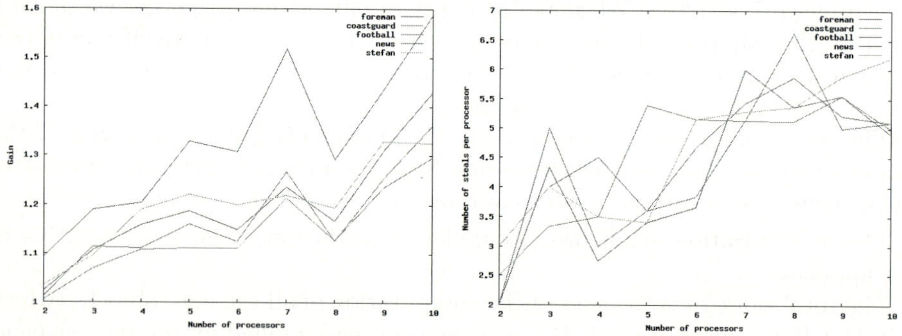

Fig. 2. (a) Gain of the work-stealing sheduling over the static placement scheduling ; (b) Number of steals per processors for the work-stealing scheduling

Figure 2(a) shows the gain in term of time of the work-stealing method over the static method (i.e. $T_{p_{sp}}/T_{p_{ws}}$) for p processors, p varying from 2 to 10 (included).

With four or more processors, the static placement scheduling is at least 10% slower. In the worse case, it can even be 50% slower than the work-stealing scheduling.

We can notice that there is a big improvement with 7 processors. This can be explained: in the case of the static placement, a total of 22 strips are shared among 7 processors, which means that all processors receives 3 strips except the last one which receives 4 strips. While the last processor computes its fourth strip, the other one are simply waiting. That's where the dynamic scheduling is far more efficient, allowing the idling processors to help the last processor.

Another static placement could have been chosen. More efficient static placement will be used and compared to in the future. But this shows that in a real case, the static placement can be penalizing.

In addition, for the work-stealing scheduling, we calculated the number of steals per processor. Figure 2(b) shows the number of steals per processor for p processors, p varying from 2 to 10 (included).

Figure 2(b) proves that the number of steals per processor does not grow too much and then remains constant. Further measures with a higher number of processors confirm this. This totally sticks to the theory (see 2.2): the amount of communications between processors remains quite low, whatever the number of processors.

These results must take into account that the overhead of the work-stealing scheduling does not have much impact as we have been able to make a very optimized version. A more general implementation would have more weight. Moreover, this results are based on simulations and the simulations must be improved in order to have more precise results.

6 Conclusion and Perspectives

In this paper, we demonstrated that the work-stealing scheduling, that was proved efficient for distributed systems, is worth being considered for MPSoC. We made some experiments on a MPSoC simulation platform based on SystemC with a MPEG-4 encoding algorithm that gave us a gain of 10% at least for four processors or more.

This is currently a proof of concept. We aim at improving the implementation of the work-stealing scheduler, and to use better MPSoC simulations so that we can have even more reliable results.

References

1. Blumofe, R.D., Leiserson, C.E.: Space-efficient scheduling of multithreaded computations. SIAM Journal on Computing **27**(1) (1998) 202–229
2. Revire, R.: Ordonnancement de graphe dynamique de tâches sur architecture de grande taille. Régulation par dégénaration séquentielle et distribuée. PhD thesis, Institut National Polytechnique de Grenoble (2004)
3. Frigo, M., Leiserson, C.E., Randall, K.H.: The implementation of the Cilk-5 multithreaded language. In: Proceedings of the ACM SIGPLAN Conference on Programming Language Design and Implementation (PLDI'98). (1998)
4. Blumofe, R.D., Leiserson, C.E.: Scheduling multithreaded computations by work stealing. In: Proceedings of the 35th Symposium on Foundations of Computer Science, Santa-Fe, New Mexico (1994) 356–368

5. Assayad, I., Gerner, P., Yovine, S., Bertin, V.: Modelling, analysis and parallel implementation of an on-line video encoder. In: 1st International Conference on Distributed Frameworks for Multimedia Applications. (2005) 295–302
6. Galilée, F., Roch, J.L., Cavalheiro, G., Doreille, M.: Athapascan-1: On-line Building Data Flow Graph in a Parallel Language. In IEEE, ed.: International Conference on Parallel Architectures and Compilation Techniques, PACT'98, Paris, France (1998) 88–95
7. Roch, J.L., Gautier, T., Revire, R.: Athapascan: API for Asynchronous Parallel Programming. Technical Report RT-0276, INRIA Rhône-Alpes, projet APACHE (2003)
8. Bender, M.A., Rabin, M.O.: Scheduling Cilk multithreaded parallel programs on processors of different speeds. In: ACM Symposium on Parallel Algorithms and Architectures. (2000) 13–21
9. Mohr, E., Kranz, D.A., Robert H. Halstead, J.: Lazy task creation: A technique for increasing the granularity of parallel programs. IEEE Transactions on Parallel and Distributed Systems **2**(3) (1991) 263–280
10. Blumofe, R.D., Papadopoulos, D.: HOOD: A user-level threads library for multiprogrammed multiprocessors. Technical report, The University of Texas at Austin (1998)
11. Willebeek-Le-Mair, M., Reeves, P.: Strategies for dynamic load-balancing on highly parallel computers. IEEE Transactions on Parallel and Distributed Systems **4**(9) (1993) 979–993
12. Chretienne, P., Coffman, E.J., Lenstra, J.K., Liu, Z.: Scheduling Theory and its Applications. John Wiley and Sons, England (1995)
13. Graham, R.: Bounds on multiprocessing timing anomalies. SIAM J. Appl. Math. **17**(2) (1969) 416–426
14. Dijkstra, E.W.: Solution of a problem in concurrent programming control. Communications of the ACM **8**(9) (1965) 569
15. Roch, J.L.: Ordonnancement de programmes parallèles sur grappes : théorie versus pratique. In: Actes du Congrès International ALA 2001, Université Mohammmed V, Rabat, Maroc (2001) 131–144
16. Pazos, N., Maxiaguine, A., Ienne, P., Leblebici, Y.: Parallel modelling paradigm in multimedia applications: Mapping and scheduling onto a multi-processor system-on-chip platform. In: Proceedings of the International Global Signal Processing Conference, Santa Clara, California (2004)

A Mechanism to Make Authorization Decisions in Open Distributed Environments Without Complete Policy Information

Chiu-Man Yu and Kam-Wing Ng

Department of Computer Science and Engineering,
The Chinese University of Hong Kong
{cmyu, kwng}@cse.cuhk.edu.hk

Abstract. To enable an open Grid environment to support organized resource sharing between multiple heterogeneous Virtual Organizations (VOs), we need to tackle the challenges of dynamic membership of VOs and trust relationships between the VOs. We propose a Dynamic Policy Management Framework (DPMF), a Conflict Analysis with Partial Information (CAPI) mechanism, and a heterogeneous authorization policy management mechanism to resolve the problems. DPMF groups VOs deploying the same model of authorization systems together to form a virtual cluster. Policy management is divided into inter-cluster heterogeneous policy management, and intra-cluster homogeneous policy management. In an open Grid environment, some VOs may prefer to keep their policy information private. The Conflict Analysis with Partial Information (CAPI) mechanism is developed to provide an approach of policy conflict analysis in open environments without complete policy information. The basis of CAPI is to generate substitution policies to replace the unknown policy information.

1 Introduction

Traditional security policy frameworks [1][2][4][5] deal with security policy management inside a VO. There is still little research on policy management for multiple heterogeneous VOs. We have proposed an authorization policy management framework, Dynamic Policy Management Framework (DPMF) [6], to handle this problem. DPMF is a hierarchical framework which aims to support dynamic, distributive, and heterogeneous authorization policy management for Grid environments of multiple VOs. Each VO consists of a number of nodes which can be service providers, or service requesters. Each VO has a policy server (or an agent which can access the policy repository of the VO). The policy server is a PDP (Policy Decision Point). The service providers and service requesters on the VOs are PEPs (Policy Enforcement Points) [3].

Figure 1 illustrates the DPMF's hierarchical architecture. There are three kinds of agents in the framework: Policy Agents (PA), Policy Processing Units (PPU), and Principal Policy Processing Unit (P-PPU). In DPMF, each VO

V.N. Alexandrov et al. (Eds.): ICCS 2006, Part IV, LNCS 3994, pp. 1007–1014, 2006.
© Springer-Verlag Berlin Heidelberg 2006

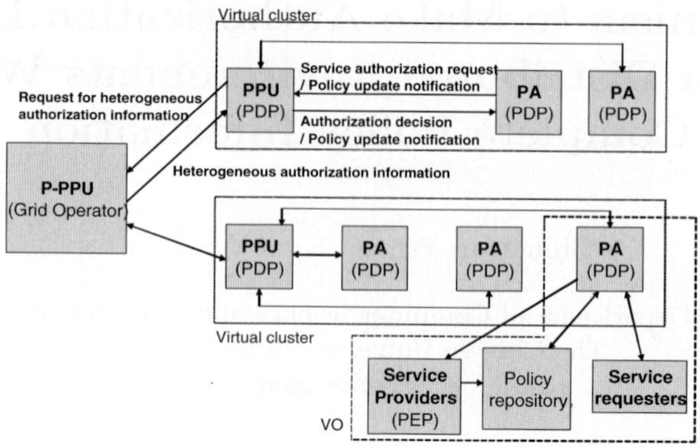

Fig. 1. DPMF architecture

needs to have a PA. The Grid Operator has a P-PPU. The DPMF groups VOs of same model of authorization system (in terms of access control model and policy model) to form a virtual cluster. Inside a virtual cluster, the workloads of policy management can be distributed among the VOs according to their trust relationships. The Conflict Analysis with Partial Information (CAPI) mechanism is developed to provide an approach to make authorization decisions in open environments without complete policy information. Section 2 and Section 3 will present the CAPI mechanism and its performance resulting from simulation.

2 Conflict Analysis with Partial Information (CAPI)

In an open environment, some VOs may prefer to keep their policy information private. In DPMF, a trust relationship between VOs (representing by PAs) means the willingness of disclosing their authorization policies. Since a trust relationship is not compulsory in DPMF, some PAs may not be trusted by some other PAs. During the manipulation of service authorization requests, if some of the involved PAs do not trust the PPU, the PPU is not able to retrieve certain policy information from the PAs. For example, a user from a VO plans to perform a task which involves inter-operation with service providers on several other VOs. So the user requests the PPU for authorization for the task. If some of the target service providers' PAs trust the PPU but some do not, then the PPU cannot retrieve the policy information from the untrustful PA(s). The PPU can only perform policy conflict analysis and make authorization decision with policy information from the trustful PAs. To handle the problem, we have developed a conflict analysis mechanism which requires only partial policy information. It is called "Conflict Analysis with Partial Information" (CAPI).

The main idea of our current approach of CAPI is that the PPU performs a conflict analysis with the known authorization policies, generates substitutions for unknown policies, finds out a set of conditions which can cause conflicts, then sends the set of conditions to the untrustful VO for its checking.

A policy template consists of a *Condition Set*, an *Action Set*, a *Target Identity*, an *Extension Set*, an **Evaluation Element Set**, and a corresponding **Priority Set**. The *Condition Set*, *Action Set*, *Target Identity* and *Extension Set* are learnt from policy information provided by trustful PAs. The **Evaluation Element Set** stores evaluation elements which are attributes of the policy owner. The attributes are defined by the PPU. They probably include the type of service providers, the type of Virtual Organizations, security levels, etc. The **Priority Set** stores the weights of importance of the *evaluation elements.*

$(Policy)_A \rightarrow\leftarrow (Policy)_B$
`if` $(Condition)_A \cap (Condition)_B \neq \emptyset$
`and` $(Action)_A$ `is: [Permit] to access` *resource* `for (Target Identity)`
`and` $(Action)_B$ `is: [Deny] to access` *resource* `for (Target Identity)`

The above expression shows a generic *conflict model* for policy conflict between two policies. The symbol $\rightarrow\leftarrow$ represents the conflict relationship, which is symmetric. The expression states that $(Policy)_A$ and $(Policy)_B$ conflict if they have an intersection of conditions which results in opposite authorization action on the *resource* (target services) for the `Target Identity` (service requester).

During conflict detection in DPMF, the PPU (or delegated PA) also needs to conclude permission conditions. An authoirzation is granted if only if there exists a permission condition set for the task in the authorization request. The permission conditions will be listed in an authorization token if the authorization for the request is positive. The method of concluding permission conditions is to *intersect* (i.e., **AND** operation) the conditions of the policies of the target services, where the action of the policies is to permit the service authorization requester to access the services. Therefore, the concluded conditions are the largest condition set which allows the requester to access the multiple target services simultaneously. The requester can then perform its task by making an access to the target services under the conditions.

2.1 Conflict Analysis with Partial Information (CAPI) Mechanism

The CAPI mechanism is divided into three phases. The *Pre-detection phase* is to prepare the "policy template database". The *Detection phase* is to generate substitution policies to perform conflict detection and to conclude permission conditions. The *Post-detection phase* is to communicate with untrustful hosts to check the conflict detection results, and finally to make authorization decisions.

In the **Pre-detection phase**, the PPU collects policy information from trustful PAs to generate policy templates. The policy templates are used for generation of substitution policies during the CAPI detection phase.

Regarding the selection of a policy template, the PPU defines two control factors: a *similarity value threshold*, and the *maximum number of selected policy templates.* A *similarity value* is the similarity of attributes of the policy owner of the policy template to that of the unknown policies.

Service similarity value = $Pr_1Ev_1 + Pr_2Ev_2 + Pr_3Ev_3 + \ldots + Pr_nEv_n$

where Pr: priority value; Ev: distance of evaluation element values of the policy template to untrustful PA.

A lower *service similarity value* (SS value) means higher similarity of policy owners in terms of the evaluation elements. A PPU would define a *similarity value threshold* and a *maximum number of substitution policies* for generation of substitution policies in the CAPI detection phase. A policy template would be selected if its similarity value is smaller than the *similarity value threshold.*

In the CAPI pre-detection phase, after generating policy templates, the PPU searches for the dynamically optimal Priority values. The PPU uses **policy similarity value** for the calculation. The **policy similarity value** (PS value) measures the similarity of two sets of policies:

$$\texttt{Policy similarity value} = \frac{\texttt{Number of matched policies}}{\texttt{Total number of policies of target policy owner}}$$

The optimal Priority values result into the highest correlation of SS value and PS value, i.e.: the lower the SS value is, the higher the PS value is. The searching method may be a complete search or a genetic algorithm [7]. The evaluation function of the searching can be the weighed PS value of a number of policy template pairs with lowest SS values.

After receiving an authorization request which requires CAPI (i.e., some of the involved PAs are keeping their policy information private), the PPU will enter the **Detection phase**. To substitute policies for an untrustful PA, the PPU generates substitution policies by selecting from the policy templates. The PPU would then use the substitution policies with the policies from trustful PAs to perform conflict detection. During conflict detection, when a conflict occurs, the corresponding substitution policy would be added into a "**Conflict Policy Set**". The "Conflict Policy Set" will be used in the CAPI post-detection phase.

After generating the "Conflict Policy Set", the PPU will enter the **Post-detection phase**. The Conflict Policy Set stores the substitution policies which result into conflicts during the CAPI detection phase. The untrustful PA needs to traverse the policies related to target service to see if any of them is a subset of a policy in the Conflict Policy Set.

For a policy $(Policy)_{untrust}$ in the untrustful PA compared to a policy $(Policy)_{conflict}$ in the Conflict Policy Set, $(Policy)_{original}$ and $(Policy)_{conflict}$ are considered to be matched if the following three criteria are satisfied:

1. $(Condition)_{conflict}$ is a subset of $(Condition)_{untrust}$
2. $(Action)_{conflict}$ is a subset of $(Action)_{untrust}$
3. $(Identity)_{conflict}$ is a subset of $(Identity)_{untrust}$

The PA sends the checking result to the PPU to state the number of "certified conflict policies". The PPU finally makes an authorization decision according to the result. The decision can be based on whether the number of matched conflict policies is within a threshold value which depends on the Grid environment.

3 Performance of CAPI

Since the detection results of CAPI are obtained through estimation, we would like to look into the accuracy of CAPI regarding several different factors. We have done simulation of CAPI based on the following environmental factors:

1. **"Correlation of service similarity and policy similarity"** ($CoSP$): It is a real value between 0 and 1. $CoSP$ represents the likeliness of the statement that "the more similar the two services are (in terms of evaluation element set), the more similar their policy sets are". It is an objective factor to the Grid environment. $CoSP$ can be expressed in the following expression: $CoSP = \frac{d(\texttt{Policy similarity})}{d(\texttt{Service similarity})}$
2. **"Maximum number of substitution policies"** and **"Similarity value threshold"**: They are used in the selection of substitution policies. They are configurable by the PPU. The similarity value threshold is represented by a percentage value to the range of service evaluation values.
3. **"Occurrence rate of opposite policy"**: It represents the occurrence probability of an opposite policy in the Grid environment. An opposite policy is a policy in conflict to the target policy. This is also an objective factor to the Grid environment.

To evaluate the accuracy of CAPI, we deploy three performance indexes: "Average PS" measures the accuracy of CAPI regarding conflict detection results.

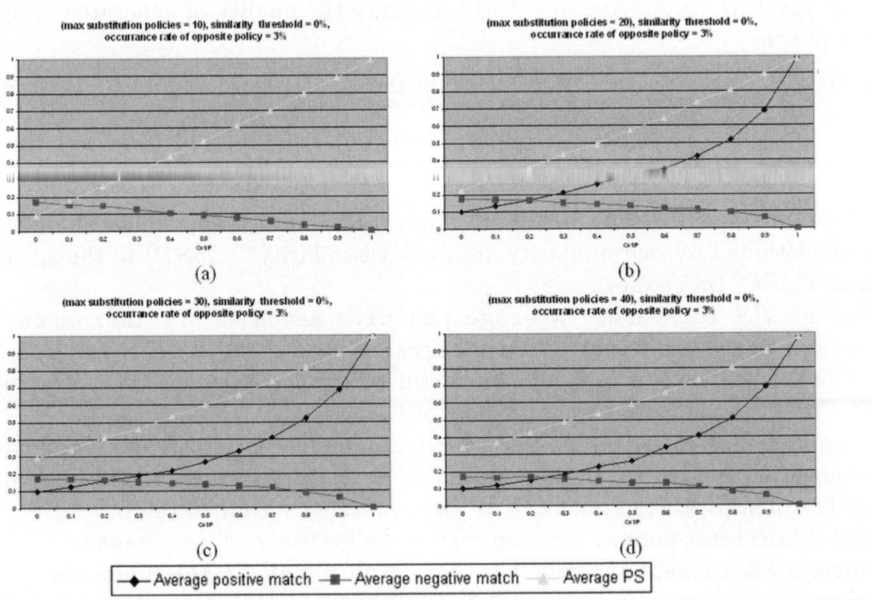

Fig. 2. Simulation results of different *maximum number of substitution policies*: (a)10; (b)20; (c)30; (d)40

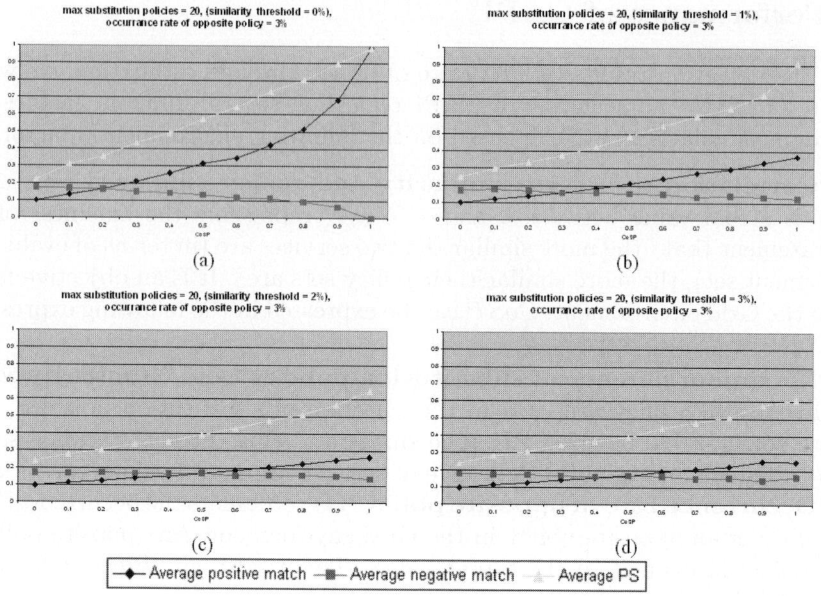

Fig. 3. Simulation results of different *similarity value threshold*: (a)0%; (b)1%; (c)2%; (d)3%

“Average PM” and ”Average NM” measure the quality of generated substitution policies.

1. $PS = \frac{\texttt{Number of matched policies}}{\texttt{Total number of policies of the untrustful service owner}}$
2. $PM = \frac{\texttt{Number of matched policies}}{\texttt{Total number of substitution policies}}$
3. $NM = \frac{\texttt{Number of opposite policies}}{\texttt{Total number of substitution policies}}$

From a single diagram in Figure 2 (e.g., Figure 2b) we can see the effect of ”Correlation of service similarity and policy similarity” ($CoSP$) to the accuracy.
`When` $CoSP$ `increases:`
`Average` PS `increases; Average positive match (`PM`) increases;`
`Average negative match (`NM`) decreases.`

Obviously, $CoSP$ is a significant factor to the accuracy of CAPI. The higher the $CoSP$ is, the higher the accuracy of CAPI is.

Figure 2 illustrates the effect of ”maximum number of substitution policies” to the accuracy of CAPI. Number of policies for each service is 10. The ”maximum number of substitution policies” in Figure 2 ranges from 10 to 40.
`When ‘‘maximum number of substitution policies" increases:`
`Average PS increases; Average positive match (`PM`) does not change; Average negative match (`NM`) does not change significantly.`

We can see that a larger “maximum number of substitution policies” can slightly increase the accuracy of the conflict detection result.

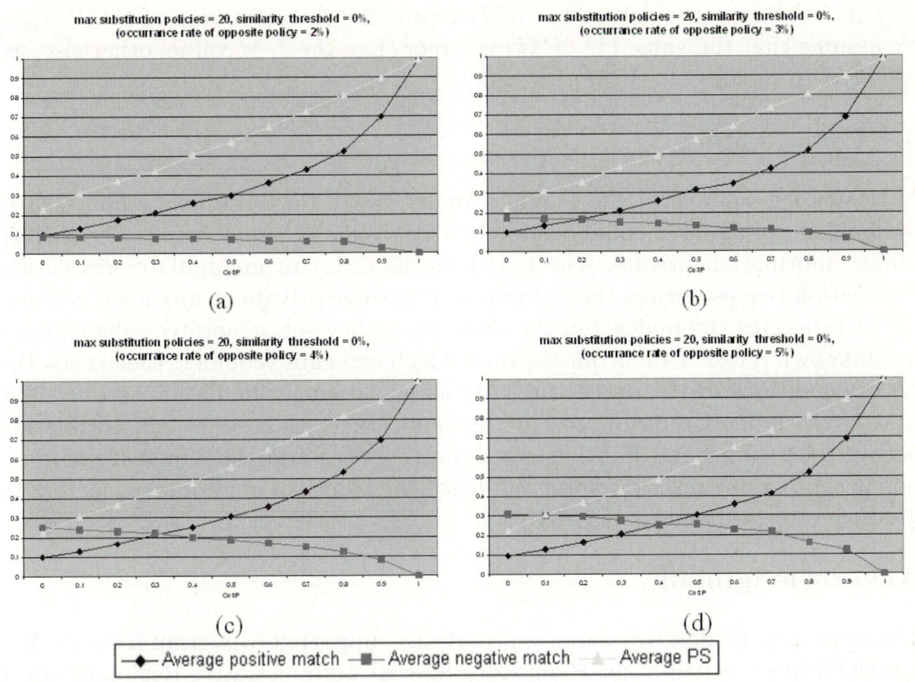

Fig. 4. Simulation results of different *occurrence rate of opposite policy*: (a)2%; (b)3%; (c)4%; (d)5%

Figure 3 illustrates the effect of "similarity value threshold" to the accuracy of CAPI. The "similarity value threshold" in Figure 3 ranges from 0% to 3%.

```
When ``similarity value threshold" increases:
Average PS decreases; Average positive match (PM) decreases;
Average negative match (NM) increases.
```

We can see that the "similarity value threshold" is a significant factor to the accuracy of CAPI. The higher the "similarity value threshold" is, the lower the accuracy of CAPI is.

According to Figure 4, a higher "occurrence rate of opposite policy" results in a higher average negative match (NM) but does not significantly affect average PS and average positive match (PM). Therefore the "occurrence rate of opposite policy" affects the error rate of conflict detection result in CAPI.

A PPU can learn the values of the environment factors during the pre-detection phase of CAPI. By knowing the environment factors, the PPU can set the threshold value for decision making in the post-detection phase. The PPU makes service authorization decision according to: number of "certified conflict policies" (CCP), negative match (NM) value, positive match (PM) value, and number of substitution policies. Positive authorization would be made if:

$$\texttt{Number of CCP} \leq (1 - PM) \times \texttt{(Number of substitution policies)}$$

This is because the value $(1 - PM)$ represents the error rate of CAPI. Here we assume that the value $(1 - PM)$ is larger than the NM value; otherwise, we use NM to replace $(1 - PM)$ instead.

4 Conclusions

In this paper, we present the Conflict Analysis with Partial Information (CAPI) mechanism. In an open collaboration of VOs, some VOs may want to keep their policy information private. The CAPI mechanism can manipulate service authorization requests when the requests may involve private or unknown policies. CAPI estimates the policies in the unknown policy set, generates substitutions for unknown policies, finds out the ones which can cause conflict, then sends the set of conditions to the untrustful VO for its checking. Simulation results show that CAPI is more suitable to environments where there is a high correlation between the similarity of policy owners' properties and the similarity of their policy sets. Policy decision point can learn the values of environment factors to improve the effectiveness of CAPI.

Acknowledgments

The work described in this paper was partially supported by a grant from the Research Grants Council of the Hong Kong Special Administrative Region (Project no. CUHK4220/04E).

References

1. B.E. Carpenter, and P. A. Janson. Abstract Interdomain Security Assertions: A Basis for extra-grid virtual organizations, IBM Systems Journal, Vol. 43, No. 4, 2004, pp. 689-701.
2. Gary N. Stone, Bert Lundy, and Geoffery G. Xie, U.S Department of Defense. Network Policy Languages: A Survey and a New Approach, in IEEE Network, Jan/Feb 2001.
3. J. Strassner and E. Ellesson. Terminology for Describing Network Policy and Services, Internet draft draft-strasner-policy-terms-01.txt, 1998.
4. Dinesh Verma, Sambit Sahu, Seraphin Calo, Manid Beigi, and Isabella Chang: A Policy Service for GRID Computing, M. Parashar(Ed.): GRID 2002, LNCS 2536, pp. 243-255.
5. Von Welch, Frank SiebenSet, Ian Foster, John Bresnahan, Karl Czajkowski, Jarek Gawor, Carl Kesselman, Sam Meder, Laura Pearlman, and Steven Tuecke. Security for Grid Services, in Proceedings of the 12th IEEE International Symposium on High Performance Distributed Computing (HPDC'03).
6. Chiu-Man Yu and Kam-Wing Ng. Dynamic Policy Management Framework for Partial Policy Information, Advances in Grid Computing - EGC 2005. European Grid Conference, Amsterdam, The Netherlands, February 14-16, 2005, Revised Selected Papers, Lecture Notes in Computer Science, vol. 3470, pages 578-588, Springer, June 2005.
7. M. Srinivas and L.M. Patnaik. Genetic Algorithms: A Survey, in IEEE Computer, vol. 27, Issue 6, June 1994, pp. 17-26.

A Reputation-Based Grid Information Service

J. H. Abawajy and A. M. Goscinski

Deakin University,
School of Engineering & Information Technology,
Geelong, Victoria 3217, Australia

Abstract. In a large-scale wide-area system such as the Grid, trust is a prime concern. The current generation of grid information services lack the ability to determine how trustworthy a particular grid service provider or grid customer is likely to be. In this paper, we propose a grid information service with reputation management facility and its underlying algorithm for computing and managing reputation in service-oriented grid computing. Our reputation management service is based on the concept of dynamic trust and reputation adaptation based on community experiences. The working model and functionality offered by the proposed reputation management service is discussed.

1 Introduction

Grid computing aims to enable resource sharing and coordinated problem solving in dynamic, multi-institutional virtual organizations (VO) [10]. In Grids, federation is typically motivated by a need to access resources or services that cannot easily be replicated locally [9]. Grids are currently evolving towards a service-oriented computing by adopting Web services-based and WSRF technologies [14] [4]. An example is the the latest Globus Toolkit (i.e., GT4) [9] [3], which is based on the Open Grid Services Architecture (OGSA) [11]. A service in a service-oriented computing can be computers, storage, data, networks, or sensors.

The Grid Information Service (GIS) is one of the main services offered by Grids. GIS maintains information about hardware, software, services and people participating in a VO. Moreover, GIS provides fundamental mechanisms for service advertisement and service discovery [10]. It allows service providers to publish their offerings while grid customers can use services provided by GIS to discover services that meet their QoS requirements. However, when integrating services across multiple partners that form a VO, trust issues become significant. For example, how can grid service customers choose a trustworthy service provider without prior interaction? Similarly, service providers are commonly concerned if the customer is legitimate and able to pay for services rendered. As customers and service providers are increasingly demanding trustworthy paid services in the digital economy [12], the answer to these questions are important. However, the current generation of grid information services lack the ability to assist service providers (users) in determining how trustworthy a specific customer (provider) is likely to be.

V.N. Alexandrov et al. (Eds.): ICCS 2006, Part IV, LNCS 3994, pp. 1015–1022, 2006.
© Springer-Verlag Berlin Heidelberg 2006

In this paper, we address this problem and propose a *reputation-enabled grid information service* that maintains a dynamic reputation metric such that grid customers can select a reliable service provider to transact with. Similarly, the proposed approach enables the service providers to tailor payment methods to a particular client based on credit rating of customers. The proposed reputation management service is based on the concept of dynamic trust and reputation adaptation based on community experiences. The feedback from the participants is analyzed, aggregated with feedback received from other members and made publicly available to the community in the form of service provider and customer reputation profiles. We present the underlying algorithm for computing and managing reputation in service-oriented grid computing environments.

The rest of the paper is organized as follows. In Section 2, we provide a short overview of the current research efforts that form the basis of our work. The working model and functionality offered by grid directory services and their limitations with respect to the questions at hand are briefly discussed. In Section 3, we propose a new framework for managing reputation in service-oriented grid computing and discuss its underlying architecture. In Section 4, we discuss how the reputation and credit values for service providers and consumers are computed. We summarize future work and conclude our work in Section 5.

2 Problem Statement and Related Work

2.1 Problem Statement

Reputation is of core importance when consumers and providers engage in situations which they perceive as risky. In this paper, *reputation* refers to the value we attribute to a specific entity including agents (e.g., brokers), services, and persons in the service grid, based on the trust exhibited by it in the past. In a typical Grid scenario users are interested in identifying possible candidate services through grid information service in a similar manner to an online shopping site [5].

There are many information services enabling service providers to publish their services while at the same time allow customers discover services of interest from among a large pool of potentially interesting services (e.g., [1] [8] and [14]). As far as we know, existing GIS have not yet addressed the issue of integrating reputation into their resource discovery algorithms. Thus, GIS users optimistically assume all trading partners are trustworthy. As a result, both service providers and consumers are forced to assume all the risks associated with a particular service and resources obtained through the existing GISs.

However, customers are interested in selecting a reliable service provider to transact with. Similarly, the service providers want to tailor payment methods to a particular client based on credit rating of customers. Reputation-enabled information service can address these concerns. Reputations hold entities accountable for their actions and deter bad behavior [12]. Entities that engage in good behavior build a positive reputation and people will continue to interact with it. Such a system encourages individuals to act in a trustworthy manner [7]. Since GISs provide automated resource and services discovery, we believe that the reputation service should be loosely integrated with the GIS such that desired services are automatically discovered by GIS and then ranked based on the reputation they obtain by the reputation services.

2.2 Related Work

Numerous information services that enabling service providers to advertise their services as well as allow customers discover services of interest. For example, the Universal Description, Discovery and Integration (UDDI) [1] defines a standard for enabling businesses to publish or discover Web services. Several efforts to extend UDDI for grid service discovery is underway (e.g., [13]). The Monitoring and Discovery System (MDS) [8] is the information services component of the Globus Toolkit [9] and provides information about the available resources on the Grid and their status. The latest version of MDS includes Web Service Resource Framework (WSRF) based implementations of the Index Service, a Trigger Service, WebMDS (formerly known as the Web Service Data Browser) and the underlying framework, the Aggregator Framework. A market-oriented grid directory service (GDS) for publication and discovery of grid service providers and their services is discussed in [14].

Several examples of feedback mechanisms are already being used in a number of well-known online communities, such as eBay [2]. A growing body of empirical evidence seems to indicate that these systems have managed to provide remarkable stability in otherwise very risky trading environments [12]. We envision a reputation service similar to the online shopping [2] for service-oriented grid computing. However, reputation requirements of the grid systems differ from that of the online shopping reputation system. For example, in eBay [2], buyers reputations matter substantially less, since sellers can hold goods until they are paid. Moreover, even if sellers wished to rely on buyers reputations it would do little good, since it is not possible to exclude buyers with bad reputations from ones auction. Also, there is a potential difficulty in aggregating and displaying feedback so that it is truly useful in influencing future decisions about who to trust. Finally, the reputation systems are for single context. In contrast, contexts in Grids can be numerous.

Recently, trust has been receiving increasing attention in grid computing community [5] [6]. The focus, however, is on grid security system to formulate trust enhanced security solutions. In contrast, we are interested in reputation enhanced grid information services. To the best of our knowledge, there is a lack of reputation information about services, resources, service providers and service users in existing grid information services. This motivated us to design a reputation service for augmenting grid information services.

3 Reputation-Based Grid Information Service

In this section, the proposed reputation-enabled grid information service infrastructure is discussed. We assume that once registered, both service providers and customers will have unforgeable identities. Unforgeable identities are usually generated by a trusted system entity and given to new users as they join. We

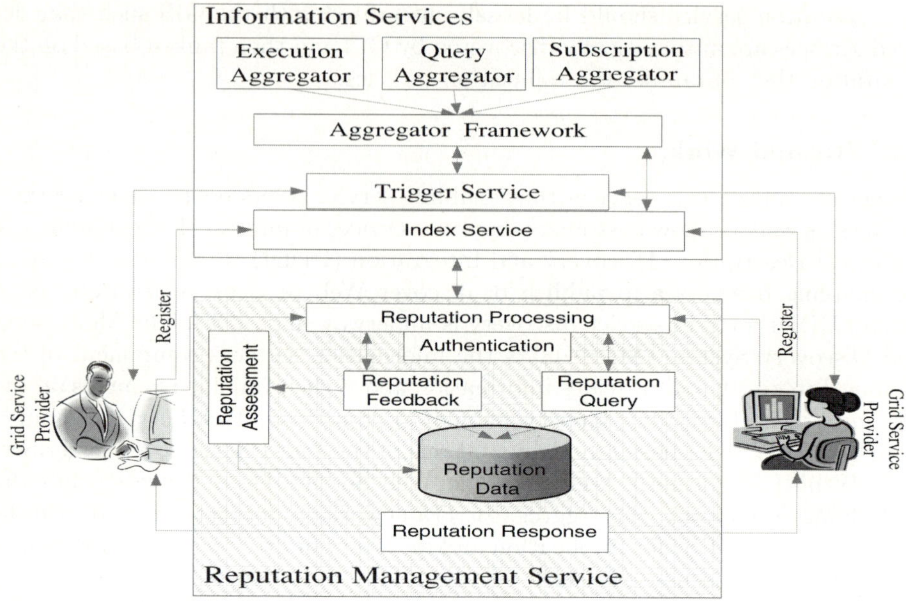

Fig. 1. Reputation-based Grid Information Service

also assume that there is no way a single user can generate multiple identities to pretend as several distinct service provider or customer or both in the network.

The overall design of the proposed reputation-based grid information service is shown in Fig. 1. The system is composed of *information service* and *reputation management service* components. Note that upon registration, the service providers (consumers) are informed to maintain a minimum trust level (Φ_{min}) at all times. By approximately automating word-of-mouth reputation for grid market environments, the proposed reputation-enabled grid information service infrastructure can provide a service that mitigates misbehavior while imposing a minimal cost on the well-behaved users.

The proposed approach uses *credit score* and *reputation score* to manage creditworthiness and trustworthnes of the GSC and GSP respectively. Reputation score have proved to be a great asset for both online shoppers and sellers to establish trust. Similarly, credit scoring is an important tool used by todays financial institutions to assess the creditworthiness of prospective customers and to monitor customer credit ratings. Credit scoring enables service providers to offer differentiated products and benefits based on credit information from customers.

3.1 Information Service

The information service component is the same as the conventional grid information services such as the MDS [8]. It is composed of a suite of web services including Index service, Trigger service and Aggregator framework, which is used by the Index and Trigger services. The central role of *information services* is to provide efficient resource registration, monitoring and resource discovery including subscription/notification. It maintains information about the resources available within grid computing such as hosts, clusters, switches, routers, links, services, sensors, available software, and services. As the focus of this paper is on the reputation management service, we will not discuss the information service components in the rest of the paper.

3.2 Reputation Management Service

The main goal of the *reputation management service (RMS)* component is to assist grid service providers and customers to confidently select trading partners. It collects and processes reputation data from the service providers and consumers. It also maintains a dynamic reputation metric for its community and provides several mechanisms for manipulating the reputation data in the repository with the objective of automating word-of-mouth reputation for grid market environments. The functionality of the RMS is encapsulated in five components: *Reputation Processing*, *Reputation Query*, *Reputation Feedback*, *Reputation Response*, and *Reputation Assessment*.

The *Reputation Feedback* allows service provider and consumer to rate each other only after they are engaged in service trading Each feedback consists of a rating: *positive*, *negative*, or *neutral*. Once the raw reputation data is collected from the service providers and consumers, the *Reputation Assessment* component calculates a reputation score for the service provider and consumer taking into account several factors including the reputation of the feedback provider and the cost of the service. The *reputation query* subcomponent allows service providers and consumers to access the reputation information after being authenticated to have been registered users.

The main responsibility of a response manager is to sanction entities in the service grid that are not behaving consistently and who break trust relations. The action that the the response manager takes when a service provider (consumer) reputation level falls below Φ_{min} is configurable. For example, it can be configured to send the service provider (consumer) a simple warning or immediately eject it from the network for a period of time or permanently banned. To reenter the system, the service providers (consumers) would need to acquire a new valid identifier, which may be costly or impossible. The shaded part of Fig. 1 shows the interaction between the RMS subcomponents as well as with the other system components (e.g., service providers and consumers).

4 Reputation Assessment

Reputation and credit data are collected from the grid service customer (GSC) and the grid service providers (GSP) after the completion of the service. The data is then processed to be represented as *credit score* and *reputation score* respectively. Reputation score provides a way of assigning quality or value to a service. In this work, reputation and credit of an entity is represented using a value in the interval between -1 and 1. As this value approaches -1, the entity becomes increasingly distrusting and conversely, as it approaches 1 the entity has complete or blind trust. We now discuss how the *reputation score* is computed for an entity P_i at time t.

Note that *service contexts (i.e., c)* in grid computing can be numerous, varying from executing jobs, storing information, downloading data, and using the network. The overall cumulative reputation score, $R(P_i, t, c)$, for context c of P_i service provider at time t is defined as follows:

$$R(P_i, t, c) = \min\Big((R_{new}(P_i, t, c), 1.0\Big) \tag{1}$$

where t is the current time and $R_{exist}(P_i, t-1, c) \geq 0$ is the reputation score of an entity P_i for context c at time $t-1$.

The parameter $R_{new}(P_i, t, c)$ in Eq. 1 is the sum of the previous and the current reputation score and is computed as follows:

$$R_{new}(P_i, t, c) = R_{current}(P_i, t, c) + R_{exist}(P_i, t-1, c) \tag{2}$$

where the parameter $R_{current}(P_i, t, c)$ is computed as follows:

$$R_{current}(P_i, t, c) = Feedback(c, t) + Price_t(c) + Credibility_t(P_i, t, c) \tag{3}$$

From Eq. 3, we observe that there are three factors contributing to the current reputation score of a given service provider (consumer): (1) feedback provided by the rater (i.e., Θ_f); (2) the price of the service (i.e., $Price_t(c)$); and (3) the credibility of the rater (i.e., $Credibility_t(P_i, t, c)$).

To determine the contribution of the *Rater Feedback* to the current reputation score, let $f \in \{positive, negative, neutral\}$ be P_i's feedback at time t (i.e., right after the completion of the latest service) for context c. Note that a service provider, P_i, will only be allowed to supply the reputation data when the payment method used involves credit (e.g., *pay-as-you-go* and *pay-after-use*). If *pay-before-use* is used, then the consumer will be automatically given $f = positive$).

$$Feedback(c, t) = 1 - \alpha^f \tag{4}$$

Following the completion of the service, both the service provider and the consumer rate the quality of service as *positive (i.e.,* $f = 1$*)* or as *negative (i.e.,* $f = -1$*)* or as *neutral (i.e.,* $f = 0$*)*. Based on the value of *f*, the contribution of the feedback to the current reputation score is computed as in Eq. 4 where $0 \leq \alpha_t \leq 1$.

The $Price_t(c)$ in Eq. 3 denotes the contribution of the service cost to the current reputation score and is given as follows:

$$Price_t(c) = 1 - e_t^{-\lambda\aleph} \tag{5}$$

where $\aleph$ is the price paid for the service. Intuitively, a GSP that defects on one \$100 transaction should have a lower reputation than one who defects on two or three \$1 transactions. It is also important that one large transaction does not elevate the rate of an entity from nothing to complete or blind trust level. Thus, we constrain the range λ can take as follows:

$$Feedback = \begin{cases} negative\ \lambda \geq 0.01 \\ positive\ \ \lambda \geq 0.0001 \\ nuetral\ \ \ \lambda \geq 0.00001 \end{cases} \tag{6}$$

This avoid one of the main problems with existing reputation systems in which service providers build up a high positive reputation based on many small transactions and then defraud one or more buyers on a high-priced item.

The *trustworthiness* of the feedback provider (i.e., $Credibility_t(P_i, t, c)$) contribution in the calculation of the current reputation score is given as follows:

$$Credibility_t(P_i, t, c) = \begin{cases} 0 & 0 \leq T \leq 1 \\ \bar{N}^{-(\frac{\bar{N}}{\beta \cdot T})} - 1 & \text{Otherwise} \end{cases} \tag{7}$$

where $\bar{N}$ denotes the total number of negative feedbacks so far received by P_i and T denotes the total number of feedbacks a grid service provider has so far accumulated and $\beta = \frac{\bar{N}}{T}$ while T is given as follows:

$$T = \sum Feedback(P_i, c, \pm) + \mu \sum Feedback(P_i, c, \sim) \tag{8}$$

where $\sum Feedback(P_i, c, \pm)$ is the sum of positive and negative feedbacks while $\sum Feedback(P_i, c, \sim)$ is the sum of neutral feedbacks received so far by P_i respectively. We believe that the neutral feedbacks should only contribute a portion to the reputation score of P_i. This is because, in real life, humans tend to take neutral position when faced with two alternatives for which they do not want to cast votes to. Thus, we use μ such that $= 0 \leq \mu \leq 1$ to capture the real life scenario.

5 Conclusion and Future Direction

Although there are variety of information services that could potentially enable grid users discover services and resources, none of these middleware services provide a direct or indirect means of discovering reputable grid services and resources. In this paper, we have introduced a reputation management support in grid information service. We believe that reputation serves as an important metric to avert the usage of under provisioned and malicious resources with the help of community feedback. As a future work, an important issue to be addressed is that if the resource selection decisions are contingent only on the reputation, severe load imbalance can occur in a large-scale grid computing with some dominant resources. Therefore, a mechanism for balancing performance against reliability will be investigated. Another future issue to be looked at is how to elicit feedback from the participants. We are currently performing analysis of reputation assessment algorithm. Finally, in the proposed approach, a service provider (consumer) can be immediately ejected it from the network for a period of time or permanently banned. A mechanism for allowing service providers (consumers) to reenter the system would be needed, which we are currently working on.

Acknowledgments. The first author would like to thank Maliha Omar for all the helps.

References

1. http://www.uddi.org.
2. http://www.ebay.com.
3. Globus toolkit. http://www.globus.org.
4. J. H. Abawajy, editor. *Grid Accounting Service Infrastructure for Service-Oriented Grid Computing Systems.* Springer-Verlag, 2005.
5. B. K. Alunkal, I. Veljkovic, G. von Laszewski, and K. Amin. Reputation-based grid resource selection. In *Proceedings of AGridM 2003*, 2003.
6. F. Azzedin and M. Maheswaran. Towards trust-aware resource management in grid computing systems. In *Proceedings of CCGRID'02*, page 452, 2002.
7. S. Ba and P. Pavlou. Evidence of the effect of trust building technology in electronic markets: Price premiums and buyer behavior. *MIS Quarterly*, 11(26(3)), 2002.
8. K. Czajkowski, S. Fitzgerald, I. Foster, and C. Kesselmand. Grid information services for distributed resource sharing. In *Proceedings of the HPDC-10*, 2001.
9. I. Foster. Globus toolkit version 4: Software for service-oriented systems. In *Proceedings of IFIP International Conference on Network and Parallel Computing*, pages 2–13, 2005.
10. I. Foster and C. Kesselman, editors. *The grid: blueprint for a new computing infrastructure.* Morgan Kaufmann Publishers Inc., San Francisco, CA, USA, 1999.
11. I. Foster, C. Kesselman, J. Nick, and S. Tuecke. The physiology of the grid: An open grid services architecture for distributed systems integration. *IEEE Computer*, 35(6), 2002.
12. P. Resnick and R. Zeckhauser. Trust among strangers in internet transactions: Empirical analysis of ebay's reputation system. In M. R. Baye, editor, *The Economics of the Internet and E-Commerce. Advances in Applied Microeconomics*, volume 11. JAI Press, Greenwich, CT, 2002.
13. B. Sinclair, A. Goscinski, and R. Dew. Enhancing uddi for grid service discovery by using dynamic parameters. In *ICCSA (3)*, pages 49–59, 2005.
14. J. Venugopal and R. Buyya. A market-oriented grid directory service for publication and discovery of grid service providers and their services. *Journal of Supercomputing*, 2005.

Transparent Resource Management with Java RM API

Arkadiusz Janik and Krzysztof Zieliński

Institute of Computer Science, AGH, al. Mickiewicza 30, 30-059 Kraków, Poland

Abstract. The Multitasking Virtual Machine has been provided with many useful features like Isolation API or Resource Consumption Management API. The latter one can be used to help in managing resources in Java applications. However, using RM API does not guarantee separation between the resource management activity and the business activity. In this paper we present the concept of The Transparent Resource Management (TRM) system. The system can be used to run Java applications with resource management policies added dynamically, as a separate aspect. The policy for a given resource is dynamic which means that it may change during the runtime, depending on the state of the application, the state of the whole system, as well as the utilization of different resources in different applications.

1 Introduction

The new Sun's Java Multitasking Virtual Machine ([2], [5]) has been provided with many interesting features like a resource consumption management ([3], [4], [7]) and an isolation of Java applications ([1], [8], [6]). Although the Resource Consumption Management API provides useful features for writing resource-aware applications in Java, the application's developer using RM API has to include parts of the code responsible for managing resources directly into the application's code. In other words, the RM API provides features which can be used to build the highest-level of management policies. The main difference between the original RM API and the proposed extension is the *transparency* of the resource management in our solution. The RM API weakest point is the fact that much of the source code has to be written every time a new application is prepared or policy is changed. The *transparency* of the proposed architecture means that the resource management activity of the application is separated from its business activity and specified on the abstract level. The resource consumption management is invisible for the application. The proposed architecture makes management simple, understandable and transparent both from the application's and developer's point of view. Different management policies can be saved for later use by different developers and by different applications. The resource management functionality can be added to the target application in different cycles of the application's life and in different ways. When considering different cycles, we can add functionality either during the deployment process or during the runtime. When considering different ways of adding the

V.N. Alexandrov et al. (Eds.): ICCS 2006, Part IV, LNCS 3994, pp. 1023–1030, 2006.
© Springer-Verlag Berlin Heidelberg 2006

RM, such possibilities as an aspect oriented programming ([9]) or wrappers seem to be useful. The fostered approach can be used to implement specific resource consumption management policy, to guarantee quality of service (QoS) and to add self-adaptability to Java applications ([11], [10]). In this article we present the overall, high-level architecture of The Transparent Resource Management (TRM) system. The RM API has been originally provided with Multitasking Virtual Machine; therefore presented approach bases on features of MVM.

The remainder of this paper is organized as follows. Section 2 introduces the the Resource Consumption Management API. Section 3 presents the architecture of Transparent Resource Management whereas different ways of using the Aspect Oriented Programming with the proposed architecture are discussed in Section 4. Conclusions and further research are summarized in Section 5.

2 Overview of RM API and Isolation API

The Resource Consumption Management API (RM API [7], [3], [4]) is provided as an experimental part of MVM. It is a framework used to supplement Java with the resource and consumption management as well as with control and reservation features. Each resource in the RM API is described as a set of *resource attributes.* In the RM API resource policies are encapsulated in *resource domains.* All *isolates* ([8]) that are bound to the same resource domain share the same policy. The isolate can be bound to many domains as long as each domain is responsible for a different resource. The policy is implemented by *consume callbacks* and *triggers* that specify when callbacks are called. Callbacks can act either as *constraints* (when called before the resource *consumption* is done) or as *notifications* (called after a consumption). An important element of the RM API is a *dispenser*, the bridge between the resource abstraction and its implementation. The dispenser monitors the amount of resource available for resource domains as well as it decides how much of the resource can be granted when the consumption is done. In order to stress that there is only one dispenser per resource system we should describe it as *global.*

The most important concept of The Application Isolation API ([8]) is an *isolate* - the abstraction of a single Java task running in a Virtual Machine. In the Sun's MVM various Java applications may be started in isolates using the same virtual machine.

3 Transparent Resource Management Architecture

The goal of the proposed system is to make resource management invisible from the application's point of view. The system is built on top of the RM API. What differs the TRM architecture from RM API is that in the TRM resource management is hidden from the application, not included into the application's source code and applied to it during deployment. The resource management policy is dynamic and the application's developer may specify it as a concern of the application. Dynamic means that the policy may change during the runtime, depending on the state of the application and the whole system. What is more, in the TRM architecture various policies may be easily applied. In contrast with the TRM architecture, the RM API can be used to build rather static policies, which are separated from the policies of different resources while in the TRM architecture the policies of different resources may be dependent. As a result, the TRM can be used to build self-adaptable applications in Java.

3.1 High-Level Architecture

The high-level architecture of the Transparent RM is presented in Figure 1. The architecture specifies different elements (rectangles) and different phases (arrows). The more detailed description of the elements of the system is presented below.

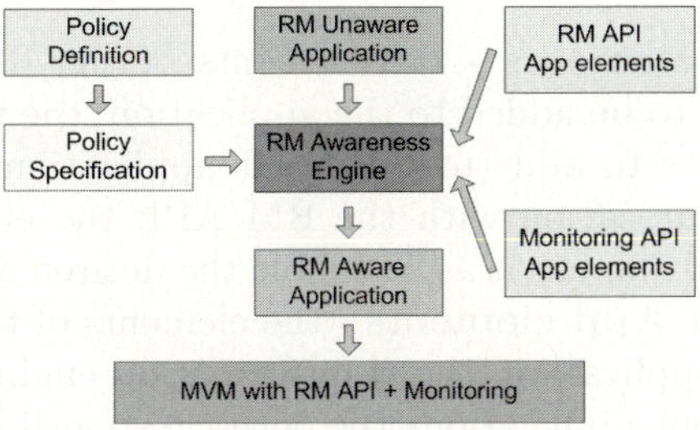

Fig. 1. The architecture of The Transparent Resource Management system

Policy definition is a high-level, user-readable description of the desired application's behavior, resource management logic and quality of service etc. This definition may limit the CPU usage or network bandwidth whereas another definition may stress that the number of concurrently running threads should not increase a given threshold. A policy definition depends on a type of the application and/or specific requirements for its current deployment. The same application running on different hardware configurations may have completely different policy definitions.

Policy specification is a formalized, more detailed presentation of the policy definition. The specialized policy specification language can be used to prepare a specification. The exemplary policy specification requires that the application should not use more than 10% of the CPU time nor to exceed 56kB of the network bandwidth. The limitations of different resources can be also composed to specify more complex policies like not to exceed 56kb of the network bandwidth as long as CPU load is less than 90%. If the CPU load increases, the higher network bandwidth is allowed (because of the switching off the CPU consuming socket compression module and sending raw data). The policies of different applications can be also merged as to treat previously seperate applications as a group. The exemplary policy can specify an upper limit of a number of threads that can be running concurrently in all started applications. The policy specification can be more or less detailed as if there was a conceptual line between the formal high-level specification and the informal low-level programming realization.

RM Unaware application is an input application which is not aware of the resource consuming policies and resource management. The developer of the application can focus on business logic. Moreover, isolating business logic from the the RM logic makes reusability possible. Each time an application is deployed/started different resource policy and different QoS can be applied.

Transformation engine is the central point of the proposed system. The engine is responsible for transforming the RM unaware application into the aware one. The transformation techniques are not presented here but will be discussed later in this paper (see Section 3.3). The transformation engine has four inputs:

- **RM unaware application** - the input application to be transformed;
- **policy specification** - policy to be applied to the application (presented above);
- **RM API App elements** - the elements of the Resource Consumption Management API to be added to the application; the transformation engine uses these elements to add (to the input application) the entity responsible for the communication with the RM API; the entity knows RM API, understands it and can use it as to enable the desired application's behavior;
- **Monitoring API App elements** - the elements of the Monitoring API to be added to the application; the transformation engine uses these elements to add (to the input application) the entity responsible for the communication with monitoring part of the system; the entity uses Monitoring API to provide information about the state of the whole system, as well as feedback on the application's behavior; the information from the Monitoring entity can be used by the RM entity as to modify the application's behavior and to guarantee the RM policy or the QoS;

RM Aware application is the only output from the transformation engine. The application has been modified in such a way that it can use the RM API and the Monitoring API to apply a given RM policy. More detailed description of the application after transformation is presented in Section 3.2.

3.2 Architecture of RM Aware Application

An abstract view on the RM aware application (which is the result of the transformation process) is presented in Figure 2. The RM aware application is built on top of the input, the RM unaware application. Two entities have been added to the original application. There exist three types of connections between elements in the system:

- connection between entities and external environment - each entity can exchange information with the external parts of the system such as entities belonging to other applications or specialized modules common for all of the elements in the system. E.g. the monitoring entity can communicate with the monitoring module which gathers and analyzes information about the state of the whole system (see Figure 3). Alternatively, the monitoring entity can communicate directly with its opposite number in the different application.

- connection between entities and the RM unaware application - the connection that can be used to get information about resource utilization and/or the application's state.
- connection between different entities within the single application - this type of a connection can be used for the effective exchange of information about different aspects of the same application. As a result, the decisions made by entities are more precise.

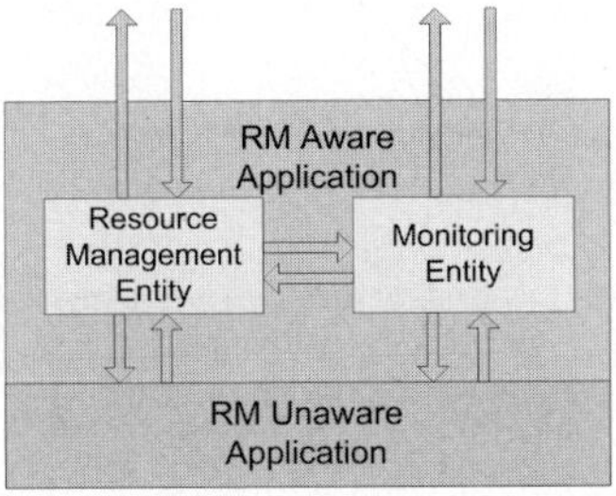

Fig. 2. The architecture of the RM aware application. The RM unaware application is instrumented with entities to communicate or juxtapose application with RM Module and Monitoring Module.

As for entities, there are a few ways of applying them. The Aspect Oriented Programming can be used to make communication between entities and original application possible.

Figure 3 illustrates four exemplary applications in the TRM system. Each application has two entities built into. Each entity communicates with the proper module (common for the whole system). The communication is bidirectional which means that modules use entities to gather information and to perform adequate actions (e.g. modify the application's behavior, definitions of RM API triggers and consume-actions etc.).

Entities of different applications can communicate with each other. This type of communication may be necessary when the strict cooperation between two applications is needed.

3.3 Transformation Engine

Transformation engine is the central point of the TRM system, responsible for transforming the RM unaware application into the RM aware one. The RM API is just a framework which does not imply any resource management policies. One can use it to write its own resources and resource management strategies. The desired result can be achieved by defining proper resources, resource domains and limitations (via consume-action callbacks and triggers). The obvious solution of the transformation process is to implement different resource policies by providing the specialized API (built on top of RM API) with a set of

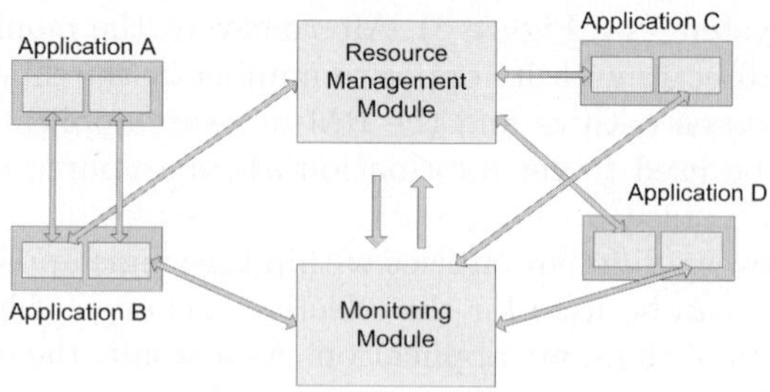

Fig. 3. Management and monitoring modules in the TRM system

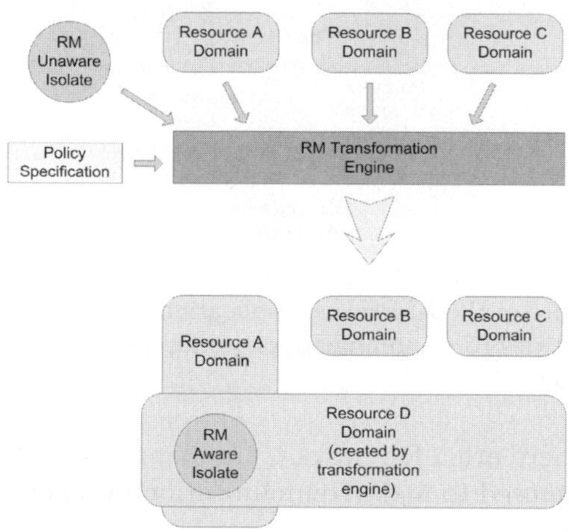

Fig. 4. The process of transforming the RM unaware isolate into the RM aware one. Resource Domain D is created by the transformation engine to meet the conditions specified in the policy specification.

ready-to-use triggers and actions. The simplest way of connecting the RM unaware application to the RM API is to use the specialized manager (isolate or program) to create and start the input application. The manager is responsible for recognition (and optionally creation) of proper resource domains, binding the original application to them and applying resource management policies (by the creation of proper triggers, actions etc.). The application's source code is not modified. However, some of the resource management limitations may result in improper behavior of the application. Let's imagine the application bound to the resource domain with limits on a number of started threads. If a maximum number of threads is present and the application wants to start another thread, resource exception is thrown and the application starts behaving in an incorrect way. The conclusion is that some code modifications might need to be done. The scheme of the transformation process is presented in Figure 4.

As it was mentioned before, the RM aware application has the resource management and monitoring entities built into. Entities can be added either by the use of threads "injected" into the original application or started in separated isolates. Specialized monitoring/managing modules can be used to change resource management policies dynamically (e.g. if network policy in the application A depends on CPU usage policy in the application B).

The presence of entities in the running application means that each application provides a well defined interface which can be used by other applications or by the TRM system to collect information about the application's behavior and resource utilization. Collected information may then be analyzed to check whether the conditions specified by resource policies of all applications are met. Proper activities can then be performed to guarantee that resource policy constraints are not violated. In particular, triggers and consume-action callbacks can be modified. Moreover, the architecture can be also extended to make changes of the application's behavior possible. E.g. if the CPU load increases, the application may change network traffic compression algorithm into the simplest one (less CPU usage), some of the application's features can be switched off, and some rescue features can be enabled. It is a simple way to introduce reflection and self-adaption to the RM aware applications.

4 RM API vs Aspect Oriented Programming

In Section 3.2 we have presented different techniques which can be used to add the resource management and monitoring modules into the RM unaware application. Aspect Oriented Programming methods can be used to support this operation.

Most applications using RM API does not see dispensers directly [3]. Usually, only JRE will create and register dispensers. In this term, the user's application is unaware of the resource management. However, the resource domain's policy has to be specified as a set of pre-consume and post-consume actions or included into the trigger's code. It means that the application's developer is involved into resource management problem and has to consider it during code development.

As it is presented before, the specialized manager can be used to start the application's isolate, and bind it to proper resource domains (either the exiting or the new ones). The manager can also create and start resource managing and monitoring entities and the code responsible for changing the currently active state. The AOP can be used to let the RM unaware applications behave properly when resource exceptions are thrown. The AOP can also define jointpoints between the application and entities. The AOP may also be used to enable the self-adaptation of the running application via changing parts of the code whenever the application's state is changed.

5 Conclusion and Future Work

In this paper we have presented the high-level architecture of the Transparent Resource Management system. We have shown that the RM API can be used to build an additional layer which helps to separate resource management logic from business logic. In the proposed architecture the RM unaware application is being transformed into the RM aware one, considering a given resource policy. Different policies can be defined and stored for later use with different applications. The policies in the TRM system are dynamic, they use information about different resources in different applications. We have presented the concept of the RM and monitoring modules. The modules are used to collect the information about the state of the whole system. The information is analyzed to check whether the resource policies constraints are met. Specialized entities are used as the bridge between the system and the application. We have shown that the Aspect Oriented Programming can be used to support the process of transforming the RM unaware application into the aware one. The next goal of our research is to define the more detailed architecture of the system.

References

1. Czajkowski, G.: Application Isolation in the Java Virtual Machine. ACM OOPSLA (2000), Minneapolis, MN.
2. Czajkowski, G., Daynes, L.: Multitasking without Compromise: A Virtual Machine Evolution. ACM OOPSLA (2001), Tampa, FL.
3. Czajkowski, G., Hahn, S., Skinner, G. and Soper, P., Bryce C. A Resource Management Interface for the Java Plaftorm. Sun Microsystems Laboratories Technical Raport, SMLI TR-2003-124, May 2003.
4. Czajkowski, G., Wegiel, M., Daynes, L., Palacz, K., Jordan, M., Skinner, G., Bryce, C., Resource Management for Clusters of Virtual Machines.
5. Heiss, J.: The Multi-Tasking Virtual Machine: Building a Highly Scalable JVM. Java developers forum. March 2005.
6. Czajkowski, G., Daynes, L., Wolczko, Mario., Automated and portable Native Code Isolation. April 2001.
7. Java Community Process. JSR-284: Resource Consumption Management API, http://www.jcp.org/en/jsr/detail?id=284
8. Java Community Process. JSR-121: The Application Isolation API Specification, Java Community Process, http://www.jcp.org/en/jsr/detail?id=121
9. JBoss Aspect-Oriented Programming home page. http://www.jboss.org/products/aop
10. Sullivan, G., Aspect-Oriented Programming using Reflection and Metaobject Protocols. Arifical Intelligence Laboratory. Massachusetts Institute of Technology. April, 2001.
11. Sadjadi, S., McKinley, P., Stirewalt, R., Cheng, B., TRAP: Transparent Reflective Aspect Programming. Proceedings of the International Symposium on Distributed Objects and Applications. 2004.

Resource Discovery in Ad-Hoc Grids*

Rafael Moreno-Vozmediano

Dept. de Arquitectura de Computadores y Automática
Universidad Complutense de Madrid. 28040 - Madrid, Spain
rmoreno@dacya.ucm.es
Tel.: (+34) 913947615; Fax: (+34) 913947527

Abstract. The extension of grid computing technology to ad-hoc mobile environments is giving rise to the development of ad-hoc grids, which enable wireless and mobile users to share computing resources, services, and information. However, the adaptation of grid technology to ad-hoc networks is not straightforward, and exhibits numerous difficulties (resource discovery, security, power consumption, QoS, etc.). This paper is focussed on the problem of resource discovery in ad-hoc grids, we study the existing resource and service discovery architectures, analyzing the main limitations of these systems (scalability, discovery delay, adaptation to changing conditions, etc.), and we propose a hybrid mechanism that overcomes these limitations.

1 Introduction

Grid technology enables organizations to share geographically distributed computing and information resources in a secure and efficient manner [1]. Shared resources can be computers, storage devices, data, software applications, or dedicated devices like scientific instruments, sensors, etc. Traditional grid infrastructures are mostly based on wired network resources owned by various individuals and/or institutions, structured in Virtual Organizations, which are subjected to specific sharing policies. Grid middleware provides basic services for resource discovery, resource management, data management, security and communication.

With the proliferation of wireless mobile devices (laptops, PDAs, mobile phones, wireless sensors, etc.), and the development of efficient protocols for communication, routing, and addressing in mobile ad-hoc networks (MANETs), wireless or ad-hoc grids are emerging as a new computing paradigm [2] [3] [4] [5] [6] [7], enabling innovative applications through the efficient sharing of information, computing resources, and services among devices in ad-hoc networks.

However, the development of ad-hoc grids entails new challenges, compared to traditional wired grids. Resource discovery, power consumption, QoS, security, etc. are problems that have still to be solved [3] [4].

In this paper we study in-depth the problem of resource discovery in ad-hoc grids. We classify the existing discovery architectures and we analyze their

* This research was supported by the Ministerio de Educación y Ciencia of Spain through the research grant TIC 2003-01321.

V.N. Alexandrov et al. (Eds.): ICCS 2006, Part IV, LNCS 3994, pp. 1031–1038, 2006.
© Springer-Verlag Berlin Heidelberg 2006

main limitations, such as scalability, discovery delays, bandwidth consumption, adaptation to changing conditions, management complexity, etc. In view of these limitations, we propose a new resource discovery mechanism, based on a hybrid peer-to-peer approach and on the concept of discovery zone, which overcomes the main shortcomings of existing approaches.

2 Classification of Service/Resource Discovery Architectures

Existing service/resource discovery mechanisms can be classified in two main categories:

2.1 Peer-to-Peer Architectures

Peer-to-peer (P2P) architectures use fully distributed mechanisms for resource or service discovery, where networks entities (providers and clients) negotiate on-to-one with each other to discover the available services and their attributes, and to find those services that meet the user requirements. Two basics mechanisms can be used to service or resource discovery in peer-to-peer systems: query mechanisms and advertising mechanisms.

P2P Query-Based Systems (P2P-Query). In P2P-Query, also called active or pull P2P mechanisms, clients send a discovery message to the network, by broadcasting or multicasting, asking for services or resources that match same specific requirements or attributes. Providers respond to the client query by sending a description of the service or resource attributes. Examples of P2P query-based systems are the Service Discovery Protocol (SDP) used in Bluetooth [8], the service discovery mechanism proposal for on-demand ad-hoc networks [9] [10], and the Konark active pull protocol [11].

P2P Advertisement-Based Systems (P2P-Adv). In P2P-Adv, also called passive or push P2P mechanisms, providers advertise periodically, by broadcasting or multicasting the location and attributes of resources and services, so that clients can build a local database with all the resources available on the network. Examples of P2P Advertising mechanisms are the Universal Plug and Play (UPnP) discovery service [12] developed by Microsoft, and the Konark passive push protocol [11].

Peer-to-peer architectures are useful for very dynamic ad-hoc environments, were network infrastructure is unpredictable, and the presence of permanent dedicated directories can not be guaranteed. However, these mechanisms, which are based on broadcasting (flooding) or multicasting, suffer from huge bandwidth usage and very low scalability, so they only suit well for small networks. Advertising mechanisms use much more bandwidth and scale worst than query mechanisms, since unsolicited information is issued periodically to the network. However, they reduce the lookup time, since every client holds updated information about all the resources and services that are available in the network.

2.2 Directory-Based Architectures

Discovery architectures based on directory use a centralized or distributed repository, which aggregates and indexes the information about resources and services offered in the network. Providers register their resources and services with this directory, and clients query the directory to obtain information about resources or services. There are three different general schemes of directory-based systems: centralized directory, distributed flat directory, distributed hierarchical directory.

Central Directory Architecture (CD). CD architecture is based on a central directory that aggregates information from every provider, and respond to queries from every client. Central directory architecture is a simple solution, easy to administrate, but directory can represent a bottleneck and a single point of failure, which causes the whole system's failure. Therefore, this solution does not scale well and is only suitable for small networks. Some examples of discovery mechanisms based on centralized architecture are the Service Location Protocol (SLP) [13] standardized by the Internet Engineering Task Force (IETF, RFC 2608), the Jini system [14], which is a platform-independent service discovery mechanism based on Java and developed by Sun Microsystems, and the Agent-Based Service Discovery proposal [15].

Distributed Flat Directory Architecture (DFD). In DFD architecture several directories cooperate in a peer-to-peer fashion, to maintain a distributed repository of information about resources and services. Flat distributed directories can work in two different ways. Directories can exchange information with all other directories, usually by multicasting, so that each directory maintains a complete database about all resources and services in the network. The Intentional Naming Service (INS) [16] and the Salutation protocol [17] are two examples of discovery mechanisms based on this technique. It is obvious that this solution generates high communication traffic level, and hence it is not scalable. The second alternative is to divide the network in clusters or domains, so that each directory maintains a repository with information about services and resources within the cluster or domain. Information exchange between directories in different clusters can be achieved using a peer-to-peer scheme, but using a lower advertising frequency than within the cluster, like for example the INS/Twine system [18], or can be achieved on-demand, like for example the service locating system based on Virtual Backbone [19]. Although clustered solutions are more scalable and suitable for large networks, they must use complex algorithms to manage clusters (cluster formation, selection of directories, addition and removal of nodes to/from the cluster, etc.), and guarantee cluster stability.

Distributed Hierarchical Directory Architecture (DHD). With DHD architecture, the network is divided in domains with a hierarchical structure (like DNS) and directories have parent and child relationship. This solution is fully scalable, but it enforces a rigid hierarchical network organization, which does not fit well in ad-hoc environments. Some examples of distributed hierarchical directory architectures are the Monitoring and Discovery Service (MDS) used in Globus [20] [21] and the Secure Service Discovery Service (SDS) developed at UC Berkeley [22].

3 A Hybrid Mechanism for Resource Discovery in Ad-Hoc Grids

In view of the advantages and limitations of the existing discovery architectures, summarized in table 1, we propose a hybrid discovery mechanism, which combines the advantages of peer-to-peer mechanisms (high adaptability for changing conditions, and low management complexity), and the advantage of clustered solutions (high scalability).

This hybrid approach is based on the idea of zone, similar to the concept introduced by the Zone Routing Protocol (ZRP) for ad-hoc networks [23]. A **discovery zone** is defined for each grid node individually, and is composed by all the neighbor nodes whose distance to the node in question does not exceed a certain number of hops, R, where R is the **zone radius**. It is obvious that the discovery zone of neighbor nodes can overlap.

Within the discovery zone of a given node, we can distinguish two kinds of nodes: the interior nodes, whose distance to the central node is lower than R; and the peripheral nodes, whose distance to the central node is exactly equal to R. Example in Fig. 1 shows the discovery zone of node node A with R=2.

The resource discovery mechanism uses a mixed peer-to-peer approach: to discover grid nodes within the zone it uses an advertisement mechanism, and to

Table 1. Main features of discovery mechanisms

	P2P-Query	P2P-Adv	CD	DFD	DHD
Suitability for changing conditions	High	High	Low	Low	Low
Scalability	Low	Low	Low	High	High
Bandwidth consumption	Medium	High	Low	Low	Low
Discovery delay	High	Low	Low	Low	Low
Management complexity	Low	Low	Medium	High	High

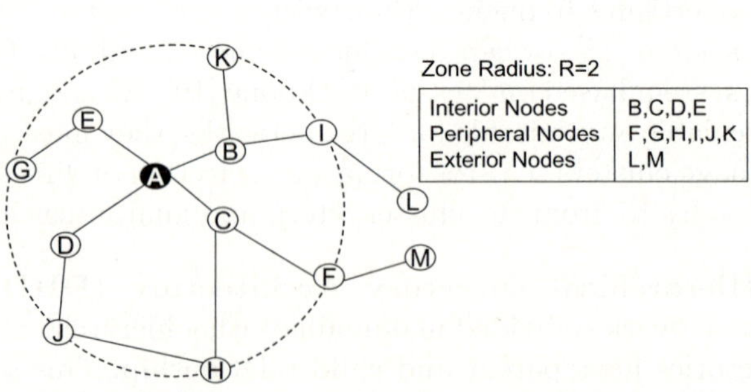

Fig. 1. Example of discovery zone for node A, with R=2

discover grid nodes out of the zone it uses a query mechanism. Each grid node periodically multicasts advertisement packets with a hop limit of R hops, so that these packets only reach those nodes within the discovery zone. Using this mechanism, every node constructs a database with detailed information about all the neighbors within its zone. If no advertisement messages are received from a given neighbor within a specific period, this node is removed from the database. This restricted multicast technique reduces the bandwidth consumption and provides a low delay mechanism for discovering grid nodes within the zone.

If the number of resources within the discovery zone is not enough to meet the client application requirements, a query mechanism is initiated. In this case, the client's node sends a query message to the peripheral nodes, to obtain information about the grid nodes existing in the adjacent zones. This procedure can be repeated several times by the client's node, to obtain information about grid nodes existing two zones away, three zones away, etc., until the number of discovered resources is enough, or until a maximum discovery delay is exceeded. To implement this behavior, each query message includes a parameter called forwarding distance, which specifies how many times the message must be forwarded by peripheral nodes to the next adjacent peripheral nodes. Figure 2 shows how a query message with Forwarding Distance = 1 is forwarded to peripheral nodes of the client's zone, and the query message with Forwarding Distance = 2 is forwarded to peripheral nodes of adjacent zones.

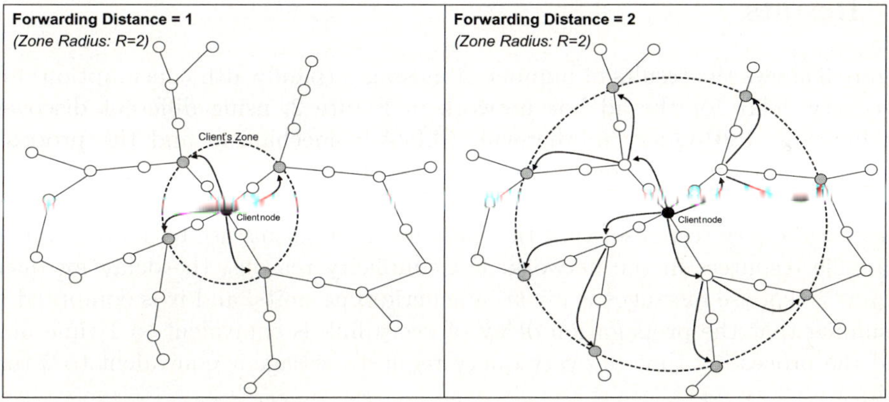

Fig. 2. forwarding of query messages

The three main messages involved in this discovery mechanism are the **Advertisement message**, which is used by a grid resource to multicast its presence and characteristics to the rest of nodes within its discovery zone. This message can contain static and dynamic information about the resource (CPU type and architecture, CPU count, processor load, OS, total and free memory, total and free disk space, bandwidth network links, software, services, etc.). Advertisement procedure is controlled by two main parameters: the Advertisement Period and the TTL period. The Advertisement Period specifies how often a grid node multicasts an Advertisement message to the discovery zone. The TTL period specifies how long a node should keep the information advertised by a neighbor, if no Advertisement messages are received from it. The **Query Request message** is sent by the client node to the peripheral nodes to discover resources out of the client's discovery zone. This message must contain the Forwarding Distance parameter, and the client application requirements, i.e., a list of static or dynamic characteristics that the remote grid nodes should meet. During the discovery process, the client node can send different query messages with increasing values of Forwarding Distance, to discover nodes further away. Finally, the **Query Response message** is used by the peripheral nodes to return to the client node a list of resources that meet the user requirements.

The hybrid method proposed is scalable, since multicast advertisement messages are restricted to the client's zone, and query messages do not use flooding, but they are propagated only by peripheral nodes of successive neighboring zones. Furthermore, discovering delays are much lower than pure peer-to-peer query mechanisms, since a peripheral node can provide information about all grid nodes within its zone. This mechanism is very suitable for changing environments, since information in node databases is updated automatically by the advertisement procedure, and it does not require any administration or management effort.

4 Results

Figure 3 shows the results of number of messages (bandwidth consumption) and discovery delay for the ad hoc network in Figure 2, using different discovery mechanisms: P2P-Query mechanism, P2P-Adv mechanism and the proposed hybrid mechanism with zone radius R=1 and R=2.

The number of messages includes all the messages (advertisement, query request, and query response) that the different mechanisms use to discover all the available resources in the network. For simplicity reasons, the delay for query request/response messages is given in generic time units, and it is computed by assuming that the propagation delay of every link is equivalent to 1 time unit, and the processing time of every query request message is equivalent to 2 time units.

We can observe that the discovery delay of P2P-Adv mechanisms is zero, since every node in the network maintains its own complete database with information about all the resources. However, because this mechanism is based on broadcasting, the number of messages (and hence the bandwidth consumption) is extremely high. On the other hand, the number of messages of the P2P-Query mechanisms is very much lower, but the discovery delay increases significantly. In the middle of these two extremes, the hybrid mechanism exhibits a good tradeoff between these two parameters, since it can reduce appreciably the discovery delay, maintaining a low bandwidth consumption.

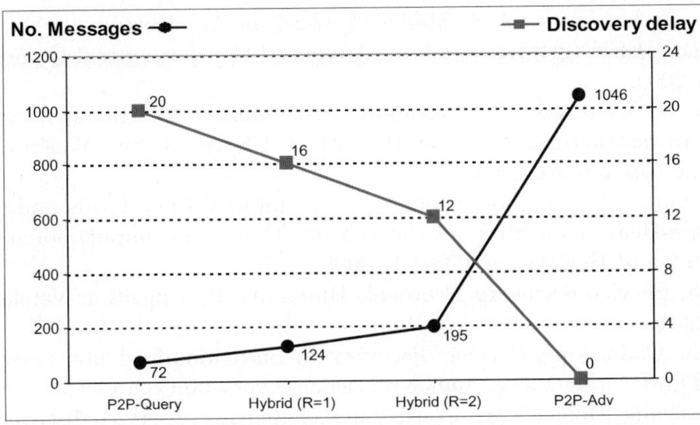

Fig. 3. Bandwidth consumption and discovery delay results

5 Conclusions and Future Work

Efficient resource discovery is a major challenge in ad-hoc grids. Most of the existing mechanisms for resource and service discovery can be classified in peer-to-peer architectures, and directory-based architectures. While peer-to-peer architectures do not scale well, directory-based systems are too rigid and could not be suitable for mobile ad-hoc environments. In this paper we propose a hybrid resource discovery mechanism that is based in the concept of discovery zone. Although it uses peer-to-peer communication, multicasting is restricted to the discovery zone, and queries are forwarded by peripheral nodes, avoiding flooding. This mechanism is scalable, exhibits low discovery delays, is adaptable to changing conditions, and does not require any management effort.

In future work we plan to introduce query control mechanisms, which try to avoid that a given node could forward the same query request several times, and to prevent query requests from being forwarded to zones already visited.

References

1. I. Foster, C. Kesselman, and S. Tuecke, The Anatomy of the Grid: Enabling Scalable Virtual Organizations, Proc. Euro-Par 2001 Parallel Processing, LCNS 2150, Springer-Verlag, 2001, pp. 1-4.
2. M. Gaynor, M. Welsh, S. Moulton, A. Rowan, E. LaCombe, J. Wynne, Integrating Wireless Sensor Networks with the Grid, IEEE Internet Computing, special issue on the wireless grid, July/Aug 2004, pp. 32-39.
3. L.W. McKnight, J.Howison, S. Bradner, Wireless Grids: Distributed Resource Sharing by Mobile, Nomadic, and Fixed Devices. IEEE Internet Computing, special issue on the wireless grid, Jul./Aug. 2004, pp. 24-31
4. D.C. Marinescu, G.M. Marinescu, Y. Ji, L. Boloni, H.J. Siegel. Ad Hoc Grids: Communication and Computing in a Power Constrained Environment. Proc. 22nd Int. Performance, Computing, and Communications Conf. (IPCCC) 2003, pp. 113-122.

5. K. Amin, G. Laszewski, A.R. Mikler, Toward an Architecture for Ad Hoc Grids, Proc. of the IEEE 12th Int. Conf. on Advanced Computing and Communications, ADCOM 2004.
6. K. Amin, G. Laszewski, M. Sosonkin, A.R. Mikler, M. Hategan, Ad Hoc Grid Security Infrastructure. Proc. of the 6th IEEE/ACM Int. Workshop on Grid Computing, pp. 69- 76, 2005.
7. Z. Li, L. Sun, E.C. Ifeachor, Challenges of Mobile ad-hoc Grids and their Applications in e-Healthcare. Proc. of the 2nd Int. Conf. on Computational Intelligence in Medicine and Healthcare, CIMED 2005.
8. Bluetooth, Service discovery Protocol, Bluetooth Specification Version 1.1, Part E, Feb. 2001.
9. R. Koodli, C. Perkins, Service discovery in on-demand ad hoc networks, IETF Internet Draft (draft-koodli-manet-servicediscovery-00.txt), October 2002.
10. Christian Frank, Holger Karl, Consistency challenges of service discovery in mobile ad hoc networks, Proc. of the 7th ACM Int. symposium on modelling, analysis and simulation of wireless and mobile systems (MSWiM), 2004, pp. 105-114.
11. S. Helal, N. Desai, V. Verma, C. Lee, Konark - A Service Discovery and Delivery Protocol for Ad-hoc Networks, Proc.of the Third IEEE Conference on Wireless Communication Networks (WCNC), New Orleans, March 2003
12. B.A. Miller, T. Nixon, C. Tai, M. D. Wood, Home Networking with Universal Plug and Play, IEEE Communications Magazine, vol. 39, no. 12 (2001), pp. 104-109.
13. E. Guttman, Service Location Protocol: Automatic Discovery of IP Network Services, IEEE Internet Computing. Vol. 3, No. 4 (1999), pp. 71-80
14. Sun Microsystems, Jini. Architecture Specification, Technical Report, version 1.2, 2001.
15. Y. Lu , A. Karmouch, M.Ahmed, R.Impey, Agent-Based Service Discovery in Ad-Hoc Networks, 22nd B. Symp. on Communications, Kingston, Ontario, Canada, 2004.
16. W. Adjie-Winoto, E. Schwartz, H. Balakrshnan, J. Lilley, The design and implementation of an intentional naming system, Proc. of the 17th ACM Symposium on Operating Systems Principles, 34(5), 1999, pp. 186-201.
17. The Salutation Consortium, Salutation Architecture Specification, Technical Report, version 2.0c, 1999.
18. M. Balazinska, H. Balakrishnan, D. Karger, INS/Twine: A Scalable Peer-to-Peer architecture for Intentional Resource Discovery, Int. Conf. on Pervasive Computing, Zurich, Switzerland, 2002.
19. J. Liu, Q. Zhang, W. Zhu, B. Li, Service Locating for Large-Scale Mobile Ad Hoc Network, International Journal of Wireless Information Networks, Vol. 10, No. 1 (2003), pp. 33-40.
20. The Globus Alliance. Globus Toolkit 2.2 MDS Technology Brief, Draft 4, http://www.globus.org/toolkit/docs/2.4/mds/mdstechnologybrief_draft4.pdf, 2003.
21. The Globus Alliance, MDS Functionality in GT3. OGSA Technical Resources, http://www.globus.org/toolkit/docs/3.0/mds/MDS.html, 2003.
22. S. Czerwinski, B. Zhao, T. Hodes, A. Joseph, R. Katz, An Architecture for a secure Service Discovery Service, Proc. of the ACM/IEEE MOBICOM, 1999, pp. 24-35.
23. Z.J. Haas, M.R. Pearlman, The performance of query control schemes for the zone routing protocol, IEEE/ACM Transactions on Networking, vol. 9, no. 4 (2001), pp. 427-438.

JIMS Extensions for Resource Monitoring and Management of Solaris 10

Krzysztof Zieliński, Marcin Jarząb, Damian Wieczorek, and Kazimierz Balos

Institute of Computer Science, University of Science and Technology,
Al. Mickiewicza 30, 30-059 Krakow, Poland
{kz, mj, dwieczor, kbalos}@agh.edu.pl
http://www.ics.agh.edu.pl

Abstract. This paper describes extensions of the JIMS system with selected mechanisms of Solaris 10 resources monitoring and management. The proposed extensions together with JIMS functionality create a foundation for the adaptive layer implementation of large distributed computer systems. The proposed solution addresses a heterogeneous self-configuring execution environment typical of Grids system. The paper presents not only the innovative concept of the proposed extensions but their existing implementation.

1 Introduction

Resource monitoring and management is one of most sought-after aspects of functionality of large computer systems, supporting their adaptability to changing workload and end-user requirements. Modern operating systems, like Solaris10, offer very rich and complex mechanisms of resource control. Being a part of large heterogeneous distributed systems, i.e. Grids, the computers running Solaris10 should expose their resource management mechanisms regarding projects and zones [10, 11] in a uniform way so as to be accessible from the integrated management platform.

This paper describes an extension of the JIMS [1, 2] platform with the concepts of Solaris10 *projects* and *zones* monitoring and management. The proposed extension has been implemented using JMX [3, 4] and expands the rich functionality of JIMS from Linux and Solaris 8, 9 to the most advanced features of Solaris10. The proposed extensions make these features accessible to applications written in Java and open a wide space for adaptive policy-driven software layer implementation. These extensions, integrated with JIMS, create a very advanced and powerful tool, beyond the capabilities of comparable existing solutions.

The structure of the paper is as follows. In Section 2 a very brief description of JIMS is presented. Next, in Section 3, the Solaris10 resource management mechanisms exploited in this work are summarized. The idea of JIMS extensions and exposition of requested mechanisms is specified in Section 4. The implementation of this extension as JIMS MBeans is presented in Section 5. The paper finishes with conclusions.

V.N. Alexandrov et al. (Eds.): ICCS 2006, Part IV, LNCS 3994, pp. 1039 – 1046, 2006.
© Springer-Verlag Berlin Heidelberg 2006

2 JIMS Overview

JIMS – the JMX-based Infrastructure Monitoring System assumes that monitored resources are represented as MBeans (Managed Beans), the simple Java objects installed in MBean Servers, key components of JMX (Java Management Extensions). The main purpose of the JIMS is to provide an integrated infrastructure (computational nodes, memory usage, network utilization) and application (Grid engines, J2EE application servers) monitoring management platform. Its features, such as adaptation to operating systems, kernels, and IP protocols, auto-configuration facility (cluster level auto-configuration, Grid level auto-configuration), and dynamic deployment of proper monitoring sensors from one common modules repository, makes it well suited for Grid environments. The JIMS communication model depicted in Fig. 1 can be split into a four-layer architecture:

- The Instrumentation Layer provides infrastructure and Java application monitoring information using specially-designed sensor modules installed in the JIMS Monitoring Agent (JIMS MA).
- The Interoperability Layer provides a common point of communication with computational nodes (also called Worker Nodes or WNs) in clusters through dedicated access nodes (Access Node -AN).
- The Integration Layer enables discovery of all accessible clusters and provides an overall view of the Grid. This layer consists of Global Discovery modules installed in chosen JIMS agents running on Access Nodes.
- The Client Application Layer consists of applications connected to the JIMS system, which are consumers of the information produced by the system.

The extensions proposed in this paper cover the instrumentation layer. The layer is equipped with Mbeans for Solaris 10 *projects* and *zones* monitoring and management.

JIMS has been deployed as a monitoring infrastructure at many Grid installations, including the EU IST CrossGrid project [1], the Polish national project Clusterix, and

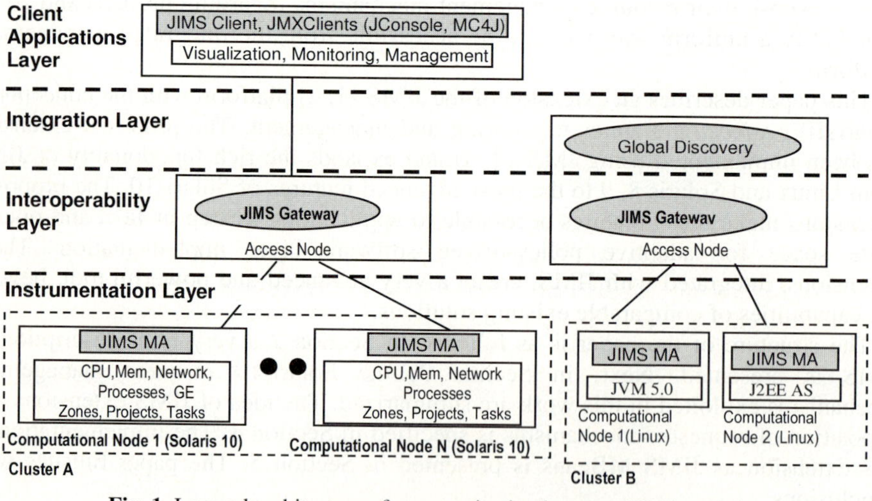

Fig. 1. Layered architecture of communication between JIMS modules

the internal installation of the N1 Sun Fire 1600 Blade System at the Department of Computer Science AGH/UST in Kraków. These JIMS applications show that features such as hierarchical information representation, fast response time (~200[ms]), scalable architecture, open source code (AXIS, JMX, SNMP, MX4J), WS interface and JSR003, 160, 255 compatibility, allow for constructing services which are easy to administer and maintain, and which allow easy integration with other components of the Grid. This is why JIMS was selected for proposed extensions regarding Solaris 10 resource monitoring and management.

3 Resource Monitoring and Management of Solaris 10 Resources

Solaris 10 [7, 8, 9, 10, 11] provides control resource usage based on workloads. A *workload* is an aggregation of all processes of an application, or group of applications, which makes sense from a business perspective. The Solaris OS provides a facility called *projects* to identify workloads. The project serves as an administrative tag used for group-related work in a manner deemed useful by the system administrator.

Once workloads are identified and labeled using projects, the next step in managing resource usage involves measuring workload resource consumption. Current consumption can be measured using the *prstat* command to obtain real-time snapshot of resource usage. This command reads data stored in */proc* VFS (Virtual File System), where each process in the zone has its own subdirectory (in the global zone there are also entries for processes from local zones). The most important information is contained in *psinfo, usage* and *status* files. By summarizing information about resource usage of processes it's possible to calculate resource usage of *projects* and *zones*. The capability to look at historical data is provided by the *Extended Accounting* facility which allows collection of statistics at the process level, task level or both. A *task* is a group of related processes executing in the same project.

Resource controls are configured through the project database. The last field of the project entry is used to set resource controls. The resource controls can be administered by two commands: *prctl* (get or set resource controls on a running process, task or project) and *rctladm* (display or modify the global state of system resource controls). This state is stored in the */etc/project* file whose changes can be monitored directly. Resource controls can define actions which should be performed when resource usage is crosses some predetermined upper bound. These actions can be observed through messages in the */var/admin/messages* log file or catching sent signals to those processes.

Solaris Zones provide a means to create one or more virtual environments on a single operating system instance, shielding applications from details of the underlying hardware. Applications in a single zone run in isolation from applications in other zones. They cannot see, monitor or affect processes running in another zone. The exception is the *global zone* which encompasses the entire system and is comparable to a normal Solaris OS instance. It has access to the physical hardware and can see and control all processes. The administrator of the global zone can control the system as a whole. The global zone always exists, even when no other zones are configured. Inside the global zone are *local zones*. These zones are isolated from the physical hardware characteristics of the machine by the virtual platform layer. This layer provides zones with a virtual network interface, one or more file systems and a virtual console. The virtual console can be accessed from the global zone using the *zlogin* command.

Zone administration tasks can be divided into two parts, *global* zone administration tasks such as creating a zone, and *local* zone administration tasks such as performing configuration *within* a zone. The four primary global zone administration tasks are: configuration, installation, virtual platform management, zone login. The configuration task defines zone resource controls which are enabled when the zone is booted.

Regular resource management facilities such as resource controls, projects, and more are available inside zones. Because projects are also virtualized inside a zone, each zone has its own project database. This allows a local zone administrator to configure projects and resource controls local to that zone. Furthermore, the Extended Accounting framework has been extended for zones. Each zone has its own extended accounting files for task-and process-based accounting that contain accounting records exclusively for that zone.

4 JIMS Extension Concept

The presented advanced resource management techniques provided by the Solaris 10 introduce a number of logical elements that must be effectively managed and configured. Such a system must be able to collect and publish various kinds of information about configured elements on many hardware virtualized nodes. The main assignments to be performed by the proposed JIMS extensions are:

- Collecting and publishing monitoring information about *zones*, *projects* and *tasks*; the system must be able to expose real-time information about the current resource usage e.g. CPU, memory, resource controls. Because Solaris 10 also introduces Extended Accounting facilities, such information might be stored in a database installed on the JIMS Access Node responsible for the management of Grid Computational Nodes. This will enable e.g. billing the customers for consumed resources or even calculating long-time resource usage trends. Since the system administrator may perform some operations using native system tools, the Monitoring Agent must also reflect these changes.
- Management of *zones*; adding or removing a given zone, lifecycle-related operations (shutdown, halt, boot, ready), modifying zone values (devices, file systems, attributes, resource controls).
- Management of *projects*; adding or removing projects, also modifying existing projects' attributes (name, comment, assigned users and groups, resource controls).

Non-functional requirements include ease of use and security. The first requirement implies that running various system components should be relatively simple (minimal need for human interaction) and also installation and configuration of JIMS on many Computing Nodes should be automatic. This may be performed by using N1 Service Provisioning Systems (N1 SPS) [5] which can install JIMS on many nodes. Automatic installation of modules containing MBeans for monitoring and management is performed by JIMS using the JMX Mlet-Service.

The security aspect must take into account many things related to a broad variety of managed resources which means that some operations such as creating zones, modifying the projects database etc. may be allowed only for restricted groups of users. The privileged position of operations performed in the global zone should be very carefully analyzed. It is necessary to take into account situations when a security domain consists only of local zones distributed over different global zones (physical computers). An example of such a case is represented by Cluster B in Fig. 2.

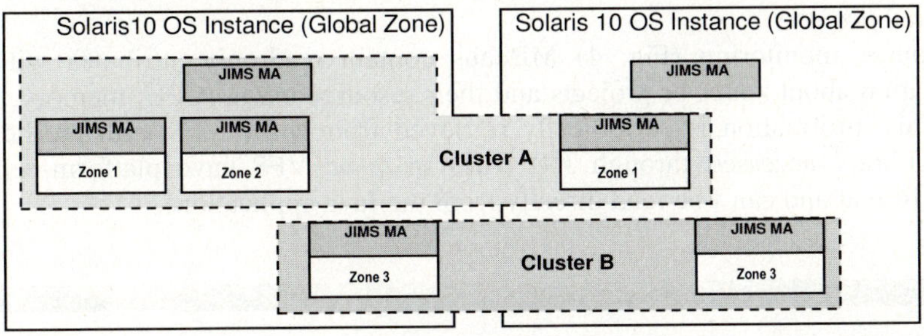

Fig. 2. Intersected clusters which consist of zones installed on different Solaris 10 instances

This is why two deployment scenarios: *Global Access* and *Restricted Access*, have to be considered. Clients in the *Global Access* scenario (see Fig. 3a) connect only to JIMS MA installed in the global zone which has permission to perform operations on all local zones in the current Solaris 10 OS instance. In the *Restricted Access* scenario (see Fig. 3b) MAs are installed in Global and local zones. The MA installed in a local zone is able to perform only operations on the given zone depending on the assigned privileges. The JIMS client – e.g. Client2 in Fig. 3b -has to connect to specific JIMS MAs to perform monitoring and management activities.

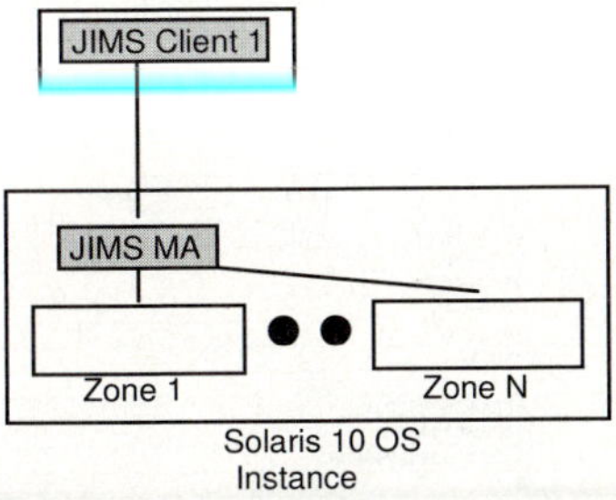

Fig. 3a. "Global Access" deployment sc-enario with only one JIMS MA installed in the global zone

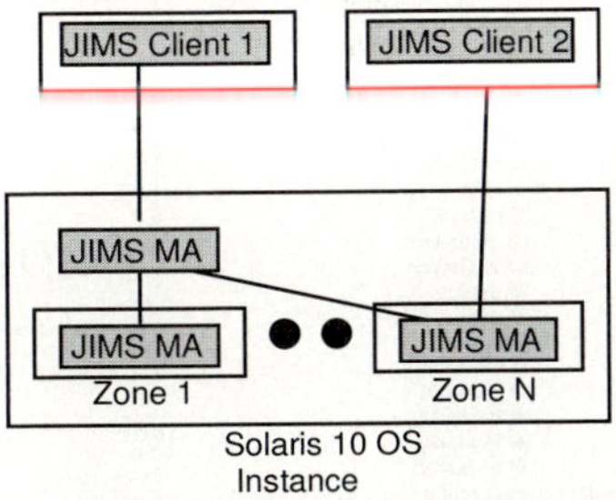

Fig. 3b. "Restricted Access" deployment scen-ario with JIMS MA installed in the global and local zones

The functionality for the monitoring and management of Solaris 10 resources will be exposed via an appropriate API which may be used by some external components. Because JIMS is based on JMX architecture, the MBeans which implement the required functionality suit this end perfectly.

5 JIMS Extension JMX-Based Implementation

The JIMS extension is a set of MBeans that expose an interface for Solaris 10 resource monitoring and management. These MBeans are divided into three groups: monitoring, management and accounting. Inside each group there are MBeans for *zones* and *projects*.

Resource monitoring (Fig. 4) MBeans contain read-only attributes with basic information about zones or projects and their resource usage (CPU, memory, threads etc.) This information is periodically retrieved from */proc* VFS with the help of a native library accessed through JNI (files in */proc* VFS have platform-dependent binary format and can't be read directly from the Java application).

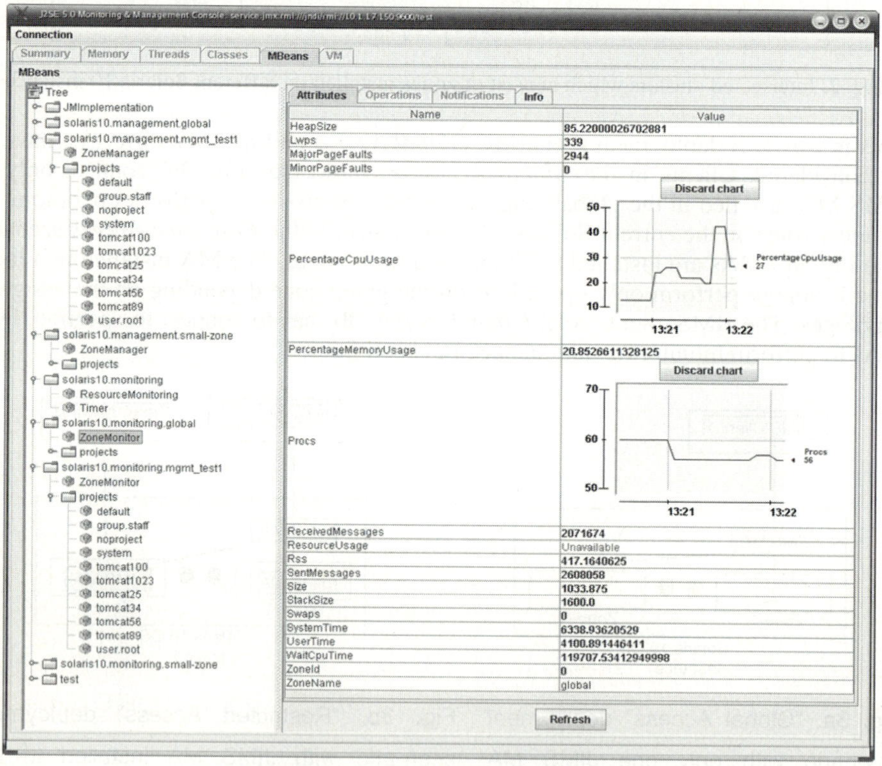

Fig. 4. Solaris 10 resource monitoring MBeans presented in JConsole

Solaris 10 management (Fig. 5) MBeans enable reading and changing properties of *zones* and *projects* (i.e. resource controls, member users etc.) and also operations that change zone state. They also allow creating or removing *zones* and *projects*. MBean interfaces depend on the zone in which the JIMS MA is running (different for global and local *zones*), i.e. local zone management MBeans expose the *boot* operation only in the global zone. Management MBeans use various methods to interact with the OS: to collect information about *zones* and *projects,* MBeans read configuration files or use JNI. Changes in configuration are applied by executing shell scripts and system commands (with `Runtime.exec()` invocation). MBeans are able also to emit JMX notifications to inform interested parties about changes in the system (i.e. concerning added projects or changed resource usage).

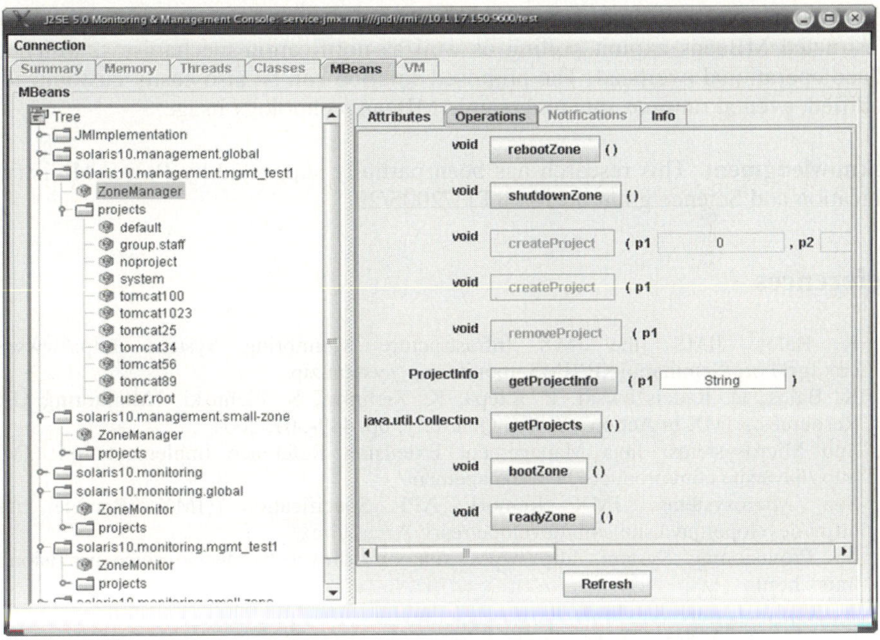

Fig. 5. Solaris 10 management MBeans accessible via JConsole

The accounting group of MBeans exposes long-time historical resource usage information. This information is collected periodically from extended accounting log files and stored in a database. We have decided to use a Java-based DBMS and integrate it within our application (to limit performance overhead). Log files are switched periodically to avoid unnecessary duplicate processing of the same information. Logs can be read only with the use of the *libexacct* API, available only for C and Perl (not Java), therefore MBeans access it through a native library and JNI.

6 Conclusions

The proposed extension of the JIMS monitoring system with selected Solaris 10 resource monitoring and management mechanisms creates a very coherent and easy to use environment. The provided functionality can be automatically discovered by the

JIMS self-configuration protocol, following which it may be accessed via a standard web browser using http adapter support, by Web Services (with the JIMS SOAP gateway implementation) or directly by any application written in Java using the standard RMI protocol. The diversity of potential access channels makes the proposed extensions easy to integrate with adaptation layers implemented in any technology.

The proposed concept of JMX-based OS mechanism exposition is rather general and can be applied to any operating system. Therefore the proposed solution can be used for adaptable reflective middleware construction, requiring access to underlying OS functionality. It is also useful for any resource-aware applications or services.

An important feature of the proposed extension is the ability to use various mechanisms for collecting operational data from the operating system. The constructed MBeans exploit polling as well as notification mechanisms, which can reduce operational overhead. The proposed solution can be also easily customized or modified, even at runtime, due to dynamic MBean technology usage.

Acknowledgment. This research has been partially supported by Polish Ministry of Education and Science grant no.1583/T11/2005/29.

References

1. K. Balos: JIMS -the JMX Infrastructure Monitoring System, http://www.eu-crossgrid.org/Seminars-INP/JIMS_monitoring_system.zip
2. K. Balos, D. Radziszowski, P. Rzepa, K. Zielinski, S. Zielinski, "Monitoring GRID Resources – JMX in Action", TASK Quarterly, pp. 487-501, 2004
3. Sun Microsystems: Java Management Extension Reference Implementation (JMX), http://java.sun.com/products/JavaManagement/
4. Sun Microsystems: JMX Remote API Specification (JMX Remote API), http://developer.java.sun.com/developer/earlyAccess/jmx/
5. N1 Provisioning System, http://wwws.sun.com/software/products/service_provisioning/index.html
6. Autonomic Computing, http://www.research.ibm.com/autonomic/
7. Daniel Price, Andrew Tucker "Solaris Zones: Operating System Support for Consolidating Commercial Workloads" (http://www.sun.com/bigadmin/content/zones/zones_lisa.pdf)
8. Amy Rich "Spotlight on Solaris Zones Feature" (http://www.sun.com/bigadmin/features/articles/solaris_zones.html)
9. Menno Lageman "Solaris Containers – What They Are and How to Use Them"
10. "System Administration Guide: Solaris Containers-Resource Management and Solaris Zones" (http://docs.sun.com/app/docs/doc/817-1592)
11. OpenSolaris project website (http://opensolaris.org)

An Agent Based Semi-informed Protocol for Resource Discovery in Grids

Agostino Forestiero, Carlo Mastroianni, and Giandomenico Spezzano

ICAR-CNR 87036 Rende (CS), Italy
{forestiero, mastroianni, spezzano}@icar.cnr.it

Abstract. A Grid information system should rely upon two basic features: the replication and dissemination of information about Grid resources, and an intelligent logical distribution of such information among Grid hosts. This paper examines an approach based on multi agent systems to build an information systems in which metadata related to Grid resources is disseminated and logically organized according to a semantic classification of resources. Agents collect resources belonging to the same class in a restricted region of the Grid, so decreasing the system entropy. A semi-informed resource discovery protocol exploits the agents' work: query messages issued by clients are driven towards "representative peers" which maintain information about a large number of resources having the required characteristics. Simulation analysis proves that the combined use of the resource mapping protocol (ARMAP) and the resource discovery protocol (ARDIP) allows users to find many useful results in a small amount of time.

1 Introduction

A Grid information system provides resource discovery and browsing services which are invoked by Grid clients when they need to use hardware or software resources belonging to a given *class*, i.e. matching given criteria and characteristics.

An agent-based protocol (i.e. ARMAP, *Ant-based Replication and MApping Protocol*) was proposed in [5] to spatially sort (or "map") resources according to their semantic classification [7]. ARMAP exploits the features of (i) epidemic mechanisms tailored to the dissemination of information in distributed systems [8] and (ii) self adaptive systems in which "swarm intelligence" emerges from the behavior of a high number of agents which interact with the environment [1]. By mapping metadata documents on Grid hosts, a *logical reorganization* of resources is achieved.

In this paper, a semi-informed discovery protocol (namely ARDIP, *Ant-based Resource DIscovery Protocol*) is proposed to exploit the logical resource organization achieved by ARMAP. The rationale is the following: if a large number of resources of a specific class are accumulated in a restricted region of the Grid, it is convenient to drive search requests (issued by hosts to search for resources of that class) towards that region, in order to maximize the number of discovered resources and minimize the response time. An ARDIP discovery operation is performed in two phases. In the first phase a *blind* mechanism, specifically the random walks technique [6], is

V.N. Alexandrov et al. (Eds.): ICCS 2006, Part IV, LNCS 3994, pp. 1047 – 1054, 2006.
© Springer-Verlag Berlin Heidelberg 2006

adopted: a number of query messages are issued by the requesting host and travel the Grid through the peer-to-peer (P2P) interconnections among Grid hosts. In the second phase, whenever a query gets close enough to a Grid region which is collecting the needed class of resources, the search becomes *informed*: the query is driven towards this Grid region and will easily discover a large number of useful resources. Simulation analysis shows that the ARMAP and the ARDIP protocol, if used together, allow for achieving a very high effectiveness in discovery operations.

The semi-informed ARDIP protocol aims to combine the benefits of both blind and informed resource discovery approaches which are currently used in P2P networks [11]. In fact, a pure blind approach (e.g. using flooding or random walks techniques) is very simple and scalable but has limited performance and can cause an excessive network load, whereas a pure informed approach (e.g. based on routing indices [2] or adaptive probabilistic search [10]) generally requires a very structured resource organization which is impractical in a large, heterogeneous and dynamic Grid.

The remainder of the paper is organized as follows. Section 2 summarizes the key points of the ARMAP protocol and describes the ARDIP protocol. Section 3 analyzes the performance of the ARDIP protocol. Section 4 concludes the paper.

2 A Multi-agent Protocol for Resource Discovery on Grids

This Section is organized as follows. In Section 2.1, the key points of the ARMAP protocol are briefly summarized (more details can be found in [5]). In Section 2.2, the new ARDIP protocol is introduced.

2.1 ARMAP Basics

The aim of the ARMAP protocol is to achieve a logical organization of Grid resources by spatially sorting them on the Grid according to their semantic classification. It is assumed that the resources have been previously classified into a number of classes `Nc`, according to their semantics and functionalities (see [7]). In the following, an information document describing a Grid resource will be referred to as a *logical resource*, or simply *resource*. When distinction is important, the actual resources will be named *physical resources*. ARMAP exploits the random movements and operations of a number of mobile agents that travel the Grid using P2P interconnections. This approach is inspired by ant systems [1, 3], in which swarm intelligence emerges from the collective behavior of very simple mobile agents (ants).

Once an agent gets to a Grid host (or *peer*), for each resource class `Ci`, it must decide whether or not to *pick* the resources of class `Ci` that are managed by that host, unless the agent already carries some resources of that class. In order to achieve the replication and mapping of resources, a pick random function `Ppick` is defined with the intention that the probability of picking the resources of a given class decreases as the local region of the Grid accumulates such resources. The ARMAP protocol can work in either the *copy* modality or the *move* modality; with the copy modality, an agent that picks some (logical) resources of class `Ci` will leave a copy of them in the current host; with the move modality, such resources are removed from the current host. A self-organization approach based on ants' pheromone [4] enables each agent to perform the modality switch (from the copy to the move modality) only on the basis of local information. Analogously, whenever an agent gets to a new Grid host, it must decide whether or not to *drop* the resources of class `Ci`, if it is carrying any of them. As opposed to the pick probability, the dropping probability is directly proportional to the relative accumulation of resources of class `Ci` in the local region. A spatial entropy function was defined in [5] to evaluate the effectiveness of the ARMAP protocol in the spatial ordering of resources. Figure 1 gives a graphical description of the mapping process performed by ARMAP. The values of system parameters are set as specified in Section 3, except for the number of resource classes which is set to 3 in order to facilitate the graphical illustration of the process.

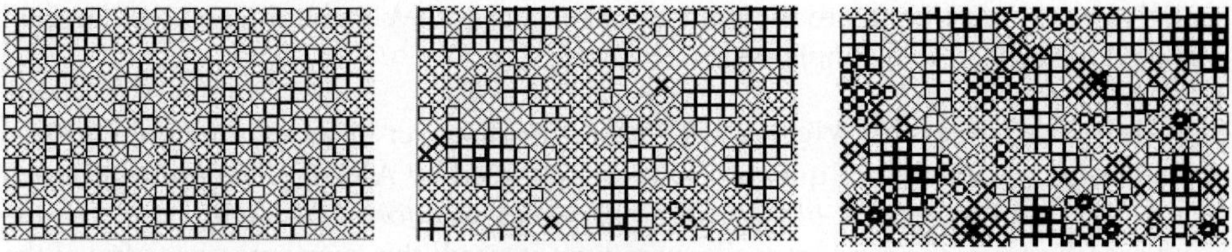

Fig. 1. Gradual mapping of logical resources in a Grid with 3 resource classes. Each peer contains a symbol (circle, square or cross) that corresponds to the most numerous class of resources contained in the peer. The symbol thickness is proportional to the number of resources belonging to the dominant class.

2.2 The ARDIP Protocol

The ARDIP (*Ant-based Resource DIscovery Protocol*) protocol is used by clients to discover Grid resources belonging to a given class. The objective of ARDIP is to drive a query message towards a region of the Grid in which the needed class of resources is being accumulating. Because ARDIP fully exploits the replication and spatial sorting of resources achieved by ARMAP, the two protocols should be used together: as ARMAP agents perform the logical reorganization of resources and build the Grid information system, it is more and more likely that ARDIP queries can find a remarkable number of useful resources in a small amount of time.

The ARDIP protocol is based upon three modules: (a) a module for the identification of representative peers which work as attractors for query messages; (b) a module which defines the semi-informed search algorithm; (c) a stigmergy mechanism that allow query messages to take advantage of the positive outcome of previous search requests. These modules are described in the following.

Identification of Representative Peers. As a class of resources `Ci` is accumulated in a Grid region, the peer that, within this region, collects the maximum number of resources belonging to the class `Ci` is elected as a *representative peer* for this class. The objective of a search operation is to let a query message get to a representative peer, since such a peer, as well as its neighbors, certainly manages a large number of useful resources. The ARDIP protocol assumes that a peer `p` is a representative peer of class `Ci` if at least one of the two following conditions are verified: (a) the peer `p` maintains a number of logical resources of class Ci that exceeds f1 times the mean number of physical resources belonging to class Ci which are offered by a generic peer; (b) the peer p maintains a number of logical resources of class Ci that exceeds f2 times (with f2<f1) the number of logical resources of the same class maintained by its neighbors.

Condition (a) is satisfied by a peer that holds a very high number of logical resources; in general it can be satisfied only when the clustering process of ARMAP is already in an advanced stage. Conversely, condition (b) can be satisfied when clustering is still in progress. Moreover, to limit the number of representative peers in the same region, each representative peer periodically checks if other representative peers are present in its neighborhood, within the *comparison radius* Rc: two neighbor representative peers must compare the number of resources they maintain, and the "looser" will be downgraded to a simple peer.

Semi-informed Search. When a user needs to discover resources belonging to a given class Ci, a number of query messages are issued by ARDIP. The semi-informed search algorithm includes a *blind* search phase and an *informed* search phase. For the blind search phase, the random walks paradigm is used: the query messages travel the Grid through the P2P interconnections by following a random path. The network load is limited with the use of a TTL parameter, which is equivalent to the maximum number of hops that can be performed by a query message before being discarded.

The blind search procedure is switched to an informed one as soon as one of the issued query messages approaches a representative peer of class Ci, i.e. when such a message is delivered to a peer which knows the existence of a representative peer and knows a route to it (see the description of the stigmergy module below). During the informed search phase, the query is driven towards the representative peer, and the TTL parameter is ignored so that the query cannot be discarded until it actually reaches the representative peer. Therefore, the semi-informed walk of a query message ends in one of two cases: (i) when the TTL is decremented to 0 during the blind phase; (ii) when the query reaches a representative peer. In both cases a queryHit message is created, and all the resources of class Ci, which are found in the current peer, are put in this message. The queryHit follows the same path back to the requesting peer and, along the way, collects all the resources of class Ci that are managed by the peers through which it passes.

Stigmergy Mechanism. The stigmergy mechanism is a mechanism, often observed in biological systems, through which elementary entities exploit the environment to communicate with each other. For example, in ant colonies, an ant that finds a food source leaves a *pheromone* along its way back to the nest, and such a pheromone will signal to other ants the presence of the food source. The ARDIP protocol exploits a similar mechanism: when a query accidentally gets to a representative peer for the first time, the returning queryHit will deposit an amount of pheromone in the peers that it encounters as it retreats from the representative peer. In this paper, it is assumed the pheromone is deposited only in the first two peers of the queryHit path.

When a query gets to a peer along its blind search, it checks the amount of pheromone which has been deposited there; if the pheromone exceeds a threshold Tf, it means that a representative peer is close, so the search becomes informed. An evaporation mechanism assures that the pheromone deposited on a peer does not drive queryHits to ex-representative peers. The pheromone level at each peer is computed every time interval of 5 minutes. The amount of pheromone Φi, computed after the i-th time interval, is given by formula (1).

$$(1)\ \Phi_i = E \cdot \Phi_{i-1} + \varphi_i$$

The evaporation rate `E` is set to 0.9; φi is equal to 1 if a pheromone deposit has been made in the last time interval by at least one agent, otherwise it is equal to 0. The threshold `Tf` is set to 2.

3 Simulation Analysis

In this section, we discuss some relevant simulation results which demonstrate the effectiveness of the ARDIP resource discovery protocol in a Grid environment, if it is used in conjunction with the ARMAP resource mapping protocol. In particular, the present section introduces the main system and protocol parameters, while Section 3.1 focuses on the performance of the ARDIP protocol.

A wide set of simulation runs were performed by exploiting the libraries and visual facilities offered by Swarm [9], a software package for multi-agent simulation of complex systems developed at the Santa Fe Institute. Table 1 reports the network, ARMAP and ARDIP parameters used in the simulation analysis.

Table 1. Environment and protocol parameters

Parameter	**Value**
Grid size (number of peer), **Np**	2500
Maximum number of neighbor peers of a Grid peer	8
Mean number of resources published by a peer	15
Number of resource classes, **Nc**	[illegible]
Number of ARMAP agents, **Na**	Np/2
Mean amount of time between two successive movements of an ARMAP agent	60 s
Maximum number of hops for each ARMAP agent's movement, **Hmax**	3
Number of query messages issued by the requesting peer	4
Time to live, **TTL**	3-7
Factor **f1**, for the identification of representative peers	5
Factor **f2**, for the identification of representative peers	2.5
Comparison radius, **Rc**	2
Mean generation frequency with which a Grid peer issues query messages	1/300 (1/s)
Mean message elaboration time at a Grid peer	100 ms

3.1 Performance of the ARDIP Protocol

The performance of the ARDIP protocol was analyzed by evaluating the performance indices defined and explained in Table 2. Figure 2(1) depicts the value of the `Nrep` index evaluated at different times while the ARMAP mapping process proceeds. This figure confirms that representative peers are selected almost exclusively with condition (a) in the first phase, but thereafter the weight of condition (b) becomes predominant, as discussed in Section 2.2.

The index `Fsq` is essential to evaluate how many search requests are actually delivered to a representative peer. Figure 2(2) proves the valuable effect caused by the combined use of ARMAP and ARDIP protocols. In fact, after a very small amount of time, the logical reorganization of resources produces a significant increase in `Fsq`. Moreover, as the `TTL` value increases, `Fsq` increases as well, since a search request extends the blind search phase and has more chances to get to a representative peer.

Table 2. Performance indices

Performance Index	Definition
Number of representative peers, **`Nrep`**	Mean number of representative peers of all classes that are selected by ARDIP to attract query messages (see Section 2.2)
Fraction of striking queries, **`Fsq`**	Fraction of queries that are actually driven towards a representative peer
Mean number of results, **`Nres`**, **`Nres(rep)`**, **`Nres(norep)`**	Mean number of resources that a node discovers after its query (computed for all the requests, for the requests that are actually delivered to a representative peer, for the requests that are not delivered to a representative peer)
Response times, **`Tr`**, **`Tr(rep)`**, **`Tr(norep)`**	Mean amount of time (s) that elapses between the generation of a query and the reception of a corresponding queryHit (computed for all the requests, for the requests that are actually delivered to a representative peer, for the requests that are not delivered to a representative peer)

The most important performance measure is `Nres`, the mean number of results that are discovered after a query request. Indeed, it is generally argued that the satisfaction of the query depends on the number of discovered resources returned to the user that issued the query [12]. The trend of `Nres` is depicted in Figure 3(1), which shows that the mean number of results is larger and larger as resources are being organized by ARMAP. Furthermore, even when the probability of reaching a representative peer begins to become stable (Figure 2(2)), the number of results continues to increase, meaning that representative peers are able to collect more and more resources of the class in which they are specialized. It is also noted that the queries which successfully get to a representative peer can discover considerably more results than the rest of the queries, and such a difference increases with the value of the `TTL` parameter.

The semi-informed discovery protocol not only increases the number of results, but also allows users to discover them in a shorter amount of time. Figure 3(2) shows that the response time decreases as the ARMAP work proceeds, and also that the response time is notably smaller if the query reaches a representative peer; in this case, in fact, the discovery operation is stopped even if the `TTL` value is still greater than 0, and a queryHit is immediately issued (see Section 2.2). However, this performance improvement is achieved only when the `TTL` value is sufficiently high.

Finally, performance results not reported in this paper show that the logical reorganization of resources, and the use of the ARDIP protocol, allows for decreasing the traffic load experienced by a single peer. Indeed, when a query messages is driven towards a representative peers, on average it has to make a lower number of hops with respect to a completely blind search.

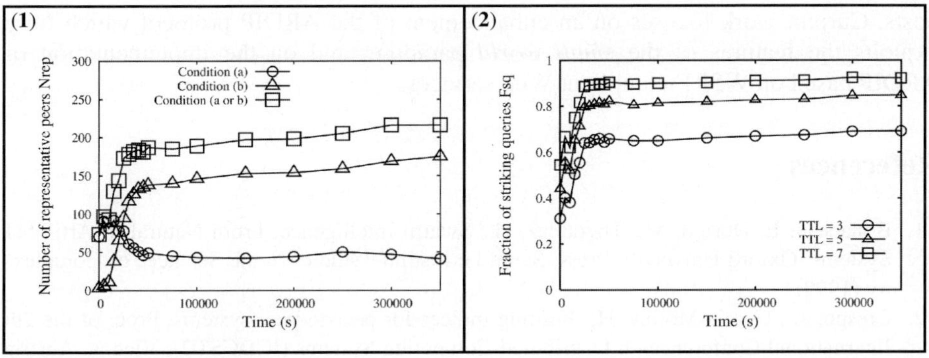

Fig. 2. Selection and actual use of representative peers as the mapping process proceeds. (1): number of representative peers selected with condition *a*, with condition *b*, and overall number of them. (2): fraction of search requests that are successfully driven to a representative peer, for different values of TTL.

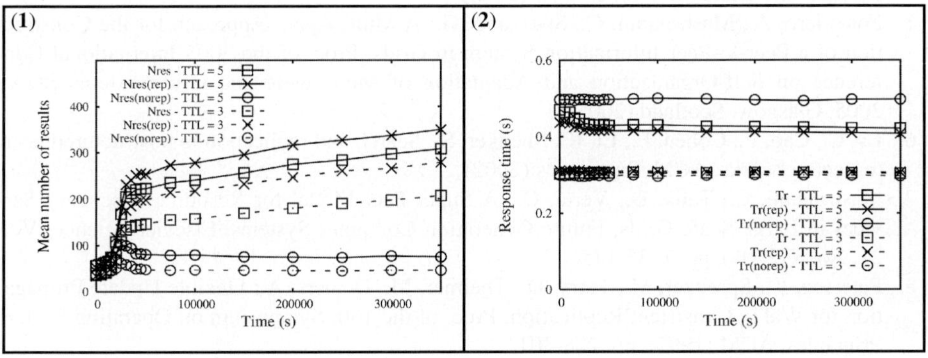

Fig. 3. Performance of search requests as the mapping process proceeds. (1): mean number of results. (2): response time. Values of performance indices, calculated for different values of TTL, are reported for queries that reach a representative peers, for queries that do not reach a representative peers, and for all the queries.

4 Conclusions and Future Work

This paper introduces an approach based on multi agent systems for building an efficient information system in Grids. A number of self-organizing agents travel the network by exploiting P2P interconnections; agents replicate and gather information related to resources having similar characteristics in restricted regions of the Grid. Such a logical reorganization of resources is exploited by a semi-informed resource discovery protocol, namely the ARDIP protocol, which is tailored to route a query message towards a "representative peer" that collects a large number of resources having the desired characteristics. Simulation analysis shows that, as the reorganization of resources proceeds, ARDIP allows users to discover more and more resources in a shorter amount of time, without increasing the traffic load experienced by Grid hosts. Current work focuses on an enhancement of the ARDIP protocol which fully exploits the features of the *small world* paradigm and on the implementation of ARDIP based on WSRF-compliant Web services.

References

1. Bonabeau, E, Dorigo, M., Theraulaz, G.: Swarm Intelligence: From Natural to Artificial Systems, Oxford University Press, Santa Fe Institute Studies in the Sciences of Complexity (1999)
2. Crespo, A., Garcia-Molina, H.: Routing indices for peer-to-peer systems. Proc. of the 2th International Conference on Distributed Computing Systems (ICDCS'02), Vienna, Austria (2002), pp. 23-33
3. Dasgupta, P.: Intelligent Agent Enabled P2P Search Using Ant Algorithms, Proc. of the 8th International Conference on Artificial Intelligence, Las Vegas, NV (2004), pp. 751-757
4. Van Dyke Parunak, H., Brueckner, S. A., Matthews, R., Sauter, J.: Pheromone Learning for Self-Organizing Agents, IEEE Transactions on Systems, Man, and Cybernetics, Part A: Systems and Humans, vol. 35, no. 3 (2005)
5. Forestiero, A., Mastroianni, C., Spezzano, G.: A Multi Agent Approach for the Construction of a Peer-to-Peer Information System in Grids, Proc. of the 2005 International Conference on Self-Organization and Adaptation of Multi-agent and Grid Systems SOAS 2005, Glasgow, Scotland (2005)
6. Lv, C., Cao, P., Cohen, E., Li, K., Shenker, S.: Search and replication in unstructured peer-to-peer networks, ACM, Sigmetrics (2002)
7. Mastroianni, C., Talia, D., Verta, O.: A Super-Peer Model for Resource Discovery Services in Large-Scale Grids, Future Generation Computer Systems, Elsevier Science, Vol. 21, No. 8 (2005), pp. 1235-1456
8. Petersen, K., Spreitzer, M., Terry, D., Theimer, M., Demers, A.: Flexible Update Propagation for Wakly Consistent Replication, Proc. of the 16th Symposium on Operating System Principles, ACM (1997), pp. 288-301
9. The Swarm environment, Swarm Development Group of Santa Fe University, New Mexico, http://www.swarm.org
10. Tsoumakos, D., Roussopoulos, N.: Adaptive probabilistic search for peer-to-peer networks. In: Third International Conference on Peer-to-Peer Computing P2P'03, Linkoping, Sweden (2003), pp. 102-110
11. Tsoumakos, D., Roussopoulos, N.: A Comparison of Peer-to-Peer Search Methods. Proc. of the Sixth International Workshop on the Web and Databases WebDB, San Diego, CA (2003), pp. 61-66
12. Yang, B., Garcia-Molina, H.: Efficient search in peer-to-peer networks, Proc. of ICDCS, Wien, Austria (2002)

Replica Based Distributed Metadata Management in Grid Environment*

Hai Jin, Muzhou Xiong, Song Wu, and Deqing Zou

Cluster and Grid Computing Lab,
School of Computer Science and Technology,
Huazhong University of Science and Technology, Wuhan, 430074, China
{hjin, mzxiong, wusong, deqingzou}@hust.edu.cn

Abstract. Metadata management is one of the key techniques in data grid. It is required to achieve two goals: high efficiency and availability. This paper presents a *Replication Based Metadata Management System* (RBMMS) as metadata server implemented in *Global Distributed Storage System* (GDSS). To address the above two goals RBMMS maintains a spares strongly connected graph to describe replica structure and relations among the replicas. The graph is used to propagate updating information and replica discovery in the process of replica addition and removal. Cache module is also implemented in RBMMS to further improve the performance of metadata access. The evaluation demonstrates that RBMMS attains high availability and efficiency of metadata management system.

1 Introduction

Metadata is the descriptive data and all the metadata in data grid [1, 2] compose the metadata catalog [3], which adopts some common structures to express metadata. All the metadata catalogs must satisfy the following requirements: 1) it should be a distributed and hierarchical structure system, such as LDAP [4]; 2) it should not breach current method of metadata expression. This paper does not discuss the expression of metadata, but focuses on the first requirement and presents a metadata management system RBMMS (*Replica based Distributed Metadata Management System*). Utilizing the solution of data replication [5-8] and web cache [9], RBMMS presents the concept of metadata replication and metadata cache, which uses LDAP directory server to store metadata and their replicas, also a cache module is added into metadata catalog. RBMMS achieves the following goals: 1) it provides multiple metadata replicas to support high availability; and 2) replica can be added or deleted dynamically. In order to achieve the above goals, RBMMS uses sparse strongly connected graph to express the metadata replica structure and the relationship among the metadata replicas.

* This paper is supported by National Science Foundation of China under grant 60125208 and 90412010, ChinaGrid project from Ministry of Education of China, and Hubei Natural Science Foundation under grant 2004ABA053.

V.N. Alexandrov et al. (Eds.): ICCS 2006, Part IV, LNCS 3994, pp. 1055 – 1062, 2006.
© Springer-Verlag Berlin Heidelberg 2006

Metadata management servers must work cooperatively to adapt to the dynamic environment in order to improve system performance in grid environment. Metadata catalogs should be organized hierarchically and symmetrically. Replica can be read and written anytime, and updating operation is to be propagated in the order of the graph. The replicating strategy always selects a nearest metadata server to the client to serve the access request and achieve high performance. The replica sparse strongly connected graph not only provides the sequence of replica updating, but also discovers the state of replica when inserting or deleting replica. When probing some replicas unavailable, system deletes it from the graph and reorganizes the graph. The above mechanism can improve the system availability and avoiding system disordered when several replicas out of work. In addition, metadata cache further improves the system performance. We have implemented RBMMS in *Global Distributed Storage System* (GDSS) [14], which is a storage middleware integrating distributed heterogeneous storage resources and providing global unified data view for users in gird environment.

This paper is organized as follows. Section 2 introduces related work. Section 3 discusses the RBMMS architecture design. Section 4 introduces the work mechanism of replica and cache in RBMMS. Section 5 analyzes the performance and the paper concludes in section 6.

2 Related Works

High performance and reliable metadata service becomes the goal of many projects. The *Storage Resource Broker* (SRB) [10] from the San Diego Supercomputing Center and its associated *Metadata Catalog* [11] provide metadata and data management services. SRB supports a logical name space that is independent of physical name space.

The *Replica Metadata* (RepMec) [13] catalog developed by the European Data Grid's Reptor project is built upon the Spitfire database service. The RepMec catalog stores logical and physical metadata. Among other functions, this catalog is used within the EDG project to map user-provided logical names of data items to unique identifiers called GUIDs. RepMec is used in the Reptor system in cooperation with a replica location service.

The Legion project provides an object-oriented middleware infrastructure for distributed computing environments [12]. The *Legion File System* (LegionFS) provides a Basic File Object with object methods that resemble UNIX read, write and seek system calls. Replication provided by LegionFS by classes maps object identifiers to multiple physical object addresses.

3 RBMMS Architecture Design

To achieve high availability and efficiency of metadata service, this paper presents RBMMS that is based on a special replica structure. Each domain has several local metadata servers that replicate metadata information for each other. System creates cache for metadata to improve the access performance. There exists a global metadata server which provides metadata index to support metadata operations across domains.

The granularity of metadata replication in RBMMS is partition, which is a subtree in LDAP directory server. Figure 1 describes a tree in LDAP directory server divided into 4 partitions, that is *P-A*, *P-B*, *P-C*, and *P-D*. When creating metadata replica, RBMMS replicates one or more partitions.

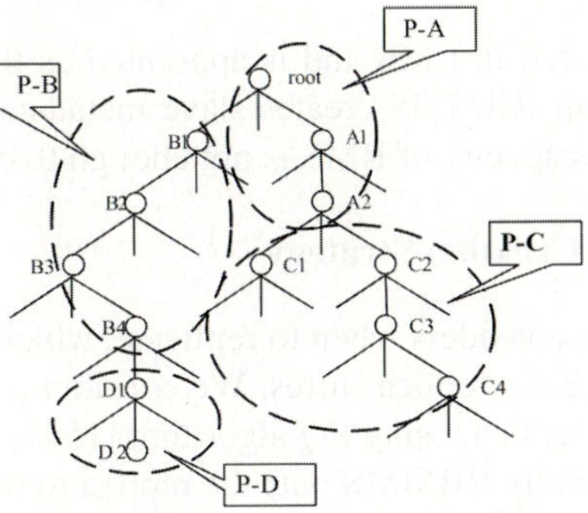

Fig. 1. Relationship between complete directory tree and partitions. A directory tree is divided into several partitions which is the smallest unit in replicating.

The relationship among the replicas in RBMMS is described as a sparse strongly connected graph. The verge of the graph means that the two replicas are connected and the updating order is according to the direction of the verge. The partition replicas build up a strongly connected graph described in Figure 2 in which each node presents a partition replica. There are two types of replicas: one is master replica which is shadowed in Fig. 2 (P-A), and the others are slave replicas. All the replicas can be read or written anytime and can be operated as the same propagating protocol for replica updating. However the master replica is the starting point of replica discovery. In RBMMS the initial partition is its master replica. The number of master replica for each partition is one and is fixed in RBMMS.

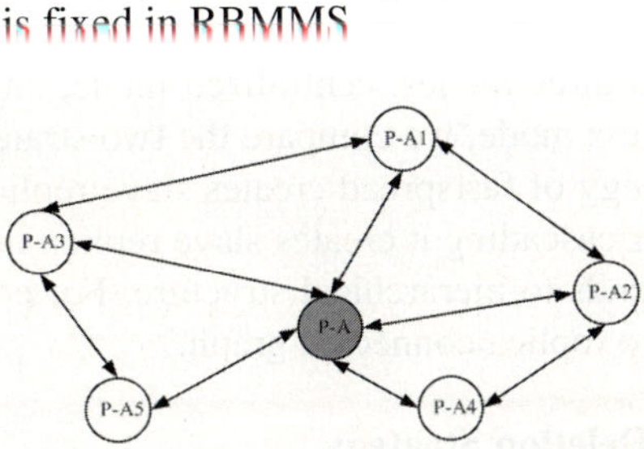

Fig. 2. Replica sparse strongly connected graph for Partition A. The shadowed node *P-A* is the master replica.

RBMMS maintains replica strongly connected graph to manage distributed replicas, which is composed by the following components. GMS (*Global Metadata Server*) maintains the index information for local metadata servers in all the domains. LMS (*Local Metadata Server*) stores master metadata replica that contains the whole metadata directory tree. There is only one LMS in a domain. RMS (*Replica Metadata*

Server) maintains the slave metadata replicas and the number of RMS can be one or more. MC (*Metadata Cache*) caches hot metadata to improve the read/write efficiency.

4 Metadata Replica Management in RBMMS

Initially metadata is created in LMS and is appointed as the master replica. Through replica creation algorithm RBMMS creates slave metadata replicas in RMS. Replica can be deleted when the capacity of RMS is not enough to contain more replicas.

4.1 Metadata Replica Creation Strategy

Replica creation strategy considers when to replicate, which metadata to replicate, and to which storage resource the replica stores. We consider partition as the smallest unit when creating metadata replica, using PQ algorithm [15] to determine when to run the replicating engine and finally RBMMS puts the replica to the place nearest to users.

RBMMS divides the whole directory tree into several partitions dynamically and records partition index table in LMS, including root node of the partition, partition size, partition response time, access number and slave replica root node. Partition response time (T_{resp}) is the sum of partition service time (T_s), waiting time (T_w), and communication cost (T_c), that is $T_{resp}= T_s+ T_w+ T_c$. A threshold $T_{threshold}$ is set for partition response time. When $T_{resp} > T_{threshold}$, the RBMMS creates slave replica for this partition.

RBMMS is free to select two parameters, P and Q ($Q>P$). Time is divided into two parts: P and Q alternatively. The concrete algorithm description is as follows: (i)if $T_{resp} < T_{threshold}$, abort replica creation, else go to (ii); (ii) $\forall \varepsilon t \in$ P, if $T_{resp} > T_{threshold}$ and $\mathrm{d}T_{resp}/\mathrm{d}t > 0$, go to (iv), else go to (iii); (iii) $\forall \varepsilon t \in$ Q, if $T_{resp} > T_{threshold}$, then go to (iv), else abort replica creation; (iv) Create slave replica.

When the time creating slave replica is determined, according to the use's access mode LMS chooses an RMS which does not contain replica of this metadata. The access mode is limited in three modes: centralized mode, uniform mode, and random mode. With different access mode, we compare the two strategies: fastspread [16] and cascading [16]. The strategy of fastspread creates slave replica at all nodes that are in metadata access path. For cascading it creates slave replica in the upper level of metadata access path and extends to hierarchical structure. For consistency issue, the new replica must be added into replica connected graph.

4.2 Metadata Replica Deletion Strategy

This subsection introduces how to delete slave replica. In RBMMS the master replica will never be deleted except that user deletes the related data. The reason of replica deletion includes no enough disk capacity and expensive cost of maintaining replicas. RBMMS randomly selects 10 slave replicas periodically and checks the access number of them. The slave replica with the least access number will be deleted and the three with the least access number will be considered as the replicas that will be deleted in the next operation. The deletion operation performs until system gets enough disk capacity.

The neighbors of the deleted replica are notified to find a new replica as their neighbor from the slave replica and create new verges, so that the strong connectivity is ensured.

4.3 Management of Metadata Cache

Metadata Cache utilizing access locality is stored in MC, containing the hot metadata to improve the metadata read/write efficiency. Different from metadata replica, metadata cache may contain incomplete information. When receiving the read/write request from user, RBMMS first checks whether the required metadata is in cache. If does, it directly reads cache; if not, it reads from a LMS or RMS and also caches the metadata. In order to confirm the metadata consistency, all the write operation is performed in MC and later updated to metadata server, which can reduce the cost of network communication.

5 Experiments

This section evaluates the performance and availability of RBMMS. We first consider the performance of RBMMS in structure of a fixed single domain; and then evaluate the cost of dynamic reorganization of replica strongly connected graph when some of the replicas are unavailable. The experiment result includes disk capacity consumption for replica storing, response time of metadata service, time of replica creation and updating under the two strategies of replica creation with different access modes. Through the experiment results, RBMMS can choose the best strategy to manage metadata.

There is no cache in the following experiments. When the access mode is centralized mode, which means all the access in RMS3, the replica distribution under the two strategies are described in Fig. 3(i). The result is in Table 1, in which the response time of fastspread is 18% faster than that of cascading and also the capacity consumption of fastspread is 2 times larger than that of cascading. The distribution of metadata replica in Fig. 3(ii) describes the access mode in uniform mode in which access to Partition A is performed in RMS3 and RMS4. Under this access mode, the response time of fastspread is 21% faster than cascading and the capacity consumption of fastspread is 3 times larger than that of cascading, shown in Table 1. The distribution of metadata replica in Fig. 3(iii) describes the access mode in random mode where access to Partition A is performed randomly. Under this access mode, the response time of fastspread is 8% faster than cascading and the capacity consumption of fastspread is 2 times larger than that of cascading, shown in Table 1.

From the above three results we can see that the response time of fastspread is faster than cascading in all cases, but also costs more disk capacity. Therefore, we decide that RBMMS utilizes fastspread under centralized access mode and uses cascading under uniform or random access mode. According to this decision, we compare the performance of RBMMS with and without MC, respectively. We conclude that adding cache into RBMMS, performance improves about 20%, shown in Fig. 4.

Table 1. Performance comparison for two strategies with centralized, uniform and random access mode

Strategy	Capacity consumption (MB)			Response Time (ms)			Replica creation Time (min)			Replica updating Time (min)		
	Central	Uniform	Random	Central	Uniform	Random	Central	Uniform	Random	Central	Uniform	Random
Fast-spread	4.4	6.5	8.7	10.4	10.8	10.4	6	6	6	40.4	60.7	80.9
Cacading	2.2	2.2	4.4	12.6	13.7	11.3	6	6	6	40.4	20.2	40.5

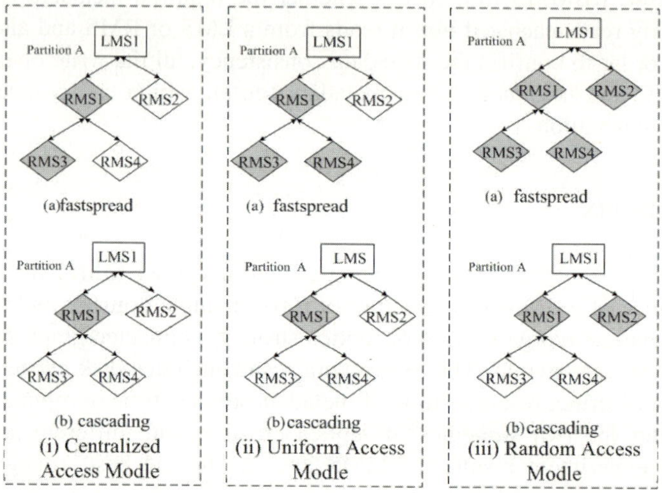

Fig. 3. The performance comparison of two strategies under the centralized, uniform and random access modes

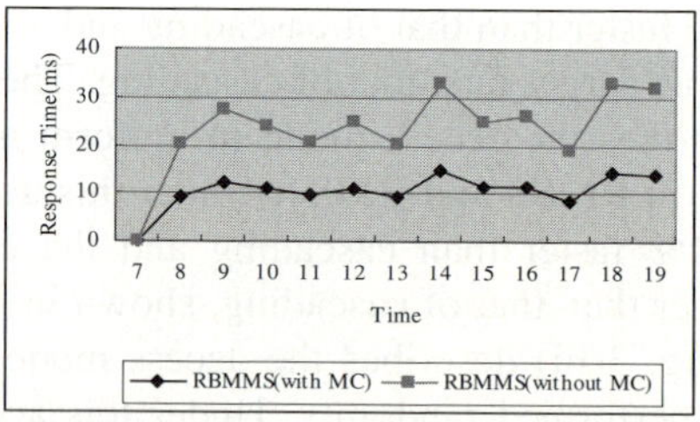

Fig. 4. Performance comparison of RBMMS with cache and without cache

When a RMS is unavailable for a long time (for example 0.5 weeks), replica stored in it is shipped to other RMS to substitute the unavailable replica and reorganize replica strongly connected graph. Because of the size of metadata is much less than those of data, the performance of reorganization can achieve high efficiency. In the following experiment, we still configure RBMMS with a single domain structure and cache works. The initial system is shown in Fig. 5(i), in which *P-A* is the master replica in

LMS1 and *P-A1~P-A4* are the slave replicas stored in RMS1~RMS4 respectively. If RMS1 is unavailable, RMS3 and RMS4 get information about RMS2 from LMS1 because RMS2 is nearest to them. RMS3 and RMS4 update their information of RMS and set RMS2 as their father node. LMS1 is notified about all the update information. LMS1 notifies RMS2 that sets RMS3 and RMS4 as its son nodes. The new reorganized graph is shown in Fig. 5(ii). In our experiment, the size of metadata in RMS1 is about 6MB (with metadata number 33900), and it takes about 7 minutes to complete the whole process.

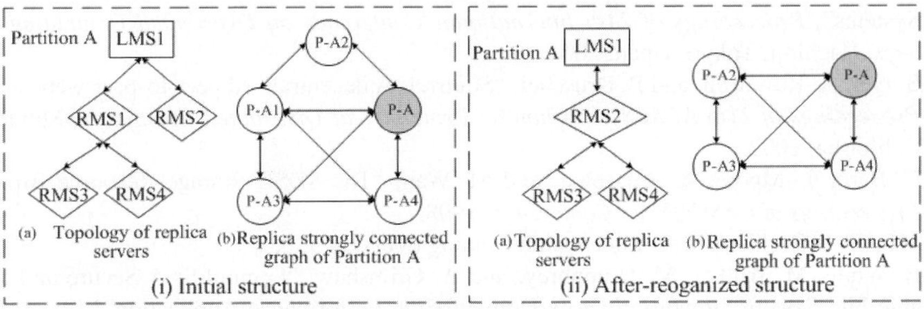

Fig. 5. (i) Initial structure of RBMMS system, (ii) Structure of RBMMS after reorganization when RMS1 is out-of-work

6 Conclusion and Future Work

In this paper, we present a replica based solution to implement metadata management in grid environment. We use a spares strongly connected graph to construct and maintain the relationship among replicas. Through that, the RBMMS system achieves a good result of availability and efficiency. In addition, we implement cache component that can farther improve the performance of RBMMS.

It is a challenge to test RBMMS in dynamic wide area network. In the future work we will present solutions to make RBMMS satisfied in such environment. How to utilize sparse strongly connected graph to maintain the consistency of replica is another issue in the system, which will also be resolved in the future.

References

1. I. Foster, "The Grid: A New Infrastructure for 21st Century Science", *Physics Today*, 2002, 55(2): 42.
2. B. Allcock, J. Bester, J. Bresnahn, A. Chervenak, I. Foster, and C. Kesselman, "Data Management and Transfer in High Performance computational Grid Environments", *Parallel Computing*, 2002, 28(5): 749-771.
3. I. Foster and C. Kesselman, "Computational Grids", *The Gird – Blueprint for a new Computing Infrastructure*, San Francisco: Morgan Kaufmann Publisher, 1999, pp.15-51.
4. G. von Laszewski and I. Foster, "Usage of LDAP in Globus", ftp://ftp.globus.org/pub/globus/papers/ldap_in_globus.pdf

5. C. Ferdean and M. Makpangou, "A Scalable Replica Selection Strategy based on Flexible Contracts", *Proceedings of Third IEEE Workshop on Internet Applications*, San Jose, California, 2003.
6. H. Lamehamedi and B. Szymanski, "Data replication strategies in grid environments", *Proceedings of Fifth International Conference on Algorithms and Architectures for Parallel Processing*, Beijing, 2002.
7. K. Ranganathan and I. Foster, "Design and Evaluation of Replication Strategies for a High Performance Data Grid", *Proceedings of International Conference on Computing in High Energy and Nuclear Physics*, Beijing, 2001.
8. M. Karlsson and C. Karamanolis, "Choosing Replica Placement Heuristics for Wide-Area Systems", *Proceedings of 24th International Conference on Distributed Computing Systems*, Hachioji, Tokyo, Japan, 2004.
9. S. Iyer, A. Rowstron, and P. Druschel, "Squirrel: A decentralized peer-to-peer web cache", *Proceedings of 21th ACM Symposium on Principles of Distributed Computing*, Monterey, California, 2002.
10. C. Baru, R. Moore, A. Rajasekar, and M. Wan, "The SDSC Storage Resource Broker", *Proceedings of CASCON'98 Conference*, 1998.
11. MCAT, MCAT – A Meta Information Catalog (Version 3.0).
12. B. White, M. Walker, M. Humphrey, and A. Grimshaw, "LegionFS: A Secure and Scalable File System Supporting Cross-Domain High-Performance Application", *Proceedings of SC'2001*, November 2001.
13. L. Guy, P. Kunszt, E. Laure, H. Stockinger, and K. Stockinger, "Replica Management in Data Grids", *Global Grid Forum 5*, 2002.
14. H. Jin, L. Ran, Z. Wang, C. Huang, Y. Chen, R. Zhou, and Y. Jia, "Architecture Design of Global Distributed Storage System for Data Grid", *High Technology Letters*, Vol.9, No.4, December 2003, pp.1-4.
15. B.-D. Lee and J. B. Weissman, "An Adaptive Service Grid Architecture Using Dynamic Replica Management", *Proceedings of 2nd International Workshop on Grid Computing*, Denver, Colorado, 2001.
16. K. Ranganathan and I. Foster, "Identifying Dynamic Replication Strategies for a High-Performance Data Grid", *Proceedings of the International Workshop on Grid Computing*, Springer-Verlag, 2001.

Data Replication Techniques for Data-Intensive Applications

Jaechun No[1], Chang Won Park[2], and Sung Soon Park[3]

[1] Dept. of Computer Software,
College of Electronics and Information Engineering,
Sejong University, Seoul, Korea
[2] Intelligent IT System Research Center,
Korea Electronics Technology Institute,
Bundang-gu, Seongnam-si, Korea
[3] Dept. of Computer Science & Engineering,
College of Science and Engineering,
Anyang University, Anyang, Korea

Abstract. Several data replication techniques have been developed to support high-performance data accesses to the remotely produced scientific data. Most of those techniques, however, do not provide the replica consistency because the data replica is just periodically updated through the remote clients. We have developed two kinds of data replication techniques, called owner-initiated replication and client-initiated replication. Our replication techniques do not need to use file system-level locking functions so that they can easily be ported to any of file systems. In this paper we describe the design and implementation of our two replication techniques and present performance results on Linux clusters.

1 Introduction

Many large-scale scientific experiments and simulations generate very large amounts of data [1, 2, 3](on the order of several hundred gigabytes to terabytes) in the geographically distributed storages. These data are shared between the researchers and colleagues for data analysis, data visualization, and so forth.

In order to reduce the communication and I/O cost, the remotely located users replicate the data sets needed for their research in the local storage. The usual way of replicating data sets is to periodically update the remotely located data replicas, like implemented in Globus toolkit [4, 5]. This method, however, does not provide the replica consistency in such a case that a data replica stored in more than one remote site is modified by a remote client and accessed by another client located at the other remote site before the modification to the replica is applied to the remote locations.

In order to provide the data replica consistency, we have developed two kinds of data replication techniques, called owner-initiated replication and client-initiated replication. We integrated those replication techniques to GEDAS (Grid

V.N. Alexandrov et al. (Eds.): ICCS 2006, Part IV, LNCS 3994, pp. 1063–1070, 2006.
© Springer-Verlag Berlin Heidelberg 2006

Environment-based Data Management System) [6, 7] that is a grid toolkit providing a high-level, user-friendly interface to share the remotely produced data among the grid communities.

In the owner-initiated replication, the data owner who owns the application data sets starts the data replication to share the data sets with remote clients. In the client-initiated replication, the client who needs the data sets starts the data replication. Moreover, our replication techniques do not need to use file system-level locking functions so that they can easily be ported to any of file systems.

The rest of this paper is organized as follows. In Section 2, we discuss an overview of GEDAS architecture to integrate our replication techniques with GEDAS. In Section 3, we present the design and implementation of our two kinds of replication techniques. Performance results on the Linux cluster located at Sejong University are presented in Section 4. We conclude in Section 5.

2 GEDAS Architecture

2.1 Overview of GEDAS

In GEDAS, users on the data owner run large-scale, data-intensive applications while writing large amounts of data to the local storage. The remote clients who want to share the application data sets with the data owner for the purpose of data visualization or data analysis are grouped into several client groups, according to the data sets replicated on the local storage. In other words, the remote clients sharing the same replicas belong to the same client group.

Figure 1 shows an overview of GEDAS. The six metadata database tables and the application data sets generated by users are located at the data owner and n

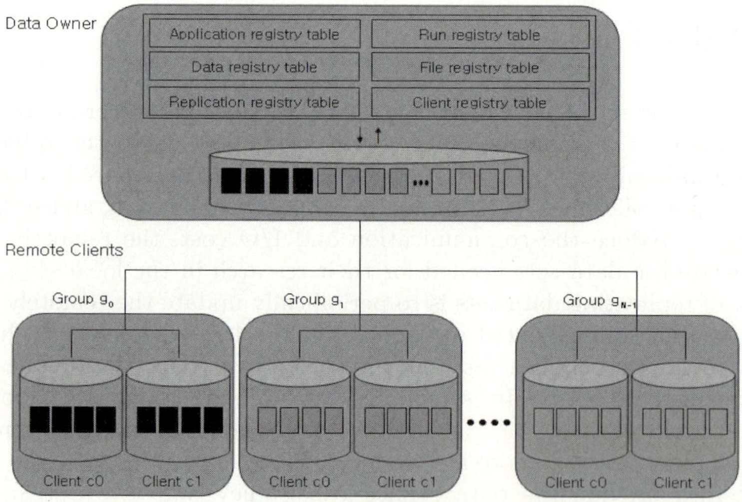

Fig. 1. An Overview of GEDAS

remote client groups are created based on the data replicas shared among them. Each client in a group is identified with groupID and clientID, such as (g_0, c_0) for the first client and (g_0, c_1) for the second client in Group g_0.

The reason for making the client groups is that if a client modifies the data replicas stored in its local storage, GEDAS can easily detect the other clients who share the same data replicas and can let them take the new copy of the modified data, without affecting other clients.

3 Data Replication

3.1 Owner-Initiated Data Replication

In order to maintain the replica consistency among the remote clients, we developed two replication approaches, called owner-initiated replication and client-initiated replication.

In the owner-initiated replication, when user generates data sets on the data owner, GEDAS replicates them to the remote clients who registered to GEDAS to share the data sets with the data owner. When a remote client changes the data replicas stored in its local storage, it broadcasts the modifications to the members in the same group where it belongs to and to the data owner for replica consistency. Figure 2 shows the steps taken in the owner-initiated replication.

Suppose that a client belonging to Group g_0 modifies the data replicas at time t_i. The client first sends the request for the IP address of other clients in g_0 to the data owner. After the client receives the IP address, it sends the modified data to the other clients in g_0 and to the data owner and waits for the acknowledgement.

When the data owner receives the modified data, it updates them to the local storage and sets the status field of the replication_registry_table to "holding"

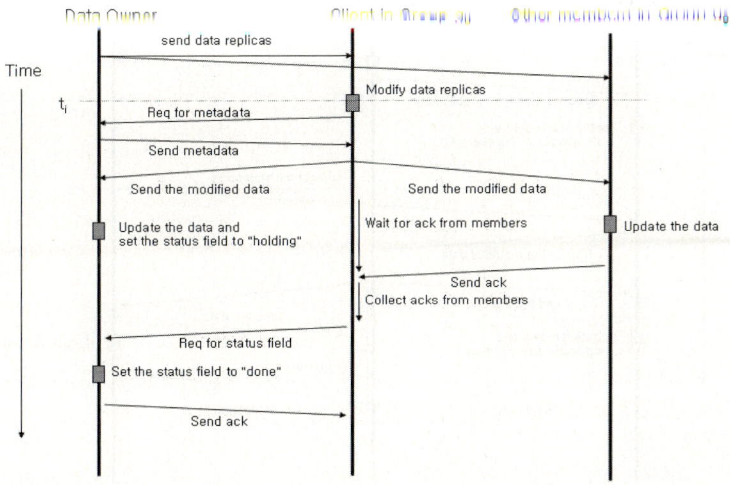

Fig. 2. Owner-Initiated Replication

to prevent another client from accessing the data sets while being updated. When the data owner receives the signal from the client who initiated the data modification, it sets the status field to "done", allowing another client to use the data replica.

The owner-initiated replication approach allows remote clients to share the data replicas safely, provided that they find out the corresponding status field is set to "done". Moreover, a remote client crash doesn't affect to the data consistency because as soon as the data replicas are modified the change is immediately reflected to the data owner and the other clients in the same group. However, if the data modification to the data replicas frequently happens, the heavy communication bottleneck then incurs even if no one else would use the data sets modified.

3.2 Client-Initiated Data Replication

Figure 3 shows the client-initiated replication where only when the modified data replicas are needed by users are those data replicas sent to the requesting client and to the data owner. Unlike in the owner-initiated replication, there is no data communication when users on the data owner produce the application data sets. If a client needs to access the remote data sets stored in the data owner, he will then get the data replica while registering to GEDAS.

Suppose that client A belonging to Group g_0 modifies a data replica at time t_i. He just sends a signal to the data owner to update the corresponding status field of the data set to the IP address of client A.

At time t_j, let client B access the data replica stored in its local storage but not been updated by client A's data modification. In order to check the replica

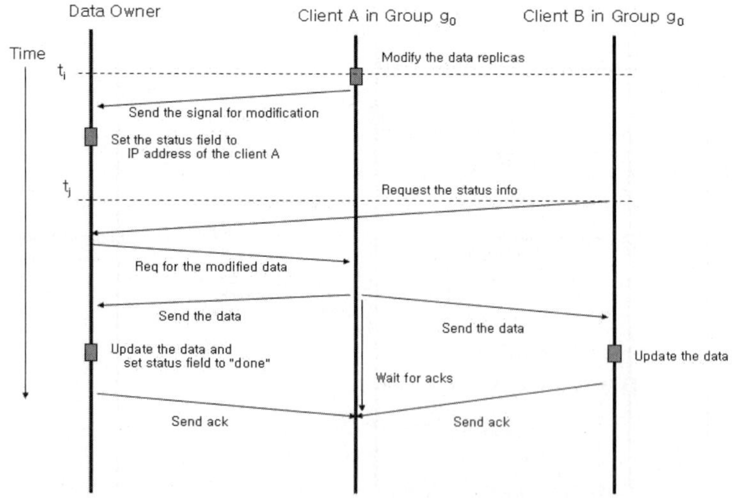

Fig. 3. Client-Initiated Replication

consistency, client B first requests the status information of the data replica to the data owner. The data owner finds out that the data set has been modified by client A and requests the data to client A. Client A sends the modified data to the data owner and to the client B, and then waits for the acknowledgement from both. After the data owner updates the modified data set and sets the status field to "done", it sends back an acknowledgement to the client A.

In the client-initiated replication approach, the data replicas are sent to the remote clients only when the data sets are actually needed by them. Therefore, unlike in the owner-initiated replication approach, the client-initiated replication does not incur unnecessary data communication. However, if a client who keeps the modification of the data replica is crashed before the data modification is updated to the data owner and to the other members of the same group, a significant data loss will be happened.

4 Performance Evaluation

In order to measure the performance, we used two Linux clusters located at Sejong university. Each cluster consists of eight nodes having Pentium3 866MHz CPU, 256 MB of RAM, and 100Mbps of Fast Ethernet each. The operating system installed on those machines was RedHat 9.0 with Linux kernel 2.4.20-8.

The performance results were obtained using the template implemented based on the three-dimensional astrophysics application, developed at the University of Chicago. The total data size generated was about 520MB and among them, 400MB of data were generated for data analysis and data restart, and then the remaining 120MB of data were generated for data visualization. The data sets produced for visualization are used by the remote clients, thus requiring the data replication to minimize the data access time.

In order to evaluate two replication approaches, a randomly chosen remote client modified 30MB of replicas at time steps 5, 10, and 15 and spread those replicas to the data owner and to the clients, according to the owner-initiated replication and to the client-initiated replication. A maximum execution time for the data replication measured among the remote clients was selected as a performance result. This time includes the cost for metadata access on the data owner, real data communication, and I/O operations.

In Figure 4, we made two client groups, while each group consisting of two nodes. At each time step, a client accesses either 30MB of replicas stored in the local storage, or 30MB of remote data sets stored on the data owner in such a case that the data sets required are not replicated.

In the owner-initiated replication, as soon as an application produces data sets at time step 0, all the remote clients receive the necessary visualization data sets to replicate them to the local storage. These replicas are used until the modification to the replicas happens at time steps 5, 10, and 15, respectively.

When the replicas stored in the local storage are used, the execution time for accessing visualization data sets drops to almost about 3 seconds needed for communicating the corresponding status information with the data owner.

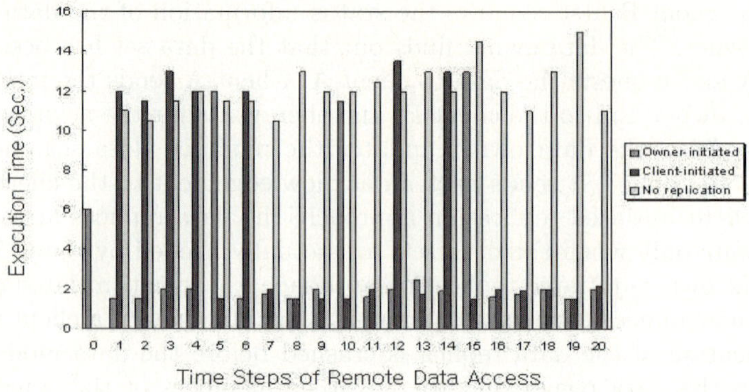

Fig. 4. Execution time for replicating visualization data on the remote clients as a function of time steps for accessing remote data sets. Two client groups were made, while each group consisting of two nodes.

If the modification to the replicas happens, like occurred at time steps 5, 10, and 15, respectively, the modified replicas are then broadcast to the data owner and to the clients in the sample group, thereby increasing the execution time for accessing remote data sets.

In the client-initiated replication, since there is no replica stored in the client side until time step 4, each remote client should communicate with the data owner to receive the data sets needed. Because each client can use the data replicas stored in its local storage from time step 5, the execution time for accessing data sets dramatically drops to almost 3 seconds.

When the replicas are modified at time steps 5, 10, and 15, the client-initiated approach just sends the IP address of the client modifying the replicas, and thus

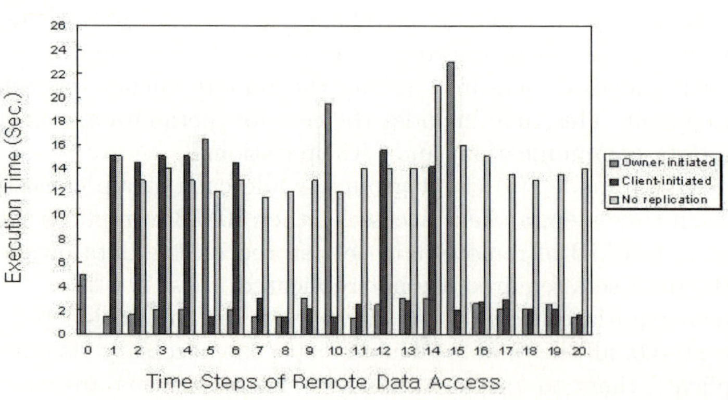

Fig. 5. Execution time for replicating visualization data on the remote clients as a function of time steps for accessing remote data sets. Two client groups were made, while each group consisting of eight nodes.

it takes no more than 3 seconds. However, we can see that in Figure 4, at time steps 6, 12, and 14, another client tries to access the modified replicas, thus occurring the data communication and I/O cost to update the replicas to the requesting client and to the data owner.

Without data replication, the data communication for accessing the remote data sets consistently happens on the clients, affecting the performance.

In Figure 5, we increased the number of nodes in each group to eight. With this configuration, when the replicas are modified at a remote client, the execution time for accessing data sets in the owner-initiated replication is significantly increased, compared to Figure 4, because of the increment in the communication cost to broadcast the replicas to the data owner and to the other clients.

On the other hand, as can be seen in Figure 5, the client-initiated replication shows not much difference in the execution time to receive the modified data sets because less number of nodes than in the own-initiated replication is involved in the communication. However, we believe that more performance evaluations should be conducted to present some valuable conclusions with these two replication approaches.

5 Conclusion

We have developed two data replication approaches to maintain the replica consistency in case of replica modifications or updates. In the owner-initiated replication, the replication occurs when applications generate the data sets in the data owner location. Whenever a remote client modifies or updates its replica, in order to maintain the data consistency, it broadcasts the replica to the other members of the same group, as well as to the data owner retaining the entire application data sets. In the client-initiated replication, only when the data sets are needed by a remote client are the necessary data sets replicated on the requesting client. If a client modifies its replica, it just sends the signal to the data owner in order for the other client to recognize the recently modified replica. Due to the data broadcast, the owner-approach shows the increased communication overhead when the number of nodes in a group becomes large. On the other hand, the client-initiated replication shows the constant communication cost even with the increased number of nodes in a group. In the future, we plan to use our replication techniques with more applications and evaluate both the usability and performance.

References

1. B. Allcock, I. Foster, V. Nefedova, A. Chervenak, E. Deelman, C. Kesselman, J. Leigh, A. Sim, A. Shoshani, B. Drach, and D. Williams. High-Performance Remote Access to Climate Simulation Data: A Challenge Problem for Data Grid Technologies. SC2001, November 2001
2. R. Moore, A. Rajasekar. Data and Metadata Collections for Scientific Applications. High Performance Computing and Networking (HPCN 2001), Amsterdam, NL, June 2001

3. A. Chervenak, E. Deelman, C. Kesselman, L. Pearlman, and G. Singh. A Metadata Catalog Service for Data Intensive Applications. GriPhyN technical report, 2002
4. I. Foster, C. Kesselman, J. Nick, and S. Tuecke. The Physiology of the Grid: An Open Grid Services Architecture for Distributed Systems Integration WG, Global Grid Forum, June 22, 2002
5. A. Chervenak, I. Foster, C. Kesselman, C. Salisbury, S. Tuecke. The Data Grid: Towards an Architecture for the Distributed Management and Analysis of Large Scientific Datasets. Journal of Network and Computer Applications, 23:187-200, 2001
6. J. No, H. Park. GEDAS: A Data Management System for Data Grid Environments. In Proceedings of International Conference on Computational Science, 2005, pages 485–492
7. J. No, R. Thakur, and A. Choudhary. High-Performance Scientific Data Management System. Journal of Parallel and Distributed Computing, (64)4:434-447, April 2003

Managing Data Using Neighbor Replication on Triangular-Grid Structure

Ali Mamat[1], M. Mat Deris[2], J.H. Abawajy[3], and Suhaila Ismail[1]

[1] Faculty of Computer Science and Information technology,
University Putra Malaysia,
43400 Serdang, Selangor, Malaysia.
ali@fsktm.upm.edu.my
[2] University College of Tun Hussein Onn,
Faculty of Information technology and Multimedia,
P.O.Box 101, 86400 Parit Raja, Batu Pahat, Johor.
mmustafa@kuittho.edu.my
[3] Deakin University, School of Information Technology, Geelong, VIC, Australia
jemal@deakin.edu.au

Abstract. Data is one of the domains in grid research that deals with the storage, replication, and management of large data sets in a distributed environment. The all-data-to-all sites replication scheme such as read-one write-all and tree grid structure (TGS) are the popular techniques being used for replication and management of data in this domain. However, these techniques have its weaknesses in terms of data storage capacity and also data access times due to some number of sites must 'agree' in common to execute certain transactions. In this paper, we propose the all-data-to-some-sites scheme called the neighbor replication on triangular grid (NRTG) technique by considering only neighbors have the replicated data, and thus, minimizes the storage capacity as well as high update availability. Also, the technique tolerates failures such as server failures, site failure or even network partitioning using remote procedure call (RPC).

1 Introduction

With the proliferation of computer networks, PCs and workstations, new models for workplaces are emerging [1]. In particular, organizations need to provide current data to users who may be geographically remote and to handle a volume of requests of data distributed around multiple sites in distributed database environment. Therefore, the storage, availability, accessibility and consistency are important issues to be addressed in order to allow distributed users efficiently and safely access data from many different sites. One way to provide access to such data is through replication. In particular, there is a need for minimal replication in order to provide an efficient way to manage the data and minimize the storage capacity.

One of the solutions is based on synchronous replication [2,3]. It is based on quorum to execute the operations with high degree of consistency [4] and also to ensure serializability. Synchronous replication can be categorized into several schemes, i.e., all data to all sites (full replication), all data to some sites and some data to all sites.

V.N. Alexandrov et al. (Eds.): ICCS 2006, Part IV, LNCS 3994, pp. 1071 – 1077, 2006.
© Springer-Verlag Berlin Heidelberg 2006

However, full replication causes high update propagation and also needs high storage capacity [4,5,6]. Several studies model partial replication in a way that each data object is replicated to some of the sites. However, it is still undefined which copies are placed on which site, such that different degrees of quality of a replication scheme can be modeled. A few studies have been done on partial replication techniques based on some data items to all sites using tree structure technique [7,8]. This technique will cause high update propagation overhead. Thus, some-data-items-to-all-sites scheme is not realistic. The European DataGrid Project [9,10] implemented this model to manage the file-based replica. It is based on the sites that have previously been registered for replication. This will cause the inconsistence number of replication occurs in the model. Also, the data availability has very high overhead as all registered replicas must be updated simultaneously. Thus, an optimum number of sites to replicate the data are required with non-tolerated availability of the system.

The focus of this paper is on modeling a technique based on synchronous solution to minimize the storage and optimize the system availability of the replicated data in a data grid. We describe the neighbor replication on triangular grid (NRTG) technique by considering only neighbors have the replicated data. This assignment provides a higher availability of executing write operations in replicated database due to the minimum number of quorum size required.

This paper is organized as follows: In Section 2 the model and the technique of the NRTG is presented. In Section 3, the replica management is discussed. In Section 4, the performance of the proposed technique is analyzed in terms of availability, and a comparison with tree grid structure technique is given.

2 Neighbor Replication on Triangular Grid (NRTG) Technique

2.1 Model

Briefly, a site i initiates a NRTG transaction to update its data object. For all accessible data objects, a NRTG transaction attempts to access a NRTG quorum. If a NRTG transaction gets a NRTG write quorum without non-empty intersection, it is accepted for execution and completion, otherwise it is rejected. We assume for the read quorum, if two transactions attempt to read a common data object, read operations do not change the values of the data object. Since read and write quorums must intersect and any two NRTG quorums must also intersect, then all transaction executions are one-copy serializable.

2.2 The NRTG Technique

In NRTG, all sites are logically organized in the form of triangular grid structure. It is inherited from binary-tree structure with the inner leaves of the tree linked together as shown in Fig. 1.

Each site has a master data file. In the remainder of this paper, we assume that replica copies are data files. A site is either operational or failed and the state (operational or failed) of each site is statistically independent to the others. When a site is operational, the copy at the site is available; otherwise it is unavailable.

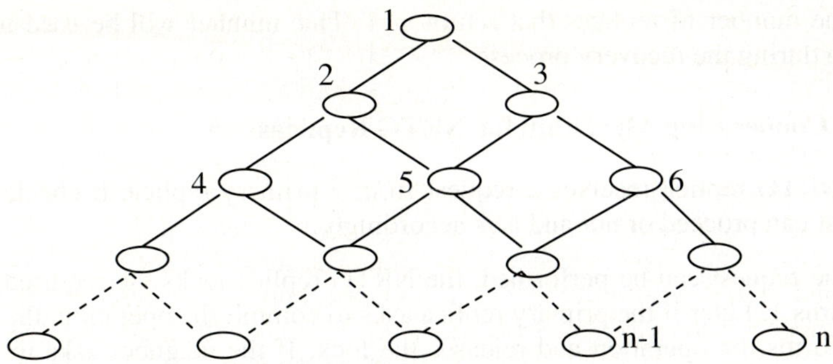

Fig. 1. Triangular-neighbor organization of n sites

For example, from Fig. 1, data from site 1 will replicate to site 2 and site 3 which are neighbors to it. Site 5 has four neighbors, which are sites 2, 3, 8, and 9. As such, site 5 has five replicas. From this property, it is clear that each site required with the maximum number of replications is 5, thus, minimizing the storage capacity as compared with full replication scheme.

3 Managing Replication Under NRTG

The quorum for an operation is defined as a set of sites (called *replicas*) whose number is sufficient to execute that operation. In other words, it is modeled as a set of replicas: $C = \{C_1, C_2, \ldots, C_n\}$, where i=1,2,…,n, are called the sequence numbers of these replicas. Each replica, C_i manage a set of data: $D_i = \{d_1^i, d_2^i, \ldots, d_m^i\}$. The consistency constraint requires that $D_i \equiv D_j$, for i,j = 1,2,…,n. The replication management of NRTG is analogous with the one that proposed in [11].

Each replica in the system provides a number of remote procedures that can be called by users for processing the data managed by the replica.

3.1 The Coordinating Algorithm for the Primary Replica

When a primary replica receives an NRTG-Procedure-Call from a transaction manager, it uses the coordinating algorithm to maintain the consistency of all replicas in terms of NRTG-Procedure-Call. This section describes the coordinating algorithm.

In the coordinating algorithm, the primary replica uses the 2PC protocol to ensure the replication consistency. In the first phase, the primary replica asks the NRTG quorum whether it can be formed or not. If the NRTG quorum can be formed (replicas under NRTG quorum return a Successful (1) for such execution), the primary replica returns a SUC to the transaction manager. If the transaction manager requests a commit, then in the second phase the primary replica asks all replicas to commit the NRTG-Procedure-Call execution. If NRTG quorum cannot be formed, then the primary replica returns a (0) to the transaction manager and asks all replicas to abort the operation in the second phase. If operation returns a Partial-Commit (-1) (the primary replica returns a PC and other non-primary replicas may return a -1), the primary replica returns a -1 to the transaction manager. The primary replica also records the number of replicas that return a -1. This number will be used in conflict resolution during the recovery process.

3.2 The Cooperating Algorithm for NRTG Replicas

When a NRTG replica receives a request from a primary replica, it checks whether the request can proceed or not and acts accordingly.

- If the request can be performed, the NRTG replica locks the required data and returns 1. Later if the primary replica asks to commit the operation, the neighbor performs the operation and releases the lock. If the neighbor asks to abort the operation, it will release the lock.
- If the neighbor finds that the operation cannot be executed (the required data is not free), it then returns an 0.
- If the NRTG replica finds that the data is in a partially committed state, then it returns a PC to the primary replica. The primary replica will then partially commits the operation and record the event when the primary replica asks to partially commit the operation. If the primary replica asks to abort the operation, then the non-primary replica aborts the operation.

The 2PC protocol used by NRTG-Procedure-Call transaction manager guarantees that the replicas will be in a consistent state if a transaction returns a 1 or 0. However, to guarantee all replicas do the same thing to the transactions with -1 pending, a majority consensus should be reached among all replicas.

The order of each partially committed NRTG-Procedure-Call over a data is also important when a re-join of network partitions carries out of these NRTG-Procedure-Calls.

If the network is partitioned into two disconnecting parts, they will eventually be re-united again. In this case, replicas in both partitions have partially committed updates. The conflict resolution algorithms during recovery process are responsible to make replicas in these two parts consistent. When recovering from a network partition, replicas of each part of the partition have to send a 're-uniting' message to replicas of the other part.

The NRTG-RPC returns -1 only when the network is partitioned. If the network is partitioned into two disconnecting parts, the two parts will eventually be re-united again. In this case, replicas in both partitions may have some partially committed updates. The conflict resolution algorithm is responsible to make replicas in these two parts consistent.

4 Performance Analysis and Comparisons

4.1 Availability Analysis

In estimating the availability, all copies are assumed to have the same availability *p*. $C_{X,Y}$ denotes the communication cost with X technique for Y operation, which is R(read) or W(write).

4.1.1 TGS Technique

Let h denotes the height of the tree, D is the degree of the copies in the tree, and $M = \lceil (D+1)/2 \rceil$ is the majority of the degree of copies. The availability of the read and write operations in the TGS can be estimated by using recurrence equations based on the tree height h. Let AR_{h+1} and AW_{h+1} be the availability of the read and the write operations with a tree of height h respectively. Then the availability of a read operation for a tree of height h+1 can be represented as:

$$AR_{h+1} = p + (1-p) \sum_{i=M}^{D} \binom{D}{i} AR_h^i (1 - AR_h)^{D-i}$$

and the availability of a write operation for a tree of height h+1 is given as:

$$AW_{h+1} = p \sum_{i=M}^{D} \binom{D}{i} AW_h^i (1 - AW_h)^{D-i} \quad (1)$$

where p is the probability that a copy is available, and $AR_0 = AW_0 = p$.

Thus system availability for TGS is

$$Av(TGS,r,w) = f AR_{h+1} + (1-f) AW_{h+1} \quad \ldots(2)$$

where f and $(1-f)$ are the probability that an arriving operation of read and write for data file x, respectively.

4.1.2 NRTG Technique

Let p_i denote the availability of site i. Read operations on the replicated data are executed by acquiring a read quorum and write operations are executed by acquiring a write quorum. For simplicity, we choose the read quorum equals to the write quorum. Thus, the quorum size for read and write operations equals to $\lfloor L_{Bx}/2 \rfloor$, that is,

$A_{NRTG,R} = A_{NRTG,W} = \lfloor L_{Bx}/2 \rfloor$. For example, if the primary site has four neighbors, each of which has vote one, then $C_{NRTG,R} = C_{NRTG,W} == \lfloor 5/2 \rfloor = 3$.

For any assignment B and quorum q for the data file x, define $\varphi(B_x,q)$ to be the probability that at least q sites in $S(B_x)$ are available, then

$\varphi(B_x,q)$ = Pr{at least q sites in $S(B_x)$ are available }

$$= \sum_{G \in Q(B_x,q)} \left(\prod_{j \in G} p_j \prod_{j \in S(B_x)-G} (1-p_j) \right) \quad \ldots (3)$$

Thus, the availability of read and write operations for the data file x, are $\varphi(B_x,r)$ and $\varphi(B_x,w)$, respectively. Let $Av(B_x,r,w)$ denote the system availability corresponding to the assignment B_x, read quorum r and write quorum w. If the probability that an arriving operation of read and write for data file x are f and $(1-f)$, respectively, then

$$Av(B_x,r,w) = f\,\varphi(B_x,r) + (1-f)\,\varphi(B_x,w).$$

Since in this paper, we consider the read and write operations are equal, then the system availability,

$$Av(B_x,r,w) = \varphi(B_x,r) = \varphi(B_x,w). \quad \ldots (4)$$

4.2 Performance Comparisons

4.2.1 Comparisons of Write Availabilities

In this section, we will compare the performance on the write availability and system availability of the TGS technique based on equations (1) and (2), and our NRTG technique based on equations (3) and (4) for the case of n=13, 40 and 121. In estimating the availability of operations, all copies are assumed to have the same availability.

Fig. 2 and Table 1, show that the NRTG technique outperform the TGS technique. When an individual copy has availability 88%, write availability in the NRTG is approximately 99% whereas write availability in the TGS is approximately 82% for n=13. Moreover, write availability in the TGS decreases as n increases. For example, when an individual copy has availability 86%, write availability is approximately 78% for n=13 whereas write availability is approximately 73% for n = 121.

Table 1. Comparison of the write availability between TGS and NRTG under the different set of copies p=0.8,…,0.98

Techniques	Write Availability								
	0.82	0.84	0.86	0.88	0.9	0.92	0.94	0.96	0.98
NRTG(R/W)	.956	.968	.978	.986	.991	.996	.998	.999	1.00
TGS(W),n=13	.692	.738	.782	.823	.861	.896	.927	.955	.979
TGS(W),n=40	.634	.697	.755	.807	.853	.892	.926	.954	.979
TGS(W),n=121	.571	.656	.731	.794	.847	.890	.925	.954	.979

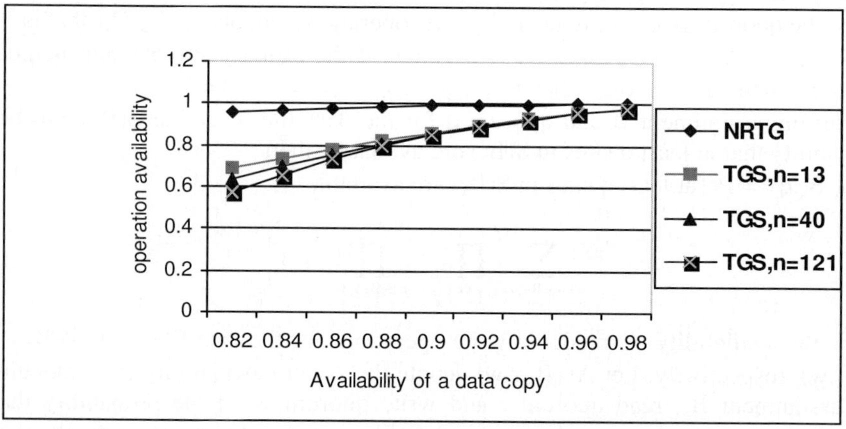

Fig. 2. Comparison of the write availability between TGS and NRTG

5 Conclusions

In this paper, a new technique, called neighbor replication on grid (NRTG) structure has been proposed to manage the data in distributed database environment. It showed that, the NRTG technique provides a convenient approach to minimize the storage, high availability for update-frequent operations by imposing a neighbor binary vote assignment to the logical grid structure on data copies, and also data access time. This is due to the minimum number of replication sites required. The fault-tolerant programs developed in the environment are able to tolerate failures such as server failure, a site failure or even a network partitioning. In comparison to the TGS structure, NRTG provides higher system availability, which is preferred for large systems.

References

1. O. Wolfson, S. Jajodia, and Y. Huang,"An Adaptive Data Replication Algorithm,"*ACM Transactions on Database Systems*, vol. 22, no 2 (1997), pp. 255-314.
2. J. Holliday, R. Steinke, D. Agrawal, and A. El-Abbadi,"Apidemic Algorithms for Replicated Databases", *IEEE Trans. On Know. and Data Engineering*, vol.15, No.3 (2003), pp.1-21.
3. H. Stockinger,"Distributed Database Management Systems and The Data Grid",IEEE-NASA Symposium, April 2001, pp. 1-12.
4. Budiarto, S. Noshio, M. Tsukamoto,"Data Management Issues in Mobile and Peer-to-Peer Environment", *Data and Knowledge Engineering, Elsevier*, 41 (2002), pp. 183-204.
5. L.Gao, M. Dahlin, A. Nayate, J. Zheng, and A. Lyengar,"Improving Availability and Performance with Application-Specific Data Replication", *IEEE Trans. On Knowledge and Data Engineering*, vol.17, no. 1, (2005), pp 106-120.
6. M.Mat Deris, J.H. Abawajy, H.M. Suzuri,"An Efficient Replicated Data Access Approach for Large Scale Distributed Systems", *IEEE/ACM Conf. On Cluster Computing and Grid (CCGRID2004)*, April 2004, Chicago, USA.
7. M. Nicola and M. Jarke," Performance Modeling of Distributed and Replicated Databases", *IEEE Trans. On Knowledge and Data Engineering*, vol.12, no. 4, (2000), pp 645-671.
8. [illegible] Huang, Qingfeng Fan, Qiongli Wu and YangXiang He,"Improved Grid Information Service Using The Idea of File-Parted Replication", *Lecture Notes in Computer Science Springer*, vol. 3584 (2005), pp. 704-711.
9. P. Kunszt, Erwin Laure, Heinz Stockinger, Kurt Stockinger,"File-based Replica Management", *International Journal of Future Generation of Computer Systems, Elsevier,* 21 (2005), pp. 115-123.
10. H. Stockinger, F Donno, E. Laure, S. Muzaffar, P. Kunszt, "Grid Data Management in Action: Experience in Running and Supporting Data Management Services in the EU Data Grid Project", Int'l. Conf. On Computing in High Energy and Nuclear Physics, March 2003, La Jolla, California. http://gridpp.ac.uk/papers/chap3-TUAT007.pdf
11. W. Zhou and A. Goscinski,"Managing Replicated Remote Procedure Call Transactions", *The Computer Journal*, Vol. 42, No. 7, pp 592-608, 1999.

Author Index

Aalto, Samuli IV-420
Abawajy, J.H. IV-1015, IV-1071
Abbas, Ghulam I-802
Abbaspour, Amin III-578
Abbod, M.F. I-993
Abed, E.H. III-448
Abramson, David I-720
Absil, P.-A. I-210
Agarwal, Pankaj K. III-409
Aguado González, S. II-350
Aguilar-Saborit, Josep I-156
Ahmad, Jamil II-887
Ahn, Chunsoo I-952
Ahn, Hye-Young IV-894
Ahn, Hyun Gi I-989
Akbal, Ayhan IV-631, IV-638
Akbar, Ali Hammad II-1073
Akcelik, Volkan III-481
Akdim, B. I-372
Akioka, Sayaka I-242
Akritas, Alkiviadis II-486
Aksyonov, Konstantin A. III-879
Aldana-Montes, Jose F. III-936
Alemani, Davide II-70
Alexandrov, Vassil I-868, II-595, II-603, III-632, III-640
Alfonsi, Giancarlo I-465
Alípio, Pedro III-240
Al-khalifah, Ali I-868
Almeida, Francisco I-872
Alonso, J. III-313
Alonso, Pedro I-348
Altinakar, Mustafa Siddik II-58
Altintas, Ilkay III-69, III-920
Alves, Domingos I-1005, III-297
Amghar, T. I-603
Amorim, Ronan M. I-68
Anagnostou, Miltiades I-579, I-892
Anai, Hirokazu II-462
Ang, Chee-Wei IV-260
Angelova, Donka III-624
Anguita, M. II-518
Antaki, James IV-855
Aramudhan, M. IV-388
Argyrakis, Panos III-1048
Arora, Nitin I-16
Arrowsmith, J. Ramon III-920
Arslan, Ahmet II-247
Artoli, Abdel Monim II-78
Assous, Franck IV-623
Asthana, A. III-161
Atanassov, Emanouil III-616
Ausloos, Marcel III-1114
Avci, Mutlu IV-615
Avila, Andres I-856
Awan, Asad III-465

Babik, Marian III-980
Babuška, I. III-530
Bachman, Timothy IV-855
Baddeley, B. II-871
Bader, M. I-673
Bae, Jungsook II-1033
Bae, Sung Eun I-595
Baek, Myung-Sun II-969, II-1058
Bagheri, Ebrahim I-588
Bajaj, C. III-530
Bajuelos, António Leslie II-255
Baker, C.G. I-210
Baker, Mark II-953
Balas, Lale I-814
Balcan, Duygu III-1083
Baldridge, Kim III-69
Balicki, Jerzy III-863
Balik, Hasan H. IV-631, IV-638
Baliś, Bartosz II-542
Balla, Sudha II-822
Balogh, Zoltan III-980
Balos, Kazimierz IV-1039
Banaś, K. III-743
Bandini, Stefania III-289
Banicescu, Ioana II-430
Bao, Yukun I-728, IV-308, IV-517
Barabási, Albert-László III-417
Barbosa, Ciro B. I-68
Baronikhina, Anastasia A. III-879
Barrientos, Ricardo J. I-611
Barsky, Prof. Brian A. II-215

Bartlett, Roscoe A. IV-525
Bartocci, Ezio III-1012
Baru, Chaitan III-920
Bass, J. III-530
Battiato, Sebastiano II-334
Baumgartner, Gerald I-267
Bauschlicher Jr., C.W. III-128
Beagley, N. II-871
Becker, Michael F. I-443
Beezley, Jonathan D. III-522
Bell, Ron I-7
Belmonte Fernández, Ó. II-350
Benavent, Xaro III-13
Bergamaschi, L. IV-685
Bernabeu, Miguel O. I-348
Bernard, Julien IV-999
Bernhard, Fabrice IV-236
Bernholdt, David E. I-267
Bernier, J.L. II-518
Bernot, Gilles II-887
Bernreuther, Martin II-161
Bernstein, D.S. III-489
Bertozzi, Luigi IV-831
Berzins, M. II-147
Bhana, Ismail II-581
Bhattacharjee, Apurba K. I-387
Bianchini, Germán I-539
Bie, Rongfang II-775, IV-781
Biely, Christoly III-1067
Bieniasz, Sławomir III-759
Biros, George III-481
Biscay, R.J. I-132
Bishop, Marvin III-608
Blais, J.A.R. III-48
Blelloch, Guy E. II-799
Blower, J.D. III-996
Bonizzoni, Paola II-622
Borowski, Stefan I-196
Botana, Francisco II-470
Bourchtein, Andrei I-258
Bourchtein, Ludmila I-258
Bourgeois, Anu II-678
Branford, Simon III-632, III-640
Brendel, Ronny II-526
Breuer, Peter T. IV-765
Brill, Downey III-401
Brinza, Dumitru II-767
Briseid, Sverre IV-204
Brogan, David III-570
Brown, Martin IV-773
Browne, J.C. III-530
Brunst, Holger II-526
Bu, Jiajun I-449
Bubak, Marian II-542
Buendia, Patricia II-807
Buffle, Jacques II-70
Bunde, Armin III-1048
Bungartz, Hans-Joachim II-161
Burguillo-Rial, J.C. III-815
Bustos, Benjamin IV-196
Byun, Yanga I-276, I-284

Caballero-Gil, P. III-337
Cai, Guoyin I-292, I-876, III-1
Cai, Yang I-1001, IV-870
Caliari, M. IV-685
Calle, Eusebi IV-136
Camahort, Emilio II-287, II-310
Campos, Celso II-310
Campos, Fernando Otaviano I-68, I-76
Cannarozzi, Gina M. II-630
Cao, Chunxiang III-9
Cao, Jian III-948
Cao, Wuchun III-9
Cao, Yuanda IV-81
Čapkovič, František III-176
Cappello, Angelo IV-831
Carbonell, F. I-132
Cariño, Ricolindo L. II-430
Carnahan, Joseph III-570
Caron, David III-514
Carpenter, Bryan II-953
Carr, Nathan A. IV-228
Carvalho, Luis Alfredo V. de I-842
Carvalho, Paulo III-240
Castelló, Pascual II-263
Čepulkauskas, Algimantas II-407
Cetnarowicz, Krzysztof III-775, III-823, III-855
Chae, Kijoon II-1024
Chai, Zhilei I-1043
Chakravarty, Manuel M.T. II-920
Chambarel, André II-50
Chandrasekar, J. III-489
Chang, Chun-Hyon IV-280
Chang, F.K. III-456
Chang, Hao-Li IV-878
Chang, Kungyen I-226
Chang, Kyungbae IV-987
Chantzara, Maria I-579

Charoy, François III-976
Chatterjee, Abhijit III-77
Chaturvedi, Alok III-433
Chauve, Cedric II-783
Chen, Chun I-449
Chen, Guihai IV-404
Chen, Jianming IV-501
Chen, Juan II-646, II-904
Chen, Junliang IV-104
Chen, Lei IV-938
Chen, Ling II-646
Chen, Ming-Jen IV-184
Chen, Shudong III-1004
Chen, Su-Shing II-830
Chen, Yangzhou II-478
Chen, Yibing I-851
Chen, Yixin II-646
Chen, Yongqiang I-896
Chen, Yu-Sheng I-1026
Chen, Zhanglong I-1043
Chen, Zhengxin IV-476, IV-485
Cheng, Guang IV-144
Cheng, Haiying I-1047
Cheng, Jingde IV-797
Cheng, Junbo I-851
Cheng, Junxia I-851
Cheng, Ruixing IV-87
Cheng, Shiduan IV-128
Cheng, Wang-Cho IV-260
Chi, Hongmei IV-773
Chiang, Tzu-Chiang II-1008
Chinnasarn, Krisana I-403
Chinnasarn, Sirima I-403
Cho, Choongho IV-168
Cho, Dong-Jun II-1058
Cho, Han Wook IV-244
Cho, Insook III-1040
Cho, Jin-Woong II-1041
Cho, Keumwon III-972
Cho, KumWon IV-293
Cho, Kyu Bong II-587
Cho, Sung-Jin I-1067
Cho, Yongyun I-965, II-510
Cho, Yookun IV-946
Choi, Bumghi I-63
Choi, Hyoung-Kee II-961
Choi, Jaeyoung I-965, I-1059, III-972
Choi, Jeong-Yong IV-25
Choi, Jin-Ghoo IV-160
Choi, Jin-Hee IV-160, IV-172
Choi, Kee-Hyun III-895, III-899
Choi, Min-Hyung I-308, I-490
Choi, Seung-Hyuk I-969
Choi, Un-Sook I-1067
Chong, Kiwon IV-902
Choo, Hyunseung I-948, I-960, I-989, II-1089
Chopard, Bastien II-70, IV-653
Choppella, Venkatesh I-267
Chou, Chien-Lung I-900
Chover, Miguel II-263
Chow, Peter II-34
Chowdhury, A.K. III-161
Christov, N.D. I-697
Chu, Yuan-Sun IV-184
Chung, Min Young I-969, I-989
Chunguo, Wu I-547
Chuyi, Song I-547
Ciarlet Jr., Patrick IV-623
Cięciwa, Renata III-823
Cirak, Fehmi II-122
Ciszewski, Stanisław III-759
Čivilis, Alminas I-1034
Ciuffo, Leandro N. I-68
Clark, James S. III-409
Clarke, Bill I-218
Cocu, Adina I-172
Coen, Janice III-522
Cohen, Reuven III-1048
Cokuslu, Deniz I-571
Cole, Martin J. III-393
Cole, Murray II-929
Collier, R. III-727
Colling, D.J. III-956
Collins, Timothy M. II-807
Combes, P. IV-653
Comet, Jean-Paul II-887
Conner, Jeffery III-920
Cornford, Dan III-586
Corradini, Flavio III-1012
Cortés, Ana I-539
Costa-Montenegro, E. III-815
Cotoi, I. II-26
Cox, Simon J. III-928
Crosby, Christopher J. III-920
Crăciun, Marian Viorel I-172
Cuadrado, J.J. IV-789
Cui, Pingyuan II-478
Culebras, R. I-395
Curley, Martin I-4

Cycon, Hans L. II-1050
Czarnul, Pawel III-944
Czekierda, Łukasz III-940

Dagdeviren, Orhan I-571
da Silva, Fabrício A.B. I-1005, III-297
Dai, Wenchao I-1047
Dai, Yafei IV-412, IV-428
Dailyudenko, Victor F. I-846
Dal Negro, Marco III-264
Danelutto, M. II-937
Danilecki, Arkadiusz I-753
Darema, Frederica III-375
Darlington, John III-964
Das, Abhimanyu III-514
Das, A.K. III-161
Dauvergne, Benjamin IV-566
Davila, Jaime II-822
Dazzi, P. II-937
Degond, Pierre II-1
Deiterding, Ralf II-122
De la Cruz, H. I-132
de la Encina, Alberto II-207
Delaplace, Franck III-1056
Della Vedova, Gianluca II-622
Del Vecchio, David I-681
De Paoli, Serge IV-999
Demkowicz, L. III-530
Deng, Hui IV-17
Deussen, Oliver IV-196
Deville, Michel II-58
Dhamdhere, Kedar II-799
Dhariwal, Amit III-514
Díaz, Manuel II-912
Dickens, L.W. III-956
Dieci, Luca IV-677
Dieter, Bill I-226
Dietz, Hank I-226
Dikshit, Anupam II-830
Diller, K.R. III-530
Dimov, Ivan III-632, III-640
Ding, Koubao I-482
Ding, Shifei I-777
Ding, Wei IV-112, IV-120, IV-144
Dobnikar, Andrej III-345
Dogan, Kaan II-996
Dokken, Tor IV-204
Dondi, Riccardo II-622
Donetti, Luca III-1075
Dong, Hongbin III-216
Dong, Jin-xiang IV-839
Dong, Yabo IV-57
Dornaika, Fadi I-563
Dou, Wen-hua I-1030
Douglas, Craig C. III-393, III-522
Draganescu, Andrei III-481
Dreżewski, Rafał III-871, III-908
Drummond, Arielle IV-855
Duan, X. I-372
Dumitriu, Luminiţa I-172, II-199
Dyshlovenko, Pavel IV-599

Easterday, Scott IV-582
Efendiev, Yalchin III-393
Eleftheriou, Maria II-846
Ellis, Carla III-409
El Yacoubi, Samira III-360
Engelmann, Christian II-573
Ensan, Faezeh I-588
Eom, Young Ik IV-356
Erciyes, Kayhan I-571
Erlebacher, Gordon II-177
Erzan, Ayşe III-1083
Escrivá, Miguel II-287
Evans, Deidre W. IV-773
Ewing, Richard III-393

Falzon, Chantal T. III-82
Fan, Liangzhong II-367
Fan, Zhiping III-601
Fang, Liqun III-9
Fang, Wei III-847
Fangohr, Hans II-139
Fantozzi, Silvia IV-831
Farhat, C. III-456
Farinella, Giovanni Maria II-334
Farinelli, Simone IV-324
Fathy, M. I-744
Feixas, Miquel II-263
Felici, G. IV-460
Feller, Scott II-846
Feng, Dan I-1063, III-671, IV-396
Feng, Gang IV-645
Feng, Y. III-530
Feng, Yi IV-468
Fernández, A. III-313
Fernández de Vega, F. III-281
Fernández, J. II-518
Fernando, Terrence III-60
Ferrari, T. III-956

Fertin, Guillaume II-622, II-783
Fesehaye Kassa, Debessay IV-65
Fidanova, Stefka I-1009
Figueiredo, Renato III-546
Filatyev, Sergei III-433
Fisher, Randy I-226
Fitch, Blake G. II-846
Fjeldly, T.A. IV-607
Fladmark, Gunnar II-102
Flikkema, Paul G. III-409
Forestiero, Agostino IV-1047
Fort, H. III-313
Fortes, José III-546
Forth, Shaun A. IV-558
Fougère, Dominique II-50
Freundl, C. II-185
Fritz, Nicolas IV-200
Froelich, Wojciech III-839
Fu, Chong I-826
Fu, Karl I-1001
Fu, Qing-hua IV-878
Fu, Xuezheng II-678
Fuentes, D. III-530
Fujimoto, Richard II-41, III-425
Funika, Włodzimierz II-534, II-549
Furlan, Luciana B. I-1005
Fürlinger, Karl II-494
Fúster-Sabater, A. III-337

Gaaloul, Khaled III-976
Gagliardi, Henrique F. I-1005, III-297
Gaitan, Rafa II-107
Galante, M.A. IV-460
Galceran, Josep II-70
Gallivan, K.A. I-210
Gallos, Lazaros K. III-1048
Gálvez, A. II-414
Gamalielsson, Jonas II-879
Gan, Honghua I-204
Ganzha, Maria III-208
Gao, Lijun IV-501
Gao, Wenzhong II-430
Gao, Xiaodong III-601
Gao, Xiaoyang I-267
Gao, Zhigang IV-918
García, Víctor M. I-324
Garšva, Gintautas IV-364
Garzón, E.M. II-106
Gashkov, Igor I-912
Gasparo, Maria Grazia IV-677
Gastineau, Mickaël II-446
Gavrilenko, Vladimir I. III-89
Gay, David M. IV-525
Geist, Al II-573
Gelfand, Alan III-409
George, E. Olusegun II-694
Germain, Robert S. II-846
Gerndt, Michael II-494
Ghattas, Omar III-481
Giampapa, Mark II-846
Giannoutakis, Konstantinos M. I-506
Giering, Ralf IV-591
Gill, Ofer H. II-638, II-654
Gilmore, Stephen II-929
Girdzijauskas, Stasys IV-364
Głut, Barbara II-302
Godart, Claude III-976
Golubchik, Leana III-514
Gomes, André Severo Pereira III-97
Goncharova, Natalia V. III-879
Gong, Jian I-1022, IV-112, IV-120, IV-144
Gonnet, Gaston H. II-630
González, Daniel I-872
González, J. I-649
González, Luis III-305
González-Castaño, F.J. III-815
González de-la-Rosa, Juan-José I-316
Gopalan, B. II-871
Gopinath, K. III-679
Gore, Jay III-433
Górriz, J.M. I-234, I-316, I-356, I-395, I-649
Goscinski, A.M. IV-1015
Goualard, Frédéric I-332
Govaerts, W. II-391
Govindan, Ramesh III-514
Grabska, Ewa III-883
Graham, Richard L. II-945
Grama, Ananth III-465
Gravvanis, George A. I-506
Grivel, E. I-697
Grochowski, M. III-783
Grossfield, Alan II-846
Grunewaldt, Lars II-565
Grzech, Adam III-224
Grzesiak-Kopeć, Katarzyna III-883
Guan, Ximeng I-250
Guensler, R. III-425
Guibas, L.J. III-456

Guisado, J.L. III-281
Guitart, Jordi I-84
Guodong, Yuan I-864
Guo, Jianping I-292, I-876, III-1, III-9
Guo, Wu II-223
Gurel, Guray II-996
Gurov, Todor III-616
Gusfield, Dan II-618
Gutsev, G.L. III-128
Guzy, Krzysztof II-542

Habala, Ondrej III-980
Haffegee, Adrian II-595, II-603
Hagen, Trond Runar IV-204, IV-220
Hager, Svenja IV-340
Haines, K. III-996
Hajiaghayi, M.T. II-758
Halperin, Eran II-799
Ham, Eunmi IV-894
Han, Chang-Wook IV-862
Han, Joohyun I-965
Han, JungHyun III-40
Han, Kijun I-940, IV-180
Han, Kyungsook I-276, I-284
Han, SeungJae II-1101
Han, Sunyoung I-936, II-1081, IV-260
Han, Wook-Shin III-648
Handzlik, Piotr II-549
Harman, Mark IV-740
Harris, J. Clay III-393
Harrison, A.B. III-996
Harrison, Ken III-401
Harrison, Robert II-710
Harrison, Rodrigo III-1091
Hartell, Mark G. I-387
Hartono, Albert I-267
Hascoët, Laurent IV-566
Hassoun, Y. III-956
Havlin, Shlomo III-1048
Hayashi, Yukio III-1106
Hazle, J. III-530
He, Gaiyun I-822
He, Jieyue II-710
He, Jing II-1069, IV-509
He, Jingwu II-750
He, Yulan II-718
Hegeman, Kyle IV-228
Heo, Junyoung IV-946
Hermer-Vazquez, Linda III-546
Hernandez, Gonzalo I-856, III-1091
Hernández Encinas, L. II-438
Hicks, Rickey P. I-387
Hidalgo-Herrero, Mercedes II-207
Hill, Judith III-481
Hiller, Stefan IV-196
Hincapie, Doracelly I-920
Hiroaki, Deguchi II-490
Hirokazu, Hashiba II-490
Hluchy, Ladislav III-980
Honavar, Vasant G. III-440
Hong, Choong Seon II-1016
Hong, Jiman IV-946, IV-970, IV-991
Hong, Kwang-Seok IV-886
Hong, Min I-308, I-490
Hong, Tzung-Pei I-1026
Horie, Daisuke IV-797
Horng, Gwoboa I-1026
Hovland, Paul IV-574, IV-582
Howard, I.C. I-993
Hrach, Rudolf I-806
Hsiao, Chieh-Ming IV-757
Hsu, Chung-Chian IV-757
Hsu, Ji-Hsen II-295
Hu, Dewen I-689
Hu, Hongyu IV-81
Hu, Xiaoping I-689
Hu, Xiaoyong I-838
Hu, Yincui I-292, I-876, III-1
Huang, Changqin I-838
Huang, Guangyan II-1069
Huang, Houkuan III-216
Huang, Jin I-1051
Huang, Wayne IV-188
Huang, Wei I-728, IV-308, IV-493, IV-517
Huang, Xianglin I-997
Huang, Xin I-411
Huang, Y. III-554
Huang, Yan-Ping IV-757
Huang, Yueh-Min II-1008
Huerta, Joaquin II-310
Huh, Eui-Nam I-960
Humphrey, Marty I-681
Hunke, Elizabeth C. IV-533
Hunter, M. III-425
Huo, Mingxu I-482
Hurtado, Pablo I. III-1075
Hwang, Ho Seon I-944
Hwang, Ho-Yon IV-264
Hwang, In-Yong I-1018

Hwang, Jae-Hyun IV-172
Hwang, Tae Jin I-944
Hwang, Yoon-Hee I-1067

Iavernaro, Felice IV-724
Iglesias, A. II-383, II-414
Im, Sungbin II-992
Imielinska, Celina IV-822
Iñiguez, B. IV-607
Ipanaqué, R. II-383
Irwin, Mary Jane I-242
Iskandarani, Mohamed III-393
Iskra, K.A. III-281
Ismail, Suhaila IV-1071
Issakova, Marina I-928

Jackson, Steven Glenn II-422
Jaeger-Frank, Efrat III-920
Jain, K. II-758
Jakop, Yanto I-522
Jakubowska, Malgorzata I-498
Jaluria, Y. III-473
Jamieson, Ronan II-595, II-603
Janicki, Aleksander IV-301
Janik, Arkadiusz IV-1023
Janik, Pawel I-300
Jankowski, Robert III-56
Jarząb, Marcin IV-1039
Javaheri Javid, Mohammad Ali III-367
Jędrzejowicz, Piotr III-719
Jeon, Jun-Cheol III-329, IV-661
Jeon, Segil IV-279, IV-297
Jeong, Karpjoo IV-264, IV-293
Jeong, Taek Sang III-40
Jeong, Taikyeong T. I-443, I-761, III-105, III-113
Jeong, Yoon-Seok IV-280
Jeong, Yoon-Su I-908
Jeras, Iztok III-345
Jermann, Christophe I-332
Jhang, Seong Tae IV-979
Jhon, Chu Shik IV-979
Jia, Jinyuan II-342
Jia, Zupeng I-851
Jiang, Gangyi II-367
Jiang, Hong IV-396
Jiang, Peng I-794, IV-693
Jie, Min Seok I-108
Jimenez, J.C. I-132
Jiménez-Morales, F. III-281
Jin, Guang IV-57
Jin, Hai I-1051, IV-380, IV-1055
Jin, Shiyao I-769
Jin, Xiaogang I-140, III-1032
Jin, Xin II-775, IV-781
Jin, Yu Xuan IV-264
Jin, Yuehui IV-128
Jingqing, Jiang I-547
Johnson, Chris R. I-3, II-147, III-393
Johnson, David II-581
Johnson, John IV-188
Joneja, Ajay II-342
Jones, Anthony III-586
Jones, Edward L. IV-773
Joo, Bok-Gyu IV-260
José L. Valcarce II-470
Juhasz, Zoltan I-830
Julià, Carme I-555
June, Sun-Do II-1041
Jung, Bokrae IV-152
Jung, Eui Bong IV-244
Jung, Jason J. III-244
Jung, Myoung-Hee I-969
Jung, Seunho IV-293
Jung, Ssang-Bong II-1041
Jung, Sungwon II-1065
Jurczyk, Tomasz II-302
Juszczyszyn, Krzysztof III-224

Kaczmarek, Paweł L. I-904
Kaihuai, Qin I-864
Kaiser, Tim I-379
Kamci, A. Kerim II-996
Kaminski, Thomas IV-591
Kaminski, Wieslaw A. II-94
Kanaujia, A. III-554
Kang, Byung-Heon III-329
Kang, Byung-Su II-969, II-1058
Kang, Dazhou IV-95
Kang, Eui Chul II-371
Kang, Lishan I-340
Kang, Mikyung IV-962, IV-970
Kang, Minho IV-152
Kang, Minhyung II-977
Kar, T. I-372
Karaivanova, Aneta III-616
Karakaya, Ziya II-375
Karam, Noha Makhoul I-148
Karcanias, Nicos I-798, II-399
Karimabadi, Homa II-41

Karl, Wolfgang II-502
Karniadakis, G.E. III-538
Karpowicz, Michał III-791
Kasperska, Elżbieta I-24
Kasprzak, Andrzej I-100
Katarzyniak, Radosław III-224, III-891
Kawasaki, Yohei IV-436
Keim, Daniel IV-196
Keller, Gabriele II-920
Kempe, David III-514
Kennedy, Catriona III-562
Keriven, Renaud IV-212, IV-236
Kernan, Warnick I-179
Kharche, Rahul V. IV-558
Khattri, Sanjay Kumar I-860, II-102, II-239
Khonsari, A. I-744
Khrenov, Alexey A. III-879
Khrustaleva, Ekaterina Yu. I-117
Kim, ByungChul II-1033
Kim, Byunggi III-1040
Kim, Chang-Hun II-279
Kim, Choelmin IV-962
Kim, Dae Sun II-1016
Kim, Daehee IV-922
Kim, Dongryung III-807
Kim, Duck Bong II-371
Kim, Han-Doo I-1067
Kim, Hanil IV-962
Kim, H.-K. III-425
Kim, Hwa-sung IV-954
Kim, Hyeoncheol II-830
Kim, HyoJin II-1101
Kim, Hyung-Jun III-895, III-899
Kim, Hyunmyung IV-293
Kim, Hyunsook I-940
Kim, Hyun-Sung I-634
Kim, I.S. III-489
Kim, In-Young I-164
Kim, Jaegwan IV-152
Kim, JangSub I-956
Kim, Jinbae II-977
Kim, Jin Ho II-1016
Kim, Jinhwan IV-970
Kim, Jong G. IV-533
Kim, Jong Hwa IV-264
Kim, Juil IV-902
Kim, Jung-Hyun IV-886
Kim, Kee-Won III-329, IV-661
Kim, Ki-Hyung II-1073
Kim, LaeYoung I-1013
Kim, Mihui II-1024
Kim, Mingon IV-152
Kim, Moonseong I-30, I-38
Kim, Myungho I-1059
Kim, Sanghun II-961
Kim, Seki I-30, I-38
Kim, Seongbaeg IV-962
Kim, Soo-Kyun II-279
Kim, Sung-Ryul IV-289
Kim, Sun I. I-164
Kim, Sun-Jeong II-279, II-326
Kim, Sun Yong II-961
Kim, Taeil I-969
Kim, Tae-Sun IV-862
Kim, Tae-Wan IV-280
Kim, Ung Mo IV-356
Kim, Won-Sik III-648
Kirby, R.M. II-147
Kirley, Michael III-248
Kisiel-Dorohinicki, Marek III-831, III-839, III-908
Kitowski, Jacek IV-252
Kleijn, C.R. II-10
Klein, Dino III-608
Klie, Hector III-384
Knight, D. III-473
Knüpfer, Andreas II-526
Ko, Young-Bae II-1097
Kobayashi, Hidetsune I-924
Kolaczek, Grzegorz III-224
Kolberg, S. IV-607
Kong, Xiaohong I-514
Koo, Jahwan I-948
Korkhov, Vladimir V. I-530
Korpeoglu, Ibrahim II-996
Kosloff, Todd J. II-215
Köstler, H. II-185
Kotsokalis, C.A. III-956
Kotulski, Leszek III-887
Kou, Gang IV-476, IV-485
Kowalczyk, W. I-1071
Kowarz, Andreas IV-541
Kožaný, Jan III-711
Koźlak, Jarosław III-703
Kozlowski, Alex II-215
Kramer, Robin II-855
Kranzlmüller, Dieter II-557
Krawczyk, Henryk I-904
Krawczyk, M.J. I-665

Kreaseck, Barbara IV-582
Krefft, Bogdan I-904
Kriksciuniene, Dalia IV-316
Krishnamoorthy, Sriram I-267
Kroc, Jiří IV-847
Kryza, Bartosz IV-252
Krzhizhanovskaya, Valeria V. I-530
Krznaric, M. III-956
Kubota, Tetsuyuki II-34
Küçükosmanoğlu, Alp I-814
Kujawski, Bernard III-1024
Kułakowski, K. I-665
Kulikov, Gennady Yu. I-117, I-781
Kulikova, Maria V. I-473
Kulvietienė, Regina II-407
Kulvietis, Genadijus II-407
Kurc, Tahsin III-384
Kurz, Haymo II-86
Küster, Uwe I-196
Kwak, Jong Wook IV-979
Kwoh, Chee Keong II-718
Kwon, Jeong Ok I-977
Kwon, Tai-Gil II-1041
Kwon, Younggoo I-973

Labatut, Patrick IV-212
Laclavik, Michal III-980
Lai, Kin Keung I-790, IV-493
Lai, Poh-Chin I-884
Laidlaw, D.H. III-538
Lambert, T. I-641
Lambiotte, Renaud III-1111
Lane, Terran II-895
Lang, E.W. I-234, I-649
Langer, Malgorzata I-498
Larriba-Pey, Josep-L. I-156
Laskar, Jacques II-446
Lastovetsky, Alexey III-1008
Lau, L.C. II-758
Lavergne, Christian IV-372
Laws, Joseph IV-870
Lazarov, Raytcho III-393
Ledoux, Veerle IV-716
Lee, Bong-Keun I-908
Lee, Chilgee II-1089
Lee, Dongeun IV-922
Lee, Dong Hoon I-977
Lee, Doo-Soo I-164
Lee, Hakjoo II-1065
Lee, Hanku IV-272, IV-293
Lee, Hongseok I-960
Lee, Hyeon-Seok II-1041
Lee, Hyewon K. I-932
Lee, Hyongwoo IV-168
Lee, Hyukjoon IV-930
Lee, Hyungkeun IV-954
Lee, Hyung Su IV-910
Lee, Inbok IV-289
Lee, J. III-425
Lee, Jaemyoung I-443, I-761
Lee, Jae-Woo IV-264
Lee, Jae Yong II-1033
Lee, Joo-Haeng II-362
Lee, Joowan II-977
Lee, Ju-Hong I-63
Lee, Junghoon I-985, IV-962, IV-970
Lee, Junguck II-1065
Lee, Jysoo I-1059
Lee, Kang Woong I-108
Lee, Keon-Myung I-908
Lee, Kwangyong IV-902
Lee, Kwan H. II-371
Lee, Kyu Min III-895, III-899
Lee, Sang-Ho I-908
Lee, Sangkeon I-1059, III-972
Lee, SeoungYoung I-1018
Lee, Seung-Heon III-200
Lee, Seung-Jun I-952
Lee, Seung-Que IV-168
Lee, Sooyoung II-1065
Lee, SuKyoung I-1013, II-985
Lee, Sung-Hee II-1097
Lee, Sung-Woon I-994
Lee, Tae-Jin I-989, II-1041
Lee, Tong-Yee II-295
Lee, William III-964
Lee, Woojin IV-902
Lee, Yeung-Hak IV-862
Lee, Yuqin IV-805
Lees, Janet M. I-834
Lefeuve-Mesgouez, Gaëlle II-50
Leonard II, J. III-425
Lepp, Dmitri I-928
Leshchinskiy, Roman II-920
Leupi, Célestin II-58
Levrat, B. I-603
Lew, A.J. III-456
Lewis, Andrew I-720
Li, Bigang III-656, III-687
Li, Deng III-522

Li, Dongdong I-435
Li, Donghai IV-645
Li, Guoqing I-880, III-17
Li, Hong IV-918
Li, Hongxing IV-404
Li, Jianping IV-501
Li, Jianyu I-997
Li, Jianzhong II-662
Li, Jingtao I-896
Li, Kuan-Ching I-900
Li, Liang III-988
Li, Minglu III-948
Li, Tao I-250
Li, Wei III-522
Li, Wen-hui II-223
Li, Xiaowei II-1069
Li, Xiaowen III-9
Li, Xing IV-176
Li, Yanhui IV-95
Li, Yiming IV-599
Li, Yin I-818
Li, Ying I-1055
Li, Yong IV-73, IV-87
Li, Yuan III-440
Li, Z. III-554
Li, Zhenhua IV-404
Li, Zhong II-358
Liao, Sheng-hui IV-839
Liao, Xiaofei IV-380
Liatsis, P. III-767
Lie, Knut-Andreas IV-220
Liljeros, Fredrik III-1048
Lim, Azman Osman IV-9
Lin, Chao-Hung II-295
Lin, Chuang IV-41
Lin, Po-Feng IV-184
Lin, Woei IV-49
Lin, Yongmin III-216
Lin, Yu IV-128
Ling, Yun III-184
Linkens, D.A. I-993
Lipscomb, William H. IV-533
Lisik, Zbigniew I-498
Little, L. II-169
Liu, Bo III-593
Liu, Chia-Lung IV-49
Liu, Dingsheng I-880, III-17
Liu, Fei I-818, III-1004
Liu, Feng II-686
Liu, Jia I-449
Liu, Jing I-514
Liu, Kun III-695
Liu, Ming I-1030
Liu, Wei II-646
Liu, Weijiang I-1022, IV-120, IV-144
Liu, Xiaojian I-769
Liu, Yang IV-188
Liu, Zhaodong IV-781
Liu, Zhiyu IV-404
Lloret, I. I-316
Lluch, Javier II-287, II-310
Lo, Shih-Ching I-1038
Lodder, Robert J. III-393
Loh, Woong-Kee III-648
Loitière, Yannick III-570
López, Antonio I-555
López-Ruiz, R. III-353
Lu, Dongming IV-57
Lu, Feng I-884
Lu, Huimei IV-81
Lu, Jianjiang IV-95
Lu, Qingda I-267
Lu, Ssu-Hsuan I-900
Lu, Yijuan II-686
Lucas, Philipp IV-200
Lumbreras, Felipe I-555
Luo, Fei IV-380
Luo, Ying I-292, I-876, III-1
Luque, Emilio I-539
Lursinsap, Chidchanok II-838
Lv, Rui I-997
Lv, Song IV-396

Ma, Fanyuan I-818, III-1004
Ma, Guangsheng IV-645
Ma, Jixin II-775, IV-781
Ma, Lizhuang II-358
Ma, Min I-769
Ma, Xi-min I-826
Ma, Yan I-880
Madey, Gregory R. III-417
Mahfouf, M. I-993
Mahinthakumar, Kumar III-401
Maik, Vivek IV-922
Makowiec, Danuta III-256
Malaschonok, Gennadi II-486
Maliekal, J. II-169
Malinowski, Krzysztof III-791
Malkowski, Konrad I-242
Mamat, Ali IV-1071

Manceny, Matthieu III-1056
Mandel, Alan K. III-522
Mandel, Jan III-522
Măndoiu, I.I. II-758
Măndoiu, Ion I. II-742
Manfroi, Fairus I-68
Manzoni, Sara III-289
Mao, Zhihong II-358
Marchese, Fabio M. III-264
Marchiori, E. I-1071
Marcjan, Robert III-775
Margalef, Tomàs I-539
Marín, Mauricio I-611
Marroquín-Alonso, Olga II-207
Martínez, A. IV-685
Martins, Ana Mafalda II-255
Martyniak, J. III-956
Marzo, Jose L. IV-136
Maskell, Douglas L. I-522
Masteika, Saulius IV-332
Mastroianni, Carlo IV-1047
Mat Deris, M. IV-1071
Mateja-Losa, Elwira I-24
Matéo-Vélez, Jean-Charles II-1
Matossian, Vincent III-384
Matsukubo, Jun III-1106
Matsumoto, Noriko IV-436
Mauch, Sean P. II-122
Maurizio, Marchese I-547
Mavridis, Pavlos II-271
McCalley, James D. III-440
McCourt, Frederick R.W. II-103
McCrindle, Rachel I-868
McGough, A.S. III-956
McGough, A. Stephen III-964
McInnes, Lois Curfman I-242
Meeker, William Q. III-440
Mei, Jian IV-669
Meiron, Daniel I. II-122
Mellema, Angela III-433
Melnik, Roderick V.N. II-114
Memon, Ashraf III-920
Meng, Yu II-223
Merelli, Emanuela III-1012
Merkevicius, Egidijus IV-364
Merkulov, Arkadi I. I-117
Merschmann, Luiz II-863
Merzky, Andre III-97
Mesgouez, Arnaud II-50
Metaxas, D. III-554
Michener, William K. III-912
Michopoulos, John G. II-131, III-456
Miehe, Philipp III-120
Mihaylova, Lyudmila III-624
Miklaszewski, Wiesław III-256
Milhous, Wilbur K. I-387
Milledge, Tom II-694, II-702
Miller, Gavin S.P. IV-228
Milthorpe, Josh I-218
Min, Jun-Ki I-364
Min, Yong I-140, III-1032
Mingarelli, Angelo B. III-360
Mishra, Bud II-638, II-654
Mitrouli, Marilena II-399
Mix, Hartmut II-526
Mo, Hongwei I-997
Mochena, M.D. III-128
Möller, Kim II-565
Monfroy, E. I-641
Monnier, J. II-26
Moon, Jongbae I-1059
Moon, Sanghoon I-276, I-284
Moreira, José E. I-2
Moreno, A. I-316
Moreno-Vozmediano, Rafael IV-1031
Morimoto, Shoichi IV-797
Morisse, Karsten II-565
Morley, Chris T. I-834
Morvan, Michel III-321
Mou, Tai-yong IV-452
Mould, David II-318
Mu, Fei III-687
Mukherjee, Joy I-46
Muldoon, C. III-727
Mun, Sung-Gon I-960
Munagala, Kamesh III-409
Muñoz Masqué, J. II-438
Muñoz, Miguel A. III-1075
Munoz, Roberto III-1091
Muntés-Mulero, Victor I-156
Murugesan, K. I-457
Murugesh, V. I-457

Nagar, Atulya I-802
Nagel, Wolfgang E. II-526
Najim, M. I-697
Namachchivaya, N.S. III-448
Namiki, Takefumi II-34
Nandigam, Viswanath III-920
Narasimhan, Giri II-694, II-702, II-807

Narayanan, Babu I-16
Nassif, Nabil R. I-148
Natvig, Jostein R. IV-220
Naumov, Maxim I-258
Navarro, Gonzalo I-611
Navas-Delgado, Ismael III-936
Nawarecki, Edward III-839
Nedjalkov, Mihail III-616
Nenortaitė, Jovita I-1034
Neves, José III-240
Ng, Kam-Wing IV-1007
Nguyen, Ngoc Thanh III-208, III-224
Nichols, Daniel A. I-387
Nicole, Denis A. III-928
Nilsson, Patric II-879
No, Jaechun IV-1063
Noël, Alfred G. II-422
Nooijen, Marcel I-267
Norris, Boyana I-242
Nou, Ramon I-84
Novák, Stanislav I-806
Nowak, Leszek I-300
Nutaro, James J. IV-814
Nygaard, Jens Olav IV-204

Oberg, Carl III-514
Oden, J.T. III-530
O'Grady, M.J. III-727
O'Hare, G.M.P. III-727
Oh, Donsung II-977
Oh, Hyukjun IV-991
Oh, Jai-Boo IV-661
Oh, Seungtak II-1089
Oladunni, Olutayo O. I-188
Olanda, R. III-13
Oliveira, Rafael Sachetto I-68, I-76
Oliveira, S. II-726
Oliver, Timothy F. I-522
Olman, Victor II-855
Olsson, Björn II-879
Osguthorpe, David I-308
Ospina, Juan I-920
Ould-Khaoua, M. I-744
Overbye, T.J. III-448
Ozaki, T. I-132

Pace, Brigida IV-724
Pachter, R. I-372
Pai, M.A. III-448
Paik, Joonki IV-922
Paik, Juryon IV-356
Paik, Woojin IV-894
Pajarola, Renato B. II-371
Pak, Jinsuk IV-180
Palekar, M. III-425
Palkow, Mark II-1050
Pan, Gang I-435
Pan, Xuezeng IV-156
Pan, Yi II-646, II-710
Pang, Wei II-223
Papaioannou, Georgios II-271
Papavassiliou, Dimitrios V. I-188
Papini, Alessandra IV-677
Paprzycki, Marcin III-208
Parashar, Manish III-384
Parasuk, Vudhichai III-136
Parasuk, Waraporn III-136
Park, Chang Won IV-1063
Park, DongGook III-232
Park, Geunyoung IV-946
Park, Gwitae IV-987
Park, Gyung Leen I-985, II-587, IV-962
Park, Hong-Shik I-1018
Park, Hyungjun II-362
Park, Jaehyung I-969
Park, Kisoeb I-30, I-38
Park, Kiyong I-936, II-1081
Park, Namhun IV-168
Park, Neungsoo IV-244
Park, Sangjoon III-1040
Park, Sang Soon I-944
Park, SeongHoon I-736
Park, Sung Soon IV-1063
Park, Taehyung II-992
Park, Taesoon III-807
Park, Tae-Su I-63
Parker, Steven G. III-393
Passalis, Georgios II-271
Pasztor, Egon II-215
Paszynski, Maciej III-751
Paternoster, Beatrice IV-700
Patist, J.P. I-1071
Paul, Samit I-16
Paventhan, A. III-928
Pazo-Robles, M.E. III-337
Peachey, Tom I-720
Pecheanu, Emilia II-199
Pei, Pengjun II-734
Peláez, Ignacio I-872
Peng, Bo I-140, III-1032

Peng, Yanbing I-1022, IV-120
Peng, Yi IV-476, IV-485
Pennington, Deana D. III-912
Pereira, António II-454
Perelman, Alex II-215
Pérez, Mariano III-13
Perumalla, Kalyan II-41
Peterson, Janet II-177
Phipps, Eric T. IV-525
Pickin, Simon IV-765
Pieczykolan, Jan IV-252
Pieczynska, Agnieszka III-224, III-891
Ping, Lingdi IV-156
Pisa, Ivan T. I-1005
Pitera, Jed II-846
Pitman, Michael C. II-846
Pitzer, Russell M. I-267
Pivkin, I.V. III-538
Plastino, Alexandre II-863
Plaza, Antonio I-888, III-24
Plaza, Javier III-24
Pokrywka, Rafał III-855
Politi, T. IV-708, IV-732
Pons, Jean-Philippe IV-212
Popolizio, M. IV-708
Posse, C. II-871
Pota, Szabolcs I-830
Pouchard, Line C. IV-814
Prank, Rein I-928
Primavera, Leonardo I-465
Príncipe, José III-546
Prăjescu, Claudia II-742
Prudhomme, S. III-530
Przekwas, Andrzej IV-822
Przytycka, Teresa M. II-620
Pugliese, A. IV-732
Puglisi, Giovanni II-334
Puntonet, Carlos G. I-234, I-316, I-356, I-649
Pyle, David Leo I-403

Qian, Jixin III-593
Qian, Liang II-904
Qiao, Daji III-440
Qin, Guan III-393, III-522
Qiu, Shibin II-895
Qu, Youli III-216
Queiruga Dios, A. II-438
Quirós, Ricardo II-310

Rafe, Vahid III-578
Raghavan, Padma I-242
Rahmani, Adel Torkaman III-578
Rajasekaran, Sanguthevar II-822
Rajasethupathy, K. II-169
Ramakrishnan, Naren I-46
Ramalingam, M. III-143
Ramanujam, J. I-267
Ramasami, K. III-143
Ramasami, Ponnadurai III-153
Ramírez, J. I-234, I-356, I-395, I-649
Ramos, J.I. II-106
Ramos, Luis IV-582
Ramsamy, Priscilla II-595, II-603
Rangel-Kuoppa, Risto II-318
Ranjithan, Ranji III-401
Rasmussen, Craig E II-945
Rasúa, Rafael A. Trujillo I-324
Ravela, Sai III-497
Ravi, R. II-799
Raychaudhuri, Dipankar IV-930
Rayshubskiy, Aleksandr II-846
Redaelli, Stefano III-289
Regensburg, Henrik II-1050
Reis, Artur E. I-842
Rendell, Alistair P. I-218, II-155
Reynolds, Paul III-570
Rhee, Seung Hyong IV-930
Rhymend Uthariaraj, V. IV-388
Richard, Adrien II-887
Richardson, P.D. III-538
Ridge, Oak II-41
Ridley, A. III-489
Riensche, R. II-871
Rigopoulos, Stelios II-18
Rizzi, Romeo II-783
Roberts, Ronald A. III-440
Roch, Jean-Louis IV-999
Rodgers, G.J. III-1024
Rodrigues, Rosália II-454
Rodríguez, D. IV-789
Rogier, Francois II-1
Roh, Yong-Wan IV-886
Rojek, Gabriel III-823, III-855
Roman, Eric II-215
Romero, L.F. II-106
Romero, Sergio II-912
Ronald, Nicole III-248
Ronchieri, E. III-956
Rong, Chunming I-794, IV-693

Ros, E. II-518
Rossello, Damiano IV-324
Rossman, T. III-473
Rountev, Atanas I-267
Rouquier, Jean-Baptiste III-321
Roux, Olivier II-887
Roy Mahapatra, Debiprosad II-114
Rubio, Bartolomé II-912
Rüde, U. II-185
Ruiz, R. IV-789
Russell, A. II-758
Ryan, J. II-66
Ryan, Sarah M. III-440
Ryba, Przemyslaw I-100
Rylander, M.N. III-530
Ryu, Jeha II-610
Ryu, Seungwan II-977, II-1033, IV-168

Sadayappan, P. I-267
Safaei, F. I-744
Sagianos, E. I-798
Sahingoz, Ozgur Koray III-192, III-903
Saidane, Mohamed IV-372
Sakalauskas, Virgilijus IV-316
Sakurai, Kouichi I-977
Salinas, Luis I-856
Saltenis, Vydunas I-704
Saltz, Joel III-384
Sameh, Ahmed III-465
Sánchez, J.R. III-353
Sanchez, Justin C. III-546
Sanchez, Maribel II-686
Sancho Chust, S. II-350
Sandu, Adrian I-712, III-120, IV-550
Sanfilippo, A. II-871
Sankoff, David II-791
Santana, Miguel IV-999
Santini, Cindy I-379
Sappa, Angel I-555
Sappa, Angel D. I-563
Sarin, Vivek I-92
Saubion, F. I-603, I-641
Sauer, P.W. III-448
Sautois, B. II-391
Sbert, Mateu II-263
Schaefer, Robert III-783, III-799
Schaubschläger, Christian II-557
Schloissnig, Siegfried II-502
Schmidt, Bertil I-522
Schmidt, Thomas C. II-1050
Schneider, Adrian II-630
Schöbel, Rainer IV-340
Schwan, K. III-425
Schwartz, Jacob T. II-654
Schwartz, Russell II-799
Seber, Dogan I-379
Sedighian, Saeed III-578
Segal, Cristina I-172, II-199
Segura, J.C. I-356, I-395
Semoushin, Innokenti V. I-473
Seok, S.C. II-726
Sequeira, Adélia II-78
Serrat, Joan I-555
Seshasayee, B. III-425
Sethuraman, V. III-143
Sevinc, Bahadir IV-638
Sfarti, Dr. Adrian II-215
Shafi, Aamir II-953
Shahinpoor, Moshen II-131
Shakhov, Vladimir V. I-948
Shang, Wenqian III-216
Sharifi, Mohsen I-981
Sharma, Abhishek III-514
Shi, Hanxiao III-184
Shi, Yong IV-452, IV-476, IV-485, IV-509
Shi, Zhong-ke IV-878
Shi, Zhongzhi I-777
Shiffler, D.A. I-372
Shigezumi, Takeya II-815
Shih, Wen-Chung I-810
Shim, Choon-Bo III-232
Shin, Chang-Sun III-232
Shin, Dongchun II-977
Shin, DongRyeol I-956, III-895, III-899
Shin, HoJin I-956
Shin, In-Hye I-985
Shin, Jeongho IV-922
Shin, Jitae I-952, IV-25
Shindin, Sergey K. I-781
Shirayama, Susumu III-1063
Shu, Jiwu III-663, III-687, III-695
Sicilia, M.A. IV-789
Simha, Rahul III-679
Simutis, Rimvydas IV-332
Siwik, Leszek III-831, III-871
Słota, Damian I-24, I-786
Ślusarczyk, Grażyna III-883
Smetek, Marcin II-549
Smoliy, Elena F. III-879

Smołka, Maciej III-799
Śnieżyński, Bartłomiej III-703, III-759
So, Won-Ho III-232
Soler, Enrique II-912
Son, Hyung-Jin III-506
Son, Jeongho IV-180
Song, Byunghun IV-910
Song, Chang-Geun II-326
Song, Hyoung-Kyu II-969, II-1058
Song, In-Ho I-164
Song, JooSeok I-1013, II-1101
Song, Jungwook I-936, II-1081
Song, Mingli I-449
Song, Minseok I-1075
Song, Sung Keun II-587
Song, Yong Ho IV-244
Song, Young Seok III-105, III-113
Song, Zhanjie I-427, I-822
Sonmez, A. Coskun III-192, III-903
Soofi, M.A. III-48
Sosonkina, Masha I-54
Sottile, Matthew J. II-945
Soukiassian, Yeran I-148
Spezzano, Giandomenico IV-1047
Spiegel, Michael III-570
Sreepathi, Sarat III-401
Sridhar, Srinath II-799
Srinivasan, Kasthuri I-92
Srovnal, Vilém III-711
Stadlthanner, K. I-234
Stafford, R.J. III-530
Stagni, Rita IV-851
Stavrou, Pavlos II-271
Stauffer, Beth III-514
Stefanescu, Diana II-199
Strout, Michelle Mills IV-574, IV-582
Su, Fanjun IV-156
Su, Hui-Kai IV-184
Su, Sen IV-73, IV-87, IV-104, IV-164
Su, Xianchuang I-140, III-1032
Sudholt, Wibke III-69
Suh, Jonghyun II-1065
Suits, Frank II-846
Sukhatme, Gaurav III-514
Sun, Bing II-654
Sun, Jizhou I-419
Sun, Jun III-847
Sunderam, Vaidy I-1
Sung, Hocheol IV-260
Susitaival, Riikka IV-420
Sussman, A. III-448
Škvor, Jiří I-806
Šuvakov, Milovan III-1098
Švec, Martin I-806
Swiecicki, Mariusz I-300
Swierszcz, Pawel II-534
Swope, William II-846
Sygkouna, Irene I-892
Sykas, Efstathios I-892
Szabo, Gabor III-417
Szczerba, Dominik II-86
Székely, Gábor II-86
Szychowiak, Michał I-753

Tabakow, Iwan III-168
Tabik, S. II-106
Tadić, Bosiljka III-1016, III-1024, III-1098
Tahar, M. II-169
Taherkordi, Amirhosein I-981
Tai, Phang C. II-710
Takahashi, Tadashi I-924
Takaoka, Tadao I-595
Takeda, Kenji III-928
Taleghan, Majid Alkaee I-981
Tan, Feng II-678
Tan, X.G. IV-822
Tang, Kai II-342
Tao, Jie II-502
Tawfik, H. III-767
Tawfik, Hissam I-802, III-60
te Boekhorst, René III-307
Tembe, B.L. III-161
Theodoropoulos, Georgios III-562
Theoharis, Theoharis II-271
Therón, Roberto III-32
Thivet, Frédéric II-1
Thomas, Sunil G. III-384
Thurner, Stefan III-1016, III-1067
Tian, Jing IV-412, IV-428
Tian, Qi II-686
Tibiletti, Luisa IV-324
Tinnungwattana, Orawan II-838
Tiyyagura, Sunil R. I-196
Tong, Ruo-feng IV-839
Toporkiewicz, W. III-783
Torosantucci, L. IV-460
Torres, Jordi I-84
Tošić, Predrag T. III-272

Trafalis, Theodore B. I-188, III-506
Trappe, Wade IV-930
Trapp, John I-490
Trebacz, Lechoslaw II-549
Trelles, Oswaldo III-936
Triantafyllou, Dimitrios II-399
Trinh, Thanh Hai IV-1
Troya, José M. II-912
Tsai, Ming-Hui II-1008
Tsechpenakis, G. III-554
Tseng, Shian-Shyong I-810
Tsukerman, Igor I-54
Tu, Shiliang I-1043
Turek, Wojciech III-775
Turias, I. I-649
Tučník, Petr III-711
Tuzun, R. II-169

Uber, Jim III-401
Uchida, Makoto III-1063
Ufuktepe, Ünal I-916
Uhruski, P. III-783
Ülker, Erkan II-247
Uribe, Roberto I-611
Urra, Anna IV-136

Vaidya, Sheila IV-188
Vaiksaar, Vahur I-928
Valaitytė, Akvilina IV-348
Valakevičius, Eimutis IV-348
Valencia, David I-888, III-24
van Bloemen Waanders, Bart III-481
Van Daele, Marnix IV-716
Vanden Berghe, Guido IV-716
Van Hamont, John E. I-387
Vanneschi, Leonardo III-289
van Veldhuizen, S. II-10
Varadarajan, Srinidhi I-46
Vazirani, V.V. II-758
Venuvanalingam, P. III-143
Versteeg, Roelof III-384
Vialette, Stéphane II-622, II-783
Vianello, M. IV-685
Vianna, Gizelle Kupac I-842
Vidal, Antonio M. I-324, I-348
Virtamo, Jorma IV-420
Visscher, Lucas III-97
Viswanathan, M. III-200
Vivó, Roberto II-310
Vodacek, Anthony III-522
Volkert, Jens II-557
von Laszewski, Gregor III-401
Vuik, C. II-10

Wählisch, Matthias II-1050
Wais, Piotr I-300
Wajs, Wieslaw I-300
Walkowiak, Krzysztof I-618, I-626
Walther, Andrea IV-541
Wan, Wei I-292, I-876, III-1
Wan, Zheng IV-156
Wang, Chaokun II-662
Wang, Chien-Lung I-1026
Wang, Chunshan IV-468
Wang, Di III-695
Wang, Fang I-1063, III-671, IV-396
Wang, Feng III-82
Wang, Guoping I-411
Wang, Hanpin III-988
Wang, Hao II-678, IV-428
Wang, Heng I-411
Wang, Hongbo IV-128
Wang, Hsiao-Hsi I-900
Wang, Jian I-880
Wang, Jianqin III-1
Wang, Kuang-Jui I-900
Wang, Lin I-728, IV-308, IV-517
Wang, Pei-rong I-826
Wang, Qijia IV-938
Wang, Shaowei I-340
Wang, Shouyang I-790, IV-308, IV-444, IV-493, IV-517
Wang, Shuanghu I-851
Wang, Xin IV-9
Wang, Xueping I-896
Wang, Yan II-223
Wang, Yang III-663
Wang, Yueming I-435
Wang, Yufeng II-686
Wang, Zhengfang I-292
Ward, Richard C. IV-814
Ward, T.J. Christopher II-846
Wasson, Glenn I-681
Weber dos Santos, Rodrigo I-68, I-76
Wedemann, Roseli S. I-842
Wei, Anne IV-17
Wei, Guiyi III-184
Wei, Guozhi IV-17
Wei, Hu I-864
Wei, Huang I-790

Wei, Wei III-948, IV-57
Wei, Xue III-656
Weihrauch, Christian III-632, III-640
Welch, Samuel I-490
Wen, Wanzhi I-851
Wheeler, Mary F. III-384
Whitlock, P.A. III-608
Wieczorek, Damian IV-1039
Wierzbowska, Izabela III-719
Wilhelm, Reinhard IV-200
Wittevrongel, Sabine IV-65
Wojcik, Grzegorz M. II-94
Wojtowicz, Hubert I-300
Wong, Kim-Sing IV-260
Wozny, Janusz I-498
Wu, Chaolin I-292, I-876, III-1
Wu, Chin-Chi IV-49
Wu, Haiping III-656
Wu, Jianping IV-33
Wu, Kun IV-33
Wu, Meiping I-689
Wu, Song IV-1055
Wu, Yuanxin I-689
Wu, Zhaohui I-435, I-1055, IV-918, IV-938
Wu, Zhongqiang IV-748
Wu, Zhauhui IV-468

Xia, Peng III-671
Xia, Yu I-8, I-124
Xia, Yunni III-988
Xiang, Quan[illegible] IV-[illegible]
Xiao, Rui I-1030
Xie, Jiang IV-669
Xie, Wen IV-444
Xie, Xia I-1051
Xiong, Muzhou IV-1055
Xu, Anbang II-775
Xu, Baowen IV-95, IV-748
Xu, Chunxiang III-988
Xu, Dong II-855
Xu, Fuyin I-838
Xu, HaiGuo IV-293
Xu, Ke IV-17, IV-33
Xu, Shanying IV-444
Xu, Weixuan IV-501
Xu, Wenbo I-514, I-1043, III-847
Xu, Xian II-670
Xu, Ying II-855
Xu, Zhe I-826
Xue, Wei I-250, III-663, III-695
Xue, Xiangyang IV-9
Xue, Yong I-292, I-876, III-1, III-9

Yaşar, O. II-169
Yaghmaee, Mohammad Hossien I-588
Yaikhom, Gagarine II-929
Yan, Chung-Ren II-295
Yang, Chao-Tung I-810
Yang, Fangchun IV-73, IV-87, IV-164
Yang, Geng I-794, IV-693
Yang, Jun III-409
Yang, Kun II-662
Yang, Mao IV-412, IV-428
Yang, Mijeong I-969
Yang, Shouyuan I-427
Yang, Shuzhong I-997
Yang, Wang I-1022
Yang, Xin-She I-834
Yang, Xuejun II-904
Yang, Yingchun I-435
Yang, Young-Kyu III-200
Yang, Zhanxin I-997
Yanami, Hitoshi II-462
Yangchun, Liang I-547
Yantır, Ahmet I-916
Yao, Jialiang III-60
Yao, Yonglei IV-164
Yazici, Ali II-375
Ye, Chaoqun I-769
Yegül, Umut I-814
Yélamos, P. I-356
Yeo, So-Young II-969
Yeon, Eunja IV-894
Yi, Huizhan II-904
Yi, Sangho IV-946
Yildirim, Tulay IV-615
Yim, Soon-Bin II-1041
Yoo, Chae-Woo I-965, II-510
Yoo, Chuck IV-160, IV-172
Yoo, Kee-Young II-1000, III-329, IV-661
Yoon, Eun-Jun II-1000
Yoon, Jungwon II-610
Yoon, Seokho II-961
Yoon, Tae-Sun II-830
Yoon, Won-Sik II-1073
Yoshida, Norihiko IV-436
You, L.H. II-231
You, Mingyu I-449

You, Young-Hwan II-969, II-1058
Youn, Choonhan I-379
Youn, Hee Yong II-587, IV-1, IV-910
Youn, Jaehwan III-807
Yu, Chiu-Man IV-1007
Yu, Dongjin I-1055
Yu, Hongliang III-656
Yu, Jun III-184
Yu, Lean I-790, IV-308, IV-444, IV-493, IV-517
Yu, Mei II-367
Yu, Shao-Ming I-1038
Yu, Shaokai II-1073
Yu, Zhiping I-250
Yuan, Ding I-179
Yuan, Ruifeng IV-404
Yue, Wuyi IV-509
Yun, Hyunho IV-152

Żabińska, Małgorzata III-735
Zechman, Emily III-401
Zelikovsky, Alex II-750
Zelikovsky, Alexander II-767
Zeng, Lingfang I-1063, III-671, IV-396
Zeng, Yurong I-728
Zenger, Ch. I-673
Zhang, Aidong II-670, II-734
Zhang, Bin III-514
Zhang, Chengwen IV-104
Zhang, Shunda IV-396
Zhang, Gendu I-896
Zhang, Guang-Yan III-663
Zhang, Guiling I-419
Zhang, Hao IV-380
Zhang, Jian J. II-231
Zhang, Jianting III-912
Zhang, Jinlong I-728
Zhang, Liguo II-478
Zhang, Qianli IV-176
Zhang, Qin I-1051, IV-380
Zhang, Shensheng III-948
Zhang, Shijia I-411
Zhang, Shunda IV-396
Zhang, Wanjun III-17
Zhang, Weide I-681
Zhang, Wenju I-818, III-1004
Zhang, Wenyi III-17
Zhang, Wu I-1047, IV-669
Zhang, Y. III-530
Zhang, Yanqing II-678
Zhang, Yingzhou IV-748
Zhang, Yiying II-830
Zhao, Bo I-657
Zhao, Chen I-204
Zhao, Guiping I-851
Zhao, Guoxing II-775, IV-781
Zhao, Jun III-593
Zhao, Liqiang IV-9
Zhao, Mingxi II-358
Zhao, Peng III-1008
Zhao, Wenyun IV-805
Zheng, Bo IV-41
Zheng, Chunfang II-791
Zheng, Gaolin II-694, II-702
Zheng, Lei I-292, I-876, III-1
Zheng, Nenggan IV-938
Zheng, Weimin III-656, III-687
Zhestkov, Yuri II-846
Zhong, Shaobo I-292, I-876, III-1, III-9
Zhong, Wei II-710
Zhou, Deyu II-718
Zhou, Mingzhong IV-112
Zhou, Ruhong II-846
Zhou, Xingwei I-427, I-822
Zhou, Yanmiao IV-938
Zhou, Yi I-1055
Zhou, Yu II-367
Zhou, Zhongmei IV-468
Zhou, Zong-fang IV-452
Zhu, Haibin III-216
Zhu, Qiuping I-340
Zhu, Wei-yong I-826
Zhuang, Yueting I-204
Zieliński, Krzysztof III-940, IV-1023, IV-1039
Zory, J. IV-653
Zou, Deqing IV-1055
Zuo, Xin III-1008
Zuzarte, Calisto I-156